Physical Science

Physical Science

Bill W. Tillery
Arizona State University

 Wm. C. Brown Publishers

Book Team

Editor *Jeffrey L. Hahn*
Developmental Editor *Lynne M. Meyers*
Production Editor *Renée Menne*
Designer *Mark Elliot Christianson*
Art Editor *Jess Schaal*
Photo Editor *Carol M. Smith*
Visuals Processor *Amy L. Saffran*

 Wm. C. Brown Publishers

President *G. Franklin Lewis*
Vice President, Publisher *George Wm. Bergquist*
Vice President, Publisher *Thomas E. Doran*
Vice President, Operations and Production *Beverly Kolz*
National Sales Manager *Virginia S. Moffat*
Advertising Manager *Ann M. Knepper*
Marketing Manager *John W. Calhoun*
Editor in Chief *Edward G. Jaffe*
Managing Editor, Production *Colleen A. Yonda*
Production Editorial Manager *Julie A. Kennedy*
Production Editorial Manager *Ann Fuerste*
Publishing Services Manager *Karen J. Slaght*
Manager of Visuals and Design *Faye M. Schilling*

Cover photo by © Chris A. Wilton/The Image Bank

The credits section for this book begins on page 599, and is considered
an extension of the copyright page.

Printed in the United States of America by Wm. C. Brown Publishers,
2460 Kerper Boulevard, Dubuque, IA 52001

10 9 8 7 6 5 4 3 2 1

"To Patricia, Tonya, Lisa, and Elizabeth"

Contents

Contents

Contents

Preface

Physical Science is a straightforward, easy-to-read but substantial introduction to the fundamental behavior of matter and energy. It is intended to serve the needs of nonscience majors who are required to complete one or more physical science courses. It introduces basic concepts and key ideas while providing opportunities for students to learn reasoning skills and a new way of thinking about their environment. No prior work in science is assumed. The language as well as the mathematics is as simple as can be practical for a college-level science course.

The book is divided into four sections: physics, chemistry, astronomy, and the earth sciences. With laboratory studies, Physical Science contains enough material for a two-semester course. The chapters and sections are flexible: the instructor can determine topic sequence and depth of coverage. Physical Science can also serve as a text in a one-semester course; the materials are complete enough to be used in a one-semester physics and chemistry course, a one-semester astronomy and earth science course, or in other combinations.

To make room for a wide spectrum of physical science focused on student needs, interests, and abilities, the technical details and terminology that are of interest to the science major are not included. In addition, special interest areas such as environmental concerns are not isolated in an arbitrary section or chapter; they are discussed where they occur naturally. After all, environmental concerns are a part of everyday life, not some isolated part to be considered at times and ignored other times.

Each chapter presents historical background where appropriate, uses everyday examples in developing concepts, and follows a logical flow of presentation. The historical chronology, of special interest to the humanistically inclined nonscience major, serves to humanize the science being presented. The use of everyday examples appeals to the nonscience major, typically accustomed to reading narration, not scientific technical writing, and also tends to bring relevancy to the material being presented. The logical flow of presentation is helpful to students not accustomed to thinking about relationships between what is being read and previous knowledge learned, a useful skill in understanding the physical sciences. Worked examples help students integrate concepts and understand the use of relationships called equations. They also serve as a problem solving model; consequently, special attention is given to *complete* unit work and to the clear, fully expressed use of mathematics. Where appropriate, chapters contain one or more activities that use everyday materials rather than specialized laboratory equipment. These activities are intended to bring the science concepts closer to the world of the student. The activities are supplemental and can be done as optional student activities or as demonstrations.

Physical Science also contains a number of innovative learning aids. Each chapter begins with an *introductory overview* and a brief *outline* that help students to organize their thoughts for the coming chapter materials. Each chapter ends with a brief *summary* that organizes the main concepts presented, a *summary of equations* (where appropriate) written both with words and

with symbols, a list of page-referenced *key terms*, a set of *multiple-choice questions* with nearby answers for immediate correction or reinforcement of major understandings, a set of *thought questions* for discussion or essay answers, and, *two sets of problem exercises* with complete solutions for one set provided in the appendix. The two sets are nearly parallel in early chapters, but they become progressively less so in successive chapters. The set with the solutions provided is intended to be a model to help students through assigned problems in the other set. In trial classroom testing, this approach proved to be a tremendous improvement over the traditional "odd problem answers." The "odd answer only" approach provided students little help in learning problem solving skills, unless it was how to work a problem backward.

Finally, each chapter of Physical Science also includes a boxed feature that discusses topics of special human or environmental concern (the use of seat belts, acid rain, and air pollution, for example), topics concerning interesting technological applications (passive solar homes, solar cells, and catalytic converters, for example), or topics on the cutting edge of scientific research (quarks, superstrings, and deep-ocean exploration, for example). All boxed features are informative materials that are supplementary in nature.

Physical Science is accompanied by an instructor's manual, a laboratory manual, a student study guide, WCB TestPak—a computer bank of multiple-choice test items, WCB QuizPak—a computer quizzing program, and a set of overhead transparencies. The laboratory manual, written and classroom tested by the author, presents a selection of traditional laboratory exercises specifically written for the interest and abilities of nonscience majors. When the laboratory manual is used with *Physical Science,* students will have an opportunity to master basic scientific principles and concepts, learn new problem solving and thinking skills, and understand the nature of scientific inquiry from the perspective of hands-on experiences.

The instructor's manual, also written by the text author, provides a chapter outline, an introduction/summary of each chapter, suggestions for discussion and demonstrations, additional multiple choice questions (with answers), and answers and solutions to all end-of-chapter questions and exercises not provided in the text.

The student study guide, cowritten with John Grant, was designed after extensive tutoring experiences with beginning physical science students. The guide provides a solid foundation for nonscience students by stressing conceptual understanding as well as techniques for successful problem solving. All the examples and illustrations are new and different from the examples and illustrations of the text, which tends to maintain interest as it adds a new dimension to student understanding of course concepts and skills.

The author has attempted to present an interesting, helpful program that will be useful to both students and instructors. Comments and suggestions about how to do a better job of reaching this goal are welcome. Any comments about the text or other parts of the program? Write to:

Bill W. Tillery
Department of Physics
Arizona State University
Tempe, AZ 85287–1504

Acknowledgments

Many constructive suggestions, new ideas, and invaluable advice were provided by reviewers through several stages of manuscript development. Special thanks and appreciation to those who reviewed all or part of the manuscript, often under tight time deadlines. Thanks is extended to Lawrence H. Adams, *Polk Community College;* Charles L. Bissell, *Northwestern State University of Louisiana;* W. H. Breazeale, Jr., *Francis Marion College;* William Brown, *Montgomery College;* Stan Celestian, *Glendale Community College;* Carl G. Davis, *Danville Area Community College;* Dennis Englin, *The Master's College;* Floretta Haggard, *Rogers State College;* Robert G. Hamerly, *University of Northern Colorado;* Eric Harms, *Brevard Community College;* L. D. Hendrick, *Francis Marion College;* Abe Korn, *New York City Tech College;* Robert Larson, *St. Louis Community College;* William Luebke, *Modesto Junior College;* Douglas L. Magnus, *St. Cloud State University;* L. Whit Marks, *Central State University;* Jesse C. Moore, *Kansas Newman College;* Michael D. Murphy, *Northwest Alabama Community College;* David L. Vosburg, *Arkansas State University.* Special thanks and appreciation is also extended to my wife, Patricia Northrop Tillery, who assisted directly with typing and photograph acquisition as well as indirectly with her ongoing support. Finally, thanks is extended to Jeffrey L. Hahn and Lynne Meyers of Wm. C. Brown for their understanding of student and instructor needs, commitment to quality, and assistance in the publication of this edition.

Chapter 1

The World Around You

Figure 1.1
Your physical surroundings include naturally occurring as well as human-made objects such as sidewalks, trash receptacles, and buildings.

HAVE you ever thought about your thinking and what you know? On a very simplified level, you could say that everything you know came to you through your senses. You see, hear, and touch things of your choosing and you can also smell and taste things in your surroundings. Information is gathered and sent to your brain by your sense organs. Somehow, your brain processes all this information in an attempt to find order and make sense of it all. Finding order helps you understand the world and what may be happening at a particular place and time. Finding order also helps you predict what may happen next, which can be very important in a lot of situations.

This is a book on thinking about and understanding your physical surroundings. These surroundings range from the obvious, such as the landscape and the day-to-day weather, to the not so obvious, such as how atoms are put together. Your physical surroundings include natural things as well as things that people have made and used (figure 1.1). You will learn how to think about your surroundings, whatever your previous experience with thought-demanding situations. This first chapter is about "tools and rules" that you will use in the thinking process.

Figure 1.2
What is your concept of a chair? Are all of these pieces of furniture chairs? Most people have concepts, or ideas of what things in general should be, that are loosely defined. The concept of a chair is one example of a loosely defined concept.

Objects and Properties

Physical science is concerned with making sense out of the physical environment. The early stages of this "search for sense" usually involve **objects** in the environment, things that can be seen or touched. These could be objects you see every day, such as a glass of water, a moving automobile, or a blowing flag. They could be quite large, such as the sun, the moon, or even the Solar System, or invisible to the unaided human eye. Objects can be any size, but people are usually concerned with objects that are larger than a pinhead and smaller than a house. Outside these limits, the actual size of an object is difficult for most people to comprehend.

As you were growing up, you learned to form a generalized mental image of objects called a **concept.** Your concept of an object is an idea of what it is, in general, or what it should be according to your idea (figure 1.2). You usually have a word stored away in your mind that represents a concept. The word "chair," for example, probably evokes an idea of "something to sit on." Your generalized mental image for the concept that goes with the word "chair" probably includes a four-legged object with a backrest. Upon close inspection, most of your (and everyone else's) concepts are found to be somewhat vague. For example, if the word "chair" brings forth a mental image of something with four legs and a backrest (the concept), what is the difference between a "high chair" and a "bar stool"? When is a chair a chair and not a stool? Thinking about this question is troublesome for most people.

Not all of your concepts are about material objects. You also have concepts about intangibles such as time, motion, and relationships between events. As was the case with concepts of material objects, words represent the existence of intangible concepts. For example, the words "second," "hour," "day," and "month" represent concepts of time. A concept of the pushes and pulls that come with changes of motion during an airplane

flight might be represented with such words as "accelerate" and "falling." Intangible concepts might seem to be more abstract since they do not represent material objects.

By the time you reach adulthood you have literally thousands of words to represent thousands of concepts. But most, you would find on inspection, are somewhat ambiguous and not at all clear-cut. That is why you find it necessary to talk about certain concepts for a minute or two to see if the other person has the same "concept" for words as you do. That is why when one person says, "Boy, was it hot!" the other person may respond, "How hot was it?" The meaning of "hot" can be quite different for two people, especially if one is from Arizona and the other from Alaska!

The problem with words, concepts, and mental images can be illustrated by imagining a situation involving you and another person. Suppose that you have found a rock that you believe would make a great bookend. Suppose further that you are talking to the other person on the telephone and you wish to discuss the suitability of the rock as a bookend. But you do not know the name of the rock. If you knew the name, you would simply state that you found a "_____." Then you would probably discuss the rock for a minute or so to see if the other person really understood what you were talking about. But not knowing the name of the rock, and wishing to communicate about the suitability of the object to serve as a bookend, what would you do? You would probably describe the characteristics, or **properties,** of the rock. Properties are the qualities or attributes that, taken together, are usually peculiar to an object. Since you commonly determine properties with your senses (smell, sight,

Figure 1.3

Could you describe this rock to another person over the telephone so that person knows *exactly* what you see? This is not probable with everyday language, which is full of implied comparisons, assumptions, and imprecise descriptions.

can look rough and feel smooth. Thus, here is another assumption, and probably all of the properties lead to implied comparisons, assumptions, and a not very accurate communication. This is the nature of your everyday language and the nature of most attempts at communication.

ACTIVITIES

1. Find out how people communicate about the properties of objects. Ask several friends to describe a paper clip while their hands are behind their backs. Perhaps they can do better describing a goatee-type beard? Try to make a sketch that represents each description.
2. Ask two classmates to sit back to back. Give one of them a sketch or photograph that shows an object in some detail, perhaps a guitar or airplane. This person is to describe the properties of the object *without naming it.* The other person is to make a scaled sketch from the description. Compare the sketch to the description, then see how the use of measurement would improve the communication.

Quantifying Properties

Typical day-to-day communications are often vague and leave much to be assumed. A communication between two people, for example, could involve one person describing some person, object, or event to a second person. The description is made by using referents and comparisons that the second person may or may not have in mind. Thus such attributes as "long" fingernails or "short" hair may have entirely different meanings to different people involved in a conversation. Assumptions and vagueness can be avoided by using **measurement** in a description. Measurement is a process of comparing a property to a well-defined and agreed-upon referent. The well-defined and agreed-upon referent is used as a standard called a **unit.** The measurement process involves three steps: (1) *comparing* the referent unit to the property being described, (2) following a *procedure,* or operation, which specifies how the comparison is made, and (3) *counting* how many standard units describe the property being considered.

As an example of how the measurement process works, consider the property of *length.* Most people are familiar with the concept of the length of something ("long" or "short"), the use of length to describe distances ("close" or "far"), and the use of length to describe heights ("tall" or "short"). The referent units used for measuring length are the familiar inch, foot, and mile from the English system and the centimeter, meter, and kilometer of the metric system. These systems and specific units will be discussed later. For now, imagine that these units do not exist but that you need to measure the length and width of this book. This imaginary exercise will illustrate how the measurement process eliminates vagueness and assumption in communication.

The first requirement in the measurement process is to choose some referent unit of length. You could arbitrarily choose something that is handy, such as the length of a standard paper

hearing, touch, and taste), you could say that the properties of an object are the effect the object has on your senses. For example, you might say that the rock is a "big, yellow, smooth rock with shiny gold cubes." But consider the mental image that is formed in the mind of the other person on the telephone when these properties are described. It is entirely possible that the other person is thinking of something very different from what you are describing (figure 1.3)!

As you can see, the example of describing a proposed bookend by listing its properties in everyday language leaves much to be desired. The description does not really help the other person form an accurate mental image of the rock. One problem with the attempted communication is that the description of any property implies some kind of **referent.** Referent means that you *refer to,* or think of, a given property in terms of another, more familiar object. Colors, for example, are sometimes stated with a referent. Examples are "sky blue," "grass green," or "lemon yellow." The referents for the colors blue, green, and yellow are, respectively, the sky, living grass, and a ripe lemon.

Referents for properties are not always explicit as with colors, but a comparison is always implied. Since the comparison is implied, it often goes unspoken and leads to assumptions in comunications. For example, when you stated that the rock was "big," you assumed that the other person knew that you did not mean as big as a house or even as big as a bicycle. You assumed that the other person knew that you meant that the rock was about as large as a book, perhaps a bit larger.

Another problem with the listed properties of the rock is the use of the word "smooth." The other person would not know if you meant that the rock *looked* smooth or *felt* smooth. After all, some objects can look smooth and feel rough. Other objects

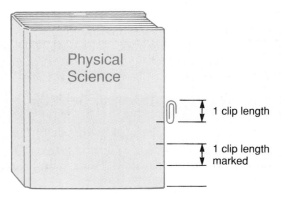

Figure 1.4

As an example of the measurement process a standard paper clip length is selected as a referent unit. The unit is compared to the property that is being described. In this example, the property of the book length is measured by counting how many clip lengths describe the length.

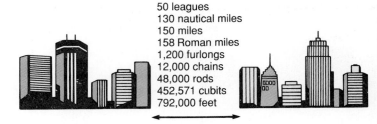

50 leagues
130 nautical miles
150 miles
158 Roman miles
1,200 furlongs
12,000 chains
48,000 rods
452,571 cubits
792,000 feet

Figure 1.5

All of these units and values have been used at some time or another to describe the same distance between two towns. Any unit could be used for this purpose, but when one particular unit is officially adopted, the unit becomes known as the standard unit.

clip, and you could call this length a "clip." Now you must decide on the procedure, which specifies how you will use the clip unit. You could define some specific procedures. For example:

1. Place a clip parallel to and on the long edge, or length, of the book so the end of the referent clip is lined up with the bottom edge of the book. Make a small pencil mark at the other end of the clip as shown in figure 1.4.
2. Move the outside end of the clip to the mark and make a second mark at the other end. Continue doing this until you reach the top edge of the book.
3. Compare how may clip replications are in the book length by counting.
4. Record the length measurements by writing (a) how many clip replications were made and (b) the name of the clip length.

If the book length did not measure to a whole number of clips, you might need to divide the clip length into smaller subunits to be more precise. You could develop a *scale* of the basic clip unit and subunits. In fact, you could use multiples of the basic clip unit for an extended scale, using the scale for measurement rather than moving an individual clip unit. You could call the scale a "clipstick" (as in yardstick or meterstick).

The measurement process thus uses a defined referent unit, which is compared to a property being measured. The *value* of the property is determined by counting the number of referent units. The name of the unit implies the procedure that results in the number. A measurement statement always contains a *number* and *name* for the referent unit. The number answers the question of "how much?" and the name answers the question "of what?" Thus a measurement always tells you "how much of what." You will find that using measurements will sharpen your communications. You will also find that using measurements is one of the first steps in understanding your physical environment.

Measurement Systems

Measurement is a process that brings precision to a description by specifying the "how much" and "of what" of a property in a particular situation. A number expresses the value of the property and the name of a unit tells you what the referent is as well as implying the procedure for obtaining the number. Referent units must be defined and established, however, if others are to understand and reproduce a measurement. It would be meaningless, for example, for you to talk about a length in "clips" if other people did not know what you meant by a "clip" unit. When standards are established the referent unit is called a **standard unit** (figure 1.5). The use of standard units makes it possible to communicate and duplicate measurements. Standard units are usually defined and established by governments and their agencies that are created for that purpose. In the United States, the agency concerned with measurement standards is appropriately named the National Bureau of Standards.

There are two major *systems* of standard units in use today, the English system and the metric system. The metric system is used throughout the world except in the United States, where both systems are in use. The continued use of the English system in the United States presents problems in international trade, so there is pressure for a complete conversion to the metric system. More and more metric units are being used in everyday measurements, but a complete conversion will involve an enormous cost. Both the metric system and the English system will be used in this text. You can therefore use units that are more familiar to you as you learn to use the metric system. Appendix A contains a simple method to convert from one system to the other easily. Consult this section if you need to convert from one metric unit to another metric unit or to convert from English to metric units or vice versa.

People have used referents to communicate about properties of things throughout human history. The ancient Greek civilization, for example, used units of *stadia* to communicate about distances and elevations. The "stadium" was a referent unit based on the length of the race track at the local stadium ("stadia" is the plural of stadium). Later civilizations, such as the ancient Romans, adopted the stadia and other referent units from the ancient Greeks. Some of these very same referent units were later adopted by the early English civilization, which eventually led to the **English system** of measurement. Some adopted units of the English system were originally based on parts of the human body, presumably because you always had these referents with you (figure 1.6). The inch, for example, used the end joint of the thumb for a referent. A foot, naturally, was the length

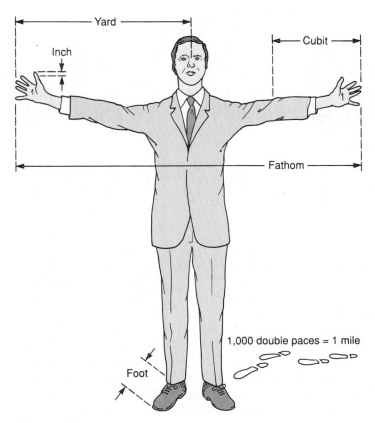

Figure 1.6

Many early units for measurement were originally based on the human body. Some of the units were later standardized by governments to become the basis of the English system of measurement.

Table 1.1

Early conversion table for English units of volume

2 mouthfuls	= 1 jigger	2 quarts	= 1 pottle
2 jiggers	= 1 jack	2 pottles	= 1 gallon
2 jacks	= 1 jill	2 gallons	= 1 pail
2 jills	= 1 cup	2 pails	= 1 peck
2 cups	= 1 pint	2 pecks	= 1 bushel
2 pints	= 1 quart		

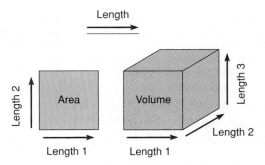

Figure 1.7

Area, or the extent of a surface, can be described by two length measurements. Volume, or the space that an object occupies, can be described by three length measurements. Length, however, can be described only in terms of how it is measured, so it is called a fundamental property.

of a foot, and a yard was the distance from the tip of the nose to the end of the fingers on an arm held straight out. A cubit was the distance from the end of an elbow to the fingertip, and a fathom was the distance between the fingertips of two arms held straight out. As you can imagine there were problems with these early units because everyone was not the same size. Beginning in the 1300s the size of the units were gradually standardized by various English kings (table 1.1). In 1879, the United States, along with sixteen other countries, signed the *Treaty of the Meter,* defining the English units in terms of the metric system. The United States thus became officially metric but not entirely metric in everyday practice.

The **metric system** was established by the French Academy of Sciences in 1791. The academy created a measurement system that was based on invariable referents in nature, not human body parts. These referents have been redefined over time to make the standard units more reproducible. In 1960, six standard metric units were established by international agreement. The *International System of Units,* abbreviated *SI,* is a modernized version of the metric system. Today, the SI system has seven units that define standards for the properties of length, mass, time, electric current, temperature, amount of substance, and light intensity. The standard units for the properties of

length, mass, and time are introduced in this chapter. The remaining units will be introduced in later chapters as the properties the units measure are discussed.

Standard Units for the Metric System

If you consider all the properties of all the objects and events in your surroundings, the number seems overwhelming. Yet, close inspection of how properties are measured reveals that some properties are combinations of other properties (figure 1.7). Volume, for example, is described by the three length measurements of length, width, and height. Area, on the other hand, is described by just the two length measurements of length and width. Length, however, cannot be defined in simpler terms of any other property. There are four properties that cannot be described in simpler terms, and all other properties are combinations of these four. For this reason they are called the **fundamental properties.** A fundamental property cannot be defined in simpler terms other than to describe how it is measured. These four fundamental properties are (1) *length,* (2) *mass,* (3) *time,* and (4) *charge.* Used individually or in combinations, these four properties will describe or measure what you observe in nature. Metric units for measuring the fundamental properties of length, mass, and time will be described next. The fourth fundamental property, charge, is associated with electricity, and a unit for this property will be discussed in a future chapter.

Length

The standard unit for length in the metric system is the **meter** (the symbol, or abbreviation, is m). The meter was originally defined, back in 1793, as one ten-millionth of the distance between the geographic North Pole and the equator of the earth. In order to make this standard accessible the length of a meter was determined and a one-meter metal bar was made as a prototype. This prototype was used to make copies for the countries of the world. The United States received its prototype meter in 1890. Beginning in 1893, the yard was legally defined in terms of the meter. Metal bars, however, tend to expand and contract with changes in temperature, so every precise measurement with the bar required a correction for the temperature. In 1960 the definition of a meter was changed to one that used the wavelength of a certain color of light given off from a particular element. In 1983 the definition was again changed, this time in terms of the distance that light travels in a vacuum during a certain time period, 1/299,792,458 second. The important thing to remember, however, is that the meter is the metric *standard unit* for length. A meter is slightly longer than a yard, 39.3 inches. It is approximately the distance from your left shoulder to the tip of your right hand when your arm is held straight out. Many door knobs are about one meter above the floor. Think about these distances when you are trying to visualize a meter length.

Mass

The standard unit for mass in the metric system is the **kilogram** (kg). The kilogram is defined as the mass of a certain metal cylinder kept by the International Bureau of Weights and Measures in France. This is the only standard unit that is still defined in terms of an object. The property of mass is sometimes confused with the property of weight since they are directly proportional to each other at a given location on the surface of the earth. They are, however, two completely different properties and are measured with different units. All objects tend to maintain their state of rest or straight-line motion, and this property is called inertia. The *mass* of an object is a measure of the inertia of an object. The *weight* of the object is a measure of the force of gravity on it. This distinction between weight and mass will be discussed in detail in chapter 3. For now, remember that weight and mass are not the same property.

Time

The standard unit for time is the **second** (sec). (The second is usually represented by the symbol "s," but "sec" will be used to avoid confusion with plurals.) The second was originally defined as 1/86,400 of a solar day (1/60 × 1/60 × 1/24). The earth's spin was found not to be as constant as thought, so the second was redefined in 1967 to be the duration required for a certain number of vibrations of a certain cesium atom. A special spectrometer called an atomic clock measures these vibrations and keeps time with an accuracy of several millionths of a second per year.

Table 1.2

Some metric prefixes

Prefix	Symbol	Meaning
Giga-	G	1,000,000,000. times the unit
Mega-	M	1,000,000. times the unit
Kilo-	k	1,000. times the unit
Hecto-	h	100. times the unit
Deka-	da	1.0 times the unit
Deci-	d	0.1 of the unit
Centi-	c	0.01 of the unit
Milli-	m	0.001 of the unit
Micro-	μ	0.000001 of the unit
Nano-	n	0.000000001 of the unit

Metric Prefixes

The metric system uses prefixes to represent larger or smaller amounts by factors of ten. Some of the more commonly used prefixes, their abbreviations, and their meanings are listed in table 1.2. Figure 1.8 illustrates how these prefixes are used. Suppose you wish to measure something smaller than the standard unit of length, the meter. The meter is subdivided into ten equal-sized subunits called *decimeters*. The prefix "deci-" has a meaning of "one tenth of," and it takes ten decimeters to equal the length of one meter. For even smaller measurements, each decimeter is divided into ten equal-sized subunits call *centimeters*. It takes ten centimeters to equal one decimeter and one hundred to equal one meter. In a similar fashion, each prefix up or down the metric ladder represents a simple increase or decrease by a factor of ten.

When the metric system was established in 1791, the standard unit of mass was defined in terms of the mass of a certain volume of water. A cubic decimeter (dm^3) of pure water at 4° C was *defined* to have a mass of one kilogram (kg). This definition was convenient because it created a relationship between length, mass, and volume. As illustrated in figure 1.9, a cubic decimeter is 10 cm on each side. The volume of this cube is therefore 10 cm × 10 cm × 10 cm, or 1,000 cubic centimeters (abbreviated as cc or cm^3). So a volume of 1,000 cm^3 of water has a mass of 1 kg. Since 1 kg is 1,000 g, 1 cm^3 of water has a mass of 1 g.

The volume of 1,000 cm^3 also defines a metric unit that is commonly used to measure liquid volume, the **liter** (L). For smaller amounts of liquid volume the milliliter (mL) is used. The relationship between liquid volume, volume, and mass of water is therefore

$$1.0 \text{ L} = 1.0 \text{ dm}^3 \text{ and has a mass of } 1.0 \text{ kg}$$

or, for smaller amounts,

$$1.0 \text{ mL} = 1.0 \text{ cm}^3 \text{ and has a mass of } 1.0 \text{ g}$$

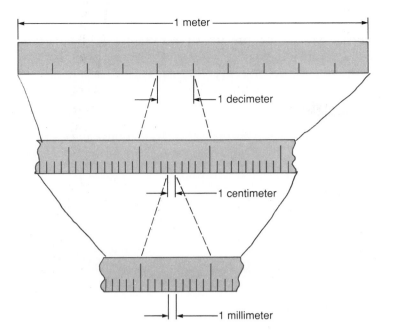

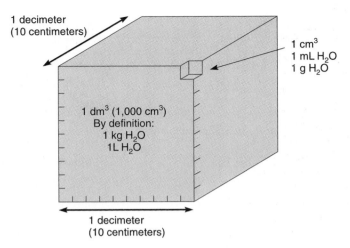

Figure 1.9

A cubic decimeter of water (1,000 cm³) has a liquid volume of 1 L (1,000 mL) and a mass of 1 kg (1,000 g). Therefore, 1 cm³ of water has a liquid volume of 1 mL and a mass of 1 g.

Figure 1.8

Prefixes are used with the standard units of the metric system to represent larger or smaller amounts by factors of ten. Measurements somewhat smaller than the standard unit of the meter, for example, are measured in decimeters. The prefix "deci-" means "one tenth of," and it takes ten decimeters to equal the length of one meter. For even smaller measurements, the decimeter is divided into ten centimeters. Continuing to even smaller measurements, the centimeter is divided into ten millimeters. There are many prefixes that can be used (table 1.2), but all are related by multiples of ten.

Understandings from Measurements

One of the more basic uses of measurement is to *describe* something in an exact way that everyone can understand. For example, if a friend in another city tells you that the weather has been "warm," you might not understand what temperature is being described. A statement that the air temperature is 70° F carries more exact information than a statement about "warm weather." The statement that the air temperature is 70° F contains two important concepts, (1) the numerical value of 70 and (2) the referent unit of degrees Fahrenheit. Note that both a numerical value and a unit are necessary to communicate a measurement correctly. Thus weather reports describe weather conditions with numerically specified units, for example, 70° Fahrenheit for air temperature, 5 miles per hour for wind speed, and 0.5 inches for rainfall (figure 1.10). When such numerically specified units are used in a description, or a weather report, everyone understands *exactly* the condition being described.

Data

Measurement information used to describe something is called **data.** Data can be used to describe objects, conditions, events, or changes that might be occurring. You really do not know if the weather is changing much from year to year until you compare the yearly weather data. The data will tell you, for example, if the weather is becoming hotter or dryer or is staying about the same from year to year.

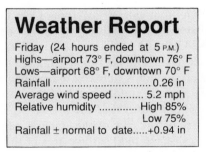

Weather Report

Friday (24 hours ended at 5 P.M.)
Highs—airport 73° F, downtown 76° F
Lows—airport 68° F, downtown 70° F
Rainfall .... 0.26 in
Average wind speed .... 5.2 mph
Relative humidity .... High 85%
                                       Low 75%
Rainfall ± normal to date.....+0.94 in

Figure 1.10

A weather report gives exact information, data that describes the weather by reporting numerically specified units for each condition being described.

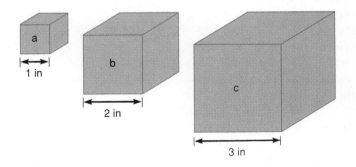

Figure 1.11

Cube a is 1 in on each side, cube b is 2 in on each side, and cube c is 3 in on each side. These three cubes can be described and compared with data, or measurement information, but some form of analysis is needed to find patterns or meaning in the data.

Let's see how data can be used to describe something and how the data can be analyzed for further understanding. The cube illustrated in figure 1.11 will serve as an example. Each cube can be described by measuring the properties of size and surface area.

7

First, consider the size of each cube. Size can be described by **volume,** which means *how much space something occupies.* The volume of a cube can be obtained by measuring and multiplying the length, width, and height. The smallest cube in figure 1.11, cube a, is 1 in on each side, so the volume of this cube is 1 in $\times$ 1 in $\times$ 1 in, or 1 in³. Note that both the numbers and the units were treated mathematically and $1 \times 1 \times 1 = 1$ and in $\times$ in $\times$ in $=$ in³. The cubic unit is thus a result of multiplying units, and a "cubic inch" does not necessarily mean that the volume is in the shape of a cube. The middle-sized cube in figure 1.11, cube b, is 2 in on each side so the volume of this cube is 2 in $\times$ 2 in $\times$ 2 in, or 8 in³. The largest cube, cube c, is 3 in on each side so the volume of this cube is 27 in³. The data so far is

volume of cube a:	1 in³
volume of cube b:	8 in³
volume of cube c:	27 in³

Now consider the surface area of each cube. **Area** means *the extent of a surface* and each cube has six surfaces, or faces (top, bottom, and four sides). The area of any face can be obtained by measuring and multiplying length and width. The smallest cube in figure 1.11, cube a, is 1 in on each side, so the area of one face of this cube is 1 in $\times$ 1 in, or 1 in². Again, note that the numbers and units were treated mathematically and in $\times$ in $=$ in². The square unit is a result of the multiplication and does not necessarily mean that the area is in the shape of a square. The area of one face of the smallest cube is 1 in² and there are 6 faces, so the total surface area of the smallest cube is 6 $\times$ 1 in², or 6 in². The middle-sized cube, cube b, is 2 in on each edge so the area of one face on this cube is 2 in $\times$ 2 in $=$ 4 in². The total surface area is 6 $\times$ 4 in², or 24 in². The largest cube, cube c, is 3 in on each edge, so each face has an area of 9 in² and the total surface area is 6 $\times$ 9 in² or 54 in². The data for the three cubes thus describes them as follows:

	Volume	Surface Area
cube a	1 in³	6 in²
cube b	8 in³	24 in²
cube c	27 in³	54 in²

Ratios and Generalizations

Data on the volume and surface area of the three cubes in figure 1.11 describes the cubes, but if it says anything about a relationship between the volume and surface area of a cube is difficult to say. Nature seems to have a tendency to camouflage relationships, making it difficult to extract meaning from raw data. Seeing through the camouflage requires the use of mathematical techniques to expose patterns. You have spent your time in mathematics classes "doing" mathematics, but here is a chance to *use* it in the real world to understand something. The key is to reduce the data to something manageable that permits you to make comparisons. Such a use of mathematics as a vehicle to understand your surroundings can be very exciting and satisfying. Let's see how such operations can be applied to the data on the three cubes and what the pattern means.

One mathematical technique for reducing data to a more manageable form is to expose patterns through a **ratio.** A ratio is a relationship between two numbers. You could think of a ratio as a *rate* obtained when one number is divided by another number. Suppose, for example, that an instructor has 50 sheets of graph paper for a laboratory group of 25 students. The relationship, or ratio, between the number of sheets and the number of students is 50 papers to 25 students and this can be written as 50 papers/25 students. This ratio is *simplified* by dividing 25 into 50 and the ratio becomes 2 papers/1 student. The 1 is usually understood (not stated) and the ratio is written as simply 2 papers/student. It is read as 2 papers "for each" student or 2 papers "per" student. The concept of simplifying with a ratio is an important one and you will see it time and time again throughout science. It is important that you understand the meaning of "per" and "for each" when used with numbers and units.

Applying the ratio concept to the three cubes in figure 1.11, the ratio of surface area to volume for the smallest cube, cube a, is 6 in² to 1 in³, or

$$\frac{6 \text{ in}^2}{1 \text{ in}^3} = 6 \frac{\text{in}^2}{\text{in}^3}$$

meaning there are 6 square inches of area *for each* cubic inch of volume.

The middle-sized cube, cube b, had a surface area of 24 in² and a volume of 8 in³. The ratio of surface area to volume for this cube is therefore

$$\frac{24 \text{ in}^2}{8 \text{ in}^3} = 3 \frac{\text{in}^2}{\text{in}^3}$$

meaning there are 3 square inches of area *for each* cubic inch of volume.

The largest cube, cube c, had a surface area of 54 in² and a volume of 27 in³. The ratio is

$$\frac{54 \text{ in}^2}{27 \text{ in}^3} = 2 \frac{\text{in}^2}{\text{in}^3}$$

or 2 square inches of area *for each* cubic inch of volume. Summarizing the ratio of surface area to volume for all three cubes,

small cube a—6:1

middle cube b—3:1

large cube c—2:1

Now that you have simplified the data through ratios, you are ready to generalize about what the information means. You can generalize that the surface area to volume ratio of a cube *decreases* as the volume of a cube becomes larger. Reasoning from this generalization will provide an explanation for a number of related observations. For example, why does crushed ice melt faster than a single large block of ice with the same volume? The explanation is that the crushed ice has a larger surface area to volume ratio than the large block so more surface is exposed to warm air. If the generalization is found to be true for shapes other than cubes, you could explain why a log chopped into small chunks burns faster than the whole log. Further generalizing

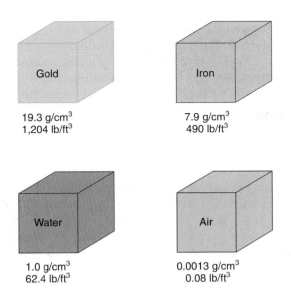

Gold
19.3 g/cm³
1,204 lb/ft³

Iron
7.9 g/cm³
490 lb/ft³

Water
1.0 g/cm³
62.4 lb/ft³

Air
0.0013 g/cm³
0.08 lb/ft³

Figure 1.12
Equal volumes of different substances do not have the same mass. The ratio of mass to volume is defined as a property called mass density, which is identified with the Greek symbol ρ. The mass density of these substances is given in g/cm³. The weight density (D) is given in lb/ft³.

Table 1.3
Densities of some common substances

Substance	Mass Density	Weight Density
	(g/cm³)	(lb/ft³)
Aluminum	2.70	169
Copper	8.96	555
Iron	7.87	490
Lead	11.40	705
Water	1.00	62.4
Seawater	1.03	64
Mercury	13.60	850
Gasoline	0.68	42

might enable you to predict if 10 lb of large potatoes would require more or less peeling than 10 lb of small potatoes. When generalized explanations result in predictions that can be verified by experience, you gain confidence in the explanation. Finding patterns of relationships is a satisfying intellectual adventure that leads to understanding and generalizations that are frequently practical.

A Ratio Called Density

The power of using a ratio to simplify things, making explanations more accessible, is evident when you compare the simplified ratio 6 to 3 to 2 with the hodgepodge of numbers that you would have to consider without using ratios. The power of using the ratio technique is also evident when considering other properties of matter. Volume is a property that is sometimes confused with mass. Larger objects do not necessarily contain more matter than smaller objects. A large balloon, for example, is much larger than this book but the book is much more massive than the balloon. The simplified way of comparing the mass of a particular volume is to find the ratio of mass to volume. This ratio is called mass **density,** which is defined as *mass per unit volume.* The "per" means "for each" as previously discussed and "unit" means one, or each. Thus "mass per unit volume" literally means the "mass of one volume" (figure 1.12). The relationship can be written as

$$\text{mass density} = \frac{\text{mass}}{\text{volume}}$$

or

$$\rho = \frac{m}{V} \qquad \textbf{equation 1.1}$$

As with other ratios, density is obtained by dividing one number and unit by another number and unit. Thus the density of an object with a volume of 17 cm³ and a mass of 16 g is

$$\text{density} = \frac{16 \text{ g}}{17 \text{ cm}^3} = 0.94 \frac{\text{g}}{\text{cm}^3}$$

The density in this example is the ratio of 16 g to 17 cm³, or 16g/17cm³, or 0.94 g to 1 cm³. Thus the density of the example object is the mass of *one* volume (a unit volume), or 0.94 g *for each* cm³.

Any unit of mass and any unit of volume may be used to express density. The densities of solids, liquids, and gases are usually expressed in grams per cubic centimeter (g/cm³), but the densities of liquids are sometimes expressed in grams per milliliter (g/mL). When density is expressed in terms of mass it is known as *mass density* and is given the symbol ρ (which is the Greek symbol for the letter "rho"). Density expressed in terms of weight is known as *weight density* and is given the symbol D (table 1.3). The units for weight and mass will be discussed fully in chapter 3. The "weight per unit volume" relationship can be written as

$$\text{weight density} = \frac{\text{weight}}{\text{volume}}$$

or

$$D = \frac{W}{V} \qquad \textbf{equation 1.2}$$

If matter is distributed the same throughout a volume, the *ratio* of mass to volume will remain the same no matter what mass and volume are being measured. Thus the density of a teaspoonful, a cup, or a lake full of fresh water at the same temperature will all have a density of about 1 g/cm³ or 1 kg/L.

Example 1.1

▼

Two blocks are on a table. Block A has a volume of 30.0 cm³ and a mass of 81.0 g. Block B has a volume of 50.0 cm³ and a mass of 135 g. Which block has the greatest density? If the two blocks have the same density, what material are they? (See table 1.3.)

Solution

Density is defined as the ratio of the mass of a substance per unit volume. Assuming the mass to be distributed equally throughout each volume, you would assume the density to be the same for both blocks. Specifically, use equation 1.1 to determine the density of each block:

Block A

mass (m) = 81.0 g

volume (V) = 30.0 cm³

density = ?

$$\rho = \frac{m}{V}$$

$$= \frac{81.0 \text{ g}}{30.0 \text{ cm}^3}$$

$$= \boxed{2.70 \frac{g}{cm^3}}$$

Block B

mass (m) = 135 g

volume (V) = 50.0 cm³

density = ?

$$\rho = \frac{m}{V}$$

$$= \frac{135 \text{ g}}{50.0 \text{ cm}^3}$$

$$= \boxed{2.70 \frac{g}{cm^3}}$$

Example 1.2

A rock with a volume of 4.50 cm³ has a mass of 15.0 g. What is the density of the rock? (Answer: 3.33 g/cm³)

▲

ACTIVITIES

1. What is the mass density of this book? Measure the length, width, and height of this book in cm, then multiply to find the volume in cm³. Use a balance to find the mass of this book in grams. Compute the density of the book by dividing the mass by the volume. Compare the density in g/cm³ with other substances listed in table 1.3.
2. Compare the densities of some common liquids. Pour a cup of vinegar in a large bottle. Carefully add a cup of corn syrup, then a cup of cooking oil. Drop a coin, tightly folded pieces of aluminum foil, and toothpicks into the bottle. Explain what you observe in terms of density.

Symbols and Equations

In the previous section, the relationship of density, mass, and volume was written with symbols. Mass density was represented by ρ, the lowercase letter rho in the Greek alphabet, mass was represented by m, and volume by V. The use of such symbols is established and accepted by convention, and these symbols are like the vocabulary of a foreign language. You learn what the symbols mean by use and practice with the understanding that *each symbol stands for a very specific property or concept*. The symbols actually represent **quantities,** or *measured properties*. The symbol m thus represents a quantity of mass that is specified by a number and a unit, for example, 16 g. The symbol V represents a quantity of volume that is specified by a number and a unit, such as 17 cm³.

Symbols usually provide a clue about which quantity they represent, such as m for mass and V for volume. However, in some cases two quantities start with the same letter, such as volume and velocity, so the uppercase letter is used for one (V for volume) and the lowercase letter is used for the other (v for velocity). Nonetheless there are more quantities than upper- and lowercase letters, so letters from the Greek alphabet are also used, for example ρ for mass density. Sometimes a subscript is used to identify a quantity in a particular situation, such as v_i for initial, or beginning, velocity and v_f for final velocity. Some symbols are also used to carry messages, for example the Greek letter delta (Δ) is a message that means "the change in" a value. Other message symbols are the symbol $\therefore$, which means "therefore," and the symbol $\propto$, which means "is proportional to."

Symbols are used in an **equation,** a statement that describes a relationship where *the quantities on one side of the equal sign are identical to the quantities on the other side*. Identical refers to both the numbers and the units. Thus in the equation describing the property of density, $\rho = m/V$, the numbers on both sides of the equal sign are identical (e.g., $5 = 10/2$). The units on both sides of the equal sign are also identical (e.g., $g/cm^3 = g/cm^3$).

Equations are used to (1) *describe a property,* (2) *define a concept,* or (3) *describe how quantities change together.* Understanding how equations are used in these three classes is basic to successful problem solving and comprehension of physical science. Each class of uses is considered separately in the following discussion.

Describing a property. You have already learned that the compactness of matter is described by the property called density. Density is a ratio of mass to a unit volume, or $\rho = m/V$. The key to understanding this property is to understand the meaning of a ratio and what "per" or "for each" means. Other examples of properties that will be defined by ratios are how fast something is moving (speed) and how rapidly a speed is changing (acceleration).

Defining a concept. A physical science concept is sometimes defined by specifying a measurement procedure. This is called an *operational definition* because a procedure is established that defines a concept as well as telling you how to measure it. Concepts of what is meant by force, mechanical work, and mechanical power and concepts involved in electrical and magnetic interactions will be defined by measurement procedures.

Describing how quantities change together. Nature is full of situations where one or more quantities change in value, or vary in size, in response to changes in other quantities. Changing quantities are called **variables.** Your weight, for example, is a variable that changes in size in response to changes in another variable, for example, the amount of food you eat. You already know about the pattern, or relationship, between these two variables. With all other factors being equal, an increase in the amount of food you eat results in an increase in your weight. When two variables increase (or decrease) together in the same ratio the variables are said to be in **direct proportion.** When two variables are in direct proportion *an increase or decrease in one variable results in the same relative increase or decrease in a second variable.* Recall that the symbol ∝ means "is proportional to," so the relationship is

amount of food consumed ∝ weight gain

Variables do not always increase or decrease together in direct proportion. Sometimes one variable *increases* while a second variable *decreases* in the same ratio. This is an **inverse proportion** relationship. Other common relationships include one variable increasing in proportion to the *square* or to the *inverse square* of a second variable. Here are the forms of these four different types of proportional relationships:

Direct: a ∝ b

Inverse: a ∝ 1/b

Square: a ∝ b²

Inverse square: a ∝ 1/b²

Proportionality statements describe in general how two variables change together, but a proportionality statement is *not* an equation. For example, consider the last time you filled your gas tank at a service station (figure 1.13). You could say that the amount of time that the gas pump was running was directly proportional to the volume of gasoline in an empty tank you are filling, or

time ∝ volume

This is not an equation because the numbers and units are not identical on both sides. Considering the units, for example, it should be clear that minutes do not equal gallons; they are two different quantities. To make a statement of proportionality into an equation, you need to apply a **proportionality constant,** which is sometimes given the symbol k. For the gas pump example the equation is

(time)(constant) = volume

or

tk = V

In the example, the constant is the flow of gasoline from the pump in gal/min (a ratio). Assume the rate of flow is 10 gal/min. In units, you can see why the statement is now an equality.

$$(min)\left(\frac{gal}{min}\right) = gal$$

$$\frac{min \times gal}{min} = gal$$

$$gal = gal$$

Figure 1.13
The amount of time a gas pump runs is directly proportional to the volume of gasoline you have added to the gas tank. This relationship can be described with an equation by using a proportionality constant.

A proportionality constant in an equation might be a **numerical constant,** a constant that is without units. Such numerical constants are said to be dimensionless, such as 2 or 3. Some of the more important numerical constants have their own symbols, for example, the ratio of the circumference of a circle to its diameter is known as π (pi). The numerical constant of π does not have units because the units cancel when the ratio is simplified by division (figure 1.14). The value of π is usually rounded to 3.14, and an example of using this numerical constant in an equation is the area of a circle equals π times the radius squared (A = πr²).

The flow of gasoline from a pump is an example of a constant that has dimensions (10 gal/min). Of course the value of this constant will vary with other conditions, such as the particular gas pump used and how far the handle on the pump hose is depressed, but it can be considered to be a constant under the same conditions for any experiment.

Students are sometimes apprehensive about equations and problem-solving activities because they appear to be highly mathematical. You *will* need an elementary knowledge of ninth-grade algebra to manipulate equations, but this use of algebra

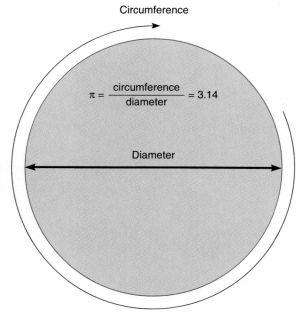

Circumference

$$\pi = \frac{\text{circumference}}{\text{diameter}} = 3.14$$

Diameter

Figure 1.14

The ratio of the circumference of *any* circle to the diameter of that circle is always π, a numerical constant that is usually rounded to 3.14. Pi does not have units because they cancel in the ratio.

will be demonstrated throughout this text in example problems. You should try to work through the example problems first, then refer to the worked solutions. If you need help with the simple operations required, refer to the mathematic review in appendix A. You could consider an equation as a *set of instructions*. The density equation, for example, is $\rho = m/V$. The equation tells you that mass density is a ratio of mass to volume, and you can find the density by dividing the mass by the volume. If you have difficulty, you either do not know the instructions or do not know how to follow the instructions (algebraic rules). The key to success is to figure out what you are having trouble with, then seek help. Appendix D and worked examples in this text can help you with both the instructions and how to follow them. There is also a Feature on problem solving at the end of this chapter.

Example 1.3

Use the equation describing the variables in the gas pump example, $tk = V$, to predict how long it will take you to fill an empty 20 gal tank. Assume $k = 10$ gal/min.

Solution

Until you are comfortable in a problem-solving situation, it is suggested that you follow a formatting procedure that will help you organize your thinking. Here is an example of such a formatting procedure:

Step 1: *List* the quantities involved together with their symbols on the left side of the page, including the unknown quantity with a question mark.

Step 2: Inspect the given quantities and the unknown quantity as listed, and *identify* the equation that expresses a relationship between these quantities. A list of the equations that express the relationships discussed in each chapter is found at the end of that chapter. *Write* the identified equation on the right side of your paper, opposite the list of symbols and quantities.

Step 3: If necessary, *solve* the equation for the variable in question. This step must be done before substituting any numbers or units in the equation. This simplifies things and keeps down the confusion that may otherwise result. If you need help solving an equation, see the section on this topic in appendix A.

Step 4: If necessary, *convert* any unlike units so they are all the same. If the equation involves a time in seconds, for example, and a speed in kilometers per hour, you should convert the km/hr to m/sec. Again, this step should be done at this point to avoid confusion and incorrect operations in a later step. If you need help converting units, see the section on this topic in appendix A.

Step 5: *Substitute* the known quantities in the equation, replacing each symbol with both the number value and units represented by the symbol.

Step 6: *Perform* the required mathematical operations on the numbers and on the units. This performance is less confusing if you first separate the numbers and units, as shown in the following example and in the examples throughout this text, and then perform the mathematical operations on the numbers and units as separate steps, showing all work.

Step 7: *Draw a box* around your answer (numbers and units) to communicate that you have found what you were looking for. The box is a signal that you have finished your work on this problem.

Here is what the solution looks like for the example problem when you follow the seven-step procedure. First, recall that the problem asks for the time (t) needed to pump a volume (V) of 20 gal when the proportionality constant (k) is 10 gal/min. Thus,

Step 1

$V = 20$ gal	$tk = V$ **Step 2**
$k = 10$ gal/min	$\dfrac{t\cancel{k}}{\cancel{k}} = \dfrac{V}{k}$ **Step 3**
$t = ?$	

$$t = \frac{V}{k}$$

$$t = \frac{20 \text{ gal}}{10 \dfrac{\text{gal}}{\text{min}}} \quad \textbf{Step 5}$$

$$= \frac{20 \text{ gal}}{10 \dfrac{\text{gal}}{\text{min}}} \quad \textbf{Step 6}$$

$$= \frac{20 \text{ \cancel{gal}}}{10} \times \frac{\text{min}}{1} \times \frac{\text{min}}{\cancel{gal}}$$

$$= \boxed{2 \text{ min}} \quad \textbf{Step 7}$$

Note that procedure step 4 was not required in this solution. See the Feature at the end of this chapter for more information on and helpful hints about problem solving.

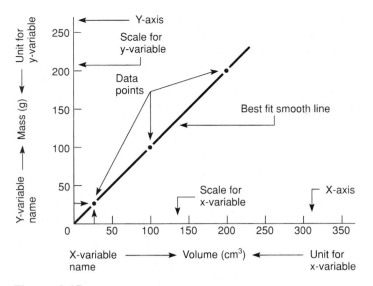

Figure 1.15

The parts of a graph. On this graph, volume is placed on the x-axis and mass on the y-axis.

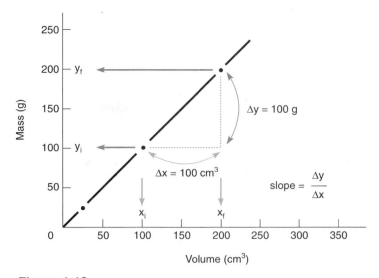

Figure 1.16

The slope is a ratio between the changes in the y-variable and the changes in the x-variable, or $\Delta y/\Delta x$.

The Simple Line Graph

An equation describes a relationship between variables, and a graph helps you "picture" this relationship. A line graph pictures how changes in one variable correspond with changes in a second variable, that is, how the two variables change together. Usually one variable can be easily manipulated. The other variable is caused to change in value by the manipulation of the first variable. The **manipulated variable** is known by various names (*independent, input,* or *cause variable*) and the **responding variable** is known by various related names (*dependent, output,* or *effect variable*). The manipulated variable is usually placed on the horizontal or x-axis of the graph so you could also identify it as the *x-variable.* The responding variable is placed on the vertical or y-axis. This variable is identified as the *y-variable.*

Figure 1.15 shows the mass of different volumes of water at room temperature. Volume is placed on the x-axis because the volume of water is easily manipulated and the mass values change as a consequence of changing the values of volume. Note that both variables are named and that the measuring unit for each is identified on the graph.

Figure 1.15 also shows a number *scale* on each axis that represents changes in the values of each variable. The scales are usually, but not always, linear. A **linear scale** has equal intervals that represent equal increases in the value of the variable. Thus a certain distance on the x-axis to the right represents a certain increase in the value of the x-variable. Likewise, certain distances up the y-axis represent certain increases in the value of the y-variable. The **origin** is the only point where both the x- and y-variables have a value of zero at the same time.

Figure 1.15 shows three **data points.** A data point represents measurements of two related variables that were made at the same time. For example, a volume of 25.0 cm³ of water was found to have a mass of 25.0 g. Locate 25.0 cm³ on the x-axis and imagine a line moving straight up from this point on the scale. Locate 25.0 g on the y-axis and imagine a line moving straight out from this point on the scale. Where the lines meet is the data point for the 25.0 cm³ and 25.0 g measurements. A data point is usually indicated with a small dot or x (dots are used in the graph illustrated in figure 1.15).

A "best fit" smooth line is drawn through all the data points as closely to them as possible. If it is not possible to draw the straight line *through* all the data points, then a straight line is drawn that has the same number of data points on both sides of the line. Such a line will represent a "best approximation" of the relationship between the two variables. The origin is also used as a data point in this example because a volume of zero will have a mass of zero.

The smooth line tells you how the two variables get larger together. A 45° line means that they are increasing in an exact direct proportion. A more flat or more upright line means that one variable is increasing faster than the other. The more you work with graphs, the easier it will become for you to analyze what the "picture" means. There are more exact ways to extract information from a graph and one of these techniques is discussed next.

The Slope of a Straight Line

One way to determine the relationship between two variables that are graphed with a straight line is to calculate the **slope.** The slope is a *ratio* between the changes in one variable and the changes in the other. The ratio is between the change in the value of the x-variable and the change in the value of the y-variable. Recall that the symbol Δ (Greek letter delta) means "change in" so the symbol Δx means the "change in x." The first step in calculating the slope is to find out how much the x-variable is changing (Δx) in relation to how much the y-variable is changing (Δy). You can find this relationship by first drawing a dashed line to the right of the slope (the straight *line,* not the data points) so that the x-variable has increased by some convenient unit (see figure 1.16). Where you start or end

The World Around You

this dashed line will not matter since the ratio between the variables will be the same everywhere on the graph line. The Δx is determined by subtracting the final value of the x-variable on the dashed line (x_f) from the initial value of the x-variable on the dashed line (x_i), or $\Delta x = x_f - x_i$. In figure 1.16, the dashed line has a x_f of 200.0 cm³ and a x_i of 100.0 cm³, so Δx is 200.0 cm³ − 100.0 cm³, or 100.0 cm³. Note that Δx has both a number value and a unit.

Now you need to find Δy. The example in figure 1.16 shows a dashed line drawn back up to the graph line from the x-variable dashed line. The value of Δy is $y_f - y_i$. In the example, $\Delta y = 200.0$ g − 100.0 g, or 100.0 g.

The slope of a straight graph line is the ratio of Δy to Δx, or

$$\text{slope} = \frac{\Delta y}{\Delta x}$$

In the example,

$$\text{slope} = \frac{100.0 \text{ g}}{100.0 \text{ cm}^3}$$

$$= 1.0 \frac{\text{g}}{\text{cm}^3}, \text{ or } 1.0 \text{ g/cm}^3$$

Thus the slope is 1.0 g/cm³ and this tells you how the variables change together. Since g/cm³ is also the definition of density, you know that you have just calculated the density of water from a graph.

Note that the slope can be calculated only for two variables that are increasing together, that is, for variables that are in direct proportion and have a line that moves upward and to the right. If variables change in any other way, mathematical operations must be performed to change the variables *into* this relationship. Examples of such necessary changes include taking the inverse of one variable, squaring one variable, taking the inverse square, and so forth.

The Nature of Science

Most humans are curious, at least when they are young, and are motivated to understand their surroundings. These traits have existed since antiquity and have proven to be a powerful motivation. In recent times the need to find out has motivated the launching of space probes to learn what is "out there," and humans have visited the moon to satisfy their curiosity. Curiosity and the motivation to understand nature were no less in the past than today. Over two thousand years ago the ancient Greeks lacked the tools and technology of today and could only make conjectures about the workings of nature. These early seekers of understanding are known as *natural philosophers,* and they thought and wrote about the workings of all of nature. They are called philosophers because their understandings come from reasoning only, without experimental evidence. Nonetheless, some of their ideas were essentially correct and are still in

use today. For example, the idea of matter being composed of *atoms* was first reasoned by certain ancient Greeks in the fifth century B.C. The idea of *elements,* basic components that make up matter, was developed much earlier but refined by the ancient Greeks in the fourth century B.C. The concept of what the elements are and the concept of the nature of atoms have changed over time, but the idea first came from ancient natural philosophers.

Some historians identify the time of Galileo and Newton, approximately three hundred years ago, as the beginnings of modern science. Like the ancient Greeks, Galileo and Newton were interested in studying all of nature. Since the time of Galileo and Newton the content of physical science has increased in scope and specialization, but the basic means of acquiring understanding, the scientific investigation, has changed little. A *scientific investigation* provides understanding through *experimental evidence* as opposed to the conjectures based on thinking only of the ancient natural philosophers. In the next chapter, for example, you will learn how certain ancient Greeks described how objects fall toward the earth with a thought-out, or reasoned, explanation. Galileo, on the other hand, changed how people thought of falling objects by developing explanations from both creative thinking and precise measurement of physical quantities, providing experimental evidence for his explanations. Experimental evidence provides explanations today, much as it did for Galileo, as relationships are found from precise measurements of physical quantities. Thus scientific knowledge about nature has grown as measurements and investigations have led to understandings that lead to further measurements and investigations.

Investigations, Data, and Explanations

What is a scientific investigation and what methods are used to conduct one? Attempts have been made to describe scientific methods, sometimes in a series of steps, but no single description has ever been satisfactory to all concerned. Scientists do similar things in investigations, but there are different approaches and different ways to evaluate what is found. One thing is certain, however, and that is scientists do not follow a routine of certain steps in an investigation. The approach depends on the individual doing the investigation as well as the particular field of science being studied. One way to understand the nature of this divergent approach to scientific investigations is to consider what all the separate fields have in common, the gathering of facts and data to provide explanations and the development of scientific laws and theories.

Recall that data is measurement information, numbers and units that describe objects and events. Data is a recording that describes *individual* objects and events, not generalizations. The data must be recorded with two criteria in mind, (1) reliability and (2) precision. *Reliability* means that everyone agrees on the meaning of the data and that others could replicate the same measurement if desired. *Precision* means accurate and specific information that differentiates the object or event as specifically as possible from similar objects or events. The more accurate and specific that reliable data is, the more acceptable the data becomes.

Data represents the facts about objects and events, and it is used to develop and test *explanations*. Explanations go by various names, depending on their use and stage of development. In its early stages an explanation is sometimes called a **hypothesis,** a tentative thought-derived or experiment-derived explanation that is compatible with the data and provides a framework for understanding. In general, a hypothesis is tested by looking for the consequences that would be observed if the hypothesis is true. In the physical sciences this test takes place through controlled experiments. If *any* experiment reveals facts that disagree with the hypothesis, the explanation is modified or rejected. Once the physical scientist arrives at an acceptable hypothesis, the process is reversed to see if new data can be found that are consistent with the explanation.

Principles and Laws

When a hypothesis is found to be acceptable through many, many precise experiments, without exception, the explanation might now be known as a *principle,* or *law*. The difference between a principle and a law is usually one of the extent of the phenomena covered by the explanation. There is not a clear division between the two, but in general, a **scientific principle** is an explanation concerned with a specific range of observations. Archimedes's principle, for example, explains the buoyancy of objects floating in liquids or gases. Thus this principle is concerned with buoyancy, a specific phenomenon.

An explanation called a **scientific law** describes a wider range of phenomena than a principle, and the term "law" is generally reserved for explanations of greater importance. A law is often described with an equation and is often identified with the name of a person associated with the formulation of the law. Newton's laws of motion, for example, were formulated some three hundred years ago by Isaac Newton and describe the relationships involved in any object moving at ordinary speeds. The name of each scientific law means something about a relationship between quantities and is usually described by an equation, but it also identifies a most interesting human story about a scientist.

Scientific laws can be expressed or described in different ways. They can be (1) expressed in verbal form as a conceptual statement; (2) summarized by an equation representing the relationship; or (3) described by a graph (figure 1.17). All three ways show that the variables are related in a certain way. Each method of presenting a relationship has its advantages. The conceptual method requires little or no formal introduction to such things as the meaning and manipulation of symbols. The equation method carries with it the precision of mathematics and the opportunity to manipulate the symbols, gaining new insights. The graphical method is very useful in the analysis of data and in finding trends and noticing points of special interest.

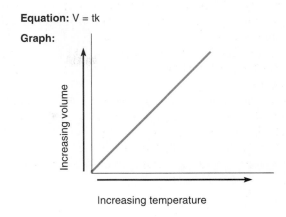

Verbal: If the pressure of a gas is constant, an increase in temperature is directly proportional to an increase in volume.

Equation: $V = tk$

Graph:

Increasing volume (vertical axis)

Increasing temperature (horizontal axis)

Figure 1.17
A relationship between variables can be described at least three different ways: (1) verbally, (2) with an equation, and (3) with a graph. This figure illustrates the three ways of describing the relationship known as Charles's law.

Have you ever heard someone state that something behaved a certain way *because* of a scientific principle or law? For example, a big truck accelerated slowly *because* of Newton's laws of motion. Perhaps this person misunderstands the nature of scientific principles and laws. Scientific principles and laws do not dictate the behavior of objects, they simply describe it. They do not say how things ought to act but rather how things *do* act. A scientific principle or law is *descriptive;* it describes how things act.

Models and Theories

Scientific principles and laws describe how nature behaves. Most people are curious beings, however, and want to know more than just *what* happens. They want to know *why* things happen as they do. Consider, for example, the relationship between the variables of pressure, volume, and the temperature of gases. You may have noticed that a balloon left outside becomes smaller as the air temperature decreases and becomes larger as the air temperature increases. This relationship was first discovered independently by A. C. Charles and Joseph Gay-Lussac. They discovered that an increase in the temperature of a gas results in a directly proportional increase in its volume if the pressure of the gas remains constant. This relationship is sometimes called *Charles's Law*. But why does nature behave this way? Why does this relationship happen?

Answers to "why" questions often concern parts of nature that cannot be directly observed. When something cannot be directly observed, it is sometimes represented mentally or physically by something that is familiar and can be observed. Such a representation is called a **model** (figure 1.18). You might notice,

Figure 1.18

A model helps you to visualize something that cannot be directly observed. Here is a physical model that helps you to visualize changes in the relationship between the pressure, volume, and speed of tiny particles when one of the variables is changed. The tiny red particles are a model for the gas particles that make up the air.

Problem Solving

Students are sometimes apprehensive when assigned problem exercises. This apprehension comes from a lack of experience in problem solving and not knowing how to proceed. This Feature is concerned with a basic approach and procedures that you can follow to simplify problem-solving exercises. Thinking in terms of quantitative ideas is a skill that you can learn. Actually, no mathematics beyond addition, subtraction, multiplication, and division is needed. What is needed is knowledge of certain techniques. If you follow the suggested formatting procedures (see example 1.3) and seek help from the appendix as needed, you will find that problem solving is a simple, fun activity that helps you to learn to think in a new way. Here are some more considerations that will prove to be helpful.

1. Read the problem carefully, perhaps several times, to understand the problem situation. If possible, make a sketch to help you visualize and understand the problem in terms of the "real world."

2. Be alert for information that is not stated directly. For example, if a moving object "comes to a stop" you know that the final velocity is zero even though this was not stated outright. Likewise, questions about "how far?" are usually asking a question about a distance unit, and questions about "how long?" are usually asking a question about a time unit. Such information can be very important in procedure step 1 (example 1.3), the listing of quantities and their symbols. Overlooked or missing quantities and symbols can make it difficult to identify the appropriate equation.

3. Understand the meaning and concepts that an equation represents. An equation represents a relationship that exists between variables. Understanding the relationship helps you to identify the appropriate equation or equations by inspection of the list of known and unknown quantities (procedure step 2). You will find a list of the equations being

considered at the end of each chapter. Information about the meaning and the concepts that an equation represents is found within each chapter.

4. Solve the equation *before* substituting numbers and units for symbols (procedure step 3). A helpful discussion of the mathematical procedures required, with examples, is in appendix A.

5. Note if the quantities are in the same units. A mathematical operation requires the units to be the same; for example, you cannot add nickles, dimes, and quarters until you first convert them all to the same unit of money. Likewise, you cannot correctly solve a problem if one time quantity is in seconds and another time quantity is in hours. The quantities must be converted to the same units before anything else is done (procedure step 4). There is a helpful section on how to use conversion ratios in appendix A.

6. Perform the required mathematical operations on the numbers and the units as if they were two separate problems (procedure step 6). You will find that following this step will facilitate problem-solving activities because the units you obtain will tell you if you have worked the problem correctly. If you just write the units that you *think* should appear in the answer, you have missed this valuable self-check.

7. Be aware that not all learning takes place in a given time frame and that solutions to problems are not necessarily arrived at "by the clock." If you have spent a half an hour or so unsuccessfully trying to solve a particular problem, move on to another problem or do something entirely different for awhile. Problem solving often requires time for "something" to happen in your brain. If you move on to some other activity, you might find that the answer to a problem that you have been stuck on will come to you "out of the blue" when you are not

even thinking about the problem. This behavior of unexpected problem solutions is common to many "real-world" professions and activities that involve thinking.

Example Problem

Mercury is a liquid metal with a mass density of 13.6 g/cm³. What is the mass of 50.0 cm³ of mercury?

Solution

The problem gives two known quantities, the mass density (ρ) of mercury and a known volume (V), and identifies an unknown quantity, the mass (m) of that volume. Make a list of these quantities:

$$\rho = 13.6 \text{ g/cm}^3$$
$$V = 50.0 \text{ cm}^3$$
$$m = ?$$

The appropriate equation for this problem is the relationship between mass density (ρ), mass (m), and volume (V):

$$\rho = \frac{m}{V}$$

But the unknown in this case is the mass, m. Solving the equation for m, by multiplying both sides by V, gives:

$$V\rho = \frac{m\cancel{V}}{\cancel{V}}$$
$$V\rho = m, \text{ or}$$
$$m = V\rho$$

Now you are ready to substitute the known quantities in the equation and perform the mathematical operations on the numbers and on the units:

$$m = \left(13.6\frac{\text{g}}{\text{cm}^3}\right)(50.0 \text{ cm}^3)$$
$$= (13.6)(50.0)\left(\frac{\text{g}}{\text{cm}^3}\right)(\text{cm}^3)$$
$$= 680 \frac{\text{g} \cdot \cancel{\text{cm}^3}}{\cancel{\text{cm}^3}}$$
$$= \boxed{680. \text{ g}}$$

for example, that a container of small bouncing rubber balls acts much like the air in a balloon. When the balls bounce faster they take up more room. The bouncing rubber balls provide a *model* for the relationship between the temperature and volume of air. This model helps you see what happens to invisible particles of air as the temperature and volume of the gas change.

Many different kinds of models are used as analogies when something cannot be directly observed. One kind of model is a *physical model* such as a model of the Solar System. You cannot directly observe the whole Solar System, and a physical model of the system helps you understand, for example, planetary motion.

Another kind of model is a *mental model*. You cannot see what is inside the earth or all of the earth's moving surfaces, and the plate tectonic model helps you understand why earthquakes occur more often in some locations than others.

A third kind of model is an *equation* that describes how some variable changes in response to changes in other variables. You often cannot directly see these changes and the use of an equation is another way of visualizing how nature behaves.

Is a bunch of bouncing rubber balls a good model for the relationship between the temperature and volume of a gas? The answer to this question would come from first identifying assumptions about the similarities and differences between the model and the actual gas, including the assumption that a gas is made up of small, rapidly moving particles. Then experiments would be conducted to determine if the same mathematics described the bouncing rubber ball model and the container of gas. If close agreement is found, it may lead to an explanation in the form of a hypothesis that suggests further experiments about the nature of matter. These experiments may lead to a new hypothesis, more experiments and hypotheses, and eventually a general **theory** is developed about the nature of matter. A theory is a broad, detailed explanation that guides the development of hypotheses and interpretation of experiments in a field of study. The term *scientific theory* is usually reserved for historic schemes of thought with explanatory and predictive values that have survived the test of long, detailed examination. Examples are the kinetic molecular theory, the atomic theory, and theories about light. These theories, with others, form the framework of scientific thought and experimentation.

Theories do not just spring from scientists' minds in a short period of time. Some theories, such as the kinetic molecular theory, take hundreds of years of work by many different people. Nor are theories always successful. Any book on the history of science is replete with examples of discarded theories. Some of these were simply wrong. Others have been replaced by a better, more detailed theory. In any case, even the theories of today are not considered to be the final word. There is always some degree of uncertainty, and the degree of acceptance or rejection as to the adequacy of a given theory varies from person to person. One scientist might have 90 percent acceptance of a certain theory, but another might accept it with a 15 percent confidence level. Generally, the degree of acceptance of a theory is the acceptance of it by a majority of those working in that field.

Scientific theories point to new ideas about the behavior of nature, and these ideas lead to more experiments, more data, and more explanations. All of this may lead to a slight modification of an existing theory, a major modification, or perhaps the creation of an entirely new theory. These activities cycle on in an ongoing attempt to understand nature and our place in it.

Summary

Physical science is a search for order in our physical surroundings. People have *concepts,* or mental images, about material *objects* and intangible *events* in their surroundings. Concepts are used for thinking and communicating. Concepts are based on *properties,* or attributes that describe a thing or event. Every property implies a *referent* that describes the property. Referents are not always explicit and most communications require assumptions. Measurement brings precision to descriptions by using numbers and standard units for referents to communicate "exactly how much of exactly what."

Measurement is a process that uses a well-defined and agreed-upon *referent* to describe a *standard unit.* The unit is compared to the property being defined by an *operation* that determines the *value* of the unit by *counting.* Measurements are always reported with a *number,* or value, and a *name* for the unit.

The two major *systems* of standard units are the *English system* and the *metric system.* The English system uses standard units that were originally based on human body parts, and the metric system uses standard units based on referents found in nature. The metric system also uses a system of prefixes to express larger or smaller amounts of units. The metric standard units for length, mass, and time are the *meter, kilogram,* and *second.*

Measurement information used to describe something is called *data.* One way to extract meanings and generalizations from data is to use a *ratio,* a simplified relationship between two numbers. Density is a ratio of mass to volume, or $\rho = m/v$.

Symbols are used to represent *quantities,* or measured properties. Symbols are used in *equations,* which are shorthand statements that describe a relationship where the quantities (both number values and units) are identical on both sides of the equal sign. Equations are used to (1) *describe* a property, (2) *define* a concept, or (3) *describe* how *quantities change* together.

Quantities that can have different values at different times are called *variables.* Variables that increase or decrease together in the same ratio are said to be in *direct proportion.* If one variable increases while the other decreases in the same ratio the variables are in *inverse proportion.* Proportionality statements are not necessarily equations. A *proportionality constant* can be used to make such a statement into an equation. Proportionality constants might have numerical value only, without units, or they might have both value and units.

A line *graph* helps you "picture" how two variables change together. The relationship can be determined by calculating the *slope,* or a ratio between changes in one variable to changes in the other. The slope can be calculated for variables that are in direct proportion. If they are not, some mathematical operation must be performed to change them to this relationship.

Modern science began about three hundred years ago during the time of Galileo and Newton. Since that time the *scientific investigation* has been used to provide *experimental evidence* about nature. The investigations provide *accurate, specific,* and *reliable* data that is used to develop and test *explanations.* A *hypothesis* is a tentative explanation that is accepted or rejected from experimental data. An accepted hypothesis may result in a *principle,* an explanation concerned with a specific range of phenomena, or a *scientific law,* an explanation concerned with important, wider-ranging phenomena. Laws are sometimes identified with the name of a scientist and can be expressed verbally, with an equation, or with a graph.

A *model* is used to help understand something that cannot be observed directly, explaining the unknown in terms of things already understood. Physical models, mental models, and equations all are examples of models that explain how nature behaves. A *theory* is a broad, detailed explanation that guides development and interpretations of experiments in a field of study.

Summary of Equations

1.1

$$\text{mass density} = \frac{\text{mass}}{\text{volume}}$$

$$\rho = \frac{m}{V}$$

1.2

$$\text{weight density} = \frac{\text{weight}}{\text{volume}}$$

$$D = \frac{W}{V}$$

Key Terms

area (p. 8)
concept (p. 2)
data (p. 7)
data points (p. 13)
density (p. 9)
direct proportion (p. 11)
English system (p. 4)
equation (p. 10)
fundamental properties (p. 5)
hypothesis (p. 15)
inverse proportion (p. 11)
kilogram (p. 6)
linear scale (p. 13)
liter (p. 6)
manipulated variable (p. 13)
measurement (p. 3)
meter (p. 6)
metric system (p. 5)
model (p. 15)

numerical constant (p. 11)
objects (p. 2)
origin (p. 13)
properties (p. 2)
proportionality constant (p. 11)
quantities (p. 10)
ratio (p. 8)
referent (p. 3)
responding variable (p. 13)
scientific law (p. 15)
scientific principle (p. 15)
second (p. 6)
slope (p. 13)
standard unit (p. 4)
theory (p. 18)
unit (p. 3)
variables (p. 10)
volume (p. 8)

1. The process of comparing a property of an object to a well-defined and agreed-upon referent is called
   a. generalizing.
   b. measurement.
   c. graphing.
   d. scientific investigation.

2. As the volume of a cube becomes larger and larger, the surface area to volume ratio
   a. increases.
   b. decreases.
   c. remains the same.
   d. sometimes increases and sometimes decreases.

3. If you consider a very small portion of a material that is the same throughout, the density of the small sample will be
   a. less.
   b. more.
   c. the same as the larger sample.

4. Symbols that are used in equations represent
   a. a message.
   b. a specific property.
   c. quantities, or measured properties.
   d. all of the above.

5. An equation is composed of symbols in such a way that
   a. the numbers and units on both sides are always equal.
   b. the units are equal but the numbers are not because one is unknown.
   c. the numbers are equal but the units are not equal.
   d. neither the numbers nor units are equal because of the unknown.

6. Quantities, or measured properties, that are capable of changing are called
   a. data.
   b. variables.
   c. proportionality constants.
   d. dimensionless constants.

7. A proportional relationship that is represented by the symbols a ∝ 1/b represents which of the following relationships?
   a. direct proportion
   b. inverse proportion
   c. direct square proportion
   d. inverse square proportion

8. A hypothesis that is concerned with a specific phenomena is found to be acceptable through many experiments over a long period of time. This hypothesis becomes known as a
   a. scientific law.
   b. theory.
   c. scientific principle.
   d. model.

9. The symbol ∝ has a meaning of
   a. almost infinity.
   b. the change in.
   c. is proportional to.
   d. therefore.

The World Around You

10. Which of the following symbols represents a measured property of the compactness of matter?
   a. ρ
   b. m
   c. V
   d. Δ

Answers

1. b 2. b 3. c 4. d 5. a 6. b 7. b 8. c 9. c
10. a

Questions for Thought

1. What is a concept?

2. Describe the process of measurement.

3. What two things does a measurement statement always contain? What do the two things tell you?

4. What is the basic difference between a referent, a measurement unit, and a standard unit of measure?

5. Compare the advantages of the English system of measurement with the advantages of the metric system of measurement.

6. Name the fundamental properties. Why are these properties called fundamental?

7. Describe the metric standard units for length, mass, and time.

8. What is data? Why is it sometimes necessary to use mathematical techniques on data?

9. What is a ratio? Of what value is it to calculate a ratio?

10. Does the density of a liquid change with the shape of a container? Explain.

11. What is the mass of 0.75 L of water?

12. Does a flattened pancake of clay have the same density as the same clay rolled into a ball? Explain.

13. What is a quantity? What two things are necessary in reporting a quantity?

14. What is an equation? How are equations used in the physical sciences?

15. What is the basic difference between a proportionality statement and an equation? How can you make a proportionality statement into an equation?

16. How does a line graph "picture" how two variables are changing together? What is the slope of a line graph? How does a slope describe the relationship between two changing variables?

17. What is the difference between the understandings about nature developed by the ancient natural philosophers and the understandings about nature developed by modern scientists?

18. Describe the role of a hypothesis in a scientific investigation.

19. Compare and contrast a scientific principle and a scientific law.

20. What is a model? How are models used?

21. What are the differences between a civil law and a scientific law?

22. Are all theories always completely accepted or completely rejected? Explain.

Exercises

Group A—Solutions Provided in Appendix D

1. What is your height in meters? In centimeters?

2. What is the mass density of mercury if 20.0 cm³ has a mass of 272.0 g?

3. What is the volume of a rock with a mass density of 3.00 g/cm³ and a mass of 1.500 kg?

4. (a) Use a ratio to compare the size of a 1,800.0 cm³ motorcycle engine to a 3.6 L automobile engine. (b) Does this ratio have units? Explain.

5. What is the mass of water in a 40.0 L aquarium?

6. A 2.0 kg pile of aluminum cans is melted, then cooled into a solid cube. What is the volume of the cube?

7. A cubic box contains 1,000 g of water. What is the length of one side of the box in meters? Explain your reasoning.

8. A ratio of the density of a substance to the density of water is called specific gravity. Specific gravity is dimensionless. Explain why this is so.

9. The weight density of gasoline is 42.0 lb/ft³. If 1.00 gal has a volume of 0.134 ft³, how much additional weight is added to a car when the 25.0 gal gas tank is filled from empty?

10. Fresh water in a container weighs 31.2 lb. How much will seawater in the same container weigh? (Consider the weight density of seawater as 64.0 lb/ft³ and fresh water as 62.4 lb/ft³.)

Group B—Solutions Not Given

1. What is your mass in kilograms? In grams?

2. What is the mass density of iron if 5.0 cm³ has a mass of 39.5 g?

3. If ice has a mass density of 0.92 g/cm³, what is the volume of 5.0 kg of ice?

4. Use a ratio to compare the mass of a cm³ of brass ($\rho = 8.7$ g/cm³) to the mass of a cm³ of gold ($\rho = 19.3$ g/cm³).

5. What is the mass of gasoline ($\rho = 0.68$ g/cm³) in a 94.6 L tank?

6. What is the volume of a 2.0 kg pile of iron cans that are melted, then cooled into a solid cube?

7. A cubic tank holds 1,000.0 kg of water. What are the dimensions of the tank in meters? Explain your reasoning.

8. If the specific gravity of alcohol is 0.79 (a) what is the density of the alcohol? and (b) what is the mass of a m³ of the alcohol?

9. The weight density of alcohol is 49.0 lb/ft³ and the weight density of gasoline is 42.0 lb/ft³. (a) Which weighs more, 5.0 gal of alcohol or 5.0 gal of gasoline? (b) How much more? (1 gal has a volume of 0.134 ft³.)

10. A 1.0 ft³ container is half filled with seawater and half filled with fresh water. What is the total weight of the combined liquids?

Section I

The concepts of motion, energy, heat, and fluid temperatures will explain what is happening in this hot air balloon race. These concepts are just a few of the connected and related concepts that you will study in this section.

Physics

This section is an introduction to the behavior of matter, that is, how the world around you moves and changes. This topic of physical science is usually called physics, and one objective of this section is to impart a useful knowledge of the physics laws, principles, and models that describe your physical environment. A second objective, and one that is just as important as the first, is to give you an opportunity to learn some of the thinking and reasoning skills that are usually identified with the subject of physics. Both of these objectives are approached from everyday experiences, designed to help you understand a little of your science-oriented culture.

You will know that you are beginning to reach the objectives of this section when you begin to discover connections between the concepts being studied. Considering the bigger picture, for example, the order of topics in the chapters are motion, energy, heat, sound, electricity, and then light. Motion comes first because the concepts of motion are basic to the other topics. Energy, for example, is best understood through the concepts of motion and position. Heat is understood through the concepts of energy and the motion of molecules. The study of sound is also based on motion, this time on the motion of

mechanical vibrations that move through materials. In addition to motion, some interesting new concepts are needed for understanding electricity and light. These new concepts are presented in a way to take some of the mystery out of these two topics. The section concludes with an introduction to some quantum ideas, concepts that will be useful in later sections of this text.

As you can see, the topics are related and build in a cumulative process. What you learn in the first chapters will be needed in later chapters. It is therefore important to review and consolidate what you have learned from time to time. In this way you will find that the world of physics is not alien, isolated bits of information but rather each concept is connected to another. This is, in fact, the nature of your physical environment since everything is related, and changes in one part will affect another part. Thus the study of physics is a study of the essence of your surroundings. Understanding that nature works according to basic natural laws helps you to find stability and order in what otherwise appears as mystery and chaos. The foundation for all this edifying and stabilizing knowledge is the topic of the first chapter in this section, motion.

Chapter 2

Motion

Figure 2.1

Some objects that move from time to time sit still for a great deal of the time. At the present time, these bicycles are not moving.

IN chapter 1, you learned some "tools and rules" and some techniques for finding order in your physical surroundings. Order is often found in the form of patterns, or relationships between quantities that are expressed as equations. Recall that equations can be used to (1) define properties, (2) define concepts, and (3) describe how quantities change together. In all three uses, patterns are quantified, conceptualized, and used to gain a general understanding about what is happening in nature.

In the study of physical science, certain parts of nature are often considered and studied together for convenience. One of the more obvious groupings involves *movement*. Most objects around you spend a great deal of time sitting quietly without motion. Buildings, rocks, utility poles, and trees rarely, if ever, move from one place to another. Even things that do move from time to time sit still for a great deal of time. This includes you, automobiles, and bicycles (figure 2.1). On the other hand, the sun, moon, and starry heavens seem to always move, never standing still. Why do things stand still? Why do things move?

Questions about motion have captured the attention of people for thousands of years. But the ancient people answered questions about motion with stories of mysticism and spirits that lived in objects. It was during the classic Greek culture, between 600 B.C. and 300 B.C., that people began to look beyond magic and spirits. One particular Greek philosopher, Aristotle, wrote a theory about the universe that offered not only explanations about things such as motion but also offered a sense of beauty, order, and perfection. The theory seemed to fit with other ideas that people had and was held to be correct for nearly two thousand years after it was written. It was not until the work of Galileo and Newton during the 1600s that a new, correct understanding about motion was developed. The development of ideas about motion is an amazing and absorbing story and is the subject of this chapter. You will learn how equations are used to (1) define the properties of motion and (2) describe how quantities of motion change together. These are basic understandings that will be used to define some concepts of motion in the next chapter.

Describing Motion

Motion is one of the more common events in your surroundings. You can see motion in natural events such as clouds moving, rain and snow falling, and streams of water moving in a never-ending cycle. Motion can also be seen in the activities of people who walk, jog, or drive various machines from place to place. Motion in nature and motion in the things people do is so common that you would think everyone would intuitively understand the concepts of motion. But history indicates that it was only during the past three hundred years or so that people began to understand motion correctly. Perhaps the correct concepts are subtle and contrary to common sense, requiring a search for simple, clear concepts in an otherwise complex situation. The process of finding such order in a multitude of sense impressions by taking measurable data, and then inventing a concept to describe what is happening, is the activity called science. We will now apply this process to motion.

What is motion? Consider a ball that you notice one morning in the middle of a lawn. Later, in the afternoon, you notice that the ball is at the edge of the lawn, against a fence, and you wonder if the wind or some person moved the ball. You do not know if the wind blew it at a steady rate, if many gusts of wind moved it, or even if some children kicked it all over the yard. All you know for sure is that the ball has been moved because it is in a different position after some time passed. These are the two important aspects of motion: (1) a change of position and (2) the passage of time.

If you did happen to see the ball rolling across the lawn in the wind, you would see more than the ball at just two locations. You would see the ball moving continuously. You could consider, however, the ball in continuous motion to be a series of individual locations with very small time intervals. Moving involves a change of position during some time period. Motion is the act or process of something changing position.

The motion of an object is usually described with respect to something else that is considered to be not moving. (Such a stationary object is said to be "at rest.") Imagine that you are traveling in an automobile with another person. You know that you are moving across the land outside the car since your location on the highway changes from one moment to another. Observing your fellow passenger, however, reveals no change of position. You are in motion relative to the highway outside the car. You are not in motion relative to your fellow passenger. Your motion, and the motion of any other object or body, is the process of a change in position *relative* to some reference object or location. Thus **motion** can be defined as the act or process of changing position relative to some reference during a period of time.

Measuring Motion

You have learned that objects can be described by measuring certain fundamental properties such as mass and length. Since motion involves (1) a change of position, or *displacement*, and (2) the passage of *time*, the motion of objects can be described by using combinations of the fundamental properties of length and time. These combinations of measurement describe the motion properties of *speed, velocity,* and *acceleration*.

Speed

Suppose you are in a car that is moving over a straight road. How could you describe your motion? Motion was defined as the process of a change of position. You need at least two measurements to describe the motion. These are (1) how much the position was changed, or the *displacement* between the two positions, and (2) how long it took for the change to take place, or the *time* that elapsed while the object moved between the two positions. Displacement and time can be combined as a ratio to define the *rate* at which a distance is covered. This rate is a property of motion called **speed,** a measurement of how fast you are moving. Speed is defined as distance per unit time, or

$$\text{speed} = \frac{\text{distance (how much the position changed)}}{\text{time (elapsed time during change)}}$$

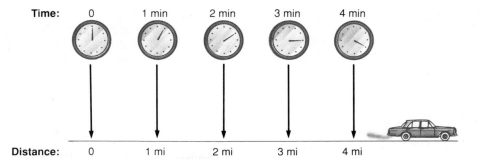

Distance: 0   1 mi   2 mi   3 mi   4 mi

Figure 2.2

This car is moving in a straight line over a distance of 1 mi each min. The speed of the car, therefore, is 60 mi each 60 min, or 60 mi/hr.

Suppose your car is moving over a straight highway and you are covering equal distances in equal periods of time (figure 2.2). If you use a stopwatch to measure the time required to cover the distance between highway mile markers (those little signs with numbers along major highways), the time intervals will all be equal. You might find, for example, that one minute lapses between each mile marker. Such a uniform straight-line motion that covers equal distances in equal periods of time is the simplest kind of motion.

If your car were moving over equal distances in equal periods of time, it would have a *constant speed*. This means that the car is neither speeding up nor slowing down. It is usually difficult to maintain a constant speed. Other cars and distractions such as interesting scenery cause you to reduce your speed. Other times you increase your speed. If you calculate your speed over an entire trip, you are considering a large distance between two places and the total time that elapsed. The increases and decreases in speed would be averaged. Therefore most speed calculations are for an *average speed*. The speed at any specific instant is called the *instantaneous speed*. To calculate the instantaneous speed you would need to consider a very short time interval—one that approaches zero. An easier way would be to use the speedometer, which shows the speed at any instant.

It is easier to study the relationships between quantities if symbols are used instead of writing out the whole word. The letter v can be used to stand for speed when dealing with straight-line motion, which is the only kind of motion that will be considered in the problems in this text. The letter d can be used to stand for distance and the letter t to stand for time. The relationship between average speed, distance, and time is therefore

$$\bar{v} = \frac{d}{t} \qquad \textbf{equation 2.1}$$

This is one of the three types of equations that were discussed earlier and in this case the equation defines a motion property. The bar over the v ($\bar{v}$) is a symbol that means *average* (it is read "v-bar" or "v-average"). You can use this relationship to

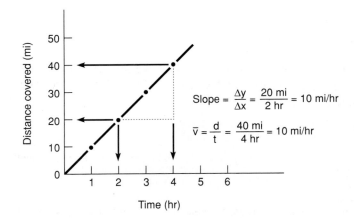

$$\text{Slope} = \frac{\Delta y}{\Delta x} = \frac{20 \text{ mi}}{2 \text{ hr}} = 10 \text{ mi/hr}$$

$$\bar{v} = \frac{d}{t} = \frac{40 \text{ mi}}{4 \text{ hr}} = 10 \text{ mi/hr}$$

Figure 2.3

Speed is defined as a ratio of the displacement, or the distance covered, in straight-line motion for the time elapsed during the change, or $\bar{v}$ = d/t. This ratio is the same as that found by calculating the slope when time is placed on the x-axis. The answer is the same as shown on this graph because both the speed and the slope are ratios of distance per unit of time. Thus you can find a speed by calculating the slope from the straight line, or "picture," of how fast distance changes with time.

find average speed (figure 2.3). For example, suppose a car travels 150 mi in 3 hr. What was the average speed? Since d = 150 mi and t = 3 hr, then

$$\bar{v} = \frac{150 \text{ mi}}{3 \text{ hr}}$$

$$= 50 \frac{\text{mi}}{\text{hr}}$$

As with other equations, you can mathematically solve the equation for any term as long as two variables are known (figure 2.4). For example, suppose you know the speed and the time but want to find the distance traveled. This can be solved by first writing the relationship

$$\bar{v} = \frac{d}{t}$$

then multiplying both sides of the equation by t (to get d on one side by itself),

$$\bar{v} \times t = \frac{d \times \cancel{t}}{\cancel{t}}$$

and the t's on the right cancel, leaving

$$\bar{v}t = d \quad \text{or} \quad d = \bar{v}t$$

If the $\bar{v}$ is 50 mi/hr and the time traveled is 2 hr, then

$$d = \left(50 \frac{mi}{hr}\right)(2 \text{ hr})$$

$$= (50 \times 2)\left(\frac{mi}{hr} \times hr\right)$$

$$= 100 \frac{mi \times \cancel{hr}}{\cancel{hr}}$$

$$= 100 \text{ mi}$$

Notice how both the numerical values and the units were treated mathematically. See the Feature in chapter 1 for more information on problem solving.

Example 2.1

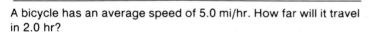

A bicycle has an average speed of 5.0 mi/hr. How far will it travel in 2.0 hr?

Solution
The bicycle has an average speed ($\bar{v}$) of 5.0 mi/hr and the time (t) is 2.0 hr. The problem asked for the distance (d). The relationship between the three variables of $\bar{v}$, t, and d is in equation 2.1, $\bar{v} = d/t$.

$$\bar{v} = 5.0 \frac{mi}{hr} \qquad \bar{v} = \frac{d}{t}$$

$$t = 2.0 \text{ hr} \qquad t\bar{v} = \frac{d \cdot \cancel{t}}{\cancel{t}}$$

$$d = ? \qquad d = \bar{v}t$$

$$d = \left(5.0 \frac{mi}{hr}\right)(2.0 \text{ hr})$$

$$= (5.0 \times 2.0)\left(\frac{mi}{hr} \times hr\right)$$

$$= 10. \frac{mi \cdot \cancel{hr}}{\cancel{hr}}$$

$$= \boxed{10. \text{ mi}}$$

Example 2.2

The driver of a car moving at 72.0 km/hr drops a road map on the floor. It takes him 3.0 sec to locate and pick up the map. How far did he travel during this time? (Answer: 60. m) (Hint: 1 hr = 3,600 sec and 1 km = 1,000 m. See appendix A on unit conversions for help with this type of problem.)

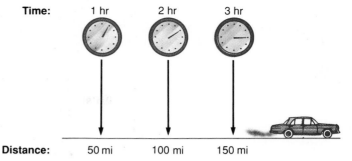

Time: 1 hr   2 hr   3 hr

Distance: 50 mi   100 mi   150 mi

Figure 2.4

If you know the value of any two of the three variables of distance, time, or speed, you can find the third. What is the average speed of this car?

Constant, instantaneous, or average speeds can be measured with any distance and time units. Common units in the English system are miles/hour and feet/second. Metric units for speed are commonly kilometers/hour and meters/second. The ratio of any distance/time is usually read as distance per time such as miles per hour. The per means "for each." Sometimes you will hear or read about the **magnitude** of speed. Magnitude is defined as the numerical value and the unit such as 30 mi/hr. It simply means the size or extent of a quantity. For example, a speed of 60 mi/hr is twice the magnitude of a speed of 30 mi/hr.

Velocity
The word "velocity" is sometimes used interchangeably with the word "speed," but there is a difference. **Velocity** describes the *speed and direction* of a moving object. For example, a speed might be described as 60 mi/hr. A velocity might be described as 60 mi/hr to the west. To produce a change in velocity, either the speed or the direction is changed (or both are changed). A satellite moving with a constant speed in a circular orbit around the earth does not have a constant velocity since its direction of movement is constantly changing. Measurements that have magnitude only, such as speed, are called scalar quantities, or **scalars.** Measurements that have both magnitude and direction, such as velocity, are called vector quantities, or **vectors.** Speed is a scalar quantity and velocity is a vector quantity. Vector quantities can be represented graphically with arrows. The length of the arrows are proportional to the magnitude and the arrowheads indicate the direction (see figure 2.5).

Acceleration
Motion can be changed three different ways: (1) by changing the speed, (2) by changing the direction of travel, or (3) by changing both the speed and direction of travel. Since velocity describes both the speed and the direction of travel, any of these three changes will result in a change of velocity. You need at least one additional measurement to describe a change of motion, which is how much time elapsed while the change was taking

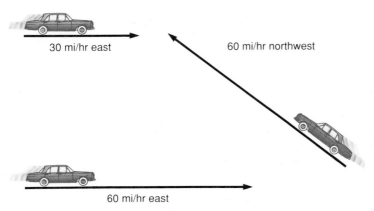

30 mi/hr east

60 mi/hr northwest

60 mi/hr east

Figure 2.5

Velocity is a vector that can be represented graphically with arrows. Here are three different velocities represented by three different arrows, with the length of each arrow proportional to the speed and the direction of travel shown.

place. The change of velocity and time can be combined to define the *rate* at which the motion was changed. This rate is called **acceleration.** Acceleration is defined as a change of velocity per unit time, or

$$\text{acceleration} = \frac{\text{change of velocity}}{\text{time elapsed}}$$

Another way of saying "change in velocity" is the final velocity minus the initial velocity, so the relationship can also be written as

$$\text{acceleration} = \frac{\text{final velocity} - \text{initial velocity}}{\text{time}}$$

Consider a car that is moving with a constant, straight-line velocity of 60 km/hr when the driver accelerates to 80 km/hr. Suppose it takes 4 sec to increase the velocity of 60 km/hr to 80 km/hr. The change in velocity is therefore 80 km/hr minus 60 km/hr, or 20 km/hr. The acceleration was

$$\text{acceleration} = \frac{80\,\frac{km}{hr} - 60\,\frac{km}{hr}}{4\,sec}$$

$$= \frac{20\,\frac{km}{hr}}{4\,sec}$$

$$= 5\,\frac{\frac{km}{hr}}{sec} \quad \text{or,}$$

$$= 5\ km/hr/sec$$

The average acceleration of the car was 5 km/hr for each ("per") second. This is another way of saying that the velocity increases an average of 5 km/hr in each second. The velocity of the car was 60 km/hr when the acceleration began (initial velocity). At the end of 1 sec, the velocity was 65 km/hr. At the end of 2 sec,

70 km/hr. At the end of 3 sec, 75 km/hr. And at the end of 4 sec (total time elapsed) the velocity was 80 km/hr (final velocity). Note how fast the velocity is changing with time. In summary,

start (initial velocity)	60 km/hr
first second	65 km/hr
second second	70 km/hr
third second	75 km/hr
fourth second (final velocity)	80 km/hr

As you can see, acceleration is really a description of how fast the velocity is changing (figure 2.6), in this case, increasing 5 km/hr each second.

Notice that the relationship between the initial and final velocities over a period of time results in an *average acceleration*. It is an average acceleration because it is calculated from the initial and the final velocities over some time period (the acceleration could have changed during this period but you would not know this). The acceleration at any specific instant, the *instantaneous acceleration*, would be calculated from a very short time interval. If the acceleration were known to be uniform throughout the time period, then the instantaneous acceleration would be the same as the average acceleration.

Usually, you would want all the units to be the same so you would convert km/hr to m/sec. You convert to m/sec because, as you will see later, many other quantities are defined in terms of meters in the form of the metric system you will be using (meter-kilogram-second). The other form of the metric system (centimeter-gram-second) will not be used. If you were given English units of mi/hr for velocity and sec for time, you would convert the mi/hr to ft/sec.

A change in velocity of 5 km/hr converts to 1.4 m/sec and the acceleration would be 1.4 m/sec/sec. The units m/sec per sec mean what change of velocity (1.4 m/sec) is occurring every second. The combination m/sec/sec is rather cumbersome, so it is typically treated mathematically to simplify the expression (to simplify a fraction, invert the divisor and multiply, or m/sec $\times$ 1/sec = m/sec^2). Remember that the expression 1.4 m/sec^2 means the same as 1.4 m/sec per sec, a change of velocity in a given time period.

The relationship between the quantities involved in acceleration can be represented with the symbols $\bar{a}$ for average acceleration, v_f for final velocity, v_i for initial velocity, and t for time. The relationship is

$$\bar{a} = \frac{v_f - v_i}{t} \qquad \textbf{equation 2.2}$$

As is the case for other such equations, any one of these quantities can be found if the others are known. For example, solving the equation for the final velocity, v_f, yields:

$$v_f = \bar{a}t + v_i$$

In problems where the initial velocity is equal to zero (starting from rest), the equation simplifies to

$$v_f = \bar{a}t$$

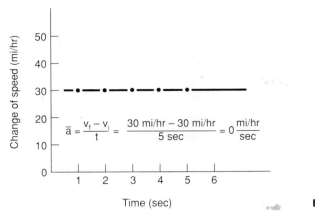

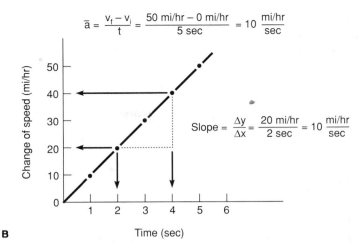

$$\overline{a} = \frac{v_f - v_i}{t} = \frac{50 \text{ mi/hr} - 0 \text{ mi/hr}}{5 \text{ sec}} = 10 \frac{\text{mi/hr}}{\text{sec}}$$

$$\text{Slope} = \frac{\Delta y}{\Delta x} = \frac{20 \text{ mi/hr}}{2 \text{ sec}} = 10 \frac{\text{mi/hr}}{\text{sec}}$$

A

B

Figure 2.6

(*a*) This graph shows how the speed changes per unit of time while driving at a constant 30 mi/hr in a straight line. As you can see, the speed is constant, and for straight-line motion, the acceleration is 0. (*b*) This graph shows the speed increasing 50 mi/hr when moving in a

straight line for 5 sec. The acceleration, or change of velocity per unit of time, can be calculated from either the equation for acceleration or by calculating the slope of the straight-line graph. Both will tell you how fast the motion is changing with time.

Example 2.3

A bicycle moves from rest to 5 m/sec in 5 sec. What was the average acceleration?

Solution

$$v_i = 0 \text{ m/sec} \qquad \overline{a} = \frac{v_f - v_i}{t}$$

$$v_f = 5 \text{ m/sec}$$

$$t = 5 \text{ sec} \qquad = \frac{5 \text{ m/sec} - 0 \text{ m/sec}}{5 \text{ sec}}$$

$$\overline{a} = ? \qquad = \frac{5}{5} \frac{\text{m/sec}}{\text{sec}}$$

$$= 1 \left(\frac{\text{m}}{\text{sec}}\right)\left(\frac{1}{\text{sec}}\right)$$

$$= \boxed{1 \frac{\text{m}}{\text{sec}^2}}$$

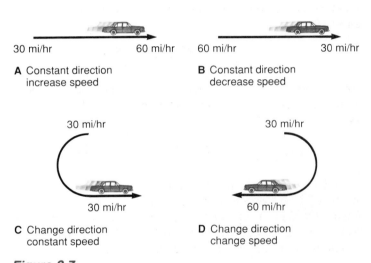

A Constant direction increase speed

B Constant direction decrease speed

C Change direction constant speed

D Change direction change speed

Figure 2.7

Four different ways (*a–d*) to accelerate a car.

Example 2.4

An automobile uniformly accelerates from rest at 15.0 ft/sec² for 6.0 sec. What is the final velocity in ft/sec? (Answer: 90. ft/sec) In mi/hr? (Answer: 61 mi/hr)

So far, you have learned only about straight-line, uniform acceleration that results in an increased velocity. There are also other changes in the motion of an object that are associated with acceleration. One of the more obvious is a change that results in a decreased velocity. Your car's brakes, for example, can slow your car or bring it to a complete stop. This is sometimes called *negative acceleration,* or *deceleration.* Another change in the

motion of an object is a change of direction. Velocity encompasses both the rate of motion as well as direction, so a change of direction is an acceleration. The satellite moving with a constant speed in a circular orbit around the earth is constantly changing its direction of movement. It is therefore constantly accelerating because of this constant change in its motion. Your automobile has three devices that could change the state of motion of the automobile. Your automobile therefore has three accelerators—the gas pedal (which can increase magnitude of velocity), the brakes (which can decrease magnitude of velocity), and the steering wheel (which can change direction of velocity). (See figure 2.7.) The important thing to remember is that acceleration results from any *change* in the motion of an object.

Motion

The final velocity (v_f) and the initial velocity (v_i) are different variables than the average velocity ($\bar{v}$). You cannot use an initial or final velocity for an average velocity. You may, however, calculate an average velocity ($\bar{v}$) from the other two variables as long as the acceleration taking place between the initial and final velocities is uniform. An example of such a uniform change would be an automobile during a constant, straight-line acceleration. This is done just as in finding other averages, such as the average weight of your class. The average class weight is the sum of all the weights divided by the number of people. To find an average velocity *during* a uniform acceleration, you add the initial velocity and the final velocity and divide by 2. This averaging can be done for a uniform acceleration that is increasing the velocity or for one that is decreasing the velocity. In symbols,

$$\bar{v} = \frac{v_f + v_i}{2} \qquad \textbf{equation 2.3}$$

Example 2.5 ▼

An automobile moving at 60.0 mi/hr (88.0 ft/sec) comes to a stop in 10.0 sec when the driver slams on the brakes. How far did the car travel while stopping?

Solution

The car has an initial velocity of 88.0 ft/sec (v_i) and the final velocity of 0 ft/sec (v_f) is implied. The time of 10.0 sec (t) is given. The problem asked for the distance (d). The relationship given between $\bar{v}$, t, and d is given in equation 2.1, $\bar{v} = d/t$, which can be solved for d. The average velocity ($\bar{v}$), however, is not given but can be found from equation 2.3, $\bar{v} = \frac{v_f + v_i}{2}$

$v_i = 88.0$ ft/sec

$v_f = 0$ ft/sec

$t = 10.0$ sec

$\bar{v} = ?$

$d = ?$

$\bar{v} = \frac{d}{t} \therefore d = \bar{v} \cdot t$

Since $\bar{v} = \frac{v_f + v_i}{2}$,

you can substitute $\left(\frac{v_f + v_i}{2} \right)$ for $\bar{v}$, and

$d = \left(\frac{v_f + v_i}{2} \right)(t)$

$= \left(\frac{0 \frac{ft}{sec} + 88.0 \frac{ft}{sec}}{2} \right)(10.0 \text{ sec})$

$= 44.0 \times 10.0 \frac{ft}{sec} \times sec$

$= 440. \frac{ft \cdot sec}{sec}$

$= \boxed{440. \text{ ft}}$

Example 2.6

What was the deceleration of the automobile in example 2.5? (Answer: 8.80 ft/sec²)

▲

Figure 2.8
Aristotle summarized all of the ancient Greek understandings when he wrote a grand theory of the universe. This theory included explanations of two kinds of motion as they were understood at the time.

ACTIVITIES

1. Is the speedometer in your car accurate? Take a friend and a stopwatch to a highway with mile markers. Drive at a constant speed while your friend measures the time required to drive one mile. Do this several times, then compare the indicated speed with the calculated speed. Which do you believe is more accurate, the speedometer or the calculated speed?
2. As long as you are driving with a friend and a stopwatch, measure the time required for your car to accelerate from one speed to another. Calculate the average acceleration. Compare your finding to the acceleration of other cars.

Aristotle's Theory of Motion

Some of the first ideas about the causes of motion were recorded back in the fourth century B.C. by the Greek philosopher Aristotle (figure 2.8). Aristotle was a former student of Plato and founded a new school, the Lyceum, in Athens in 337 B.C. As head of the Lyceum, Aristotle taught his students while walking within the school grounds. He also wrote an encyclopedia of classified knowledge that organized all knowledge known at that time as a unified theory of nature. He produced a philosophic theory of the physical world that the Greeks used to interpret and understand nature. The theory is said to be philosophic because it was developed from reasoning only.

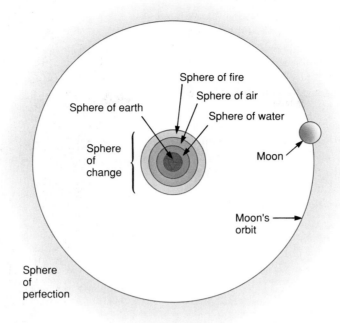

Figure 2.9

The shape of a sphere represented an early Greek idea of perfection. Aristotle's grand theory of the universe was made up of spheres, which explained the "natural motion" of objects.

Aristotle's theory presented a grand scheme of the universe, which was considered to have two main parts: (1) the *sphere of change* within the orbit of the moon and (2) the *sphere of perfection* outside the orbit of the moon. The earth was seen to be fixed and unmoving at the center of the sphere of change. All other heavenly bodies (moon, sun, planets, stars, etc.) were seen to be in the sphere of perfection. Everything in the sphere of perfection was thought to be perfect in every respect. Everything was perfectly round and smooth and moved in perfect circles at perfectly uniform speeds. These perfect and unchanging bodies moved in "ether," a rarefied material believed to fill the sphere of perfection. The ether was a necessary part of the theory because Aristotle did not consider a vacuum possible.

The part of the universe inside the sphere of change was considered to be made of four elements: earth, air, fire, and water. These were not your ordinary earth, air, fire, and water but were idealized elements that did not exist in their pure forms. They always existed in combination with each other and the different, varying combinations made up the materials of the earth (figure 2.9). Rocks, for example, were thought to be mostly the element earth in combination with lesser amounts of air, water, and fire. The various amounts of the other elements accounted for the different kinds of rocks. Other materials, such as steam, were considered combinations of two elements, perhaps fire and water. Wood was a combination of all four elements of earth, air, fire, and water.

Natural Motion

Ideally, the four elements would exist alone in their own spheres. The center sphere would consist of the heaviest element, earth. This would be surrounded by a sphere of water, which in turn would be surrounded by a sphere of air. Fire, the lightest element, would surround the sphere of air (figure 2.9). But this region of the universe was thought not to be perfect and mixing did occur. When the mixing occurred, each element would seek out its ideal, or natural, place. Thus a rock, being made up of mostly the heavy element earth, would tumble down a hill through the air because the element earth was seeking its natural home. Water would flow over the rock and bubbles of air would rise through the water as water and air moved toward their natural homes. Aristotle called this spontaneous movement of materials seeking their natural place a *natural motion*. In natural motion, heavy bodies moved down and light bodies moved up according to the four elements making them up. Aristotle is supposed to have said that objects fall at speeds proportional to their weight but at a constant speed. In other words, a ten-pound ball would fall twice as fast as a five-pound ball. Furthermore, the speed was seen to be acquired immediately and remain constant during the fall. This belief was never checked by measurement, as reasoning, not observation, was used to develop the concept.

Forced Motion

Aristotle recognized that not all motion was in response to objects seeking their "natural places." People could throw rocks and horses could move carts, and the motion of the rocks and the carts was not from these objects seeking their natural places. This motion was imposed on the rocks and the carts. He called imposed motion a *forced motion*. A forced motion required a push or pull, such as a person pushing on a rock or a horse pulling a cart. But Aristotle believed that the cause must be continuously acting or else the motion would stop. If a horse stops pulling on a cart, for example, the cart will come to rest. A rock thrown through the air continues to move, he believed, because air moved in behind the moving rock to force it along. If it were not for the air forcing the rock to move, it would stop and fall straight down according to its natural motion.

Aristotle's work did offer explanations for the motion of falling objects and for horizontal motion, but they were disastrously wrong. His physical science theory was based on thinking, not measurement, and no one checked to see if his beliefs were accurate. He should not be blamed, however, for the fact that his writings later became part of Middle Ages theology, acquiring the status of priestly authority. With such authority, a person who questioned Aristotle's reasoned beliefs was viewed as a devious devil's advocate. It therefore took about two thousand years to correct his beliefs about motion.

Forces

Aristotle recognized the need for a force to produce forced motion. He was partly correct, because a force is closely associated with *any* change of motion, as you will see. This section introduces the concept of a force, which will be developed more fully when the relationship between forces and motion is explained in the next chapter.

A **force** is usually considered as a push or a pull. These pushes and pulls can result from two kinds of interaction, (1) *contact interaction* and (2) *interaction at a distance*. An example of a contact interaction is people exerting a force by contact with a stalled automobile. They push on it directly until it moves. Likewise, a tow truck pulls through the contact of a tow cable on the automobile until it moves. A common example of interaction at a distance is gravitational attraction. A meteor is pulled toward the immediate surface of the earth by this gravitational force. Raindrops and autumn leaves are also pulled by interaction at a distance, the gravitational force.

Through contact or interaction at a distance, a force is a push or a pull that is *capable* of changing the state of motion of an object. A force may be capable of changing the state of motion of an object but there are other considerations. Consider, for example, a stalled automobile. Suppose you find that two people can push with sufficient force to move the car. Does it matter where on the car they push? It should be obvious that the people can push with a certain strength and determine the direction that the force is directed. Since a force has magnitude, or strength, as well as direction, it is a vector.

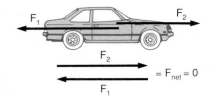

A When two people push in opposite directions with forces of equal magnitude

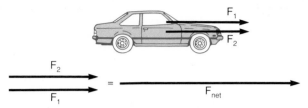

B When two people push parallel with forces of equal magnitude

Figure 2.10

(*a*) Two opposite forces parallel and equal in magnitude result in a net force of 0. (*b*) Two parallel forces of equal magnitude in the same direction result in an unbalanced, or net force, of twice the magnitude.

What effect does direction have on two force vectors? If they act in exactly opposite directions, the force on the object is the difference between the strength of the two forces. If they have the same force, the overall effect is to cancel each other without producing any motion. The forces balance each other and there is not an *unbalanced* force. The **net force** is said to be zero. "Net" means final, after the vector forces are added (figure 2.10).

When two parallel forces act in the same direction, their forces can be simply added together. In this case, there is a net force that is the equivalent to the vector sum of the two forces (figure 2.10).

When two vector forces act in a way different from exactly together or exactly opposite each other, the result will be like a new, different net force having a new direction. The new net force is called a resultant and can be calculated by drawing vector arrows to scale (length and direction) and using geometry. The tail of the vector arrow is placed on the object that feels the force, and the arrowhead points in the direction that the force is exerted. The length of the arrow is proportional to the magnitude of the force (figure 2.11). In summary, there are four things to understand and remember about vector arrows that represent forces:

1. The tail of the arrow is placed on the object that *feels* the force.
2. The arrowhead points in the *direction* that the force is applied.
3. The *length* of the arrow is proportional to the magnitude of the force.
4. The *net force* is the sum of the vector forces.

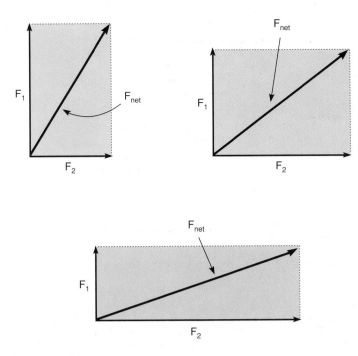

Figure 2.11
Vector addition results in a net force that has a different magnitude and a different direction.

Horizontal Motion on Land

Everyday experience seems to indicate that Aristotle's idea about horizontal motion on the earth's surface is correct. After all, moving objects that are not pushed or pulled do come to rest in a short period of time. It would seem that an object keeps moving only if a force continues to push it. A moving automobile will slow and come to rest if you turn off the ignition. Likewise, a ball that you roll along the floor will slow until it comes to rest. Is the natural state of an object to be at rest and is a force necessary to keep an object in motion? This is exactly what people thought until Galileo (figure 2.12) published his book *Two New Sciences* in 1638, which described his findings about motion.

Consider the era when Galileo lived, from 1564 to 1642. Aristotle's ideas about motion had become joined with theology some three hundred years earlier and to challenge Aristotle's idea was considered to be heresy. Indeed, people were burned at the stake at that time for questioning the current theologic dogma. But this was a time of change. Europe was in the middle of the Renaissance movement, and it was the time of Columbus, da Vinci, Shakespeare, and Michelangelo among others. Galileo was a key figure in this change as he challenged the Aristotelian views and focused attention on the concepts of distance, time, velocity, and acceleration.

Galileo was a professor at the University of Pisa and studied a variety of physical science topics. The telescope was invented about 1608 and Galileo built his own in 1609, focusing on many things that had not been previously observed. For the first time he could see mountains on the Moon, spots on the Sun, and a bulging Saturn. These were not the perfectly round and

Figure 2.12
Galileo challenged the Aristotelian view on motion and focused attention on the concepts of distance, time, velocity, and acceleration.

perfectly smooth celestial objects that Aristotle's theory called for. Furthermore, Galileo observed moons revolving around Jupiter, an impossibility if Earth was the center of rotation for the universe. Galileo was probably best known, however, for the widespread story of his experiments of dropping objects from the Leaning Tower of Pisa to study the motion of falling objects.

Galileo's constant challenge of the ancient authority of Aristotle created many enemies and eventually led to a trial by the Inquisition. Galileo was a prisoner of the Inquisition when he wrote *Two New Sciences,* which was secretly published in Holland in 1638. The book had three parts that dealt with uniform motion, accelerated motion, and projectile motion. Galileo described details of simple experiments, measurements, calculations, and thought experiments as he developed definitions and concepts of motion. In one of his thought experiments Galileo presented an argument against Aristotle's view that a force is needed to keep an object in motion. Galileo imagined an object (such as a ball) moving over a horizontal surface without the force of friction. He concluded that the object would move forever with a constant velocity as long as there was no unbalanced force acting to change the motion.

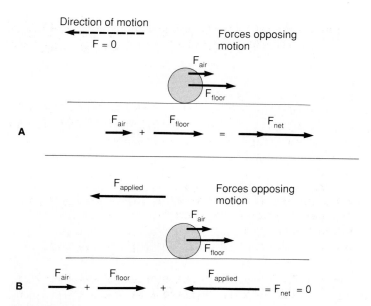

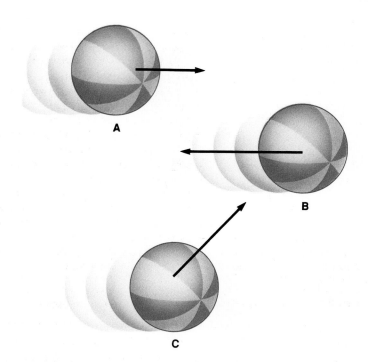

Figure 2.13

(*a*) This ball is rolling to your left with no forces in the direction of motion. The vector sum of the force of floor friction (F_{floor}) and the force of air friction (F_{air}) result in a net force opposing the motion, so the ball slows to a stop. (*b*) A force is applied to the moving ball, perhaps by a hand that moves along with the ball. The force applied ($F_{applied}$) equals the vector sum of the forces opposing the motion, so the ball continues to move with a constant velocity.

Figure 2.14

Galileo concluded that objects persist in their states of motion in the absence of an unbalanced force, a property of matter called inertia. Thus an unbalanced force, as shown by the vector arrows, is needed to (*a*) speed up, (*b*) slow, or (*c*) change the direction of a moving object.

Why does a rolling ball slow to a stop? You know that a ball will roll farther across a smooth, waxed floor such as a bowling lane than it will across floor covered with carpet. The rough carpet offers more resistance to the rolling ball. The resistance of the floor friction is shown by a force vector in figure 2.13 as is the resistance produced by air friction. Now imagine what force you would need to exert by pushing with your hand, moving along with the ball to keep it rolling at a uniform rate. The answer is the same force as the combined force from the floor and air resistance. Therefore, the ball continues to roll at a uniform rate when you *balance* the force opposing its motion. The force that you applied with your hand simply balanced the sum of the two opposing forces. It is reasonable, then, to imagine if the opposing forces were removed you would not need to apply *any* force to the ball to keep it moving; it would roll forever on the frictionless floor. This was the kind of reasoning that Galileo did when he discredited the Aristotelian view that a force was necessary to keep an object moving. Galileo concluded that a moving object would continue moving with a constant velocity if no unbalanced forces were applied, that is, if the net force were zero.

It could be argued that the difference in Aristotle's and Galileo's views of "forced" motion is really a degree of analysis. After all, moving objects on earth do come to rest unless continuously pushed or pulled. But Galileo's conclusion describes *why* they must be pushed or pulled and reveals the true nature of the motion of objects. Aristotle argued that the natural state of objects is to be at rest and attempted to explain why objects

move. Galileo, on the other hand, argued that it is just as natural for an object to be moving and attempted to explain why they come to rest. Galileo called the behavior of matter to persist in its state of motion **inertia.** Inertia is the *tendency of an object to remain in unchanging motion or at rest in the absence of an unbalanced force* (friction, gravity, or whatever). The development of this concept changed the way people viewed the natural state of an object and opened the way for further understandings about motion.

Falling Objects

Return now to the other kind of motion described by Aristotle, natural motion. Recall that Aristotle viewed natural motion as the motion produced when objects return to their natural places. For example, if you were to hold a rock above the ground and let it go, the rock would fall toward the ground, its natural place. Today, you know that the rock falls because the gravitational force of the earth pulls the rock downward. But the concern here is what happens during the fall. Aristotle is supposed to have said that objects fall at speeds proportional to their weight. In other words, suppose you drop a one-hundred-pound iron ball at the same time a one-pound iron ball is dropped from a height of one hundred feet. According to Aristotle, the one-hundred-pound ball would reach the ground before the one-pound ball had fallen one foot. As stated in a popular story, Galileo discredited Aristotle's conclusion by dropping a solid iron ball and a solid wooden ball simultaneously from the top of the Leaning

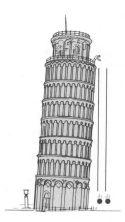

Figure 2.15
According to a widespread story, Galileo dropped two objects with different weights from the Leaning Tower of Pisa. They were supposed to have hit the ground at about the same time, discrediting Aristotle's view that the time of the fall is proportional to weight.

Tower of Pisa. Both balls, according to the story, hit the ground nearly at the same time. To do this, they would have to fall with the same velocity. In other words, the velocity of a falling object does not depend on its weight. Any difference in freely falling bodies is explainable by air resistance. Soon after the time of Galileo the air pump was invented. The air pump could be used to remove the air from a glass tube. The affect of air resistance on falling objects could then be demonstrated by comparing how objects fall in the air with how they fall in an evacuated glass tube. You know that a coin falls faster when dropped with a feather in the air. A feather and heavy coin will fall together in the near vacuum of an evacuated glass tube because the affect of air resistance on the feather has been removed. When objects fall toward the earth without considering air resistance, they are said to be in **free fall.** Free fall considers only gravity and neglects air resistance.

Galileo versus Aristotle's Natural Motion

Galileo concluded that light and heavy objects fall together in free fall, but he also wanted to know the details of what was going on while they fell. He now knew that the velocity of an object in free fall was *not* proportional to the weight of the object. He observed that the velocity of an object in free fall *increased* as the object fell and reasoned from this that the velocity of the falling object would have to be (1) somehow proportional to the *time* of fall and (2) somehow proportional to the *distance* the object fell. If the time and distance were both related to the velocity of a falling object, how were they related to one another? To answer this question Galileo made calculations involving distance, velocity, and time and, in fact, introduced the concept of acceleration. The relationships between these variables are found in the same three equations that you have already learned. Let's see how the equations can be rearranged to incorporate acceleration, distance, and time for an object in free fall.

Step 1: Equation 2.1 gives a relationship between average velocity ($\bar{v}$), distance (d), and time (t). Solving this equation for distance gives

$$d = \bar{v}t$$

Step 2: An object in free fall should have uniformly accelerated motion, so the average velocity could be calculated from equation 2.3,

$$\bar{v} = \frac{v_f + v_i}{2}$$

Substituting this equation in the rearranged equation 2.1, the distance relationship becomes

$$d = \left(\frac{v_f + v_i}{2}\right)(t)$$

Step 3: The initial velocity of a falling object is always zero just as it is dropped, so the v_i can be eliminated,

$$d = \left(\frac{v_f}{2}\right)(t)$$

Step 4: Now you want to get acceleration into the equation in place of velocity. This can be done by solving equation 2.2 for the final velocity (v_f), then substituting. The initial velocity (v_i) is again eliminated because it equals zero.

$$\bar{a} = \frac{v_f - v_i}{t}$$

$$v_f = \bar{a}t$$

$$d = \left(\frac{\bar{a}t}{2}\right)(t)$$

Step 5: Simplifying, the equation becomes

$$d = \frac{1}{2}\bar{a}t^2 \qquad \text{equation 2.4}$$

Thus, Galileo reasoned that a freely falling object should cover a distance *proportional to the square of the time of the fall* ($d \propto t^2$). In other words the object should fall 4 times as far in 2 sec as in 1 sec ($2^2 = 4$), 9 times as far in 3 sec ($3^2 = 9$), and so on. Compare this prediction with figure 2.16.

Galileo checked this calculation by rolling balls on an inclined board with a smooth groove in it. He used the inclined board to slow the motion of descent in order to measure the distance and time relationships, a necessary requirement since he lacked the accurate timing devices that exist today. He found, as predicted, that the falling balls moved through a distance proportional to the square of the time of falling. This also means that the *velocity of the falling object increased at a constant rate,* as shown in figure 2.17. Recall that a change of velocity during some time period is called acceleration. In other words, a falling object *accelerates* toward the surface of the earth.

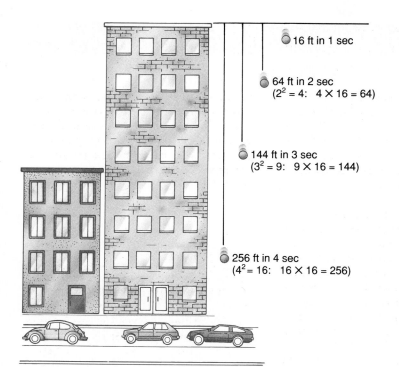

16 ft in 1 sec

64 ft in 2 sec
($2^2 = 4$: $4 \times 16 = 64$)

144 ft in 3 sec
($3^2 = 9$: $9 \times 16 = 144$)

256 ft in 4 sec
($4^2 = 16$: $16 \times 16 = 256$)

Figure 2.16

An object dropped from a tall building covers increasing distances with every successive second of falling. The distance covered is proportional to the square of the time of falling ($d \propto t^2$).

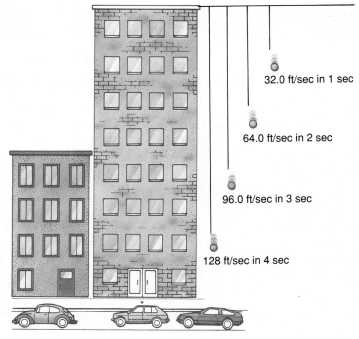

32.0 ft/sec in 1 sec

64.0 ft/sec in 2 sec

96.0 ft/sec in 3 sec

128 ft/sec in 4 sec

Figure 2.17

The velocity of a falling object increases at a constant rate, 32.0 ft/sec².

Acceleration Due to Gravity

Since the velocity of a falling object increases at a constant rate, this must mean that falling objects are *uniformly accelerated* by the force of gravity. *All objects in free fall do so with a constant acceleration.* During each second of fall the object gains a velocity of 9.80 m/sec (32.0 ft/sec). This gain is the acceleration of the falling object, 9.80 m/sec² (32.0 ft/sec²).

The acceleration of objects falling toward the earth varies slightly from place to place on the earth's surface because of the earth's shape and spin. The acceleration of falling objects decreases from the poles to the equator and also varies from place to place because the earth's mass is not distributed equally. The value of 9.80 m/sec² (32.0 ft/sec²) is an average value that is fairly close to, but not exactly, the acceleration due to gravity in any particular location. The acceleration due to gravity is important in a number of situations, so the acceleration from this force is given a special symbol, **g**.

Example 2.7

A rock that is dropped into a well hits the water in 3.00 sec. Ignoring air resistance, how far is it to the water?

Solution

The problem concerns a rock in free fall. The time of fall (t) is given and the problem asks for a distance (d). Since the rock is in free fall, the acceleration due to the force of gravity (g) is implied. The metric value and unit for g is 9.80 m/sec² and the English value and unit is 32.0 ft/sec². You would use the metric g to obtain an answer in meters and the English unit to obtain an answer in feet.

Equation 2.4, $d = 1/2\bar{a}t^2$, gives a relationship between distance (d), time (t), and average acceleration ($\bar{a}$). The acceleration in this case is the acceleration due to gravity (g), so

$t = 3.00$ sec

$g = 9.80$ m/sec²

$d = ?$

$d = \dfrac{1}{2} g t^2$ ($\bar{a} = g = 9.80$ m/sec²)

$d = \dfrac{1}{2} (9.80 \text{ m/sec}^2)(3.00 \text{ sec})^2$

$= (4.90 \text{ m/sec}^2)(9.00 \text{ sec}^2)$

$= 44.1 \dfrac{\text{m} \cdot \text{sec}^2}{\text{sec}^2}$

$= \boxed{44.1 \text{ m}}$

or, you could do each step separately. Check this solution by a three-step procedure:

1. find the final velocity, v_f, of the rock from $v_f = \bar{a}t$;
2. calculate the average velocity ($\bar{v}$) from the final velocity

$$\bar{v} = \frac{v_f + v_i}{2}$$

then;

3. use the average velocity ($\bar{v}$) and the time (t) to find distance (d), $d = \bar{v}t$.

Note that the one-step procedure is preferred over the three-step procedure because fewer steps mean fewer possibilities for mistakes and mathematical blunders.

Deceleration and Seat Belts

Do you "buckle up" when you ride in a car? As you know, seat belts and shoulder harnesses are designed to protect people if an accident occurs. Clearly, many accidents are survivable if such restraints are worn (box figure 2.1). This Feature is about the role of seat belts and air bags during the rapid deceleration of a car collision.

Car collisions range from the relatively simple, straightforward fender-bender to the complex interactions that occur when a car is crunched into a lump of deformed metal. Injuries occur when passengers are thrown violently forward by the rapid deceleration of the car. But it is not the deceleration that hurts people. Injuries occur from the tendency of the passengers to retain their states of motion as the car rapidly decelerates. As a result, passengers hit the steering wheel, the dash, or the windshield because of their inertia.

Deceleration is measured the same as acceleration except it is a decrease of velocity. The acceleration due to gravity, g, is 9.80 m/sec^2 or 32.0 ft/sec^2. This value of g is used as a base unit when discussing the acceleration or deceleration of cars, aircraft, and rockets. A normally braking car might decelerate with a velocity reduction of 16.0 ft/sec every second. Since this is one-half of 32.0 ft/sec^2, this car is decelerating at 0.5 g. A hard-braking car might decelerate at 1 g, or 32.0 ft/sec^2. A 1 g deceleration is equivalent to reducing the velocity about 22 mi/hr each second.

The human body is surprisingly hardy when dealing with deceleration. People have survived a deceleration of 35 g's without serious injury and have a good chance of surviving a car collision without injury if the deceleration does not exceed 25 g's. Compare this level of acceptable deceleration—25 g's—to the deceleration experienced when coming to a dead stop in a quarter of a second from the velocities shown in box table 2.1.

As you can see in box table 2.1, you can survive a deceleration of even 70 mi/hr in a quarter of a second. In real crashes a number of factors influence the time involved in the deceleration. The front end of most cars is designed to collapse, extending the time variable. An air bag is a safety feature designed to inflate in the passenger compartment the instant of a collision. The air bag protects passengers from hitting the steering wheel, windshield, and so forth but also increases the stopping time as well. This reduces the deceleration as it helps prevent injury. Most injuries that occur in a car collision are not from the rapid deceleration but occur because people fail to wear seat belts.

Box Table 2.1
Deceleration to a complete stop in a quarter of a Second

Speed		Deceleration	g's
mi/hr	ft/sec	ft/sec^2	
10	15	59	1.8
20	29	117	3.7
30	44	176	5.5
40	59	235	7.3
50	73	293	9.1
60	88	352	11.0
70	103	411	12.8

A

B

C

D

Box Figure 2.1
This Pontiac Grand Am, equipped with automatic belts for front seat passengers, hits a barrier (a–d) in a crash that is equivalent to a car moving at 60 mi/hr when it slams into a parked car. General Motors conducts such tests to evaluate the crashworthiness of their vehicle designs.

1. Place three or four coins of different masses such as a dime, nickle, and quarter flat on the edge of a table. Place a notebook or book flat on the table behind the coins so it is parallel to the edge of the table. Keeping the book parallel to the table, push the coins over the edge at the same time. Do this several times. Does the way they hit the floor support the view of Aristotle?
2. What is your reaction time? Have a friend hold a meter stick vertically from the top while you position your thumb and index finger at the 50 cm mark. Your friend will drop the stick (unannounced) and you will catch it with your thumb and finger. Measure how far the stick drops. Use this distance in the equation,

$$t = \sqrt{\frac{2d}{g}}$$

to calculate your reaction time. (This is equation 2.4 solved for t.)

Compound Motion

So far we have considered two types of motion, (1) the horizontal, straight-line motion of objects moving on the surface of the earth and (2) the vertical motion of dropped objects that accelerate toward the surface of the earth. A third type of motion occurs when an object is thrown, or projected, into the air. Essentially, such a projectile (rock, football, bullet, golfball, or whatever) could be directed straight upward as a vertical projection, directed straight out as a horizontal projection, or directed at some angle between the vertical and the horizontal. Basic to understanding such compound motion is the observation that (1) gravity acts on objects *at all times,* no matter where they are, and (2) the acceleration due to gravity (*g*) is *independent of any motion* that an object may have.

Vertical Projectiles

Consider first a ball that you throw straight upward, a vertical projection. The ball has an initial velocity but then reaches a maximum height, stops for an instant, then accelerates back toward the earth. Gravity is acting on the ball throughout its climb, stop, and fall. As it is climbing, the force of gravity is accelerating it back to the earth. The overall effect during the climb is deceleration, which continues to slow the ball until the instantaneous stop. The ball then accelerates back to the surface just like a ball that has been dropped. If it were not for air resistance, the ball would return with the same velocity that it had initially. The velocity vectors for a ball thrown straight up are shown in figure 2.18a.

Horizontal Projectiles

Horizontal projections are easier to understand if you split the complete motion into vertical and horizontal parts. Consider, for example, a bullet that is fired horizontally from a rifle over perfectly level ground. The force of gravity accelerates the bullet

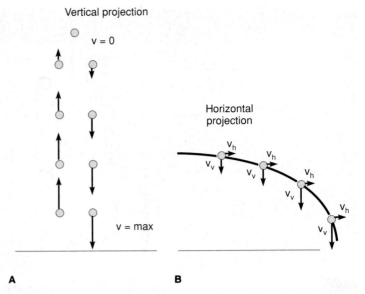

Figure 2.18

(*a*) On its way up, a vertical projectile is slowed by the force of gravity until an instantaneous stop; then it accelerates back to the surface just like a dropped object. (*b*) A horizontal projectile has the same horizontal velocity throughout the fall as it accelerates toward the surface. The combined effect results in a curved path.

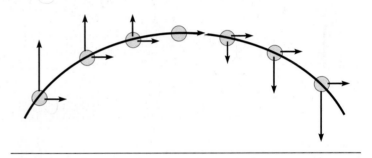

Figure 2.19

A projectile is launched at 45° to the horizon. The horizontal velocity is a constant (neglecting air resistance) and the vertical velocity decreases, then increases, just as a vertical projection. The combined motions produce a parabolic path.

downward, giving it an increasing velocity as it falls vertically toward the earth. This increasing downward velocity is the same as that of a dropped bullet and is represented by the vector arrows in figure 2.18b. There are no forces in the horizontal direction (if you ignore air resistance), so the horizontal velocity remains the same as shown by the v_h arrows. The combination of the vertical (v_v) motion and the horizontal (v_h) motion causes the bullet to follow a curved path until it hits the ground. An interesting prediction that can be made from this analysis is that a second bullet dropped from the same height as the horizontal rifle the same time the rifle is fired will hit the ground at the same time as the bullet fired from the rifle.

Golfballs, footballs, and baseballs are usually projected upward at some angle to the horizon. The horizontal motion of these projectiles is constant as before because there are no horizontal forces involved. The vertical motion is the same as that

Figure 2.20

This softball player is probably aware of the relationship between the projection angle and the maximum distance acquired for a given projection velocity.

of a ball projected directly upward. The combination of these two motions causes the projectile to follow a curved path called a *parabola* as shown in figure 2.19. The next time you have the opportunity, observe the path of a ball that has been projected at some angle (figure 2.20). Note that the second half of the path is almost a reverse copy of the first half. If it were not for air resistance, the two values of the path would be exactly the same. Also note the distance that the ball travels as compared to the angle of projection. An angle of projection of 45° results in the maximum distance of travel.

Summary

Motion is defined as the act or process of changing position. The change is usually described by comparing the moving object to something that is not moving. Motion can be measured by speed, velocity, and acceleration. *Speed* is a measure of how fast something is moving. It is a ratio of the distance covered between two locations to the time that elapsed while moving between the two locations. The *average speed* considers the distance covered during some period of time while the *instantaneous speed* is the speed at some specific instant. *Velocity* is a measure of the speed and direction of a moving object. *Acceleration* is a change of velocity per unit of time. The *average acceleration* is the change of velocity during some period of time while the *instantaneous acceleration* is the acceleration at some specific instant. Speed has magnitude only and is a *scalar*. Velocity and acceleration have both magnitude and direction and are *vectors*.

Early ideas about motion were recorded in the fourth century B.C. by Aristotle. Aristotle's theory of the universe considered two types of motion, *natural motion* and *forced motion*. Natural motion was the motion of falling objects. Aristotle's theory explained the motion of falling objects as the result of objects seeking their natural places. During the fall, an object was seen to acquire an immediate and constant velocity that was proportional to its weight. Forced motion required an imposed force to move an object. The motion was understood to stop unless the force was continuously acting.

A *force* is a push or a pull that can change the motion of an object. A force results from *contact interactions* or *interactions at a distance*. Force is a vector and *net force* is the vector sum of all the forces acting on an object.

Galileo challenged the authority of Aristotle's views, which had become a part of Middle Ages theology. Galileo determined that a continuously applied force is unnecessary for motion and defined the concept of *inertia*, that an object remains in unchanging motion in the absence of an unbalanced force. Galileo also determined that falling objects accelerate toward the earth's surface and that the acceleration is independent of the weight of the object. He found the acceleration due to gravity, *g*, to be 9.80 m/sec² (32.0 ft/sec²), and the distance an object falls is proportional to the square of the time of free fall ($d \propto t^2$).

Compound motion occurs when an object is projected into the air. Compound motion can be described by splitting the motion into vertical and horizontal parts. The acceleration due to gravity, *g*, is a constant that is acting at all times and acts independent of any motion that an object has. The path of an object that is projected at some angle to the horizon is therefore a parabola.

Summary of Equations

2.1

$$\text{average speed or velocity} = \frac{\text{distance}}{\text{time}}$$

$$\bar{v} = \frac{d}{t}$$

2.2

$$\text{average acceleration} = \frac{\text{change of velocity}}{\text{time}}$$

$$= \frac{\text{final velocity} - \text{initial velocity}}{\text{time}}$$

$$\bar{a} = \frac{v_f - v_i}{t}$$

2.3

$$\text{average velocity} = \frac{\text{final velocity} + \text{initial velocity}}{2}$$

$$\bar{v} = \frac{v_f + v_i}{2}$$

2.4

$$\text{distance} = \frac{1}{2}(\text{acceleration})(\text{time})^2$$

$$d = \frac{1}{2}\bar{a}t^2$$

Key Terms

acceleration (p. 28)
force (p. 32)
free fall (p. 35)
g (p. 36)
inertia (p. 34)
magnitude (p. 27)

motion (p. 25)
net force (p. 32)
scalars (p. 27)
speed (p. 25)
vectors (p. 27)
velocity (p. 27)

Applying the Concepts

1. For an object traveling in a straight line, the symbol $\bar{v}$ represents
   a. instantaneous speed.
   b. average speed.
   c. constant speed.
   d. total speed.

2. If $\bar{v} = d/t$, then d =
   a. $\bar{v}/t$.
   b. $t/\bar{v}$.
   c. $\bar{v}t$.
   d. $(\bar{v})(1/t)$.

3. A quantity of 5 m/sec² is a measure of
   a. velocity.
   b. acceleration.
   c. speed.

4. A ratio of $\Delta v/\Delta t$ is a measure of motion that is known as (recall that the symbol Δ means "a change in")
   a. speed.
   b. velocity.
   c. acceleration.
   d. none of the above.

5. A ratio of $\Delta d/\Delta t$ is a measure of straight-line motion that is known as
   a. speed.
   b. acceleration.
   c. mass density.
   d. none of the above.

6. Ignoring air resistance, an object falling toward the surface of the earth has a *velocity* that is
   a. constant.
   b. increasing.
   c. decreasing.
   d. acquired instantaneously but dependent on the weight of the object.

7. Ignoring air resistance, an object falling toward the surface of the earth has an *acceleration* that is
   a. constant.
   b. increasing.
   c. decreasing.
   d. dependent on the weight of the object.

8. A ball rolling across the floor slows to a stop because
   a. there are unbalanced forces acting on it.
   b. the force that started it moving wears out.
   c. the forces are balanced.
   d. the net force equals zero.

9. If there are no unbalanced forces acting on a moving object it will
   a. slow to a stop.
   b. continue to move.

10. When you throw a ball directly upward, it is accelerated
    a. only as it falls.
    b. during the instantaneous stop and during the fall.
    c. at all times.

Answers

1. b   2. c   3. b   4. c   5. a   6. b   7. a   8. a   9. b
10. c

Questions for Thought

1. What is the difference between speed and velocity?

2. What is acceleration?

3. Can you assume that an object moving at a constant speed is not accelerating? Explain.

4. A weather forecast predicts the possibility of wind with a velocity of 40 mi/hr. What is technically wrong with this forecast?

5. What is the difference between an acceleration of 9.80 m/sec² and a speed of 9.80 m/sec?

6. An insect inside a bus flies from the back towards the front at 5.0 mi/hr. The bus is moving in a straight line at 50.0 mi/hr. What is the speed of the insect?

7. Give several examples of objects that keep moving without an unbalanced force. Explain how this is possible.

8. Explain what happens to the velocity and acceleration of an object in free fall.

9. In the equation $d = 1/2\bar{a}t^2$, if g is 9.80 m/sec² and t is in sec, what is the unit for d?

10. What is the major difference between Aristotle's and Galileo's concept of forced motion?

11. What is inertia?

12. In the unit of acceleration 9.80 m/sec² what is the meaning of a "square second"?

13. Disregarding air friction, describe all the forces acting on a bullet shot from a rifle into the air.

14. Can gravity act in a vacuum? Explain.

15. Does the force of gravity acting on a ball change as the ball is thrown straight up from the surface of the earth? Explain.

16. What is the difference between a constant speed, an instantaneous speed, and an average speed?

17. Explain the difference between the variables represented by the symbols v_f, v_i, and $\bar{v}$.

18. How does the velocity of a ball thrown straight up compare to the velocity the ball has when it falls to the same place on the way down? Explain.

19. What are the major differences between Aristotle's and Galileo's understanding of what happens to an object in free fall?

20. Are there any unbalanced forces acting on an object that falls a short distance? Explain.

Exercises

Group A—Solutions Provided in Appendix D

1. What is the average speed of a truck that makes a 250.0 mi trip in 5.0 hr?

2. A bicyclist averages 18.0 km/hr for 20.0 min. How many km did the bicyclist travel?

3. What average speed must you maintain to make a (a) 400.0 mi trip in 8.0 hr? (b) The same 400.0 mi trip in 7.0 hr?

4. Suppose you average 10.0 mi/hr going 10.0 mi up a hill on a bicycle and you average 20.0 mi/hr covering the same distance down the other side of the hill. What was your average speed? (Hint: The answer is *not* 15.0 mi/hr.)

5. (a) What is the average acceleration of a car that moves from rest to 60.0 ft/sec in 10.0 sec? (b) At a constant acceleration, what will the velocity be after 15.0 sec?

6. What is the velocity of a car that accelerates from rest at 9.0 ft/sec² for 8.0 sec?

7. A sports car moving at 88.0 ft/sec is able to stop in 100.0 ft. (a) How much time was required for the stop? (b) What was the deceleration in g's?

8. A ball thrown straight up climbs for 3.0 sec before falling. Neglecting air resistance, with what velocity was the ball thrown?

9. A ball dropped from a building falls for 4.00 sec before it hits the ground. (a) What was its final velocity just as it hits the ground? (b) What was the average velocity during the fall? (c) How high was the building?

10. You drop a rock from a cliff and 5.00 sec later you see it hit the ground. How high is the cliff?

11. How long must a car accelerate at 4.0 m/sec² to go from 10.0 m/sec to 50.0 m/sec?

12. What is the velocity of a rock in free fall 3.0 sec after it is released from rest?

Group B—Solutions Not Given

1. What is the average speed of a car that travels 270.0 mi in 4.50 hr?

2. A ferry boat requires 45.0 min to travel 9.0 mi across a bay. What was the average speed of the boat?

3. A car with an average speed of 60.0 mi/hr will (a) require how much time to cover 300.0 mi? (b) travel how far in 8.0 hr?

4. A car travels 60.0 mi/hr for 1.0 hr, at 50.0 mi/hr for the next 2.0 hr, then at 70.0 mi/hr for the last hour of a trip. What was the average velocity for the entire trip?

5. (a) What is the acceleration of a car that moves from rest to 30.0 mi/hr (44.0 ft/sec) in 8.0 sec? (b) At the same acceleration, how fast will the car be moving after 9.0 sec?

6. What is the velocity of a car that accelerates from rest at 6.0 m/sec² for 5.0 sec?

7. A car with an initial velocity of 30.0 m/sec is able to come to a stop over a distance of 100.0 m when the brakes are applied. (a) How much time was required for the stopping process? (b) How many g's did the driver experience?

8. A rock thrown straight up climbs for 2.50 sec, then falls to the ground. Neglecting air resistance, with what velocity did the rock strike the ground?

9. An object is observed to fall from a bridge, striking the water below 2.50 sec later. (a) With what velocity did it strike the water? (b) What was its average velocity during the fall? (c) How high is the bridge?

10. A ball dropped from a window strikes the ground 2.00 sec later. How high is the window above the ground?

11. What is the resulting velocity if a car moving at 10.0 m/sec accelerates at 4.00 m/sec² for 10.0 sec?

12. If a rock in free fall is moving at 96.0 ft/sec, how long has it been falling?

Chapter 3

Patterns of Motion

Figure 3.1

In a moving airplane you feel forces in many directions when the plane changes its motion. You cannot help but notice the forces involved when there is a change of motion.

In the previous chapter you learned how to describe motion in terms of distance, time, velocity, and acceleration. In addition, you learned about different kinds of motion such as straight-line motion, the motion of falling objects, and the compound motion of objects projected up from the surface of the earth. You were also introduced, in general, to two concepts closely associated with motion: (1) that objects have inertia, a tendency to resist a change in motion, and (2) that forces are involved in a change of motion.

The relationship between forces and a change of motion is obvious in many everyday situations (figure 3.1). When a car, bus, or plane starts moving you feel a force on your back. Likewise, you feel a force on the bottoms of your feet when an elevator starts moving upward. On the other hand, you seem to be forced toward the dashboard if a car stops quickly, and it feels as if the floor pulls away from your feet when an elevator drops rapidly. These examples all involve patterns between forces and motion, patterns that can be quantified, conceptualized, and used to answer questions about why things move or stand still. These patterns are the subject of this chapter.

Laws of Motion

Isaac Newton was born on Christmas Day in 1642, the same year that Galileo died. Newton was a quiet farm boy who seemed more interested in mathematics and tinkering than farming. He entered Trinity College of Cambridge University at the age of eighteen, where he enrolled in mathematics. He graduated four years later, the same year that the university was closed because the bubonic plague, or "Black Death," was ravaging Europe. During this time, Newton returned to his boyhood home, where he thought out most of the ideas that would later make him famous. Here, between the ages of twenty-three and twenty-four, he invented the field of mathematics called calculus and clarified his ideas on motion and gravitation (figure 3.2). After the plague he returned to Cambridge, where he was appointed professor of mathematics at the age of twenty-six. He lectured and presented papers on optics. One paper on his theory about light and colors caused such a controversy that Newton resolved never to publish another line. Newton was a shy, introspective person who was too absorbed in his work for such controversy. In 1684, Edmund Halley (of Halley's comet fame) asked Newton to resolve a dispute involving planetary motions. Newton had already worked out the solution to this problem in addition to other problems on gravity and motion. Halley persuaded the reluctant Newton to publish the material. Two years later, in 1687, Newton published *Principia,* which was paid for by Halley. In spite of his fear of controversy, the book was accepted almost at once and established Newton as one of the greatest thinkers who ever lived.

Newton built his theory of motion on the previous work of Galileo and others. In fact, Newton's first law is similar to the concept of inertia presented earlier by Galileo. Newton acknowledged the contribution of Galileo and others to his work, stating that if he had seen further than others "it was by standing upon the shoulders of giants."

Figure 3.2

Among other accomplishments, Isaac Newton invented calculus, developed the laws of motion, and developed the law of gravitational attraction.

Newton's First Law of Motion

Newton's first law of motion is also known as the *law of inertia* and is very similar to one of Galileo's findings about motion. Recall that Galileo used the term *inertia* to describe the tendency of an object to resist changes in motion. Newton's first law describes this tendency more directly. In modern terms (not Newton's words), the **first law of motion** is as follows:

Every object retains its state of rest or its state of uniform straight-line motion unless acted upon by an unbalanced force.

This means that an object at rest will remain at rest unless it is put into motion by an unbalanced force, that is, the vector sum of the forces must be greater than zero if more than one force is involved. Likewise, an object moving with uniform straight-line motion will retain that motion unless an unbalanced force causes it to speed up, slow down, or change its direction of travel. Thus, Newton's first law describes the tendency of an object to resist *any* change in its state of motion, a property defined by inertia.

Some objects have greater inertia than other objects. For example, it is easier to push a compact car into motion than to push a heavy truck into motion. The truck has greater inertia than the compact car. It is also more difficult to stop the heavy truck from moving than it is to stop a compact car. Again the heavy truck has greater inertia. The amount of inertia an object

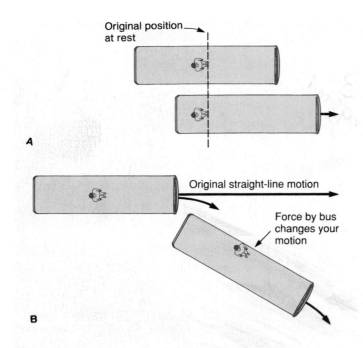

Original position at rest

A

Original straight-line motion

Force by bus changes your motion

B

Figure 3.3
Top view of a person standing in the aisle of a bus. (*a*) The bus is at rest, then starts to move forward. Inertia causes the person to remain in the original position, appearing to fall backward. (*b*) The bus turns to the right but inertia causes the person to retain the original straight-line motion until forced by the side of the bus into a new direction.

has describes the **mass** of the object. Mass is a *measure of inertia*. The more inertia an object has, the greater its mass. Thus the heavy truck has more mass than the compact car. You know this because the truck has greater inertia. Newton originally defined mass as the "quantity of matter" in an object, and this definition is intuitively appealing. However, Newton needed to measure inertia because of its obvious role in motion and redefined mass as a measure of inertia. Thinking of mass in terms of a resistance to a change of motion may seem strange at first, but it will begin to make more sense as you explore the relationships between mass, forces, and acceleration.

Think of Newton's first law of motion when you ride standing in the aisle of a bus. The bus begins to move and you, being an independent mass, tend to remain at rest. You take a few steps back as you tend to maintain your position relative to the ground outside. You reach for a seat back or some part of the bus. Once you have a hold on some part of the bus it supplies the forces needed to give you the same motion as the bus and you no longer find it necessary to step backward. You now have the same motion as the bus and no forces are involved. Until, that is, the bus goes around a curve. You now feel a tendency to move to the side of the bus. The bus has changed its straight-line motion but you, again being an independent mass, tend to move straight ahead. The side of the seat forces you into following the curved motion of the bus. The forces you feel when the bus starts moving or turning are a result of your inertia. You tend to remain at rest or follow a straight path until your motion is corrected by forces to be the same as the bus (figure 3.3).

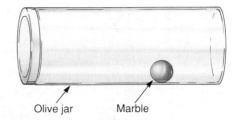

Olive jar   Marble

Figure 3.4
This marble can be used to demonstrate inertia. See the activities below.

Consider a second person standing next to you in the aisle of the moving bus. The bus stops and the two of you tend to retain your straight-line motion and you move forward in the aisle. If it is more difficult for the other person to stop moving, you know that the other person has greater inertia. Since mass is a measure of inertia you know that the other person has more mass than you. But do not confuse this with the other person's weight. Weight is a different property than mass and is explained by Newton's second law of motion.

ACTIVITIES

1. Place a marble inside any cylindrical container such as an olive bottle or pill vial. Place the container on its side on top of a table and push it along with the mouth at the front end (figure 3.4). Note the position of the marble inside the container. Stop the container immediately. Explain the reaction of the marble.
2. Place the marble inside the container and push it along as before. This time push the container over a carpeted floor, a smooth floor, a waxed tabletop, and so forth. Notice the distance that the marble rolls over the different surfaces. Repeat the procedure with large and small marbles of various masses as you use the same velocity each time. Explain the differences you observe.

Newton's Second Law of Motion

Newton had successfully used Galileo's ideas to describe the nature of motion. Newton's first law of motion explains that any motion, once started, will continue with a constant velocity in a straight line unless a force acts on the moving object. This law not only describes motion but establishes the role of a force as well. A change of motion is therefore *evidence* of the action of some unbalanced force or forces. The association of forces and a change of motion is common in your everyday experience. You have felt forces on your back in an accelerating automobile, and you have felt other forces as the automobile turns or stops. You have also learned about gravitational forces that accelerate objects toward the surface of the earth. Forces and acceleration are involved in any change of motion. The amount of inertia, or mass, is also involved since inertia is a resistance to a change of motion. Newton's second law of motion is a relationship between *force, acceleration,* and *mass* that describes the cause of a change of motion (figure 3.5).

Figure 3.5

These bicycle riders know the relationship between force, acceleration, and mass.

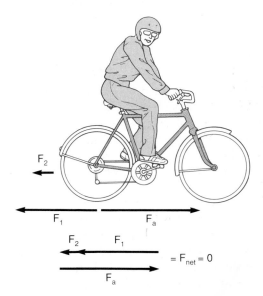

Figure 3.6

The force of tire friction (F_1) and the force of air resistance (F_2) have a vector sum that equals the force applied (F_a). The net force is therefore 0.

Consider the motion of you and a bicycle you are riding. Suppose you are riding your bicycle over level ground in a straight line at three miles per hour. Newton's first law tells you that this motion will continue with a constant velocity in a straight line as long as no external, unbalanced force acts on you and the bicycle. The force that you *are* exerting on the pedals seems to equal some external force that moves you and the bicycle along (more on this later). The force exerted as you move along is needed to *balance* the resisting forces of tire friction and air resistance. If these resisting forces were removed you would not need to exert any force at all to continue moving at a constant velocity. The net force is thus the force you are applying minus the forces from tire friction and air resistance. The *net force* is therefore zero (figure 3.6).

If you now apply a force on the pedals that is greater than the net force your velocity will increase. You will accelerate to an increasing velocity during, and *only* during, the time that applied force is unbalanced. Likewise, you will slow down if you apply a force to the brakes, another kind of resisting friction. A third way to change your velocity is to apply a force on the handlebars, changing the direction of your velocity. Thus, *unbalanced forces* on you and your bicycle produce an *acceleration*.

Starting a bicycle from rest suggests a relationship between force and acceleration. You observe that the harder you push on the pedals, the greater your acceleration. If you exert

45

Figure 3.7
More mass results in less acceleration when the same force is applied. With the same force applied, the riders and bike with twice the mass will have half the acceleration.

a net force on the pedals you will accelerate to a certain velocity. If you double the net force you will double the acceleration, reaching the same velocity in half the time. Likewise, if you triple the net force you will increase the acceleration threefold. Recall that when quantities increase or decrease together in the same ratio, they are said to be *directly proportional*. The acceleration is therefore directly proportional to the net force applied. Recall also that the symbol ∝ means "is proportional to." The relationship between acceleration (a) and the unbalanced force (F) can thus be abbreviated as

$$\text{acceleration} \propto \text{force}$$
or in symbols,
$$a \propto F$$

Suppose that your bicycle has two seats and you have a friend who will ride with you. Suppose also that the addition of your friend on the bicycle will double the mass of the bike and riders. If you use the same net force as before, the bicycle will undergo a much smaller acceleration. In fact, doubling the mass and applying the same force will produce an acceleration of only half as much (figure 3.7). An even more massive friend would reduce the acceleration even more. If you triple the mass and apply the same net force, the acceleration will be one-third as

much. Recall that when a relationship between two quantities shows that one quantity increases as another decreases, in the same ratio, the quantities are said to be *inversely proportional*. The acceleration (a) of an object is therefore inversely proportional to its mass (m). This relationship can be abbreviated as

$$\text{acceleration} \propto \frac{1}{\text{mass}}$$

or in symbols,

$$a \propto \frac{1}{m}$$

Since the mass (m) is in the denominator, doubling the mass will result in one-half the acceleration, tripling the mass will result in one-third the acceleration, and so forth.

Now the relationships can be combined to give

$$\text{acceleration} \propto \frac{\text{force}}{\text{mass}}$$

or

$$a \propto \frac{F}{m}$$

The acceleration of an object therefore depends on *both* the *net force applied* and the *mass* of the object. The **second law of motion** is as follows:

The acceleration of an object is directly proportional to the net force acting on it and inversely proportional to the mass of the object.

The proportion between acceleration, force, and mass is not yet an equation because the units have not been defined. The unit of force is *defined* as a force that will produce an acceleration of 1.0 m/sec² when applied to an object with a mass of 1.0 kg. This unit of force is called the **newton** (N) in honor of Isaac Newton. Now that a unit of force has been defined the equation for Newton's second law of motion can be written. First, rearrange

$$a \propto \frac{F}{m} \qquad \text{to} \qquad F \propto ma$$

then replace the proportionality with the more explicit equal sign and Newton's second law becomes

$$F = ma \qquad\qquad \textbf{equation 3.1}$$

The newton unit of force is a derived unit that is based on the three fundamental units of mass, length, and time. This is readily observed if the units are placed in the second law equation

$$F = ma$$

$$1N = (1 \text{ kg})\left(1 \ \frac{m}{\text{sec}^2}\right)$$

or

$$1 \text{ N} = 1 \ \frac{\text{kg} \cdot \text{m}}{\text{sec}^2}$$

Until now, equations were used to *describe properties* of matter such as density, velocity, and acceleration. This is your first example of an equation that is used to *define a concept,* specifically the concept of what is meant by a force. Since the concept is defined by specifying a measurement procedure, it is also an example of an *operational definition.* You are not only told what a newton of force is but also how to go about measuring it. Notice that the newton is defined in terms of mass measured in kg and acceleration measured in m/sec². Any other units must be converted to kg and m/sec² before a problem can be solved for newtons of force.

Example 3.1

A 60.0 kg bicycle and rider accelerate at 1.5 m/sec². How much force was applied?

Solution

The mass (m) of 60.0 kg and the acceleration (a) of 1.5 m/sec² are given. The problem asked for the applied force (F) needed to give the mass the acquired acceleration. The relationship is found in equation 3.1, F = ma.

$$m = 60.0 \text{ kg} \qquad F = ma$$

$$a = 1.5 \frac{m}{sec^2} \qquad = (60.0 \text{ kg})\left(1.5 \frac{m}{sec^2}\right)$$

$$F = ? \qquad = (60.0 \times 1.5) \text{ kg} \times \frac{m}{sec^2}$$

$$= 90 \frac{kg \cdot m}{sec^2}$$

$$= \boxed{90. \text{ N}}$$

(Note that the units kg · m/sec² is the definition of a newton of force, so the symbol N is used.)

Example 3.2

What is the acceleration of a 20 kg cart if the net force on it is 40 N? (Answer: 2 m/sec²)

The difference between mass and weight can be confusing since they are proportional to one another. If you double the mass of an object, for example, its weight is also doubled. But there is an important distinction between mass and weight and, in fact, they are different concepts. *Weight is a downward force,* the gravitational force acting on an object. Mass, on the other hand, refers to the amount of matter in the object and is *independent* of the force of gravity. Mass is measured by inertia, the extent to which the object resists a change of motion. Recall that the force of gravity varies from place to place on the surface of the earth. It is slightly less, for example, in Colorado than in Florida. Imagine that you could be instantly transported from Florida to Colorado. You would find that you weigh less in Colorado than you did in Florida even though the amount of matter in you has not changed. Now imagine that you could be instantly transported to the moon. You would weigh one-sixth

Table 3.1

Units of mass and weight in the metric and English systems of measurement

	Mass	×	Acceleration	=	Force
Metric System	kg	×	$\frac{m}{sec^2}$	=	$\frac{kg \cdot m}{sec^2}$ (newton)
English System	$\left(\frac{lb}{32.0 \text{ ft/sec}^2}\right)$	×	ft/sec²	=	lb (pound)

as much on the moon because the force of gravity on the moon is one-sixth of that on the earth. Yet, your mass would be the same in both locations.

Weight is a measure of the force of gravity acting on an object and this force can be calculated from Newton's second law of motion,

$$F = ma$$

or

downward force = (mass)(acceleration due to gravity)

or

weight = (mass)(*g*)

or $W = mg$ **equation 3.2**

In the metric system, *mass* is measured in kilograms. The acceleration due to gravity, *g*, is 9.80 m/sec². According to equation 3.2, weight is mass times acceleration. A kilogram multiplied by an acceleration measured in m/sec² results in kg · m/sec², a unit you now recognize as a force called a newton. The *unit of weight* in the metric system is therefore the *newton (N).*

In the English system the pound is the unit of *force.* The acceleration due to gravity, *g*, is 32.0 ft/sec². The force unit of a pound is defined as the force required to accelerate a unit of mass called the *slug.* Specifically, a force of 1.0 lb will give a 1.0 slug mass an acceleration of 1.0 ft/sec². Since W = mg, the *mass* in slugs can be found by simply dividing the *weight* of an object by *g*. Thus a 100 lb person has a mass of $\frac{100 \text{ lb}}{32.0 \text{ ft/sec}^2}$, or 3.1 slugs. It is not necessary to use the term slug if you use the ratio of $\frac{lb}{32.0 \text{ ft/sec}^2}$ for a mass unit in calculations. These units are compared in table 3.1.

The important thing to remember is that *pounds* and *newtons* are units of *force.* A *kilogram,* on the other hand, is a measure of *mass.* Thus the English unit of 1.0 lb is comparable to the metric unit of 4.5 N (or 0.22 lb is equivalent to 1.0 N). Conversion tables sometimes show how to convert from pounds (a unit of weight) to kilograms (a unit of mass). This is possible because weight and mass are proportional on the surface of the earth. It can also be confusing since some variables depend on weight and others depend on mass. To avoid confusion, it is important to remember the distinction between weight and mass and that a kilogram is a unit of mass. Newtons and pounds are units of force that can be used to measure weight.

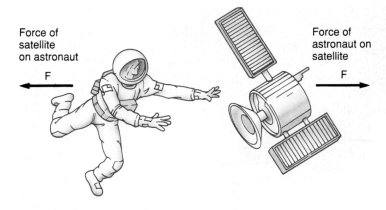

Force of satellite on astronaut

F ←

Force of astronaut on satellite

→ F

Figure 3.8
Forces occur in matched pairs that are equal in magnitude and opposite in direction.

Example 3.3

What is the weight of a 60.0 kg person on the surface of the earth?

Solution

A mass (m) of 60.0 kg is given and the acceleration due to gravity (*g*) 9.80 m/sec² is implied. The problem asked for the weight (W). The relationship is found in equation 3.2, W = m*g*, which is a form of F = ma.

$$m = 60.0 \text{ kg}$$

$$g = 9.80 \frac{m}{sec^2}$$

$$W = ?$$

$$W = mg$$

$$= (60.0 \text{ kg})\left(9.80 \frac{m}{sec^2}\right)$$

$$= 60.0 \times 9.80 \text{ kg} \times \frac{m}{sec^2}$$

$$= 588 \frac{kg \cdot m}{sec^2}$$

$$= \boxed{588 \text{ N}}$$

Example 3.4

A 60.0 kg person weighs 100.0 N on the moon. What is the value of *g* on the moon? (Answer: 1.67 m/sec²)

ACTIVITIES

1. Stand on a bathroom scale in an operating elevator and note the reading when the elevator accelerates, decelerates, or has a constant velocity. See if you can figure out how to use the readings to calculate the acceleration of the elevator.
2. Calculate the acceleration of your car (see chapter 2). Check the owner's manual of the car (or see the dealer) to obtain the weight of the car. Convert this to mass (for English units, mass equals weight in pounds divided by 32.0 ft/sec²), then calculate the force, mass, and acceleration relationships.

Newton's Third Law of Motion

Newton's first law of motion states that an object retains its state of motion when the net force is zero. The second law states what happens when the net force is *not* zero, describing how the object moves when a given force is applied with a known mass. The two laws give one aspect of the concept of a force, that is, if you observe an object start moving, speed up, slow down, or change its direction of travel, you can conclude that an unbalanced force is acting on the object. Thus any change in the state of motion of an object is *evidence* that an unbalanced force has been applied.

Newton's third law of motion is also concerned with forces and considers how a force is produced. First, consider where a force comes from. A force is always produced by the interaction of two or more objects. There is always a second object pushing or pulling on the first object to produce a force. To simplify the many interactions that occur on earth, consider a satellite freely floating in space. According to Newton's second law (F = ma), a force must be applied to change the state of motion of the satellite. What is a possible source of such a force? Perhaps an astronaut pushes on the satellite for one second. The satellite would accelerate *during* the application of the force, then move away from the original position at some constant velocity. But the astronaut, too, would move away from the original position but in the opposite direction (figure 3.8). A *single* force *does not exist* by itself. There is always a matched and opposite force that occurs at the same time. Thus the astronaut exerted a momentary force on the satellite. But the satellite evidently exerted a momentary force back on the astronaut as well, for the astronaut moved away from the original position in the opposite direction. Newton did not have astronauts and satellites to think about, but this is the kind of reasoning he did when he concluded that forces always occur in matched pairs that are equal and opposite. Thus the **third law of motion** is as follows:

> **Whenever two objects interact, the force exerted on one object is equal in size and opposite in direction to the force exerted on the other object.**

The third law states that forces always occur in matched pairs that act in opposite directions and on two *different* bodies. You could express this law with symbols as

$$F_{A \text{ due to } B} = F_{B \text{ due to } A} \qquad \text{equation 3.3}$$

where the force on the astronaut, for example, would be "A due to B" and the force on the satellite would be "B due to A."

Sometimes the third law of motion is expressed, "For every action there is an equal and opposite reaction," but this can be misleading. Neither force is the cause of the other. The forces

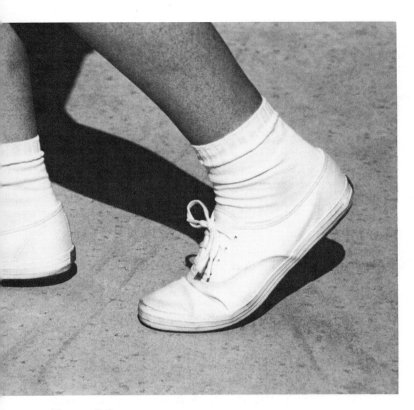

Figure 3.9
This foot is pushing against the ground, but it is the ground pushing against the foot that accelerates the person forward.

the same mass as you, however, it would accelerate backward at the same rate that you were accelerated forward. The earth is much more massive than you, however, so any acceleration of the earth is a vanishingly small amount. The overall effect is that you are accelerated forward by the force the ground exerts on you.

Return now to the example of riding a bicycle that was discussed previously. What is the source of the *external* force that accelerates you and the bike? Pushing against the pedals is not external to you and the bike so that force will *not* accelerate you and the bicycle forward. This force is transmitted through the bike mechanism to the rear tire, which pushes against the ground. It is the ground exerting an equal and opposite force against the system of you and the bike that accelerates you forward. You must consider the forces that act on the system of the bike and you before you can apply F = ma. The only forces that will affect the forward motion of the bike system are the force of the ground pushing it forward and the frictional forces that oppose the forward motion. This is another example of the third law.

Example 3.5

A 60.0 kg astronaut is freely floating in space and pushes on a freely floating 120.0 kg satellite with a force of 30.0 N for 1.50 sec. (a) Compare the force exerted on the astronaut and the satellite, and (b) compare the acceleration of the astronaut to the acceleration of the satellite.

Solution
(a) According to Newton's third law of motion (equation 3.3),

$$F_{A \text{ due to } B} = F_{B \text{ due to } A}$$

$$30.0 \text{ N} = 30.0 \text{ N}$$

Both feel a 30.0 N for 1.50 sec but in opposite directions.

(b) Newton's second law describes a relationship between force, mass, and acceleration, F = ma.
For the astronaut:

$$m = 60.0 \text{ kg}$$
$$F = 30.0 \text{ N}$$
$$a = ?$$

$$F = ma \therefore a = \frac{F}{m}$$

$$a = \frac{30.0 \frac{\text{kg} \cdot \text{m}}{\text{sec}^2}}{60.0 \text{ kg}}$$

$$= \frac{30.0}{60.0} \frac{\text{kg} \cdot \text{m}}{\text{sec}^2} \times \frac{1}{\text{kg}}$$

$$= 0.500 \frac{\text{kg} \cdot \text{m}}{\text{kg} \cdot \text{sec}^2}$$

$$= \boxed{0.500 \frac{\text{m}}{\text{sec}^2}}$$

are at every instant the cause of each other and they appear and disappear at the same time. If you are going to describe the force exerted on a satellite by an astronaut, then you must realize that there is a simultaneous force exerted on the astronaut by the satellite. The forces (astronaut on satellite and satellite on astronaut) are equal in magnitude but opposite in direction.

Perhaps it would be more common to move a satellite with a small rocket. A satellite is maneuvered in space by firing a rocket in the direction opposite to the direction someone wants to move the satellite. Exhaust gases (or compressed gases) are accelerated in one direction and the expelled gases exert an equal but opposite force on the satellite that accelerates it in the opposite direction. This is another example of the third law.

Consider how the pairs of forces work on the earth's surface. You walk by pushing your feet against the ground (figure 3.9). Of course you could not do this if it were not for friction. You would slide as on slippery ice without friction. But since friction does exist, you exert a backward horizontal force on the ground and, as the third law explains, the ground exerts an equal and opposite force on you. You accelerate forward from the unbalanced force as explained by the second law. If the earth had

Patterns of Motion

For the satellite:

$$m = 120.0 \text{ kg}$$
$$F = 30.0 \text{ N}$$
$$a = ?$$

$$F = ma \therefore a = \frac{F}{m}$$

$$a = \frac{30.0 \dfrac{\text{kg} \cdot \text{m}}{\text{sec}^2}}{120.0 \text{ kg}}$$

$$= \frac{30.0}{120.0} \frac{\text{kg} \cdot \text{m}}{\text{sec}^2} \times \frac{1}{\text{kg}}$$

$$= 0.250 \frac{\text{kg} \cdot \text{m}}{\text{kg} \cdot \text{sec}^2}$$

$$= \boxed{0.250 \frac{\text{m}}{\text{sec}^2}}$$

Example 3.6

After the interaction and acceleration between the astronaut and satellite described previously, they both move away from their original positions. What is the new speed for each? (Answer: Astronaut $v_f = 0.750$ m/sec. Satellite $v_f = 0.375$ m/sec) (Hint: $v_f = \bar{a}t + v_i$)

Momentum

Sportscasters often refer to the "momentum" of a team, and newscasters sometimes refer to an election where one of the candidates has "momentum." Both situations describe a competition where one side is moving toward victory and it is difficult to stop. It seems appropriate to borrow this term from the physical sciences because momentum is a property of movement. It takes a longer time to stop something from moving when it has a lot of momentum. The physical science concept of momentum is closely related to Newton's laws of motion. **Momentum** (p) is defined as the product of the mass (m) of an object and its velocity (v),

$$\text{momentum} = \text{mass} \times \text{velocity}$$

or

$$p = mv \qquad\qquad \textbf{equation 3.4}$$

The astronaut in example 3.5 had a mass of 60.0 kg and a velocity of 0.750 m/sec as a result of the interaction with the satellite. The resulting momentum was therefore (60.0 kg) (0.750 m/sec), or 45.0 kg·m/sec. As you can see, the momentum would be greater if the astronaut had acquired a greater velocity or if the astronaut had a greater mass and acquired the same velocity. Momentum involves both the inertia and the velocity of a moving object.

Notice that the momentum acquired by the satellite in example 3.5 is *also* 45.0 kg·m/sec. The astronaut gained a certain momentum in one direction and the satellite gained the *very*

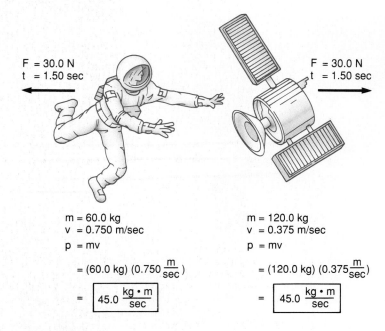

$$F = 30.0 \text{ N}$$
$$t = 1.50 \text{ sec}$$

$$F = 30.0 \text{ N}$$
$$t = 1.50 \text{ sec}$$

$$m = 60.0 \text{ kg}$$
$$v = 0.750 \text{ m/sec}$$
$$p = mv$$

$$= (60.0 \text{ kg}) (0.750 \frac{\text{m}}{\text{sec}})$$

$$= \boxed{45.0 \frac{\text{kg} \cdot \text{m}}{\text{sec}}}$$

$$m = 120.0 \text{ kg}$$
$$v = 0.375 \text{ m/sec}$$
$$p = mv$$

$$= (120.0 \text{ kg}) (0.375 \frac{\text{m}}{\text{sec}})$$

$$= \boxed{45.0 \frac{\text{kg} \cdot \text{m}}{\text{sec}}}$$

Figure 3.10

Both the astronaut and the satellite received a force of 30.0 N for 1.50 sec when they pushed on each other. Both then have a momentum of 45.0 kg·m/sec in the opposite direction. This is an example of the law of conservation of momentum.

same momentum in the opposite direction. Newton originally defined the second law in terms of a rate of change of momentum being proportional to the net force applied to an object. Since the third law explains that the forces exerted on both the astronaut and satellite were equal and opposite, you would expect both objects to acquire equal momentum in the opposite direction. This result is observed any time a system of objects interact and the only forces involved are those between the interacting objects (figure 3.10). This statement leads to a particular kind of relationship called a *law of conservation.* In this case the law applies to momentum and is called the **law of conservation of momentum:**

> **The total momentum of a group of interacting objects remains the same in the absence of external forces.**

The law of conservation of momentum is useful in analyzing motion in simple systems of collisions such as billiard balls, automobiles, or railroad cars. It is also useful in measuring action and reaction interactions such as rocket propulsion, where the backward momentum of the expelled gases equals the momentum given to the rocket in the opposite direction (figure 3.11). When this is done, momentum is found to be conserved each and every time.

Figure 3.11
The momentum of the expelled gases in one direction equals the
momentum of the rocket and space shuttle in the other direction.

Patterns of Motion

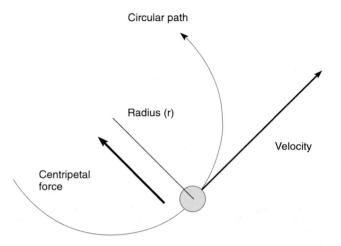

Figure 3.12
Centripetal force on the ball causes it to change direction continuously, or accelerate into a circular path. Without the unbalanced force acting on it, the ball would continue in a straight line.

Compared to the other concepts of motion, there are two aspects of momentum that are unusual: (1) the symbol for momentum (p) does not give a clue about the quantity it represents, and (2) the combination of metric units that results from a momentum calculation (kg·m/sec) does not have a name of its own. So far, you have been introduced to one combination of units with a name. The units kg·m/sec² are known as a unit of force called the newton (N). More combinations of metric units, all with names of their own, will be introduced later.

Forces and Circular Motion

Consider a communications satellite that is moving in a uniform speed around the earth in a circular orbit. According to the first law of motion there *must be* forces acting on the satellite since it does *not* move off in a straight line. The second law of motion also indicates forces since an unbalanced force is required to change the motion of an object.

Recall that acceleration is defined as a change in velocity and that velocity is a vector quantity, having both magnitude and direction. A vector quantity is changed by a change in speed, direction, or both speed and direction. The satellite in a circular orbit is continuously being accelerated. This means that there is a continuously acting unbalanced force acting on the satellite that pulls it out of a straight-line path.

The force that pulls an object out of its straight-line path and into a circular path is called a **centripetal** (center-seeking) **force.** Perhaps you have revolved a ball on the end of a string in a horizontal circle over your head. Once you have the ball moving, the only unbalanced force (other than gravity) acting on the ball is the centripetal force your hand exerts on the ball through the string. This centripetal force pulls the ball from its

natural straight-line path into a circular path. There are no outward forces acting on the ball. The force that you feel on the string is a consequence of the third law; the ball exerts an equal and opposite force on your hand. If you were to release the string, the ball would move away from the circular path in a *straight line* that has a right angle to the radius at the point of release (figure 3.12). When you release the string the centripetal force ceases and the ball then follows its natural straight-line motion. If other forces were involved it would follow some other path. Nonetheless, the apparent outward force has been given a name just as if it were a real force. The outward tug is called a **centrifugal force.**

The magnitude of the centripetal force required to keep an object in a circular path depends on the inertia, or mass, of the object and the acceleration of the object just as you learned in the second law of motion. The acceleration of an object moving in a circle can be shown by geometry or calculus to be directly proportional to the square of the speed around the circle (v^2) and inversely proportional to the radius of the circle (r). (A smaller radius requires a greater acceleration.) Therefore the acceleration of an object moving in uniform circular motion (a_c) is

$$a_c = \frac{v^2}{r} \qquad \text{equation 3.5}$$

The magnitude of the centripetal force of an object with a mass (m) that is moving with a velocity (v) in a circular orbit of a radius (r) can be found by substituting equation 3.5 in F = ma, or

$$F = \frac{mv^2}{r} \qquad \text{equation 3.6}$$

Example 3.7

A 0.25 kg ball is attached to the end of a 0.5 m string and moved in a horizontal circle at 2.0 m/sec. What net force is needed to keep the ball in its circular path?

Solution

m = 0.25 kg

r = 0.5 m

v = 2.0 m/sec

F = ?

$$F = \frac{mv^2}{r}$$

$$= \frac{(0.25 \text{ kg})(2.0 \text{ m/sec})^2}{0.5 \text{ m}}$$

$$= \frac{(0.25 \text{ kg})(4.0 \text{ m}^2/\text{sec}^2)}{0.5 \text{ m}}$$

$$= \frac{(0.25)(4.0)}{0.5} \frac{\text{kg} \cdot \text{m}^2}{\text{sec}^2} \times \frac{1}{\text{m}}$$

$$= 2 \frac{\text{kg} \cdot \text{m}^2}{\text{m} \cdot \text{sec}^2}$$

$$= 2 \frac{\text{kg} \cdot \text{m}}{\text{sec}^2}$$

$$= \boxed{2 \text{ N}}$$

Example 3.8

Suppose you make the string in example 3.7 half as long, 0.25 m. What force is now needed? (Answer: 4.0 N)

▲

Newton's Law of Gravitation

You know that if you drop an object it always falls to the floor. You define "down" as the direction of the object's movement and "up" as the opposite direction. Objects fall because of the force of gravity, which accelerates objects at $g = 9.80$ m/sec² (32.0 ft/sec²) and gives them weight, $W = mg$.

Gravity is an attractive force, a pull that exists between all objects in the universe. It is a mutual force that, just like all other forces, comes in matched pairs. Since the earth attracts you with a certain force, you must attract the earth with an exact opposite force. The magnitude of this force of mutual attraction depends on several variables. These variables were first described by Newton in *Principia,* his famous book on motion that was printed in 1687. Newton had, however, worked out his ideas much earlier by the age of twenty-four, along with ideas about his laws of motion and the formula for centripetal acceleration. In a biography written by a friend in 1752, Newton stated that the notion of gravitation came to mind during a time of thinking that "was occasioned by the fall of an apple." He was thinking about why the moon stays in orbit around the earth rather than moving off in a straight line as would be predicted by the first law of motion. Perhaps the same force that attracts the moon toward the earth, he thought, attracts the apple to the earth. Newton developed a theoretical equation for gravitational force that explained not only the motion of the moon but the motion of the whole solar system. Today, this relationship is known as the **universal law of gravitation:**

> **Every object in the universe is attracted to every other object with a force that is directly proportional to the product of their masses and inversely proportional to the square of the distances between them.**

In symbols, m_1 and m_2 can be used to represent the masses of two objects, d the distance between their centers, and G a constant of proportionality. The equation for the law of universal gravitation is therefore

$$F = G \frac{m_1 m_2}{d^2}$$

equation 3.7

The constant G is a universal constant since the law applies to all objects in the universe. It was first measured experimentally by Henry Cavendish in 1798. The accepted value today is $G = 6.67 \times 10^{-11}$ N·m²/kg². Do not confuse G, the universal constant, with g, the acceleration due to gravity on the surface of the earth.

Thus the magnitude of the force of gravitational attraction is determined by the mass of the two objects and the distance between them (figure 3.13). The law also states that *every*

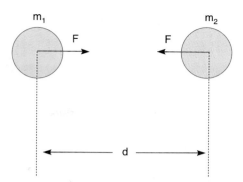

Figure 3.13

The variables involved in gravitational attraction. The force of attraction (F) is proportional to the product of the masses (m_1, m_2) and inversely proportional to the square of the distance (d) between the centers of the two masses.

object is attracted to every other object. You are attracted to all the objects around you—chairs, tables, other people, and so forth. Why don't you notice the forces between you and other objects? The answer is in example 3.9.

Example 3.9

▼

What is the force of gravitational attraction between two 60.0 kg (132 lb) students who are standing 1.00 m apart?

Solution

$G = 6.67 \times 10^{-11}$ N · m²/kg²

$m_1 = 60.0$ kg

$m_2 = 60.0$ kg

$d = 1.00$ m

$F = ?$

$$F = G \frac{m_1 m_2}{d^2}$$

$$= \frac{(6.67 \times 10^{-11} \text{ N} \cdot \text{m}^2/\text{kg}^2)(60.0 \text{ kg})(60.0 \text{ kg})}{(1.00 \text{ m})^2}$$

$$= (6.67 \times 10^{-11})(3.6 \times 10^3) \frac{\text{N} \cdot \text{m}^2 \cdot \cancel{\text{kg}^2}}{\cancel{\text{kg}^2}}$$

$$= 2.4 \times 10^{-7} \text{ N} \cdot \text{m}^2 \times \frac{1}{\text{m}^2}$$

$$= 2.4 \times 10^{-7} \frac{\text{N} \cdot \cancel{\text{m}^2}}{\cancel{\text{m}^2}}$$

$$= \boxed{2.4 \times 10^{-7} \text{ N}}$$

(Note: A force of 2.4×10^{-7} (0.00000024) N is equivalent to a force of 3.0×10^{-6} (0.000003) lb, a force that you would not notice. In fact, it would be difficult to measure such a small force.)

▲

Patterns of Motion

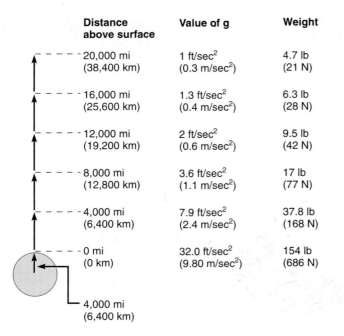

Distance above surface	Value of g	Weight
20,000 mi (38,400 km)	1 ft/sec² (0.3 m/sec²)	4.7 lb (21 N)
16,000 mi (25,600 km)	1.3 ft/sec² (0.4 m/sec²)	6.3 lb (28 N)
12,000 mi (19,200 km)	2 ft/sec² (0.6 m/sec²)	9.5 lb (42 N)
8,000 mi (12,800 km)	3.6 ft/sec² (1.1 m/sec²)	17 lb (77 N)
4,000 mi (6,400 km)	7.9 ft/sec² (2.4 m/sec²)	37.8 lb (168 N)
0 mi (0 km)	32.0 ft/sec² (9.80 m/sec²)	154 lb (686 N)

4,000 mi
(6,400 km)

Figure 3.14

The force of gravitational attraction decreases inversely with the square of the distance from the earth's center. Note the weight of a 686 N (154 lb) person at various distances above the earth's surface.

As you can see in example 3.9, one or both of the interacting objects must be quite massive before a noticeable force results from the interaction. That is why you do not notice the force of gravitational attraction between you and objects that are not very massive compared to the earth. The attraction between you and the earth overwhelmingly predominates and that is all you notice.

Newton was able to show that the distance used in the equation is the distance from the center of one object to the center of the second object. This does not mean that the force originates at the center, but that the overall effect is the same as considering all the mass concentrated at a center point. The weight of an object, for example, can be calculated by using a form of Newton's second law, $F = ma$. This general law shows a relationship between *any* force, the mass of a body, and the resulting acceleration. When the acceleration is due to gravity, the equation becomes $F = mg$. The law of gravitation deals *specifically with the force of gravity* and how it varies with distance and mass. Since weight is a force, then $F = mg$. You can write the two equations together,

$$mg = G\frac{mm_e}{d^2}$$

where m is the mass of some object on earth, m_e is the mass of the earth, g is the acceleration due to gravity, and d is the distance between the centers of the masses. Canceling the m's in the equation leaves

$$g = G\frac{m_e}{d^2}$$

which tells you that the acceleration due to gravity, 9.80 m/sec², is a constant because the other three variables are constant. Since the m's canceled you also know that the mass of an object does not affect the rate of free fall; all objects fall at the same rate, with the same acceleration, no matter what their masses are.

Example 3.10

The surface of the earth is approximately 6,400 km from its center. If the mass of the earth is 6.0×10^{24} kg, what is the acceleration due to gravity, g, near the surface?

$G = 6.67 \times 10^{-11} N \cdot m^2/kg^2$

$m_e = 6.0 \times 10^{24}$ kg

$d = 6400$ km

$g = ?$

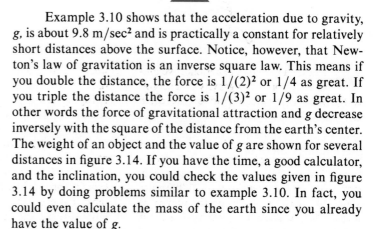

(Note: In the unit calculation, remember that a newton is a kg · m/sec².)

Example 3.10 shows that the acceleration due to gravity, g, is about 9.8 m/sec² and is practically a constant for relatively short distances above the surface. Notice, however, that Newton's law of gravitation is an inverse square law. This means if you double the distance, the force is $1/(2)^2$ or $1/4$ as great. If you triple the distance the force is $1/(3)^2$ or $1/9$ as great. In other words the force of gravitational attraction and g decrease inversely with the square of the distance from the earth's center. The weight of an object and the value of g are shown for several distances in figure 3.14. If you have the time, a good calculator, and the inclination, you could check the values given in figure 3.14 by doing problems similar to example 3.10. In fact, you could even calculate the mass of the earth since you already have the value of g.

Using reasoning similar to that found in example 3.10, Newton was able to calculate the acceleration of the moon toward the earth, about 0.0027 m/sec². The moon "falls" toward the earth because it is accelerated by the force of gravitational attraction. This attraction acts as a *centripetal force* that keeps the moon from following a straight-line path as would be predicted from the first law. Thus the acceleration of the moon keeps it in a somewhat circular orbit around the earth. Figure 3.15 shows that the moon would be in position A if it followed a

Space Station Weightlessness

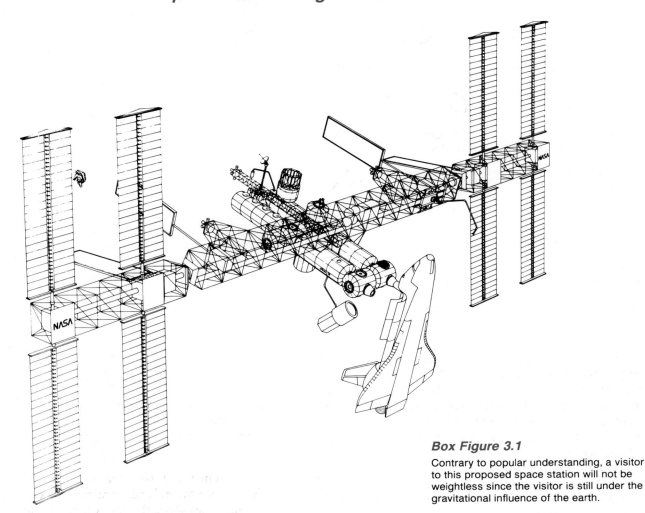

Box Figure 3.1

Contrary to popular understanding, a visitor to this proposed space station will not be weightless since the visitor is still under the gravitational influence of the earth.

When do astronauts experience weightlessness, or "zero gravity"? Theoretically, the gravitational field of the earth extends to the whole universe. You know that it extends to the moon, and indeed, even to the sun some 93 million miles away. There is a distance, however, at which the gravitational force must become immeasurably small. But even at an altitude of 20,000 miles above the surface of the earth, gravity is measurable. At 20,000 miles the value of g is about 1 ft/sec^2 (0.3 m/sec^2) compared to 32.0 ft/sec^2 (9.80 m/sec^2) on the surface. Since gravity does exist at these distances, how can an astronaut experience "zero gravity"?

Gravity does act on astronauts in spacecraft that are in orbit around the earth. The spacecraft stays in orbit, in fact, because of the gravitational attraction and because it has the correct tangential speed. If the tangential speed were less than 5 mi/sec, the spacecraft

would return to the earth. Astronauts fire their "retro-rockets," which slow the tangential speed, causing the spacecraft to fall down to the earth. If the tangential speed were more than 7 mi/sec, the spacecraft would fly off into space. The spacecraft stays in orbit because it has the right tangential speed to continuously "fall" around and around the earth. Gravity provides the necessary centripetal force that causes the spacecraft to "fall" out of its natural straight-line motion.

Since gravity is acting on the astronaut and spacecraft, the term "zero gravity" is not an accurate description of what is happening. The astronaut, spacecraft, and everything in it are experiencing *apparent weightlessness* because they are continuously "falling" toward the earth (box figure 3.1). Everything seems to float because everything is falling together. But, strictly speaking, everything still has weight because weight is defined

as a gravitational force acting on an object ($W = mg$).

Whether an apparent weightlessness or a real weightlessness, however, the effects on people are the same. Long-term orbital flights have provided evidence that the human body changes from the effect of weightlessness. Bones lose calcium and minerals, the heart shrinks to a much smaller size, and leg muscles shrink so much on prolonged flights that astronauts cannot walk when they return to the earth. These and other problems resulting from prolonged weightlessness must be worked out before long-term weightless flights can take place. One solution to these problems might be a large, uniformly spinning spacecraft. The astronauts tend to move in a straight line and the side of the turning spacecraft (now the "floor") exerts a force on them to make them go in a curved path. This force would act as an artificial gravity.

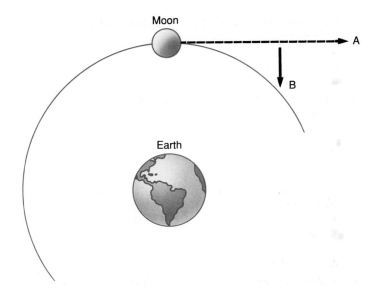

Figure 3.15

Gravitational attraction acts as a centripetal force that keeps the moon from following a straight-line path as shown by position A. It was pulled to position B by gravity (0.0027 m/sec²) and thus "fell" toward the earth the distance A to B, resulting in a somewhat circular path.

straight-line path instead of "falling" to position B as it does. The moon thus "falls" around the earth. Newton was able to analyze the motion of the moon quantitatively as evidence that it is gravitational force that keeps the moon in its orbit. The law of gravitation was extended to the sun, other planets, and eventually the universe. The quantitative predictions of observed relationships among the planets was strong evidence that all objects obey the same law of gravitation. In addition, the law provided a means to calculate the mass of the earth, the moon, the planets, and the sun. Newton's law of gravitation, laws of motion, and work with mathematics formed the basis of most physics and technology for the next two centuries as well as accurately describing the world of everyday experience.

Summary

Isaac Newton developed a complete explanation of motion with three laws of motion. The laws explain the role of a *force* and the *mass* of an object involved in a *change of motion.*

Newton's *first law of motion* is concerned with the motion of an object and the *lack* of an unbalanced force. Also known as the *law of inertia*, the first law states that an object will retain its state of straight-line motion (or state of rest) unless an unbalanced force acts on it. The amount of resistance to a change of motion, *inertia*, describes the *mass* of an object.

The *second law of motion* describes a relationship between *force, mass,* and *acceleration.* The relationship is $F = ma$. A *newton* of force is the force needed to give a 1.0 kg mass an acceleration of 1.0 m/sec². *Weight* is the downward force that results from the earth's gravity acting on the mass of an object. Weight can be calculated from $F = ma$, or $W = mg$. Weight is measured in *newtons* in the metric system and *pounds* in the English system.

Newton's *third law of motion* states that forces are produced by the interaction of *two different* objects and that these forces *always* occur in *matched pairs* that are *equal* and *opposite*. These forces are capable of producing an acceleration in accord with the second law of motion.

Momentum is the product of the mass of an object and its velocity. In the absence of external forces, the momentum of a group of interacting objects always remains the same. This relationship is the *law of conservation of momentum.*

An object moving in a circular path must have a force acting on it since it does not move off in a straight line. The force that pulls an object out of its straight-line path is called a *centripetal force*. The centripetal force needed to keep an object in a circular path depends on the mass (m) of the object, its velocity (v), and the size of the circle (r), or

$$F = \frac{mv^2}{r}$$

The *universal law of gravitation* is a relationship between the masses of two objects, the distance between the objects, and a proportionality constant. The relationship is

$$F = G \frac{m_1 m_2}{r^2}$$

Newton was able to use this relationship to show that gravitational attraction provides the centripetal force that keeps the moon in its orbit. This relationship was found to explain the relationship between all parts of the Solar System.

Summary of Equations

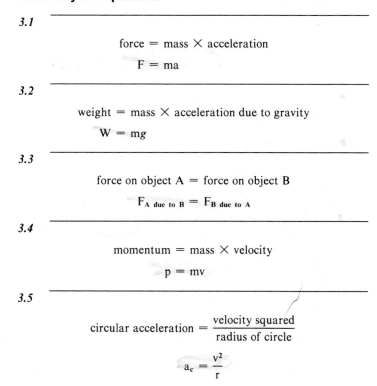

3.1

force = mass × acceleration

$$F = ma$$

3.2

weight = mass × acceleration due to gravity

$$W = mg$$

3.3

force on object A = force on object B

$$F_{A \text{ due to } B} = F_{B \text{ due to } A}$$

3.4

momentum = mass × velocity

$$p = mv$$

3.5

$$\text{circular acceleration} = \frac{\text{velocity squared}}{\text{radius of circle}}$$

$$a_c = \frac{v^2}{r}$$

3.6

$$\text{centripetal force} = \frac{\text{mass} \times \text{velocity squared}}{\text{radius of circle}}$$

$$F = \frac{mv^2}{r}$$

3.7

$$\text{gravitational force} = \text{constant} \times \frac{\text{one mass} \times \text{another mass}}{\text{distance squared}}$$

$$F = G\,\frac{m_1 m_2}{d^2}$$

Key Terms

centrifugal force (p. 52)
centripetal force (p. 52)
first law of motion (p. 43)
law of conservation of
   momentum (p. 50)
mass (p. 44)

momentum (p. 50)
newton (p. 46)
second law of motion (p. 46)
third law of motion (p. 48)
universal law of gravitation
   (p. 53)

Applying the Concepts

1. The extent of resistance to a change of motion is determined by an object's
   a. weight.
   b. mass.
   c. density.
   d. all of the above.

2. Mass is a measure of
   a. inertia.
   b. how difficult it is to stop a moving object.
   c. how difficult it is to change the direction of travel of a moving object.
   d. all of the above.

3. A change in the state of motion is evidence of
   a. a force.
   b. an unbalanced force.
   c. a force that has been worn out after an earlier application.
   d. any of the above.

4. Considering the forces on the system of you and a bicycle as you pedal the bike at a constant velocity in a straight line,
   a. all forces are in balance with the net force equal to zero.
   b. the force you are exerting on the pedal is greater than the resisting forces.
   c. the resisting forces of air and tire friction are less than the force you are exerting.
   d. the resisting forces are greater than the force you are exerting.

5. Which of the following is a unit that can be used for a measure of weight?
   a. kilogram
   b. newton
   c. kg·m/sec
   d. none of the above

6. A newton of force has what combination of units?
   a. kg·m/sec
   b. kg/m²
   c. kg/m²/sec²
   d. kg·m/sec²

7. Which of the following is *not* a unit or combination of units for a downward force?
   a. lb
   b. kg
   c. N
   d. kg·m/sec²

8. Which of the following is a unit for a measure of resistance to a change of motion?
   a. lb
   b. kg
   c. N
   d. all of the above

9. Doubling the distance between an orbiting satellite and the earth will result in what change in the gravitational attraction between the two?
   a. one-half as much
   b. one-fourth as much
   c. twice as much
   d. four times as much

10. If a ball swinging in a circle on a string is moved twice as fast the force on the string will be
    a. twice as great.
    b. four times as great.
    c. one-half as much.
    d. one-fourth as much.

Answers

1. b 2. d 3. b 4. a 5. b 6. d 7. b 8. b 9. b
10. b

Questions for Thought

1. Define a newton of force.

2. Can a small car have the same momentum as a heavy truck? Explain.

3. What evidence tells you about the existence of an unbalanced force?

4. What net force is needed to maintain the constant velocity of a car moving in a straight line? Explain.

5. How can there ever be an unbalanced force on an object if every action has an equal and opposite reaction?

6. Describe the relationship between force and motion when one or the other is varied when the mass remains constant.

7. Describe the relationship between mass and acceleration when the mass is varied as a constant force is applied.

8. You exert a force (push) on a table against a wall and the table does not move. Explain why the table does not move if F = ma.

9. Compare the force of gravity on an object at rest to the force of gravity on the very same object in free fall.

10. Does an object in free fall have weight? Explain.

11. What path would the moon take if it were possible to "turn off" gravity? Explain.

12. An operational definition describes a concept in terms of a measurement. Give operational definitions for (a) force, (b) mass, and (c) weight.

13. How is it possible for your weight to change if your mass remains constant?

14. What maintains the speed of the moving earth in its orbit around the sun?

15. What is wrong with listing the "net weight" of potato chips in grams or kilograms? What does "net" mean?

16. Suppose you are standing on the ice of a frozen lake and there is no friction whatsoever. How can you get off the ice? (Hint: Friction is necessary to crawl or walk, so that will not get you off the ice.)

17. A rocket blasts off from a platform on a space station. An identical rocket blasts off from free space. Considering everything else to be equal, will the two rockets have the same acceleration? Explain.

18. Is a constant force necessary for a constant acceleration? Explain.

19. Is an unbalanced force necessary to maintain a constant speed? Explain.

20. An astronaut leaves a space ship moving through free space to adjust an antenna. Will the space ship move off and leave the astronaut behind? Explain.

21. What problems would an astronaut in free space have while trying to tighten a bolt on a satellite with a wrench? Suggest a solution to the problem.

22. How can an artificial satellite stay in orbit around the earth without firing its rockets?

23. An object always moves in the same direction as the direction of an applied force. With this in mind, explain where the force comes from that moves you when you begin walking.

24. What is the centripetal force that keeps a car moving around a curve on the road?

25. Use Newton's laws of motion to explain why water leaves the wet clothes when a washer is in the spin cycle.

Exercises

Group A—Solutions Provided in Appendix D

1. (a) What is the weight of a 1.25 kg book? (b) What is the acceleration when a net force of 10.0 N is applied to the book?

2. What net force is needed to accelerate a 1.25 kg book 5.00 m/sec²?

3. What net force does the road exert on a 70.0 kg bicycle and rider to give them an acceleration of 2.0 m/sec²?

4. A 1,500 kg car accelerates uniformly from 44.0 km/hr to 80.0 km/hr in 10.0 sec. What was the net force exerted on the car?

5. A net force of 5,000.0 N accelerates a car from rest to 90.0 km/hr in 5.0 sec. (a) What is the mass of the car? (b) What is the weight of the car?

6. What is the weight of a 70.0 kg person?

7. What net force is needed to accelerate a 320.0 lb motorcycle from 30.0 mi/hr to 60.0 mi/hr in 4.0 sec? (Note: 60.0 mi/hr = 88.0 ft/sec, a conversion that is handy to remember.)

8. A jet plane accelerates at 23.0 ft/sec² during takeoff. What is the force of the back of the seat on a 160.0 lb person during this acceleration?

9. An elevator and its occupants have a total mass of 750.0 kg. What total force must the supporting cable be able to withstand while the elevator accelerates upward at 2.00 m/sec²?

10. A 1,000.0 kg car at rest experiences a net force of 1,000.0 N for 10.0 sec. What is the final speed of the car?

11. What is the momentum of a 50.0 kg person walking at a speed of 2.00 m/sec?

Group B—Solutions Not Given

1. (a) What is the weight of a 5.00 kg backpack? (b) What is the acceleration of the backpack if a net force of 10.0 N is applied?

2. What net force is required to accelerate a 20.0 kg object to 10.0 m/sec²?

3. What forward force must the ground apply to the foot of a 60.0 kg person to result in an acceleration of 1.00 m/sec²?

4. A 1,000.0 kg car accelerates uniformly to double its speed from 36.0 km/hr in 5.00 sec. What net force acted on this car?

5. A net force of 3,000.0 N accelerates a car from rest to 36.0 km/hr in 5.00 sec. (a) What is the mass of the car? (b) What is the weight of the car?

6. How much does a 60.0 kg person weigh?

7. What net force is required to accelerate a 3,200.0 lb car from rest to a velocity of 30.0 mi/hr in 6.00 sec?

8. What braking force is needed to decelerate a 2,400.0 lb car at 10.0 ft/sec²?

9. (a) A 75.0 kg person standing on a scale in a motionless elevator would exert what force on the scale? (b) What total force is exerted on the scale as the elevator accelerates upward at 2.00 m/sec?

10. A 60.0 gram tennis ball is struck by a racket with a force of 425.0 N for 0.01 sec. What is the speed of the tennis ball in km/hr as a result?

11. Compare the momentum of (a) a 2,000.0 kg car moving at 25.0 m/sec and (b) a 1,250 kg car moving at 40.0 m/sec.

12. How much centripetal force is needed to keep a 0.20 kg ball on a 1.50 m string moving in a circular path with a speed of 3.0 m/sec?

13. What is the velocity of a 100.0 g ball on a 50.0 cm string moving in a horizontal circle that requires a centripetal force of 1.0 N?

14. A 1,000.0 kg car moves around a curve with a 20.0 m radius with a velocity of 10.0 m/sec. (a) What centripetal force is required? (b) What is the source of this force?

15. On earth, an astronaut and equipment weighs 1,960.0 N. While weightless in space, the astronaut fires a 100 N rocket backpack for 2.0 sec. What is the resulting velocity of the astronaut and equipment?

12. What tension must a 50.0 cm length of string support in order to whirl an attached 1,000.0 gram stone in a circular path at 5.00 m/sec?

13. What is the maximum speed a 1,000.0 kg car can move around a curve with a radius of 30.0 m if the tires provide a maximum frictional force of 2,700.0 N?

14. How much centripetal force is needed to keep a 60.0 kg person and skateboard moving at 6.0 m/sec in a circle with a 10.0 m radius?

15. A 200.0 kg astronaut and equipment move with a velocity of 2.00 m/sec toward an orbiting spacecraft. How long will the astronaut need to fire a 100.0 N rocket backpack to stop the motion relative to the spacecraft?

60.0 Kg M

r
10 m

6 m/sec = V

$\dfrac{(60)(6)^2}{10}$

$\dfrac{6^2}{10}\quad\dfrac{36}{10}$

Fc

$Fc = \dfrac{mv^2}{r}$

Energy

Figure 4.1
Matter in motion exerts forces that can be used as an energy source. Compare the size of this stream of moving water to the pickup truck.

THE term *energy* is closely associated with the concepts of force and motion. Naturally moving matter, such as the wind or moving water, exerts forces. You have felt these forces if you have ever tried to walk against a strong wind or stand in one place in a stream of rapidly moving water. The motion and forces of moving air and moving water are used as *energy sources* (figure 4.1). The wind is an energy source as it moves the blades of a windmill performing useful work. Moving water is an energy source as it forces the blades of a water turbine to spin, turning an electric generator. Thus, moving matter exerts forces that are used as energy sources.

Matter does not have to be moving to supply energy; matter *contains* energy. Food supplied the energy for the muscular exertion of the humans and animals that accomplished most of the work before this century. Today, machines do the work that was formerly accomplished by muscular exertion. Machines also use the energy contained in matter. They use gasoline, for example, as they supply the forces and motion to accomplish work.

Moving matter and matter that contains energy can be used as energy sources to perform work. The concepts of work and energy and the relationship to matter are the topics of this chapter. You will learn how energy flows in and out of your surroundings as well as a broad, conceptual view of energy that will be developed more fully throughout the course.

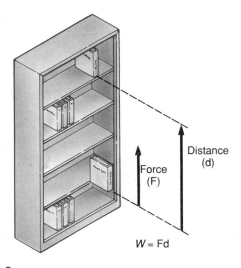

Figure 4.2

The force on the book moves it through a vertical distance from the second shelf to the fifth shelf and work is done, $W = Fd$.

Work

You learned earlier that the term *force* has a special meaning in science that is different from your everyday concept of force. In everyday use you use the term in a variety of associations such as police force, economic force, or the force of an argument. Earlier, force was discussed in a general way as a push or pull. Then a more precise scientific definition of force was developed from Newton's laws of motion—a force is a result of an interaction that is capable of changing the state of motion of an object.

The word *work* represents another one of those concepts that has a special meaning in science that is different from your everyday concept. In everyday use "work" is associated with a task to be accomplished or the time spent in performing the task. You might "work" at understanding physical science, for example, or you might tell someone that physical science is a lot of "work." You also probably associate physical work, such as lifting or moving boxes, with how tired you become from the effort. The scientific definition of work is not concerned with tasks, time, or how tired you become from doing a task. It is concerned with the application of a force to an object and the distance the object moves as a result of the force. The **work** done on the object is defined as *the magnitude of the applied force multiplied by the distance through which the force acts:*

$$\text{work} = \text{force} \times \text{distance}$$

$$W = Fd \qquad \qquad \textbf{equation 4.1}$$

Work, in the scientific sense, is the product of a force and the distance an object moves as a result of the force. There are two important considerations to remember about this definition: (1) something *must move* whenever work is done, and (2) the movement must be in the *same direction* as the direction of the force. When you move a book to a higher shelf in a bookcase you are doing work on the book. You apply a vertically upward force equal to the weight of the book as you move it in the same direction as the direction of the applied force. The work done on the book can therefore be calculated by multiplying the weight of the book by the distance it was moved (figure 4.2).

If you simply stand there holding the book, however, you are doing no work on the book. Your arm may become tired from holding the book since you must apply a vertically upward force equal to the weight of the book. But this force is not acting through a distance since the book is not moving. According to equation 4.1, a distance of zero results in zero work (figure 4.3). Only a force that results in motion in the same direction results in work.

Suppose you walk across the room while holding the book. You are exerting a vertically upward force equal to the weight of the book as before. But the direction of movement is perpendicular (90°) to the upward force on the book. Since the movement is perpendicular to the direction of the applied force, no work is done on the book. In this case there is no *relationship* between the applied force and the direction of movement. Since the upward force has nothing to do with the horizontal movement, no work is done (figure 4.4). This may seem odd at this point, but it will make sense after you learn some concepts of energy.

Energy

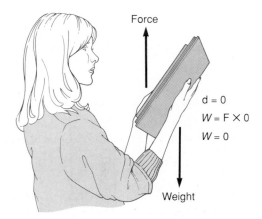

Force

d = 0
W = F × 0
W = 0

Weight

Figure 4.3
It is true that a force is exerted simply to hold a book, but the book does not move through a distance. Therefore the distance moved is 0 and the work accomplished is also 0.

The applied force does not have to be exactly parallel to the direction of movement. A force is a vector that can be resolved into the component force that acts in the same direction as the movement (figure 4.5). The force vector, however, cannot be perpendicular (90°) to the direction of movement.

Units of Work

The units of work are defined by the definition of work, Fd = W. In the metric system a force is measured in newtons (N) and distance is measured in meters (m) so the unit of work is

$$Fd = W$$

$$(newton)(meter) = newton\text{-}meter$$

$$(N)(m) = N \cdot m$$

The newton-meter is therefore the unit of work. This derived unit has a name. The newton-meter is called a **joule** (J) (pronounced "jool").

$$1 \text{ joule} = 1 \text{ newton-meter}$$

The units for a newton are kg·m/sec² and the unit for a meter is m. It therefore follows that the units for a joule are kg·m²/sec².

In the English system the force is measured in pounds (lb) and the distance is measured in feet (ft). The unit of work in the English system is therefore the ft·lb. The ft·lb does not have a name of its own as the N·m does (figure 4.6).

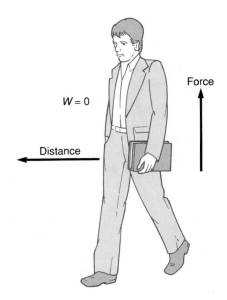

W = 0

Force

Distance

Figure 4.4
If the direction of movement (the distance moved) is perpendicular to the direction of the force, no work is done. This person is thus doing no work by carrying a book across a room. *on the book, He is doing work on himself.*

Example 4.1

How much work is needed to lift a 5.0 kg backpack to a shelf 1.0 m above the floor?

Solution

The backpack has a mass (m) of 5.0 kg and the distance (d) is 1.0 m. To lift the backpack requires a vertically upward force equal to the weight of the backpack. Weight can be calculated from W = mg:

m = 5.0 kg $W = mg$

g = 9.80 m/sec² $= (5.0 \text{ kg})\left(9.80 \dfrac{m}{sec^2}\right)$

W = ?

$$= (5.0 \times 9.80) \text{ kg} \times \frac{m}{sec^2}$$

$$= 49 \frac{kg \cdot m}{sec^2}$$

$$= \boxed{49 \text{ N}}$$

The definition of work is found in equation 4.1,

F = 49 N $W = Fd$

d = 1.0 m $= (49 \text{ N})(1.0 \text{ m})$

W = ? $= (49 \times 1.0)(N \cdot m)$

$$= 49 \text{ N} \cdot \text{m}$$

$$= \boxed{49 \text{ J}}$$

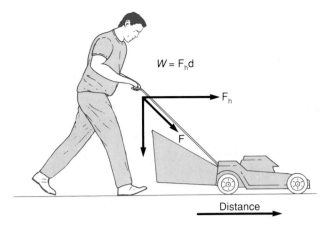

$W = F_h d$

Distance

Figure 4.5

A force at some angle (not 90°) to the direction of movement can be resolved into the horizontal component to calculate the work done.

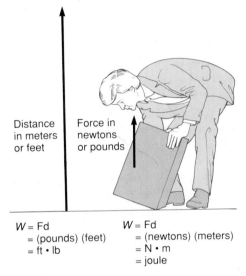

Distance in meters or feet

Force in newtons or pounds

$W = Fd$
= (pounds) (feet)
= ft • lb

$W = Fd$
= (newtons) (meters)
= N • m
= joule

Figure 4.6

Work is done against gravity when lifting an object. Work is measured in joules or in foot-pounds.

Example 4.2

How much work is required to lift a 50 lb box vertically a distance of 2 ft? (Answer: 100 ft·lb)

Simple Machines

Simple machines are tools that people use to help them do work. These machines help by (1) reducing the magnitude of the force needed to do work or (2) changing the direction of the applied force. The lever and inclined plane are simple machines that are often used to reduce the magnitude of a force needed to do work,

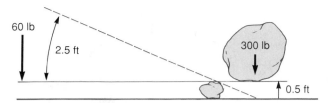

60 lb

2.5 ft

300 lb

0.5 ft

Figure 4.7

A 60 lb force through a distance of 2.5 ft can move a 300 lb force through a distance of 0.5 ft, disregarding friction. Thus the work input on a simple machine equals the work output.

and the simple pulley is often used to change the direction of a force. Figure 4.7 shows how a lever can be used to reduce the force needed to do work and change the direction of a force.

Suppose you want to lift a 300 lb rock a vertical distance of 0.5 ft. You would do 150 ft·lb of work—if you could lift that much weight. Consider using a lever as shown in figure 4.7. You would push down with a force of only 60 lb to raise the 300 lb rock! Note, however, that you would push down with a force of 60 lb through a distance of 2.5 feet. You did 60 × 2.5 ft, or 150 ft·lb of work. The 300 lb rock, on the other hand, was raised a distance of 0.5 ft. The work done on the rock was therefore 300 lb × 0.5 ft, or 150 ft·lb of work. The work done by you is the same as the work done *on* the rock, so

work input = work output

The lever enabled you to trade force for distance. Ignoring friction, the work you get out of any simple machine is the same as the work you put into it. There are six kinds of simple machines: the inclined plane, the wedge, the screw, the lever, the wheel and axle, and the pulley. All complex machines are combinations of these simple machines.

Power

You are doing work when you walk up a stairway since you are lifting yourself through a distance. You are lifting your weight (force exerted) the *vertical* height of the stairs (distance through which the force is exerted). Consider a person who weighs 120 lbs and climbs a stairway with a vertical distance of 10 ft. This person will do (120 lb)(10 ft) or 120 ft·lb of work. Will the amount of work change if the person were to run up the stairs? The answer is no, the same amount of work is accomplished. Running up the stairs, however, is more tiring than walking up the stairs. You use up your energy at a greater *rate* when running. The rate at which energy is transformed or the rate at which work is done is called **power** (figure 4.8). Power is defined as work per unit of time,

$$power = \frac{work}{time}$$

$$P = \frac{W}{t}$$

equation 4.2

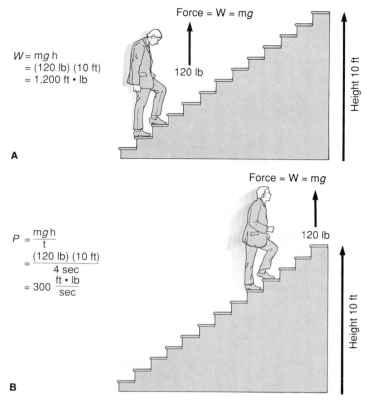

$W = mg\,h$
$= (120\ \text{lb})\ (10\ \text{ft})$
$= 1{,}200\ \text{ft} \cdot \text{lb}$

Force = W = mg

120 lb

Height 10 ft

A

Force = W = mg

$P = \dfrac{mg\,h}{t}$
$= \dfrac{(120\ \text{lb})\ (10\ \text{ft})}{4\ \text{sec}}$
$= 300\ \dfrac{\text{ft} \cdot \text{lb}}{\text{sec}}$

120 lb

Height 10 ft

B

Figure 4.8

(a) The work accomplished climbing a stairway is the person's weight times the vertical distance. (b) The power level is the work accomplished per unit of time.

The 120 lb person who ran up the 10 ft height of stairs in 4 sec would have a power rating of

$$P = \frac{W}{t} = \frac{(120\ \text{lb})(10\ \text{ft})}{4\ \text{sec}} = 300\ \frac{\text{ft} \cdot \text{lb}}{\text{sec}}$$

If the person had a time of 3 sec on the same stairs, the power rating would be greater, 400 ft·lb/sec. This is a greater rate of energy use, or greater power.

When the steam engine was first invented there was a need to describe the rate at which the engine could do work. Since people at this time were familiar with using horses to do their work, the steam engines were compared to horses. James Watt, who designed a workable steam engine, defined **horsepower** as a power rating of 550. ft·lb/sec (figure 4.9a). To convert a power rating in the English units of ft·lb/sec to horsepower, divide the power rating by 550. ft·lb/sec. For example, the 120.0 lb person who had a power rating of 400 ft·lb/sec had a horsepower of 400 ft·lb/sec ÷ 550. ft·lb/sec, or 0.7.

In the Metric system, power is measured in joules per second. The unit J/sec, however, has a name. A J/sec is called a **watt** (w). The watt (figure 4.9b) is used with metric prefixes for large numbers: 1,000 w = 1 kilowatt (kw) and 1,000,000 w = 1 megawatt (mw). It takes 746 w to equal 1 horsepower. One

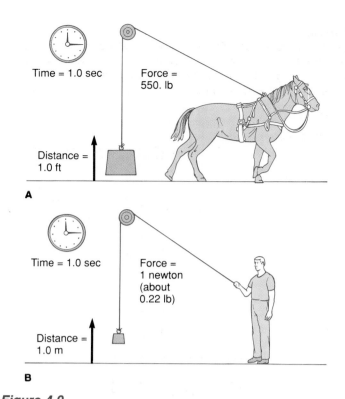

Time = 1.0 sec

Force = 550. lb

Distance = 1.0 ft

A

Time = 1.0 sec

Force = 1 newton (about 0.22 lb)

Distance = 1.0 m

B

Figure 4.9

(a) A horsepower is defined as a power rating of 550 ft·lb/sec. (b) A watt is defined as a newton-meter, or joule per second.

kilowatt is equal to about 1-1/3 horsepower. The electric utility company charges you for how much power you have used with your electrical appliances. But they also want to know how long you have used that power, so electrical energy is measured by power (kw) times the time of use (hr). Thus electrical energy is measured in kwhr. The kwhr is a unit of *work,* not power. Since power is

$$P = \frac{W}{t}$$

then it follows that

$$W = P \times t$$

So power times time equals a unit of work, kwhr. We will return to kilowatts and kilowatt-hours later when electricity is discussed. (Note that the symbol *W* is used for work, the symbol w is used for watt, and the symbol W is used for weight. Do not confuse the symbols for these three "W concepts"; they represent very different quantities.)

Example 4.3

An electric lift can raise a 500. kg mass a distance of 10.0 m in 5.0 sec. What is the power of the lift?

Solution

Power is work per unit time ($P = W/t$), and work is force times distance ($W = F/d$). The vertical force required is the weight lifted and $W = mg$. Therefore the work accomplished would be $W = mgd$ and the power would be $P = mgd/t$.

$m = 500.$ kg

$g = 9.80$ m/sec²

$d = 10.0$ m

$t = 5.0$ sec

$P = ?$

$$P = \frac{mgd}{t}$$

$$= \frac{(500. \text{ kg})(9.80 \text{ m/sec}^2)(10.0 \text{ m})}{5.0 \text{ sec}}$$

$$= \frac{(500.)(9.80)(10.0)}{5.0} \frac{\text{kg} \cdot \frac{\text{m}}{\text{sec}^2} \cdot \text{m}}{\text{sec}}$$

$$= 9,800 \frac{\text{N} \cdot \text{m}}{\text{sec}}$$

$$= 9,800 \frac{\text{J}}{\text{sec}}$$

$$= 9,800 \text{ w}$$

$$= 9.8 \text{ kw}$$

The power in horsepower (hp) units would be

$$9,800 \text{ w} \times \frac{\text{hp}}{746 \text{ w}} = \boxed{13 \text{ hp}}$$

Example 4.4

A 150 lb person runs up a 15.0 ft stairway in 10.0 sec. What is the horsepower rating of the person? (Answer: 0.41 horsepower)

ACTIVITIES

1. Find your horsepower rating. Measure the vertical height of a stairway that is more than 3 m (10 ft) high. Have someone time you while you walk up the stairs. Find your walking power in both watts and horsepower.
2. Repeat number 1 but this time go up the stairs as fast as you can. Compare your exerted power rating to your walking power in both watts and horsepower.

Motion, Position, and Energy

Closely related to the concept of work is the concept of **energy.** Energy can be defined as the *ability to do work*. This definition of energy seems consistent with everyday ideas about energy and physical work. After all, it takes a lot of energy to do a lot of work. In fact, one way of measuring the energy of something is to see how much work it can do. Likewise, when work is done *on* something, a change occurs in its energy level. The following examples will help clarify this close relationship between work and energy.

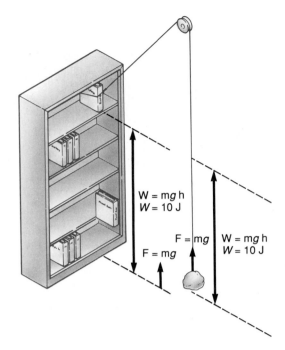

Figure 4.10

If moving a book from the floor to a high shelf requires 10 J of work, then the book will do 10 J of work on the object of the same mass when the book falls from the shelf.

Potential Energy

Consider a book on the floor next to a bookcase. You can do work on the book by vertically raising it to a shelf. You can measure this work by multiplying the vertical upward force applied times the distance that the book is moved. You might find, for example, that you did an amount of work equal to 10 J on the book (see example 4.1).

Suppose that the book has a string attached to it as shown in figure 4.10. The string is threaded over a frictionless pulley and attached to an object on the floor. If the book is caused to fall from the shelf, the object on the floor will be vertically lifted through some distance by the string. The falling book exerts a force on the object through the string and the object is moved through a distance. In other words, the *book* did work on the object, $W = Fd$.

The book can do more work on the object if it falls from a higher shelf since it will move the object a greater distance. The higher the shelf the greater the *potential* for the book to do work. The ability to do work is defined as energy. The energy that an object has because of its position is called **potential energy** (PE). Potential energy is defined as *energy due to position*. This type of potential energy is called *gravitational potential energy* since it is a result of gravitational attraction. There are other types of potential energy such as that in a compressed or stretched spring.

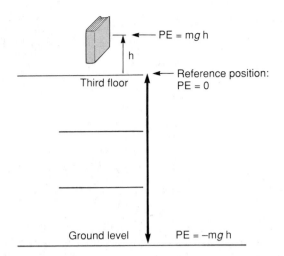

Figure 4.11
The 0 reference level for potential energy is chosen for convenience. Here the reference position chosen is the third floor, so the book will have a negative potential energy when at ground level.

Note the relationship between work and energy in the example. You did 10 J of work to raise the book to a higher shelf. In so doing, you increased the potential energy of the book by 10 J. The book now has the *potential* of doing 10 J of additional work on something else, therefore,

$$
\begin{array}{c}
\text{Work done on} \\
\text{an object to} \\
\text{change position}
\end{array}
=
\begin{array}{c}
\text{Increase in} \\
\text{potential energy}
\end{array}
=
\begin{array}{c}
\text{Increase in} \\
\text{work the object} \\
\text{can do}
\end{array}
$$

$$
\begin{array}{ccc}
\text{work on book} & = & \text{potential energy of book} & = & \text{work by book} \\
(10\ \text{J}) & & (10\ \text{J}) & & (10\ \text{J})
\end{array}
$$

As you can see, a joule is a measure of work accomplished on an object. A joule is also a measure of potential energy. And, a joule is a measure of how much work an object can do. Both work and energy are measured in joules (or ft·lbs).

The potential energy of an object can be calculated, as described previously, from the work done *on* the object to change its position. You exert a force equal to its weight as you lift it some height above the floor, and the work you do is the product of the weight and height. Likewise, the amount of work the object *could* do because of its position is the product of its weight and height. For the metric unit of mass, weight is the product of the mass of an object times *g*, the acceleration due to gravity, so

$$\text{potential energy} = \text{weight} \times \text{height}$$

$$PE = mgh \qquad \qquad \textbf{equation 4.3}$$

For English units, the pound *is* the gravitational unit of force, or weight, so equation 4.3 becomes PE = (W)(h).

When does an object have zero potential energy? Considering the book in the bookcase, you could say that the book has zero potential energy when it is flat on the floor. It can do no work when it is on the floor. But what if that floor happens to

be the third floor of a building? You could, after all, drop the book out of a window. The answer is that it makes no difference. The same results would be obtained in either case since it is the *change of position* that is important in potential energy. The zero reference position for potential energy is therefore arbitrary. A zero reference point is chosen as a matter of convenience. Note that if the third floor of a building is chosen as the zero reference position, a book on ground level would have negative potential energy. This means that you would have to do work on the book to bring it back to the zero potential energy position (figure 4.11). You will learn more about negative energy levels later in the chapters on chemistry.

Example 4.5

What is the potential energy of a 2.0 lb book that is on a bookshelf 4.0 ft above the floor?

Solution
Equation 4.3, PE = mgh, shows the relationship between potential energy (PE), weight (mg), and height (h).

$$
\begin{array}{ll}
W = 2.0\ \text{lb} & PE = mgh = (W)(h) \\
h = 4.0\ \text{ft} & \quad = (2.0\ \text{lb})(4.0\ \text{ft}) \\
PE = ? & \quad = (2.0 \times 4.0)\ \text{ft} \cdot \text{lb} \\
& \quad = \boxed{8.0\ \text{ft} \cdot \text{lb}}
\end{array}
$$

Example 4.6

How much work can a 5.00 kg mass do if it is 5.00 m above the ground? (Answer: 245 J)

Kinetic Energy
Moving objects have the ability to do work on other objects because of their motion. A rolling bowling ball exerts a force on the bowling pins and moves them through a distance (figure 4.12). A moving car has the ability to exert a force on a tree and knock it down. Objects in motion have the ability to do work, so they have energy. The energy of motion is known as **kinetic energy.** Kinetic energy can be measured (1) in terms of the work done to put the object in motion or (2) in terms of the work the moving object will do in coming to rest. Consider objects that you put into motion by throwing. You exert a force on a football or baseball as you move them through a distance before they leave your hand. The kinetic energy that the ball now has is equal to the work (force times distance) that you did on the ball. You exert a force on a baseball through a distance as the ball leaves your hand. The kinetic energy that the ball now has is equal to the work that you did on the ball. The ball exerts a force on the hand of the person catching the ball and moves it through a distance. The net work done on the hand is equal to the kinetic energy that the ball had. Therefore,

$$
\begin{array}{c}
\text{Work done to} \\
\text{put an object} \\
\text{in motion}
\end{array}
=
\begin{array}{c}
\text{Increase in} \\
\text{kinetic energy}
\end{array}
=
\begin{array}{c}
\text{Increase in} \\
\text{work the object} \\
\text{can do}
\end{array}
$$

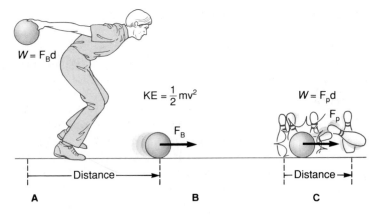

$W = F_Bd$

$KE = \frac{1}{2}mv^2$

F_B

$W = F_pd$

F_p

—Distance—

—Distance—

A

B

C

Figure 4.12

(a) Work is done on the bowling ball as a force (F_B) moves it through a distance. (b) This gives the ball a kinetic energy equal in amount to the work done on it. (c) The ball does work on the pins and has enough remaining energy to crash into the wall behind the pins.

A baseball and a bowling ball moving with the same velocity do not have the same kinetic energy. You cannot knock down many bowling pins with a slowly rolling baseball. Obviously the more massive bowling ball can do much more work than a less massive baseball with the same velocity. Is it possible for the bowling ball and the baseball to have the same kinetic energy? The answer is yes, if you can give the baseball a sufficiently greater velocity. This might require shooting the baseball from a cannon, however. Kinetic energy is proportional to the mass of a moving object but velocity has a greater influence. Consider two balls of the same mass, but one is moving twice as fast as the other. The ball with twice the velocity will do *four* times as much work as the slower ball. A ball with three times the velocity will do *nine* times as much work as the slower ball. Kinetic energy is proportional to the square of the velocity ($2^2 = 4$; $3^2 = 9$). The kinetic energy (KE) of an object is

$$\text{kinetic energy} = \frac{1}{2}(\text{mass})(\text{velocity})^2$$

$$KE = \frac{1}{2}mv^2 \qquad \textbf{equation 4.4}$$

The 1/2 comes from the fact that an average velocity is used to calculate kinetic energy (adding the initial and final velocities and dividing by 2). The unit of mass is the kg and the unit of velocity is m/sec. Therefore the unit of kinetic energy is

$$KE = (kg)\left(\frac{m}{sec}\right)^2$$

$$= (kg)\left(\frac{m^2}{sec^2}\right)$$

$$= \frac{kg \cdot m^2}{sec^2}$$

which is the same thing as

$$\left(\frac{kg \cdot m}{sec^2}\right)(m)$$

or

$$N \cdot m$$

or

$$\text{joule (J)}$$

Kinetic energy is measured in joules as is work ($F \times d$ or $N \cdot m$) and potential energy (mgh or $N \cdot m$).

Example 4.7

A 7.00 kg bowling ball is moving in a bowling lane with a velocity of 5.00 m/sec. What is the kinetic energy of the ball?

Solution

The relationship between kinetic energy (KE), mass (m), and velocity (v) is found in equation 4.4, $KE = 1/2\ mv^2$:

$m = 7.00$ kg

$v = 5.00$ m/sec

$KE = ?$

$$KE = \frac{1}{2}mv^2$$

$$= \frac{1}{2}(7.00 \text{ kg})\left(5.00 \frac{m}{sec}\right)^2$$

$$= \frac{1}{2}(7.00 \times 25.0) \text{ kg} \times \frac{m^2}{sec^2}$$

$$= \frac{1}{2}\ 175\ \frac{kg \cdot m^2}{sec^2}$$

$$= 87.5\ \frac{kg \cdot m}{sec^2} \cdot m$$

$$= 87.5\ N \cdot m$$

$$= \boxed{87.5 \text{ J}}$$

Example 4.8

A 100.0 kg football player moving with a velocity of 6.0 m/sec runs into a stationary quarterback. How much work was done on the quarterback? (Answer: 1,800 J)

Energy Flow

The key to understanding the individual concepts of work and energy is to understand the close relationship between the two. When you do work on something you give it energy of position (potential energy) or you give it energy of motion (kinetic energy). In turn, objects that have kinetic or potential energy can now do work on something else as the transfer of energy continues. Where does all this energy come from and where does it go? The answer to this question is the subject of this section on energy flow.

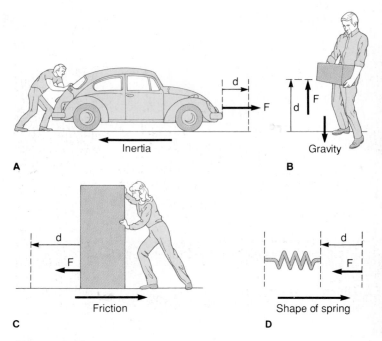

Figure 4.13

Examples of working against (*a*) inertia, (*b*) gravity, (*c*) friction, and (*d*) shape.

Figure 4.14

Mechanical energy is the energy of motion, or the energy of position, of many familiar objects. This boat has energy of motion.

Work and Energy

Energy is used to do work on an object, exerting a force through a distance. This force is usually *against* something (figure 4.13) and there are five main groups of resistance:

1. *Work against inertia*. A net force that changes the state of motion of an object is working against inertia. According to the laws of motion, a net force acting through a distance is needed to increase, decrease, or change the direction of the velocity of an object.
2. *Work against fundamental forces*. There are four fundamental forces that result from interactions between matter. Of the four, three are more common to everyday experiences concerned with work and energy. These three are gravitational, electromagnetic, and nuclear forces. Consider the force from gravitational attraction. A net force that changes the position of an object is a downward force from the acceleration due to gravity acting on a mass, $W = mg$. To change the position of an object a force opposite to mg is needed to act through the distance of the position change. Thus lifting an object straight up requires doing work against the fundamental force of gravity.
3. *Work against friction*. The force that is needed to maintain the motion of an object is working against friction. Friction is always present when two surfaces in contact move over each other. Friction resists motion.
4. *Work against shape*. The force that is needed to stretch or compress a spring is working against the shape of the spring. Other examples of work against shape include compressing or stretching elastic materials. If the elastic limit is reached, then the work goes into deforming or breaking the material.
5. *Work against any combination of inertia, fundamental forces, friction, and/or shape*. It is a rare occurrence on the earth that work is against only one type of resistance. Pushing on the back of a stalled automobile to start it moving up a slope would involve many resistances. This is complicated, however, so a single resistance is usually singled out for discussion.

Work is done against the main groups of resistances, but what is the result? The result is that some kind of *energy change* has taken place. Among the possible energy changes are the following:

1. *Increased kinetic energy*. Work against inertia results in an increase of kinetic energy, the energy of motion.
2. *Increased potential energy*. Work against fundamental forces and work against shape result in an increase of potential energy, the energy of position.
3. *Increased temperature*. Work against friction results in an increase in the temperature. Temperature is a manifestation of the kinetic energy of the particles making up an object, as you will learn in the next chapter.
4. *Increased combinations of kinetic energy, potential energy, and/or temperature*. Again, isolated occurrences are more the exception than the rule. In all cases, however, the sum of the total energy changes will be equal to the work done.

Work was done *against* various resistances and energy was *increased* as a result. The object with increased energy can now do work on some other object or objects. A moving object has kinetic energy so it has the ability to do work. An object with potential energy has energy of position and it, too, has the ability to do work. You could say that energy *flowed* into and out of

Figure 4.15
Chemical energy is a form of potential energy that is released during a chemical reaction. The trees in the background store chemical energy through the process of photosynthesis. The crew uses chemical energy to propel the shell.

an object during the entire process. The following energy scheme is intended to give an overall conceptual picture of energy flow. Use it to develop a broad view of energy. You will learn the details later throughout the course.

Energy Forms

Energy comes in various forms, and different terms are used to distinguish one form from another. Although energy comes in various *forms,* this does not mean that there are different *kinds* of energy. The forms are the result of the more common fundamental forces—gravitational, electromagnetic, and nuclear—and objects that are interacting. There are five forms of energy, (1) *mechanical,* (2) *chemical,* (3) *radiant,* (4) *electrical,* and (5) *nuclear.* The following is a brief discussion of each of the five forms of energy.

Mechanical energy is the form of energy of familiar objects and machines (figure 4.14). A car moving on a highway has kinetic mechanical energy. Water behind a dam has potential mechanical energy. The spinning blades of a steam turbine have kinetic mechanical energy. The form of mechanical energy is usually associated with the kinetic energy of everyday-sized objects and the potential energy that results from gravity. There are other possibilities (e.g., sound), but this description will serve the need for now.

Chemical energy is the form of energy involved in chemical reactions (figure 4.15). Chemical energy is released in the chemical reaction known as oxidation. The fire of burning wood is an example of rapid oxidation. A slower oxidation releases energy from food units in your body. As you will study in the chemistry unit, chemical energy involves electromagnetic forces between the parts of atoms. Until then, consider the following comparison. Photosynthesis is carried on in green plants. The plants use the energy of sunlight to rearrange carbon dioxide and water into plant materials and oxygen. This reaction could be represented by the following word equation:

energy + carbon dioxide + water = wood + oxygen

Energy

Figure 4.16
Radiant energy is energy that travels through space. This solar cell array converts radiant energy to electrical energy, producing an average of 200,000 watts alternating current.

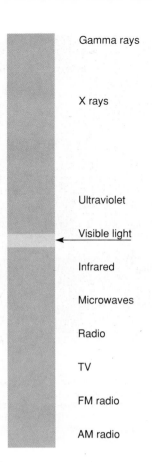

Gamma rays

X rays

Ultraviolet

Visible light ←

Infrared

Microwaves

Radio

TV

FM radio

AM radio

Figure 4.17
The electromagnetic spectrum includes many forms of radiant energy. Note that visible light occupies only a tiny part of the entire spectrum.

The plant took energy and two substances and made two different substances. This is similar to raising a book to a higher shelf in a bookcase. That is, the new substances have more energy than the original ones did. Consider a word equation for the burning of wood:

$$wood + oxygen = carbon\ dioxide + water + energy$$

Notice that this equation is exactly the reverse of photosynthesis. In other words, the energy used in photosynthesis was released during oxidation. Chemical energy is a kind of potential energy that is stored and later released during a chemical reaction.

Radiant energy is energy that travels through space (figure 4.16). Most people think of light or sunlight when considering this form of energy. Visible light, however, occupies only a small part of the complete electromagnetic spectrum, as shown in figure 4.17. Radiant energy includes light and all other parts of the spectrum. Infrared radiation is sometimes called "heat radiation" because of the association with heating when this type of radiation is absorbed. For example, you feel the interaction of infrared radiation when you hold your hand near a warm range element. However, infrared radiation is another type of radiant energy. In fact, some snakes, such as the sidewinder, can "see" infrared radiation emitted from warm animals where you see total darkness. Microwaves are another type of radiant energy that are used in cooking. As with other forms of energy,

light, infrared, and microwaves will be considered in more detail later. For now, consider all types of radiant energy to be forms of energy that travel through space.

Electrical energy is another form of energy from electromagnetic interactions that will be considered in detail later. You are familiar with electrical energy that travels through wires to your home from a power plant (figure 4.18), electrical energy that is generated by chemical cells in a flashlight, and electrical energy that can be "stored" in a car battery.

Nuclear energy is a form of energy often discussed because of its use as an energy source in power plants. Nuclear energy is another form of energy from the atom, but this time the energy involves the nucleus, the innermost part of an atom, and nuclear interactions.

Energy Conversion

Potential energy can be converted to kinetic energy and vice versa. The simple pendulum offers a good example of this conversion. A simple pendulum is an object, called a bob, suspended by a string or wire from a support. If the bob is moved to one side then released, it will swing back and forth in an arc. At the moment that the bob reaches the top of its swing, it stops

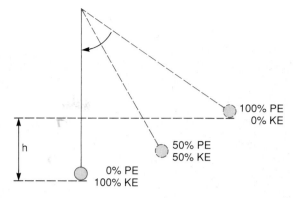

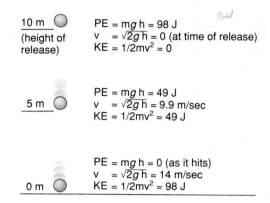

Figure 4.19

This pendulum bob loses potential energy (PE) and gains an equal amount of kinetic energy (KE) as it falls through a distance h. The process reverses as the bob moves up the other side of its swing.

10 m (height of release)
$PE = mgh = 98$ J
$v = \sqrt{2gh} = 0$ (at time of release)
$KE = 1/2mv^2 = 0$

5 m
$PE = mgh = 49$ J
$v = \sqrt{2gh} = 9.9$ m/sec
$KE = 1/2mv^2 = 49$ J

0 m
$PE = mgh = 0$ (as it hits)
$v = \sqrt{2gh} = 14$ m/sec
$KE = 1/2mv^2 = 98$ J

Figure 4.20

The ball trades potential energy for kinetic energy as it falls. Notice that the ball had 98 J of potential energy when dropped and a kinetic energy of 98 J just as it hits the ground.

Figure 4.18

The blades of a steam turbine. In a power plant, chemical or nuclear energy is used to heat water to steam, which is directed against the turbine blades. The mechanical energy of the turbine turns an electric generator. Thus, a power plant converts chemical or nuclear energy to mechanical energy, which is then converted to electrical energy.

for an instant, then begins another swing. At the instant of stopping, the bob has 100 percent potential energy and no kinetic energy. As the bob starts back down through the swing it is gaining kinetic energy and losing potential energy. At the instant the bob is at the bottom of the swing, it has 100 percent kinetic energy and no potential energy. As the bob now climbs through the other half of the arc it is gaining potential energy and losing kinetic energy until it again reaches an instantaneous stop at the top and the process starts over. The kinetic energy of the bob at the bottom of the arc is equal to the potential energy it had at the top of the arc (figure 4.19). Disregarding friction, the sum of the potential energy and the kinetic energy remains constant throughout the swing. The potential energy lost during a fall equals the kinetic energy gained (figure 4.20). In other words,

$$PE_{lost} = KE_{gained}$$

Substituting the values from equations 4.3 and 4.4,

$$mgh = \frac{1}{2}mv^2$$

Canceling the m and solving for v,

$$v = \sqrt{2gh} \qquad \text{equation 4.5}$$

Equation 4.5 tells you the speed of a falling object as its potential energy is converted to kinetic energy. This assumes, however, that the object is in free fall since the effect of air resistance is ignored. Note that the m's cancel, showing again that the mass of an object has no effect on its final speed.

Example 4.9

A 1.0 kg book falls from a height of 1.0 m. What is its velocity just as it hits the floor?

Solution

The relationships involved in the velocity of a falling object are given in equation 4.5.

$h = 1.0$ m
$g = 9.80$ m/sec²
$v = ?$

$v = \sqrt{2gh}$
$= \sqrt{(2)(9.80 \text{ m/sec}^2)(1.0 \text{ m})}$
$= \sqrt{2 \times 9.80 \times 1.0 \frac{m}{sec^2} \cdot m}$
$= \sqrt{29.6 \frac{m^2}{sec^2}}$
$= \boxed{4.4 \text{ m/sec}}$

Example 4.10

What is the kinetic energy of a 1.0 kg book just before it hits the floor after a 1.0 m fall? (Answer: 9.7 J)

Energy

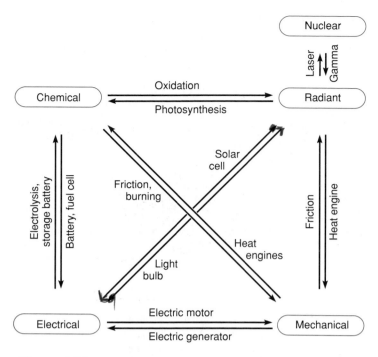

Figure 4.21

The energy forms and some conversion pathways.

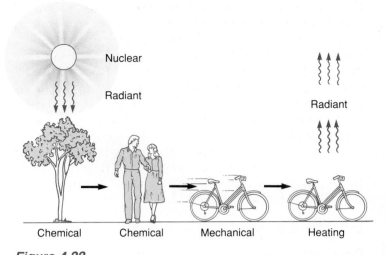

Figure 4.22

Energy arrives from the sun, goes through a number of conversions, then radiates back into space. The total sum leaving eventually equals the original amount that arrived.

Any *form* of energy can be converted to another form. In fact, most technological devices that you use are nothing more than *energy form converters* (figure 4.21). A light bulb, for example, converts electrical energy to radiant energy. A car converts chemical energy to mechanical energy. A solar cell converts radiant energy to electrical energy, and an electric motor converts electrical energy to mechanical energy. Each technological device converts some form of energy (usually chemical or electrical) to another form that you desire (usually mechanical or radiant).

It is interesting to trace the *flow of energy* that takes place in your surroundings. Suppose, for example, that you are riding a bicycle. The bicycle has kinetic mechanical energy as it moves along. Where did the bicycle get this energy? From you, as you use the chemical energy of food units to contract your muscles and move the bicycle along. But where did your chemical energy come from? It came from your food, which consists of plants, animals who eat plants, or both plants and animals. In any case, plants are at the bottom of your food chain. Plants convert radiant energy from the sun into chemical energy. Radiant energy comes to the plants from the sun because of the nuclear reactions that took place in the core of the sun. Your bicycle is therefore powered by nuclear energy that has undergone a number of form conversions!

Energy Conservation

Energy can be transferred from one object to another, and it can be converted from one form to another form. If you make a detailed accounting of all forms of energy before and after a transfer or conversion, the total energy will be *constant*. Consider your bicycle coasting along over level ground when you apply the brakes. What happened to the kinetic mechanical energy of the bicycle? It went into heating the rim and brakes of your bicycle, then eventually radiated to space as infrared radiation. All radiant energy that reaches the earth is eventually radiated back to space (figure 4.22). Thus, throughout all the form conversions and energy transfers that take place, the total sum of energy remains constant.

The total energy is constant in every situation that has been measured. This consistency leads to another one of the conservation laws of science, the **law of conservation of energy:**

Energy is never created or destroyed. Energy can be converted from one form to another but the total energy remains constant.

You may be wondering about the source of nuclear energy. Does a nuclear reaction create energy? Albert Einstein answered this question back in the early 1900s when he formulated his now-famous relationship between mass and energy, $E = mc^2$. This relationship will be discussed in detail in chapter 17. Basically, the relationship states that mass *is* a form of energy, and this has been experimentally verified many times.

Energy Transfer

Earlier it was stated that when you do work on something, you give it energy. The result of work could be increased kinetic mechanical energy, increased gravitational potential energy, or an increase in the temperature of an object. You could summarize this by stating that either *working* or *heating* is always involved anytime energy is transformed. This is not unlike your financial situation. In order to increase or decrease your financial status, you need some mode of transfer, such as cash or checks, as a means of conveying assets. Just as with cash flow from one individual to another, energy flow from one object to another requires a mode of transfer. In energy matters the mode of transfer is working or heating. Anytime you see working or heating occurring you know that an energy transfer is taking place. The next time you see heating, think about what energy form is being converted to what new energy form. (The final form is usually radiant energy.) Heating is the topic of the next chapter, where you will consider the role of heat in energy matters.

Energy Sources Today

Prometheus, according to ancient Greek mythology, stole fire from heaven and gave it to humankind. Fire has propelled human advancement ever since. All that was needed was something to burn—fuel for Prometheus's fire.

Any substance that burns can be used to fuel a fire, and various fuels have been used over the centuries as humans advanced. First, wood was used as a primary source for heating. Then coal fueled the industrial revolution. Eventually, humankind roared into the twentieth century burning petroleum. Today, petroleum is the most widely used source of energy. It provides about 40 percent of the total energy used by the nation, but this dependence has been dropping since the 1970s. Natural gas also contributes about 23 percent of the total energy used today. The use of coal has been increasing and today provides about 23 percent of the total. Note that petroleum, coal, and natural gas are all chemical sources of energy, sources that are mostly burned for their energy. These chemical sources supply about 86 percent of the total energy consumed. About a third of this is burned for heating, and the rest is burned to drive engines or generators.

Nuclear energy and hydropower are the nonchemical sources of energy. These sources are used to generate electrical energy. The alternative sources of energy such as solar and geothermal provide less than .5 percent of the total energy consumed in the United States today.

The energy source mix has changed from past years and it will change in the future. Wood supplied 90 percent of the energy until the 1850s, when the use of coal increased. Then, by 1910, coal was supplying about 75 percent of the total energy needs. Then petroleum began making increased contributions to the energy supply. Now increased economic constraints and a decreasing supply of petroleum are producing another supply shift. The present petroleum-based energy era is about to shift to a new energy era.

Over 99 percent of the total energy consumed today is provided by four sources, (1) petroleum (including natural gas), (2) coal, (3) hydropower, and (4) nuclear (figure 4.23). The following is a brief introduction to these four sources.

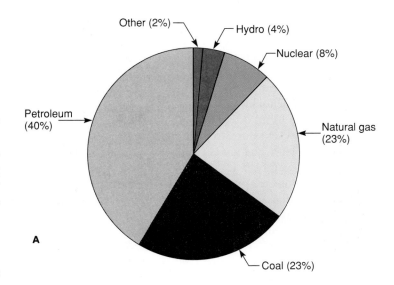

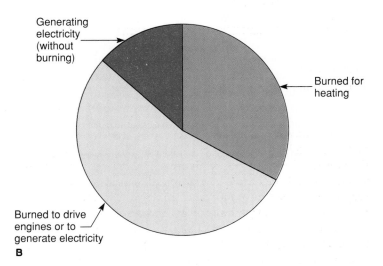

Figure 4.23
(*a*) The sources of energy and (*b*) the uses of energy during the 1990s.

Petroleum

The word *petroleum* is derived from the word "petra," meaning rock, and the word "oleum," meaning oil. Petroleum is oil that comes from oil-bearing rock. Natural gas is universally associated with petroleum and has similar origins. Both petroleum and natural gas form from organic sediments, materials that have settled out of bodies of water. Sometimes a local condition permits the accumulation of sediments that are exceptionally rich in organic material. This could occur under special conditions in a freshwater lake or it could occur on shallow ocean basins. In either case, most of the organic material is from plankton—tiny free-floating animals and plants such as algae. It is from such accumulations of buried organic material that petroleum and natural gas are formed.

The exact process by which buried organic material becomes petroleum and gas is not understood. It is believed that bacteria, pressure, appropriate temperatures, and time are all important. Natural gas is formed at higher temperatures than is petroleum. Varying temperatures over time may produce a mixture of petroleum and gas or natural gas alone.

Petroleum forms a thin film around the grains of the rock where it formed. Pressure from the overlying rock and water move the petroleum and gas through the rock until it reaches a rock type or structure that stops it. If natural gas is present it occupies space above the accumulating petroleum. It is from such accumulations of petroleum and natural gas that these energy sources are obtained.

Discussions about the petroleum supply and the cost of petroleum usually refer to a "barrel of oil." The "barrel" is an accounting device of forty-two United States gallons. Such a forty-two-gallon barrel does not exist. When or if oil is shipped in barrels, each drum holds fifty-five United States gallons. The various uses of petroleum products are discussed in chapter 16.

The supply of petroleum and natural gas is limited. Most of the continental drilling prospects appear to be exhausted and the search for new petroleum supplies is now offshore. In the early 1990s, over 25 percent of our nation's petroleum is estimated to come from offshore wells. The amount of imported petroleum has ranged from 30 to 50 percent over the years with most imported oil coming from Mexico, Canada, Great Britain, Indonesia, and Saudi Arabia.

Petroleum is used for gasoline (about 45 percent), diesel (about 40 percent), and heating oil (about 15 percent). Petroleum is also used in making medicine, clothing fabrics, plastics, and ink.

Coal

Petroleum and natural gas formed from the remains of tiny plants and animals that lived millions of years ago. Coal, on the other hand, formed from an accumulation of plant materials that collected under special conditions millions of years ago. Thus, petroleum, natural gas, and coal are called **fossil fuels.** Fossil fuels contain the stored radiant energy of plants that lived millions of years ago.

Fossil plants found in coal are similar to plants that grow today in the swamps of Florida, Georgia, and Louisiana. When plants die in such a swamp they fall into the water, become waterlogged, and sink. There, in the stagnant water, they are protected from consumption by termites, fungi, and bacteria. The plants rot somewhat and layers of loose plant materials collect at the bottom of swampy lakes. This carbon-rich decayed plant material is called peat (not to be confused with peat moss). Peat can be used as a fuel. It is also the earliest stage of coal. Pressure, compaction, and heating brought about by movements of the earth's crust changed the water content and the free carbon in the material as it gradually changed the *rank*, or stage of development of the coal. The lowest rank is lignite (brown coal), then subbituminous, then bituminous (soft coal), and finally the highest rank of anthracite (hard coal).

Each rank of coal has different burning properties and a different energy content. Coal also contains impurities of clay, silt, iron oxide, and sulfur. The mineral impurities leave an ash when the coal is burned and the sulfur produces sulfur dioxide, a pollutant.

Most of the coal mined today is burned by utilities to generate electricity (about 80 percent). The coal is ground to a face-powder consistency and blown into furnaces. This greatly increases efficiency but produces fly ash, ash that "flies" up the chimney. Industries and utilities are required by the Federal Clean Air Act to remove sulfur dioxide and fly ash from plant emissions. About 20 percent of the cost of a new coal-fired power plant goes into air pollution control equipment. Coal is an abundant but dirty energy source.

Water Power

Moving water has been used as a source of energy for thousands of years. It is considered a renewable energy source, inexhaustible as long as the rain falls. Today, hydroelectric plants generate about 4 percent of the nation's *total* energy consumption at about 1,200 power generating dams across the nation. Hydropower furnished about 40 percent of the nation's electric power in 1940. Today, dams furnish 13 percent of the electric power. It is projected that this will drop even lower, perhaps to 7 percent in the near future. Energy consumption has increased but hydropower production has not kept pace because geography limits the number of sites that can be built.

Water from a reservoir is conducted through large pipes called penstocks to a powerhouse, where it is directed against turbine blades that turn a shaft on an electric generator. A rough approximation of the power that can be extracted from the falling water can be made by multiplying the depth of the water (in feet) by the amount of water flowing (in cubic feet per second), then dividing by ten. The result is roughly equal to the horsepower.

ACTIVITIES

Compare amounts of energy sources needed to produce electric power. Generally, 1 mw (1,000,000 w) will supply the electrical needs of 1,000 people.

1. Use the population of your city to find how many megawatts of electricity are required for your city.
2. Use the following equivalencies to find out how much coal, oil, gas, or uranium would be consumed in one day to supply the electrical needs.

$$1.0 \text{ kw-hr of electricity} = \begin{cases} 1.0 \text{ lb of coal} \\ 0.08 \text{ gal of oil} \\ 9 \text{ cubic ft of gas} \\ 0.00013 \text{ g of uranium} \end{cases}$$

Example
Assume your city has 36,000 people. Then 36 mw of electricity will be needed. How much oil is needed to produce this electricity?

$$36 \text{ mw} \times \frac{1{,}000 \text{ kw}}{\text{mw}} \times \frac{24 \text{ hr}}{\text{day}} \times \frac{0.08 \text{ gal}}{\text{kwhr}} = 69{,}120 \text{ gal/day}$$

Since there are 42 gallons in a barrel,

$$\frac{69{,}120 \text{ gal}}{42 \text{ gal/barrel}} = \frac{69{,}120 \text{ gal}}{42} \times \frac{\text{barrel}}{1 \text{ gal}} = 1{,}646 \text{ barrel}$$

Nuclear Power

Nuclear power plants use nuclear energy to produce electricity. Energy is released as the nuclei of uranium and plutonium atoms split, or fission. The fissioning takes place in a large steel vessel called a reactor. Water is pumped through the reactor to produce steam, which is used to produce electrical energy just as in the fossil fuel power plants. The nuclear processes are described in detail in chapter 17, and the process of producing electrical energy is described in detail in chapters 8 and 9. Nuclear power plants use nuclear energy to produce electricity, but there are opponents to this process. The electric utility companies view nuclear energy as *one* energy source used to produce electricity. They state that they have no allegiance to any one energy source but are seeking to utilize the most reliable and dependable of several energy sources. Petroleum, coal, and hydropower are also presently utilized as energy sources for electric power production. The electric utility companies are concerned that petroleum and natural gas are becoming increasingly expensive and there are questions about long-term supplies. Hydropower has limited potential for growth, and solar is prohibitively expensive today. Utility companies see two major energy sources that are available for growth: coal and nuclear. There are problems and advantages to each, but the utility companies feel they must use coal and nuclear power until the new technologies such as solar are economically feasible.

Summary

Work is defined as the product of an applied force and the distance through which the force acts. Work is measured in newton-meters, a metric unit called a *joule*. Simple machines reduce the magnitude of a force by increasing the distance through which the force acts. Ignoring friction, the work input to a simple machine equals the work output of the machine. *Power* is work per unit of time. Power is measured in *watts*. One watt is one joule per second. Power is also measured in *horsepower*. One horsepower is 550. ft·lb/sec.

Energy is defined as the ability to do work. An object that is elevated against gravity has a potential to do work. The object is said to have *potential energy*, or *energy of position*. Moving objects have the ability to do work on other objects because of their motion. The *energy of motion* is called *kinetic energy*.

Work is usually done *against inertia, fundamental forces, friction, shape*, or *combinations of these*. As a result there is a gain of *kinetic energy, potential energy, an increased temperature*, or *any combination of these*. Energy comes in the *forms* of *mechanical, chemical, radiant, electrical*, or *nuclear*. Potential energy can be converted to kinetic and kinetic can be converted to potential. Any form of energy

can be *converted* to any other form. Most technological devices are *energy form converters* that do work for you. Energy flows into and out of the surroundings but the amount of energy is always constant. The *law of conservation of energy* states that *energy is never created or destroyed*. Energy conversion always takes place through *heating* or *working*.

The basic energy sources today are the chemical *fossil fuels* (petroleum, natural gas, and coal), *nuclear energy,* and *hydropower*. Petroleum and *natural gas* were formed from organic material of plankton, tiny free-floating plants and animals. A barrel of petroleum is forty-two United States gallons, an actual container that does not exist. *Coal* formed from plants that were protected from consumption by falling into a swamp. The decayed plant material, *peat,* was changed into the various *ranks* of coal by pressure and heating over some period of time. Coal is a dirty fuel that contains impurities and sulfur. Controlling air pollution from coal burning is costly. Water power and nuclear energy are used for the generation of electricity.

Summary of Equations

4.1

$$\text{work} = \text{force} \times \text{distance}$$
$$W = Fd$$

4.2

$$\text{power} = \frac{\text{work}}{\text{time}}$$
$$P = \frac{W}{t}$$

4.3

$$\text{potential energy} = \text{weight} \times \text{height}$$
$$PE = mgh$$

4.4

$$\text{kinetic energy} = \frac{1}{2}(\text{mass})(\text{velocity})^2$$
$$KE = \frac{1}{2}mv^2$$

4.5

$$\text{velocity} = \text{square root of } 2 \times \frac{\text{acceleration due}}{\text{to gravity}} \times \text{height of fall}$$
$$v = \sqrt{2gh}$$

Key Terms

chemical energy (p. 69)
electrical energy (p. 70)
energy (p. 65)
fossil fuels (p. 74)
horsepower (p. 64)
joule (p. 62)
kinetic energy (p. 66)
law of conservation of energy
  (p. 72)

mechanical energy (p. 69)
nuclear energy (p. 70)
potential energy (p. 65)
power (p. 63)
radiant energy (p. 70)
watt (p. 64)
work (p. 61)

Solar Technologies

An alternative source of energy is one that is different from the typical sources used today. The sources used today are the fossil fuels (coal, petroleum, and natural gas), nuclear, and falling water. Alternative sources could be solar, geothermal, hydrogen gas, fusion, or any other energy source that a new technology could utilize.

The term "solar energy" is used to describe a number of technologies that directly or indirectly utilize sunlight as an alternative energy source (box figure 4.1). There are eight main categories of these solar technologies:

1. *Solar cells.* A solar cell is a thin crystal of silicon, gallium, or some polycrystalline compound that generates electricity when exposed to light. Also called photovoltaic devices, solar cells have no moving parts and produce electricity directly, without the need for hot fluids or intermediate conversion states. Solar cells have been used extensively in space vehicles and satellites. Here on the earth, however, use has been limited to demonstration projects, remote site applications, and consumer specialty items such as solar-powered watches and calculators. The problem with solar cells today is the manufacturing cost is too high (they are essentially handmade). Research is continuing on the development of highly efficient, affordable solar cells that could someday produce electricity for the home.

2. *Power tower.* This is another solar technology designed to generate electricity. One type of planned power tower will have a 171 m (560 ft) tower surrounded by some 9,000 special mirrors called heliostats. The heliostats will focus sunlight on a boiler at the top of the tower where salt (a mixture of sodium nitrate and potassium nitrate) will be heated to about 566° C (about 1,050° F). This molten salt will be pumped to a steam generator, and the steam will be used to drive a generator, just like other power plants. Water could be heated directly in the power tower boiler. Molten salt is used because it can be stored in an insulated storage tank for use when the sun is not shining, perhaps for up to twenty hours.

Box Figure 4.1

These solar cells produce a yearly average of 4,000 watts alternating current, providing about a third of the electrical energy needs of the household. Over twenty years of savings will be required, however, to pay for the cost of the solar cell installation.

3. *Passive application.* Passive means that energy flows by natural means, without mechanical devices such as motors, pumps, and so forth. A passive solar house would include considerations like orientation of a house to the sun, the size and positioning of windows, and a roof overhang that lets sunlight in during the winter but keeps it out during the summer. There are different design plans to capture, store, and distribute solar energy throughout a house.

4. *Active application.* An active solar application requires a solar collector in which sunlight heats air, water, or some liquid. The liquid or air is pumped through pipes in a house to generate electricity or it is used directly for hot water. Solar water heating makes more economic sense today than the other applications.

5. *Wind energy.* The wind has been used for centuries to move ships, grind grain into flour, and pump water. The wind blows, however, because of radiant energy from the sun, so it is another form of solar energy. Wind turbines are used to generate electrical energy or mechanical energy. The biggest problem with wind energy is the inconsistency of the wind. Some times the wind speed is too great and other times it is not great enough. Several methods of solving this problem are being researched.

6. *Biomass.* Biomass is any material formed by photosynthesis including plants, trees, and crops and any garbage, crop residue, or animal waste. Biomass can be burned directly as a fuel, converted into a gas fuel (methane), or converted into liquid fuels such as alcohol. The problems with using biomass include the energy expended in gathering the biomass as well as the energy used to convert it to a gaseous or liquid fuel.

7. *Agriculture and industrial heating.* This is a technology that simply uses sunlight to dry grains, cure paint, or do anything that can be done with sunlight rather than using traditional energy sources.

8. *Ocean thermal energy conversion (OTEC).* This is an electric generating plant that would take advantage of the approximately 22° C (about 40° F) temperature difference between the surface and the depths of tropical, subtropical, and equatorial ocean waters.

Basically, warm water is drawn into the system to vaporize a fluid, which expands through a turbine generator. Cold water from the depths condenses the vapor back to a liquid form, which is then cycled back to the warm water side. The concept has been tested several times and was found to be technically successful. The greatest interest seems to be islands that have warm surface waters (and cold depths) such as Hawaii, Puerto Rico, Guam, the Virgin Islands, and others.

Applying the Concepts

1. The metric unit of a joule (J) is a unit of
   a. potential energy.
   b. work.
   c. kinetic energy.
   d. any of the above.

2. A N·m/sec is a unit of
   a. work.
   b. power.
   c. energy.
   d. any of the above.

3. A kilowatt-hour is actually a unit of
   a. power.
   b. work.
   c. time.

4. In calculating the upward force required to lift an object, it is necessary to use g if the mass is given in kg. The quantity of g is not needed if the weight is given in lb because
   a. the rules of measurement are different in the English system.
   b. the symbol for metric mass has a "g" in it and the symbol for pound does not.
   c. a pound is defined as a measure of force and a kilogram is not.
   d. a kilogram is a unit of weight.

5. The potential energy of a box on a shelf, relative to the floor, is a measure of the
   a. work that was required to put the box on the shelf from the floor.
   b. weight of the box times the distance above the floor.
   c. energy the box has because of its position above the floor.
   d. all of the above.

6. A rock on the ground had zero potential energy, then fell into a well. In the well, the rock now has
   a. zero potential energy as before it fell.
   b. negative potential energy.
   c. positive potential energy.
   d. zero potential energy but will require work to bring it back to ground level.

7. Which quantity has the greatest influence on the amount of kinetic energy that a large truck has while moving down the highway?
   a. mass
   b. weight
   c. velocity
   d. size

8. Electrical energy can be converted to
   a. chemical energy.
   b. mechanical energy.
   c. radiant energy.
   d. any of the above.

9. Almost all energy comes to and leaves the earth in the form of
   a. nuclear energy.
   b. chemical energy.
   c. radiant energy.
   d. kinetic energy.

10. The law of conservation of energy is basically that
    a. energy must not be used up faster than it is created or the supply will run out.
    b. energy should be saved because it is easily destroyed.
    c. energy is never created or destroyed.
    d. you are breaking a law if you needlessly destroy energy.

ANSWERS

1. d   2. b   3. b   4. c   5. d   6. b   7. c   8. d   9. c
10. c

Questions for Thought

1. How is work related to energy?

2. Why is work always accomplished *against* something?

3. Is it possible for any of the things that work is typically done *against* (inertia, basic forces, friction, shape) to do work on something else? Explain.

4. What is the relationship between the work done while moving a book to a higher bookshelf to the potential energy that the book has on the higher shelf?

5. Describe how you could measure the work done on a person standing in the aisle of a bus when the bus starts to move from rest.

6. Does a person standing motionless in the aisle of a moving bus have kinetic energy? Explain.

7. A lamp bulb is rated at 100 w. Why is a time factor not included in the rating?

8. Is a kwhr a unit of work, energy, power, or more than one of these? Explain.

9. Which forms of energy must be used at once? Which forms can be saved for future use? Give examples.

10. If energy cannot be destroyed, why do some people worry about the energy supplies?

11. A spring clamp exerts a force on a stack of papers it is holding together. Is the spring clamp doing work on the papers? Explain.

12. What is the meaning of a positive change in potential energy? A negative change?

13. What happens to the potential energy of an orbiting space vehicle as the vehicle returns to the surface of the earth?

14. Compare the effect of doubling the mass to the effect of doubling the velocity on the energy of a moving object.

15. One person calculates the potential energy of an object to be 10.0 J, a second person calculates it to be 0, and a third calculates it to be −10.0 J. Assuming no errors, how can this be?

16. Why are petroleum, natural gas, and coal called "fossil fuels"?

17. From time to time people claim to have invented a machine that will run forever without energy input and develops more energy than it uses (perpetual motion). Why would you have reason to question such a machine?

18. Does a higher power rating mean that more work is done than would be done with a lower power rating? Explain.

19. Define a joule. What is the difference between a joule of work and a joule of energy?

20. Explain why you do the same work, disregarding friction, rolling a cart up a long incline with a gradual slope as you do pushing the same cart up a short incline with a steep slope.

21. Compare the energy needed to raise a mass 10 m on the earth to the energy needed to raise the same mass 10 m on the moon. Explain the difference, if any.

22. What happens to the kinetic energy of a falling book when the book hits the floor?

23. Why could you consider the potential energy of water behind a dam to be a form of stored solar energy?

24. The braking force of two identical cars is the same, yet it takes four times the distance to stop one of the cars. Explain what this means in terms of kinetic energy and work.

Exercises

▼

Group A—Solutions Provided in Appendix D

1. A horizontal force of 10.0 lb is needed to push a bookcase 15.0 ft across the floor. (a) How much work was done on the bookcase? (b) How much did the gravitational potential energy change as a result?

2. (a) How much work is done in moving a 2.0 kg book to a shelf 2.00 m high? (b) What is the potential energy of the book as a result? (c) How much kinetic energy will the book have after it falls and hits the ground?

3. A 150.0 g baseball has a velocity of 30.0 m/sec. What is its kinetic energy in joules?

4. (a) What is the kinetic energy of a 1,000.0 kg car that is traveling at 90.0 km/hr? (b) How much work was done to give the car this kinetic energy? (c) How much work must be done now to stop the car?

5. A 60.0 kg jogger moving at 3.00 m/sec decides to double the jogging speed. How did this change in speed change the kinetic energy?

6. A bicycle and rider have a combined mass of 70.0 kg and are moving at 6.00 m/sec. A 70.0 kg person is now given a ride on the bicycle. How did the addition of the new rider change the kinetic energy at the same speed?

7. A 170.0 lb student runs up a stairway to a classroom 25.0 ft above ground level in 10.0 sec. (a) How much work did the student do? (b) What was the average power output in horsepower?

8. (a) How many seconds will it take a 20.0 hp motor to lift a 2,000.0 lb elevator a distance of 20.0 ft? (b) What was the average velocity of the elevator?

9. A 10.0 lb brick falls from a ledge that is 15.0 ft above the ground to a ledge that is 10.0 ft above the ground. (a) How much potential energy did the brick lose? (b) How much does it still have?

Group B—Solutions Not Given

1. A force of 50.0 lb is used to push a box 10.0 ft across a level floor. (a) How much work was done on the box? (b) What is the change of potential energy as a result of this move?

2. (a) How much work is done in raising a 50.0 kg crate a distance of 1.5 m above a storeroom floor? (b) What is the change of potential energy as a result of this move? (c) How much kinetic energy will the crate have as it falls and hits the floor?

3. What is the kinetic energy, in joules, of a 60.0 g tennis ball approaching a tennis racket at 20.0 m/sec?

4. (a) What is the kinetic energy of a 1,500.0 kg car with a velocity of 72.0 km/hr? (b) How much work must be done on this car to bring it to a complete stop?

5. The driver of a 800.0 kg car decides to double the speed from 20.0 m/sec to 40.0 m/sec. What effect would this have on the amount of work required to stop the car, that is, on the kinetic energy of the car?

6. Compare the kinetic energy of a 800.0 kg car moving at 20.0 m/sec to the kinetic energy of a 1,600.0 kg car moving at an identical speed.

7. A 175.0 lb hiker is able to ascend a 1,980.0 ft high slope in 1 hr 45 min. (a) How much work did the hiker do? (b) What was the average power output in horsepower?

8. (a) What distance will a 10.0 hp motor lift a 2,000.0 lb elevator in 30.0 sec? (b) What was the average velocity of this elevator during the lift?

9. An accomplished skateboard rider uses a ramp to jump to a ledge 6.00 ft above the ground, then to a lower ledge 3.00 ft above the ground. If the rider and skateboard have a mass of 60.0 kg (a) what was the potential energy on the ledge 6.0 ft above the ground? (b) on the ledge that is 3.00 ft above the ground? (c) on the ground?

10. Too heavy and bulky for one person to lift, a piece of furniture weighing 6,500.0 N must be somehow moved to the top of a stairway 400.0 cm above ground level. It could be pushed up a 20.0 m ramp by workers pushing with a force of 1,300.0 N, or it could be lifted straight up by workers pulling on a rope over a pulley. (a) How much work would the workers do by pushing the furniture up the 20.0 m ramp? (b) How much work would the workers do by pulling the furniture straight up to the 400. cm height with a rope and pulley? (c) Evaluate the two methods of moving the furniture.

11. A 200. g ball is thrown straight up with 40.0 J of energy. How high did the ball go?

12. A ball is dropped from 9.80 ft above the ground. Using energy considerations only, find the velocity of the ball just as it hits the ground.

13. What is the velocity of a 1,000.0 kg car if its kinetic energy is 200. kJ?

14. A Foucault pendulum swings to 3.0 in above the ground at the highest points and is practically touching the ground at the lowest point. What is the maximum velocity of the pendulum?

15. An electric hoist is used to lift a 250.0 kg load to a height of 80.0 m in 39.2 sec. (a) What is the power of the hoist motor in kw? (b) in hp?

10. A 200.0 kg crate must be raised 5.00 m to a loading dock. (a) How much work is performed in lifting the crate straight up to the dock? (b) How much work is performed in pushing it up a 20.0 m ramp with a force of 490.0 N? (c) Evaluate the advantages and disadvantages of each method of raising the crate.

11. How much work would you need to perform on a 60.0 g tennis ball in order to throw it 10.0 m high?

12. A ball is dropped from 20.0 ft above the ground. (a) At what height is half of its energy kinetic and half potential? (b) Using energy considerations only, what is the velocity of the ball just as it hits the ground?

13. What is the velocity of a 60.0 kg jogger with a kinetic energy of 1,080.0 J?

14. A small sports car and a pickup truck start coasting down a 10.0 m hill together, side by side. Assuming no friction, what is the velocity of each vehicle at the bottom of the hill?

15. A 70.0 kg student runs up the stairs of a football stadium to a height of 10.0 m above the ground in 10.0 sec. (a) What is the power of the student in kw? (b) in hp?

1. $U = Fd$ $(10.0\ lb)(15.0)$ a. $150\ lb/ft$

b. none

2. $W = mgd$ $(2.0\ kg)(9.80\ m/sec^2)(2.00\ m) = 39.2\ kg\cdot m^2/sec^2$ b. 39 J c. 39 J

3. $Ke = \frac{1}{2}mv^2$ $\frac{1}{2}(.15\ kg)(30.0)\ m/sec^2$ $(1/2)(.15)(900\ kg\cdot m^2/sec^2) = 67.5\ J$

6. $K = \frac{1}{2}mv^2 = (1/2)(70\ kg)(6100\ m/sec) = 1,260\ J$

$L = \frac{1}{2}mv^2 = \frac{1}{2}(70\ kg)(36.0\ m/sec^2) = 2520\ J$

10. $U = Fd$ $(1,300\ \frac{kg\cdot m}{sec^2})(20.0\ m) = 26,000\ \frac{kg\cdot m}{sec^2}$

$(6500)(4.00) = 26,000\ \frac{kg\cdot m^2}{sec^2}$

Heat and Temperature

Figure 5.1
Heat and modern technology are inseparable. These glowing steel
slabs, at over 1,100° C (about 2,000° F), are cut by an automatic flame
torch. The slab caster converts three hundred tons of molten steel into
slabs in about forty-five minutes. The slabs are converted to sheet steel
for use in the automotive, appliance, and building industries.

HEAT has been closely associated with the comfort and support of people throughout history. You can imagine the appreciation when your earliest ancestors first discovered fire and learned to keep themselves warm and cook their food. You can also imagine the wonder and excitement about 3000 B.C., when people put certain earthlike substances on the hot, glowing coals of a fire and later found metallic copper, lead, or iron. The use of these metals for simple tools followed soon afterwards. Today, metals are used to produce complicated engines that use heat for transportation and that do the work of moving soil and rock, construction, and agriculture. Devices made of heat-extracted metals are also used to control the temperature of structures, heating or cooling the air as necessary. Thus the production and control of heat gradually built the basis of civilization today (figure 5.1).

The sources of heat are the energy forms that you learned about in chapter 4. The fossil fuels are *chemical* sources of heat. Heat is released when oxygen is combined with these fuels. Heat also results when *mechanical* energy does work against friction such as in the brakes of a car coming to a stop. Heat also appears when *radiant* energy is absorbed. This is apparent when solar energy heats water in a solar collector or when sunlight melts snow. The transformation of *electrical* energy to heat is apparent in toasters, heaters, and ranges. *Nuclear* energy provides the heat to make steam in a nuclear power plant. Thus all the energy forms can be converted to heat.

The relationship between energy forms and heat appears to give an order to nature, revealing patterns that you will want to understand. All that you need is some kind of explanation for the relationships—a model or theory that helps make sense of it all. This chapter is concerned with heat and temperature and their relationship to energy. It begins with a simple theory about the structure of matter, then uses the theory to explain the concepts of heat, energy, and temperature changes.

The Caloric Theory of Heat

Heat results when mechanical energy does work against friction, when radiant energy is absorbed, or when electrical energy warms a wire. Heat is also produced when stored chemical energy or nuclear energy is released. These are all very different means of producing heat, but each form of energy disappeared as heat appeared. What is heat and where does it come from? The answer to this question has not been an easy one for humankind and, in fact, was only worked out in the last one hundred years or so. Here is part of this most interesting story.

Recall that Aristotle considered matter to be composed of the four idealized elements of earth, air, fire, and water. Recall also that it was not until the early 1600s that Galileo challenged the Aristotelian view of motion, which lead to Newton's laws of motion by the late 1600s. During this time people held the belief that heat was a fluid form of matter that could flow into and out of objects, "flowing" from a hot object to a cooler one. Not everyone accepted a fluid theory of heat, however, including Bacon, Galileo, Newton, and Boyle. They tended to think of heat

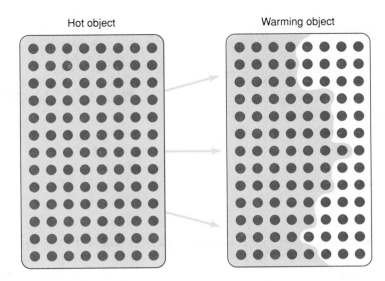

Figure 5.2

The caloric theory of heat viewed heat as an invisible, weightless fluid that flowed from hot objects into a cooler object. The fluid, called *caloric*, was thought to surround the particles making up the object and force them farther apart, expanding the object.

as movement, or vibrations, of the particles that made up bodies. Both the "fluid substance" and "vibration of particles" ideas were vague, however, so neither viewpoint of heat became firmly established during the 1600s and 1700s.

Then, in the 1780s, a Scotsman named Joseph Black made quantitative measurements of heat and defined several key concepts in the study of heat. He found, for example, that hot and cold objects brought together cooled and warmed to some equilibrium, but that the total amount of heat remained *constant*. This seemed convincing evidence that heat was indeed a fluid substance. The fluid was given a name, *caloric*, and more and more properties were assigned to it as the **caloric theory of heat** was developed and refined. Caloric was supposed to be an invisible, weightless substance that was attracted to ordinary matter. According to the theory, caloric fluid flowed into a cool object from a warmer object or objects. The fluid was thought to flow inside the mass of the cooler object, surrounding all the particles making up the object much as air surrounds the earth. The caloric fluid was considered to be self-repulsive, causing the engulfed particles making up an object to move further apart. If enough caloric flowed into a solid, it would force the gravity-attracted particles (the accepted explanation at that time) completely apart, melting the solid. Thus the caloric theory supposedly explained why warmer solids expand and even warmer ones melt. As a melted substance cooled, the caloric supposedly flowed out and the gravitational attraction pulled the particles closer together, forming a solid. Further cooling, and supposed further loss of caloric, allowed the solid to contract. That was the caloric explanation (figure 5.2).

Heat and Temperature

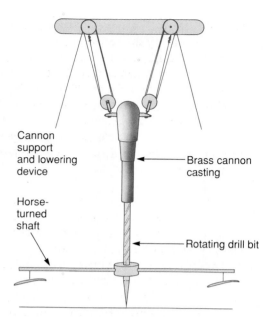

Figure 5.3

During the time of Count Rumford, cannons were made by lowering a brass cannon casting on a rotating drill bit turned by horses. Rumford observed that the casting, the bit, and the metal chips became very hot during the process and that heat was generated as long as there was motion. He therefore doubted that the heat could come from the release of a fluid called caloric.

Another assigned property of caloric was that the fluid came in two forms, *latent* (hidden) and *sensible*. Latent caloric was supposedly hidden from the senses so you could not identify its presence by a warmer temperature. Latent caloric was somehow squeezed out of hiding inside of objects by friction, hammering, or pressure to become sensible caloric, that is, heat that could be felt and measured. For example, a paper clip feels cool to the touch, but if you bend the wire back and forth the wire becomes warm to the touch. The warming was explained as the transformation of latent caloric to sensible caloric. The caloric theory seemed to explain this and many other observations concerning heat and temperature changes logically.

The caloric theory of heat was widely accepted during the late 1700s and into the early 1800s because it did seem to explain many observations about heat and the apparent movement of heat. The existence of an invisible, weightless fluid was not a problem since other phenomena, such as electricity and light, were also considered to involve invisible, weightless fluids during this period of history. There was one part of the caloric theory that did present a problem, however, and that part was the requirement that the amount of caloric had to remain constant through all heating and cooling events. If more or less of the caloric fluid was observed after such an event it would mean that the theory was flawed. The work of Count Rumford and others revealed this flaw with some convincing experiments and logical arguments.

Count Rumford was born Benjamin Thompson in Woburn, Massachusetts, in 1753. He was a Tory who took refuge in England during the American Revolution. Thompson became Count Rumford when he was knighted by King George III, taking the original name of Concord, New Hampshire, which was at one time his home. One of Count Rumford's many duties as Minister of War was to supervise the production of brass cannons in Germany. During that time, cannons were made by a horse-driven lathe drilling out a barrel in a metal casting (figure 5.3). Rumford noted that the drill, the cannon casting, and the metal chips became very hot during the drilling process. He conducted a number of experiments concerning the mechanical work and the heat produced. Rumford found that more heating resulted from a dull drilling bit than from a sharp one and that the heating was proportional to the drilling motion, not to the amount of metal actually drilled. More important, he found that heat was generated *as long as there was motion*. He therefore doubted that heat could come from the release of a material substance (caloric) since the source of heat seemed to be inexhaustible. Rumford rejected the caloric theory and proposed that the heating was produced by mechanical energy, not a material substance called caloric. He had struck a blow against the caloric theory by showing the unlimited creation of heat by friction. But this failed to discredit the caloric theory, however, since there was not an alternative theory. It would take the work of other scientists through the early 1800s to convince the caloric theory proponents of the relationships between work, energy, and heat.

Part of the work of disproving the caloric theory was done by James P. Joule, the son of a Manchester, England, brewer. Joule was a pupil of John Dalton, a famous chemist who will be discussed later. By his early twenties Joule was using the brewery's workshops to perform careful, repeated experiments that would occupy most of his life. He was determined to prove the relationship that Rumford (and others) had suggested, that an exact amount of heat is created whenever mechanical work is done. One of his experiments compared the work needed to operate an electric generator with the heat generated by the current. Another compared the work needed to pump water through thin tubes with the heat produced by the friction of the water flowing through the pipes. A third experiment compared the work needed to compress a gas to the heat created by the compression. In his most famous experiment, Joule compared the mechanical energy of falling weights to the temperature increase that occurred from the friction of turning paddle wheels (figure 5.4). By the year 1849, Joule was able to summarize all his experiments and show that a measured amount of work *always* produces a certain amount of heat. Joule found that heat is not a substance and can be converted from energy forms into a predictable amount of heat. Joule's work established the basis for the law of conservation of energy, the basis for understanding the nature of heat, and established the groundwork for the kinetic molecular theory of matter. It would be up to a group

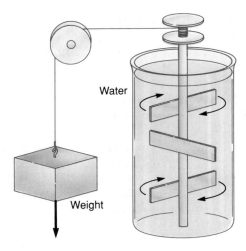

Figure 5.4
This experiment, one of Joule's many experiments concerning the forms of energy and heat, compared the mechanical energy of a falling weight to the temperature increase that resulted from the fluid friction of the turning paddles.

of theoreticians (Kelvin, Clausius, Boltzmann, and others) to develop the collection of mathematical equations used in the study of heat and the transformation of energy forms. This resulted, by 1860, in the kinetic molecular theory, a major development of nineteenth-century physical science.

The caloric theory was finally shown to be an inadequate explanation for heat by the 1850s. Yet, a few traces of the discarded theory can be found in today's language, such as statements that food "contains Calories" and that "heat flows." Heat is no longer properly thought of as a fluid that flows into and out of objects. Nonetheless, even people who understand the nature of heat continue to use such terms because these analogies can be used to communicate clearly about what is being observed. You will learn an exact, modern definition of heat after first considering the kinetic molecular theory, a necessary explanation about the relationships between energy and matter.

The Kinetic Molecular Theory

The idea that substances are composed of very small particles can be traced back to certain early Greek philosophers. The earliest record of this idea was written by Democritus during the fifth century B.C. He wrote that matter was empty space filled with tremendous numbers of tiny, indivisible particles called *atoms*. This idea, however, was not acceptable to most of the ancient Greeks because matter seemed continuous and empty space was simply not believable. The idea of atoms was rejected by Aristotle as he formalized his belief in continuous matter composed of the earth, air, fire, and water elements. Aristotle's belief about matter, as his beliefs about motion, predominated through the 1600s. Some people, such as Galileo and Newton,

believed the ideas about matter being composed of tiny particles, or atoms, since this theory seemed to explain the behavior of matter. Widespread acceptance of the particle model did not occur, however, until strong evidence was developed through chemistry in the late 1700s and early 1800s. The experiments finally led to a collection of assumptions about the small particles of matter and the space around them. Collectively, the assumptions could be called the **kinetic molecular theory**. The following is a general description of some of these assumptions.

Molecules

The basic assumption of the kinetic molecular theory is that all matter is made up of tiny, basic units of structure called *atoms*. Atoms are neither divided, created, nor destroyed during any type of chemical or physical change. There are similar groups of atoms that make up the pure substances known as chemical *elements*. Each element has its own kind of atom, which is different from the atoms of other elements. For example, hydrogen, oxygen, carbon, iron, and gold are chemical elements and each has its own kind of atom.

In addition to the chemical elements, there are pure substances called *compounds* that have more complex units of structure (figure 5.5). Such pure substances, such as water, sugar, and alcohol, are composed of atoms of two or more elements that join together in definite proportions. Water, for example, has structural units that are made up of two atoms of hydrogen tightly bound to one atom of oxygen (H_2O). These units are not easily broken apart and stay together as small physical particles of which water is composed. Each is the smallest particle of water that can exist, a molecule of water. A *molecule* is generally defined as a tightly bound group of atoms in which the atoms maintain their identity. How atoms become bound together to form molecules is discussed in chapters 11–13.

Some elements exist as gases at ordinary temperatures, and all elements are gases at sufficiently high temperatures. At ordinary temperatures the atoms of oxygen, nitrogen, and other gases are paired in groups of two to form *diatomic molecules*. Other gases, such as helium, exist as single, unpaired atoms at ordinary temperatures. At sufficiently high temperatures, iron, gold, and other metals vaporize to form gaseous, single, unpaired atoms. In the kinetic molecular theory the term *molecule* has the additional meaning of the smallest, ultimate particles of matter that can exist. Thus the ultimate particle of a gas, whether it is made up of two or more atoms bound together or of a single atom, is conceived of as a molecule. A single atom of helium, for example, is known as a *monatomic molecule*. For now, a **molecule** is defined as the smallest particle of a compound, or a gaseous element, that can exist and still retain the characteristic properties of that substance.

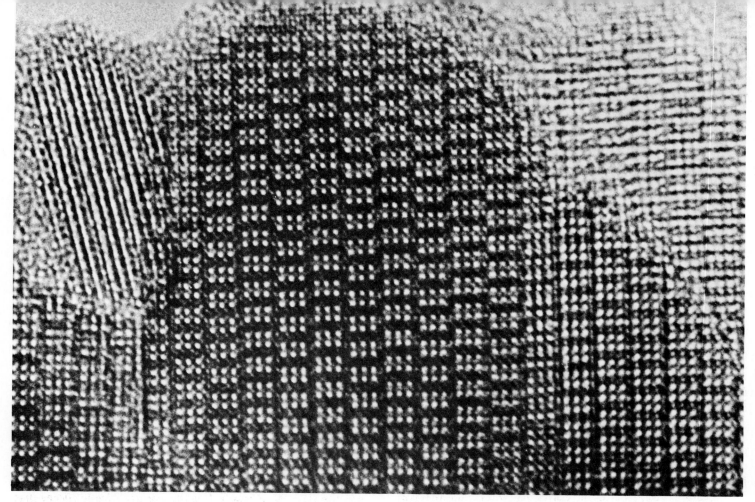

Figure 5.5
Metal atoms appear as black dots in this micrograph of a crystal of titanium niobium oxide, magnified ten million times by an electron microscope.

Molecules Interact

Some molecules of solids and liquids interact, strongly attracting and clinging to each other. When this attractive force is between the same kind of molecules, it is called *cohesion*. It is a stronger cohesion that makes solids and liquids different from gases, and without cohesion all matter would be in the form of gases. Sometimes one kind of molecule attracts and clings to a different kind of molecule. The attractive force between unlike molecules is called *adhesion*. Water wets your skin because the adhesion of water molecules and skin is stronger than the cohesion of water molecules. Some substances, such as glue, have a strong force of adhesion when they harden from a liquid state and they are called "adhesives."

States of Matter

Different states of matter have different molecular arrangements. You know that matter can occur in three states of solid, liquid, and gas. The different characteristics of these states can be attributed to the molecular arrangements and the strength of attraction between the molecules (figure 5.6).

Solids have definite shapes and volumes because they have molecules that are fixed distances apart with relatively strong cohesive forces. Each molecule is a nearly fixed distance from the next, but it does vibrate and move around an equilibrium position. The masses of these molecules and the spacing between them determine the density of the solid. The hardness of a solid is the resistance of a solid to forces that tend to push its molecules further apart.

Liquids have molecules that are not confined to an equilibrium position as in a solid. The molecules of a liquid are close together and have cohesive forces, but they are not as strong as in a solid. This permits the molecules to move from place to place within the liquid. The molecular forces are strong enough to give the liquid a definite volume but not strong enough to give it a definite shape. Thus a pint of milk is always a pint of milk (unless it is under tremendous pressure) and takes the shape of the container holding it. Because the forces between the molecules of a liquid are weaker than the forces between the molecules of a solid, a liquid cannot support the stress of a rock placed on it as a solid does. The liquid molecules *flow,* rolling

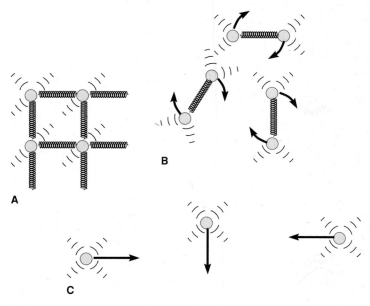

Figure 5.6

(a) In a solid, molecules vibrate around a fixed equilibrium position and are held in place by strong molecular forces. (b) In a liquid, molecules can rotate and roll over each other because the molecular forces are not as strong. (c) In a gas, molecules move rapidly in random, free paths.

over each other as the rock pushes its way between the molecules. Yet, the molecular forces are strong enough to hold the liquid together so it keeps the same volume.

Gases are composed of molecules with weak cohesive forces acting between them. The gas molecules are relatively far apart and move freely in a constant, random motion that is changed often by collisions with other molecules. Gases therefore have neither fixed shapes nor fixed volumes.

There are other distinctions between the states of matter. The term **vapor** is sometimes used to describe a gas that is usually in the liquid state. Water vapor, for example, is the gaseous state of liquid water. Liquids and gases are collectively called **fluids** because of their ability to flow, a property that is lacking in most solids. A not-so-ordinary state of matter on the earth is called **plasma.** Plasma occurs at extremely high temperatures and is made up of charged parts of atoms, which will be discussed later.

Molecules Move

Suppose you are in an evenly heated room with no air currents. If you open a bottle of ammonia, the odor of ammonia is soon noticeable everywhere in the room. According to the kinetic molecular theory, molecules of ammonia leave the bottle and bounce around among the other molecules making up the air until they are everywhere in the room, slowly becoming more evenly distributed. The ammonia molecules *diffuse,* or spread, throughout the room. The ammonia odor diffuses throughout the room faster

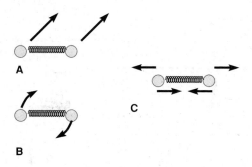

Figure 5.7

The basic forms of kinetic energy of molecules. (a) Translational motion is the motion of a molecule as a whole moving from place to place. (b) Rotational motion is the motion of a turning molecule. (c) Vibrational motion is the back-and-forth movement of a vibrating molecule.

if the air temperature is higher and slower if the air temperature is lower. This would imply a relationship between the temperature and the speed at which molecules move about.

The mathematical relationship between the temperature of a gas and the motion of molecules was formulated in 1857 by Rudolf Clausius. He showed that the temperature of a gas is proportional to the average kinetic energy of the gas molecules. This means that ammonia molecules have a greater average velocity at a higher temperature and a slower average velocity at a lower temperature. This explains why gases diffuse at a greater rate at higher temperatures. Recall, however, that kinetic energy involves the mass of the molecules as well as their velocity ($KE = 1/2\ mv^2$). It is the *average kinetic energy* that is proportional to the temperature, which involves the molecular mass as well as the molecular velocity.

The kinetic energy of molecules can be in three basic forms: vibrational, rotational, or translational (involving the motion of a molecule as a whole) (figure 5.7). The kinetic energy of the molecules of a solid is mostly in the form of vibrational kinetic energy. The molecules of a liquid can have vibrational, rotational, and some translational kinetic energy. The molecules of a liquid have been compared to dancers on a packed dance floor; they are free to move about, but it is difficult to move across the room. Gas molecules, on the other hand, can move about easily since they are far apart. Gas molecules can have all three forms of kinetic energy—translational, rotational, and vibrational. The *total* kinetic energy of the molecules of a substance can be complicated combinations of the three basic forms plus pulsing shape changes, twisting, and other forms of motion. In general, the type of motion represented by kinetic energy varies with the state of matter. Whether the kinetic energy is jiggling, vibrating, rotating, or moving from place to place, the **temperature** of a substance is *a measure of the average kinetic energy of the molecules making up the substance* (figure 5.8).

Heat and Temperature

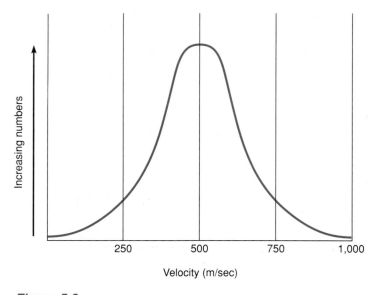

Figure 5.8

The number of oxygen molecules with certain velocities that you might find in a sample of air at room temperature. Notice that a few are barely moving and some have velocities over 1,000 m/sec at a given time, but the *average* velocity is somewhere around 500 m/sec.

The kinetic molecular theory explains why matter generally expands with increased temperatures and contracts with decreased temperatures. At higher temperatures the molecules of a substance move faster, with increased agitation, and therefore they move a little further apart, thus expanding the substance. As the substance cools the motion slows, and the molecular forces are able to pull the molecules closer together, thus contracting the substance.

These assumptions do not provide a complete nor rigorous description of the kinetic molecular theory. They are, however, sufficient for an introduction to energy, heat, and some of the more interesting related behaviors of matter.

ACTIVITIES

1. Obtain a glass of very hot water and a glass of very cold water. When the water is still, place a drop of food coloring in each. Describe how this activity provides evidence that substances with a higher temperature have molecules with more kinetic energy.
2. Again obtain a glass of very hot water and a glass of very cold water. Carefully lower a sugar cube into each glass and observe what happens. Describe how this provides evidence that substances with a higher temperature have molecules with more kinetic energy.

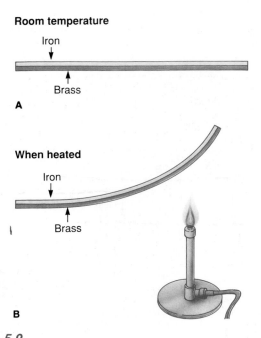

Figure 5.9

(a) A bimetallic strip is two different metals, such as iron and brass, bonded together as a single unit, shown here at room temperature. (b) Since one metal expands more than the other, the strip will bend when it is heated. In this example the brass expands more than the iron, so the bimetallic strip bends away from the brass.

Temperature

If you ask people about the temperature, they usually respond with a referent ("hotter than the summer of '89") or a number ("68° F or 20° C"). Your response, or feeling, about the referent or number depends on a number of factors, including a *relative* comparison. A temperature of 20° C (68° F), for example, might seem cold during the month of July but warm during the month of January. The 20° C temperature is compared to what is expected at the time even though 20° C is 20° C, no matter what month it is.

When people ask about the temperature they are really asking *how hot or how cold something is*. Without a thermometer, however, most people can do no better than *hot* or *cold*, or perhaps *warm* or *cool*, in describing a relative temperature. Even then there are other factors that confuse people about temperature. Your body judges temperature on the basis of the net *direction* of energy flow. You call energy flowing into your body "warm" and energy flowing out "cool." Perhaps you have experienced having your hands in snow for some time, then washing your hands in cold water. The cold water "feels" warm. Your hands are colder than the water, energy flows into your hands, and they communicate "warm."

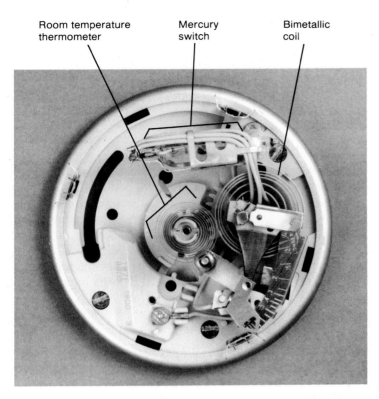

Room temperature thermometer   Mercury switch   Bimetallic coil

Figure 5.10

This thermostat has a coiled bimetallic strip that expands and contracts with changes in the room temperature. The attached vial of mercury is thus tilted one way or the other, and the mercury completes or breaks an electric circuit that turns the heating or cooling system on or off.

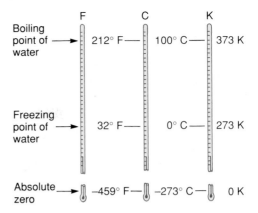

Figure 5.11

The Fahrenheit, Celsius, and Kelvin temperature scales.

Thermometers

The human body is a poor sensor of temperature, so a device called a **thermometer** is used to measure the hotness or coldness of something. Most thermometers are based on the relationship between some property of matter and changes in temperature. Almost all materials expand with increasing temperatures. A strip of metal is slightly longer when hotter and slightly shorter when cooler, but the change of length is too small to be useful in a thermometer. A more useful, larger change is obtained when two metals that have different expansion rates are bonded together in a strip. The bimetallic strip will bend toward the metal with less expansion when the strip is heated (figure 5.9). Such a bimetallic strip is formed into a coil and used in thermostats and dial thermometers (figure 5.10).

The common glass thermometer is a glass tube with a small bore connected to a relatively large glass bulb. The bulb contains mercury or colored alcohol, which expands up the bore with increases in temperature and contracts back toward the bulb with decreases in temperature. The height of this liquid column is used with a referent scale to measure temperature. Some thermometers, such as a "fever thermometer," have a small constriction in the bore so the liquid cannot normally return to the bulb. This results in the thermometer showing the highest reading even if the temperature it measures has fluctuated up and down during the reading. The liquid must be forced back into the bulb by a small swinging motion, bulb-end down, then sharply stopping the swing in a "snap." The inertia of the mercury in the bore forces it past the constriction and into the bulb. The fever thermometer is then ready to use again.

Thermometer Scales

There are several referent scales used to define numerical values for measuring temperatures (figure 5.11). The **Fahrenheit scale** was developed by the German physicist Gabriel D. Fahrenheit in about 1715. Fahrenheit invented a mercury-in-glass thermometer with a scale based on two arbitrarily chosen reference points. The original Fahrenheit scale was based on the temperature of an ice and salt mixture for the lower reference point (0°) and the temperature of the human body as the upper reference point (about 100°). Thus the original Fahrenheit scale was a centigrade scale with 100 divisions between the high and the low reference points. The distance between the two reference points was then divided into equal intervals called *degrees*. There were problems with identifying a "normal" human body temperature as a reference point, since body temperature naturally changes during a given day and from day to day. The only consistent thing about the human body temperature is constant change. The standards for the Fahrenheit scale were eventually changed to something more consistent, the freezing point and the boiling point of water at normal atmospheric pressure. The original scale was retained with the new reference points, however, so the "odd" numbers of 32° F (freezing point of water) and 212° F (boiling point of water under normal pressure) came to be the reference points. There are 180 equal intervals, or degrees, between the freezing and boiling points on the Fahrenheit scale.

The **Celsius scale** was invented by Anders C. Celsius, a Swedish astronomer, in about 1735. The Celsius scale uses the freezing point and the boiling point of water at normal atmospheric pressure, but it has different arbitrarily assigned values.

The Celsius scale identifies the freezing point of water as 0° C and the boiling point as 100° C. There are 100 equal intervals, or degrees, between these two reference points, so the Celsius scale is sometimes called the **centigrade** scale.

There is nothing special about either the Celsius scale or the Fahrenheit scale. Both have arbitrarily assigned numbers and one is no more accurate than the other. The Celsius scale is more convenient because it is a decimal scale and because it has a direct relationship with a third scale to be described shortly, the Kelvin scale. Both scales have arbitrarily assigned reference points and an arbitrary number line that indicates *relative* temperature changes. Zero is simply one of the points on each number line and does *not* mean that there is no temperature. Likewise, since the numbers are relative measures of temperature change, 2° is not twice as hot as a temperature of 1° and 10° is not twice as hot as a temperature of 5°. The numbers simply mean some measure of temperature *relative to* the freezing and boiling points of water under normal conditions.

You can convert from one temperature to the other by considering two differences in the scales: (1) the difference in the degree size between the freezing and boiling points on the two scales, and (2) the difference in the values of the lower reference points.

The Fahrenheit scale has 180° between the boiling and freezing points (212° F − 32° F) and the Celsius scale has 100° between the same two points. Therefore each Celsius degree is 180/100 or 9/5 as large as a Fahrenheit degree. Each Fahrenheit degree is 100/180 or 5/9 of a Celsius degree. You know that this is correct since there are more Fahrenheit degrees than Celsius degrees between freezing and boiling. The relationship between the degree sizes is 1° C = 9/5° F and 1° F = 5/9° C. In addition, considering the difference in the values of the lower reference points (0° C and 32° F) gives the equations for temperature conversion.

$$T_F = \frac{9}{5} T_C + 32° \qquad \text{equation 5.1}$$

$$T_C = \frac{5}{9} (T_F - 32°) \qquad \text{equation 5.2}$$

Example 5.1

The average human body temperature is 98.6° F. What is the equivalent temperature on the Celsius scale?

Solution

$$T_C = \frac{5}{9} (T_F - 32°)$$

$$= \frac{5}{9} (98.6° - 32°)$$

$$= \frac{5}{9} (66.6°)$$

$$= \frac{333°}{9}$$

$$= \boxed{37° C}$$

Example 5.2

A bank temperature display indicates 20° C (room temperature). What is the equivalent temperature on the Fahrenheit scale? (Answer: 68° F)

There is a temperature scale that does not have arbitrarily assigned reference points and zero *does* mean nothing. This is not a relative scale but an absolute temperature scale called the **absolute scale,** or **Kelvin scale.** The zero point on the absolute scale is thought to be the lowest limit of temperature. **Absolute zero** is the *lowest temperature attainable,* occurring when all random motion of molecules has ceased. Absolute zero is written as 0 K. A degree symbol is not used and the K stands for the scale unit, Kelvin. The absolute scale uses the same degree size as the Celsius scale and −273° C = 0 K. Note in figure 5.11 that 273 K is the freezing point of water and 373 K is the boiling point. You could think of the absolute scale as a Celsius scale with the zero point shifted by 273°. Thus the relationship between the absolute and Celsius scales is

$$T_K = T_C + 273 \qquad \text{equation 5.3}$$

Example 5.3

A science article refers to a temperature of 300.0 K. (a) What is the equivalent Celsius temperature? (b) the equivalent Fahrenheit temperature?

Solution

(a) The relationship between the absolute scale and Celsius scale is found in equation 5.3, $T_K = T_C + 273$. Solving this equation for Celsius yields $T_C = T_K - 273$.

$$T_C = T_K - 273$$

$$= 300.0 - 273$$

$$= \boxed{27.0° C}$$

(b)

$$T_F = \frac{9}{5} T_C + 32°$$

$$= \frac{9}{5} 27.0° + 32°$$

$$= \frac{243°}{5} + 32°$$

$$= 48.6° + 32°$$

$$= \boxed{80.6° F}$$

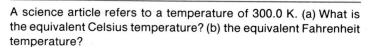

Heat

Temperature is a numerical measure of how hot or how cold an object is, but what is heat? From the time of the early Greek philosophers to the middle of the nineteenth century, heat was assumed to be a substance. The kinetic molecular theory offers

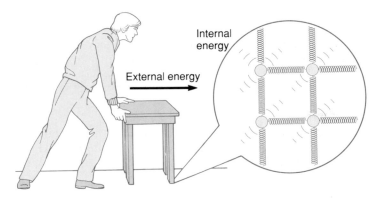

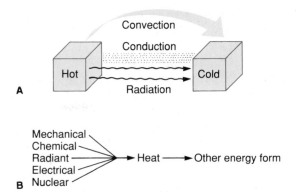

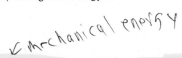

Figure 5.12

The external energy of the table is the kinetic and potential energy that you can see. The internal energy of the table is the kinetic and potential energy of the molecules making up the table. When you push the table across the floor, you work against friction, and the external mechanical energy goes into internal kinetic and potential energy.

Figure 5.13

Heat is the total internal kinetic and potential energy of an object. The object gains heat through a transfer of energy that takes place (*a*) from a temperature difference or (*b*) through an energy form conversion.

another interpretation, that heat is associated with the average kinetic energy *and* the potential energy of the molecules of matter. The findings of Count Rumford and the very careful experiments of Joule showed that the caloric theory could not explain the heat generated by the expenditure of mechanical energy.

When you push a desk, sliding it across a level floor, for example, you do mechanical work on the desk. Yet, the desk apparently has no more kinetic or potential energy after you pushed it than it did before. Where did the mechanical energy go? It went into the system of molecules making up the surface of the floor and the system of molecules making up the bottom surface of the desk legs. Both the surfaces were warmed from the addition of this energy. A similar heating takes place anytime that mechanical energy exerts a force against friction, including fluid friction. The heat resulting from a baseball, bicycle, or automobile moving through the air is widely distributed, and the resulting temperature changes are usually not noticed. On the other hand, a meteor or space vehicle moves at a high velocity as it enters the earth's atmosphere. Such high-velocity objects have more localized fluid friction next to the immediate surface, which produces some very high temperatures. In all cases the expenditure of mechanical energy against friction results in the generation of heat. For common objects moving at ordinary speeds on the earth's surface, this production of heat goes unnoticed.

Internal and External Energy

Joule found in his experiments that the amount of heat generated is *proportional* to the amount of mechanical energy used. The more mechanical kinetic energy you give a desk by pushing on it, for example, the more heating you will notice between the surfaces in contact. The external energy of the moving desk went into the internal vibrational energy of the molecules. The sum of all the energy of all the molecules of an object is called the

internal energy. The *external* energy of an object is its observable kinetic energy and gravitational potential energy. The *internal* energy is the kinetic and potential energy of molecules and is characterized by the properties of temperature, density, heat, volume, pressure of a gas, and so forth. When you push a desk across the floor the observable *external* kinetic energy of the desk was transferred to the *internal* kinetic energy of the molecules (figure 5.12). This explains why the heat generated is proportional to the amount of mechanical energy used. Remember, however, that temperature is a measure of the average kinetic energy of the molecules of an object. Heat refers to the *total* internal energy of the molecules involved in an energy transfer.

Heat as Energy Transfer

Heat is the total internal energy of the molecules, which is increased by two methods: (1) by gaining energy from a temperature difference or (2) by gaining energy from an energy form conversion (figure 5.13). When a temperature difference occurs, energy is transferred from a region of higher temperature to a region of lower temperature. Energy flows from a hot range element, for example, to a pot of cold water on the element. It is a natural process for energy to flow from a region of higher temperature to a region of a lower temperature just as it is natural for a ball to roll downhill. During a form conversion, energy that has been absorbed from some form (mechanical, radiant, electrical, etc.) results in heat and may then come out in some other form. Electrical energy, for example, is dissipated in the element of an electric range. The element becomes hotter as a result and gives off radiant energy (there are other possibilities). Energy transfer took place from electrical energy to heating to radiant energy. On a molecular level, the energy forms are doing work on the molecules that can result in an increase of the average kinetic energy, which is indicated by a temperature increase. Thus heating by energy form conversion is actually a

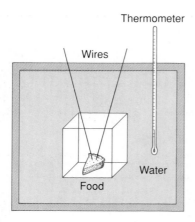

Figure 5.14

The Calorie value of food is determined by measuring the heat released from burning the food. If there is 10.0 kg of water and the temperature increased from 10° to 20° C, the food contained 100 Calories (100,000 calories). The food illustrated here would release much more energy than this.

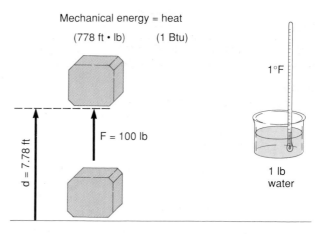

Figure 5.15

Joule worked with the English system of measurement used during his time and found that 778 ft·lb of mechanical energy produced 1 Btu of heat.

transfer of energy by *working*. All energy transfers therefore take place through (1) heating from a temperature difference or (2) working that occurs during an energy form conversion.

These *methods* of energy transfer through heating will be considered in more detail after considering some measures of energy in transit known as heating.

Measures of Heat

Since heating is a method of energy transfer, a quantity of heat can be measured just like any quantity of energy. The metric unit for measuring work, energy, or heat is the **joule.** However, the separate historical development of the concepts of heat (Black, Rumford, Joule) and the concepts of motion (Galileo, Newton) resulted in separate units, some based on temperature differences.

The metric unit of heat is called the **calorie** (cal), a left-over term from the caloric theory of heat. A calorie is defined as the *amount of energy (or heat) needed to increase the temperature of one gram of water one degree Celsius.* A more precise definition specifies the degree interval from 14.5° C to 15.5° C because the energy required varies slightly at different temperatures. This precise definition is not needed for a general discussion. **A kilocalorie** (kcal) is the *amount of energy (or heat) needed to increase the temperature of one kilogram of water one degree Celsius.* The measure of the energy released by the oxidation of food is the kilocalorie, but it is called the Calorie (with a capital C) by nutritionists (figure 5.14). This results in much confusion, which can be avoided by making sure that the scientific calorie is never capitalized (cal) and the dieter's Calorie is always capitalized (Cal). The best solution would be to call the Calorie what it is, a kilocalorie (kcal).

The English system's measure of heating is called the **British thermal unit** (Btu). A Btu is *the amount of energy (or heat) needed to increase the temperature of one pound of water*

one degree Fahrenheit. The Btu is commonly used to measure the heating or cooling rates of furnaces, air conditioners, water heaters, and so forth. The rate is usually expressed or understood to be in Btu per hour. A much larger unit is sometimes mentioned in news reports and articles about the national energy consumption. This unit is the **quad,** which is 1 quadrillion Btu (a million billion or 10^{15} Btu). It is predicted that the United States will be using 91 quads of energy per year by the mid 1990s.

The relationship between the three units of heat is

$$252 \text{ cal} = 1.0 \text{ Btu} = 0.252 \text{ kcal}$$

The amount of heat generated by mechanical work was first investigated by Count Rumford, then quantified by Joule in 1849. Using English units, Joule determined the *mechanical equivalent of heat* (figure 5.15) to be

$$778 \text{ ft} \cdot \text{lb} = 1 \text{ Btu}$$

In metric units the mechanical equivalent of heat is

$$4.184 \text{ J} = 1 \text{ cal}$$

or

$$4,184 \text{ J} = 1 \text{ kcal}$$

The establishment of this precise proportionality means that, fundamentally, mechanical energy and heat are different forms of the same thing.

Specific Heat

There are at least three variables that influence the energy transfer that takes place during heating: (1) the *temperature change,* (2) the *mass* of the substance, and (3) the nature of the *material* being heated. In some normal range of temperature (between 0° C and 100° C) you would expect a temperature change that is proportional to the amount of heating or

Example 5.4

A 1,000.0 kg car is moving at 90.0 km/hr (25.0 m/sec). How many kilocalories are generated when the car brakes to a stop?

Solution

The kinetic energy of the car is

$$KE = \frac{1}{2}mv^2$$

$$= \frac{1}{2}(1,000.0 \text{ kg})(25.0 \text{ m/sec})^2$$

$$= (500.0)(625)\frac{\text{kg} \cdot \text{m}^2}{\text{sec}^2}$$

$$= 312,500 \text{ J} = 313,000 \text{ J}$$

You can convert this to kcal by using the relationship between mechanical energy and heat:

$$(313,000 \text{ J})\left(\frac{1 \text{ kcal}}{4,184 \text{ J}}\right)$$

$$\frac{313,000}{4,184} \frac{\text{J} \cdot \text{kcal}}{\text{J}}$$

$$\boxed{74.8 \text{ kcal}}$$

(Note: The temperature increase from this amount of heating could be calculated from equation 5.4.)

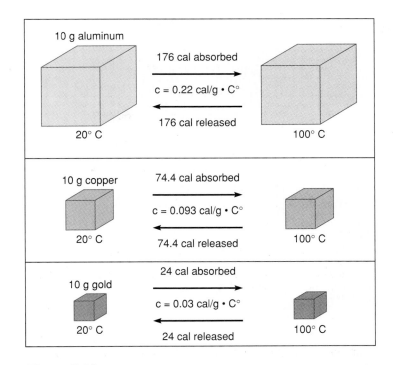

Figure 5.16
Of these three metals, aluminum needs the most heat per gram per degree when warmed and releases the most heat when cooled.

cooling. The quantity of heat (Q) needed to increase the temperature of a pot of water from an initial temperature (T_i) to a final temperature (T_f) is therefore proportional to ($T_f - T_i$), or $Q \propto (T_f - T_i)$. Recalling that the symbol Δ means "a change in," this relationship could also be written as $Q \propto \Delta T$. This simply means that the temperature change of a substance is proportional to the quantity of heat that is added. The quantity ΔT is found from $T_f - T_i$.

The quantity of heat (Q) absorbed or given off during a certain change of temperature is also proportional to the mass (m) of the substance being heated or cooled. A larger mass requires more heat to go through the same temperature than a smaller mass. In symbols, $Q \propto m$.

Different materials require different quantities of heat (Q) to go through the same temperature range when their masses are equal (figure 5.16). If the units for this relationship are the same units used to measure the quantity of heat (Q), then all the relationships can be written in equation form,

$$Q = mc\Delta T \qquad \text{equation 5.4}$$

where c is called the **specific heat**. The specific heat is *the amount of energy (or heat) needed to increase the temperature of one gram of a substance one degree Celsius*. Specific heat is related to the internal structure of a substance, and some of the energy goes into the internal potential energy of the molecules and some goes into the internal kinetic energy of the molecules. The different values for the specific heat of different substances are related to the number of molecules in the one gram sample and to the way they form a molecular structure. If you could isolate a single molecule of solid metal, for example, you would find that it would take the same amount of energy for each molecule to produce the same kinetic energy, that is, the same temperature increase. Table 5.1 lists some specific heats of some common substances.

Note that equation 5.4 can be used for problems of heating *or* cooling. A negative result would mean that the energy is leaving the material, or cooling. When two materials of different temperatures are involved in heat transfer and are perfectly insulated from the surroundings, the heat lost by one will equal the heat gained by the other,

heat lost = heat gained

(by warmer material)(by cooler material)

or

$$Q_{lost} = Q_{gained}$$

or

$$(mc\Delta T)lost = (mc\Delta T)gained$$

Table 5.1

The specific heat of selected substances

Substance	Specific Heat (cal/g·C° or kcal/kg·C°)
Air	0.17
Aluminum	0.22
Concrete	0.16
Copper	0.093
Glass (average)	0.16
Gold	0.03
Ice	0.50
Iron	0.11
Lead	0.03
Mercury	0.033
Seawater	0.93
Silver	0.056
Soil (average)	0.20
Steam	0.48
Water	1.00

Note: To convert to specific heat in J/kg·C°, multiply each value by 4,184.

Example 5.5

How much heat must be supplied to a 500.0 g pan to raise its temperature from 20.0° C to 100.0° C if the pan is made of (a) iron and (b) aluminum?

Solution

The relationship between the heat supplied (Q), the mass (m), and the temperature change (ΔT) is found in equation 5.4. The specific heats (c) of iron and aluminum can be found in table 5.1.

(a)

$m = 500.0$ g

$c = 0.11$ cal/g · °C

$T_f = 100.0°$ C

$T_i = 20.0°$ C

$Q = ?$

$Q = mc\Delta T$

$= (500.0 \text{ g})\left(0.11 \frac{\text{cal}}{\text{g} \cdot °\text{C}}\right)(80.0° \text{ C})$

$= (500.0)(0.11)(80.0) \text{ g} \times \frac{\text{cal}}{\text{g} \cdot °\text{C}} \times °\text{C}$

$= 4,400 \frac{\text{g} \cdot \text{cal} \cdot °\cancel{\text{C}}}{\cancel{\text{g}} \cdot °\cancel{\text{C}}}$

$= 4,400$ cal

$= \boxed{4.4 \text{ kcal}}$

(b)

$m = 500.0$ g

$c = 0.22$ cal/g · °C

$T_f = 100.0°$ C

$T_i = 20.0°$ C

$Q = ?$

$Q = mc\Delta T$

$= (500.0 \text{ g})\left(0.22 \frac{\text{cal}}{\text{g} \cdot °\text{C}}\right)(80.0° \text{ C})$

$= (500.0)(0.22)(80.0) \text{ g} \times \frac{\text{cal}}{\text{g} \cdot °\text{C}} \times °\text{C}$

$= 8,800 \frac{\text{g} \cdot \text{cal} \cdot °\cancel{\text{C}}}{\cancel{\text{g}} \cdot °\cancel{\text{C}}}$

$= 8,800$ cal

$= \boxed{8.8 \text{ kcal}}$

It takes twice as much heat energy to warm the aluminum pan through the same temperature range as an iron pan. Thus, with equal rates of energy input, the iron pan will warm twice as fast as an aluminum pan.

Example 5.6

What is the specific heat of a 2.0 kg metal sample if 1.2 kcal are needed to increase the temperature from 20.0° C to 40.0° C? (Answer: 0.03 kcal/kg·°C)

Heat Flow

In a previous section you learned that heating is a transfer of energy that involves (1) a temperature difference or (2) energy form conversions. Heat transfer that takes place because of a temperature difference takes place three different ways: by conduction, convection, or radiation.

Conduction

Anytime there is a temperature difference there is a natural transfer of heat from the region of higher temperature to the region of lower temperature. In solids, this transfer takes place as heat is *conducted* from a warmer place to a cooler one. Recall that the molecules in a solid vibrate in a fixed equilibrium position and that molecules in a higher temperature region have more kinetic energy, on the average, than those in a lower temperature region. When a solid, such as a metal rod, is held in a flame the molecules in the warmed end vibrate violently. Through molecular interaction, this increased energy of vibration is passed on to the adjacent, slower-moving molecules, which also begin to vibrate more violently. They, in turn, pass on more vibrational energy to the molecules next to them. The increase in activity thus moves from molecule to molecule, causing the region of increased activity to extend along the rod. This is called **conduction,** the transfer of energy from molecule to molecule (figure 5.17).

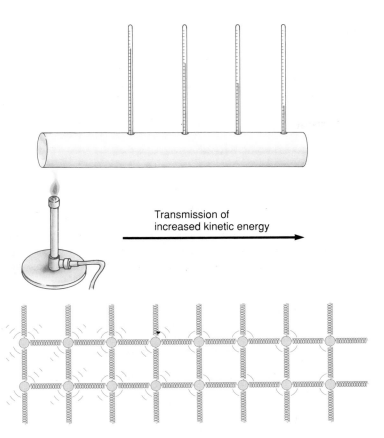

Figure 5.17

Thermometers placed in holes drilled in a metal rod will show that heat is conducted from a region of higher temperature to a region of lower temperature. The increased molecular activity is passed from molecule to molecule in the process of conduction.

The *rate* of conduction depends on the temperature difference between the two regions, the area and the thickness of the substance, and the nature of the material. Table 5.2 shows the conduction of heat through various materials when the temperature difference, area, and thickness are the same for each. The materials with the higher values are good conductors since heat flows through them quickly. The materials with the smallest values are poor conductors and are called heat **insulators** since heat flows through them slowly.

Most insulating materials are good insulators because they contain many small air spaces (figure 5.18). The small air spaces are poor conductors because the molecules of air are far apart, compared to a solid, making it more difficult to pass the increased vibrating motion from molecule to molecule. Styrofoam, glass wool, and wool cloth are good insulators because they have many small air spaces, not because of the material they are made of. The best insulator is a vacuum since there are no molecules to pass on the vibrating motion.

Table 5.2

Rate of conduction of materials*

Silver	0.97
Copper	0.92
Aluminum	0.50
Iron	0.11
Lead	0.08
Concrete	4.0×10^{-3}
Glass	2.5×10^{-3}
Tile	1.6×10^{-3}
Brick	1.5×10^{-3}
Water	1.3×10^{-3}
Wood	3.0×10^{-4}
Cotton	1.8×10^{-4}
Styrofoam	1.0×10^{-4}
Glass wool	9.0×10^{-5}
Air	6.0×10^{-5}
Vacuum	0

*Based on temperature difference of 1° C per cm. Values are cal/sec through a square centimeter of the material.

Wooden and metal parts of your desk have the same temperature but the metal parts will feel cooler if you touch them. Metal is a better conductor of heat than wood and feels cooler because it conducts heat from your finger faster. This is the same reason that a wood or tile floor feels cold to your bare feet. You use an insulating rug to slow the conduction of heat from your feet.

ACTIVITIES

1. Objects that have been in a room with a constant temperature for some time should all have the same temperature. Touch metal, plastic, and wooden parts of a desk or chair to sense their temperature. Explain your findings.
2. Fill one pan with a mixture of ice and water, a second pan with lukewarm water, and a third with the hottest water you can stand. Place one hand in the hot water and the other hand in the cold water for several minutes, then quickly dry your hands and place them both in the lukewarm water. Explain what each hand feels.

Convection

Convection is the transfer of heat by a large-scale displacement of groups of molecules with relatively higher kinetic energy. In conduction, increased kinetic energy is passed from molecule to

Heat and Temperature

Figure 5.18

Fiberglass insulation installed in the walls of a house is rated in terms of R-value, a ratio of the conductivity of the material to its thickness. The fiberglass shown here has a value of R-11.

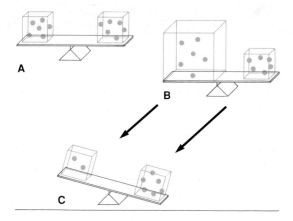

Figure 5.19

(*a*) Two identical volumes of air are balanced since they have the same number of molecules and the same mass. (*b*) Increased temperature causes one volume to expand from the increased kinetic energy of the gas molecules. (*c*) The same volume of the expanded air now contains fewer gas molecules and is less dense, and it is buoyed up by the cooler, more dense air.

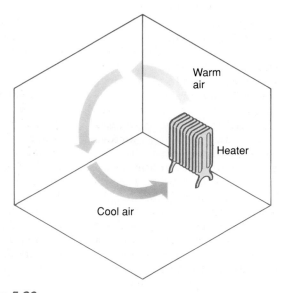

Figure 5.20

Convection currents move warm air throughout a room as the air over the heater becomes warmed, expands, and is moved upward by cooler air.

molecule. In convection, molecules with higher kinetic energy are moved from one place to another place. Conduction happens primarily in solids, but convection only happens in liquids and gases, where fluid motion can carry molecules with higher kinetic energy over a distance. When molecules gain energy, they move more rapidly and push more vigorously against their surroundings. The result is an expansion as the region of heated molecules pushes outward and increases the volume. Since the same amount of matter now occupies a larger volume, the overall density has been decreased (figure 5.19).

In fluids, expansion sets the stage for convection. Warm, less dense fluid is pushed upward by the cooler, more dense fluid around it. In general, cooler air is more dense and sinks and flows downhill. Cold air, being more dense, flows out near the bottom of an open refrigerator. You can feel the cold, dense air pouring from the bottom of a refrigerator to your toes on the floor. On the other hand, you hold your hands *over* a heater because the warm, less dense air is pushed upward. In a room, warm air is pushed upward from a heater. The warm air spreads outward along the ceiling and is slowly displaced as newly warmed air is pushed upward to the ceiling. As the air cools it sinks over another part of the room, setting up a circulation pattern known as a *convection current* (figure 5.20). Convection currents can also be observed in a large pot of liquid that is heating on a range. You can see the warmer liquid being forced upward over the warmer parts of the range element, then sink over the cooler parts. Overall, convection currents give the pot of liquid an appearance of turning over as it warms.

Passive Solar Design

Passive solar application is an economically justifiable use of solar energy today. Passive solar design uses a structure's construction to heat a living space with solar energy. There are few electric fans, motors, or other energy sources used. The passive solar design takes advantage of free solar energy; it stores and then distributes this energy through natural conduction, convection, and radiation.

Sunlight that reaches the earth's surface is mostly absorbed. Buildings, the ground, and objects become warmer as the radiant energy is absorbed. Nearly all materials, however, reradiate the absorbed energy at longer wavelengths, wavelengths too long to be visible to the human eye. The short wavelengths of sunlight pass readily through ordinary window glass, but the longer, reemitted wavelengths cannot. Therefore, sunlight passes through a window and warms objects inside a house. The reradiated longer wavelengths cannot pass readily back through the glass but are absorbed by certain molecules in the air. The temperature of the air is thus increased. This is called the "greenhouse effect." Perhaps you have experienced the effect when you left your car windows closed on a sunny, summer day.

In general, a passive solar home makes use of the materials from which it is constructed to capture, store, and distribute solar energy to its occupants. Sunlight enters the house through large windows facing south and warms a thick layer of concrete, brick, or stone. This energy "storage mass" then releases energy during the day, and, more important, during the night. This release of energy can be by direct radiation to occupants, by conduction to adjacent air, or by convection of air across the surface of the storage mass. The living space is thus heated without special plumbing or forced air circulation. As you can imagine, the key to a successful passive solar home requires considering every detail of natural energy flow, including the materials of which floors and walls are constructed, convective air circulation patterns, and the size and placement of windows. In addition, a passive solar home requires a different life-style and living patterns. Carpets, for example, would defeat the purpose of a storage mass floor since carpeting would insulate the storage mass from sunlight. Glass is not a good insulator so windows must have curtains or movable insulation panels to slow energy loss at night. This requires the daily activity of closing curtains

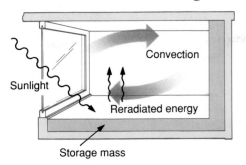

Box Figure 5.1
The direct solar gain design collects and stores solar energy in the living space.

or moving insulation panels at night and then opening curtains and moving panels in the morning. Passive solar homes therefore require a high level of personal involvement by the occupants.

There are three basic categories of passive solar design: (1) direct solar gain, (2) indirect solar gain, and (3) isolated solar gain.

A *direct solar gain* home is one in which solar energy is collected in the actual living space of the home (box figure 5.1). The advantage of this design is the large, open window space with a calculated overhang, which admits maximum solar energy in the winter but prevents solar gain in the summer. The disadvantage is that the occupants are living in the collection and storage components of the design and can place nothing (such as carpets and furniture) that would interfere with warming the storage mass in the floors and walls.

An *indirect solar gain* home uses a massive wall inside a window that serves as a storage mass. This wall, called a *Trombe wall* is shown in box figure 5.2. The Trombe wall collects and stores solar energy, then warms the living space with radiant energy and convection currents. The disadvantage to the indirect solar gain design is that large windows are blocked by the Trombe wall. The advantage is that the occupants are not in direct contact with the solar collection and storage area so they can place carpets and furniture as they wish. Controls to prevent energy loss at night are still necessary with this design.

An *isolated solar gain* home uses a structure that is separated from the living space to collect and store solar energy. Examples of an isolated gain design are an attached greenhouse or sun porch (box figure 5.3). Energy flow between the attached structure and the living space can be by con-

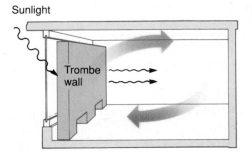

Box Figure 5.2
The indirect solar gain design uses a Trombe wall to collect, store, and distribute solar energy.

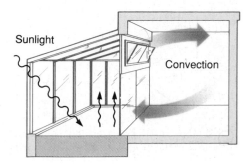

Box Figure 5.3
The isolated solar gain design uses a separate structure to collect and store solar energy.

duction, convection, and radiation, which can be controlled by opening or closing off the attached structure. This design provides the best controls since it can be completely isolated, opened to the living space as needed, or directly used as living space when the conditions are right. Additional insulation is needed for the glass at night, however, and for sunless winter days.

It has been estimated that building a passive solar home would cost about 10 percent more than building a traditional home of the same size. Considering the possible energy savings, you might believe that most homes would now have a passive solar design. They do not, however, as most new buildings require technology and large amounts of energy to maximize comfort. Yet, it would not require too much effort to consider where to place windows in relation to the directional and seasonal intensity of the sun and where to plant trees. Perhaps in the future you will have an opportunity to consider using the environment to your benefit through the natural processes of conduction, convection, and radiation.

Convection currents will ventilate a room if the top *and* bottom parts of a window are opened. In the winter, cool and denser air will pour in the bottom opening, forcing the warmer, stale air out the top opening. An open fireplace functions because of convection currents. Smoke (and much of the heat) goes up the chimney as fresh air from the room furnishes oxygen for the fire. One way to start smoldering wood burning in a fireplace is to open a window to let cool, dense air into the room. The resulting convection has the same effect as fanning the smoldering wood and it will burst into flames.

Radiation

The third way that heat transfer takes place because of a temperature difference is called **radiation.** Radiation involves the form of energy called radiant energy, energy that moves through space. As you will learn in chapter 9, radiant energy includes visible light and many other forms as well. All objects with a temperature above absolute zero give off radiant energy. The absolute temperature of the object determines the rate, intensity, and kinds of radiant energy emitted. You know that visible light is emitted if an object is heated to a certain temperature. A heating element on an electric range, for example, will glow with a reddish-orange light when at the highest setting, but it produces no visible light at lower temperatures although you feel warmth in your hand when you hold it near the element. Your hand absorbs the nonvisible radiant energy being emitted from the element. The radiant energy does work on the molecules of your hand, giving them more kinetic energy. You sense this as an increase in temperature, that is, warmth.

All objects above absolute zero emit radiant energy, but all objects also absorb radiant energy. A hot object, however, emits more radiant energy than a cold object. The hot object will emit more energy than it absorbs from the cold object, and the cold object will absorb more energy from the hot object than it emits. There is therefore a net energy transfer that will take place by radiation as long as there is a temperature difference between the two objects.

Summary

Heat was considered to be a *substance* from at least the time of the ancient Greek civilization through the 1800s. During the late 1700s this concept led to the formulation of the *caloric theory of heat,* which considered heat to be an invisible, weightless and self-repulsive fluid. The caloric theory was widely accepted until the work of Rumford showed that heat was produced by mechanical motion. Joule measured and quantified the heat produced from different energy forms and established a mechanical equivalent of heat. The *kinetic molecular theory of matter* was developed by 1860 from this pioneering work.

The kinetic theory of matter assumes that all matter is made up of tiny, ultimate particles of matter called *molecules*. A molecule is defined as the smallest particle of a compound, or a gaseous element, that can exist and still retain the characteristic properties of that substance. Molecules interact, attracting each other through a force of *cohesion*. Liquids, solids, and gases are the states of matter that are explained by the molecular arrangements and forces of attraction between their molecules. A *solid* has a definite shape and volume because it has molecules that vibrate in a fixed equilibrium position with strong cohesive forces. A *liquid* has molecules that have cohesive forces strong enough to give it a definite volume but not strong enough to give it a definite shape. The molecules of a liquid can flow, rolling over each other. A *gas* is composed of molecules that are far apart with weak cohesive forces. Gas molecules move freely in a constant, random motion.

Molecules can have *vibrational, rotational,* or *translational* kinetic energy. The *temperature* of an object is related to the *average kinetic energy* of the molecules making up the object. A measure of temperature tells how hot or cold an object is on two arbitrary scales, the *Fahrenheit scale* and the *Celsius scale*. The *absolute scale,* or *Kelvin scale,* has the coldest temperature possible ($-273°$ C) as zero (0 K).

The observable potential and kinetic energy of an object is the *external energy* of that object, while the potential and kinetic energy of the molecules making up the object is the *internal energy* of the object. Heat refers to the total internal energy and is a transfer of energy that takes place because of (1) a *temperature difference* between two objects or (2) an *energy form conversion*. An energy form conversion is actually an energy conversion involving work at the molecular level, so all energy transfers involve *heating* and *working*.

A quantity of heat can be measured in *joules* (a unit of work or energy) or *calories* (a unit of heat). A *kilocalorie* is 1,000 calories, another unit of heat. A *Btu,* or *British thermal unit,* is the English system unit of heat. The *mechanical equivalent of heat* is 4,184 J = 1 kcal.

The *specific heat* of a substance is the amount of energy (or heat) needed to increase the temperature of one gram of a substance one degree Celsius. The specific heats of various substances are not the same because the molecular structure of each substance is different.

Energy transfer that takes place because of a temperature difference does so through conduction, convection, or radiation. *Conduction* is the transfer of increased kinetic energy from molecule to molecule. Substances vary in their abilities to conduct heat, and those that are poor conductors are called *insulators*. Gases, such as air, are good insulators. The best insulator is a vacuum. *Convection* is the transfer of heat by the displacement of large groups of molecules with higher kinetic energy. Convection takes place in fluids, and the fluid movement that takes place because of density differences is called a *convection current*. *Radiation* is radiant energy that moves through space. All objects with an absolute temperature above zero give off radiant energy, but all objects absorb it as well. Energy is transferred from a hot object to a cold one through radiation.

Summary of Equations

5.1

$$T_F = \frac{9}{5} T_C + 32°$$

5.2

$$T_C = \frac{5}{9} (T_F - 32°)$$

5.3

$$T_K = T_C + 273$$

5.4

quantity of heat = (mass)(specific heat)(temperature change)

$$Q = mc\Delta T$$

Key Terms

absolute scale (p. 88)	joule (p. 90)
absolute zero (p. 88)	Kelvin scale (p. 88)
British thermal unit (p. 90)	kilocalorie (p. 90)
caloric theory of heat (p. 81)	kinetic molecular theory (p. 83)
calorie (p. 90)	liquids (p. 84)
Celsius scale (p. 87)	molecule (p. 83)
centigrade (p. 88)	plasma (p. 85)
conduction (p. 92)	quad (p. 90)
convection (p. 93)	radiation (p. 96)
Fahrenheit scale (p. 87)	solids (p. 84)
fluids (p. 85)	specific heat (p. 91)
gases (p. 85)	temperature (p. 85)
heat (p. 89)	thermometer (p. 87)
insulators (p. 93)	vapor (p. 85)
internal energy (p. 89)	

Applying the Concepts

1. Joule founded the beginnings of a new theory of heat when he noted that
   a. caloric was absorbed in a beer-producing process.
   b. a measured amount of energy always produced a predictable amount of heat.
   c. there was no relationship between mechanical work and heat.
   d. mechanical work could be converted to heat, but heat could not be converted to mechanical work.

2. The Fahrenheit temperature scale
   a. was originally a centigrade scale.
   b. has 180 divisions between two reference points.
   c. has an arbitrarily assigned number line.
   d. all of the above.

3. A two degree interval on the Celsius temperature scale is
   a. equal to 3.6 Fahrenheit degrees.
   b. equal to 35.6 Fahrenheit degrees.
   c. twice as hot as 1 Celsius degree.
   d. all of the above.

4. A temperature reading of 2° C is
   a. equivalent to 3.6° F.
   b. equivalent to 35.6° F.
   c. twice as hot as 1° C.
   d. all of the above.

5. The metric unit of heat called a calorie is
   a. the specific heat of water.
   b. the energy needed to increase the temperature of one gram of water one degree Celsius.
   c. equivalent to a little over 4 joules of mechanical work.
   d. all of the above.

6. Which of the following is a shorthand way of stating that "the temperature change of a substance is directly proportional to the quantity of heat that is added"?
   a. $Q \propto m$
   b. $m \propto T_f - T_i$
   c. $Q \propto \Delta T$
   d. $Q = T_f - T_i$

7. Table 5.1 lists the specific heat of soil as 0.20 kcal/kg·C° and the specific heat of water as 1.00 kcal/kg·C°. This means that if equal masses of soil and water receive the same 1 kcal of energy,
   a. the water will be warmer than the soil by 0.8° C.
   b. the soil will be 5° C warmer than the water.
   c. the water will be 5° C warmer than the soil.
   d. the water will warm by 1° C and the soil will warm by 0.2° C.

8. The energy transfer that takes place by moving directly from molecule to molecule is called
   a. conduction.
   b. convection.
   c. radiation.

9. The energy transfer that does not require the presence of matter to take place is called
   a. conduction.
   b. convection.
   c. radiation.
   d. impossible, for matter is always required.

10. The transfer of heat that takes place because of density difference in fluids is
    a. conduction.
    b. convection.
    c. radiation.

Answers

1. b   2. d   3. a   4. b   5. d   6. c   7. b   8. a   9. c
10. b

Questions for Thought

1. Can you accurately sense the temperature of an object? Explain.

2. What is a molecule?

3. Describe the properties of a solid, a liquid, and a gas according to the kinetic molecular theory.

4. What is temperature? What is heat?

5. Describe the main understandings and features of the caloric theory of heat. What observation or observations eventually led to the rejection of this theory?

6. Does heat flow out of a hot object or does cold flow in? Explain.

7. What is the basic difference in the Fahrenheit and the Celsius temperature scales? What is the meaning of zero (0°) on these two temperature scales?

8. If the average kinetic energy of the molecules of a gas were to double, the temperature would double on one temperature scale but not the other two scales. Identify these scales and explain.

9. Why is radiation considered to be the transmission of radiant energy rather than the transmission of heat?

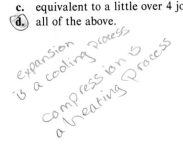

expansion is a cooling process

compression is a heating process

evaporation is a cooling process

Heat and Temperature

10. Insulating windows have two glass panes separated by a small air space. Why is this better than a single glass pane? How could you improve the insulating ability of the double-paned window?

11. Conduction and convection require matter to take place but radiation does not. Explain.

12. Can convection occur in a solid? Explain.

13. If a container of hydrogen gas has the same absolute temperature as a container of oxygen gas, would the average speed of the molecules in the two containers be the same? Explain. (Hint: Consider the mass of an oxygen molecule as compared to the mass of a hydrogen molecule.)

14. Explain why most materials become less dense as their temperature is increased.

15. Explain why different substances have different specific heats.

16. Explain how you know that solar energy travels from the sun to the earth by radiation and not by conduction or convection.

17. Would the tight packing of more insulation, such as glass wool, in an enclosed space increase or decrease the insulation value? Explain.

18. If the specific heat of soil is about 0.2 cal/g° C, explain why bodies of water heat and cool more slowly than nearby land.

19. Give an example of converting each of the five energy forms into heat. What does this mean about the nature of heat?

20. What does a thermometer actually measure? Would this depend on the mass of an object being measured? Explain.

21. Describe how you could find the specific heat of an unidentified substance.

22. A true vacuum bottle has a double-walled, silvered bottle with the air removed from the space between the walls. Describe how this design keeps food hot or cold by dealing with conduction, convection, and radiation.

23. Why is cooler air found in low valleys on calm nights?

24. Why is air a good insulator?

25. Explain the meaning of the mechanical equivalent of heat.

26. What do people really mean when they say that a certain food "has a lot of Calories"?

27. A piece of metal feels cooler than a piece of wood at the same temperature. Explain why.

28. Using the kinetic molecular theory, explain why gases and liquids flow but solids do not.

Exercises

Group A—Solutions Provided in Appendix D

1. What is the Celsius equivalent of the following temperatures?
   (a) 20.0° F
   (b) 68.0° F
   (c) 98.6° F
   (d) 118.0° F

   $T_C = \frac{5}{9} - 32°$

2. What is the Fahrenheit equivalent of the following temperatures?
   (a) 20.0° C
   (b) 100.0° C
   (c) 50.0° C
   (d) 0.0° C

   $T_F = \frac{9}{5}(T_C + 32°)$

3. What is the value of absolute zero on the (a) Celsius scale? (b) Fahrenheit scale?

4. What is the Fahrenheit equivalent of the following temperatures?
   (a) 4.00 K
   (b) 100.0 K
   (c) 300.0 K
   (d) 6,000.0 K

5. An electric current heats a 200.0 g copper wire from 20.0° C to 40.0° C. How much heat was generated by the current? ($c_{copper} = 0.093$ kcal/kg° C)

6. A bicycle and rider have a combined mass of 100.0 kg. How many calories of heat are generated in the brakes when the bicycle comes to a stop from a speed of 36.00 km/hr?

Group B—Solutions Not Given

1. What is the Celsius equivalent of the following temperatures?
   (a) 50.0° F
   (b) 80.0° F
   (c) −40.0° F
   (d) 90.0° F

2. What is the Fahrenheit equivalent of the following temperatures?
   (a) 37.0° C
   (b) 10.0° C
   (c) −40.0° C
   (d) 40.0° C

3. The freezing point of oxygen is about −362° F. (a) What is the Celsius equivalent of this temperature? (b) What is the Kelvin equivalent of this temperature?

4. What is the Fahrenheit equivalent of the following temperatures?
   (a) 3 K
   (b) 400 K
   (c) 10,000 K
   (d) 100,000 K

5. A 0.25 kg length of aluminum wire is warmed 10.0° C by an electric current. How much heat was generated by the current? ($c_{aluminum} = 0.22$ kcal/kg° C)

6. A 1,000.0 kg car with a speed of 90.0 km/hr brakes to a stop. How many calories of heat are generated by the brakes as a result?

7. A 10.0 kg loose bag of soil falls 5.00 m at a construction site. If all the energy is retained by the bag of soil, how much will its temperature increase? (c_{soil} = 0.250 kcal/kg° C)

8. A 75.0 kg person consumes a small order of french fries (250.0 Cal) and wishes to "work off" the energy by climbing a 10.0 m stairway. How many vertical climbs are needed to use all the energy?

9. A 0.5 kg glass bowl (c_{glass} = 0.2 kcal/kg° C) and a 0.5 kg iron pan (c_{iron} = 0.11 kcal/kg° C) have a temperature of 68° F when placed in a freezer. How much heat will the freezer have to remove from each to cool them to 32° F?

10. A sample of silver at 20.0° C is warmed to 100.0° C when 896 cal is added. What is the mass of the silver? (c_{silver} = 0.056 kcal/kg° C)

11. A 300.0 w immersion heater is used to heat 250.0 g of water from 10.0° C to 70.0° C. About how many minutes did this take?

12. A 100.0 g sample of metal is warmed 20.0° C when 60 cal is added. What is the specific heat of this metal?

7. A 1.0 kg metal head of a geology hammer strikes a solid rock with a velocity of 5.0 m/sec. Assuming all the energy is retained by the hammerhead, how much will its temperature increase? (c_{head} = 0.11 kcal/kg° C)

8. In general, how many times will a 60.0 kg person have to climb a 10.0 m stairway to "work off" each excess Calorie (kcal) consumed?

9. A 50.0 g silver spoon at 20.0° C is placed in a cup of coffee at 90.0° C. How much heat does the spoon absorb from the coffee to reach a temperature of 89.0° C?

10. If the silver spoon placed in the coffee in problem 9 causes it to cool 0.75° C, what is the mass of the coffee? (assume c_{coffee} = 1.0 cal/g° C)

11. How many minutes would be required for a 300.0 w immersion heater to heat 250.0 g of water from 20.0° C to 100.0° C?

12. A 200.0 g china serving bowl is warmed 65.0° C when it absorbs 2.6 kcal of heat from a serving of hot food. What is the specific heat of the china dish?

1. a $\boxed{-6.67°C}$ $\boxed{b. \ 20°C}$ c. $\boxed{37.0°C}$ D. $\boxed{47.78°C}$

2. a. $\boxed{68.0°F}$ $\boxed{b. \ 212°F}$ C. $\boxed{122°F}$ D. $\boxed{32°F}$

3. $\boxed{-273°C}$ B. $\boxed{-459°F}$

5. $Q = mc\Delta T \ (200.g)(.093 kcal/kg°c)(4c-20)$

$\boxed{372 \ cal}$

10. $Q = mc\Delta T \ \therefore \ m = \dfrac{Q}{c\Delta T} \ \dfrac{896}{(.056 \ kcal/kg°c)(80c)}$

$200 g = \boxed{.200 kg}$

Heat and Technology

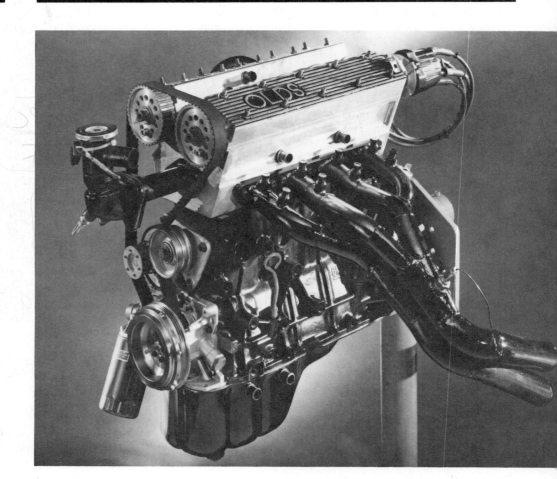

Figure 6.1

This is the Quad 4, a technologically advanced 150 horsepower, dual-overhead cam engine with 4 valves per cylinder. The General Motors engine was designed by computer for low friction, increased power, and increased fuel economy.

THINK about how you can control your environment through modern technology. You can live comfortably during a winter of cold northern climates and during a summer of hot southern climates. You can make it day or night as you change the temperature with a thermostat setting controlling the environment of a building. Furthermore, you can rapidly travel from place to place, again in comfort, using modern technology. Technology is, in fact, defined as the *controlling and shaping of the environment for the support and comfort of people.*

The first central heating system and the steam-heated radiator were invented in the late 1790s by a man you will recognize from his association with early studies of heat, Count Rumford. In addition, Rumford was one of the first to describe convection currents and to explain correctly how clothing insulates the human body. Other heat-related technological inventions of Count Rumford include pots and pans of modern design, an efficient cooking stove, the drip coffee pot, and the vacuum thermos bottle.

Transportation, in the meantime, was being transformed by the invention of the *heat engine,* a device that produces mechanical energy from heat. The first practical heat engine was the steam engine. The steam engine was improved dramatically through the work of James Watt during the years of the American Revolution, and by 1790 steam-driven boats were moving up and down rivers along the Atlantic Coast. Experimental steam-driven carriages were moving on land about this time. It was not until about one hundred years later that the first patent was granted for an internal combustion engine that burned petroleum. Soon after, the Ford Motor Company started mass producing the automobile and people roared into the twentieth century on four wheels propelled by heat engines (figure 6.1).

The first heat engines were invented before people understood energy, heat, or the kinetic molecular theory. This was a case of technology developing before the understanding that comes from science. This chapter is concerned with these understandings and the relationships between energy, heat, and mechanical energy. The first section considers a mental model to gain some insight about energy flow and the states of matter. The second section uses this insight as human comfort and technological devices used to support this comfort are explored. Then the process of converting energy to mechanical work is considered as you learn about heat engines. Finally, the laws of thermodynamics are discussed, along with an interesting implication for the ultimate fate of the universe.

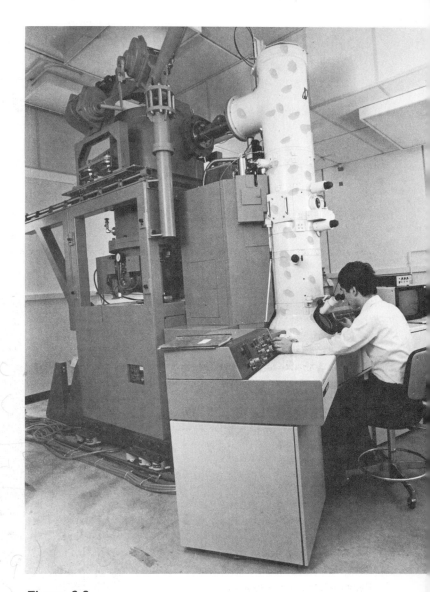

Figure 6.2

This is a 400,000 volt transmission electron microscope. It is so powerful it can magnify as much as 1,500,000 times, revealing patterns of molecules and atoms.

Energy, Heat, and Molecular Theory

The kinetic molecular theory of matter is based on evidence from different fields of physical science, not just one subject area. Chemists and physicists developed some convincing conclusions about the structure of matter over the past 150 years, using carefully designed experiments and mathematical calculations that explained observable facts about matter. Step by step the detailed structure of this submicroscopic, invisible world of particles became firmly established. Today, an understanding of this particle structure is basic to physics, chemistry, biology, geology, and practically every other science subject. This understanding has also resulted in present-day technology.

The molecular structure of matter is invisible, beyond what you can see or touch (figure 6.2). This means that you will need to use your imagination and develop a mental model in order to understand this invisible structure. The following section is about some observable properties of matter that will help you build such a model of the relationships between molecules, heat, and energy. First, recall the structure of matter according to the kinetic molecular theory discussed in chapter 5. This theory pictured a solid object as composed of molecules that are each

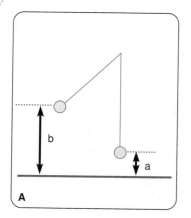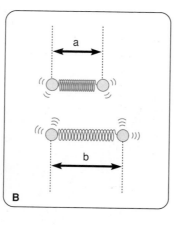

Figure 6.3

(a) As a pendulum swings back and forth, it trades kinetic energy for potential energy and potential energy for kinetic energy. (b) As a molecule vibrates back and forth, it trades molecular kinetic energy for molecular potential energy and molecular potential energy for molecular kinetic energy.

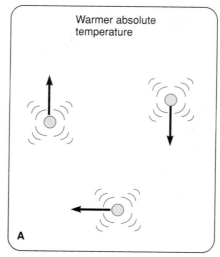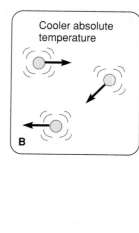

Figure 6.4

(a) The random, chaotic motion of gas molecules at a higher absolute temperature is greater than (b) the random, chaotic motion of gas molecules at some cooler absolute temperature above absolute zero. The arrows represent velocities, not forces.

locked in an equilibrium position by intermolecular forces. Each molecule vibrates back and forth and in all directions in its equilibrium position, with some value of kinetic energy. However, the molecules are coupled by intermolecular forces so they influence one another. In addition, as the molecules vibrate they move a little closer and a little further apart, trading kinetic energy for potential energy and vice versa, much as a moving pendulum trades potential energy for kinetic energy as it swings back and forth (figure 6.3). Thus the average values of molecular kinetic energy and potential energy are nearly equal. The total internal energy of a solid is therefore energy that is about half potential energy and half kinetic energy.

There are three related parts to your model so far: (1) internal energy is the *total* energy of the molecules of an object, (2) molecules have both kinetic *and* potential energy, and (3) the internal energy is made up of both molecular potential energy and molecular kinetic energy.

Another important part of the model is understanding the effect of the addition of energy, or heat, to the internal energy. Adding heat to a solid causes the molecules to vibrate faster and move increasingly further apart, causing the solid to expand. If this vibratory motion becomes too great, the equilibrium position of the molecules becomes disorganized. Molecules begin to wander about, but they are still held together by intermolecular forces. Thus, a solid is warmed to the melting point and is now in a liquid state. Further heating gives the molecules in the liquid state sufficient energy to escape, becoming a gas.

The molecules of solids, liquids, and gases are in a state of motion as long as the temperature is above absolute zero. This motion increases in activity with increases in temperature,

and the energy entering the material is called *heat* (Q), a quantity of energy from the absorption of one of the five forms of energy or from heat flow caused by a temperature difference (figure 6.4).

The production of heat from some energy form is satisfactorily explained by the kinetic molecular theory. Consider, for example, the conversion of mechanical energy into heat. Heat is produced by friction when one surface is rubbed across another, such as the surface of the floor and the bottom of table legs as you slide a table across the floor. If you think of the solid surface as vibrating molecules held together by elastic forces, you can imagine that one surface moving over another might pull molecules aside, then release them into a state of increased vibration, much like a plucked guitar string. The result of this increased motion is an increase in surface temperature. The production of heat from the other four forms of energy can likewise be explained by the kinetic interpretation of heat. Each will be explained as electricity, radiant energy, chemical energy, and nuclear energy are discussed in future chapters.

Change of State

Solids, liquids, and gases are the three common states of matter, and each state is characterized by different molecular arrangements. The motion of the molecules in any of the three states can be increased by (1) adding heat through a temperature difference or (2) the absorption of one of the five forms of energy, which results in heating. In either case the temperature of the solid, liquid, or gas increases according to the specific heat of the substance, and generally more heating means higher temperatures.

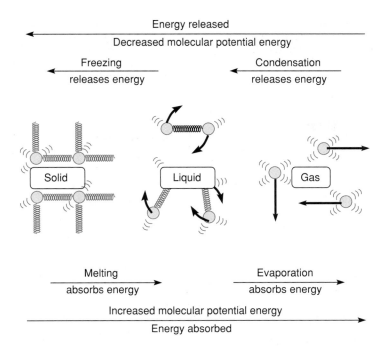

Figure 6.5

Each phase change absorbs or releases a quantity of latent heat, which goes into or is released from molecular potential energy.

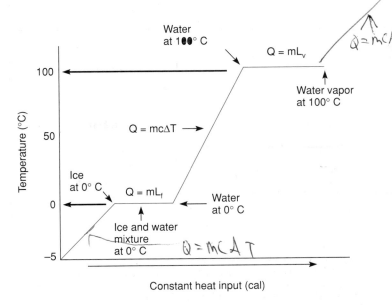

Figure 6.6

This graph shows two warming sequences and two phase changes with a constant input of heat. The ice warms to the melting point, then absorbs heat during the phase change as the temperature remains constant. When all the ice has melted, the now-liquid water warms to the boiling point, where the temperature again remains constant as heat is absorbed during this second phase change.

More heating, however, does not always result in increased temperatures. When a solid, liquid, or gas changes from one state to another, the transition is called a **phase change.** A phase change always absorbs or releases energy, *a quantity of heat that is not associated with a temperature change.* Since the quantity of heat associated with a phase change is not associated with a temperature change, it is called **latent heat.** Latent heat is a term borrowed from the old caloric theory of heat and means "hidden heat." Today, latent heat refers to the "hidden" energy of phase changes, which is energy (heat) that goes into or comes out of *internal potential energy* (figure 6.5).

There are three kinds of phase changes that can occur: (1) *solid-liquid,* (2) *liquid-gas,* and (3) *solid-gas.* In each case the phase change can go in either direction. For example, the solid-liquid phase change occurs when a solid melts to a liquid or when a liquid freezes to a solid. Ice melting to water and water freezing to ice is a common example of this phase change and its two directions. Both occur at a temperature called the **freezing point** or the **melting point** depending on which direction the phase change is occurring. In either case, however, the freezing and melting points are the same temperature.

The liquid-gas phase change also occurs in two different directions. The temperature at which a liquid boils and changes to a gas (or vapor) is called the **boiling point.** The temperature at which a gas or vapor changes back to a liquid is called the **condensation point.** The boiling and condensation points are the same temperature. There are conditions other than boiling under which liquids may undergo liquid-gas phase changes, and these conditions are discussed in the next section.

You probably are not as familiar with solid-gas phase changes, but they are common. A phase change that takes a solid directly to a gas or vapor is called **sublimation.** Moth balls and dry ice (solid CO_2) are common examples of materials that undergo sublimation, but frozen water, meaning common ice, also sublimates under certain conditions. Perhaps you have noticed ice cubes in a freezer become smaller with time as a result of sublimation. The frost that forms in a freezer, on the other hand, is an example of a solid-gas phase change that takes place in the other direction. In this case water vapor forms the frost without going through the liquid state, a solid-gas phase change that takes place in an opposite direction to sublimation.

For a specific example of a phase change and latent heat, consider the temperature changes that occur when ice is subjected to a constant source of heat. The temperature changes that occur with a constant input of heat are shown in figure 6.6. Starting at the left side of the graph, you can see that the temperature of the ice increases from the constant input of heat. The ice warms according to $Q = mc\Delta T$, where c is the specific heat of ice. When the temperature reaches the melting point (0° C), it stops increasing as the ice begins to melt. More and more liquid water appears as the ice melts, but the temperature *remains* at 0° C even though heat is still being added at a constant rate. Finally, when all the ice is completely melted, the

Heat and Technology

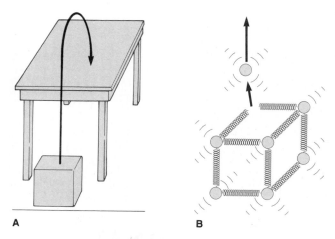

Figure 6.7

(a) Work is done against gravity to lift an object, giving the object gravitational potential energy. (b) Work is done against intermolecular forces in separating a molecule from a solid, giving the molecule potential energy.

temperature again increases at a constant rate. The quantity of heat involved in this phase change was used in doing the work of breaking the molecule-to-molecule bonds in the solid, making a liquid with molecules that are now free to move about and roll over one another. Since the quantity of heat (Q) was absorbed without a temperature change, it is called the **latent heat of fusion** (L_f). The latent heat of fusion is *the heat involved in a solid-liquid phase change in melting or freezing.* You learned in a previous chapter that when you do work on something you give it energy. In this case, the work done in breaking the molecular bonds in the solid gave the molecules more *potential* energy (figure 6.7). This energy is "hidden" since heat was absorbed but a temperature increase did not take place. This same potential energy is given up when the molecules of the liquid return to the solid state. A melting solid absorbs energy and a freezing liquid releases this *same amount* of energy, warming the surroundings. Thus you put ice in a cooler because the melting ice absorbs the latent heat of fusion from the beverage cans, cooling them. Citrus orchards are flooded with water when freezing temperatures are expected because freezing water releases the latent heat of fusion, which warms the air around the trees. For water, the latent heat of fusion is 80.0 cal/g (144.0 Btu/lb). This means that every gram of ice that melts in your cooler *absorbs* 80.0 cal of heat. Every gram of water that freezes *releases* 80.0 cal. The total heat involved in a solid-liquid phase change depends on the mass of the substance involved, so

$$Q = mL_f \qquad \text{equation 6.1}$$

where L_f is the latent heat of fusion for the substance involved.

Refer again to figure 6.6. After the solid-liquid phase change is complete, the constant supply of heat increases the temperature of the water according to $Q = mc\Delta T$, where c is

now the specific heat of liquid water. When the water reaches the boiling point the temperature again remains constant even though heat is still being supplied at a constant rate. The quantity of heat involved in the liquid-gas phase change again goes into doing the work of overcoming the attractive molecular forces. This time the molecules escape from the liquid state to become single, independent molecules of gas. The quantity of heat (Q) absorbed or released during this phase change is called the **latent heat of vaporization** (L_v). The latent heat of vaporization is *the heat involved in a liquid-gas phase change where there is evaporation or condensation.* The latent heat of vaporization is the energy gained by the gas molecules as work is done in overcoming molecular forces. Thus the leaving molecules absorb energy from the surroundings and a condensing gas (or vapor) releases this *exact same amount of energy.* For water, the latent heat of vaporization is 540.0 cal/g (970.0 Btu/lb). This means that every gram of water vapor that condenses on your bathroom mirror releases 540.0 cal, which warms the bathroom. The total heating depends on how much water vapor condensed, so

$$Q = mL_v \qquad \text{equation 6.2}$$

where L_v is the latent heat of vaporization for the substance involved.

Example 6.1

How much energy does a refrigerator remove from 100.0 g of water at 20° C to make ice at −10° C?

Solution

This type of problem is best solved by subdividing it into smaller steps that consider (1) the heat added or removed and the resulting temperature changes *for each state* of the substance and (2) the heat flow resulting from any *phase change* that occurs within the ranges of changes as identified by the problem. The heat involved in each phase change and the heat involved in the heating or cooling of each state are identified as Q_1, Q_2, and so forth. Temperature readings are calculated with *absolute values,* so you ignore any positive or negative signs.

1. Water in the liquid state cools from 20° C to 0° C (the freezing point) according to the relationship $Q = mc\Delta T$, where c is the specific heat of water, and

$$Q_1 = mc\Delta T$$

$$= (100.0 \text{ g})\left(1.0 \frac{\text{cal}}{\text{g} \cdot °\text{C}}\right)(0° - 20° \text{ C})$$

$$= (100.0)(1.0)(20) \frac{\text{g} \cdot \text{cal} \cdot °\cancel{\text{C}}}{\cancel{\text{g}} \cdot °\cancel{\text{C}}}$$

$$= 2,000 \text{ cal}$$

$$Q_1 = \boxed{2 \text{ kcal}}$$

2. The latent heat of fusion must now be removed as water at 0° C becomes ice at 0° C through a phase change, and

$$Q_2 = mL_F$$

$$= (100.0 \text{ g})\left(80.0 \ \frac{\text{cal}}{\text{g}}\right)$$

$$= (100.0)(80.0) \ \frac{\text{g} \cdot \text{cal}}{\text{g}}$$

$$= 8,000 \text{ cal}$$

$$Q_2 = \boxed{8.00 \text{ kcal}}$$

3. The ice is now at 0° C and is cooled to −10° C as specified in the problem. The ice cools according to $Q = mc\Delta T$, where c is the specific heat of ice. The specific heat of ice is 0.50 cal/g·°C, and

$$Q_3 = mc\Delta T$$

$$= (100.0 \text{ g})\left(0.50 \ \frac{\text{cal}}{\text{g} \cdot °\text{C}}\right)(-10° - 0° \text{ C})$$

$$= (100.0)(0.50)(10) \ \frac{\text{g} \cdot \text{cal} \cdot °\text{C}}{\text{g} \cdot °\text{C}}$$

$$= 500 \text{ cal}$$

$$Q_3 = \boxed{0.5 \text{ kcal}}$$

The total energy removed is then

$$Q_T = Q_1 + Q_2 + Q_3$$

$$= (2 \text{ kcal}) + (8.00 \text{ kcal}) + (0.5 \text{ kcal})$$

$$Q_T = \boxed{10.5 \text{ kcal}}$$

Evaporation and Condensation

Liquids do not have to be at the boiling point to change to a gas and, in fact, tend to undergo a phase change at any temperature when left in the open. The phase change occurs at any temperature but does occur more rapidly at higher temperatures. The temperature of the water is associated with the *average* kinetic energy of the water molecules. The word average implies that some of the molecules have a greater energy and some have less (refer back to figure 5.8). If a molecule of water that has an exceptionally high energy is near the surface, and is headed in the right direction, it may overcome the attractive forces of the other water molecules and escape the liquid to become a gas. This is the process of *evaporation*. Evaporation reduces a volume of liquid water as water molecules leave the liquid state to become water vapor in the atmosphere (figure 6.8).

Evaporation occurs at any temperature, but the energy (heat) required per gram of liquid water changing to water vapor is the same in evaporation as in boiling. A gram of water at

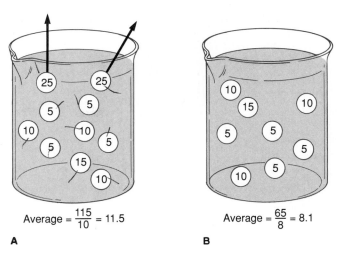

Average $= \dfrac{115}{10} = 11.5$ **A**

Average $= \dfrac{65}{8} = 8.1$ **B**

Figure 6.8

Temperature is associated with the average kinetic energy of the molecules of a substance. These numbered circles represent arbitrary levels of molecular kinetic energy that, in turn, represent temperature. The two molecules with the higher kinetic energy values [25 in (a)] escape, which lowers the average values from 11.5 to 8.1(b). Thus evaporation of water molecules with more kinetic energy contributes to the cooling effect of evaporation in addition to the absorption of latent heat.

20° C (room temperature) will need 80.0 cal for the equivalent energy of boiling water and an additional 540.0 cal for the latent heat of vaporization. Thus each gram of room-temperature water requires 620.0 cal to change to a vapor. This supply of energy must be present to maintain the process of evaporation, and the water robs this energy from its surroundings. This explains why water at a higher temperature evaporates more rapidly than water at a cooler temperature. More energy is available at higher temperatures to maintain the process, so the water evaporates more rapidly. It also explains why evaporation is a cooling process. Consider, for example, how perspiring cools your body. Water from sweat glands runs onto your skin and evaporates, removing heat from your body. Since your body temperature is about 37.0° C (98.6° F), each gram of water will require 63.0 cal to give it sufficient energy to evaporate. Each gram of water will then remove 540.0 cal (the latent heat of vaporization) as it changes to water vapor. Each gram of water evaporated therefore removed about 603 cal (540.0 + 63.0) from your body, or about 603 kcal per liter of water.

Water molecules that evaporate move about in all directions and some will return, striking the liquid surface. The same forces that they escaped from earlier capture the molecules, returning them to the liquid state. This is called the process of condensation. Condensation is the opposite of evaporation. In **evaporation,** more molecules are leaving the liquid state than are returning. In **condensation,** more molecules are returning to the liquid state than are leaving. This is a dynamic, ongoing

Figure 6.9

The inside of this closed bottle is isolated from the environment, so the space above the liquid becomes saturated. While it is saturated, the evaporation rate equals the condensation rate. When the bottle is cooled, condensation exceeds evaporation and droplets of liquid form on the inside surfaces.

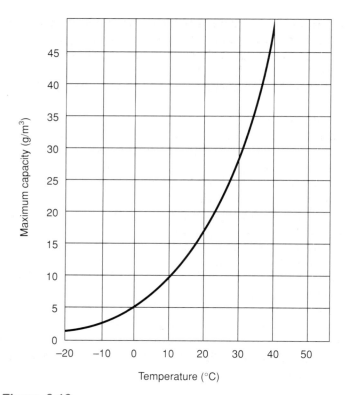

Figure 6.10

The curve shows the *maximum* amount of water vapor in g/m³ that can be in the air at various temperatures.

process with molecules leaving and returning continuously. The net number leaving or returning determines if evaporation or condensation is taking place (figure 6.9).

When the condensation rate *equals* the evaporation rate the air above the liquid is said to be **saturated.** The air immediately next to a surface may be saturated, but the condensation of water molecules is easily moved away with air movement. There is no net energy flow when the air is saturated since the heat carried away by evaporation is returned by condensation. This is why you fan your face when you are hot. The moving air from the fanning action pushes away water molecules from the air near your skin, preventing the adjacent air from becoming saturated, thus increasing the rate of evaporation. Think about this process the next time you see someone fanning his or her face.

There are four ways to increase the rate of evaporation. (1) An increase in the temperature of the liquid will increase the average kinetic energy of the molecules and thus increase the number of high energy molecules able to escape from the liquid state. (2) Increasing the surface area of the liquid will also increase the likelihood of molecular escape to the air. This is why you spread out wet clothing to dry or spread out a puddle you want to evaporate. (3) Removal of water vapor from near the surface of the liquid will prevent the return of the vapor molecules to the liquid state and thus increase the net rate of evaporation. This is why things dry more rapidly on a windy day. (4) **Pressure** is defined as *force per unit area,* which can be measured in lb/in² or N/m². Gases exert a pressure, which is interpreted in terms of the kinetic molecular theory later in this chapter. Atmospheric pressure is discussed in detail in chapter 23. For now, consider that the atmosphere exerts a pressure of about 14.7 lb/in² at sea level. The atmospheric pressure, as well as the intermolecular forces, tend to hold water molecules in the liquid state. Thus reducing the atmospheric pressure will reduce one of the forces holding molecules in a liquid state. Perhaps you have noticed that wet items dry more quickly at higher elevations, where the atmospheric pressure is less.

Relative Humidity

There is a relationship between evaporation-condensation and the air temperature. If the air temperature is decreased, the average kinetic energy of the molecules making up the air is decreased. Water vapor molecules condense from the air when they slow enough that molecular forces can pull them into the liquid state. Fast-moving water vapor molecules are less likely to be captured than slow-moving ones. Thus, as the air temperature increases, there is less tendency for water vapor molecules to return to the liquid state. Warm air can therefore hold more water vapor than cool air. In fact, air at 38° C (100° F) can hold five times as much vapor as air at 10° C (50° F) (figure 6.10).

The ratio of how much water vapor *is* in the air to how much water vapor *could be* in the air at a certain temperature is called **relative humidity.** This ratio is usually expressed as a percent, and

$$\text{relative humidity} = \frac{\text{water vapor in air}}{\text{capacity at present temperature}} \times 100\%$$

$$\text{R.H.} = \frac{\text{g/m}^3 \text{(present)}}{\text{g/m}^3 \text{(max)}} \times 100\% \qquad \textbf{equation 6.3}$$

Figure 6.10 shows the maximum amount of water vapor that can be in the air at various temperatures. Suppose that the air contains 10 g/m³ of water vapor at 10° C (50° F). According to figure 6.10, the maximum amount of water vapor that *can be* in the air when the air temperature is 10° C (50° F) is 10 g/m³. Therefore the relative humidity is (10 g/m³) ÷ (10 g/m³) × 100%, or 100 percent. This air is therefore saturated. If the air held only 5 g/m³ of water vapor at 10° C, the relative humidity would be 50 percent, and 2 g/m³ of water vapor in the air at 10° C is 20 percent relative humidity.

As the air temperature increases, the capacity of the air to hold water vapor also increases. This means that if the *same amount* of water vapor is in the air during a temperature increase, the relative humidity will decrease. Thus the relative humidity increases every night because the air temperature decreases, not because water vapor has been added to the air. The relative humidity is important because it is one of the things that controls the rate of evaporation. And the evaporation rate is one of the variables involved in how well you can cool yourself in hot weather.

Example 6.2

The air temperature is 30° C and the relative humidity is 33 percent. At what cooler temperature will saturation (100 percent relative humidity) occur if no water vapor is added to or removed from the air?

Solution

First, find the 30° C line on the graph in figure 6.10. Tracing this line up and down, you see that it intersects the maximum slope line at about 30 g/m³ of water vapor. This means that saturated air, with a relative humidity of 100 percent, will contain 30 g/m³ of water vapor at 30° C. Since the relative humidity at 30° C is given as 33 percent, this means that the air contains 33 percent of the moisture it can hold at that temperature, or about 10 g/m³ (30 g/m³ × 0.33). Tracing the 10 g/m³ line from the 30° C temperature line to the left, you can see that it intersects the maximum slope line at about 10° C. This means that air containing 10 g/m³ that is cooled from 30° C will reach saturation at an air temperature of 10° C. Therefore, the relative humidity will be 100 percent if the air is cooled to 10° C.

Example 6.3

The air temperature is 40° C and the relative humidity is 50 percent. At what approximate air temperature will the relative humidity become 100 percent? (Answer: about 25° C)

Boiling

If a pot of cold water is heated on a stove, the temperature of the water increases until the boiling point is reached. At this temperature you first see small points of bubbles that appear, then disappear, on some flaw or imperfection in the bottom of the pot directly over the heat source. Then the bubbles begin to grow larger, breaking free and moving to the surface. Soon the surface is churning from the bubbles as they break through the surface, and rapid boiling has begun.

The bubbles observed in boiling are water vapor. While evaporation is the process of water vapor escaping from the *surface* of the liquid, **boiling** is evaporation that occurs *throughout* the liquid (figure 6.11). When a liquid is not boiling, the intermolecular forces as well as the atmospheric pressure on the liquid tend to keep the molecules in a liquid state. Increased temperature means increased kinetic energy of the molecules in a liquid state. Increased kinetic energy enables the molecules in the liquid state to overcome the intermolecular forces and pressure on the liquid, and not just at the surface. Bubbles of water vapor therefore form throughout the liquid and rise to the top. More bubbles are actually formed at the bottom of the pot because it is hotter there.

Heat and Technology

Figure 6.11

This pot of water is boiling as evaporation is occurring throughout the liquid. The bubbles are water vapor, which escapes only from the surface during normal evaporation.

There are *two* variables involved in boiling: (1) the energy of the molecules in the liquid state and (2) the pressure on the molecules that tends to hold them in the liquid state. Thus there are two ways that a liquid can be made to boil: (1) by *increasing* the temperature and therefore the kinetic energy of the molecules and (2) by *reducing* the pressure on the liquid. Either method will produce boiling by causing the pressure of the vapor inside the bubbles to equal or exceed the atmospheric pressure on the liquid.

Since the boiling point of water changes with pressure, you can find the approximate elevation of where you live by measuring the boiling point of water. Atmospheric pressure decreases with increases in elevation, and the boiling point of water decreases about 1.0° C for each 305 m increase of elevation (1.8° F per 1,000 ft). If you find, for example, that water boils at 203° F (95° C) you can determine your approximate elevation as follows:

a. 212° F − 203° F = 9° F (boiling point reduction)
b. 9° F ÷ 1.8° F = 5 (units of 1.8° F in boiling point reduction)
c. 5 × 1,000 ft = 5,000 ft (elevation)

Water boils at about 90° C (194° F) at an elevation of 3,048 m (10,000 ft). At what elevation will water boil at room temperature, 20° C (68° F)? The answer is about 24,400 m (about

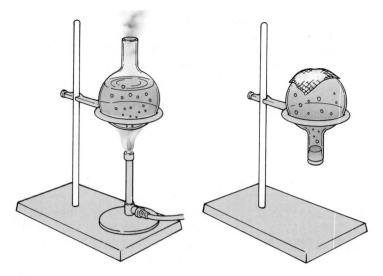

Figure 6.12

This demonstrates how to boil water without heating. See the activities below to find out how.

80,000 ft). According to these calculations, water should boil and freeze at the same time at an altitude of about 30,500 m (about 100,000 ft).

Cooking changes the molecular structure of foods and these changes occur more slowly at lower temperatures. Since water boils at a lower temperature at increased elevations, you must cook food longer at increased elevations. For example, at a high elevation in the mountains five minutes might be required to boil an egg to the same degree of doneness that is accomplished in three minutes at a lower elevation. Increasing the rate of heating will not cook the food any faster, however. Since evaporation is a cooling process and since boiling is evaporation that occurs throughout the liquid, boiling is a cooling process. Increasing the rate of heating simply increases the evaporation rate. Thus increasing the rate of heating will not increase the temperature, and the food takes as long to cook as before. The temperature of water will go no higher, under normal conditions, than its boiling point. Cooking food at a rapid boil is therefore a waste of energy (and water).

ACTIVITIES

1. Half fill a thick-walled flask and heat it to a rapid boil. Let it boil for several minutes then remove it from the heat source and carefully place a tight fitting stopper in the flask. Observe the water, and continue observing as you turn the flask upside down and cool it with a wet cloth (figure 6.12). What happens to the water in the flask? Turn the flask right side up and carefully remove the stopper. Does the water continue to boil? Use the kinetic molecular theory to explain what happened.
2. Since the temperature of water will go no higher than its boiling point, you can boil water in a paper container. The sea level boiling temperature of water at normal

atmospheric pressure is 100° C (212° F), and the temperature at which paper will burn (the kindling point) is about 234° C (454° F). Since the boiling process does not allow the temperature of water to go above 100° C, you can boil water in a paper container without burning the paper. (Note: Do *not* use a waxed paper container. The kindling point of wax is considerably less than that of paper.)

Increased heating will not change the boiling point of a liquid but it can be changed by varying the pressure on the liquid. Boiling at increased pressure takes place in a pressure cooker. Water vapor is confined by the pressure cooker and this increases the pressure, which increases the boiling temperature. As a consequence, foods cook at a higher temperature and therefore require less time for cooking. Conversely, boiling at reduced pressure avoids high temperatures. Vacuum cookers are used in the production of sugar to boil solutions without scorching the sugar. Vacuum chambers are also used to prepare freeze-dried foods. Water readily boils and evaporates from foods placed in a vacuum chamber. Evaporation from the resulting low-temperature boiling cools the food until it is frozen and evaporation continues until no water remains. The process does not change the molecular arrangement of the food and the original food is later reconstituted when water is added. Since most foods are from 50 percent to 90 percent water, freeze-dried foods are light and easy to store.

Gas Molecules and Working

A gas is not very dense, compared to solids and liquids, and diffuses throughout any container in which it is placed. This diffusion is spontaneous and against gravity, and only the movement of highly mobile molecules can account for it. The gas molecules must be rapidly moving about, colliding often with other gas molecules and the walls of the container. Since it is *not* necessary to supply heat to maintain the temperature of a perfectly insulated sample of gas, all of these collisions must be perfectly elastic. On the average, there is no loss of kinetic energy when molecules collide. Extending this assumption, the pressure (force per unit area) a gas exerts on the container walls is the result of the continuous bombardment of gas molecules on the container walls (figure 6.13). This explains why adding air to a tire increases the air pressure in the tire. More air means more bombardment by the added molecules. Likewise, when you drive the air pressure increases in your tires because friction with the road increases the absolute temperature of the air in your tires. Increasing the absolute temperature increases the average molecular kinetic energy and therefore increases the frequency of impact on the tire walls as well as the force of impact. Increased frequency and force of impact result in greater pressure. This is why you should check the air pressure in your tires when they are cool, before you have driven very far.

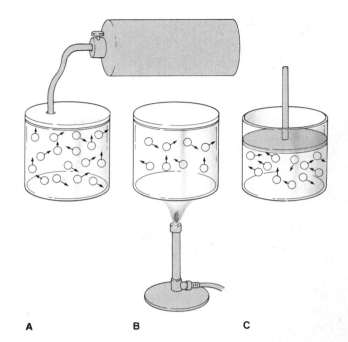

Figure 6.13
The pressure of a confined gas can be increased by (*a*) adding more gas molecules, (*b*) increasing the temperature of the gas, or (*c*) reducing the volume. Arrows represent velocities, not forces.

Air must be compressed by an air pump before it can be added to a tire. It is compressed by a moving piston that pushes the molecules closer together. This increases the absolute temperature because the moving piston gives the molecules additional kinetic energy. By analogy, this is like throwing a perfectly elastic ball against the back of a truck backing *toward* you. The moving truck will give the ball additional energy and it will bounce back with a greater velocity. Compressing a gas thus *increases* the temperature of the gas.

Conversely, expanding a gas will *reduce* the temperature of the gas because the molecules are doing work on their surroundings, giving up some of their kinetic energy in the process. Again by analogy, this is like throwing a perfectly elastic ball against the back of a truck pulling *away* from you. The ball will give some of its energy to the truck and it will bounce back with less velocity. Expanding a gas thus *decreases* the temperature of the gas. You have probably noticed this cooling effect when releasing a compressed gas from a tire, pressurized spray can, or carbon dioxide fire extinguisher. The CO_2 from a fire extinguisher cools by expansion to such an extent that solid carbon dioxide (dry ice) is produced with a temperature of −78° C (−172° F).

Human Comfort and Technology

The human body maintains an average internal core temperature of about 37.0° C (98.6° F). This temperature is slightly less at night and slightly more during the day but usually not more than a degree or so. In addition, hormonal changes are

Heat and Technology

associated with monthly temperature fluctuations, but in general, the core body temperature is maintained independent of the temperature of the immediate environment. The hypothalamus at the base of the brain acts as a thermostat, conserving body heat at low environmental temperatures and getting rid of body heat at high environmental temperatures. In general, a sense of comfort is felt in a range of environmental temperature between about 20° C (68° F) and 28° C (about 82° F). The actual comfort range is influenced by the level of activity, age, gender, the relative humidity, and air movement. Temperature, however, is the most important variable influencing a sense of comfort.

Suitable clothing is worn to assist the human body in maintaining its core temperature by controlling conduction, convection, and radiation. We also use structures to limit outside temperature extremes and modify the interior temperature above or below the temperature extremes. This modification requires technology and energy to run the technology. The following describes how three common technologies are used to help people keep cool: fans, evaporative coolers, and heat pumps and refrigerators.

Fans

Suppose you place a thermometer in front of an electric fan where the air blows on it and place a second thermometer behind the fan. Would the thermometer in front of the fan with the air blowing on it show a cooler temperature than the thermometer behind the fan? The thermometer in front of the fan will *not* show a cooler temperature and, in fact, may show a slightly warmer temperature because of the motor friction. A fan does not make the air cooler. It cools you because it moves water vapor away from your skin. In *still air* some water evaporates from the surface of your skin but some also condenses back on your skin. This is an ongoing process that continues until a layer of air around you becomes saturated. When the air is saturated the rate of evaporation equals the rate of condensation. Heat is removed from your skin by evaporating water molecules through the latent heat of vaporization. But the very same amount of heat is returned to your skin by condensation. A fan cools you because it moves the concentration of water vapor away from the surface of your skin. Evaporation then exceeds condensation and each liter (about a quart) of evaporated condensation carries more than 450 kcal away from your body.

How much can a fan cool you? It depends on the relative humidity. With 100 percent relative humidity (saturated air) there can be no net evaporation, and no net evaporation means no cooling. With extremely low humidity there is maximum cooling. Thus the saying, "It's not the heat, it's the humidity," actually means that you cannot become much cooler by fanning yourself when the humidity is high.

Table 6.1

Cooling power of an 80 percent efficient evaporative cooler when the outside air temperature is 38° C (100° F)

At This Humidity	Air Will Be Cooled to
10%	22° C (72° F)
20%	24° C (76° F)
30%	27° C (80° F)
40%	29° C (84° F)
50%	31° C (88° F)
60%	33° C (91° F)
70%	34° C (93° F)
80%	35° C (95° F)
90%	37° C (98° F)

Note: The cooling power at other outside air temperatures can be determined from the relative humidity table in appendix C.

Evaporative Coolers

Evaporative cooling is one of the earliest forms of mechanical air conditioning. An evaporative cooler (sometimes called a "swamp cooler") works by cooling outside air by evaporation, then blowing the cooler, but now more humid air through a house. Usually, an evaporative cooler moves a sufficient amount of air through a house to completely change the air every two minutes.

An evaporative cooler is a metal or fiberglass box with louvers. Inside the box is a fan and motor, and there might be a small pump to recycle water. Behind the louvers are loose pads made of wood shavings (excelsior) or some other material. Water is pumped from the bottom of the cooler and trickles down through the pads, thoroughly wetting them. The fan forces outside air through the wet pads and into the house. Windows and/or doors must be partly open to allow air to flow through the house.

The water in the pads evaporates, robbing heat from the air moving through them. Each liter of water that evaporates could take over 540 kcal of energy from the air. The actual amount removed depends on the temperature of the water, the efficiency of the cooler, and the relative humidity of the outside air.

An electric fan cools you by evaporation, but an evaporative cooler cools the air that blows around you. Relative humidity is the main variable that determines how much an evaporative cooler will cool the air. Table 6.1 shows the cooling power of a typical evaporative cooler at various relative humidities when the outside air is 38° C (about 100° F).

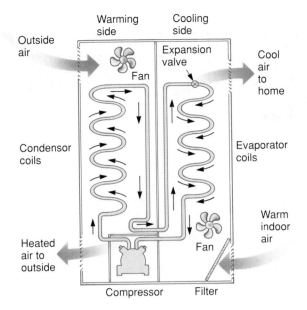

Figure 6.14

A heat pump moves heat from the inside of a house during the summer outside against the natural flow of heat. Some are reversible, moving heat to the inside during the winter.

The advantage of an evaporative cooler is the low operating cost compared to refrigeration. The disadvantages are that it does not cool the air that much when the humidity is high, it adds even more humidity to the air, and outside air with its dust, pollen, and pollution is continually being forced through the house. Another disadvantage is that mineral deposits left by the evaporating water require a regular maintenance schedule.

Heat Pumps and Refrigerators

A refrigeration system is a mechanical device that uses mechanical energy to move heat from a region of low temperature to a region of high temperature. The same basic components are used in refrigerators, freezers, home air conditioners, and car air conditioners. Each device could be called a **heat pump** since it moves, or "pumps" heat from one place to another. The term *heat pump,* however, is also used to describe a reversible home air conditioner that cools the home in summer and warms it in the winter.

All heat pumps use a **refrigerant** to absorb and transport heat. A refrigerant is a volatile fluid that undergoes a liquid-gas phase change at a low temperature. The most common refrigerants are fluorocarbons and ammonia. Of some sixteen existing fluorocarbons, Freon-12 (CCl_2F_2) is commonly used in home heat pumps and ammonia is used in commercial heat pumps. Freon-12 has a boiling point of $-5.6°$ C ($22°$ F) and

a latent heat of vaporization of about 40 cal/g (71 Btu/lb). Ammonia has a boiling point of $-2.2°$ C ($28°$ F) and a latent heat of vaporization of about 304 cal/g (547 Btu/lb). Both of these refrigerants thus have boiling points less than the freezing point of water.

A heat pump uses evaporation and condensation of the refrigerant to move heat. Figure 6.14 illustrates the basic parts of a heat pump system that accomplishes this process: the compressor, condenser, expansion valve, evaporator, and tubes that connect these different parts into a closed system. The system is charged with a refrigerant such as Freon, which is cycled through the gaseous and liquid phases inside the tubes.

When the heat pump is started, Freon vapors from the evaporator (cooling) section are compressed by the compressor, which increases the temperature and pressure of the vapor. The vapors are then pushed into the condenser, where they lose heat by conduction to cooler air moving across the condenser coils. Home refrigerators radiate this heat into air in the kitchen through the condenser coils on the back of the refrigerator. Newer models blow air across the condenser coils at the bottom of the refrigerator. In any case, the process lowers the temperature sufficiently to allow the compressed Freon to condense, forming a liquid inside the coils. As the vapors condense they release the latent heat of vaporization, which is also conducted through the condenser coils.

The liquid Freon is now forced by the compressor through a high-pressure tube to the evaporator. Here, the liquid Freon is forced through an expansion valve to the low-pressure side of the system. The expansion lowers the pressure on the liquid Freon, which now begins to boil inside the evaporation coils, absorbing the latent heat of vaporization. This "energy robbing" section of the heat pump is used in different ways depending on the application. A fan may blow air across the evaporation coils, for example, for cooling the air in a car, room, or house. In refrigerators, the evaporation coils are located in the freezer compartment.

After removing heat in the evaporation coils, the Freon vapors are now pushed back to the compressor through the low-pressure tube. The overall process is repeated as often as needed to reach the desired temperature.

Transportation and Technology

Recall that mechanical energy can be converted to heat and that the heat produced by a definite quantity of mechanical energy is always the same. A **heat engine** accomplishes the opposite process, converting heat into mechanical energy. The operation of any heat engine can be explained by the kinetic molecular theory. Consider the simple apparatus shown in figure 6.15. A cylinder, much like a large can, has a closely fitting piston that traps a sample of air in the cylinder. The piston is like a slightly

Heat and Technology

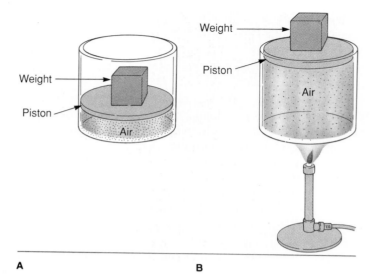

A   **B**

Figure 6.15

A very simple heat engine. The air in (a) has been heated, increasing the molecular motion and thus the pressure in (b). Some of the heat is transferred to the increased gravitational potential energy of the weight as it is converted to mechanical energy.

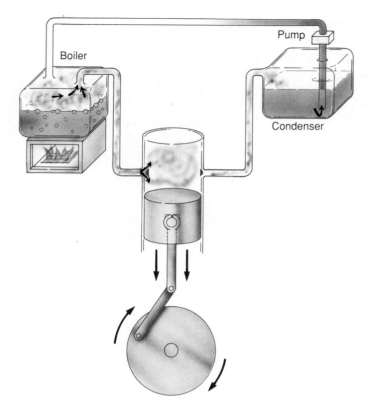

Figure 6.16

A reciprocating steam engine. Steam enters the cylinder through a valve and forces the piston down, turning the wheel. The exhaust valve then opens and the motion of the wheel returns the piston upward, exhausting the spent steam into the condenser.

smaller cylinder and has a weight resting on it, supported by the trapped air. If the air in the large cylinder is now heated, the molecules of the air will increase in motion and produce an increased pressure. This expands the air, pushing the piston and weight upward as shown in figure 6.15b. Thus, some of the heat has now been transformed to the increased gravitational potential energy of the weight.

When a heat engine operates, a burning fuel gives additional motion to the molecules of gas in the engine. The molecules make violent collisions with all surfaces they contact and produce an increased pressure. When they strike a piston that is moving away, they bounce back with less energy, losing energy to the piston. In other words, they do work on the piston, transferring energy to it. The *random* motion of thousands of molecules of gas are thus converted into *directed* and therefore useful motion of the piston.

There are two main types of heat engines: (1) those that have combustion external to the engine, such as a steam engine, and (2) those that have combustion inside the engine, such as gasoline, diesel, and jet engines. The first working steam engine was developed by Thomas Newcomen in England about 1711. This engine was primarily used to pump water out of mines. Because it was bulky and inefficient, however, it was never extensively used. James Watt improved the engine about fifty years later and opened the way for its use in industry and transportation.

Watt's steam engine had a piston connected by a crank to a wheel. Steam from an external boiler entered and left the engine through valves as the steam pushed the piston, then the

motion of the wheel returned the piston through a cycle of back-and-forth movements. Belts transferred the circular motion of the wheel to do useful work. Such an engine that has alternating back-and-forth motion of a piston or pistons is called a *reciprocating engine* (figure 6.16).

Today, most steam is used in a steam *turbine*. A steam turbine does not have pistons, cylinders, or valves. High-pressure steam is directed against sets of rotor blades attached to a shaft (figure 6.17). The steam gives some of its internal kinetic energy to the blades by colliding with them. It is then redirected by a row of stationary blades to a larger set of blades in the next chamber. The steam expands against the second set of blades and is redirected to a larger set and so on until, at the opposite end of the turbine, the now low-pressure, cooler steam is exhausted to be condensed, heated again to steam, and recycled. The steam turbine is more efficient than the steam reciprocating engine and is used in large electric power plants. The major difference in a coal, oil, or nuclear power plant is the source of energy for making steam. Once steam is made it is directed to a steam turbine, which turns the shaft on a large electric generator.

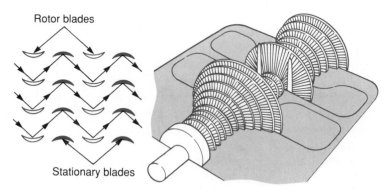

Figure 6.17

A steam turbine spins as high-pressure, high-temperature steam is directed toward the rotor blades. This steam engine is used in large electric power plants.

The steam turbine is the most efficient of the heat engines, converting about 40 percent of the generated heat to mechanical energy. The major problem of the steam turbine is that it operates best in large units that run at high speeds continuously. This is no problem for an electric power plant, but it limits the application of the steam turbine elsewhere. There are comparatively small, lightweight engines that convert heat to mechanical energy over a wide range of operating speeds. These are the *internal combustion engines* of which there are two main types: the reciprocating designs and the Wankel rotary engine.

Thermodynamics

The branch of physical science called **thermodynamics** is concerned with the study of heat and its relationship to mechanical energy. It is the science of heat pumps, heat engines, and the transformation of energy in all its forms. The *laws of thermodynamics* describe what happens as energy is transformed to work and the reverse, and they also serve as useful intellectual tools in meteorology, chemistry, and biology.

Thermodynamics is concerned with the *internal energy* (U), the total internal potential and kinetic energies of molecules making up an object or set of objects. The variables of temperature, gas pressure, volume, heat, and so forth characterize the total internal energy, which is called the *state* of the system. Once the system is identified everything else is called the *surroundings*. A system can exist in a number of states since the variables that characterize a state can have any number of values and combinations of values. However, any two systems that have the same values of variables that characterize internal energy are said to be in the same state.

The First Law of Thermodynamics

Any thermodynamic system has a unique set of properties that will identify the internal energy of the system. This state can be changed two ways: (1) by heat flowing into (Q_{in}) or out of

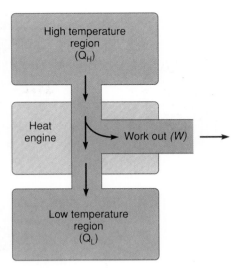

Figure 6.18

The heat supplied (Q_H) to a heat engine goes into the mechanical work (*W*) and the remainder is expelled in the exhaust (Q_L). The work accomplished is therefore the difference in the heat input and output ($Q_H - Q_L$), so the work accomplished represents the heat used, $W = J(Q_H - Q_L)$.

(Q_{out}) the system or (2) by the system doing work (W_{out}) or by work being done on the system (W_{in}). Thus work (*W*) and heat (Q) can change the internal energy of a thermodynamic system according to

$$Q - W = U_2 - U_1 \qquad \text{equation 6.4}$$

where ($U_2 - U_1$) is the internal energy difference between two states. In other words, the heat *added* to a thermodynamic system minus the work done *by* the system is equal to the change in internal energy. This equation represents the **first law of thermodynamics,** which you will recognize as another way of stating the law of conservation of energy, that *energy is conserved*. The law of conservation of energy is a general law, applying to all energy matters. The first law of thermodynamics, on the other hand, is concerned specifically with a thermodynamic system. Consider, for example, energy conservation in the thermodynamic system of a heat engine (figure 6.18). As the engine cycles back to the original state of internal energy ($U_2 - U_1 = 0$), the net external work accomplished must be equal to the net heat absorbed in the cycle. The heat supplied to the engine from a high temperature source (Q_H) is partly converted to work (*W*) and the rest is rejected in the lower temperature exhaust (Q_L). The work accomplished is therefore the difference in the heat input and the heat output ($Q_H - Q_L$), so the work accomplished represents the heat used,

$$W = J(Q_H - Q_L) \qquad \text{equation 6.5}$$

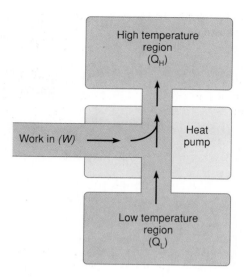

Figure 6.19

A heat pump uses work (W) to move heat from a low-temperature region (Q_L) to a high-temperature region (Q_H). The heat moved (Q_L) requires work (W), so $JQ_L = W$.

where J is the mechanical equivalent of heat (1 kcal = 4,184 J or 1 Btu = 778 ft·lb). A schematic diagram of this relationship is shown in figure 6.18.

Example 6.4

A heat engine operates with 65.0 kcal of heat supplied and exhausts 40.0 kcal of heat. How much work did the engine do?

Solution

The heat input (Q_H) is 65.0 kcal and the heat rejected in the exhaust (Q_L) is 40.0 kcal. The mechanical equivalent of heat is 1 kcal = 4,184 J, so

$Q_H = 65.0$ kcal

$Q_L = 40.0$ kcal

$J = 4,184$ J/kcal

$W = ?$

$$W = J(Q_H - Q_L)$$

$$= 4,184 \frac{J}{kcal} (65.0 \text{ kcal} - 40.0 \text{ kcal})$$

$$= 4,184 \frac{J}{kcal} (25.0 \text{ kcal})$$

$$= 4,184 \times 25.0 \frac{J \cdot kcal}{kcal}$$

$$= 104,600 \text{ J} = \boxed{105 \text{ kJ}}$$

The Second Law of Thermodynamics

A heat pump is the *opposite* of a heat engine as shown schematically in figure 6.19. The heat pump does work (W) in compressing the refrigerant vapors in the compressor and moves heat from a region of low temperature (Q_L) to a region of high temperature (Q_H). This is in accord with the observation that heat *naturally flows,* or moves from a region of higher temperature to a region of lower temperature. Energy, or work, is *required* to move heat from a cooler region to a warmer region. Heat has never been observed to move spontaneously from a cold region to a warmer region on its own. A statement about the direction of this natural process is called the **second law of thermodynamics.** The second law can be stated several different ways. One way is that *heat does not flow spontaneously from a colder object to a warmer object*. In other words, if you want heat to flow from a colder region to a warmer one you must *cause* it to do so. And if you do, such as with the use of a heat pump, you cause changes elsewhere, particularly in the energy sources used in the generation of electricity. Another statement of the second law is that *it is impossible to convert heat completely into mechanical energy*. This does not say that you cannot convert mechanical energy completely into heat, for example, in the brakes of a car when they bring the car to a stop. This statement says that the reverse process is not possible, that you cannot convert 100 percent of a heat source into mechanical energy. Both of these statements of the second law are concerned with a *direction* of thermodynamic processes, and the implications of this direction will be discussed shortly.

The foundations of thermodynamics were laid in 1824 when Sadi Carnot (pronounced "car-no") summarized his theoretical analysis of the heat engine. Carnot was a French engineer who recognized that steam power could replace human and animal labor and thus improve the human condition. He therefore set out to analyze and improve the efficiency of the steam engine. The only problem was that Carnot, as most people during his time, was a believer in the caloric theory of heat. Carnot thus based his analysis on the assumption that a steam engine performed work as the caloric fluid ran from a high-temperature source to a low-temperature receiver, much as water turns a waterwheel as it runs downhill. The misconceptions in Carnot's work were corrected by the next generation of scientists, including Joule, Lord Kelvin, and Clausius. Carnot's work, however, had planted the ideas that would lead others to understanding some of the innermost workings of nature.

Efficiency of Heat Engines

Carnot recognized that no mechanical energy could be obtained from a heat engine unless there was a *temperature difference* between the operating cycles of the engine. Today, the kinetic molecular theory explains why a temperature difference is necessary. Consider the steam engine illustrated in figure 6.16. Steam moves the piston because of the steam pressure, which is produced by the crowded, high-energy motion of the water vapor molecules. If the temperature, and thus the pressure of the steam, were the same after doing work on the piston, then the piston would have to do the *same* work on the steam to push it out of the cylinder! Thus a temperature difference is required for the heat engine to extract mechanical work from heat. It follows then, that the greater the temperature difference the greater the amount of work that a heat engine can extract. A temperature difference is attained by using the hottest source possible, but it could also be attained by cooling the exhaust as much as possible.

Today, there are many kinds of heat engines including the steam turbine, the jet engine, diesel engines, and gasoline engines of various designs. The operation of all these engines can be described in terms of an ideal cycle called the **Carnot cycle.** The Carnot cycle is an idealized way of describing the transformations that take place between heat and work in the heat engine. It considers the working substance, which could be steam, jet fuel and air, diesel fuel and air, or gasoline and air, to move through an operation called a *cycle*. The working substance starts the cycle with a high temperature, which does work through the engine. Not all the heat is converted to work, however, and the leftover heat is delivered to a *heat sink*. A heat sink is the opposite of a heat source. The surroundings that absorb the leftover heat without an overall increase in temperature, for example, is the heat sink for a heat engine.

The idealized cycle of the operation of a heat engine is shown schematically in figure 6.18. A quantity of heat, Q_H, is supplied to the engine to begin the cycle. The engine accomplishes an amount of work, W, and the leftover heat, Q_L, is exhausted to a lower temperature heat sink. The relationship between the work accomplished (W), the heat input (Q_H), and leftover heat (Q_L) was given in equation 6.5,

$$W = J(Q_H - Q_L)$$

The ratio of the work accomplished by an engine to the heat it absorbs is called the **thermal efficiency** (e) of the engine. During one cycle this ratio is

$$e = \frac{\text{work accomplished}}{\text{heat supplied}} \times 100\%$$

or

$$e = \frac{W}{JQ_H} \times 100\% \qquad \text{equation 6.6}$$

Substituting W from equation 6.5,

$$e = \frac{J(Q_H - Q_L)}{JQ_H} \times 100\%$$

or

$$e = 1 - \frac{Q_L}{Q_H} \times 100\% \qquad \text{equation 6.7}$$

Ignoring any friction and any forces between the molecules of gases, it can be shown that the ratio of Q_L/Q_H is equal to the ratio of the *absolute*, or Kelvin temperature (T), so

$$e = 1 - \frac{T_L}{T_H} \times 100\% \qquad \text{equation 6.8}$$

Equation 6.8 represents the *ideal maximum efficiency* of a heat engine without losses due to friction, conduction, or radiation. Any real engine has these losses, of course, which reduce the efficiency to a range of 60 to 80 percent of the ideal maximum value. In other words, even if it were possible to convert 100 percent of the heat to mechanical energy, the actual efficiency would be only 60 to 80 percent because of heat losses and friction.

Example 6.5

A steam turbine is run by steam entering at 590° F and has a condenser temperature of 126° F. What is the ideal maximum efficiency of the turbine?

Solution
First the temperatures must be converted to the absolute scale. Recall from chapter 5 that $T_C = 5/9(T_F - 32°)$ and that $T_K = T_C + 273$, and

$$T_K = \frac{5}{9}(T_F - 32°) + 273 \therefore T_H = \frac{5}{9}(590 - 32) + 273 = 583K$$

$$T_L = \frac{5}{9}(126 - 32) + 273 = 325K$$

From equation 6.8, the maximum ideal efficiency is

$$e = 1 - \frac{T_L}{T_H} \times 100\%$$

$$= 1 - \frac{325K}{583K} \times 100\%$$

$$= 0.443 \times 100\%$$

$$= \boxed{44.3\%}$$

However, the actual efficiency of this turbine could be as low as 26 percent (60 percent of 44 percent) due to heat losses and friction. Therefore, this turbine probably converts at least 26 percent of the heat supplied into mechanical energy.

Modern gas, oil, and coal power plants use steam at about 1,000° F (811 K) to run the steam turbine and condense the steam of the turbine exhaust at the condensing temperature of water, 212° F (373 K). The *maximum* ideal efficiency of this turbine is therefore about 54 percent. Higher steam temperatures would improve the turbine efficiency, but the durability of materials limits the higher temperatures. Is there a theoretical lower limit to the condensing temperature? The coldest temperature possible is *absolute zero*. Would a heat sink temperature of absolute zero result in an efficiency of 100 percent? The answer is no, and for several reasons. First, from inspection of equation 6.8 you can see that 1 minus the ratio of the low temperature divided by the higher temperature would require the

Cooling with Heat

There are two sources of energy that occur on the surface of the earth that can be used to provide hot water naturally. These sources are geothermal energy and solar energy. Geothermal energy is energy from beneath the earth's surface that usually appears in the form of hot springs, geysers, and naturally venting steam. But the springs, geysers, and steam only provide a hint about the far greater energy resource that is underground. Solar energy also provides hot water. A solar collector is a miniature greenhouse or a collector in which water or some other liquid is heated by sunlight.

Hot water from geothermal resources or from solar collectors can be used to heat *or* cool buildings. The technology for using hot water to cool buildings is not new. Called *absorption refrigeration,* the technology dates back to the old gas flame refrigerators. The absorption refrigerator utilizes the same physical science concepts as the electric heat pump, that evaporation *takes* heat from the surroundings, cooling them, and that condensation *gives* heat to the surroundings, warming them. The absorption refrigerator applies these concepts in a way different than a heat pump, however, using hot water and cold water rather than an electric-run compressor. This Feature is about how the absorption refrigerator is able to provide cooling without a compressor.

An absorption refrigerator utilizes a mixture of two liquids, a *refrigerant* and an *absorber*. The refrigerant has a much lower boiling point than the absorber. For example, ammonia can be used as a refrigerant and water as an absorber. Ammonia has a boiling point of $-2.2°$ C ($28°$ F) and water has a boiling point of $100°$ C ($212°$ F). Ammonia also has a relatively high heat of vaporization, about 304 cal/g (547 Btu/lb).

Box figure 6.1 illustrates the basic components of an absorption refrigeration unit. Basically, there must be a source of hot water (solar, geothermal, or other) and a source of cooling water (perhaps an evaporative cooler). The hot water runs through a heat exchanger coil in the *generator tank*. The generator tank

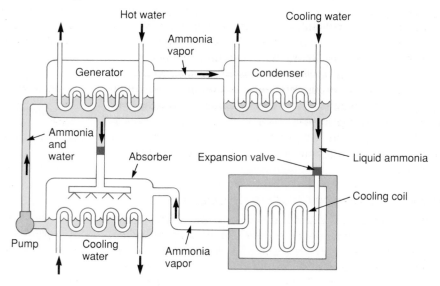

Box Figure 6.1

This absorption refrigeration system runs on hot water, cool water, and a small amount of electricity. The hot water can be provided from solar or geothermal sources and the cool water from an evaporative cooler.

holds a mixture of the working fluids, for example, a mixture of ammonia and water. The heat exchanger heats the ammonia-water mixture, causing the lower boiling point ammonia to vaporize. The function of the generator tank is to separate the ammonia from the water.

The ammonia vapor moves through a pipe to the *condenser tank*. Here, cool water runs through a heat exchanger coil, cooling the ammonia vapors and condensing them into a liquid. The cooling water could be provided from an evaporative cooler or from some natural source. The function of the condenser tank is to condense the ammonia vapor into liquid ammonia.

The liquid ammonia moves through an *expansion valve* inside the low-pressure side of the system. It boils and evaporates inside the cooling coil, absorbing the latent heat of vaporization. A fan blows air across the outside of the coil, which takes heat from the air. The ammonia vapor, containing the latent heat of vaporization it took from the air

through the coil, now moves into the *absorber tank*. The vapor is dissolved (absorbed) by water being cycled from the generator tank as shown in box figure 6.1. As the ammonia vapor dissolves, it gives up the heat removed from the cooling coils. Cooling water is circulated through a heat exchanger coil in the tank to remove the released heat. Again, the cooling water can be provided by the evaporative cooler or some natural source. The cooled ammonia-water solution is now pumped to the generator tank to continue this ongoing cycle.

Water and ammonia are used in high-pressure absorption refrigeration systems that produce very cold temperatures suitable for freezer applications. A lower-pressure, not-as-cold-system uses lithium bromide and water. This system can be used to cool a structure, such as a large building.

The advantage to the absorption refrigeration is the low, fixed cost for the energy to run the system. All you need is a source of hot water, a source of cooling water, and a small amount of electrical energy to run the small pumps.

heat source and the heat sink to be the same temperature in order to achieve an efficiency of 100 percent. Since a heat engine cannot operate unless there is a difference in the two temperatures, the efficiency of a heat engine must always be less than 100 percent. Second, the second law of thermodynamics implies that heat always flows from objects with a higher temperature to objects with a lower temperature. One interpretation of the second law is that it is *impossible* to attain a temperature of absolute zero. There is no place to transfer the heat in order to reach a temperature of absolute zero, according to this interpretation, so it is impossible to cool an object to exactly 0 K. Low-temperature experiments have produced temperatures as cold as 10^{-6} K, but absolute zero has never been reached. The impossibility of absolute zero is sometimes called the *third law of thermodynamics*.

The Second Law and Natural Processes

The consequences and implications of the second law of thermodynamics are numerous and widespread, making this law one of the major developments in the physical sciences. Before considering the consequences of this major law, return for a moment to some of the concepts of energy that were introduced in the previous two chapters. Energy can be viewed from two considerations of scale: (1) the observable *external energy* of an object and (2) the *internal energy* of the molecules, or particles that make up an object. A ball, for example, has kinetic energy after it is thrown through the air, and the entire system of particles making up the ball acts like a single massive particle as the ball moves. The motion and energy of the single system can be calculated from the laws of motion and from the equations representing the concepts of work and energy. All of the particles are moving together, in *coherent motion,* when the external kinetic energy is considered.

But the particles making up the ball have another kind of kinetic energy, with the movements and vibrations of internal kinetic energy. In this case the particles are not moving uniformly together but are vibrating with motions in many different directions. Since there is a lack of net motion and a lack of correlation, the particles have a jumbled *incoherent motion,* which is often described as chaotic. This random, chaotic motion is sometimes called *thermal motion.*

Thus there are two kinds of motion that the particles of an object can have: (1) a coherent motion where they move together, in step, and (2) an incoherent, chaotic motion of individual particles. These two types of motion are related to the two modes of energy transfer that were introduced in chapter 4: working and heating. The relationship is that *work* on an object is associated with its *coherent motion,* while *heating* an object is associated with its internal *incoherent motion.*

The second law of thermodynamics implies a direction to the relationship between work (coherent motion) and heat (incoherent motion), and this direction becomes apparent as you analyze what happens to the motions during energy conversions.

Some forms of energy, such as electrical and mechanical, have a greater amount of order since they involve particles moving together in a coherent motion. The amount of coherent motion is also referred to as the *quality* of energy. Energy with high order and coherence is called a *high-quality energy.* Energy with less order and less coherence, on the other hand, is called *low-quality energy.* In general, high-quality energy can be easily converted to work but low-quality energy is less able to do work.

High-quality electrical and mechanical energy can be used to do work, but then they become dispersed as heat through energy form conversions and friction. The resulting heat can be converted to do more work only if there is a sufficient temperature difference. The temperature differences do not last long, however, as conduction, convection, and radiation quickly disperse the energy even more. Thus, a natural process of the environment is the transformation of high-quality energy into lower-quality energy. Energy tends to disperse, both from the conversion of an energy form to heat and from the heat flow processes of conduction, convection, and radiation. Both processes flow in one direction only and cannot be reversed. This transformation of high-quality energy to lower-quality energy is called the **degradation of energy.** In every known example it is a natural process of energy to degrade, becoming less and less available to do work. The process is *irreversible* even though it is possible to temporarily transform heat to mechanical energy through a heat engine or to upgrade the temperature through the use of a heat pump. Eventually the upgraded mechanical energy will degrade to heat and the increased heat will disperse through the processes of heat flow. Furthermore, the upgrading of energy in both of these examples is always accompanied by the degrading of energy someplace else. The electrical energy used to run the heat pump, for example, was produced by the downgrading of chemical or nuclear energy at an electrical power plant. The overall result is that the *total* energy was degraded toward a more disorderly state.

A thermodynamic measure of disorder is called **entropy.** Order means patterns and coherent arrangements. Disorder means dispersion, no patterns, and a randomized, or spread out, arrangement. Entropy is therefore a measure of chaos, and this leads to another statement about the second law of thermodynamics and the direction of natural change:

The total entropy of the universe continually increases.

Note the use of the words *total* and *universe* in this statement of the second law. The entropy of a system can decrease (more order), for example, when a heat pump cools and condenses the random, chaotically moving water vapor molecules into the more ordered state of liquid water. However, when the energy source for the production, transmission, and use of electrical energy is considered, the *total* entropy will be seen as an increase. Likewise the total entropy increases during the growth of a plant or animal. When all the food, waste products, and products of metabolism are considered, there is again an increase in *total* entropy.

Thus the *natural* process is for a state of order to degrade into a state of disorder with a corresponding increase in entropy. This means that all the available energy of the universe is gradually diminishing, and over time, the universe should therefore approach some limit of maximum disorder. At this theoretical limit all the molecules will be spread far, far apart, vibrating slowly with a uniform low temperature. This limit of disorder is called the **heat death** of the universe. It seems to be a logical consequence of the second law of thermodynamics, but scientists are not certain if the second law should apply to the whole universe. Do you think the universe, with all its complexities, will end with such simplicity?

Summary

The kinetic molecular theory provides a *model* that explains the *expansion and contraction* of matter with temperature changes, *melting and vaporization, temperature,* and how mechanical energy is *converted* into heat.

The transition from one state of matter to another is called a *phase change*. A phase change always absorbs or releases a quantity of *latent heat* not associated with a temperature change. Latent heat is energy that goes into or comes out of *internal potential energy*. The *latent heat of fusion* is absorbed or released at a solid-liquid phase change. The latent heat of fusion for water is 80.0 cal/g (144.0 Btu/lb). The *latent heat of vaporization* is absorbed or released at a liquid-gas phase change. The latent heat of vaporization for water is 54.0 cal/g (970.0 Btu/lb).

Molecules of liquids sometimes have a high enough velocity to escape the surface through the process of *evaporation*. Evaporation is a cooling process since the escaping molecules remove the latent heat of vaporization in addition to their high molecular energy. Vapor molecules return to the liquid state through the process of *condensation*. Condensation is the opposite of evaporation and is a warming process. When the condensation rate equals the evaporation rate, the air is said to be *saturated*. The rate of evaporation can be *increased* by (1) increased temperature, (2) increased surface area, (3) removal of evaporated molecules, and (4) reduced atmospheric pressure.

Warm air can hold more water vapor than cold air and the ratio of how much water vapor is in the air to how much could be in the air at that temperature (saturation) is called *relative humidity*.

Boiling is evaporation that occurs throughout a liquid. It is influenced by (1) the *energy* of the molecules in the liquid state and (2) the atmospheric *pressure* on the liquid. Thus pressure increases will *raise* the boiling point and pressure reductions will *lower* the boiling point.

Gases are highly mobile molecules that easily diffuse. Since additional heat is not necessary to maintain the temperature of a sample of gas, the molecules must have perfectly *elastic collisions*. The *pressure* of a gas is the result of molecular bombardment, and the pressure is increased by *adding molecules* or by *increasing the absolute temperature*. *Compressing* a gas increases the temperature of the gas since work is done on the gas molecules. *Expanding* a gas reduces the temperature since the gas molecules do work on the surroundings.

Humans use the technology of fans, evaporative coolers, or heat pumps to help them maintain a constant core temperature. *Fans* cool by moving saturated air away from the skin. *Evaporative coolers* evaporate water, which removes the latent heat of vaporization from the air. A *heat pump* uses a refrigerant to move heat from a cool area to a warm area through liquid-gas phase changes.

Humans use the technology of *heat engines* for transporting and doing work. A heat engine converts heat to mechanical energy. The *steam turbine* is a highly efficient means of producing mechanical energy from the fossil fuels or nuclear energy sources. *Internal combustion engines* are smaller and run at any speed on gasoline, diesel, or other liquid fuels.

The study of heat and its relationships to mechanical energy is called *thermodynamics*. The *laws of thermodynamics* describe these relationships. The *first law of thermodynamics* describes the relationship between the change of internal energy state of a system (U), the heat added (Q), and the work done by the system (W), or $Q - W = U_2 - U_1$, where $U_2 - U_1$ is the internal energy difference between two states of the system. The *second law of thermodynamics* states that (1) heat cannot flow on its own from a cold object to a warmer object and (2) it is impossible to convert heat completely into mechanical energy. Both of these statements describe the *direction* of a thermodynamic process.

A *temperature difference* between the operating cycles of a heat engine is necessary if the engine is to function. The *efficiency* (e) of a heat engine is the ratio of the work the engine does to the heat it absorbs. Ignoring any friction and any forces between the molecules of gases, the ratio of the heat absorbed (Q_L/Q_H) is equal to the ratio of the absolute temperature, so the efficiency can be calculated from $e = 1 - T_L/T_H \times 100\%$. This is the *ideal maximum efficiency* of a heat engine without consideration of losses due to friction, conduction, or radiation.

The second law of thermodynamics describes the natural process as a movement towards a greater *state of disorder*. Energy is degraded from higher quality to lower quality as energy form transformations and heat flow disperse the energy, resulting in increased disorder. A thermodynamic measure of this disorder is called *entropy*, and the second law of thermodynamics means that the *total* entropy of the *universe* is continually increasing. The theoretical limit to this growing disorder is called the *heat death* of the universe.

Summary of Equations

6.1

$$\frac{\text{heat absorbed}}{\text{or released}} = (\text{mass})(\text{latent heat of fusion})$$

$$Q = mL_f$$

6.2

$$\frac{\text{heat absorbed}}{\text{or released}} = (\text{mass})(\text{latent heat of vaporization})$$

$$Q = mL_v$$

6.3

$$\frac{\text{relative}}{\text{humidity}} = \frac{\text{water vapor in air}}{\text{capacity at present temperature}} \times 100\%$$

$$\text{R.H.} = \frac{\text{g/m}^3(\text{present})}{\text{g/m}^3(\text{max})} \times 100\%$$

6.4

$$\text{heat} - \text{work} = \text{change of internal energy}$$

$$Q - W = U_2 - U_1$$

6.5

$$\text{work} = \left(\begin{array}{c}\text{mechanical}\\\text{equivalence}\\\text{of heat}\end{array}\right)\left(\begin{array}{c}\text{heat supplied}\\\text{at high}\\\text{temperatures}\end{array}\right) - \left(\begin{array}{c}\text{heat rejected}\\\text{at low}\\\text{temperatures}\end{array}\right)$$

$$W = J(Q_H - Q_L)$$

6.6

$$\frac{\text{maximum}}{\text{efficiency}} = \frac{\text{work accomplished}}{\text{heat supplied}} \times 100\%$$

$$e = \frac{W}{JQ_H} \times 100\%$$

6.7

$$\frac{\text{maximum}}{\text{efficiency}} = 1 - \frac{\text{heat rejected}}{\text{heat supplied}} \times 100\%$$

$$e = 1 - \frac{Q_L}{Q_H} \times 100\%$$

6.8

$$\frac{\text{maximum}}{\text{efficiency}} = 1 - \frac{\begin{array}{c}\text{absolute temperature}\\\text{of heat rejected}\end{array}}{\begin{array}{c}\text{absolute temperature}\\\text{of heat supplied}\end{array}} \times 100\%$$

$$e = 1 - \frac{T_L}{T_H} \times 100\%$$

Key Terms

boiling (p. 107)
boiling point (p. 103)
Carnot cycle (p. 115)
condensation (p. 105)
condensation point (p. 103)
degradation of energy (p. 117)
entropy (p. 117)
evaporation (p. 105)
first law of thermodynamics
 (p. 113)
freezing point (p. 103)
heat death (p. 118)
heat engine (p. 111)
heat pump (p. 111)
latent heat (p. 103)

latent heat of fusion (p. 104)
latent heat of vaporization
 (p. 104)
melting point (p. 103)
phase change (p. 103)
pressure (p. 106)
refrigerant (p. 111)
relative humidity (p. 107)
saturated (p. 106)
second law of thermodynamics
 (p. 114)
sublimation (p. 103)
thermal efficiency (p. 115)
thermodynamics (p. 113)

Applying the Concepts

1. Latent heat is "hidden" because it
   a. goes into or comes out of internal potential energy.
   b. is a fluid (caloric) that cannot be sensed.
   c. does not actually exist.
   d. is a form of internal kinetic energy.

2. The melting point of a solid is the same temperature as the solid's
   a. condensation point.
   b. freezing point.
   c. sublimation point.
   d. none of the above.

3. As a solid undergoes a phase change to a liquid state, it
   a. releases heat while remaining at a constant temperature.
   b. absorbs heat while remaining at a constant temperature.
   c. releases heat as the temperature decreases.
   d. absorbs heat as the temperature increases.

4. The condensation of water vapor actually
   a. warms the surroundings.
   b. cools the surroundings.
   c. sometimes warms and sometimes cools the surroundings, depending on the relative humidity at the time.
   d. neither warms nor cools the surroundings.

5. No water vapor is added to or removed from a sample of air that is cooling, so the relative humidity of this sample of air will
   a. remain the same.
   b. be lower.
   c. be higher.

6. Compared to cooler air, warm air can hold
   a. more water vapor.
   b. less water vapor.
   c. the same amount of water vapor.

7. With increasing elevation the boiling point of water is
   a. higher.
   b. lower.
   c. constant.

8. Foods cooked in a pot of boiling water will cook faster if you
   a. increase the rate of heating.
   b. increase the pressure on the water.
   c. take it to a higher elevation.
   d. any of the above.

9. A gas released from a pressurized container feels cold because
   a. liquids under pressure always have a colder temperature.
   b. compression squeezes out the latent heat of fusion.
   c. compression removes kinetic energy from the molecules because of the closeness resulting from the compression.
   d. an expanding gas does work on the surroundings.

10. The work that a heat engine is able to accomplish is ideally equivalent to the
    a. difference in the heat supplied and the heat rejected.
    b. heat that was produced in the cycle.
    c. heat that appears in the exhaust gases.
    d. sum total of the heat input and the heat output.

Answers

1. a   2. b   3. b   4. a   5. c   6. a   7. b   8. b   9. d
10. a

Questions for Thought

1. Explain how latent heat of fusion and latent heat of vaporization are "hidden."

2. How are evaporation and boiling similar? How are they different?

3. Explain what happens to the boiling point of water with increases in elevation. Does this mean that it should take more time or less time to bring a pot of water to a boil at a higher elevation? Explain your reasoning.

4. What is relative humidity? Explain why the relative humidity should change with changes in the air temperature; that is, what does the air temperature have to do with how much water vapor can be in the air?

5. What is condensation? Explain, on a molecular level, how the condensation of water vapor on a bathroom mirror warms the bathroom.

6. Explain how a fan is able to cool you if it does not provide cooler air.

7. What is the meaning of the statement, "It's not the heat, it's the humidity?"

8. Using your understanding of relative humidity and condensation, think of an explanation for the observation, "It's too cold to snow."

9. Briefly describe, using sketches as needed, how a heat pump is able to move heat from a cooler region to a warmer region.

10. Which provides more cooling for a styrofoam cooler, one with 10 lb of ice at 0° C or one with 10 lb of ice water at 0° C? Explain your reasoning.

11. Explain why a glass filled with a cold beverage seems to "sweat." Would you expect more sweating inside a house during the summer or during the winter? Explain.

12. A diesel engine operates at a higher temperature than a gasoline engine. Why would this fact make a diesel engine more efficient?

13. Which has more entropy—ice, liquid water, or water vapor? Explain your reasoning.

14. Explain why a burn from 100° C steam is more severe than a burn from water at 100° C.

15. Give an example of adding heat to a solid, a liquid, and a gas when no temperature change occurs in spite of the added heat. Explain each example.

16. Using a molecular point of view, describe how mechanical energy is converted into heat by friction.

17. The relative humidity increases most every evening after sunset. Explain how this is possible if no additional water vapor is added to or removed from the air.

18. Should a pot of boiling soup cook faster with a lid or without a lid? Explain.

19. Explain why air released from a high-pressure can is cold.

20. Fans and evaporative coolers work better some days than other days. Offer an explanation for the variable cooling power.

21. Suppose you use a heat engine to do the work to drive a heat pump. Could the heat pump be used to provide the temperature difference to run the heat engine? Explain.

22. Briefly describe why it is considered impossible to have a 100 percent efficient heat engine.

23. What would it mean about the efficiency of an automobile engine if the temperature of the exhaust gases were the same temperature as the air going into the engine? Explain.

24. Describe how the second law of thermodynamics implies a direction to a number of natural processes involving energy.

Exercises

Group A—Solutions Provided in Appendix D

1. How much heat is needed to change 250.0 g of ice at 0° C to water at 0° C?

2. How much heat is needed to change 250.0 g of water at 80.0° C to steam at 100.0° C?

3. A 100.0 g sample of water at 20.0° C is heated in a pressure cooker to steam at 125.0° C. How much heat was absorbed?

4. In an electric freezer, 400.0 g of water at 18.0° C is cooled, frozen, and the ice is chilled to −5.00° C. (a) How much total heat was removed from the water? (b) If the latent heat of vaporization of the Freon refrigerant is 40.0 cal/g, how many grams of Freon must be evaporated to absorb this heat?

Group B—Solutions Not Given

1. A 1.00 kg block of ice at 0° C is added to a picnic cooler. How much heat will the ice remove as it melts to water at 0° C?

2. A 500.0 g pot of water at room temperature (20.0° C) is placed on a stove. How much heat is required to change this water to steam at 100.0° C?

3. Spent steam from a certain electric generating plant leaves the turbines at about 120.0° C and is cooled by water from a cooling tower in a heat exchanger called the condenser. How much heat is removed by the cooling tower water for each kilogram of spent steam?

4. Lead is a soft, dense metal with a specific heat of 0.028 kcal/kg·°C, a melting point of 328.0° C, and a heat of fusion of 5.5 kcal/kg. How much heat must be provided to melt a 250.0 kg sample of lead with a temperature of 20.0° C?

5. An automobile gasoline engine produces a temperature of 4,937° F and the exhaust temperature is 2,832° F. What is the ideal maximum efficiency of this heat engine?

6. A heat engine is supplied with 5.00 kcal as it does 2,092 J of mechanical work. What is the ideal efficiency of this engine?

7. A heat engine is supplied with 300.0 cal and rejects 200.0 cal in the exhaust. (a) How many joules of mechanical work was done? (b) What is the ideal maximum efficiency of this engine?

8. An ocean thermal energy conversion (OTEC) power plant would use the 80.0° F surface water and 40.0° F water from the depths to run a heat engine. What is the ideal maximum efficiency of this OTEC engine?

9. A refrigerator removes 40.0 kcal of heat from the freezer and releases 55.0 kcal through the condenser on the back. How much work was done by the compressor?

10. During the winter, outside air with a temperature of 32° F and a relative humidity of 30 percent is warmed inside a home to 68° F. What is the relative humidity of the air inside the house?

5. A small steam engine has steam going in at 200.0° C and steam leaving at 100.0° C. What is the ideal maximum efficiency of this engine?

6. A heat engine does 8,368 J of work when it is supplied with 10,000.0 cal of heat. What is the ideal maximum efficiency of this engine?

7. A heat engine converts 100.0 cal from a supply of 400.0 cal into work. (a) How much mechanical work was done? (b) What is the ideal maximum efficiency of the engine?

8. An ocean thermal energy conversion (OTEC) power plant uses 80.0° F surface water and 40.0° F water from the depths to run a heat engine. How much would the ideal efficiency of this engine be increased by using solar water heaters to increase the surface water temperature to 100.0° F?

9. A heat pump releases 60.0 kcal as it removes 40.0 kcal at the evaporator coils. How much work does this heat pump ideally accomplish?

10. The outside air temperature is 30.0° C with a relative humidity of 33.33 percent. At what temperature will the air become saturated if no water vapor is added to or removed from the air?

1. $Q = ML_f$ $(250.0 g)(80.0 \, cal/g) = 20,000 \, cal$ $\boxed{20.0 \, kcal}$

2. $m = 250.0 g$ $\quad Q_1 = mc\Delta T = (250.0)(1.0 \, cal/g°C)(20.0°C) = 5000 \, cal$
$L_v = 540.0$ $\hspace{8cm} 5.0 \, kcal$

$Q_2 = ML_v$ b. $(250.0 g)(540.0 \, cal/g)$ $135,000 \, cal$ $= 135.0 \, kcal$

$Q \, total = Q_1 + Q_2 = 5.0 + 135.0 = \boxed{140.0 \, kcal}$

5. $T_K = \frac{5}{9}(T_f - 32) + 273$
$T_H = \frac{5}{9}(4,937 - 32) + 273 = 2,998 \, K =$
$T_L = \frac{5}{9}(2,832 - 32) + 273 = 1829 \, K$

$e = 1 - \frac{T_L}{T_H} \quad \frac{1829}{2998} \times 100\%$
$= 1 - .6100 \times 100\%$
$= .3400 \times 100\% = \boxed{39\%}$

8. $T_H = \frac{5}{9}(80.0 - 32) + 273 = 300 \, K$
$T_L = \frac{5}{9}(40.0 - 32) + 273 = 277 \, K$

$e = 1 - \frac{T_L}{T_H} = \frac{277}{300} \times 100$
$= 1 - .923 \times 100\% = \cancel{.03\%}$
$.077 \times 100\% = \boxed{7.7\%}$

9. $U = J(Q_A - Q_C)$
$= 4,184 \frac{J}{kcal}(55.0 \, kcal - 40 kcal)$
$= 4,184 \times 15 \, kcal = 62,760 \, J = 62.8 \, kJ$

Wave Motion and Sound

Figure 7.1
Vibrations are common in many elastic materials, and you can see and hear the results of many in your surroundings. Other vibrations in your surroundings, such as those involved in heat, electricity, and light, are invisible to the senses.

SOMETIMES you can feel the floor of a building shake for a moment when something heavy is dropped. You can also feel prolonged vibrations in the ground when a nearby train moves by. The floor of a building and the ground are solids that transmit vibrations from a disturbance. Vibrations are common in most solids because the solids are elastic, having a tendency to rebound, or snap back, after a force or an impact deforms them. Usually you cannot see the vibrations in a floor or the ground, but you sense they are there because you can feel them.

There are many examples of vibrations that you can see. You can see the rapid blur of a vibrating guitar string (figure 7.1). You can see the vibrating up-and-down movement of a bounced-upon diving board. Both the vibrating guitar string and the diving board set up a vibrating motion of air that you identify as a sound. You cannot see the vibrating motions of the air, but you sense it is there because you hear sounds.

There are many kinds of vibrations that you cannot see but can sense. Heat, as you have learned, is associated with molecular vibrations that are too rapid and too tiny for your senses to detect other than an increase in temperature. Other invisible vibrations include electrons that vibrate, generating spreading electromagnetic radio waves or visible light. Thus vibrations take place as an observable motion of objects but are also involved in sound, heat, electricity, and light. The vibrations involved in all these phenomena are fundamentally alike in many ways and all involve energy. Therefore, many topics of physical science are concerned with vibrational motion. In this chapter you will learn about the nature of vibrations and how they produce waves in general. These concepts will be applied to sound in this chapter and to electricity, light, and radio waves in later chapters.

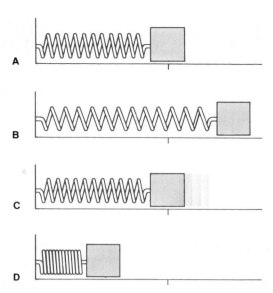

Figure 7.2

A mass on a frictionless surface is at rest at an equilibrium position (a) when undisturbed. When the spring is stretched (b) or compressed (d), then released (c), the mass vibrates back and forth because restoring forces pull opposite to and proportional to the displacement.

Forces and Elastic Materials

An *elastic* material is one that is capable of recovering its shape and form after some force has deformed it. A spring, for example, can be stretched or compressed, two changes that deform the shape of the spring. As you can imagine, there is a direct relationship between the extent of stretching or compression of a spring and the amount of force applied to it. As long as the applied force does not exceed the elastic limit of the spring, the spring always returns to its original shape when the applied force is removed. There are three important considerations about the applied force and deformation relationship: (1) the greater the applied force, the greater the compression or stretch of the spring from its original shape, (2) the spring appears to have an *internal restoring force,* which returns it to its original shape, and (3) the farther the spring is pushed or pulled the *stronger* the restoring force that returns the spring to its original shape.

Forces and Vibrations

A **vibration** is a back-and-forth motion that repeats itself. Almost any solid can be made to vibrate if it is elastic. To see how forces are involved in vibrations, consider the spring and mass in figure 7.2. The spring and mass are arranged so that the mass can

freely move back and forth on a frictionless surface. When the mass has not been disturbed, it is at rest at an *equilibrium position* (figure 7.2a). At the equilibrium position the spring is not compressed or stretched so it applies no force on the mass. If, however, the mass is pulled to the right (figure 7.2b) the spring is stretched and applies a restoring force on the mass towards the left. The farther the mass is displaced the greater the stretch of the spring and thus the greater the restoring force. The restoring force is proportional to the displacement and is in the opposite direction of the applied force.

If the mass is now released, the restoring force is the only force acting (horizontally) on the mass so it accelerates back toward the equilibrium position. This force will continuously decrease until the moving mass arrives back at the equilibrium position, where the force is zero. But the mass will have a maximum velocity when it arrives, so it overshoots the equilibrium position and continues moving to the left (figure 7.2c). As it moves to the left of the equilibrium position it compresses the spring, which exerts an increasing force on the mass. The moving mass comes to a temporary halt, but now the restoring force again starts it moving back toward the equilibrium position. The whole process repeats itself again and again as the mass moves back and forth over the same path.

The periodic vibration, or oscillation, of the mass is similar to many vibrational motions found in nature called **simple harmonic motion.** Simple harmonic motion is defined as the vibratory motion that occurs when there is a restoring force opposite to and proportional to a displacement.

Wave Motion and Sound

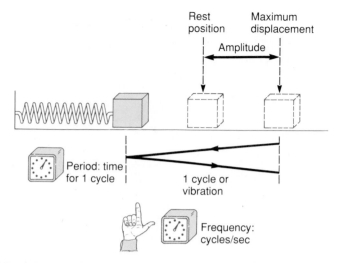

Figure 7.3

A vibrating mass attached to a spring is displaced from the rest, or equilibrium position, then released. The maximum displacement is called the amplitude of the vibration. A cycle is one complete vibration. The period is the time required for one complete cycle. The frequency is how many cycles it completes in 1 sec.

The vibrating mass and spring system will continue to vibrate for a while, slowly decreasing with time until the vibrations stop completely. The slowing and stopping is due to air resistance and internal friction. If these could be eliminated or compensated for with additional energy, the mass would continue to vibrate with a repeating, or *periodic,* motion.

Describing Vibrations

A vibrating mass is described by measuring several variables (figure 7.3). The extent of displacement from the equilibrium position is called the **amplitude.** A vibration that has a mass displaced a greater distance from equilibrium thus has a greater amplitude than a vibration with less displacement.

A complete vibration is called a **cycle.** A cycle is the movement from some point, say the far left, all the way to the far right, and back to the same point again, the far left in this example. The **period** (T) is simply the time required to complete one cycle. For example, suppose 0.1 sec is required for an object to move through one complete cycle, to complete the back-and-forth motion from one point, then back to that point. The period of this vibration is 0.1 sec.

Sometimes it is useful to know how frequently a vibration completes a cycle every second. The number of cycles per second is called the **frequency** (f). For example, a vibrating object moves through 10 cycles in 1 sec. The frequency of this vibration is 10 cycles per second. Frequency is measured in a unit called a **hertz** (hz). The unit for a hertz is 1/sec since a cycle does not have dimensions. Thus a frequency of 10 cycles per second is referred to as 10 hertz or 10 1/sec.

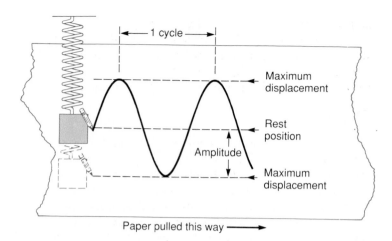

Figure 7.4

A graph of simple harmonic motion is described by a sinusoidal curve.

The period and frequency are two ways of describing the time involved in a vibration. Since the period (T) is the total time involved in one cycle and the frequency (f) is the number of cycles per second, the relationship is

$$T = \frac{1}{f}$$

equation 7.1

or

$$f = \frac{1}{T}$$

equation 7.2

Example 7.1

A vibrating system has a period of 0.1 sec. What is the frequency in hz?

Solution

$$T = 0.1 \text{ sec}$$
$$f = ?$$

$$f = \frac{1}{T}$$

$$= \frac{1}{0.1 \text{ sec}}$$

$$= \frac{1}{0.1}\frac{1}{\text{sec}}$$

$$= 10\,\frac{1}{\text{sec}}$$

$$= \boxed{10 \text{ hz}}$$

You can obtain a graph of a vibrating object, which makes it easier to measure the amplitude, period, and frequency. If a pen is fixed to a vibrating mass and a paper is moved beneath it at a steady rate, it will draw a curve as shown in figure 7.4. The greater the amplitude of the vibrating mass the greater the height of this curve. The greater the frequency, the closer together the peaks and valleys. Note the shape of this curve. This shape is characteristic of simple harmonic motion and is called a *sinusoidal*, or sine, graph. It is so named because it is the same shape as a graph of the sine function in trigonometry.

Waves

A vibration is a repeating, or *periodic*, type of motion that can disturb the surroundings. A *pulse* is a disturbance of a single event of short duration. Both pulses and periodic vibrations can create a physical **wave** in the surroundings. A wave is a disturbance that moves through a medium such as a solid or the air. A heavy object dropped on the floor, for example, makes a pulse that sends a mechanical wave that you feel. It might also make a sound wave in the air that you hear. In either case, the medium that transported a wave (solid floor or air) returns to its normal state after the wave has passed. The medium does not travel from place to place, the wave does. Two major considerations about a wave are that (1) a wave is a traveling disturbance and (2) a wave transports energy.

You can observe waves when you drop a rock into a still pool of water. The rock pushes the water into a circular mound as it enters the water. Since it is forcing the water through a distance, it is doing work to make the mound. The mound starts to move out in all directions, in a circle, leaving a depression behind. Water moves into the depression and a circular wave—mound and depression—moves from the place of disturbance outward. Any floating object in the path of the wave, such as a leaf, exhibits an up-and-down motion as the mound and depression of the wave passes. But the leaf merely bobs up and down and after the wave has passed it is much in the same place as before the wave. Thus it was the disturbance that traveled across the water, not the water itself. If the wave reaches a leaf floating near the edge of the water, it may push the leaf up and out of the water, doing work on the leaf. Thus the wave is a moving disturbance that transfers energy from one place to another.

Kinds of Waves

If you could see the motion of an individual water molecule near the surface as a water wave passed, you would see it trace out a circular path as it moves up and over, down and back. This circular motion is characteristic of the motion of a particle reacting to a water wave disturbance. There are other kinds of waves, and each involves particles in a characteristic motion.

A **longitudinal wave** is a disturbance that causes particles to move closer together or farther apart in the same direction that the wave is moving. If you attach one end of a coiled spring to a wall then pull it tight, you will make longitudinal waves in the spring if you grasp the spring, then move your hand back and forth parallel to the spring. Each time you move your hand toward the length of the spring a pulse of closer together coils will move across the spring. Each time you pull your hand back

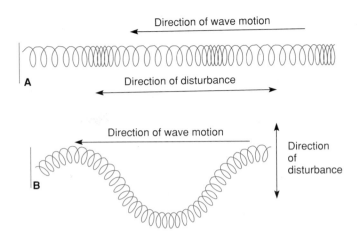

Figure 7.5

(*a*) Longitudinal waves are created in a spring when the free end is moved back and forth parallel to the spring. (*b*) Transverse waves are created in a spring when the free end is moved up and down.

a pulse of farther apart coils will move across the spring. The coils move back and forth in the same direction that the wave is moving, which is the characteristic movement in reaction to a longitudinal wave.

You will make a different kind of wave in the stretched spring if you now move your hand up and down perpendicular to the length of the spring. This creates a **transverse wave.** A transverse wave is a disturbance that causes motion perpendicular to the direction that the wave is moving. Particles responding to a transverse wave do not move closer together or farther apart in response to the disturbance, but rather they vibrate back and forth or up and down in a direction perpendicular to the direction of the wave motion (figure 7.5).

Whether you make longitudinal or transverse waves depends on the nature of the disturbance creating the waves but also on the nature of the medium. Transverse waves can move through a material only if there is some interaction, or attachment, between the molecules making up the medium. In a gas, for example, the molecules move about freely without attachments to one another. A pulse can cause these molecules to move closer together or farther apart, so a gas can carry a longitudinal wave. But if a gas molecule is caused to move up or down, there is no reason for other molecules to do the same since they are not attached. Thus a gas will carry longitudinal waves but not transverse waves. Likewise a liquid will carry longitudinal waves but not transverse waves since the liquid molecules simply slide past one another. The surface of a liquid, however, is another story because of surface tension. A surface water wave is, in fact, a combination of longitudinal and transverse wave patterns that produce the circular motion of a disturbed particle. Solids can and do carry both longitudinal and transverse waves because of the strong attachments between the molecules.

Wave Motion and Sound

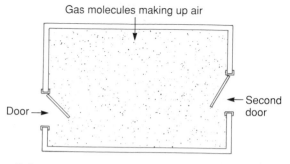

Figure 7.6

When you open one door into this room, the other door closes. Why does this happen? The answer is that the first door creates a pulse of compression that moves through the air like a sound wave. The pulse of compression pushes on the second door, closing it.

Waves in Air

Waves that move through the air are longitudinal, so sound waves must be an example of a longitudinal wave. A familiar situation will be used to describe the nature of a longitudinal wave moving through air before considering sound specifically. The situation concerns a small room with no open windows and two doors that open into the room (figure 7.6). When you open one door into the room the other door closes. Why does this happen? According to the kinetic molecular theory, the room contains many tiny randomly moving gas molecules that make up the air. As you opened the door, it pushed on these gas molecules creating a "jammed together" zone of molecules immediately adjacent to the door. This "jammed together" zone of air now has a greater density and pressure, which immediately spreads outward from the door as a pulse. The disturbance is rapidly passed from molecule to molecule and the pulse of compression spreads through the room. It is not unlike a pulse of movement that can sometimes be seen in a swarm of flying insects. A momentary disturbance will cause nearby individual insects to momentarily move away from the disturbance and toward another flying insect, then back toward their original position. The movement is passed on through the swarm. During the movement of the disturbance, each insect maintains its own random motion. The overall effect, however, is that of a pulse moving through the swarm. In the example of the closing door, the pulse of greater density and pressure of air reached the door at the other side of the room, and the composite effect of the molecules impacting with the door, that is, the increased pressure, caused it to close.

If the door at the other side of the room does not latch, you can probably cause it to open again by pulling on the first door quickly. By so doing, you send a pulse of "thinned out" molecules of lowered density and pressure. The door you pulled quickly pushed some of the molecules out of the room. Other molecules quickly move into the region of less pressure, then back to their normal positions. The overall effect is the movement of a "thinned out" pulse that travels through the room.

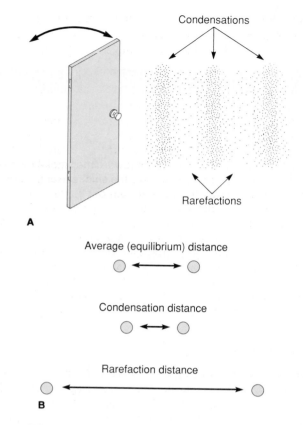

Figure 7.7

(a) Swinging the door inward produces pulses of increased density and pressure called condensations. Pulling the door outward produces pulses of decreased density and pressure called rarefactions. (b) In a condensation, the average distance between gas molecules is momentarily decreased as the pulse passes. In a rarefaction, the average distance is momentarily increased.

When the pulse of slightly reduced pressure reaches the other door, molecules exerting their normal pressure on the other side of the door cause it to move. After a pulse has passed a particular place, the molecules are very soon homogeneously distributed again due to their rapid, random movement.

If you were to swing a door back and forth it would be a vibrating object. As it vibrates back and forth it would have a certain frequency in terms of the number of vibrations per second. As the vibrating door moves toward the room, it creates a pulse of "jammed together" molecules called a **condensation** (or compression) that quickly moves throughout the room. As the vibrating door moves away from the room a pulse of "thinned out" molecules called a **rarefaction** quickly moves throughout the room. The vibrating door sends repeating pulses of condensation (increased density and pressure) and rarefaction (decreased density and pressure) through the room as it moves back and forth (figure 7.7). You know that the pulses transmit energy because they produce movement, or do work on, the other door.

Individual molecules execute a harmonic motion about their equilibrium position and can do work on a moveable object. Energy is thus transferred by this example of longitudinal waves.

In a very still room with no air movement whatsoever, place a smoking incense, punk, or appropriate smoke source in an ash tray on a table. It should make a thin stream of smoke that moves straight up. Hold one hand flat, fingers together, and parallel to the stream of smoke. Quickly move it toward the smoke for a very short distance as if pushing air toward the smoke. Repeat but this time pull it quickly away from the stream of smoke. You should be able to see the smoke stream move away from, then towards your hand. What is the maximum distance from the smoke that you can still make the smoke stream move? There are at least two explanations for the movement of the smoke stream: (1) pulses of condensation and rarefaction or (2) movement of a mass of air such as occurs when you splash water. How can you prove one explanation or the other to be correct without a doubt?

Hearing Waves in Air

You cannot hear a vibrating door because the human ear normally hears sounds originating from vibrating objects with a frequency between 20 to 20,000 hz. Longitudinal waves with frequencies less than 20 hz are called **infrasonic.** You usually *feel* sounds below 20 hz rather than hear them, particularly if you are listening to a good sound system. Longitudinal waves above 20,000 hz are called **ultrasonic.** Although 20,000 hz is usually considered the upper limit of hearing, the actual limit varies from person to person and becomes lower and lower with increasing age. Humans do not hear infrasonic nor ultrasonic sounds, but various animals have different limits. Dogs, cats, rats, and bats can hear higher frequencies than humans. Dogs can hear an ultrasonic whistle when a human hears nothing, for example. Some bats make and hear sounds of frequencies up to 100,000 hz as they navigate and search for flying insects in total darkness.

A tuning fork that vibrates at 260 hz makes longitudinal waves much like the swinging door, but these longitudinal waves are called *sound waves* because they are within the frequency range of human hearing. The prongs of a struck tuning fork vibrate, moving back and forth. This is more readily observed if the prongs of the fork are struck, then held against a sheet of paper or plunged into a beaker of water. In air, the vibrating prongs first move toward you, pushing the air molecules into a condensation of increased density and pressure. As the prongs then move back, a rarefaction of decreased density and pressure is produced. The alternation of increased and decreased pressure pulses moves from the vibrating tuning fork and spreads outward in all directions, much like the surface of a rapidly expanding balloon (figure 7.8). When the pulses reach your eardrum, it is forced in and out by the pulses. It now vibrates with

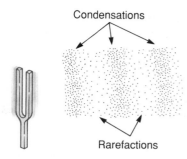

Figure 7.8

A vibrating tuning fork produces a series of condensations and rarefactions that move away from the tuning fork. The pulses of increased and decreased pressure reach your ear, vibrating the eardrum. The ear sends nerve signals to the brain about the vibrations and the brain interprets the signals as sounds.

the same frequency as the tuning fork. The vibrations of the eardrum are transferred by three tiny bones to a fluid in a coiled chamber. Here, tiny hairs respond to the frequency and size of the disturbance, activating nerves that transmit the information to the brain. The brain interprets a frequency as a sound with a certain **pitch.** High-frequency sounds are interpreted as high-pitched musical notes, for example, and low-frequency sounds are interpreted as low-pitched musical notes. The brain then selects certain sounds from all you hear and you "tune" to certain ones, enabling you to listen to whatever sounds you want while ignoring the background noise, which is made up of all the other sounds.

Wave Terms

A tuning fork vibrates with a certain frequency and amplitude, producing a longitudinal wave of alternating pulses of increased pressure condensations and reduced pressure rarefactions. A graph of the frequency and amplitude of the vibrations is shown in figure 7.9a, and a representation of the condensations and rarefactions is shown in figure 7.9b. The wave pattern can also be represented by a graph of the changing air pressure of the traveling sound wave as shown in figure 7.9c. This graph can be used to define some interesting concepts associated with sound waves. Note the correspondence between the (1) amplitude, or displacement, of the vibrating prong, (2) the pulses of condensations and rarefactions, and (3) the changing air pressure. Note also the correspondence between the frequency of the vibrating prong and the frequency of the wave cycles.

Figure 7.10 shows the terms commonly associated with waves from a continuously vibrating source. The wave *crest* is the maximum disturbance from the undisturbed (rest) position. For a sound wave, this would represent the maximum increase of air pressure. The wave *trough* is the maximum disturbance in the opposite direction from the rest position. For a sound wave, this would represent the maximum decrease of air pressure. The *amplitude* of a wave is the displacement from rest to the crest

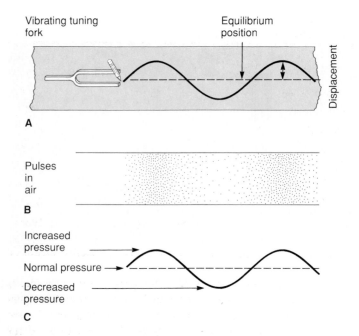

A

B

Increased
pressure

Normal pressure →

Decreased
pressure

C

Figure 7.9

Compare the (a) back-and-forth vibrations of a tuning fork with (b) the resulting condensations and rarefactions that move through the air and (c) the resulting increases and decreases of air pressure on a surface that intercepts the condensations and rarefactions.

or from rest to the trough. The time required for a wave to repeat itself is the *period* (T). To repeat itself means the time required to move through one full wave, such as from the crest of one wave to the crest of the next wave. This length in which the wave repeats itself is called the **wavelength** (the symbol is λ, which is the Greek letter lambda). Wavelength is measured in centimeters or meters just like any other length.

There is a relationship between the wavelength, period, and speed of a wave. Recall that speed is

$$v = \frac{\text{distance}}{\text{time}}$$

Since it takes one period (T) for a wave to move one wavelength (λ), then the speed of a wave can be measured from

$$v = \frac{\text{one wavelength}}{\text{one period}} = \frac{\lambda}{T}$$

The *frequency,* however, is more convenient than the period for dealing with waves that repeat themselves rapidly. Recall the relationship between frequency (f) and the period (T) is

$$f = \frac{1}{T}$$

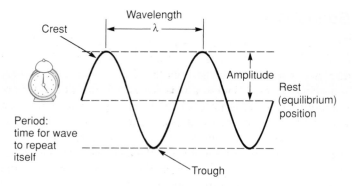

Figure 7.10

Here are some terms associated with periodic waves. The wavelength is the distance from a part of one wave to the same part in the next wave, such as from one crest to the next. The amplitude is the displacement from the rest position. The period is the time required for a wave to repeat itself, that is, the time for one complete wavelength to move past a given location.

Substituting f for 1/T yields

$$v = \lambda f \qquad\qquad \textbf{equation 7.3}$$

which is known as the **wave equation.** This equation tells you that the velocity of a wave can be obtained from the product of the wavelength and the frequency. Note that it also tells you that the wavelength and frequency are inversely proportional at a given velocity.

Example 7.2

A sound wave with a frequency of 260.0 hz has a wavelength of 1.27 m. With what speed would you expect this sound wave to move?

Solution

$$f = 260.0 \text{ hz} \qquad v = \lambda f$$

$$\lambda = 1.27 \text{ m}$$

$$v = ? \qquad\qquad = (1.27 \text{ m})\left(260.0\frac{1}{\text{sec}}\right)$$

$$= 1.27 \times 260.0 \text{ m} \times \frac{1}{\text{sec}}$$

$$= \boxed{330.\frac{\text{m}}{\text{sec}}}$$

Example 7.3

In general, the human ear is most sensitive to sounds at 2,500. hz. Assuming sound to move at 330. m/sec, what is the wavelength of sounds to which people are most sensitive? (Answer: 13.2 cm)

Sound Waves

The transmission of a sound wave requires a medium, that is, a solid, liquid, or gas to carry the disturbance. Therefore sound does not travel through the vacuum of outer space since there is nothing to carry the vibrations from a source. The nature of the molecules making up a solid, liquid, or gas determines how well or how rapidly the substance will carry sound waves. The two variables, (1) the inertia of the molecules and (2) the strength of the interaction of the molecules, are attached to one another. Thus hydrogen gas, with the least massive molecules with no interaction or attachments, will carry a sound wave at 1,284 m/sec (4,213 ft/sec) when the temperature is 0° C. More massive helium gas molecules have more inertia and carry a sound wave at only 965 m/sec (3,166 ft/sec) at the same temperature. A solid, however, has molecules that are strongly attached so vibrations are passed rapidly from molecule to molecule. Steel, for example, is highly elastic and sound will move through a steel rail at 5,940 m/sec (19,488 ft/sec). Thus there is a reason for the old saying, "Keep your ear to the ground" because sounds move through solids more rapidly than through a gas (table 7.1).

Velocity of Sound in Air

Most people have observed that sound takes some period of time to move through the air. If you watch a person hammering on a roof a block away, the sounds of the hammering are not in sync with what you see. Light travels so rapidly that you can consider what you see to be instantaneous with what is actually happening for all practical purposes. Sound, however, travels much more slowly and the sounds arrive late for what you are seeing. This is dramatically illustrated by seeing a flash of lightning, then hearing thunder seconds later. Perhaps you know of a way to estimate the distance to a lightning flash by timing the time interval between the flash and boom. You will learn a precise way to measure this distance shortly.

The air temperature influences how rapidly sound moves through the air. The gas molecules in warmer air have a greater kinetic energy than those of cooler air. The molecules of warmer air therefore transmit an impulse from molecule to molecule more rapidly. More precisely, the speed of a sound wave increases 0.60 m/sec (2.0 ft/sec) for *each* Celsius degree increase in temperature. In *dry* air at sea-level density (normal pressure) and 0° C (32° F), the velocity of sound is about 331 m/sec (1,087 ft/sec). Therefore the velocity of sound at different temperatures can be calculated from the following relationships:

$$v_{T_p} \text{ (m/sec)} = v_0 + \left(\frac{0.60 \text{ m/sec}}{°C}\right)(T_p) \quad \textbf{equation 7.4}$$

Table 7.1
Speed of sound in various materials

Medium	m/sec	ft/sec
Carbon dioxide (0° C)	259	850
Dry air (0° C)	331	1,087
Helium (0° C)	965	3,166
Hydrogen (0° C)	1,284	4,213
Water (25° C)	1,497	4,911
Seawater (25° C)	1,530	5,023
Lead	1,960	6,430
Glass	5,100	16,732
Steel	5,940	19,488

where v_{T_p} is the velocity of sound at the present temperature, v_0 is the velocity of sound at 0° C, and T_p is the present temperature. This equation tells you that the velocity of a sound wave increases 0.6 m/sec for each °C above 0° C. For units of ft/sec,

$$v_{T_p} \text{ (ft/sec)} = v_0 + \left(\frac{2.0 \text{ ft/sec}}{°C}\right)(T_p) \quad \textbf{equation 7.5}$$

Equation 7.5 tells you that the velocity of a sound wave increases 2.0 ft/sec for each degree Celsius above 0° C.

Example 7.4

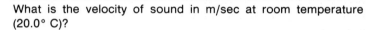

What is the velocity of sound in m/sec at room temperature (20.0° C)?

Solution
$v_0 = 331$ m/sec

$T_p = 20.0°$ C

$v_{T_p} = ?$

$$v_{T_p} = v_0 + \left(\frac{0.60 \text{ m/sec}}{°C}\right)(T_p)$$

$$= 331 \text{ m/sec} + \left(\frac{0.6 \text{ m/sec}}{°C}\right)(20.0° C)$$

$$= 331 + (0.60 \times 20.0) \text{ m/sec} + \frac{\text{m/sec}}{°C} \times °C$$

$$= 331 + 12 \text{ m/sec} + \frac{\text{m/sec}/°\cancel{C}}{°\cancel{C}}$$

$$= \boxed{343 \text{ m/sec}}$$

Example 7.5

The air temperature is 86.0° F. What is the velocity of sound in ft/sec? (Note that °F must be converted to °C for equation 7.5.) (Answer: 1,147 ft/sec)

Wave Motion and Sound

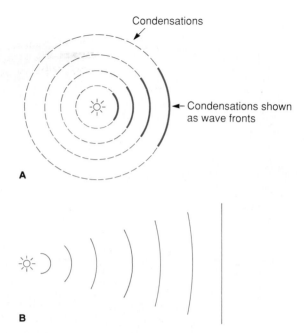

Figure 7.11

(*a*) Spherical waves move outward from a sounding source much as a rapidly expanding balloon. This two-dimensional sketch shows the repeating condensations as spherical wave fronts. (*b*) Some distance from the source, a spherical wave front is considered a linear, or plane, wave front.

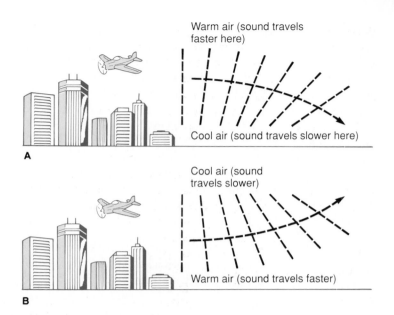

Figure 7.12

(*a*) Since sound travels faster in warmer air, a wave front becomes bent, or refracted, toward the earth's surface when the air is cooler near the surface. (*b*) When the air is warmer near the surface, a wave front is refracted upward, away from the surface.

Refraction and Reflection

When you drop a rock into a still pool of water circular patterns of waves move out from the disturbance. These water waves are on a flat, two-dimensional surface. Sound waves, however, move in three-dimensional space like a rapidly expanding balloon. Sound waves are *spherical waves* that move outward from the source. Spherical waves of sound move as condensations and rarefactions from a continuously vibrating source at the center. If you identify the same part of each wave in the spherical waves, you have identified a **wave front.** For example, the crests of each condensation could be considered as a wave front. From one wave front to the next, therefore, identifies one complete wave or wavelength. At some distance from the source a small part of a spherical wave front can be considered a *linear wave front* (figure 7.11).

Waves move within a homogeneous medium such as a gas or a solid at a fairly constant rate but gradually lose energy to friction. When a wave encounters a different condition, however, drastic changes may occur rapidly. The division between two physical conditions is called a **boundary.** Boundaries are usually encountered (1) between different materials or (2) between the same materials with different conditions. An example of a wave moving between different materials is a sound made in the next room that moves through the air to the wall

and through the wall to the air in the room where you are. The boundaries are air-wall and wall-air. If you have ever been in a room with "thin walls" it is obvious that sound moved through the wall and air boundaries.

An example of sound waves moving through the same material with different conditions is found when a wave front moves through air of different temperatures. Since sound travels faster in warm air than in cold air the wave front becomes bent. The bending of a wave front between boundaries is called **refraction.** Refraction changes the direction of travel of a wave front. Consider, for example, that on calm, clear nights the air near the earth's surface is cooler than air further above the surface. Air at rooftop height above the surface might be four or five degrees warmer under such ideal conditions. Sound will travel faster in the higher, warmer air than it will in the lower, cooler air close to the surface. A wave front will therefore become bent, or refracted, toward the ground on a cool night and you will be able to hear sounds from farther away than on warm nights (figure 7.12a). The opposite process occurs during the day as the earth's surface becomes warmer from sunlight (figure 7.12b). Wave fronts are refracted upward because part of the wave front travels faster in the warmer air near the surface. Thus sound does not seem to carry as far in the summer as it does in the winter. What is actually happening is that during the summer the wave fronts are refracted away from the ground before they travel very far.

When a wave front strikes a boundary that is parallel to the front the wave may be absorbed, transmitted, or undergo **reflection** depending on the nature of the boundary medium. Or, the wave may be partly absorbed, partly transmitted, and partly reflected or any combination thereof. Some materials, such as hard, smooth surfaces, reflect sound waves more than they are absorbed. Other materials, such as soft, ruffly curtains, absorb sound waves more than they are reflected. If you have ever been in a room with smooth, hard walls and with no curtains, carpets, or furniture, you know that sound waves may be reflected several times before they are finally absorbed. Sounds seem to increase in volume because of all the reflections and the apparent increase in volume due to **reverberation.** Reverberation is the mixing of sound with reflections and is one of the factors that determines the acoustical qualities of a room, lecture hall, or auditorium.

If a reflected sound arrives after 0.10 sec, the human ear can detect the reflected sound distinct from the original sound. Such a reflected sound that can be distinguished from the original is called an **echo.** Thus a reflected sound that arrives before 0.10 sec is perceived as an increase in volume and is called a reverberation, but a sound that arrives after 0.10 sec is perceived as an echo.

Example 7.6

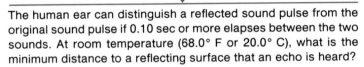

The human ear can distinguish a reflected sound pulse from the original sound pulse if 0.10 sec or more elapses between the two sounds. At room temperature (68.0° F or 20.0° C), what is the minimum distance to a reflecting surface that an echo is heard?

Solution

t = 0.10 sec
(minimum)

$$v = \frac{d}{t} \therefore d = vt$$

v = 343 m/sec
(from example
7.4)

$$= \left(343 \frac{m}{sec}\right)(0.10 \ sec)$$

d = ?

$$= 343 \times 0.10 \frac{m}{sec} \times sec$$

$$= 34.3 \frac{m \cdot sec}{sec}$$

$$= 34 \ m$$

Since the sound pulse must travel from the source to the reflecting surface, then back to the source,

$$34 \ m \times 1/2 = \boxed{17 \ m}$$

The minimum distance to a reflecting surface to hear an echo when the air is at room temperature is therefore 17 m (about 56 ft).

Example 7.7

An echo is heard exactly 1.00 sec after a sound when the air temperature is 30.0° C. How many feet away is the reflecting surface? (Answer: 574 ft)

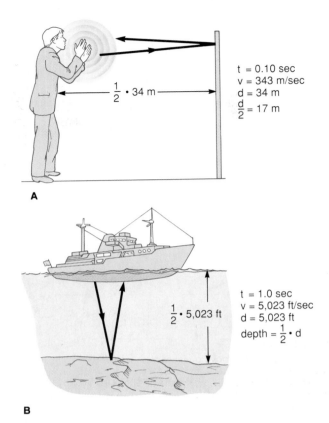

Figure 7.13

(a) At room temperature sound travels at 343 m/sec. In 0.10 sec, sound would travel 34 m. Since the sound must travel to a surface and back in order for you to hear an echo, the distance to the surface is one half the total distance. (b) Sonar measures a depth by measuring the elapsed time between an ultrasonic sound pulse and the echo. The depth is one half the round trip.

Sound wave echoes are measured to determine the depth of water or to locate underwater objects by a *sonar* device. The word "sonar" is taken from *sound navigation ranging.* The device generates an underwater ultrasonic sound pulse, then measures the elapsed time for the returning echo. Sound waves travel at about 1,531 m/sec (5,023 ft/sec) in seawater at 25° C (77° F). A 1 sec lapse between the ping of the generated sound and the echo return would mean that the sound traveled 5,023 ft for the round trip. The bottom would be half this distance below the surface (figure 7.13).

Interference

Waves interact with a boundary much as a particle would, reflecting or refracting because of the boundary. A moving ball, for example, will bounce from a surface at the same angle it strikes the surface just as a wave does. A particle or a ball, however, can be only in one place at a time, but waves can be spread over a distance at the same time. You know this since many different people in different places can hear the same sound at the same time.

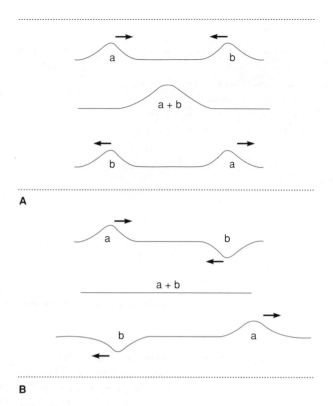

A

B

Figure 7.14

(*a*) Constructive interference occurs when two equal in-phase waves meet. (*b*) Destructive interference occurs when two equal out-of-phase waves meet. In both cases, the wave displacements are superimposed when they meet, but they then pass through one another and return to their original amplitudes.

Another difference between waves and particles is that two or more waves can exist in the same place at the same time. When two patterns of waves meet, they pass through each other without refracting or reflecting. However, at the place where they meet the waves interfere with each other, producing a *new* disturbance. This new disturbance has a different amplitude, which is the algebraic sum of the amplitudes of the two separate wave patterns. If the wave crests or wave troughs arrive at the same place at the same time, the two waves are said to be *in phase*. The result of two waves arriving in phase is a new disturbance with a crest and trough that has greater displacement than either of the two separate waves. This is called **constructive interference** (figure 7.14a). If the trough of one wave arrives at the same place and time as the crest of another wave, the waves are completely *out of phase*. When two waves are completely out of phase, the crest of one wave (positive displacement) will cancel the trough of the other wave (negative displacement) and the result is zero total disturbance, or no wave. This is called **destructive interference** (figure 7.14b). If the two sets of wave patterns do not have the exact same amplitudes or wavelengths, they will be neither completely in phase nor completely out of phase. The result will be partly constructive or destructive interference, depending on the exact nature of the two wave patterns.

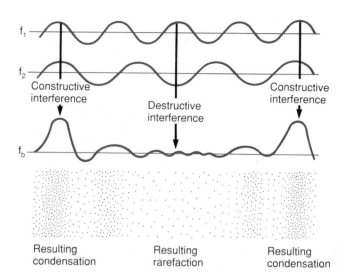

Figure 7.15

Two waves of equal amplitude but slightly different frequencies interfere destructively and constructively. The result is an alternation of loudness called a beat.

Suppose that two vibrating sources produce sounds that are equal in amplitude and equal in frequency. The resulting sound will be increased in volume because of constructive interference. But suppose the two sources are slightly different in frequency, for example, 350 and 352 hz. You will hear a regularly spaced increase and decrease of sound known as **beats.** Beats occur because the two sound waves experience alternating constructive and destructive interferences (figure 7.15). The phase relationship changes because of the difference in frequency as you can see in the illustration. These alternating constructive and destructive interference zones are moving from the source to the receiver, and the receiver hears the results as a rapidly rising and falling sound level. The beat frequency is the difference between the frequencies of the two sources. A 352 hz source and 350 hz source sounded together would result in a beat frequency of 2 hz. Thus, the frequencies are closer and closer together, and fewer beats will be heard per second. You may be familiar with the phenomenon of beats if you have ever flown in an airplane with two engines. If one engine is running slightly faster than the other, you hear a slow beat. The same phenomena can sometimes be heard from two or more engines of a diesel locomotive and from two snow tires on a car driving down a highway. The beat frequency (f_b) is equal to the absolute difference in frequency of two interfering waves with slightly different frequencies, or

$$f_b = f_2 - f_1 \qquad \text{equation 7.6}$$

Energy and Sound

All waves involve the transportation of energy, including sound waves. The vibrating mass and spring in figure 7.2 vibrates with an amplitude that depends on how much work you did on the mass in moving it from its equilibrium position. More work on

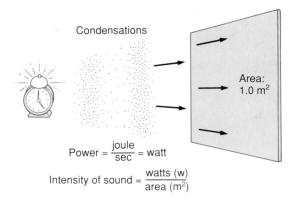

Condensations

Area: 1.0 m²

$$\text{Power} = \frac{\text{joule}}{\text{sec}} = \text{watt}$$

$$\text{Intensity of sound} = \frac{\text{watts (w)}}{\text{area (m}^2)}$$

Figure 7.16

The intensity, or energy, of a sound wave is the rate of energy transferred to an area perpendicular to the waves. Intensity is measured in watts per square meter, w/m².

Table 7.2

Comparison of noise levels in decibels with intensity

Example	Response	Decibels	Intensity
Least Needed for Hearing	Barely perceived	0	1×10^{-12}
Calm Day in Woods	Very, very quiet	10	1×10^{-11}
Whisper (15 ft)	Very quiet	20	1×10^{-10}
Library	Quiet	40	1×10^{-8}
Talking	Easy to hear	65	3×10^{-6}
Heavy Street Traffic	Conversation difficult	70	1×10^{-5}
Pneumatic Drill (50 ft)	Very loud	95	3×10^{-3}
Jet Plane (200 ft)	Discomfort	120	1

the mass results in a greater displacement and a greater amplitude of vibration. A vibrating object that is producing sound waves will produce more intense condensations and rarefactions if it has a greater amplitude. The **intensity** of a sound wave is a measure of the energy the sound wave is carrying. Intensity is directly proportional to the square of the amplitude. It is defined as the power (in watts) transmitted by a wave to a unit area (in square meters) that is perpendicular to the waves. Intensity is therefore measured in watts per square meter (w/m²) (figure 7.16).

Loudness

The **loudness** of a sound is a subjective interpretation that varies from person to person. Loudness is also related to (1) the energy of a vibrating object, (2) the condition of the air the sound wave travels through, and (3) the distance between you and the vibrating source. Furthermore, doubling the amplitude of the vibrating source will quadruple the *intensity* of the resulting sound wave, but the sound will not be perceived as four times as loud. The relationship between perceived loudness and the intensity of a sound wave is not a linear relationship. In fact, a sound that is twice as loud requires ten times the intensity, and quadrupling the loudness requires a one-hundred-fold increase in intensity.

The human ear is very sensitive, capable of hearing sounds with intensities as low as 10^{-12} w/m² and is not made uncomfortable by sound until the intensity reaches about 1 w/m². The second intensity is a million million (10^{12}) times greater than the first. Within this range the subjective interpretation of loudness seems to vary by powers of ten. This observation led to the development of the **decibel scale** to measure relative loudness. The scale is a ratio of the intensity level of a given sound to the threshold of hearing, which is defined as 10^{-12} w/m² at 1,000 hz. In keeping with the power-of-ten subjective interpretations of loudness, a logarithmic scale is used rather than a linear scale.

Originally the scale was the logarithm of the ratio of the intensity level of a sound to the threshold of hearing. This definition set the zero point at the threshold of human hearing. The unit was named the *bel* in honor of Alexander Graham Bell. This unit was too large to be practical, so it was reduced by one-tenth and called a *decibel*. The intensity level of a sound is therefore measured in decibels (table 7.2). Compare the decibel noise level of familiar sounds listed in table 7.2 and note that each increase of ten on the decibel scale is matched by a *multiple* of ten on the intensity level. For example, moving from a decibel level of ten to a decibel level of twenty requires *ten times* more intensity. Likewise, moving from a decibel level of twenty to forty requires a one-hundred-fold increase in the the intensity level. As you can see, the decibel scale is not a simple linear scale.

Resonance

You know that sound waves transmit energy when you hear a thunderclap rattle the windows. In fact, the sharp sounds from an explosion have been known not only to rattle but also break windows. The source of the energy is obvious when thunderclaps or explosions are involved. But sometimes energy transfer occurs through sound waves when it is not clear what is happening. A truck drives down the street, for example, and one window rattles but the others do not. A singer shatters a crystal water glass by singing a single note, but other objects remain undisturbed. A closer look at the nature of vibrating objects and the transfer of energy will explain these phenomena.

Almost any elastic object can be made to vibrate and will vibrate freely at a constant frequency after being sufficiently disturbed. Entertainers sometimes discover this fact and appear on late-night talk shows playing saws, wrenches, and other odd objects as musical instruments. All material objects have a **natural frequency** of vibration determined by the materials and shape of the objects. The natural frequencies of different

Wave Motion and Sound

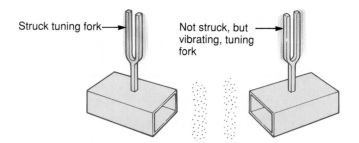

Struck tuning fork →

Not struck, but vibrating, tuning fork →

Figure 7.17

When the frequency of an applied force, including the force of a sound wave, matches the natural frequency of an object, energy is transferred very efficiently. The condition is called resonance.

wrenches enable an entertainer to use the suspended tools as if they were the bars of a xylophone.

If you have ever pumped a swing, you know that small forces can be applied at any frequency. If the frequency of the applied forces matches the natural frequency of the moving swing, there is a dramatic increase in amplitude. When the two frequencies match energy is transferred very efficiently. This condition, when the frequency of an external force matches the natural frequency, is called **resonance.** The natural frequency of an object is thus referred to as the *resonant frequency,* that is, the frequency at which resonance occurs.

A silent tuning fork will resonate if a second tuning fork with the same frequency is struck and vibrates nearby (figure 7.17). You will hear the previously silent tuning fork sounding if you stop the vibrations of the struck fork by touching it. The waves of condensations and rarefactions produced by the struck tuning fork produce a regular series of impulses that match the natural frequency of the silent tuning fork. This illustrates that at resonance, relatively little energy is required to start vibrations.

A truck causing vibrations as it is driven past a building may cause one window to rattle while others do not. Vibrations caused by the truck have matched the natural frequency of this window but not the others. The window is undergoing resonance from the sound wave impulses that matched its natural frequency. It is also resonance that enables a singer to break a water glass. If the tone is at the resonant frequency of the glass, the resulting vibrations may be large enough to shatter it.

Resonance considerations are important in engineering. A large water pump, for example, was designed for a nuclear power plant. Vibrations from the electric motor matched the resonant frequency of the impeller blades and they shattered after a short period of time. The blades were redesigned to have a different natural frequency when the problem was discovered. Resonance vibrations are particularly important in the design of buildings.

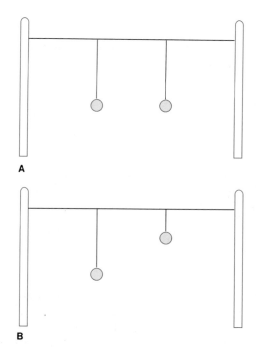

A

B

Figure 7.18

The activities below will demonstrate that one of these pendulum arrangements will show resonance and the other will not. Can you predict which one will show resonance?

ACTIVITIES

Set up two identical pendulums as shown in figure 7.18a. The bobs should be identical and suspended from identical strings of the same length attached to a tight horizontal string. Start one pendulum vibrating by pulling it back, then releasing it. Observe the vibrations and energy exchange between the two pendulums for the next several minutes. Now change the frequency of vibrations of *one* of the pendulums by shortening the string (figure 7.18b). Again start one pendulum vibrating and observe for several minutes. Compare what you observe when the frequencies are matched and when they are not. Explain what happens in terms of resonance.

Sources of Sounds

All sounds have a vibrating object as their source. The vibrations of the object send pulses or waves of condensations and rarefactions through the air. These sound waves have physical properties that can be measured, such as frequency and intensity. Subjectively, your response to frequency is to identify a certain pitch. A high-frequency sound is interpreted as a high-pitched sound and a low-frequency sound is interpreted as a low-

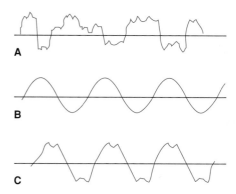

Figure 7.19
Different sounds that you hear include (a) noise, (b) pure tones, and (c) musical notes.

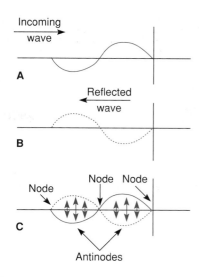

Figure 7.20
An incoming wave (a) meets a reflected wave (b) with the same amplitude and frequency, producing a standing wave (c). Note that a standing wave of one wavelength has three nodes and two antinodes.

pitched sound. Likewise, a greater intensity is interpreted as increased loudness, but there is not a direct relationship between intensity and loudness as there is between frequency and pitch.

There are other subjective interpretations about sounds. Some sounds are bothersome and irritating to some people but go unnoticed by others. In general, sounds made by brief, irregular vibrations such as those made by a slamming door, dropped book, or sliding chair are called **noise.** Noise is characterized by sound waves with mixed frequencies and jumbled intensities (figure 7.19). On the other hand, there are sounds made by very regular, repeating vibrations such as those made by a tuning fork. A tuning fork produces a **pure tone** with a sinusoidal curved pressure variation and regular frequency. Yet a tuning fork produces a tone that most people interpret as bland. You would not call a tuning fork sound a musical note! Musical sounds from instruments have a certain frequency and loudness as do noise and pure tones. But you can readily identify the source of the very same musical note made by two different instruments. You recognize it as a musical note, not noise and not a pure tone. You also recognize if the note was produced by a violin or a guitar. The difference is in the wave form of the sounds made by the two instruments, and the difference is called the **sound quality.** How does a musical instrument produce a sound of a characteristic quality? The answer may be found by looking at the two broad categories of instruments, (1) those that make use of vibrating strings and (2) those that make use of vibrating columns of air. These two categories will be considered separately.

Vibrating Strings

A stringed musical instrument, such as a guitar, has strings that are stretched between two fixed ends. When a string is plucked, waves of many different frequencies travel back and forth on the string, reflecting from the fixed ends. Many of these waves

quickly fade away but certain frequencies resonate, setting up patterns of waves. Before considering these resonate patterns in detail, keep in mind that (1) two or more waves can be in the same place at the same time, traveling through one another from opposite directions; (2) a confined wave will be reflected at a boundary and the reflected wave will be inverted (a crest becomes a trough); and (3) reflected waves interfere with incoming waves of the same frequency to produce **standing waves.** Figure 7.20 is a graphic "snapshot" of what happens when reflected wave patterns meet incoming wave patterns. The incoming wave is shown as a solid line and the reflected wave is shown as a dotted line. The result is (1) places of destructive interference, called **nodes,** which show no disturbance, and (2) loops of constructive interference, called **antinodes,** which take place where the crests and troughs of the two wave patterns produce a disturbance that rapidly alternates upward and downward. This pattern of alternating nodes and antinodes does not move along the string and is thus called a "standing wave." Note that the standing wave for *one wavelength* will have a node at both ends and in the center and also two antinodes. Standing waves occur at the natural, or resonant, frequencies of the string, which are a consequence of the nature of the string, the string length, and the tension in the string. Since the standing waves are resonant vibrations, they continue as all other waves quickly fade away.

Since the two ends of the string are not free to move, the ends of the string will have nodes. The *longest* wave that can make a standing wave on such a string has a wavelength (λ) that is twice the length (L) of the string. Since frequency (f) is inversely proportional to wavelength (f = vλ from equation 7.3),

$$F = v/\lambda$$

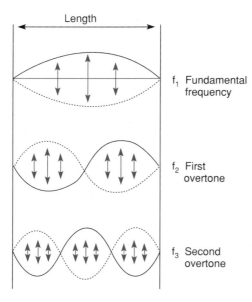

Figure 7.21

A stretched string of a given length has a number of possible resonant frequencies. The lowest frequency is the fundamental, f_1, and the next higher frequencies, or overtones, shown are f_2 and f_3.

this longest wavelength has the lowest frequency possible, called the **fundamental frequency.** The fundamental frequency has one antinode, which means that the length of the string has one-half a wavelength. The fundamental frequency (f_1) determines the pitch of the *basic* musical note being sounded. Other resonant frequencies occur at the same time, however, since other standing waves can also fit onto the string. A higher frequency of vibration (f_2) could fit two half-wavelengths between the two fixed nodes. An even higher frequency (f_3) could fit three half-wavelengths between the two fixed nodes (figure 7.21). Any whole number of halves of the fundamental wavelength will permit a standing wave to form. The higher frequencies (f_2, f_3, etc.) of these wavelengths are called the **overtones, or harmonics,** beyond the fundamental. It is the presence and strength of various overtones that give a musical note from a certain instrument its characteristic quality. The fundamental and the overtones add together to produce the characteristic *sound quality,* which is different for the same-pitched note produced by a violin and by a guitar (figure 7.22).

Since nodes must be located at the ends, only half-wavelengths ($1/2 \lambda$) can fit on a string of a given length (L), so the fundamental frequency of a string is $1/2\lambda = L$, or $\lambda = 2L$. Substituting this value in the wave equation (solved for frequency, f) will give the relationship for finding the fundamental frequency and the overtones when the string length and velocity of waves on the string are known. The relationship is

$$f_n = \frac{nv}{2L}$$ **equation 7.7**

where $n = 1, 2, 3, 4 \ldots$, and $n = 1$ is the fundamental frequency and $n = 2$, $n = 3$, and so forth are the overtones.

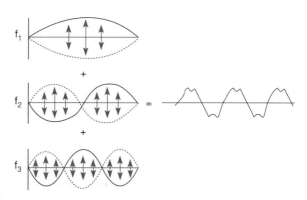

Figure 7.22

A combination of the fundamental and overtone frequencies produces a composite wave form with a characteristic sound quality.

Example 7.8

What is the fundamental frequency of a 0.500 m string if wave speed on the string is 400.0 m/sec?

Solution

The length (L) and the velocity (v) are given. The relationship between these quantities and the fundamental frequency ($n = 1$) is given in equation 7.7, and

$$L = 0.500 \text{ m}$$
$$v = 400.0 \text{ m/sec}$$
$$f_1 = ?$$

$$f_n = \frac{nv}{2L} \quad \text{where } n = 1 \text{ for the fundamental frequency,}$$

$$f_1 = \frac{1 \times 400.0 \text{ m/sec}}{2 \times 0.500 \text{ m}}$$

$$= \frac{400.0}{1.00} \frac{\text{m}}{\text{sec}} \times \frac{1}{\text{m}}$$

$$= 400. \frac{\text{m}}{\text{sec} \cdot \text{m}}$$

$$= 400. \frac{1}{\text{sec}}$$

$$= \boxed{400. \text{ hz}}$$

Example 7.9

What is the frequency of the first overtone in a 0.500 string when the wave speed is 400.0 m/sec? (Answer: 800. hz)

Vibrating Air Columns

A musical instrument that makes use of a vibrating air column, such as a wind instrument or a pipe organ, produces standing waves in air in a tube that is open at both ends (*open tube*) or open at one end and closed at the other end (*closed tube*). First, consider a tube that is closed at one end and open at the other, which is called a *closed tube.*

Closed Tube

The air in a closed tube is made to vibrate by a reed, turbulence, or some other means, depending on the instrument. The air at the closed end of the tube is not free to vibrate but reflects the

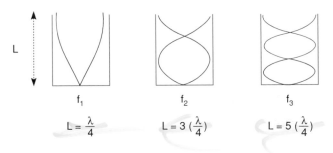

$$L = \frac{\lambda}{4} \qquad L = 3\left(\frac{\lambda}{4}\right) \qquad L = 5\left(\frac{\lambda}{4}\right)$$

Figure 7.23

Standing wave patterns of air vibrating in a closed tube. Note the node at the closed end and the antinode at the open end. Only odd multiples of the fundamental are thus possible.

condensations and rarefactions of the vibrating disturbance. The closed end is therefore a node and the open end is an antinode, where the air is free to resonate freely. Thus the bottom of the tube is a node and the top is an antinode for a longitudinal standing wave in the air. The *longest* wave that can make a standing wave in such a vibrating air column has a wavelength (λ) that is four times the length (L) of the air column (figure 7.23). Thus the longest wavelength that will fit into an air column that is closed at one end is $L = 1/4\lambda$, or $\lambda = 4L$. Substituting this value in the wave equation (solved for the frequency, f) will give the relationship for finding the fundamental frequency and the overtones when the length of the closed air column and the velocity are known. The relationship for a *closed tube* is

$$f_n = \frac{nv}{4L} \qquad \text{equation 7.8}$$

where n = 1 is the fundamental and 3, 5, 7, and so on are the possible harmonics.

Open Tube

An open tube can have an antinode at both ends since the air is free to vibrate at both places (figure 7.24). The *longest* wavelength that can fit into the open tube is therefore the distance between two antinodes, or $1/2\lambda$. Therefore the wavelength of the fundamental is twice the length of the tube, or $\lambda_1 = 2L$. The relationship between wavelength and frequency is $f = v/\lambda$, so substituting 2L for λ will give the relationship between the frequency, velocity, and length of an air column for an *open tube*,

$$f_n = \frac{nv}{2L} \qquad \text{equation 7.9}$$

where n = 1, 2, 3, etc. The overtone frequencies are therefore equal to whole number multiples (2, 3, 4, etc) of the fundamental. Note that this is the same relationship as that for a vibrating string (equation 7.7). Both have a relationship of $1/2\lambda = L$, but for different reasons.

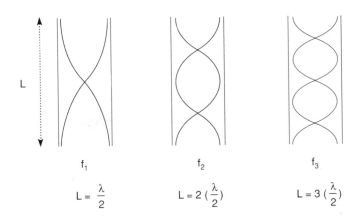

$$L = \frac{\lambda}{2} \qquad L = 2\left(\frac{\lambda}{2}\right) \qquad L = 3\left(\frac{\lambda}{2}\right)$$

Figure 7.24

Standing wave patterns of air vibrating in an open tube. Note that both ends have antinodes. Any whole number of multiples of the fundamental are thus possible.

Example 7.10

A tuning fork is found to resonate with an air column closed at one end when the tube is 60.0 cm long. If the air is 20.0° C, the speed of sound in air is 343.0 m/sec. What is the frequency of the tuning fork?

Solution

The relationship between frequency, velocity, and wavelength of a sound formed in a closed tube is found in equation 7.8, $f_n = nv/4L$, where n = 1 is the fundamental.

$$n = 1 \qquad\qquad f_n = \frac{nv}{4L}$$

$$L = 60.0 \text{ cm}$$

$$= 0.600 \text{ m} \qquad = \frac{1 \times 343.0 \dfrac{m}{sec}}{4 \times 0.600 \text{ m}}$$

$$v = 343.0 \text{ m/sec}$$

$$= \frac{343.0}{4 \times 0.600} \frac{m}{sec} \times \frac{1}{m}$$

$$= \frac{343.0}{2.40} \frac{m}{sec \cdot m}$$

$$= 143 \frac{1}{sec}$$

$$= \boxed{143 \text{ hz}}$$

This problem can also be solved by using a two-step method:

1. Find the wavelength of the longest wave possible, the fundamental from $L = \frac{1}{4}\lambda$

$$L = 0.600 \text{ m} \qquad L = \frac{1}{4}\lambda \therefore \lambda = 4L$$

$$\lambda = ? \qquad\qquad\qquad = (4)(0.600 \text{ m})$$

$$= 2.40 \text{ m}$$

Wave Motion and Sound

2. Find the frequency from the wave equation, $v = f\lambda$

$v = 343.0$ m/sec $\qquad v = f\lambda \therefore f = \dfrac{v}{\lambda}$

$\lambda = 2.40$ m

$f = ?$

$$= \dfrac{343.0 \, \dfrac{m}{sec}}{2.40 \, m}$$

$$= \dfrac{343.0}{2.40} \, \dfrac{m}{sec} \times \dfrac{1}{m}$$

$$= 143 \, \dfrac{m}{sec \cdot m}$$

$$= 143 \, \dfrac{1}{sec}$$

$$= \boxed{143 \, hz}$$

Example 7.11

What is the fundamental frequency for a 60.0 cm *open* tube when the velocity of sound is 343.0 m/sec? (Answer: 286 hz)

Sounds from Moving Sources

When the source of a sound is stationary, waves of condensation and rarefaction expand from the source like the surface of an expanding balloon. These concentric waves have equal spacing in all directions, and an observer standing in any location will hear the same pitch (frequency) as another observer standing at any other location. However, if the source of the sound moves then the center of each successive wave will be displaced in the direction of movement. The successive crests are therefore crowded closer together in the direction of the motion and spread farther apart in the opposite direction. An observer in front of the moving source will encounter more wave crests per second than are being generated by the source. The frequency of the sound will therefore seem higher than it really is, and the observer will interpret this as a higher pitch. As the moving source passes the observer, the source is moving away from the waves sent backward. The frequency will then seem lower than it really is, and the observer will interpret this as a lower pitch. The overall effect is that of a higher pitch as the source approaches, then a lower pitch as it moves away. This apparent shift of frequency of a sound from a moving source is called the **Doppler effect.** The Doppler effect is evident if you stand by a street and an approaching car sounds its horn as it drives by you. You will hear a higher-pitched horn as the car approaches, which shifts to a lower-pitched horn as the waves go by you. The driver of the car, however, will hear the continual, true pitch of the horn since the driver is moving with the source (figure 7.25).

A Doppler shift is also observed if the observer is moving and the source of sound is stationary. When the observer moves toward the source, the wave fronts are encountered more frequently than if the observer were standing still. As the observer moves away from the source, the wave fronts are encountered

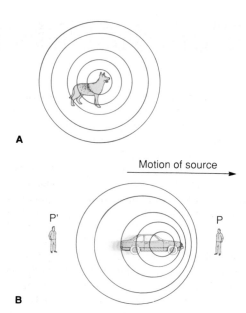

A

Motion of source

P'　　　　　　　　　　P

B

Figure 7.25

(*a*) Spherical sound waves from a stationary source spread out evenly in all directions. (*b*) If the source is moving, an observer at position P will encounter more wave crests per second than an observer at position P'. The observer at P interprets this as a higher pitch, and the phenomena is called the Doppler effect.

less frequently than would be the case if the observer were not moving. An observer on a moving train approaching a crossing with a sounding bell thus hears a high-pitched bell that shifts to a lower-pitched bell as the train whizzes by the crossing.

When an object moves through the air at the speed of sound, it keeps up with its own sound waves. All the successive wave fronts pile up on one another, creating a large wave disturbance called a **shock wave.** The shock wave from a supersonic airplane is a cone-shaped shock wave of intense condensations trailing backward at an angle dependent on the speed of the aircraft. Wherever this cone of superimposed crests passes, a **sonic boom** occurs. The many crests have been added together, each contributing to the pressure increase. The human ear cannot differentiate between such a pressure wave created by a supersonic aircraft and a pressure wave created by an explosion.

Summary

Elastic objects *vibrate,* or move back and forth, in a repeating motion when disturbed by some external force. They are able to do this because they have an *internal restoring force* that returns them to their original positions after being deformed by some external force. If the internal restoring force is opposite to and proportional to the deforming displacement, the vibration is called a *simple harmonic motion.* The extent of displacement is called the *amplitude,* and one complete back-and-forth motion is one *cycle.* The time required for one cycle is a *period.* The *frequency* is the number of cycles per second and the unit of frequency is the *hertz.* A *graph* of the displacement as a function of time for a simple harmonic motion produces a *sinusoidal* graph.

Ultrasonics

Ultrasonic waves are mechanical waves that have frequencies above the normal limit of hearing of the human ear. The arbitrary upper limit is about 20,000 hz, so an ultrasonic wave has a frequency of 20,000 hz or greater. Intense ultrasonic waves are used in many ways in industry and medicine.

Industrial and commercial applications of ultrasound utilize lower-frequency ultrasonic waves in the 20,000 to 60,000 hz range. Commercial devices that send ultrasound through the air include burglar alarms and rodent repellers. An ultrasonic burglar alarm sends ultrasonic spherical waves through the air of a room. The device is adjusted to ignore echoes from the contents of the room. The presence of a person provides a new source of echoes, which activates the alarm. Rodents emit ultrasonic frequencies up to 150,000 hz that are used in rodent communication when they are disturbed or during aggressive behavior. The ultrasonic rodent repeller generates ultrasonic waves of similar frequency. Other commercial applications of ultrasound include sonar and depth measurements, cleaning and drilling, welding plastics and metals, and material flaw detection. Ultrasonic cleaning baths are used to remove dirt and foreign matter from solid surfaces, usually within a liquid solvent. The ultrasonic waves create vapor bubbles in the liquid, which vibrate and emit audible and ultrasonic sound waves. The audible frequencies are often heard as a hissing or frying sound.

Medical applications of ultrasound use frequencies in the 1,000,000 to 20,000,000 hz range. Ultrasound in this frequency range cannot move through the air because the required displacement amplitudes of the gas molecules in the air are less than the average distance between the molecules. Thus a gas molecule that is set into motion in this frequency range cannot collide with other gas molecules to transmit the energy of the wave. Intense ultrasound is used for cleaning teeth and disrupting kidney stones. Less intense ultrasound is used for therapy (heating and reduction of pain). The least intense ultrasounds are used for diagnostic imaging. The largest source of exposure of humans to ultrasound is for the purpose of ultrasonic diagnostic imaging, particularly in fertility and pregnancy cases. In the United States, it has been estimated that more than half of the children born in the 1980s were scanned at least once by ultrasound before birth.

The ultrasonic medical scanner uses a transducer probe that is held next to the skin. The transducer emits an ultrasonic pulse that passes through the skin and into the body. Echoes from the internal tissues and organs are reflected back to the transducer, which sends the signals to a computer. Another pulse of ultrasound is then sent out after the echoes from the first pulse have returned. The strength of and number of pulses per second vary with the application, ranging from hundreds to thousands of pulses per second. The computer constructs a picture from the returning echoes, showing an internal view without the use of more dangerous X rays.

Ultrasonic scanners have been refined to the point that the surface of the ovaries can now be viewed, showing the number and placement of developing eggs. The ovary scan is typically used in conjunction with fertility-stimulating drugs, where multiple births are possible, and to identify the exact time of ovulation. After conception, the ultrasonic scanner is used to identify and monitor the fetus as early as four weeks. The scan is usually performed on a patient with a full bladder, which displaces the uterus from the pelvic bones and possibly gas-filled bowel, providing a clear view of the uterus. By the thirteenth week an ultrasonic scan can show the fetus heart movement, bone and skull size, and internal organs. The ultrasonic scanner is often used to show the position of the fetus and placenta for the purpose of amniocentesis, which involves withdrawing a sample of amniotic fluid from the uterus for testing.

Periodic, or repeating, vibrations or the *pulse* of a single disturbance can create *waves,* disturbances that carry energy through a medium. A wave that disturbs particles in a back-and-forth motion in the direction of the wave travel is called a *longitudinal wave.* A wave that disturbs particles in a motion perpendicular to the direction of wave travel is called a *transverse wave.* The nature of the medium and the nature of the disturbance determine the type of wave created.

Waves that move through the air are longitudinal and cause a back-and-forth motion of the molecules making up the air. A zone of molecules forced closer together produce a *condensation,* a pulse of increased density and pressure. A zone of reduced density and pressure is a *rarefaction.* A vibrating object produces condensations and rarefactions that expand outward from the source. If the frequency is between 20 and 20,000 hz, the human ear perceives the waves as *sound.* High frequency is interpreted as high-*pitched* sound and low frequency as low-pitched sound.

A graph of pressure changes produced by condensations and rarefactions can be used to describe sound waves. The condensations produce *crests* and the rarefactions produce *troughs.* The *amplitude* is the maximum change of pressure from the normal. The *wavelength* is the distance between any two successive places on a wave train, such as the distance from one crest to the next crest. The *period* is the time required for a wave to repeat itself. The *velocity* of a wave is how frequently a wavelength passes. The *frequency* can be calculated from the *wave equation,* $v = \lambda f$.

Sound waves can move through any medium but not a vacuum. The velocity of sound in a medium depends on the molecular inertia and strength of interactions. Sound therefore travels most rapidly through a solid, then a liquid, then a gas. In air, sound has a greater velocity in warmer air than in cooler air because the molecules of air are moving about more rapidly, therefore transmitting a pulse more rapidly.

Sound waves are *reflected* or *refracted* from a *boundary,* which means a change in the transmitting medium. Reflected waves that are *in phase* with incoming waves undergo *constructive interference* and waves that are *out of phase* undergo *destructive interference.* Two waves that are otherwise alike but with slightly different frequencies produce an alternating increasing and decreasing of loudness called *beats.*

The *energy* of a sound wave is called the wave *intensity*, which is measured in watts per square meter. The intensity of sound is expressed on the *decibel scale*, which relates it more to changes in loudness as perceived by the human ear.

All elastic objects have *natural frequencies* of vibration that are determined by the materials they are made of and their shapes. When energy is transferred at the natural frequencies, there is a dramatic increase of amplitude called *resonance*. The natural frequencies are also called *resonant frequencies*.

Sounds are compared by pitch, loudness, and *quality*. The quality is determined by the instrument sounding the note. Each instrument has its own characteristic quality because of the resonant frequencies that it produces. The basic, or *fundamental, frequency* is the longest standing wave that it can make. The fundamental frequency determines the basic note being sounded and other resonant frequencies, or standing waves called *overtones* or *harmonics* combine with the fundamental to give the instrument its characteristic quality.

A moving source of sound or a moving observer experiences an apparent shift of frequency called the *Doppler effect*. If the source is moving as fast or faster than the speed of sound, the sound waves pile up into a *shock wave* called a *sonic boom*. A sonic boom sounds very much like the pressure wave from an explosion.

Summary of Equations

7.1

$$\text{period} = \frac{1}{\text{frequency}}$$

$$T = \frac{1}{f}$$

7.2

$$\text{frequency} = \frac{1}{\text{period}}$$

$$f = \frac{1}{T}$$

7.3

$$\text{velocity} = (\text{wavelength})(\text{frequency})$$

$$v = \lambda f$$

7.4

$$\begin{array}{l}\text{velocity of} \\ \text{sound (m/sec)} \\ \text{at present} \\ \text{temperature}\end{array} = \begin{array}{l}\text{velocity} \\ \text{of sound} \\ \text{at } 0°\text{ C}\end{array} + \begin{array}{l}0.60\text{ m/sec} \\ \text{increase per} \\ \text{degree Celsius}\end{array} \times \begin{array}{l}\text{present} \\ \text{temperature} \\ \text{in }°\text{C}\end{array}$$

$$v_{T_p}\text{ (m/sec)} = v_0 + \left(\frac{0.60\text{ m/sec}}{°\text{C}}\right)(T_p)$$

7.5

$$\begin{array}{l}\text{velocity of} \\ \text{sound (ft/sec)} \\ \text{at present} \\ \text{temperature}\end{array} = \begin{array}{l}\text{velocity} \\ \text{of sound} \\ \text{at } 0°\text{ C}\end{array} + \begin{array}{l}2.0\text{ ft/sec} \\ \text{increase per} \\ \text{degree Celsius}\end{array} \times \begin{array}{l}\text{present} \\ \text{temperature} \\ \text{in }°\text{C}\end{array}$$

$$v_{T_p}\text{ (ft/sec)} = v_0 + \left(\frac{2.0\text{ ft/sec}}{°\text{C}}\right)(T_p)$$

7.6

$$\text{beat frequency} = \text{one frequency} - \text{other frequency}$$

$$f_b = f_2 - f_1$$

7.7

$$\text{resonant frequency} = \frac{\text{number} \times \text{velocity on string}}{2 \times \text{length of string}}$$

where number 1 = fundamental frequency and numbers 2, 3, and so on = overtones.

$$f_n = \frac{nv}{2L}$$

7.8

$$\text{resonant frequency in closed tube} = \frac{\text{number} \times \text{velocity of sound}}{4 \times \text{length of air column}}$$

where number = 1 for fundamental frequency and numbers 2, 3, and so on = frequency of overtones

$$f_n = \frac{nv}{4L}$$

7.9

$$\text{resonant frequency in open tube} = \frac{\text{number} \times \text{velocity of sound}}{2 \times \text{length of air column}}$$

where number = 1 for fundamental frequency and numbers 2, 3, and so on = frequency of overtones

$$f_n = \frac{nv}{2L}$$

Key Terms

amplitude (p. 124)
antinodes (p. 135)
beats (p. 132)
boundary (p. 130)
condensation (p. 126)
constructive interference (p. 132)
cycle (p. 124)
decibel scale (p. 133)
destructive interference (p. 132)
Doppler effect (p. 138)
echo (p. 131)
frequency (p. 124)
fundamental frequency (p. 136)
harmonics (p. 136)
hertz (p. 124)
infrasonic (p. 127)
intensity (p. 133)
longitudinal wave (p. 125)
loudness (p. 133)
natural frequency (p. 133)
nodes (p. 135)

noise (p. 135)
overtones (p. 136)
period (p. 124)
pitch (p. 127)
pure tone (p. 135)
rarefaction (p. 126)
reflection (p. 130)
refraction (p. 130)
resonance (p. 134)
reverberation (p. 131)
shock wave (p. 138)
simple harmonic motion (p. 123)
sonic boom (p. 138)
sound quality (p. 135)
standing waves (p. 135)
transverse wave (p. 125)
ultrasonic (p. 127)
vibration (p. 123)
wave (p. 125)
wave equation (p. 128)
wave front (p. 130)
wavelength (p. 128)

Applying the Concepts

1. The characteristic of a wave that is responsible for what you interpret as pitch is the wave
   a. amplitude.
   b. shape.
   c. frequency.
   d. height.

2. The extent of displacement of a vibrating tuning fork is related to the resulting sound wave characteristic of
   a. frequency.
   b. amplitude.
   c. wavelength.
   d. period.

3. The number of cycles that a vibrating tuning fork experiences each second is related to the resulting sound wave characteristic of
   a. frequency.
   b. amplitude.
   c. wave height.
   d. quality.

4. From the wave equation of $v = \lambda f$, you know that the wavelength and frequency at a given velocity are
   a. directly proportional.
   b. inversely proportional.
   c. not related.

5. The difference between an echo and a reverberation is
   a. an echo is a reflected sound; reverberation is not.
   b. the time interval between the original sound and the reflected sound.
   c. the amplitude of an echo is much greater.
   d. reverberation comes from acoustical speakers; echoes come from cliffs and walls.

6. Sound interference is necessary to produce the phenomena known as
   a. resonance.
   b. decibels.
   c. beats.
   d. reverberation.

7. The efficient transfer of energy that takes place at a natural frequency is known as
   a. resonance.
   b. beats.
   c. the Doppler effect.
   d. reverberation.

8. The fundamental frequency of a standing wave on a string has
   a. one node and one antinode.
   b. one node and two antinodes.
   c. two nodes and one antinode.
   d. two nodes and two antinodes.

9. The fundamental frequency of a standing wave in an air column of a *closed* tube has
   a. one node and one antinode.
   b. one node and two antinodes.
   c. two nodes and one antinode.
   d. two nodes and two antinodes.

10. An observer on the ground will hear a sonic boom from an airplane traveling faster than the speed of sound
    a. only when the plane breaks the sound barrier.
    b. as the plane is approaching.
    c. when the plane is directly overhead.
    d. after the plane has passed by.

Answers
1. c 2. b 3. a 4. b 5. b 6. c 7. a 8. c 9. a
10. d

Questions for Thought

1. What is a wave?

2. What does the amplitude of a wave represent?

3. What happens to two waves traveling in opposite directions when they meet? After they meet?

4. How is a transverse wave different from a longitudinal wave?

5. Is it possible for a transverse wave to move through air? Explain.

6. A piano tuner hears three beats per second when a tuning fork and a note are sounded together and six beats per second after the string is tightened. What should the tuner do next, tighten or loosen the string? Explain.

7. Why do astronauts on the moon have to communicate by radio even when close to one another?

8. What is resonance?

9. Explain why sounds travel faster in warm air than in cool air.

10. Do all frequencies of sound travel with the same velocity? Explain your answer by using the wave equation.

11. What eventually happens to a sound wave traveling through the air?

12. Considering condensations and rarefactions alone, what is one wavelength? What is frequency? What is intensity?

13. What gives a musical note its characteristic quality?

14. Does a supersonic aircraft make a sonic boom only when it cracks the sound barrier? Explain.

15. Is the Doppler effect a real or apparent shift of frequency? Explain.

16. Explain why a standing wave on a string that consists of two nodes and one antinode is equivalent to one-half a wavelength.

17. Explain how you know that a sound wave is a longitudinal wave and not a transverse wave.

18. Explain how overtones are generated in stringed instruments and instruments with vibrating air columns.

19. What is an echo?

20. Will the air temperature affect the pitch of a musical instrument with a vibrating air column? Explain.

21. How are noise, a pure tone, and a musical note different?

22. What is a decibel?

23. If you could filter out all higher-frequency overtones, could you tell which instrument the same musical note came from? Explain.

24. Explain how interference produces a standing wave.

25. Is there a loss of energy when two waves destructively interfere? Explain.

26. How must the length of an open tube compare to the wavelength of a sound to produce a fundamental resonance?

27. What is a fundamental frequency?

28. Why are fundamental frequencies and overtones also called resonance frequencies?

Exercises

▼

Group A—Solutions Provided in Appendix D

1. A vibrating object produces periodic waves with a wavelength of 50.0 cm and a frequency of 10.0 hz. How fast do these waves move away from the object?

2. The distance between the center of a condensation and the center of an adjacent rarefaction is 1.50 m. If the frequency is 112.0 hz, what is the speed of the wave front?

3. Water waves are observed to pass under a bridge at a rate of one complete wave every 4.00 sec. (a) What is the period of these waves? (b) What is the frequency?

4. A sound wave with a frequency of 260. hz moves with a velocity of 330. m/sec. What is the distance from one condensation to the next?

5. The following sound waves have what velocity?
   (a) Middle C, or 256 hz and 1.34 m λ.
   (b) Note A, or 440. hz and 78.0 cm λ.
   (c) A siren at 750. hz and λ of 45.7 cm.
   (d) Note from a stereo at 2,500. hz and λ of 13.72 cm.

6. What is the speed of sound, in ft/sec, if the air temperature is
   (a) 0.0° C
   (b) 20.0° C
   (c) 40.0° C
   (d) 80.0° C

7. An echo is heard from a cliff 4.80 sec after a rifle is fired. If the air temperature is 43.7° F, how many feet away is the cliff?

8. The air temperature is 80.00° F during a thunderstorm and thunder was timed 4.63 sec after lightning was seen. How many feet away did the lightning strike?

9. A 340.0 hz tuning fork resonates with an air column in a closed tube that is 25 cm long. What was the speed of sound in the tube?

10. If the velocity of a 440 hz sound is 1,125 ft/sec in the air and 5,020 ft/sec in seawater, find the wavelength of this sound (a) in air and (b) in seawater.

11. What is the frequency of a tuning fork that resonates with an air column in a 24.0 cm closed tube with an air temperature of 20.0° C?

12. What is the fundamental frequency and the frequency of the first overtone of a 70.0 cm closed organ pipe?

Group B—Solutions Not Given

1. A tuning fork vibrates 440.0 times a second, producing sound waves with a wavelength of 78.0 cm. What is the velocity of these waves?

2. The distance between the center of a condensation and the center of an adjacent rarefaction is 65.23 cm. If the frequency is 256.0 hz, how fast are these waves moving?

3. A warning buoy is observed to rise every 5.0 sec as a crest of waves passes by it. What is the period of these waves? What is the frequency?

4. The siren of an emergency vehicle has a frequency of 750.0 hz and moves with a velocity of 343.0 m/sec. What is the distance from one condensation to the next?

5. The following sound waves have what velocity?
   (a) 20.0 hz, λ of 17.15 m
   (b) 200.0 hz, λ of 1.72 m
   (c) 2,000.0 hz, λ of 17.15 cm
   (d) 20,000.0 hz, λ of 1.72 cm

6. How much time is required for a sound to travel 1 mi (5,280.0 ft) if the air temperature is
   (a) 0.0° C
   (b) 20.0° C
   (c) 40.0° C
   (d) 80.0° C

7. A ship at sea sounds a whistle blast and an echo returns from the coastal land 10.0 sec later. At an air temperature of 10.0° C, how many km is it to the coastal land?

8. If the air temperature is 90.0° F, how many seconds will elapse between seeing lightning and hearing the thunder if the lightning strikes 1 mi (5,280.0 ft) away?

9. A 440.0 hz sound resonates with a 19.5 cm air column in a closed tube. What was the speed of sound in the tube?

10. A 600.0 hz sound has a velocity of 1,087.0 ft/sec in the air and a velocity of 4,920.0 ft/sec in water. Find the wavelength of this sound in the air and in the water.

11. What is the shortest length of a closed air column in which a 440.0 hz tuning fork can produce resonance at 20.0° C?

12. The air temperature increases from 20.0° C to 25.0° C. What effect will this have on the fundamental frequency produced by a 70.0 cm closed organ pipe?

Chapter 8

Electricity

$1.\ v = f \lambda\quad (10.0^{1}/\text{sec})(.500\,m)$

$$\boxed{5.00\ m/\text{sec}}$$

$2.\ v = f \lambda\ (112.0^{1}/\text{sec})(3.00\,m)$

$$\boxed{336\ m/\text{sec}}$$

$3.\ f = \frac{1}{T}\quad \frac{1}{4.00\,\text{sec}} = \boxed{1250\,\text{hz}}$

$4.\ v = f \lambda \therefore \lambda\ \frac{v}{f}$

$$\frac{330\ m/\text{sec}}{260^{1}/\text{sec}}\qquad \boxed{1.27\,m}$$

Figure 8.1
The importance of electrical power seems obvious in a modern industrial society. What is not so obvious is the role of electricity in magnetism, light, chemical change, and the very basis for the structure of matter. All matter, in fact, is electrical in nature, as you will see.

$9.\ L = \frac{1}{4}\lambda \therefore \lambda = 4L$

$$= 4(.025)$$

$$= 1.0$$

$$(340-0)(1.0)^{1}/\text{sec} \times m$$

$$\boxed{340\ m/\text{sec}}$$

IN previous chapters we have been concerned with *mechanical* concepts, explanations of the motion of objects that exert forces on one another. These concepts were used to explain straight-line motion, the motion of free fall, and the circular motion of objects on the earth as well as the circular motion of planets and satellites. The mechanical concepts were based on Newton's laws of motion and are sometimes referred to as Newtonian physics. The mechanical explanations were then extended into the submicroscopic world of matter through the kinetic molecular theory. The objects of motion were now particles, molecules that exert forces on one another, and concepts associated with heat were interpreted as the motion of these particles. In a further extension of Newtonian concepts, mechanical explanations were made for concepts associated with sound, a mechanical disturbance that follows the laws of motion as it moves through the molecules of matter.

You might wonder, as did the scientists of the 1800s, if mechanical interpretations would also explain other natural phenomena such as electricity, chemical reactions, and light. A mechanical model would be very attractive since it already explained so many other facts of nature and scientists have always looked for basic, unifying theories. Mechanical interpretations were tried; electricity was considered as a moving fluid, and light was considered as a mechanical wave moving through a material fluid. But there were many unsolved puzzles with such a model, and gradually it was recognized that electricity, light, and chemical reactions could not be explained by mechanical interpretations. Gradually, the point of view changed from a study of particles to a study of the properties of the *space* around the particles. In this chapter you will learn about electric charge in terms of the space around particles. This model of electric charge, called the field model, will be used to develop understandings about electric current, the electric circuit, and electrical work and power (figure 8.1). These concepts and understandings will form the foundation for the following chapter, which considers magnetism, how electricity is produced, and electrical applications.

Electric Charge

You are familiar with the use of electricity in many electrical devices such as lights, toasters, radios, and calculators. You are also aware that electricity is used for transportation and for heating and cooling places where you work and live. Many people accept electrical devices as part of their surroundings with only a hazy notion of how they work. To many people electricity is seemingly magical. Electricity is not magical, and it can be understood just as any other natural phenomenon. There are theories that explain observations, quantities that can be measured, and relationships between these quantities, or laws, that lead to understanding. All of the observations, measurements, and laws begin with an understanding of *electric charge*.

Electron Theory of Charge

As early as 600 B.C. the ancient Greeks marveled about the observation that amber, a fossilized tree resin, would attract hair, threads, and other bits of light material when rubbed with wool.

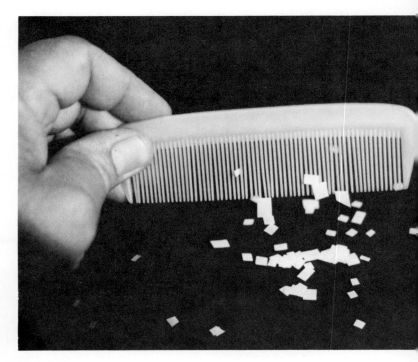

Figure 8.2

Many objects will attract small bits of light materials after being rubbed; this effect has been known for thousands of years. In the past, the effect was considered to be from invisible fluid or fluids released or moved by the rubbing. Today, the effect is understood to be from electrical charges transferred by the friction of rubbing.

Early explanations about these mysterious attractions considered a "sympathy" to exist between the objects or an invisible "aura" that pulled the objects together. In general, the amber effect was considered a mystical curiosity for the next two thousand years. Then in 1600, William Gilbert, a court physician to Elizabeth I and James I, published the results of his extensive investigations about the amber effect and magnetism. Gilbert identified many different substances, in addition to amber, that would attract objects when rubbed. He called these substances *electrics* after the Greek word for amber, "elektron." Gilbert supposed that the attraction of electrics was produced by an invisible fluid that he called "effluvium," which resulted from the rubbing. The fluid was thought to spread out and "draw in" light objects (figure 8.2).

The amber effect and other related effects that repel as well as attract were discovered over the next three hundred years. During that period of time the effects were explained as the attraction and repulsion of invisible fluids that were squeezed out of matter by rubbing. Some believed that there were two fluids that attracted and repelled each other. Others, including Benjamin Franklin, believed that there was only one fluid that resulted in attractions and repulsions because objects had either a surplus or deficiency of the fluid. The "fluid theories" of electric charge were the common understanding until experiments

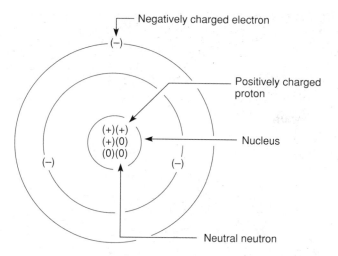

Negatively charged electron

Positively charged proton

Nucleus

Neutral neutron

Figure 8.3
A highly simplified model of an atom has most of the mass in a small dense center called the nucleus. The nucleus has positively charged protons and neutral neutrons. Negatively charged electrons move around the nucleus at some distance. Ordinary atoms are neutral because there is a balance between the number of positively charged protons and negatively charged electrons.

concerning the structure of matter took place, beginning about 1900. These experiments, and the resulting understanding about atomic structure, will be presented in some detail in chapter 11. The following is only a brief overview that is intended to serve the purpose of understanding electricity.

One of the first understandings about the structure of matter was discovered in 1897 by the physicist Joseph J. Thomson. From his experiments with cathode ray tubes, Thomson concluded that negatively charged particles, now called **electrons,** are present in all matter. He thought that matter was made up of electrons embedded in a positive "fluid." This model of the atom was modified by Ernest Rutherford in 1911. As a result of his work with radioactivity, Rutherford concluded that most of the mass of an atom is concentrated in a small dense center of the atom he called the **nucleus** and that the nucleus contains positively charged particles called **protons.** In 1932 James Chadwick discovered another particle in the nucleus, the **neutron.** Neutrons have no charge and are slightly more massive than protons (figure 8.3).

Today, the atom is considered to be made up of a number of subatomic particles with the electron, proton, and neutron being the most stable. The electron belongs to a family of elementary particles of which at least six are known. The proton and neutron belong to a family of composite particles of which hundreds are known. These composite particles are believed to be made up of even more elementary particles. For your information, you can read about all of the families of particles and their members in the Feature at the end of chapter 11. For understanding electricity you need only to consider the positively charged protons in the nucleus and the negatively charged electrons that move around the nucleus. Electrons can be moved

from an atom and caused to move through a material, but the forces required to do this vary from one substance to another. Basically, the electrical, light, and chemical phenomena involve the *electrons* and not the more massive nucleus. The massive nuclei remain in a relatively fixed position in a solid, but some of the electrons can move about from atom to atom.

Electric Charge and Electrical Forces

Electrons have a **negative electric charge** and protons have a **positive electric charge.** The negative charge on an electron and the positive charge on a proton describe the way that electrons and protons *are* and how they behave. Charge is as fundamental to these subatomic particles as gravitational attraction is fundamental to masses. This means that you cannot separate gravity from masses, and you cannot separate charge from electrons and protons.

There are only two kinds of electric charge, and the charges interact to produce a force that is called the **electrical force.** *Like charges produce a repulsive electrical force* as positive repels positive and negative repels negative. *Unlike charges produce an attractive electrical force* as positive and negative charges attract each other. The electrical *force* is as fundamental to subatomic particles as the force of gravitational attraction is between two masses. The electrical force is billions and billions of times stronger than the gravitational force between the tiny particles with their tiny masses. Thus when dealing with subatomic particles such as the electron and the proton the electrical force is the force of consequence and the gravitational force is ignored for all practical purposes.

Ordinary atoms are usually neutral because there is a balance between the number of positively charged protons and the number of negatively charged electrons. But when there is an imbalance, say by an electron being torn away by friction, the remaining atom has a net positive charge and is now called a **positive ion** (figure 8.4). The removed electron might continue to be a free negative charge or it might become attached to a neutral atom to make a **negative ion.** Of course, the electron could also become attached to a positive ion to make a neutral atom.

Electrostatic Charge

Electrons can be moved from atom to atom to create ions. They can also be moved from one object to another by friction and by other means that will be discussed soon. Since electrons are negatively charged, an object that acquires an excess of electrons becomes a negatively charged body. The loss of electrons by another body results in a deficiency of electrons, which results in a positively charged object. Thus, *electric charges on objects result from the gain or loss of electrons.* Because the electric charge is confined to an object and is not moving, it is called an **electrostatic charge.** You probably call this charge *static electricity.* Static electricity is an accumulated electric charge at rest, that is, one that is not moving. When you comb your hair with a hard rubber comb, the comb becomes negatively charged because electrons are transferred *from* your hair

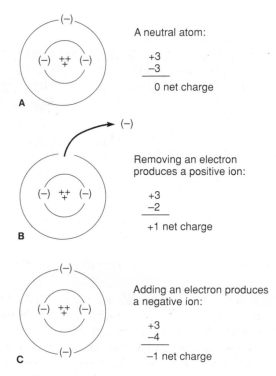

Figure 8.4

(*a*) A neutral atom has no net charge because the number of electrons and protons are balanced. (*b*) Removing an electron produces a net positive charge; the charged atom is called a positive ion. (*c*) The addition of an electron produces a net negative charge and a negative ion.

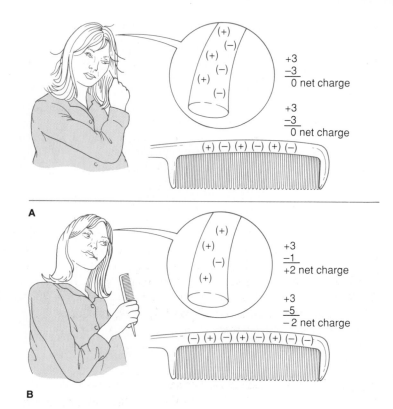

Figure 8.5

Arbitrary numbers of protons (+) and electrons (−) on a comb and in hair (*a*) before and (*b*) after combing. Combing transfers electrons from the hair to the comb by friction, resulting in a negative charge on the comb and a positive charge on the hair.

to the comb. Your hair becomes positively charged with a charge equal in magnitude to the charge gained by the comb (figure 8.5). Both the negative charge on the comb from an excess of electrons and the positive charge on your hair from a deficiency of electrons are charges that are momentarily at rest, so they are electrostatic charges.

Once charged by friction, objects such as a rubber comb soon return to a neutral or balanced state by the movement of electrons. This happens more quickly on a humid day because water vapor assists with the movement of electrons to or from charged objects. Thus static electricity is more noticeable on dry days than on humid ones.

An object can become electrostatically charged (1) by *friction,* which transfers electrons from one object to another, (2) by *contact* with another charged body, which results in the transfer of electrons, or (3) by *induction.* Induction produces a charge by a redistribution of charges in a material. When you comb your hair, for example, the comb removes electrons from your hair and acquires a negative charge. When the negatively charged comb is held near small pieces of paper, it repels some electrons in the paper to the opposite side of the paper. This leaves the side of the paper closest to the comb with a positive charge, and there is an attraction between the pieces of paper and the comb since unlike charges attract. Note that no transfer of electrons takes place in induction; the attraction results from a reorientation of the charges in the paper (figure 8.6).

A C T I V I T I E S

1. This activity works best on a day with low humidity. Tie a string around the lip of a small glass test tube. Have a partner hold one end of the tube with a cloth while rubbing the tube with a silk cloth. You hold a second test tube with a cloth and also rub it with a silk cloth for several minutes. As your partner allows the tube to hang freely from the string, bring your tube near by, but not touching, and observe any interactions. (If nothing happens, try rubbing the tubes longer.)
2. Your partner should again rub the test tube with the silk cloth while you rub a comb with fur or flannel for several minutes. Bring the comb near the hanging test tube and observe any interactions.
3. Tie a string on a second comb. Your partner should rub this comb with fur or flannel as you rub your comb again for several minutes. Bring your comb near the hanging comb and observe any interactions.
4. Describe what you observe during this procedure and give evidence that two kinds of electric charge exist.

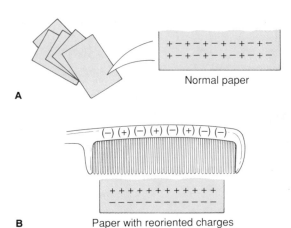

A Normal paper

B Paper with reoriented charges

Figure 8.6

Charging by induction. The comb has become charged by friction, acquiring an excess of electrons. The paper (a) normally has a random distribution of (+) and (−) charges. (b) When the charged comb is held close to the paper, there is a reorientation of charges from the repulsion of like charges. This leaves a net positive charge on the side close to the comb, and since unlike charges attract, the paper is attracted to the comb.

Electrical Conductors and Insulators

When you slide across a car seat or scuff your shoes across a carpet, you are rubbing some electrons from the materials and acquiring an excess of negative charges. Because the electric charge is confined to you and is not moving, it is an electrostatic charge. The electrostatic charge is produced by friction between two surfaces and will remain until the electrons can move away because of their mutual repulsion. This usually happens when you reach for a metal doorknob, and you know when it happens because the electron movement makes a spark. Materials like the metal of a doorknob are good **electrical conductors** because they have electrons that are free to move throughout the metal. If you touch plastic or wood, however, you will not feel a shock. Materials like plastic and wood do not have electrons that are free to move throughout the material, and they are called **electrical nonconductors.** Nonconductors are also called **electrical insulators.** Electrons do not move easily through an insulator, but electrons can be added to or removed and the charge tends to remain. In fact, your body is a poor conductor, which is why you become charged by friction in the first place (table 8.1).

Materials vary in their ability to conduct charges, and this ability is determined by how tightly or loosely the electrons are held to the nucleus. Metals have millions of free electrons that can take part in the conduction of an electric charge. Materials such as rubber, glass, and plastics hold tightly to their electrons and are good insulators. Thus metal wires are used to conduct an electric current from one place to another, and rubber, glass, and plastics are used as insulators to keep the current from going elsewhere.

Table 8.1

Electrical conductors and insulators

Conductors	Insulators
Silver	Rubber
Copper	Glass
Gold	Plastics
Aluminum	Wood
Carbon	
Tungsten	
Iron	
Lead	
Nichrome	

There is a third class of materials, such as silicon and germanium, that sometimes conduct and sometimes insulate depending on the conditions and how pure they are. These materials are called **semiconductors,** and their special properties make possible a number of technological devices such as the electrostatic copying machine, solar cells, and so forth. A Feature on solar cells appears in chapter 9.

Measuring Electrical Charges

As you might have experienced, sometimes you receive a slight shock after walking across a carpet but other times you are really zapped. You receive a greater shock when you have accumulated a greater electric charge. Since there is less electric charge one time and more another, it should be evident that charge is a measurable quantity. The magnitude of an electric charge is identified with the number of electrons that have been transferred onto or away from an object. The quantity of such a charge (q) is measured in a unit called a **coulomb** (C). A coulomb unit is equivalent to the charge resulting from the transfer of 6.24×10^{18} of the charge carried by particles such as the electron. The coulomb is a fundamental metric unit of measure like the meter, kilogram, and second. There is not, however, a direct way of measuring coulombs as exists for measuring meters, kilograms, or seconds because of the difficulty of measuring tiny charged particles.

The coulomb is a *unit* of electric charge that is used with other metric units such as meters for distance and newtons for force. Thus a quantity of charge (q) is described in units of coulomb (C). This is just like the process of a quantity of mass (m) being described in units of kilogram (kg). The concepts of charge and coulomb may seem less understandable than the concepts of mass and kilogram since you cannot *see* charge or how it is measured. But charge does exist and it can be measured, so you can understand both the concept and the unit by working with them. Consider, for example, that an object has a net electric charge (q) because it has an unbalanced number (n) of electrons (e^-) and protons (p^+). The net charge on you after walking across a carpet depends on how many electrons you rubbed from the carpet. The net charge in this case would be the excess of electrons, or

$$\text{quantity of charge} = (\text{number of electrons})(\text{electron charge})$$

or

$$q = ne \qquad \textbf{equation 8.1}$$

Since 1.00 coulomb is equivalent to the transfer of 6.24×10^{18} particles such as the electron, the charge on one electron must be

$$e = \frac{q}{n}$$

where q is 1.00 C and n is 6.24×10^{18} electrons,

$$e = \frac{1.00 \text{ coulomb}}{6.24 \times 10^{18} \text{ electron}}$$

$$= 1.60 \times 10^{-19} \frac{\text{coulomb}}{\text{electron}}$$

This charge, 1.60×10^{-19} coulomb, is the *smallest* common charge known (more exactly $1.6021892 \times 10^{-19}$ C). It is the **fundamental charge** of the electron ($e^- = 1.60 \times 10^{-19}$ C) and the proton ($p^+ = 1.60 \times 10^{-19}$ C). All charged objects have multiples of this fundamental charge.

Example 8.1

Combing your hair on a day with low humidity results in a comb with a negative charge on the order of 1.00×10^{-8} coulomb. How many electrons were transferred from your hair to the comb?

Solution
The relationship between the quantity of charge on an object (q), the number of electrons (n), and the fundamental charge on an electron (e^-) is found in equation 8.1, q = ne.

$$q = 1.00 \times 10^{-8} \text{ C} \qquad q = ne \therefore n = \frac{q}{e}$$

$$e = 1.60 \times 10^{-19} \frac{C}{e}$$

$$n = \frac{1.00 \times 10^{-8} \text{ C}}{1.60 \times 10^{-19} \frac{C}{e}}$$

$$n = ?$$

$$= \frac{1.00 \times 10^{-8}}{1.60 \times 10^{-19}} \cancel{C} \times \frac{e}{\cancel{C}}$$

$$= \boxed{6.25 \times 10^{10} \text{ e}}$$

Thus the comb acquired an excess of approximately 62.5 billion electrons. (Note the convention in scientific notation is to express an answer with one digit to the left of the decimal. See appendix A for further information on scientific notation.)

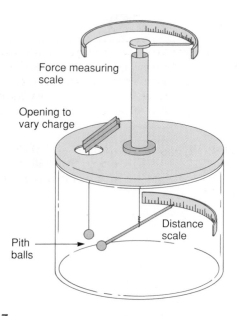

Figure 8.7
Coulomb constructed a torsion balance to test the relationships between a quantity of charge, the distance between the charges, and the electrical force produced. He found the inverse square law held accurately for various charges and distances.

Measuring Electrical Forces
Recall that two objects with like charges, $(-)$ and $(-)$ or $(+)$ and $(+)$, produce a repulsive force and two objects with unlike charges, $(+)$ and $(-)$, produce an attractive force. These forces were investigated by Charles Coulomb, a French military engineer, in 1785. Coulomb invented a sensitive balance to measure the forces between two pith balls (figure 8.7). Pith is a light material made from the dried inside of the stem of a plant or from dried potatoes. Coulomb found that the force between two electrically charged pith balls was (1) directly proportional to the product of the electric charge and (2) inversely proportional to the square of the distance between them. In symbols, q_1 represents the quantity of electric charge on object 1 and q_2 represents the quantity of electric charge on object 2, d is the distance between objects 1 and 2, and k is a constant of proportionality. The magnitude of the electrical force between object 1 and object 2, F, is either attractive (unlike charges) or repulsive (like charges) depending on the charges on the two objects. This relationship, known as **Coulomb's law,** is

$$F = k \frac{q_1 q_2}{d^2} \qquad \textbf{equation 8.2}$$

where k has the value of 9.00×10^9 newton-meters^2/coulomb^2 (9.00×10^9 N·m^2/C^2).

Example 8.2

Electrons carry a negative electric charge and revolve about the nucleus of the atom, which carries a positive electric charge from the proton. The electron is held in orbit by the force of electrical attraction at a typical distance of 1.00×10^{-10} m. What is the force of electrical attraction between an electron and proton?

Solution

The fundamental charge of an electron (e^-) is 1.60×10^{-19} C and the fundamental charge of the proton (p^+) is 1.60×10^{-19} C. The distance is given and the force of electrical attraction can be found from equation 8.2:

$$q_1 = 1.60 \times 10^{-19} \text{ C}$$
$$q_2 = 1.60 \times 10^{-19} \text{ C}$$
$$d = 1.00 \times 10^{-10} \text{ m}$$
$$k = 9.00 \times 10^9 \text{ N} \cdot \text{m}^2/\text{C}^2$$
$$F = ?$$

$$F = k \frac{q_1 q_2}{d^2}$$

$$= \frac{\left(9.00 \times 10^9 \frac{\text{N} \cdot \text{m}^2}{\text{C}^2}\right)(1.60 \times 10^{-19} \text{ C})(1.60 \times 10^{-19} \text{ C})}{(1.00 \times 10^{-10} \text{ m})^2}$$

$$= \frac{(9.00 \times 10^9)(1.60 \times 10^{-19})(1.60 \times 10^{-19})}{1.00 \times 10^{-20}} \frac{\left(\frac{\text{N} \cdot \text{m}^2}{\text{C}^2}\right)(\text{C}^2)}{\text{m}^2}$$

$$= \frac{2.30 \times 10^{-28} \text{ N} \cdot \text{m}^2}{1.00 \times 10^{-20}} \times \frac{\text{C}^2}{\text{C}^2} \times \frac{1}{1} \times \frac{1}{\text{m}^2}$$

$$= \boxed{2.30 \times 10^{-8} \text{ N}}$$

The electrical force of attraction between the electron and proton is 2.30×10^{-8} newton.

Force Fields

Does it seem odd to you that gravitational forces and electrical forces can act on objects that are not touching? How can gravitational forces act through the vast empty space between the earth and the sun? How can electrical forces act through a distance to pull pieces of paper to your charged comb? Such questions have bothered people since the early discovery of small, light objects being attracted to rubbed amber. There was no mental model of how such a force could act through a distance without touching. The idea of "invisible fluids" was an early attempt to develop a mental model that would help people visualize how a force could act over a distance without physical contact. Then Newton developed the law of universal gravitation, which correctly predicted the magnitude of gravitational forces acting through space. Coulomb's law of electrical forces had similar success in describing and predicting electrostatic forces acting through space. "Invisible fluids" were no longer needed to explain what was happening because the two laws seemed to explain the results of such actions. But it was still difficult to visualize what was happening physically when forces acted through a distance, and there were a few problems with the concept of action at a distance. Not all observations were explained by the model.

The work of Michael Faraday and James Maxwell in the early 1800s finally provided a new mental model for interaction at a distance. This new model did *not* consider the force that one object exerts on another one through a distance. Instead, it considered *the condition of space* around an object. The condition of space around an electric charge is considered to be changed by the presence of the charge. The charge produces a **force field** in the space around it. Since this force field is produced by an electrical charge, it is called an **electric field**. Imagine a second electric charge, called a *test charge*, that is far enough away from the electric charge that no forces are experienced. As you move the test charge closer and closer, it will experience an increasing force as it enters the electric field. The test charge is assumed not to change the field that it is entering and can be used to identify the electric field that spreads out and around the space of an electric charge.

All electric charges are considered to be surrounded by an electric field. All *masses* are considered to be surrounded by a *gravitational field*. The earth, for example, is considered to change the condition of space around it because of its mass. A spaceship far, far from the earth does not experience a measurable force. But as it approaches the earth, it enters the earth's gravitational field and thus it experiences a measurable force. Likewise, a magnet creates a *magnetic field* in the space around it. You can visualize a magnetic field by moving a magnetic compass needle around a bar magnet. Far from the bar magnet the compass needle does not respond. Moving it closer to the bar magnet, you can see where the magnetic field begins. Another way to visualize a magnetic field is to place a sheet of paper over a bar magnet, then sprinkle iron filings on the paper. The filings will clearly identify the presence of the magnetic field.

Another way to visualize a field is to make a map of the field. Consider a small positive test charge that is brought into an electric field. A *positive* test charge is always used by convention. As shown in figure 8.8, a positive test charge is brought near a negative charge and a positive charge. The vector arrow points in the direction of the force that the *test charge experiences*. Thus, when brought near a negative charge, the test charge is attracted toward the unlike charge and the arrow points that way. When brought near a positive charge the test charge is repelled so the arrow points away from the positive charge.

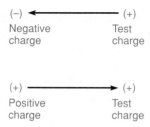

(−) ◄────────── (+)
Negative   Test
charge   charge

(+) ──────────► (+)
Positive   Test
charge   charge

Figure 8.8

A *positive test charge* is used by convention to identify the properties of an electric field. The vector arrow points in the direction of the force that the test charge would experience.

An electric field is defined by using a positive test charge as a measuring device. The test charge will experience a different force in different directions when located at various points in the field. The strength of the field is represented by the symbol E, which is a vector quantity since it has direction as well as magnitude.

The magnitude of E is defined as the force on a positive test charge divided by the magnitude of the test charge, or

$$\text{electric field strength} = \frac{\text{force}}{\text{quantity of charge}}$$

$$E = \frac{F}{q} \qquad \textbf{equation 8.3}$$

In units, the electric force (F) is measured in newtons and the test charge (q) is measured in coulombs, so the **electric field strength** is measured in force per unit charge, or newtons per coulomb.

Note that an electric field is *defined* in terms of a measurement on a test charge. Using a measurement procedure to define a concept is another operational definition; it is the same procedure that was used to define a newton of force.

An electric field is represented by drawing **lines of force** or **electric field lines** that show the direction of the field. The vector arrows in figure 8.9 show field lines that could extend outward forever from isolated charges since there is always some force on a distant test charge (review Coulomb's law; the force ideally never reaches zero). The field lines between pairs of charges in figure 8.9 show curved field lines that originate on positive charges and end on negative charges. By convention, the field lines are closer together where the field is stronger and farther apart where the field is weaker.

The field concept explains some observations that were not explained with the Newtonian concept of action at a distance. Suppose, for example, that a charge produces an electric field. This field is not instantaneously created all around the charge, but it is seen to build up and spread into space. If the charge is suddenly neutralized, the field that it created continues to spread outward, then collapses back at some speed even though the source of the field no longer exists. Consider an example with the gravitational field of the sun. If the mass of the sun were to instantaneously disappear, would the earth notice this instantaneously? Or would the gravitational field of the sun collapse

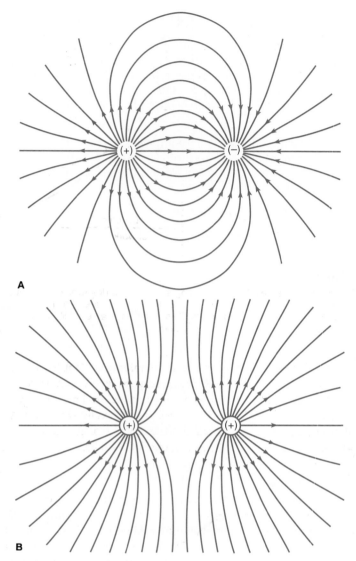

A

B

Figure 8.9

Lines of force diagrams for (a) a negative charge and (b) a positive charge when the charges have the same magnitude as the test charge.

at some speed, say the speed of light, to be noticed by the earth some eight minutes later? The Newtonian concept of action at a distance did not consider any properties of space, so according to this concept the gravitational force from the sun would disappear instantly. The field concept, however, explains that the disappearance would be noticed after some period of time, about eight minutes. This time delay agrees with similar observations and objects interacting with fields, so the field concept is more useful than a mysterious action at a distance concept, as you will see.

Actually there are three models for explaining how gravitational, electrical, and magnetic forces operate at a distance. (1) The *action at a distance model* recognizes that masses are attracted gravitationally and that electric charges and magnetic poles attract and repel each other through space but gives no further explanation; (2) the *field model* considers a field to be a condition of space around a mass, electric charge, or magnet, and the properties of fields are described by field lines; and

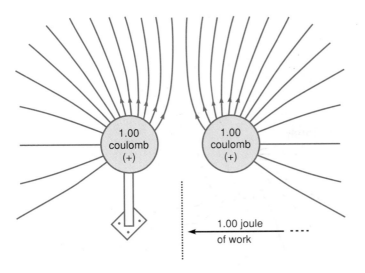

Figure 8.10

Electric potential results from moving a positive coulomb of charge into the electric field of a second positive coulomb of charge. When 1.00 joule of work is done in moving 1.00 coulomb of charge, 1.00 volt of potential results. A volt is a joule/coulomb.

(3) the *field particle model* is a complex and highly mathematical explanation of attractive and repulsive forces as the rapid emission and absorption of subatomic particles. This model explains electrical and magnetic forces as the exchange of *virtual photons,* gravitational forces as the exchange of *gravitons,* and strong nuclear forces as the exchange of *mesons.*

Electric Potential

Recall from chapter 4 that work is accomplished as you move an object to a higher location on the earth, say by moving a book from the first shelf of a bookcase to a higher shelf. By virtue of its position, the book now has gravitational potential energy that can be measured by mgh (the force of the book's weight × distance), so many joules of gravitational potential energy. Using the field model, you could say that this work was accomplished against the gravitational field of the earth. Likewise, an electric charge has an electric field surrounding it, and work must be done to move a second charge into or out of this field. Bringing a like charged particle *into* the field of another charged particle will require work since like charges repel, and separating two unlike charges will also require work since unlike charges attract. In either case, the **electric potential energy** (E_p) is changed just as the gravitational potential energy is changed by moving a mass in the earth's gravitational field.

One useful way to measure electrical potential energy (E_p) is to consider the *potential difference* that occurs when a certain amount of work (W) is used to move a certain quantity of charge (q). For example, suppose there is a firmly anchored and insulated metal sphere that has a positive charge (figure 8.10). The sphere will have a positive electric field in the space around it. Suppose also that you have a second sphere and that this sphere has exactly 1.00 coulomb of positive charge. You begin moving the coulomb of positive charge toward the anchored sphere. As you enter the electric field you will have to push

harder and harder to overcome the increasing repulsion. If you stop moving when you have done exactly 1.00 joule of work, the repulsion will *do* one joule of work if you now release the sphere. The sphere has potential energy in the same way that a compressed spring possesses potential energy. In electrical matters, the *potential difference that is created by doing 1.00 joule of work in moving 1.00 coulomb of charge is defined to be 1.00 volt.* The **volt** (V) is a measure of potential difference between two points, or

$$\text{electric potential} = \frac{\text{work to create potential}}{\text{charge moved}}$$

$$E_p = \frac{W}{q} \qquad \text{equation 8.4}$$

In units,

$$1.00 \text{ volt (V)} = \frac{1.00 \text{ joule (J)}}{1.00 \text{ coulomb (C)}}$$

The voltage of any electric charge, either static or moving, is the energy transfer per coulomb. The energy transfer can be measured by the *work that is done to move the charge* or by the *work that the charge can do* because of its position in the field. This is perfectly analogous to the work that must be done to give an object gravitational potential energy or to the work that the object can potentially do because of its new position. Thus when a 12 volt battery is charged, 12.0 joules of work are done to transfer 1.00 coulomb of charge from an outside source against the electric field of the battery terminal. When the 12 volt battery is used, it does 12.0 joules of work for each coulomb of charge transferred from one terminal of the battery through the electrical system and back to the other terminal.

Electric Current

So far, we have considered electric charges that have been instantaneously moved by friction but then generally stayed in one place. Experiments with static electricity played a major role in the development of the understanding of electricity by identifying charge, the attractive and repulsive forces between charges, and the field concept. The work of Franklin, Coulomb, Faraday, and Maxwell thus increased studies of electrical phenomena and eventually led to insights into connections between electricity and magnetism, light, and chemistry. These connections will be discussed later, but for now, consider the sustained flowing or moving of charge, an **electric current** (I). Electric current means a flow of charge in the same way that "water current" means a flow of water. Since the word "current" *means* flow, you are being redundant if you speak of "flow of current." It is the *charge* that flows, and the current is defined as the flow of charge. Note that the symbol for a quantity of current (I) is not an abbreviation for the word "current."

The Electric Circuit

When you slide across a car seat, you are acquiring electrons on your body by friction. Through friction, you did *work* on the electrons as you removed them from the seat covering. You now

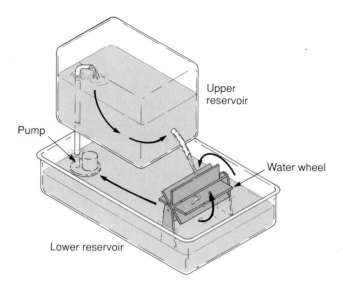

Figure 8.11

The falling water can do work in turning the waterwheel only as long as the pump maintains the potential difference between the upper and lower reservoirs.

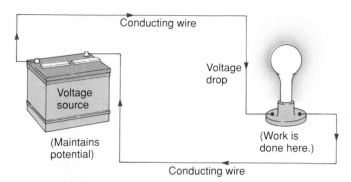

Figure 8.12

A simple electric circuit has a voltage source (such as a generator or battery) that maintains the electrical potential, some device (such as a lamp or motor) where work is done by the potential, and continuous pathways for the current to follow.

have a net negative charge from the imbalance of electrons, which tend to remain on you because you are a poor conductor. But the electrons are now closer than they want to be, within a repulsive electric field, and there is an electrical potential difference (E_p) between you and some uncharged object, say a metal door handle. When you touch the handle, the electrons will flow, creating a momentary current in the form of a spark, which lasts only until the charge on you is neutralized.

In order to keep an electric current going, you must maintain the separation of charges and therefore maintain the electric field (or potential difference), which can push the charges through a conductor. This might be possible if you could somehow continuously slide across the car seat, but this would be a hit-and-miss way of maintaining a separation of charges and would probably result in a series of sparks rather than a continuous current. This is how electrostatic machines work.

A useful analogy for understanding the requirements for a sustained electric current is the decorative waterwheel device (figure 8.11). Water in the upper reservoir has a greater gravitational potential energy than water in the lower reservoir. As water flows from the upper reservoir it can do work in turning the waterwheel, but it can continue to do this only as long as the pump does the work to maintain the potential difference between the two reservoirs. This "water circuit" will do work in turning the waterwheel as long as the pump returns the water to a higher potential continuously as the water flows back to the lower potential.

So, by a water circuit analogy, a steady electric current is maintained by pumping charges to a higher potential, and the charges do work as they move back to a lower potential. An **electric circuit** contains some device, such as a battery or electric generator, that acts as a source of energy as it gives charges

a higher potential against an electric field. The charges do work in another part of the circuit as they light bulbs, run motors, or provide heat. The charges flow through connecting wires to make a continuous path along which the current flows, maintaining the potential difference. An electric switch is a means of interrupting or completing this continuous path.

The potential difference between the two connecting wires shown in figure 8.12 is one factor in the work done *by* the device that creates a higher potential (battery, for example) and the work done *in* some device (lamp, for example). Disregarding any losses, the work done in both places would be the same. Recall that work done per unit of charge is joules/coulomb, or volts (equation 8.4). The source of the potential difference is therefore referred to as a **voltage source,** and the device where the charges do their work uses the potential difference and causes a **voltage drop.** Potential difference is measured in volts so the term *voltage* is often used for it. Household circuits usually have a difference of potential of 120 or 240 volts. A voltage of 120 means that each coulomb of charge that moves through the circuit can do 120 joules of work in some electrical device.

Voltage describes the potential difference, in joules/coulomb, between two places in an electric circuit. By way of analogy to pressure on water in a circuit of water pipes, this potential difference is sometimes called an "electrical force" or "electromotive force" (emf). Note that in electrical matters, however, the potential difference is the *source* of a force rather than being a force such as water under pressure. Nonetheless, just as you can have a small water pipe and a large water pipe under the same pressure, the two pipes would have a different rate of water flow in gallons per minute. The *rate* at which an electric current (I) flows is the quantity of charge (q) that moves through a cross section of a conductor in a unit of time (t), or

$$\text{electric current} = \frac{\text{quantity of charge}}{\text{time}}$$

$$I = \frac{q}{t} \qquad \textbf{equation 8.5}$$

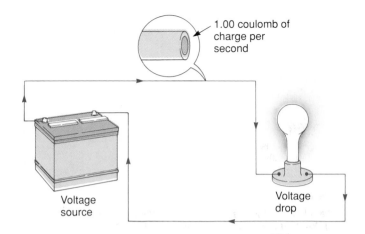

Figure 8.13

A simple electric circuit carrying a current of 1.00 coulomb per second through a cross section of a conductor has a current of 1.00 amp.

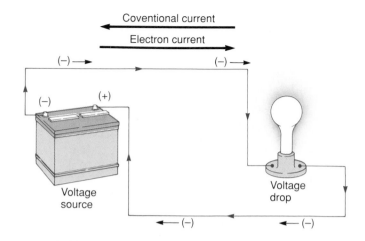

Figure 8.14

A conventional current describes positive charges moving from the positive terminal (+) to the negative terminal (−). An electron current describes negative charges (−) moving from the negative terminal (−) to the positive terminal (+).

The units of current are thus coulombs/second. A coulomb/second is called an **ampere** (A), or **amp** for short. In units, current is therefore

$$1.00 \text{ amp (A)} = \frac{1.00 \text{ coulomb (C)}}{1.00 \text{ second (sec)}}$$

A 1.00 amp current is 1.00 coulomb of charge moving through a conductor each second, a 2.00 amp current is 2.00 coulombs per second, and so forth (figure 8.13).

Using the water circuit analogy, you would expect a greater rate of water flow (gallons/minute) when the water pressure is produced by a greater gravitational potential difference. The rate of water flow is thus directly proportional to the difference in gravitational potential energy. In an electric circuit the rate of current (coulombs/second, or amps) is directly proportional to the difference of electrical potential (joules/coulombs, or volts) between two parts of the circuit, $I \propto E_p$.

The Nature of Current

There are two ways to describe the current that flows outside the power source in a circuit, (1) a historically based description called **conventional current** and (2) a description based on a flow of charges called **electron current**. The *conventional* current describes current as positive charges that flow from the positive to the negative terminal of a battery. This description has been used by convention ever since Ben Franklin first misnamed the charge of an object based on an accumulation, or a positive amount, of "electrical fluid." Conventional current is still used in circuit diagrams. The *electron* current description is in an opposite direction to the conventional current. The electron current describes current as the drift of negative charges that flow from the negative to the positive terminal of a battery. Today, everyone understands the role of electrons in a current, something that was unknown to Franklin. But conventional current is still used by tradition. It actually does not make any difference which description is used since positive charges moving

from the positive terminal is mathematically equivalent to negative charges moving from the negative terminal (figure 8.14).

The description of an electron current also retains historical traces of the earlier fluid theories of electricity. Today, people understand that electricity is not a fluid but still speak of current, rate of flow, and resistance to flow (figure 8.15). Fluid analogies can be helpful because they describe the overall electrical effects. But they can also lead to incorrect concepts such as: (1) an electric current is the movement of electrons through a wire just as water flows through a pipe; (2) electrons are pushed out one end of the wire as more electrons are pushed in the other end; and (3) electrons must move through a wire at the speed of light since a power plant failure hundreds of miles away results in an instantaneous loss of power. Perhaps you have held one or more of these misconceptions from fluid analogies.

What is the exact nature of an electric current? First, consider the nature of a metal conductor without a current. The atoms making up the metal have unattached electrons that are free to move about, much as the molecules of a gas in a container. They randomly move at high speed in all directions, often colliding with each other and stationary positive ions of the metal. This motion is chaotic, and there is no net movement in any one direction, but the motion does increase with increases in the absolute temperature of the conductor.

When a potential difference is applied to the wire in a circuit an electric field is established everywhere in the circuit. The *electric field* travels through the conductor near the speed of light as it is established. A force is exerted on each electron by the field, which accelerates the free electrons in the direction of the force. The resulting increased velocity of the electrons is superimposed on their existing random, chaotic movement. This added motion is called the *drift velocity* of the electrons. The drift velocity of the electrons is a result of the imposed electric field. The electrons do not drift straight through the conductor, however, because they undergo countless collisions with other

Figure 8.15

What is the nature of the electric current carried by these conducting lines? It is an electric field that moves at near the speed of light. The field causes a net motion of electrons that constitutes a flow of charge, a current.

electrons and the stationary positive ions. This results in a random zigzag motion with a net motion in one direction. *This net motion constitutes a current,* a flow of charge (figure 8.16).

When the voltage across a conductor is zero the drift velocity is zero and there is no current. The current that occurs when there is a voltage depends on (1) the number of free electrons per unit volume of the conducting material, (2) the charge on each electron (the fundamental charge), (3) the drift velocity, which depends on the electronic structure of the conducting material and the temperature, and (4) the cross-sectional area of the conducting wire.

The relationship between the number of free electrons, charge, drift velocity, area, and current can be used to determine the drift velocity when a certain current flows in a certain size wire made of copper. A 1.0 amp current in copper bell wire (#18), for example, has an average drift velocity on the order

of 0.01 cm/sec. At that rate, it would take over 5 hr for an electron to travel the 200 cm from your car battery to the brake light of your car (figure 8.17). Thus it seems clear that it is the *electric field,* not electrons, that causes your brake light to come on almost instantaneously when you apply the brake. The electric field gives energy to the electrons already in the filament of the light bulb in the brake light and the bulb glows.

Conclusions about the nature of an electric current are that (1) a potential difference establishes an electric field that travels through a circuit at near the speed of light, (2) the field causes a net motion that constitutes a flow of charge, or current, and (3) the average velocity of the electrons moving as a current is very slow even though the electric field that moves them travels with a speed close to the speed of light.

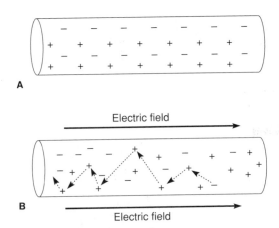

A

Electric field →

B

Electric field →

Figure 8.16

(*a*) A metal conductor without a current has immovable positive ions surrounded by a swarm of chaotically moving electrons. (*b*) An electric field causes the electrons to shift positions, creating a separation charge as the electrons move with a zigzag motion from collisions with stationary positive ions and other electrons.

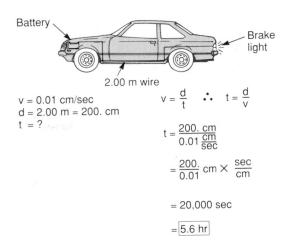

$v = 0.01$ cm/sec
$d = 2.00$ m $= 200.$ cm
$t = ?$

$v = \dfrac{d}{t} \quad \therefore \quad t = \dfrac{d}{v}$

$t = \dfrac{200.\text{ cm}}{0.01\frac{\text{cm}}{\text{sec}}}$

$= \dfrac{200.}{0.01}$ cm $\times \dfrac{\text{sec}}{\text{cm}}$

$= 20,000$ sec

$= \boxed{5.6\text{ hr}}$

Figure 8.17

Electrons move very slowly in a direct current circuit. With a drift velocity of 0.01 cm/sec, more than 5 hr would be required for an electron to travel 200 cm from a car battery to the brake light. It is the electric field, not the electrons, that moves at near the speed of light in an electric circuit.

Another aspect of the nature of an electric current is the direction the charge is flowing. A circuit like the one described with your car battery has a current that always moves in one direction, a **direct current** (dc). Chemical batteries, fuel cells, and solar cells produce a direct current, and direct currents are utilized in electronic devices. Electric utilities and most of the electrical industry, on the other hand, use an **alternating current** (ac). An alternating current, as the name implies, moves the electrons alternately one way then the other way. Since the electrons are simply moving back and forth, there is no electron drift along a conductor in an alternating current. Since household electric circuits use alternating current, there is no movement of electrons from the electrical outlets through the circuits. The electric field moves back and forth through the circuit near the speed of light, moving electrons back and forth. This movement constitutes a current that flows one way, then the other with the changing field. The current changes like this 120 times a second in a 60 hertz alternating current.

Electrical Resistance

Recall the natural random and chaotic motion of electrons in a conductor and their frequent collisions with each other and with the stationary positive ions. When these collisions occur, electrons lose energy that they gained from the electric field. The stationary positive ions gain this energy, and their increased energy of vibration results in a temperature increase. Thus there is a resistance to the movement of electrons being accelerated by an electric field and a resulting energy loss. Materials have a property of opposing or reducing a current, and this property is called **electrical resistance** (R).

Recall that the current (I) through a conductor is directly proportional to the potential difference (E_p) between two points in a circuit. If a conductor offers a small resistance, less voltage would be required to push an amp of current through the circuit. If a conductor offers more resistance, then more voltage will be required to push the same amp of current through the circuit. Resistance (R) is therefore a *ratio* between the potential difference (E_p) between two points and the resulting current (I). This ratio is

$$\text{resistance} = \frac{\text{electrical potential}}{\text{current}}$$

$$R = \frac{E_p}{I}$$

In units, this ratio is

$$1.00 \text{ ohm } (\Omega) = \frac{1.00 \text{ volt (V)}}{1.00 \text{ amp (A)}}$$

The ratio of volts/amps is the unit of resistance called an **ohm** (Ω) after the German physicist who discovered the relationship. The resistance of a conductor is therefore 1.00 ohm if 1.00 volt is required to maintain a 1.00 amp current. The ratio of volt/amp is *defined as* an ohm. Therefore,

$$\text{ohm}\left(\frac{\text{volt}}{\text{amp}}\right) = \frac{\text{volt}}{\text{amp}}$$

Another way to show the relationship between the voltage, current, and resistance is

$$V = IR \qquad\qquad \textbf{equation 8.6}$$

which is known as **Ohm's law.** This is one of three ways to show the relationship, but this way (solved for V) is convenient for easily solving the equation for other unknowns.

Electricity

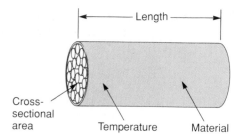

Figure 8.18
The four factors that influence the resistance of an electrical conductor are the length of the conductor, the cross-sectional area of the conductor, the material the conductor is made of, and the temperature of the conductor.

The magnitude of the electrical resistance of a conductor depends on four variables: (1) the length of the conductor, (2) the cross-sectional area of the conductor, (3) the material the conductor is made of, and (4) the temperature of the conductor (figure 8.18). Different kinds of material have different resistances as shown by the list of conductors in table 8.1. Silver, for example, is at the top of the list because it offers the least resistance, followed by copper, gold, then aluminum. Of the materials listed in table 8.1, nichrome is the conductor with the greatest resistance. By definition, conductors have less electrical resistance than insulators, which have a very large electrical resistance.

The length of a conductor varies directly with the resistance, that is, a longer wire has more resistance and a shorter wire has less resistance. In addition, the cross-sectional area of a conductor varies inversely with the resistance. A thick wire has a greater area and therefore has less resistance than a thin wire. For most materials the resistance increases with increases in temperature. As previously discussed, this is a consequence of the increased motion of electrons and ions at higher temperatures, which offers increased interference to the current. At very low temperatures the resistance of some materials approaches zero, and the materials are said to be **superconductors.** Recent discoveries and possible uses of superconductors are discussed in the Feature at the end of this chapter.

Example 8.3

A light bulb in a 120. V circuit is switched on and a current of 0.500 A flows through the filament. What is the resistance of the bulb?

Solution

The current (I) of 0.500 is given with a potential difference (V) of 120. V. The relationship to resistance (R) is given by Ohm's law (equation 8.6)

$$I = 0.500 \text{ A} \qquad V = IR \therefore R = \frac{V}{I}$$

$$V = 120. \text{ V}$$

$$R = ? \qquad \qquad = \frac{120. \text{ V}}{0.500 \text{ A}}$$

$$= 240. \frac{V}{A}$$

$$= 240. \text{ ohm}$$

$$= \boxed{240. \ \Omega}$$

Example 8.4

What current would flow through an electrical device in a circuit with a potential difference of 120 V and a resistance of 30 Ω? (Answer: 4 A)

Electrical Power and Electrical Work

All electric circuits have three parts in common: (1) a *voltage source* such as a battery or electric generator that uses some nonelectric source of energy to do work on electrons, moving them *against* an electric field to a higher potential; (2) an *electric device* such as a light bulb or electric motor where work is done *by* the electric field; and, (3) *conducting wires* that maintain the potential difference across the electrical device. In a direct current circuit, the electric field moves from one terminal of a battery to the electric device through one wire. The second wire from the device carries the now low potential field back to the other terminal, maintaining the potential difference. In an alternating current circuit, such as a household circuit, one wire supplies the alternating electric field from the electric generator of a utility company. The second wire from the device is connected to a pipe in the ground and is at the same potential as the earth. This wire is analogous to the cold reservoir in a heat engine, and its purpose is to maintain the potential difference. The observation that a bird can perch on a current-carrying wire without harm is explained by the fact that there is no potential difference across the bird's body. If the bird were to come into contact with the earth through a second, grounded wire a potential difference would be established and a current would flow through it.

The work done by a voltage source (battery, electric generator) is equal to the work done by the electric field in an electric device (light bulb, electric motor) *plus* the energy lost to resistance. Resistance is analogous to friction in a mechanical device, so low resistance conducting wires are used in a way to reduce this loss. Disregarding losses to resistance, electrical work can therefore be measured where the voltage source creates a potential difference by doing work (W) to move charges (q) to a higher potential (E_p). From equation 8.4, this relationship is

$$\text{work} = (\text{potential})(\text{charge})$$

or

$$W = E_p q$$

In units, the electrical potential is measured in joules/coulomb and a quantity of charge is measured in coulombs. Therefore the unit of electrical work is the *joule*,

$$E_p \times q = W$$

$$\frac{\text{joules}}{\text{coulomb}} \times \text{coulomb} = \text{joules}$$

Recall that a joule is a unit of work in mechanics (a newton-meter). In electricity a joule is also a unit of work, but it is derived from moving a quantity of charge (coulomb) to higher potential difference (joules/coulomb). In mechanics the work put into a simple machine equals the work output when you disregard *friction*. In electricity the work put into an electric

A

B

Figure 8.19

(a) This light bulb is designed to operate on a potential difference of 120 volts and will do work at the rate of 75 watts. (b) This strobe light does work at the rate of 2 amp × 240 volt, or 480 watts.

circuit equals the work output when you disregard *resistance*. Thus the work done by a voltage source is ideally equal to the work done by electrical devices in the circuit.

Recall also that mechanical power (P) was defined as work (W) per unit time (t), or

$$P = \frac{W}{t} \qquad \text{equation 8.7}$$

Since electrical work is $W = E_p q$, then electrical power must be

$$P = \frac{E_p q}{t} \qquad \text{equation 8.8}$$

Equation 8.5 defined a quantity of charge (q) per unit time (t) as a current (I), or $I = q/t$. Therefore electrical power is

$$P = \left(\frac{q}{t}\right)(E_p)$$

or, in units

$$P = IV \qquad \text{equation 8.9}$$

In units, you can see that multiplying the current ($I = C/\text{sec}$) by the potential ($V = J/C$) yields

$$\frac{\text{coulombs}}{\text{second}} \times \frac{\text{joules}}{\text{coulombs}} = \frac{\text{joules}}{\text{second}}$$

A joule/second is a unit of power called the **watt.** Therefore electrical power is measured in units of watts and

$$I \times V = P$$

or

current (in amps) × potential (in volts) = power (in watts)

or

amps × volts = watts

Household electrical devices are designed to operate on a particular voltage, usually 120 or 240 volts (figure 8.19). They therefore draw a certain current to produce the designed power. Information about these requirements is usually found somewhere on the device. A light bulb, for example, is usually stamped with the designed power such as 100 watts. Other electrical devices may be stamped with amp and volt requirements. You can determine the power produced in these devices by using equation 8.9, that is, amps × volts = watts. Another handy conversion factor to remember is that 746 watts are equivalent to 1.00 horsepower.

Example 8.5

A 1,300. w hair dryer is designed to operate on 120. V. How much current does the dryer require?

Solution

The power (P) produced is given in watts with a potential difference of 120. V across the dryer. The relationship between the units of amps, volts, and watts is found in equation 8.9, $P = IV$

$$P = 1,300 \text{ w} \qquad P = IV \ \therefore \ I = \frac{P}{V}$$

$$V = 120. \text{ V}$$

$$I = ? \text{ A}$$

$$= \frac{1,300. \ \dfrac{\text{joule}}{\text{second}}}{120. \ \dfrac{\text{joule}}{\text{coulomb}}}$$

$$= \frac{1,300.}{120.} \ \frac{J}{\text{sec}} \times \frac{C}{J}$$

$$= 10.8 \ \frac{J \cdot C}{\text{sec} \cdot J}$$

$$= 10.8 \ \frac{C}{\text{sec}}$$

$$= \boxed{10.8 \text{ A}}$$

Figure 8.20

This meter measures the amount of electric *work* done in the circuits, usually over a time period of a month. The work is measured in kwhr.

Example 8.6

An electric fan is designed to draw 0.500 A in a 120. V circuit. What is the power rating of the fan? (Answer: 60.0 w)

An electric utility company measures the electrical work done in a household with a meter located near where the wires enter the house. The meter measures kilowatt-hours (kwhr) of work. Since household electrical devices are designed to operate at a particular potential (V), the utility company wants to know the current (I) you used for what time period. When you multiply volts × amps × time, you are really multiplying power × time and the answer is the work done, $W = P \times t$ (from equation 8.7). Since $P = IV$ (equation 8.9), then an expression of work can be obtained by combining the two equations, or

$$IV \times t = W \qquad \textbf{equation 8.10}$$

In units,

$$\frac{\text{coulomb}}{\text{second}} \times \frac{\text{joules}}{\text{coulomb}} \times \text{second} = \text{joules}$$

Thus, the utility meter shows the amount of work that was done in a household for a month. The joule is a small unit of work and its use would result in some very large numbers. The electric utility therefore uses a kilowatt- (1,000 × amp × volt)

hour (3,600 sec) to measure the electrical work done (figure 8.20). One kilowatt-hour of work is thus equivalent to 3,600,000 joules. A typical monthly electric bill shows the charge for 1,000 kwhr, not 3,600,000,000 joules.

The electric utility charge for the electrical work done is at a rate of cents per kwhr. The rate varies from place to place across the country depending on the cost of producing the power. You can predict the cost of running a particular electrical appliance by first finding the work done in kwhr from equation 8.10. In units,

$$\text{kwhr} = \frac{(\text{volts})(\text{amps})(\text{time})}{1{,}000 \, \frac{\text{w}}{\text{kw}}}$$

Note that volts × amps = watts, so a watt power rating can be substituted for amps × volts. Also note that the time unit is in hours, so if you want to know the cost of running an appliance for a number of minutes, this must be converted to the decimal equivalent of a hour. Once the work in kwhr is found the cost of running the appliance is determined by multiplying the rate by the kwhr used, or

$$(\text{work})(\text{rate}) = \text{cost} \qquad \textbf{equation 8.11}$$

Example 8.7

What is the cost of operating a 100. w light bulb for 1.00 hr if the utility rate is $0.10 per kwhr?

Solution

The power rating is given as 100. w so the volt and amp units are not needed. Therefore,

$$IV = P = 100. \, \text{w} \qquad\qquad \text{kwhr} = \frac{(\text{volts})(\text{amps})(\text{time})}{1{,}000 \, \frac{\text{w}}{\text{kw}}}$$

$$t = 1.00 \, \text{hr}$$

$$\text{rate} = \$0.10/\text{kwhr} \qquad\qquad = \frac{(100. \, \text{w})(1.00 \, \text{hr})}{1{,}000 \, \frac{\text{w}}{\text{kw}}}$$

$$\text{cost} = ?$$

$$= \frac{100. \times 1.00}{1{,}000} \, \text{w} \cdot \text{hr} \times \frac{\text{kw}}{\text{w}}$$

$$= 0.100 \, \text{kwhr}$$

At a rate of $0.10/kwhr the cost is

$$\text{cost} = (\text{work})(\text{rate})$$

$$= (0.100 \, \text{kwhr})\left(\frac{\$0.10}{\text{kwhr}}\right)$$

$$= 0.100 \times 0.10 \, \frac{(\$)(\text{kwhr})}{\text{kwhr}}$$

$$= \boxed{\$0.01}$$

The cost of operating a 100 w light bulb at a rate of 10¢/kwhr is 1¢/hr.

Superconductivity

Most common metals have a property of electrical resistance that decreases uniformly as the temperature drops. At normal operating temperatures the resistance decreases about 0.004 for each 1° C of cooling. At very low temperatures, below 20 K (about −423° F), some pure metals lose virtually all resistance to an electric current (less than $4 \times 10^{-25} \, \Omega \cdot m$) and thus have the property known as *superconductivity*. Superconductivity was discovered in 1911 by Heike Kamerlingh Onnes, a Dutch physicist. Onnes had developed a method for liquefying the normally gaseous element helium. By accident, he discovered that frozen mercury submerged in liquid helium at 4.2 K (about −452° F) lost its electrical resistance and became a superconductor. There were no practical applications of this curiosity, however, since the superconducting property could be maintained only by continued submersion in the expensive, difficult to handle helium. Furthermore, there was no theory or adequate explanation of superconductivity to guide development.

A theory of superconductivity was not developed until 1957, some forty-six years after Onnes's initial discovery. The theory is known as the BCS theory of superconductivity after the names of the developers, John Bardeen, Leon Cooper, and John Schrieffer. The BCS theory explains that electrons in a superconductor form pairs. These pairs, called Cooper pairs, interact with surrounding atoms in the superconducting material, overcoming their normal repulsive force between the electrons. The interaction also permits the Cooper pairs to move between the atoms without collisions, and, without collisions, there is no electrical resistance. Once started, a current in a closed loop of superconducting material continues for days or weeks, continuously creating a magnetic field.

Superconductivity research during the next several decades found that niobium alloys become superconductors at temperatures as high as 23 K (about −418° F) and could carry currents in the kiloamps without excessive heating losses. Liquid helium was still needed as a coolant, however, and applications were restricted to specialized uses such as (1) superstrong magnetic fields for research in particle physics, (2) superconducting magnetic imaging devices for use in medical diagnosis, and (3) a prototype train with superconducting magnets to magnetically levitate and propel the train above its tracks at a speed of 480 km/hr (about 300 mi/hr).

The pace of research in superconductivity increased dramatically when, in 1986, George Bednorz and Alex Muller at IBM's Zurich laboratory discovered that certain ceramic materials called perovskites became superconductors at 30. K (about 405° F). Researchers began testing other ceramic superconductors, oxides or metals mixed with rare earth elements, for example, Lanthenum strontium copper oxide, Yttrium barium copper oxide, and Gadolinium barium copper oxide. They found that some of the perovskites become superconductors at temperatures as high as 100. K (about −279° F). This relatively high-temperature superconductivity was not consistent with the BCS theory. Scientists still believe that electron pairs move through the perovskites, but they do not understand why the electrons form pairs. Some scientists believe that the development of a new theory to explain the high-temperature superconductivity will result in the eventual development of materials that superconduct at room temperatures. The possibilities of such a development are staggering: (1) superconducting switches that process information about ten times faster than silicon chips presently used in computers; (2) high-efficiency superconducting electric generators, superconducting transmission lines without losses, and direct storage of electricity with superconducting magnets; (3) expanded efforts and applications of magnetically levitated and propelled trains; (4) new industrial processes to purify, separate materials, and grow crystals; and (5) the development of new magnetic sensing devices with exceptional sensitivity. Much research and development remains to be done, however, and there are technical hurdles that must be overcome before the scientific breakthrough in high-temperature superconductivity results in these applications.

Example 8.8

An electric fan draws 0.500 A in a 120. V circuit. What is the cost of operating the fan if the rate is 10¢/kwhr? (Answer: $0.006, which is 0.6 of a cent per hour)

Summary

The first electrical phenomenon recognized was charges produced by friction, which today is called *static electricity*. At first, the charge was thought to be produced by a *fluid* that was released by rubbing. By the early 1900s the *electron theory of charge* was developed from studies of the *atomic nature of matter*. These studies led to the understanding that matter is made of *atoms* that are composed of *negatively charged electrons* moving about a central *nucleus,* which contains *positively charged protons*. The two kinds of charges interact as *like charges produce a repulsive force* and *unlike charges produce an attractive force.*

Table 8.2

Summary of electrical quantities and units

Quantity	Definition*	Units
Charge	$q = ne$	1.00 coulomb (C) = charge equivalent to 6.24×10^{18} particles such as the electron
Electrical potential difference	$E_p = \dfrac{W}{q}$	$1.00 \text{ volt (V)} = \dfrac{1.00 \text{ joule (J)}}{1.00 \text{ coulomb (C)}}$
Electric current	$I = \dfrac{q}{t}$	$1.00 \text{ amp (A)} = \dfrac{1.00 \text{ coulomb (C)}}{1.00 \text{ second (sec)}}$
Electrical resistance	$R = \dfrac{E_p}{I}$	$1.00 \text{ ohm } (\Omega) = \dfrac{1.00 \text{ volt (V)}}{1.00 \text{ amp (A)}}$
Electrical power	$P = IV$	$1.00 \text{ watt (w)} = \dfrac{C}{sec} \times \dfrac{J}{C}$

*See Summary of Equations for more information.

Electrical *conductors* have electrons that are easily moved, and *nonconductors,* which are called *electrical insulators,* hold tightly to their electrons. An object acquires an *electric charge* when it has an excess or deficiency of electrons, which is called an *electrostatic charge.* An object becomes electrostatically charged by *friction,* which transfers electrons, by *contact* with a charged object and by *induction,* in which a redistribution of charges takes place.

A *quantity of charge* (q) is measured in units of *coulombs* (C), the charge equivalent to the transfer of 6.24×10^{18} charged particles such as the electron. The *fundamental charge* of an electron or proton is 1.60×10^{-19} coulomb. The *electrical forces* between two charged objects can be calculated from the relationship between the quantity of charge and the distance between two charged objects. The relationship is known as *Coulomb's law.*

A *force field* is a model for explaining the interaction of electrical charges. This model considers the *electric field* to be a condition of space surrounding a charge that can be mapped with electric field *lines of force.* The field is understood to build up and move through space, independent of the charge that created it.

A charged object in an electric field has electrical *potential energy* that is related to the charge on the object and the work done to move it into a field of like charge. The resulting *electrical potential* (E_p) is a ratio of the work done (W) to move a quantity of charge (q). In units, a joule of work to move a coulomb of charge is called a *volt.*

A flow of electric charge is called an *electric current* (I). A current requires some device, such as a generator or battery, to maintain a potential difference. The device is called a *voltage source.* An *electric circuit* contains (1) a voltage source, (2) a *continuous path* along which the current flows, and (3) a device such as a lamp or motor where work is done called a *voltage drop.* Current (I) is measured as the *rate* of flow of charge, that is, the quantity of charge (q) through a conductor in a period of time (t). The unit of current in coulomb/second is called an *ampere* or *amp* for short (A).

Current occurs in a conductor when a *potential difference* is applied and an *electric field* travels through the conductor near the speed of light. The electrons move *very slowly,* accelerated by the electric field. The field moves the electron in one direction in a *direct current* (dc) and moves them back and forth in an *alternating current* (ac).

Materials have a property of opposing or reducing an electric current called *electrical resistance* (R). Resistance is a ratio between the potential difference (V) between two points and the resulting current (I), or R = V/I. The unit is called the *ohm* (Ω) and $1.00\ \Omega = 1.00$ volt/1.00 amp. The relationship between voltage, current, and resistance is called *Ohm's law.*

Disregarding the energy lost to resistance, the *work* done by a voltage source is equal to the work accomplished in electrical devices in a circuit. The *rate* of doing work is *power,* or *work per unit time,* $P = W/t$. *Electrical power* can be calculated from the relationship of $P = IV$, which gives the power unit of *watts.*

Summary of Equations

8.1

quantity of charge = (number of electrons)(electron charge)

$$q = ne$$

8.2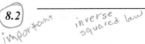

important   inverse squared law

$$\text{electrical force} = (\text{constant}) \times \frac{\text{charge on one object} \times \text{charge on second object}}{\text{distance between objects squared}}$$

$$F = k\frac{q_1 q_2}{d^2}$$

where $k = 9.00 \times 10^9$ newton-meters²/coulomb²

8.3

$$\text{electric field strength} = \frac{\text{force}}{\text{quantity of charge}}$$

$$E = \frac{F}{q}$$

8.4

$$\text{electric potential} = \frac{\text{work to create potential}}{\text{charge moved}}$$

$$E_p = \frac{W}{q}$$

8.5   Know

Definition of electric current

$$\text{electric current} = \frac{\text{quantity of charge}}{\text{time}}$$

$$I = \frac{q}{t}$$

8.6   Know

$$\text{volts} = \text{current} \times \text{resistance}$$

$$V = IR$$

8.7   Know

$$\text{power} = \frac{\text{work}}{\text{time}}$$

$$P = \frac{W}{t}$$

$$\text{electric power} = \frac{(\text{electric potential})(\text{charge})}{\text{time}}$$

$$P = \frac{E_p q}{t}$$

 yes

$$\text{electric power} = (\text{amps})(\text{volts})$$

$$P = IV$$

$$\text{electric current} \times \text{electric potential} \times \text{time} = \text{electric work}$$

$$IV \times t = W$$

$$(\text{work})(\text{rate}) = \text{cost}$$

Key Terms

alternating current (p. 155)
amp (p. 153)
ampere (p. 153)
conventional current (p. 153)
coulomb (p. 147)
Coulomb's law (p. 148)
direct current (p. 155)
electrical conductors (p. 147)
electrical force (p. 145)
electrical insulators (p. 147)
electrical nonconductors (p. 147)
electrical resistance (p. 155)
electric circuit (p. 152)
electric current (p. 151)
electric field (p. 149)
electric field lines (p. 150)
electric field strength (p. 150)
electric potential energy (p. 151)
electron current (p. 153)
electrons (p. 145)

electrostatic charge (p. 145)
force field (p. 149)
fundamental charge (p. 148)
lines of force (p. 150)
negative electric charge (p. 145)
negative ion (p. 145)
neutron (p. 145)
nucleus (p. 145)
ohm (p. 155)
Ohm's law (p. 155)
positive electric charge (p. 145)
positive ion (p. 145)
protons (p. 145)
semiconductors (p. 147)
superconductors (p. 156)
volt (p. 151)
voltage drop (p. 152)
voltage source (p. 152)
watt (p. 157)

Applying the Concepts

1. Of the three more stable subatomic particles that make up an atom two are understood to remain fixed in the atom while one can be forced away from the atom. Which is the one more easily moved?
   a. electron
   b. proton
   c. neutron
   d. all are equally difficult to move away from an atom

2. How does a proton acquire a positive charge?
   a. by removing electrons from an atom
   b. by adding protons to an atom
   c. by a charging process
   d. Charge is a fundamental property of a proton.

3. A quantity of electric charge is measured in a unit called a
   a. coulomb.
   b. volt.
   c. amp.
   d. watt.

4. The unit that describes the potential difference that occurs when a certain amount of work is used to move a certain quantity of charge is called a
   a. ohm.
   b. volt.
   c. amp.
   d. watt.

5. Which of the following units are measures of *rates?*
   a. amp and volt
   b. coulomb and joule
   c. volt and watt
   d. amp and watt

6. An electric current is measured in units of
   a. coulomb.
   b. volt.
   c. amp.
   d. watt.

7. In which of the following currents is there no electron movement from one end of a conducting wire to the other end?
   a. electron current
   b. direct current
   c. alternating current
   d. none of the above

8. If you multiply amps × volts the answer will be in units of
   a. resistance.
   b. work.
   c. current.
   d. power.

9. Which of the following is a measure of electrical work?
   a. kilowatt
   b. C
   c. kwhr
   d. C/sec

10. If an electric charge is somehow suddenly neutralized, the electric field that surrounds it will
    a. immediately cease to exist.
    b. collapse inward at some speed.
    c. continue to exist until neutralized.
    d. move off into space until it finds another charge.

Answers
1. a 2. d 3. a 4. b 5. d 6. c 7. c 8. d 9. c
10. b

Questions for Thought

1. Describe how the electron theory explains that objects can have two kinds of electrostatic charge.

2. In what unit is the electrostatic charge of an object measured? Describe how you could measure an electrostatic charge using this unit.

3. Explain why a balloon that has been rubbed sticks to a wall for a while.

4. Name the three basic parts of a simple electric circuit and describe what happens in each part.

5. Explain how you could use a variable power source, an ammeter, and a voltmeter to measure the resistance of a length of wire.

6. When you multiply amps times volts the answer is in watts. Explain.

7. Explain what is happening when you walk across a carpet, then receive a shock when you touch a metal object.

8. Why does a positively or negatively charged object have multiples of the fundamental charge?

9. If the electrical force is billions of times stronger than the gravitational force, explain why the universe is not pulled together or forced apart by this electrical force.

10. Compare and contrast gravitational potential difference and the potential difference in electrical matters, including how each is measured.

11. In what unit is the flow of electric charge measured? By analogy, what is a comparable measure in a water circuit?

12. What is a conventional current and an electron current? Why are there two ways to describe an electric current?

13. Explain how you know that it is an electric field, not electrons, that moves rapidly through a circuit.

14. How is it possible for a bird to perch on a high-voltage wire without harm?

15. Is a kwhr a unit of power or a unit of work? Explain.

16. What is the difference between ac and dc?

17. What happens to the electrical force between two charges (a) if you double the distance between them? (b) if you halve the distance between them?

18. An operational definition defines a concept by specifying a measurement procedure. Explain what would be measured to describe the following concepts: (a) electrical potential, (b) electrical current, (c) electrical resistance, (d) electric field.

19. A battery is connected to a lamp bulb with one wire from the positive terminal to the base of the bulb and a second wire from the negative terminal to the bottom button on the bulb. (a) Make a sketch of the circuit showing the direction of the *electron current*. (b) Make a sketch of the circuit showing the *conventional current*. In both sketches show how the current flows through the inside of the bulb.

20. Describe as many ways as you can that you could make a copper wire have the same electrical resistance as an iron wire.

21. Describe the following and the role that each plays in an electric current: (a) voltage, (b) electron drift, (c) electric field, (d) amperage, (e) the conducting wire.

Exercises

Group A—Solutions Provided in Appendix D

1. A rubber balloon has become negatively charged from rubbing with a wool cloth, and the charge is measured as 1.00×10^{-14} C. According to this charge, the balloon contains an excess of how many electrons?

2. One rubber balloon with a negative charge of 3.00×10^{-14} C is suspended by a string and hangs 2.00 cm from a second rubber balloon with a negative charge of 2.00×10^{-12} C. (a) What is the direction of the force between the balloons? (b) What is the magnitude of the force?

3. A dry cell does 7.50 J of work through chemical energy to transfer 5.00 C to the negative terminal. What is the electric potential between the two terminals?

4. An electric current through a wire is 6.00 C every 2.00 sec. What is the magnitude of this current?

5. A 1.00 A electric current corresponds to the charge of how many electrons flowing through a wire per second?

6. A current of 4.00 A flows through a toaster connected to a 120. V circuit. What is the resistance of the toaster?

7. What is the current in a 60.0 Ω resistor when the potential difference across it is 120. V?

Group B—Solutions Not Given

1. An inflated rubber balloon is rubbed with a wool cloth until an excess of a billion electrons are on the balloon. What is the magnitude of the charge on the balloon?

2. What is the force between two balloons, each with a negative charge of 1.6×10^{10} C if the balloons are 5.0 cm apart?

3. How much energy is available from a 12.0 V storage battery that can transfer a total charge equivalent to 100,000 C?

4. A wire carries a current of 2.0 A. At what rate is the current flowing?

5. What is the magnitude of the least possible current that could theoretically exist?

6. A current of 0.83 amp flows through a light bulb in a 120.0 V circuit. What is the resistance of this light bulb?

7. What is the voltage across a 60.0 Ω resistor with a current of 3 1/3 A?

8. A light bulb with a resistance of 10.0 Ω allows a 1.20 A current to flow when connected to a battery. (a) What is the voltage of the battery? (b) What is the power of the light bulb?

9. A small radio operates on 3.00 V and has a resistance of 15.0 Ω. At what rate does the radio use electric energy?

10. A 1,200. w hair dryer is operated on a 120. V circuit for 15 min. If electricity costs $0.10/kwhr, what was the cost of using the hair dryer?

11. An automobile starter rated at 2.00 hp draws how many amps from a 12.0 V battery?

12. An average-sized home refrigeration unit has a 1/3 hp fan motor for blowing air over the inside cooling coils, a 1/3 hp fan motor for blowing air over the outside condenser coils, and an 3.70 hp compressor motor. (a) All three motors use electric energy at what rate? (b) If electricity costs $0.10/kwhr, what is the cost of running the unit per 1.00 hr? (c) What is the cost for running the unit 12 hr a day for a 30. day month?

13. A 15.0 Ω toaster is turned on in a circuit that already has a 0.20 hp motor, three 100. w light bulbs, and a 600. w electric iron that are on. Will this trip a 15.0 A circuit breaker?

8. A 10.0 Ω light bulb is connected to a 12.0 V battery. (a) What current flows through the bulb? (b) What is the power of the bulb?

9. A light bulb designed to operate in a 120.0 V circuit has a resistance of 192.0 Ω. At what rate does the bulb use electric energy?

10. What is the monthly energy cost of leaving a 60.0 w bulb on continuously if electricity costs 10¢ per kwhr?

11. An electric motor draws a current of 11.50 A in a 240.0 V circuit. (a) What is the power of this motor in watts? (b) How many horsepower is this?

12. A swimming pool requiring a 2.0 hp motor to filter and circulate the water runs for eighteen hours a day. What is the monthly electrical cost for running this pool pump if electricity costs 10¢ per kwhr?

13. Is it possible for two people to simultaneously operate 1,300.0 w hair dryers on the same 120.0 V circuit without tripping a 15.0 A circuit breaker? Explain.

Chapter 9

Electricity and Magnetism

Figure 9.1

This nuclear power plant generator converts mechanical energy into electrical energy, producing 1,300 megawatts. In general, a capacity of 1 kilowatt of power is needed for every person served. One megawatt is 1 million watts, so this one generator meets the electrical needs of 1,300,000 people.

In chapter 8 you learned about electric charge, electric current, and some electrical quantities and units. You were also introduced to the field concept, a conceptual way of explaining how charges are able to produce attractive and repulsive forces at a distance. Patterns were found between these forces and electrical energy and the patterns were quantified, conceptualized, and used to answer questions about electric currents.

But many questions remain about how a current is produced by mechanical energy and how electricity is used in technological applications. This chapter begins with the fascinating topic of magnetism—what it is and how it is produced. A relationship is found between electricity and magnetism, and this relationship is used to explain the mechanical production of electricity, how it is measured, and how many everyday devices use electrical energy (figure 9.1).

The chapter concludes with a discussion of light and other forms of radiant energy. You will learn why light could be considered a subtopic of magnetism and electricity. This concluding section will also serve as an introduction to the next chapter on light.

Figure 9.2

Every magnet has ends, or poles, about which the magnetic properties seem to be concentrated. As this photo shows, more iron filings are attracted to the poles, revealing their location.

Magnetism

The ability of a certain naturally occurring rock to attract iron has been known since at least 600 B.C. The early Greeks called this rock "Magnesian stone" since it was discovered near the ancient city of Magnesia in western Turkey. Knowledge about the iron-attracting properties of the "Magnesian stone" grew slowly. About A.D. 100, the Chinese learned to magnetize a piece of iron with a "Magnesian stone," and sometime before A.D. 1000 they learned to use the magnetized iron or stone as a direction finder (compass). Today, the rock that attracts iron is known to be the black iron oxide mineral named *magnetite* after the city of Magnesia. If a sample of magnetite acts as a natural magnet, it is called *lodestone* after its use as a "leading stone" or compass.

A lodestone is a *natural magnet* and strongly attracts iron and steel but also attracts cobalt and nickel. Such substances that are attracted to magnets are said to have **ferromagnetic properties,** or simply *magnetic* properties. Iron, cobalt, and nickel are considered to have magnetic properties, and most other common materials are considered not to have magnetic properties. Most of these nonmagnetic materials, however, are slightly attracted or slightly repelled by a strong magnet. In addition, certain rare earth elements (chapter 12) as well as certain metal oxides exhibit strong magnetic properties.

Magnetic Poles

The ancient Chinese were the first to discover that a lodestone has two **magnetic poles,** or ends, about which the force of attraction seems to be concentrated. Iron filings or other small pieces of iron are attracted to the poles of a lodestone, for example, revealing their location (figure 9.2). The Chinese also

discovered that when a lodestone is free to turn, such as a lodestone floating on a piece of wood, one pole moves toward the north as the other pole moves toward the south. Today, the north-seeking pole is simply called the **north pole** and the south-seeking pole is called the **south pole.** Such floating lodestones were used as early magnetic compasses. A modern magnetic compass is a magnetic needle on a pivot, usually with the north-seeking end colored blue or black.

You are probably familiar with the fact that two magnets exert forces on each other. For example, if you move the north pole of one magnet near the north pole of a second magnet resting on a table top, each experiences a repulsive force. If you move your magnet slowly, you can push the second magnet across the table without the magnets ever touching. A repulsive force also occurs if two south poles are moved close together. But if the north pole of one magnet is brought near the south pole of a second magnet, an attractive force occurs. Moving two like poles together usually causes a magnet resting on a table to repel, rotate, then move toward the magnet you are holding. This occurs because *like magnetic poles repel* and *unlike magnetic poles attract.*

Magnetic Fields

A magnet moved into the space near a second magnet experiences a magnetic force as it enters the **magnetic field** of the second magnet. Recall that the electric field in the space near a charged particle was represented by electric field lines of force. A magnetic field can be represented by *magnetic field lines*. By convention, magnetic field lines are drawn to indicate how the *north pole* of a tiny imaginary magnet would point when in various places in the magnetic field. Arrowheads indicate the direction that the north pole would point, thus defining the direction of the magnetic field. The strength of the magnetic

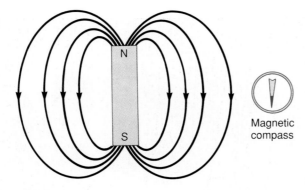

Figure 9.3

These lines are a map of the magnetic field around a bar magnet. The needle of a magnetic compass will follow the lines, with the north end showing the direction of the field.

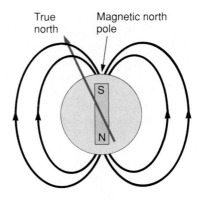

Figure 9.4

The earth's magnetic field. Note that the magnetic north pole and the geographic North Pole are not in the same place. Note also that the magnetic north pole acts as if the south pole of a huge bar magnet were inside the earth. You know that it must be a magnetic south pole since the north end of a magnetic compass is attracted to it and opposite poles attract.

field is greater where the lines are closer together and weaker where they are further apart. Figure 9.3 shows the magnetic field lines around the familiar bar magnet. Note that magnetic field lines emerge from the magnet at the north pole and enter the magnet at the south pole. Magnetic field lines always form closed loops.

The north end of a magnetic compass needle points north because the earth has a magnetic field (figure 9.4). Since the north, or north-seeking, pole of the needle is attracted to the geographic North Pole, it must be a magnetic *south* pole since unlike poles attract. However, it is conventionally called the *north magnetic pole* since it is located near the geographic North Pole. This can be confusing. Also, the two poles are not in the same place. The north magnetic pole is presently about 2,000 km (1,200 mi) south of the geographic North Pole. Thus, depending on your location, the north pole of a compass needle does not always point to the exact geographic north, or true north. The angle between the magnetic north and the geographic true north is called the *magnetic declination*. The magnetic declination map in figure 9.5 shows how many degrees east or west of true north a compass needle will point in different locations.

The typical compass needle pivots in a horizontal plane, moving to the left or right without up or down motion. Inspection of figure 9.4, however, shows that the earth's magnetic field is horizontal to the surface only at the magnetic equator. A compass needle that is pivoted in such a way to move only up and down will be horizontal only at the magnetic equator. Elsewhere it shows the angle of the field from the horizontal, called the *magnetic dip*. The angle of dip is the vertical component of the earth's magnetic field. As you travel from the equator the angle of magnetic dip increases from zero to a maximum of 90° at the magnetic poles.

The Source of Magnetic Fields

The observation that like magnetic poles repel and unlike magnetic poles attract might remind you of the forces involved with like and unlike charges. Recall that electric charges exist as single isolated units of positive protons and units of negative electrons. An object becomes electrostatically charged when charges are separated and the object acquires an excess or deficiency of negative charges. You might wonder, by analogy, if the poles of a magnet are similarly made up of an excess or deficiency of magnetic poles. The answer is no; magnetic poles are different from electric charges. Positive and negative charges *can* be separated and isolated. But suppose that you try to separate and isolate the poles of a magnet by cutting a magnet into two halves. Cutting a magnet in half will produce two new magnets, each with north and south poles. You could continue cutting each half into new halves, but each time the new half will have its own north and south poles (figure 9.6). It seems that no subdivision will ever separate and isolate a single magnetic pole, called a *monopole*. Magnetic poles always come in matched pairs of north and south and a monopole has never been found. Scientists continue to search for monopoles, seeking a symmetry between magnetism and electricity. The two poles are always found to come together and as it is understood today, magnetism is thought to be produced by *electric currents*, not an excess of monopoles. The modern concept of magnetism is electric in origin, and magnetism is understood to be a secondary property of electricity.

The key discovery about the source of magnetic fields was reported in 1820 by a Danish physics professor named Hans Christian Oersted. Oersted found that a wire conducting an electric current caused a magnetic compass needle below the

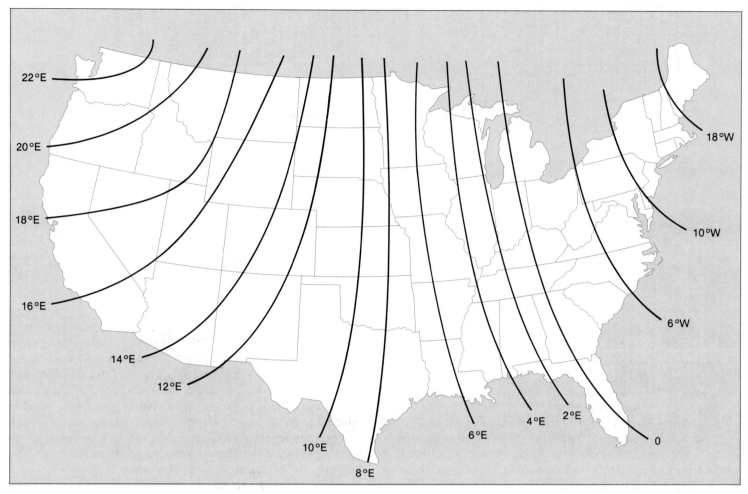

Figure 9.5

This magnetic declination map shows the approximate number of degrees east or west of the true geographic north that a magnetic compass will point in various locations.

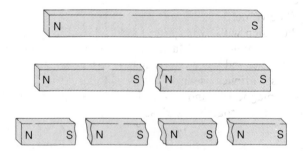

Figure 9.6

A bar magnet cut into halves always makes new, complete magnets with both a north and a south pole. The poles always come in pairs and the separation of a pair into single poles, called monopoles, has never been accomplished.

wire to move. When the wire was not connected to a battery the needle of the compass was lined up with the wire and pointed north as usual. But when the wire was connected to a battery, the compass needle moved perpendicular to the wire (figure 9.7). When he reversed the current in the wire the magnetic needle swung 180°, again perpendicular to the wire. Oersted had discovered that an electric current produces a magnetic field. An electric current is understood to be the movement of electric charges, so Oersted's discovery suggested that magnetism is a property of charges in motion. Recall that every electric charge is surrounded by an electric field. If the charge is moving, it is surrounded by an electric field *and* a magnetic field. The electric field and magnetic field are different, and the differences all point to the understanding that magnetism is a secondary property of electricity. The electric field of a charge, for example, is fixed according to the fundamental charge of the particle. The magnetic field, however, changes with the velocity of

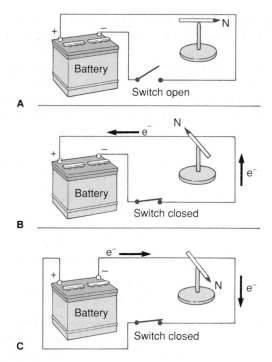

Figure 9.7

Oersted discovered that a compass needle below a wire (*a*) pointed north when there was not a current, (*b*) moved at right angles when a current flows one way, and (*c*) moved at right angles in the opposite direction when the current was reversed.

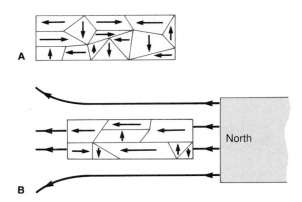

Figure 9.8

(*a*) In an unmagnetized piece of iron, the magnetic domains have a random arrangement that cancels any overall magnetic effect. (*b*) When an external magnetic field is applied to the iron, the magnetic domains parallel to the field grow in size at the expense of the other domains, and the iron is magnetized.

the moving charge. The magnetic field does not exist at all if the charge is not moving, and the strength of the magnetic field increases with increases in velocity. It seems clear that magnetic fields are produced by the motion of charges, or electric currents. Thus a magnetic field is a *property* of the space around a moving charge.

Permanent Magnets

The magnetic fields of bar magnets, horseshoe magnets, and other so-called permanent magnets are explained by the relationship between magnetism and moving charges. According to modern atomic theory, all matter is made up of atoms. An extremely simplified view of an atom pictures electrons moving about the nucleus of the atom. Since electrons are charges in motion, they produce magnetic fields. In most materials these magnetic fields cancel one another and neutralize the overall magnetic effect. In other materials, such as iron, cobalt, and nickel, the electrons are arranged and oriented in a complicated way that imparts a magnetic property to the atomic structure. These atoms are grouped in a tiny region called a **magnetic domain.** A magnetic domain is roughly 0.01 to 1 mm in length or width and does not have a fixed size. The atoms in each domain are magnetically aligned, contributing to the polarity of the domain. Each domain becomes essentially a tiny magnet with a north and south pole. In an unmagnetized piece of iron the domains are oriented in all possible directions and effectively cancel any overall magnetic effect. The net magnetism is therefore zero or near zero (figure 9.8).

When an unmagnetized piece of iron is placed in a magnetic field, the orientation of the domain changes to align with the magnetic field and the size of aligned domains may grow at the expense of unaligned domains. This explains why a "string" of iron paper clips is picked up by a magnet. Each paper clip has domains that become temporarily and slightly aligned by the magnetic field, and each paper clip thus acts as a temporary magnet while in the field of the magnet. In a strong magnetic field the size of the aligned domains grows to such an extent that the paper clip becomes a "permanent magnet." The same result can be achieved by repeatedly stroking a paper clip with the pole of a magnet. The magnetic effect of a "permanent magnet" can be reduced or destroyed by striking, dropping, or heating the magnet to a sufficiently high temperature. These actions randomize the direction of the magnetic domains and the overall magnetic field disappears.

Earth's Magnetic Field

The earth's magnetic field is believed to originate deep within the earth. Like all other magnetic fields, earth's magnetic field is believed to originate with moving charges. Earthquake waves and other evidence suggest that the earth has a solid inner core with a radius of about 1,200 km (about 750 mi) surrounded by a fluid outer core some 2,200 km (about 1,400 mi) thick. This core is probably composed of iron and nickel, which flows as the earth rotates, creating electric currents that result in the earth's magnetic field. How the electric currents are generated is not yet understood.

Other planets have magnetic fields, and there seems to be a relationship between the rate of rotation and the strength of the planet's magnetic field. Jupiter and Saturn rotate faster than the Earth and have stronger magnetic fields than the Earth. Venus and Mercury rotate more slowly than the Earth and have weaker magnetic fields. This is indirect evidence that the rotation of a planet is associated with internal fluid movements, which somehow generate electric currents and produce a magnetic field.

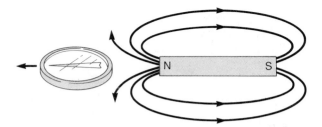

Figure 9.9

As shown by a compass needle, the magnetic force is at a maximum when at right angles to the field lines and at zero when parallel. This is why the compass needle lines up with the field lines and not in some other orientation.

In addition to questions about how the electric current is generated, there are puzzling questions from geologic evidence. Lava contains magnetic minerals that act like tiny compasses that are oriented to the earth's magnetic field when the lava is fluid but become frozen in place as the lava cools. Studies of these rocks by geologic dating and studies of the frozen magnetic mineral orientation show that the earth's magnetic field has undergone sudden reversals in polarity, where the north magnetic pole becomes the south magnetic pole and vice versa. This has happened many times over the distant geologic past. The cause of such magnetic field reversals is unknown, but it must be related to changes in the flow patterns of the earth's fluid outer core of iron and nickel.

The Strength of Magnetic Fields

The space around a moving charge and the space around a current flowing in a wire is changed by the movement of charges, creating a magnetic field. As with gravitational and electrical fields, a magnetic field can be described and measured with a hypothetical test object. Since a charged particle in motion produces its own magnetic field, it will interact with another magnetic field with some magnetic force. The strength of a magnetic field can be described in terms of this force experienced by a moving test charge. The magnetic force on a test charge in any given magnetic field depends on the magnitude of the charge and on its velocity. The magnetic force (F) varies directly with the magnitude of charge (q) and its velocity (v), or $F \propto qv$. The ratio of $F \propto qv$ could be used to measure the strength of a magnetic field at some point. However, the force is also found to vary with the *direction* of test charge movement within the magnetic field. The *force is zero if the charge is moving parallel to the field and at a maximum if the charge is moving at a right angle to the field*. At some other angle between these two extremes, the force is between zero and its maximum value as you might expect. The test charge is therefore used to identify the *direction and strength* of a given magnetic field with vectors identified with the symbol B. The compass needle in figure 9.9 is aligned parallel to the magnetic field lines because the maximum forces on it were greatest when it was at a right angle, but they are zero when the needle is parallel to the field lines. The direction of B is therefore the direction in which a test charge experiences zero force. The magnetic field represented by B vectors is known as the *magnetic field strength* or the *magnetic induction*. Remember that the maximum force experienced by a test charge is always at right angles to the field lines.

The maximum magnetic force (F_{max}) a test charge experiences is proportional to its charge (q) and velocity (v) when the particle is moving in a direction perpendicular to the magnetic field vectors (B), or $F_{max} = qvB$. Thus the strength of a given magnetic field can be defined as

$$B = \frac{F_{max}}{qv} \qquad \textbf{equation 9.1}$$

In units,

$$B = \frac{newton}{(coulomb)(meters/second)}$$

Since a coulomb/sec is an amp, this can be written as

$$B = \frac{newton}{amp \cdot meter}$$

which is called a **tesla** (T). The tesla is a measure of the strength of a magnetic field. Near the surface, the earth's horizontal magnetic field in some location is about 2×10^{-5} tesla. A small bar magnet produces a magnetic field of about 10^{-2} tesla but large, strong magnets can produce magnetic fields of 2 tesla. Superconducting magnets have magnetic fields as high as 30 tesla. Another measure of magnetic field strength is called the **gauss** (G) (1 tesla $= 10^4$ gauss). Thus the process of demagnetizing something is sometimes referred to as "degaussing."

Example 9.1

An electron used as a test charge to measure the strength of a magnetic field is deflected by a force of 3.20×10^{-13} N when moving at 4.00×10^6 m/sec at right angles to the field. What is the strength of the magnetic field in tesla?

Solution

Recall from chapter 8 that the fundamental charge of an electron is 1.60×10^{-19} C. The other quantities needed to find the strength of the magnetic field in tesla are given in the problem. According to equation 9.1,

$q = 1.60 \times 10^{-19}$ C

$F_{max} = 3.20 \times 10^{-13}$ N

$v = 4.00 \times 10^6$ m/sec

$B = ?$

$$B = \frac{F_{max}}{qv}$$

$$= \frac{3.20 \times 10^{-13} \text{ N}}{(1.60 \times 10^{-19} \text{ C})\left(4.00 \times 10^6 \dfrac{m}{sec}\right)}$$

$$= \frac{3.20 \times 10^{-13}}{(1.60 \times 10^{-19})(4.00 \times 10^6)} \frac{N}{C \cdot \dfrac{m}{sec}}$$

$$= \frac{3.20 \times 10^{-13}}{6.40 \times 10^{-13}} \frac{N}{\dfrac{C}{sec} \cdot m}$$

$$= 0.500 \frac{N}{A \cdot m}$$

$$= \boxed{0.500 \text{ tesla}}$$

Electricity and Magnetism

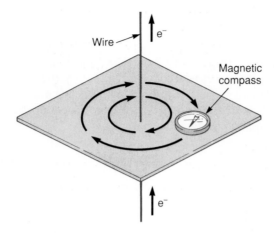

Figure 9.10
A magnetic compass shows the presence and direction of the magnetic field around a straight length of current-carrying wire.

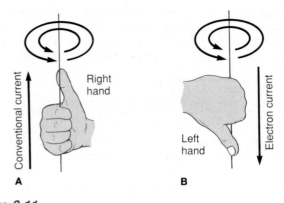

Figure 9.11
Use (a) a right-hand rule of thumb to determine the direction of a conventional current and (b) a left-hand rule of thumb to determine the direction of an electron current.

Electric Currents and Magnetism

As Oersted discovered, electric charges in motion produce a magnetic field around the charges. You can "map" the magnetic field established by the current in a wire by running a straight wire vertically through a sheet of paper. The wire is connected to a battery and iron filings are sprinkled on the paper. The filings will become aligned as the domains in each tiny piece of iron are forced parallel to the field. Overall, filings near the wire form a pattern of concentric circles with the wire in the center.

The direction of the magnetic field around a current-carrying wire can be determined by using the common device for finding the direction of a magnetic field, the magnetic compass. The north-seeking pole of the compass needle will point in the direction of the magnetic field lines (by definition). If you move the compass around the wire the needle will always move to a position that is tangent to a circle around the wire. Evidently the magnetic field lines are closed concentric circles that are at right angles to the length of the wire (figure 9.10).

If you *reverse* the direction of the current in the wire and again move a compass around the wire, the needle will again move to a position that is tangent to a circle around the wire. But this time the north pole direction is reversed. Thus the magnetic field around a current-carrying wire has closed concentric field lines that are perpendicular to the length of the wire. The direction of the magnetic field is determined by the direction of the current.

You can also determine the direction of a magnetic field around a current-carrying wire by using a "rule of thumb." If you are considering a *conventional* (positive) current use your *right* hand to grasp a current-carrying wire with your thumb pointing in the direction of the conventional current. Your curled fingers will point in the circular direction of the magnetic field (figure 9.11). If you are considering an *electron* (negative) current use your *left* hand to grasp the wire with your thumb

pointing in the direction of the electron current. This rule is easily understood when you realize that a conventional current runs in the opposite direction of an electron current. When you point upward with your right thumb and downward with your left thumb, the curled fingers of both hands will point in the same direction.

The strength of the magnetic field (B) around a long, straight, current-carrying wire is directly proportional to the current (I) in the wire and inversely proportional to some distance (d) from the wire, or

$$B = k\frac{2I}{d} \qquad \text{equation 9.2}$$

where the proportionality constant k is 1.00×10^{-7} newton/amp². Note that the magnetic field strength varies with the distance from the wire (not the square of the distance) and that the unit is newton/amp·meter, or tesla.

Example 9.2

A vertical straight wire carries a 5.00 A current. Is the magnetic field produced sufficiently strong to attract a magnetic compass needle 5.00 cm from the wire? Assume the earth's horizontal magnetic field to be 2.00×10^{-5} tesla.

Solution
The magnetic field of a straight current-carrying wire can be found from equation 9.2. The current (I) of 5.00 A is given, and a distance (d) of 5.00 cm (0.050 m) from the wire is given.

I = 5.00 A

d = 0.050 m

$k = 1.00 \times 10^{-7}$ N/A²

B = ?

$$B = k\frac{2I}{d}$$

$$= \left(1.00 \times 10^{-7}\,\frac{N}{A^2}\right)\frac{2 \times 5.00\ A}{0.050\ m}$$

$$= (1.00 \times 10^{-7})\frac{2 \times 5.00}{0.050}\,\frac{N}{A^2} \times \frac{A}{m}$$

$$= (1.00 \times 10^{-7})(200)\,\frac{N}{A \cdot m}$$

$$= \boxed{2.00 \times 10^{-5}\ \text{tesla}}$$

The magnetic field produced by the current is the *same* strength as the earth's magnetic field, so the compass needle will point in a direction that is the sum of the two magnetic fields. In the proper orientation, the needle will point between the two fields.

Example 9.3

What is the magnitude of a magnetic field 5.00 cm from the wire if a 10.0 A current is carried by the wire? (Answer: 4.00×10^{-5} tesla. Note that twice the current at the same distance as in example 9.2 produces a magnetic field with twice the strength since $B \propto I$).

Single Charges Interact with a Magnetic Field

Cosmic rays are subatomic charged particles such as protons, electrons, and other parts of atoms that move at tremendous velocities through space. They are believed to be particles that were ejected in the explosion of a dying star, a supernova, and have traveled through space since they were ejected. Cosmic rays are ionizing radiation and can cause biological damage that can lead to mutations. This might be an ongoing problem on the earth, except the magnetic field of the earth exerts a force on many of these charged particles, deflecting them from the surface. Let's see how the earth's magnetic field protects you this way.

A moving charge produces a magnetic field of its own. This magnetic field will interact with a second magnetic field, just as the magnetic fields of two bar magnets interact. The force of the interactions between a moving charge and a magnetic field depends on the velocity and magnitude of charge on the moving charge and on the strength of the second magnetic field. A maximum force occurs when the moving charge is moving at a right angle to the magnetic field. A charged particle moving directly toward the earth's equator would be moving at a right angle to the earth's magnetic field and would thus experience a maximum deflecting force as it approaches the earth.

Some of the charged cosmic ray particles are deflected away from earth and others may become trapped, following a looping circular path as they move along magnetic field lines between the poles. This creates belts of radiation known as the **Van Allen belts.** In addition, charged particles sometimes enter the earth's atmosphere at the poles, causing the atmosphere to glow as a colorful shimmering curtain. The glowing atmosphere is called the aurora borealis (northern lights) in the Northern Hemisphere and the aurora australis in the Southern Hemisphere.

Current-Carrying Wires Interact with a Magnetic Field

Just as a single moving charge interacts with a second magnetic field, a current-carrying wire interacts with a second magnetic field. The force on the wire is directly proportional to the current (I), (coulombs/second), and the length of wire within the

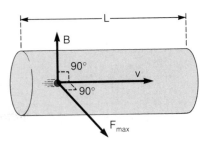

Figure 9.12
A length of wire (L) carrying a current (I) at right angles to a magnetic field (B) experiences a maximum force (F_{max}) that is at right angles to both the direction of the current and the magnetic field.

second magnetic field (L). If the wire is at a right angle to the vector B of the second magnetic field, the force is

$$F_{max} = ILB \qquad \text{equation 9.3}$$

As with a single moving charge, the force on a current-carrying wire is at a maximum when the current direction makes a right angle to the second magnetic field. The *direction* of the force on the current-carrying wire is at a right angle to the direction of the current *and* at a right angle to the direction of the second magnetic field (figure 9.12).

Example 9.4

A 10.0 cm length of wire carrying a current of 1.00 A is placed perpendicular to the field lines of a magnetic field B of 0.5 tesla. What is the force on the wire?

Solution
The force on a current-carrying wire is calculated from equation 9.3, $F_{max} = ILB$. The current (I) of 1.00 A, wire length (L) of 10.0 cm, and strength (B) of the second field of 0.5 tesla is given.

$$I = 1.00 \text{ A}$$
$$L = 10.0 \text{ cm} = 0.100 \text{ m}$$
$$B = 0.5 \text{ tesla}$$
$$F_{max} = ?$$

$$F_{max} = ILB$$
$$= (1.00 \text{ A})(0.100 \text{ m})\left(0.5 \frac{N}{A \cdot m}\right)$$
$$= (1.00)(0.100)(0.5) \text{ A} \times \text{m} \times \frac{N}{A \cdot m}$$
$$= 0.05 \frac{A \cdot m \cdot N}{A \cdot m}$$
$$= \boxed{0.05 \text{ N}}$$

A force of 0.05 N is equivalent to a force of about 0.01 lb, or about 0.16 oz.

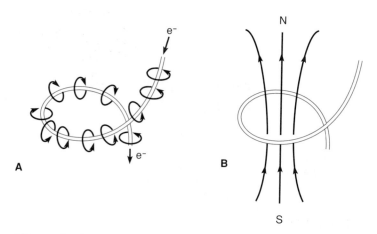

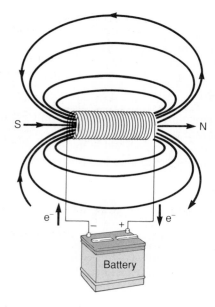

Figure 9.13

(*a*) Forming a wire into a loop causes the magnetic field to pass through the loop in the same direction. (*b*) This gives one side of the loop a north pole and the other side a south pole.

Figure 9.14

When a current is run through a cylindrical coil of wire, a solenoid, it produces a magnetic field like the magnetic field of a bar magnet.

Current Loops

As you can see from example 9.4, the force on a 10.0 cm straight wire carrying 1.00 amp is only about 0.160 oz., which is not a very useful force. The force could be increased by increasing the current, but there is a more efficient way to obtain a larger force. A current-carrying wire that is formed into a loop has perpendicular, circular field lines that pass through the inside of the loop in the same direction. This has the effect of concentrating the field lines, which increases the magnetic field intensity. Since the field lines all pass through the loop in the same direction, one side of the loop will have a north pole and the other side a south pole (figure 9.13).

Many loops of wire formed into a cylindrical coil is called a **solenoid.** When a current is in a solenoid, each loop contributes field lines along the length of the cylinder (figure 9.14). The overall effect is a magnetic field around the solenoid that acts just like the magnetic field of a bar magnet. This magnet, called an **electromagnet,** can be turned on or off by turning the current on or off. In addition, the strength of the electromagnet depends on the magnitude of the current and the number of loops (ampere-turns). The strength of the electromagnet can also be increased by placing a piece of soft iron in the coil. The domains of the iron become aligned by the influence of the magnetic field. This induced magnetism increases the overall magnetic field strength of the solenoid as the magnetic field lines are gathered into a smaller volume within the core.

Applications of Electromagnets

The discovery of the relationship between an electric current, magnetism, and the resulting forces created much excitement in the 1820s and 1830s. This excitement was generated because it was now possible to explain some seemingly separate phenomena in terms of an interrelationship and because people began to see practical applications almost immediately. Within

a year of Oersted's discovery, Andre Ampere had fully explored the magnetic effects of currents, combining experiments and theory to find the laws describing these effects. Soon after Ampere's work, the possibility of doing mechanical work by sending currents through wires was explored. The electric motor, similar to motors in use today, was invented in 1834, only fourteen years after Oersted's momentous discovery.

The magnetic field produced by an electric current is used in many practical applications, including electrical meters, electromagnetic switches that make possible the remote or programmed control of moving mechanical parts, and electric motors. In each of these applications, an electric current is applied to an electromagnet.

Electric Meters

Since you cannot measure electricity directly, it must be measured indirectly through one of the effects that it produces. The strength of the magnetic field produced by an electromagnet is proportional to the electric current in the electromagnet. Thus one way to measure a current is to measure the magnetic field that it produces. A device that measures currents from their magnetic fields is called a *galvanometer* (figure 9.15). A galvanometer has a coil of wire that can rotate on pivots in the magnetic field of a permanent magnet. The coil has an attached pointer that moves across a scale and control springs that limit its motion and return the pointer to zero when there is no current. When there is a current in the coil the electromagnetic field is attracted and repelled by the field of the permanent

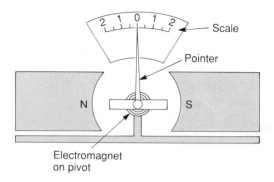

Figure 9.15

A galvanometer measures the direction and relative strength of an electric current from the magnetic field it produces. A coil of wire wrapped around an iron core becomes an electromagnet that rotates in the field of a permanent magnet. The rotation moves a pointer on a scale.

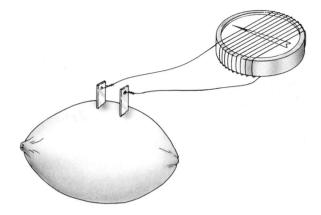

Figure 9.16

You can use these materials to create and detect an electric current. See the activities below.

magnet. The larger the current the greater the force and the more the coil will rotate until it reaches an equilibrium position with the control springs. The amount of movement of the coil (and thus the pointer) is proportional to the current in the coil. With certain modifications and applications, the galvanometer can be used to measure current (ampmeter), potential difference (voltmeter), and resistance (ohmmeter).

ACTIVITIES

1. You can make a simple compass galvanometer that will detect a small electric current (figure 9.16). All you need is a magnetic compass and some thin insulated wire (the thinner the better).

2. Wrap the thin insulated wire in parallel windings around the compass. Make as many parallel windings as you can but leave enough room to see both ends of the compass needle. Leave the wire ends free for connections.

3. To use the galvanometer, first turn the compass so the needle is parallel to the wire windings. When a current passes through the coil of wire the magnetic field produced will cause the needle to move from its north-south position, showing the presence of a current. The needle will deflect one way or the other depending on the direction of the current.

4. Test your galvanometer with a "lemon battery." Roll a soft lemon on a table while pressing on it with the palm of your hand. Cut two slits in the lemon about 1 cm apart. Insert a 1 cm by 8 cm (approximate) copper strip in one slit and a same-sized strip of zinc in the other slit, making sure the strips do not touch inside the lemon. Connect the galvanometer to the two metal strips. Try the two metal strips in other fruits, vegetables, and liquids. Can you find a pattern?

Electromagnetic Switches

A relay is an electromagnetic switch device that makes possible the use of a low-voltage control current to switch a larger, high-voltage circuit on and off (figure 9.17). A thermostat, for example, utilizes two thin, low-voltage wires in a glass tube of mercury. The glass tube of mercury is attached to a metal coil that expands and contracts with changes in temperature, tipping the attached glass tube. When the temperature changes enough to tip the glass tube, the mercury flows to the bottom end, which makes or breaks contact with the two wires, closing or opening the circuit. When contact is made, a weak current activates an electromagnetic switch, which closes the circuit on the large current furnace or heat pump motor.

A solenoid is a coil of wire with a current. Some solenoids have a spring-loaded movable piece of iron inside. When a current flows in such a coil the iron is pulled into the coil by the magnetic field, and the spring returns the iron when the current is turned off. This device could be utilized to open a water valve, turning the hot or cold water on in a washing machine or dishwasher, for example. Solenoids are also used as mechanical switches on VCRs, automobile starters, and signaling devices such as bells and buzzers. The "pin printer" used by many computer printers is actually a group of small solenoids that are activated by electric currents from the computer. Seven or more of these small solenoids work together to strike a print ribbon, forming the letters or images as they rapidly move in and out.

Electricity and Magnetism

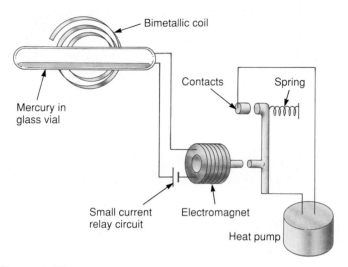

Figure 9.17

A schematic of a relay circuit. The mercury vial turns as changes in temperature expand or contract the coil, moving the mercury and making or breaking contact with the relay circuit. When the mercury moves to close the relay circuit, a small current activates the electromagnet, which closes the contacts on the large current circuit.

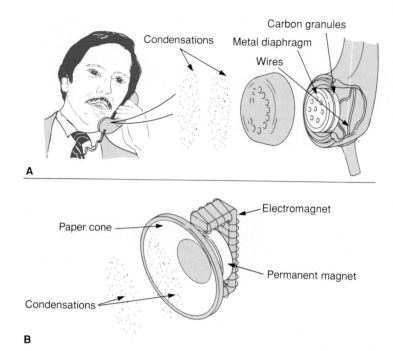

Figure 9.18

(*a*) Sound waves are converted into a changing electrical current in a telephone. (*b*) Changing electrical current can be changed to sound waves in a speaker by the action of an electromagnet pushing and pulling on a permanent magnet. The permanent magnet is attached to a stiff paper cone or some other material that makes sound waves as it moves in and out.

Telephones and Loudspeakers

The mouthpiece of a telephone contains a cylinder of carbon granules with a thin metal diaphragm facing the front. When someone speaks into the telephone the diaphragm moves in and out with the condensations and rarefactions of the sound wave (figure 9.18). This movement alternately compacts and loosens the carbon granules, increasing and decreasing the electric current that increases and decreases with the condensations and rarefactions of the sound waves.

The moving electric current is fed to the "earphone" part of a telephone at another location. The current runs through a coil of wire that attracts and repels a permanent magnet attached to a speaker cone. When repelled forward the speaker cone makes a condensation, and when attracted back the cone makes a rarefaction. The overall result is a series of condensations and rarefactions that, through the changing electric current, accurately match the sounds made by the other person.

The loudspeaker in a radio or stereo system works from changes in an electric current in a similar way, attracting and repelling a permanent magnet attached to the speaker cone. You can see the speaker cone in a large speaker moving back and forth as it creates condensations and rarefactions.

Electric Motors

An electric motor is an electromagnetic device that converts electrical energy to mechanical energy. Basically, a motor has two working parts, a stationary electromagnet called a *field magnet* and a cylindrical, movable electromagnet called an *armature*. The armature is on an axle and rotates in the magnetic field of the field magnet. The axle turns fan blades, compressors, drills, pulleys, or many other devices where mechanical work is desired.

There are different designs of electric motors that are used for various applications, but the simple demonstration motor shown in figure 9.19 can be used as an example of the basic operating principle. Both the field coil and the armature are connected to an electric current. The armature turns and it receives the current through a *commutator* and *brushes*. The brushes are contacts that "brush" against the commutator, maintaining contact. When the current is turned on, the field coil and the armature become electromagnets and the unlike poles attract, rotating the armature. If the current is dc, the armature would turn no farther, stopping as it does in a galvanometer. But the commutator has insulated segments so when it turns halfway, the commutator segments switch brushes and the current flows through the armature in the *opposite* direction. This switches the armature poles, which are now repelled for another half-turn. The commutator again reverses the polarity and the motion continues in one direction. An actual motor has many coils (called "windings") in the armature to obtain a useful force and many commutator segments. This gives the motor a smoother operation with a greater turning force.

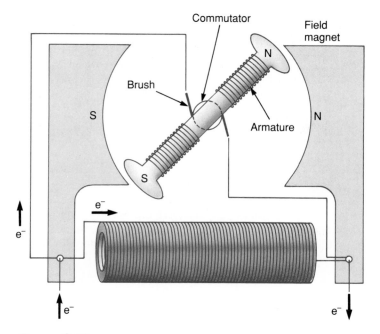

Figure 9.19

A schematic of a simple electric motor.

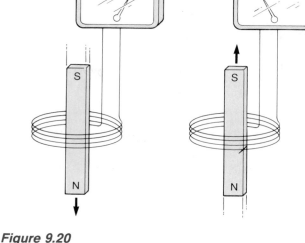

Figure 9.20

A current is induced in a coil of wire moved through a magnetic field, and the direction of the current depends on the direction of motion.

Electromagnetic Induction

So far, you have learned that (1) a moving charge and a current-carrying wire produce a magnetic field; (2) a second magnetic field exerts a force on a moving charge and exerts a force on a current-carrying wire as their magnetic fields interact; and (3) the direction of the maximum force produced on a moving charge or moving charges is at right angles to their velocity and to the interacting magnetic field lines.

Soon after the discovery of these relationships by Oersted and Ampere, people began to wonder if the opposite effect was possible, that is, would a magnetic field produce an electric current. The discovery was made independently in 1831 by Joseph Henry in the United States and by Michael Faraday in England. They found that *if a loop of wire is moved in a magnetic field, or if the magnetic field is changed, a voltage is induced in the wire.* The voltage is called an *induced voltage,* and the resulting current in the wire is called an *induced current.* The overall interaction is called **electromagnetic induction.**

One way to produce electromagnetic induction is to move a bar magnet into or out of a coil of wire (figure 9.20). A galvanometer shows that the induced current flows one way when the bar magnet is moved toward the coil but flows the other way when the bar magnet is moved away from the coil. The same effect occurs if you move the coil back and forth over a stationary magnet. Furthermore, no current is detected when the magnetic field and the coil of wire are not moving. Thus electromagnetic induction depends on the relative motion of the magnetic field and the coil of wire. It does not matter which moves or changes, but one must move or change relative to the other for electromagnetic induction to occur.

Electromagnetic induction occurs when the loop of wire cuts across magnetic field lines or when magnetic field lines cut across the loop. The magnitude of the induced voltage is proportional to (1) the number of wire loops cutting the magnetic field lines, (2) the strength of the magnetic field, and (3) the rate at which magnetic field lines are cut by the wire.

ACTIVITIES

1. Make a coil of wire from insulated bell wire (#18 copper wire) by wrapping fifty windings around a narrow jar. Tape the coil several places so it does not come apart.
2. Connect the coil to a compass galvanometer (see previous activity).
3. Move a strong bar magnet into and out of the stationary coil of wire and observe the galvanometer. Note the magnetic pole, direction of movement, and direction of current for both in and out movements.
4. Move the coil of wire back and forth over the stationary bar magnet.

Generators

Soon after the discovery of electromagnetic induction the **electric generator** was developed. This development began the tremendous advance of technology that followed soon after. The generator is essentially an axle with many wire loops that rotates in a magnetic field. The axle is turned by some form of mechanical energy such as a water turbine or a steam turbine, which uses steam generated from fossil fuels or nuclear energy. The use of the electric generator began in the 1890s, when George Westinghouse designed and built a waterwheel-powered

Electricity and Magnetism

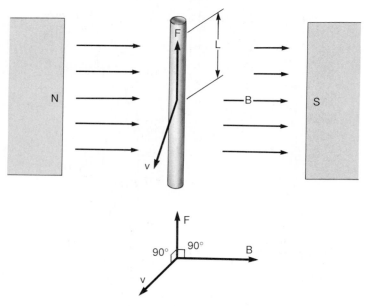

Figure 9.21

A wire moved at right angles to a magnetic field causes electrons to experience a force one way along the wire, establishing a potential between two points on the wire.

75 kilowatt generator near Ouray, Colorado. Thomas Edison built a dc power plant in New York City about the same time. In 1896 a 3.7 megawatt power plant at Niagara Falls sent electrical power to factories in Buffalo, New York. The factories no longer had to be near a source of energy such as a waterfall since electric energy could now be transmitted by wires.

To understand electromagnetic induction, and thus how the electric generator works, consider the length of copper wire (L) in the magnetic field (B) as shown in figure 9.21. Think of this wire as part of one of the many wire loops that rotate in a magnetic field inside a generator. Imagine that as some form of mechanical energy turns the axle of the generator, the wire moves at right angles to and across the magnetic field lines with a certain velocity (v). Copper contains many electrons that are free to move, and these electrons with a charge of q experience a maximum force of $F_{max} = qvB$ (equation 9.1). Since this force is at right angles to both B and v, it is directed one way along the length of the wire.

Since work is force times distance, the work done on the electrons moved across the wire of L length is $W = FL$. From equation 9.1, the maximum force the electrons experienced is $F_{max} = qvB$. Substituting this F in the expression of work gives $W = qvBL$. Voltage is work/charge, so the voltage can be found by dividing by q, or

$$\text{induced voltage} = BLv \qquad \textbf{equation 9.4}$$

In units,

$$\text{induced voltage} = \left(\frac{\text{newton}}{\text{amp} \cdot \text{meter}}\right)(\text{meter})\left(\frac{\text{meter}}{\text{second}}\right)$$

$$= \frac{\text{newton} \cdot \text{meter}}{\text{amp} \cdot \text{second}}$$

$$= \frac{\text{joule}}{\text{coulomb}}$$

$$= \text{volts}$$

Example 9.5

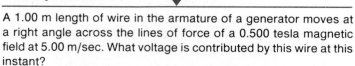

A 1.00 m length of wire in the armature of a generator moves at a right angle across the lines of force of a 0.500 tesla magnetic field at 5.00 m/sec. What voltage is contributed by this wire at this instant?

Solution

The induced voltage can be found from equation 9.4, induced voltage = BLv.

B = 0.500 tesla

L = 1.00 m

v = 5.00 m/sec

V = ?

$$\text{induced voltage} = BLv$$

$$= \left(0.500 \frac{N}{A \cdot m}\right)(1.00 \text{ m})\left(5.00 \frac{m}{\text{sec}}\right)$$

$$= (0.500)(1.00)(5.00) \frac{N}{A \cdot m} \times m \times \frac{m}{\text{sec}}$$

$$= 2.50 \frac{N \cdot m}{A \cdot \text{sec}}$$

$$= 2.50 \frac{J}{\frac{C}{\text{sec}} \times \text{sec}}$$

$$= 2.50 \frac{J}{C}$$

$$= \boxed{2.50 \text{ V}}$$

Transformers

In the 1890s the production of electrical power from electromagnetic induction began in the United States with generators built by George Westinghouse and Thomas Edison. A controversy arose, however, because Edison built dc generators and Westinghouse built ac generators. Edison believed that alternating current was dangerous and argued for the use of direct current only. Alternating current eventually won because (1) the voltage of ac could be easily changed to meet different applications, but the voltage of dc could *not* be easily changed, and (2) dc transmission suffered from excessive power losses in transmission while ac power was made possible by a **transformer,** a device that uses electromagnetic induction to increase or decrease ac voltage.

A transformer has two basic parts: (1) a **primary coil,** which is connected to a source of alternating current, and (2) a **secondary coil,** which is close by. Both coils are often wound on a single iron core but are always fully insulated from each other. When an alternating current flows through the primary coil a magnetic field grows around the coil to a maximum size, collapses to zero, then grows to a maximum size with an opposite polarity. This happens 120 times a second as the alternating current oscillates at 60 hertz. The magnetic field is strengthened and directed by the iron core. The growing and collapsing magnetic field moves across the wires in the secondary coil, inducing a voltage in the secondary coil. The growing and collapsing magnetic field from the primary coil thus induces a voltage in the secondary coil, just as an induced voltage occurs in the wire loops of a generator.

The transformer increases or decreases the voltage in an alternating current because the magnetic field grows and collapses past the secondary coil, inducing a voltage. If a direct current is applied to the primary coil the magnetic field grows around the primary coil as the current is established but then becomes stationary. Recall that electromagnetic induction occurs when there is relative motion between the magnetic field lines and a wire loop. Thus an induced voltage occurs from a direct current (1) only for an instant when the current is established and the growing field moves across the secondary coil and (2) only for an instant when the current is turned off and the field collapses back across the secondary coil. In order to use dc in a transformer the current must be continually interrupted to produce a changing magnetic field.

When an alternating current or a continually interrupted direct current is applied to the primary coil, the magnitude of the induced voltage in the secondary coil is proportional to the ratio of wire loops in the two coils. If they have the same number of loops, the primary coil produces just as many magnetic field lines as are intercepted by the secondary coil. In this case the induced voltage in the secondary coil will be the same as the voltage in the primary coil. Suppose, however, that the secondary coil has one-tenth as many loops as the primary coil. This means that the secondary loops will cut one-tenth as many field lines as the primary coil produces. As a result, the induced voltage in the secondary coil will be one-tenth the voltage in the primary coil. This is called a **step-down transformer** since the voltage was stepped down in the secondary coil. On the other hand, more wire loops in the secondary coil will intercept more magnetic field lines. If the secondary coil has ten times *more* loops than the primary coil, then the voltage will be *increased* by a factor of ten. This is a **step-up transformer.** How much the voltage is stepped up or stepped down depends on the ratio of wire loops in the primary and secondary coils (figure 9.22). Note that the *volts per wire loop* are the same in each coil. The relationship is

$$\frac{\text{volts}_{\text{primary}}}{(\text{number of loops})_{\text{primary}}} = \frac{\text{volts}_{\text{secondary}}}{(\text{number of loops})_{\text{secondary}}}$$

or

$$\frac{V_p}{N_p} = \frac{V_s}{N_s} \qquad \textbf{equation 9.5}$$

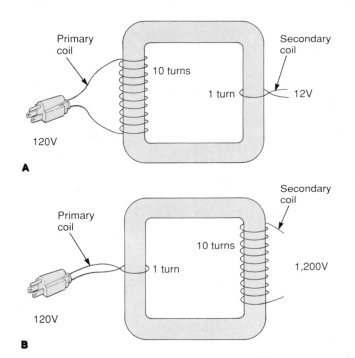

A

B

Figure 9.22

(a) This step-down transformer has 10 turns on the primary for each turn on the secondary and reduces the voltage from 120 V to 12 V. (b) This step-up transformer increases the voltage from 120 V to 1,200 V since there are 10 turns on the secondary to each turn on the primary.

Example 9.6

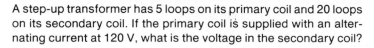

A step-up transformer has 5 loops on its primary coil and 20 loops on its secondary coil. If the primary coil is supplied with an alternating current at 120 V, what is the voltage in the secondary coil?

Solution

$N_p = 5 \text{ loops}$

$N_s = 20 \text{ loops}$

$V_p = 120 \text{ V}$

$V_s = ?$

$$\frac{V_p}{N_p} = \frac{V_s}{N_s} \therefore V_s = \frac{V_p N_s}{N_p}$$

$$V_s = \frac{(120 \text{ V})(20 \text{ loops})}{5 \text{ loops}}$$

$$= \frac{120 \times 20}{5} \frac{\text{V} \cdot \text{loops}}{\text{loops}}$$

$$= \boxed{480 \text{ V}}$$

A step-up or step-down transformer steps up or steps down the *voltage* of an alternating current according to the ratio of wire loops in the primary and secondary coils. Assuming no losses in the transformer the *power input* on the primary coil equals the *power output* on the secondary coil. Since P = IV, you can see that when the voltage is stepped up the current is correspondingly decreased, as

$$\text{power input} = \text{power output}$$

$$\text{watts input} = \text{watts output}$$

$$(\text{amps} \times \text{volts})_{\text{in}} = (\text{amps} \times \text{volts})_{\text{out}}$$

or

$$V_p I_p = V_s I_s \qquad \textbf{equation 9.6}$$

Electricity and Magnetism

Figure 9.23

This transformer receives current from the uppermost high-voltage lines and steps it down to a lower voltage in the lower set of lines.

Example 9.7

The step-up transformer in example 9.6 is supplied with an alternating current at 120 V and a current of 10.0 A in the primary coil. What current flows in the secondary circuit?

Solution

$$V_p = 120 \text{ V} \qquad V_p I_p = V_s I_s \therefore I_s = \frac{V_p I_p}{V_s}$$

$$I_p = 10.0 \text{ I}$$

$$V_s = 480 \text{ V} \qquad I_s = \frac{120 \text{ V} \times 10.0 \text{ A}}{480 \text{ V}}$$

$$I_s = ? \qquad = \frac{120 \times 10.0}{480} \frac{\text{V} \cdot \text{A}}{\text{V}}$$

$$= \boxed{2.5 \text{ A}}$$

Energy losses in transmission are reduced by stepping up the voltage. Recall that electrical resistance results in an energy loss and a corresponding absolute temperature increase in the conducting wire. If the current is large, there are many collisions between the moving electrons and positive ions of the wire resulting in a large energy loss. Each collision takes energy from the electric field, diverting it into increased kinetic energy of the positive ions and thus increased temperature of the conductor. The energy lost to resistance is therefore reduced by *lowering* the current, which is what a transformer does by increasing the voltage. Hence electric power companies step up the voltage of generated power for economical transmission. A step-up transformer at a power plant, for example, might step up the voltage from 22,000 volts to 500,000 volts for transmission across the country to a city. This step-up in voltage correspondingly reduces the current, lowering the resistance losses to a more acceptable 4 or 5 percent over long distances. A step-down transformer at a substation near the city reduces the voltage to several thousand volts for transmission around the city. Step-down transformers reduce this voltage to 120 volts for transmission to three or four houses (figure 9.23).

Electromagnetic Waves

The topic of *radiant energy* was briefly introduced in chapter 5 and defined as energy that moves through space. You know, for example, that sunlight carries energy since you are warmed by sunlight. You also know that there is nothing between the earth and the sun other than the vacuum or near-vacuum of outer space. Since there are no molecules between the earth and the sun neither conduction nor convection can explain this transfer of energy from the sun to the earth. The energy is said to move through space as radiant energy.

Radiant energy is also known as *electromagnetic radiation*. Visible light is one kind of electromagnetic radiation and other kinds include infrared, ultraviolet, microwaves, radio waves, X rays, and gamma rays. All of the kinds of electromagnetic radiation have the properties of waves and travel through space at the speed of light. They also have the property of carrying energy, which can be delivered to a surface.

The idea of waves traveling through a vacuum, carrying energy, sounds very baffling. What, for example, does the moving? What is the *it*, or material, that transports energy from place to place through empty space? If you feel baffled, you are not alone. People have wondered about questions such as these since at least the time of Aristotle. It was not until 1865 that James Clerk Maxwell found the answer in a relationship between electric and magnetic fields. Maxwell developed a highly theoretical and mathematical theory that showed the relationships with four equations, known today as Maxwell's equations. These equations require calculus and differential equations and are beyond the scope of this book. It is possible to develop an intuitive, qualitative idea of the physical process involved in electromagnetic waves by using your imagination. If you have been successful in imagining an electric field in the space around a charged particle and imagining a magnetic field that is produced by a moving charged particle, then you should be able to combine the two to imagine electromagnetic radiation that

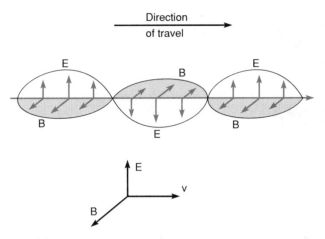

Figure 9.24

An electromagnetic wave is a pulse of electric fields (E) and magnetic fields (B). The fields regenerate each other in an endless cycle as they move through space at the speed of light. The E field, B field, and direction of travel of the wave are at right angles.

Table 9.1

The electromagnetic spectrum

Name of Wave	Approximate Frequency Range (hz)	Approximate Wavelength Range (m)
Gamma	5×10^{19} and up	6×10^{-12} and smaller
X Ray	$3 \times 10^{17} - 5 \times 10^{19}$	$1 \times 10^{-9} - 6 \times 10^{-12}$
Ultraviolet	$8 \times 10^{14} - 3 \times 10^{17}$	$3.8 \times 10^{-7} - 1 \times 10^{-9}$
Visible Light	$3.8 \times 10^{14} - 7.7 \times 10^{14}$	$7.9 \times 10^{-7} - 3.9 \times 10^{-7}$
Infrared	$3 \times 10^{11} - 4 \times 10^{14}$	$1 \times 10^{-3} - 7.5 \times 10^{-7}$
Microwaves	$1 \times 10^{9} - 3 \times 10^{11}$	$0.001 - 0.3$
Radar	$3 \times 10^{9} - 3 \times 10^{10}$	$0.01 - 0.1$
UHF	$3 \times 10^{8} - 3 \times 10^{9}$	$0.1 - 1$
VHF	$3 \times 10^{7} - 3 \times 10^{8}$	$1 - 10$
Radio	$3 \times 10^{5} - 3 \times 10^{7}$	$10 - 1,000$
CB	$3 \times 10^{4} - 3 \times 10^{5}$	$1,000 - 10,000$

moves as a wave. The key to all this imagining is to understand that (1) a changing electric field generates a magnetic field, (2) a changing magnetic field generates an electric field, and (3) both magnetic and electric fields carry energy.

When an electric charge is at rest it has a constant, uniform electric field around it. But what happens if you somehow *accelerate* the charge? As it begins to move the electric field is changed, and this changing field moves away from the charge into space at some speed. The changing electric field generates a magnetic field. But since the magnetic field grows from zero to a maximum, it is a changing magnetic field. The changing magnetic field now generates a new electric field, and the process continues on with the electric and magnetic fields coupled together in the form of a pulse. The two fields are now locked together, separated from the charge that originated them. As one field changes it generates a new field a little farther out. So the pulse moves out through space, exchanging energy back and forth between the electric and magnetic fields as they regenerate each other in an endless cycle. This pulse of rapidly regenerating electric and magnetic fields is called an **electromagnetic wave** (figure 9.24). The electric and magnetic fields are two parts of the wave, which is a single entity. Once generated by an accelerated charge, the wave disturbance moves from its source and is independent of it. It will move through space at the speed of light until it is absorbed, giving up its energy.

Maxwell's equations predicted that electromagnetic waves would be produced whenever charges are accelerated but also gave the velocity of electromagnetic waves as 3.00×10^8 m/sec in a vacuum (about 186,000 mi/sec), which is the same as the velocity of light. Maxwell concluded that light is an electromagnetic wave. Today, visible light is understood to comprise a very small part of the **electromagnetic spectrum.** All parts of this spectrum are similar to light but have higher or lower frequencies, or numbers of oscillations per second of an electromagnetic wave. The relationship between the frequency

of a wave (f) and wavelength (λ) was given by the wave equation in chapter 7 (equation 7.3), $v = \lambda f$. In the case of electromagnetic waves, v is given the symbol c for the speed of light, or $c = \lambda f$. In general, the frequency and wavelength are inversely proportional ($f = c/\lambda$), so a higher frequency means a shorter wavelength.

There is not a sharp dividing point between the different regions as table 9.1 suggests, and some regions overlap. Notice that the visible light region of the spectrum is the only part that can be detected by the human eye. Visible light is the topic of the next chapter.

Summary

Magnets have two *poles* about which their attraction is concentrated. When free to turn, one pole moves to the north and the other to the south. The north-seeking pole is called the *north pole* and the *south-seeking* pole is called the south pole. *Like poles repel* one another and *unlike poles attract.*

The property of magnetism is *electric in origin,* produced by charges in motion. *Permanent magnets* have tiny regions called *magnetic domains,* each with its own north and south pole. An *unmagnetized* piece of iron has randomly arranged domains. When *magnetized,* the domains become aligned and contribute to the overall magnetic effect.

The *strength* of a magnetic field is measured in terms of the force experienced by a *moving test charge.* The *maximum force* experienced is proportional to the *charge* (q) and *velocity* (v) of the charge when it is moving at right angles to the magnetic field (B), or $F_{max} = qvB$. The strength of B can therefore be defined as $B = F_{max}/qv$, having units of newton/amp · meter, which is called a *tesla.*

A *current-carrying wire* has magnetic field lines of closed *concentric circles* that are at right angles to the length of wire. The *direction* of the magnetic field depends on the direction of the current. The *strength* of the magnetic field is directly proportional to the *current* and inversely proportional to the *distance* from the wire. When placed in a *second magnetic field* at right angles, the magnetic field of the

Electricity and Magnetism

Box Figure 9.1

Solar cells generate electricity when exposed to light, without moving parts or hot fluids of any kind. The solar cells on this patio roof furnish an average of 500 to 600 kwhr of electrical energy per month for a home in Arizona.

You may be familiar with many solid-state devices such as calculators, computers, word processors, digital watches, VCRs, digital stereos, and camcorders. All of these are called solid-state devices because they use a solid material, such as the semiconductor silicon, in an electric circuit in place of vacuum tubes. Solid-state technology developed from breakthroughs in the use of semiconductors during the 1950s, and the use of thin pieces of silicon crystal is common in many electric circuits today.

A related technology also uses thin pieces of a semiconductor such as silicon but not as a replacement for a vacuum tube. This technology is concerned with photovoltaic devices, also called *solar cells,* that generate electricity when exposed to light (box figure 9.1). A solar cell is unique in generating electricity since it produces electricity directly, without moving parts or chemical reactions, and potentially has a very long lifetime. This Feature is concerned with how a solar cell generates electricity.

Light is an electromagnetic wave in a certain wavelength range. This wave is a pulse of electric and magnetic fields that are locked together, moving through space as they exchange energy back and forth, regenerating each other in an endless cycle until the wave is absorbed, giving up its energy. Light also has the ability to give its energy to an electron, knocking it out of a piece of metal. This phenomenon, called the *photoelectric effect,* is described in the next chapter.

Solid-state technology is mostly concerned with crystals and electrons in crystals. A *crystal* is a solid with an ordered, three-dimensional arrangement of atoms. A crystal of common table salt, sodium chloride, has an ordered arrangement that produces a cubic structure. You can see this cubic structure of table salt if you look closely at a few grains. The regular geometric arrangement of atoms (or ions) in a crystal is called the *lattice,* or framework, of the crystal structure.

current-carrying wire interacts with a maximum force that is proportional to the *current* (I) in the wire, the *length* (L) of the wire, and the *strength* of the second magnetic field (B), or $F_{max} = ILB$. This force can be *increased* by making wire loops from the straight wire, which concentrates the magnetic field lines. A coil of many loops is called a *solenoid,* or *electromagnet.* The electromagnet is the working part in electrical meters, electromagnetic switches, and the electric motor.

When a loop of wire is moved in a magnetic field or if a magnetic field is moved past a wire loop a voltage is induced in the wire loop. The interaction is called *electromagnetic induction.* An electric generator is a rotating coil of wire in a magnetic field. The coil is rotated by mechanical energy and electromagnetic induction induces a voltage, thus converting mechanical energy to electrical energy. A *transformer* steps up or steps down the voltage of a current through electromagnetic induction in a secondary coil. The ratio of input and output voltage is determined by the number of loops in the primary and secondary coils. Increasing the voltage decreases the current, which makes long-distance transmission of electrical energy economically feasible.

The answer about the nature of light was found by Maxwell as he found mathematical relationships between electric and magnetic fields. His mathematical theory showed that light and other forms of radiant energy are electromagnetic waves. An electromagnetic wave is a pulse consisting of electric and magnetic fields that continually regenerate each other as the pulse moves through space. The pulse travels at the speed of light until it is absorbed, giving up its energy. Electromagnetic waves can have any frequency, and different ranges of frequencies are known as radio waves, infrared, visible light, ultraviolet, X rays, and gamma rays.

A highly simplified picture of a crystal of silicon has each silicon atom bonded with four other silicon atoms, with each pair of atoms sharing two electrons. Normally, this ties up all the electrons and none are free to move and produce an electric current. This happens in the dark in silicon crystals and the silicon crystal is an insulator. Light, however, can break electrons free in a silicon crystal so they can move in a current. Silicon is therefore a semiconductor.

The conducting properties of silicon can be changed by *doping*, that is, artificially forcing atoms of other elements into the crystal lattice. Phosphorus, for example, has five electrons in its outermost shell compared to the four in a silicon atom. When phosphorus atoms replace silicon atoms in the crystal lattice, there are extra electrons not tied up in the two electron bonds. The extra electrons move easily through the crystal lattice, carrying a charge. Since the phosphorus-doped silicon carries a negative charge it is called a *n-type* semiconductor. The n means negative charge carrier.

A silicon crystal doped with boron will have atoms in the lattice with only three electrons in the outermost shell. This results in a deficiency, that is, electron "holes" that act as positive charges. A hole can move as an electron is attracted to it, but it leaves another hole elsewhere, where it moved from. Thus a flow of electrons in one direction is equivalent to a flow of holes in the opposite

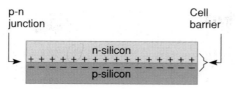

Box Figure 9.2

The cell barrier forms at the p-n junction between the n-silicon and the p-silicon. The barrier creates a "one-way" door that accumulates negative charges in the n-silicon.

direction. A hole therefore behaves as a positive charge. Since the boron-doped silicon carries a positive charge it is called a *p-type* semiconductor. The p means positive charge carrier.

The basic operating part of a silicon solar cell is typically an 8 cm wide and 3×10^{-3} mm (about one-hundredth of an inch) thick wafer cut from a silicon crystal. One side of the wafer is doped with boron to make p-silicon and the other side is doped with phosphorus to make n-silicon. The place of contact between the two is called the p-n junction, which creates a *cell barrier*. The cell barrier forms as electrons are attracted from the n-silicon to the holes in the p-silicon. This creates a very thin zone of negatively charged p-silicon and positively charged n-silicon (box figure 9.2). Thus an internal electric field is established at the p-n junction and the field is the cell barrier. The barrier is about

3×10^{-7} mm (about one-millionth of an inch) thick.

A metal base plate is attached to the p-silicon side of the wafer and a grid of metal contacts to the n-silicon side for electrical contacts. The grid is necessary to allow light into the cell. The entire cell is then coated with a transparent plastic covering.

The cell is thin and light can penetrate through the p-n junction. Light impacts the p-silicon, freeing electrons. Low-energy free electrons might combine with a hole, but high-energy electrons cross the cell barrier into the n-silicon. The electron loses some of its energy and the barrier prevents it from returning, creating an excess negative charge in the n-silicon and a positive charge in the p-silicon. This establishes a potential of about 0.5 volt, which will drive a 2 amp current (8 cm cell) through a circuit connected to the electrical contacts.

Solar cells are connected different ways for specific electrical energy requirements in about a 1 meter arrangement called a *module*. Several modules are used to make a solar *panel*, and panels are the units used to design a solar cell *array*.

Today, solar cells are essentially handmade and are economical only in remote power uses (navigational aids, communications, or irrigation pumps) and in consumer specialty items (solar-powered watches and calculators). Research continues on finding methods of producing highly efficient, highly reliable solar cells that are affordably priced.

Summary of Equations

9.1

$$\text{magnetic field strength} = \frac{\text{maximum force on test charge}}{\left(\begin{array}{c}\text{charge on}\\\text{test charge}\end{array}\right)\left(\begin{array}{c}\text{velocity of}\\\text{test charge}\end{array}\right)}$$

$$B = \frac{F_{max}}{qv}$$

9.2

$$\begin{array}{c}\text{magnetic field strength}\\\text{(straight wire)}\end{array} = (\text{constant})\frac{2 \times \text{current}}{\text{distance from wire}}$$

$$B = k\frac{2I}{d}$$

where $k = 1.00 \times 10^{-7}$ newton/amp^2

9.3

$$\begin{array}{c}\text{maximum force}\\\text{on current-}\\\text{carrying wire}\end{array} = \left(\begin{array}{c}\text{electric}\\\text{current}\\\text{in wire}\end{array}\right)\left(\begin{array}{c}\text{length of}\\\text{wire in}\\\text{magnetic}\\\text{field}\end{array}\right)\left(\begin{array}{c}\text{strength of}\\\text{magnetic}\\\text{field}\end{array}\right)$$

$$F_{max} = ILB$$

9.4

$$\begin{array}{c}\text{induced}\\\text{voltage}\end{array} = \left(\begin{array}{c}\text{strength of}\\\text{magnetic}\\\text{field}\end{array}\right)\left(\begin{array}{c}\text{length of}\\\text{wire in}\\\text{magnetic}\\\text{field}\end{array}\right)\left(\begin{array}{c}\text{velocity}\\\text{of moving}\\\text{wire}\end{array}\right)$$

$$\text{induced voltage} = BLv$$

Charge moving in a magnetic field feels a force.

Electricity and Magnetism

9.5

$$\frac{volts_{primary}}{(number\ of\ loops)_{primary}} = \frac{volts_{secondary}}{(number\ of\ loops)_{secondary}}$$

$$\frac{V_p}{N_p} = \frac{V_s}{N_s}$$

9.6

$$(volts_{primary})(current_{primary}) = (volts_{secondary})(current_{secondary})$$

$$V_p I_p = V_s I_s$$

Key Terms

electric generator (p. 175)
electromagnet (p. 172)
electromagnetic induction (p. 175)
electromagnetic spectrum (p. 179)
electromagnetic wave (p. 179)
ferromagnetic properties (p. 165)
gauss (p. 169)
magnetic domain (p. 168)
magnetic field (p. 165)

magnetic poles (p. 165)
north pole (p. 165)
primary coil (p. 177)
secondary coil (p. 177)
solenoid (p. 172)
south pole (p. 165)
step-down transformer (p. 177)
step-up transformer (p. 177)
tesla (p. 169)
transformer (p. 176)
Van Allen belts (p. 171)

Applying the Concepts

1. The north pole of an ordinary bar magnet that is suspended will point
   a. generally toward the geographic north.
   b. to the geographic south since opposite poles attract.
   c. in random directions since it is a bar magnet.
   d. in the direction of the largest regional magnetite deposit.

2. The earth's north magnetic pole
   a. is located exactly in the same place as the geographic North Pole.
   b. is a magnetic south pole.
   c. has always had the same orientation.
   d. none of the above.

3. A permanent magnet has magnetic properties because
   a. the magnetic fields of its electrons are balanced.
   b. of an accumulation of monopoles in the ends.
   c. the magnetic domains are aligned.
   d. all of the above.

4. The force that exists between a second magnetic field and the magnetic field of a current-carrying wire can be increased by
   a. increasing the current in the wire.
   b. forming the wire into a series of loops.
   c. wrapping many turns of the wire around a soft iron rod.
   d. all of the above.

5. A device that measures an electric current, such as a galvanometer, actually measures
   a. the flow of individual electrons past a given point in the circuit.
   b. the force produced by the moving electrons.
   c. the magnetic field produced by the current.
   d. groups of electrons called coulombs.

6. The operation of a stereo speaker requires the use of
   a. one electromagnet.
   b. two electromagnets.
   c. one electromagnet and one permanent magnet.
   d. two permanent magnets.

7. The operation of an electric motor requires the use of
   a. an electromagnet on an axle that maintains its magnetic orientation.
   b. two electromagnets, one of which has a changing magnetic orientation as it spins on an axle.
   c. two electromagnets with constant magnetic field orientations in order to do a specified amount of work.

8. The induced voltage produced in a generator is directly proportional to the
   a. length of wire rotating inside the magnetic field.
   b. strength of the magnetic field inside the generator.
   c. velocity of the moving wire that is rotating.
   d. all of the above.

9. A step-up transformer steps up the
   a. power.
   b. current.
   c. voltage.
   d. all of the above.

10. Electric power companies use step-up transformers at power plants because
    a. the current would not otherwise have enough power to move the long distances to the cities.
    b. step-up transformers produce more current, therefore resulting in more electricity to sell and greater profits for the company.
    c. this reduces the current, which in turn reduces the energy lost to resistance.
    d. the current is increased enough to overcome the large resistance encountered in transmitting electricity over long distances.

Answers

1. a 2. b 3. c 4. d 5. c 6. c 7. b 8. d 9. c
10. c

Questions for Thought

1. What is a magnetic pole? How are magnetic poles named?

2. Describe a model that would explain how the magnetic poles of magnets are able to exert forces on one another without touching.

3. What is magnetic declination? What occurrence is responsible for magnetic declination?

4. When does an electric charge (a) have only an electric field? (b) have both an electric field and a magnetic field? (c) produce an electromagnetic wave?

5. How is an unmagnetized piece of iron different from the same piece of iron when it is magnetized?

6. Two identical pieces of iron are attracted to each other. Without the use of other materials to test the pieces on, describe how you could find out (a) if both are permanent magnets or not and (b) which is the magnet if only one is a permanent magnet.

7. Describe how the strength of a magnetic field is measured. What are the units of magnetic field strength and what do they mean?

8. Suppose there is a vertical wire directly in front of you with the electron current running from the floor to the ceiling. Would you expect a compass held in the magnetic field around this wire to point the north end of the needle toward you on the right or the left side of the wire? Explain.

9. What are the northern lights? Why do they occur at the North Pole and not elsewhere?

10. Explain why an electric utility company increases the voltage of electricity for long-distance transmission.

11. How does a galvanometer measure an electric current?

12. Compare and contrast an electric motor and an electric generator.

13. Why does a moving electric charge interact with a magnetic field?

14. What factors determine the induced voltage produced by electromagnetic induction in an electric generator? Explain.

15. Describe as many ways as you can to increase the strength of a magnetic field around a current-carrying wire.

16. Describe how an electric generator and a transformer use electromagnetic induction.

17. Why does the north pole of a magnet point to the geographic North Pole if like poles repel?

18. A maximum force is experienced by a moving electron nearing the earth's magnetic field while a second, identical electron moving toward the earth with an identical velocity experiences zero net force. Describe as many reasons as you can that explain how this could occur.

19. Describe the units for measuring the strength of a magnetic field.

20. Describe how you could use a magnetic compass to determine the direction of a current in a straight wire.

21. What does an electric meter (ampmeter, voltmeter, or ohmmeter) really measure? How can this be justified as a measure of electrical units?

22. Does a telephone transmit sound waves? Explain.

23. A step-up transformer increases the voltage of a current. Are you getting something for nothing? Explain.

24. What is an electromagnetic wave? How does it move through space until it interacts with a surface?

25. A certain magnetic compass is constructed so the needle moves only up and down. Where on the earth would this compass needle point straight out, parallel to the ground? Where would it point straight down? Where would it point straight up?

26. Explain what causes an electron to move toward one end of a wire when the wire is moved across a magnetic field.

Exercises

▼

Group A—Solutions Provided in Appendix D

1. A 150. cm wire moves at a right angle to a 0.400 tesla magnetic field at 30.0 cm/sec. What voltage is induced in the wire?

2. A power plant generator produces a 1,200. V, 40.0 A alternating current that is fed to a step-up transformer before transmission over the high lines. The transformer has a ratio of 200. to 1 wire loops. (a) What is the voltage of the transmitted power? (b) What is the current?

3. Assuming the earth's magnetic field to be 2.00×10^{-5} tesla, what voltage is induced in a 5.00 m length of wire that is moved 1.00 m/sec as a jump rope at right angles to the earth's magnetic field?

4. A 100. m length of transmission line is perpendicular to the earth's 2.00×10^{-5} tesla magnetic field. If the line carries 48.0 kw of electric power with a potential of 240,000. V, what magnetic force does the wire experience?

5. A 500. cm wire at right angles to a magnetic field experiences a force of 0.400 N when it carries a current of 8.00 A. What is the strength of the magnetic field?

6. An electron is moving at right angles to a 0.500 tesla magnetic field at 6.00×10^5 m/sec. If the mass of the electron is about 9.00×10^{-32} kg (a) what force acts on the particle? (b) what acceleration does the electron experience as a result of this force acting on it?

Group B—Solutions Not Given

1. A small generator used to power a bicycle light has a total length of 50.0 cm of wire that moves across a 0.500 tesla field from a permanent magnet. How many volts are induced into the wire when the bicycle moves fast enough to move the wire at 24.0 m/sec?

2. A step-up transformer has a primary coil with 100 loops and a secondary coil with 1,500 loops. If the primary coil is supplied with a household current of 120. V and 15.0 A (a) what voltage is produced in the secondary circuit? (b) what current flows in the secondary circuit?

3. Assuming the earth's magnetic field to have a strength of 2.00×10^{-5} tesla, at what velocity would you need to move a 5.00 m length of wire at right angles to the field in order to induce a voltage of 1.00 V?

4. A straight 50.0 cm wire in a car carries 24.0 w of electrical power with a potential of 12.0 V. What force acts on the wire if a magnetic field with a strength of 1.00 tesla moves across it at right angles?

5. A 1.50 m wire is carrying a current of 5.00 A at right angles to a magnetic field when it experiences a force of 0.5 N. What is the magnetic field strength?

6. A beam of electrons enters a magnetic field at right angles with a velocity of 4.00×10^6 m/sec. If the magnetic field strength is 0.600 tesla (a) what force acts on the electrons? (b) what acceleration do the electrons undergo as a result?

1.00 x 10⁻⁷

7. A high-voltage transmission line carries 765 kilovolt and 0.260 A. What is the magnetic field strength 60.0 m from this line?

8. A straight horizontal wire carries a 5.00 A electron current towards the east. At a distance of 1.00 m directly below the wire on the floor (a) what is the direction of the magnetic field? (b) what is the magnitude of the magnetic field?

9. What current do you need in a straight horizontal wire to produce a magnetic field 10.0 cm directly below the wire with a similar strength to the earth's 2.00×10^{-5} tesla field?

10. A step-down transformer has an output of 12.0 V and 0.500 A when connected to a 120. V line. Assuming no losses (a) what is the ratio of primary to secondary loops? (b) what current does the transformer draw from the line? (c) what is the power output of the transformer?

11. A step-up transformer on a 120. V line has 50 loops on the primary coil and 150 loops on the secondary coil and draws a 5.00 A current. Assuming no losses (a) what is the voltage from the secondary coil? (b) what is the current from the secondary coil? (c) What is the power output?

12. An electron moves at right angles through a magnetic field with a strength of 6.00×10^{-2} tesla. What force acts on the electron if it has a velocity of 3.00×10^7 m/sec?

7. A wire in a wall carries a 120. V, 15.0 A current. What is the magnetic field strength experienced by a person 1.00 m from the wire?

8. If the earth's magnetic field has a strength of 2.00×10^{-5} tesla (a) what current is needed in a wire to produce an equal-sized magnetic field at a distance of 1.50 m? (b) how should this wire be oriented to oppose the earth's magnetic field directly below the wire?

9. What current is required in a straight wire to produce a magnetic field strength of a permanent bar magnet of 1.00×10^{-2} tesla? Assume the wire is 10.0 cm away.

10. The step-down transformer in a local neighborhood reduces the voltage from a 7,200. V line to 120. V. (a) If there are 125 loops on the secondary coil, how many are on the primary coil? (b) What current does the transformer draw from the line if the current in the secondary coil is 36.0 A? (c) What is the power input and output?

11. A step-up transformer connected to a 120. V electric generator has 30 loops on the primary coil to each loop on the secondary coil. (a) What is the voltage of the secondary coil? (b) If the transformer has a 90.0 A current in the primary coil, what is the current in the primary coil? (c) What is the power input and output?

12. A beta particle, which is a high-speed electron, moves with a velocity of 3.00×10^7 m/sec across a magnetic field with a strength of 4.00×10^{-2} tesla. What maximum force acts on the beta particle?

Chapter 10

Light

Figure 10.1

Light provides the most information about your surroundings but seems more dramatic and mysterious than solid matter. Light seems more mysterious because it can only be studied indirectly, that is, in terms of how it behaves. As a result, the behavior of light has stimulated thinking, scientific investigations, and debate for hundreds of years. Perhaps you have wondered about light and its behaviors. What is light?

You use light and your eyes more than any other sense to learn about your surroundings. All of your other senses—touch, taste, sound, and smell—involve matter, but the most information is provided by light. Yet, light seems more mysterious than matter. You can study matter directly, measuring its dimensions, taking it apart, and putting it together to learn about it. Light, on the other hand, can only be studied indirectly in terms of how it behaves (figure 10.1). Once you understand its behavior, you know everything there is to know about light. Anything else is thinking about what the behavior means.

The behavior of light has stimulated thinking, scientific investigations, and debate for hundreds of years. The investigations and debate have occurred because light cannot be directly observed, which makes the exact nature of light very difficult to pin down. For example, you know that light moves energy from one place to another place. You can feel energy from the sun as sunlight warms you, and you know that light has carried this energy across millions of miles of empty space. The ability of light to move energy like this could be explained (1) as energy transported by waves, just as sound waves carry energy from a source, or (2) as the kinetic energy of a stream of moving particles, which give up their energy when they strike a surface. The movement of energy from place to place could be explained equally well by a wave model of light or by a particle model of light. When two possibilities exist like this in science, experiments are designed and measurements are made to support one model and reject the other. Light, however, presents a baffling dilemma. Some experiments provide evidence that light consists of waves and not a stream of moving particles. Yet other experiments provide evidence of just the opposite, that light is a stream of particles and not a wave. Evidence for accepting a wave or particle model seems to depend on which experiments are considered.

The purpose of using a model is to make new things understandable in terms of what is already known. When these new things concern light, three models are useful in visualizing separate behaviors. Thus the electromagnetic wave model that was introduced in chapter 9 will be used to describe how light is created at a source. Another model, a model of light as a ray, a small beam of light, will be used to discuss some common properties of light such as reflection and the refraction, or bending, of light. Finally, properties of light that provide evidence for a particle model will be discussed before ending with a discussion of the present understanding of light.

Sources of Light

The sun, stars, light bulbs, and burning materials all give off light. When something produces light it is said to be **luminous.** The sun is a luminous object that provides almost all of the *natural* light on the earth. A small amount of light does reach the earth from the stars but not really enough to see by on a moonless night. The moon and planets shine by reflected light and do not produce their own light, so they are not luminous.

Burning has been used as a source of *artificial* light for thousands of years. A wood fire and a candle flame are luminous because of their high temperatures. When visible light is given off as a result of high temperatures the light source is said to be **incandescent.** A flame from any burning source, an ordinary light bulb, and the sun are all incandescent sources because of high temperatures.

How do incandescent objects produce light? One explanation is given by the electromagnetic wave model that was introduced in chapter 9. This model describes a relationship between electricity, magnetism, and light. The model pictures an electromagnetic wave as forming whenever an electric charge is *accelerated* by acquiring additional energy. The acceleration produces a wave consisting of electrical and magnetic fields that become isolated from the accelerated charge, moving off into space. As the wave moves through space the two fields exchange energy back and forth, continuing on until they are absorbed by matter and give up their energy.

The frequency of an electromagnetic wave depends on the acceleration of the charge, and the greater the acceleration the higher the frequency of the wave that is produced. The complete range of frequencies is called the *electromagnetic spectrum* (see table 9.1). The spectrum ranges from radio waves at the low-frequency end of the spectrum to gamma rays at the high-frequency end. Visible light occupies only a small part of the middle portion of the complete spectrum.

Visible light is emitted from incandescent sources at high temperatures, but actually electromagnetic radiation is given off from matter at *any* temperature. This radiation is called **blackbody radiation,** which refers to an idealized material (the "blackbody") that perfectly absorbs and perfectly emits electromagnetic radiation. From the electromagnetic wave model, the radiation originates from the acceleration of charged particles near the surface of an object. The frequency of the blackbody radiation, and if it is in the range of visible light or not, is determined by the energy available for accelerating charged particles, that is, the temperature of the object. Near absolute zero there is little energy available and no radiation is given off. As the temperature of an object is increased more energy is available and this energy is distributed over a range of values, so more than one frequency of radiation is emitted. A graph of the frequencies emitted from the range of available energy is thus somewhat bell-shaped. The steepness of the curve and the position of the peak depends on the temperature (figure 10.2). As the temperature of an object increases there is an increase in the *amount* of radiation given off, and the peak radiation emitted progressively *shifts* toward higher and higher frequencies.

At room temperature the radiation given off from an object is in the infrared region, invisible to the human eye. When the temperature of the object reaches about 700° C (about 1,300° F) the peak radiation is still in the infrared region, but the peak has shifted enough toward the higher frequencies that a little visible light is emitted as a dull red glow. As the temperature of the object continues to increase the amount of radiation increases and the peak continues to shift toward shorter wavelengths. Thus the object begins to glow brighter and the color changes from red to orange, to yellow, and eventually to

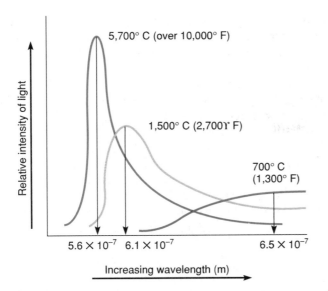

Figure 10.2

The intensity of blackbody radiation increases with increasing temperature, and the peak wavelength emitted shifts toward shorter wavelengths.

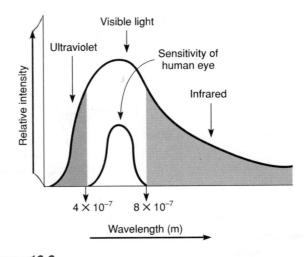

Figure 10.3

Sunlight is about 9 percent ultraviolet radiation, 40 percent visible light, and 51 percent infrared radiation before it travels through the earth's atmosphere.

white. The association of this color change with temperature is noted in the referent description of an object being "red hot," "white hot," and so forth.

The incandescent flame of a candle or fire results from the blackbody radiation of carbon particles in the flame. At a blackbody temperature of 1,500° C (about 2,700° F), the carbon particles emit visible light in the red to yellow frequency range. The tungsten filament of an incandescent light bulb is heated to about 2,200° C (about 4,000° F) by an electric current. At this temperature the visible light emitted is in the reddish, yellow-white range.

The radiation from the sun, or sunlight, comes from the sun's surface, which has a temperature of about 5,700° C (about 10,000° F). As shown in figure 10.3, the sun's radiation has a broad spectrum centered near the yellow-green wavelength. Your eye is most sensitive to this wavelength of sunlight. The spectrum of sunlight before it travels through the earth's atmosphere is infrared (about 51 percent), visible light (about 40 percent), and ultraviolet (about 9 percent). Sunlight originated as energy released from nuclear reactions in the sun's core. This energy requires about a million years to work its way up to the surface. At the surface, the energy from the core accelerates charged particles, which then emit light like tiny antennas. The sunlight requires about eight minutes to travel the distance from the sun's surface to the earth.

Properties of Light

You can see luminous objects from the light they emit, and you can see nonluminous objects from the light they reflect, but you cannot see the path of the light itself. For example, you cannot see a flashlight beam unless you fill the air with chalk dust or

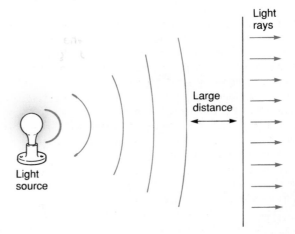

Figure 10.4

Light rays are perpendicular to a wave front. A wave front that has traveled a long distance is a plane wave front, and its rays are parallel to each other. The rays show the direction of the wave motion.

smoke. The dust or smoke particles reflect light, revealing the path of the beam. This simple observation must be unknown to the makers of science fiction movies since they always show visible laser beams zapping through the vacuum of space.

Some means of representing the invisible travels of light are needed in order to discuss some of its properties. Throughout history a **light ray model** has been used to describe the travels of light. The meaning of this model has changed over time, but it has always been used to suggest that "something" travels in *straight-line paths*. The light ray is a line that is drawn to represent the straight-line travel of light. It is often used with a discussion of waves since many properties of light can be explained in terms of the behavior of waves. The light ray is, nonetheless, a line that represents an imaginary thin beam of light (figure 10.4). A line is drawn to represent this imaginary beam

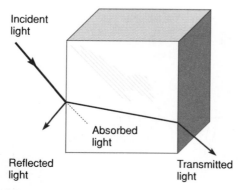

Figure 10.5

Light that interacts with matter can be reflected, absorbed, or transmitted through transparent materials. Any combination of these interactions can take place, but a particular substance is usually characterized by what it mostly does.

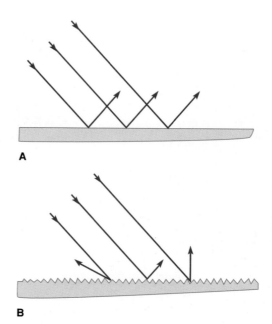

Figure 10.6

(a) Rays reflected from a perfectly smooth surface are parallel to each other. (b) Diffuse reflection from a rough surface causes rays to travel in many random directions.

to illustrate the law of reflection (as from a mirror) and the law of refraction (as through a lens). There are limits to using a light ray for explaining some properties of light, but it works very well in explaining mirrors, prisms, and lenses.

Light Interacts with Matter

A ray of light travels in a straight line from a source until it encounters some object or particles of matter (figure 10.5). What happens next depends on several factors, including (1) the smoothness of the surface, (2) the nature of the material, and (3) the angle at which it strikes the surface.

The *smoothness* of the surface of an object can range from perfectly smooth to extremely rough. If the surface is perfectly smooth rays of light undergo **reflection,** leaving the surface parallel to each other. A mirror is a good example of a very smooth surface that reflects light this way (figure 10.6a). If a surface is not smooth the light rays are reflected in many random directions as **diffuse reflection** takes place (figure 10.6b). Rough and irregular surfaces and dust in the air make diffuse reflections. It is diffuse reflection that provides light in places not in direct lighting, such as under a table or under a tree. Such shaded areas would be very dark without diffuse reflection of light.

Some *materials* allow much of the light that falls on them to move through the material without being reflected. Materials that allow transmission of light through them are called **transparent.** Glass and clear water are examples of transparent materials. Many materials do not allow transmission of any light and are called **opaque.** Opaque materials reflect light, absorb light, or some combination of partly absorbing and partly reflecting light. The light that is reflected varies with wavelength and gives rise to the perception of color, which will be discussed shortly. Absorbed light gives up its energy to the material and may be reemitted at a different wavelength or it may simply show up as a temperature increase.

The *angle* of the light ray to the surface and the nature of the material determine if the light is absorbed, transmitted through a transparent material, or reflected. Vertical rays of light, for example, are mostly transmitted through a transparent material with some reflection and some absorption. If the rays strike the surface at some angle, however, much more of the light is reflected, bouncing off the surface. Thus the glare of reflected sunlight is much greater around a body of water in the late afternoon than when the sun is directly overhead.

Light that interacts with matter is reflected, transmitted, or absorbed, and all combinations of these interactions are possible. Materials are usually characterized by which of these interactions they *mostly* do, but this does not mean that other interactions are not occurring too. For example, a window glass is usually characterized as a transmitter of light. Yet, the glass always *reflects* about 4 percent of the light that strikes it. The reflected light usually goes unnoticed during the day because of the bright light that is transmitted from the outside. When it is dark outside you notice the reflected light as the window glass now appears to act much like a mirror. A one-way mirror is another example of both reflection and transmission occurring (figure 10.7). A mirror is usually characterized as a reflector of light. A one-way mirror, however, has a very thin silvering that reflects most of the light but still transmits a little. In a lighted room a one-way mirror appears to reflect light just as any other mirror. But a person behind the mirror in a dark room can see into the lighted room by means of the transmitted light. Thus you know that this mirror transmits as well as reflects light. One-way mirrors are used to unobtrusively observe for shoplifters in many businesses.

Figure 10.7

A one-way mirror reflects most of the light that strikes it but still transmits some light to a person behind the mirror in a darkened room.

Reflection

Most of the objects that you see are visible from diffuse reflection. For example, consider some object such as a tree that you see during a bright day. Each *point* on the tree must reflect light in all directions since you can see any part of the tree from any angle (figure 10.8). As a model, think of bundles of light rays entering your eye, which enable you to see the tree. This means

Figure 10.8

Bundles of light rays are reflected diffusely in all directions from every point on an object. Only a few light rays are shown from only one point on the tree in this illustration. The light rays that move to your eyes enable you to see the particular point from which they reflected.

Figure 10.9

Adjacent light rays spread farther and farther apart after reflecting from a point. Close to the point the rate of spreading is great. At a far distance, the rays are almost parallel. The rate of ray spreading carries information about distances.

that you can see any part of the tree from any angle because different bundles of reflected rays will enter your eye from different parts of the tree.

Groups of rays move from each point on the tree, traveling in all directions. Figure 10.9 shows a two-dimensional ray representation of this diffuse reflection. If you consider just two of the light rays you will notice that they are spreading farther and farther apart as they travel. You are not aware of it, but at a relatively close distance the rate of spreading carries information that your eyes and brain interpret about distance. At some great distance from a source of light only rays that are almost parallel will reach your eye. Light rays from great distances, such as the distance to the sun or more, are almost parallel. Overall, the rate at which groups of rays are spreading or if they are essentially parallel carries information about distances.

Light rays that are diffusely reflected move in all possible directions, but rays that are reflected from a smooth surface, such as a mirror, leave the mirror in a definite direction. Suppose you look at a tree in a mirror. There is only one place on the mirror where you look to see any one part of the tree. While

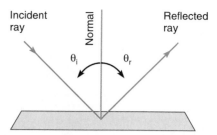

Figure 10.10

The law of reflection states that the angle of incidence (θ_i) is equal to the angle of reflection (θ_r). Both angles are measured from the normal, a reference line drawn perpendicular to the surface at the point of reflection.

light is reflecting off the mirror from all parts of the tree, the only rays that reach your eye are the rays that are reflected at a certain angle from the place where you look. The relationship between the light rays moving from the tree and the direction they are reflected from the mirror to reach your eyes can be understood by drawing three lines: (1) a line representing an original ray from the tree, called the **incident ray,** (2) a line representing a reflected ray, called the **reflected ray,** and (3) a reference line that is perpendicular to the reflecting surface and is located at the point where the incident ray struck the surface. This line is called the **normal.** The angle between the incident ray and the normal is called the **angle of incidence,** θ_i, and the angle between the reflected ray and the normal is called the **angle of reflection,** θ_r (figure 10.10). The *law of reflection,* which was known to the ancient Greeks, is that the *angle of incidence equals the angle of reflection,* or

$$\theta_i = \theta_r \qquad \textbf{equation 10.1}$$

Figure 10.11 shows how the law of reflection works when you look at a flat mirror. Light is reflected from all points on the box, and of course only the rays that reach your eyes are detected. These rays are reflected according to the law of reflection, with the angle of reflection equaling the angle of incidence. If you move your head slightly, then a different bundle of rays reaches your eyes. Of all the bundles of rays that reach your eyes, only two rays from a point are shown in the illustration. After these two rays are reflected, they continue to spread apart at the same rate that they were spreading before reflection. Your eyes and brain do not know that the rays have been reflected, and the diverging rays appear to come from behind the mirror as the dashed lines show. The image therefore appears to be the same distance *behind* the mirror as the box is from the front of the mirror. Thus a mirror image is formed where the rays of light *appear* to originate. This is called a **virtual image.** A virtual image is the result of your eyes' and brain's interpretations of light rays, not actual light rays originating from an image. Light rays that do originate from the other kind of image are called a **real image.** A real image is like the one displayed on a movie screen, with light originating from the image. A virtual image cannot be displayed on a screen since it results from an interpretation.

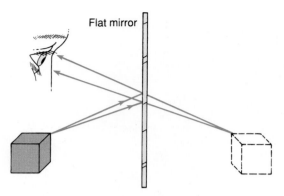

Figure 10.11

Light rays leaving a point on the block are reflected according to the law of reflection, and those reaching your eye are seen. After reflecting, the rays continue to spread apart at the same rate. You interpret this to be a block the same distance behind the mirror. You see a virtual image of the block because light rays do not actually move from the image.

Curved mirrors are either *concave,* with the center part curved inward, or *convex,* with the center part bulging outward. A concave mirror can be used to form an enlarged virtual image, such as a shaving or makeup mirror, or it can be used to form a real image as in a reflecting telescope. The convex mirror is often used to increase the field of vision, for example, the mirrors on the sides of trucks and vans. Convex mirrors are also used above an aisle in a store to show a wide area.

Refraction

You may have observed that an object that is partly in the air and partly in water appears to be broken, or bent, where the air and water meet. When a light ray moves from one transparent material to another, such as from water through air, the ray undergoes a change in the direction of travel at the boundary between the two materials. This change of direction of a light ray at the boundary is called **refraction.** The amount of direction change can be measured as an angle from the normal just as it was for the angle of reflection. The incoming ray is called the incident ray as before, and the new direction of travel is called the *refracted ray.* The angles of both rays are measured from the normal (figure 10.12).

Refraction results from a *change in speed* when light passes from one transparent material into another. The speed of light in a vacuum is 3.00×10^8 m/sec, but it is slower when moving through a transparent material. In water, for example, the speed of light is reduced to about 2.30×10^8 m/sec. The speed of light has a magnitude that is specific for various transparent materials.

When light moves from a transparent material to a new transparent material with a *slower* speed of light, the ray is refracted *toward* the normal (figure 10.13a). For example, light travels through air faster than through water. Light traveling from air into water is therefore refracted toward the normal as it enters the water. On the other hand, if light has a *faster* speed in the new material it is refracted *away* from the normal. Thus light traveling from water into the air is refracted away from the normal as it enters the air (figure 10.13b).

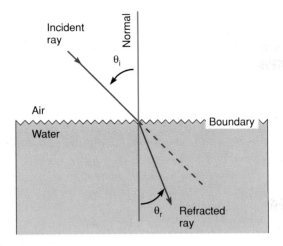

Figure 10.12

A ray diagram shows refraction at the boundary as a ray moves from air through water.

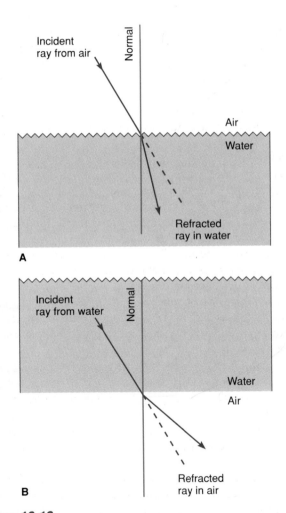

A

B

Figure 10.13

(a) A light ray moving to a new material with a slower speed of light is refracted toward the normal. (b) A light ray moving to a new material with a faster speed is refracted away from the normal.

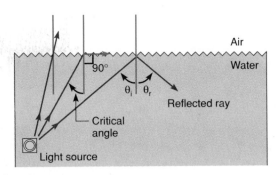

Figure 10.14

When the angle of incidence results in an angle of refraction of 90°, the refracted light ray is refracted along the water surface. The angle of incidence for a material that results in an angle of refraction of 90° is called the *critical angle*. When the incident ray is at this critical angle or greater, the ray is reflected internally. The critical angle for water is about 49° and for a diamond about 25°.

The magnitude of refraction depends on (1) the angle at which light strikes the surface and (2) the ratio of the speed of light in the two transparent materials. An incident ray that is perpendicular (90°) to the surface is not refracted at all. As the angle of incidence is increased the angle of refraction is also increased. There is a limit, however, that occurs when the angle of refraction reaches 90°, or along the water surface. Figure 10.14 shows rays of light traveling from water to air at various angles. When the incident ray is about 49°, the angle of refraction that results is 90°, along the water surface. This limit to the angle of incidence that results in an angle of refraction of 90° is called the **critical angle** for a water to air surface. At any incident angle greater than the critical angle the light ray does not move from the water to the air but is *reflected* back from the surface as if it were a mirror. This is called **total internal reflection** and implies that the light is trapped inside if it arrived at the critical angle or beyond. Faceted transparent gemstones such as the diamond are brilliant because they have a small critical angle and thus reflect much light internally. Total internal reflection is also important in fiber optics as discussed in the Feature at the end of this chapter.

As was stated earlier, refraction results from a change in speed when light passes from one transparent material into another. The ratio of the speeds of light in the two materials determines the magnitude of refraction at any given angle of incidence. The greatest speed of light possible, according to current theory, occurs when light is moving through a vacuum. The speed of light in a vacuum is accurately known to nine decimals but is usually rounded to 3.00×10^8 m/sec for general discussion. The speed of light in a vacuum is a very important constant in physical science, so it is given a symbol of its own, c. The ratio of c to the speed of light in some transparent material, v, is called the **index of refraction,** n, of that material or

$$n = \frac{c}{v}$$

equation 10.2

Light

Table 10.1
Index of refraction

Substance	$n = c/v$
Glass	1.50
Diamond	2.42
Ice	1.31
Water	1.33
Benzene	1.50
Carbon tetrachloride	1.46
Ethyl alcohol	1.36
Air (0° C)	1.00029
Air (30° C)	1.00026

The indexes of refraction for some substances are listed in table 10.1. The values listed are constant physical properties and can be used as a means to identify a specific substance. Note that a larger value means a greater angle of refraction at a given angle. Of the materials listed, diamond thus refracts light the most and air the least. The index for air is nearly one, which means that light is slowed only slightly in air.

Example 10.1

What is the speed of light in a diamond?

Solution
The relationship between the speed of light in a material (v), the speed of light in a vacuum (c = 3.00 × 10⁸ m/sec), and the index of refraction is given in equation 10.2. The index of refraction of a diamond is found in table 10.1 (n = 2.42).

$$n_{diamond} = 2.42$$
$$c = 3.00 \times 10^8 \text{ m/sec}$$
$$v = ?$$

$$n = \frac{c}{v} \therefore v = \frac{c}{n}$$

$$v = \frac{3.00 \times 10^8 \text{ m/sec}}{2.42}$$

$$= \boxed{1.24 \times 10^8 \text{ m/sec}}$$

Note that table 10.1 shows that colder air at 0° C (32° F) has a higher index of refraction than warmer air at 30° C (86° F), which means that light travels faster in warmer air. This difference explains the "wet" highway that you sometimes see at a distance in the summer. In the summer the air near the road is hotter on a clear, calm day. Light rays traveling toward you in this hotter air are refracted upwards as they enter the cooler air. Your brain interprets this refracted light as *reflected* light. Light traveling downwards from other cars is also refracted upwards toward you, and you think you are seeing cars "reflected" from the wet highway. When you reach the place where the "water" seemed to be, it disappears only to appear again farther down the road.

Sometimes convection currents produce a mixing of warmer air near the road with the cooler air just above. This mixing refracts light one way, then the other as the warmer and cooler air mix. This produces a shimmering or quivering that some people call "seeing heat." They are actually seeing changing refraction, which is a *result* of heating and convection. In addition to causing distant objects to quiver, the same effect causes the point source of light from stars to appear to twinkle. The light from closer planets does not twinkle because the many light rays from the disklike sources are not refracted together as easily as the fewer rays from the point sources of stars. The light from planets will appear to quiver, however, if the atmospheric turbulence is great.

Dispersion and Color
Recall that Maxwell had determined that electromagnetic waves travel with the speed of light and that there is a whole spectrum of waves with various frequencies and wavelengths (see table 9.1). The speed of electromagnetic waves (c) is related to the wavelength (λ) and the frequency (f) by a form of the wave equation, or

$$c = \lambda f \qquad \text{equation 10.3}$$

Visible light is the part of the electromagnetic spectrum that your eyes can detect, a narrow range of wavelength from about 7.90×10^{-7} m to 3.90×10^{-7} m. In general, this range of visible light can be subdivided into ranges of wavelengths that you perceive as colors. These are the colors of the rainbow, and there are six distinct colors that blend one into another. These colors are *red, orange, yellow, green, blue,* and *violet.* The corresponding ranges of wavelengths and frequencies of these colors are given in table 10.2.

In general, light is interpreted to be white if it has the same mixture of colors as the solar spectrum. That sunlight is made up of component colors was first investigated in detail by Issac Newton. While a college student, Newton became interested in grinding lenses, light, and color. At the age of twenty-three, Newton visited a local fair and bought several triangular glass prisms and proceeded to conduct a series of experiments with a beam of sunlight in his room. From the beginning, Newton intended to learn from direct observation and avoid speculation about the nature of light. In 1672, he reported the results of his experiments with prisms and color, concluding that white light is a mixture of all the independent colors. Newton found that a beam of sunlight falling on a glass prism in a darkened room produced a band of colors he called a *spectrum.* Further, he found that a second glass prism would not subdivide each separate color but would combine all the colors back into white sunlight. Newton concluded that sunlight consists of a mixture of the six colors. Today, light that is composed, or made up of, several colors of light is called *polychromatic* light. Light that is one wavelength only is called *monochromatic* light, but each color of the spectrum is a range of wavelengths.

Table 10.2

Range of wavelengths and frequencies of the colors of visible light

Color	Wavelength (in meters)	Frequency (in hertz)
Red	7.9×10^{-7} to 6.2×10^{-7}	3.8×10^{14} to 4.8×10^{14}
Orange	6.2×10^{-7} to 6.0×10^{-7}	4.8×10^{14} to 5.0×10^{14}
Yellow	6.0×10^{-7} to 5.8×10^{-7}	5.0×10^{14} to 5.2×10^{14}
Green	5.8×10^{-7} to 4.9×10^{-7}	5.2×10^{14} to 6.1×10^{14}
Blue	4.9×10^{-7} to 4.6×10^{-7}	6.1×10^{14} to 6.6×10^{14}
Violet	4.6×10^{-7} to 3.9×10^{-7}	6.1×10^{14} to 7.7×10^{14}

Example 10.2

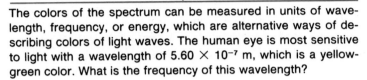

The colors of the spectrum can be measured in units of wavelength, frequency, or energy, which are alternative ways of describing colors of light waves. The human eye is most sensitive to light with a wavelength of 5.60×10^{-7} m, which is a yellow-green color. What is the frequency of this wavelength?

Solution

The relationship between the wavelength (λ), frequency (f), and speed of light in a vacuum (c), is found in equation 10.3, $c = \lambda f$.

$$c = 3.00 \times 10^8 \text{ m/sec}$$

$$\lambda = 5.60 \times 10^{-7} \text{ m}$$

$$f = ?$$

$$c = \lambda f \therefore f = \frac{c}{\lambda}$$

$$f = \frac{3.00 \times 10^8 \ \frac{m}{sec}}{5.60 \times 10^{-7} \text{ m}}$$

$$= \frac{3.00 \times 10^8}{5.60 \times 10^{-7}} \ \frac{m}{sec} \times \frac{1}{m}$$

$$= 5.40 \times 10^{14} \ \frac{1}{sec}$$

$$= \boxed{5.40 \times 10^{14} \text{ hz}}$$

A glass prism separates sunlight into a spectrum of colors because the index of refraction is different for different wavelengths of light. The same processes that slow the speed of light in a transparent substance have a greater effect on short wavelengths than they do on longer wavelengths. As a result violet light is refracted most, red light is refracted least, and the other colors are refracted between these extremes. This results in a beam of white light being separated, or dispersed, into a spectrum when it is refracted. Any transparent material in which the index of refraction varies with wavelength has the property of **dispersion.** The dispersion of light by ice crystals sometimes produces a colored halo around the sun and the moon.

Evidence for Waves

So far we have considered reflection and refraction by using light rays to explain the behavior of light. A light ray successfully accounts for the path of light after reflection and during refraction. The light ray represents the straight-line path of light, but what exactly *is* the ray? During the early 1600s many people believed that all space in the universe was filled with "ether," a thin, invisible, and mysterious fluid that was highly elastic. The name "ether" was borrowed from the old Greek idea that was advanced by Aristotle (see chapter 2), but this ether had different properties. The ether concept of the 1600s was very vague and mysterious, but it did offer a mechanical model for things that were not otherwise understood. Electrostatic attraction for small bits of paper, for example, occurred when a glass tube was rubbed. This attraction at a distance, without touching, was difficult to comprehend. With the ether concept, however, you could convince yourself that some of the invisible ether must become condensed on the glass by the rubbing action. Other ether would move to the rarefied zone produced by the condensation, carrying the bits of paper with it. Light, on the other hand, could be considered as some kind of wave in the ether that moved outward from a light source. Or, perhaps light was uncountable numbers of tiny, invisible particles of matter that accelerated from the light source but were slowed to some terminal speed by the ether. Both of these ideas had been around since the time of the early Greeks.

The nature of light became a topic of debate toward the end of the 1600s as Issac Newton published his *particle theory* of light. He believed that the straight-line travel of light could be better explained as small particles of matter that traveled at great speed from a source of light. Particles, reasoned Newton, should follow a straight line according to the laws of motion. Waves, on the other hand, should bend as they move, much as water waves on a pond bend into circular shapes as they move away from a disturbance.

About the same time that Newton developed his particle theory of light, Christian Huygens (pronounced "hi-ganz") was concluding that light is not a stream of particles but rather a longitudinal wave that moves through the "ether." Huygens's *wave theory* proposed that waves were created by a light source. The wave front, according to this theory, did not simply move off through the ether. Each point on the wave front was pictured as impacting an "ether molecule." Each "ether molecule" produced a small *wavelet* as a result. Each wavelet was a small hemisphere that spread outward in the direction of the movement. A line connecting all the leading surfaces of the wavelets described a new wave front, which now impacted other "ether molecules," which made new wavelets, which formed a new wave front, and so on. The forming and reforming of the wave front resulted in a plane wave that moved through space in a straight line. Thus Huygens's wave theory explained the straight-line travel of light as the continual formation of new waves (figure 10.15).

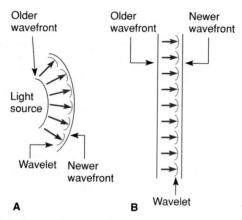

Figure 10.15

(a) Huygens's wave theory described wavelets that formed from an older wave front from a light source. The wavelets then moved out to form a new wave front. (b) As the wavelets spread, the wave front eventually becomes a plane wave, or straight wave, at some distance. This continuous process explained how waves could travel in a straight line.

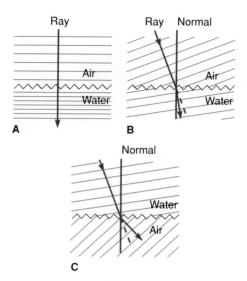

Figure 10.16

Light waves move from one material to another where the speed of light changes. The incident ray is perpendicular to the wave fronts and shows the direction of movement. (a) Waves arriving perpendicular to the boundary are not refracted. (b) Waves moving to a material where they have a slower speed are refracted toward the normal. (c) Waves moving to a material where they have a faster speed are refracted away from the normal.

Both theories had advocates during the 1700s, but the majority favored Newton's particle theory. By the beginning of the 1800s new evidence was found that favored the wave theory, evidence that could not be explained in terms of anything but waves. The remainder of this section describes this evidence, which led to the acceptance of the wave theory during the 1800s.

Reflection and Refraction

The laws of reflection and refraction were known when Newton developed his particle theory and when Huygens developed his wave theory. Both theories successfully accounted for reflection. Waves that can be observed, such as water waves, will reflect from a boundary, and the angle of reflection always equals the incidence angle. But particles also exhibited the same behavior. You can predict the path of a bounced basketball or banked pool ball as long as the ball does not have spin ("English") because the angle or reflection always equals the angle of incidence.

Newton had used the laws of mechanics to verify how particles of light could be reflected. Since these laws could be observed directly with everyday-sized objects his analysis seemed convincing. At the same time Newton was thought to have had a similar success using the idea of particles to explain refraction. Newton had to assume, however, that an attractive force acted on the particles of light as they entered a new material such as glass. The force was necessary to accelerate the particles, changing their direction of travel toward the perpendicular. Thus Newton had to assume that the speed of light *increased* as it traveled across a boundary, for example, from air to glass. It was not possible to measure the speed of light through glass during Newton's time, and this measurement was not made until 1850. The laws of mechanics could be observed at the time, however, so the particle model seemed to be a valid explanation of refraction.

The wave theory explains that the speed of light is *decreased* as light moves through a transparent material. The speed of light is at a maximum in a vacuum and is slower in different transparent materials. The slowing correlates with the electrical properties of the transparent material the electromagnetic wave is traveling through. Why a change of speed results in a change of the direction of travel can be understood by imagining light as a series of waves (figure 10.16). When the waves strike the surface of a transparent material at some angle, part of each wave will enter the material before the other parts. If they have a slower speed in the new material, the part of the wave that reached the surface first are slowed first and thus fall behind the rest of the wave front. The result is that the light is deflected toward the normal (figure 10.16b). If light has a faster speed in the new material the parts of the waves that reach the surface first are speeded up first and thus get in front of the rest of the wave front. The result is that light is deflected away from the normal (figure 10.16c). Thus the wave theory successfully accounts for refraction, but the particle theory does not. By the time the speed of light in different transparent materials was measured, the wave theory was already accepted from the evidence to be presented next. Finding that light slowed in transparent materials, not increasing in speed as the particle theory predicted, thus offered strong evidence to reject Newton's particle theory.

Diffraction

One of Newton's arguments for a particle nature of light concerned the observation that light moved in straight lines through openings such as a window. He felt that if light were waves they

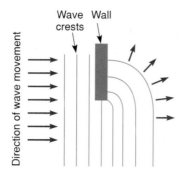

Figure 10.17

Water waves are observed to bend around an obstacle such as a wall in the water. Light appears to move straight past an obstacle, forming a shadow. This is one reason that Newton thought light must be particles, not waves.

would "bend around" obstacles such as the edge of the window. By analogy, the water waves in figure 10.17 strike the edge of a wall, an obstacle, and noticeably bend around the corner, some bending enough to travel parallel to the wall. Light does not bend around an obstacle like this and appears to make a sharp shadow behind an obstacle. So Newton argued for a particle nature of light since particles would move straight past an obstacle and make a sharp shadow.

In 1665, Francesco Grimaldi reported that there was not a sharp edge to a shadow moving through a small pinhole and the light formed a larger spot than would be expected by light traveling in a straight line. Grimaldi had made the first description of **diffraction,** the bending of light around the edge of an opaque object. Any small opening or a sharp-edged object was found to produce diffraction. Newton could not imagine the extremely small wavelengths of light that would be required to produce this result, so he dismissed the observation as an interaction between particles of light and the edges of the pinhole. As shown in figure 10.18, the determining factor is the size of an opening (or obstacle) as *compared to the wavelength*. Everyday-sized openings and obstacles are much larger than the wavelengths of light, so little diffraction occurs and the light travels in straight lines.

Diffraction can be explained by using Huygens's wave theory. Recall that Huygens described a light wave as a disturbance in the form of waves moving away from a light source. Each point on a crest is considered to be a source of *small wavelets* that spread in the direction of movement. The wavelets are small hemispheres and a line connecting all the surfaces at the leading edge forms a new wave front that is made up of all the little wavelets. Arrows perpendicular to the wave front are called rays, and the rays show the direction of movement. From a distant source such as the sun the wave front is considered to be in a plane, which means that the perpendicular rays are parallel to each other.

According to the Huygens model, a wave front passing through a large opening will continue to generate wavelets so the original shape of the wave front moves straight through the

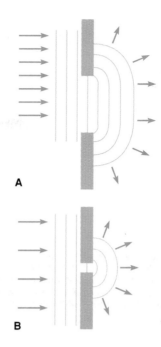

Figure 10.18

(a) When an opening is large as compared to the wavelength, waves appear to move mostly straight through the opening. (b) When the opening is about the same size as the wavelengths, the waves diffract, spreading into an arc.

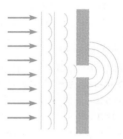

Figure 10.19

Huygens's wave theory explains diffraction of light. When the opening is large as compared to the wavelength, the wavelets form a new wave front as usual, and the front continues to move in a straight line. When the opening is the size of the wavelength, a single wavelet moves through, expanding in an arc.

opening. If the opening is very small, about the same size as the wavelength, a *single* wavelet will move out in all directions as an expanding arc from the opening. This explains diffraction from an opening (figure 10.19).

When a wave front strikes an obstacle the same size as the wavelength, a wavelet will move off in all directions from the corners, including into the shadow behind the obstacle. Thus the use of light rays is restricted to describing the behavior of light when obstacles and openings are large compared to the wavelength of light, where little diffraction occurs.

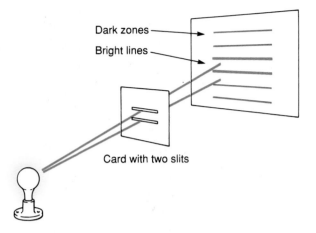

Figure 10.20

Young's double-split experiment produced a pattern of bright lines and dark zones when light from a single source passed through the slits.

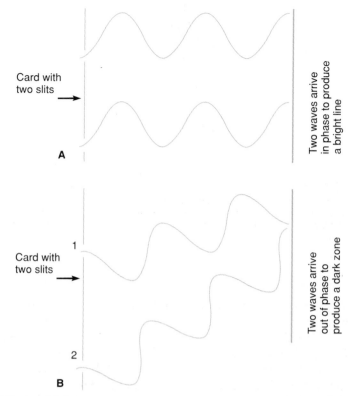

Figure 10.21

An interference pattern of bright lines and dark zones is produced because of the different distances that waves must travel from the two slits. (*a*) When they arrive together, a bright line is produced. (*b*) Light from slit 2 must travel farther than light from slit 1, so they arrive out of phase.

Interference

In 1801, Thomas Young published evidence of a behavior of light that could only be explained in terms of a wave model of light. Young's experiment is illustrated in figure 10.20. Light from a single source is used to produce two beams of light that are in phase, that is, having their crests and troughs together as they move away from the source. This light falls on a card with two slits, each less than a millimeter in width. The light is diffracted from each slit, moving out from each as an expanding arc. Beyond the card the light from one slit crosses over the light from the other slit to produce a series of bright lines on a screen. Young had produced a phenomenon of light called **interference,** and interference can only be explained by waves.

The pattern of bright lines and dark zones is called an *interference pattern* (figure 10.21). The light moved from each slit in phase, crest to crest and trough to trough. Light from both slits traveled the same distance directly across to the screen, so they arrived in phase. The crests from the two slits are superimposed here, and constructive interference produces a bright line in the center of the pattern. But for positions above and below the center, the light from the two slits must travel different distances to the screen. At a certain distance above and below the bright center line light from one slit had to travel a greater distance and arrives one-half wavelength after light from the other slit. Destructive interference produces a zone of darkness at these positions. Continuing up and down the screen, a bright line of constructive interference will occur at each position where the distance traveled by light from the two slits differs by any whole number of wavelengths. A dark zone of destructive interference will occur at each position where the distance traveled by light from the two slits differs by any half number wavelength. Thus bright lines occur above and below the center bright line at positions representing differences in paths of 1, 2, 3, 4, and so on wavelengths. Similarly, zones of

darkness occur above and below the center bright line at a position representing differences in paths of ½, 1½, 2½, 3½, and so on wavelengths. Young found all of the experimental data such as this in full agreement with predictions from a wave theory of light. About fifteen years later, A. J. Fresnel (pronounced "fray-nel") demonstrated mathematically that diffraction as well as other behaviors of light could be fully explained with the wave theory. In 1821 Fresnel determined that the wavelength of red light was about 4×10^{-7} m and of violet light about 8×10^{-7} m, with other colors in between these two extremes. The work of Young and Fresnel seemed to resolve the issue of considering light to be a stream of particles or a wave, and it was generally agreed that light must be waves. The waves, however, were considered to be mechanical waves in the "ether" until after the work of Maxwell, Lorentz, and Einstein.

Polarization

Huygens's wave theory and Newton's particle theory could explain some behaviors of light satisfactorily, but there were some behaviors that neither (original) theory could explain. Both theories failed to explain some behaviors of light, such as light moving through certain transparent crystals. For example, a slice

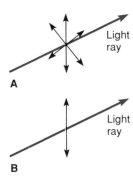

Figure 10.22

(a) Unpolarized light has transverse waves vibrating in all possible directions perpendicular to the direction of travel. (b) Polarized light vibrates only in one plane. In this illustration, the wave is vibrating in a vertical direction only.

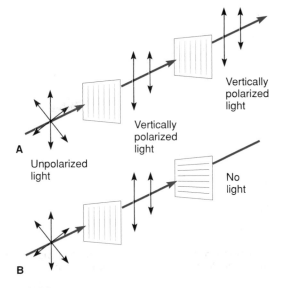

Figure 10.23

(a) Two crystals that are aligned both transmit vertically polarized light that looks like any other light. (b) When the crystals are crossed, no light is transmitted.

of the mineral tourmaline transmits what appears to be a low-intensity greenish light. But if a second slice of tourmaline is placed on the first and rotated, the transmitted light passing through both slices begins to dim. The transmitted light is practically zero when the second slice is rotated 90°. Newton suggested that this behavior had something to do with "sides" or "poles" and introduced the concept of what is now called the "polarization" of light.

The waves of Huygens's wave theory were longitudinal, moving like sound waves with wave fronts moving in the direction of travel. A longitudinal wave could not explain the polarization behavior of light. In 1817, Young modified Huygens's theory by describing the waves as *transverse,* vibrating at right angles to the direction of travel. This modification helped explain the polarization behavior of light transmitted through the two crystals and provided firm evidence that light is a transverse wave. As shown in figure 10.22a, **unpolarized light** is assumed to consist of transverse waves vibrating in all conceivable random directions. Polarizing materials, such as the tourmaline crystal, transmit light that is vibrating in one direction only, such as the vertical direction in figure 10.22b. Such a wave is said to be **polarized,** or *plane-polarized,* since it vibrates only in one plane. The single crystal polarized light by transmitting only waves that vibrate parallel to a certain direction while selectively absorbing waves that vibrate in all other directions. Your eyes cannot tell the difference between unpolarized and polarized light, so the light transmitted through a single crystal looks just like any other light. When a second crystal is placed on the first, the amount of light transmitted depends on the alignment of the two crystals (figure 10.23). When the two crystals are *aligned,* the polarized light from the first crystal passes through the second with little absorption. When the crystals are *crossed* at 90°, the light transmitted by the first is vibrating in a plane that is absorbed by the second crystal and practically all the light is absorbed. At some other angle only a fraction of the polarized light from the first crystal is transmitted by the second.

You can verify whether or not a pair of sunglasses is made of polarizing material by rotating a lens of one pair over a lens of a second pair. Light is transmitted when the lenses are aligned but mostly absorbed at 90° when the lenses are crossed.

Light is completely polarized when all the waves are removed except those vibrating in a single direction. Light is partially polarizing when some of the waves are in a particular orientation and any amount of polarization is possible. There are several means of producing partially or completely polarized light, including (1) selective absorption, (2) reflection, and (3) scattering.

Selective absorption is the process that takes place in certain crystals, such as tourmaline, where light in one plane is transmitted as all the other planes are absorbed. A method of manufacturing a polarizing film was developed in the 1930s by Edwin H. Land. The film is called **Polaroid.** Today Polaroid is made of long chains of hydrocarbon molecules that are aligned in a film. The long-chain molecules ideally absorb all light waves that are parallel to their lengths and transmit light that is perpendicular to their lengths. The direction that is *perpendicular* to the oriented molecular chains is thus called the polarization direction or the *transmission axis.*

Reflected light with an angle of incidence between 1° and 89° is partially polarized as the waves parallel to the reflecting surface are reflected more than other waves. Complete polarization, with all waves parallel to the surface, occurs at a particular angle of incidence. This angle depends on a number of variables, including the nature of the reflecting material. Figure 10.24 illustrates polarization by reflection. Polarizing sunglasses reduce the glare of reflected light because they have vertically oriented transmission axes. This absorbs the horizontally

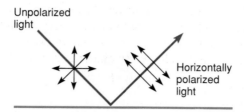

Figure 10.24

Light that is reflected becomes partially or fully polarized in a horizontal direction depending on the incident angle and other variables.

oriented reflected light. If you turn your head from side to side so as to rotate your sunglasses while looking at a reflected glare, you will see the intensity of the reflected light change. This means that the reflected light is partially polarized.

The phenomenon called *scattering* occurs when light is absorbed and reradiated by particles about the size of gas molecules that make up the air. Sunlight is initially unpolarized. When it strikes a molecule, electrons are accelerated and vibrate horizontally and vertically. The vibrating charges reradiate polarized light. Thus if you look at the blue sky with a pair of polarizing sunglasses and rotate them, you will observe that light from the sky is polarized. Bees are believed to be able to detect polarized skylight and use it to orient the direction of flights. Violet and blue light have the shortest wavelengths of visible light and red and orange light the largest. The violet and blue rays of sunlight are scattered the most. However, sunlight has more blue than violet light (figure 10.3), and this is why the sky appears blue when the sun is high in the sky. At sunset the path of sunlight through the atmosphere is much longer than when the sun is more directly overhead. Much of the blue and violet have been scattered away as a result of the longer path through the atmosphere at sunset. The remaining light that comes through is mostly red and orange and these are the colors you see at sunset.

Evidence for Particles

The evidence from diffraction, interference, and polarization of light was very important in the acceptance of the wave theory because there was simply no way to explain these behaviors with a particle theory. Then, in 1850, J. L. Foucault was able to prove that light travels much slower in transparent materials than it does in air. This was in complete agreement with the wave theory and completely opposite of the particle theory explanation. By the end of the 1800s, Maxwell's theoretical concept of electric and magnetic fields changed the concept of light from mechanical waves to waves of changing electric and magnetic fields. Further evidence removed the necessity for ether, the "material" supposedly needed for waves to move through. Light was now seen as electromagnetic waves that could move through empty space. By this time it was possible to explain all behaviors of light moving through empty space or through matter with a wave theory. Yet, there were nagging problems that the wave theory could not explain. In general, these problems concerned light that is absorbed or emitted from matter.

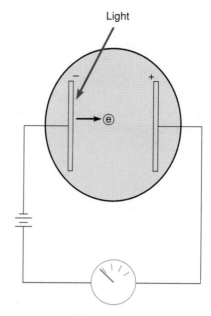

Figure 10.25

A setup for observing the photoelectric effect. Light strikes the negatively charged plate and electrons are ejected. The ejected electrons move to the positively charged plate and can be measured as a current in the circuit.

Photoelectric Effect

Light is a form of energy, and it gives its energy to matter when it is absorbed. Usually the energy of absorbed light results in a temperature increase such as the warmth you feel from absorbed sunlight. Sometimes, however, the energy from absorbed light results in other effects. In some materials the energy is acquired by electrons, and some of the electrons acquire sufficient energy to jump out of the material. The movement of electrons as a result of energy acquired from light is known as the **photoelectric effect.** The photoelectric effect is put to a practical use in a solar cell, which transforms the energy of light into an electric current (figure 10.25).

The energy of light can be measured with great accuracy. The kinetic energy of electrons after they absorb light can also be measured with great accuracy. When these measurements were made of light and electrons involved in the photoelectric effect some unexpected results were observed. Monochromatic light, that is, light of a single, fixed frequency, was used to produce the photoelectric effect. First, a low-intensity, or dim, light was used and the numbers and energy of the ejected electrons were measured. Then a high-intensity light was used and the numbers and energy of the ejected electrons were again measured. Measurement showed that (1) low-intensity light caused fewer electrons to be ejected, and high-intensity light caused many to be ejected, and (2) all electrons ejected from low- or high-intensity light ideally had the *same* kinetic energy. Surprisingly, the kinetic energy of the ejected electrons was found to be *independent* of the light intensity. This is contrary to what the wave theory of light would predict since a stronger light should mean that waves with more energy have more energy to give to the electrons. Here is a behavior involving light that the wave theory could not explain.

Quantization of Energy

In addition to the problem of the photoelectric effect, there were problems with blackbody radiation, light emitted from hot objects. The experimental measurements of light emitted through blackbody radiation did not match predictions made from theory. In 1900, Max Planck (pronounced "plonk"), a German physicist, found that he could fit the experimental measurements and theory together by assuming that the vibrating molecules that emitted the light could only have a *fixed amount* of energy. Instead of energy existing through a continuous range of amounts, Planck found that the vibrating molecules could only have energy in multiples of energy in certain amounts, or **quanta** (meaning fixed amounts; quantum is singular and quanta plural). Planck thus developed the concept of quantization of energy, that is, vibrating molecules involved in blackbody radiation vibrate with quantized energy E according to the relationship

$$E = nhf \qquad \text{equation 10.4}$$

where n is a positive whole number, f is the frequency of vibration of the molecules, and h is a proportionality constant known today as **Planck's constant.** The value of Planck's constant is 6.63×10^{-34} J·sec.

Planck's discovery of quantized energy states was a radical, revolutionary development and most scientists, including Planck, did not believe it at the time. Planck, in fact, spent considerable time and effort trying to disprove his own discovery. It was, however, the beginning of the quantum theory, which was eventually to revolutionize physics.

Five years later, in 1905, Albert Einstein applied Planck's quantum concept to the problem of the photoelectric effect. Einstein described the energy in a light wave as quanta of energy called **photons.** Each photon has an energy E that is related to the frequency f of the light through Planck's constant h, or

$$E = hf \qquad \text{equation 10.5}$$

This relationship says that higher-frequency light (e.g., blue light at 6.50×10^{14} hz) has more energy than lower-frequency light (e.g., red light at 4.00×10^{14} hz). The energy of such high- and low-frequency light can be verified by experiment.

The photon theory also explained the photoelectric effect. According to this theory, light is a stream of moving photons. It is the number of photons in this stream that determines if the light is dim or has a high intensity. A high-intensity light has many, many photons and a low-intensity light has only a few photons. At any particular fixed frequency, all the photons would have the same energy, the product of the frequency and Planck's constant (hf). When a photon interacts with matter it is absorbed and gives up all of its energy. In the photoelectric effect, this interaction takes place between photons and electrons. When an intense light is used this means that there are more photons to interact with the electrons, so more electrons are ejected. The energy given up by each photon is a function of the frequency of the light, so at a fixed frequency the energy of each photon, hf, is the same and the acquired kinetic energy of each ejected electron is the same. Thus the photon theory explains the measured experimental results of the photoelectric effect.

Example 10.3

What is the energy of a photon of red light with a frequency of 4.00×10^{14} hz?

Solution

The relationship between the energy of a photon (E) and its frequency (f) is found in equation 10.5. Planck's constant is given as 6.63×10^{-34} J · sec.

$f = 4.00 \times 10^{14}$ hz

$h = 6.63 \times 10^{-34}$ j · sec

$E = ?$

$$E = hf$$
$$= (6.63 \times 10^{-34} \text{ J} \cdot \text{sec})\left(4.00 \times 10^{14} \frac{1}{\text{sec}}\right)$$
$$= (6.63 \times 10^{-34})(4.00 \times 10^{14}) \text{ J} \cdot \text{sec} \times \frac{1}{\text{sec}}$$
$$= 2.65 \times 10^{-19} \frac{\text{J} \cdot \text{sec}}{\text{sec}}$$
$$= \boxed{2.65 \times 10^{-19} \text{ J}}$$

Example 10.4

What is the energy of a photon of violet light with a frequency of 7.00×10^{14} hz? (Answer: 4.64×10^{-19} J)

The photoelectric effect is explained by considering light to be photons with quanta of energy, not a wave of continuous energy. This is not the only evidence about the quantum nature of light, and more will be presented in the next chapter. But, as you can see, there is a dilemma. The electromagnetic wave theory and the photon theory seem incompatible. Some experiments cannot be explained by the wave theory and seem to support the photon theory. Other experiments are contradictions, providing seemingly equal evidence to reject the photon theory in support of the wave theory.

The Present Theory

Today, light is considered to have a dual nature, sometimes acting like a wave and sometimes acting like a particle. A wave model is useful in explaining how light travels through space and how it exhibits such behaviors as refraction, interference, and diffraction. A particle model is useful in explaining how light is emitted from and absorbed by matter, exhibiting such behaviors such as blackbody radiation and the photoelectric effect. Together, both of these models are part of a single theory of light, a theory that pictures light as having both particle and wave properties. Some properties are more useful when explaining some observed behaviors and other properties are more useful when explaining other behaviors.

Frequency is a property of a wave and the energy of a photon is a property of a particle. Both frequency and the energy of a photon are related in equation 10.5, $E = hf$. It is thus possible to describe light in terms of a frequency (or wavelength) or in terms of a quantity of energy. Any part of the electromagnetic spectrum can thus be described by units of frequency, wavelength, or energy, which are alternative means of describing light. The radio radiation parts of the spectrum are low-frequency, low-energy, and long-wavelength radiations. Radio radiations have more wave properties and practically no particle properties since the energy levels are low. Gamma radiation, on the other hand, is high-frequency, high-energy, and short-wavelength radiation. Gamma radiation has more particle properties since the extremely short wavelengths have very high energy levels. The more familiar part of the spectrum, visible light, is between these two extremes and exhibits both wave and particle properties, but it never exhibits both properties at the same time in the same experiment.

Part of the problem in forming a concept or mental image of the exact nature of light is understanding this nature in terms of what is already known. The things you already know about are observed to be particles, or objects, or they are observed to be waves. You can see objects that move through the air such as baseballs or footballs, and you can see waves on water or in a field of grass. There is nothing that acts like a moving object in some situations but acts like a wave in other situations. Objects are objects and waves are waves, but objects do not become waves and waves do not become objects. If this dual nature did exist it would seem very strange. Imagine, for example, holding this book at a certain height above a lake (figure 10.26). You can make measurements and calculate the kinetic energy the book will have when dropped into the lake. When it hits the water, the book disappears and water waves move away from the point of impact in a circular pattern that moves across the water. When the waves reach another person across the lake a book identical to the one you dropped pops up out of the water as the waves disappear. As it leaves the water across the lake the book has the same kinetic energy that your book had when it hit the water in front of you. You and the other person could measure things about either book, and you could measure things about the waves, but you could not measure both at the same time. You might say that this behavior is not only strange but impossible. Yet, it is an analogy to the observed behavior of light.

As stated, light has a dual nature, sometimes exhibiting the properties of a wave and sometimes exhibiting the properties of moving particles but never exhibiting both properties at the same time. Both the wave and the particle nature are accepted as being part of one model today, with the understanding that the exact nature of light is not describable in terms of anything that is known to exist in the everyday-sized world. Light is an extremely small-scale phenomena that must be different,

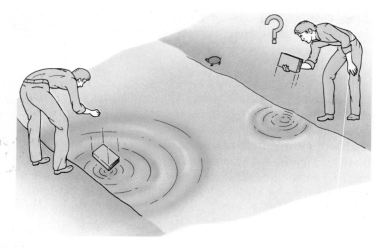

Figure 10.26

It would seem very strange if there were not a sharp distinction between objects and waves in our everyday world. Yet this appears to be the nature of light.

without a sharp distinction between a particle and a wave. Evidence about this strange nature of extremely small-scale phenomena will be considered again in the next chapter as a basis for introducing the quantum theory of matter.

Summary

Electromagnetic radiation is emitted from all matter with a temperature above absolute zero, and as the temperature increases more radiation and shorter wavelengths are emitted. Visible light is emitted from matter hotter than about 700° C, and this matter is said to be *incandescent*. The sun, a fire, and the ordinary light bulb are incandescent sources of light.

The behavior of light is shown by a light ray model that uses straight lines to show the straight-line path of light. Light that interacts with matter is *reflected* with parallel rays, moves in random directions by *diffuse reflection* from points, or is *absorbed*, resulting in a temperature increase. Matter is *opaque*, reflecting light, or *transparent*, transmitting light.

In reflection, the incoming light, or *incident ray*, has the same angle as the *reflected ray* when measured from a perpendicular from the point of reflection, called the *normal*. That the two angles are equal is called the *law of reflection*. The law of reflection explains how a flat mirror forms a *virtual image*, one from which light rays do not originate. Light rays do originate from the other kind of image, a *real image*.

Light rays are bent, or *refracted*, at the boundary when passing from one transparent media to another. The amount of refraction depends on the *incident angle* and the *index of refraction*, a ratio of the speed of light in a vacuum to the speed of light in the media. When the refracted angle is 90° to the incident angle *total internal reflection* takes place. This limit to the angle of incidence is called the *critical angle*, and all light rays with an incident angle at or beyond this angle are reflected internally.

Each color of light has a range of wavelengths that forms the *spectrum* from red to violet. A glass prism has the property of *dispersion*, separating a beam of white light into a spectrum. Dispersion occurs because the index of refraction is different for each range of colors, with short wavelengths refracted more than larger ones.

Fiber Optics

The communication capacities of fiber optics are enormous. When fully developed and exploited a single fiber the size of a human hair could in principle carry all the telephone conversations and all the radio and television broadcasts in the United States at one time without any interfering with another. This enormous capacity is possible because of a tremendous range of frequencies in a beam of light. This range determines the information-carrying capacity, which is called *bandwidth*. The traditional transmission of telephone communication depends on the movement of electrons in a metallic wire, and a single radio station usually broadcasts on a single frequency. Light has a great number of different frequencies, as you can imagine when looking at table 10.2. An optic fiber has a bandwidth about a million times greater than the frequency used by a radio station.

Physically, an optical fiber has two main parts. The center part, called the *core,* is practically pure silica glass with an index of refraction of about 1.6. Larger fibers, about the size of a human hair or larger, transmit light by total internal reflection. Light rays internally reflect back and forth about a thousand times or so in each 15 cm length of such an optical fiber core. Smaller fibers, about the size of a light wavelength, act as waveguides, with rays moving straight through the core. A layer of transparent material called the *cladding* is used to cover the core. The cladding has a lower index of refraction, perhaps 1.50, and helps keep the light inside the core.

Presently fiber optics systems are used to link telephone offices across the nation and in a submarine cable that links the United States and Europe. The electrical signals of a telephone conversation are converted to light signals, which travel over optical fibers. The light signals are regenerated by devices called repeaters until the signal reaches a second telephone office. Here, the light signals are reformed to electrical signals, which are then transmitted by more traditional means.

Currently, new optical fiber systems have increased the number of conversations almost tenfold that can be carried through the much smaller, lighter-weight fiber optics. Extension of the optical fiber to the next generation of computers will be the next logical step, with hundreds of available television channels, two-way videophone conversations, and unlimited data availability. The use of light for communications will have come a long way since the first use of signaling mirrors and flashing lamps from mountain tops and church steeples.

A wave model of light can be used to explain reflection and refraction, and the properties of diffraction, interference, and polarization provide strong evidence for the wavelike nature of light. *Diffraction* is the bending of light around the edge of an object or spreading of light in an arc after passing through a tiny opening. *Interference* occurs when light passes through two small slits or holes and produces an *interference pattern* of bright lines and dark zones. *Polarized light* vibrates in one direction only, in a plane. Light can be polarized by certain materials, by reflection, or by scattering. Polarization can only be explained by a transverse wave model.

A wave model fails to explain observations of light behaviors in the *photoelectric effect* and *blackbody radiation.* Max Planck found that he could modify the wave theory to explain blackbody radiation by assuming that vibrating molecules could only have fixed amounts, or *quanta,* of energy and found that the quantized energy is related to the frequency and a constant known today as *Planck's constant.* Albert Einstein applied Planck's quantum concept to the photoelectric effect and described a light wave in terms of quanta of energy called *photons.* Each photon has an energy that is related to the frequency and Planck's constant.

Today, the properties of light are explained by a model that incorporates both the wave and the particle nature of light. Light is considered to have both wave and particle properties and is not describable in terms of anything known in the everyday-sized world.

Summary of Equations

10.1

$$\text{angle of incidence} = \text{angle of reflection}$$
$$\theta_i = \theta_r$$

10.2

$$\frac{\text{index of}}{\text{refraction}} = \frac{\text{speed of light in vacuum}}{\text{speed of light in material}}$$
$$n = \frac{c}{v}$$

10.3

$$\text{speed of light in vacuum} = (\text{wavelength})(\text{frequency})$$
$$c = \lambda f$$

10.4

$$\frac{\text{quantized}}{\text{energy}} = \begin{pmatrix} \text{positive} \\ \text{whole} \\ \text{number} \end{pmatrix} \begin{pmatrix} \text{Planck's} \\ \text{constant} \end{pmatrix} (\text{frequency})$$
$$E = nhf$$

10.5

$$\frac{\text{energy of}}{\text{photon}} = \begin{pmatrix} \text{Planck's} \\ \text{constant} \end{pmatrix} (\text{frequency})$$
$$E = hf$$

Key Terms

angle of incidence (p. 190)
angle of reflection (p. 190)
blackbody radiation (p. 186)
critical angle (p. 191)
diffraction (p. 195)

diffuse reflection (p. 188)
dispersion (p. 193)
incandescent (p. 186)
incident ray (p. 190)
index of refraction (p. 191)

Applying the Concepts

1. According to the electromagnetic wave model, electromagnetic radiation is produced
   a. when an electric charge is accelerated.
   b. by an object heated to a high temperature.
   c. by any object with a temperature above absolute zero.
   d. all of the above.

2. An object is hot enough to emit a dull red glow. When this object is heated even more, it will emit
   a. more shorter-wavelength, higher-frequency radiation.
   b. longer-wavelength, lower-frequency radiation.
   c. the same wavelengths as before, but with more energy.
   d. more of the same wavelengths with more energy.

3. You are able to see in shaded areas, such as under a tree, because light has undergone
   a. refraction.
   b. incident bending.
   c. a change in speed.
   d. diffuse reflection.

4. Light traveling at some angle as it moves from water into the air is refracted away from the normal as it enters the air, so the fish you see under water is actually (draw a sketch if needed)
   a. above the refracted image.
   b. below the refracted image.
   c. beside the refracted image.
   d. in the same place as the refracted image.

5. When viewed straight down (90° to the surface), a fish under water is
   a. above the image (away from you).
   b. below the image (closer to you).
   c. beside the image.
   d. in the same place as the image.

6. Which of the following can only be explained by a wave model of light?
   a. reflection
   b. refraction
   c. interference
   d. photoelectric effect

7. The photoelectric effect proved to be a problem for a wave model of light because
   a. the number of electrons ejected varied directly with the intensity of the light.
   b. the light intensity had no effect on the energy of the ejected electrons.
   c. the number of electrons ejected varied inversely with the intensity of the light.

8. Max Planck made the revolutionary discovery that the energy of vibrating molecules involved in blackbody radiation existed only in
   a. multiples of certain fixed amounts.
   b. amounts that smoothly graded one into the next.
   c. the same, constant amount of energy in all situations.
   d. amounts that were never consistent from one experiment to the next.

9. Einstein applied Planck's quantum discovery to light and found
   a. a direct relationship between the energy and frequency of light.
   b. that the energy of a photon divided by the frequency of the photon always equaled a constant known as Planck's constant.
   c. that the energy of a photon divided by Planck's constant always equaled the frequency.
   d. all of the above.

10. Today, light is considered to have a nature of
    a. tiny particles of matter that move through space.
    b. electromagnetic waves.
    c. a small-scale phenomena without a sharp distinction between particle and wave properties.
    d. something that is completely unknown.

Answers
1. a   2. a   3. d   4. b   5. d   6. c   7. b   8. a   9. d
10. c

Questions for Thought

1. What determines if an electromagnetic wave emitted from an object is a visible light wave or a wave of infrared radiation?

2. What is the difference between a luminous object and an object that is illuminated? Is it possible for an object to be both luminous and illuminated? Explain.

3. Does the reflection of light support a particle model of light or a wave model of light? Explain.

4. What is the basic difference between reflection of light from a smooth, shiny surface and diffuse reflections from a rougher surface?

5. What is the basic difference between a real image and a virtual image?

6. What model of light does diffraction of light support? Why does diffraction not occur through a large opening?

7. Why was "ether" a necessary part of Huygens's wave model of light?

8. How did Huygens's wave model explain the straight-line travel of light when water waves are observed to bend as they travel?

9. What model of light does the polarization of light support? Explain.

10. What is the physical difference between the different colors of light?

11. What observation supports the model of light as transverse waves? Explain.

12. Which carries more energy, red light or blue light? Should this have meaning about the preferred color of warning and stop lights? Explain.

13. What model of light is supported by the photoelectric effect? Explain.

14. What is the angle of incidence if a light ray is not refracted at the boundary of two different transparent materials?

15. What happens to light that is absorbed by matter?

16. One star is reddish and another is bluish. Do you know anything about the relative temperatures of the two stars? Explain.

17. Explain the meaning of the "critical angle" of incidence.

18. When is light refracted (a) toward the normal? (b) away from the normal?

19. When does total internal reflection occur? Why does this occur in the diamond more than other gemstones?

20. Why does a highway sometimes appear wet on a hot summer day when it is not wet?

21. What is a "ray" of light?

22. Explain how sunlight is dispersed by a prism.

23. Explain how frequency, wavelength, and energy could all be used to describe parts of the electromagnetic spectrum.

24. Does light interference support a wave or particle model of light? Explain.

25. How can you tell if a pair of sunglasses is polarizing or not?

26. What conditions are necessary for two light waves to form an interference pattern of bright lines and dark areas?

27. Explain why the intensity of reflected light appears to change if you tilt your head from side to side while wearing polarizing sunglasses.

28. Why do astronauts in orbit around the earth see a black sky with stars that do not twinkle but see a blue earth?

29. What was so unusual about Planck's findings about blackbody radiation? Why was this revolutionary?

30. Why are both the photon model and the electromagnetic wave model accepted today as a single theory? Why was this so difficult for people to accept at first?

Exercises

Group A—Solutions Provided in Appendix D 3.00×10^8 m/s

1. What is the speed of light while traveling through (a) water and (b) ice? 1.33 Ice 1.31

2. If the sun is 1.50×10^8 km from the earth, how many minutes are required for sunlight to reach the earth?

3. How many hours are required for a radio signal from a space probe near the planet Pluto, 6.00×10^9 km away, to reach the Earth?

4. A light ray is reflected from a mirror with an angle $10°$ to the normal. What was the angle of incidence?

5. Light travels through a transparent substance at 2.20×10^8 m/sec. What is the substance?

6. The wavelength of a monochromatic light source is measured to be 6.00×10^{-7} m in a diffraction experiment. (a) What is the frequency? (b) What is the energy of a photon of this light?

7. At a particular location and time, sunlight is measured on a one square meter solar collector with an intensity of 1,000.0 w. If the peak intensity of this sunlight has a wavelength of 5.60×10^{-7} m, how many photons are arriving each second?

8. A light wave has a frequency of 4.90×10^{14} cycles per second. (a) What is the wavelength? (b) What color would you observe? 6.63×10^{-34}

9. What is the energy of a gamma photon of frequency 5.00×10^{20} hz?

10. What is the energy of a microwave photon of wavelength 1.00 mm?

Group B—Solutions Not Given

1. (a) What is the speed of light while traveling through a vacuum? (b) While traveling through air at $30°$ C? (c) While traveling through air at $0°$ C?

2. If the distance between the earth and the moon is 3.85×10^2 km, how much time is required for reflected sunlight to travel from the moon to the earth?

3. If the planet Mars is 7.83×10^7 km from the Earth, how many minutes are required for a radio signal to travel from the Earth to a space station on Mars?

4. An incident light ray strikes a mirror with an angle of $30°$ to the surface of the mirror. What is the angle of the reflected ray?

5. The speed of light through a transparent substance is 2.00×10^8 m/sec. What is the substance?

6. A monochromatic light source used in a diffraction experiment has a wavelength of 4.60×10^{-7} m. What is the energy of a photon of this light?

7. In black and white photography, a photon energy of about 4.00×10^{-19} J are needed to bring about the changes in the silver compounds used in the film. Explain why a red light used in a darkroom does not affect the film during developing.

8. The wavelength of light from a monochromatic source is measured to be 6.80×10^{-7} m. (a) What is the frequency of this light? (b) What color would you observe?

9. How much greater is the energy of a photon of ultraviolet radiation ($\lambda = 3.00 \times 10^{-7}$ m) than the energy of an average photon of sunlight ($\lambda = 5.60 \times 10^{-7}$)?

10. At what rate must electrons in a wire vibrate to emit microwaves with a wavelength of 1.00 mm?

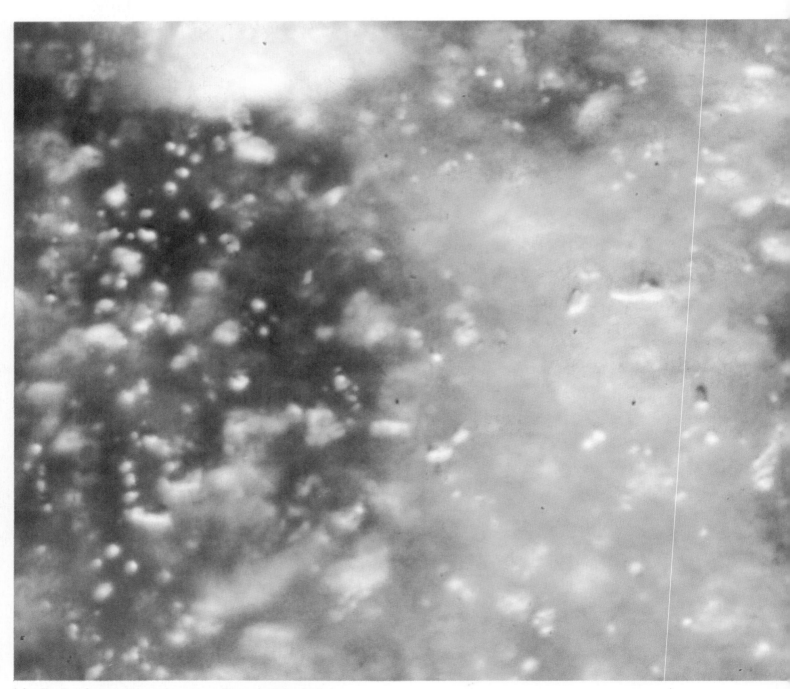

A familiar item from your everyday surroundings, photographed at a
very close range. Can you reach agreement with other students about
the identity of the item?

Chemistry

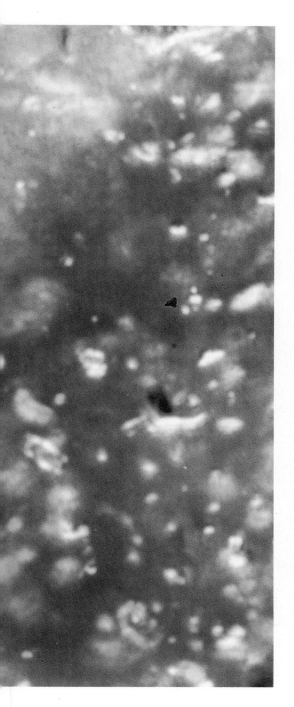

This section is an introduction to the structure of matter and the changes it undergoes. This is the field of science known as chemistry, a broad area of study concerned with the form and properties of matter on a large scale and the atoms, molecules, and electrons of matter on a much smaller scale. You might wonder about the boundary between physics (the behavior of matter) and chemistry (how matter is put together and how it changes). Scientific fields of study, and the boundaries between the fields, are purely arbitrary and human-made definitions made for convenience of study or made out of habit. Today the fields overlap, and it is often difficult to tell where one field ends and another begins. This is especially true with some of the more complex and vast modern-day fields such as biochemistry, geophysics, physical chemistry, and physical geochemistry. Space travel and the modern-day problems of a weakening ozone shield, increasing air pollution, acid rain, and the greenhouse effect all require the combined efforts of scientists from many fields of study.

The study of chemistry gives you an opportunity to learn thinking and reasoning skills as you study the forms, the structure, and the changes that matter undergoes. The photograph, for example, is a very close-range picture of a common item from the world around you. To identify what the item is requires a knowledge of some structures and properties of the overwhelming variety of matter in your surroundings. With this knowledge you are able to reason, thinking out what the item might or might not be. In this section you will examine matter at a much closer range than shown in the photograph. You will consider the smallest, discreet unit of matter called a molecule and in some substances an atom. You will learn about (1) the structure of an atom and how people learned about the structure, (2) patterns that exist between matter with similar atomic structures, (3) how and why certain kinds of matter interact, forming new chemical substances, (4) ways of expressing the reactions of matter and what can be learned from these expressions, (5) water solutions of acids, bases, and salts and how they are used in your everyday life, (6) petroleum, petroleum products, and the chemicals of life, and (7) reactions that can take place involving the innermost part of an atom, the nucleus, and why these reactions involve tremendous amounts of energy. These new concepts are presented from a viewpoint of understanding your environment, removing the apparent mystery and helping you to understand a little bit more of your science-oriented culture.

Atomic Structure

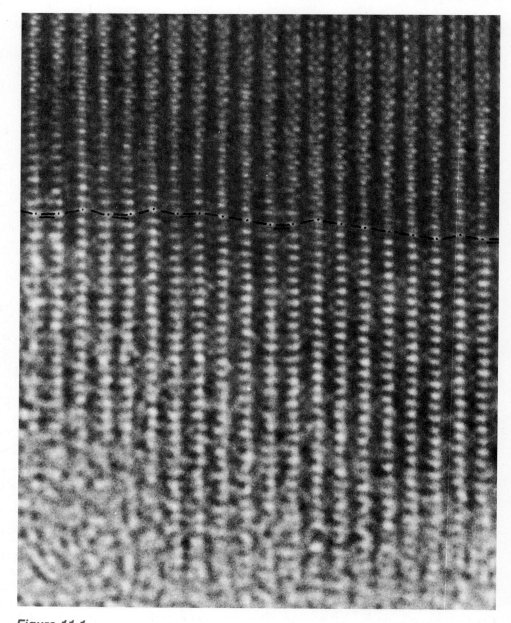

Figure 11.1
This electron microscope high-resolution image shows magnification of the thin edge of a piece of mica. The white dots are "empty tunnels" between layers of silicon-oxygen tetrahedrons, and the black dots are potassium atoms that bond the tetrahedrons together.

MANY materials used today are relatively new, created in the last few decades. These new materials are the result of modern chemical research, produced and manufactured through controlled chemical reactions. The new materials include synthetic fibers, from nylon to polyesters, and plastics, from polyethylene to Teflon. They also include water-based paints and super adhesives used in construction. The manufactured materials are lighter, stronger, and have special properties not found in natural materials. Today, such synthetic materials are used extensively in buildings, clothing, automobiles, and airplanes. The packaging, preserving, and marketing of many convenience foods are also made possible by the products of chemical research as are manufactured vitamins and drugs that help keep you healthy. From synthetic fibers to synthetic drugs, there are millions of products today that are the direct result of chemical research.

The countless numbers of new products resulting from chemical research demonstrate understandings about matter and how it is put together. These understandings start with the most basic unit of matter, the atom. Perhaps you have wondered how incredibly tiny atoms were discovered and how they can be studied. Atoms are so tiny that they are invisible to any optical device. Even more incredible is the study of the innermost parts of these invisible atoms and the development of knowledge of how they are put together. You will soon know the answer to questions about how atoms were discovered and studied (figure 11.1). This chapter contains the essence of the fascinating story of how the atomic concept was discovered and developed.

The development of the modern atomic model illustrates how modern scientific understanding comes from many different fields of study. For example, you will learn how studies of electricity led to the discovery that atoms have subatomic parts called electrons. The discovery of radioactivity led to the discovery of more parts, a central nucleus that contains protons and neutrons. Information from the absorption and emission of light was used to construct a model of how these parts are put together, a model resembling a miniature solar system with electrons circling the nucleus. The "solar system" model had initial, but limited, success and was inconsistent with other understandings about matter and energy. Modifications of this model were attempted, but none solved the problems. Then the discovery of wave properties of matter led to an entirely new model of the atom.

The atomic model will be put to use in later chapters to explain the countless varieties of matter and the changes that matter undergoes. In addition, you will learn how these changes can be manipulated to make new materials, from drugs to ceramics. In short, you will learn how understanding the atom and all the changes it undergoes not only touches your life directly but shapes and affects all parts of civilization. Basic to all of this is understanding the atom.

First Definition of the Atom

The atom concept is very old, dating back to ancient Greek philosophers some 2,500 years ago. The ancient Greeks thought, reasoned, and speculated about the basis for their surroundings. Consider, as an example, that you have observed water in puddles, ponds, lakes, rivers, and perhaps an ocean. You recognize rain, snow, sleet, dew, and ice to be water. You have also observed that plants and animals contain water, and that water can be produced by heating wood and other substances. If you think about all the forms, places, and things that water is a part of, you might begin to get the idea that water is a basic, fundamental substance that makes up other materials. Over time, this kind of reasoning led the ancient Greeks to the concept that everything is made up of four basic, fundamentally different substances, or "elements." These elements were earth, air, fire, and water. From logical considerations of what they observed, the ancient Greeks considered all substances to be composed of these four elements in varying proportions.

The ancient Greeks also reasoned about the way that pure substances are put together. A glass of water, for example, appears to be the same throughout. Is it the same throughout? Two plausible, but conflicting, ideas were possible as an intellectual exercise. The water could have a continuous structure, that is, it could be completely homogeneous throughout. The other idea was that the water only appears to be continuous but is actually *discontinuous*. This means that if you continue to divide the water into smaller and smaller volumes, you would eventually reach a limit to this dividing, a particle that could not be further subdivided. This model was developed by Leucippus and Democritus in the fourth century B.C. Democritus called the indivisible particle an *atom* from a Greek word meaning "uncuttable."

Democritus speculated that the atoms of a substance were separated by *empty space* and that the atoms of the four elements had different shapes. Since this is an intellectual exercise you can imagine anything you wish—as long as it is logically consistent with what is observed. For example, you might imagine that atoms of water might be round and smooth since water pours and flows. Atoms of earth might have cubic shapes that would prevent them from pouring and flowing. Democritus speculated that one substance changed to another by separation or combination of atoms.

When Aristotle organized and recorded all the ancient Greek knowledge that was available during this time, it was about one hundred years after Democritus had proposed the existence of atoms. Aristotle adopted the continuous, four element model because of perceived logical problems with the empty space between the atoms—he did not believe a vacuum could exist in nature. The idea of four elements—earth, air, fire, and water—with a continuous structure became the accepted model for the next thousand years. If anyone during this period did support the idea of atoms, and a few did, they were definitely in the minority.

Atomic Structure Discovered

In the 1600s, Robert Boyle's work with gases provided evidence to reject Aristotle's idea that a vacuum could not exist. In 1661, Boyle published *The Skeptical Chemist,* in which he rejected Aristotle's four element theory, too. Boyle defined an element as a "simple substance" that could not be broken down into anything simpler. Today, an **element** is defined as a pure substance

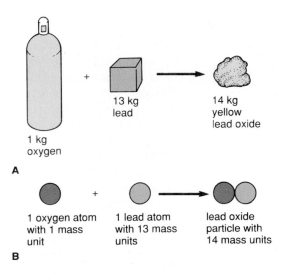

A

B

Figure 11.2

(a) Oxygen and lead combine to form yellow lead oxide in a ratio of 1:13 for a total of 14. (b) If 1 atom of oxygen combines with 1 atom of lead, the fixed ratio in which oxygen and lead combine must mean that 1 atom of lead is 13 times more massive than 1 atom of oxygen.

A

B

Figure 11.3

Reasoning the existence of atoms from the way elements combine in fixed weight ratios. (a) If matter were a continuous, infinitely divisible material, there would be no reason for one amount to go with another amount. (b) If matter is made up of discontinuous, discrete units (atoms), then the units would combine in a fixed weight ratio that equals the weight of the combination.

that cannot be broken down to anything simpler by chemical or physical means. Water is a pure substance, but it can be broken down into oxygen and hydrogen so water is not an element. Oxygen and hydrogen are pure substances that cannot be broken down into anything simpler, so they are elements. Oxygen, silicon, iron, gold, and aluminum are common elements, and over one hundred elements are known today. Elements will be considered in the next chapter. The development of a model of the atom is the topic of interest in this chapter, a model that will explain how elements are different.

It was information about how elements combine that led John Dalton in the early 1800s to bring back the ancient Greek idea of hard, indivisible atoms. In general, he had noted, as others had, that certain elements always combined with other elements in fixed ratios. For example, 1 gram of oxygen always combined with 13 grams of lead to produce 14 grams of a yellow compound called lead oxide (figure 11.2). Dalton reasoned that this must mean that the oxygen and lead were made up of individual particles called *atoms,* not a form of matter that was completely homogeneous in structure. If elements were continuous, there would be no reason for the oxygen and lead to combine in fixed ratios. If elements were composed of atoms, however, then whole atoms would combine to make the compound. By way of analogy, on a macroscopic scale you could consider both peanut butter and jelly to be homogeneous, continuous substances. If you combine peanut butter and jelly there is no reason for a particular amount of peanut butter to go with a particular amount of jelly. Crackers and uniform slices of cheese, however, could both be considered on a macroscopic scale to be individual units. If you combine a slice of cheese and a cracker, the combination would have a total weight with part contributed by a cracker and part contributed by the cheese slice.

A fixed ratio of the weight of a slice of cheese to the weight of an individual cracker would result in the same total weight each time. Using this kind of reasoning, Dalton theorized that elements must be composed of atoms (figure 11.3).

During the 1800s Dalton's concept of hard, indivisible atoms was familiar to most scientists. Yet, the existence of atoms was not generally accepted by all scientists. There was skepticism about something that could not be observed directly. Strangely, full acceptance of the atom came in the early 1900s with the discovery that the atom was not indivisible after all. The atom has parts that give it an internal structure. The first part to be discovered was the *electron,* a part that was discovered through studies of electricity.

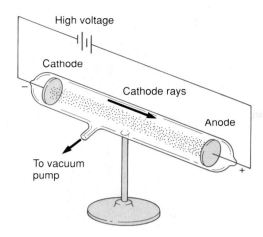

Figure 11.4

A vacuum tube with metal plates attached to a high-voltage source produces a greenish beam called cathode rays. These rays move from the cathode (negative charge) to the anode (positive charge).

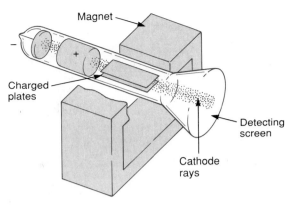

Figure 11.5

A cathode ray passed between two charged plates is deflected toward the positively charged plate. The ray is also deflected by a magnetic field. By measuring the deflection by both, J. J. Thomson was able to calculate their ratio of charge to mass. He was able to measure the deflection because the detecting screen was coated with zinc sulfide, a substance that produces a visible light when struck by a charged particle.

Discovery of the Electron

Scientists of the 1800s were interested in understanding the nature of electricity, but it was impossible to see anything but the effects of a current as it ran through a wire. To observe a current directly, they tried to produce a current by itself, away from matter, by pumping the air from a tube and then running a current through the empty space. By 1885 a good air pump was invented, and this pump could evacuate the air from a glass tube until the pressure was 1/10,000 of normal air pressure. When metal plates inside such a tube were connected to the negative and positive terminals of a high-voltage source (figure 11.4), a greenish beam was observed that seemed to move from the cathode (negative terminal) through the empty tube and collect at the anode (positive terminal). Since this mysterious beam seemed to come out of the cathode it was said to be made of **cathode rays.**

Just what the cathode rays were became a source of controversy. Some scientists suggested that cathode rays were atoms of the cathode material. But the properties of cathode rays were found to be the same when the cathode was made of any material. Perhaps the rays were a form of light? But the rays were observed to be deflected by a magnet, not a behavior observed with light. What *were* the cathode rays?

The mystery of the cathode rays was finally solved by the English physicist J. J. Thomson, who provided the answer in 1897. Thomson had placed a positively charged metal plate on one side of the beam and a negatively charged metal plate on the other side (figure 11.5). The beam was deflected toward the positive plate and away from the negative plate. Since it was known that unlike charges attract and like charges repel, this meant that the beam must be composed of *negatively charged particles.*

The cathode ray was also deflected when caused to pass between the poles of a magnet. Thomson knew that moving charges are deflected by a magnetic field. He found more information by adjusting the electric charge on the plates above and below the beam, then measuring the deflection. The same procedure was then used with a measured magnetic field. A greater charge on a particle would result in a greater deflection by an electric field, and a greater moving mass would be more difficult to deflect than a lesser moving mass. By balancing the deflections made by the magnet with the deflections made by the electric field, Thomson could thus determine the *ratio of the charge to mass* for an individual particle. Today, the charge to mass ratio is considered to be 1.7584×10^{11} coulomb/kilogram. The significant part of Thomson's experiments was finding that the charge to mass ratio was the *same* no matter what gas was in the tube and of what materials the electrodes were made. Thomson was convinced that he had discovered a fundamental particle, the stuff of which atoms are made.

Thomson did not propose any special name for the particle. Some time earlier, the term **electron** had been proposed as a name for the unit of charge gained or lost when atoms became ions. Not long after Thomson reported his findings in 1897, the existence of the fundamental particle was generally accepted and everyone started calling the particles electrons.

A method for measuring the charge and mass of the electron was worked out by an American physicist, Robert A. Millikan, around 1906. Millikan used an apparatus such as the one illustrated in figure 11.6 to measure the charge indirectly. Small droplets of mineral oil sprayed into the apparatus could be observed with a magnifier, and measurements could be made as the droplets drifted downwards. With a vertical electric field turned on, the droplets would drift upwards at a rate depending on the electric charge on each droplet. Measuring the rise and fall of the droplets as the electric field is turned on and off enabled Millikan to deduce the charge from the speed of fall and rise.

Atomic Structure

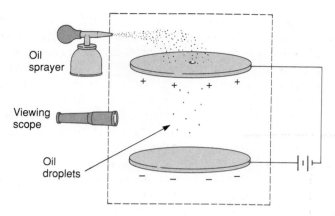

Figure 11.6

Millikan measured the charge of an electron by balancing the pull of gravity on oil droplets with an upward electrical force. Knowing the charge to mass ratio that Thomson had calculated, Millikan was able to calculate the charge on each droplet. He found that all the droplets had a charge of 1.60×10^{-19} coulomb or multiples of that charge. The conclusion was that this had to be the charge of an electron.

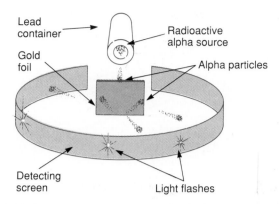

Figure 11.7

Rutherford and his co-workers studied alpha particle scattering from a thin metal foil. The alpha particles struck the detecting screen, producing a flash of visible light. Measurements of the angles between the flashes, the metal foil, and the source of the alpha particles showed that the particles were scattered in all directions, including straight back toward the source.

Millikan found that none of the droplets had a charge less than one particular value (1.60×10^{-19} coulomb) and that larger charges on various droplets were always multiples of this unit of charge. Since all of the droplets carried the single unit of charge or multiples of the single unit, the unit of charge was assumed to be the charge of a single electron.

Knowing the charge of a single electron and knowing the charge to mass ratio that Thomson had measured now made it possible to calculate the mass of a single electron. The mass of an electron was thus determined to be about 9.11×10^{-31} kg, or about 1/1,840 of the mass of the lightest atom, hydrogen.

Thomson had discovered the negatively charged electron, and Millikan had measured the charge and mass of the electron. But atoms themselves are electrically neutral. If an electron is part of an atom, there must be something else that is positively charged, canceling the negative charge of the electron. The next step in the sequence of understanding atomic structure would be to find what is neutralizing the negative charge and to figure out how all the parts are put together.

About 1900, Thomson proposed a model for what was known about the atom at the time. He suggested that an atom could be a blob of positively charged matter in which electrons were stuck like "raisins in plum pudding." If the mass of a hydrogen atom is due to the electrons embedded in a positively charged matrix, then 1,840 electrons would be needed together with sufficient positive matter to make the atom electrically neutral.

About this same time, 1896, radioactivity was discovered by Antonie Becquerel, which was soon described in terms of alpha, beta, and gamma rays. The details of radioactivity are considered in chapter 17. For now, all you need to know is what was known in 1907, that alpha particles are very fast, massive, and positively charged particles that are spontaneously given off from radioactive elements. The experimental application of alpha particles would lead to the discovery of the nucleus, then other parts of the atom.

The Nucleus

The nature of radioactivity and matter were the research interests of an English physicist, Ernest Rutherford. In 1907, Rutherford was studying the scattering of alpha particles directed toward a thin sheet of metal. As shown in figure 11.7, alpha particles from a radioactive source were allowed to move through a small opening of a lead container, so only a narrow beam of the massive, fast-moving particles would penetrate a very thin sheet of gold. The alpha particles were then detected by plates covered with zinc sulfide, which produced a small flash of light when struck by the positively charged alpha particle.

A C T I V I T I E S

The luminous dial of a watch or clock contains a mixture of zinc sulfide and a radioactive substance. Obtain a watch or clock with a luminous dial and a magnifying glass. Wait in a completely dark room about ten minutes until your eyes adjust to the darkness. Observe the glowing parts of the dial with the magnifying glass. Is the glow continuous or is it made up of tiny flashes of light? What would flashes of light mean?

Rutherford and his co-workers found that most of the alpha particles went straight through the foil, and just as expected, none were scattered by more than a few degrees. Then he suggested that one of his young co-workers check to see if any alpha particles were scattered through very large angles. He really did not expect any great amount of scattering since alpha particles were very fast, massive particles with a great deal of energy and electrons were not very massive in comparison. When a young co-worker reported that alpha particles were deflected at very large angles and some were even reflected

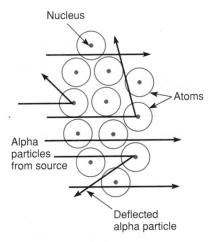

Figure 11.8
Rutherford's nuclear model of the atom explained the alpha scattering experiment as positive alpha particles experiencing a repulsive force from the positive nucleus. Measurements of the various angles of scattering gave Rutherford a means of estimating the size of the nucleus.

backwards, Rutherford was astounded. Rutherford could account for the large deflections and backward scattering only by assuming that the massive, positively charged particles were repelled by a massive positive charge concentrated in a small region of the atom. He concluded that an atom must have a tiny, massive, and positively charged **nucleus** surrounded by electrons. Since the electrons had an opposite charge than the nucleus, the electrons must be moving or else they would be attracted to it (figure 11.8).

From measurements of the scattering, Rutherford estimated the radius of the nucleus to be approximately 10^{-13} cm. Other researchers had estimated the radius of the atom to be on the order of 10^{-8} cm, so the electrons must be moving around the nucleus at a distance 100,000 times the radius of the nucleus. To visualize this spatial relationship, think of the thickness of a dime, which is about 1 mm thick. A distance 100,000 times the thickness of a dime is 100,000 mm, or 100 m. Thus if the radius of a nucleus were the thickness of a dime, the electrons would be moving around the nucleus at a distance of about 100 m away, or about the length of a football field. Thus if the radius of a nucleus were about the same size as the thickness of a dime, the atom would be about two football fields wide (figure 11.9). As you can see, the volume of an atom is mostly empty space.

Rutherford announced his conclusions and evidence for the existence of the atomic nucleus in 1911. This was a revolutionary development that would soon lead to more developments and more evidence about atomic structure. In 1917, Rutherford was able to break up the nucleus of a nitrogen atom by using alpha particles and identified the discrete unit of positive charge he called a **proton.** Rutherford also speculated about the existence of a neutral particle in the nucleus, a **neutron.** The neutron was eventually identified in 1932 by James Chadwick.

Today, the number of *protons* in the nucleus of an atom is called the **atomic number.** An element is made up of atoms that all have the same number of protons in their nucleus, so all atoms of an element have the same atomic number. Hydrogen has an atomic number of 1, so any atom that has one proton in its nucleus is an atom of the element hydrogen. Today, there are 109 different kinds of elements with a different number of protons. The *neutrons* of the nucleus, along with the protons, contribute to the mass of an atom. Atomic mass (and atomic weight) will be considered in the next chapter.

So the atom has a tiny, massive nucleus containing positively charged protons and neutral neutrons. Negatively charged electrons, equal in number to the protons, are moving around at a distance of about 100,000 times the radius of the nucleus. How are these electrons moving? It might occur to you, as it did to Rutherford and others, that an atom might be similar to a miniature solar system. This idea would picture the nucleus in the role of the sun, electrons in the role of moving planets in their orbits, and electrical attractions between the nucleus and electrons in the role of gravitational attraction. There are, however, significant problems with this idea. If electrons were moving in circular orbits, they would continually change their direction of travel and would therefore be accelerating. According to the Maxwell model of electromagnetic radiation, an accelerating electric charge emits electromagnetic radiation such as light. If an electron gave off light, it would lose energy. The energy loss would mean that the electron could not maintain its orbit, and it would be pulled into the oppositely charged nucleus. The atom would collapse as electrons spiraled into the nucleus. Since atoms do not collapse like this, there is a significant problem with the "solar system" model of the atom.

The Bohr Model

Niels Bohr was a young Danish student who visited Rutherford's laboratory in 1912 and became very interested in questions about the solar system model of the atom. He wondered what determined the size of the electron orbits and the energies of the electrons. He wanted to know why orbiting electrons did not give off electromagnetic radiation. Seeking answers to questions such as these led Bohr to incorporate the *quantum concept* of Planck and Einstein with Rutherford's model to describe the electrons in the outer part of the atom. This quantum concept will be briefly reviewed before proceeding with the development of Bohr's model of the hydrogen atom.

The Quantum Concept

In the year 1900, Max Planck introduced the idea that matter emits and absorbs energy in discrete units that he called **quanta.** Planck had been trying to match data from experiments with data that could be predicted from the theory of electromagnetic radiation. In order to match the experimental findings with the theory he had to assume that specific, discrete amounts of energy were associated with different frequencies of radiation. In 1905,

Figure 11.9

From measurements of alpha particle scattering, Rutherford estimated the radius of an atom to be 100,000 times greater than the radius of the nucleus. This is analogous to a ratio comparing the thickness of a dime to the length of a football field.

Albert Einstein extended the quantum concept to light, stating that light consists of discrete units of energy that are now called **photons**. The energy of a photon is directly proportional to the frequency of vibration, and the higher the frequency of light the greater the energy of the individual photons. In addition, the interaction of a photon with matter is an "all-or-none" affair, that is, matter absorbs an entire photon or none of it. The relationship between frequency (f) and energy (E) is

$$E = hf \qquad \textbf{equation 11.1}$$

where h is the proportionality constant known as *Planck's constant* (6.63×10^{-34} J·sec). This relationship means that higher-frequency light, such as ultraviolet, has more energy than lower-frequency light, such as red light.

Example 11.1

What is the energy of a photon of red light with a frequency of 4.60×10^{14} hz?

Solution

$f = 4.60 \times 10^{14}$ hz

$h = 6.63 \times 10^{-34}$ J · sec

$E = ?$

$$E = hf$$

$$= (6.63 \times 10^{-34} \text{ J} \cdot \text{sec})\left(4.60 \times 10^{14} \, \frac{1}{\text{sec}}\right)$$

$$= (6.63 \times 10^{-34})(4.60 \times 10^{14}) \text{ J} \cdot \text{sec} \times \frac{1}{\text{sec}}$$

$$= \boxed{3.05 \times 10^{-19} \text{ J}}$$

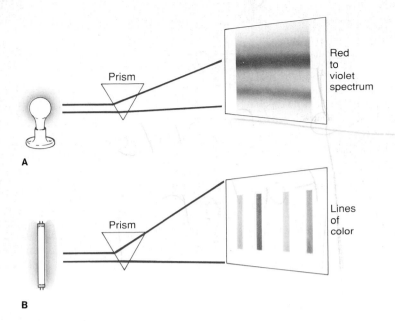

A

B

Figure 11.10

(a) Light from incandescent solids, liquids, or dense gases produces a continuous spectrum as atoms interact to emit all frequencies of visible light. (b) Light from an incandescent gas produces a line spectrum as atoms emit certain frequencies that are characteristic of each element.

Example 11.2

What is the energy of a photon of violet light with a frequency of 7.30×10^{14} hz? (Answer: 4.84×10^{-19} J)

▲

Atomic Spectra

Planck was concerned with hot solids that emit electromagnetic radiation. The nature of this radiation, called blackbody radiation, depends on the temperature of the source. When this light is passed through a prism it is dispersed into a *continuous spectrum*, with one color gradually blending into the next as in a rainbow. Today, it is understood that a continuous spectrum comes from solids, liquids, and dense gases because the atoms interact and all frequencies within a temperature-determined range are emitted. Light from an incandescent gas, on the other hand, is dispersed into a **line spectrum,** narrow lines of colors with no light between the lines (figure 11.10). The atoms in the incandescent gas are able to emit certain characteristic frequencies, and each frequency is a line of color that represents a definite value of energy. The line spectra are specific for a substance, and increased or decreased temperature changes only the intensity of the lines of colors. Thus hydrogen always produces the same colors of lines in the same position. Helium has

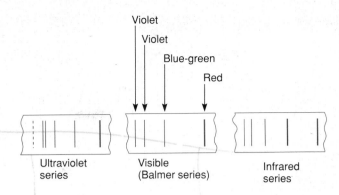

Figure 11.11

Atomic hydrogen produces a series of characteristic line spectra in the ultraviolet, visible, and infrared parts of the total spectrum. The visible light spectra always consist of two violet lines, a blue-green line, and a bright red line.

its own specific set of lines, as do other substances. Line spectra are a kind of "fingerprint" that can be used to identify a gas. A line spectrum might also extend beyond visible light into ultraviolet, infrared, and other electromagnetic regions.

In 1885, a Swiss mathematics teacher named J. J. Balmer was studying the regularity of spacing of the hydrogen line spectra. Balmer was able to develop an equation that fit all the visible lines. By assigning a value of n of 3, 4, 5, and 6 to the four lines he found the wavelengths fit the equation

$$\frac{1}{\lambda} = R \left(\frac{1}{2^2} - \frac{1}{n^2} \right) \qquad \textbf{equation 11.2}$$

when R is a constant of 1.097×10^7 1/m.

Balmer's findings were:

Violet line	(n = 6)	$\lambda = 4.1 \times 10^{-7}$ m
Violet line	(n = 5)	$\lambda = 4.3 \times 10^{-7}$ m
Blue-green line	(n = 4)	$\lambda = 4.8 \times 10^{-7}$ m
Red line	(n = 3)	$\lambda = 6.6 \times 10^{-7}$ m

These four lines became known as the **Balmer series.** Other series were found later, outside the visible part of the spectrum (figure 11.11). The equations of the other series were different only in the value of n and what number was squared.

Such regularity of observable spectral lines must reflect some unseen regularity in the atom. At this time it was known that hydrogen had only one electron. How could one electron produce series of spectral lines with such regularity?

Example 11.3

Calculate the wavelength of the violet line in the hydrogen line spectra according to Balmer's equation.

Solution

$n = 6$

$R = 1.097 \times 10^7 \ 1/m$

$\lambda = ?$

$$\frac{1}{\lambda} = R\left(\frac{1}{2^2} - \frac{1}{n^2}\right)$$

$$= 1.097 \times 10^7 \ \frac{1}{m}\left(\frac{1}{2^2} - \frac{1}{6^2}\right)$$

$$= 1.097 \times 10^7 \left(\frac{1}{4} - \frac{1}{36}\right)\frac{1}{m}$$

$$= 1.097 \times 10^7 \ (0.222)\frac{1}{m}$$

$$\frac{1}{\lambda} = 2.44 \times 10^6 \ \frac{1}{m}$$

$$\lambda = \boxed{4.11 \times 10^{-7} \ m}$$

Bohr's Theory

An acceptable model of the hydrogen atom would have to explain the characteristic line spectra and their regularity described by Balmer. In fact, a successful model should be able to *predict* the occurrence of each color line as well as account for its origin. By 1913, Bohr was able to do this by applying the quantum concept to a "solar system" model of the atom. He began by considering the single hydrogen electron to be a single "planet" revolving in a circular orbit around the nucleus. He assumed that the electron could occupy more than one circular orbit and that it would have a different state of energy, which depended on the radius of the orbit it was in at any particular time. Each orbit was assigned a number *n,* which could be any whole number 1, 2, 3, and so on out from the nucleus. These were known as the orbit **quantum numbers.** The following describes how orbit quantum numbers were used with definitions of (1) allowed orbits, (2) radiationless orbits, and (3) quantum jumps to describe the **Bohr model** of the atom.

Allowed Orbits

An electron can revolve around an atom only in specific allowed orbits. Bohr considered the electron to be a particle with a known mass in motion around the nucleus. Rotational motion is measured by *angular momentum,* a product of the mass (m), velocity (v), and radius of the orbit (r), or mvr. Conservation of angular momentum requires a greater velocity for a smaller orbit and less velocity for a larger orbit. It was Bohr's assumption

that the allowed orbits are those for which the angular momentum of the electron (mvr) equaled the orbit quantum number ($n = 1, 2, 3, \ldots$) times Planck's constant (h) divided by 2π, or

$$mvr = n\frac{h}{2\pi} \qquad \textbf{equation 11.3}$$

(Note that $2\pi r$ describes a circumference and that mv describes momentum.) Bohr used this relationship to determine the allowed orbits because differences between the energy levels it describes fit exactly with the differences between the frequencies (and thus energies) of the line spectra described by Balmer. Bohr did not have a reason for this assumption but used it simply because it worked.

Bohr also assumed that the force of electrical attraction between the electron (q_1) and proton (q_2) must be equal to the centripetal force if the electron moves in a circular orbit. This means that the force according to Coulomb's law of electrical attraction must be equal to the centripetal force described by Newton's second law of motion, or

$$\frac{kq_1q_2}{r^2} = \frac{mv^2}{r} \qquad \textbf{equation 11.4}$$

This describes a relationship that is very similar to the relationship between the gravitational forces between the earth and moon and the centripetal force that keeps the moon in its orbit around the earth (see chapter 3). The earth and moon are attracted by the force of gravity, however, not electrical forces.

Equations 11.3 and 11.4 can be solved for r, which will yield the distances, or radii, of the allowed orbits. When r is obtained for the first quantum number ($n = 1$), the radius of the closest orbit is found to be 0.529×10^{-10} m, which is known as the calculated *Bohr radius.* The next allowed orbit has a radius of 2.12×10^{-10} m, the next a radius of 4.46×10^{-10} m, and so on. According to the Bohr model, electrons can exist *only* in one of these allowed orbits and nowhere else.

Radiationless Orbits

An electron in an allowed orbit does not emit radiant energy as long as it remains in the orbit. It had been understood since the development of Maxwell's theory of electromagnetic radiation that an accelerating electron should emit an electromagnetic wave, such as light, which would move off into space from the electron. Bohr recognized that electrons moving in a circular orbit are accelerating, since they are changing direction continuously. Yet, light was not observed to be emitted from hydrogen atoms in their normal state. Bohr decided that the situation must be different for orbiting electrons, and that electrons could stay in their allowed orbits and *not* give off light. Thus he postulated this rule as a way to make his theory consistent with other scientific theories.

Quantum Jumps

An electron gains or loses energy only by moving from one allowed orbit to another (figure 11.12). The reference level for the potential energy of an electron is considered to be zero when

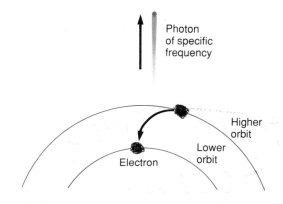

Figure 11.12

Each time an electron makes a "quantum leap," moving from a higher energy orbit to a lower energy orbit, it emits a photon of a specific frequency and energy value.

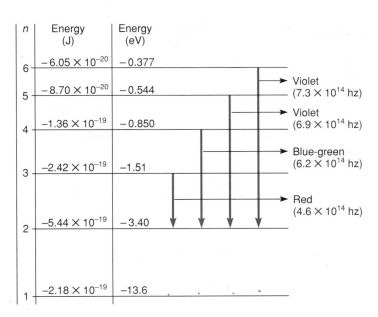

Figure 11.13

An energy level diagram for a hydrogen atom, not drawn to scale. The energy levels (*n*) are listed on the left side, followed by the energies of each level in J and eV. The color and frequency of the visible light photons emitted are listed on the right side, with the arrow showing the orbit moved from and to.

the electron is *removed* from an atom. The electron therefore has a lower and lower potential energy at closer and closer distances to the nucleus and has a negative value when it is in some allowed orbit. By way of analogy, you could consider ground level as a reference level where the potential energy of some object equals zero. But suppose there are two basement levels below the ground. An object on either basement level would have a gravitational potential energy less than zero, and work would have to be done on each object to bring each back to the zero level. Thus each object would have a negative potential energy. The object on the lowest level would have the largest negative value of energy since more work would have to be done on it to bring it back to the zero level. Therefore the object on the lowest level would have the *least* potential energy, and this would be expressed as the *largest negative value.*

Just as the objects on different basement levels, the electron has a definite negative potential energy in each of the allowed orbits. Bohr calculated the energy of an electron in the orbit closest to the nucleus to be -2.17×10^{-18} J, which is called the energy of the lowest state. The energy of electrons is expressed in units of the **electron volt** (eV). An electron volt is defined as the energy of an electron moving through a potential of one volt. Since this energy is charge times voltage (from $V = w/q$), 1.00 eV is equivalent to 1.60×10^{-19} J. Therefore the energy of an electron in the innermost orbit is its energy in joules divided by 1.60×10^{-19} J/eV, or -13.6 eV.

Bohr found that the energy of each of the allowed orbits could be found from the simple relationship of

$$E_n = \frac{E_1}{n^2} \qquad \text{equation 11.5}$$

where E_1 is the energy of the innermost orbit (-13.6 eV), and *n* is the quantum numbers for the orbits, or 1, 2, 3, and so on. Thus the energy for the second orbit ($n = 2$) is $E_2 = -13.6$ eV$/4 = -3.40$ eV. The energy for the third orbit out ($n = 3$) is $E_3 = -13.6$ eV$/9 = -1.51$ eV, and so forth (figure 11.13). Thus the energy of each orbit is *quantized,* occurring only as definite values.

In the Bohr model, the energy of the electron is determined by which allowable orbit it occupies. The only way that an electron can change its energy is to jump from one allowed orbit to another in quantum "jumps." An electron must *acquire* energy to jump from a lower orbit to a higher one. Likewise an electron *gives up* energy when jumping from a higher orbit to a lower one. Such jumps must be all at once, not part way and not gradual. By way of analogy, this is very much like the gravitational potential energy that you have on the steps of a staircase. You have the lowest potential on the bottom step and the greatest amount on the top step. Your potential energy is quantized because you can increase or decrease it by going up or down a number of steps, but you cannot stop between the steps.

An electron acquires energy from high temperatures or from electrical discharges to jump to a higher orbit. An electron jumping from a higher to a lower orbit gives up energy in the form of light. A single photon is emitted when a downward jump occurs, and the *energy of the photon is exactly equal to the difference in the energy level* of the two orbits. If E_1 represents the lower energy level (closest to the nucleus) and E_h represents a higher energy level (farthest from the nucleus), the energy of the emitted photon is

$$hf = E_h - E_1 \qquad \text{equation 11.6}$$

where h is Planck's constant and f is the frequency of the emitted light.

The energy level diagram in figure 11.13 shows the energy states for the orbits of a hydrogen atom. The lowest energy state, $n = 1$, is known as the **ground state** (or normal state). The higher states, $n = 2$, $n = 3$, and so on, are known as the **excited states.** The electron in a hydrogen atom would, under normal conditions, be located in the ground state ($n = 1$). But high temperatures or electric discharge can give the electron sufficient energy to jump to one of the excited states. Once in an excited state, the electron immediately jumps back to a lower state as shown by the arrows in the figure. The length of the arrow represents the frequency of the photon that the electron emits in the process. A hydrogen atom can give off only one photon at a time, and the many lines of a hydrogen line spectrum come from many atoms giving off many photons at the same time.

As you can see, the energy level diagram in figure 11.13 shows how the change of known energy levels from known orbits results in the exact energies of the color lines in the Balmer series. Bohr's theory did offer an explanation for the lines in the hydrogen spectrum with a remarkable degree of accuracy. However, the model did not have much success with larger atoms. Larger atoms had spectra lines that could not be explained by the Bohr model with its single quantum number. A German physicist, A. Sommerfield, tried to modify Bohr's model by adding elliptical orbits in addition to Bohr's circular orbits. It soon became apparent that the "patched up" model, too, was not adequate. Bohr had made the rule that there were radiationless orbits without an explanation, and he did not have an explanation for the quantized orbits. There was something fundamentally incomplete about the model.

Example 11.4

An electron in a hydrogen atom jumps from the excited energy level $n = 4$ to $n = 2$. What is the frequency of the emitted photon?

Solution

The frequency of an emitted photon can be calculated from equation 11.6, $hf = E_h - E_1$. The values for the two energy levels can be obtained from figure 11.13. (Note: E and E_1 must be in joules. If the values are in electron volts, they can be converted to joules by multiplying by the ratio of joules per electron volt, or $(eV)(1.6 \times 10^{-19} \text{ J/eV}) = \text{joules.}$)

$E_h = -1.36 \times 10^{-19}$ J

$E_1 = -5.44 \times 10^{-19}$ J

$h = 6.63 \times 10^{-34}$ J · sec

$f = ?$

$$hf = E_h - E_1 \therefore f = \frac{E_h - E_1}{h}$$

$$f = \frac{(-1.36 \times 10^{-19} \text{ J}) - (-5.44 \times 10^{-19} \text{ J})}{6.63 \times 10^{-34} \text{ J} \cdot \text{sec}}$$

$$= \frac{4.08 \times 10^{-19} \text{ J}}{6.63 \times 10^{-34} \text{ J} \cdot \text{sec}}$$

$$= 6.15 \times 10^{14} \frac{1}{\text{sec}}$$

$$= \boxed{6.15 \times 10^{14} \text{ hz}}$$

This is approximately the blue-green line in the hydrogen line spectrum.

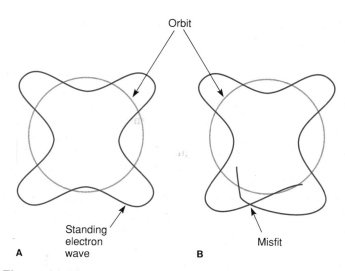

Figure 11.14

(a) Schematic of de Broglie wave, where the standing wave pattern will just fit in the circumference of an orbit. This is an allowed orbit. (b) This orbit does not have a circumference that will match a whole number of wavelengths; it is not an allowed orbit.

Quantum Mechanics

The Bohr model of the atom successfully accounted for the line spectrum of hydrogen and provided an understandable mechanism for the emission of photons by atoms. However, the model did not predict the spectra of any atom larger than hydrogen, and there were other limitations. A new, better theory was needed. The roots of a new theory would again come from experiments with light. Experiments with light had established that sometimes light behaves like a stream of particles and other times it behaves like a wave (see chapter 10). Eventually scientists began to accept that light has both wave properties and particle properties, which is now referred to as the *wave-particle duality of light.* This dual nature of light was recognized in 1905 when Einstein applied Planck's quantum concept to the energy of a photon with the relationship found in equation 11.1, $E = hf$, where E is the energy of a photon particle and f is the frequency of the associated wave, and h is Planck's constant.

Matter Waves

In 1923 Louis de Broglie, a French physicist, reasoned that symmetry is usually found in nature, so if a particle of light has a dual nature then particles such as electrons should too. De Broglie reasoned further that if this is true, an electron in its circular path around the nucleus would have to have a particular wavelength that would fit into the circumference of the orbit (figure 11.14). The circumference of the orbit must be a whole number of wavelengths long or

circumference = (number)(wavelength)

or

$$2\pi r = n\lambda \qquad \text{equation 11.7}$$

where $n = 1, 2, 3, \ldots$

De Broglie derived a relationship from equations concerning light and energy, which was

$$\lambda = \frac{h}{mv} \qquad \textbf{equation 11.8}$$

where λ is the wavelength, m is mass, v is velocity, and h is again Planck's constant. This equation means that any moving particle has a wavelength that is associated with its mass and velocity. In other words, de Broglie was proposing a wave-particle duality of matter, the existence of **matter waves.** According to equation 11.8, *any* moving object should exhibit wave properties. However, an ordinary-sized object would have wavelengths so small that they could not be observed. This is different for electrons because they have such a tiny mass.

Example 11.5

What is the de Broglie wavelength associated with a 0.150 kg baseball with a velocity of 50.0 m/sec?

Solution
m = 0.150 kg

v = 50.0 m/sec

h = 6.63 × 10⁻³⁴ J · sec

$\lambda = ?$

$$\lambda = \frac{h}{mv}$$

$$= \frac{6.63 \times 10^{-34} \text{ J} \cdot \text{sec}}{(0.150 \text{ kg})\left(50.0 \dfrac{m}{sec}\right)}$$

$$= \frac{6.63 \times 10^{-34}}{(0.150)(50.0)} \frac{\text{J} \cdot \text{sec}}{\text{kg} \times \dfrac{m}{sec}}$$

$$= \frac{6.63 \times 10^{-34}}{7.50} \frac{\dfrac{\text{kg} \cdot m^2}{sec^2} \cdot \text{sec}}{\dfrac{\text{kg} \cdot m}{sec}}$$

$$= \boxed{8.84 \times 10^{-35} \text{ m}}$$

What is the de Broglie wavelength associated with an electron with a velocity of 6.00 × 10⁶ m/sec?

Solution
m = 0.150 kg

v = 50.0 m/sec

h = 6.63 × 10⁻³⁴ J · sec

$\lambda = ?$

$$\lambda = \frac{h}{mv}$$

$$= \frac{6.63 \times 10^{-34} \text{ J} \cdot \text{sec}}{(9.11 \times 10^{-31} \text{ kg})\left(6.00 \times 10^6 \dfrac{m}{sec}\right)}$$

$$= \frac{6.63 \times 10^{-34}}{5.47 \times 10^{-24}} \frac{\text{J} \cdot \text{sec}}{\text{kg} \times \dfrac{m}{sec}}$$

$$= 1.21 \times 10^{-10} \frac{\dfrac{\text{kg} \cdot m^2}{sec^2} \cdot \text{sec}}{\dfrac{\text{kg} \cdot m}{sec}}$$

$$= \boxed{1.21 \times 10^{-10} \text{ m}}$$

The baseball wavelength of 8.84 × 10⁻³⁵ m is much too small to be detected or measured. The electron wavelength of 1.21 × 10⁻¹⁰ m, on the other hand, is comparable to the distances between atoms in a crystal, so a beam of electrons through a crystal should produce diffraction.

The idea of matter waves was soon tested after de Broglie published his theory. Experiments with a beam of light passing through a very small opening or by the edge of a sharp-edged obstacle were described in chapter 10. In these experiments the beam of light produces diffraction and interference patterns. This was part of the evidence for the wave nature of light since such results could only be explained by waves, not particles. When similar experiments were performed with a beam of electrons, *identical* wave property behaviors were observed. This and many related experiments showed without doubt that electrons have both wave properties and particle properties. And, as was the case with light waves, measurements of the electron interference patterns provided a means to measure the wavelength of electron waves.

Recall that waves confined on a fixed string establish resonate modes of vibration called *standing waves* (see chapter 7). Only certain fundamental frequencies and harmonics can exist on a string, and the combination of the fundamental and overtones gives the stringed instrument its particular quality. The same result of resonate modes of vibrations is observed in *any* situation where waves are confined to a fixed space. Characteristic standing wave patterns depend on the wavelength and wave velocity for waves formed on strings, in enclosed columns of air, or for any kind of wave in a confined space. Electrons are confined to the space near a nucleus, and electrons have wave properties, so an electron in an atom must be a confined wave. Does an electron form a characteristic wave pattern? This was the question being asked in about 1925 when Heisenberg, Schrödinger, Dirac, and others applied the wave nature of the electron to develop a new model of the atom based on the mechanics of electron waves. The new theory is now called **wave mechanics,** or **quantum mechanics.**

Obtain a long spiral spring such as a "slinky." Connect one end to the other, forming a circle. Suspend the circular spring with rubber bands so that the spring makes a horizontal circle. Wiggle one part of the spring at various frequencies until standing waves are formed around the circle. From your observations, explain why n is a whole number in the equation $2\pi r = n\lambda$ (equation 11.7).

Wave Mechanics

Erwin Schrödinger, an Austrian physicist, treated the atom as a three-dimensional system of waves to derive what is now called the *Schrödinger equation.* Instead of the simple circular planetary orbits of the Bohr model, solving the Schrödinger equation results in a description of three-dimensional shapes of the patterns that develop when electron waves are confined by a nucleus. Schrödinger first considered the hydrogen atom, calculating the states of vibration that would be possible for an electron wave confined by a nucleus. He found that the frequency of these vibrations, when multiplied by Planck's constant, matched exactly, to the last decimal point, the observed energies of the quantum states of the hydrogen atom ($E = hf$). The conclusion is that the wave nature of the electron is the important property to consider for a successful model of the atom.

The quantum mechanics theory of the atom proved to be very successful as it confirmed all the known experimental facts and predicted new discoveries. The theory does have some of the same quantum ideas as the Bohr model; for example, an electron emits a photon when jumping from a higher state to a lower one. The Bohr model, however, considered the particle nature of an electron moving in a circular orbit with a definitely assigned position at a given time. Quantum mechanics considers the wave nature, with the electron as a confined wave with well-defined shapes and frequencies. A wave is not localized like a particle and is spread out in space. The quantum mechanics model is therefore a series of orbitlike smears, or fuzzy statistical representations, of where the electron might be found.

The Quantum Mechanics Model

The quantum mechanics model is a highly mathematical treatment of the mechanics of matter waves. In addition, the wave properties are considered as three-dimensional problems, and three quantum numbers are needed to describe the fuzzy electron cloud. The mathematical detail will not be presented here. The following is a qualitative description of the main ideas in the quantum mechanics model. It will describe the results of the mathematics and will provide a mental visualization of what it all means.

First, understand that the quantum mechanical theory is not an extension or refinement of the Bohr model. The Bohr model considered electrons as particles in circular orbits that could be only certain distances from the nucleus. The quantum mechanical model, on the other hand, considers the electron as a wave and considers the energy of its harmonics, or modes, of standing waves. In the Bohr model the location of an electron was certain—in an orbit. In the quantum mechanical model the electron is a spread-out wave.

Quantum mechanics describes the energy state of an electron wave with four *quantum numbers,* in terms of its (1) distance from the nucleus, (2) energy sublevel, (3) orientation in space, and (4) direction of spin.

The **principal quantum number,** called *n,* describes the *main energy level* of an electron in terms of its most probable distance from the nucleus. The lowest energy state possible is closest to the nucleus and is assigned the principal quantum number of 1 ($n = 1$). Higher states are assigned progressively higher positive whole numbers of $n = 2$, $n = 3$, $n = 4$, and so on. Electrons with higher principal quantum numbers have higher energies and are located farther from the nucleus.

The **angular momentum quantum number** defines energy sublevels within the main energy levels. Each sublevel is identified with a letter. The first four of these letters, in order of increasing energy, are s, p, d, and f. The choice of these letters goes back to spectral studies when the spectral lines were described as *s*harp, *p*rincipal, *d*iffuse, and *f*ine. The letter s represents the lowest sublevel and the letter f represents the highest sublevel. A principal quantum number and a letter indicating the angular momentum quantum number are combined to identify the main energy state and energy sublevel of an electron. For an electron in the lowest main energy level, n = 1, and in the lowest sublevel, s, the number and letter are 1s (read as "one-s"). Thus 1s indicates an electron that is as close to the nucleus as possible in the lowest energy sublevel possible.

As stated in the Bohr model, the location of an electron was certain, that is, in an orbit. In the quantum mechanical model the electron is spread out and knowledge of its location is very uncertain. The **Heisenberg uncertainty principle** states that you cannot measure both the momentum and the exact position of an electron at the same time. The location of the electron can only be described in terms of *probabilities* where it might be at a given instant. The probability of location is described by a fuzzy region of space called an **orbital.** An orbital defines the space where an electron is likely to be found. Orbitals have characteristic three-dimensional shapes and sizes and are identified with electrons of characteristic energy levels (figure 11.15).

An orbital shape represents where an electron could probably be located at any particular instant. This "probability cloud" could likewise have any particular orientation in space, and the direction of this orientation is uncertain. On the other hand, an external magnetic field applied to an atom produces different energy levels that are related to the orientation of the orbital to the magnetic field. The orientation of an orbital in space is described by the **magnetic quantum number.** This number is related to the energies of orbitals as they are oriented in space relative to an external magnetic field, a kind of energy sub-sublevel. In general, the lowest energy sublevel (s) has only

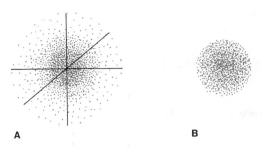

A **B**

Figure 11.15

(*a*) An electron distribution sketch representing probability regions where an electron is most likely to be found. (*b*) A boundary surface, or contour, that encloses about 90 percent of the electron distribution shown in (*a*). This three-dimensional space around the nucleus, where there is the greatest probability of finding an electron, is called an orbital.

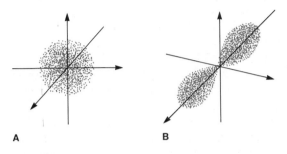

A **B**

Figure 11.16

(*a*) A contour representation of an s orbital. (*b*) A contour representation of a p orbital.

Table 11.1

Quantum numbers and electron distribution to $n = 4$

Main Energy Level	Energy Sublevels	Maximum Number of Electrons	Maximum Number of Electrons per Main Energy Level
$n = 1$	s	2	2
$n = 2$	s	2	
	p	6	8
$n = 3$	s	2	
	p	6	
	d	10	18
$n = 4$	s	2	
	p	6	
	d	10	
	f	14	32

one orbital orientation. The next higher energy sublevel (p) can have three orbital orientations (figure 11.16). The d sublevel can have five orbital orientations, and the highest sublevel of f can have a total of seven different orientations (table 11.1).

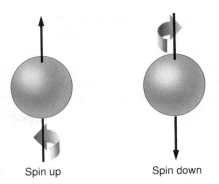

Spin up   Spin down

Figure 11.17

Experimental evidence supports the concept that electrons can be considered to spin one way or the other as they move about an orbital under an external magnetic field.

High-resolution studies of the hydrogen line spectra revealed details that were not known when the Bohr model of the atom was first developed. These studies showed, for example, that what was previously believed to be a single red line was actually two lines that were very close together. The only way to explain the splitting was to consider the electron to be spinning on its axis like a top. Such a spinning movement would cause the electron to produce a magnetic field. The energy of the electron would depend on which way the electron magnetic field was aligned with an external magnetic field. Thus an electron spinning one way (say clockwise) would have a different energy than one spinning the other way (say counterclockwise). These two spin orientations are described by the **spin quantum number** (figure 11.17).

Electron spin is an important property of electrons that helps determine the electronic structure of an atom. As it turns out, two electrons spinning in opposite directions produce unlike magnetic fields that are attractive, balancing some of the normal repulsion from two like charges. Two electrons of opposite spin, called an **electron pair,** can thus occupy the same orbital. This was summarized in 1924 by Wolfgang Pauli, a German physicist. His summary, now known as the **Pauli exclusion principle,** states that *no two electrons in an atom can have the same four quantum numbers.* This provides the key for understanding the electron structure of atoms.

Electron Configuration

Recall that the energy of electrons is measured by considering an unattached electron to have an energy value of zero. It takes energy to remove an electron from an atom, so a bound electron must have *less* energy than a free one. The arrangement of electrons in orbitals is called the **electron configuration.** When in the ground state, electrons always adopt the lowest possible energies consistent with the Pauli exclusion principle, arranging themselves in the lowest energy orbitals as close to the nucleus as possible. The lowest possible energy level is $n = 1$ and the lowest sublevel is s. *The number of electrons that can occupy* *this orbital is limited.* According to the Pauli exclusion principle, no two electrons in an atom can have all four of their quantum numbers exactly the same. If one electron is in the first energy level ($n = 1$) in the orbital of the lowest energy (s), a second electron can occupy this same $n = 1$, s orbital only if it has a different spin orientation. Thus the exclusion principle states there can only be two electrons in the $n = 1$, s orbital and, as it works out, there can only be *a maximum of two electrons in any given orbital.* An atom of helium has two electrons and both can occupy the $n = 1$, s orbital because they have opposite spins. In general, electrons occupy the orbitals in an order starting with the lowest energy. Before you can describe the electron arrangement, you need to know how many electrons are present in an atom.

An atom is electrically neutral, so the number of protons (positive charge) must equal the number of electrons (negative charge). The atomic number therefore identifies the number of electrons as well as the number of protons:

atomic number = number of protons = number of electrons

Now that you have a means of finding the number of electrons, consider the various energy levels to see how the electron configuration is determined. There are four things to consider: (1) the main energy level, (2) the energy sublevel, (3) the number of orbital orientations, and (4) the electron spin. Recall that the lowest energy level is $n = 1$, and successive numbers identify progressively higher energy levels. Recall also that the energy sublevels, in order of increasing energy, are s, p, d, and f. This electron configuration is written in shorthand with 1s standing for the lowest energy sublevel of the first energy level. A superscript gives the number of electrons present in a sublevel. Thus the electron configuration for a helium atom, which has two electrons, is written as

$$1s^2$$

This combination of symbols has the following meaning:

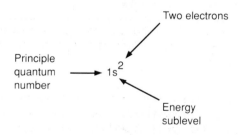

The symbols mean an atom with two electrons in the s sublevel of the first main energy level.

Table 11.2 gives the electron configurations for the first twenty elements. The configurations of the p energy sublevel have been condensed in this table. There are three possible orientations of the p orbital, each with two electrons. This is shown

Table 11.2

Electron configuration for the first twenty elements

Atomic Number	Element	Electron Configuration
1	Hydrogen	$1s^1$
2	Helium	$1s^2$
3	Lithium	$1s^2 2s^1$
4	Beryllium	$1s^2 2s^2$
5	Boron	$1s^2 2s^2 2p^1$
6	Carbon	$1s^2 2s^2 2p^2$
7	Nitrogen	$1s^2 2s^2 2p^3$
8	Oxygen	$1s^2 2s^2 2p^4$
9	Fluorine	$1s^2 2s^2 2p^5$
10	Neon	$1s^2 2s^2 2p^6$
11	Sodium	$1s^2 2s^2 2p^6 3s^1$
12	Magnesium	$1s^2 2s^2 2p^6 3s^2$
13	Aluminum	$1s^2 2s^2 2p^6 3s^2 3p^1$
14	Silicon	$1s^2 2s^2 2p^6 3s^2 3p^2$
15	Phosphorus	$1s^2 2s^2 2p^6 3s^2 3p^3$
16	Sulfur	$1s^2 2s^2 2p^6 3s^2 3p^4$
17	Chlorine	$1s^2 2s^2 2p^6 3s^2 3p^5$
18	Argon	$1s^2 2s^2 2p^6 3s^2 3p^6$
19	Potassium	$1s^2 2s^2 2p^6 3s^2 3p^6 4s^1$
20	Calcium	$1s^2 2s^2 2p^6 3s^2 3p^6 4s^2$

as p^6, which is a condensation of the three possible p orientations. Note that the number of electrons in all the orbitals add up to the atomic number. Note also that as you proceed from a lower atomic number to a higher one, the higher element has the same configuration as the element before it with the addition of one more electron. In general, it is then possible to begin with the simplest atom, hydrogen, and add one electron at a time to the order of energy sublevels and obtain the electron configuration for all the elements. The exclusion principle limits the number of electrons in any orbital, and allowances will need to be made for the more complex behavior of atoms with many electrons.

The energies of the orbital are not fixed as you progress through the atomic numbers, and there are several factors that influence their energies. The first orbitals are filled in a straightforward 1s, 2s, 2p, 3s, then 3p order. Then the order becomes contrary to what you might expect. One useful way of figuring out the order in which orbitals are filled is illustrated in figure 11.18. Each row of this matrix represents a principal energy level with possible energy sublevels increasing from left to right. The order of filling is indicated by the diagonal arrows. There are exceptions to the order of filling as shown by the matrix, but it works for most of the elements.

The Quark

Some understanding about how matter is put together came with the discovery of the electron, proton, and neutron—three elementary particles that make up an atom. In the early 1900s a particle outside the atom, the *photon,* was verified by experimental evidence. Two other particles were verified in the 1930s, the *neutrino* ("little neutral one") and the *positron* (a positively charged electron). By the mid-1930s a total of six elementary particles were known. Since that time, high-energy accelerator experiments made it possible to collide particles with great violence, probing the inner parts of atoms and how they are put together. A multitude of elementary particles are now known to exist. This feature is about these elementary particles.

There are now thought to be twelve elementary particles that make up matter. The elementary particles can be divided into three main groups: (1) *leptons,* which exist independently, (2) *quarks,* which exist together, making up a third group, (3) *hadrons.*

Leptons are a group of fundamental particles that include the familiar electron, the muon (an overweight relative of the electron), and three types of neutrinos. In radioactive decay an electron (beta particle) is emitted with an electron neutrino. For each lepton there is a corresponding antiparticle, or antilepton, with the same mass but opposite electric charge.

Hadrons are a group of composite particles with an internal structure, so they are not elementary particles. There are two subgroups of hadrons: (1) the *mesons* (meaning "intermediate mass," between electrons and protons) and (2) the *baryons* (meaning "greater mass"). Hundreds of short-lived hadrons have been identified that exist briefly after high-energy collisions. Among the more stable are the baryons named protons and neutrons.

Hadrons are composed of different combinations of fundamental particles called quarks. Five kinds of quarks, fancifully called *flavors,* have been identified and a sixth is believed to exist. The existing quarks are called up, down, sideways (or strange), charm, and bottom (or beauty). The sixth flavor yet to be discovered will be named top (or truth). Each flavor carries a fractional charge that is either $-1/3$ or $+2/3$. Antiquarks have equal but opposite charges. In order to explain how identical quarks could combine as observed, each flavor was assigned three quantum states that are called *color.* Each flavor can carry a charge of red, green, or blue. Antiquarks carry a corresponding anticolor, for example, a red quark has an antiquark of the color cyan, a green quark has an antiquark of the color magenta, and a blue quark has an antiquark of the color yellow. The idea of quark color was designed to follow the allowable combination of quarks and antiquarks according to the exclusion principle. Hadrons do not have a color charge, so the sum of the quark colors making up the hadron must result in a white hadron. Baryons are made up of three quarks, so a combination of a red, a green, and a blue quark would be acceptable since this would result in a white baryon. Mesons are made up of a quark and an antiquark so a combination of a blue quark and a yellow antiquark would be acceptable since this would result in a white meson.

The story of subnuclear elementary particles is by no means complete. There is no explanation at present for why quarks and leptons exist. Are there more fundamental particles? Answers to this and more questions await further research. Eventually the studies of fundamental particles will reveal the fundamental laws of nature, explaining why matter is put together the way it is. The next chapter in this story will likely be written at the Superconducting Supercollider.

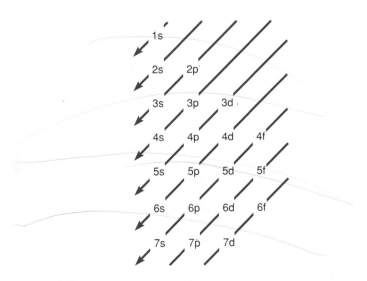

Figure 11.18

A matrix showing the order in which the orbitals are filled. Start at the top left, then move from the head of each arrow to the tail of the one immediately below it. This sequence moves from the lowest energy level to the next higher level for each orbital.

Example 11.6

Certain strontium (atomic number 38) compounds are used to add the pure red color to flares and fireworks. Write the electron configuration for strontium.

Solution

First, note that an atomic number of 38 means a total of thirty-eight electrons. Second, refer to the order-of-filling matrix in figure 11.18. Remember that only two electrons can occupy an orbital, but there are three orientations of the p orbital for a total of six electrons. There are likewise five possible orientations of the d orbital for a total of ten electrons. Starting at the lowest energy level, two electrons go in 1s making $1s^2$, then two go in 2s making $2s^2$. That is a total of four electrons so far. Next $2p^6$ and $3s^2$ use eight more electrons for a total of twelve so far. The $3p^6$, $4s^2$, $3d^{10}$, and $4p^6$ use up twenty-four more electrons for a total of thirty-six. The remaining two go into the next orbital $5s^2$ and the complete answer is

Strontium: $1s^2\ 2s^2\ 2p^6\ 3s^2\ 3p^6\ 4s^2\ 3d^{10}\ 4p^6\ 5s^2$

Summary

Attempts at understanding matter date all the way back to ancient Greek philosophers, who viewed matter as being composed of *elements*, or simpler substances. Two models were developed that considered matter to be (1) *continuous*, or infinitely divisible, or (2) *discontinuous*, made up of particles called *atoms*.

During the 1600s Robert Boyle provided experimental evidence to reject the ancient Greek ideas of continuous matter made up of earth, air, fire, and water. Boyle reasoned there were *elements*, which could not be broken down to anything simpler, and compounds, which were made up of combinations of elements.

In the early 1800s Dalton published an *atomic theory*, reasoning that matter was composed of hard, indivisible atoms that were joined together or dissociated during chemical change.

When a good air pump was invented in 1885, *cathode rays* were observed to move from the negative terminal in an evacuated glass tube. The nature of cathode rays was a mystery. The mystery was solved in 1887 when Thomson discovered they were negatively charged particles now known as *electrons*. Thomson had discovered the first elementary particle of which atoms are made and measured their charge to mass ratio.

Rutherford developed a "solar system" model based on experiments of alpha particles scattered from a thin sheet of metal. This model had a small, massive, and positively charged *nucleus* surrounded by moving electrons. These electrons were calculated to be at a distance 100,000 times the radius of the nucleus, so the volume of an atom is mostly empty space. Later, Rutherford proposed that the nucleus contained two elementary particles, *protons* with a positive charge and *neutrons* with no charge. The *atomic number* is the number of protons in an atom.

Bohr developed a model of the hydrogen atom to explain the characteristic *line spectra* emitted by hydrogen. His model specified that (1) electrons can move only in allowed orbits, (2) electrons do not emit radiant energy when they remain in an orbit, and (3) electrons move from one allowed orbit to another when they gain or lose energy. When an electron jumps from a higher orbit to a lower one it gives up energy in the form of a single photon. The energy of the photon corresponds to the difference in energy between the two levels. The Bohr model worked well for hydrogen but not other atoms.

De Broglie proposed that moving particles of matter (electrons) should have wave properties like moving particles of light (photons). His derived equation, $\lambda = h/mv$, showed that these *matter waves* were only measurable for very small particles such as electrons. De Broglie's proposal was tested experimentally, and the experiments confirmed that electrons do have wave properties.

Schrödinger and others used the wave nature of the electron to develop a new model of the atom called *wave mechanics*, or *quantum mechanics*. This model was found to confirm exactly all the experimental data as well as predict new data. The quantum mechanical model describes the energy state of the electron in terms of quantum numbers based on the wave nature of the electron. The quantum numbers defined the *probability* of the location of an electron in terms of a fuzzy region of space called *orbitals*.

Summary of Equations

11.1

$$\text{energy} = (\text{Planck's constant})\,(\text{frequency})$$

$$E = hf$$

where $h = 6.63 \times 10^{-34}$ J · sec

11.2

$$\frac{1}{\text{wavelength}} = \text{constant}\left(\frac{1}{2^2} - \frac{1}{\text{number}^2}\right)$$

$$\frac{1}{\lambda} = R\left(\frac{1}{2^2} - \frac{1}{n^2}\right)$$

11.3

$$\text{angular momentum} = \left(\begin{array}{c}\text{orbit}\\\text{quantum}\\\text{number}\end{array}\right)\left(\frac{\text{Planck's constant}}{2\pi}\right)$$

$$mvr = n\frac{h}{2\pi}$$

where $h = 6.63 \times 10^{-34}$ J · sec and $n = 1, 2, 3, \ldots$ for an orbit

11.4

$$\text{electrical force} = \text{centripetal force}$$

$$\text{Coulomb's law} = \text{Newton's second law for circular motion}$$

$$\frac{kq_1q_2}{r^2} = \frac{mv^2}{r}$$

11.5

$$\text{energy state of orbit number} = \frac{\text{energy state of innermost orbit}}{\text{number squared}}$$

$$E_n = \frac{E_1}{n^2}$$

where $E_1 = -13.6$ eV and $n = 1, 2, 3, \ldots$

11.6

$$\begin{array}{c}\text{energy}\\\text{of}\\\text{photon}\end{array} = \left(\begin{array}{c}\text{energy state}\\\text{of}\\\text{higher orbit}\end{array}\right) - \left(\begin{array}{c}\text{energy state}\\\text{of}\\\text{lower orbit}\end{array}\right)$$

$$hf = E_h - E_1$$

where $h = 6.63 \times 10^{-34}$ J · sec, E_h and E_1 must be in joules

11.7

$$\text{circumference of orbit} = (\text{whole number})(\text{wavelength})$$

$$2\pi r = n\lambda$$

where $n = 1, 2, 3, \ldots$

11.8

$$\text{wavelength} = \frac{\text{Planck's constant}}{(\text{mass})(\text{velocity})}$$

$$\lambda = \frac{h}{mv}$$

where $h = 6.63 \times 10^{-34}$ J · sec

Key Terms

angular momentum quantum number (p. 218)
atomic number (p. 211)
Balmer series (p. 213)
Bohr model (p. 214)
cathode rays (p. 209)
electron (p. 209)
electron configuration (p. 219)
electron pair (p. 219)
electron volt (p. 215)
element (p. 207)
excited states (p. 216)
ground state (p. 215)
Heisenberg uncertainty principle (p. 218)
line spectrum (p. 213)
magnetic quantum number (p. 218)

matter waves (p. 217)
neutron (p. 211)
nucleus (p. 211)
orbital (p. 218)
Pauli exclusion principle (p. 219)
photons (p. 212)
principal quantum number (p. 218)
proton (p. 211)
quanta (p. 211)
quantum mechanics (p. 217)
quantum numbers (p. 214)
spin quantum number (p. 219)
wave mechanics (p. 217)

Applying the Concepts

1. John Dalton reasoned that atoms exist from the evidence that
   a. elements could not be broken down into anything simpler.
   b. water pours and flows when in the liquid state.
   c. elements always combined in certain fixed ratios.
   d. peanut butter and jelly could be combined in any ratio.

2. According to Rutherford's calculations, the volume of an atom is mostly
   a. occupied by protons and neutrons.
   b. filled with electrons.
   c. occupied by tightly bound protons, electrons, and neutrons.
   d. empty space.

3. The main problem with a "solar system" model of the atom is that
   a. electrons move in circular, not elliptical, orbits.
   b. the electrons should lose energy since they are accelerating.
   c. opposite charges should attract one another.
   d. the mass ratio of the nucleus to the electrons is wrong.

4. The lines of color in a line spectrum from a given element
   a. change colors with changes in the temperature.
   b. are always the same with a regular spacing pattern.
   c. are randomly spaced, having no particular pattern.
   d. have the same colors with a spacing pattern that varies with the temperature.

5. Hydrogen, with its one electron, produces a line spectrum in the visible light range with
   a. one color line.
   b. two color lines.
   c. three color lines.
   d. four color lines.

6. According to the Bohr model, an electron gains or loses energy only by
   a. moving faster or slower in an allowed orbit.
   b. jumping from one allowed orbit to another.
   c. being completely removed from an atom.
   d. jumping from one atom to another atom.

7. When an electron in a hydrogen atom jumps from an orbit farther from the nucleus to an orbit closer to the nucleus, it emits
   a. a single photon with an energy equal to the energy difference of the two orbits.
   b. four photons, one for each of the color lines observed in the line spectrum of hydrogen.
   c. a number of photons dependent on the number of orbit levels jumped over.

8. The Bohr model of the atom
   a. explained the color lines in the hydrogen spectrum.
   b. could not explain the line spectrum of atoms larger than hydrogen.
   c. had some made-up rules without explanations.
   d. all of the above.

9. The quantum mechanics model of the atom is based on
   a. the quanta, or measured amounts, of energy of a moving particle.
   b. the energy of a standing electron wave that can fit into an orbit.
   c. calculations of the energy of the three-dimensional shape of a circular orbit of an electron particle.
   d. Newton's laws of motion, but scaled down to the size of electron particles.

10. Two electrons can occupy the same orbital because they have different
    a. principal quantum numbers.
    b. angular momentum quantum numbers.
    c. magnetic quantum numbers.
    d. spin quantum numbers.

Answers

1. c 2. d 3. b 4. b 5. d 6. b 7. a 8. d 9. b
10. d

Questions for Thought

1. What reason did Dalton have for bringing back the ancient Greek idea of matter being composed of hard, indivisible atoms?

2. What was the experimental evidence that Thomson had discovered the existence of a subatomic particle when working with cathode rays?

3. Describe the experimental evidence that led Rutherford to the concept of a nucleus in an atom.

4. What was Thomson's model of the atom and how was it shown to be wrong?

5. What is the main problem with a "solar system" model of the atom?

6. Compare the size of an atom to the size of its nucleus.

7. What does atomic number mean? How is the atomic number related to what an atom is an element of? How is atomic number related to the number of electrons in an atom?

8. An atom has eleven protons in the nucleus. What is the atomic number? What is the name of this element? What is the electron configuration of this atom?

Atomic Structure

9. Describe the three main points in the Bohr model of the atom.

10. Why do the energies of electrons in an atom have negative values? (Hint: It is *not* because of the charge of the electron.)

11. Which has the lowest energy, an electron in the first energy level ($n = 1$) or an electron in the third energy level ($n = 3$)? Explain.

12. Which has the lowest energy, an electron in the s sublevel or an electron in a p sublevel of the same main energy level?

13. How did the Bohr model of the atom account for the Balmer spectral series of the hydrogen atom?

14. Explain how a hydrogen atom, with its single electron, can produce four separate spectral lines of visible light.

15. Considering the second energy level ($n = 2$), how many electrons can occupy (a) the 2s orbital and (b) the entire second energy level? Explain.

16. Using the matrix in figure 11.18, which orbital would fill with electrons first: (a) 2s or 3s, (b) 3d or 4s, (c) 3d or 4p?

17. How did Bohr use Planck's quantum theory of energy in his model of the atom? What was Bohr's reason for using the quantum theory?

18. What is similar about the Bohr model of the atom and the quantum mechanical model? What are the fundamental differences?

19. What is the difference between a hydrogen atom in the ground state and one in the excited state?

20. Explain how the Bohr model accounted for the line spectra of hydrogen.

Exercises

Group A—Solutions Provided in Appendix D

1. A neutron with a mass of 1.68×10^{-27} kg moves from a nuclear reactor with a velocity of 3.22×10^3 m/sec. What is the de Broglie wavelength of the neutron?

2. Calculate the energy (a) in eV and (b) in joules for the sixth energy level ($n = 6$) of a hydrogen atom.

3. How much energy is needed to move an electron in a hydrogen atom from $n = 2$ to $n = 6$? Give the answer (a) in joules and (b) in eV. (See figure 11.13 for needed values.)

4. What frequency of light is emitted when an electron in a hydrogen atom jumps from $n = 6$ to $n = 2$?

5. How much energy is needed to remove the electron from a hydrogen atom in the ground state completely?

6. Thomson determined the charge to mass ratio of the electron to be -1.76×10^{11} C/kg. Millikan determined the charge on the electron to be -1.60×10^{-19} C. According to these findings, what is the mass of an electron?

7. Assume that an electron wave making a standing wave in a hydrogen atom has a wavelength of 1.67×10^{-10} m. Considering the mass of an electron to be 9.11×10^{-31} kg, use the de Broglie equation to calculate the velocity of an electron in this orbit.

8. Using any reference you wish, write the complete electron configurations for (a) boron, (b) aluminum, and (c) potassium.

9. Explain how you know that you have the correct *total* number of electrons in your answer for 8a, 8b, and 8c.

10. Refer to figure 11.18 *only* and write the complete electron configurations for (a) argon, (b) zinc, and (c) bromine.

Group B—Solutions Not Given

1. An electron with a mass of 9.11×10^{-31} kg has a velocity of 4.3×10^6 m/sec in the innermost orbit of a hydrogen atom. What is the de Broglie wavelength of the electron?

2. Calculate the energy (a) in eV and (b) in joules of the third energy level ($n = 3$) of a hydrogen atom.

3. How much energy is needed to move an electron in a hydrogen atom from the ground state ($n = 1$) to $n = 3$? Give the answer (a) in joules and (b) in eV.

4. What frequency of light is emitted when an electron in a hydrogen atom jumps from $n = 2$ to the ground state ($n = 1$)?

5. How much energy is needed to remove an electron from $n = 2$ in a hydrogen atom completely?

6. If the charge to mass ratio of a proton is 9.58×10^7 C/kg and the charge is 1.60×10^{-19} C, what is the mass of the proton?

7. An electron wave making a standing wave in a hydrogen atom has a wavelength of 8.33×10^{-11} m. If the mass of the electron is 9.11×10^{-31} kg, what is the velocity of the electron according to the de Brolgie equation?

8. Using any reference you wish, write the complete electron configurations for (a) nitrogen, (b) phosphorus, and (c) chlorine.

9. Explain how you know that you have the correct *total* number of electrons in your answer for 8a, 8b, and 8c.

10. Referring to Figure 11.18 *only,* write the complete electron configuration for (a) neon, (b) sulfur, and (c) calcium.

Elements and the Periodic Table

Figure 12.1

Classification is the arrangement into groups or categories according to some criteria. The act of classifying creates a pattern that helps you to recognize and understand the behavior of fish, chemicals, or any matter in your surroundings. These fish, for example, are classified as salmon because they ascend rivers from the sea for breeding.

IN chapter 11 you learned how the concept of the atom was developed, evolving into the modern model of a fuzzy cloud of matter with an internal structure. We considered this internal structure and how the electrons make up the fuzzy cloud in atoms of different elements. But there was no discussion about the meaning or implications of the different structures or how they could be used to understand the properties of matter. That is the goal of this chapter, to understand matter based on the electron structure of atoms.

The behavior of matter seems bewildering when you consider all of its different kinds, forms, and shapes and all the changes it undergoes. Things seem bewildering and confusing because you cannot make connections between behaviors, that is, there are no apparent patterns. Often in such situations the act of grouping or classifying things by similar properties is helpful. Classifying helps you to find patterns of similarities and identities of groups. Once you have a pattern you will want to find a reason for its existence. You are now on your way to an understanding that not only accounts for the patterns but also provides a means of organizing all the information as well.

This chapter presents an example of gaining understanding through grouping, which is also known as classifying (figure 12.1). We will consider several different ways that matter can be classified, for example, into classes of mixtures and pure substances. Pure substances can be further subdivided into groups of elements and compounds. Some of the more interesting elements will be described, along with how they were named and their symbols. Matter can be classified other ways, including changes in matter. All changes are either physical or chemical, and chemical changes are the key to grouping according to the electron structure of atoms. After an introduction of a few new properties of atoms, the periodic table will be discussed in terms of its systematic classification of elements. This periodic classification can be used to predict how elements react with one another, making the study of matter much easier.

Classifying Matter

The universe is made up of just two basics, matter and energy. **Matter** is usually defined as anything that occupies space and has mass. It is the substance of any particle or object, from the parts that make up atoms to the bulk of a giant star. Between these two extremes of size there is an overwhelming and complex variety of observable sizes, shapes, and kinds of matter on the earth. One way to make thinking about such a variety a little less complex is to *classify*. Classifying is the act of mentally making groups based on similar properties. Classifying helps to organize your thinking and to find patterns that might otherwise go unnoticed. The following considers several different ways of classifying matter by considering different properties.

Metals and Nonmetals

Humans learned thousands of years ago that certain earthlike materials placed on a bed of glowing coals produced a new substance, a metal. The smelting of copper and tin dates back to

about 3500 B.C., the beginning of the bronze age. Iron smelting dates back to about 1500 B.C., and steel was first made about 1200 B.C. These metals all have very different properties than the earthlike materials they were extracted from. Today a **metal** is recognized as a kind of matter having the following physical properties:

1. metallic *luster,* the way a shiny metal reflects light to have the look of gold, copper, and other metals;
2. high heat and electrical *conductivity;*
3. *malleability,* which means that you can roll it or pound it into a thin sheet; and
4. *ductility,* which means that you can pull it into a wire.

All metals are solids at room temperature except mercury, which is a liquid (figure 12.2).

A **nonmetal** is a kind of matter that does not have a metallic luster, is a poor conductor of heat and electricity, and when solid, is a brittle material that cannot be pounded or pulled into new shapes. Nonmetals occur as solids, liquids, or gases at room temperature.

Most unprotected metals do not last long when exposed to air. Iron, for example, rusts to a reddish, earthlike material with nonmetallic properties. In time, most metals seem to return to the same kind of nonmetallic, earthlike materials from which they were extracted. Eventually humans began to wonder about the transformation of nonmetallic, earthlike materials to metals and why they returned to nonmetallic forms. Today, such questions are considered in **chemistry,** the science concerned with the study of matter and the changes it undergoes.

Solids, Liquids, and Gases

The Greek philosophers of some 2,500 years ago thought of all matter as being made up of four elements that were identified as earth, air, water, and fire. It is interesting that, as we know now, the ancient Greeks were not describing elements. They were describing the general forms in which all matter exists and one form of energy. Earth, air, and water are the most common examples of the solid, gaseous, and liquid states of matter. Fire is the most common example of heat, the energy involved in changes of matter such as nonmetallic ores to metals. Thus the ancient Greeks were actually describing the basics that make up the universe, matter and energy.

Solid, liquid, and gas are called the **states of matter,** and these states represent another way to classify matter. On the earth, all matter generally belongs to one of the three groups, or states, which are defined by two general properties. These properties are how well a sample of matter maintains (1) its shape and (2) its volume (figure 12.3). For example, the *gaseous* state of matter is not able to maintain a definite shape or a definite volume. A sample of gas released into a completely evacuated, rigid container will disperse throughout the entire space inside the container, taking the shape and volume of the container. Since density is mass per unit volume, the density of the gas will depend on how much gas is placed in the container as well as the volume of the container.

Figure 12.2

Most matter can be classified as metals or nonmetals according to physical properties. Aluminum, for example, is a lightweight kind of matter that can be melted and rolled into a thin sheet or pulled into a wire. Here you see aluminum soft-drink cans that have been compressed into 1,600 lb bales for recycling, destined to again be formed into a new soft-drink can, aluminum foil, or aluminum wire.

Liquids, like gases, will flow and can be poured from one container to another. Both liquids and gases are fluids. Both have an indefinite shape and take the shape of their container. A liquid, however, has a definite volume and does not fill any container as a gas does. A 500 cm³ sample of gas will disperse to fill a 1,000 cm³ container completely. A 500 cm³ sample of liquid will still have that volume when placed into a 1,000 cm³ container. Liquids have a much greater density than gases and are not very compressible in comparison. Most common liquids have a density not much different from water at 1 g/cm³ (1 kg/L) at 4° C.

Solids, like liquids, have a definite volume. Unlike gases and liquids, a solid has a definite shape. A solid with a cubic shape placed in a container of some other shape maintains both its volume and its original shape. Most solids are more dense when in the solid state than in the liquid state. Ice is the common

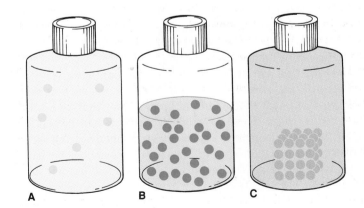

Figure 12.3

(*a*) A gas disperses throughout a container, taking the shape and volume of the container. (*b*) A liquid takes the shape of the container but retains its own volume. (*c*) A solid retains its own shape and volume.

A   B   C

exception to this generalization as ice floats on water, which is more dense. This exception will be explained when water is discussed in a later chapter.

Mixtures and Pure Substances

Still another way to classify matter is to consider its composition, that is, how it is put together (figure 12.4). In this classification scheme all common matter occurs in one of two groups, mixtures or pure substances. **Mixtures** are made of unlike parts that have a variable composition. Common sand, for example, is a mixture of bits of rocks, minerals, and perhaps hard parts from plants and animals. Since sand is a mixture, any given sample might have more of some things and less of others. The black sand of Hawaii is mostly tiny pieces of volcanic rocks. The white sand of Florida is mostly tiny pieces of ground-up seashells. The composition and proportions of a sand mixture depend on the source of the components and how far the sand has been transported. Solutions, gases, and alloys are also mixtures since they are made of unlike parts with a variable composition.

Mixtures can be separated into their component parts by physical means. For example, you can physically separate the parts making up a sand mixture by using a magnifying glass and tweezers to move and isolate each part. A solution of salt in water is a mixture since the amount of salt dissolved in water can have a variable composition. But how do you separate the parts of a solution? One way is to evaporate the water, leaving the salt behind. There are many methods for separating mixtures but all involve a **physical change.** A physical change does not alter the *identity* of matter. When water is evaporated from the salt solution, for example, it changes to a different state of matter (water vapor) but is still recognized as water. Physical changes involve physical properties only and no new substances are formed. Examples of physical changes include evaporation, condensation, melting, freezing, and dissolving.

Mixtures can be physically separated into **pure substances,** materials that are the same throughout and have a fixed, definite composition. If you closely examine a sample of table salt you will see that it is made up of hundreds of tiny cubes. Any one of these cubes will have the identical properties as any other cube including a salty taste. Sugar, like table salt, has all of its parts alike. Unlike salt, sugar grains have no special shape or form. But each grain has the same sweet taste and other properties as any other sugar grain.

If you heat salt and sugar in separate containers, you will find very different results. Some pure substances, such as salt, undergo a physical change and melt, changing back to a solid upon cooling with the same properties that it had originally. Sugar, however, changes to a black material upon heating while it gives off water vapor. The black material does not change back to sugar upon cooling. The sugar has *decomposed* to a new substance while the salt did not. The new substance is carbon and it has properties completely different from sugar. The original substance (sugar) and its properties have changed to a new substance (carbon) with new properties. The sugar has gone through

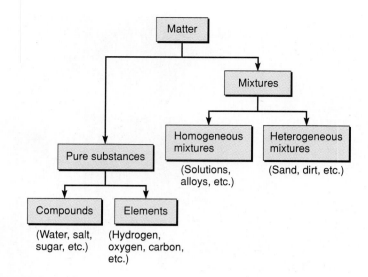

Figure 12.4

A classification scheme for matter.

a **chemical change.** A chemical change alters the identity of matter, producing new substances with different properties. In this case the chemical change was one of decomposition. Heat produced a chemical change by decomposing the sugar into carbon and water vapor.

The decomposition of sugar always produces the same mass ratio of carbon to water, so sugar has a fixed, definite composition. A **compound** is a pure substance that can be decomposed by a chemical change into simpler substances with a fixed mass ratio. This means that sugar is a compound (figure 12.5).

A pure substance that cannot be broken down into anything simpler by chemical or physical means is an **element.** Sugar is decomposed by heating into carbon and water vapor. Carbon cannot be broken down further, so carbon is an element. It has been known since about 1800 that water is a compound that can be broken down by electrolysis into hydrogen and oxygen, two gases that cannot be broken down to anything simpler. So sugar is a compound made from the elements of carbon, hydrogen, and oxygen.

But what about the table salt? Is table salt a compound? Table salt is a stable compound that is not decomposed by heating. It melts at a temperature of about 1,000° C, then returns to the solid form with the same salty properties upon cooling. Electrolysis was used in the early 1800s to decompose table salt into the elements sodium and chlorine, positively proving that it is a compound. Heat brings about a chemical change and decomposes some compounds, such as sugar. Heat will not bring about a chemical change in table salt, so other means are needed.

Pure substances are either compounds or elements. Decomposition through heating and decomposition through electrolysis are two means of distinguishing between compounds and elements. If a substance can be decomposed into something simpler, you know for sure that it is a compound. If the substance cannot be decomposed it might be an element or it might be a stable compound that resists decomposition. More testing would

Figure 12.5
Sugar is a compound that can be easily decomposed to simpler substances by heating. One of the simpler substances is the black element carbon, which cannot be further decomposed by chemical or physical means.

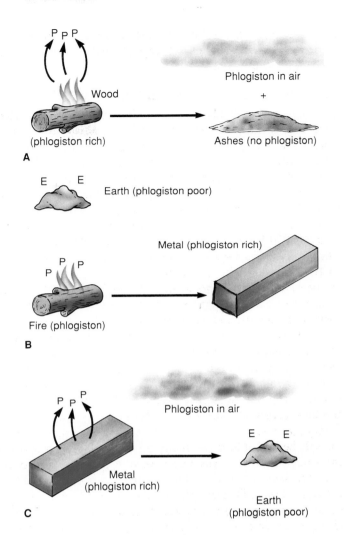

Figure 12.6
The phlogiston theory. (*a*) In this theory, burning was considered to be the escape of phlogiston into the air. (*b*) Smelting combined phlogiston-poor earth with phlogiston from a fire to make a metal. (*c*) Metal rusting was considered to be the slow escape of phlogiston from a metal into the air.

be necessary before you can be confident that you have identified an element. Most pure substances are compounds. There are millions of different compounds but only 109 known elements at the present time. These elements are the fundamental materials of which all matter is made.

Elements

Modern science is usually identified as beginning just after the time of Galileo and Newton, about three hundred years ago. Understandings about matter at this time were still based on the writings of Aristotle of some two thousand years earlier. There were alternative ideas about the number of elements, but consideration of earth, air, fire, and water as the basic elements was generally accepted.

Reconsidering the Fire Element

During the early 1700s the use of fuels such as coal, the process of burning, and the production of steam became topics of general interest. The first working steam engine was invented by Thomas Newcomen in 1711, and the time was ripe for new ideas, new theories, and answers to questions about burning and what was going on during the burning process.

About 1700 a new theory about burning was introduced by a German physician. This theory considered all burnable materials to contain a substance called *phlogiston,* a word from the Greek meaning "fire." Burning was considered to be the escape of phlogiston from fuels into the air. Materials that did not burn were considered not to contain phlogiston, either because they never had it or because they had already lost it (such as ashes). So far, phlogiston might remind you of Aristotle's "fire element" with a different name. But this theory continued on in detail to explain other observations. For example, a candle

in a closed container would burn for a few minutes, then go out. The explanation for this observation was that the air could hold only so much phlogiston, which was released during burning. When the air was saturated it could accept no more phlogiston. The escape of more phlogiston was thus prevented and the fire would go out.

Other observations explained by the theory involved (1) the conversion of nonmetallic, earthlike materials into metals by fire and (2) the rusting of metals, returning them to nonmetallic, earthlike matter over a period of time. The earthlike materials were considered to be phlogiston poor. Placing the materials in a fire permitted them to absorb phlogiston, becoming metals. The metals could not hold on to the phlogiston and, over time, it leaked away, returning the metals to their nonmetallic form (figure 12.6). The phlogiston theory, with its convincing explanations, became the accepted understanding of burning, metal smelting, rusting, and the role of air in these processes during the 1700s.

Elements and the Periodic Table

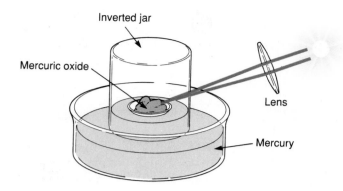

Figure 12.7

Priestley produced a gas (oxygen) by using sunlight to heat mercuric oxide kept in a closed container. The oxygen forced some of the mercury out of the jar as it was produced, increasing the volume about five times.

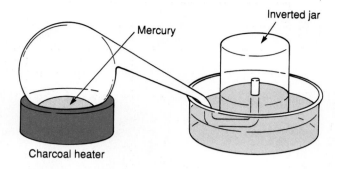

Figure 12.8

Lavoisier heated a measured amount of mercury to form the red oxide of mercury. He measured the amount of oxygen removed from the jar and the amount of red oxide formed. When the reaction was reversed, he found the original amounts of mercury and oxygen.

Discovery of Modern Elements

A series of events would lead to the downfall of the phlogiston theory and the discovery of modern elements in the 1770s. The first of these events was an experiment in gas chemistry conducted by an English minister named Joseph Priestley. Priestley was an amateur chemist with a natural flair for experimental research. One of his first discoveries was a method for producing carbon dioxide by reacting chalk and sulfuric acid, then forcing the gas into a container of water. Priestley had invented soda water, which soon would become modern-day cola. During Priestley's time, a navy officer won approval to supply ships with soda water, believing it might help with scurvy. The officer, coincidentally, was named Lord Sandwich.

Priestley is recognized today for the discovery of oxygen. In one of his experiments, Priestley heated a nonmetallic red powder that was then called mercurius calcinatus (mercuric oxide) in a jar inverted in a bowl of mercury (figure 12.7). The powder gave off a gas, forcing some of the mercury out of the inverted jar. From this, Priestley could tell that the red powder gave off about five times its own volume of gas. He found that this gas caused a candle to burn vigorously, a glowing piece of wood to sparkle, and a mouse to jump about with vigor.

Priestley had discovered oxygen, but he interpreted his findings in terms of the then-current phlogiston theory. He believed that the red powder combined with phlogiston from the air. That was why things burned vigorously in the air. The air was depleted of phlogiston, Priestley thought, which rapidly "pulled" it from flaming candles and glowing pieces of wood. The phlogiston theory provided an explanation for what was observed, and Priestley was not much concerned with measurements.

Antoine Lavoisier was already a recognized French chemist in 1772 when, at the age of twenty-nine, he began to experiment with the role of phlogiston in burning and the rusting of metals. He had observed that sulfur and phosphorus *gained* weight when they burned, not lost weight as the phlogiston theory predicted. He supposed that the sulfur and phosphorus

combined with air to make the additional weight. When he read about Priestley's "dephlogisticated air," Lavoisier repeated Priestley's work with a carefully measured analysis. Following through with his earlier observations about burning and weight gain, Lavoisier proved that Priestley's "dephlogisticated air" was actually a substance he called *oxygen* (figure 12.8). He replaced the phlogiston theory with a theory that burning is a chemical combination of some material with oxygen.

In the meantime, Henry Cavendish had isolated a very light, highly flammable gas by reacting metals with acids. Cavendish found that when this gas burned, pure water was produced. Lavoisier repeated the Cavendish experiment and concluded that the light gas combined with oxygen to produce water. Lavoisier named the gas *hydrogen* from Greek words meaning "water former." This proved to be the final blow to the ancient Greek concepts of water and air as elements as well as the end of the phlogiston theory. Lavoisier recognized the need for a whole new concept of elements, compounds, and chemical change. The concept of elements was reconsidered, and by working with other leading chemists of the time, a whole new list of elements was developed.

Lavoisier published the results of his experiments and a list of the known elements in 1789. Lavoisier's list of thirty-three elements included twenty-three that are recognized as elements today and eight that are recognized as compounds along with light and heat. This was the beginning of modern chemistry and of understanding elements as they are known today.

Figure 12.9 shows the number of known elements since the time of Lavoisier. New elements were discovered as new chemical and analytical techniques were developed. The increase just before 1800 resulted from an interest in rocks, minerals, and the materials of the earth. Just after the 1800s the invention of the electric battery was followed by experiments in which compounds were decomposed into elements by electrolysis. These experiments resulted in the discovery of more elements. By 1809, about half of the natural elements had been

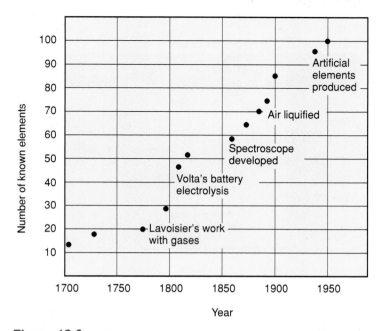

Figure 12.9

The number of known elements increased as new chemical and analytical techniques were developed.

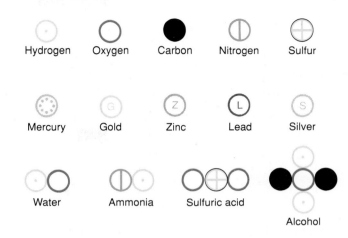

Figure 12.10

Here are some of the symbols Dalton used for atoms of elements and molecules of compounds. He probably used a circle for each because he thought of atoms as the ancient Greeks did, as tiny, round, hard spheres.

discovered. The development of the spectroscope in the late 1850s led to the identification of more elements by their spectral lines. Mendeleev found patterns in elements in 1869, which resulted in even more discoveries. By the 1890s, technology was developed that made it possible to produce very cold, liquified gases, and this resulted in even more discoveries. All the natural elements were discovered by 1940. The naturally occurring elements range from hydrogen (atomic number 1) to uranium (atomic number 92). However, technetium (atomic number 43) and promethium (atomic number 61) are artificially produced elements that are not found anywhere on the earth. In addition, small amounts of plutonium (atomic number 94) have been found occurring naturally. The elements that are not natural ones were artificially made during nuclear physics experiments and are not found on the earth. Thus there are 91 naturally occurring elements and 18 artificially prepared ones that are known at the present time for a total of 109.

Names of the Elements

Elements such as sulfur, zinc, tin, and iron have been known since ancient times but were not recognized as elements. The original meanings of their names are ancient and have become obscured with time. More recently the right to name an element belonged to the discoverer, who could call it anything as long as the name of a metal ended with "-um." The first 103 elements have internationally accepted names. Sources of these names have included the following:

1. the compound or substance in which the element was discovered;
2. an unusual or identifying property of the element;
3. places, cities, and countries;
4. famous scientists;
5. Greek mythology or some other mythology;
6. astronomical objects.

Chemical Symbols

When John Dalton introduced his atomic theory in the early 1800s he introduced a system of symbols to represent atoms and how they formed compounds. Each element had its own special symbol, which stood for one atom of that element. Some of the symbols Dalton used to explain his atomic theory are shown in figure 12.10. Each atom was a circle, probably because Dalton thought of atoms as tiny, indivisible spheres. He indicated the different elements by symbols or letters within each circle, for example, a dot for hydrogen and a "G" for gold. Dalton represented compounds with combinations of element symbols as shown in the figure.

Letter symbols came into use about 1913 after an earlier recommendation by Jons Berzelius, a Swedish chemist. Berzelius recommended that the letter symbol should be the capitalized first letter of the name of an element, for example, the symbol H for hydrogen. Today, there are about a dozen common elements that have single capitalized first letters as their symbols. All the rest have two letters with the first letter *always* capitalized and the second letter *never* capitalized. The second, lowercase letter is either (1) the second letter in the name of the element or (2) a letter representing a strong consonant heard when the name of the element is spoken. Examples of using the first two letters of the name are Ca for calcium and Si for silicon. Examples of using the first letter and the letter that is heard when the name is spoken are Cl for chlorine and Cr for chromium.

Some elements have symbols from the earlier use of Latin names for the elements. For example, the symbol Au is used for gold because the metal was earlier known by its Latin name of *aurum,* meaning "shining dawn." There are ten elements with symbols from Latin names and one with a symbol from a German name. These eleven elements are listed in table 12.1, together with the sources of their names and their symbols.

Table 12.1

Elements with symbols from Latin or German names

Atomic Number	Name	Source of Symbol	Symbol
11	Sodium	Latin: Natrium	Na
19	Potassium	Latin: Kalium	K
26	Iron	Latin: Ferrum	Fe
29	Copper	Latin: Cuprum	Cu
47	Silver	Latin: Argentum	Ag
50	Tin	Latin: Stannum	Sn
51	Antimony	Latin: Stibium	Sb
74	Tungsten	German: Wolfram	W
79	Gold	Latin: Aurum	Au
80	Mercury	Latin: Hydrargyrum	Hg
82	Lead	Latin: Plumbum	Pb

Table 12.2

Elements making up 99 percent of earth's crust

Element (Symbol)	Percent by Weight
Oxygen (O)	46.6
Silicon (Si)	27.7
Aluminum (Al)	8.1
Iron (Fe)	5.0
Calcium (Ca)	3.6
Sodium (Na)	2.8
Potassium (K)	2.6
Magnesium (Mg)	2.1

Chemical symbols are like the vocabulary of a new language and are used to describe chemical changes with clarity and precision. The first key to understanding this language is to understand that a chemical symbol identifies a specific element and represents one atom of that element. Thus the symbol H means one atom of the element hydrogen. The symbol Hg means one atom of the element mercury and so on.

You usually learn the symbols by using them and looking up a symbol as needed from a table such as the one located on the inside back cover of this text. This is really not as big a task as it might seem at first, because the elements are not equally abundant and only a few are common. In table 12.2, for example, you can see that only eight elements make up about 99 percent of the solid surface of the earth. Oxygen is most abundant, making up about 50 percent of the weight of the earth's crust. Silicon makes up more than 25 percent, so these two nonmetals alone make up about 75 percent of the earth's solid surface. Almost all the rest is made up of just six metals as shown in the table.

The number of common elements is limited elsewhere, too. Only two elements make up about 99 percent of the atmospheric air around the earth. Air is mostly nitrogen (about 78 percent) and oxygen (about 21 percent) with traces of five other elements and compounds. Water on the earth is hydrogen and oxygen, of course, but seawater also contains elements in solution. These elements are chlorine (55 percent), sodium (31 percent), sulfur (8 percent), and magnesium (4 percent). Only three elements make up about 97 percent of your body. These elements are hydrogen (60 percent), oxygen (26 percent), and carbon (11 percent). Generally, all of this means that the elements are not equally distributed or equally abundant in nature.

Symbols and Atomic Structures

When Dalton developed his atomic theory he pictured atoms as tiny, hard spheres. He considered these spheres to differ only in their masses, with all the atoms of one element having the same mass, which was different from the atomic masses of the other elements. An important contribution of Dalton's theory was the attempt to determine a comparison of the atomic masses. He used the fixed mass ratios of combining elements to determine the relative atomic masses. For example, hydrogen always combines with oxygen in a mass ratio of 1:8 to make water. This means that one gram of hydrogen combines with eight grams of oxygen to form nine grams of water. Dalton assumed that one atom of hydrogen combined with one atom of oxygen to form a particle of water. Today, the particle of water is called a **molecule.** For now, consider a molecule to be a particle composed of two or more atoms held together by an attractive force called a chemical bond.

From the assumption that one atom of hydrogen joined with one atom of oxygen to form a molecule of water, Dalton reasoned that the mass ratio of 1:8 could be explained if one atom of oxygen is eight times more massive than one atom of hydrogen. Measuring mass ratios of hydrogen, oxygen, and other elements, Dalton was able to construct the table of relative atomic masses (figure 12.11).

Dalton's relative atomic mass table was a step in the right direction, but his findings were all wrong. They were wrong because of measurement errors and the assumption that one atom of an element combines with one atom of another element. Today, for example, it is known that two atoms of hydrogen combine with one atom of oxygen to form water. Chemists eventually worked out techniques for determining the number of atoms of elements that combine to form compounds, and these techniques will be discussed in later chapters. It took about one hundred years of carefully measured chemical experimentation before accurate values of relative atomic masses were established.

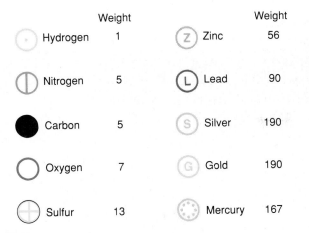

		Weight				Weight
	Hydrogen	1		Zinc	Zinc	56
	Nitrogen	5		Lead	Lead	90
	Carbon	5		Silver	Silver	190
	Oxygen	7		Gold	Gold	190
	Sulfur	13		Mercury	Mercury	167

Figure 12.11

Using information from the fixed mass ratios of combining elements, Dalton was able to calculate the relative atomic masses of some of the elements. Many of his findings were wrong, as you can see from this sample of his table.

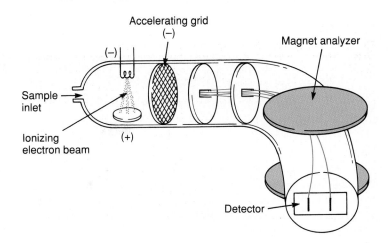

Figure 12.12

A schematic of a mass spectrometer. The atoms of a sample of gas become positive ions after being bombarded by a beam of electrons. The ions are deflected into a curved path by a magnetic field, which separates them according to their charge to mass ratio. Less massive ions are deflected the most, so the device identifies different groups of particles with different masses.

Radioactivity was discovered in the late 1890s, and before long new but puzzling data were observed. Recall that Dalton considered all atoms of an element to have the same mass and all to act the same when combining with other elements. That is, all the atoms of an element were considered to have the same physical properties and the same chemical properties. Analysis of radioactive elements found that the masses of atoms were *not* always the same. One sample of the metal lead, for example, was radioactive but another sample was not radioactive. Further analysis found that the two samples had different masses even though the two samples were the same element with identical chemical behaviors. In 1910 Frederic Soddy called such varieties of an element an **isotope** because they were in the same place chemically (from the Greek *iso* meaning "same" and *tope* meaning "place"). So isotopes are atoms of an element with identical chemical properties but with different masses.

The masses and abundance of isotopes were measured by Francis William Aston in the 1910s. Aston had been an assistant to J. J. Thompson and was familiar with his method of measuring the charge to mass ratio of the electron. But Aston now had a new device called a *mass spectrometer* (figure 12.12). This instrument uses a beam of electrons to ionize, for example, vaporized atoms of an element. Positive ions are formed, then accelerated by high voltage through a tube. In the tube they are deflected by a magnetic field, and the amount of deflection depends on their masses. A wide range of mass values is scanned by varying the magnetic field and accelerating voltage. A detector is thus not only able to measure the presence of various masses but also the abundance of each mass from the intensity. A plot of the intensity measured by the detector of the mass is called a *mass spectrum*.

Aston was able to provide the first clear, undisputed evidence for the existence of isotopes with the mass spectrograph. He confirmed that suspected isotopes existed for some elements and discovered new ones as well. Figure 12.13, for example,

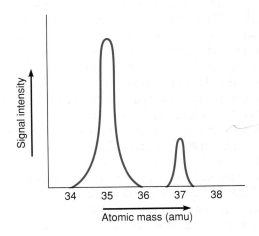

Figure 12.13

A mass spectrum of chlorine from a mass spectrometer. Note that two separate masses of chlorine atoms are present, and their abundance can be measured from the signal intensity. The greater the signal intensity, the more abundant the isotope.

shows that a sample of chlorine gas has two chlorine isotopes in different proportions, and this was unknown before Aston analyzed chlorine in the mass spectrograph. In 1919 Aston reported that the atomic mass of each substance was very close to, but not exactly, a whole number. Chlorine, for example, has one isotope with an atomic mass very close to 35 (34.969 units). Thus the atomic mass of an element with more than one isotope must be the *average* of the atomic masses of the isotopes. The contribution of a particular isotope to this average value depends on its abundance. For example, if the chlorine isotope with

Elements and the Periodic Table

an atomic mass of 34.969 units makes up 75.53 percent of a sample of chlorine gas, and the chlorine isotope with an atomic mass of 36.966 units makes up the other 24.47 percent, then

$$\text{average atomic mass} = (75.53\% \times 34.969 \text{ units})$$
$$+ (24.47\% \times 36.966 \text{ units})$$
$$= 26.41 \text{ units} + 9.05 \text{ units}$$
$$= 35.46 \text{ units}$$

The average atomic mass of chlorine in this example is 35.46 units, which is a *weighted average* of the masses of the chlorine isotopes as they occur. **Atomic weight** is the name given to the weighted average of the masses of stable isotopes of an element as they occur in nature.

Dalton had originally assigned hydrogen a mass of one unit, then compared the mass of other atoms to this standard. Today, the mass of any isotope is compared to the mass of a particular isotope of carbon. This isotope is *assigned* a mass of exactly 12.00 units called **atomic mass units** (u). This isotope, called carbon-12, provides the standard to which the masses of all other isotopes are compared. The mass of any isotope is based on the mass of a carbon-12 isotope. A table of atomic weights, such as the one on the inside back cover of this text, lists the atomic weight of carbon as slightly more than 12 u. Carbon occurs naturally as two isotopes, one with an atomic mass of exactly 12.00 u (98.9 percent) and one with an atomic mass of about 13 u (1.11 percent). Thus the weighted average, or *atomic weight,* of carbon is slightly greater than about 12.01 u. Other examples are illustrated in table 12.3.

The abundance of each isotope making up a naturally occurring element is always the same, no matter where the sample is measured. While isotopes have different masses, they have the *same chemical behavior.* Isotopes occur because of varying numbers of neutrons in the nuclei of atoms with the same atomic number. The atomic number is the number of protons in the nucleus and the number of electrons in the atom. What element an atom belongs to is identified by the number of protons, and the chemical nature is identified by the number of electrons. Since the isotopes of an element all have the same number of protons and electrons with the same configuration, they all have the same chemical properties. Since they have different numbers of neutrons they have different atomic masses.

A neutron has about the same mass as a proton, but an electron has a comparatively trivial mass (about 1/1,840 the mass of a proton). Thus neutrons and protons make up almost all the mass of an atom. The sum of the number of protons and neutrons in a nucleus is called the **mass number** of an atom. Mass numbers are used to identify isotopes. A chlorine atom with 17 protons and 20 neutrons has a mass number of 17 + 20, or 37, and is referred to as "chlorine-37." A chlorine atom with 17 protons and 18 neutrons has a mass number of 17 + 18, or 35, and is referred to as "chlorine-35." Using symbols, chlorine-37 is written as

$$^{37}_{17}\text{Cl}$$

Table 12.3

Selected atomic weights calculated from mass and abundance of isotopes

Stable Isotopes	Mass of Isotope Compared to C-12	Abundance	Atomic Weight
$^{1}_{1}\text{H}$	1.007	99.985%	
$^{2}_{1}\text{H}$	2.0141	0.015%	1.0079
$^{9}_{4}\text{Be}$	9.01218	100.%	9.01218
$^{14}_{7}\text{N}$	14.00307	99.63%	
$^{15}_{7}\text{N}$	15.00011	0.37%	14.0067
$^{16}_{8}\text{O}$	15.99491	99.759%	
$^{17}_{8}\text{O}$	16.99914	0.037%	
$^{18}_{8}\text{O}$	17.00016	0.204%	15.9994
$^{19}_{9}\text{F}$	18.9984	100.%	18.9984
$^{20}_{10}\text{Ne}$	19.99244	90.92%	
$^{21}_{10}\text{Ne}$	20.99395	0.257%	
$^{22}_{10}\text{Ne}$	21.99138	8.82%	20.179
$^{27}_{13}\text{Al}$	26.9815	100.%	26.9815

where Cl is the chemical symbol for chlorine, the subscript to the bottom left is the atomic number, and the superscript to the top left is the mass number. Consider the following rules when you use a chemical symbol to identify isotopes with a mass number:

1. The mass number is the closest whole number to the atomic mass of an isotope. Only carbon-12 has an atomic mass with a whole number (by definition).
2. The number of protons in the nucleus of an atom equals the atomic number.
3. The number of neutrons in the nucleus of an atom equals the mass number minus the atomic number.

Atomic numbers are always whole numbers because they represent the number of protons in the nucleus of an atom. *Mass numbers* are always whole numbers because they represent the number of protons and neutrons in the nucleus of an isotope. The *atomic mass of an isotope* is *not* a whole number, with the exception of carbon-12, which was assigned a mass of exactly 12 units to set up a relative scale from which the masses of all other atoms are compared. Why are the atomic masses of other isotopes not whole numbers? The answer is found in a mass contribution from the internal energy of the nucleus. There is a relationship between energy and mass, and this relationship will be explained in a future chapter on nuclear energy. *Atomic weights* are also *not* whole numbers, but you might expect this because atomic weight is a weighted average of the masses of the isotopes. Compare these values in table 12.3.

Example 12.1

Identify the number of protons, neutrons, and electrons in an atom of $^{16}_{8}O$.

Solution

The subscript to the bottom left is the atomic number. Atomic number is defined as the number of protons in the nucleus, so this number identifies the number of protons as 8. Any atom with 8 protons is an atom of oxygen, which is identified with the symbol O. The superscript to the top left identifies the mass number of this isotope of oxygen, which is 16. The mass number is defined as the sum of the number of protons and the number of neutrons in the nucleus. Since you already know the number of protons is 8 (from the atomic number), then the number of neutrons is 16 minus 8, or 8 neutrons. Since a neutral atom has the same number of electrons as protons, an atom of this oxygen isotope has 8 protons, 8 neutrons, and 8 electrons.

Example 12.2

How many protons, neutrons, and electrons are found in an atom of $^{17}_{8}O$? (Answer: 8 protons, 9 neutrons, and 8 electrons)

The Periodic Law

Many new elements were discovered in a short period of time in the early 1800s, nearly doubling the number of known elements to 60 by the 1840s. As the list grew, information about the behavior of elements and their compounds created a large body of apparently unrelated facts. Seeking to organize and make sense out of this mass of information, chemists made attempts to classify the elements according to their properties. They were looking for some underlying order, a pattern that would not only organize the large body of facts but also perhaps form the basis of a theory that would account for the facts. The process began with attempts to relate somehow the behaviors of elements to their atomic weights.

As information about the behavior of elements increased, chemists began to recognize certain patterns of similarities. Lithium, sodium, and potassium were found to be shiny, soft metals that reacted vigorously with water to form an alkaline solution. Calcium, strontium, and barium were found to be another group of soft metals with similar properties, but properties that were different from the sodium group. Iron, cobalt, and nickel were similar hard metals. Chlorine, bromine, and iodine were similar nonmetals. By 1829 a German chemist named

Table 12.4

Examples of element triads

Element	Atomic Weight (u)	Average Atomic Weight (First and Third)	Density (g/cm³)
Lithium (Li)	6.9		0.53
Sodium (Na)	23.0	23.0	0.97
Potassium (K)	39.1		0.86
Calcium (Ca)	40.1		1.55
Strontium (Sr)	87.6	88.7	2.54
Barium (Ba)	137.3		3.50
Chlorine (Cl)	35.5		1.56
Bromine (Br)	79.9	81.2	3.12
Iodine (I)	126.9		4.93

Johann Dobereiner described the existence of *triads,* groups of three elements with similar chemical properties. Dobereiner found that when the elements of a triad were listed in order of atomic weights, the atomic weight of the middle element was almost equal to the average atomic weight of the other two elements. In most cases the values for density, melting point, and other properties of the middle element were also midway between values for the other two elements. Three of these triads are identified in table 12.4. There was no explanation that would account for the occurrence of triads at this time, and they were considered an interesting curiosity.

A workable classification scheme of the elements was developed independently by two scientists. The Russian chemist Dmitri Mendeleev and the German physicist Lothar Meyer published similar classification schemes in 1869. Mendeleev's scheme was based mostly on the chemical properties of elements and their atomic weight, and Meyer's was based on physical properties and atomic weight. But both arranged the elements in order of increasing atomic weight and observed that the properties recur periodically. Both had devised schemes that would systematize the study of chemistry and lead to the modern periodic table.

Mendeleev and Meyer both arranged the elements in rows of increasing atomic weights, arranged so that elements with similar properties made vertical columns. These vertical columns contained Dobereiner's triads of elements with similar properties. Mendeleev and Meyer were not restricted by trying to fit their individual schemes to some perceived "law" as others before them. Both left blank spaces if the element with the next

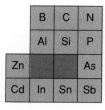

Figure 12.14

Mendeleev left blank spaces in his table when the properties of the elements above and below did not seem to match. The existence of unknown elements was predicted by Mendeleev on the basis of the blank spaces. When the unknown elements were discovered, it was found that Mendeleev had closely predicted the properties of the elements as well as their discovery.

highest atomic weight did not fit with a vertical family. For example, the next known element following zinc (Zn) in the sequence of increasing atomic weight was arsenic (As). But placing As after Zn in the table would place it in the same vertical column with aluminum (Al). The properties of As suggested that it belonged in the column with phosphorus (P), not Al. So two blank spaces were left after Zn, suggesting undiscovered elements.

Mendeleev demonstrated his understanding and daring by predicting that elements would be discovered to fill the gaps in his table and predicting the physical and chemical properties of the yet to be discovered elements. The gaps were directly below boron, aluminum, and silicon in the 1871 version of his table. The gaps below aluminum and silicon are illustrated in figure 12.14. Mendeleev named the elements eka-boron, eka-aluminum, and eka-silicon ("eka" is Sanskrit for "one" and in this context means "next one"). Eka-aluminum was discovered in 1875 and is now called gallium (Ga), eka-boron was discovered in 1879 and is now called scandium (Sc), and eka-silicon was discovered in 1886 and is now called germanium (Ge). There was an impressive correspondence between Mendeleev's predicted properties of these elements and the properties that were observed after their discovery. Mendeleev is usually given credit for developing the periodic table, probably because of his dramatic and highly publicized predictions about the unknown elements.

There were some problems with Mendeleev's original periodic table because not all atomic weights were quite right for the proper placement. And there was no theoretical accounting for similar chemical and physical properties in terms of atomic weight. After the work of Rutherford and Moseley on the atomic nucleus, it became clear that the *atomic number,* not the atomic weight, was the fundamental factor. The atomic number, that is, the number of protons in the nucleus and the number of electrons around the nucleus, is the significant, essential basis for the modern periodic table. Thus the modern **periodic law** is:

Similar physical and chemical properties recur periodically when the elements are listed in order of increasing atomic number.

The Modern Periodic Table

The periodic table is made up of squares, with each element having its own square in a specific location. The squares are arranged in rows and columns but not symmetrically. The arrangement has a meaning, both about atomic structure and about chemical behaviors. The key to meaningful, satisfying use of the table is to understand the code of this structure. The following explains some of what the code means. It will facilitate your understanding of the code if you refer frequently to a periodic table during the following discussion (see the inside back cover of this text and figure 12.20).

An element is identified in each square with its chemical symbol. The number above the symbol is the atomic number of the element, and the number below the symbol is the rounded atomic weight of the element. Horizontal *rows* of elements run from left to right with increasing atomic numbers. Each row is called a **period.** The periods are numbered from 1 to 7 on the left side. Period 1, for example, has only two elements, H (hydrogen) and He (helium). Period 2 starts with Li (lithium) and ends with Ne (neon). The two rows at the bottom of the table are actually part of periods 6 and 7 (between atomic numbers 57 and 72, 89 and 104). They are moved so the table is not so wide.

A vertical *column* of elements is called a **family** (or *group*) of elements. Elements in families have similar properties, but this is true of some families more than others. Note that the families are identified with Roman numerals and letters at the top of each column. Group IIA, for example, begins with Be (beryllium) at the top and has Ra (radium) at the bottom. The A families are in sequence from left to right. The B families are not in sequence, and one group contains more elements than the others (figure 12.15).

Example 12.3

Identify the periodic table (a) period and (b) family of the element silicon.

Solution

According to the list of elements on the inside back cover of this text, silicon has the symbol Si and an atomic number of 14. The square with the symbol Si and the atomic number 14 is located in the third period (third row) and in the column identified as IVA.

Example 12.4

Identify the (a) period and (b) family of the element iron. (Answer: Iron has the symbol Fe and is located in period 4, family VIIIB)

Periodic Patterns

Hydrogen is a colorless, odorless, light gas that burns explosively, combining with oxygen to form water. Helium is a colorless, odorless, light gas that does not burn and, in fact, will not react with other elements at all. Why are hydrogen and

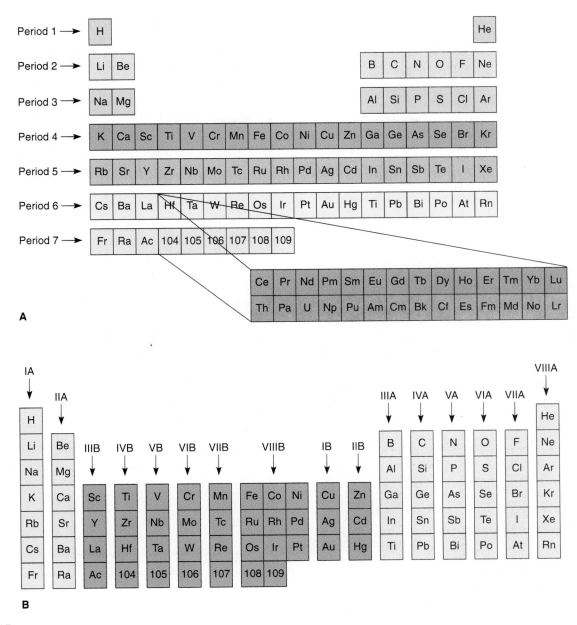

Figure 12.15

(a) Periods of the periodic table, and (b) families of the periodic table.

helium different? Why does one have atoms that are very re-active while the other has atoms that do not react at all? The answer to these questions is found in the atomic structures of the two gases. Recall that the number of protons determines the identity of an atom. Any atom that has one proton is an atom of hydrogen. Any atom that has two protons is an atom of helium. The chemical behavior of the two gases is determined by their electron configuration, which, in turn, is determined by the number of electrons they have.

Since the energy levels of electrons are quantized, they roughly correspond to a series of distances that resemble the spherical layers of an onion. These spherical layers are some-times referred to as **shells.** The term *shell* is used to represent all the electrons with the same value of *n*. The shell concept first came from studies of X-ray spectra, and X-ray vocabulary is used to identify the shells. The shortest X rays are called K X rays, and they were experimentally found to be produced by electrons closest to the nucleus. This energy level closest to the nucleus, *n* = 1, came to be called the K shell, and the shells further out were called the L shell, M shell, N shell, and so on. As noted in chapter 11, the number of electrons that can occupy a given orbital is limited, so *there is a limit to how many electrons can occupy a given shell.* We will return to this important key concept shortly.

Elements and the Periodic Table

As already noted, the first period contains just hydrogen (H) and helium (He). Hydrogen has an atomic number of one, so it has one proton and one electron. This electron is the lowest possible energy level, in the K shell. Helium has two protons in the nucleus so it has two electrons. Both electrons can occupy the K shell, which means that the helium atom has a filled outside shell since a maximum of two electrons is allowed in the K shell. Note that period 1 ends with the filling of the orbital in the K shell.

After helium, lithium (Li) is the next element, and it begins the second period. Li has three electrons since it has an atomic number of three. Two of these electrons are accommodated in the K shell, but the third must go to the next higher level, in the L shell. Lithium is followed by beryllium, boron, carbon, nitrogen, oxygen, fluorine, and neon in order. Each element adds one additional electron to the outermost shell, in this case, the L shell. Note that period 2 ends with the filling of the orbitals in the L shell (table 12.5).

The third period also contains eight elements, from sodium (Na) to argon (Ar). The outer shell is filled just as the elements in the second period, but this time at the third energy level. The third period ends with the filling of the orbitals in the M shell.

By now a couple of patterns are becoming apparent. First, note that the first three periods contain just A families. Each period *begins* with a single electron in a new outer shell. Second, each period *ends* with the filling of an orbital in an outer shell, completing the maximum number of electrons that can occupy that shell. Since the first A family is identified as IA, this means that all the atoms of elements in this family have one electron in their outer shells. All the atoms of elements in family IIA have two electrons in their outer shells. This pattern continues on to family VIIIA, in which all the atoms of elements have eight electrons in their outer shells except helium. Thus the number identifying the A families *also identifies the number of electrons in the outer shells* with the exception of helium. Helium is nonetheless similar to the others in this family since all have filled orbitals in their outer shells. The electron theory of chemical bonding, which is discussed in the next chapter, states that only the electrons in the outermost shell of an atom are involved in chemical reactions. Thus *the outer shell electrons are mostly responsible for the chemical properties of an element.* Therefore, since the members of a family all have similar outer configurations, you would expect them to have similar chemical behaviors, and they do.

The members of the "A-group" families are called the *main-group* or **representative elements.** The members of the "B-group" families are called the **transition elements** (or metals). How the members of the families (figure 12.16) in the representative elements resemble one another will be discussed next.

Table 12.5

Electron configurations for periods 2 and 3

Period 2 From the End of Period 1 Where He: 1s²

Element (Atomic Number and Symbol)	Electron Configuration	Number of Electrons in K (First) Shell	Number of Electrons in L (Second) Shell
Lithium ($_3$Li)	[He] 2s¹	2	1
Beryllium ($_4$Be)	[He] 2s²	2	2
Boron ($_5$B)	[He] 2s²2p¹	2	3
Carbon ($_6$C)	[He] 2s²2p²	2	4
Nitrogen ($_7$N)	[He] 2s²2p³	2	5
Oxygen ($_8$O)	[He] 2s²2p⁴	2	6
Fluorine ($_9$F)	[He] 2s²2p⁵	2	7
Neon ($_{10}$Ne)	[He] 2s²2p⁶	2	8

Period 3 From the End of Period 2 Where Ne: 1s²2s²2p⁶

Element (Atomic Number and Symbol)	Electron Configuration	Electrons in K Shell	Electrons in L Shell	Electrons in M Shell
Sodium ($_{11}$Na)	[Ne] 3s¹	2	8	1
Magnesium ($_{12}$Mg)	[Ne] 3s²	2	8	2
Aluminum ($_{13}$Al)	[Ne] 3s²3p¹	2	8	3
Silicon ($_{14}$Si)	[Ne] 3s²3p²	2	8	4
Phosphorus ($_{15}$P)	[Ne] 3s²3p³	2	8	5
Sulfur ($_{16}$S)	[Ne] 3s²3p⁴	2	8	6
Chlorine ($_{17}$Cl)	[Ne] 3s²3p⁵	2	8	7
Argon ($_{18}$Ar)	[Ne] 3s²3p⁶	2	8	8

Example 12.5

How many outer shell electrons are found in an atom of (a) oxygen, (b) calcium, and (c) aluminum?

Solution

(a) According to the list of elements on the inside back cover of this text, oxygen has the symbol O and an atomic number of 8. The square with the symbol O and the atomic number 8 is located in the column identified as VIA. Since the A family number is the same as the number of electrons in the outer shell, oxygen therefore has six outer shell electrons. (b) Calcium has the symbol Ca (atomic number 20) and is located in column IIA, so a calcium atom has two outer shell electrons. (c) Aluminum has the symbol Al (atomic number 13) and is located in column IIIA, so an aluminum atom has three outer shell electrons.

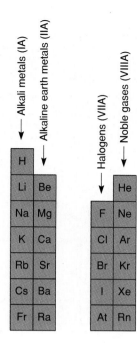

Figure 12.16

Four chemical families of the periodic table: the alkali metals (IA), alkaline earth metals (IIA), halogens (VIIA), and the noble gases (VIIIA).

Chemical Families

As shown in table 12.6, all of the elements in group IA have an outside electron configuration of one electron. With the exception of hydrogen, the IA elements are shiny, low-density metals that are so soft you can cut them easily with a knife. These IA metals are called the **alkali metals** because they react violently with water to form an alkaline solution. The alkali metals do not occur in nature as free elements because they are so reactive. Hydrogen is a unique element in the periodic table. It is not an alkali metal and is placed in the IA group because it seems to fit there with one outer shell electron.

The elements in group IIA all have an outside configuration of two electrons and are called the **alkaline earth metals.** The alkaline earth metals are soft, reactive metals but not as reactive or soft as the alkali metals. Calcium and magnesium are familiar examples of this group.

The elements in group VIIA all have an outside configuration of seven electrons, needing only one more electron to completely fill the outer shell. These elements are called the **halogens.** The halogens are very reactive nonmetals. The halogens fluorine and chlorine are greenish colored gases. Bromine is a reddish brown liquid and iodine is a dark purple solid. Halogens are used as disinfectants, bleaches, and combined with a metal, as a source of light in halogen headlights. Halogens react with metals to form a group of chemicals called salts, such as sodium chloride. In fact, the word halogen is Greek, meaning "salt former."

Table 12.6

Electron structures of the alkali metal family

Element	Electron Configuration	Number of Electrons in Shell						
		1st	2nd	3rd	4th	5th	6th	7th
Lithium (Li)	[He] 2s^1	2	1	—	—	—	—	—
Sodium (Na)	[Ne] 3s^1	2	8	1	—	—	—	—
Potassium (K)	[Ar] 4s^1	2	8	8	1	—	—	—
Rubidium (Rb)	[Kr] 5s^1	2	8	18	8	1	—	—
Cesium (Cs)	[Xe] 6s^1	2	8	18	18	8	1	—
Francium (Fr)	[Rn] 7s^1	2	8	18	32	18	8	1

Table 12.7

Electron structures of the noble gas family

Element	Electron Configuration	Number of Electrons in Shell						
		1st	2nd	3rd	4th	5th	6th	7th
Helium (He)	1s^2	2	—	—	—	—	—	—
Neon (Ne)	[He] 2s^{2}2p^8	2	8	—	—	—	—	—
Argon (Ar)	[Ne] 3s^{2}3p^6	2	8	8	—	—	—	—
Krypton (Kr)	[Ar] 4s^{2}3d^{10}4p^6	2	8	18	8	—	—	—
Xenon (Xe)	[Kr] 5s^{2}4d^{10}5p^6	2	8	18	18	8	—	—
Radon (Rn)	[Xe] 6s^{2}4f^{14}5d^{10}6p^6	2	8	18	32	18	8	—

As shown in table 12.7, the elements in group VIIIA have orbitals that are filled to capacity in the outside shells. These elements are colorless, odorless gases that almost never react with other elements to form compounds. Sometimes they are called the **noble gases** because they are chemically inert, perhaps indicating they are above the other elements. They have also been called the *rare gases* because of this scarcity and *inert gases* because they are mostly chemically inert, not forming compounds. The noble gases are inert because they have filled outer electron configurations, a particularly stable condition.

Example 12.6

(a) To what chemical family does chlorine belong? (b) How many electrons does an atom of chlorine have in its outer shell?

Solution

According to the list of elements on the inside back cover of this text, chlorine has the symbol Cl and an atomic number of 17. The square with the symbol Cl and the atomic number 17 is located in the third period and in the column identified as VIIA. (a) Column VIIA is the chemical family known as the halogens. (b) Each A family number is the same as the number of electrons in the outer shell, so an atom of chlorine has seven electrons in its outer shell.

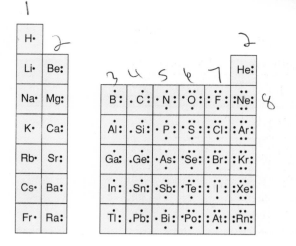

Figure 12.17

Electron dot notation for the representative elements.

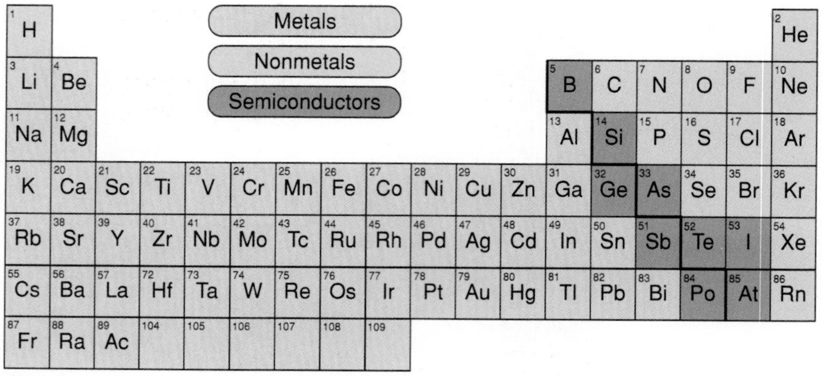

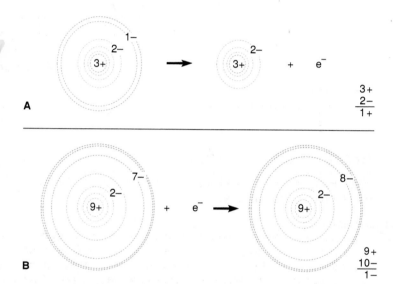

Figure 12.18

The location of metals, nonmetals, and semiconductors in the periodic table.

Metals, Nonmetals, and Semiconductors

As indicated earlier, chemical behavior is mostly concerned with the outer shell electrons. The outer shell electrons, that is, the highest energy level electrons, are conveniently represented with an **electron dot notation.** This notation is made by writing the chemical symbol with dots around it indicating the number of outer shell electrons. Electron dot notations are shown for the representative elements in figure 12.17. Again, note the pattern in figure 12.17—all the noble gases are in group VIIIA and all (except helium) have eight outer electrons. All the group IA elements (alkali metals) have one dot, all the IIA elements have two dots, and so on. This pattern will explain the difference in metals, nonmetals, and a third group of in-between elements called semiconductors.

This chapter began with a discussion of several ways to classify matter according to properties. One example given was to group substances according to the physical properties of metals and nonmetals—luster, conductivity, malleability, and ductility. Metals and nonmetals also have certain chemical properties that are related to their positions in the periodic table. Figure 12.18 shows where the *metals, nonmetals,* and *semiconductors* are located. Note that about 80 percent of all the elements are metals.

The noble gases have completely filled outer orbitals in their highest energy levels, and this is a particularly stable arrangement. Other elements react chemically, either *gaining or losing electrons to attain a filled outermost energy level like the noble gases.* When an atom loses or gains electrons it acquires an unbalanced electron charge called an **ion.** An atom of lithium, for example, has three protons (plus charges) and three electrons (negative charges). If it loses the outermost electron it now has an outer filled orbital structure like helium, a noble gas. It is also now an ion since it has three protons (3+) and two electrons (2−), for a net charge of 1+. A lithium ion thus has a 1+ charge.

Figure 12.19

(*a*) Metals lose their outer electrons to acquire a noble gas structure and become positive ions. Lithium becomes a 1+ ion as it loses its one outer electron. (*b*) Nonmetals gain electrons to acquire an outer noble gas structure and become negative ions. Fluorine gains a single electron to become a 1− ion.

Elements with one, two, or three outer electrons tend to lose electrons to form positive ions. The metals lose electrons like this and the *metals are elements that lose electrons to form positive ions* (figure 12.19). Nonmetals, on the other hand, are elements with five to seven outer electrons that tend to acquire electrons to fill their outer orbitals. *Nonmetals are elements that gain electrons to form negative ions.* In general, elements located in the left two-thirds or so of the periodic table are metals. The nonmetals are on the right side of the table (figure 12.20).

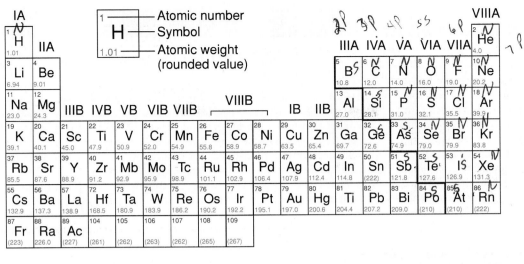

Figure 12.20

The periodic table of the elements.

( ) represents an isotope

Example 12.7

(a) Is strontium a metal, nonmetal, or semiconductor? (b) What is the charge on a strontium ion?

Solution

(a) The list of elements inside the back cover identifies the symbol for strontium as Sr (atomic number 38). In the periodic table, Sr is located in family IIA, which means that an atom of strontium has two electrons in its outer shell. For several reasons, you know that strontium is a metal: (1) An atom of strontium has two electrons in its outer shell and atoms with one, two, or three outer electrons are identified as metals; (2) Strontium is located in the IIA family, which is named the alkaline earth metals, and; (3) Strontium is located on the left side of the periodic table and, in general, elements located in the left two-thirds of the table are metals. (b) Elements with one, two, or three outer electrons tend to lose electrons to form positive ions. Since strontium has an atomic number of 38, you know that it has thirty-eight protons (38+) and thirty-eight electrons (38−). When it loses its two outer shell electrons, it has 38+ and 36− for a charge of 2+.

Example 12.8

(a) Is iodine a metal, nonmetal, or semiconductor? (b) What is the charge on an iodine ion? (Answer: Nonmetal with a charge of 1−)

The dividing line between the metals and nonmetals is a steplike line from the left top of group IIIA down to the bottom left of group VIIA. This is not a line of sharp separation between the metals and nonmetals, and elements *along* this line sometimes act like metals, sometimes act like nonmetals, and sometimes act like both. These ill-defined elements are called **semiconductors** (or *metalloids*). Silicon, germanium, and arsenic have physical properties of nonmetals, for example, they are brittle materials that cannot be hammered into a new shape. Yet these elements conduct electric currents under certain conditions. The ability to conduct an electric current is a property of a metal, and nonmalleability is a property of nonmetals, so as you can see these semiconductors have the properties of both metals and nonmetals.

Some of the elements near the semiconductors exhibit **allotropic forms,** which means the same element can have several different structures with very different physical properties. Carbon atoms, for example, are joined differently in diamond, graphite, and charcoal. All three are made of the element carbon, but all three have different physical properties. They are allotropic forms of carbon. Phosphorus, sulfur, and tin are other elements near the dividing line that also have several allotropic forms.

ACTIVITIES

Make a periodic table. List the elements in boxes on a roll of adding tape. Spiral the tape so that the noble gases appear one below the other. Use cellophane tape to hold the noble gas family together, then cut the tape just to the right of this group. When this cut spiral is spread flat, a long form of the periodic table will be the result. Moving the inner transition elements will produce the familiar short form of the periodic table.

Elements and the Periodic Table

The Rare Earths

Compounds of the rare earths were first identified when they were isolated from uncommon minerals in the late 1700s. The elements are very reactive and have similar chemical properties, so they were not recognized as elements until some fifty years later. Thus they were first recognized as earths, that is, nonmetal substances, when in fact they are metallic elements. They were also considered to be rare since, at that time, they were known to occur only in uncommon minerals. Today, these metallic elements are known to be more abundant in the earth than gold, silver, mercury, or tungsten. The rarest of the rare earths, thulium, is twice as abundant as silver. The rare earth elements are neither rare nor earths, and they are important materials in glass, electronic, and metallurgical industries.

You can identify the rare earths in the two lowermost rows of the periodic table. These rows contain two series of elements that actually belong in the periods 6 and 7, but they are moved below so the entire table is not so wide. Together, the two series are called the inner transition elements. The top series is fourteen elements wide from elements 58 through 71. Since this series belongs next to element 57, lanthanum, it is sometimes called the *lanthanide series*. This series is also known as the rare earths. The second series of fourteen elements is likewise called the *actinide series*. These are mostly the artificially prepared elements that do not occur naturally.

You may have never heard of the rare earth elements, but they are key materials in many advanced or high-technology products. Lanthanum, for example, gives glass special refractive properties and is used in optic fibers and expensive camera lenses. Samarium, neodymium, and dysprosium are used to manufacture crystals used in lasers. Samarium, ytterbium, and terbium have special magnetic properties that have made possible new electric motor designs, magnetic-optical devices in computers, and the creation of a ceramic superconductor. Other rare earth metals are also being researched for use in possible high-temperature superconductivity materials. Many rare earths are also used in metal alloys, for example, an alloy of cerium is used to make heat resistant jet engine parts. Erbium is also used in high-performance metal alloys. Dysprosium and holmium have neutron absorbing properties and are used in control rods to control nuclear fission. Europium should be mentioned because of its role of making the red color of color television screens. The rare earths are relatively abundant metallic elements that play a key role in many common and high-technology applications. They may also play a key role in superconductivity research (see the Feature in chapter 8).

The transition elements, which are all metals, are located in the B group families. Unlike the representative elements, which form vertical families of similar properties, the transition elements tend to form horizontal groups of elements with similar properties. Iron (Fe), cobalt (Co), and nickel (Ni) in group VIIIB, for example, are three horizontally arranged metallic elements that show magnetic properties.

A family of representative elements all form ions with the same charge. Alkali metals, for example, all lose an electron to form a 1+ ion. The transition elements have *variable charges*. Some transition elements, for example, lose their one outer electron to form 1+ ions (copper, silver, gold). Copper, because of its special configuration, can also lose an additional electron to form a 2+ ion. Thus copper can form either a 1+ ion or a 2+ ion. Some transition elements have two outer electrons and lose them both to form 2+ ions (iron, cobalt, nickel). But some of these elements also have special configurations that permit them to lose more of their electrons. Thus iron and cobalt, for example, can form either a 2+ ion or a 3+ ion. Much more can be interpreted from the periodic table, and more generalizations will be made as the table is used in the following chapters.

Summary

Matter can be *classified,* that is, mentally grouped into sets with common properties to make thinking about the wide variety of matter a little less complex. One classification scheme identifies matter as *metals* or *nonmetals* according to the *metallic properties* of *luster, conductivity, malleability, and ductility.* A second classification scheme identifies matter as *solids, liquids, and gases* according to how well it maintains its shape and volume. A third scheme identifies groups of matter according to how it is put together. *Mixtures* are made up of *unlike parts* with a *variable composition. Pure substances* are the *same throughout* and have a *definite composition.* Mixtures can be separated into their components by *physical changes,* changes that do not alter the identity of matter. Some pure substances can be broken down into simpler substances by a *chemical change,* a change that *alters the identity of matter as it produces new substances with different properties.* A pure substance that can be decomposed by chemical change into simpler substances with a definite composition is a *compound.* A pure substance that cannot be broken down into anything simpler is an *element.* There are ninety-one naturally occurring elements, eighteen that have been made artificially, and millions of known compounds.

Elements are identified by *letter symbols,* and each symbol for an element identifies the element and *represents one atom* of that element. Studies of radioactive elements found that elements have identical chemical behaviors with different masses. The name *isotope* was given to atoms of the same element with different masses. The masses of isotopes were accurately measured with a *mass spectroscope,* and new isotopes were discovered for many elements. These isotopes were found to occur in different proportions with a mass very near a whole number.

The mass of each isotope was compared to the mass of the carbon-12 isotope, which was assigned a mass of exactly 12.00 *atomic mass units* (u). The mass contribution of isotopes according to their abundance is used to determine a *weighted average* of all the isotopes of an element. The weighted average is called the *atomic weight* of that element. Isotopes are identified by their *mass number,* defined as the sum of the number of protons and neutrons in the nucleus. The mass number is the closest whole number to the actual atomic mass of an isotope. Isotopes are identified by a chemical symbol with the atomic number (the number of protons) shown as a subscript and the mass number (the number of protons and neutrons) shown as a superscript.

The *periodic law* states that the properties of elements recur periodically when the elements are listed in order of increasing atomic number. The periodic table has horizontal rows of elements called *periods* and vertical columns of elements called *families.* Families have the same outer shell electron configurations, and it is the electron configuration that is mostly responsible for the chemical properties of an element. The *chemical families* of *alkali metals, alkaline earth metals, halogens,* and *noble gases* all have elements with similar properties and identical outer electron arrangements. The outer electrons are represented by *electron dot notations,* which use dots around a chemical symbol to represent the outer electrons.

Elements react chemically to *gain or lose electrons to attain a filled outer orbital structure like the noble gases.* When an atom gains or loses electrons it acquires an imbalanced charge and is called an *ion.* Metals *lose electrons to form positive ions.* Nonmetals *gain electrons to form negative ions.* Elements on the dividing line between metals and nonmetals are *semiconductors.* Elements near the dividing line exhibit *allotropic forms* with different physical properties.

Key Terms

alkaline earth metals (p. 239)	matter (p. 226)
alkali metals (p. 239)	metal (p. 226)
allotropic forms (p. 241)	mixtures (p. 228)
atomic mass units (p. 234)	molecule (p. 232)
atomic weight (p. 234)	noble gases (p. 239)
chemical change (p. 228)	nonmetal (p. 226)
chemistry (p. 226)	period (p. 236)
compound (p. 228)	periodic law (p. 236)
electron dot notation (p. 240)	physical change (p. 228)
element (p. 228)	pure substances (p. 228)
family (p. 236)	representative elements (p. 238)
halogens (p. 239)	semiconductors (p. 241)
ion (p. 240)	shells (p. 237)
isotope (p. 233)	states of matter (p. 226)
mass number (p. 234)	transition elements (p. 238)

(handwritten note in margin: nucleods, mass# = nuetrons + protons)

Applying the Concepts

1. A sample of saltwater solution is homogeneous throughout. Is this sample a mixture or a pure substance?
   a. A pure substance since it is the same throughout.
   b. A mixture because it can have a variable composition.
   c. A pure substance because it has a definite composition.
   d. A mixture because it has a fixed, definite composition.

2. You see the symbols "CO" in a newspaper article. According to the list of elements on the inside back cover of this text, these symbols mean one atom of
   a. cobalt.
   b. copper.
   c. carbon and one atom of oxygen.
   d. copper and one atom of oxygen.

3. Two isotopes of the same element have the same number of
   a. protons, neutrons, and electrons.
   b. protons and neutrons but different numbers of electrons.
   c. protons and electrons but different numbers of neutrons.
   d. neutrons and electrons but different numbers of protons.

4. Atomic weight is
   a. the weight of an atom in grams.
   b. the average atomic mass of the isotopes as they occur in nature.
   c. the number of protons and neutrons in the nucleus.
   d. all of the above.

5. The mass of any isotope is based on the mass of
   a. hydrogen, which is assigned the number 1 since it is the lightest element.
   b. oxygen, which is assigned a mass of 16.
   c. an isotope of carbon, which is assigned a mass of 12.
   d. its most abundant isotope as found in nature.

6. The isotopes of a given element always have
   a. the same mass and the same chemical behavior.
   b. the same mass and a different chemical behavior.
   c. different masses and different chemical behaviors.
   d. different masses and the same chemical behavior.

7. If you want to know the number of protons in an atom of a given element, you would look up the
   a. mass number.
   b. atomic number.
   c. atomic weight.
   d. abundance of isotopes compared to the mass number.

8. If you want to know the number of neutrons in an atom of a given element, you would
   a. round the atomic weight to the nearest whole number.
   b. add the mass number and the atomic number.
   c. subtract the atomic number from the mass number.
   d. add the mass number and the atomic number, then divide by two.

9. Which of the following is always a whole number?
   a. atomic mass of an isotope
   b. mass number of an isotope
   c. atomic weight of an element
   d. none of the above

10. A particular element is located in column IVA of the periodic table. How many dots would be placed around the symbol of this element in its electron dot notation?
    a. 1
    b. 3
    c. 4
    d. 8

Answers
1. b 2. c 3. c 4. b 5. c 6. d 7. b 8. c 9. b
10. c

Elements and the Periodic Table

Questions for Thought

1. What is the periodic law?

2. How was Mendeleev able to predict the chemical and physical properties of elements that were not yet discovered?

3. What is an isotope? Are *all* atoms isotopes? Explain.

4. Which of the following are whole numbers and which are not whole numbers? Explain why for each.
   (a) atomic number
   (b) isotope mass
   (c) mass number
   (d) atomic weight

5. Why does the carbon-12 isotope have a whole number mass but not the other isotopes?

6. Describe how you could determine if a substance is a mixture, a compound, or an element.

7. An element is a brittle solid that cannot be pounded or pulled into a new shape but is an excellent electrical conductor. In what part of the periodic table would you expect to find this element? Explain.

8. What is chemistry?

9. Describe how you know if a substance has undergone a physical change and a chemical change.

10. What two things does a chemical symbol represent?

11. What do the members of the noble gas family have in common? What are the differences?

12. How did Dalton reason the relative atomic masses of certain elements to construct a table?

13. How are the isotopes of an element similar? How are they different?

14. What is electron dot notation? How does it identify metals or nonmetals?

15. Why do metals lose electrons to make positive ions while nonmetals gain electrons to make negative ions?

16. What are the physical properties of metals and nonmetals?

17. A glacier of ice flows downhill over time, moving trees and rocks in its path. According to the definitions of the states of matter, in what state of matter is a glacier?

18. A solution of salt in water is the same throughout, so why is it called a mixture?

19. What is the basic difference between a chemical change and a physical change? Give three examples of each.

20. How many naturally occurring elements are known to exist? How many have been artificially prepared? About how many different compounds are known to exist?

21. What is an ion? How are ions formed?

22. What patterns are noted in the electron structures of elements found in a period and a family in the periodic table?

23. What is a shell?

24. What pattern exists between the identifying number of the A families of the periodic table and the number of electrons in the outer shell of that family?

25. What is a semiconductor?

26. Why do chemical families exist?

Exercises

Group A—Solutions Provided in Appendix D

1. Write the chemical symbols for the following chemical elements:
   (a) Silicon
   (b) Silver
   (c) Helium
   (d) Potassium
   (e) Magnesium
   (f) Iron

2. Lithium has two naturally occurring isotopes, lithium-6 and lithium-7. Lithium-6 has a mass of 6.01512 relative to carbon-12 and makes up 7.42 percent of all naturally occurring lithium. Lithium-7 has a mass of 7.016 compared to carbon-12 and makes up the remaining 92.58 percent. According to this information, what is the atomic weight of lithium?

3. Identify the numbers of protons, neutrons, and electrons in the following isotopes:
   (a) $^{12}_{6}C$
   (b) $^{1}_{1}H$
   (c) $^{40}_{18}Ar$
   (d) $^{2}_{1}H$
   (e) $^{197}_{79}Au$
   (f) $^{235}_{92}U$

Group B—Solutions Not Given

1. Write the chemical symbols for the following chemical elements:
   (a) Argon
   (b) Gold
   (c) Neon
   (d) Sodium
   (e) Calcium
   (f) Tin

2. Boron has two naturally occurring isotopes, boron-10 and boron-11. Boron-10 has a mass of 10.0129 relative to carbon-12 and makes up 19.78 percent of all naturally occurring boron. Boron-11 has a mass of 11.00931 compared to carbon-12 and makes up the remaining 80.22 percent. What is the atomic weight of boron?

3. Identify the numbers of protons, neutrons, and electrons in the following isotopes:
   (a) $^{14}_{7}N$
   (b) $^{7}_{3}Li$
   (c) $^{35}_{17}Cl$
   (d) $^{48}_{20}Ca$
   (e) $^{63}_{29}Cu$
   (f) $^{230}_{92}U$

4. Identify the period and family in the periodic table for the following elements:
   (a) Radon
   (b) Sodium
   (c) Copper
   (d) Neon
   (e) Iodine
   (f) Lead

5. How many outer shell electrons are found in an atom of the following:
   (a) Li ~ 1
   (b) N ~ 5
   (c) F ~ 7
   (d) Cl ~ 7
   (e) Ra ~ 2
   (f) Be ~ 2

6. Write electron dot notations for the following elements:
   (a) Boron
   (b) Bromine
   (c) Calcium
   (d) Potassium
   (e) Oxygen
   (f) Sulfur

7. Identify the charge on the following ions:
   (a) Boron 3+
   (b) Bromine 1-
   (c) Calcium 2+
   (d) Potassium 1+
   (e) Oxygen 2-
   (f) Nitrogen 2-

8. Use the periodic table to identify if the following are metals, nonmetals, or semiconductors:
   (a) Krypton N
   (b) Cesium M
   (c) Silicon S
   (d) Sulfur N
   (e) Molybdenum M
   (f) Plutonium M

9. From their charges, predict the periodic table family number for the following ions:
   (a) Br^{-1} VII A
   (b) K^{+1} I A
   (c) Al^{+3} III A
   (d) S^{-2} VI A
   (e) Ba^{+2} II A
   (f) O^{-2} VI A

10. Use chemical symbols and numbers to identify the following isotopes:
    (a) Oxygen-16 $^{16}_{8}O$
    (b) Sodium-23 ^{23}Na
    (c) Hydrogen-3 $^{3}_{1}H$
    (d) Chlorine-35

    $^{35}_{17}Cl$

4. Identify the period and the family in the periodic table for the following elements:
   (a) Xenon
   (b) Potassium
   (c) Chromium
   (d) Argon
   (e) Bromine
   (f) Barium

5. How many outer shell electrons are found in an atom of the following:
   (a) Na
   (b) P
   (c) Br
   (d) I
   (e) Te
   (f) Sr

6. Write electron dot notations for the following elements:
   (a) Aluminum
   (b) Fluorine
   (c) Magnesium
   (d) Sodium
   (e) Carbon
   (f) Chlorine

7. Identify the charge on the following ions:
   (a) Aluminum
   (b) Chlorine
   (c) Magnesium
   (d) Sodium
   (e) Sulfur
   (f) Hydrogen

8. Use the periodic table to identify if the following are metals, nonmetals, or semiconductors:
   (a) Radon
   (b) Francium
   (c) Arsenic
   (d) Phosphorus
   (e) Hafnium
   (f) Uranium

9. From their charges, predict the periodic table family number for the following ions:
   (a) F^{-1}
   (b) Li^{+1}
   (c) B^{+3}
   (d) O^{-2}
   (e) Be^{+2}
   (f) Si^{+4}

10. Use chemical symbols and numbers to identify the following isotopes:
    (a) Potassium-39
    (b) Neon-22
    (c) Tungsten-184
    (d) Iodine-127

Chapter 13

Compounds and Chemical Change

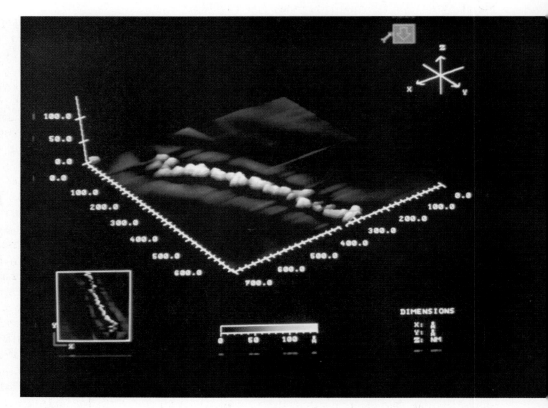

Figure 13.1

This is a scanning electron microscope image of DNA, showing the
actual grooves of the double helix. The scale is in Angstroms
(1 Angstrom = 10^{-10} m), and each twist of the helix is about 35
Angstroms. A molecule like this one determined the characteristics that
other people recognize as you.

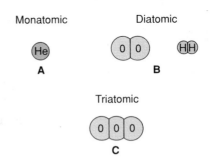

Figure 13.2

(a) The noble gases are monatomic, occurring as single atoms. (b) Many gases, such as hydrogen and oxygen, are diatomic, with two atoms per molecule. (c) Ozone is a form of oxygen that is triatomic, occurring with three atoms per molecule.

IN the previous two chapters you learned how the modern atomic theory is used to describe the structures of atoms of different elements. The electron structures of different atoms successfully account for the position of elements in the periodic table as well as for groups of elements with similar properties. On a large scale all metals were found to have a similarity in electron structure as were nonmetals. On a smaller scale, chemical families such as the alkali metals were found to have the same outer electron configurations. Thus the modern atomic theory accounts for observed similarities between elements in terms of atomic structure.

So far, only individual, isolated atoms have been discussed without considering how atoms of elements join together to produce compounds. There is a relationship between the electron structure of atoms and the reactions they undergo to produce specific compounds. Understanding this relationship will explain the changes that matter itself undergoes. For example, hydrogen is a highly flammable, gaseous element that burns with an explosive reaction. Oxygen, on the other hand, is a gaseous element that supports burning. As you know, hydrogen and oxygen combine to form water. Water is a liquid that neither burns nor supports burning. What happens when atoms of elements such as hydrogen and oxygen join to form molecules such as water? Why do such atoms join and why do they stay together? Why does water have different properties from the elements that combine to produce it? And finally, why is water H_2O and not H_3O or H_4O?

Answers to questions about why and how atoms join together in certain numbers are provided by considering the electronic structures of the atoms. Chemical substances are formed from the interactions of electrons as their structures merge, forming new patterns that result in molecules with new properties (figure 13.1). It is the new electron pattern of the water molecule that gives water different properties than the oxygen or hydrogen from which it formed. Understanding how electron structures of atoms merge to form new patterns is understanding the changes that matter itself undergoes, the topic of this chapter.

Compounds and Chemical Change

The air you breath, the liquids you drink, and all the things around you are elements, compounds, or mixtures. Most are compounds, however, and very few are pure elements. Water, sugar, gasoline, and chalk are examples of compounds. Each can be broken down into the elements that make it up. Recall that elements are basic substances that cannot be broken down into simpler substances. Examples of elements are hydrogen, carbon, and calcium. Why and how these elements join together in different ways to form different compounds is the subject of this chapter.

You have already learned that elements are made up of atoms that can be described by the modern atomic theory. You can also consider an **atom** to be *the smallest unit of an element that can exist alone or in combination with other elements*. Compounds are formed when atoms are held together by an attractive force called a *chemical bond*. The chemical bond binds individual atoms together in a compound. A molecule is generally thought of as a tightly bound group of atoms that maintains its identity. More specifically, a **molecule** is defined as *the smallest particle of a compound, or a gaseous element, that can exist and still retain the characteristic properties of a substance*. Compounds with one type of chemical bond, as you will see, have molecules that are electrically neutral groups of atoms held together strongly enough to be considered independent units. For example, water is a compound. The smallest unit of water that can exist alone is an electrically neutral unit made up of two hydrogen atoms and one oxygen atom held together by chemical bonds. The concept of a molecule will be expanded as chemical bonds are discussed.

Compounds occur naturally as gases, liquids, and solids. Many common gases occur naturally as molecules made up of two or more atoms. For example, at ordinary temperatures hydrogen gas occurs as molecules of two hydrogen atoms bound together. Oxygen gas also usually occurs as molecules of two oxygen atoms bound together. Both hydrogen and oxygen occur naturally as *diatomic molecules* ("di-" means "two"). Oxygen sometimes occurs as molecules of three oxygen atoms bound together. These *triatomic molecules* ("tri-" means "three") are called *ozone*. The noble gases are unique, occurring as single atoms called *monatomic* ("mon-" or "mono-" means "one") (figure 13.2). These monatomic particles are sometimes called monatomic molecules since they are the smallest unit of the noble gases that can exist alone. Helium and neon are examples of the monatomic noble gases.

When multiatomic molecules of any size are formed or broken down into simpler substances, new materials with new properties are produced. This kind of a change in matter is called a chemical change, and the process is called a chemical reaction. A **chemical reaction** is defined as

a change in matter where different chemical substances are created by forming or breaking chemical bonds.

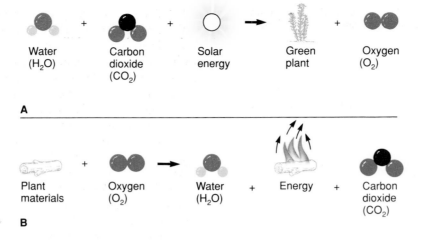

Figure 13.3

(a) New chemical bonds are formed as a green plant makes new materials and stores solar energy through the photosynthesis process. (b) The chemical bonds are later broken and the same amount of energy and the same original materials are released. The same energy and the same materials are released rapidly when the plant materials burn, and they are released slowly when the plant decomposes.

In general, chemical bonds are formed when atoms of elements are bound together to form compounds. Chemical bonds are broken when a compound is decomposed into simpler substances.

Chemical reactions happen all the time, all around you. A growing plant, burning fuels, and your body's utilization of food all involve chemical reactions. These reactions produce different chemical substances with greater or smaller amounts of internal potential energy (see chapter 4 for a discussion of internal potential energy). Energy is *absorbed* to produce new chemical substances with more internal potential energy. Energy is *released* when new chemical substances are produced with less internal potential energy. In general, changes in internal potential energy are called **chemical energy**. For example, new chemical substances are produced in green plants through the process called photosynthesis. A green plant uses radiant energy (sunlight), carbon dioxide, and water to produce new chemical materials and oxygen. These new chemical materials, the stuff that leaves, roots, and wood are made of, contain more chemical energy than the carbon dioxide and water they were made from.

A **chemical equation** is a way of describing what happens in a chemical reaction. Later, you will learn how to use formulas in a chemical reaction. For now, the chemical reaction of photosynthesis will be described by using words in an equation:

$$\text{energy (sunlight)} + \text{carbon dioxide molecules} + \text{water molecules} \rightarrow \text{plant material molecules} + \text{oxygen molecules}$$

The substances that are changed are on the left side of the word equation and are called *reactants*. The reactants are carbon dioxide molecules and water molecules. The equation also indicates that energy is absorbed since the term *energy* appears on the left side. The arrow means *yields*. The new chemical substances are on the right side of the word equation and are called *products*. Reading the photosynthesis reaction as a sentence, "Carbon dioxide and water use energy to react, yielding plant materials and oxygen."

The plant materials produced by the reaction have more internal potential energy, also known as chemical energy, than the reactants. You know this from the equation because the term *energy* appears on the left side but not the right. This means that the energy on the left went into internal potential energy on the right. You also know this because the reaction can be reversed to release the stored energy (figure 13.3). When plant materials (such as wood) are burned, the materials react with oxygen and chemical energy is released in the form of radiant energy (light) and high kinetic energy of the newly formed gases and vapors. In words,

$$\text{plant material molecules} + \text{oxygen molecules} \rightarrow \text{carbon dioxide molecules} + \text{water molecules} + \text{energy}$$

If you compare the two equations, you will see that burning is the opposite process of photosynthesis! The energy released in burning is the exact same amount of solar energy that was stored as internal potential energy by the plant. Such chemical changes where chemical energy is stored in one reaction and released by another reaction are the result of the making, then the breaking, of chemical bonds. Chemical bonds were formed by utilizing energy to produce new chemical substances. Energy was released when these bonds were broken to produce the original substances. Chemical reactions and energy flow can be explained by the making and breaking of chemical bonds. Chemical bonds can be explained in terms of changes in the electron structures of atoms. Thus the place to start in seeking understanding about chemical reactions is the electron structure of the atoms themselves.

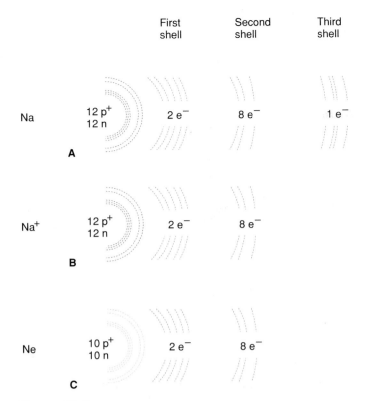

		First shell	Second shell	Third shell
Na	12 p⁺ 12 n	2 e⁻	8 e⁻	1 e⁻
	A			
Na⁺	12 p⁺ 12 n	2 e⁻	8 e⁻	
	B			
Ne	10 p⁺ 10 n	2 e⁻	8 e⁻	
	C			

Figure 13.4

(*a*) A sodium atom has two electrons in the first energy level, eight in the second energy level, and one in the third level. (*b*) When it loses its one outer, or valence, electron, it becomes a sodium ion with the same electron structure as an atom of neon (*c*).

Valence Electrons and Ions

As discussed in chapter 12, it is the number of electrons in the outermost shell that usually determines the chemical properties of an atom. These outer electrons are called **valence electrons,** and it is the valence electrons that participate in chemical bonding. The inner electrons are in stable, fully occupied orbitals and do not participate in chemical bonds. The representative elements have valence electrons in the outermost orbitals, which contain from one to eight valence electrons. Recall that you can easily find the number of valence electrons by referring to a periodic table. The number at the top of each representative family is the same as the number of outer shell electrons (with the exception of helium).

The noble gases have filled outer orbitals and do not normally form compounds. Apparently half-filled and filled orbitals are particularly stable arrangements. Atoms have a tendency to seek such a stable, filled outer orbital arrangement such as the one found in the noble gases. For the representative elements, this tendency is called the **octet rule.** The octet rule states that *atoms attempt to acquire an outer orbital with eight electrons* through chemical reactions. This rule is a generalization, and a few elements do not meet the requirement of "eight" but do seek the same general trend of stability. There are a few other exceptions, and the octet rule should be considered a generalization that helps keep track of the valence electrons in most representative elements.

The periodic table representative element family number tells you the number of valence electrons and what the atom must do to reach the stability suggested by the octet rule. For example, consider sodium (Na). Sodium is in family IA, so sodium has one valence electron. If the sodium atom can get rid of this outer valence electron through a chemical reaction, it will have the same outer electron configuration as an atom of the noble gas neon (Ne) (compare figures 13.4b and 13.4c).

When a sodium atom (Na) loses an electron to form a sodium ion (Na^+) it has the same, stable outer electron configuration as a neon atom (Ne). The sodium ion (Na^+) is still a form of sodium since it still has eleven protons. But it is now a sodium *ion,* not a sodium *atom,* since it has eleven protons (eleven positive charges) and now has ten electrons (ten negative charges) for a total of

$$\begin{array}{l} 11+ \text{ (protons)} \\ \underline{10- \text{ (electrons)}} \\ 1+ \text{ (net charge on sodium ion)} \end{array}$$

This charge is shown on the chemical symbol of Na^+ and is called a *sodium ion.* Note that the sodium nucleus and the inner orbitals do not change when the sodium atom is ionized. The sodium ion is formed when a sodium atom loses its valence electron, and the process can be described by

$$\text{energy} + Na\cdot \rightarrow Na^+ + e^- \qquad \textbf{equation 13.1}$$

where Na· is the electron dot symbol for sodium and the e^- is the electron that has been pulled off the sodium atom.

Example 13.1

What is the symbol and charge for a calcium ion?

Solution

From the list of elements on the inside back cover, the symbol for calcium is Ca and the atomic number is 20. The periodic table tells you that Ca is in family IIA, which means that calcium has 2 valence electrons. According to the octet rule, the calcium ion must lose 2 electrons to acquire the stable outer arrangement of the noble gases. Since the atomic number is 20 a calcium atom has 20 protons (20+) and 20 electrons (20−). When it is ionized, the calcium ion will lose 2 electrons for a total charge of (20+) + (18−), or 2+. The calcium ion is represented by the chemical symbol for calcium and the charge shown as a superscript: Ca^{2+}.

Example 13.2

What is the symbol and charge for an aluminium ion? (Answer: Al^{3+})

Chemical Bonds

Atoms gain or lose electrons through a chemical reaction to achieve a state of lower energy, the stable electron arrangement of the noble gas atoms. Such a reaction results in a **chemical**

Compounds and Chemical Change

bond, an *attractive force that holds atoms together in a compound.* There are three general classes of chemical bonds: (1) ionic bonds, (2) covalent bonds, and (3) metallic bonds.

Ionic bonds are formed when atoms *transfer* electrons to achieve the noble gas electron arrangement. Electrons are given up or acquired in the transfer, forming positive and negative ions. The electrostatic attraction between oppositely charged ions forms ionic bonds, and ionic compounds are the result. In general, ionic compounds are formed when a metal from the left side of the periodic table reacts with a nonmetal from the right side.

Covalent bonds result when atoms achieve the noble gas electron structure by *sharing* electrons. Covalent bonds are generally formed between the nonmetallic elements on the right side of the periodic table.

Metallic bonds are formed in solid metals such as iron, copper, and the other metallic elements that make up about 80 percent of all the elements. The atoms of metals are closely packed and share many electrons in a "sea" that is free to move throughout the metal, from one metal atom to the next. Metallic bonding accounts for metallic properties such as high electrical conductivity.

Ionic, covalent, and metallic bonds are attractive forces that hold atoms or ions together in molecules and crystals. There are two ways to describe what happens to the electrons when one of these bonds is formed, by considering (1) the new patterns formed when atomic orbitals overlap to form a combined orbital called a *molecular orbital* or (2) the atoms in a molecule as *isolated atoms* with changes in their outer shell arrangements. The molecular orbital description considers that the electrons belong to the whole molecule and form a molecular orbital with its own shape, orientation, and energy levels. The isolated atom description considers the electron energy levels as if the atoms in the molecule were alone, isolated from the molecule. The isolated atom description is less accurate than the molecular orbital description, but it is less complex and more easily understood. Thus the following details about chemical bonding will mostly consider individual atoms and ions in compounds.

Ionic Bonds

An **ionic bond** is defined as the *chemical bond of electrostatic attraction* between negative and positive ions. Ionic bonding forms when an atom of a metal reacts with an atom of a nonmetal. The reaction results in a transfer of one or more valence electrons from the metal atom to the valence shell of the nonmetal atom. The atom that loses electrons becomes a positive ion, and the atom that gains electrons becomes a negative ion. Oppositely charged ions attract one another, and when pulled together, they form an ionic solid with the ions arranged in an orderly geometric structure. This results in a crystalline solid that is typical of salts such as sodium chloride (figure 13.5).

As an example of ionic bonding, consider the reaction of sodium (a soft reactive metal) with chlorine (a pale yellow-green gas). When an atom of sodium and an atom of chlorine collide,

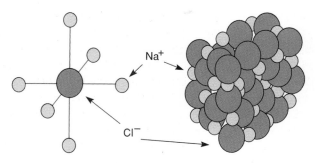

Figure 13.5

Sodium chloride crystals are composed of sodium and chlorine ions held together by electrostatic attraction. Each sodium ion is surrounded by six chlorine ions, and each chlorine ion is surrounded by six sodium ions. A crystal builds up like this, giving the sodium chloride crystal a cubic structure.

they react violently as the valence electron is transferred from the sodium to the chlorine atom. This produces a sodium ion and a chlorine ion. The reaction can be illustrated with electron dot symbols as follows:

equation 13.2

$$\text{Na} \cdot \ + \ \cdot \ddot{\underset{\cdot\cdot}{\text{Cl}}} \colon \ \longrightarrow \ \text{Na}^+ \ \left(\colon \ddot{\underset{\cdot\cdot}{\text{Cl}}} \colon \right)^-$$

As you can see, the sodium ion transferred its valence electron and the resulting ion now has a stable electron configuration. The chlorine atom accepted the electron in its outer orbital to acquire a stable electron configuration. Thus a stable positive ion and a stable negative ion are formed. Because of opposite electrical charges, the ions attract each other to produce an ionic bond. When many ions are involved, each Na^+ ion is surrounded by six Cl^- ions and each Cl^- ion is surrounded by six Na^+ ions. This gives the resulting solid NaCl its crystalline cubic structure as shown in figure 13.5. In the solid state all the sodium ions and all the chlorine ions are bound together in one giant unit. Thus the term "molecule" is not really appropriate for ionic solids such as sodium chloride. But the term is sometimes used anyway since any given sample will have the same number of Na^+ ions as Cl^- ions.

Energy and Electrons in Ionic Bonding

The sodium-chloride reaction can be represented with electron dot notation as occurring in three steps:

1. $\text{energy} \ + \ \text{Na} \cdot \ \longrightarrow \ \text{Na}^+ \ + \ e^-$

2. $\cdot \ddot{\underset{\cdot\cdot}{\text{Cl}}} \colon \ + \ e^- \ \longrightarrow \ \left(\colon \ddot{\underset{\cdot\cdot}{\text{Cl}}} \colon \right)^- \ + \ \text{energy}$

3. $\text{Na}^+ \ + \ \left(\colon \ddot{\underset{\cdot\cdot}{\text{Cl}}} \colon \right)^- \ \longrightarrow \ \text{Na}^+ \ \left(\colon \ddot{\underset{\cdot\cdot}{\text{Cl}}} \colon \right)^- \ + \ \text{energy}$

The energy released in steps 2 and 3 is greater than the energy absorbed in step 1, and an ionic bond is formed. The energy released is called the **heat of formation.** It is also the amount of energy required to decompose the compound (sodium chloride) into its elements. The reaction does not take place in steps as described, however, but occurs all at once. Note again, as in the photosynthesis-burning reactions described earlier, that the total amount of chemical energy is conserved. The energy released by the formation of the sodium chloride compound is the *same* amount of energy needed to decompose the compound.

Ionic bonds are formed by electron transfer, and electrons are conserved in the process. This means that electrons are not created or destroyed in a chemical reaction. The same total number of electrons exists after a reaction that existed before the reaction. There are two rules you can use for keeping track of electrons in ionic bonding reactions:

1. Ions are formed as atoms gain or lose valence electrons to achieve the stable noble gas structure.
2. There must be a balance between the number of electrons lost and the number of electrons gained by atoms in the reaction.

The sodium-chlorine reaction follows these two rules. The loss of one valence electron from a sodium atom formed a stable sodium ion. The gain of one valence electron by the chlorine atom formed a stable chlorine ion. Thus both ions have noble gas configurations (rule 1), and one electron was lost and one was gained, so there is a balance in the number of electrons lost and the number gained (rule 2).

Ionic Compounds and Formulas

The **formula** of a compound *describes what elements are in the compound and in what proportions*. Sodium chloride contains one positive sodium ion for each negative chlorine ion. The formula of the compound sodium chloride is NaCl. If there are no subscripts at the lower right part of each symbol, "1" is understood. Thus NaCl indicates a compound made up of the elements sodium and chlorine, and there is one sodium atom for each chlorine atom.

Calcium (Ca) is an alkaline metal in family IIA, and fluorine (F) is a halogen in family VIIA. Since calcium is a metal and fluorine is a nonmetal, you would expect calcium and fluorine atoms to react, forming a compound with ionic bonds. Calcium has two valence electrons to lose to acquire a noble gas configuration. Fluorine needs one valence electron to acquire a noble gas configuration. So calcium needs to lose two electrons and fluorine needs to gain one electron to achieve a stable configuration (rule 1). Two fluorine atoms, each acquiring one electron, are needed to balance the number of electrons lost and the number of electrons gained. The compound formed from the reaction, calcium fluoride, will therefore have a calcium ion with a charge of plus two for every fluorine ion with a charge of minus one. Recalling that electron dot symbols show only the outer valence electrons, you can see that the reaction is

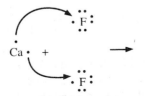

which shows that a calcium atom transfers two electrons, one each to two fluorine atoms. Now showing the results of the reaction, a calcium ion is formed from the loss of two electrons (charge 2+) and two fluorine ions are formed by gaining one electron each (charge 1−):

equation 13.3

The formula of the compound is therefore CaF_2, with the subscript 2 for fluorine and the understood subscript 1 for calcium. This means that there are two fluorine atoms for each calcium atom in the compound.

Sodium chloride (NaCl) and magnesium fluoride (MgF_2) are examples of compounds held together by ionic bonds. Compounds held together by ionic bonds are called **ionic compounds.** Ionic compounds of the representative elements are generally white, crystalline solids that form colorless solutions. Sodium chloride, the most common example, is common table salt. Many of the transition elements form colored compounds that make colored solutions. Ionic compounds dissolve in water, producing a solution of ions that can conduct an electric current.

In general, the elements in families IA and IIA of the periodic table tend to form positive ions by losing electrons. The ion charge for these elements equals the family number of these elements. The elements in families VIA and VIIA tend to form negative ions by gaining electrons. The ion charge for these elements equals their family number minus eight. The elements in families IIIA and VA have less of a tendency to form ionic compounds except those in higher periods. Common ions of representative elements are given in table 13.1. The transition elements form positive ions of several different charges. Some common ions of the transition elements are listed in table 13.2.

The single-charge representative elements and the variable-charge transition elements form single, monatomic negative ions. There are also many polyatomic ("poly" means "many") negative ions, charged groups of atoms that act like a single unit in ionic compounds. Polyatomic ions are held together by covalent bonds, which will be discussed in the next section.

Example 13.3

Use electron dot notation to predict the formula of a compound formed when aluminum (Al) combines with fluorine (F).

Solution

Aluminum, atomic number 13, is in family IIIA so it has three valence electrons and an electron dot notation of

$$\dot{Al}\cdot$$

According to the octet rule, the aluminum atom would need to lose three electrons to acquire the stable noble gas configuration. Fluorine, atomic number 9, is in family VIIA so it has seven valence electrons and an electron dot notation of

$$\cdot \ddot{F} \colon$$

Table 13.1

Common ions of representative elements

Element	Symbol	Ion
Lithium	Li	1+
Sodium	Na	1+
Potassium	K	1+
Magnesium	Mg	2+
Calcium	Ca	2+
Barium	Ba	2+
Aluminum	Al	3+
Oxygen	O	2−
Sulfur	S	2−
Hydrogen	H	1+, 1−
Fluorine	F	1−
Chlorine	Cl	1−
Bromine	Br	1−
Iodine	I	1−

Table 13.2

Common ions of transition elements

Single-Charge Ions		
Element	Symbol	Charge
Zinc	Zn	2+
Tungsten	W	6+
Silver	Ag	1+
Cadmium	Cd	2+

Variable-Charge Ions		
Element	Symbol	Charge
Chromium	Cr	2+, 3+, 6+
Manganese	Mn	2+, 4+, 7+
Iron	Fe	2+, 3+
Cobalt	Co	2+, 3+
Nickel	Ni	2+, 3+
Copper	Cu	1+, 2+
Tin	Sn	2+, 4+
Gold	Au	1+, 3+
Mercury	Hg	1+, 2+
Lead	Pb	2+, 4+

Fluorine would acquire a noble gas configuration by accepting one electron. Three fluorine atoms, each acquiring one electron, are needed to balance the three electrons lost by aluminum. The reaction can be represented as

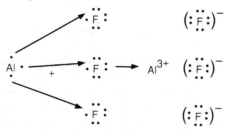

The ratio of aluminum atoms to fluorine atoms in the compound is 1:3. The formula for aluminum fluoride is therefore AlF_3.

Example 13.4

Predict the formula of the compound formed between aluminum and oxygen using electron dot notation. (Answer: Al_2O_3)

Covalent Bonds

Most substances do not have the properties of ionic compounds since they are not composed of ions. Most substances are molecular, composed of electrically neutral groups of atoms that are tightly bound together. As noted earlier, many gases are diatomic, occurring naturally as two atoms bound together as an electrically neutral molecule. Hydrogen, for example, occurs as molecules of H_2 and no ions are involved. The hydrogen atoms are held together by a covalent bond. A **covalent bond** is a *chemical bond formed by the sharing of a pair of electrons.* In the

diatomic hydrogen molecule each hydrogen atom contributes a single electron to the shared pair. Both hydrogen atoms count the shared pair of electrons in achieving their noble gas configuration. Hydrogen atoms both share one pair of electrons, but other elements might share more than one pair to achieve a noble gas structure.

Consider how the covalent bond forms between two hydrogen atoms by imagining two hydrogen atoms moving toward one another. Each atom has a single electron. As the atoms move closer and closer together, their orbitals begin to overlap. Each electron is attracted to the oppositely charged nucleus of the other atom and the overlap tightens. Then the repulsive forces from the like-charged nuclei will halt the merger. A state of stability is reached between the two nuclei and two electrons, and a H_2 molecule has been formed. The two electrons are now shared by both atoms, and the attraction of one nucleus for the other electron and vice versa holds the atoms together (figure 13.6).

Covalent Compounds and Formulas

Electron dot notation can be used to represent the formation of covalent bonds. For example, the joining of two hydrogen atoms to form a H_2 molecule can be represented as:

$$H\cdot \; + \; H\cdot \longrightarrow H:H$$

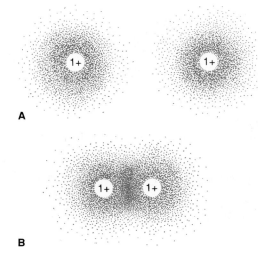

A

B

Figure 13.6

(a) Two hydrogen atoms, each with its own probability distribution of electrons about the nucleus. (b) When the hydrogen atoms bond, a new electron distribution pattern forms around the entire molecule and both electrons occupy the molecular orbital.

Since an electron pair is *shared* in a covalent bond the two electrons move throughout the entire molecular orbital. Since each hydrogen atom now has both electrons on an equal basis, each can be considered to now have the noble gas configuration of helium. A dashed circle around each symbol shows that both atoms have two electrons:

$$\text{H} \cdot \ + \ \text{H} \cdot \ \longrightarrow \ (\text{H} \colon \text{H}) \qquad \textbf{equation 13.4}$$

Hydrogen and fluorine react to form a covalent molecule (how this is known will be discussed shortly), and this bond can be represented with electron dots. Fluorine is in the VIIA family, so you know an atom of fluorine has seven valence electrons in the outermost energy level. The reaction is

$$\text{H} \cdot \ + \ \cdot \overset{\cdot\cdot}{\underset{\cdot\cdot}{\text{F}}} \colon \ \longrightarrow \ (\text{H} \colon \overset{\cdot\cdot}{\underset{\cdot\cdot}{\text{F}}} \colon) \qquad \textbf{equation 13.5}$$

Each atom shares a pair of electrons to achieve a noble gas configuration. Hydrogen achieves the helium configuration and fluorine achieves the neon configuration. All the halogens have seven valence electrons like this, and all need to gain one electron (ionic bond) or share an electron pair (covalent bond) to achieve a noble gas configuration. This also explains why the halogen gases occur as diatomic molecules. Two fluorine atoms can achieve a noble gas configuration by sharing a pair of electrons:

$$\cdot \overset{\cdot\cdot}{\underset{\cdot\cdot}{\text{F}}} \colon \ + \ \cdot \overset{\cdot\cdot}{\underset{\cdot\cdot}{\text{F}}} \colon \ \longrightarrow \ (\overset{\cdot\cdot}{\underset{\cdot\cdot}{\text{F}}} \colon \overset{\cdot\cdot}{\underset{\cdot\cdot}{\text{F}}} \colon) \qquad \textbf{equation 13.6}$$

Each fluorine atom thus achieves the neon configuration by bonding together. Note that there are two types of electron pairs: (1) orbital pairs and (2) bonding pairs. Orbital pairs are not shared since they are the two electrons in an orbital, each with

Table 13.3

Structures and compounds of nonmetal elements combined with hydrogen

Nonmetallic Elements	Element (E Represents Any Element of Family)	Compound
Family IVA: C, Si, Ge	$\cdot \overset{\cdot}{\underset{\cdot}{\text{E}}} \cdot$	$\text{H} \colon \overset{\text{H}}{\underset{\text{H}}{\text{E}}} \colon \text{H}$
Family VA: N, P, As, Sb	$\cdot \overset{\cdot\cdot}{\underset{\cdot}{\text{E}}} \cdot$	$\text{H} \colon \overset{\cdot\cdot}{\underset{\text{H}}{\text{E}}} \colon \text{H}$
Family VIA: O, S, Se, Te	$\cdot \overset{\cdot\cdot}{\underset{\cdot\cdot}{\text{E}}} \cdot$	$\text{H} \colon \overset{\cdot\cdot}{\underset{\cdot\cdot}{\text{E}}} \colon \text{H}$
Family VIIA: F, Cl, Br, I	$\cdot \overset{\cdot\cdot}{\underset{\cdot\cdot}{\text{E}}} \colon$	$\text{H} \colon \overset{\cdot\cdot}{\underset{\cdot\cdot}{\text{E}}} \colon$

a separate spin. Orbital pairs are also called *lone pairs* since they are not shared. *Bonding pairs,* as the name implies, are the electron pairs shared between two atoms. Considering again the F_2 molecule,

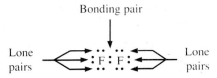

Bonding pair

Lone pairs ← :F: F: → Lone pairs

Often, the number of bonding pairs that are formed by an atom is the same as the number of single, *unpaired* electrons in the atomic electron dot notation. For example, hydrogen has one unpaired electron and oxygen has two unpaired electrons. Hydrogen and oxygen combine to form an H_2O molecule as

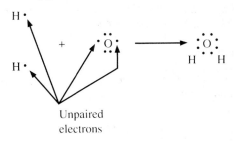

Unpaired electrons

The diatomic hydrogen (H_2) and fluorine (F_2), hydrogen fluoride (HF), and water (H_2O) are examples of compounds held together by covalent bonds. A compound held together by covalent bonds is called a **covalent compound.** In general, covalent compounds form from nonmetal elements on the right side of the periodic table. For elements in families IVA through VIIA, the number of unpaired electrons (and thus the number of covalent bonds formed) is eight minus the family number. You can get a lot of information from the periodic table from generalizations like this one. For another generalization, compare table 13.3 with the periodic table. The table gives the structures of nonmetals combined with hydrogen and the resulting compounds.

Compounds and Chemical Change

Multiple Bonds

Two dots can represent a lone pair of valence electrons or it can represent a bonding pair, a single pair of electrons being shared by two atoms. Bonding pairs of electrons are often represented by a simple line between two atoms. For example,

$$H : H \qquad \text{is shown as} \qquad H - H$$

and

$$\overset{..}{\underset{..}{:}} \overset{..}{O} \overset{..}{:} \qquad \text{is shown as} \qquad \overset{O}{\underset{H \quad H}{/ \ \backslash}}$$

Note that the line between the two hydrogen atoms represents an electron pair, so each hydrogen atom has two electrons in the outer shell like helium. In the water molecule each hydrogen atom has two electrons as before. The oxygen atom has two lone pairs (a total of four electrons) and two bonding pairs (a total of four electrons) for a total of eight electrons. Thus oxygen has acquired a stable octet of electrons.

A covalent bond in which a single pair of electrons is shared by two atoms is called a *single covalent bond* or simply a **single bond.** Some atoms have two unpaired electrons and can share more than one electron pair. A **double bond** is a covalent bond formed when *two pairs* of electrons are shared by two atoms. This happens mostly in compounds involving atoms of the elements C, N, O, and S. Ethylene, for example, is a gas given off from ripening fruit. The electron dot formula for ethylene is

$$\overset{H}{\underset{H}{}} : C :: C : \overset{H}{\underset{H}{}} \qquad \text{or} \qquad \overset{H}{\underset{H}{>}} C = C \overset{H}{\underset{H}{<}}$$

The ethylene molecule has a double bond between two carbon atoms. Since each line represents two electrons, you can simply count the lines around each symbol to see if the octet rule has been satisfied. Each H has one line, so each H atom is sharing two electrons. Each C has four lines so each C atom has eight electrons, satisfying the octet rule.

A **triple bond** is a covalent bond formed when *three pairs* of electrons are shared by two atoms. Triple bonds occur mostly in compounds with atoms of the elements C and N. Acetylene, for example, is a gas often used in welding torches. The electron dot formula for acetylene is

$$H : C ::: C : H \qquad \text{or} \qquad H - C \equiv C - H$$

The acetylene molecule has a triple bond between two carbon atoms. Again, note that each line represents two electrons. Each C atom has four lines, so the octet rule is satisfied.

Coordinate Covalent Bonds

The single, double, and triple covalent bonds are formed when an atom shares one, two, or three pairs of electrons. For each pair each atom contributes one electron and another atom shares one of its electrons in return. There is another type of covalent bond that is a "hole and plug" kind of sharing. A **coordinate covalent bond** is formed when *the shared electron pair is contributed by one atom.* The coordinate covalent bond is like the other covalent bonds since a pair of electrons is shared between two atoms. The difference is that both of these electrons come from one atom, not one each from two atoms. Ammonia, for example, has an electron dot structure of

$$\begin{array}{c} \text{Lone} \\ \text{pair} \end{array} \longrightarrow \overset{H}{\underset{H}{: N : H}}$$

which appears to be a stable structure since the octet rule is satisfied. However, notice the lone pair of electrons on the ammonia molecule. Ammonia can contribute its pair of nonbonding electrons to, say, a hydrogen ion, which shares the pair with ammonia

$$H^+ \quad + \quad \overset{H}{\underset{H}{: N : H}} \quad \longrightarrow \quad \left[\overset{H}{\underset{H}{H : N : H}} \right]^+ \qquad \textbf{equation 13.7}$$

forming an ammonium ion. All of the covalent bonds with hydrogen are now identical, with each hydrogen atom sharing a pair of electrons. The molecule is now an ion, however, since it has a net charge. An ion made up of many atoms is called a **polyatomic ion.** Coordinate covalent bonding is common in many polyatomic ions. These ions are important in many common chemicals, which will be discussed later. Some of the common polyatomic ions are listed in table 13.4. Note that (1) all have negative charges except the ammonium ion, (2) all are formed exclusively of nonmetals except three that contain metals, and (3) some are similar with different "-ite" and "-ate" endings. The "-ate" ion always has one more oxygen than the "-ite" ion.

Bond Polarity

How do you know if a bond between two atoms will be ionic or covalent? In general, ionic bonds form between metal atoms and nonmetal atoms, especially those from the opposite sides of the periodic table. Also in general, covalent bonds form between the atoms of nonmetals. If an atom has a much greater electron-pulling ability than another atom, the electron is pulled completely away from the atom with lesser pulling ability and an ionic bond is the result. If the electron-pulling ability is more even between the two atoms, the electron is shared and a covalent bond results. As you can imagine, all kinds of reactions

Table 13.4

Some common polyatomic ions

Ion Name	Formula
Acetate	$(C_2H_3O_2)^{1-}$
Ammonium	$(NH_4)^{1+}$
Borate	$(BO_3)^{3-}$
Carbonate	$(CO_3)^{2-}$
Chlorate	$(ClO_3)^{1-}$
Chromate	$(CrO_4)^{2-}$
Cyanide	$(CN)^{1-}$
Dichromate	$(Cr_2O_7)^{2-}$
Hydrogen carbonate (or bicarbonate)	$(HCO_3)^{1-}$
Hydrogen sulfate (or bisulfate)	$(HSO_4)^{1-}$
Hydroxide	$(OH)^{1-}$
Hypochlorite	$(ClO)^{1-}$
Nitrate	$(NO_3)^{1-}$
Nitrite	$(NO_2)^{1-}$
Perchlorate	$(ClO_4)^{1-}$
Permanganate	$(MnO_4)^{1-}$
Phosphate	$(PO_4)^{3-}$
Phosphite	$(PO_3)^{3-}$
Sulfate	$(SO_4)^{2-}$
Sulfite	$(SO_3)^{2-}$

are possible between atoms with different combinations of electron-pulling abilities. The result is that it is possible to form many gradations of bonding between completely ionic and completely covalent bonding. Which type of bonding will result can be found by comparing the electronegativity of the elements involved. **Electronegativity** is the *comparative ability of atoms of an element to attract bonding electrons*. The assigned numerical values for electronegativities are given in figure 13.7. Elements with higher values have the greatest attraction for bonding electrons, and elements with the lowest values have the least attraction for bonding electrons.

The absolute ("absolute" means without plus or minus signs) difference in the electronegativity of two bonded atoms can be used to predict if a bond is ionic or covalent. A large difference means that one element has a much greater attraction for bonding electrons than the other element. *If the absolute difference in electronegativity is 1.7 or more,* one atom pulls the bonding electron completely away and *an ionic bond results*. For example, sodium (Na) has an electronegativity of 0.9. Chlorine (Cl) has an electronegativity 3.0. The difference is 2.1, so you can expect sodium and chloride to form ionic bonds. *If the absolute difference in electronegativity is 0.5 or less,* both atoms have about the same ability to attract bonding electrons.

The result is that the electron is shared and *a covalent bond results*. A given hydrogen atom (H) has an electronegativity of another hydrogen atom, so the difference is 0. Zero is less than 0.5 so you can expect a molecule of hydrogen gas to have a covalent bond.

An ionic bond can be expected when the difference in electronegativity is 1.7 or more, and a covalent bond can be expected when the difference is less than 0.5. What happens when the difference is between 0.5 and 1.7? A covalent bond is formed, but there is an inequality since one atom has a greater bonding electron attraction than the other atom. Thus the bonding electrons are shared unequally. **A polar covalent bond** is *a covalent bond in which there is an unequal sharing of bonding electrons*. Thus the bonding electrons spend more time around one atom than the other. The term "polar" means "poles," and that is what forms in a polar molecule. Since the bonding electrons spend more time around one atom than the other, one end of the molecule will have a negative pole and the other end will have a positive pole. Since there are two poles the molecule is sometimes called a *dipole*. Note that the molecule as a whole still contains an equal number of electrons and protons, so it is overall electrically neutral. The poles are created by an uneven charge distribution, not an imbalance of electrons and protons. Figure 13.8 shows this uneven charge distribution for a polar covalent compound. The bonding electrons spend more time near the atom on the right, giving this side of the molecule a negative pole.

Figure 13.8 also shows a molecule that has an even charge distribution. The electron distribution around one atom is just like the charge distribution around the other. This molecule is thus a *nonpolar molecule* with a *nonpolar bond*. Thus a polar bond can be viewed as an intermediate type of bond between a nonpolar covalent bond and an ionic bond. Many gradations are possible between the transition from a purely nonpolar covalent bond and a purely ionic bond.

Example 13.5

Predict if the following bonds are nonpolar covalent, polar covalent, or ionic: (a) H-O; (b) C-Br; and, (c) K-Cl

Solution

From the electronegativity values in figure 13.7, the absolute differences are

(a) H-O, 1.4
(b) C-Br, 0.3
(c) K-Cl, 2.2

Since an absolute difference of less than 0.5 means nonpolar covalent, between 0.5 and 1.7 means polar covalent, and greater than 1.7 means ionic, then

(a) H-O, polar covalent
(b) C-Br, nonpolar covalent
(c) K-Cl, ionic

Compounds and Chemical Change

Figure 13.7

Electronegativities of the elements. These values are comparative only, assigned an arbitrary scale to indicate the relative tendency of atoms to attract shared electrons.

Electron distribution and kinds of bonding

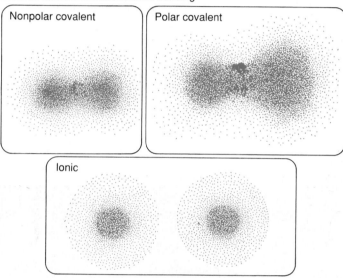

Figure 13.8

The absolute difference in electronegativities determines the kind of bond formed.

Example 13.6

Predict if the following bonds are nonpolar covalent, polar covalent, or ionic: (a) Ca-O; (b) H-Cl, and; (c) C-O.
(Answer: (a) ionic; (b) polar covalent; (c) polar covalent)

Composition of Compounds

As you can imagine, there are literally millions of different chemical compounds from all the possible combinations of over ninety natural elements held together by ionic or covalent bonds. Each of these compounds has its own name, so there are millions of names and formulas for all the compounds. In the early days, compounds were given *common names* according to how they were used, where they came from, or some other means of identifying them. Thus sodium carbonate was called "soda," and closely associated compounds were called baking soda (sodium bicarbonate), washing soda (sodium carbonate), caustic soda (sodium hydroxide), and the bubbly drink made by reacting "soda" with acid was called "soda water," later called "soda pop" (figure 13.9). Potassium carbonate was extracted from charcoal by soaking in water and came to be called "potash." Such common names are colorful, and some are descriptive, but it was impossible to keep up with the names as the number of known compounds grew. So a systematic set of rules was developed to determine the name and formula of each compound. Once you know the rules you can write the formula when you hear the name. Conversely, seeing the formula will tell you the systematic name of the compound. This can be an interesting intellectual activity and can also be important when reading the list of ingredients to understand the composition of a product.

There is a different set of systematic rules to be used with ionic compounds and covalent compounds, but there are a few rules in common. For example, a compound made of only two different elements always ends with the suffix "-ide." So when you hear the name of a compound ending with "-ide" you automatically know that the compound is made up of only two elements. Sodium chlor*ide* is an ionic compound made up of sodium and chlorine ions. Carbon diox*ide* is a covalent compound with carbon and oxygen atoms. Thus the systematic name tells you what elements are present in a compound with an "-ide" ending.

Ionic Compound Names

Ionic compounds formed by representative metal ions are named by stating the name of the metal (positive ion) first, then the name of the nonmetal (negative ion). Ionic compounds formed by variable-charge ions of the transition elements have an additional rule to identify which variable-charged ion is involved. There was an old way of identifying the charge on the ion by

Figure 13.9

These substances all have common names with the term "soda" because they now, or at one time, were made from sodium and some form of a carbonate ion. Soda water (or "soda pop") was first made by reacting soda (sodium carbonate) with an acid.

adding either "-ic" or "-ous" to the name of the metal. The suffix "-ic" meant the higher of two possible charges, and the suffix "-ous" meant the lower of two possible charges. For example, iron has two possible charges, $2+$ or $3+$. The old system used the Latin name for the root. The Latin name for iron is ferrum, so a higher charged iron ion ($3+$) was named a "ferric ion." The lower charged iron ion ($2+$) was called a "ferrous ion."

You still hear the old names sometimes, but chemists now have a better way to identify the variable-charge ion. The newer system uses the English name of the metal with Roman numerals in parentheses to indicate the charge number. Thus an iron ion with a charge of $2+$ is called an "iron (II) ion" and an iron ion with a charge of $3+$ is an "iron (III) ion." Table 13.5 gives some of the modern names for variable-charge ions. These names are used with the name of a nonmetal ending in "-ide" just like the single-charge ions in ionic compounds made up of two different elements.

Some ionic compounds are more complex than a combination of a metal and a nonmetal ion, containing three or more elements. This is possible because they have *polyatomic ions,* groups of two or more atoms that are bound together tightly and behave very much like a single monatomic ion. For example, the OH^{1-} ion is an oxygen atom bound to a hydrogen atom with a net charge of $1-$. This polyatomic ion is called a *hydroxide ion.* The hydroxide compounds make up one of the

Table 13.5

Modern names of some variable-charged ions

Ion	Name of Ion
Fe^{2+}	Iron (II) ion
Fe^{3+}	Iron (III) ion
Cu^{+}	Copper (I) ion
Cu^{2+}	Copper (II) ion
Pb^{2+}	Lead (II) ion
Pb^{4+}	Lead (IV) ion
Sn^{2+}	Tin (II) ion
Sn^{4+}	Tin (IV) ion
Cr^{2+}	Chromium (II) ion
Cr^{3+}	Chromium (III) ion
Cr^{6+}	Chromium (VI) ion

main groups of ionic compounds, the *metal hydroxides.* A metal hydroxide is an ionic compound consisting of a metal with the hydroxide ion. Another main group consists of the salts with polyatomic ions.

The metal hydroxides are named by identifying the metal first and the term *hydroxide* second. Thus NaOH is named sodium hydroxide and KOH is potassium hydroxide. The salts

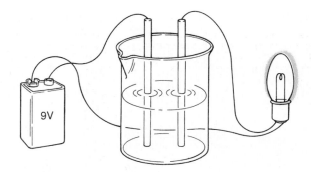

Figure 13.10
A battery and bulb will tell you if a solution contains ion. See activities below.

are similarly named with the metal (or ammonium ion) identified first, then the name of the polyatomic ion. So $NaNO_3$ is named sodium nitrate and $NaNO_2$ is sodium nitrite. Note that the suffix "-ate" means the polyatomic ion with one more oxygen atom than the "-ite" ion. For example, the chlor*ate* ion is $(ClO_3)^{1-}$ and the chlor*ite* ion is $(ClO_2)^{1-}$. Sometimes more than two possibilities exist, and more oxygen atoms are identified with the prefix "per-" and less with the prefix "hypo-". Thus the *per*chlor*ate* ion is $(ClO_4)^{1-}$ and the *hypo*chlor*ite* ion is $(ClO)^{1-}$.

ACTIVITIES

Dissolving an ionic compound in water results in ions being pulled from the crystal lattice to form free ions. A solution that contains ions will conduct an electric current, so electrical conductivity is therefore one way to test a dissolved compound to see if it is an ionic compound or not.

Make a conductivity tester like the one illustrated in figure 13.10. This one is a 9 V radio battery and a miniature Christmas tree bulb with two terminal wires. Sand these wires to make a good electrical contact. Try testing dry (1) baking soda, (2) table salt, and (3) sugar. Test solutions of these substances dissolved in distilled water. Test vinegar and rubbing alcohol. Explain which contains ions and why.

Ionic Compound Formulas

The formulas for ionic compounds are easy to write. There are two rules: (1) the symbol for the positive element is written first, followed by the symbol for the negative element just as in the order in the name, and (2) subscripts are used to indicate the numbers of ions needed to produce an electrically neutral compound. For example, the name "calcium chloride" tells you that this compound consists of positive calcium ions and negative chlorine ions. Again, "-ide" means only two elements are present. The calcium ion is Ca^{2+} and the chlorine ion is Cl^- (you know this by applying the atomic theory, knowing their positions in the periodic table, or by using a table of ions and their charges). To be electrically neutral, the compound must have an equal number of pluses and minuses. Thus two negative chlorine ions are needed for every calcium ion with its $2+$ charge. The formula is $CaCl_2$. The total charge of two chlorines is thus $2-$, which balances the $2+$ charge on the calcium ion.

One easy way to write a formula showing that a compound is electrically neutral is to cross over the absolute charge numbers (without plus or minus signs) and use them as subscripts. For example, the symbols for the calcium ion and the chlorine ion are

$$Ca^{2+} \quad Cl^{1-}$$

Crossing the absolute numbers as subscripts,

$$Ca_1^{2+} \quad Cl_2^{1-}$$

then dropping the charge numbers gives

$$Ca_1Cl_2$$

No subscript is written for 1; it is understood. The formula for calcium chloride is thus

$$CaCl_2$$

When using the crossover technique it is sometimes necessary to reduce the ratio to the lowest common multiple. Thus Mg_2O_2 means an equal ratio of magnesium and oxygen ions, so the correct formula is MgO. The crossover technique works because ionic bonding results from a transfer of electrons and the net charge is conserved. A calcium ion has a $2+$ charge because the atom lost two electrons and two chlorine atoms gain one electron each for a total of two electrons gained. Two electrons lost equals two electrons gained, and the net charge on calcium chloride is zero as it has to be.

The formulas for variable-charge ions are easy to write since the Roman numeral tells you the charge number. The formula for tin (II) fluoride is written by crossing over the charge numbers (Sn^{2+}, F^{1-}) and the formula is SnF_2.

Example 13.7

Name the following compounds: (a) LiF and (b) PbF_2. Write the formulas for the following compounds: (c) potassium bromide and (d) copper (I) sulfide.

Solution

(a) The formula LiF means that the positive metal ions are lithium, the negative nonmetal ions are fluorine, and there are only two elements in the compound. Lithium ions are Li^{1+} (family IA) and fluorine ions are F^{1-} (family VIIA). The name is lithium fluoride.

(b) Lead is a variable-charge transition element (table 13.5) and fluorine ions are F^{1-}. The lead ion must be Pb^{2+} because the compound PbF_2 is electrically neutral. Therefore, the name is lead (II) fluoride.

(c) The ions are K^{1+} and Br^{1-}. Crossing over the charge numbers and dropping the signs gives the formula KBr.

(d) The Roman numeral tells you the charge on the copper ion so the ions are Cu^{1+} and S^{2-}. The formula is Cu_2S.

The formulas for ionic compounds with polyatomic ions are written from combinations of positive metal ions or the ammonium ion with the polyatomic ions as listed in table 13.4. Since the polyatomic ion is a group of atoms that has a charge and stays together in a unit, it is sometimes necessary to indicate this with parentheses. For example, magnesium hydroxide is composed of Mg^{2+} ions and $(OH)^{1-}$ ions. Using the crossover technique to write the formula, you get

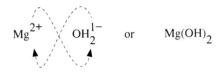

The parentheses are used and the subscript is written *outside* the parenthesis to show that the entire hydroxide unit is taken twice. The formula $Mg(OH)_2$ means

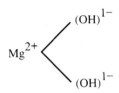

which shows that the pluses equal the minuses. Parentheses are not used, however, when only one polyatomic ion is present. Sodium hydroxide is NaOH, not $Na(OH)_1$.

Example 13.8

Name the following compounds: (a) Na_2SO_4 and (b) $Cu(OH)_2$. Write formulas for the following compounds: (c) calcium carbonate and (d) calcium phosphate.

Solution

(a) The ions are Na+ (sodium ion) and $(SO_4)^{2-}$ (sulfate ion). The name of the compound is sodium sulfate.

(b) Copper is a variable-charge transition element (table 13.5) and the hydroxide ion $(OH)^{1-}$ has a charge of 1−. Since the compound $Cu(OH)_2$ must be electrically neutral, the copper ion must be Cu^{2+}. The name is copper (II) hydroxide.

(c) The ions are Ca^{2+} and $(CO_3)^{2-}$. Crossing over the charge numbers and dropping the signs gives the formula $Ca_2(CO_3)_2$. Reducing the ratio to the lowest common multiple gives the correct formula of $CaCO_3$.

(d) The ions are Ca^{2+} and $(PO_4)^{3-}$ (from table 13.4). Using the crossover technique gives the formula $Ca_3(PO_4)_2$. The parentheses indicate that the entire phosphate unit is taken twice.

Covalent Compound Names

Covalent compounds are molecular and the molecules are composed of two *nonmetals,* as opposed to the metal and nonmetal elements that make up ionic compounds. The combinations of nonmetals alone do not present simple names as the ionic compounds did, so a different set of rules for naming and formula writing is needed.

Table 13.6

Prefixes and element stem names

Prefixes		Stem Names	
Prefix	Meaning	Element	Stem
Mono-	1	Hydrogen	Hydr-
Di-	2	Carbon	Carb-
Tri-	3	Nitrogen	Nitr-
Tetra-	4	Oxygen	Ox-
Penta-	5	Fluorine	Fluor-
Hexa-	6	Phosphorus	Phosph-
Hepta-	7	Sulfur	Sulf-
Octa-	8	Chlorine	Chlor-
Nona-	9	Bromine	Brom-
Deca-	10	Iodine	Iod-

Note: the *a* or *o* ending on the prefix is often dropped if the stem name begins with a vowel, e.g., "tetroxide," not "tetraoxide."

Ionic compounds were named by stating the name of the positive metal ion, then the name of the negative nonmetal ion with an "-ide" ending. This system is not adequate for naming the covalent compounds. To begin, covalent compounds are composed of two or more nonmetal atoms that form a molecule. It is possible for some atoms to form single, double, or even triple bonds with other atoms, including atoms of the same element, and coordinate covalent bonding is also possible in some compounds. The net result is that the same two elements can form more than one kind of covalent compound. Carbon and oxygen, for example, can combine to form the gas released from burning and respiration, carbon dioxide (CO_2). Under certain conditions the very same elements combine to produce a different gas, the poisonous carbon monoxide (CO). Similarly, sulfur and oxygen can combine differently to produce two different covalent compounds. A successful system for naming covalent compounds must therefore provide a means of identifying different compounds made of the same elements. This is accomplished by using a system of Greek prefixes (see table 13.6). The rules are as follows:

1. The first element in the formula is named first with a prefix indicating the number of atoms if the number is greater than one.

2. The stem name of the second element in the formula is next. A prefix is used with the stem if two elements form more than one compound. The suffix "-ide" is again used to indicate a compound of only two elements.

For example, CO is carbon monoxide and CO_2 is carbon dioxide. The compound BF_3 is boron trifluoride and N_2O_4 is dinitrogen tetroxide. Knowing the formula and the prefix and stem information in table 13.6, you can write the name of any

Compounds and Chemical Change

Figure 13.11
Once you understand chemical names and formulas, you can figure out what chemical compounds are contained in different household products.

covalent compound made up of two elements by ending it with "-ide." Conversely, the name will tell you the formula. However, there are a few polyatomic ions that have "-ide" endings that are compounds made up of more than just two elements (hydroxide and cyanide). Compounds formed with the ammonium will also have an "-ide" ending, and these are also made up of more than two elements.

Covalent Compound Formulas

The systematic name tells you the formula for a covalent compound. The gas that dentists use as an anesthetic, for example, is dinitrogen monoxide. This tells you there are two nitrogen atoms and one oxygen atom in the molecule, so the formula is N_2O. A different molecule composed of the very same elements is nitrogen dioxide. Nitrogen dioxide is the pollutant responsible for the brownish haze of smog. The formula for nitrogen dioxide is NO_2. Other examples of formulas from systematic names are carbon dioxide (CO_2) and carbon tetrachloride (CCl_4).

Formulas of covalent compounds indicate a pattern of how many atoms of one element combine with atoms of another. Carbon, for example, combines with no more than two oxygen atoms to form carbon dioxide. Carbon combines with no more than four chlorine atoms to form carbon tetrachloride. Electron dot formulas show these two molecules as

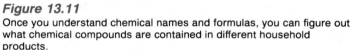

Using a dash to represent bonding pairs,

$$O - C - O \qquad\qquad \begin{matrix} & & Cl & & \\ & & | & & \\ Cl & - & C & - & Cl \\ & & | & & \\ & & Cl & & \end{matrix}$$

In both of these compounds the carbon atom forms four covalent bonds with another atom. The number of covalent bonds that an atom can form is called its **valence.** Carbon has a valence

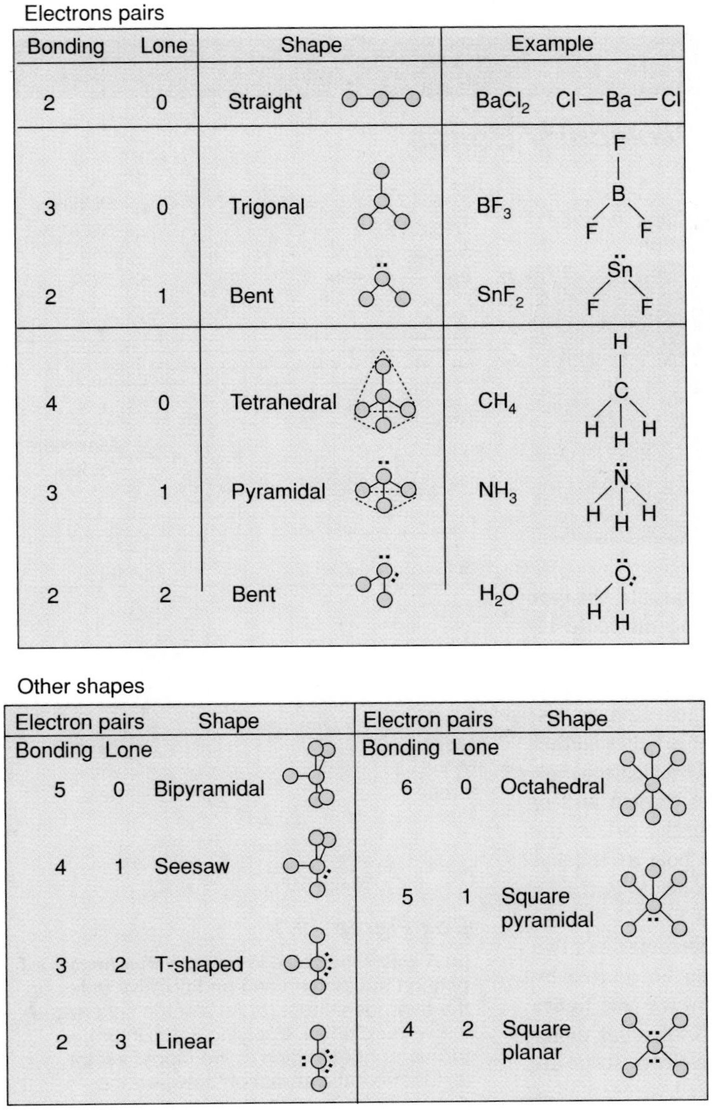

Electrons pairs

Bonding	Lone	Shape		Example	
2	0	Straight	O—O—O	$BaCl_2$	Cl—Ba—Cl
3	0	Trigonal		BF_3	F, B, F, F
2	1	Bent		SnF_2	Sn, F, F
4	0	Tetrahedral		CH_4	H, C, H, H, H
3	1	Pyramidal		NH_3	N, H, H, H
2	2	Bent		H_2O	O, H, H

Other shapes

Electron pairs		Shape		Electron pairs		Shape
Bonding	Lone			Bonding	Lone	
5	0	Bipyramidal		6	0	Octahedral
4	1	Seesaw		5	1	Square pyramidal
3	2	T-shaped				
2	3	Linear		4	2	Square planar

Figure 13.12
As you can see by studying these two charts, there is a relationship between the number of bonding electron pairs and the number of lone electron pairs and the shape of a molecule.

of four and can form single, double, or triple bonds. Here are the possibilities for a single carbon atom (combining elements not shown):

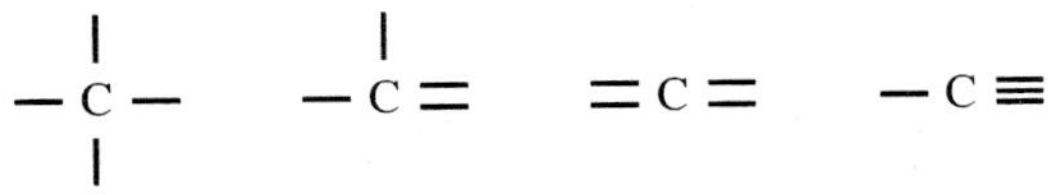

$$ -\overset{|}{\underset{|}{C}}- \qquad -\overset{|}{C}= \qquad =C= \qquad -C\equiv $$

Hydrogen has only one unshared electron so the hydrogen atom has a valence of one. Oxygen has a valence of two and nitrogen has a valence of three. Here are the possibilities for hydrogen, oxygen, and nitrogen:

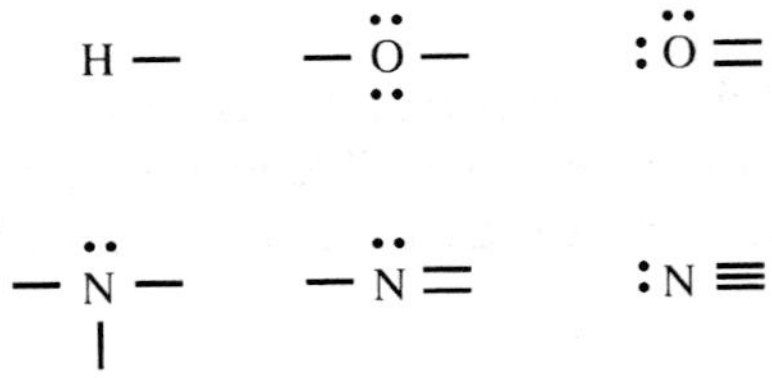

$$ H- \qquad -\overset{..}{\underset{..}{O}}- \qquad :\overset{..}{O}= $$

$$ -\overset{..}{\underset{|}{N}}- \qquad -\overset{..}{N}= \qquad :N\equiv $$

Note the lone pairs shown on the oxygen and nitrogen atoms. Such lone pairs create the possibility of forming a coordinate covalent bond with another atom. The number of bonds and the number of lone pairs also determine the *shape* of a molecule. Molecules are not flat like the formulas on paper. Molecules have three dimensional shapes as illustrated in figure 13.12. More will be said about molecular shapes later.

Summary

Elements are basic substances that cannot be broken down into anything simpler, and an *atom* is the smallest unit of an element. *Compounds* are combinations of two or more elements and can be broken down into simpler substances. Compounds are formed when atoms are held together by an attractive force called a *chemical bond*. A *molecule* is the smallest unit of a compound, or a gaseous element, that can exist and still retain the characteristic properties of a substance.

A *chemical change* produces new substances with new properties, and the new materials are created by making or breaking chemical bonds. The process of chemical change in which different chemical substances are created by forming or breaking chemical bonds is called a *chemical reaction*. During a chemical reaction different chemical substances with greater or lesser amounts of internal potential energy are produced. *Chemical energy* is the change of internal potential energy during a chemical reaction, and other reactions absorb energy. A *chemical equation* is a shorthand way of describing a chemical reaction. An equation shows the substances that are changed, the *reactants* on the left side, and the new substances produced, the *products* on the right side.

Chemical reactions involve *valence electrons,* the electrons in the outermost shell of an atom. Atoms tend to lose or acquire electrons to achieve the configuration of the noble gases with stable, filled outer orbitals. This tendency is generalized as the *octet rule,* that atoms lose or gain electrons to acquire the noble gas structure of eight electrons in the outer orbital. Atoms form negative or positive *ions* in the process.

A chemical bond is an attractive force that holds atoms together in a compound. Chemical bonds that are formed when atoms transfer electrons to become ions are *ionic bonds*. An ionic bond is an electrostatic attraction between oppositely charged ions. Chemical bonds formed when ions share electrons are *covalent bonds*.

Ionic bonds result in *ionic compounds* with a crystalline structure. The energy released when an ionic compound is formed is called the *heat of formation*. It is the same amount of energy that is required to decompose the compound into its elements. A *formula* of a compound uses symbols to tell what elements are in a compound and in what proportions. Ions of representative elements have a single, fixed charge but many transition elements have variable charges. Electrons are conserved when ionic compounds are formed and the ionic compound is electrically neutral. The formula shows this overall balance of charges.

Covalent compounds are molecular, composed of electrically neutral groups of atoms bound together by *covalent bonds*. A *single covalent bond* is formed by the sharing of a pair of electrons, with each atom contributing a single electron to the shared pair. Covalent bonds formed when two pairs of electrons are shared are called *double bonds*. A *triple bond* is the sharing of three pairs of electrons. If a shared electron pair comes from a single atom, the bond is called a *coordinate covalent bond*. Coordinate covalent bonding sometimes results in a group of atoms with a charge that acts together as a unit. The charged unit is called a *polyatomic ion*.

The electron-pulling ability of an atom in a bond is compared with arbitrary values of *electronegativity*. A high electronegative value means a greater attraction for bonding electrons. If the absolute difference in electronegativity of two bonded atoms is 1.7 or more, one atom pulls

Microwave Ovens and Molecular Bonds

A microwave oven rapidly cooks foods that contain water, but paper, glass, and plastic products remain cool in the oven. If they are warmed at all it is from the heat conducted from the food. The explanation of how the microwave oven heats water, but not most other substances, begins with the nature of the chemical bond.

A chemical bond acts much like a stiff spring, resisting both compression and stretching as it maintains an equilibrium distance between the atoms. As a result, a molecule tends to vibrate when energized or buffeted by other molecules. The rate of vibration depends on the "stiffness" of the spring, which is determined by the bond strength, and the mass of the atoms making up the molecule. Each kind of molecule therefore has its own set of characteristic vibrations, a characteristic natural frequency.

Disturbances with a wide range of frequencies can impact a vibrating system. When the frequency of a disturbance matches the natural frequency, energy is transferred very efficiently and the system undergoes a large increase in amplitude. Such a frequency match is called resonance. When the disturbance is visible light or some other form of radiant energy, a resonant match results in absorption of the radiant energy and an increase in the molecular kinetic energy of vibration. Thus a resonant match results in a temperature increase.

The natural frequency of a water molecule matches the frequency of infrared radiation, so resonant heating occurs when infrared radiation strikes water molecules. It is the water molecules in your skin that absorb infrared radiation from the sun, a fire, or some hot object, resulting in the warmth that you feel. Because of this match between the frequency of infrared radiation and the natural frequency of a water molecule, infrared is often called "heat radiation." Since infrared radiation is absorbed by water molecules, it is mostly absorbed on the surface of an object, penetrating only a short distance.

The frequency ranges of visible light, infrared radiation, and microwave radiation are given in box table 13.1. Most microwave ovens operate at the lower end of the microwave frequency range, between 1×10^9 hz to 3×10^9 hz, or 1 to 3 gigahertz. This range is too low for a resonant match with water molecules, so something else must transfer energy from the microwaves to heat the water. This something else is a result of another

Box Table 13.1

Approximate ranges of visible light, infrared radiation, and microwave radiation

Radiation	Frequency Range (hz)
Visible light	4×10^{14} to 8×10^{14}
Infrared radiation	3×10^{11} to 4×10^{14}
Microwave radiation	1×10^9 to 3×10^{11}

characteristic of the water molecule, the type of covalent bond holding the molecule together.

The difference in electronegativity between a hydrogen and oxygen atom is 1.4, meaning the water molecule is held together by a polar covalent bond. The electrons are strongly shifted toward the oxygen end of the molecule, creating a negative pole at the oxygen end and a positive pole at the hydrogen ends. The water molecule is thus a dipole as shown in figure 13.13.

The dipole of water molecules has two effects: (1) the molecule can be rotated by the electric field of a microwave (see figure 13.13b) and (2) groups of individual molecules are held together by an electrostatic attraction between the positive hydrogen ends of a water molecule with the negative oxygen end of another molecule (see figure 13.13c). This attraction is called a hydrogen bond (see chapter 15).

One model to explain how microwaves heat water involves a particular group of three molecules, arranged so that the end molecules of the group are aligned with the microwave electric field, with the center molecule not aligned. The microwave torques the center molecule, breaking its hydrogen bond. The energy of the microwave goes into doing the work of breaking the hydrogen bond, and the molecule now has increased potential energy as a consequence. The detached water molecule reestablishes its hydrogen bond, giving up its potential energy, which goes into the vibration of the group of molecules. Thus the energy of the microwaves is converted into a temperature increase of the water. The temperature increase is high enough to heat and cook most foods.

Microwave cooking is different from conventional cooking because the heating results from energy transfer in polar water molecules, not conduction and convection. The

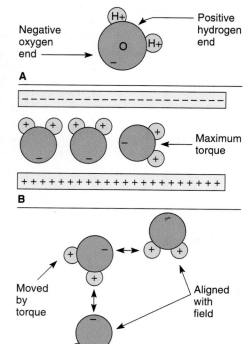

Box Figure 13.1

(a) A water molecule is polar, with a negative pole on the oxygen end and positive poles on the hydrogen ends. (b) An electric field aligns the water dipoles, applying a maximum torque at right angles to the dipole vector. (c) Electrostatic attraction between the dipoles holds groups of water molecules together.

surface of the food never reaches a temperature over the boiling point of water, so a microwave oven does not brown food (a conventional oven may reach temperatures almost twice as high). Large food items continue to cook for a period of time after being in a microwave oven as the energy is conducted from the water molecules to the food. Most recipes allow for this continued cooking by specifying a waiting period after removing the food from the oven.

Microwave ovens are able to defrost frozen foods because ice always has a thin layer of liquid water (which is what makes it slippery). To avoid "spot cooking" of small pockets of liquid water, many microwave ovens cycle on and off in the defrost cycle. The electrons in metals, like the dipole water molecules, are affected by the electric field of a microwave. A piece of metal near the wall of a microwave oven can result in sparking, which can ignite paper. Metals also reflect microwaves, which can damage the radio tube that produces the microwaves.

the bonding electron away and an ionic bond results. If the difference is less than 0.5, the electrons are equally shared in a covalent bond. Between 0.5 and 1.7, the electrons are shared unequally in a *polar covalent bond*. A polar covalent bond results in electrons spending more time around the atom or atoms with the greater pulling ability, creating a negative pole at one end and a positive pole at the other. Such a molecule is called a *dipole* since it has two poles, or centers, of charge.

Compounds are named with systematic rules for ionic and covalent compounds. Both ionic and covalent compounds that are made up of only two different elements always end with an "-ide" suffix, but there are a few "-ide" names for compounds that have more than just two elements.

The modern systematic system for naming variable-charge ions states the English name and gives the charge with Roman numerals in parentheses. Ionic compounds are electrically neutral and formulas must show a balance of charge. The *crossover technique* is an easy way to write formulas that show a balance of charge.

Covalent compounds are molecules of two or more nonmetal atoms held together by a covalent bond. The system for naming covalent compounds uses Greek prefixes to identify the numbers of atoms since more than one compound can form from the same two elements (CO and CO_2, for example).

Summary of Equations

13.1

Ionization of a metal atom:

energy + Na· ⟶ Na⁺ + e⁻

13.2

Ionic bonding reaction (single charges):

Na· + ·Cl: ⟶ Na⁺ (:Cl:)⁻

13.3

Ionic bonding reaction (single and double charges):

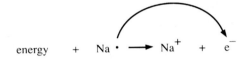

13.4

Covalent bonding (hydrogen-hydrogen):

H· + H· ⟶ (H:H)

13.5

Covalent bonding (hydrogen-fluorine):

H· + ·F: ⟶ (H:F:)

13.6

Covalent bonding (fluorine-fluorine):

·F: + ·F: ⟶ (:F:F:)

13.7

Coordinate covalent bonding (ammonium ion):

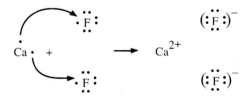

Key Terms

Applying the Concepts

1. Atoms of the representative elements have a tendency to seek stability through
   a. acquiring the noble gas structure.
   b. filling or emptying their outer orbitals.
   c. any situation that will satisfy the octet rule.
   d. all of the above.

2. An atom of an element that is in family VIA will have what charge when it is ionized?
   a. 2+
   b. 6+
   c. 6−
   d. 2−

3. If there are two bromine atoms for each barium ion in a compound the chemical formula is
   a. ₂Br₁Ba.
   b. Ba₂Br.
   c. BaBr₂.
   d. none of the above.

4. Which combination of elements forms crystalline solids that will dissolve in water, producing a solution of ions that can conduct an electric current?
   a. metal and metal
   b. metal and nonmetal
   c. nonmetal and nonmetal
   d. all of the above

Compounds and Chemical Change

5. The number of pairs of shared electrons in a covalent compound is often the same as the number of
   a. unpaired electrons in the electron dot notation.
   b. valence electrons.
   c. orbital pairs.
   d. protons in the nucleus.

6. Sulfur and oxygen are both in the VIA family of the periodic table. If element X combines with oxygen to form the compound X_2O, element X will combine with sulfur to form the compound
   a. XS_2.
   b. X_2S.
   c. X_2S_2.
   d. It is impossible to say without more information.

7. One element is in the IA family of the periodic table and a second is in the VIIA family. What type of compound will the two elements form?
   a. ionic
   b. covalent
   c. More information is needed to answer this question.

8. One element is in the VA family of the periodic table and a second is in the VIA family. What type of compound will these two elements form?
   a. ionic
   b. covalent
   c. They will not form a compound.
   d. More information is needed to answer this question.

9. A compound made of only two different elements has a name that always ends with the suffix
   a. -ite.
   b. -ate.
   c. -ide.
   d. -ous.

10. Dihydrogen monoxide is the systematic name for a compound that has the common name of
    a. laughing gas.
    b. water.
    c. smog.
    d. rocket fuel.

Answers

1. d 2. d 3. c 4. b 5. a 6. b 7. a 8. b 9. c
10. b

Questions for Thought

1. Describe how the following are alike and how they are different: (a) a sodium atom and a sodium ion and (b) a sodium ion and a neon atom.

2. Suggest a reason why solid NaCl does not conduct an electric current, but molten NaCl will conduct an electric current.

3. Why is it technically incorrect to use the term *molecule* when describing an ionic compound?

4. What is electronegativity? How does it determine the type of chemical bond?

5. What is the basic difference between a polar covalent bond and a nonpolar covalent bond?

6. What is the basic difference between an ionic and covalent bond? What do atoms forming the two bond types have in common?

7. What is the immediate source of the chemical energy released when wood burns? How do the molecules in wood store this energy?

8. What is the octet rule?

9. Describe the basic difference between a single covalent bond and a coordinate covalent bond.

10. Is there a relationship between the number of valence electrons and how many covalent bonds an atom can form? Explain.

11. Describe the differences of electronegativity of two atoms involved in (a) an ionic bond, (b) a nonpolar bond, and (c) a polar covalent bond.

12. Write electron dot formulas for molecules formed when hydrogen combines with (a) chlorine, (b) oxygen, and (c) carbon.

13. What is a lone pair? How many lone pairs exist on a molecule of hydrogen and a nonmetal from family VA? How many lone pairs exist on a molecule of hydrogen and a nonmetal from family IVA?

14. Sodium fluoride is often added to water supplies to strengthen teeth. Is sodium fluoride ionic, nonpolar covalent, or polar covalent? Explain the bases of your answer.

15. What is the modern systematic name of a compound with the formula (a) SnF_2? (b) PbS?

16. What kinds of elements are found in (a) ionic compounds with a name ending with an "-ide" suffix and (b) covalent compounds with a name ending with an "-ide" suffix?

17. Why is it necessary to use a system of Greek prefixes to name binary covalent compounds?

18. What are variable-charge ions? Explain how variable-charge ions are identified in the modern system of naming compounds.

19. What is a polyatomic ion? Give the names and formulas for several common polyatomic ions.

20. Write the formula for magnesium hydroxide. Explain what the parentheses mean.

21. What is a double bond? A triple bond?

22. Would you expect the elements in family VIIA to form double bonds? Explain.

Exercises

Group A—Solutions Provided in Appendix D

1. Use electron dot symbols in equations to predict the formula of the ionic compound formed between the following:
   (a) K and I   K I
   (b) Sr and S   SrS
   (c) Na and O   Na$_2$O
   (d) Al and O   Al$_2$O$_2$

2. Name the following ionic compounds formed from variable-charge transition elements:
   (a) CuS
   (b) Fe$_2$O$_3$
   (c) CrO
   (d) PbS

3. Name the following polyatomic ions:
   (a) $(OH)^{1-}$
   (b) $(SO_3)^{2-}$
   (c) $(ClO)^{1-}$
   (d) $(NO_3)^{1-}$
   (e) $(CO_3)^{2-}$
   (f) $(ClO_4)^{1-}$

4. Use the crossover technique to write formulas for the following compounds:
   (a) Iron (III) hydroxide
   (b) Lead (II) phosphate
   (c) Zinc carbonate
   (d) Ammonium nitrate
   (e) Potassium hydrogen carbonate
   (f) Potassium sulfite

5. Write formulas for the following covalent compounds:
   (a) Carbon tetrachloride
   (b) Dihydrogen monoxide
   (c) Manganese dioxide
   (d) Sulfur trioxide
   (e) Dinitrogen pentoxide
   (f) Diarsenic pentasulfide

6. Name the following covalent compounds:
   (a) CO
   (b) CO$_2$
   (c) CS$_2$
   (d) N$_2$O
   (e) P$_4$S$_3$
   (f) N$_2$O$_3$

7. Predict if the bonds formed between the following pairs of elements are ionic, polar covalent, or nonpolar covalent:
   (a) Si and O   P C B 1.0
   (b) O and O   N P C 0
   (c) H and Te   N P C 0
   (d) C and H   N P C
   (e) Li and F   N P C
   (f) Ba and S   P C

Group B—Solutions Not Given

1. Use electron dot symbols in equations to predict the formulas of the ionic compound formed between the following:
   (a) Li and F
   (b) Be and S
   (c) Li and O
   (d) Al and S

2. Name the following ionic compounds formed from variable-charge transition elements:
   (a) PbCl$_2$
   (b) FeO
   (c) Cr$_2$O$_3$
   (d) PbO

3. Name the following polyatomic ions:
   (a) $(C_2O_3O_2)^{1-}$
   (b) $(HCO_3)^{1-}$
   (c) $(SO_4)^{2-}$
   (d) $(NO_4)^{1-}$
   (e) $(MnO_4)^{1-}$
   (f) $(CO_3)^{2-}$

4. Use the crossover technique to write formulas for the following compounds:
   (a) Aluminum hydroxide
   (b) Sodium phosphate
   (c) Copper (II) chloride
   (d) Ammonium sulfate
   (e) Sodium hydrogen carbonate
   (f) Cobalt (II) chloride

5. Write formulas for the following covalent compounds:
   (a) Silicon dioxide
   (b) Dihydrogen sulfide
   (c) Boron trifluoride
   (d) Dihydrogen sulfide
   (e) Carbon tetrafluoride
   (f) Nitrogen trihydride

6. Name the following covalent compounds:
   (a) N$_2$O
   (b) SO$_2$
   (c) SiC
   (d) PF$_5$
   (e) SeCl$_6$
   (f) N$_2$O$_4$

7. Predict if the bonds formed between the following pairs of elements are ionic, polar covalent, or nonpolar covalent:
   (a) Si and C
   (b) Br and Cl
   (c) S and O
   (d) Sr and F
   (e) O and H
   (f) K and F

Chemical Formulas and Equations

Figure 14.1

The products of chemical research have substantially increased food supplies but have also increased the possibilities of pollution. Balancing the benefits and hazards of the use of chemicals requires a knowledge of chemistry and a knowledge of the alternatives.

W E live in a chemical world that has been partly manufactured through controlled chemical change. Consider all of the synthetic fibers and plastics that are used in clothing, housing, and cars. Consider all the synthetic flavors and additives in foods, how these foods are packaged, and how they are preserved. Consider also the synthetic drugs and vitamins that keep you healthy. There are millions of such familiar products that are the direct result of chemical research. Most of these products simply did not exist sixty years ago.

Many of the products of chemical research have remarkably improved the human condition. For example, synthetic fertilizers have made it possible to supply food in quantities that would not otherwise be possible. Chemists learned how to take nitrogen from the air and convert it into fertilizers on an enormous scale. Other chemical research resulted in products such as weed killers, insecticides, and mold and fungus inhibitors. The fertilizers and these products have made it possible to supply food for millions of people who would have otherwise starved (figure 14.1).

Yet, we also live in a world with concerns about chemical pollutants, the greenhouse effect, acid rain, and a disappearing ozone shield. The very nitrogen fertilizers that have increased food supplies also wash into rivers, polluting the waterways and bays. Such dilemmas require an understanding of chemical products and the benefits and hazards of possible alternatives. Understanding requires a knowledge of chemistry, since the benefits, and risks, are chemical in nature.

The previous chapters were about the modern atomic theory and how it explains elements and how compounds are formed in chemical change. This chapter is concerned with describing chemical changes and the different kinds of chemical reactions that occur. These reactions are explained with balanced chemical equations, which are concise descriptions of reactions that produce the products used in our chemical world.

Chemical Formulas

In chapter 13 you learned how to name and write formulas for ionic and covalent compounds, including the ionic compound of table salt and the covalent compound of ordinary water. Recall that a formula is a shorthand way of describing the elements or ions that make up a compound. There are basically three kinds of formulas that describe compounds: (1) *empirical* formulas, (2) *molecular* formulas, and (3) *structural* formulas. Empirical and molecular formulas, and their use, will be considered in this chapter. Structural formulas will be considered in chapter 16.

An **empirical formula** identifies the elements present in a compound and describes the *simplest whole number ratio* of atoms of these elements with subscripts. For example, the empirical formula for ordinary table salt is NaCl. This tells you that the elements sodium and chlorine make up this compound and there is one atom of sodium for each chlorine atom. The empirical formula for water is H_2O, meaning there are two atoms of hydrogen for each atom of oxygen.

Name	Molecular formula	Sketch	Structural formula
Water	H_2O		
Ammonia	NH_3		
Hydrogen peroxide	H_2O_2		
Carbon dioxide	CO_2		$O=C=O$

Figure 14.2

The name, molecular formula, sketch, and structural formula of some common molecules. Compare the kinds and numbers of atoms making up each molecule in the sketch to the molecular formula (see figure 13.12).

Covalent compounds exist as molecules. A chemical formula that identifies the *actual numbers* of atoms in a molecule is known as a **molecular formula.** Figure 14.2 shows the structure of some common molecules and their molecular formulas. Note that each formula identifies the elements and numbers of atoms in each molecule. The figure also indicates how molecular formulas can be written to show how the atoms are arranged in the molecule. Formulas that show the relative arrangements are called structural formulas. Compare the structural formulas in the illustration with the three-dimensional representations and the molecular formulas.

How do you know if a formula is empirical or molecular? First, you need to know if the compound is ionic or covalent. You know that ionic compounds are usually composed of metal and nonmetal atoms with an electronegativity difference greater than 1.7. Formulas for ionic compounds are *always* empirical formulas. Ionic compounds are composed of many positive and negative ions arranged in an electrically neutral array. There is no discrete unit, or molecule, in an ionic compound so it is only possible to identify ratios of atoms with an empirical formula.

Covalent compounds are generally nonmetal atoms bonded to nonmetal atoms in a molecule. You could therefore assume that a formula for a covalent compound is a molecular formula unless it is specified otherwise. You can be certain it is a molecular formula if it is not the simplest whole number ratio. Glucose, for example, is a simple sugar (also known as dextrose)

Chemical Formulas and Equations

with the formula $C_6H_{12}O_6$. This formula is divisible by six, yielding a formula with the simplest whole number ratio of CH_2O. Therefore CH_2O is the empirical formula for glucose and $C_6H_{12}O_6$ is the molecular formula.

Molecular and Formula Weights

The **formula weight** of a compound is the sum of the atomic weights of all the atoms in a chemical formula. For example, the formula for water is H_2O. Hydrogen and oxygen are both nonmetals, so the formula means that one atom of oxygen is bound to two hydrogen atoms in a molecule. From the periodic table, you know that the approximate (rounded) atomic weight of hydrogen is 1.0 u and oxygen is 16.0 u. Adding the atomic weights for *all* the atoms,

Atoms	Atomic Weight		Totals
2 of H	2 × 1.0 u	=	2.0 u
1 of O	1 × 16.0 u	=	16.0 u
	Formula weight	=	18.0 u

Thus the formula weight of a water molecule is 18.0 u.

The formula weight of an ionic compound is found the same way, by adding the rounded atomic weights of atoms (or ions) making up the compound. Sodium chloride is NaCl, so the formula weight is 23.0 u plus 35.5 u, or 58.5 u. The *formula weight* can be calculated for an ionic or molecular substance. The **molecular weight** is the formula weight of a molecular substance. The term *molecular weight* is sometimes used for all substances, whether or not they have molecules. Since ionic substances such as NaCl do not occur as molecules, this is not strictly correct. Both molecular and formula weights are calculated the same way, but formula weight is a more general term.

Example 4.1

What is the formula weight of table sugar (sucrose), which has the formula $C_{12}H_{22}O_{11}$?

Solution
The formula identifies the numbers of each atom and the atomic weights are from a periodic table:

Atoms	Atomic Weight		Totals
12 of C	12 × 12.0 u	=	144.0 u
22 of H	22 × 1.0 u	=	22.0 u
11 of O	11 × 16.0 u	=	176.0 u
	Formula weight	=	342.0 u

Example 14.2

What is the molecular weight of ethyl alcohol, C_2H_5OH? (Answer: 46.0 u)

Percent Composition of Compounds

The formula weight of a compound can provide useful information about the elements making up a compound (figure 14.3). For example, suppose you want to know how much calcium is provided by a dietary supplement. The label lists the main ingredient as calcium carbonate, $CaCO_3$. To find how much calcium is supplied by a pill with a certain mass you need to find the *mass percentage* of calcium in the compound.

Percent is simply the fractional part of the whole times 100 percent (meaning "per 100"), or

$$\left(\frac{part}{whole}\right)(100\% \text{ of whole}) = \% \text{ of part} \qquad \textbf{equation 14.1}$$

For example, if 13 students in a class of 50 are freshmen, the percentage of freshmen in the class is

$$\left(\frac{13 \text{ freshmen}}{50 \text{ classmates}}\right)(100\% \text{ of classmates})$$

$$= \left(0.26 \frac{\text{freshmen}}{\text{classmates}}\right)(100\% \text{ of classmates})$$

$$= 26\% \text{ freshmen}$$

Note the classmate units cancel, giving the answer in percent freshmen.

Since the formula weight of a compound represents all of its composition, the formula weight is the "whole" in equation 14.1, with all the atoms contributing a part of the whole weight. The "part" in equation 14.1 is the atomic weight times the number of atoms of the element in which you are interested. Thus the mass percentage of an element in a compound can be found from

$$\frac{\left(\begin{array}{c}\text{atomic weight}\\\text{of element}\end{array}\right)\left(\begin{array}{c}\text{number of atoms}\\\text{of element}\end{array}\right)}{\text{formula weight of compound}} \times \begin{array}{c}100\% \text{ of}\\\text{compound}\end{array} = \begin{array}{c}\% \text{ of}\\\text{element}\end{array}$$

$$\textbf{equation 14.2}$$

The mass percentage of calcium in $CaCO_3$ can be found in two steps:

Step 1: Determine formula weight:

Atoms	Atomic Weight		Totals
1 of Ca	1 × 40.1 u	=	40.1 u
1 of C	1 × 12.0 u	=	12.0 u
3 of O	3 × 16.0 u	=	48.0 u
	Formula weight	=	100.1 u

Step 2: Determine percentage of Ca:

$$\frac{(40.1 \text{ u Ca})(1)}{100.1 \text{ u } CaCO_3} \times 100\% \; CaCO_3 = 40.1\% \text{ Ca}$$

Knowing the percentage of the total mass contributed by the calcium, this fractional part (as a decimal) can be multiplied by the mass of the supplement pill to find the calcium supplied. The mass percentage of the other elements can also be determined with equation 14.2.

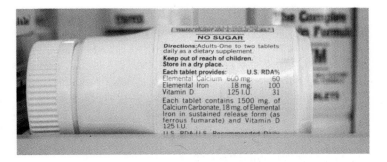

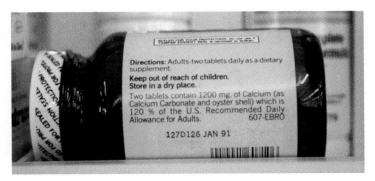

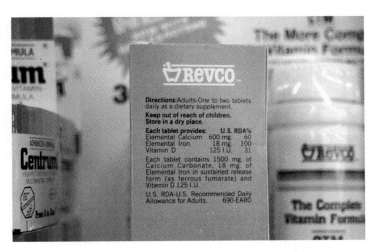

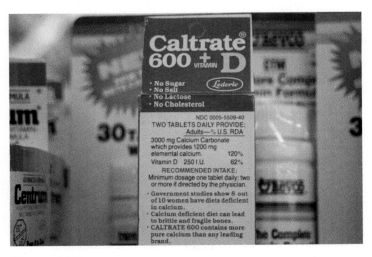

Figure 14.3

If you know the name of an ingredient, you can write a chemical formula and the percent composition of a particular substance can be calculated from the formula. This can be useful information for consumer decisions.

Example 14.3

Sodium fluoride is added to water supplies and to some toothpastes for fluoridation. What is the percentage composition of the elements in sodium fluoride?

Solution

Step 1: Write the formula for sodium fluoride, NaF.

Step 2: Determine the formula weight.

Atoms	Atomic Weight		Totals
1 of Na	1×23.0 u	=	23.0 u
1 of F	1×19.0 u	=	19.0 u
	Formula weight	=	42.0 u

Step 3: Determine the percentage of Na and F.

For Na:

$$\frac{(23.0 \text{ u Na})(1)}{42.0 \text{ u NaF}} \times 100\% \text{ NaF} = \boxed{54.7\% \text{ Na}}$$

For F:

$$\frac{(19.0 \text{ u F})(1)}{42.0 \text{ u NaF}} \times 100\% \text{ NaF} = \boxed{45.2\% \text{ F}}$$

The percentage often does not total to exactly 100 percent because of rounding.

Example 14.4

Calculate the percentage composition of carbon in table sugar, sucrose, which has a formula of $C_{12}H_{22}O_{11}$. (Answer: 42.1% C)

ACTIVITIES

Chemical fertilizers are added to the soil when it does not contain sufficient elements that are essential for plant growth. The three critical elements are nitrogen, phosphorus, and potassium, and these are the basic ingredients in most chemical fertilizers. In general, lawns require fertilizers high in nitrogen and gardens require fertilizers high in phosphorus.

Read the labels on commercial packages of chemical fertilizers sold in a garden shop. Find the name of the chemical that supplies each of these critical elements, for example, nitrogen is sometimes supplied by ammonium sulfate $(NH_4)_2SO_4$. Calculate the mass percentage of each critical element supplied according to the label information. Compare these percentages to the grade number of the fertilizer, for example, 10–20–10. Determine which fertilizer brand gives you the most nutrients for the money.

Chemical Equations

Chemical reactions occur when bonds between the outermost parts of atoms are formed or broken. Bonds are formed, for example, when a green plant uses sunlight—a form of energy—to create molecules of sugar, starch, and plant fibers. Bonds are broken and energy is released when you digest the sugars and starches or when plant fibers are burned. Chemical reactions thus involve changes in matter, the creation of new materials with new properties, and energy exchanges. So far you have considered chemical symbols as a concise way to represent elements and formulas as a concise way to describe what a compound is made of. There is also a concise way to describe a chemical reaction, the **chemical equation.**

Balancing Equations

Word equations were introduced in the previous chapter. Word equations are useful in identifying the *reactants,* the substances that existed before the reaction, and the *products,* the new substances formed as a result of the reaction. For example, the charcoal used in a barbecue grill is carbon. The carbon reacts with oxygen while burning, and the reaction (1) releases energy and (2) forms carbon dioxide. The reactants and products for this reaction can be described as

carbon + oxygen → carbon dioxide

The arrow means *yields* and the word equation is read as, "Carbon reacts with oxygen to yield carbon dioxide." This word equation describes what happens in the reaction but says nothing about the quantities of reactants or products.

Chemical symbols and formulas can be used in the place of words in an equation and the equation will have a whole new meaning. For example, the equation describing carbon reacting with oxygen to yield carbon dioxide becomes

$C + O_2 \rightarrow CO_2$ (balanced)

The new, added meaning is that one atom of carbon (C) reacts with one molecule of oxygen (O_2) to yield one molecule of carbon dioxide (CO_2) (figure 14.4). Note that the equation also shows one atom of carbon and two atoms of oxygen (recall that oxygen occurs as a diatomic molecule) as reactants on the left side and one atom of carbon and two atoms of oxygen as products on the right side. Since the same number of each kind of atom appears on both sides of the equation, the equation is said to be *balanced.*

You would not want to use a charcoal grill in a closed room because there might not be enough oxygen. An insufficient supply of oxygen produces a completely different product, the poisonous gas, carbon monoxide (CO). An equation for this reaction is

$C + O_2 \rightarrow CO$ (not balanced)

As it stands, this equation describes a reaction that violates the **law of conservation of mass,** that matter is neither created nor destroyed in a chemical reaction. From the point of view of atoms, this law states that *atoms are neither created nor destroyed in a chemical reaction.* A chemical reaction is the making or breaking of chemical bonds between atoms or groups

Figure 14.4

The charcoal used in a grill is basically carbon. The carbon reacts with oxygen to yield carbon dioxide. The chemical equation for this reaction, $C + O_2 \rightarrow CO_2$, contains the same information as the English sentence but has quantitative meaning as well.

of atoms. Atoms are not lost or destroyed in the process nor are they changed to a different kind. The equation for the formation of carbon monoxide has two oxygen atoms in the reactants (O_2) but only one in the product (in CO). An atom of oxygen has disappeared somewhere and that violates the law of conservation of mass. You cannot fix the equation by changing the CO to a CO_2, because this would change the identity of the compounds. Carbon monoxide is a poisonous gas that is different from carbon dioxide, a relatively harmless product of burning and respiration. *You cannot change the subscript in a formula* because that would change the formula. A different formula means a different composition and thus a different compound.

You cannot change the subscripts of a formula but you can place a number called a *coefficient* in *front* of the formula. Changing a coefficient changes the *amount* of a substance, not the identity. Thus 2 CO means two molecules of carbon monoxide and 3 CO means three molecules of carbon monoxide. If there is no coefficient, 1 is understood as with subscripts. The meaning of coefficients and subscripts is illustrated in figure 14.5.

Placing a coefficient of 2 in front of the C and a coefficient of 2 in front of the CO in the equation will result in the same numbers of each kind of atom on both sides:

$2 C + O_2 \rightarrow 2 CO$

Reactants:	2 C	Products:	2 C
	2 O		2 O

The equation is now balanced.

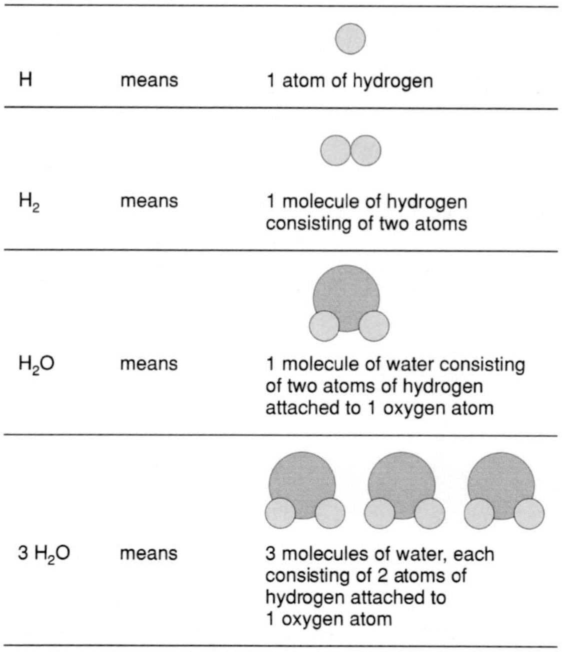

H	means	1 atom of hydrogen
H_2	means	1 molecule of hydrogen consisting of two atoms
H_2O	means	1 molecule of water consisting of two atoms of hydrogen attached to 1 oxygen atom
$3 H_2O$	means	3 molecules of water, each consisting of 2 atoms of hydrogen attached to 1 oxygen atom

Figure 14.5

The meaning of subscripts and coefficients used with a chemical formula. The subscripts tell you how many atoms of a particular element are in a compound. The coefficient tells you about the quantity, or number, of molecules of the compound.

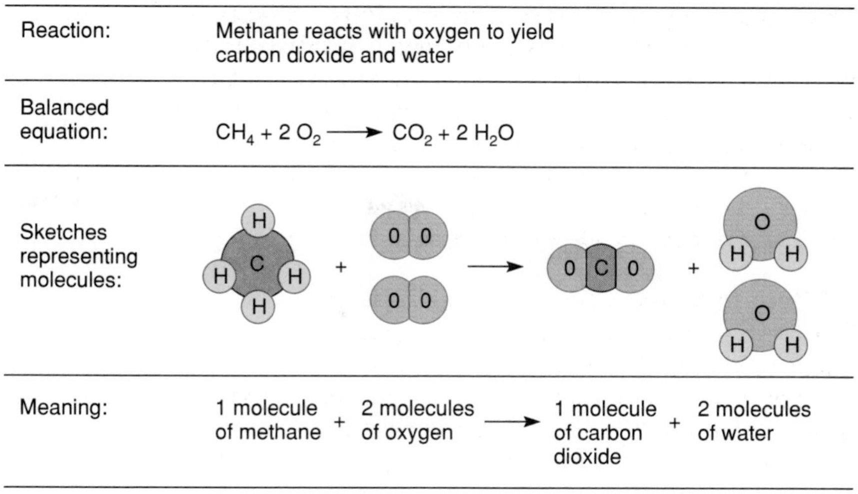

Reaction:	Methane reacts with oxygen to yield carbon dioxide and water
Balanced equation:	$CH_4 + 2 O_2 \longrightarrow CO_2 + 2 H_2O$

Sketches representing molecules:

Meaning:	1 molecule of methane	+	2 molecules of oxygen	→	1 molecule of carbon dioxide	+	2 molecules of water

Figure 14.6

Compare the numbers of each kind of atom in the balanced equation with the numbers of each kind of atom in the sketched representation. Both the equation and the sketch have the same number of atoms in the reactants and in the products.

Suppose your barbecue grill burns natural gas, not charcoal. Natural gas is mostly methane, CH_4. Methane burns by reacting with oxygen (O_2) to produce carbon dioxide (CO_2) and water vapor (H_2O). A balanced chemical equation for this reaction can be written by following a procedure of four steps.

Step 1: Write the correct formulas for the reactants and products in an unbalanced equation. The reactants and products could have been identified by chemical experiments, or they could have been predicted from what is known about chemical properties. This will be discussed in more detail later. For now, assume that the reactants and products are known and are given in words. For the burning of methane, the unbalanced, but otherwise correct, formula equation would be

$$CH_4 + O_2 \rightarrow CO_2 + H_2O \qquad \text{(not balanced)}$$

Step 2: Inventory the number of each kind of atom on both sides of the unbalanced equation. In the example there are

Reactants:	1 C	Products:	1 C
	4 H		2 H
	2 O		3 O

This shows that the H and O are unbalanced.

Step 3: Determine where to place coefficients in front of formulas to balance the equation. It is often best to focus on the simplest thing you can do with whole number ratios. The H and the O are unbalanced, for example, and there are 4 H atoms on the left and 2 H atoms on the right. Placing a coefficient 2 in front of H_2O will balance the H atoms:

$$CH_4 + O_2 \rightarrow CO_2 + 2 H_2O \qquad \text{(not balanced)}$$

Now take a second inventory:

Reactants:	1 C	Products:	1 C
	4 H		4 H
	2 O		4 O ($O_2 + 2 O$)

This shows the O atoms are still unbalanced with 2 on the left and 4 on the right. Placing a coefficient of 2 in front of O_2 will balance the O atoms.

$$CH_4 + 2 O_2 \rightarrow CO_2 + 2 H_2O \qquad \text{(balanced)}$$

Step 4: Take another inventory to determine (a) if the number of atoms on both sides are now equal and, if so, (b) if the coefficients are in the lowest possible whole number ratio. The inventory is now

Reactants:	1 C	Products:	1 C
	4 H		4 H
	4 O		4 O

(a) The number of each kind of atom on each side of the equation is the same, and (b) the ratio of $1:2 \rightarrow 1:2$ is the lowest possible whole number ratio. The equation is balanced, which is illustrated with sketches of molecules in figure 14.6.

Balancing chemical equations is mostly a trial-and-error procedure. But with practice, you will find there are a few generalized "role models" that can be useful in balancing equations for many simple reactions. The key to success at balancing equations is to think it out step-by-step while remembering the following:

1. Atoms are neither lost nor gained nor do they change their identity in a chemical reaction. The same kind and number of atoms in the reactants must appear in the products, meaning atoms are conserved.
2. A correct formula of a compound cannot be changed by altering the number or placement of subscripts. Changing subscripts changes the identity of a compound and the meaning of the entire equation.
3. A coefficient in front of a formula multiplies everything in the formula by that number.

There are also a few generalizations that can be helpful for success in balancing equations:

1. Look first to formulas of compounds with the most atoms and try to balance the atoms or compounds they were formed from or decomposed to.
2. Polyatomic ions that appear on both sides of the equation should be treated as independent units with a charge. That is, consider the polyatomic ion as a unit while taking an inventory rather than the individual atoms making up the polyatomic ion. This will save time and simplify the procedure.
3. Both the "crossover technique" and the use of "fractional coefficients" can be useful in finding the least common multiple to balance an equation. All of these generalizations are illustrated in examples 14.5, 14.6, and 14.7.

The physical state of reactants and products in a reaction is often identified by the symbols (g) for gas, (l) for liquid, (s) for solid, and (aq) for an aqueous solution (aqueous means water). If a gas escapes, this is identified with an arrow pointing up ($\uparrow$). A solid formed from a solution is identified with an arrow pointing down ($\downarrow$). The Greek symbol delta (Δ) is often used to indicate that heat is supplied.

Example 14.5

Propane is a liquified petroleum gas (LPG) that is often used as a bottled substitute for natural gas. Propane (C_3H_8) reacts with oxygen (O_2) to yield carbon dioxide (CO_2) and water vapor (H_2O). What is the balanced equation for this reaction?

Solution
Step 1: Write the correct formulas of the reactants and products in an unbalanced equation.

$$C_3H_8(g) + O_2(g) \rightarrow CO_2(g) + H_2O(g) \quad \textbf{(unbalanced)}$$

Step 2: Inventory the numbers of each kind of atom.

Reactants: 3 C   Products: 1 C
           8 H   2 H
           2 O   3 O

Step 3: Determine where to place coefficients to balance the equation. Looking at the compound with the most atoms (generalization 1), you can see that a propane molecule has 3 Cs and 8 Hs. Placing a coefficient of 3 in front of CO_2 and a 4 in front of H_2O will balance these atoms (3 of C and $4 \times 2 = 8$ H atoms on the right has the same number of atoms as C_3H_8 on the left),

$$C_3H_8(g) + O_2(g) \rightarrow 3 CO_2(g) + 4 H_2O(g) \quad \textbf{(not balanced)}$$

A second inventory shows

Reactants: 3 C
             8 H
             2 O

Products: 3 C
             8 H ($4 \times 2 = 8$)
             10 O [$(3 \times 2) + (4 \times 1) = 10$]

The O atoms are still unbalanced. Place a 5 in front of O_2 and the equation is balanced ($5 \times 2 = 10$). Remember that you cannot change the subscripts and that oxygen occurs as a diatomic molecule of O_2.

$$C_3H_8(g) + 5 O_2(g) \rightarrow 3 CO_2(g) + 4 H_2O(g) \quad \textbf{(balanced)}$$

Step 4: Another inventory shows (a) the number of atoms on both sides are now equal and (b) the coefficients are 1:5 $\rightarrow$ 3:4, the lowest possible whole number ratio. The equation is balanced.

Example 14.6

One type of water hardness is caused by the presence of calcium bicarbonate in solution, $Ca(HCO_3)_2$. One way to remove the troublesome calcium ions from wash water is to add washing soda, which is sodium carbonate, Na_2CO_3. The reaction yields sodium bicarbonate ($NaHCO_3$) and calcium carbonate ($CaCO_3$), which is insoluble. Since $CaCO_3$ is insoluble, the reaction removes the calcium ions from solution. Write a balanced equation for the reaction.

Solution
Step 1: Write the unbalanced equation

$$Ca(HCO_3)_2(aq) + Na_2CO_3(aq) \rightarrow NaHCO_3(aq) + CaCO_3\downarrow$$
$$\textbf{(not balanced)}$$

Step 2: Inventory the numbers of each kind of atom. This reaction has polyatomic ions that appear on both sides, so they should be treated as independent units with a charge (generalization 2). The inventory is

Reactants: 1 Ca   Products: 1 Ca
           2 $(HCO_3)^{1-}$ 1 $(HCO_3)^{1-}$
           2 Na   1 Na
           1 $(CO_3)^{2-}$ 1 $(CO_3)^{2-}$

Step 3: Placing a coefficient of 2 in front of $NaHCO_3$ will balance the equation,

$$Ca(HCO_3)_2(aq) + Na_2CO_3(aq) \rightarrow 2 NaHCO_3(aq) + CaCO_3\downarrow \quad \textbf{(balanced)}$$

Step 4: An inventory shows

Reactants:	1 Ca	Products:	1 Ca
	2 $(HCO_3)^{1-}$		2 $(HCO_3)^{1-}$
	2 Na		2 Na
	1 $(CO_3)^{2-}$		1 $(CO_3)^{2-}$

The coefficient ratio of 1:1 → 2:1 is the lowest whole number ratio. The equation is balanced.

Example 14.7

Gasoline is a mixture of hydrocarbons, including octane (C_8H_{18}). Combustion of octane produces CO_2 and H_2O, with the release of energy. Write a balanced equation for this reaction.

Solution

Step 1: Write the correct formulas in an unbalanced equation,

$$C_8H_{18}(g) + O_2(g) \rightarrow CO_2(g) + H_2O(g) \quad \textbf{(not balanced)}$$

Step 2: Take an inventory,

Reactants:	8 C	Products:	1 C
	18 H		2 H
	2 O		3 O

Step 3: Start with the compound with the most atoms (generalization 1) and place coefficients to balance these atoms,

$$C_8H_{18}(g) + O_2(g) \rightarrow 8\ CO_2(g) + 9\ H_2O(g) \quad \textbf{(not balanced)}$$

Redo the inventory,

Reactants:	8 C	Products:	8 C
	18 H		18 H
	2 O		25 O

The O atoms are still unbalanced. There are 2 O atoms in the reactants but 25 O atoms in the products. Since the subscript cannot be changed, it will take 12.5 O_2 to produce 25 oxygen atoms (generalization 3).

$$C_8H_{18}(g) + 12.5\ O_2(g) \rightarrow 8\ CO_2(g) + 9\ H_2O(g) \quad \textbf{(balanced)}$$

Step 4: (a) An inventory will show that the atoms balance,

Reactants:	8 C	Products:	8 C
	18 H		18 H
	25 O		25 O

(b) The coefficients are not in the lowest whole number ratio (one-half an O_2 does not exist). To make the lowest possible whole number ratio, all coefficients are multiplied by two. This results in a correct balanced equation of

$$2\ C_8H_{18}(g) + 25\ O_2(g) \rightarrow 16\ CO_2(g) + 18\ H_2O(g) \quad \textbf{(balanced)}$$

Generalizing Equations

In the previous chapters you learned that the act of classifying, or grouping, something according to some property makes the study of a large body of information less difficult. Generalizing from groups of chemical reactions also makes it possible to predict what will happen in similar reactions. For example, you have studied equations in the previous section describing the combustion of methane (CH_4), propane (C_3H_8), and octane (C_8H_{18}). Each of these reactions involves a *hydrocarbon,* a compound of the elements hydrogen and carbon. Each hydrocarbon reacted with O_2, yielding CO_2 and releasing the energy of combustion. Generalizing from these reactions, you could predict that the combustion of any hydrocarbon would involve the combination of atoms of the hydrocarbon molecule with O_2 to produce CO_2 and H_2O with the release of energy. Such reactions could be analyzed by chemical experiments, and the products could be identified by their physical and chemical properties. You would find your predictions based on similar reactions would be correct, thus justifying predictions from such generalizations. Butane, for example, is a hydrocarbon with the formula C_4H_{10}. The balanced equation for the combustion of butane is

$$2\ C_4H_{10}(g) + 13\ O_2(g) \rightarrow 8\ CO_2(g) + 10\ H_2O(g)$$

You could extend the generalization further, noting that the combustion of compounds containing oxygen as well as carbon and hydrogen also produces CO_2 and H_2O (figure 14.7). These compounds are *carbohydrates,* composed of carbon and water. Glucose, for example, was identified earlier as a compound with the formula $C_6H_{12}O_6$. Glucose combines with oxygen to produce CO_2 and H_2O and the balanced equation is

$$C_6H_{12}O_6(s) + 6\ O_2(g) \rightarrow 6\ CO_2(g) + 6\ H_2O(g)$$

Note that three molecules of oxygen were not needed from the O_2 reactant since the other reactant, glucose, contains six oxygen atoms per molecule. An inventory of atoms will show that the equation is thus balanced.

Combustion is a rapid reaction with O_2 that releases energy, usually with a flame. A very similar, although much slower reaction takes place in plant and animal respiration. In respiration, carbohydrates combine with O_2 and release energy used for biological activities. This reaction is slow compared to combustion and requires enzymes to proceed at body temperature. Nonetheless, CO_2 and H_2O are the products.

Oxidation-Reduction Reactions

The reactions involving hydrocarbons and carbohydrates with oxygen are examples of an important group of chemical reactions called *oxidation-reduction* reactions. When the term "oxidation" was first used, it specifically meant reactions involving the combination of oxygen with other atoms. But fluorine, chlorine, and other nonmetals were soon understood to have similar reactions as oxygen, so the definition was changed to one concerning the shifts of electrons in the reaction.

Chemical Formulas and Equations

Figure 14.7

Hydrocarbons are composed of the elements hydrogen and carbon. Propane (C_3H_8) and gasoline (which contains octane [C_8H_{18}]) are examples of hydrocarbons. *Carbohydrates* are composed of the elements hydrogen, carbon, and oxygen. Table sugar, for example, is the carbohydrate $C_{12}H_{22}O_{11}$. Generalizing, all hydrocarbons and carbohydrates react completely with oxygen to yield CO_2 and H_2O.

There are two broad classes of chemical reactions: (1) those in which electrons are transferred as a result of the reaction and (2) those in which electrons are not transferred. Reactions in which electrons are not transferred will be considered in the next section. For now, consider the **oxidation-reduction reaction,** a chemical reaction in which electrons are transferred from one atom to another. From the first letters, it is sometimes called a **redox reaction** for short.

When one substance loses electrons in a chemical reaction some other substance gains electrons. Substances and processes are given special names according to this loss or gain of electrons. The process of losing electrons is thus called **oxidation,** and the substance doing the losing is said to be **oxidized.** The substance oxidized loses electrons to another substance, which gains the negative charges. Gaining electrons is called **reduction,** and the substance doing the gaining is said to be **reduced.** Oxidation and reduction always occur together, and one atom's loss (oxidation) is another atom's gain (reduction). These names and definitions may seem backward (a gain is a reduction?), but they actually refer to changes in the **oxidation number,** the positive or negative charge of the atoms of an element in a compound. A gain of negative electrons makes a positive oxidation number smaller, thus the process is one of reduction of the number.

Sometimes one aspect of the redox reaction is often the focus of attention and the other is hardly mentioned. For example, foods and fuels are oxidized for energy, and the products

Table 14.1

Oxidation-reduction reaction terms

During Reaction Oxidation: Electrons are being lost Reduction: Electrons are being gained
Reactants Oxidizing agent: Gains electrons Reducing agent: Loses electrons
Products Substance oxidized: Lost electrons Substance reduced: Gained electrons

of this oxidation are CO_2 and H_2O, but little is said about what is reduced. For another example, consider that metals are separated from ores by reduction, but oxidation is seldom mentioned. Remember in both cases that (1) oxidation means a loss of electrons, not necessarily a combination with oxygen, (2) reduction means a gain of electrons, and it has nothing to do with making an ore body smaller, and (3) oxidation and reduction always occur together.

Substances that take electrons from other substances are called **oxidizing agents** (table 14.1). Oxidizing agents take electrons from the substances being oxidized. Oxygen is the most common oxidizing agent, and several examples have already been given about how it oxidizes foods and fuels. Chlorine is another commonly used oxidizing agent, often for the purposes of bleaching or killing bacteria (figure 14.8).

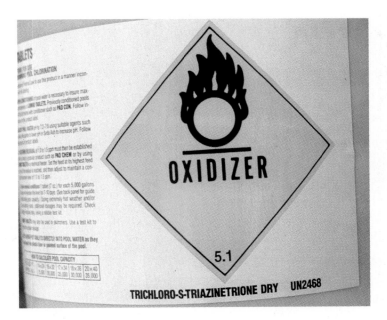

Figure 14.8
Oxidizing agents take electrons from other substances that are being oxidized. Oxygen and chlorine are commonly used strong oxidizing agents.

A **reducing agent** supplies electrons to the substance being reduced. Hydrogen and carbon are commonly used reducing agents. Carbon is commonly used as a reducing agent to extract metals from their ores. For example, carbon (from coke, which is coal that has been baked) reduces Fe_2O_3, an iron ore in the reaction

$$2 Fe_2O_3(s) + 3 C(s) \rightarrow 4 Fe(s) + 3 CO_2 \uparrow$$

The Fe in the ore gained electrons from the carbon, the reducing agent in this reaction.

ACTIVITIES

Silverware and silver-plated objects often become tarnished when the silver is oxidized by sulfur, forming Ag_2S. Commercial silver polishes often act by removing the oxidized layer with an abrasive. The silver can also be polished by reducing the Ag_2S back to metallic silver without removing a layer. Place the tarnished silver in a clean aluminum pan with about 80 g sodium bicarbonate ($NaHCO_3$) and 80 g NaCl dissolved in each liter of near boiling water. A sufficient amount should be prepared to cover the silver object or objects. The salts provide ions to help transfer electrons and facilitate the reaction. The reaction is

$$3 Ag_2S + 2 Al + 6 H_2O \rightarrow 6 Ag + 2 Al(OH)_3 + 3 H_2S$$

(Note: H_2S has a rotten egg odor and is a poisonous gas.)

Types of Chemical Reactions

Many chemical reactions can be classified as redox or nonredox reactions. Another way to classify chemical reactions is to consider what is happening to the reactants and products. This type of classification scheme leads to four basic categories of chemical reactions, which are (1) *combination,* (2) *decomposition,*

(3) *replacement,* and (4) *ion exchange* reactions. The first three categories are subclasses of redox reactions. It is in the ion exchange reactions that you will find the first example of a reaction that is not a redox reaction.

Combination Reactions

A **combination reaction** is a synthesis reaction in which two or more substances combine to form a single compound. The combining substances can be (1) elements, (2) compounds, or (3) combinations of elements and compounds. In generalized form, a combination reaction is

$$X + Y \rightarrow XY \qquad \text{equation 14.3}$$

Many redox reactions are combination reactions. For example, metals are oxidized when they burn in air, forming a metal oxide. Consider magnesium, which gives off a bright white light as it burns:

$$Mg(s) + O_2(g) \rightarrow 2 MgO(s)$$

Note how the magnesium-oxygen reaction follows the generalized form of equation 14.3.

The rusting of metals is oxidation that takes place at a slower pace than burning, but metals are nonetheless oxidized in the process (figure 14.9). Again noting the generalized form of a combination reaction, consider the rusting of iron:

$$4 Fe(s) + 3 O_2(g) \rightarrow 2 Fe_2O_3(s)$$

Nonmetals are also oxidized by burning in air, for example, when carbon burns with a sufficient supply of O_2:

$$C(s) + O_2(g) \rightarrow CO_2(g)$$

Note that all the combination reactions follow the generalized form of $X + Y \rightarrow XY$.

Decomposition Reactions

A **decomposition reaction,** as the term implies, is the opposite of a combination reaction. In decomposition reactions a compound is broken down (1) into the elements that make up the compound, (2) into simpler compounds, or (3) into elements and simpler compounds. Decomposition reactions have a generalized form of

$$XY \rightarrow X + Y + \ldots \qquad \text{equation 14.4}$$

Decomposition reactions generally require some sort of energy, which is usually supplied in the form of heat or electrical energy. An electric current, for example, decomposes water into hydrogen and oxygen:

$$2 H_2O(l) \overset{electricity}{\rightarrow} 2 H_2(g) + O_2(g)$$

Mercury (II) oxide is decomposed by heat, an observation that led to the discovery of oxygen (figure 14.10):

$$2 HgO(s) \overset{\Delta}{\rightarrow} 2 Hg(s) + O_2 \uparrow$$

Chemical Formulas and Equations

Figure 14.9
Rusting iron is a common example of a combination reaction, where two or more substances combine to form a new compound. Rust is iron (III) oxide formed from the combination of iron and oxygen. These new girders are made of unprotected iron that has begun to rust during construction of a new building.

Plaster is a building material made from a mixture of calcium hydroxide, $Ca(OH)_2$, and plaster of Paris, $CaSO_4$. The calcium hydroxide is prepared by adding water to calcium oxide (CaO), which is commonly called quicklime. Calcium oxide is made by heating limestone or chalk ($CaCO_3$), and

$$CaCO_3(s) \overset{\Delta}{\rightarrow} CaO(s) + CO_2\uparrow$$

Note that all the decomposition reactions follow the generalized form of $XY \rightarrow X + Y + \ldots$

Replacement Reactions

In a **replacement reaction,** an atom or polyatomic ion is replaced in a compound by a different atom or polyatomic ion. The replaced part can be either the negative or positive part of the compound. In generalized form, a replacement reaction is

$$XY + Z \rightarrow XZ + Y \qquad \textbf{equation 14.5}$$
(negative part replaced)

or

$$XY + A \rightarrow AY + X \qquad \textbf{equation 14.6}$$
(positive part replaced)

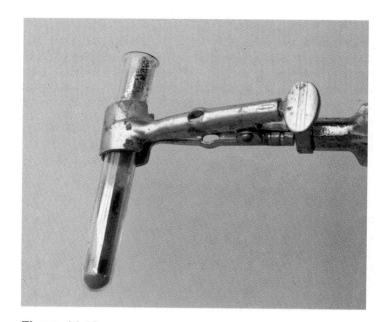

Figure 14.10
Mercury (II) oxide is decomposed by heat, leaving the silver-colored element mercury behind as oxygen is driven off. This is an example of a decomposition reaction, $2\ HgO \rightarrow 2\ Hg + O_2\uparrow$. Compare this equation to the general form of a decomposition reaction.

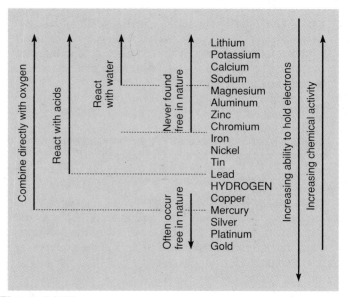

Figure 14.11

The activity series for common metals, together with some generalizations about the chemical activities of the metals. The series is used to predict which replacement reactions will take place and which reactions will not occur. (Note that hydrogen is not a metal and is placed in the series for reference to acid reactions.)

Figure 14.12

A reaction between metallic aluminum and the blue solution of copper (II) chloride. Aluminum is above copper in the activity series, and aluminum replaces the copper ions from the solution as copper is deposited as a metal. The aluminum loses electrons to the copper and forms aluminum ions in solution.

Replacement reactions occur because some elements have a stronger electron holding ability than other elements. Elements that have the least ability to hold on to their electrons are the most chemically active. Figure 14.11 shows a list of chemical activity of some metals, with the most chemically active at the top. Hydrogen is included because of its role in acids (see chapter 15). Take a few minutes to look over the generalizations listed in figure 14.11. The generalizations apply to combination, decomposition, and replacement reactions.

Replacement reactions take place as more active metals give up electrons to elements lower on the list with a greater electron holding ability. For example, aluminum is higher on the activity series than copper. When aluminum foil is placed in a solution of copper (II) chloride, aluminum is oxidized, losing electrons to the copper. The loss of electrons from metallic aluminum forms aluminum ions in solution, and the copper comes out of solution as a solid metal (figure 14.12).

$$2 \ Al(s) + 3 \ CuCl_2(aq) \rightarrow 2 \ AlCl_3(aq) + 3 \ Cu(s)$$

A metal will replace any metal ion in solution that it is above in the activity series. If the metal is listed below the metal ion in solution, no reaction occurs. For example, $Ag(s) + CuCl_2(aq) \rightarrow$ no reaction.

The very active metals (lithium, potassium, calcium, and sodium) react with water to yield metal hydroxides and hydrogen. For example,

$$2 \ Na(s) + 2 \ H_2O(l) \rightarrow 2 \ NaOH(aq) + H_2\uparrow$$

Acids yield hydrogen ions in solution, and metals above hydrogen in the activity series will replace hydrogen to form a metal salt. For example,

$$Zn(s) + H_2SO_4(aq) \rightarrow ZnSO_4(aq) + H_2\uparrow$$

In general, the energy involved in replacement reactions is less than the energy involved in combination or decomposition reactions.

Ion Exchange Reactions

An **ion exchange reaction** is a reaction that takes place when the ions of one compound interact with the ions of another compound, forming (1) a solid that comes out of solution (a precipitate), (2) a gas, or (3) water.

A water solution of dissolved ionic compounds is a solution of ions. For example, solid sodium chloride dissolves in water to become ions in solution,

$$NaCl(s) \rightarrow Na^{1+}Cl^{1-}(aq)$$

If a second ionic compound is dissolved with a solution of another, a mixture of ions results. The formation of a precipitate, a gas, or water, however, removes ions from the solution, and this must occur before you can say that an ionic exchange reaction has taken place. For example, water being treated for domestic use sometimes carries suspended matter that is removed by adding aluminum sulfate and calcium hydroxide to the water. The reaction is

$$3 \ Ca(OH)_2(aq) + Al_2(SO_4)_3(aq) \rightarrow 3 \ CaSO_4(aq) + 2 \ Al(OH)_3\downarrow$$

Chemical Formulas and Equations

The aluminum hydroxide is a jellylike solid, which traps the suspended matter for sand filtration. The formation of the insoluble aluminum hydroxide removed the aluminum and hydroxide ions from the solution, so an ion exchange reaction took place.

In general, an ion exchange reaction has the form

$$XY + AZ \rightarrow XZ + AY \qquad \text{equation 14.7}$$

where one of the products removes ions from the solution. The calcium hydroxide and aluminum sulfate reaction took place as the aluminum and calcium ions traded places. A solubility table such as the one in appendix B will tell you if an ionic exchange reaction has taken place. Aluminum hydroxide is insoluble, according to the table, so the reaction did take place. No ionic exchange reaction occurred if the new products are both soluble.

Another way for an ion exchange reaction to occur is if a gas or water molecule forms to remove ions from the solution. When an acid reacts with a base (an alkaline compound), a salt and water are formed

$$HCl(aq) + NaOH(aq) \rightarrow NaCl(aq) + H_2O(l)$$

The reactions of acids and bases are discussed in chapter 15.

Example 14.8

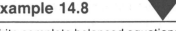

Write complete balanced equations for the following, and identify if each reaction is combination, decomposition, replacement, or ion exchange:

(a) silver(s) + sulfur(g) → silver sulfide(s)
(b) aluminum(s) + iron (III) oxide(s) → aluminum oxide(s) + iron
(c) sodium chloride(aq) + silver nitrate(aq) → ?
(d) potassium chlorate(s) $\overset{\Delta}{\rightarrow}$ potassium chloride(s) + oxygen(g)

Solution

(a) The reactants are two elements and the product is a compound, following the general form X + Y → XY of a combination reaction. Table 13.2 gives the charge on silver as Ag^{1+}, and sulfur (as the other nonmetals in family VIA) is S^{2-}. The balanced equation is

$$2 \, Ag(s) + S(g) \rightarrow Ag_2S(s)$$

Silver sulfide is the tarnish that appears on silverware.

(b) The reactants are an element and a compound that react to form a new compound and an element. The general form is XY + Z → XZ + Y, which describes a replacement reaction. The balanced equation is

$$2 \, Al(s) + Fe_2O_3(s) \rightarrow Al_2O_3(s) + 2 \, Fe(s)$$

This is known as a thermite reaction, and in the reaction aluminum reduces the iron oxide to metallic iron with the release of sufficient energy to melt the iron. The thermite reaction is sometimes used to weld large steel pieces, such as railroad rails.

(c) The reactants are water solutions of two compounds with the general form of XY + AZ →, so this must be the reactant part of an ion exchange reaction. Completing the products part of the equation by exchanging parts as shown in the general form and balancing,

$$NaCl(aq) + AgNO_3(aq) \rightarrow NaNO_3(?) + AgCl(?)$$

The solubility chart in appendix B is now consulted to find out if either of the products is insoluble. $NaNO_3$ is soluble and AgCl is insoluble. Since at least one of the products is insoluble, the reaction did take place and the equation is rewritten as

$$NaCl(aq) + AgNO_3(aq) \rightarrow NaNO_3(aq) + AgCl\downarrow$$

(d) The reactant is a compound and the products are a simpler compound and an element, following the generalized form of a decomposition reaction, XY → X + Y. The delta sign (Δ) also means that heat was added, which provides another clue that this is a decomposition reaction. The formula for the chlorate ion is in table 13.4. The balanced equation is

$$2 \, KClO_3(s) \overset{\Delta}{\rightarrow} 2 \, KCl(s) + 3 \, O_2\uparrow$$

Information from Chemical Equations

A balanced chemical equation describes what happens in a chemical reaction in a concise, compact way. The balanced equation also carries information about (1) atoms, (2) molecules, and (3) atomic weights. The balanced equation for the combustion of hydrogen, for example, is

$$2 \, H_2(g) + O_2(g) \rightarrow 2 \, H_2O(l)$$

An inventory of each kind of atom in the reactants and products shows

Reactants:		Products:	
	4 hydrogen		4 hydrogen
	2 oxygen		2 oxygen
Total:	6 atoms	Total:	6 atoms

There are six atoms before the reaction and there are six atoms after the reaction, which is in accord with the law of conservation of mass.

In terms of molecules, the equation says that two diatomic molecules of hydrogen react with one (understood) diatomic molecule of oxygen to yield one molecule of water. The number of coefficients in the equation is the number of molecules involved in the reaction. If you are concerned how two molecules plus one molecule could yield two molecules, remember that *atoms* are conserved in a chemical reaction, not molecules.

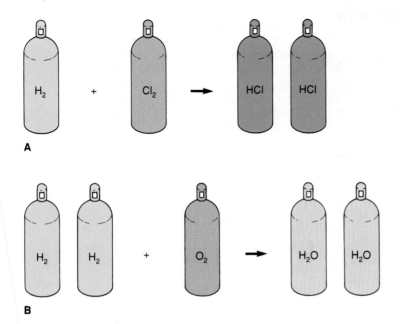

Figure 14.13

Reacting gases combine in ratios of small, whole number volumes when the temperature and pressure are the same for each volume. (a) One volume of hydrogen gas combines with one volume of chlorine gas to yield two volumes of hydrogen chloride gas. (b) Two volumes of hydrogen gas combine with one volume of oxygen gas to yield two volumes of water vapor.

Since atoms are conserved in a chemical reaction, their atomic weights should be conserved, too. One hydrogen atom has an atomic weight of 1.0 u so the formula weight of a diatomic hydrogen molecule must be 2×1.0 u, or 2.0 u. The formula weight of O_2 is 2×16.0 u, or 32 u. If you consider the equation in terms of atomic weights, then

Equation

$$2\,H_2 + O_2 \rightarrow 2\,H_2O$$

Atomic weights

$$2\,(1.0\ u + 1.0\ u) + (16.0\ u + 16.0\ u) \rightarrow 2\,(2 \times 1.0\ u + 16.0\ u)$$

$$4\ u + 32\ u \rightarrow 36\ u$$

$$36\ u \rightarrow 36\ u$$

The formula weight for H_2O is $(1.0\ u \times 2) + 16$ u, or 18 u. The coefficient of 2 in front of H_2O means there are two molecules of H_2O, so the mass of the products is 2×18 u, or 36 u. Thus the reactants had a total mass of 4 u + 32 u, or 36 u, and the products had a total mass of 36 u. Again, this is in accord with the law of conservation of mass.

The equation says that 4 u of hydrogen will combine with 32 u of oxygen. Thus hydrogen and oxygen combine in a mass ratio of 4:32, which reduces to 1:8. So one gram of hydrogen will combine with eight grams of oxygen, and, in fact, they will combine in this ratio no matter what the measurement units are (gram, kilogram, pound, etc.). They always combine in this mass ratio because this is the mass of the individual reactants.

Back in the early 1800s John Dalton attempted to work out a table of atomic weights as he developed his atomic theory (see chapter 12). Dalton made two major errors in determining the atomic weights, including (1) measurement errors about mass ratios of combining elements and (2) incorrect assumptions about the formula of the resulting compound. For water, for example, Dalton incorrectly measured that 5.5 g of oxygen combined with 1.0 g of hydrogen. He assumed that one atom of hydrogen combined with one atom of oxygen, resulting in a formula of HO. Thus Dalton concluded that the atomic mass of oxygen was 5.5 u and the atomic mass of hydrogen was 1.0 u. Incorrect atomic weights for hydrogen and oxygen led to conflicting formulas for other substances, and no one could show that the atomic theory worked.

The problem was solved during the first decade of the 1800s through the separate work of a French chemistry professor, Joseph Gay-Lussac, and an Italian physics professor, Amadeo Avogadro. In 1808, Gay-Lussac reported that reacting gases combined in small, whole number *volumes* when the temperature and pressure were constant. Two volumes of hydrogen, for example, combined with one volume of oxygen to form two volumes of water vapor. The term "volume" means any measurement unit, for example, a liter. Other reactions between gases were also observed to combine in small, whole number ratios and the pattern became known as the *law of combining volumes* (figure 14.13).

Avogadro proposed an explanation for the law of combining volumes in 1811. Avogadro proposed that equal volumes of all gases at the same temperature and pressure *contain the same number of molecules*. Avogadro's hypothesis had two important implications for the example of water. First, since two volumes of hydrogen combine with one volume of oxygen, it means that a molecule of water contains twice as many hydrogen atoms as oxygen atoms. The formula for water must be H_2O, not HO. Second, since *two* volumes of water vapor were produced, each molecule of hydrogen and each molecule of oxygen must be diatomic. Diatomic molecules of hydrogen and oxygen would double the number of hydrogen and oxygen atoms, thus producing twice as much water vapor. These two implications are illustrated in figure 14.14, along with a balanced equation for the reaction. Note that the coefficients in the equation now have two meanings, (1) the number of molecules of each substance involved in the reaction and (2) the ratios of combining volumes. The coefficient of 2 in front of the H_2, for example, means two molecules of H_2. It also means two volumes of H_2 gas when all volumes are measured at the same temperature and pressure. Recall that the volume of any gas at the same temperature and pressure contains the same number of molecules. Thus the coefficient in a balanced equation means a ratio of *any number* of molecules, from 2 of H_2 and 1 of O_2, 20 of H_2 and 10 of O_2, 2,000 of H_2 and 1,000 of O_2, or however many are found in 2 L of H_2 and 1 L of O_2.

Chemical Formulas and Equations

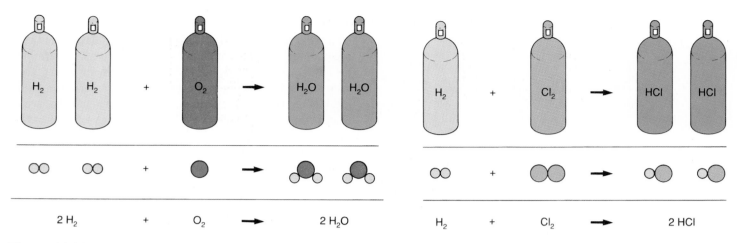

Figure 14.14
Avogadro's hypothesis of equal volumes of gas having equal numbers of molecules offered an explanation for the law of combining volumes.

Example 14.9

Propane is a hydrocarbon with the formula C_3H_8 that is used as a bottled gas. (a) How many liters of oxygen are needed to burn 1 L of propane gas? (b) How many liters of carbon dioxide are produced by the reaction? Assume all volumes to be measured at the same temperature and pressure.

Solution
The balanced equation is

$$C_3H_8(g) + 5\ O_2(g) \rightarrow 3\ CO_2(g) + 4\ H_2O(g)$$

The coefficients tell you the relative number of molecules involved in the reaction, that 1 molecule of propane reacts with 5 molecules of oxygen to produce 3 molecules of carbon dioxide and 4 molecules of water. Since equal volumes of gases at the same temperature and pressure contain equal numbers of molecules, the coefficients also tell you the relative volumes of gases. Thus 1 L of propane (a) requires 5 L of oxygen and (b) yields 3 L of carbon dioxide (and 4 L of water vapor) when reacted completely.

Units of Measurement Used with Equations

The coefficients in a balanced equation represent a ratio of any *number* of molecules involved in a chemical reaction. The equation has meaning about the atomic *weights* and formula *weights* of reactants and products. The counting of numbers and the use of atomic weights are brought together in a very important measurement unit called a *mole* (from the Latin meaning "a mass"). Here are the important ideas in the mole concept:

1. Recall that the atomic weights of elements are average relative masses of the isotopes of an element. The weights are based on a comparison to carbon-12, with an assigned mass of exactly 12.00 (see chapter 12).

2. The *number* of C-12 atoms in exactly 12.00 g of C-12 has been measured experimentally to be 6.02×10^{23}. This number is called **Avogadro's number,** named after the scientist who reasoned that equal volumes of gases contain equal numbers of molecules.

3. An amount of a substance that contains Avogadro's number of atoms, ions, molecules or any other chemical unit is defined as a **mole** of the substance. Thus a mole is 6.02×10^{23} atoms, ions, etc., just as a dozen is 12 eggs, apples, etc. The mole is the chemist's measure of atoms, molecules, or other chemical units. A mole of Na^+ ions is 6.02×10^{23} Na^+ ions.

4. A mole of C-12 atoms is defined as having a mass of exactly 12.00 g, a mass that is numerically equal to its atomic mass. So the mass of a mole of C-12 atoms is 12.00 g, or

$$\underset{(12.00\ u)}{\text{mass of one atom}} \times \underset{(6.02 \times 10^{23})}{\text{one mole}} = \underset{12.00\ g}{\text{mass of a mole of C-12}}$$

The masses of all the other isotopes are *based* on a comparison to the C-12 atom. Thus a He-4 atom has one-third the mass of a C-12 atom. An atom of Mg-24 is twice as massive as a C-12 atom. Thus

	1 Atom	×	1 Mole	=	Mass of Mole
C-12:	12.00 u	×	6.02×10^{23}	=	12.00 g
He-4:	4.00 u	×	6.02×10^{23}	=	4.00 g
Mg-24:	24.00 u	×	6.02×10^{23}	=	24.00 g

Therefore the mass of a mole of any element is numerically equal to its atomic mass. And any mass of an element that is in the same ratio as their atomic masses must contain the same number of atoms (figure 14.15).

This reasoning can be used to generalize about formula weights, molecular weights, and atomic weights since they are all based on atomic mass units relative to C-12. The **gram-atomic weight** is the mass in grams of one mole of an element that is numerically equal to its atomic weight. The atomic weight of

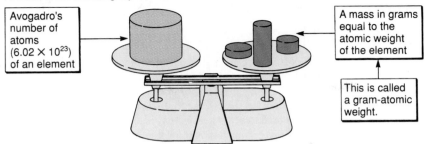

Avogadro's number of atoms (6.02×10^{23}) of an element

A mass in grams equal to the atomic weight of the element

This is called a gram-atomic weight.

B. Each of the following represents one mole of a compound:

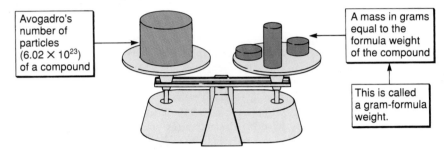

Avogadro's number of particles (6.02×10^{23}) of a compound

A mass in grams equal to the formula weight of the compound

This is called a gram-formula weight.

C. Each of the following represents one mole of a molecular substance:

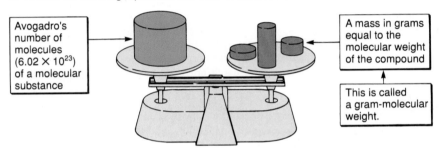

Avogadro's number of molecules (6.02×10^{23}) of a molecular substance

A mass in grams equal to the molecular weight of the compound

This is called a gram-molecular weight.

Figure 14.15

The mole concept for (a) elements, (b) compounds, and (c) molecular substances. A mole contains 6.02×10^{23} particles. Since every mole contains the same number of particles, the ratio of the mass of any two moles is the same as the ratio of the masses of individual particles making up the two moles.

carbon is 12.01 u, the gram-atomic weight of carbon is 12.01 g. The atomic weight of magnesium is 24.3 u; the gram-atomic weight of magnesium of 24.3 g. Any gram-atomic weight contains Avogadro's number of atoms. Therefore the gram-atomic weights of the elements all contain the same number of atoms.

Similarly, the **gram-formula weight** of a compound is the mass in grams of one mole of the compound that is numerically equal to its formula weight. The **gram-molecular weight** is the gram-formula weight of a molecular compound. Note that one mole of O atoms (6.02×10^{23} oxygen atoms) has a gram-atomic weight of 16.0 g, but one mole of O_2 molecules (6.02×10^{23} oxygen molecules) has a gram-molecular weight of 32.0 g. Stated the other way around, 32.0 g of O_2 and 16.0 g of O both contain the same Avogadro's number of particles.

Example 14.10

(a) A 100 percent silver chain has a mass of 107.9 g. How many silver atoms are in the chain? (b) What is the mass of one mole of sodium chloride, NaCl?

Solution

The mole concept and Avogadro's number provide a relationship between numbers and masses. (a) The atomic weight of silver is 107.9 u, so the gram-atomic weight of silver is 107.9 g. A gram-atomic weight is one mole of an element, so the silver chain contains 6.02×10^{23} silver atoms. (b) The formula weight of NaCl is 58.5 u, so the gram-formula weight is 58.5 g. One mole of NaCl has a mass of 58.5 g.

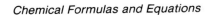

Chemical Formulas and Equations

The Catalytic Converter

The modern automobile produces two troublesome products in the form of (1) nitrogen monoxide (NO) and (2) hydrocarbons from the incomplete combustion of gasoline. These products from the exhaust enter the air to react in sunlight, eventually producing an irritating haze known as *photochemical smog*. To reduce photochemical smog, modern automobiles are fitted with a catalytic converter as part of the automobile exhaust system. This Feature is about how the catalytic converter combats smog-forming pollutants.

Chemical reactions proceed at a rate that is affected by (1) the concentration of the reactants, (2) the temperature at which a reaction occurs, and (3) the surface area of the reaction. In general, a higher concentration, higher temperatures, and greater surface area mean a faster reaction. The problem with nitrogen monoxide is that it is easily oxidized to nitrogen dioxide (NO_2), a reddish brown, damaging gas that also plays a key role in the formation of photochemical smog. Nitrogen dioxide and the hydrocarbons are oxidized slowly in the air when left to themselves. What is needed is a means to decompose nitrogen monoxide and uncombusted hydrocarbons rapidly before they are released into the air.

The rate at which a chemical reaction proceeds is affected by a *catalyst*, a material that speeds up a chemical reaction without being permanently changed by the reaction. Apparently, molecules require a certain amount of energy to change the chemical bonds that tend to keep them as they are, unreacted. This certain amount of energy is called the *activation energy*, and it represents an energy barrier that must be overcome before a chemical reaction can take place. This explains why chemical reactions proceed at a faster rate at higher temperatures. At higher temperatures, molecules have greater average kinetic energies, thus they already have part of the minimum energy needed for a reaction to take place.

A catalyst appears to speed a chemical reaction by lowering the activation energy. Molecules become temporarily attached to the surface of the catalyst, which weakens the chemical bonds holding the molecule together. Thus the weakened molecule is easier to break apart and the activation energy is lowered. Some catalysts do this better with some specific compounds than others, and extensive chemical research programs are devoted to finding new and more effective catalysts.

Automobile catalytic converters use unreactive metals, such as platinum, and transition metal oxides such as copper (II) oxide and chromium (III) oxide. Catalytic reactions that occur in the converter include the following:

$$H_2O + CO \overset{catalyst}{\rightarrow} H_2 + CO_2$$

$$2\ NO + 2\ CO \overset{catalyst}{\rightarrow} N_2 + 2\ CO_2$$

$$2\ NO + 2\ H \overset{catalyst}{\rightarrow} N_2 + 2\ H_2O$$

$$O_2 + CO* \overset{catalyst}{\rightarrow} CO_2 + H_2O$$

*or other hydrocarbons

Thus nitrogen monoxide is reduced to nitrogen gas and hydrocarbons are oxidized to CO_2 and H_2O. Sometimes small quantities of sulfur in gasoline are burned, producing SO_2 in the exhaust, which is converted to H_2S by the catalyst. H_2S has a rotten egg odor. Leaded fuels such as those containing tetraethyl lead, $Pb(C_2H_5)_4$, will poison a catalytic converter, rendering it useless. In spite of these problems, a catalytic converter can reduce or oxidize about 90 percent of the hydrocarbons, 85 percent of the carbon monoxide, and 40 percent of the nitrogen monoxide from exhaust gases. Other controls, such as exhaust gas recirculation (EGR) are used to further reduce NO formation.

Quantitative Uses of Equations

A balanced chemical equation can be interpreted in terms of (1) a *molecular ratio* of the reactants and products, (2) a *mole ratio* of the reactants and products, or (3) a *mass ratio* of the reactants and products. Consider, for example, the balanced equation for reacting hydrogen with nitrogen to produce ammonia,

$$3\ H_2(g) + N_2(g) \rightarrow 2\ NH_3(g)$$

From a *molecular* point of view, the equation says that three molecules of hydrogen combine with one molecule of N_2 to form two molecules of NH_3. The coefficients of $3:1 \rightarrow 2$ thus express a molecular ratio of the reactants and the products.

The molecular ratio leads to the concept of a *mole ratio* since any number of molecules can react as long as they are in the ratio of $3:1 \rightarrow 2$. The number could be Avogadro's number, so $(3) \times (6.02 \times 10^{23})$ molecules of H_2 will combine with $(1) \times (6.02 \times 10^{23})$ molecules of N_2 to form $(2) \times (6.02 \times 10^{23})$ molecules of ammonia. Since 6.02×10^{23} molecules is the number of particles in a mole, the coefficients therefore represent the *numbers of moles* involved in the reaction. Thus three moles of H_2 react with one mole of N_2 to produce two moles of NH_3.

The mole ratio of a balanced chemical equation leads to the concept of a *mass ratio* interpretation of a chemical equation. The gram-formula weight of a compound is the mass in grams of *one mole* that is numerically equal to its formula weight. Therefore the equation also describes the mass ratios of the reactants and the products. The mass ratio can be calculated from the mole relationship described in the equation. The three interpretations are summarized in table 14.2.

Table 14.2

Three interpretations of a chemical equation

Equation: $3 H_2 + N_2 \rightarrow 2 NH_3$
Molecular Ratio:
3 molecules H_2 + 1 molecule $N_2 \rightarrow$ 2 molecules NH_3
Mole Ratio:
3 moles H_2 + 1 mole $N_2 \rightarrow$ 2 moles NH_3
Mass Ratio:
6.0 g H_2 + 28.0 g $N_2 \rightarrow$ 34.0 g NH_3

Thus the coefficients in a balanced equation can be interpreted in terms of molecules, which leads to an interpretation of moles, mass, or any formula unit. The mole concept thus provides the basis for calculations about the quantities of reactants and products in a chemical reaction.

Summary

A chemical formula is a shorthand way of describing the composition of a compound. An *empirical formula* identifies the simplest whole number ratio of element. A *molecular formula* identifies the actual number of atoms in a molecule.

The sum of the atomic weights of all the atoms in any formula is called the *formula weight*. The *molecular weight* is the formula weight of a molecular substance. The formula weight of a compound can be used to determine the *mass percentage* of elements making up a compound.

A concise way to describe a chemical reaction is to use formulas in a *chemical equation*. A chemical equation with the same number of each kind of atom on both sides is called a *balanced equation*. A balanced equation is in accord with the *law of conservation of mass,* which states that atoms are neither created nor destroyed in a chemical reaction. To balance a chemical equation, *coefficients* are placed in front of chemical formulas. Subscripts of formulas may not be changed since this would change the formula, meaning a different compound.

One important group of chemical reactions is called *oxidation-reduction reactions,* or *redox* reactions for short. Redox reactions are reactions where shifts of electrons occur. The process of losing electrons is called *oxidation,* and the substance doing the losing is said to be *oxidized.* The process of gaining electrons is called *reduction,* and the substance doing the gaining is said to be *reduced.* The process is called reduction because gaining electrons reduces the *oxidation number,* the charge an atom appears to have while in a compound. Substances that take electrons from other substances are called *oxidizing agents.* Substances that supply electrons are called *reducing agents.*

Chemical reactions can also be classified as (1) *combination,* (2) *decomposition,* (3) *replacement,* or (4) *ion exchange.* The first three of these are redox reactions, but ion exchange is not.

A balanced chemical equation describes chemical reactions and has quantitative meaning about numbers of atoms, numbers of molecules, and conservation of atomic weights. The coefficients also describe the *volumes* of combining gases. At a constant temperature and pressure gases combine in small, whole number ratios that are given by the coefficients. Each volume at the same temperature and pressure contains the *same number of molecules.*

The number of atoms in exactly 12.00 g of C-12 is called *Avogadro's number,* which has a value of 6.02×10^{23}. Any substance that contains Avogadro's number of atoms, ions, molecules, or any chemical unit called a *mole* of that substance. The mole is a measure of a number of atoms, molecules, or other chemical units. The mass of a mole of any substance is equal to the atomic mass of that substance.

The mass, number of atoms, and mole concepts are generalized to other units. The *gram-atomic weight* of an element is the mass in grams that is numerically equal to its atomic weight. The *gram-formula weight* of a compound is the mass in grams that is numerically equal to the formula weight of the compound. The *gram-molecular weight* is the gram-formula weight of a molecular compound. The relationships between the mole concept and the mass ratios can be used with a chemical equation for calculations about the quantities of reactants and products in a chemical reaction.

Summary of Equations

14.1

$$\left(\frac{part}{whole}\right)(100\% \text{ of whole}) = \% \text{ of part}$$

14.2

$$\frac{\left(\begin{array}{c}\text{atomic weight} \\ \text{of element}\end{array}\right)\left(\begin{array}{c}\text{number of atoms} \\ \text{of element}\end{array}\right)}{\text{formula weight of compound}} \times \begin{array}{c}100\% \text{ of} \\ \text{compound}\end{array} = \begin{array}{c}\% \text{ of} \\ \text{element}\end{array}$$

14.3

Combination reaction, general form

$$X + Y \rightarrow XY$$

14.4

Decomposition reaction, general form

$$XY \rightarrow X + Y + \ldots$$

14.5

Replacement reaction, general form for negative part replaced

$$XY + Z \rightarrow XZ + Y$$

14.6

Replacement reaction, general form for positive part replaced

$$XY + A \rightarrow AY + X$$

14.7

Ion exchange reaction, general form

$$XY + AZ \rightarrow XZ + AY$$

(Note: One of the products must remove ions by forming an insoluble product, water, or a gas.)

Key Terms

Avogadro's number (p. 280)
chemical equation (p. 270)
combination reaction (p. 275)
decomposition reaction (p. 275)
empirical formula (p. 267)
formula weight (p. 268)
gram-atomic weight (p. 280)
gram-formula weight (p. 281)
gram-molecular weight (p. 281)
ion exchange reaction (p. 277)
law of conservation of mass
  (p. 270)
mole (p. 280)

molecular formula (p. 267)
molecular weight (p. 268)
oxidation (p. 274)
oxidation number (p. 274)
oxidation-reduction reaction
  (p. 274)
oxidized (p. 274)
oxidizing agents (p. 274)
redox reaction (p. 274)
reduced (p. 274)
reducing agent (p. 275)
reduction (p. 274)
replacement reaction (p. 276)

Applying the Concepts

1. The formula weight of sulfuric acid, H_2SO_4, is
   a. 49 u.
   b. 50 u.
   c. 98 u.
   d. 194 u.

2. A balanced chemical equation has the same
   a. number of molecules on both sides of the equation.
   b. kinds of molecules on both sides of the equation.
   c. number of each kind of atom on both sides of the equation.
   d. all of the above.

3. A chemical equation is balanced by changing the
   a. subscripts.
   b. superscripts.
   c. coefficients.
   d. any of the above as necessary to achieve a balance.

4. Since wood is composed of carbohydrates, you should expect what gases to exhaust from a fireplace when complete combustion takes place?
   a. carbon dioxide, carbon monoxide, and pollutants
   b. carbon dioxide and water vapor
   c. carbon monoxide and smoke
   d. it depends on the type of wood being burned

5. According to the activity series for metals, adding metallic iron to a solution of aluminum chloride should result in
   a. a solution of iron chloride and metallic aluminum.
   b. a mixed solution of iron and aluminum chloride.
   c. the formation of iron hydroxide with hydrogen given off.
   d. no reaction.

6. In a replacement reaction, elements that have the most ability to hold onto their electrons are
   a. the most chemically active.
   b. the least chemically active.
   c. not generally involved in replacement reactions.
   d. none of the above.

7. The incomplete equation of $2 KClO_3(s) \xrightarrow{\Delta}$ probably represents which type of chemical reaction?
   a. combination
   b. decomposition
   c. replacement
   d. ion exchange

8. In the equation of $2 H_2 + O_2 \rightarrow 2 H_2O$
   a. the total mass of the gaseous reactants is less than the total mass of the liquid product.
   b. the total number of molecules in the reactants is equal to the total number of molecules in the products.
   c. one volume of oxygen combines with two volumes of hydrogen to produce two volumes of water.
   d. all of the above.

9. An amount of a substance that contains Avogadro's number of atoms, ions, or molecules is a
   a. mole.
   b. gram-atomic weight.
   c. gram-formula weight.
   d. any of the above.

10. If you have 6.02×10^{23} atoms of metallic iron you will have how many grams of iron?
    a. 26
    b. 55.8
    c. 334.8
    d. 3.4×10^{25}

Answers

1. c   2. c   3. c   4. b   5. d   6. b   7. b   8. c   9. d
10. b

Questions for Thought

1. How is an empirical formula like and unlike a molecular formula?

2. How can you be sure that a formula of a covalent compound is an empirical formula or a molecular formula?

3. What, if anything, is wrong with calculating a molecular weight of sodium chloride?

4. Describe how you could calculate how much pure iron would be furnished by a 1,000 g sample of the iron ore hematite, iron (III) oxide.

5. Describe the basic parts of a chemical equation. Identify how the physical state of elements and compounds is identified in an equation.

6. Describe a general equation that can be used to find percent. Describe a general equation that can be used to find the mass percentage of an element in a compound.

7. What is the law of conservation of mass? How do you know if a chemical equation is in accord with this law?

8. What is meant by a balanced chemical equation? What, if anything, is wrong with changing subscripts in a formula to balance an equation? What is a coefficient and what does it mean?

9. Describe in your own words how a chemical equation is balanced.

10. What is a hydrocarbon? What is a carbohydrate? In general, what are the products of complete combustion of hydrocarbons and carbohydrates?

11. What is meant by oxidation? Reduction? An oxidation-reduction reaction?

12. Define and give two examples of (a) oxidizing agents and (b) reducing agents.

13. Explain why an oxidizing agent is always reduced in a redox reaction.

14. Define and give an example in the form of a balanced equation of (a) a combination reaction, (b) a decomposition reaction, (c) a replacement reaction, and (d) an ion exchange reaction.

15. What must occur in order for an ion exchange reaction to take place? What is the result if this does not happen?

16. Potassium reacts with iodine to form potassium iodide. Write a balanced chemical equation for this reaction. Identify the oxidizing agent.

17. Predict the products for the following reactions: (a) the combustion of ethyl alcohol, C_2H_5OH, (b) the rusting of aluminum, and (c) the reaction between iron and sodium chloride.

18. The formula for butane is C_4H_{10}. Is this an empirical formula or a molecular formula? Explain the reason(s) for your answer.

19. How is the activity series for metals (see figure 14.11) used to predict whether or not a replacement reaction will occur?

20. What is a gram-formula weight? How is it calculated?

21. Avogadro hypothesized that equal volumes of all gases at the same temperature and pressure contain the same number of molecules. For the equation $2 H_2 + O_2 \rightarrow 2 H_2O$ describe how this hypothesis explains (a) that the formula for water is H_2O, not HO, and (b) that both oxygen and hydrogen are diatomic.

22. What is the meaning and value of Avogadro's number? What is a mole?

Exercises

Group A—Solutions Provided in Appendix D

1. Identify the following as empirical formulas or molecular formulas and indicate any uncertainty with (?):
   (a) $MgCl_2$
   (b) C_2H_2
   (c) BaF_2
   (d) C_8H_{18}
   (e) CH_4
   (f) S_8

2. What is the formula weight for each of the following compounds?
   (a) Copper (II) sulfate
   (b) Carbon disulfide
   (c) Calcium sulfate
   (d) Sodium carbonate

3. What is the mass percentage composition of the elements in the following compounds?
   (a) Fool's gold, FeS_2
   (b) Boric acid, H_3BO_3
   (c) Baking soda, $NaHCO_3$
   (d) Aspirin, $C_9H_8O_4$

4. Write balanced chemical equations for each of the following unbalanced reactions:
   (a) $2SO_2 + O_2 \rightarrow 2SO_3$
   (b) $4 P + 5O_2 \rightarrow 2P_2O_5$
   (c) $2 Al + 6 HCl \rightarrow 2AlCl_3 + 3 H_2$
   (d) $2NaOH + H_2SO_4 \rightarrow Na_2SO_4 + 2 H_2O$
   (e) $Fe_2O_3 + 3CO \rightarrow 2Fe + 3CO_2$
   (f) $Mg(OH)_2 + H_3PO_4 \rightarrow Mg_3(PO_4)_2 + H_2O$

5. Identify the following as combination, decomposition, replacement, or ion exchange reactions:
   (a) $NaCl(aq) + AgNO_3(aq) \rightarrow NaNO_3(aq) + AgCl\downarrow$
   (b) $H_2O(l) + CO_2(g) \rightarrow H_2CO_3(l)$
   (c) $2 NaHCO_3(s) \rightarrow Na_2CO_3(s) + H_2O(g) + CO_2(g)$
   (d) $2 Na(s) + Cl_2(g) \rightarrow 2 NaCl(s)$
   (e) $Cu(s) + 2 AgNO_3(aq) \rightarrow Cu(NO_3)_2(aq) + 2 Ag(s)$
   (f) $CaO(s) + H_2O(l) \rightarrow Ca(OH)_2(aq)$

Group B—Solutions Not Given

1. Identify the following as empirical formulas or molecular formulas and indicate any uncertainty with (?):
   (a) CH_2O
   (b) $C_6H_{12}O_6$
   (c) NaCl
   (d) CH_4
   (e) F_6
   (f) CaF_2

2. Calculate the formula weight for each of the following compounds:
   (a) Dinitrogen monoxide
   (b) Lead (II) sulfide
   (c) Magnesium sulfate
   (d) Mercury (II) chloride

3. What is the mass percentage composition of the elements in the following compounds?
   (a) Potash, K_2CO_3
   (b) Gypsum, $CaSO_4$
   (c) Saltpeter, KNO_3
   (d) Caffeine, $C_8H_{10}N_4O_2$

4. Write balanced chemical equations for each of the following unbalanced reactions:
   (a) $NO + O_2 \rightarrow NO_2$
   (b) $KClO_3 \rightarrow KCl + O_2$
   (c) $NH_4Cl + Ca(OH)_2 \rightarrow CaCl_2 + NH_3 + H_2O$
   (d) $NaNO_3 + H_2SO_4 \rightarrow Na_2SO_4 + HNO_3$
   (e) $PbS + H_2O_2 \rightarrow PbSO_4 + H_2O$
   (f) $Al_2(SO_4)_3 + BaCl_2 \rightarrow AlCl_3 + BaSO_4$

5. Identify the following as combination, decomposition, replacement, or ion exchange reactions:
   (a) $ZnCO_3(s) \rightarrow ZnO(s) + CO_2\uparrow$
   (b) $2 NaBr(aq) + Cl_2(g) \rightarrow 2 NaCl(aq) + Br_2(g)$
   (c) $2 Al(s) + 3 Cl_2(g) \rightarrow 2 AlCl_3(s)$
   (d) $Ca(OH)_2(aq) + H_2SO_4(aq) \rightarrow CaSO_4(aq) + 2 H_2O(l)$
   (e) $Pb(NO_3)_2(aq) + H_2S(g) \rightarrow 2 HNO_3(aq) + PbS\downarrow$
   (f) $C(s) + ZnO(s) \rightarrow Zn(s) + CO\uparrow$

6. Write complete, balanced equations for each of the following reactions:
   (a) $C_5H_{12}(g) + O_2(g) \rightarrow$
   (b) $HCl(aq) + NaOH(aq) \rightarrow$
   (c) $Al(s) + Fe_2O_3(s) \rightarrow$
   (d) $Fe(s) + CuSO_4(aq) \rightarrow$
   (e) $MgCl(aq) + Fe(NO_3)_2(aq) \rightarrow$
   (f) $C_6H_{10}O_5(s) + O_2(g) \rightarrow$

7. Write complete, balanced equations for each of the following decomposition reactions. Include symbols for physical states, heating, and others as needed:
   (a) When solid potassium chlorate is heated, solid potassium chloride and oxygen gas are formed.
   (b) Upon electrolysis, molten bauxite (aluminum oxide) yields solid aluminum metal and oxygen gas.
   (c) Upon heating, solid calcium carbonate yields solid calcium oxide and carbon dioxide gas.

8. Write complete, balanced equations for each of the following replacement reactions. If no reaction is predicted, write "no reaction" as the product:
   (a) $Na(s) + H_2O(l) \rightarrow$
   (b) $Au(s) + HCl(aq) \rightarrow$
   (c) $Al(s) + FeCl_3(aq) \rightarrow$
   (d) $Zn(s) + CuCl_2(aq) \rightarrow$

9. Write complete, balanced equations for each of the following ion exchange reactions. If no reaction is predicted, write "no reaction" as the product:
   (a) $NaOH(aq) + HNO_3(aq) \rightarrow$
   (b) $CaCl_2(aq) + KNO_3(aq) \rightarrow$
   (c) $Ba(NO_3)_2(aq) + Na_3PO_4(aq) \rightarrow$
   (d) $KOH(aq) + ZnSO_4(aq) \rightarrow$

10. The gas welding torch is fueled by two tanks, one containing acetylene (C_2H_2) and the other pure oxygen (O_2). The very hot flame of the torch is produced as acetylene burns,

$$2\ C_2H_2(g) + O_2(g) \rightarrow 4\ CO_2(g) + H_2O(g)$$

According to this equation, how many liters of oxygen are required to burn 1 L of acetylene?

6. Write complete, balanced equations for each of the following reactions:
   (a) $C_3H_5(g) + O_2(g) \rightarrow$
   (b) $H_2SO_4(aq) + KOH(aq) \rightarrow$
   (c) $C_6H_{12}O_6(s) + O_2(g) \rightarrow$
   (d) $Na_3PO_4(aq) + AgNO_3(aq) \rightarrow$
   (e) $NaOH(aq) + Al(NO_3)_3(aq) \rightarrow$
   (f) $Mg(OH)_2(aq) + H_3PO_4(aq) \rightarrow$

7. Write complete, balanced equations for each of the following decomposition reactions. Include symbols for physical states, heating, and others as needed:
   (a) When solid zinc carbonate is heated, solid zinc oxide and carbon dioxide gas are formed.
   (b) Liquid hydrogen peroxide decomposes to liquid water and oxygen gas.
   (c) Solid ammonium nitrite decomposes to liquid water and nitrogen gas.

8. Write complete, balanced equations for each of the following replacement reactions. If no reaction is predicted, write "no reaction" as the product:
   (a) $Zn(s) + FeCl_2(aq) \rightarrow$
   (b) $Zn(s) + AlCl_3(aq) \rightarrow$
   (c) $Cu(s) + HgCl_2(aq) \rightarrow$
   (d) $Al(s) + HCl(aq) \rightarrow$

9. Write complete, balanced equations for each of the following ion exchange reactions. If no reaction is predicted, write "no reaction" as the product:
   (a) $Ca(OH)_2(aq) + H_2SO_4(aq) \rightarrow$
   (b) $NaCl(aq) + AgNO_3(aq) \rightarrow$
   (c) $NH_4NO_3(aq) + MgPO_4(aq) \rightarrow$
   (d) $Na_3PO_4(aq) + AgNO_3(aq) \rightarrow$

10. Iron (III) oxide, or hematite, is one mineral used as an iron ore. Other iron ores are magnetite (Fe_3O_4) and siderite ($FeCO_3$). Assume that you have pure samples of all three ores that will be reduced by reaction with carbon monoxide. Which of the three ores will have the highest yield of metallic iron?

Chapter 15

Water and Solutions

Figure 15.1

Water is the most abundant liquid on the earth and is necessary for all life. Because of water's great dissolving properties, any sample is a solution containing solids, other liquids, and gases from the environment. This city water processing facility removes such dissolved and suspended solids from water for domestic use.

THE previous three chapters were concerned with elements, compounds, and their chemical reactions. Elements and compounds are pure substances, materials with a definite, fixed composition that is the same throughout. Mixtures, on the other hand, have a composition that may vary from one sample to the next. A mixture contains the particles of one substance physically dispersed throughout the particles of another substance. These particles can be any size, from the size of rock particles in a gravel mixture down to the smallest size possible—particles the size of ions or molecules. The size of the particles gives a mixture its appearance. When the particles are relatively large and visible to the eye, a mixture appears to be heterogeneous. Thus a pile of gravel appears to be a heterogeneous mixture. At the other end of the size scale, a uniform mixture of ion- or molecule-sized particles appears to be homogeneous, or the same throughout. Such homogeneous mixtures are called *solutions*. This chapter is concerned with solutions, those involving water in particular.

Many common liquids are solutions. Solutions are commonly used in everyday household activities, as well as in the chemistry laboratory. The household activities of cooking, cleaning, and painting all involve solutions that will be considered in this chapter. Detergents, cleaners, and drain openers all function because a solution is a medium for rapid chemical change. Solids react slowly, if at all, because they have limited contact only at their immediate surfaces. Thus solutions are commonly used to speed reactions. The water you drink is also a solution (figure 15.1). Hard water results in certain reactions that occur between soap and the water solution. This chapter considers hard water and how it is softened, in addition to acids, bases, the pH scale, and many common solutions used in everyday activities.

Solutions

A **solution** is defined as a homogeneous mixture of ions or molecules of two or more substances. The process of making a solution is called *dissolving,* and during dissolving the different components that make up the solution become mixed. The components could be sugar and water, for example, and when sugar dissolves in water, molecules of sugar are uniformly dispersed throughout the molecules of water. The uniform taste of sweetness of any part of the sugar solution is a result of this uniform mixing. In a salt and water solution, however, the salt dissolves into sodium and chlorine ions. The components of a salt and water solution are thus sodium and chlorine ions dissolved in water.

Solutions are not limited to solids, such as sugar or salt, dissolved in liquids, such as water. There are three states of matter, and a solution can involve any combination of the ions or molecules of gases, liquids, or solids. Thus it is possible to have nine kinds of solutions. Table 15.1 gives an example of each of these nine kinds of solutions.

Table 15.1

Examples of each kind of solution

Kind of Solution	Example
Gas in gas	*Air* is O_2, CO_2, and other gases dissolved in nitrogen gas.
Liquid in gas	*Water vapor* is water molecules dissolved in air.
Solid in gas	*Smoke* is solid particles dissolved in air (many smoke particles are larger than molecules, so they are not part of the solution).
Gas in liquid	*Soda water* is CO_2 dissolved in water.
Liquid in liquid	*Alcohol* is alcohol molecules dissolved in water (unless 200 proof, which is pure alcohol).
Solid in liquid	*Seawater* is the ions of salts dissolved in water (mostly sodium chloride).
Liquid in solid	*Dental fillings* are prepared from a solution of mercury in silver.
Solid in solid	*Brass* is zinc dissolved in copper.

The amounts of the components of a solution are identified by the general terms of *solvent* and *solute*. The **solvent** is the component present in the larger amount. The **solute** is the component that dissolves in the solvent. Atmospheric air, for example, is about 78 percent nitrogen, so nitrogen is considered the solvent. Oxygen (about 21 percent), argon (about 0.9 percent), and other gases make up the solutes.

If one of the components of a solution is a liquid, it is usually identified as the solvent. An *aqueous solution* is a solution of a solid, a liquid, or a gas in water. Water is the solvent in an aqueous solution. A *tincture* is a solution of something dissolved in alcohol. Tincture of iodine, for example, is iodine dissolved in the solvent alcohol.

Concentration of Solutions

The relative amounts of solute and solvent are described by the **concentration** of a solution. In general, a solution with a large amount of solute is *concentrated* and a solution with much less solute is *dilute*. The terms "dilute" and "concentrated" are somewhat arbitrary, and it is sometimes difficult to know the difference between a solution that is "weakly concentrated" and one that is "not very diluted." More meaningful information is provided by measurement of the *amount of solute in a solution*. There are different ways to express concentration measurements, each lending itself to a particular kind of solution or to how the information will be used. For example, you read about concentrations of parts per million in an article about pollution, but most of the concentration of solutions sold in stores are reported in percent by volume or percent by weight (figure 15.2).

Figure 15.2
There are different ways to express the concentration of a solution.
How many different ways can you identify in this photograph?

Each of these concentrations are concerned with the amount of *solute* in the *solution*.

Concentration ratios that describe small concentrations of solute are sometimes reported as a ratio of **parts per million** (ppm) or **parts per billion** (ppb). This ratio could mean ppm by volume or ppm by weight, depending if the solution is a gas or a liquid. For example, a drinking water sample with 1 ppm Na^+ by weight has 1 weight measure of solute, sodium ions, *in* every 1,000,000 weight measures of the total solution. By way of analogy, 1 ppm expressed in money means 1 cent in every $10,000 (which is one million cents). A concentration of 1 ppb means 1 cent in $10,000,000. Thus the concentrations of very dilute solutions, such as certain salts in seawater, minerals in drinking water, and pollutants in water or in the atmosphere are often reported in ppm or ppb.

Sometimes it is useful to know the conversion factors between ppm or ppb and the more familiar percent concentration by weight. These factors are ppm $\div$ (1×10^4) = percent concentration and ppb $\div$ (1×10^7) = percent concentration. For example, very hard water (water containing Ca^{2+} or Mg^{2+} ions), by definition, contains more than 300 ppm of the ions. This is a percent concentration of $300 \div 1 \times 10^4$, or 0.03 percent. To be suitable for agricultural purposes, irrigation water must not contain more than 700 ppm of total dissolved salts, which means a concentration no greater than 0.07 percent salts.

Water and Solutions

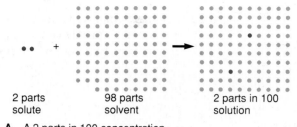

2 parts
solute

98 parts
solvent

2 parts in 100
solution

A. A 2 parts in 100 concentration

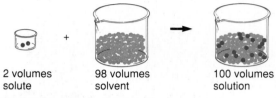

2 volumes
solute

98 volumes
solvent

100 volumes
solution

B. A 2% by volume concentration

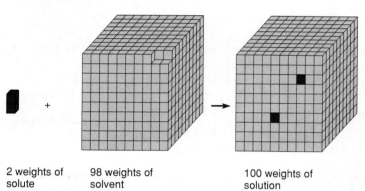

2 weights of
solute

98 weights of
solvent

100 weights of
solution

C A 2% by weight concentration

Figure 15.3

Three ways of expressing the amount of solute in a solution include (*a*) expressed as parts, as in parts per million, (*b*) expressed as percent by volume, and (*c*) expressed as percent by weight. Note that all three expressions are concerned with amounts of solute and amounts of solution.

The concentration term of **percent by volume** is defined as the *volume of solute in 100 volumes of solution*. This concentration term is just like any other percentage ratio, that is, "part" divided by the "whole" times 100 percent. The distinction is that the part and the whole are concerned with a volume of solute and a volume of solution. Knowing the meaning of percent by volume can be useful in consumer decisions. Rubbing alcohol, for example, can be purchased at a wide range of prices. The various brands range from a concentration, according to the labels, of "12% by volume" to "70% by volume." If the volume unit is mL, a "12% by volume" concentration contains 12 mL of pure isopropyl (rubbing) alcohol in every 100 mL of solution. The "70% by volume" contains 70 mL of isopropyl alcohol in every 100 mL of solution. The relationship for % by volume is

$$\frac{\text{volume solute}}{\text{volume solution}} \times 100\% \text{ solution} = \% \text{ solute}$$

or

$$\frac{V_{solute}}{V_{solution}} \times 100\% \text{ solution} = \% \text{ solute} \qquad \textbf{equation 15.1}$$

The concentration term of **percent by weight** is defined as the *weight of solute in 100 weight units of solution*. This concentration term is just like any other percentage composition, the difference being that it is concerned with the weight of solute (the part) in a weight of solution (the whole) (figure 15.3). Hydrogen peroxide, for example, is usually sold in a concentration of "3% by weight." This means that 3 oz (or other weight units) of pure hydrogen peroxide are in 100 oz of solution. Since weight is proportional to mass in a given location, mass units such as grams are sometimes used to calculate a percent by weight. The relationship for percent by weight (using mass units) is

$$\frac{\text{mass of solute}}{\text{mass of solution}} \times 100\% \text{ solution} = \% \text{ solute}$$

or

$$\frac{m_{solute}}{m_{solution}} \times 100\% \text{ solution} = \% \text{ solute} \qquad \textbf{equation 15.2}$$

Example 15.1

Vinegar that is prepared for table use is a mixture of acetic acid in water, usually 5.00% by weight. How many grams of pure acetic acid are in 25.0 g of vinegar?

Solution

The percent by weight is given (5.00%), the mass of the solution is given (25.0 g), and the mass of the solute (HCl) is the unknown. The relationship between these quantities is found in equation 15.2, which can be solved for the mass of the solute:

% solute = 5.00%

$m_{solution}$ = 25.0 g

m_{solute} = ?

$$\frac{m_{solute}}{m_{solution}} \times 100\% \text{ solution} = \% \text{ solute}$$

$$\therefore$$

$$m_{solute} = \frac{(m_{solution})(\% \text{ solute})}{100\% \text{ solution}}$$

$$= \frac{(m_{solution})(\% \text{ solute})}{100\% \text{ solution}}$$

$$= \frac{(25.0 \text{ g})(5.00)}{100} \text{ solute}$$

$$= \boxed{1.25 \text{ g solute}}$$

Example 15.2

A solution used to clean contact lenses contains 0.002% by volume of thimerosal as a preservative. How many mL of this preservative are needed to make 100,000 L of the cleaning solution? (Answer: 2.0 mL)

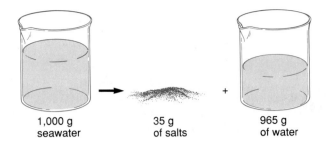

Figure 15.4

Salinity is a measure of the amount of salts dissolved in 1 kg of solution. If 1,000 g of seawater were evaporated, 35.0 g of salts would remain as 965.0 g of water leave.

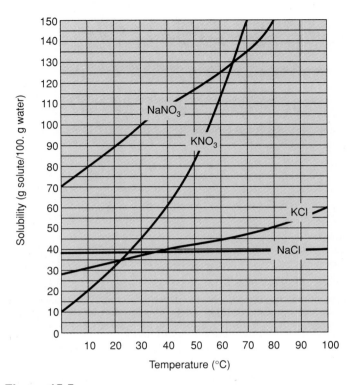

Figure 15.5

Approximate solubility curves for sodium nitrate, potassium nitrate, potassium chloride, and sodium chloride.

Both percent by volume and percent by weight are defined as the volume or weight per 100 units of solution because percent *means* parts per hundred. The measure of dissolved salts in seawater is called *salinity*. **Salinity** is defined as the mass of salts dissolved in 1,000 g of solution. As illustrated in figure 15.4, evaporation of 965 g of water from 1,000 g of seawater will leave an average of 35 g salts. Thus the average salinity of the seawater is 35‰. Note the ‰, which means parts per thousand just as % means parts per hundred. Thus, the average salinity of seawater is 35‰, which means there are 35 g of salts dissolved in every 1,000 g of seawater. The equivalent percent measure for salinity is 3.5%, which equals 35‰.

Solubility

Gases and liquids appear to be soluble in all proportions, but there is an obvious limit to how much solid can be dissolved in a liquid. You may have noticed that a cup of hot tea will dissolve several teaspoons of sugar, but the limit of solubility is reached quickly in a glass of iced tea. The limit of how much sugar will dissolve seems to depend on the temperature of the tea. More sugar added to the cold tea after the limit is reached will not dissolve, and solid sugar granules begin to accumulate at the bottom of the glass. At this limit the sugar and tea solution is said to be *saturated*. Dissolving does not actually stop when a solution becomes saturated and undissolved sugar continues to enter the solution. However, dissolved sugar is now returning to the undissolved state at the same rate. The overall equilibrium condition of sugar dissolving and sugar coming out of solution is called a **saturated solution.** A saturated solution is a *state of equilibrium that exists between dissolving solute and solute coming out of solution.* You actually cannot see the dissolving and coming out of solution that occurs in a saturated solution because the exchanges are taking place with particles the size of molecules or ions.

Not all compounds dissolve as sugar does, and more or less of a given compound may be required to produce a saturated solution at a particular temperature. In general, the difficulty of dissolving a given compound is referred to as solubility. More specifically, the **solubility** of a solute is defined as the *concentration that is reached in a saturated solution at a particular temperature.* Solubility varies with the temperature as the sodium and potassium salt examples show in figure 15.5. These solubility curves describe the amount of solute required to reach the saturation equilibrium at a particular temperature. In general, the solubilities of most ionic solids increase with temperature, but there are exceptions. In addition, some salts release heat when dissolved in water and other salts absorb heat when dissolved. The "instant cold pack" used for first aid is a bag of water containing a second bag of ammonium nitrate (NH_4NO_3). When the bag of ammonium nitrate is broken, the compound dissolves and absorbs heat.

You can usually dissolve more of a solid, such as salt or sugar, as the temperature of the water is increased. Contrary to what you might expect, gases usually become *less* soluble in water as the temperature increases. As a glass of water warms, small bubbles collect on the sides of the glass as dissolved air comes out of solution. The first bubbles that appear when warming a pot of water to boiling are also bubbles of dissolved air coming out of solution. This is why water that has been boiled usually tastes "flat." The dissolved air has been removed by the heating. The "normal" taste of water can be restored by pouring the boiled water back and forth between two glasses. The water dissolves more air during this process, restoring the usual taste.

Changes in pressure have no effect on the solubility of solids in liquids but greatly affect the solubility of gases. The fizz of an opened bottle or can of soda occurs because pressure is reduced on the beverage and dissolved carbon dioxide comes out of solution. In general, *gas solubility decreases with temperature and increases with pressure.* As usual, there are exceptions to this generalization.

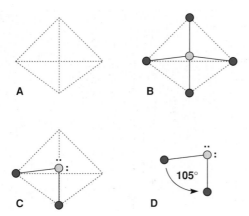

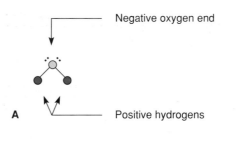

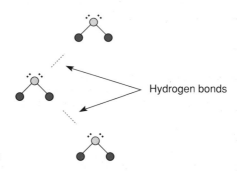

Figure 15.6

(a) A tetrahedron is a four-sided pyramid. (b) Molecules with four bonding electron pairs have a tetrahedral shape like the CCl₄ molecule illustrated here. Note the carbon atom is in the center of the pyramid, with chlorine atoms at each of the four corners. (c) A water molecule has two bonding pairs and two lone pairs in a tetrahedral arrangement. (d) The water molecule has an angular arrangement called bent. If something were attached to the two lone pairs, it would be a tetrahedral arrangement (see also figure 15.8).

Water Solutions

Water is the one chemical that is absolutely essential for all organisms, both plant and animal. Organisms use water to transport food and essential elements and to carry biological molecules in a solution. In fact, living organisms are mostly cells filled with water solutions. Your foods are mostly water, with fruits and vegetables consisting of up to 95 percent water and meat consisting of 50 percent water. Your body consists of over 70 percent water by weight. It is the specific properties of water that make it so important for life, in particular the unusual ability of water to act as a solvent. The properties of a water molecule must be considered in order to account for the solvent abilities of water.

Properties of Water Molecules

A water molecule is composed of two atoms of hydrogen and one atom of oxygen joined by a polar covalent bond. The electron dot formula for a water molecule is

$$:\!\overset{\cdot\,\cdot}{\underset{\cdot\,\cdot}{O}}\!:$$
$$H \qquad H$$

Notice the four pairs of electrons around the oxygen atom, consisting of two lone pairs and two bonding pairs. These four electron pairs are arranged in the direction of a tetrahedral arrangement (figure 15.6). Since there are only two bonding pairs, however, a water molecule has a bent molecular arrangement, represented with a structural formula as

$$\overset{O}{/\ \backslash}$$
$$H \qquad H$$

veen the two bonds is not 90° but has been ex-
sured to be about 105°, which is very close to
'e (figure 15.6).

Figure 15.7

(a) The water molecule is polar, with centers of positive and negative charges. (b) Attractions between these positive and negative centers establish hydrogen bonds between adjacent molecules.

Oxygen has a stronger electronegativity (3.5) than hydrogen (2.1), and thus it has a greater ability to attract shared electrons. This results in the bonding electrons spending more time near the oxygen atom than the hydrogen atoms, essentially leaving the hydrogen as exposed protons on one end of the molecule. Thus the water molecule has polar covalent bonds and is a *polar molecule* with centers of negative or positive charges.

The polar water molecule, with its negative oxygen end and positive hydrogen end, sets the stage for **intermolecular forces,** forces of interaction between molecules. The positive end of the water molecule can attract the negative end of another molecule, including the hydrogen end of another water molecule. The general term for weak attractive intermolecular forces is a **van der Waals force,** named after the Dutch physicist who first proposed the concept. Specifically, the intermolecular force of attraction between a hydrogen atom in a polar molecule and electrons in another molecule is called a **hydrogen bond.** A hydrogen bond is a weak to moderate bond between the hydrogen end (+) of a polar molecule and the negative end (−) of a second polar molecule (figure 15.7). In general, hydrogen bonding occurs between the hydrogen atom of one molecule and the oxygen, fluorine, or nitrogen atom of another molecule.

Hydrogen bonding accounts for the physical properties of water, including its unusually high heat of fusion and heat of vaporization as well as its unusual density changes. Figure 15.8 shows the hydrogen bonded structure of ice. Each oxygen atom is associated with four hydrogen atoms, two in the H₂O molecule and two from other water molecules, held by hydrogen bonds. The arrangement is tetrahedral, forming a six-sided hexagonal structure. The open space of the hexagonal channel in

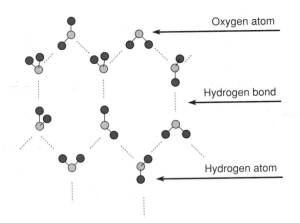

Figure 15.8
The hexagonal structure of ice. Hydrogen bonding between the oxygen atom and two hydrogen atoms of other water molecules results in a tetrahedral arrangement, which forms the open, hexagonal structure of ice.

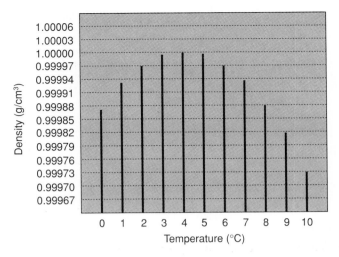

Figure 15.9
The density of water from 0° C to 10° C. The density of water is at a maximum at 4° C, becoming less dense as it is cooled or warmed from this temperature. Hydrogen bonding explains this unusual behavior.

this structure results in ice being less dense than water. The shape of the channel also suggests why snowflakes always have six sides.

When ice is warmed, the increased vibrations of the molecules begin to expand and stretch the hydrogen bond structure. When ice melts, about 15 percent of the hydrogen bonds break and the open structure collapses into the more compact arrangement of liquid water. As the liquid water is warmed from 0° C still more hydrogen bonds break down and the density of the water steadily increases. At 4° C the expansion of water from the increased molecular vibrations begins to predominate and the density decreases steadily with further warming (figure 15.9). Thus water has its greatest density at a temperature of 4° C.

The heat of fusion, specific heat, and heat of vaporization of water are unusually high when compared to other, but chemically similar, substances. These high values are accounted for by the additional energy needed to break hydrogen bonds.

The Dissolving Process

A solution is formed when the molecules or ions of two or more substances become homogeneously mixed. But the process of dissolving must be more complicated than the simple mixing together of particles because (1) solutions become saturated, meaning there is a limit on solubility, and (2) some substances are *insoluble,* not dissolving at all or at least not noticeably. In general, the forces of attraction between molecules or ions of the solvent and solute determine if something will dissolve and any limits on the solubility. These forces of attraction and their role in the dissolving process will be considered in the following examples.

First, consider the dissolving process in gaseous and liquid solutions. In a gas, the intermolecular forces are small so gases can mix in any proportion. Fluids that can mix in any proportion like this are called **miscible fluids.** Fluids that do not mix are called *immiscible fluids.* Air is a mixture of gases and vapors, so gases and vapors are miscible.

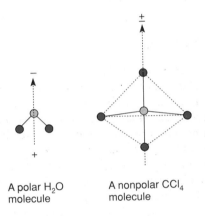

A polar H_2O molecule   A nonpolar CCl_4 molecule

Figure 15.10
Water is polar and carbon tetrachloride is nonpolar. Since like dissolves like, water and carbon tetrachloride are immiscible.

Liquid solutions can have a gas, another liquid, or a solid as a solute. Gases are miscible in liquids, and a carbonated beverage (your favorite cola) is the common example, consisting of carbon dioxide dissolved in water. Whether or not two given liquids form solutions depends on some similarities in their molecular structures. The water molecule, for example, is a polar molecule with a negative end and a positive end. On the other hand, carbon tetrachloride (CCl_4) is a molecule with polar bonds that are symmetrically arranged. Because of the symmetry, CCl_4 has no negative or positive ends so it is nonpolar. So some liquids have polar molecules and some liquids have nonpolar molecules. The general rule for forming solutions is *like dissolves like* (figure 15.10). A nonpolar compound, such as carbon tetrachloride, will dissolve oils and greases because they are nonpolar compounds. Water, a polar compound, will not dissolve the nonpolar oils and greases. Carbon tetrachloride was at one time used as a cleaning solvent because of its oil and grease dis-

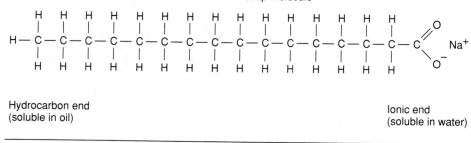

Structural formula of a soap molecule

Hydrocarbon end
(soluble in oil)

Ionic end
(soluble in water)

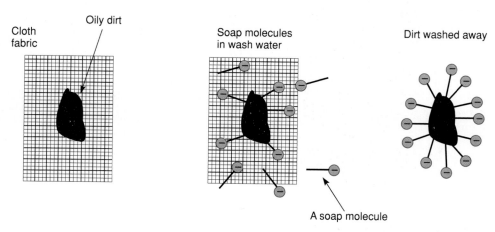

A soap molecule

Figure 15.11

Soap cleans oil and grease because one end of the soap molecule is
soluble in water and the other end is soluble in oil and grease. Thus,
the soap molecule provides a link between two substances that would
otherwise be immiscible.

solving abilities. Its use is no longer recommended because it is
also a possible health hazard (liver damage).

Some molecules, such as ethyl alcohol and soap, have a
part of the molecule that is polar and a part that is nonpolar.
Washing with water alone will not dissolve oils because water
and oil are immiscible. When soap is added to the water, however, the polar end of the soap molecule is attracted to the polar
water molecules and the nonpolar end is absorbed into the oil.
A particle (larger than a molecule) is formed and the oil is
washed away with the water (figure 15.11).

The "like dissolves like" rule applies to solids and liquid
solvents as well as liquids and liquid solvents. Polar solids, such
as salt, will readily dissolve in water, which has polar molecules,
but do not dissolve readily in oil, grease, or other nonpolar solvents. Polar water readily dissolves salt because the charged
polar water molecules are able to exert an attraction force on
the ions, pulling them away from the crystal structure. Thus
ionic compounds dissolve in water.

As noted in figure 15.5, ionic compounds vary in their solubilities in water. This difference is explained by the existence
of two different forces involved in an ongoing "tug of war." One
force is the attraction between an ion on the surface of the crystal
and a water molecule, an *ion-polar molecule force.* When solid

sodium chloride and water are mixed together, the negative ends
of the water molecules (the oxygen end) become oriented toward
the positive sodium ions on the crystal. Likewise, the positive
ends of water molecules (the hydrogen ends) become oriented
toward the negative chlorine ions. The attraction of water molecules for ions is called **hydration.** If the force of hydration is
greater than the attraction between the ions in the solid, they
are pulled away from the solid and dissolving occurs. Not considering the role of water in this dissolving process, the equation
is

$$Na^+Cl^-(s) \rightarrow Na^+(aq) + Cl^-(aq)$$

which shows that the ions were separated from the solid to
become a solution of ions. In other compounds the attraction
between the ions in the solid might be greater than the energy
of hydration. In this case, the ions of the solid would win the
"tug of war" and the ionic solid is insoluble (figure 15.12).

The saturation of soluble compounds is explained in terms
of hydration eventually occupying a large number of the polar
water molecules. Fewer available water molecules means less
attraction on the ionic solid, with more solute ions being pulled
back to the surface of the solid. The tug of war continues back
and forth as an equilibrium condition is established.

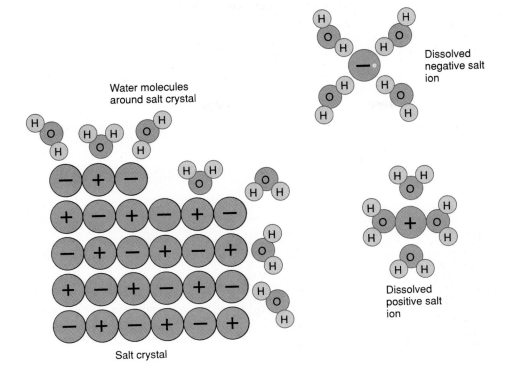

Water molecules
around salt crystal

Dissolved
negative salt
ion

Salt crystal

Dissolved
positive salt
ion

Figure 15.12

An ionic solid dissolves in water because the number of water molecules around the surface is greater than the number of other ions of the solid. This attraction between polar water molecules and a charged ion enables the water molecules to pull ions away from the crystal, a process called dissolving.

Properties of Solutions

Pure solvents have characteristic physical and chemical properties that are changed by the presence of the solute. Following are some of the more interesting changes.

Electrolytes

Water solutions of ionic substances will conduct an electric current, so they are called **electrolytes.** Ions must be present and free to move in a solution to carry the charge, so electrolytes are solutions containing ions. Pure water will not conduct an electric current as it is a covalent compound, which ionizes only very slightly. Water solutions of sugar, alcohol, and most other covalent compounds are nonconductors, so are called **nonelectrolytes.** Nonelectrolytes are covalent compounds that form molecular solutions, so they cannot conduct an electric current (figure 15.13).

Some covalent compounds are nonelectrolytes as a pure liquid but become electrolytes when dissolved in water. Pure hydrogen chloride (HCl), for example, does not conduct an electric current, so you can assume that it is a molecular substance. When dissolved in water, hydrogen chloride does conduct a current, so it must now contain ions. Evidently, the hydrogen chloride has become **ionized** by the water. The process of forming ions from molecules is called **ionization.** Hydrogen chloride, just

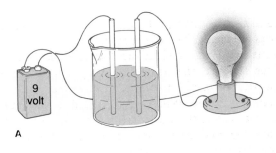

A

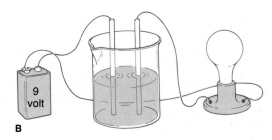

B

Figure 15.13

(a) Water solutions that conduct an electric current are called electrolytes. (b) Water solutions that do not conduct electricity are called nonelectrolytes.

Water and Solutions

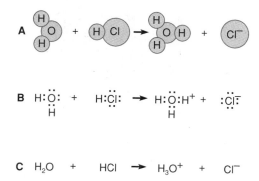

Figure 15.14

Three representations of water and hydrogen chloride in an ionizing reaction. (*a*) Sketches of molecules involved in the reaction. (*b*) Electron dot equation of the reaction. (*c*) The chemical equation for the reaction. Each of these representations show the hydrogen being pulled away from the chlorine atom to form H_3O^+, the hydronium ion.

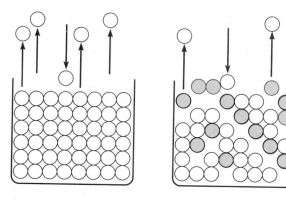

Figure 15.15

The rate of evaporation, and thus the vapor pressure, is less for a solution than for a solvent in the pure state. The greater the solute concentration, the less the vapor pressure.

as water, has polar molecules. The positive hydrogen atom on the HCl molecule is attracted to the negative oxygen end of a water molecule, and the force of attraction is strong enough to break the hydrogen-chlorine bond, forming charged particles (figure 15.14). The reaction is

$$HCl(l) + H_2O(l) \rightarrow H_3O^+(aq) + Cl^-(aq)$$

The H_3O^+ ion is called a **hydronium ion.** A hydronium ion is basically a molecule of water with an attached hydrogen ion. The presence of the hydronium ion gives the solution new chemical properties, and the solution is no longer hydrogen chloride but is *hydrochloric acid.* Hydrochloric acid, and other acids, will be discussed shortly.

Boiling Point

Boiling occurs when the pressure of the vapor escaping from a liquid is equal to the atmospheric pressure on the liquid. The *normal* boiling point is defined as the temperature at which the vapor pressure is equal to the average atmospheric pressure at sea level. For pure water, this temperature is 100° C (212° F).

During the later 1880s, a French chemist named Francois Raoult observed that the vapor pressure over a solution is *less* than the vapor pressure over the pure solvent at the same temperature. Molecules of a liquid can escape into the air only at the surface of the liquid, and the presence of molecules of a solute means that fewer solvent molecules can be at the surface to escape. Thus the vapor pressure over a solution is less than the vapor pressure over a pure solvent (figure 15.15).

Because the vapor pressure of a solution is less than over the pure solvent, the solution boils at a higher temperature. A higher temperature is required to increase the vapor pressure to that of the atmospheric pressure. Some cooks have been observed to add a "pinch" of salt to a pot of water before boiling. Is this to increase the boiling point, and therefore cook the food more quickly? How much does a pinch of salt increase the boiling temperature? The answers are found in the relationship between the concentration of a solute and the boiling point of the solution.

It is the number of solute particles (ions or molecules) at the surface of a solution that increases the boiling point. Recall from chapter 14 that a mole is a measure that can be defined as a number of particles called Avogadro's number. Since the number of particles at the surface is proportional to the ratio of particles in the solution, the concentration of the solute will directly influence the increase in the boiling point. In other words, the boiling point of any dilute solution is increased proportional to the concentration of the solute. For water, the boiling point is increased 0.521° C for every mole of solute dissolved in 1,000 g of water. Thus any water solution will boil at a higher temperature than pure water. Since it boils at a higher temperature, it also takes a longer time to reach the boiling point.

It makes no difference what the substance is that is dissolved in the water, one mole of solute in 1,000 g of water will elevate the boiling point by 0.521° C. A mole contains Avogadro's number of particles, so a mole of any solute will lower the vapor pressure by the same amount. Sucrose, or table sugar, for example, is $C_{12}H_{22}O_{11}$ and has a gram-formula weight of 342 g. So 342 g of sugar in 1,000 g of water (about a liter) will increase the boiling point by 0.521° C. Thus if you measure the boiling point of a sugar solution you can determine the concentration of sugar in the solution. For example, pancake syrup that boils at 100.261° C (sea level pressure) must contain 171 g of sugar dissolved in 1,000 g of water. You know this because the increase of 0.261° C over 100° C is one-half of 0.521° C. If the boiling point were increased by 0.521° C over 100° C, the syrup would have the full gram-formula weight (342 g) dissolved in a kg of water.

Since it is the number of particles of solute in a specific sample of water that elevates the boiling point, there is a different effect by dissolved covalent and dissolved ionic compounds (figure 15.16). Sugar is a covalent compound, and the solute is thus molecules of sugar moving between the water molecules. Sodium chloride, on the other hand, is an ionic compound and dissolves by the separation of ions, or

$$Na^+Cl^-(s) \rightarrow Na^+(aq) + Cl^-(aq)$$

This equation tells you that one mole of NaCl separates into one mole of sodium ions and one mole of chlorine ions for a total of *two* moles of solute. The boiling point elevation of a solution made from one mole of NaCl (58.5 g) is therefore multiplied by two, or $2 \times 0.521° C = 1.04° C$. The boiling point of a solution made by adding 58.5 g of NaCl to 1,000 g of water is therefore 101.04° C at normal sea level pressure.

Now back to the question of how much a pinch of salt increases the boiling point of a pot of water. Assuming the pot contains about a liter of water (about a quart), and assuming that a "pinch" of salt has a mass of about 0.2 gram, the boiling point will be increased by 0.0037° C. So, there must be some reason other than increasing the boiling point that a cook adds a pinch of salt to a pot of boiling water.

Freezing Point

Freezing occurs when the kinetic energy of molecules has been reduced sufficiently so the molecules can come together, forming the crystal structure of the solid. Reduced kinetic energy of the molecules, that is, reduced temperature, results in a specific freezing point for each pure liquid. The *normal* freezing point for pure water, for example, is 0° C (32° F) under normal pressure. The presence of solute particles in a solution interferes with the water molecules as they attempt to form the six-sided hexagonal structure. The water molecules cannot get by the solute particles until the kinetic energy of the solute particles is reduced, that is, until the temperature is below the normal freezing point. Thus the presence of solute particles lowers the freezing point, and solutions freeze at a lower temperature than the pure solvent.

The freezing-point depression of a solution has a number of interesting implications for solutions such as seawater. When seawater freezes, the water molecules must work their way around the salt particles as was described earlier. Thus the solute particles are *not* normally included in the hexagonal structure of ice. Ice formed in seawater is practically pure water. Since the solute was *excluded* when the ice formed, the freezing of seawater increases the salinity. Increased salinity means increased concentration, so the freezing point of seawater is further depressed and more ice forms only at a lower temperature. When this additional ice forms more pure water is removed, and the process goes on. Thus seawater does not have a fixed freezing point but has a lower and lower freezing point as more and more ice freezes.

The depression of the freezing point by a solute has a number of interesting applications in colder climates. Salt, for example, is spread on icy roads to lower the freezing point (and thus the melting point) of the ice. Calcium chloride, $CaCl_2$, is a salt that is often used for this purpose. Water in a car radiator would also freeze in colder climates if a solute, called antifreeze, were not added to the radiator water. Methyl alcohol has been used as an antifreeze because it is soluble in water and does not damage the cooling system. Methyl alcohol, however, has a low boiling point and tends to boil away. Ethylene glycol has a higher boiling point, so it is called a "permanent" antifreeze. As other solutes, ethylene glycol also raises the boiling point, which is an added benefit for summer driving.

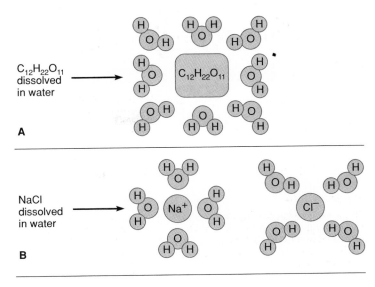

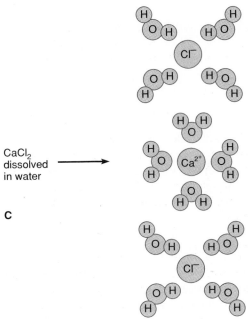

Figure 15.16
Since ionic compounds dissolve by separation of ions, they provide more particles in solution than molecular compounds. (*a*) A mole of sugar provides Avogadro's number of particles. (*b*) A mole of NaCl provides two times Avogadro's number of particles. (*c*) A mole of $CaCl_2$ provides three times Avogadro's number of particles.

Acids, Bases, and Salts

The electrolytes known as acids, bases, and salts are evident in environmental quality, foods, and everyday living. Environmental quality includes the hardness of water, which is determined by the presence of certain salts, the acidity of soils, which determines how well plants grow, and acid rain, which is a by-product of industry and automobiles. Many concerns about air and water pollution are often related to the chemistry concepts of acids, bases, and salts. These concepts, and uses of acids, bases, and salts, will be considered in this section.

Water and Solutions

Table 15.2

Some common acids

Name	Formula	Comment
Acetic acid	CH_3COOH	A weak acid found in vinegar
Boric acid	H_3BO_3	A weak acid used in eyedrops
Carbonic acid	H_2CO_3	The weak acid of carbonated beverages
Formic acid	$HCOOH$	Makes the sting of insects and certain plants
Hydrochloric acid	HCl	Also called muriatic acid; used in swimming pools, soil acidifiers, and stain removers
Lactic acid	$CH_3CHOHCOOH$	Found in sour milk, sauerkraut, and pickles; gives tart taste to yogurt
Nitric acid	HNO_3	A strong acid
Phosphoric acid	H_3PO_4	Used in cleaning solutions; added to carbonated beverages for tartness
Sulfuric Acid	H_2SO_4	Also called oil of vitriol; used as battery acid and in swimming pools

Table 15.3

Some common bases

Name	Formula	Comment
Sodium hydroxide	$NaOH$	Also called lye or caustic soda; a strong base used in oven cleaners and drain cleaners
Potassium hydroxide	KOH	Also called caustic potash; a strong base used in drain cleaners
Ammonia	NH_3	A weak base used in household cleaning solutions
Calcium hydroxide	$Ca(OH)_2$	Also called slaked lime; used to make brick mortar
Magnesium hydroxide	$Mg(OH)_2$	Solution is called milk of magnesia; used as antacid and laxative

Properties of Acids and Bases

Acids and bases are classes of chemical compounds that have certain characteristic properties. These properties can be used to identify if a substance is an acid or a base (tables 15.2 and 15.3). The following are the properties of *acids* dissolved in water:

1. Acids have a sour taste such as the taste of citrus fruits.
2. Acids change the color of certain substances, for example, litmus changes from blue to red when placed in an acid solution.
3. Acids react with active metals, such as magnesium or zinc, releasing hydrogen gas.
4. Acids *neutralize* bases, forming water and salts from the reaction.

Likewise, *bases* have their own characteristic properties. Bases are also called alkaline substances, and the following are the properties of bases dissolved in water:

1. Bases have a bitter taste, for example, the taste of caffeine.
2. Bases reverse the color changes that were caused by acids. Red litmus is changed back to blue when placed in a solution containing a base.
3. Basic solutions feel slippery on the skin. They have a *caustic* action on plant and animal tissue, converting tissue into soluble materials. A strong base, for example, reacts with fat to make soap and glycerine. This accounts for the slippery feeling on the skin.
4. Bases *neutralize* acids, forming water and salts from the reaction.

Tasting an acid or base to see if it is sour or bitter can be hazardous, since some are highly corrosive or caustic. Many organic acids are not as corrosive and occur naturally in foods. Citrus fruit, for example, contains citric acid, vinegar is a solution of acetic acid, and sour milk contains lactic acid. The stings or bites of some insects (bees, wasps, and ants) and some plants (stinging nettles) are painful because an organic acid, formic acid, is injected by the insect or plant. Your stomach contains a solution of hydrochloric acid. In terms of relative strength, the hydrochloric acid in your stomach is about ten times stronger than the carbonic acid (H_2CO_3) of carbonated beverages.

Examples of bases include solutions of sodium hydroxide ($NaOH$), which has a common name of lye or caustic soda, and potassium hydroxide (KOH), which has a common name of caustic potash. These two bases are used in products known as drain cleaners. They open plugged drains because of their caustic action, turning grease, hair, and other organic "plugs" into soap and other soluble substances that are washed away. A weaker base is a solution of ammonia (NH_3), which is often used as a household cleaner. A solution of magnesium hydroxide, $Mg(OH)_2$, has a common name of milk of magnesia and is sold as an antacid and laxative.

Many natural substances change color when mixed with acids or bases. You may have noticed that tea changes color slightly, becoming lighter, when lemon juice (which contains citric acid) is added. Some plants have flowers of one color when grown in acidic soil and flowers of another color when grown in basic soil. Vegetable dyes that change color in the presence of acids or bases can be used as an **acid-base indicator.** An indicator is simply a vegetable dye that is used to distinguish between acid and base solutions by a color change. Litmus, for example, is an acid-base indicator made from a dye extracted from certain species of lichens. The dye is applied to paper strips, which turn red in acidic solutions and blue in basic solutions.

To see how acids and bases change the color of certain vegetable dyes, consider the dye that gives red cabbage its color. Shred several leaves of red cabbage and boil them in a pan of water to extract the dye. After you have a purple solution, squeeze the juice from the cabbage into the pan and allow the solution to cool. Add vinegar in small amounts as you stir the solution, continuing until the color changes. Add ammonia in small amounts, again stirring until the color changes again. Reverse the color change again by adding vinegar in small amounts. Will this purple cabbage acid-base indicator tell you if other substances are acids or bases?

Explaining Acid-Base Properties

Comparing the lists in tables 15.2 and 15.3, you can see that acids and bases appear to be chemical opposites. Notice in table 15.2 that the acids all have an H, or hydrogen atom, in their formulas. In table 15.3, most of the bases have a hydroxide ion, OH^-, in their formulas. Could this be the key to acid-base properties?

Lavoisier was one of the first to attempt an explanation for the properties of acids. Lavoisier thought that all acids contained oxygen, so in 1777 he proposed the name *oxygen,* which means "acid former" in Greek. The formulas of acids listed in table 15.2 show that most acids do contain oxygen, with the exception of hydrochloric acid, which is HCl. During Lavoisier's time chlorine was believed to be an oxygen compound. Not until 1810 was chlorine discovered as an element. The modern understanding of acids and bases originated with the introduction of a *theory of ionization,* developed by Svante Arrhenius in 1883. Arrhenius proposed that electrolytes such as acids, bases, and salts produced equal numbers of positive and negative ions when dissolved in dilute solutions. In concentrated solutions, these ions were in equilibrium with solute molecules that were not ionized.

Chemical equilibrium occurs when two opposing reactions happen at the same time and at the same rate. Pure water, for example, ionizes very slightly to produce a hydronium ion, H_3O^+, and a hydroxide ion, OH^-, from the dissociation of a water molecule:

$$H_2O(l) + H_2O(l) \rightarrow H_3O^+(aq) + OH^-(aq)$$

This is termed the *forward reaction,* which occurs to about 1 molecule in 500 million molecules of pure water. The H_3O^+ and OH^- now react in a *reverse reaction* to produce a molecule of water. Chemical equilibrium occurs when the forward reaction occurs at the same rate as the reverse reaction. This is shown by a double arrow:

$$H_2O(l) + H_2O(l) \rightleftarrows H_3O^+(aq) + OH^-(aq)$$

In a condition of chemical equilibrium the concentration of molecules and ions remains constant, even though both reactions are always occurring.

The modern concept of an acid considers the properties of acids in terms of the hydronium ion, H_3O^+. As was mentioned earlier, the hydronium ion is a water molecule to which a H^+ ion is attached. Since a hydrogen ion is a hydrogen atom without its single electron, it could be considered as an ion consisting of a single proton. Thus the H^+ ion can be called a *proton.* An **acid** is defined as any substance that is a *proton donor* when dissolved in water, increasing the hydronium ion concentration. For example, hydrogen chloride dissolved in water has the following reaction:

$$\overset{\frown}{(H)}Cl_{(aq)} + H_2O_{(l)} \longrightarrow H_3O^+_{(aq)} + Cl^-_{(aq)}$$

The dotted circle and arrow were added to show that the hydrogen chloride donated a proton to a water molecule. The resulting solution contains H_3O^+ ions and has acid properties, so the solution is called hydrochloric acid.

The bases listed in table 15.3 all appear to have a hydroxide ion, OH^-. Water solutions of these bases do contain OH^- ions, but the definition of a base is much broader. A **base** is defined as any substance that is a *proton acceptor* when dissolved in water, increasing the hydroxide ion concentration. For example, ammonia dissolved in water has the following reaction:

$$NH_{3(g)} + \overset{\frown}{(H_2)}O_{(l)} \rightleftarrows (NH_4)^+ + OH^-$$

The dotted circle and arrow show that the ammonia molecule accepted a proton from a water molecule, providing a hydroxide ion. The resulting solution contains OH^- ions and has basic properties, so a solution of ammonium hydroxide is a base.

Carbonates, such as sodium carbonate (Na_2CO_3), form basic solutions because the carbonate ion reacts with water to produce hydroxide ions.

$$(CO_3)^{2-}(aq) + H_2O(l) \rightarrow (HCO_3)^-(aq) + OH^-(aq)$$

Thus sodium carbonate produces a basic solution.

Acids could be thought of as simply solutions of hydronium ions in water, and bases could be considered solutions of hydroxide ions in water. The proton donor and proton acceptor definition is much broader, and it does include the definition of acids and bases as hydronium and hydroxide compounds. The broader, more general definition covers a wider variety of reactions and is therefore more useful.

The modern concept of acids and bases explains why the properties of acids and bases are **neutralized,** or lost, when acids and bases are mixed together. For example, consider the hydronium ion produced in the hydrochloric acid solution and the hydroxide ion produced in the ammonia solution. When these solutions are mixed together, the hydronium ion reacts with the hydroxide ion,

$$H_3O^+(aq) + OH^+(aq) \rightarrow H_2O(l) + H_2O(l)$$

Thus a proton is transferred from the hydronium ion (an acid), and the proton is accepted by the hydroxide ion (a base). A molecule of water is produced and both the acid and base properties disappear or are neutralized.

Strong and Weak Acids and Bases

Acids and bases are classified according to their degree of ionization when placed in water. **Strong acids** ionize completely in water, with all molecules dissociating into ions. Nitric acid, for example, reacts completely in the following equation:

$$HNO_3(aq) + H_2O(l) \rightarrow H_3O^+(aq) + (NO_3)^-(aq)$$

Nitric acid, hydrochloric acid, and sulfuric acid are common strong acids.

Acids that react only partially produce fewer hydronium ions, so they are weaker acids. **Weak acids** are only partially ionized because of an equilibrium reaction with water. Vinegar, for example, contains acetic acid that reacts with water in the following reaction:

$$HC_2H_3O_2 + H_2O \rightleftarrows H_3O^+ + (C_2H_3O_2)^-$$

The double yield arrows indicate an equilibrium reaction, which means that at any given time, not many of the hydronium ions are in solution. In fact, only about 1 percent or less of the acetic acid molecules ionize, depending on the concentration.

Bases are also classified as strong or weak. A **strong base** is completely ionic in solution and has hydroxide ions. Sodium hydroxide, or lye, is the most common example of a strong base. It dissolves in water to form sodium and hydroxide ions:

$$Na^+OH^-(s) \rightarrow Na^+(aq) + OH^-(aq)$$

A **weak base** is only partially ionized because of an equilibrium reaction with water. Ammonia, magnesium hydroxide, and calcium hydroxide are examples of weak bases. Magnesium and calcium hydroxide are only slightly soluble in water, and this reduces the *concentration* of hydroxide ions in a solution. It would appear that $Ca(OH)_2$ would produce two moles of hydroxide ions. It would, if it were completely soluble and reacted completely. It is the concentration of hydroxide ions in solution that determines if a base is weak or strong, not the number of ions per mole.

The pH Scale

The strength of an acid or a base is usually expressed in terms of a range of values called a **pH scale.** The pH scale is based on the concentration of the hydronium ion (in moles/L) in an acidic or a basic solution. To understand how the scale is able to express both acid and base strength in terms of the hydronium ion, first recall that pure water is very slightly ionized in the equilibrium reaction:

$$H_2O(l) + H_2O(l) \rightleftarrows H_3O^+(aq) + OH^-(aq)$$

The amount of self-ionization by water has been determined through measurements. In pure water at 25° C or any neutral water solution at that temperature, the H_3O^+ concentration is 1×10^{-7} moles/L and the OH^- concentration is also 1×10^{-7} moles/L. Since both ions are produced in equal numbers, then the H_3O^+ concentration equals the OH^- concentration and pure water is neutral, neither acidic nor basic.

Table 15.4

The pH and hydronium ion concentration (moles/L)

Hydronium Ion Concentration (moles/L)	Reciprocal of Hydronium Ion Concentration	pH
1×10^0	1×10^0	0
1×10^{-1}	1×10^1	1
1×10^{-2}	1×10^2	2
1×10^{-3}	1×10^3	3
1×10^{-4}	1×10^4	4
1×10^{-5}	1×10^5	5
1×10^{-6}	1×10^6	6
1×10^{-7}	1×10^7	7
1×10^{-8}	1×10^8	8
1×10^{-9}	1×10^9	9
1×10^{-10}	1×10^{10}	10
1×10^{-11}	1×10^{11}	11
1×10^{-12}	1×10^{12}	12
1×10^{-13}	1×10^{13}	13
1×10^{-14}	1×10^{14}	14

In general, adding an acid substance to pure water increases the H_3O^+ concentration. Adding a base substance to pure water increases the OH^- concentration. Adding a base also *reduces* the H_3O^+ concentration as the additional OH^- ions are able to combine with more of the hydronium ions to produce unionized water. Thus at a given temperature, an increase in OH^- concentration is matched by a *decrease* in H_3O^+ concentration. Thus the concentration of the hydronium ion can be used as a measure of acidic, neutral, and basic solutions. In general, (1) acidic solutions have H_3O^+ concentrations above 1×10^{-7} moles/L, (2) neutral solutions have H_3O^+ concentrations equal to 1×10^{-7} moles/L, and (3) basic solutions have H_3O^+ concentrations less than 1×10^{-7} moles/L. These three statements lead directly to the pH scale, which is named from the French *pouvoir hydrogene*, meaning "hydrogen power." Power refers to the exponent of the hydronium ion concentration, and the pH is a *power of ten notation that expresses the H_3O^+ concentration* (table 15.4).

A neutral solution has a pH of 7.0. Acidic solutions have pH values below 7, and smaller numbers mean greater acidic properties. Increasing the OH^- concentration decreases the H_3O^+ concentration, so the strength of a base is indicated on the same scale with values greater than 7. Note that the pH scale is logarithmic, so a pH of 2 is ten times more acidic than a pH of 3. Likewise, a pH of 10 is one hundred times more basic than a pH of 8. Figure 15.17 is a diagram of the pH scale, and table 15.5 compares the pH of some common substances.

H_3O^+ concentration (moles/liters)	pH	Meaning
$1 \times 10^{-0} (=1)$	0	
1×10^{-1}	1	
1×10^{-2}	2	Increasing acidity
1×10^{-3}	3	
1×10^{-4}	4	
1×10^{-5}	5	
1×10^{-6}	6	
1×10^{-7}	7	Neutral
1×10^{-8}	8	
1×10^{-9}	9	
1×10^{-10}	10	Increasing basicity
1×10^{-11}	11	
1×10^{-12}	12	
1×10^{-13}	13	
1×10^{-14}	14	

Figure 15.17
The pH scale.

Properties of Salts

Salt is produced by a neutralization reaction between an acid and a base. A **salt** is defined as any ionic compound except those with hydroxide or oxide ions. Table salt, NaCl, is but one example of this large group of ionic compounds. As an example of a salt produced by a neutralization reaction, consider the reaction of HCl (an acid in solution) with $Ca(OH)_2$ (a base in solution). The reaction is

$$2\ HCl(aq) + Ca(OH)_2(aq) \rightarrow CaCl_2(aq) + 2\ H_2O(l)$$

This is an ionic exchange reaction that forms molecular water, leaving Ca^{2+} and Cl^- in solution. As the water is evaporated, these ions begin forming ionic crystal structures as the solution concentration increases. When the water is all evaporated, the white crystalline salt of $CaCl_2$ remains.

If sodium hydroxide had been used as the base instead of calcium hydroxide, a different salt would be produced:

$$HCl(aq) + NaOH(aq) \rightarrow NaCl(aq) + H_2O(l)$$

Salts are also produced when elements combine directly, when an acid reacts with a metal, and by other reactions. Salts are usually prepared commercially by a neutralization reaction between an acid and a base that furnishes the desired ions.

Salts are essential in the diet both as electrolytes and as a source of certain elements, usually called *minerals* in this context. Plants must have certain elements that are derived from water-soluble salts. Potassium, nitrates, and phosphate salts are often used to supply the needed elements. There is no scientific

Table 15.5

The approximate pH of some common substances

Substance	pH (or pH Range)
Hydrochloric acid (4%)	0
Gastric (stomach) solution	1.6–1.8
Lemon juice	2.2–2.4
Vinegar	2.4–3.4
Carbonated soft drinks	2.0–4.0
Grapefruit	3.0–3.2
Oranges	3.2–3.6
Acid rain	4.0–5.5
Tomatoes	4.2–4.4
Potatoes	5.7–5.8
Natural rain water	5.6–6.2
Milk	6.3–6.7
Pure water	7.0
Seawater	7.0–8.3
Blood	7.4
Sodium bicarbonate solution	8.4
Milk of magnesia	10.5
Ammonia cleaning solution	11.9
Sodium hydroxide solution	13.0

evidence that plants prefer to obtain these elements from natural sources, as compost, or from chemical fertilizers. After all, a nitrate ion is a nitrate ion, no matter what its source. Table 15.6 lists some common salts and their uses.

Hard and Soft Water

Salts vary in their solubility in water, and a solubility chart is in appendix B. Table 15.7 lists some generalizations concerning the various common salts. Some of the salts are dissolved by water that will eventually be used for domestic supply. When the salts are soluble calcium or magnesium compounds, the water will contain calcium or magnesium ions in solution. A solution of Ca^{2+} of Mg^{2+} ions is said to be **hard water** because it is hard to make soap lather in the water. "Soft" water, on the other hand, makes a soap lather easily. The difficulty occurs because soap is a sodium or potassium compound that is soluble in water. The calcium or magnesium ions, when present, replace the sodium or potassium ions in the soap compound, forming an insoluble compound. It is this insoluble compound that forms a "bathtub ring" and also collects on clothes being washed, preventing cleansing.

Table 15.6

Some common salts and their uses

Common Name	Formula	Use
Alum	$KAl(SO_4)_2$	Medicine, canning, baking powder
Baking powder	(a mixture)	Baking (makes bread rise)
Baking soda	$NaHCO_3$	Fire extinguisher, antacid, deodorizer, baking powder
Bleaching powder (chlorine tablets)	$CaOCl_2$	Bleaching, deodorizer, disinfectant in swimming pools
Borax	$Na_2B_4O_7$	Water softener
Chalk	$CaCO_3$	Antacid tablets, scouring powder
Cobalt chloride	$CoCl_2$	Hygrometer (pink in damp weather, blue in dry weather)
Chile saltpeter	$NaNO_3$	Fertilizer
Epsom salt	$MgSO_4 \cdot 7 H_2O$	Laxative
Fluorspar	CaF_2	Metallurgy flux
Gypsum	$CaSO_4 \cdot 2 H_2O$	Plaster of Paris, soil conditioner
Lunar caustic	$AgNO_3$	Germicide and cauterizing agent
Niter (or saltpeter)	KNO_3	Meat preservative, makes black gunpowder (75 parts KNO_3, 15 of carbon, 10 of sulfur)
Potash	K_2CO_3	Makes soap, glass
Rochelle salt	$KNaC_4H_4O_6$	Baking powder ingredient
TSP	Na_3PO_4	Water softener, fertilizer

Table 15.7

Generalizations about salt solubilities

Salts	Solubility	Exceptions
Sodium Potassium Ammonium	Soluble	None
Nitrate Acetate Chlorate	Soluble	None
Chlorides	Soluble	Ag and Hg (I) are insoluble
Sulfates	Soluble	Ba, Sr, and Pb are insoluble
Carbonates Phosphates Silicates	Insoluble	Na, D, and NH_4 are soluble
Sulfides	Insoluble	Na, K, and NH_4 are soluble: Mg, Ca, Sr, and Ba decompose

The key to "softening" hard water is to remove the troublesome calcium and magnesium ions (figure 15.18). If the hardness is caused by magnesium or calcium bicarbonates, the removal is accomplished by simply heating the water. Upon heating, they decompose, forming an insoluble compound that effectively removes the ions from solution. The decomposition reaction for calcium bicarbonate is

$$Ca^{2+}(HCO_3)_2(aq) \rightarrow CaCO_3(s) + H_2O(l) + CO_2\uparrow$$

The reaction is the same for magnesium bicarbonate. As the solubility chart in appendix B shows, magnesium and calcium carbonates are insoluble, so the ions are removed from solution in the solid that is formed. Perhaps you have noticed such a white compound forming around faucets if you live where bicarbonates are a problem. Commercial products to remove such deposits usually contain an acid, which reacts with the carbonate to make a new, soluble salt that can be washed away.

Water hardness is also caused by magnesium or calcium sulfate, which requires a different removal method. Certain

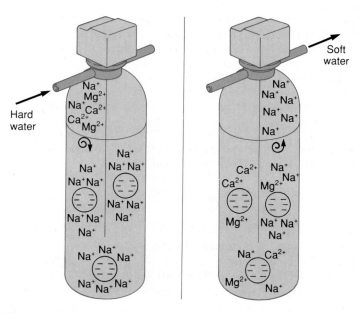

Figure 15.18

A water softener exchanges sodium ions for the calcium and magnesium ions of hard water. Thus the water is now soft, but it contains the same number of ions as before.

chemicals such as sodium carbonate (washing soda), trisodium phosphate (TSP), and borax will react with the troublesome ions, forming an insoluble solid that removes them from solution. For example, washing soda and calcium sulfate react as follows:

$$Na_2CO_3(aq) + CaSO_4(aq) \rightarrow Na_2SO_4(aq) + CaCO_3\downarrow$$

Acid Rain

Acid rain is a general term used to describe any acidic substances, wet or dry, that fall from the atmosphere. Wet acidic deposition could be in the form of rain, but snow, sleet, and fog could also be involved. Dry acidic deposition could include gases, dust, or any solid particles that settle out of the atmosphere to produce an acid condition.

Pure, unpolluted rain is naturally acidic. Carbon dioxide in the atmosphere is absorbed by rainfall, forming carbonic acid (H_2CO_3). Carbonic acid lowers the pH of pure rainfall to a range of 5.6 to 6.2. Decaying vegetation in local areas can provide more CO_2, making the pH even lower. A pH range of 4.5 to 5.0, for example, has been measured in remote areas of the Amazon jungle. Human-produced exhaust emissions of sulfur and nitrogen oxides can lower the pH of rainfall even more, to a 4.0 to 5.5 range. This is the pH range of acid rain.

The sulfur and nitrogen oxides that produce acid rain come from exhaust emissions of industries and electric utilities that burn coal and from the exhaust of cars, trucks, and buses. The emissions are sometimes called "SO_x" and "NO_x," which is read "socks" and "knox." The "x" subscript implies the variable presence of any or all of the oxides, for example, nitrogen monoxide (NO), nitrogen dioxide (NO_2), and dinitrogen tetroxide (N_2O_4) for NO_x.

SO_x and NO_x are the raw materials of acid rain and are not themselves acidic. They react with other atmospheric chemicals to form sulfates and nitrates, which combine with water vapor to form sulfuric acid (H_2SO_4) and nitric acid (HNO_3). These are the chemicals of concern in acid rain.

Many variables influence how much and how far SO_x and NO_x are carried in the atmosphere and if they are converted to acid rain or simply return to the surface as a dry gas or particles. During the 1960s and 1970s, concerns about local levels of pollution led to the replacement of short smokestacks of about 60 m (about 200 ft) with taller smokestacks of about 200 m (about 650 ft). This did reduce the local levels of pollution by dumping the exhaust higher in the atmosphere where winds could carry it away. It also set the stage for longer-range transport of SO_x and NO_x and their eventual conversion into acids.

There are two main reaction pathways by which SO_x and NO_x are converted to acids: (1) reactions in the gas phase and (2) reactions in the liquid phase, such as in water droplets in clouds and fog. In the gas phase SO_x and NO_x are oxidized to acids, mainly by hydroxyl ions and ozone, and the acid is absorbed by cloud droplets and precipitated as rain or snow. Most of the nitric acid in acid rain and about one-fourth of the sulfuric acid is formed in gas phase reactions. Most of the liquid phase reactions that produce sulfuric acid involve the absorbed SO_x and hydrogen peroxide (H_2O_2), ozone, oxygen, and particles of carbon, iron oxide, and manganese oxide particles. These particles also come from the exhaust of fossil fuel combustion.

Acid rain falls on the land, bodies of water, forests, crops, buildings, and people. The concerns about acid rain center on its environmental impact on lakes, forests, crops, materials, and human health. Lakes in different parts of the world, for example, have been increasing in acidity over the past fifty years. Lakes in northern New England, the Adirondacks, and parts of Canada now have a pH of less than 5.0, and correlations have been established between lake acidity and decreased fish populations. Trees, mostly conifers, are dying at unusually rapid rates in the northeastern United States. Red spruce in Vermont's Green Mountains and the mountains of New York and New Hampshire have been affected by acid rain as have pines in New Jersey's Pine Barrens. It is believed that acid rain leaches essential nutrients, such as calcium, from the soil and also mobilizes aluminum ions. The aluminum ions disrupt the water equilibrium of fine root hairs, and when the root hairs die, so do the trees.

Human-produced emissions of sulfur and nitrogen oxides from burning fossil fuels are the cause of acid rain. The heavily industrialized northeastern part of the United States, from the Midwest through New England, release sulfur and nitrogen emissions that result in a precipitation pH of 4.0 to 4.5. This region is the geographic center of the nation's acid rain problem. The solution to the problem is found in (1) using fuels other than fossil fuels and (2) reducing the thousands of tons of SO_x and NO_x that are dumped into the atmosphere per day when fossil fuels are used.

Calcium carbonate is insoluble, thus the calcium ions are removed from solution before they can react with the soap. Many laundry detergents have Na_2CO_3, TSP, or borax ($Na_2B_4O_7$) added to soften the water. TSP causes other problems, however, as the additional phosphates in the waste water can act as a fertilizer, stimulating the growth of algae to such an extent that other organisms in the water die.

A water softener unit is an ion exchanger. The unit contains a mineral that exchanges sodium ions for calcium and magnesium ions as water is run through it. The softener is regenerated periodically by flushing with a concentrated sodium chloride solution. The sodium ions replace the calcium and magnesium ions, which are carried away in the rinse water. The softener is then ready for use again. The frequency of renewal cycles depends on the water hardness, and each cycle can consume from four to twenty pounds of sodium chloride per renewal cycle. In general, water with less than 75 ppm calcium and magnesium ions is called soft water, and with greater concentrations is called hard water. The greater the concentration above 75 ppm, the harder the water.

Buffers

A **buffer solution** consists of a weak acid together with a salt and has the same negative ion as the acid. A buffer has the ability to resist changes in the pH when small amounts of an acid or a base are added. Acetic acid, for example, is a weak acid that forms hydronium ions and acetate ions in equilibrium:

$$HC_2H_3O_2(aq) + H_2O(l) \rightarrow H_3O^+(aq) + (C_2H_3O_2)^-(aq)$$

When sodium acetate, $NaC_2H_3O_2$, is added to the solution, it becomes a buffer solution. If a small amount of an acid is added, the hydronium ions are neutralized by reacting with the acetate ions in solution:

$$(C_2H_3O_2)^-(aq) + H_3O^+(aq) \rightarrow HC_2H_3O_2(aq) + H_2O(l)$$

If a small amount of a base is added, the hydroxide ions are neutralized by reacting with acetic acid:

$$HC_2H_3O_2(aq) + OH^-(aq) \rightarrow (C_2H_3O_2)^-(aq) + H_2O(l)$$

Thus the addition of an acid or a base does not change the pH, but it changes the ratio of $HC_2H_3O_2$ and $(C_2H_3O_2)^-$ instead. The solution will continue its buffering action as long as the number of H_3O^+ or OH^- added does not exceed the number of $HC_2H_3O_2$ molecules or acetate ions in the solution. Your blood contains buffer solutions that maintain the pH at about 7.4. Seawater is a buffer solution that maintains a pH of about 8.2. Buffers are also added to medicines and to foods. Many lemon-lime carbonated beverages, for example, contain citric acid and sodium citrate (check the label), which forms a buffer in the acid range. Sometimes the label says that these chemicals are to impart and regulate "tartness." Any acid will produce a tart taste. In this case, the tart taste comes from the citric acid and the addition of sodium citrate makes it a buffered solution.

Summary

A *solution* is a homogeneous mixture of ions or molecules of two or more substances. The substance present in the large amount is the *solvent*, and the *solute* is dissolved in the solvent. If one of the components is a liquid, however, it is called the solvent. The relative amounts of solute in a solvent is called the *concentration* of a solution. Concentrations are measured (1) in *parts per million* (ppm) or *parts per billion* (ppb), (2) *percent by volume*, the volume of a solute per 100 volumes of solution, (3) *percent by weight*, the weight of solute per 100 weight units of solution, and (4) *salinity*, the mass of salts in 1 kg of solution.

A limit to dissolving solids in a liquid occurs when the solution is *saturated*. A *saturated solution* is one with equilibrium between solute dissolving and solute coming out of solution. The *solubility* of a solid is the concentration of a saturated solution at a particular temperature.

A water molecule consists of two hydrogen atoms and an oxygen atom with bonding and electron pairs in a tetrahedral arrangement. This results in a *bent molecular arrangement*, with 105° between the hydrogen atoms. Oxygen is more electronegative than hydrogen, so electrons spend more time around the oxygen, producing a *polar molecule*, with centers of negative and positive charge. Polar water molecules interact with an *intermolecular force*, or *van der Waals force*, between the negative center of one molecule and the positive center of another. The force of attraction is called a *hydrogen bond*. The hydrogen bond accounts for the decreased density of ice, the high heat of fusion, and the high heat of vaporization of water. The hydrogen bond is also involved in the *dissolving* process.

Fluids that mix in any proportion are called *miscible fluids*, and *immiscible fluids* do not mix. Polar substances dissolve in polar solvents, but not nonpolar solvents, and the general rule is *like dissolves like*. Thus oil, a nonpolar substance, is immiscible in water, a polar substance. When a polar substance is added to a polar solvent, the substance dissolves if the *ion-polar molecular force* is greater than the *ion-ion force*. If the ion-ion force is greater, the substance is *insoluble*.

Water solutions that carry an electric current are called *electrolytes*, and nonconductors are called *nonelectrolytes*. In general, ionic substances make electrolyte solutions and molecular substances make nonelectrolyte solutions. Polar molecular substances may be *ionized* by polar water molecules, however, making an electrolyte from a molecular solution.

The *boiling point of a solution* is greater than the boiling point of the pure solvent, and the increase depends only on the concentration of the solute (at a constant pressure). For water, the boiling point is increased 0.521° C for each mole of solute in each kg of water. The *freezing point of a solution* is lower than the freezing point of the pure solvent, and the depression also depends on the concentration of the solute.

Acids, bases, and salts are chemicals that form ionic solutions in water, and each can be identified by simple properties. These properties are accounted for by the modern concepts of each. *Acids* are *proton donors* that form *hydronium ions* (H_3O^+) in water solutions. *Bases* are *proton acceptors* that form *hydroxide ions* (OH^-) in water solutions. *Strong acids* and *strong bases* ionize completely in water, and *weak acids* and *weak bases* are only partially ionized because of an *equilibrium reaction* with the solvent. The strength of an acid or base is measured on the *pH scale*, a power of ten notation of the hydronium ion concentration. On the scale, numbers from 0 up to 7 are acids, 7 is neutral, and numbers above 7 and up to 14 are bases. Each unit represent a tenfold increase or decrease in acid or base properties.

A *salt* is any ionic compound except those with hydroxide or oxide ions. Salts provide plants and animals with essential elements. The solubility of salts varies with the ions that make up the compound. Solutions of magnesium or calcium produce *hard water*, water that is hard to make soap lather in. Hard water is softened by removing the magnesium and calcium ions. A *buffer* solution is a solution of a weak acid and one of its salts. The solution resists changes in pH by reacting with acids or bases that are added.

Summary of Equations

15.1

Percent by volume

$$\frac{V_{solute}}{V_{solution}} \times 100\% \text{ solution} = \% \text{ solute}$$

15.2

Percent by weight (mass)

$$\frac{m_{solute}}{m_{solution}} \times 100\% \text{ solution} = \% \text{ solute}$$

Key Terms

acid (p. 299)
acid-base indicator (p. 298)
base (p. 299)
buffer solution (p. 303)
chemical equilibrium (p. 299)
concentration (p. 288)
electrolytes (p. 295)
hard water (p. 301)
hydration (p. 294)
hydrogen bond (p. 292)
hydronium ion (p. 296)
intermolecular forces (p. 292)
ionization (p. 295)
ionized (p. 295)
miscible fluids (p. 293)
neutralized (p. 299)
nonelectrolytes (p. 295)

parts per billion (p. 289)
parts per million (p. 289)
percent by volume (p. 290)
percent by weight (p. 290)
pH scale (p. 300)
salinity (p. 291)
salt (p. 301)
saturated solution (p. 291)
solubility (p. 291)
solute (p. 288)
solution (p. 288)
solvent (p. 288)
strong acids (p. 300)
strong base (p. 300)
van der Waals force (p. 292)
weak acids (p. 300)
weak base (p. 300)

Applying the Concepts

1. The solution concentration terms of parts per million, percent by volume, and percent by weight are concerned with the amount of
   a. solvent in the solution.
   b. solute in the solution.
   c. solute compared to solvent.
   d. solvent compared to solute.

2. According to the label a bottle of vodka has a 40 percent by volume concentration. This means the vodka contains 40 mL of pure alcohol
   a. in each 140 mL of vodka.
   b. to every 100 mL of water.
   c. to every 60 mL of vodka.
   d. mixed with 60 mL of water.

3. When a salt solution has solid salt on the bottom of the container and salt is dissolving at the same rate that it is coming out of solution, you know the solution is a(n)
   a. electrolyte.
   b. nonelectrolyte.
   c. buffered solution.
   d. saturated solution.

4. As the temperature of water *decreases* the solubility of carbon dioxide gas in the water
   a. increases.
   b. decreases.
   c. remains the same.

5. Water has the greatest density at what temperature?
   a. 100° C
   b. 20° C
   c. 4° C
   d. 0° C

6. Adding sodium chloride to water raises the boiling point of water because
   a. sodium chloride has a higher boiling point.
   b. sodium chloride ions occupy space at the water surface.
   c. sodium chloride ions have stronger ion-ion bonds than water.
   d. the energy of hydration is higher.

7. The ice that forms in freezing seawater is
   a. pure water.
   b. the same salinity as liquid seawater.
   c. more salty than liquid seawater.
   d. more dense than liquid seawater.

8. Which of the following would have a pH of *more* than 7?
   a. a solution of ammonia
   b. a solution of sodium chloride
   c. pure water
   d. carbonic acid

9. Substance "A" has a pH of 2 and substance "B" has a pH of 3. This means that
   a. substance A has more basic properties than substance B.
   b. substance B has more acidic properties than substance A.
   c. substance A is ten times more acidic than substance B.
   d. substance B is ten times more acidic than substance A.

10. A solution that is able to resist changes in the pH when small amounts of an acid or base are added is called a
    a. neutral solution.
    b. saturated solution.
    c. balanced solution.
    d. buffer solution.

Answers

1. b 2. d 3. d 4. a 5. c 6. b 7. a 8. a 9. c
10. d

Questions for Thought

1. How is a solution different from other mixtures?

2. Based on the temperature differences of the ocean water, why would the cold waters around Alaska have a greater abundance of fish than the warmer equatorial waters? (Hint: Think of a food chain and the solubility of gases.)

3. Explain why some ionic compounds are soluble while others are insoluble in water.

4. Explain why adding salt to water increases the boiling point.

5. A deep lake in Minnesota is covered with ice. What is the water temperature at the bottom of the lake? Explain your reasoning.

6. Is the freezing point depressed the same by equal amounts of NaCl and $CaCl_2$? Explain your reasoning.

7. Icebergs of frozen seawater have been considered as a source of fresh water. Explain why the ice is fresh water.

8. Why is it impossible to know the freezing point of seawater?

9. Explain why water has a greater density at 4° C than at 0° C.

10. Is a fog in the air a solution? Explain.

11. Provide an explanation for the six-sided structure of snowflakes.

12. What is hard water? How is it softened?

13. Based on the definition of an acid and the definition of a base, does the pH increase, decrease, or remain the same when NaCl is added to pure water? Explain.

14. Briefly define the properties of acids and bases. Give three examples of acids and three examples of bases.

15. Substance "A" has a pH of 4 and substance "B" has a pH of 5. Describe some general properties of the substances. Which substance has the greater display of these properties? How much greater?

16. Explain the meaning of a concentration reported as 9 ppm.

17. What is the difference between percent by weight and salinity?

18. What is a hydrogen bond? Explain how a hydrogen bond forms.

19. Explain why oil and water do not form a solution.

20. What feature of a soap molecule gives it cleaning ability?

21. A gas dissolved in another gas can form a solution of any concentration. An ionic compound dissolved in water, however, has a limit to how much can be dissolved before the solution becomes saturated. Provide an explanation for this difference.

22. What ion is basically responsible for (a) acidic properties? (b) for basic properties?

Water and Solutions

23. If the pH scale is a measure of hydronium ion concentration, how is it possible to use the scale to measure the strength of a base?

24. What is chemical equilibrium? How is equilibrium indicated in a chemical equation?

25. What is the difference between (a) a strong acid and a weak acid? (b) a strong base and a weak base?

26. Explain why a pH of 7 indicates a neutral solution—why not some other number?

27. What is a buffer solution and how does it work?

Exercises

▼

Group A—Solutions Provided in Appendix D

1. A 50.0 g sample of a saline solution contains 1.75 g NaCl. What is the percent by weight concentration?

2. A student attempts to prepare a 3.50% by weight saline solution by dissolving 3.50 g NaCl in 100 g of water. Since equation 15.2 calls for 100 g of solution, the correct amount of solvent would have been 96.5 g water ($100 - 3.5 = 96.5$). What percent by weight solution did the student actually prepare?

3. Seawater contains 30,113 ppm by weight dissolved sodium and chlorine ions. What is the percent by weight concentration of sodium chloride in seawater?

4. What is the mass of hydrogen peroxide, H_2O_2, in 250. g of a 3.0% by weight solution?

5. How many mL of pure alcohol are in a 200. mL glass of wine that is 12% alcohol by volume?

6. How many mL of pure alcohol are in a single cocktail made with 50 mL of 40% vodka? (Note: "Proof" is twice the percent, so 80 proof = 40%.)

7. If fish in a certain lake are reported to contain 5 ppm by weight DDT (a) what percent of the fish meat is DDT? (b) How much of this fish would have to be consumed to reach a poisoning accumulation of 17.0 grams of DDT?

8. For each of the following reactants, draw a circle around the proton donor and a box around the proton acceptor. Label which acts as an acid and which acts as a base.
   (a) $HC_2H_3O_2(aq) + H_2O(l) \rightleftarrows H_3O^+(aq) + C_2H_5O_2^-(aq)$
   (b) $C_6H_6NH_2(l) + H_2O(l) \rightleftarrows C_6H_5NH_3^+(aq) + OH^-(aq)$
   (c) $HClO_4(aq) + HC_2H_3O_2(aq) \rightleftarrows H_2C_2H_3O_2^+(aq) + ClO_4^-(aq)$
   (d) $H_2O(l) + H_2O(l) \rightleftarrows H_3O^+(aq) + OH^-(aq)$

Group B—Solutions Not Given

1. What is the percent by weight of a solution containing 2.19 g NaCl in 75 g of the solution?

2. What is the percent by weight of a solution prepared by dissolving 10 g of NaCl in 100 g of H_2O?

3. A concentration of 0.5 ppm SO_2 is harmful to plant life. What is the percent by volume of this concentration?

4. What is the volume of water in a 500 mL bottle of rubbing alcohol that has a concentration of 70% by volume?

5. If the legal definition of intoxication is an alcohol concentration of 0.05% by volume in blood, how much alcohol would be present in the average (155 lb) person's 6,300 mL of blood?

6. How much pure alcohol is in a 355 mL bottle of a "wine cooler" that is 5.0% alcohol by volume?

7. In the 1970s, when lead was widely used in "ethyl" gasoline, the blood level of the average American contained 0.25 ppm lead. The danger level of lead poisoning is 0.80 ppm. (a) What percent of the average person was lead? (b) How much lead would be in an average 80 kg person? (c) How much more lead would the person accumulate to reach the danger level?

8. Draw a circle around the proton donor and a box around the proton acceptor for each of the reactants and label which acts as an acid and which acts as a base.
   (a) $H_3PO_4(aq) + H_2O(l) \rightleftarrows H_3O^+(aq) + H_2PO_4^-(aq)$
   (b) $N_2H_4(l) + H_2O(l) \rightleftarrows N_2H_5^+(aq) + OH^-(aq)$
   (c) $HNO_3(aq) + HC_2H_3O_2(aq) \rightleftarrows H_2C_2H_3O_2^+(aq) + NO_3^-(aq)$
   (d) $2 NH_4^+(aq) + Mg(s) \rightleftarrows Mg^{2+}(aq) + 2 NH_3^+(aq) + H_2(g)$

Chapter 16

Organic Chemistry

Figure 16.1
Refinery and tank storage facilities, like this one in New Jersey, are
needed to change the hydrocarbons of crude oil to many different
petroleum products. The classes and properties of hydrocarbons is one
topic of study in organic chemistry.

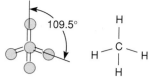

A Three-dimensional model

B An unbranched chain

$$C—C—C—C—C$$

C Simplified unbranched chain

Figure 16.2

(a) The carbon atom forms bonds in a tetrahedral structure with a bond angle of 109.5°. (b) Carbon-to-carbon bond angles are 109.5°, so a chain of carbon atoms makes a zigzag pattern. (c) The unbranched chain of carbon atoms is usually simplified in a way that looks like a straight chain, but it is actually a zigzag as shown in (b).

THE impact of ancient Aristotelian ideas on the development of understandings of motion, elements, and matter was discussed in earlier chapters. Historians also trace the "vitalist theory" back to Aristotle. According to Aristotle's idea, all living organisms are composed of the four elements (earth, air, fire, and water) and have in addition an *actuating force,* the life or soul that makes the organism different from nonliving things made of the same four elements. Plants, as well as animals, were considered to have this actuating, or vital, force in the Aristotelian scheme of things.

There were strong proponents of the vitalist theory as recent as the early 1800s. Their basic argument was that organic matter, the materials and chemical compounds recognized as being associated with life, could not be produced in the laboratory. Organic matter could only be produced in a living organism, they argued, because the organism had a vital force that is not present in laboratory chemicals. Then, in 1828, a German chemist named Fredrich Wohler reacted two chemicals that were *not organic* to produce urea (N_2H_4CO), a known *organic* compound that occurs in urine. Wohler's synthesis of an organic compound was soon followed by the synthesis of other organic substances by other chemists. The vitalist theory gradually disappeared with each new synthesis, and a new field of study, organic chemistry, emerged.

This chapter is an introductory survey of the field of study known as organic chemistry. Organic chemistry is concerned with compounds and reactions of compounds that contain carbon. You will find this an interesting, informative introduction particularly if you have ever wondered about synthetic materials, natural foods and food products, or any of the thousands of carbon-based chemicals you use every day. The survey begins with the simplest of organic compounds, those consisting of only carbon and hydrogen atoms, compounds known as hydrocarbons. Hydrocarbons are the compounds of crude oil, which is the source of hundreds of petroleum products (figure 16.1). In this section you will find information about things you may have wondered about, for example, what an octane rating is and how petroleum products differ.

Most common organic compounds can be considered derivatives of the hydrocarbons, such as alcohols, ethers, fatty acids, and esters. Some of these are the organic compounds that give flavors to foods, and others are used to make hundreds of commercial products, from face cream to oleo. The main groups, or classes, of derivatives will be briefly introduced, along with some interesting examples of each group. Some of the important organic compounds of life, including proteins, carbohydrates, and fats, are discussed next. The chapter concludes with an introduction to synthetic polymers, what they are, and how they are related to the fossil fuel supply.

Organic Compounds

Organic compounds are sensitive to increases in temperature, decomposing or burning when heated to 400° C (about 750° F) or greater. When sugar is decomposed by heating, for example, it often leaves a black residue of carbon. When burned completely, sugar and other organic materials produce carbon dioxide (and other products). Carbon is the essential element of organic matter, and today, **organic chemistry** is defined as the study of compounds in which carbon is the principal element, whether the compound was formed by living things or not. The study of all the other elements and compounds is called **inorganic chemistry.** An *organic compound* is thus a compound in which carbon is the principal element, and an *inorganic compound* is any other compound.

Organic compounds, by definition, must contain carbon while all the other compounds can contain all the other elements. Yet, there are *millions* of different organic compounds but fewer than fifty thousand inorganic compounds. It is the unique properties of carbon that allow it to form so many different compounds. A carbon atom has a simple $1s^2 2s^2 2p^2$ electron structure, and there is room for four more electrons in the outer shell. Carbon has a valence of four and can form four electron pairs, with no lone pairs. The molecular shape of a carbon compound such as CH_4 is therefore tetrahedral. The carbon atom has a valence of four, and can combine with one, two, three, or four *other carbon atoms* in addition to a wide range of other kinds of atoms (figure 16.2). The number of possible molecular combinations is almost limitless, which explains why there are so many organic compounds. Fortunately, there are patterns of groups of carbon atoms and groups of other atoms that lead to similar chemical characteristics, making the study of organic chemistry less difficult. The key to success in studying organic chemistry is to recognize patterns and to understand the code and meaning of organic chemical names. The first patterns to be discussed will be those of the simplest organic compounds, consisting of only two elements.

A Molecular formula

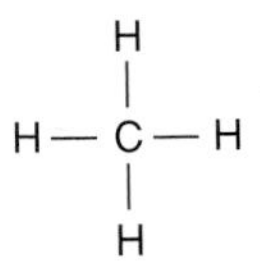

B Structural formula

Figure 16.3

Recall that a molecular formula (a) describes the numbers of different kinds of atoms in a molecule, and a structural formula (b) represents a two-dimensional model of how the atoms are bonded to each other. Each dash represents a bonding pair of electrons.

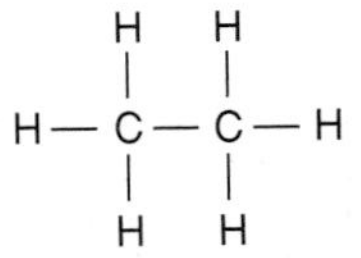

A Ethane

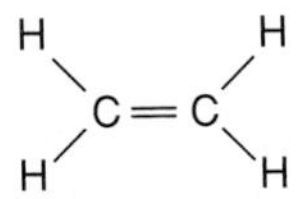

B Ethylene

C Ethyne

Figure 16.4

Carbon-to-carbon bonds can be single (a), double (b), or triple (c). Note that in each example, each carbon atom has four dashes, which represent four bonding pairs of electrons, satisfying the octet rule.

Hydrocarbons

A **hydrocarbon** is an organic compound consisting of only two elements. As the name implies, these elements are hydrogen and carbon. The simplest hydrocarbon has one carbon atom and four hydrogen atoms, CH_4, but since carbon atoms can combine with one another, there are thousands of possible structures and arrangements. The carbon-to-carbon bonds are nonpolar covalent and can be single, double, or triple (figure 16.4). Recall that the dash in a structural formula means one shared electron pair. To satisfy the octet rule, this means that each carbon atom must have a total of four dashes around it, no more and no less. Note that when the carbon atom has double or triple bonds, fewer hydrogen atoms can be attached as the octet rule is satisfied. There are four groups of hydrocarbons that are classified according to how the carbon atoms are put together, the (1) *alkanes*, (2) *alkenes*, (3) *alkynes*, and (4) *aromatic hydrocarbons*.

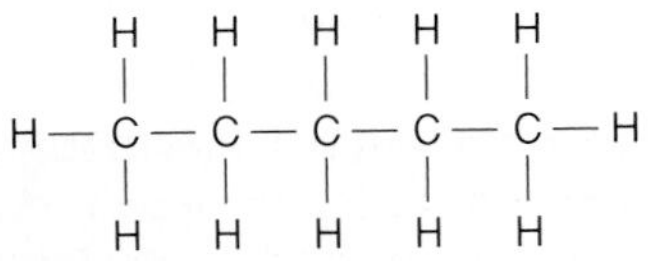

A Straight chain for C_5H_{10}

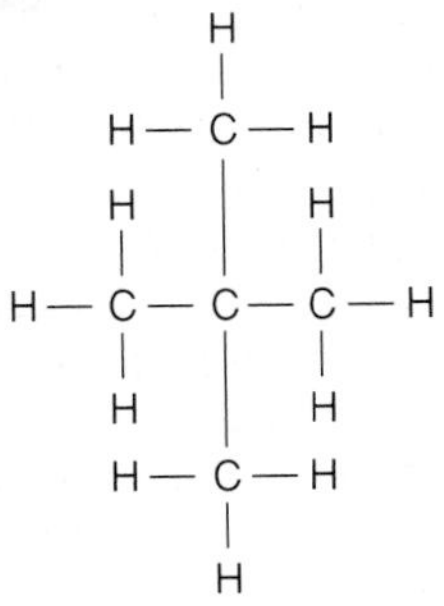

B Branched chain for C_5H_{10}

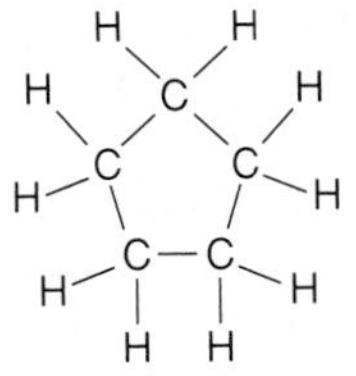

C Ring chain for C_5H_{10}

Figure 16.5

Carbon-to-carbon chains can be (a) straight, (b) branched, or (c) in a closed ring.

The **alkanes** are *hydrocarbons with single covalent bonds* between the carbon atoms. Alkanes that are large enough to form chains of carbon atoms occur with a straight structure, a branched structure, or a ring structure as shown in figure 16.5. (The "straight" structure is actually a zigzag as shown in figure 16.2.) You are familiar with many alkanes, for they make up the bulk of petroleum and petroleum products, which will be discussed shortly. The clues and codes in the names of the alkanes will be considered first.

The alkanes are also called the *paraffin series*. The alkanes are not as chemically reactive as the other hydrocarbons, and the term *paraffin* means "little affinity." They are called a series because *each higher molecular weight alkane has an additional CH_2*. The simplest alkane is methane, CH_4, and the next highest molecular weight alkane is propane, C_2H_6. As you can see, C_2H_6 is CH_4 with an additional CH_2. If you compare the first ten alkanes in table 16.1, you will find that each successive compound in the series always has an additional CH_2.

Note the names of the alkanes listed in table 16.1. After pentane the names have a consistent prefix and suffix pattern. The prefix and suffix pattern is a code that provides a clue about

Table 16.1

The first ten straight-chained alkanes

Name	Molecular Formula	Structural Formula	Name	Molecular Formula	Structural Formula
Methane	CH_4		Hexane	C_6H_{14}	
Ethane	C_2H_6		Heptane	C_7H_{16}	
Propane	C_3H_8		Octane	C_8H_{18}	
Butane	C_4H_{10}		Nonane	C_9H_{20}	
Pentane	C_5H_{12}		Decane	$C_{10}H_{22}$	

the compound. The Greek prefix tells you the *number of carbon atoms* in the molecule, for example, "oct-" means eight, so *oct*ane has eight carbon atoms. The suffix "-ane" tells you this hydrocarbon is a member of the alk*ane* series, so it has single bonds only. With the general alkane formula of C_nH_{2n+2}, you can now write the formula when you hear the name. Octane has eight carbon atoms with single bonds and n = 8. Two times 8 plus 2 (2n + 2) is 18, so the formula for octane is C_8H_{18}. Most organic chemical names provide clues like this, as you will see.

The alkanes in table 16.1 all have straight chains. A straight, continuous chain is identified with the term *normal*, which is abbreviated *n*. Figure 16.6a shows *n*-butane with a straight chain and a molecular formula of C_4H_{10}. Figure 16.6b shows a different branched structural formula that has the same C_4H_{10} molecular formula. Compounds with the same molecular

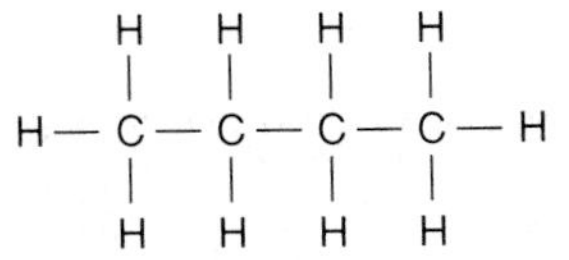

A *n*-butane, C_4H_{10}

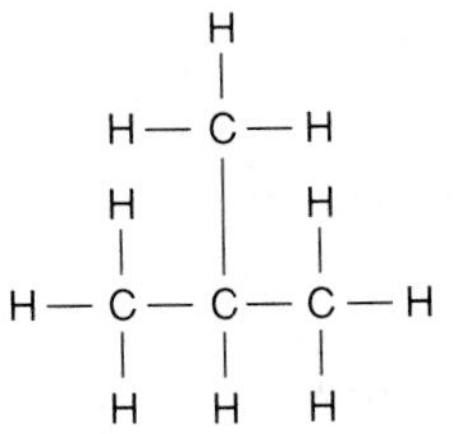

B Isobutane (2-methylpropane), C_4H_{10}

Figure 16.6

(*a*) A straight-chain alkane is identified by the prefix "*n*-" for "normal" in the common naming system. (*b*) A branched-chain alkane isomer is identified by the prefix "iso-" for "isomer" in the common naming system. In the IUPAC name, isobutane is 2-methylpropane.

Table 16.2

Alkane hydrocarbons and corresponding hydrocarbon groups

Alkane Name	Molecular Formula	Hydrocarbon Group	Molecular Formula
Methane	CH_4	Methyl	$-CH_3$
Ethane	C_2H_6	Ethyl	$-C_2H_5$
Propane	C_3H_8	Propyl	$-C_3H_7$
Butane	C_4H_{10}	Butyl	$-C_4H_9$
Pentane	C_5H_{12}	Amyl	$-C_5H_{11}$
Hexane	C_6H_{14}	Hexyl	$-C_6H_{13}$
Heptane	C_7H_{16}	Heptyl	$-C_7H_{15}$
Octane	C_8H_{18}	Octyl	$-C_8H_{17}$
Nonane	C_9H_{20}	Nonyl	$-C_9H_{19}$
Decane	$C_{10}H_{22}$	Decyl	$-C_{10}H_{21}$

Note: $-CH_3$ means $* - \overset{\displaystyle H}{\underset{\displaystyle H}{C}} - H$ where * denotes unattached. The attachment

takes place on a base chain or functional group.

formulas with different structures are called **isomers.** Since the straight-chained isomer is called *n*-butane, the branched isomer is called *isobutane.* The isomers of a particular alkane, such as butane, have different physical properties because they have different structures. Isobutane, for example, has a boiling point of $-10°$ C. The boiling point of *n*-butane, on the other hand, is $-0.5°$ C.

Methane, ethane, and propane can have only one structure each, and butane has two isomers. The number of possible isomers for a particular molecular formula increases rapidly as the number of carbon atoms increase. After butane, hexane has five isomers, octane eighteen isomers, and decane seventy-five isomers. Because they have different structures, each isomer has different physical properties. A different naming system is needed because there are just too many isomers to keep track of. The system of naming the branched-chain alkanes is described by rules agreed upon by the International Union of Pure and Applied Chemistry, or IUPAC. Here are the steps in naming the alkane isomers.

Step 1: The longest continuous chain of carbon atoms determines the *base name* of the molecule. The longest continuous chain is not necessarily straight and can take any number of right-angle turns as long as the continuity is not broken. The base name corresponds to the number of carbon atoms in this chain as in table 16.1. For example, the structure

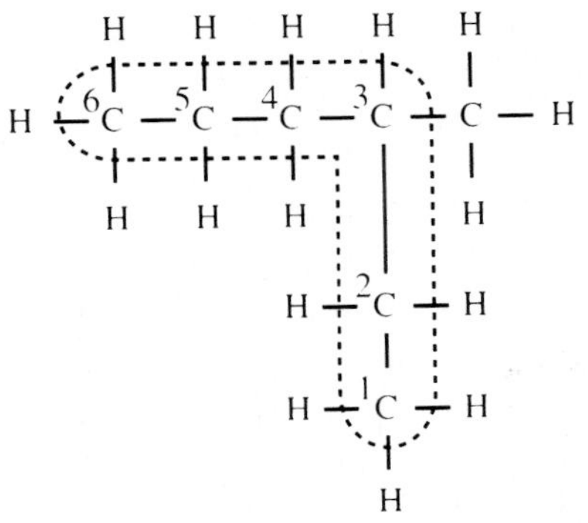

has six carbon atoms in the longest chain, so the base name is *hexane.*

Step 2: The locations of other groups of atoms attached to the base chain are identified by counting carbon atoms from either the left or from the right. The direction selected is the one that results in the *smallest* numbers for attachment locations. For example, the hexane chain has a CH_3 attached to the third or the fourth carbon atom, depending on which way you count. The third atom direction is chosen since it results in a smaller number.

Step 3: The hydrocarbon groups attached to the base chain are named from the number of carbons in the group by changing the alkane suffix "-ane" to "-yl." Thus a hydrocarbon group attached to a base chain that has one carbon atom is called methyl. Note that the "-yl" hydrocarbon groups have one less hydrogen than the corresponding alkane. Therefore methane is CH_4 and a *methyl group* is CH_3. The first ten alkanes and their corresponding hydrocarbon group names are listed in table 16.2. In the example, a methyl group is attached to the third carbon atom of the base hexane chain. The name and address of this hydrocarbon group is 3-methyl. The compound is named 3-methylhexane.

Step 4: The prefixes "di-," "tri-," and so on are used to indicate if a particular hydrocarbon group appears on the main chain more than once. For example,

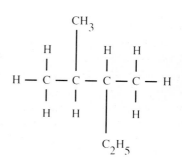

(or)

is 2,2-dimethylbutane and

(or)

is 2,3-dimethylbutane.

If hydrocarbon groups with different numbers of carbon atoms are on a main chain they are listed by order of increasing size, for example,

(or)

is named 2-methyl-3-ethylpentane.

Example 16.1 ▼

What is the name of an alkane with the following formula?

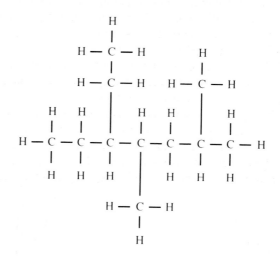

Solution

The longest continuous chain has seven carbon atoms, so the base name is heptane. The smallest numbers are obtained by counting from right to left and counting the carbons on this chain; there is a methyl group in carbon atom 2, a second methyl group on atom 4, and an ethyl group on atom 5. There are two methyl groups, so the prefix "di-" is needed, and the methyl group contains fewer carbon atoms, so it is listed first. The name of the compound is 2,4-dimethyl-5-ethylheptane.

Example 16.2

Write the structural formula for 2,2-dichloro-3-methyloctane.

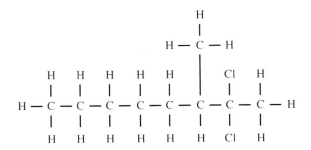

Answer

Alkenes and Alkynes

The alkanes are hydrocarbons with single carbon-to-carbon bonds. The **alkenes** are *hydrocarbons with a double covalent carbon-to-carbon bond*. To denote the presence of a double bond the "-ane" suffix of the alkanes is changed to "-ene" as in alk*ene*. Figure 16.4 shows the structural formula for (a) ethane, C_2H_6, and (b) ethylene, C_2H_4. Alkenes have room for two less hydrogen atoms because of the double bond, so the general alkene formula is C_nH_{2n}.

Ethylene is an important raw material in the chemical industry. Obtained from the processing of petroleum, about half of the commercial ethylene is used to produce the familiar polyethylene plastic. It is also produced by plants to ripen fruit, which explains why unripe fruit enclosed in a sealed plastic bag with ripe fruit will ripen more quickly. The ethylene produced by the ripe fruit acts on the unripe fruit. Commercial fruit packers sometimes use small quantities of ethylene gas to ripen fruit quickly that was picked while green.

Perhaps you have heard the terms "saturated" and "unsaturated" in advertisements for cooking oil and oleomargarine. The meaning of these terms with reference to foods will be discussed shortly. First, you need to understand the meaning of the terms. An organic molecule, such as a hydrocarbon, that does not contain the maximum number of hydrogen atoms is an **unsaturated** hydrocarbon. For example, ethylene can add more hydrogen atoms by reacting with hydrogen gas to form ethane:

The ethane molecule has all the hydrogen atoms possible, so ethane is a **saturated** hydrocarbon. Unsaturated molecules are less stable, which means that they are more chemically reactive than saturated molecules. Again, the role of saturated and unsaturated fats in foods will be discussed later.

Alkenes are named just as the alkanes except (1) the longest chain of carbon atoms must contain the double bond, (2) the base name now ends in "-ene," (3) the carbon atoms are numbered from the end nearest the double bond, and (4) the base name is given a number of its own, which identifies the address of the double bond. For example,

is named 4-methyl-1-pentene. The 1-pentene tells you there is a double bond (-ene) and the 1 tells you the double bond is after the first carbon atom in the longest chain containing the double bond. The methyl group is on the fourth carbon atom in this chain.

An **alkyne** is a *hydrocarbon with a carbon-to-carbon triple bond* and the general formula of C_nH_{2n-2}. The alkynes are highly reactive and the simplest one, ethyne, has a common name of acetylene. Acetylene is commonly burned with oxygen gas in a welding torch because the flame reaches a temperature of about 3,000° C. Acetylene is also an important raw material in the production of plastics. The alkynes are named as the alkenes, except the longest chain must contain the triple bond and the base name suffix is changed to "-yne."

Cyclo-Alkanes and Aromatic Hydrocarbons

The hydrocarbons discussed up until now have been straight or branched open-ended chains of carbon atoms. Carbon atoms can also bond to each other to form a ring, or cycle, structure. Figure 16.7 shows the structural formulas for some of these cycle structures.

The six-carbon ring structure shown in Figure 16.8a has three double bonds but does not behave like the double bonds in the alkenes. In this six-carbon ring the double bonds are not localized in one place but have many different possible electron locations. This gives the C_6H_6 molecule increased stability. As a result, the molecule does not act unsaturated, that is, it does not readily react to add hydrogen to the ring. The C_6H_6 molecule is the organic compound named *benzene*. Organic compounds that are based on the benzene ring structure are called **aromatic hydrocarbons.** To denote the six-carbon ring with delocalized electrons, benzene is represented by the symbol shown in figure 16.8b.

Organic Chemistry

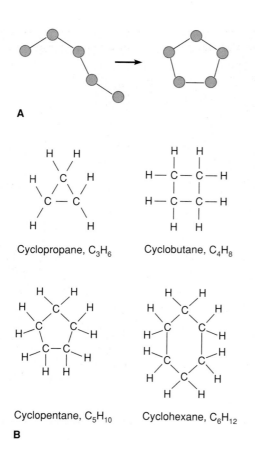

Figure 16.7

(a) The "straight" chain has carbon atoms that are able to rotate freely around their single bonds, sometimes linking up in a closed ring.
(b) Ring compounds of the first four cycloalkanes.

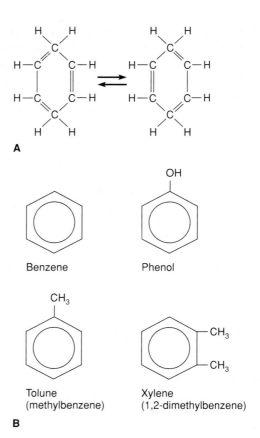

Figure 16.8

(a) The double bonds in C_6H_6 are continually shifting, which gives it different chemical properties than other double-bonded hydrocarbons.
(b) The C_6H_6 ring compound with shifting double bonds is benzene. The six-sided symbol with a circle represents the benzene ring. Organic compounds based on the benzene ring are called aromatic hydrocarbons because of their aromatic character.

The circle in the six-sided benzene symbol represents the delocalized electrons. Figure 16.8b illustrates how this benzene ring symbol is used to show the structural formula of some aromatic hydrocarbons. You may have noticed some of the names on labels of paints, paint thinners, and lacquers. Toluene and the xylenes are commonly used in these products as a solvent. A benzene ring attached to another molecule or functional group is given the name *phenyl*.

Petroleum

Petroleum is a mixture of alkanes, cyclo-alkanes, and some aromatic hydrocarbons. The origin of petroleum is uncertain, but it is believed to have formed from the slow anaerobic decomposition of buried marine life, primarily plankton and algae. Time, temperature, pressure, and perhaps bacteria are considered important in the formation of petroleum. As the petroleum formed, it was forced through porous rock until it reached a rock type or rock structure that stopped it. Here, it accumulated to saturate the porous rock, forming an accumulation called an **oil field.** The composition of petroleum varies from one oil field to the next. The oil from a given field might be dark or light in color, and it might have an asphalt base or a paraffin base. Some

oil fields contain oil with a high quantity of sulfur, referred to as "sour crude." Because of such variations, some fields have oil with more desirable qualities than oil from other fields.

In some locations an oil field occurs close to the surface and petroleum seeps to the surface, often floating on water from a spring. Such seepage is the source of petroleum that has been collected and used since about 3000 B.C. Ancient Babylonians, Egyptians, and Roman civilizations used this oil for medicinal purposes, for paving roads, and when thickened by drying, as a caulking compound in early wooden ships.

Early settlers found oil seeps in the eastern United States and collected the oil for medicinal purposes. One enterprising oil peddler tried to improve the taste by running the petroleum through a whiskey still. He obtained a clear liquid by distilling the petroleum and, by accident, found that the liquid made an excellent lamp oil. This was fortunate timing, for the lamp oil used at that time was whale oil, and whale oil production was declining. This clear liquid obtained by distilling petroleum is today known as kerosene.

The first oil well was drilled in Titusville, Pennsylvania, in 1859. The well struck oil at a depth of seventy feet and produced two thousand barrels a year. This is not much compared to the billions of barrels produced per year today, but it had an economic impact in 1859. Before the well was drilled, oil was selling for $40 a barrel. Two years later the price was 10¢ a barrel. A "barrel of oil" is an accounting measure of forty-two United States gallons. Such a barrel size does not really exist. When or if oil is shipped in barrels, each drum holds fifty-five United States gallons.

Wells were drilled and crude oil refineries were built to produce the newly discovered lamp oil. Gasoline was a by-product of the distillation process and was used primarily as a spot remover. With Henry Ford's automobile production and Edison's electric light invention, the demand for gasoline increased and the demand for kerosene decreased. The refineries were converted to produce gasoline, and the petroleum industry grew to become one of the world's largest industries.

Crude oil is petroleum that is pumped from the ground, a complex and variable mixture of hydrocarbons with one or more carbon atoms, with an upper limit of about fifty atoms. This thick, smelly black mixture is not usable until it is refined, that is, separated into usable groups of hydrocarbons called petroleum products. The petroleum products are separated by distillation, and any particular product has a boiling point range, or "cut" of the distilled vapors. Thus each product, such as gasoline, heating oil, and so forth is made up of hydrocarbons within a range of carbon atoms per molecule (figure 16.9). The products, their boiling ranges, and ranges of carbon atoms per molecule are listed in table 16.3.

The hydrocarbons that have one to four carbon atoms (CH_4 to C_4H_{10}) are gases at room temperature. They can be pumped from certain wells as a gas, but they also occur dissolved in crude oil. *Natural gas* is a mixture of hydrocarbon gases, but it is about 95 percent methane (CH_4). Propane (C_3H_8) and butane (C_4H_{10}) are liquified by compression and cooling and are sold as liquified petroleum gas, or *LPG*. LPG is used where natural gas is not available for cooking or heating and is widely used as a fuel in barbecue grills and camp stoves.

Hydrocarbons with five to seven carbon atoms per molecule are volatile liquids at room temperature. Different groups of these closely related volatile hydrocarbons are used for various commercial purposes under the general heading of *petroleum ether*, also called "petroleum distillates" and naphtha. Petroleum ether is used as a cleaning fluid by the dry cleaning industry. It is also used as a solvent. Naphtha is also present in gasoline.

Gasoline is a mixture of volatile hydrocarbons with five to twelve carbon atoms per molecule. Gasoline distilled from crude oil consists mostly of straight-chain molecules not suitable for use as an automotive fuel. Straight-chain molecules burn too rapidly in an automobile engine, producing more of an explosion than a smooth burn. You hear these explosions as a knocking or pinging in the engine, and they mean poor efficiency and they could damage the engine. On the other hand, branched chain molecules burn comparatively slower, without the pinging or knocking explosions. The burning rate of gasoline is described by the *octane number* scale. The scale is based on pure *n*-heptane, straight-chain molecules that are assigned an

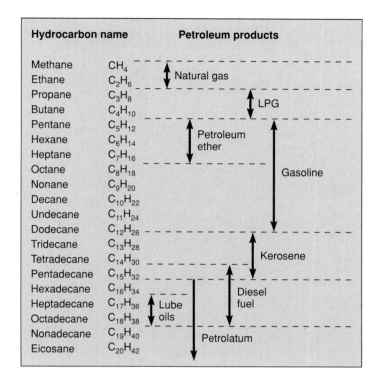

Figure 16.9

Petroleum products and the ranges of hydrocarbons in each product.

Table 16.3

Petroleum products

Name	Boiling Range (°C)	Carbon Atoms per Molecule
Natural Gas	Less than 0	C_1 to C_4
Petroleum Ether	35–100	C_5 to C_7
Gasoline	35–215	C_5 to C_{12}
Kerosene	35–300	C_{12} to C_{15}
Diesel Fuel	300–400	C_{15} to C_{18}
Motor Oil, Grease	350–400	C_{16} to C_{18}
Paraffin	Solid, melts at about 55	C_{20}
Asphalt	Boiler residue	C_{40} or more

octane number of 0, and a multiple branched isomer of octane, 2,2,4-trimethylpentane, which is assigned an octane number of 100 (figure 16.10). Most unleaded gasolines have an octane rating of 87, which could be obtained with a mixture that is 87 percent 2,2,4-trimethylpentane and 13 percent *n*-heptane. Gasoline, however, is a much more complex mixture.

The octane rating of gasoline can be improved one of two ways: (1) by adding a substance that slows the burning rate, such as tetraethyl lead, $(C_2H_5)_4Pb$ ("ethyl"), or (2) by converting some of the straight-chain hydrocarbons into branched-chain ones. The use of tetraethyl lead is less expensive but is

Organic Chemistry

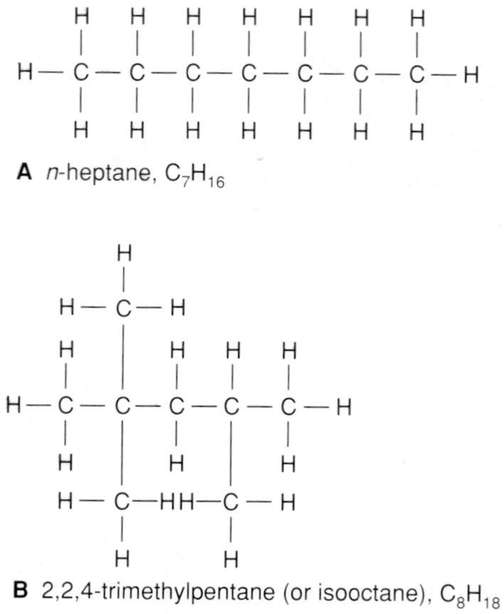

A n-heptane, C_7H_{16}

B 2,2,4-trimethylpentane (or isooctane), C_8H_{18}

Figure 16.10

The octane rating scale is a description of how rapidly gasoline burns. It is based on (a) n-heptane, with an assigned octane number of 0 and (b) 2,2,4-trimethylpentane, with an assigned number of 100.

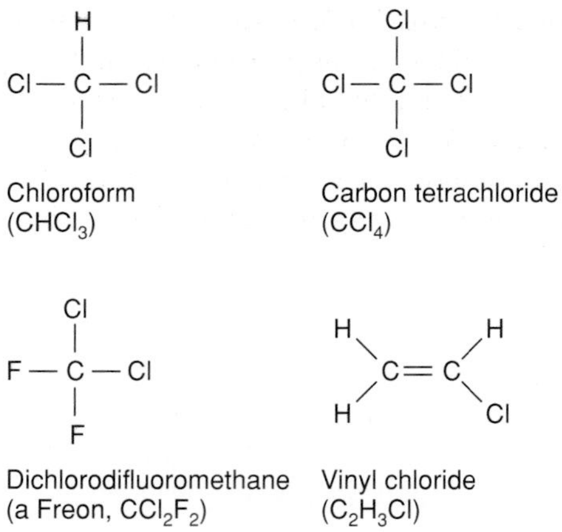

Chloroform
($CHCl_3$)

Carbon tetrachloride
(CCl_4)

Dichlorodifluoromethane
(a Freon, CCl_2F_2)

Vinyl chloride
(C_2H_3Cl)

Figure 16.11

Common examples of organic halides.

being phased out because of increased concerns over lead pollution and because lead poisons the antipollution catalytic converter (see the Feature at the end of this chapter). It is more expensive to produce unleaded gasoline because some of the straight-chain hydrocarbon molecules must be converted into branched molecules. The process is one of "cracking and reforming" some of the straight-chain molecules. First, the gasoline is passed through metal tubes heated to 500° C to 800° C (932° F to 1,470° F). At this high temperature, and in the absence of oxygen, the hydrocarbon molecules decompose by breaking into smaller carbon-chain units. These smaller hydrocarbons are then passed through tubes containing a catalyst, which causes them to reform into branched-chain molecules. Unleaded gasoline is produced by the process. Without the reforming that produces unleaded gasoline, low-numbered hydrocarbons (such as ethylene) can be produced. Ethylene is used as a raw material for many plastic materials, antifreeze, and other products. Cracking is also used to convert higher-numbered hydrocarbons, such as heating oil, into gasoline.

Kerosene is a mixture of hydrocarbons that have from twelve to fifteen carbon atoms. The petroleum product called kerosene is also known by other names, depending on its use. Some of these names are lamp oil (with coloring and odorants added), jet fuel (with a flash flame retardant added), heating oil, #1 fuel oil, and in some parts of the country, "coal oil."

Diesel fuel is a mixture of a group of hydrocarbons that have from fifteen to eighteen carbon atoms per molecule. Diesel fuel also goes by other names, again depending on its use. This group of hydrocarbons is called diesel fuel, distillate fuel oil, heating oil, or #2 fuel oil. During the summer season there is a greater demand for gasoline than for heating oil, so some of the supply is converted to gasoline by the cracking process.

Motor oil and *lubricating oils* have sixteen to eighteen carbon atoms per molecule. Lubricating grease is heavy oil that is thickened with soap. *Petroleum jelly,* also called petrolatum (or Vaseline), is a mixture of hydrocarbons with sixteen to thirty-two carbon atoms per molecule. *Mineral oil* is a light lubricating oil that has been decolorized and purified.

Depending on the source of the crude oil, varying amounts of *paraffin* wax (C_{20} or greater) or *asphalt* (C_{36} or more) may be present. Paraffin is used for candles, waxed paper, and home canning. Asphalt is mixed with gravel and used to surface roads. It is also mixed with refinery residues and lighter oils to make a fuel called #6 fuel oil or residual fuel oil. Industries and utilities often use this semisolid material that must be heated before it will pour. Number 6 fuel oil is used as a boiler fuel, costing about half as much as #2 fuel oil.

Hydrocarbon Derivatives

The hydrocarbons account for only about 5 percent of the known organic compounds, but the other 95 percent can be considered as hydrocarbon derivatives. **Hydrocarbon derivatives** are formed when *one or more hydrogen atoms on a hydrocarbon have been replaced by some element or group of elements other than hydrogen.* For example, the halogens (F_2, Cl_2, Br_2) react with an alkane in sunlight or when heated, replacing a hydrogen:

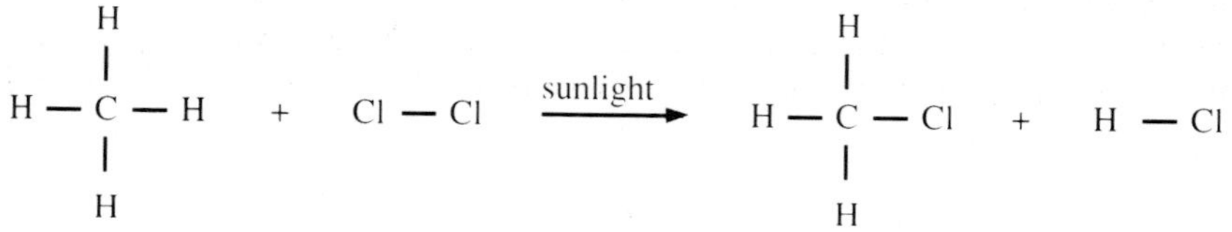

In this particular *substitution reaction* a hydrogen atom on methane is replaced by a chlorine atom to form methyl chloride. Replacement of any number of hydrogen atoms is possible, and a few *organic halides* are illustrated in figure 16.11.

If a hydrocarbon molecule is unsaturated (has a multiple bond), a hydrocarbon derivative can be formed by an *addition reaction:*

$$\begin{array}{ccccccc} & H & H & H & & & H & H & H \\ & | & | & | & & & | & | & | \\ H-&C&-C&=C&-H + Br-Br \longrightarrow H-&C&-C&-C&-H \\ & | & & & & & | & | & | \\ & H & & & & & H & Br & Br \end{array}$$

The bromine atoms add to the double bond on propene, forming 1,2-dibromopropane.

Alkene molecules can also add to each other in an addition reaction to form a very long chain consisting of hundreds of molecules. A long chain of repeating units is called a **polymer,** and the reaction is called *addition polymerization.* Ethylene, for example, is heated under pressure with a catalyst to form *polyethylene.* Heating breaks the double bond,

$$\begin{array}{ccc} H \quad\quad H & & H \quad H \\ \diagdown\;\;\diagup & & | \quad | \\ C = C & \xrightarrow{\;\Delta\;} & * - C - C - * \\ \diagup\;\;\diagdown & & | \quad | \\ H \quad\quad H & & H \quad H \end{array}$$

which provides sites for single covalent bonds to join the ethylene units together,

$$\begin{array}{cccc} H \quad H & & H \quad H & & H \quad H \quad H \quad H \\ | \quad | & & | \quad | & & | \quad | \quad | \quad | \\ * - C - C - * & + & * - C - C - * \longrightarrow & * - C - C - C - C - * \\ | \quad | & & | \quad | & & | \quad | \quad | \quad | \\ H \quad H & & H \quad H & & H \quad H \quad H \quad H \end{array}$$

which continues the addition polymerization until the chain is hundreds of units long. Synthetic polymers such as polyethylene are discussed in a later section.

The addition reaction and the addition polymerization reaction can take place because of the double bond of the alkenes, and, in fact, the double bond is the site of most alkene reactions. The atom or group of atoms in an organic molecule that is the site of a chemical reaction is identified as a **functional group.** *It is the functional group that is responsible for the chemical properties of an organic compound.* Functional groups usually have (1) multiple bonds or (2) lone pairs of electrons that cause them to be sites of reactions. Table 16.4 lists some of the common hydrocarbon functional groups. Look over this list, comparing

Table 16.4

Selected organic functional groups

Name of Functional Group	General Formula	General Structure
Organic Halide	RCl	R — C̈l:
Alcohol	ROH	R — Ö — H
Ether	ROR'	R — Ö — R'
Aldehyde	RCHO	R — C — H ‖ :O:
Ketone	RCOR'	R — C — R' ‖ :O:
Organic Acid	RCOOH	R — C — Ö — H ‖ :O:
Ester	RCOOR'	R — C — Ö — R' ‖ :O:
Amine	RNH_2	R — N̈ — H | H

the structure of the functional group with the group name. Some of the more interesting examples from a few of these groups will be considered next. Note that the R and R' stand for one or more of the hydrocarbon groups from table 16.2. For example, in the reaction between methane and chlorine, the product is methyl chloride. In this case the R in RCl stands for methyl, but it could represent any hydrocarbon group.

Alcohols

An **alcohol** is an organic compound formed by replacing one or more hydrogens on an alkane with a hydroxyl functional group (−OH). The hydroxyl group should not be confused with the

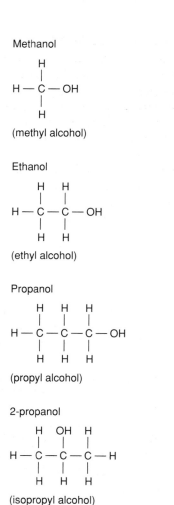

Methanol

H
|
H — C — OH
|
H

(methyl alcohol)

Ethanol

H   H
| |
H — C — C — OH
| |
H   H

(ethyl alcohol)

Propanol

H   H   H
| | |
H — C — C — C — OH
| | |
H   H   H

(propyl alcohol)

2-propanol

H   OH H
| | |
H — C — C — C — H
| | |
H   H   H

(isopropyl alcohol)

Figure 16.12

Four different alcohols. The IUPAC name is given above each structural formula and the common name is given below.

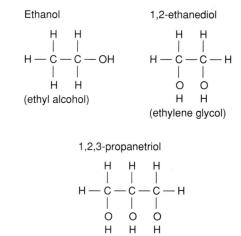

Ethanol

H   H
| |
H — C — C — OH
| |
H   H

(ethyl alcohol)

1,2-ethanediol

H   H
| |
H — C — C — H
| |
O   O
H   H

(ethylene glycol)

1,2,3-propanetriol

H   H   H
| | |
H — C — C — C — H
| | |
O   O   O
H   H   H

(glycerol or glycerin)

Figure 16.13

Common examples of alcohols with one, two, and three hydroxyl groups per molecule. The IUPAC name is given above each structural formula and the common name is given below.

hydroxide ion, OH^-. The hydroxyl group is attached to an organic compound and does not form ions in solution as the hydroxide ion does. It remains attached to a hydrocarbon group (R), giving the compound its set of properties that are associated with alcohols.

The name of the hydrocarbon group (table 16.2) determines the name of the alcohol. If the hydrocarbon group in ROH is methyl, for example, the alcohol is called *methyl alcohol.* Using the IUPAC naming rules, the name of an alcohol has the suffix "-ol." Thus the IUPAC name of methyl alcohol is *methanol.* If the molecule has a sufficient number of carbon atoms that further definition is needed, the base name is determined from the longest continuous chain of carbon atoms that has the —OH. The location of the hydroxyl group is identified with a number (figure 16.12).

All alcohols have the hydroxyl functional group and all are chemically similar. Alcohols are toxic to humans, for example, except that ethanol can be consumed in limited quantities. Consumption of other alcohols such as methanol or 2-propanol (isopropyl alcohol) can result in blindness and death.

Ethanol, C_2H_5OH, is produced by the action of yeast or by a chemical reaction of ethylene derived from petroleum refining. Yeast acts on sugars to produce ethanol and CO_2. When beer, wine, and other such beverages are the desired product, the CO_2 escapes during fermentation and the alcohol remains in solution. In baking, the same reaction utilizes the CO_2 to make the dough rise and the alcohol is evaporated during baking. Most alcoholic beverages are produced by the yeast fermentation reaction, but some are made from ethanol derived from petroleum refining.

The hydroxyl group is strongly polar, and alcohols are soluble in both alkanes and water. A solution of ethanol and gasoline is called **gasohol.** Alcoholic beverages are a solution of ethanol and water. The **proof** of such a beverage is double the ethanol concentration by volume. Thus a solution of 40 percent ethanol by volume in water is 80 proof, and wine that is 12 percent alcohol by volume is 24 proof. Distillation alone will produce a 190 proof concentration, but other techniques are necessary to obtain 200 proof absolute alcohol. *Denatured alcohol* is ethanol with acetone, formaldehyde, and other chemicals in solution that are difficult to separate by distillation. Since these denaturants make consumption impossible, denatured alcohol is sold without the consumption tax.

Methanol, ethanol, and isopropyl alcohol each has one hydroxyl group per molecule. An alcohol with two hydroxyl groups per molecule is called a **glycol.** Ethylene glycol is perhaps the best-known glycol since it is used as an antifreeze. An alcohol with three hydroxyl groups per molecule is called **glycerol** (or glycerin). Glycerol is a by-product in the making of soap. It is added to toothpastes, lotions, and some candies to retain moisture and softness. Ethanol, ethylene glycol, and glycerol are compared in figure 16.13.

Glycerol reacts with nitric acid in the presence of sulfuric acid to produce glyceryl trinitrate, commonly known as *nitroglycerine.* Nitroglycerine is a clear oil that is violently explosive,

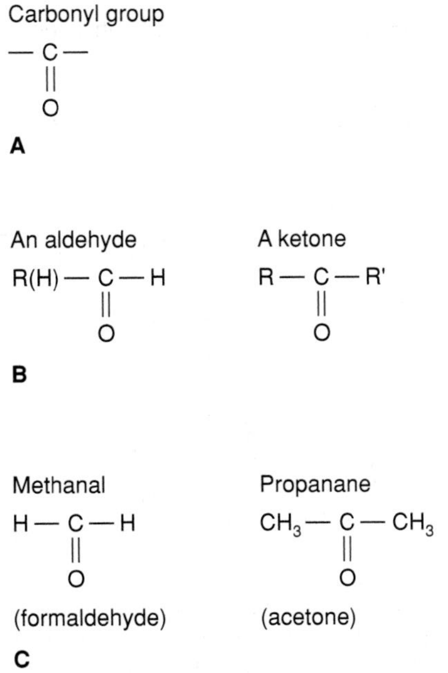

Carbonyl group

$$-C-$$
$$\|$$
$$O$$

A

An aldehyde

$$R(H)-C-H$$
$$\|$$
$$O$$

A ketone

$$R-C-R'$$
$$\|$$
$$O$$

B

Methanal

$$H-C-H$$
$$\|$$
$$O$$

(formaldehyde)

Propanane

$$CH_3-C-CH_3$$
$$\|$$
$$O$$

(acetone)

C

Figure 16.14
The carbonyl group (a) is present in both aldehydes and ketones as shown in (b). (c) The simplest example of each with the IUPAC name above and the common name below each formula.

and when warmed, it is extremely unstable. In 1867, Alfred Nobel discovered that a mixture of nitroglycerine and siliceous earth was more stable than pure nitroglycerine but was nonetheless explosive. The mixture is packed in a tube and is called *dynamite*. Old dynamite tubes, however, leak pure nitroglycerine that is again sensitive to a slight shock.

Ethers, Aldehydes, and Ketones

An **ether** has a general formula of ROR', and the best-known ether is diethylether. In a molecule of diethylether, both the R and the R' are ethyl groups. Diethylether is a volatile, highly flammable liquid that was used as an anesthetic in the past. Today, it is used as an industrial and laboratory solvent.

Aldehydes and ketones both have a functional group of a carbon atom doubly bonded to an oxygen atom called a *carbonyl group*. The **aldehyde** has a hydrocarbon group, R (or a hydrogen in one case), and a hydrogen attached to the carbonyl group. A **ketone** has a carbonyl group with two hydrocarbon groups attached (figure 16.14).

The simplest aldehyde is *formaldehyde*. Formaldehyde is soluble in water, and a 40 percent concentration called *formalin* has been used as an embalming agent and to preserve biological specimens. Formaldehyde is also a raw material used to make plastics such as Bakelite. All the aldehydes have odors, and the odors of some aromatic hydrocarbons include the odors of almonds, cinnamon, and vanilla. The simplest ketone is *acetone*. Acetone has a fragrant odor and is used as a solvent in paint removers and nail polish removers.

Organic Acids and Esters

Mineral acids, such as hydrochloric and sulfuric acid, are made of inorganic materials. Acids that were derived from organisms are called **organic acids.** Because many of these organic acids can be formed from fats, they are sometimes called *fatty acids*. Chemically, they are known as the *carboxylic acids* because they contain the carboxyl functional group, $-COOH$, and have a general formula of RCOOH.

The simplest carboxylic acid has been known since the Middle Ages, when it was isolated by the distillation of ants. The Latin word *formica* means "ant," so this acid was given the name *formic acid*. Formic acid is

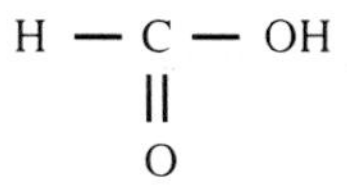

$$H-C-OH$$
$$\|$$
$$O$$

It is formic acid, along with other irritating materials, that causes the sting of bees, ants, and certain plants.

Acetic acid, the acid of vinegar, has been known since antiquity. Acetic acid forms from the oxidation of ethanol. An oxidized bottle of wine contains acetic acid in place of the alcohol, which gives the wine a vinegar taste. Before wine is served in a restaurant, the person ordering is customarily handed the bottle cork and a glass with a small amount of wine. You first break the cork in half to make sure it is dry, which tells you that the wine has been sealed from oxygen. The small sip is to taste for vinegar before the wine is served. If the wine has been oxidized, the reaction is

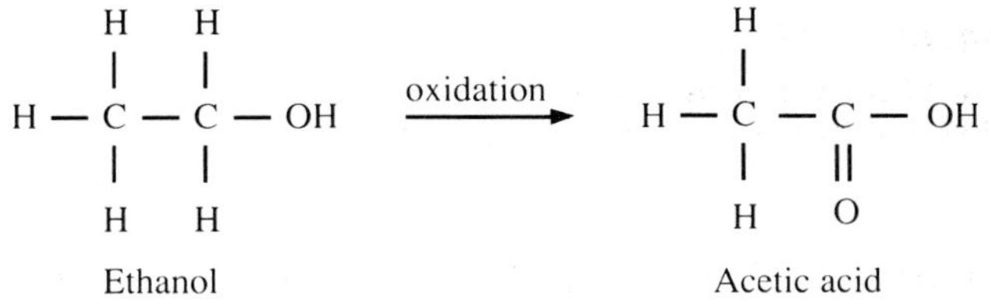

Ethanol → oxidation → Acetic acid

Organic acids are common in many foods. The juice of citrus fruit, for example, contains citric acid, which relieves a thirsty feeling by stimulating the flow of saliva. Lactic acid is found in sour milk, buttermilk, sauerkraut, and pickles. Lactic acid also forms in your muscles as a product of carbohydrate metabolism, causing a feeling of fatigue. Citric and lactic acids are small molecules compared to some of the carboxylic acids that are formed from fats. Palmitic acid, for example, is $C_{16}H_{32}O_2$ and comes from palm oil. The structure of palmitic acid is a chain of fourteen CH_2 groups with CH_3^- at one end and $-COOH$ at the other. Again, it is the functional carboxyl group, $-COOH$, that gives the molecule its acid properties. Organic acids are also raw materials used in the making of polymers of fabric, film, and paint.

Esters are common in both plants and animals, giving fruits and flowers their characteristic odor and taste. Esters are also used in perfumes and artificial flavorings. A few of the flavors that particular esters are responsible for are listed in table 16.5. These liquid esters can be obtained from natural sources or they

Table 16.5

Flavors and esters

Ester Name	Formula	Flavor
Amyl Acetate	$CH_3 - \overset{\displaystyle O}{\underset{\displaystyle \|}{C}} - O - C_5H_{11}$	Banana
Octyl Acetate	$CH_3 - \overset{\displaystyle O}{\underset{\displaystyle \|}{C}} - O - C_8H_{17}$	Orange
Ethyl Butyrate	$C_3H_7 - \overset{\displaystyle O}{\underset{\displaystyle \|}{C}} - O - C_2H_5$	Pineapple
Amyl Butyrate	$C_3H_7 - \overset{\displaystyle O}{\underset{\displaystyle \|}{C}} - O - C_5H_{11}$	Apricot
Ethyl Formate	$H - \overset{\displaystyle O}{\underset{\displaystyle \|}{C}} - O - C_2H_5$	Rum

large macromolecules. A **macromolecule** is a very large molecule that is a combination of many smaller, similar molecules joined together in a chainlike structure. Macromolecules have molecular weights of thousands or millions of atomic mass units. There are three main types of macromolecules: (1) proteins, (2) carbohydrates, and (3) nucleic acids, in addition to fats. A living organism, even a single-cell organism such as a bacterium, contains six thousand or so different kinds of macromolecules. The basic unit of an organism is called a *cell*. Cells are made of macromolecules that are formed inside the cell. The cell decomposes organic molecules taken in as food and uses energy from the food molecules to build more macromolecules. The process of breaking down organic molecules and building up macromolecules is called *metabolism*. Through metabolism, the cell grows, then divides into two cells. Each cell is an exact duplicate of the other, even down to the number and kinds of macromolecules contained within. Each new cell continues the process of growth, then reproduces again, making more cells. This is the basic process of life. The complete process is complicated and very involved, easily filling a textbook in itself, so the details will not be presented here. The following discussion will be limited to three groups of organic molecules involved in the process: proteins, carbohydrates, and fats and oils.

Proteins

Proteins are macromolecular polymers made up of smaller molecules of amino acids. These very large macromolecules have molecular weights that vary from about six thousand to fifty million. Some proteins are simple straight-chain polymers of amino acids, but others contain metal ions such as Fe^{2+} or parts of organic molecules derived from vitamins. Proteins serve as major structural and functional materials in animals. *Structurally*, proteins are major components of muscles, connective tissue, and the skin, hair, and nails. *Functionally*, some proteins are enzymes, which catalyze metabolic reactions; hormones, which regulate body activities; hemoglobin, which carries oxygen to cells; and antibodies, which protect the body.

Proteins are formed from 20 **amino acids,** which are organic acid functional groups with the general formula of

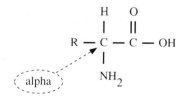

Note the carbon atom labeled "alpha" in the general formula. The amino functional group (NH_2) is attached to this carbon atom, which is next to the carboxylic group (COOH). This arrangement is called an *alpha-amino acid,* and the building blocks of proteins are all alpha-amino acids. The 20 amino acids differ in the nature of the R group, also called the *side chain*. It is the side chain that determines the properties of a protein. Figure 16.15 gives the structural formula for the 20 amino acids found in most proteins and the three-letter abbreviations of the name of each amino acid.

can be chemically synthesized. Whatever the source, amyl acetate, for example, is the chemical responsible for what you identify as the flavor of banana. Natural flavors, however, are complex mixtures of these esters along with other organic compounds. Lower molecular weight esters are fragrant-smelling liquids, but higher molecular weight esters are odorless oils and fats. These are discussed in the next section along with carbohydrates and proteins.

Organic Compounds of Life

Aristotle and the later proponents of the vitalist theory were *partly* correct in their concept that living organisms are different from inorganic substances made of the same elements. Living organisms, for example, have the ability to (1) exchange matter and energy with their surroundings and (2) transform matter and energy into different forms as they (3) respond to changes in their surroundings. In addition, living organisms can use the transformed matter and energy to (4) grow and (5) reproduce. Living organisms are able to do these things through a great variety of organic reactions that are catalyzed by enzymes, however, and not through some mysterious "vital force." These enzyme-regulated organic reactions take place because living organisms are highly organized and have an incredible number of relationships between many different chemical processes.

The chemical processes regulated by living organisms begin with relatively small organic molecules and water. The organism uses energy and matter from the surroundings to build

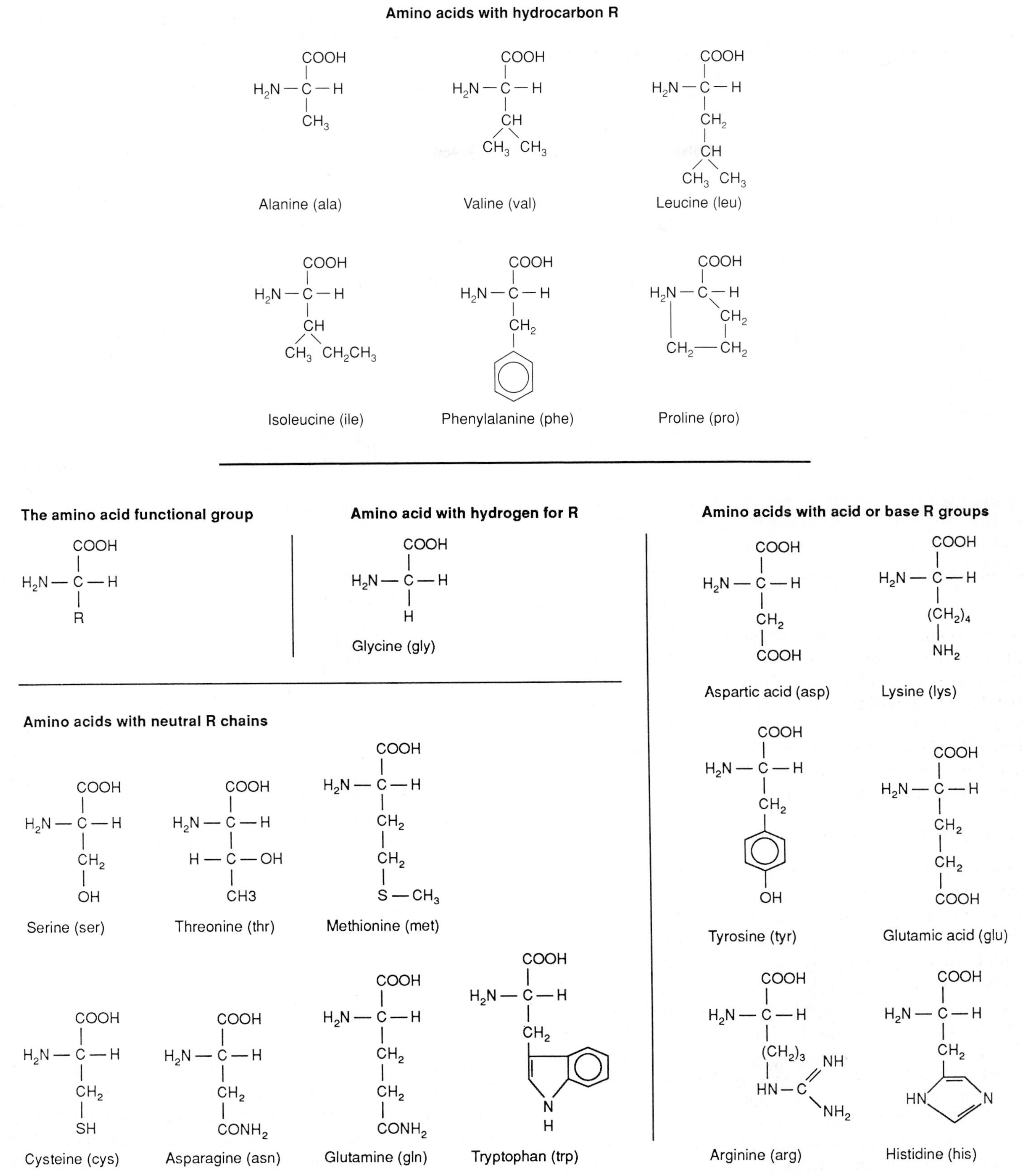

Amino acids with hydrocarbon R

Alanine (ala)

Valine (val)

Leucine (leu)

Isoleucine (ile)

Phenylalanine (phe)

Proline (pro)

The amino acid functional group

Amino acid with hydrogen for R

Glycine (gly)

Amino acids with acid or base R groups

Aspartic acid (asp)

Lysine (lys)

Amino acids with neutral R chains

Serine (ser)

Threonine (thr)

Methionine (met)

Tyrosine (tyr)

Glutamic acid (glu)

Cysteine (cys)

Asparagine (asn)

Glutamine (gln)

Tryptophan (trp)

Arginine (arg)

Histidine (his)

Figure 16.15

The twenty amino acids that make up proteins, with three-letter abbreviations. The carboxyl group of one amino acid bonds with the amino group of a second acid to yield a dipeptide and water. Proteins are polypeptides.

Organic Chemistry

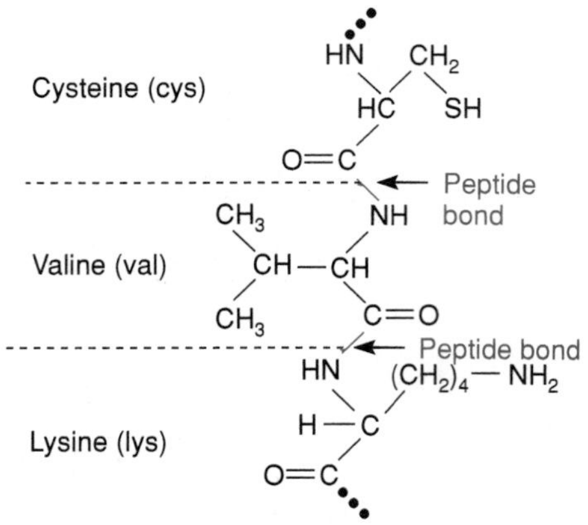

Figure 16.16

Part of a protein polypeptide made up of the amino acids cysteine (cys), valine (val), and lysine (lys). A protein can have from fifty to one thousand of these amino acid units, with each protein having its own unique sequence.

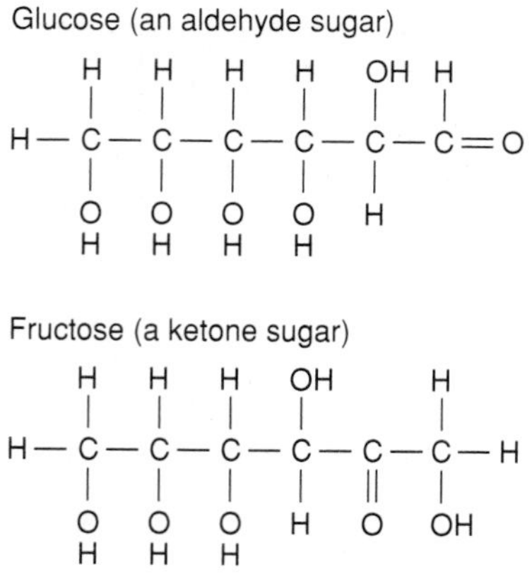

Figure 16.17

Glucose (blood sugar) is an aldehyde and fructose (fruit sugar) is a ketone. Both have a molecular formula of $C_6H_{12}O_6$.

Amino acids are linked to form a protein by a peptide bond between the amino group of one amino acid and the carboxyl group of a second amino acid. A polypeptide is a polymer formed from linking many amino acid molecules. If the polypeptide has a role as a biological structure or function, it is called a *protein*. A protein chain can consist of different combinations of the 20 amino acids with hundreds or even thousands of amino acid molecules held together with peptide bonds (figure 16.16). The arrangement or sequence of these amino acid molecules determines the structure that gives the protein its unique set of biochemical properties. Insulin, for example, is a protein hormone that biochemically regulates the blood sugar level. Insulin contains 86 amino acids that begin (at the amino group) with phenylalanine, valine, asparagine, and then 83 other amino acid molecules in the chain. Hemoglobin is the protein that carries oxygen in the bloodstream, and its biochemical characteristics are determined by its chain of 146 amino acid molecules.

Carbohydrates

Carbohydrates are an important group of organic compounds that includes sugars, starches, and cellulose, and they are important in plants and animals for structure, protection, and food. Cellulose is the skeletal substance of plants and plant materials, and chitin is a similar material that forms the hard, protective covering of insects and shellfish such as crabs and lobsters. *Glucose*, $C_6H_{12}O_6$, is the most abundant carbohydrate and serves as a food and a basic building block for other carbohydrates.

Carbohydrates were named when early studies found that water vapor was given off and carbon remained when sugar was heated. The name *carbohydrate* literally means "watered carbon," and the empirical formulas for most carbohydrates indeed indicate carbon (C) and water (H_2O). Glucose, for example, could be considered to be six carbons with six waters, or $C_6(H_2O)_6$. However, carbohydrate molecules are more complex

than just water attached to a carbon atom. They are polyhydroxyl aldehydes and ketones, two of which are illustrated in figure 16.17. The two carbohydrates in this illustration belong to a group of carbohydrates known as **monosaccharides,** or *simple sugars*. They are called simple sugars because they are mostly 6-carbon molecules such as glucose and fructose. Glucose (also called dextrose) is found in the sap of plants, and in the human bloodstream it is called *blood sugar*. Corn syrup, which is often used as a sweetener, is mostly glucose. Fructose, as its name implies, is the sugar that occurs in fruits and it is sometimes called *fruit sugar*. Both glucose and fructose have the same molecular formula, but glucose is an aldehyde sugar and fructose is a ketone sugar (figure 16.17). A mixture of glucose and fructose is found in honey. This mixture also is formed when table sugar is reacted with water in the presence of an acid, a reaction that takes place in the preparation of canned fruit and candies. The mixture of glucose and fructose is called *invert sugar*. Invert sugar is about twice as sweet to the taste as the same amount of sucrose.

Two monosaccharides are joined together to form **disaccharides** with the loss of a water molecule, for example,

$$C_6H_{12}O_6 + C_6H_{12}O_6 \rightarrow C_{12}H_{22}O_{11} + H_2O$$
$$\text{glucose} \qquad \text{fructose} \qquad \text{sucrose}$$

The most common disaccharide is *sucrose,* or ordinary table sugar. Sucrose occurs in high concentrations in sugar cane and sugar beets. It is extracted by crushing the plant materials, then dissolving the sucrose from the materials with water. The water is evaporated and the crystallized sugar is decolorized with charcoal to produce white sugar. Other common disaccharides include *lactose* (milk sugar) and *maltose* (malt sugar). All three disaccharides have similar properties, but maltose tastes only about one-third as sweet as sucrose. Lactose tastes only about one-sixth as sweet as sucrose. No matter which sugar is consumed (sucrose, lactose, or maltose), it is converted into glucose and transported by the bloodstream for use by the body.

Polysaccharides are polymers consisting of monosaccharide units joined together in straight or branched chains. Polysaccharides are the energy-storage molecules of plants and

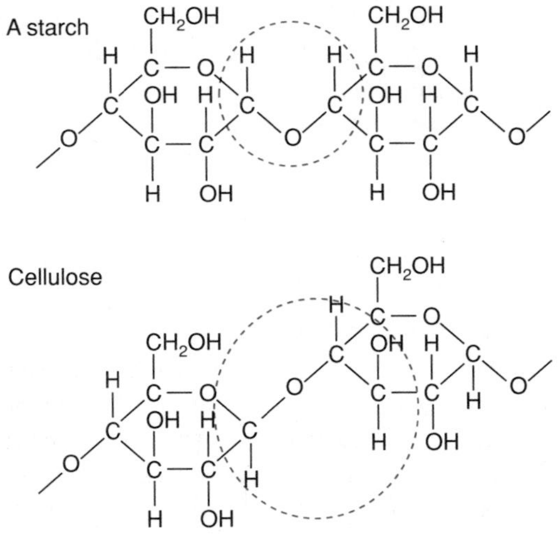

Figure 16.18

Starch and cellulose are both polymers of glucose, but humans cannot digest cellulose. The difference in the bonding arrangement might seem minor, but enzymes must fit a molecule very precisely. Thus, enzymes that break down starch do nothing to cellulose.

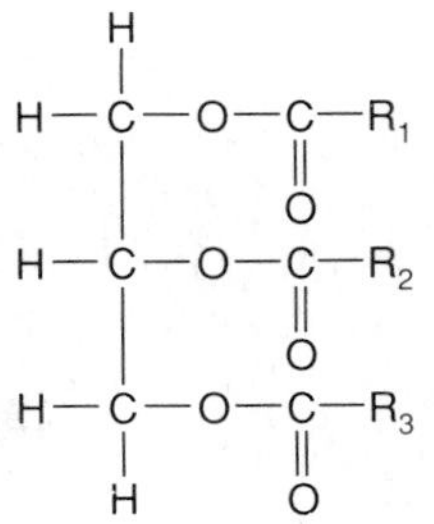

Figure 16.19

The triglyceride structure of fats and oils. Note the glycerol structure on the left and the ester structure on the right. Also notice that R_1, R_2, and R_3 are carboxylic acid (fatty acid) groups.

animals (starch and glycogen) and the structural-building molecules of plants (cellulose). **Starch** is a group of complex carbohydrates that plants use as a stored food source. Potatoes, rice, corn, and wheat contain starch granules and serve as an important source of food for humans. The human body breaks down the starch molecules to glucose, which is transported by the bloodstream and utilized just like any other glucose. This digestive process begins with enzymes secreted with saliva in the mouth. You may have noticed a result of this enzyme catalyzed reaction as you chew bread. If you chew the bread for awhile it begins to taste sweet.

Plants store sugars in the form of starch polysaccharides, and animals store sugars in the form of the polysaccharide **glycogen**. Glycogen is a starchlike polysaccharide that is synthesized by the human body and stored in the muscles and liver. Glycogen, like starch, is a very high molecular weight polysaccharide but it is more highly branched. These highly branched polysaccharides serve as a direct reserve source of energy in the muscles. In the liver, they serve as a reserve source to maintain the blood sugar level.

Cellulose is a polysaccharide that is abundant in plants, forming the fibers in cell walls that preserves the structure of plant materials. Cellulose molecules are straight chains, consisting of large numbers of glucose units. These glucose units are arranged very similar to the glucose units of starch but with differences in the bonding arrangement that holds the glucose units together (figure 16.18). This difference turns out to be an important one where humans are concerned, because enzymes that break down starches do not affect cellulose. Humans do not have the necessary enzymes to break down the cellulose chain (digest it), so humans receive no food value from cellulose. Cattle and termites that do utilize cellulose as a source of food have bacteria (with the necessary enzymes) in their digestive systems. Cellulose is still needed in the human diet, however, for fiber and bulk.

Fats and Oils

The human body can normally synthesize all of the amino acids needed to build proteins except for eight called the *essential amino acids*. An adequate diet must contain the eight essential amino acids or health problems result. Meat and dairy products usually provide the essential amino acids, but they can also be acquired by combining cereal grains (corn, wheat, rice, etc.) with a legume (beans, peanuts, etc.). Interestingly, many ethnic foods have such a combination, for example, corn and beans (Mexican), rice and soybeans (tofu) (Japanese), and rice and red beans (Cajun).

Cereal grains and legumes also provide carbohydrates, the human body's preferred food for energy. When a sufficient amount of carbohydrates are consumed, the body begins to store some of its energy source in the form of glycogen in the muscles and liver. Beyond this storage for short-term needs, the body begins to store energy in a different chemical form for longer-term storage. This chemical form is called **fats** in animals and **oils** in plants. Fats and oils are esters formed from glycerol (1,2,3-trihydroxypropane) and three long-chain carboxylic acids (fatty acids). This ester is called a **triglyceride,** and its structural formula is shown in figure 16.19. Fats are solids and oils are liquids at room temperature, but they both have this same general structure.

Fats and oils usually have two or three different fatty acids, and several are listed in table 16.6. Animal fats can be either saturated or unsaturated but most are saturated. Oils are liquids at room temperature because they contain a higher number of unsaturated units. These unsaturated oils (called "poly" unsaturated in news and advertisements), such as safflower and corn oils, are used as liquid cooking oils because unsaturated oils are believed to lead to lower cholesterol levels in the bloodstream. Saturated fats, along with cholesterol, are believed to contribute to hardening of the arteries over time.

Cooking oils from plants, such as corn and soybean oil, are hydrogenated to convert the double bonds of the unsaturated oil to the single bonds of a saturated one. As a result, the liquid oils are converted to solids at room temperature. For example, one brand of oleomargarine lists ingredients as "liquid

Organic Chemistry

Table 16.6

Some fatty acids occurring in fats

Common Name	Condensed Structure	Source
Lauric Acid	$CH_3(CH_2)_{10}COOH$	Coconuts
Palmitic Acid	$CH_3(CH_2)_{14}COOH$	Palm oil
Stearic Acid	$CH_3(CH_2)_{16}COOH$	Animal fats
Oleic Acid	$CH_3(CH_2)_7CH = CH(CH_2)_7COOH$	Corn oil
Linoleic Acid	$CH_3(CH_2)_4CH = CHCH_2 = CH(CH_2)_7COOH$	Soybean oil
Linolenic Acid	$CH_3CH_2(CH = CHCH_2)_3(CH_2)_6COOH$	Fish oils

soybean oil (nonhydrogenated) and partially hydrogenated cottonseed oil with water, salt, preservatives, and coloring." Complete hydrogenation would result in a hard solid, so the cottonseed oil is partially hydrogenated, then mixed with liquid soybean oil. Coloring is added because oleo is white, not the color of butter. Vegetable shortening is the very same product without added coloring. Reaction of a triglyceride with a strong base such as KOH or NaOH yields a fatty acid of salt and glycerol. A sodium or potassium fatty acid is commonly known as *soap.*

Excess food from carbohydrate, protein, or fat and oil sources is converted to fat for long-term energy storage in *adipose tissue,* which also serves to insulate and form a protective padding. In terms of energy storage, fats yield more than twice the energy per gram oxidized than carbohydrates or proteins.

Synthetic Polymers

Polymers are huge, chainlike molecules made of hundreds or thousands of smaller, repeating molecular units called *monomers.* Polymers occur naturally in plants and animals. Cellulose, for example, is a natural plant polymer made of glucose monomers. Wool and hair are natural animal polymers made of protein monomers. Synthetic polymers are now manufactured from a wide variety of substances, and you are familiar with these polymers as synthetic fibers such as nylon and the inexpensive light plastic used for wrappings and containers (figure 16.20).

The first synthetic polymer was a modification of the naturally existing cellulose polymer. Cellulose was chemically modified in 1862 to produce celluloid, the first *plastic.* The term "plastic" means that celluloid could be molded to any desired shape. Celluloid was produced by first reacting cotton with a mixture of nitric and sulfuric acids, which produced an ester of cellulose nitrate. This ester is an explosive compound known as "guncotton," or smokeless gunpowder. When made with ethanol and camphor, the product is less explosive and can be formed and molded into useful articles. This first plastic, celluloid, was

used to make dentures, combs, glasses frames, and photographic film. Before the discovery of celluloid, many of these articles, including dentures, were made from wood. Today, only Ping-Pong balls are made from cellulose nitrate.

Cotton reacted with acetic acid and sulfuric acid produces a cellulose acetate ester. This polymer, through a series of chemical reactions, produces viscose rayon filaments when forced through small holes. The filaments are twisted together to form viscose rayon thread. When forced through a thin slit, a sheet is formed rather than filaments, and the transparent sheet is called *cellophane.* Both rayon and cellophane, as celluloid, are manufactured by modifying the natural polymer of cellulose.

The first truly synthetic polymer was produced in the early 1900s by reacting two chemicals with relatively small molecules rather than modification of a natural polymer. Phenol, an aromatic hydrocarbon, was reacted with formaldehyde, the simplest aldehyde, to produce the polymer named *Bakelite.* Bakelite is a *thermosetting* material that forms cross-links between the polymer chains. Once the links are formed during production, the plastic becomes permanently hardened and cannot be softened or made to flow. Some plastics are *thermoplastic* polymers and soften during heating and harden during cooling because they do not have cross-links.

Polyethylene is a familiar thermoplastic polymer used for vegetable bags, dry cleaning bags, grocery bags, and plastic squeeze bottles. Polyethylene is a polymer produced by a polymerization reaction of ethylene, which is derived from petroleum. Polyethylene was invented just before World War II and was used as an electrical insulating material during the war. Today, there are many variations of polyethylene that are produced by different reaction conditions or by substitution of one or more hydrogen atoms in the ethylene molecule. When soft polyethylene near the melting point is rolled in alternating perpendicular directions or expanded and compressed as it is cooled, the polyethylene molecules become ordered in such a way to improve the rigidity and tensile strength. This change in the microstructure produces *high-density polyethylene* with a superior rigidity and tensile strength compared to *low-density polyethylene.* High-density polyethylene is used as liners in screw-on jar tops, bottle caps, and as a material for toys.

The properties of polyethylene are changed by replacing one of the hydrogen atoms in a molecule of ethylene. If the hydrogen is replaced by a chlorine atom the compound is called vinyl chloride, and the polymer formed from vinyl chloride is

$$\underset{H}{\overset{H}{\diagdown}} C = C \underset{Cl}{\overset{H}{\diagup}}$$

polyvinyl chloride (PVC). Polyvinyl chloride is used to make plastic water pipes, synthetic leather, and other vinyl products. It differs from the waxy plastic of polyethylene because of the chlorine atom that replaces hydrogen on each monomer.

Name	Chemical unit	Uses	Name	Chemical unit	Uses
Polyvinyl acetate	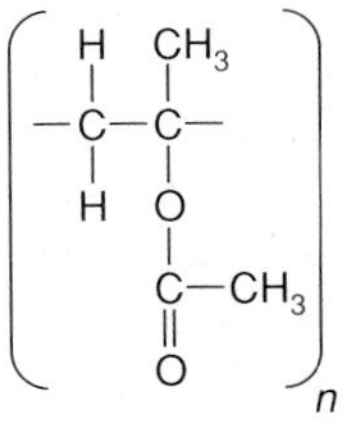	Mixed with vinyl chloride to make vinylite; used as an adhesive and resin in paint	Polyethylene		Sqeeze bottles, containers, laundry and trash bags, packaging
Styrene-butadiene rubber	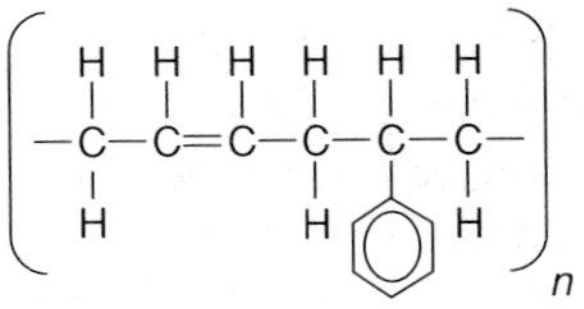	Automobile tires	Polypropylene		Indoor-outdoor carpet, pipe valves, bottles
Polychloroprene (Neoprene)	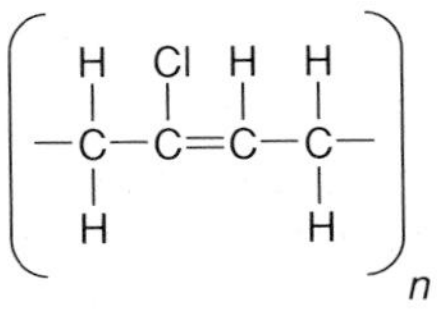	Shoe soles, heels	Polyvinylchloride (PVC)		Plumbing pipes, synthetic leather, plastic tablecloths, phonograph records, vinyl tile
Polymethyl methacrylate (Plexiglas, Lucite)	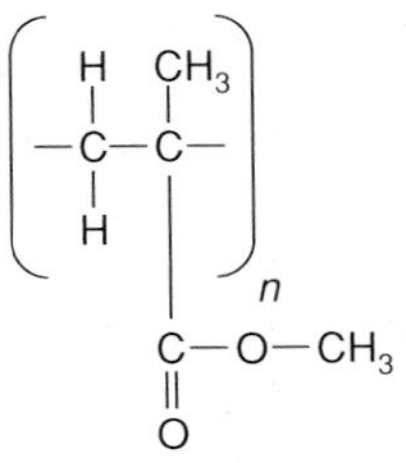	Moldings, transparent surfaces on furniture, lenses, jewelery, transparent plastic "glass"	Polystyrene		Styrofoam coolers, cups, insulating foam, shock resistant packing material, simulated wood furniture
Polyvinylidene chloride (Saran)	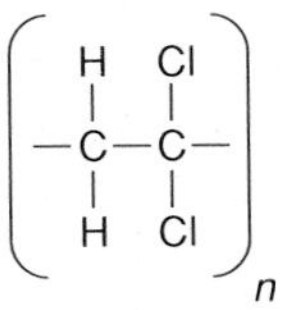	Flexible food wrap	Polytetrafluoroethylene (Teflon)		Gears, bearings, coating for nonstick surface of cooking utensils
Polyacrylonitrile (Orlon, Acrilan, Creslan)	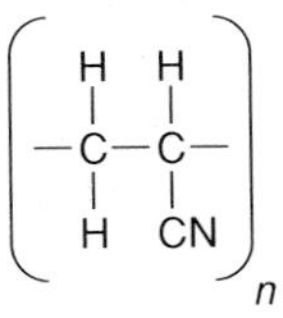	Textile fibers			
Polycarbonate (Lexan)	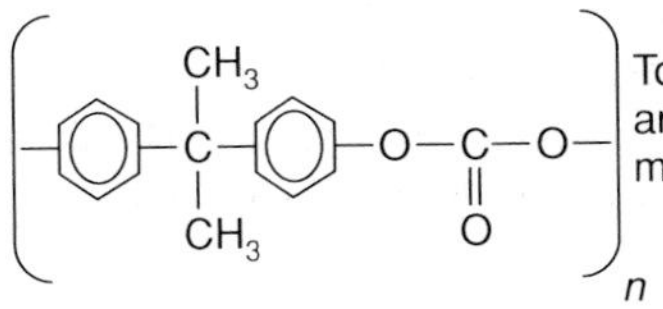	Tough molded articles such as motorcycle helmets			

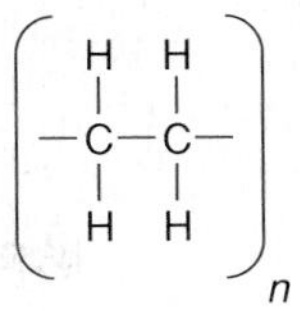

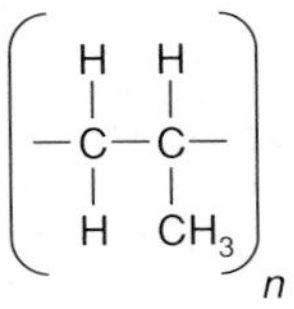

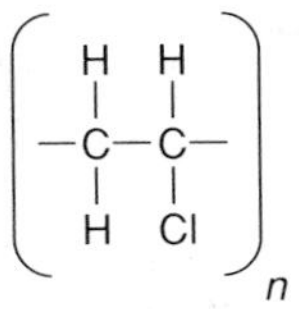

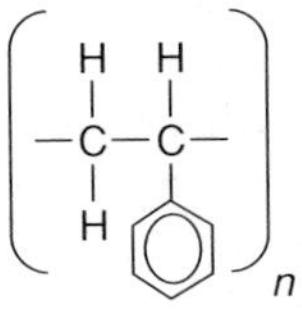

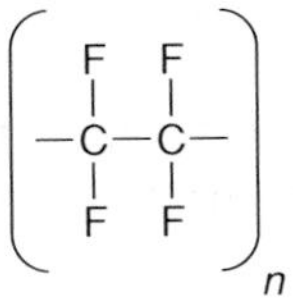

Figure 16.20

Synthetic polymers, the polymer unit, and some uses of each polymer.

Replacement of a hydrogen atom with a benzene ring makes a monomer called *styrene*. Styrene is

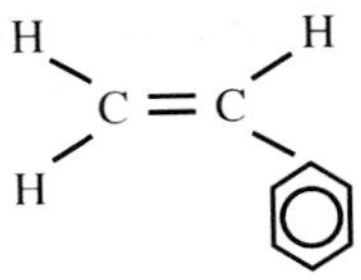

and polymerization of styrene produces *polystyrene*. Polystyrene is puffed full of air bubbles to produce the familiar Styrofoam coolers, cups, and insulating materials.

If all hydrogens of an ethylene molecule are replaced with atoms of fluorine, the product is polytetrafluorethylene, a tough plastic that resists high temperatures and acts more like a metal

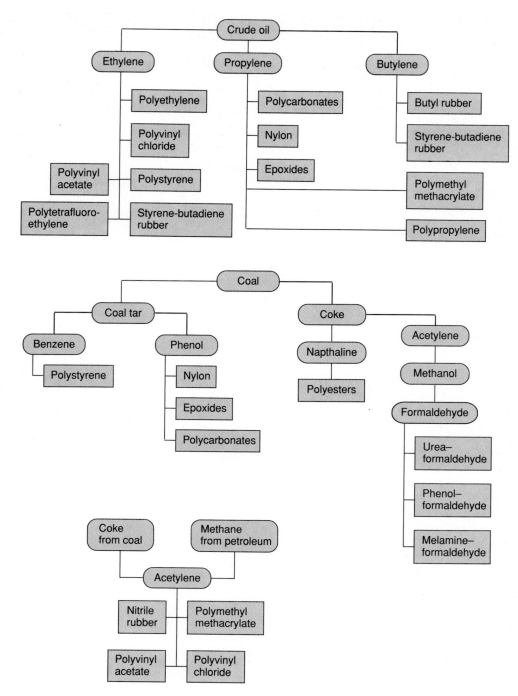

Figure 16.21

Petroleum and coal as sources of raw materials for manufacturing synthetic polymers.

than a plastic. Since it has a low friction it is used for bearings, gears, and as a nonsticking coating on frying pans. You probably know of this plastic by its trade name of *Teflon*.

There are many different polymers in addition to PVC, Styrofoam, and Teflon, and the monomers of some of these are shown in figure 16.20. There are also polymers of isoprene, or synthetic rubber, in wide use. Fibers and fabrics may be polyamides (such as nylon), polyesters (such as Dacron), or

polyacrylonitriles (Orlon, Acrilon, Creslon), which have a CN in place of a hydrogen atom on an ethylene molecule and are called acrylic materials. All of these synthetic polymers have added much to practically every part of your life. It would be impossible to list all of their uses here, however, they present problems since (1) they are manufactured from raw materials obtained from coal and a dwindling petroleum supply (figure 16.21) and (2) they do not readily decompose when dumped into

Organic Molecules and Heredity

D id you ever wonder how hereditary characteristics are passed on from parents to their children and how one protein becomes a muscle but another forms hair? Hereditary information and the synthesis of proteins are controlled and carried out by certain organic molecules. This Feature is but a brief introduction to the complicated and involved chemical processes that determine everything biological, from what kinds of proteins are formed as you grow to why you are you.

Nucleic acids are organic polymers that give the self-replicating ability to living cells (box figure 16.1). There are two types of nucleic acid polymers, deoxyribonucleic acids (DNA) and ribonucleic acids (RNA). Both of these polymers consist of repeating units called *nucleotides*. The nucleotides of DNA are made up of (1) the organic base amines of thymine, adenine, guanine, and cytosine; (2) a phosphoric acid molecule; and (3) the simple sugar deoxyribose. RNA is different in structure in that thymine is replaced by a different organic base, uracil, and the deoxyribose is replaced by ribose.

In general, DNA is found in the nucleus of a cell. The linear sequences of nucleotides are the genetic codes of *chromosomes,* the cell structures that contain the DNA. A DNA molecule is very large, with molecular weights up to sixteen million, and is configured in a double helix. The phosphate and sugar groups are on the outside of this spiral arrangement and the amines are inside, bonded in a specific way. When a cell divides, the DNA strands are separated and each single strand takes other organic molecules to build two new, identical double helixes. The sequences of amines determine the hereditary information for the synthesis of proteins that work together to produce the particular organism being replicated.

Protein synthesis is not carried out by the DNA. The DNA only carries the genetic information. Information coded in the DNA is transmitted to structures called *ribosomes,* where the protein synthesis takes place, by RNA. Actually, there are three kinds of RNA: (1) messenger RNA, (2) transfer RNA, and (3) ribosomal RNA. The transfer RNA is a complementary strand of the DNA, which produces a smaller messenger RNA

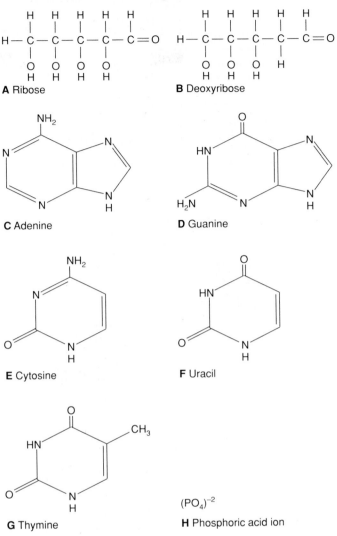

A Ribose

B Deoxyribose

C Adenine

D Guanine

E Cytosine

F Uracil

G Thymine

$(PO_4)^{-2}$

H Phosphoric acid ion

Box Figure 16.1

The building blocks of nucleotides are two simple sugars, (*a*) and (*b*), five amides, (*c*) through (*g*), and phosphoric acid molecules, (*h*).

molecule. Messenger RNA carries the pattern of protein synthesis in a sequence of three amines. The messenger RNA molecules move about the cell, attaching to the ribosome, where they serve as a template for protein synthesis. The sequence of this template determines the kind of protein formed and thus the function and purpose of that protein. There are sixty-one different arrangements, or codes, that the RNA template can carry

that determine if muscle proteins or the proteins of hormones are produced, for example. It is a complicated and involved process and any small error, from the replication of DNA to the production of proteins, can lead to mutations or genetic diseases that result from faulty protein synthesis. Research in *genetic engineering,* the manipulation of DNA molecules, may someday provide answers to how to correct such genetic errors.

Organic Chemistry

rivers, oceans, or other parts of the environment. However, research in the polymer sciences is beginning to reflect new understandings learned from research on biological tissues. This could lead to whole new molecular designs for synthetic polymers that will be more compatible with the ecosystem.

Summary

Organic chemistry is the study of compounds that have carbon as the principal element. Such compounds are called *organic compounds* and all the rest are *inorganic compounds*. There are millions of organic compounds because a carbon atom can link with other carbon atoms as well as atoms of other elements.

A *hydrocarbon* is an organic compound consisting of hydrogen and carbon atoms. The simplest hydrocarbon is one carbon atom and four hydrogen atoms, or CH_4. All hydrocarbons larger than CH_4 have one or more carbon atoms bonded to another carbon atom. The bond can be single, double, or triple, and this forms a basis for classifying hydrocarbons. A second basis is if the carbons are in a ring or not. The *alkanes* are hydrocarbons with single carbon-to-carbon bonds, the *alkenes* have a double carbon-to-carbon bond, and the *alkynes* have a triple carbon-to-carbon bond. The alkanes, alkenes, and alkynes can have straight- or branched-chain molecules. When the number of carbon atoms is greater than three, there are different arrangements that can occur for a particular number of carbon atoms. The different arrangements with the same molecular formula are called isomers. *Isomers* have different physical properties, so each isomer is given its own name. The name is determined by (1) identifying the longest continuous carbon chain as the base name, (2) locating the attachment of other atoms or hydrocarbon groups by counting from the direction that results in the smallest numbers, (3) identifying attached hydrocarbon groups by changing the "-ane" suffix of alkanes to "-ly," (4) identifying the number of these hydrocarbon groups with prefixes, and (5) identifying the location of the groups with the carbon atom number.

The alkanes have all the hydrogen atoms possible, so they are *saturated* hydrocarbons. The alkenes and the alkynes can add more hydrogens to the molecule, so they are *unsaturated* hydrocarbons. Unsaturated hydrocarbons are more chemically reactive than saturated molecules.

Hydrocarbons that occur in a ring or cycle structure are cyclohydrocarbons. A six-carbon cyclohydrocarbon with three double bonds has different properties than the other cyclohydrocarbons because the double bonds are not localized. This six-carbon molecule is *benzene*, the basic unit of the *aromatic hydrocarbons*.

Petroleum is a mixture of alkanes, cycloalkanes, and a few aromatic hydrocarbons that formed from the slow decomposition of buried marine plankton and algae. Petroleum from the ground, or *crude oil*, is distilled into petroleum products of *natural gas, LPG, petroleum ether, gasoline, kerosene, diesel fuel,* and *motor oils*. Each group contains a range of hydrocarbons and is processed according to use.

In addition to oxidation, hydrocarbons react by *substitution, addition,* and *polymerization* reactions. Reactions take place at sites of multiple bonds or lone pairs of electrons on the *functional groups*. The functional group determines the chemical properties of organic compounds. Functional group results in the *hydrocarbon derivatives* of *alcohols, ethers, aldehydes, ketones, organic acids, esters,* and *amines*.

Living organisms have an incredible number of highly organized chemical reactions that are catalyzed by *enzymes,* using food and energy to grow and reproduce. The process involves building large *macromolecules* from smaller molecules and units. The organic molecules involved in the process are proteins, carbohydrates, and fats and oils.

Proteins are macromolecular polymers of *amino acids* held together by *peptide bonds*. There are twenty amino acids that are used in various polymer combinations to build structural and functional proteins. *Structural proteins* are muscles, connective tissue, and the skin, hair, and nails of animals. *Functional proteins* are enzymes, hormones, and antibodies.

Carbohydrates are polyhydroxyl aldehydes and ketones that form three groups, the monosaccharides, disaccharides, and polysaccharides. The *monosaccharides* are simple sugars such as *glucose* and *fructose*. Glucose is *blood sugar,* a source of energy. The disaccharides are *sucrose* (table sugar), *lactose* (milk sugar), and *maltose* (malt sugar). The disaccharides are broken down (digested) to glucose for use by the body. The polysaccharides are polymers or glucose in straight or branched chains used as a near-term source of stored energy. Plants store the energy in the form of *starch,* and animals store it in the form of *glycogen*. *Cellulose* is a polymer similar to starch that humans cannot digest.

Fats and oils are esters formed from three fatty acids and glycerol into a *triglyceride*. *Fats* are usually solid triglycerides associated with animals, and *oils* are liquid triglycerides associated with plant life, but both represent a high-energy storage material.

Polymers are huge, chain-like molecules of hundreds or thousands of smaller, repeating molecular units called *monomers*. Polymers occur naturally in plants and animals, and many *synthetic polymers* are made today from variations of the ethylene-derived monomers. Among the more widely used synthetic polymers derived from ethylene are polyethylene, polyvinyl chloride, polystyrene, and Teflon. Problems with the synthetic polymers include that (1) they are manufactured from fossil fuels that are also used as the primary energy supply and (2) they do not readily decompose and tend to accumulate in the environment.

Key Terms

alcohol (p. 317)
aldehyde (p. 319)
alkanes (p. 309)
alkenes (p. 313)
alkyne (p. 313)
amino acids (p. 320)
aromatic hydrocarbons (p. 313)
carbohydrates (p. 322)
cellulose (p. 323)
crude oil (p. 315)
disaccharides (p. 322)
esters (p. 319)
ether (p. 319)
fats (p. 323)
functional group (p. 317)
gasohol (p. 318)
glycerol (p. 318)
glycogen (p. 323)
glycol (p. 318)
hydrocarbon (p. 309)

hydrocarbon derivatives (p. 316)
inorganic chemistry (p. 308)
isomers (p. 311)
ketone (p. 319)
macromolecule (p. 320)
monosaccharides (p. 322)
oil field (p. 314)
oils (p. 323)
organic acids (p. 319)
organic chemistry (p. 308)
petroleum (p. 314)
polymer (p. 317)
polysaccharides (p. 322)
proof (p. 318)
proteins (p. 320)
saturated (p. 313)
starch (p. 323)
triglyceride (p. 323)
unsaturated (p. 313)

Applying the Concepts

1. An organic compound is a compound that
   a. contains carbon and was formed only by a living organism.
   b. is a natural compound that has not been synthesized.
   c. contains carbon, no matter if it was formed by a living thing or not.
   d. was formed by a plant.

2. The organic compound 2,2,4-trimethylpentane is an isomer of
   a. propane.
   b. pentane.
   c. heptane.
   d. octane.

3. The label on a container states that the product contains "petroleum distillates." Which of the following hydrocarbons is probably present?
   a. CH_4
   b. C_5H_{12}
   c. $C_{16}H_{34}$
   d. $C_{40}H_{82}$

4. Ethylene molecules can add to each other in a reaction to form a long chain called a
   a. monomer.
   b. dimer.
   c. trimer.
   d. polymer.

5. The R in ROH represents a(n)
   a. functional group.
   b. hydrocarbon group with a name ending in "-yl."
   c. atom of an inorganic element.
   d. polyatomic ion that does not contain carbon.

6. The OH in ROH represents
   a. a functional group.
   b. a hydrocarbon group with a name ending in "-yl."
   c. the hydroxide ion, which ionizes to form a base.
   d. the site of chemical activity in a strong base.

7. What is the proof of a "wine cooler" that is 5 percent alcohol by volume?
   a. 2.5 proof
   b. 5 proof
   c. 10 proof
   d. 50 proof

8. A bottle of wine that has "gone bad" now contains
   a. CH_3OH
   b. CH_3OCH_3
   c. CH_3COOH
   d. CH_3COOCH_3

9. Liquid oils from plants can be converted to solids by adding what to the molecule?
   a. metal ions
   b. carbon
   c. polyatomic ions
   d. hydrogen

10. You know for sure that the compound named decane has
    a. more than ten isomers.
    b. ten carbon atoms in each molecule.
    c. only single bonds.
    d. all of the above.

Answers

1. c 2. d 3. b 4. d 5. b 6. a 7. c 8. c 9. d
10. d

Questions for Thought

1. What is an organic compound?

2. There are millions of organic compounds but only thousands of inorganic compounds. Explain why this is the case.

3. Define and give an example of an (a) alkane, (b) alkene, and (c) alkyne.

4. Using the same number of carbon atoms for both, draw the structural formula and give the name for any (a) straight-chain hydrocarbon and (b) branched-chain hydrocarbon.

5. Using the same number of carbon atoms for both, draw the structural formula and give the name for any (a) saturated hydrocarbon and (b) unsaturated hydrocarbon.

6. Propylene is an alkene. Draw the structural formula for propylene.

7. What is cracking and reforming? For what purposes are either or both used by the petroleum industry?

8. What is an isomer? Describe how the properties of isomers are alike and how they are different.

9. Is it possible to have an isomer of ethane? Explain.

10. Describe the differences between the following and use structural formulas to illustrate your descriptions: (a) ethane, (b) ethyl, (c) ethene, and (d) ethyne.

11. Suggest a reason why ethylene is an important raw material used in the production of plastics but ethane is not.

12. Describe how a branched-chain alkane is named.

13. What is the size of the "barrel of oil" that is described in news reports.

14. What is (a) natural gas, (b) LPG, and (c) petroleum ether?

15. What does the octane number of gasoline describe? On what is the number based?

16. What is a functional group? What is it about the nature of a functional group that makes it the site of chemical reactions?

17. Draw a structural formula for alcohol. Describe how alcohols are named.

18. A soft drink is advertised to contain "no sugar." The label lists the ingredients as carbonated water, dextrose, corn syrup, fructose, and flavorings. Evaluate the advertising and the list of ingredients.

19. What are fats and oils? What does saturated and unsaturated fats and oils mean?

20. What is a polymer? Give an example of a naturally occurring plant polymer. Give an example of a synthetic polymer.

21. Explain why a small portion of wine is customarily poured before a bottle of wine is served. Sometimes the cork is handed to the person who ordered the wine with the small portion. What is the person supposed to do with the cork and why?

Exercises

Group A—Solutions Provided in Appendix D

1. Draw the structural formulas for (a) *n*-pentane and (b) an isomer of pentane with the maximum possible branching. (c) Give the IUPAC name of this isomer.

2. Write structural formulas for all the hexane isomers you can identify and give their IUPAC names.

3. Write structural formulas for
   (a) 3,3,4-trimethyloctane
   (b) 2-methyl-1-pentene
   (c) 5,5-dimethyl-3-heptyne

4. Write the IUPAC name for each of the following:

```
              H
              |
          H — C — H

      H       |     H   Cl  H
      |       |     |   |   |
  H — C — C — C — C — C — H
      |   |   |   |   |
      H   H   H   H   H
```

```
              H
              |
          H — C — H

      H       |     H   H   H
      |       |     |   |   |
  H — C = C — C — C — C — H
                  |   |   |
                  H   H   H
```

```
              H
              |
          H — C — H

      H       |     H   H   H
      |       |     |   |   |
  H — C — C — C = C — C — H
      |   |   |   |   |
      H   H   |   H   H
              H — C — H
                  |
              H — C — H
                  |
                  H
```

Group B—Solutions Not Given

1. Write structural formulas for (a) *n*-octane and (b) an isomer of octane with the maximum possible branching. (c) Give the IUPAC name of this isomer.

2. Write the structural formulas for all the heptane isomers you can identify and give their IUPAC names.

3. Write structural formulas for
   (a) 2,3-dimethylpentane
   (b) 1-butene
   (c) 2-methyl-3-ethyl-3-hexene

4. Write the IUPAC name for each of the following:

```
              H
              |
          H — C — H

      H       |     H   H
      |       |     |   |
  H — C — C — C — C — H
      |   |   |   |
      H   H   Br  H
```

```
      H   H   H   H   H   H
      |   |   |   |   |   |
  H — C — C = C — C — C — C — H
      |           |   |   |
      H           H   |   H
                      H — C — H
                          |
                          H
```

```
      H   H   H               H
      |   |   |               |
  H — C — C = C — C ≡ C — C — H
      |   |   |               |
      H   Cl  H               H
```

```
      H   H   H   H
      |   |   |   |
 Br — C — C = C — C — H
      |           |
      H           H
```

5. Which would have the higher octane rating, 2,2,3-trimethylbutane or 2,2-dimethylpentane? Explain with an illustration.

6. Use the information in table 16.4 to classify the following as an alcohol, ether, organic acid, ester, or amide.

```
    H   H   H
    |   |   |
H — C — C — C — OH
    |   |   |
    H   H   H
```

```
    H   H   H
    |   |   |
H — C — C — C — NH2
    |   |   |
    H   H   H
```

```
    H   H           H   H   H
    |   |           |   |   |
H — C — C — O — C — C — C — C — H
    |   |       ||  |   |   |
    H   H       O   H   H   H
```

```
    H   H   H           H   H   H
    |   |   |           |   |   |
H — C — C — C — O — C — C — C — H
    |   |   |           |   |   |
    H   H   H           H   H   H
```

```
    H   H   H   H
    |   |   |   |
H — C — C — C — C — C — OH
    |   |   |   |   ||
    H   H   H   H   O
```

5. Which would have the higher octane rating, 2-methyl-butane or dimethylpropane? Explain with an illustration.

6. Classify the following as an alcohol, ether, organic acid, ester, or amide.

```
    H   H   H
    |   |   |
H — C — C — C — H
    |   |   |
    O   O   O
    H   H   H
```

```
    H   H
    |   |
H — C — C — C — OH
    |   |   ||
    H   H   O
```

```
    H   H           H   H
    |   |           |   |
H — C — C — O — C — C — H
    |   |           |   |
    H   H           H   H
```

```
    H   H   H
    |   |   |
H — C — C — C — NH2
    |   |   |
    H   H   H
```

```
    H   H           H   H   H   H   H   H   H
    |   |           |   |   |   |   |   |   |
H — C — C — O — C — C — C — C — C — C — C — H
    |   |       ||  |   |   |   |   |   |   |
    H   H       O   H   H   H   H   H   H   H
```

Nuclear Reactions

Figure 17.1

Decisions about nuclear energy require some understanding about nuclear reactions and the nature of radioactivity. This is the Palo Verde nuclear power plant in Arizona. With all three units running, this plant produces enough power to meet the electrical needs of nearly four million people.

THE ancient alchemist dreamed of changing one element into another, such as lead into gold. The alchemist was never successful, however, because such changes were attempted with chemical reactions. Chemical reactions are reactions that involve only the electrons of atoms. Electrons are shared or transferred in chemical reactions and the internal nucleus of the atom is unchanged. Elements thus retain their identity during the sharing or transferring of electrons. This chapter is concerned with a different kind of reaction, one that involves the *nucleus* of the atom. In nuclear reactions, the nucleus of the atom is often altered, changing the identity of the elements involved. The ancient alchemist's dream of changing one element into another was actually a dream of achieving a nuclear change, that is, a nuclear reaction.

Understanding nuclear reactions is important because although fossil fuels are the major source of energy today, there are growing concerns about (1) air pollution from fossil fuel combustion, (2) increasing levels of CO_2 from fossil fuel combustion, which may be warming the earth (the greenhouse effect), and (3) the dwindling fossil fuel supply itself, which cannot last forever. Energy experts see nuclear energy as a means of meeting rising energy demands in an environmentally acceptable way. However, the topic of nuclear energy is controversial, and discussions about the topic often result in strong emotional responses. Decisions about the use of nuclear energy require some understandings about nuclear reactions and some facts about radioactivity and radioactive materials (figure 17.1). These understandings and facts are the topics of this chapter.

Natural Radioactivity

Radioactivity was discovered in 1896 by Henri Becquerel, a French scientist who was very interested in the recent discovery of X rays. Becquerel was experimenting with fluorescent minerals, minerals that give off visible light after being exposed to sunlight. He wondered if fluorescent minerals emitted X rays in addition to visible light. From previous work with X rays, Becquerel knew that they would penetrate a wrapped, light-tight photographic plate, exposing it as visible light exposes an unprotected plate. Thus Becquerel decided to place a fluorescent uranium mineral on a protected photographic plate while the mineral was exposed to sunlight. Sure enough, he found a silhouette of the mineral on the plate when it was developed. Believing the uranium mineral emitted X rays, he continued his studies until the weather turned cloudy. Storing a wrapped, protected photographic plate and the uranium mineral together during the cloudy weather, Becquerel returned to the materials later and developed the photographic plate to again find an image of the mineral (figure 17.2). He concluded that the mineral was emitting an "invisible radiation" that was not induced by sunlight. Becquerel named the emission of invisible radiation *radioactivity*. Materials that have the property of radioactivity are called *radioactive* materials.

Becquerel's discovery led to the beginnings of the modern atomic theory and to the discovery of new elements. Ernest Rutherford studied the nature of radioactivity and found that there are three kinds, which are today known by the first three letters of the Greek alphabet—alpha (α), beta (β), and gamma

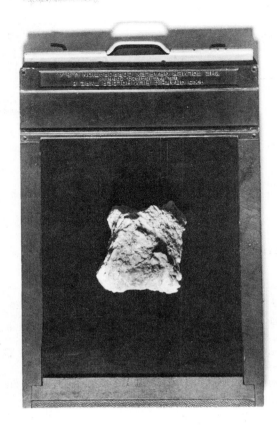

A

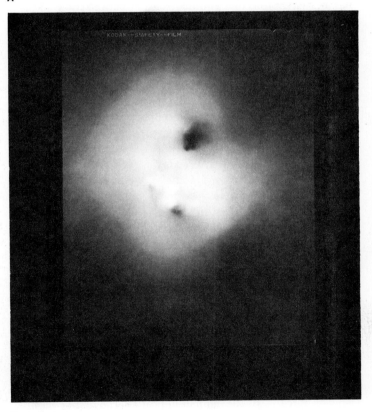

B

Figure 17.2

Radioactivity was discovered by Henri Becquerel when he exposed a light-tight photographic plate to a radioactive mineral, then developed the plate. (*a*) A photographic film is exposed to an uranite ore sample. (*b*) The film, developed normally after a four-day exposure to uranite. Becquerel found an image like this one and deduced that the mineral gave off "invisible radiation" he called radioactivity.

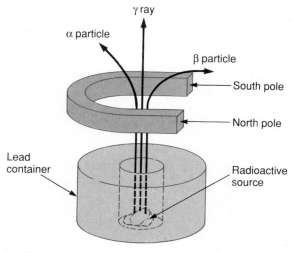

Figure 17.3

Radiation passing through a magnetic field shows that massive, positively charged alpha particles are deflected one way and less massive beta particles with their negative charge are greatly deflected in the opposite direction. Gamma rays, like light, are not deflected.

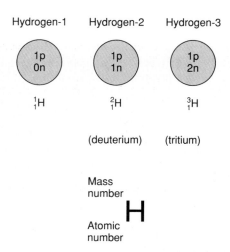

Figure 17.4

The three isotopes of hydrogen have the same number of protons but different numbers of neutrons. Hydrogen-1 is the most common isotope. Hydrogen-2, with an additional neutron, is named deuterium, and hydrogen-3 is called tritium. Neutrons and protons are called *nucleons* because they are in the nucleus.

(γ). These Greek letters were used at first before the nature of the radiation was known. Today, an **alpha particle** (sometimes called an alpha ray) is known to be the nucleus of a helium atom, that is, two protons and two neutrons. A **beta particle** (or beta ray) is a high-energy electron. A **gamma ray** is electromagnetic radiation, as is light, but of very short wavelength (figure 17.3).

It was Rutherford's work with alpha particles that resulted in the discovery of the nucleus and the proton (see chapter 11). At Becquerel's suggestion, Madame Marie Curie searched for other radioactive materials and in the process discovered two new elements, polonium and radium. More radioactive elements were discovered since that time, and, in fact, all the isotopes of all the elements with an atomic number greater than 83 (bismuth) are radioactive. Today, **radioactivity** is defined as the *spontaneous emission of particles or energy from an atomic nucleus* as it disintegrates. As a result of the disintegration the nucleus of an atom often undergoes a change of identity, becoming a simpler nucleus. The spontaneous disintegration of a given nucleus is a purely natural process and cannot be controlled or influenced. The natural spontaneous disintegration or decomposition of a nucleus is also called **radioactive decay.** Although it is impossible to know *when* a given nucleus will undergo radioactive decay, as you will see later, it is possible to deal with the *rate* of decay for a given radioactive material with precision.

Nuclear Equations

There are two main subatomic particles in the nucleus, the proton and the neutron. The proton and neutron are called **nucleons.** Recall that the number of protons, the *atomic number,* determines what element an atom is, and that all atoms of a given element have the same number of protons. The number of neutrons varies in *isotopes,* which are atoms with the same

atomic number but different numbers of neutrons (figure 17.4). The number of protons and neutrons together determines the *mass number,* so different isotopes of the same element are identified with their mass numbers. Thus the two most common, naturally occurring isotopes of uranium are referred to as uranium-238 and uranium-235, and the 238 and 235 are the mass numbers of these isotopes. Isotopes are also represented by the symbol:

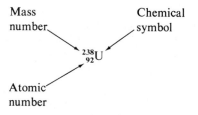

Subatomic particles involved in nuclear reactions are represented by symbols with the following form:

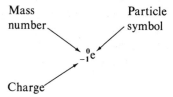

Symbols for these particles are illustrated in table 17.1.

Symbols are used in an equation for a nuclear reaction that is written much like a chemical reaction with reactants and products. When a uranium-238 nucleus emits an alpha particle ($^4_2$He), for example, it loses two protons and two neutrons. The nuclear reaction is written in equation form as

$$^{238}_{92}\text{U} \rightarrow ^{234}_{90}\text{Th} + ^4_2\text{He}$$

Table 17.1

Names, symbols, and properties of particles in nuclear equations

Name	Symbol	Mass Number	Charge
Proton	$_1^1H$ (or $_1^1p$)	1	1+
Electron	$_{-1}^0e$ (or $_{-1}^0\beta$)	0	1−
Neutron	$_0^1n$	1	0
Gamma Photon	$_0^0\gamma$	0	0

The *products* of this nuclear reaction from the decay of a uranium-238 nucleus are (1) the alpha particle ($_2^4He$) given off and (2) the nucleus, which remains after the alpha particle leaves the original nucleus. What remains is easily determined since all nuclear equations must show conservation of charge and conservation of the total number of nucleons. Therefore: (1) the number of protons (positive charge) remains the same, and the sum of the subscripts (atomic number, or numbers of protons) in the reactants must equal the sum of the subscripts in the products; and (2) the total number of nucleons remains the same, and the sum of the superscripts (atomic mass, or number of protons plus neutrons) in the reactants must equal the sum of the superscripts in the products. The new nucleus remaining after the emission of an alpha particle therefore has an atomic number of 90 (92 − 2 = 90). According to the table of atomic numbers on the inside back cover of this text, this new nucleus is thorium (Th). The mass of the thorium isotope is 238 minus 4, or 234. The emission of an alpha particle thus decreases the number of protons by 2 and the mass number by 4. From the subscripts, you can see that the total charge is conserved (92 = 90 + 2). From the superscripts, you can see that the total number of nucleons is also conserved (238 = 234 + 4). The mass numbers (superscripts) and the atomic numbers (subscripts) are *balanced* in a correctly written nuclear equation. Such nuclear equations are considered to be independent of any chemical form or chemical reaction. Nuclear reactions are independent and separate from chemical reactions, whether or not the atom is in the pure element or in a compound. The particles that are involved in nuclear reactions each has its own symbol with a superscript indicating mass number and a subscript indicating the charge. These symbols, names, and numbers are given in table 17.1.

Example 17.1

A plutonium-242 nucleus undergoes radioactive decay, emitting an alpha particle. Write the nuclear equation for this nuclear reaction.

Solution

Step 1: The table of atomic weights on the inside back cover gives the atomic number of plutonium as 94. Plutonium-242 therefore has a symbol of $_{94}^{242}Pu$. The symbol for an alpha particle is ($_2^4He$), so the nuclear equation so far is

$$_{94}^{242}Pu \rightarrow {_2^4}He + ?$$

Step 2: From the subscripts, you can see that 94 = 2 + 92, so the new nucleus has an atomic number of 92. The table of atomic weights identifies element 92 as uranium with a symbol of U.

Step 3: From the superscripts, you can see that the mass number of the uranium isotope formed is 242 − 4 = 238, so the product nucleus is $_{92}^{238}U$ and the complete nuclear equation is

$$_{94}^{242}Pu \rightarrow {_2^4}He + {_{92}^{238}}U$$

Step 4: Checking the subscripts (94 = 2 + 92) and the superscripts (242 = 4 + 238), you can see that the nuclear equation is balanced.

Example 17.2

What is the product nucleus formed when radium emits an alpha particle? (Answer: Radon-222, a chemically inert, radioactive gas)

The Nature of the Nucleus

The modern atomic theory does not picture the nucleus as a group of stationary protons and neutrons clumped together by some "nuclear glue." The protons and neutrons are understood to be held together by a **nuclear force,** a strong fundamental force of attraction that is functional only at very short distances, on the order of 10^{-15} m or less. At distances greater than about 10^{-15} m the nuclear force is negligible, and the weaker **electromagnetic force,** the force of attraction between like charges, is the operational force. Thus like charged protons experience a repulsive force when they are farther apart than about 10^{-15} m. When closer together than 10^{-15} m, the short-range, stronger nuclear force predominates and the protons experience a strong attractive force. This explains why the like charged protons of the nucleus are not repelled by their like electric charges.

Observations of radioactive decay reactants and products and experiments with nuclear stability have led to a **shell model of the nucleus.** This model considers the protons and neutrons moving in energy levels, or shells, in the nucleus analogous to the shell structure of electrons in the outermost part of the atom. As in the electron shells, there are certain configurations of nuclear shells that have a greater stability than others. Considering electrons, filled and half-filled shells are more stable than other arrangements, and maximum stability occurs with the noble gases and their 2, 10, 18, 36, 54, and 86 electrons. Considering the nucleus, atoms with 2, 8, 20, 28, 50, 82, or 126 protons or neutrons have a maximum nuclear stability. The stable numbers are not the same for electrons and nucleons because of differences in nuclear and electromagnetic forces.

Isotopes of uranium, radium, and plutonium, as well as other isotopes, emit an alpha particle during radioactive decay to a simpler nucleus. The alpha particle is a helium nucleus, $_2^4He$. The alpha particle contains two protons as well as two neutrons, which is one of the nucleon numbers of stability, so you would expect the helium nucleus (or alpha particle) to have a

Nuclear Reactions

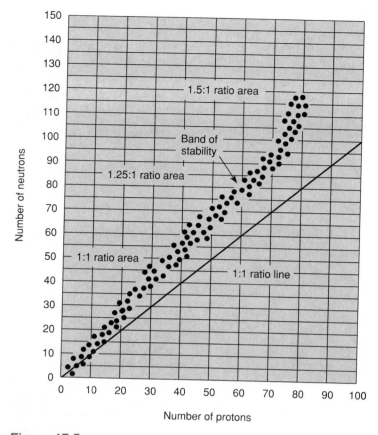

Figure 17.5

The dots indicate stable nuclei, which group in a band of stability according to their neutron-to-proton ratio. As the size of nuclei increases, so does the neutron-to-proton ratio that represents stability. Nuclei outside this band of stability are radioactive.

stable nucleus and it does. Stable means it does not undergo radioactive decay. Pairs of protons and pairs of neutrons have increased stability just as pairs of electrons in a molecule. As a result, nuclei with an *even number* of both protons and neutrons are, in general, more stable than nuclei with odd numbers of protons and neutrons. There are a little more than 150 isotopes with an even number of protons and an even number of neutrons, but there are 5 stable isotopes with odd numbers of each. Just as in the case of electrons, there are other factors that come into play as the nucleus becomes larger and larger with increased numbers of nucleons.

The results of some of these factors are shown in figure 17.5, which is a graph of the number of neutrons versus the number of protons in nuclei. As the number of protons increases, the neutron-to-proton ratio of the *stable nuclei* also increases in a **band of stability.** Within the band the neutron-to-proton ratio increases from about 1:1 at the bottom left to about 1½:1 at the top right. The increased ratio of neutrons is needed to produce a stable nucleus as the number of protons increases. Neutrons provide additional attractive *nuclear* (not electrical) forces, which counters the increased electrical repulsion from a larger number of positively charged protons. Thus more neutrons are required in larger nuclei to produce a stable nucleus. However, there is a limit to the additional attractive

forces that can be provided by more and more neutrons, and all isotopes of all elements with more than 83 protons are unstable and thus undergo radioactive decay.

The generalizations about nuclear stability provide a means of predicting if a particular nucleus is radioactive. The generalizations are as follows:

1. All isotopes with an atomic number greater than 83 have an unstable nucleus.
2. Isotopes that contain 2, 8, 20, 28, 50, 82, or 126 protons or neutrons in their nucleus are generally more stable than other numbers of protons or neutrons.
3. Pairs of protons and pairs of neutrons have increased stability, so isotopes that have nuclei with even numbers of both protons and neutrons are generally more stable than nuclei with odd numbers of both protons and neutrons.
4. Isotopes with an atomic number less than 83 are stable when the ratio of neutrons to protons in the nucleus is about 1:1 in isotopes with up to 20 protons, but the ratio increases in larger nuclei in a band of stability (figure 17.5). Isotopes with a ratio to the left or right of this band are unstable and thus will undergo radioactive decay.

Example 17.3

Would you predict the following isotopes to be radioactive or stable?

(a) $^{60}_{27}Co$

(b) $^{222}_{86}Rn$

(c) $^{3}_{1}H$

(d) $^{40}_{20}Ca$

Solution

(a) Cobalt-60 has 27 protons and 33 neutrons, both odd numbers, so you might expect $^{60}_{27}Co$ to be radioactive.
(b) Radon has an atomic number of 86, and all isotopes of all elements beyond atomic number 83 are radioactive. Radon-222 is therefore radioactive.
(c) Hydrogen-3 has an odd number of protons and an even number of neutrons, but its 2:1 neutron-to-proton ratio places it outside the band of stability. Hydrogen-3 is radioactive.
(d) Calcium-40 has an even number of protons and an even number of neutrons, containing 20 of each. The number 20 is a particularly stable number of protons or neutrons, and calcium-40 has 20 of each. In addition, the neutron-to-proton ratio is 1:1, placing it within the band of stability. All indications are that calcium-40 is stable, not radioactive.

Example 17.4

Which of the following would you predict to be radioactive?

(a) $^{127}_{53}I$

(b) $^{131}_{53}I$

(c) $^{206}_{82}Pb$

(d) $^{214}_{82}Pb$

(Answer: (b) and (d))

Types of Radioactive Decay

Through the process of radioactive decay, an unstable nucleus becomes a more stable one with less energy. The three more familiar types of radiation emitted—alpha, beta, and gamma—were introduced earlier. There are five common types of radioactive decay, and three of these involve alpha, beta, and gamma radiation.

1. *Alpha emission.* Alpha (α) emission is the expulsion of an alpha particle ($_2^4$He) from an unstable, disintegrating nucleus. The alpha particle, a helium nucleus, travels from 2 to 12 cm through the air, depending on the energy of emission from the source. An alpha particle is easily stopped by a sheet of paper close to the nucleus. As an example of alpha emission, consider the decay of a radon-222 nucleus,

$$_{86}^{222}\text{Rn} \rightarrow _{84}^{218}\text{Po} + _2^4\text{He}$$

The spent alpha particle eventually acquires two electrons and becomes an ordinary helium atom.

2. *Beta emission.* Beta (β^-) emission is the expulsion of a different particle, a beta particle, from an unstable disintegrating nucleus. A beta particle is simply an electron ($_{-1}^0$e) ejected from the nucleus at a high speed. The emission of a beta particle *increases the number of protons* in a nucleus. It is as if a neutron changed to a proton by emitting an electron, or

$$_0^1\text{n} \rightarrow _1^1\text{p} + _{-1}^0\text{e}$$

Carbon-14 is a carbon isotope that decays by beta emission:

$$_6^{14}\text{C} \rightarrow _7^{14}\text{N} + _{-1}^0\text{e}$$

Note that the number of protons increased from six to seven, but the mass number remained the same. The mass number is unchanged because the mass of the expelled electron (beta particle) is negligible.

Beta particles are more penetrating than alpha particles and may travel several hundred centimeters through the air. They can be stopped by a thin layer of metal close to the emitting nucleus, such as a 1 cm thick piece of aluminum. A spent beta particle may eventually join an ion to become part of an atom, or it may remain a free electron.

3. *Gamma emission.* Gamma (γ) emission is a high-energy burst of electromagnetic radiation from an excited nucleus. It is a burst of light (photon) of a wavelength much too short to be detected by the eye. Other types of radioactive decay, such as alpha or beta emission, sometimes leave the nucleus with an excess of energy, a condition called an *excited state.* As in the case of excited electrons, the nucleus returns to a lower energy state by emitting electromagnetic radiation. From a nucleus, this radiation is in the high-energy portion of the electromagnetic spectrum. Gamma is the most penetrating of the three common types of nuclear radiation. Like X rays, gamma rays can pass completely through a person, but all gamma radiation can be stopped by a 5 cm thick piece of lead close to the source. As with other types of electromagnetic radiation, gamma radiation is absorbed by and gives its energy to materials. Since the product nucleus changed from an

Table 17.2

Radioactive decay

Unstable Condition	Type of Decay	Emitted	Product Nucleus
More than 83 protons	Alpha emission	$_2^4$He	Lost 2 protons and 2 neutrons
Neutron-to-proton ratio too large	Beta emission	$_{-1}^0$e	Gained 1 proton, no mass change
Excited nucleus	Gamma emission	$_0^0\gamma$	No change
Neutron-to-proton ratio too small	Other emission	$_1^0$e	Lost 1 proton, no mass change

excited state to a lower energy state, there is no change in the number of nucleons. For example, radon-222 is an isotope that emits gamma radiation:

$$_{86}^{222}\text{Rn*} \rightarrow _{86}^{222}\text{Rn} + _0^0\gamma$$
(* denotes excited state)

Alpha, beta, and gamma emission radioactive decay are summarized in table 17.2, which also lists the unstable nuclear conditions that lead to the particular type of emission. Just as electrons seek a state of greater stability, a nucleus undergoes radioactive decay to achieve a balance between nuclear attractions, electromagnetic repulsions, and a low quantum of nuclear shell energy. The key to understanding the types of reactions that occur is found in the band of stable nuclei illustrated in figure 17.5. The isotopes within this band have achieved the state of stability, and other isotopes above, below, or beyond the band are unstable and thus radioactive.

Nuclei that have a neutron-to-proton ratio beyond the upper right part of the band are unstable because of an imbalance between the proton-proton electromagnetic repulsions and all the combined proton and neutron nuclear attractions. Recall that the neutron-to-proton ratio increases from about 1:1 to about 1½:1 in the larger nuclei. The additional neutron provided additional nuclear attractions to hold the nucleus together, but atomic number 83 appears to be the upper limit to this additional stabilizing contribution. Thus all nuclei with an atomic number greater than 83 are outside the upper right limit of the band of stability. Emission of an alpha particle reduces the number of protons by two and the number of neutrons by two, moving the nucleus more toward the band of stability. Thus you can expect a nucleus that lies beyond the upper right part of the band of stability to be an alpha emitter (figure 17.6).

A nucleus that has a neutron-to-proton ratio that is too large will be on the left side of the band of stability. Emission of a beta particle decreases the number of neutrons and increases the number of protons, so a beta emission will lower the

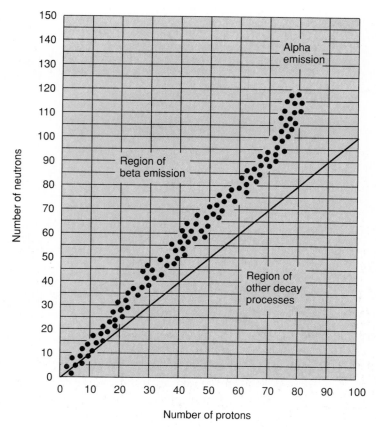

Figure 17.6
Unstable nuclei undergo different types of radioactive decay to obtain a more stable nucleus. What type of decay depends, in general, on the neutron-to-proton ratio as shown.

neutron-to-proton ratio. Thus you can expect a nucleus with a large neutron-to-proton ratio, that is, one to the left of the band of stability, to be a beta emitter.

A nucleus that has a neutron-to-proton ratio that is too small will be on the right side of the band of stability. These nuclei can increase the number of neutrons and reduce the number of protons in the nucleus by other types of radioactive decay. As is usually the situation when dealing with broad generalizations and trends, there are exceptions to the summarized relationships between neutron-to-proton ratios and radioactive decay.

Example 17.5

Refer to figure 17.6 and predict the type of radioactive decay for each of the following unstable nuclei:

(a) $^{131}_{53}I$

(b) $^{241}_{94}Pu$

Solution

(a) Iodine-131 has a nucleus with 53 protons and 131 minus 53, or 78 neutrons, so it has a neutron-to-proton ratio of 1.47:1. This places iodine-131 on the left side of the band of stability, with a high neutron-to-proton ratio that can be reduced by beta emission. The nuclear equation is

$$^{131}_{53}I \rightarrow ^{131}_{54}Xe + ^{0}_{-1}e$$

(b) Plutonium-241 has 94 protons and 241 minus 94, or 147 neutrons, in the nucleus. This nucleus is to the upper right, beyond the band of stability. It can move back toward stability by emitting an alpha particle, losing 2 protons and 2 neutrons from the nucleus. The nuclear equation is

$$^{241}_{94}Pu \rightarrow ^{238}_{92}U + ^{4}_{2}He$$

Radioactive Decay Series

A radioactive decay reaction produces a simpler, more stable nucleus than the reactant nucleus. As discussed in the previous section, large nuclei with an atomic number greater than 83 decay by alpha emission, giving up two protons and two neutrons with each alpha particle. A nucleus with an atomic number greater than 86, however, will emit an alpha particle and *still* have an atomic number greater than 83, which means the product nucleus will also be radioactive. This nucleus will also undergo radioactive decay, and the process will continue through a series of decay reactions until a stable nucleus is achieved. Such a series of decay reactions that (1) begins with one radioactive nucleus, which (2) decays to a second nucleus, which (3) then decays to a third nucleus, and so on until (4) a stable nucleus is reached is called a **radioactive decay series.** There are three naturally occurring radioactive decay series. One begins with thorium-232 and ends with lead-208, another begins with uranium-235 and ends with lead-207, and the third series begins with uranium-238 and ends with lead-206. Figure 17.7 shows the uranium-238 radioactive decay series.

As figure 17.7 illustrates, the uranium-238 begins with uranium-238 decaying to thorium-234 by alpha emission. Thorium has a new position on the graph because it now has a new atomic number and a new mass number. Thorium-234 is unstable and decays to protactinium-234 by beta emission, which is also unstable and decays by beta emission to uranium-234. The process continues with five sequential alpha emissions, then two beta-beta-alpha decay steps before the series terminates with the stable lead-206 nucleus.

Uranium-238 is radioactive and decays to thorium-234 by emitting an alpha particle. Yet not all uranium-238 has decayed to lead-206, and in fact a sample of uranium-238 will continue to give off alpha particles for millions of years. Uranium-238, as uranium-235 and thorium-232, undergoes radioactive decay very slowly, and any given nucleus may disintegrate today or it may disintegrate millions of years later. It is not possible to predict when a nucleus will decay because it is a random process. It is possible, however, to deal with nuclear disintegration statistically since the rate of decay is not changed by any external conditions of temperature, pressure, or any chemical state. When dealing with a large number of nuclei, the ratio of the rate of nuclear disintegration per unit of time to the total number of radioactive nuclei will be a constant, or

$$radioactive\ decay\ constant = \frac{decay\ rate}{number\ of\ nuclei}$$

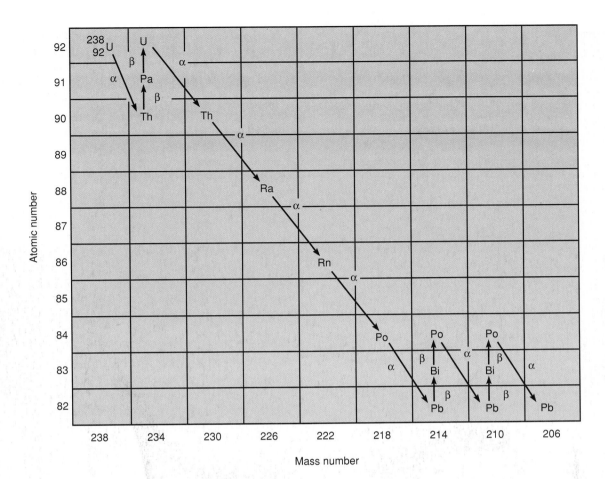

Figure 17.7

The radioactive decay series for uranium-238. This is one of three naturally occurring series.

or, in symbols,

$$k = \frac{\text{rate}}{n}$$ **equation 17.1**

The **radioactive decay constant,** k, is a specific constant for a particular isotope, and each isotope has its own decay constant. For example, a 238 g sample of uranium-238 (1 mole) that has 2.93×10^6 disintegrations per second would have a decay constant of

$$k = \frac{\text{rate}}{n} = \frac{2.93 \times 10^6 \text{ nuclei/sec}}{6.02 \times 10^{23} \text{ nuclei}}$$

$$= 4.87 \times 10^{-18}/\text{sec}$$

The rate of radioactive decay is usually described in terms of its *half-life*. The **half-life** is the time required for one-half of the unstable nuclei to decay. Since each isotope has a characteristic decay constant, then each isotope has its own characteristic half-life. Half-lives of some highly unstable isotopes are measured in fractions of seconds, and other isotopes have half-lives measured in seconds, minutes, hours, days, months, years, or billions of years. Table 17.3 lists half-lives of some of the isotopes, and the process is illustrated in figure 17.8.

Table 17.3

Half-life of some radioactive isotopes

Isotope	Half-Life	Mode of Decay
$^{3}_{1}\text{H}$ (tritium)	12.26 years	Beta
$^{14}_{6}\text{C}$	5,930 years	Beta
$^{90}_{38}\text{Sr}$	28 years	Beta
$^{131}_{53}\text{I}$	8 days	Beta
$^{133}_{54}\text{Xe}$	5.27 days	Beta
$^{238}_{92}\text{U}$	4.51×10^9 years	Alpha
$^{242}_{94}\text{Pu}$	3.79×10^5 years	Alpha
$^{240}_{94}\text{Pu}$	6,760 years	Alpha
$^{239}_{94}\text{Pu}$	24,360 years	Alpha
$^{40}_{19}\text{K}$	1.3×10^9 years	Alpha

Nuclear Reactions

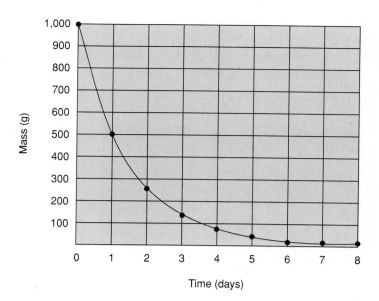

Figure 17.8

Radioactive decay of a hypothetical isotope with a half-life of one day. Each sample decays by one-half to some other element in each half-life of one day. Actual half-lives may be in seconds, minutes, or any time unit up to billions of years.

As an example of the half-life measure, consider a hypothetical isotope that has a half-life of one day. The half-life is independent of the amount of the isotope being considered, but suppose you start with a 1.0 kg sample of this element with a half-life of one day. One day later, you will have half of the original sample, or 500 g. The other half did not disappear, but it is now the decay products, that is, some new element. During the next day half of the remaining nuclei will disintegrate and only 250 g of the initial sample is still the original element. One-half of the remaining sample will disintegrate each day until the original sample no longer exists.

The half-life of a radioactive nucleus is related to its radioactive decay constant by

$$\text{half-life} = \frac{\text{a mathematical constant}}{\text{decay constant}}$$

or

$$t_{1/2} = \frac{0.693}{k} \qquad \textbf{equation 17.2}$$

For example, the radioactive decay constant for uranium-238 was determined earlier to be 4.87×10^{-18}/sec. The half-life or uranium-238 is therefore

$$t_{1/2} = \frac{0.693}{4.87 \times 10^{-18}/\text{sec}} = 1.42 \times 10^{17} \text{ sec}$$

This is the half-life of uranium-238 in seconds. There are $60 \times 60 \times 24 \times 365$, or 3.15×10^7 sec in a year, so

$$\frac{1.42 \times 10^{17} \text{ sec}}{3.15 \times 10^7 \text{ sec/yr}} = 4.5 \times 10^9 \text{ yr}$$

The half-life of uranium-238 is thus 4.5 billion years. Figure 17.9 gives the half-life for each step in the uranium-238 decay series.

As you can see from equations 17.1 and 17.2, the half-life of a radioactive isotope is directly proportional to its rate of disintegration. Thus isotopes with a shorter half-life are more active and are disintegrating at a faster rate. On the other hand, longer half-lives mean less activity and lower rates of radiation.

Measurement of Radiation

The measurement of radiation is important in determining the half-life of radioactive isotopes, as you learned in the previous section. Radiation measurement is also important in considering biological effects, which will be discussed in the next section. As is the case with electricity, it is not possible to make direct measurements on things as small as electrons and other parts of atoms. Indirect measurement methods are possible, however, by considering the effects of the radiation.

Measurement Methods

As Becquerel discovered, radiation affects photographic film, exposing it as visible light does. Since the amount of film exposure is proportional to the amount of radiation, photographic film can be used as an indirect measure of radiation. Today, people who work around radioactive materials or X rays carry light-tight film badges. The film is replaced periodically and developed. The optical density of the developed film provides a record of the worker's exposure to radiation.

There are also devices that indirectly measure radiation by measuring an effect of the radiation. An **ionization counter** is one type of device and measures ions produced by radiation. A second type of device is called a **scintillation counter.** "Scintillate" is a word meaning "sparks or flashes," and a scintillation counter measures the flashes of light produced when radiation strikes a phosphor.

The most common example of an ionization counter is known as a **Geiger counter** (figure 17.10). The working components of a Geiger counter are illustrated in figure 17.11. Radiation is received in a metal tube filled with an inert gas, such as argon, through a thin plastic window that is transparent to alpha, beta, and gamma radiation. An insulated wire inside the tube is connected to the positive terminal of a direct current source. The metal cylinder around the insulated wire is connected to the negative terminal. There is not a current between the center wire and the metal cylinder because the gas acts as an insulator. When radiation passes through the window, however, it ionizes some of the gas atoms, releasing free electrons. These electrons are accelerated by the field between the wire and cylinder, and the accelerated electrons ionize more gas molecules, which results in an *avalanche* of free electrons. The avalanche creates a pulse of current that is amplified, then measured. More radiation means more avalanches, so the pulses are an indirect means of measuring radiation. When connected to a speaker or earphone, each avalanche produces a "pop" or "click."

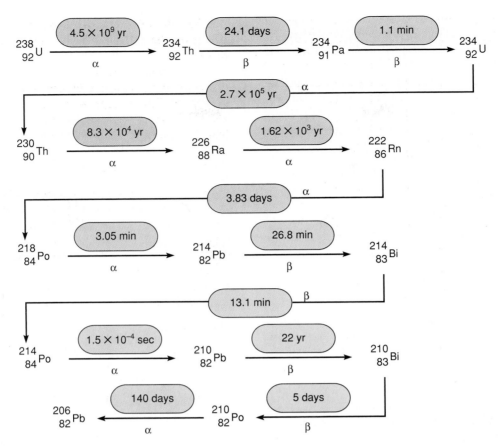

Figure 17.9

The half-life of each step in the uranium-238 radioactive decay series.

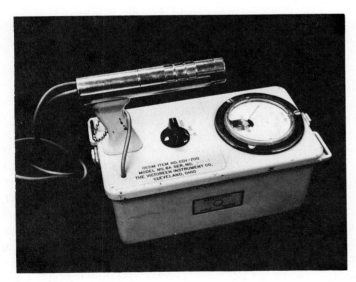

Figure 17.10

This is a portable Geiger counter that gives readings in counts per minute, millirem, or clicks in an earphone.

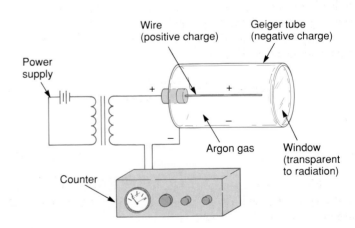

Figure 17.11

The working parts of a Geiger counter.

Nuclear Reactions

Some materials are *phosphors,* substances that emit a flash of light when excited by radiation. Zinc sulfide, for example, is used in television screens and luminous watches, and it was used by Rutherford to detect alpha particles. A luminous watch dial has a mixture of zinc sulfide and a small amount of radium sulfate. A zinc sulfide atom gives off a tiny flash of light when struck by radiation from a disintegrating radium nucleus. A scintillation counter measures the flashes of light through the photoelectric effect, producing free electrons that are accelerated to produce a pulse of current. Again, the pulses of current are used as an indirect means to measure radiation.

Radiation Units

You have learned that *radioactivity* is a property of isotopes with unstable, disintegrating nuclei and *radiation* is emitted particles (alpha or beta) or energy traveling in the form of photons (gamma). Radiation can be measured (1) at the source of radioactivity or (2) at a place of reception, where the radiation is absorbed.

The *activity* of a radioactive source is a measure of the number of nuclear disintegrations per unit of time. The unit of activity at the source is called a **curie** (Ci), which is defined as 3.70×10^{10} nuclear disintegrations per second. The radioactivity can be measured by a radiation counter, or it can be calculated from the radioactive decay rate. For example, a 238 g sample of uranium-238 has a decay rate of 2.93×10^6 disintegrations per second, so the activity in curies is

$$\frac{2.93 \times 10^6}{3.70 \times 10^{10}} = 7.92 \times 10^{-5} \text{ Ci}$$

A unit frequently mentioned is a *picocurie,* which is a millionth of a millionth of a curie.

As radiation from a source moves out and strikes a material it gives the material energy. The amount of energy released by radiation striking living tissue is usually very small, but it can cause biological damage nonetheless. Chemical bonds are broken and free polyatomic ions are produced by radiation, and the broken bonds and free polyatomic ions are the damaging results.

One measure of radiation received by a material is called the **rad.** The term *rad* is from *r*adiation *a*bsorbed *d*ose, and one rad releases 1×10^{-5} J/g (1×10^{-3} J/kg). Another measure of radiation received considers the biological effect from a rad. This unit is called a **rem,** which takes into account the possible biological damage produced by different types of radiation. The term *rem* is from *r*oentgen *e*quivalent *m*an (a roentgen is another measure of radiation). The equivalent measure is needed because alpha radiation, for example, has a greater ionizing power than beta or gamma radiation, so fewer rads of alpha are required to produce a rem. Beta and gamma radiation are the most penetrating, however, and alpha radiation barely penetrates the skin. Alpha radiation can be very damaging if the source gets inside the body, which is the reason so many people are concerned about exposure to radon gas. Radon is chemically inert and cannot be filtered or absorbed by a gas mask or any other means. Most common isotopes of radon are alpha emitters.

Overall, there are many factors and variables that affect the possible damage from radiation, including the distance from the source and what shielding materials are between a person and a source. A *millirem* is 1/1,000 of a rem and is the unit of choice when low levels of radiation are discussed.

Radiation Exposure

Natural radioactivity is a part of your environment, and you receive between 100 and 500 millirems each year from natural sources. This radiation from natural sources is called **background radiation.** Background radiation comes from outer space in the form of cosmic rays and from unstable isotopes in the ground, building materials, and foods. Many activities and situations will increase your yearly exposure to radiation. For example, the atmosphere absorbs some of the cosmic rays from space, so the less atmosphere above you, the more radiation you will receive. You are exposed to one additional millirem per year for each 100 feet you live above sea level. You receive approximately 0.3 millirem for each hour spent on a jet flight. Airline crews thus receive an additional 300 to 400 millirems per year because they spend so much time high in the atmosphere. Additional radiation exposure comes from medical X rays, television sets, and luminous objects such as watch and clock dials. In general, the background radiation exposure for the average person is about 130 millirems per year.

What are the consequences of radiation exposure? Radiation can be a hazard to living organisms because it produces ionization along its path of travel. This ionization can (1) disrupt chemical bonds in essential macromolecules such as DNA and (2) produce molecular fragments, which are free polyatomic ions that can interfere with enzyme action and other essential cell functions. Tissues with highly active cells are more vulnerable to radiation damage than others, such as blood-forming tissue. Thus one of the symptoms of an excessive radiation exposure is an altered blood count. Table 17.4 compares the estimated results of various levels of acute radiation exposure.

Radiation is not a mysterious, unique health hazard. It is a hazard that should be understood and neither ignored nor exaggerated. Excessive radiation exposure should be avoided just as you avoid excessive exposure to other hazards such as certain chemicals, electricity, or even sunlight. Everyone agrees that *excessive* radiation exposure should be avoided, but there is some controversy about long-term, low-level exposure and its possible role in cancer. Some claim that tolerable low-level exposure does not exist because that is not possible. Others point to many studies comparing high and low background radioactivity with cancer mortality data. For example, no cancer mortality differences could be found between people receiving 500 or more millirems a year and those receiving less than 100 millirems a year.

Table 17.4

Approximate single dose, whole body effects of radiation exposure

Level	Comment
0.130 rem	Average annual exposure to natural background radiation
0.500 rem	Upper limit of annual exposure to general public
25.0 rem	Threshold for observable effects such as blood count changes
100.0 rem	Fatigue and other symptoms of radiation sickness
200.0 rem	Definite radiation sickness, bone marrow damage, possibility of developing leukemia
500.0 rem	Lethal dose for 50 percent of individuals
1,000.0 rem	Lethal dose for all

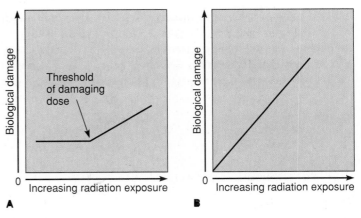

Figure 17.12

Graphic representation of the (a) threshold model and (b) linear model of low-level radiation exposure. The threshold model proposes that the human body can repair damage up to a threshold. The linear model proposes that any radiation exposure is damaging.

The controversy continues, however, because of lack of knowledge about long-term exposure. Two models of long-term, low-level radiation exposure have been proposed: (1) a linear model and (2) a threshold model. The *linear model* proposes that any radiation exposure above zero is damaging and can produce cancer and genetic damage. The *threshold model* proposes that the human body can repair damage and get rid of damaging free polyatomic ions up to a certain exposure level called the threshold (figure 17.12). The controversy over long-term, low-level radiation exposure will probably continue until there is clear evidence about which model is correct. Whichever is correct will not lessen the need for rational risks versus cost-benefit analyses of all energy alternatives.

Nuclear Energy

As discussed, some nuclei are unstable because they are too large or because they have an unstable neutron-to-proton ratio. These unstable nuclei undergo radioactive decay, forming products of greater stability. An example of this radioactive decay is the alpha emission reaction of uranium-238 to thorium-234,

$$^{238}_{92}U \rightarrow ^{234}_{90}Th + ^{4}_{2}He$$

$$238.0003 \text{ u} \rightarrow 233.9942 \text{ u} + 4.00150 \text{ u}$$

The numbers below the nuclear equation are the *nuclear* masses (u) of the reactant and products. As you can see, there seems to be a loss of mass in the reaction,

$$233.9942 + 4.00150 - 238.0003 = -0.0046 \text{ u}$$

This change in mass is related to the energy change according to the relationship that was formulated by Albert Einstein in 1905. The relationship is

$$E = mc^2 \qquad \text{equation 17.3}$$

where E is a quantity of energy, m is a quantity of mass, and c is a constant of the speed of light in a vacuum, 3.00×10^8 m/sec. According to this relationship, matter and energy are the same thing, and energy can be changed to matter and vice versa. Since the mass of a mole in grams is numerically equal to the atomic mass unit (u) of a nucleus, the mass change for a mole of decaying uranium-238 is -0.0046 g, or -4.6×10^{-6} kg. Using this mass loss (Δm) in equation 17.3, you can calculate the energy change (ΔE),

$$\Delta E = \Delta mc^2$$

$$= (-4.6 \times 10^{-6} \text{ kg})\left(3.00 \times 10^8 \frac{m}{sec}\right)^2$$

$$= (-4.6 \times 10^{-6} \text{ kg})\left(9.00 \times 10^{16} \frac{m^2}{sec^2}\right)$$

$$= (-4.6 \times 9.00) \times 10^{(-6 + 16)} \frac{kg \cdot m^2}{sec^2}$$

$$= -4.14 \times 10^{11} \text{ J}$$

Thus the products of a mole of uranium-238 decaying to more stable products (1) have a lower energy of 4.14×10^{11} J and (2) lost a mass of 4.6×10^{-6} kg. As you can see, a very small amount of matter was converted into a large amount of energy in the process, forming products of lower energy.

The relationship between mass and energy explains why the mass of a nucleus is always *less* than the sum of the masses of the individual particles of which it is made. For example, the masses of the particles making up a helium-4 nucleus are

$$2 \text{ protons } = 2(1.00728 \text{ u}) = 2.01456 \text{ u}$$

$$2 \text{ neutrons} = 2(1.00867 \text{ u}) = \underline{2.01734 \text{ u}}$$

$$4.03190 \text{ u}$$

But the mass of a helium-4 nucleus is 4.00150 u, a difference of 0.03040. The difference between (1) the mass of the individual nucleons making up a nucleus and (2) the actual mass of the nucleus is called the **mass defect** of the nucleus. The explanation for the mass defect is again found in $E = mc^2$. When nucleons join to make a nucleus, energy is released as the more stable nucleus is formed. A mole of helium-4 nuclei would release a very large amount of energy,

$$\Delta E = \Delta mc^2$$

$$= (-3.04 \times 10^{-6}\ kg)\left(3.00 \times 10^8\ \frac{m}{sec}\right)^2$$

$$= (-3.04 \times 10^{-6}\ kg)\left(9.00 \times 10^{16}\ \frac{m^2}{sec^2}\right)$$

$$= (-3.04 \times 9.00) \times 10^{(-6\ +\ 16)}\ \frac{kg \cdot m^2}{sec^2}$$

$$= -2.74 \times 10^{12}\ J$$

By comparison, hydrogen atoms coming together to form one mole of H_2 molecules release about 4.3×10^5 J of chemical energy, or about 10,000,000 times less energy per mole.

The energy equivalent released when a nucleus is formed is the same as the **binding energy,** the energy required to break the nucleus into individual protons and neutrons. The binding energy of the nucleus of any isotope can be calculated from the mass defect of the nucleus.

The ratio of binding energy to nucleon number is a reflection of the stability of a nucleus (figure 17.13). The greatest binding energy per nucleon occurs near mass number 56, with about 1.5×10^{-12} J per nucleon, then decreases for both more massive and less massive nuclei. This means that more massive nuclei can gain stability by splitting into smaller nuclei with the release of energy. It also means that less massive nuclei can gain

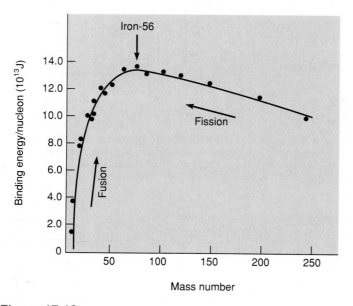

Figure 17.13

The maximum binding energy per nucleon occurs around mass number 56, then decreases in both directions. As one result, fission of massive nuclei and fusion of less massive nuclei both release energy.

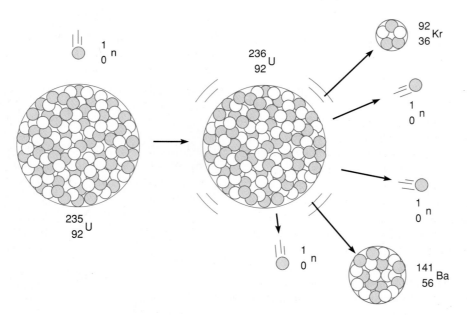

Figure 17.14

The fission reaction occurring when a neutron is absorbed by a uranium-235 nucleus. The deformed nucleus splits any number of ways into lighter nuclei, releasing neutrons in the process.

stability by joining together with the release of energy. The slope also shows that more energy is released in the coming-together process than in the splitting process.

The nuclear reaction of splitting a massive nucleus into more stable, less massive nuclei with the release of energy is **nuclear fission** (figure 17.14). Nuclear fission occurs rapidly in an atomic bomb explosion and occurs relatively slowly in a nuclear reactor. The nuclear reaction of less massive nuclei, coming together to form more stable, and more massive, nuclei with the release of energy is **nuclear fusion.** Nuclear fusion occurs rapidly in a hydrogen bomb explosion and occurs continually in the sun, releasing the energy essential for the continuation of life on the earth. Nuclear fission and nuclear fusion are the topics of the next sections.

Nuclear Fission

Nuclear fission was first accomplished in the late 1930s when researchers were attempting to produce isotopes by bombarding massive nuclei with neutrons. In 1938 two German scientists, Otto Hahn and Fritz Strassman, identified the element barium in a uranium sample that had been bombarded with neutrons. Where the barium came from was a puzzle at the time, but soon afterward Lise Meitner, an associate who had moved to Sweden, deduced that uranium nuclei had split, producing barium. The reaction might have been

$$_{0}^{1}n + _{92}^{235}U \rightarrow _{56}^{141}Ba + _{36}^{92}Kr + 3\ _{0}^{1}n$$

The phrase "might have been" is used because a massive nucleus can split in many different ways, producing different products. About thirty-five different, less massive nuclei have been identified among the fission products of uranium-235. Some of these products are fission fragments and some are produced as unstable fragments that undergo radioactive decay. These fission fragments are listed in table 17.5, together with their major modes of radioactive decay and half-lives. Some of the isotopes are the focus of concern about nuclear wastes, the topic of the Feature at the end of this chapter.

Fissioning of a uranium-235 nucleus produces two or three neutrons along with other products. These neutrons can each move to other uranium-235 nuclei where they are absorbed, causing fission with the release of more neutrons, which move to other uranium-235 nuclei to continue the process. A reaction where the products are able to produce more reactions in a self-sustaining series is called a **chain reaction.** A chain reaction is self-sustaining until all the uranium-235 nuclei have fissioned or until the neutrons fail to strike a uranium-235 nucleus (figure 17.15).

You might wonder why all the uranium in the universe does not fission in a chain reaction. Natural uranium is mostly uranium-238, an isotope that does not fission easily. Only about 0.7 percent of natural uranium is the highly fissionable uranium-235. This low ratio of readily fissionable uranium-235 nuclei makes it unlikely that a stray neutron would be able to achieve a chain reaction.

Table 17.5

Fragments and products from fission of uranium-235

Isotope	Major Mode of Decay	Half-Life
Tritium	Beta	12.26 years
Carbon-14	Beta	5,930 years
Argon-41	Beta, gamma	1.83 hours
Iron-55	Electron capture	2.7 years
Cobalt-58	Beta, gamma	71 days
Cobalt-60	Beta, gamma	5.26 years
Nickel-63	Beta	92 years
Krypton-85	Beta, gamma	10.76 years
Strontium-89	Beta	5.4 days
Strontium-90	Beta	28 years
Yttrium-91	Beta	59 days
Zirconium-93	Beta	9.5×10^5 years
Zirconium-95	Beta, gamma	65 days
Niobium-95	Beta, gamma	35 days
Technetium-99	Beta	2.1×10^5 years
Ruthenium-106	Beta	1 year
Iodine-129	Beta	1.6×10^7 years
Iodine-131	Beta, gamma	8 days
Xenon-133	Beta, gamma	5.27 days
Cesium-134	Beta, gamma	2.1 years
Cesium -135	Beta	2×10^6 years
Cesium-137	Beta	30 years
Cerium-141	Beta	32.5 days
Cerium-144	Beta, gamma	285 days
Promethium-147	Beta	2.6 years
Samarium-151	Beta	90 years
Europium-154	Beta, gamma	16 years
Lead-210	Beta	22 years
Radon-222	Alpha	3.8 days
Radium-226	Alpha, gamma	1,620 years
Thorium-229	Alpha	7,300 years
Thorium-230	Alpha	26,000 years
Uranium-234	Alpha	2.48×10^5 years
Uranium-235	Alpha, gamma	7.13×10^8 years
Uranium-238	Alpha	4.51×10^9 years
Neptunium-237	Alpha	2.14×10^6 years
Plutonium-238	Alpha	89 years
Plutonium-239	Alpha	24,360 years
Plutonium-240	Alpha	6,760 years
Plutonium-241	Beta	13 years
Plutonium-242	Alpha	3.79×10^5 years
Americium-241	Alpha	458 years
Americium-243	Alpha	7,650 years
Curium-242	Alpha	163 days
Curium-244	Alpha	18 years

Nuclear Reactions

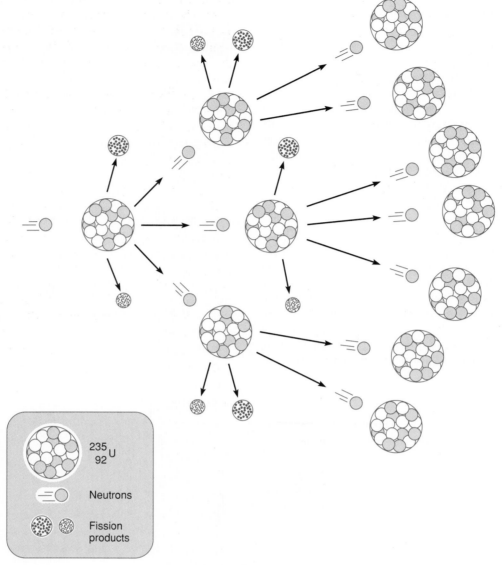

Figure 17.15

A schematic representation of a chain reaction. Each fissioned nucleus releases neutrons, which move out to fission other nuclei. The number of neutrons can increase quickly with each series.

In order to achieve a chain reaction, there must be (1) a sufficient mass with (2) a sufficient concentration of fissionable nuclei. When the mass and concentration are sufficient to sustain a chain reaction the amount is called a **critical mass.** Likewise, a mass too small to sustain a chain reaction is called a *subcritical mass.* A mass of sufficiently pure uranium-235 (or plutonium-239) that is large enough to produce a rapidly accelerating chain reaction is called a *supercritical mass.* An atomic bomb is simply a device that uses a small, conventional explosive to push subcritical masses of fissionable material into a supercritical mass. Fissioning occurs almost instantaneously in the supercritical mass, and tremendous energy is released in a violent explosion.

Nuclear Power Plants

The nuclear part of a nuclear power plant is the **nuclear reactor,** a steel vessel in which a controlled chain reaction of fissionable material releases energy (figure 17.16). In the most popular design, called a pressurized light-water reactor, the fissionable material is enriched 3 percent uranium-235 and 97 percent uranium-238 that has been fabricated in the form of small ceramic pellets. The pellets are encased in a long zirconium alloy tube called a **fuel rod.** The fuel rods are locked into a *fuel rod assembly* by locking collars, arranged in a way to permit pressurized water to flow around each fuel rod and to allow the insertion of *control rods* between the fuel rods. **Control rods** are

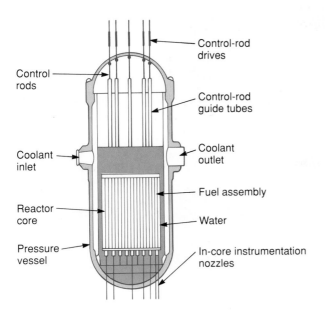

Figure 17.16

A schematic representation of the basic parts of a nuclear reactor. The largest commercial nuclear power plant reactors are nine- to eleven-inch thick steel vessels with a stainless steel liner, standing about forty feet high with a diameter of sixteen feet. Such a reactor has four pumps, which move 440,000 gallons of water per minute through the primary loop.

constructed of materials, such as cadmium, that absorb neutrons. The lowering or raising of control rods within the fuel rod assemblies slows or increases the chain reaction by varying the amount of neutrons absorbed. When lowered completely into the assembly, enough neutrons are absorbed to stop the chain reaction.

It is physically impossible for the low-concentration fuel pellets to form a supercritical mass. A nuclear reactor in a power plant can only release energy at a comparatively slow rate, and it is impossible for a nuclear power plant to produce a nuclear explosion. In a pressurized water reactor the energy released is carried away from the reactor by pressurized water in a closed pipe called the **primary loop** (figure 17.17). The water is pressurized at about 150 atmospheres (about 2,200 lb/in^2) to keep the water from boiling, which may be 350° C (about 660° F).

In the pressurized light-water (ordinary water) reactor the circulating pressurized water acts as a coolant, carrying heat away from the reactor. The water also acts as a **moderator,** a substance that slows neutrons so they are more readily absorbed by uranium-235 nuclei. Other reactor designs use heavy water (deuterium dioxide) or graphite as a moderator.

Water from the closed primary loop is circulated through a heat exchanger called a **steam generator.** The pressurized high-temperature water from the reactor moves through hundreds of

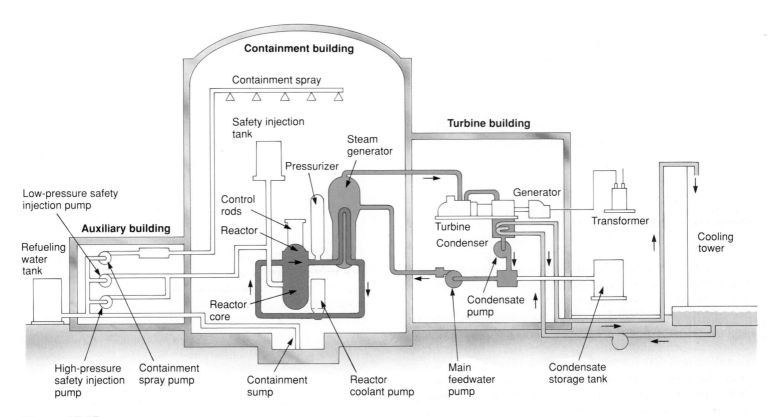

Figure 17.17

A schematic general system diagram of a pressurized water nuclear power plant, not to scale. The containment building is designed to withstand an internal temperature of 300° F at a pressure of 60 lbs/in^2 and still maintain its leak-tight integrity.

small tubes inside the generator as *feedwater* from the **secondary loop** flows over the tubes. The water in the primary loop heats feedwater in the steam generator, then returns to the nuclear reactor to become heated again. The feedwater is heated to steam at about 235° C (455° F) with a pressure of about 68 atmospheres (1,000 lb/in²). This steam is piped to the turbines, which turn an electric generator.

After leaving the turbines, the spent steam is condensed back to liquid water in a second heat exchanger receiving water from the cooling towers. Again, the cooling water does not mix with the closed secondary loop water. The cooling tower water enters the condensing heat exchanger at about 32° C (90° F) and leaves at about 50° C (about 120° F) before returning to a cooling tower, where it is cooled by evaporation. The feedwater is preheated, then recirculated back to the steam generator to start the cycle over again. The steam is condensed back to liquid water because of the difficulty of pumping and reheating steam.

After a period of time the production of fission products in the fuel rods begins to interfere with effective neutron transmission, so the reactor is shut down annually for refueling. During refueling about one-third of the fuel that had the longest exposure in the reactor is removed as "spent" fuel. New fuel rod assemblies are inserted to make up for the part removed. However, only about 4 percent of the "spent" fuel is unusable waste, about 94 percent is uranium-238, 0.8 percent is uranium-235, and about 0.9 percent is plutonium (figure 17.18). Thus "spent" fuel rods contain an appreciable amount of usable uranium and plutonium. For now, spent reactor fuel rods are mostly stored in cooling pools at the nuclear plant sites. In the future, a decision will be made either to reprocess the spent fuel, recovering the uranium and plutonium through chemical reprocessing, or put the fuel in terminal storage. Concerns about reprocessing are based on the fact that plutonium-239 and uranium-235 are fissionable and could possibly be used by terrorist groups to construct nuclear explosive devices. Six other countries do have reprocessing plants, however, and the spent fuel rods represent an energy source that will accumulate by the year 2000 to an amount equivalent to more than 25 billion barrels of petroleum. Some energy experts say that it would be inappropriate to dispose of such an energy source.

The technology to dispose of fuel rods exists if the decision is made to do so. The longer half-life waste products are mostly alpha emitters. These metals could be converted to oxides, mixed with powdered glass (or a ceramic), melted, and then poured into stainless steel containers. The solidified canisters would then be buried in a stable geologic depository. The glass technology is used in France for disposal of high-level wastes. Buried at two-thousand- to three-thousand-foot depths in solid granite, the only significant means of the radioactive wastes reaching the surface would be through groundwater dissolving the stainless steel, glass, and waste products and then transporting it back to the surface. Many experts believe that if such groundwater

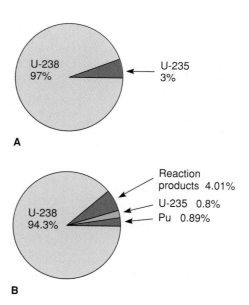

Figure 17.18

The composition of the nuclear fuel in a fuel rod (*a*) before and (*b*) after use over a three-year period in a nuclear reactor.

dissolving were to take place it would require thousands of years. The radioactive isotopes would thus undergo natural radioactive decay by the time they could reach the surface. Nonetheless, research is continuing on nuclear waste and its disposal. In the meantime, the question of whether it is best to reprocess fuel rods or place them in permanent storage remains unanswered.

What is the volume of nuclear waste under question? If all the spent fuel rods from all the commercial nuclear plants accumulated up to the year 2000 were reprocessed, then mixed with glass, the total amount of glassified waste would make a pile on one football field an estimated 4 m (about 13 ft) high.

Nuclear Fusion

As the graph of nuclear binding energy versus mass numbers shows (see figure 17.13), nuclear energy is released when (1) massive nuclei such as uranium-235 undergo fission and (2) when less massive nuclei come together to form more massive nuclei through nuclear fusion. Nuclear fusion is responsible for the energy released by the sun and other stars. The sun is composed of about 73 percent helium, 26 percent hydrogen, and about 1 percent other elements. Through fusion, the sun converts about 600 million tons of hydrogen to 596 million tons of helium every second. The other 4 million tons of matter are converted into energy. Even at this rate, the sun has enough hydrogen to continue the process for an estimated 5 billion years. There are several fusion reactions that take place between hydrogen and helium isotopes including the following:

$$\,^1_1H + \,^1_1H \rightarrow \,^2_1H + \,^0_1e$$

$$\,^2_1H + \,^2_1H \rightarrow \,^3_2He + \,^1_0n$$

$$\,^3_2He + \,^3_2He \rightarrow \,^4_2He + 2\,^1_1H$$

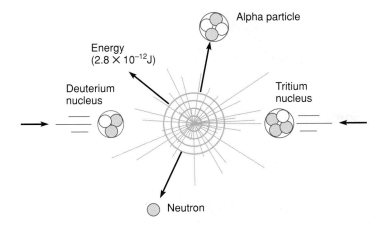

Figure 17.19
A fusion reaction between a tritium nucleus and a deuterium nucleus requires a certain temperature, density, and time of containment to take place.

The fusion process would seem to be a desirable energy source on earth because (1) two isotopes of hydrogen, deuterium ($_1^2H$) and tritium ($_1^3H$), undergo fusion at a relatively low temperature; (2) the supply of deuterium is practically unlimited, with each gallon of seawater containing about a teaspoonful of deuterium dioxide; and (3) enormous amounts of energy are released with no radioactive by-products.

The oceans contain enough deuterium to generate electricity for the entire world for million of years, and tritium can be constantly produced by a fusion device. Researchers know what needs to be done to tap this tremendous energy source. The problem is *how* to do it in an economical, continuous energy-producing fusion reactor. The problem, one of the most difficult engineering tasks ever attempted, is meeting three basic fusion reaction requirements of (1) temperature, (2) density, and (3) time (figure 17.19):

1. *Temperature.* Nuclei contain protons and are positively charged, so they experience the electromagnetic repulsion of like charges. This force of repulsion can be overcome, moving the nuclei close enough to fuse together, by giving the nuclei sufficient kinetic energy. The fusion reaction of deuterium and tritium, which has the lowest temperature requirements of any fusion reaction known at the present time, requires temperatures on the order of 100 million °C.
2. *Density.* There must be a sufficiently dense concentration of heavy hydrogen nuclei, on the order of $10^{14}/cm^3$, so many reactions occur in a short time.
3. *Time.* The nuclei must be confined at the appropriate density up to a second or longer at pressures of a least 10 atmospheres to permit a sufficient number of reactions to take place.

The temperature, density, and time requirements of a fusion reaction are interrelated. A short time of confinement, for example, requires an increased density, and a longer confinement time requires less density. The primary problems of fusion research are the high-temperature requirements and confinement. No material in the world can stand up to a temperature of 100 million °C, and any material container would be instantly vaporized. Thus research has centered on meeting the fusion reaction requirements without a material container. Two approaches are being tested, *magnetic confinement* and *inertial confinement.*

Magnetic confinement utilizes a **plasma,** a very hot gas consisting of atoms that have been stripped of their electrons because of the high kinetic energies. The resulting positively and negatively charged particles respond to electrical and magnetic forces, enabling researchers to develop a "magnetic bottle," that is, magnetic fields that confine the plasma and avoid the problems of material containers that would vaporize. A magnetically confined plasma is very unstable, however, and researchers have compared the problem to trying to carry a block of jello on a pair of rubber bands. Different magnetic field geometries and magnetic "mirrors" are the topics of research in attempts to stabilize the hot, wobbly plasma. Electric currents, injection of fast ions, and radio frequency (microwave) heating methods are also being studied.

Inertial confinement is an attempt to heat and compress small frozen pellets of deuterium and tritium with energetic laser beams or particle beams, producing fusion. The focus of this research is new and powerful lasers, light ion and heavy ion beams. If successful, magnetic or inertial confinement will provide a long-term solution for future energy requirements.

The Source of Nuclear Energy

When elements undergo the natural radioactive decay process, energy is released and the decay products have less energy than the original reactant nucleus. When massive nuclei undergo fission, much energy is rapidly released along with fission products that continue to release energy through radioactive decay. What is the source of all this nuclear energy? The answer to this question is found in current theories about how the universe started and in theories about the life cycle of the stars. Theories about the life cycle of stars are discussed in chapters 18 and 19. For now, consider just a brief introduction to the life cycle of a star in order to understand the ultimate source of nuclear energy.

The current universe is believed to have started with a "big bang" of energy, which created a plasma of protons and neutrons. This primordial plasma cooled rapidly and, after several minutes, began to form hydrogen nuclei. Throughout the newly formed universe massive numbers of hydrogen atoms—on the order of 10^{57} nuclei—were gradually pulled together by gravity into masses that would become the stars. As the hydrogen atoms fell toward the center of each mass of gas they accelerated, just like any other falling object. As they accelerated the contracting mass began to heat up because the average kinetic energy of the atoms increased from acceleration. Eventually, after say ten million years or so of collapsing and heating, the mass of hydrogen condensed to a sphere with a diameter of 1.5 million miles or so, or about twice the size of the sun today. At the same time the temperature increased to millions of degrees,

Nuclear Waste

There are two general categories of nuclear wastes: (1) low-level wastes and (2) high-level wastes. The *low-level wastes* are produced by the normal operation of a nuclear reactor. Radioactive isotopes sometimes escape from fuel rods in the reactor and in the spent fuel storage pools. These isotopes are removed from the water by ion-exchange resins and from the air by filters. The used resins and filters will contain the radioactive isotopes and will become low-level wastes. In addition, any contaminated protective clothing, tools, and discarded equipment also become low-level wastes.

Low-level liquid wastes are evaporated, mixed with cement, then poured into fifty-five-gallon steel drums. Solid wastes are compressed and placed in similar drums. The drums are currently disposed of by burial in government-licensed facilities. In general, low-level waste has an activity of less than 1.0 curie per cubic foot. Contact with the low-level waste could expose a person to up to 20 millirems per hour of contact.

High-level wastes from nuclear power plants are spent nuclear fuel rods. At the present time most of the commercial nuclear power plants have these rods in temporary storage at the plant site. These rods are "hot" in the radioactive sense, producing about 100,000 curies per cubic foot. They are also hot in the thermal sense, continuing to generate heat for months after removal from the reactor. The rods are cooled by heat exchangers connected to storage pools; they could otherwise achieve an internal temperature as high as 800° C for several decades. In the future, these spent fuel rods will be reprocessed or disposed of through terminal storage.

Agencies of the United States federal government have also accumulated millions of gallons of high-level wastes from the manufacture of nuclear weapons and nuclear research programs. These liquid wastes are stored in million-gallon stainless steel containers that are surrounded by concrete. The containers are located in the states of Washington, Idaho, and South Carolina. The future of this large amount of high-level wastes may be evaporation to a solid form or mixture with a glass or ceramic matrix, which is melted and poured into stainless steel containers. These containers would be buried in solid granite rock in a stable geologic depository. Such high-level wastes must be contained for thousands of years as they undergo natural radioactive decay. Burial at a depth of two thousand to three thousand feet in solid granite would provide protection from exposure by explosives, meteorite impact, or erosion. One major concern about this plan is that a hundred generations later, people might lose track of what is buried in the nuclear garbage dump.

---

reaching the critical points of density, temperature, and containment for a fusion reaction to begin. Thus a star was born as hydrogen nuclei fused into helium nuclei, releasing enough energy that the star began to shine.

Hydrogen nuclei in the newborn star had a higher energy per nucleon than helium nuclei, and helium nuclei had more energy per nucleon than other nuclei up to around iron. The fusion process continued for billions of years, releasing energy as heavier and heavier nuclei were formed. Eventually, the star materials were fused into nuclei around iron, the elements with the lowest amount of energy per nucleon, and the star used up its energy source. Larger, more massive dying stars explode into a supernova (discussed in chapter 18). Such an explosion releases a flood of neutrons, which bombard medium-weight nuclei and build them up to more massive nuclei, all the way from iron up to uranium. Thus the more massive elements were born from an exploding supernova, then spread into space as dust. In a process to be discussed later, this dust became the materials of which planets were made, including the earth. The point for the present discussion, however, is that the energy of naturally radioactive elements, and the energy released during fission, can be traced back to the force of gravitational attraction, which provided the initial energy for the whole process.

Summary

Radioactivity is the spontaneous emission of particles or energy from an unstable atomic nucleus. The modern atomic theory pictures the nucleus as protons and neutrons held together by a short-range *nuclear force* that has moving *nucleons* (protons and neutrons) in *energy shells* analogous to the shell structure of electrons. A graph of the number of neutrons to the number of protons in a nucleus reveals that stable nuclei have a certain neutron-to-proton ratio in a *band of stability*. Nuclei that are above or below the band of stability, and nuclei that are beyond atomic number 83, are radioactive and undergo *radioactive decay*.

Three common examples of radioactive decay involve the emission of an *alpha particle*, a *beta particle*, and a *gamma ray*. An alpha particle is a helium nucleus, consisting of two protons and two neutrons. A beta particle is a high-speed electron that is ejected from the nucleus. A gamma ray is a short wavelength electromagnetic radiation from an excited nucleus. In general, nuclei with an atomic number of 83 or larger become more stable by alpha emission. Nuclei with a neutron-to-proton ratio that is too large become more stable by beta emission. Gamma ray emission occurs from a nucleus that was left in a high-energy state by the emission of an alpha or beta particle.

Each radioactive isotope has its own specific *radioactive decay constant* (k), a ratio of the rate of nuclear disintegration to the total number of nuclei (n), or $k = $ rate/n. The rate is usually described in terms of *half-life*, the time required for one-half the unstable nuclei to decay. Half-life is related to the decay constant by half-life $= 0.693/k$, where 0.693 is a mathematical constant for exponential decay (the natural log of 2).

Radiation is measured by (1) its effects on photographic film, (2) the number of ions it produces, or (3) the flashes of light produced on a phosphor. It is measured at a source in units of a *curie*, defined as 3.70×10^{10} nuclear disintegrations per second. It is measured where received in units of a *rad*, defined as 1×10^{-5} J. A *rem* is a measure of radiation that takes into account the biological effectiveness of different types of radiation damage. In general, the natural environment exposes everyone to 100 to 500 millirems per year, an exposure called *background radiation*. Life-style and location influence the background radiation received, but the average is 130 millirems per year.

Energy and mass are related by Einstein's famous equation of $E = mc^2$, which means that *matter can be converted to energy and energy to matter*. The mass of a nucleus is always less than the sum of the masses of the individual particles of which it is made. This *mass defect* of a nucleus is equivalent to the energy released when the nucleus was formed according to $E = mc^2$. It is also the *binding energy*, the energy required to break the nucleus apart into nucleons.

When the binding energy is plotted against the mass number, the greatest binding energy per nucleon is seen to occur for an atomic number near iron. More massive nuclei therefore release energy by fission, or splitting to more stable nuclei. Less massive nuclei release energy by fusion, the joining of less massive nuclei to produce a more stable, more massive nucleus. Nuclear fission provides the energy for atomic explosions and nuclear power plants. Nuclear fusion is the energy source of the sun and other stars and also holds promise as a future energy source. The source of the energy of a nucleus can be traced back to the gravitational attraction that formed a star.

Summary of Equations

17.1

$$\text{radioactive decay constant} = \frac{\text{decay rate}}{\text{number of nuclei}}$$

$$k = \frac{\text{rate}}{n}$$

17.2

$$\text{half-life} = \frac{\text{a mathematical constant}}{\text{decay constant}}$$

$$t_{1/2} = \frac{0.693}{k}$$

17.3

$$\text{energy} = \text{mass} \times \text{the speed of light squared}$$

$$E = mc^2$$

Key Terms

alpha particle (p. 334)
background radiation (p. 342)
band of stability (p. 336)
beta particle (p. 334)
binding energy (p. 344)
chain reaction (p. 345)
control rods (p. 346)
critical mass (p. 346)
curie (p. 342)

electromagnetic force (p. 335)
fuel rod (p. 346)
gamma ray (p. 334)
Geiger counter (p. 340)
half-life (p. 339)
ionization counter (p. 340)
mass defect (p. 344)
moderator (p. 347)
nuclear fission (p. 345)

nuclear force (p. 335)
nuclear fusion (p. 345)
nuclear reactor (p. 346)
nucleons (p. 334)
plasma (p. 349)
primary loop (p. 347)
rad (p. 342)
radioactive decay (p. 334)
radioactive decay constant (p. 339)

radioactive decay series (p. 338)
radioactivity (p. 334)
rem (p. 342)
scintillation counter (p. 340)
secondary loop (p. 348)
shell model of the nucleus (p. 335)
steam generator (p. 347)

Applying the Concepts

1. An atom of radon-222 loses an alpha particle to become a more stable atom of
   a. radium.
   b. bismuth.
   c. polonium.
   d. radon.

2. Which of the following isotopes is more likely to be radioactive?
   a. magnesium-24
   b. calcium-40
   c. astatine-210
   d. ruthenium-101

3. Hydrogen-3 is a radioactive isotope of hydrogen. Which type of radiation would you expect an atom of this isotope to emit?
   a. an alpha particle
   b. a beta particle
   c. either of the above
   d. neither of the above

4. The rate of radioactive decay can be increased by increasing the
   a. temperature.
   b. pressure.
   c. size of the sample.
   d. none of the above.

5. Isotope A has a half-life of seconds and isotope B has a half-life of millions of years. Which isotope is more radioactive?
   a. it depends on the sample size
   b. isotope A
   c. isotope B
   d. unknown from the information given

6. A measure of radioactivity at the *source* is the
   a. curie.
   b. rad.
   c. rem.
   d. any of the above.

7. A measure of radiation received that considers the biological effect resulting from the radiation is the
   a. curie.
   b. rad.
   c. rem.
   d. any of the above.

8. The mass of a nucleus is always _____ the sum of the masses of the individual particles of which it is made.
   a. equal to
   b. less than
   c. more than

9. When protons and neutrons join together to make a nucleus, energy is
   a. released.
   b. absorbed.
   c. neither released nor absorbed.

10. The energy released by radioactive decay and the energy released by nuclear reactions can be traced back to the energy that isotopes acquired from
    a. fusion.
    b. the sun.
    c. gravitational attraction.

Answers

1. c   2. c   3. b   4. d   5. b   6. a   7. c   8. b   9. a
10. c

Questions for Thought

1. How is a radioactive material different from a material that is not radioactive?

2. What is radioactive decay? Describe how the radioactive decay rate can be changed, if this is possible.

3. What is a nucleon?

4. Describe three kinds of radiation emitted by radioactive materials. Describe what eventually happens to each kind of radiation after it is emitted.

5. How does a nuclear equation show that mass and charge are conserved in radioactive decay?

6. Describe how the atomic number and the mass number of a radioactive nucleus is changed by emission of (a) an alpha particle, (b) a beta particle, and (c) a gamma ray.

7. How are positively charged protons able to stay together in a nucleus since like charges repel?

8. Describe the structure of the nucleus of an atom according to modern atomic theory.

9. Give four generalizations about the stability of a nucleus and how you can use this information to predict if a particular isotope is radioactive or not.

10. What is half-life? Give an example of the half-life of an isotope, describing the amount remaining and the time elapsed after five half-life periods.

11. In general, would you expect an isotope with a long half-life to be more, the same, or less radioactive than an isotope with a short half-life? Explain.

12. Does the radioactive decay constant depend on the amount of radioactive materials? Explain why or why not.

13. Does the radioactive decay constant depend on the kind of radioactive materials? Explain why or why not.

14. Describe three methods that can be used to measure radioactivity.

15. What is (a) a curie? (b) a rad? (c) a rem?

16. What is meant by background radiation? What is the normal radiation dose for the average person from background radiation?

17. Why is radiation a hazard to living organisms? Describe harmful levels of radiation that occur from a single dose, whole body exposure.

18. Why is there some controversy about the effects of long-term, low levels of radiation exposure?

19. How do you know that radioactive decay produces products of a lower energy?

20. What is a mass defect? How is it related to the binding energy of a nucleus? How can both be calculated?

21. Describe how the binding energy to nucleon ratio changes as nuclei become more massive from hydrogen up. What is the approximate mass number for nuclei with the greatest amount of binding energy? What does this mean about nuclear stability?

22. Compare and contrast nuclear fusion and nuclear fission. Why do both reactions yield energy?

23. What is a chain reaction?

24. Is it possible for a nuclear power plant to produce a nuclear explosion? Explain.

25. How much of a used nuclear fuel rod is really waste? Describe several possible futures for used nuclear fuel rods.

26. What conditions are necessary to produce a fusion reaction in a reactor? What problems do these conditions present? What approaches are being tested to solve these problems?

27. Where does a nucleus get its energy that is released during radioactive decay?

Exercises

Group A—Solutions Provided in Appendix D

1. Give the number of protons and the number of neutrons in the nucleus of each of the following isotopes:
   (a) cobalt-60
   (b) potassium-40
   (c) neon-24
   (d) lead-204

2. Write the nuclear symbols for each of the nuclei in exercise 1.

3. Predict if the nuclei in exercise 1 are radioactive or stable, giving your reasoning behind each prediction.

4. Write a nuclear equation for the decay of the following nuclei as they give off a beta particle:
   (a) $^{56}_{26}Fe$
   (b) $^{7}_{4}Be$
   (c) $^{64}_{29}Cu$
   (d) $^{24}_{11}Na$
   (e) $^{214}_{82}Pb$
   (f) $^{32}_{15}P$

5. Write a nuclear equation for the decay of the following nuclei as they undergo alpha emission:
   (a) $^{235}_{92}U$
   (b) $^{226}_{88}Ra$
   (c) $^{239}_{94}Pu$
   (d) $^{214}_{83}Bi$
   (e) $^{230}_{90}Th$
   (f) $^{210}_{84}Po$

6. The half-life of iodine-131 is 8 days. How much of a 1.0 oz sample of iodine-131 will remain after 32 days?

7. If the half-life of strontium-90 is 27.6 years, what is the decay constant for strontium-90?

8. Using the decay constant for strontium-90 obtained in exercise 7, find the number of nuclear disintegrations over a period of time for a molar mass of strontium-90.

9. What is the activity in curies of the molar mass of strontium-90 described in exercise 8?

10. How much energy must be supplied to break a single iron-56 nucleus into separate protons and neutrons? (The mass of an iron-56 nucleus is 55.9206 u, one proton is 1.00728 u, and one neutron is 1.00867 u.)

Group B—Solutions Not Given

1. Give the number of protons and the number of neutrons in the nucleus of each of the following isotopes:
   (a) aluminum-25
   (b) technetium-95
   (c) tin-120
   (d) mercury-200

2. Write the nuclear symbols for each of the nuclei in exercise 1.

3. Predict if the nuclei in exercise 1 are radioactive or stable, giving your reasoning behind each prediction.

4. Write a nuclear equation for the beta emission decay of each of the following:
   (a) $^{14}_{6}C$
   (b) $^{60}_{27}Co$
   (c) $^{24}_{11}Na$
   (d) $^{241}_{94}Pu$
   (e) $^{131}_{53}I$
   (f) $^{210}_{82}Pb$

5. Write a nuclear equation for each of the following alpha emission decay reactions:
   (a) $^{241}_{95}Am$
   (b) $^{232}_{90}Th$
   (c) $^{223}_{88}Ra$
   (d) $^{234}_{92}U$
   (e) $^{242}_{96}Cm$
   (f) $^{237}_{93}Np$

6. If the half-life of cesium-137 is 30 years, how much time will be required to reduce a 1.0 kg sample to 1.0 g?

7. The half-life of tritium ($^{3}_{1}H$) is 12.26 years. What is the radioactive decay constant for tritium?

8. What is the number of disintegrations per unit of time for a molar mass of tritium? The decay constant is obtained from exercise 7.

9. Calculate the activity in curies of the molar mass of tritium described in exercise 8.

10. How much energy is needed to separate the nucleus in a single lithium-7 nucleus? (The mass of a lithium-7 nucleus is 7.01435 u, one proton is 1.00728 u, and one neutron is 1.00867 u.)

Section III

Space probes have revealed some exciting and fascinating surprises about the solar system. This photograph of Saturn, from an early 1970s *Pioneer* flyby, shows bright and dark zones in Saturn's clouds. The dark zone in a plane with the rings is the rings' shadow on the surface of Saturn.

Astronomy

History tells us that people have always been intrigued with the night sky and its moving moon, stars, and planets. In recent times this fascination has flourished with space missions and with each exciting new finding about space. In part, this fascination comes from a growing awareness of what is "out there" and from the idea of space travel. It also comes from the opportunity to find stability and order in a universe that otherwise appears as mystery and chaos. One objective of this section is to impart the conceptual background that is needed to understand astronomical ideas and to appreciate the meaning of each new exciting discovery. As in other sections of this text, there are thinking and reasoning skills associated with the topic under study. Astronomy requires you to use your imagination, both in understanding the vast distances and spatial relationships and in imagining the past and the future. You will also learn the humanistic side of astronomy, that is, how it is ingrained in the human culture in a way that goes beyond a night view of the sky.

This interesting field of physical science begins with a chapter on how humans developed a big picture of the cosmos. Today, it is understood that the sun is but one of billions and billions of stars that move in galactic islands. You will learn why the sun is considered to be an average, middle-aged star and how it is different from other stars. How stars and galaxies

form and what eventually happens to them are the two big ideas introduced in this chapter.

Moving in closer to home, the local part of our galaxy, the Solar System, is considered next. Some of the humanistic perspectives of the Solar System are first considered as you learn how people grappled with ideas about the sun and its planets. The present-day understandings of the Solar System are presented next, based on data and inferences from data collected from space probes. You will learn why the earth and other planets are considered to have formed from the remains of previously existing stars. You will also learn how the planets are different and some of the fascinating findings of instrumented space probes.

Your home planet, earth, is considered next in terms of its size, shape, and motions. Earth's motions have cultural, as well as practical, significance since all the present-day concepts of time and location are based on them. Finally, recent understanding about earth's moon and where it came from are considered. Presently the moon is the only place outside the earth to be visited by humans. Many of the present-day understandings about the moon are a result of visits by humans and studies of the moon rocks that were brought back to the earth. The earth itself will be considered in more detail in the following section on the earth sciences.

The Universe

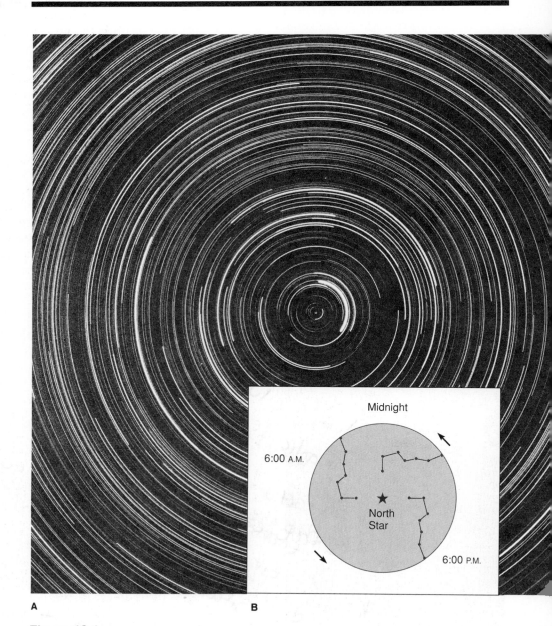

A   B

Figure 18.1
Ancient civilizations used celestial cycles of motion as clocks and
calendars. (a) The path of stars around the North Star. (b) The position
of the Big Dipper is shown over a twenty-four hour period as it turns
around the North Star. This shows how the Big Dipper can be used as
a clock.

ASTRONOMY is an exciting and mind-expanding physical science that has fascinated and intrigued people since the beginnings of recorded history. Ancient civilizations searched the heavens in wonder, some recording on clay tablets what they observed. Many religious and philosophical beliefs were originally based on interpretations of these ancient observations. Today, we are still awed by the heavens and space, but now we are fascinated with ideas of space travel, black holes, and the search for extraterrestrial life. Throughout history, people have speculated about the universe and their place in it and watched the sky and wondered (figure 18.1). What is out there and what does it all mean? Are there other people such as ourselves on other planets, looking at the star in their sky that is our sun, wondering if we exist?

Until about thirty years ago progress in astronomy was limited to what could be observed and photographed. Developments in astronomy, technology, and other branches of science then began to provide the details of what is happening in the larger expanses of space away from the earth. These developments included understandings about nuclear reactions and implications about what must be going on inside a star and the nature of the light emitted; new data made available from the development of infrared telescopes, radio telescopes, and X-ray telescopes; and detailed spectral analysis of light from the stars. All of these developments and the discovery and theoretical meaning of pulsars, neutron stars, and black holes began to fit together like the pieces of a puzzle. Theoretical models emerged about how stars evolve, about what galaxies are and how they evolve, and eventually, about the explosive beginnings of the universe and the chain of events that led to the formation of the sun and the earth. This chapter is concerned with these topics, beginning with historical attempts to understand how the stars are arranged in space. The chapter concludes with theoretical models of how the universe began and what may happen to it in the future.

Ancient Ideas

Early civilizations had a much better view of the night sky before city lights, dust, and pollution obscured much of the sky. Today, you must travel far from the cities, perhaps to a remote mountaintop, to see a clear night sky as early people observed it. Back then, people could clearly see the motion of the moon and stars night after night, observing recurring cycles of motion. These cycles became important as they became associated with the timing of certain events. Thus watching the sun, moon, and star movements became a way to identify when to plant crops, when to harvest, and when it was time to plan for other events. Observing the sky was an important activity, and many early civilizations built observatories with sighting devices to track and record astronomical events. Stonehenge, for example, was an ancient observatory built in England around 2600 B.C. by Neolithic people (figure 18.2).

There were many different centers of ancient civilizations, but historians identify those of the Euphrates River valleys (Babylon and Chaldea), the Nile River valleys (Egypt), and the Greek and Roman countries as civilizations that contributed the first known astronomical knowledge. As early as 2000 B.C., for example, the Babylonians kept track of long periods of time by dividing the year into 12 months, with 7 days to a week and 360 days to a year. Later, the Babylonians maintained observatories, compiled star catalogs, and were able to predict certain eclipses. They did not, however, attempt to explain any of the observed motions or cycles. The Babylonian concept of nature did not include the notion of physical cause and effect. To them, individual gods created and controlled the different parts of nature, so any explanation beyond a religious myth was not possible. They conceived of the universe as being the valley where they

Figure 18.2
The stone pillars of Stonehenge were positioned so that the movement of the sun and moon could be followed with the seasons of the year.

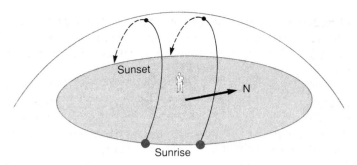

Figure 18.3

The people of ancient civilizations observed that the sun, the moon, and the five planets that were known at that time all moved within a narrow band across the sky. This shows the sun's path when it is lower on the horizon in the winter and when it is higher on the horizon in the summer. These two extreme paths identify the outer edges of the band of travel.

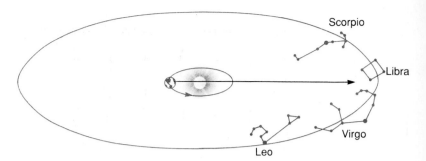

Figure 18.4

The sun, moon, and planets move across the constellations of the zodiac, with the sun moving around all twelve constellations during a year. From the earth, the sun will appear to be "in" Libra at sunrise in this sketch. As the earth revolves around the sun, the sun will seem to move from Libra into Scorpio, then through each constellation in turn.

lived, with other celestial objects doing the moving. This concept of a **geocentric,** or earth-centered, universe was typical of ancient concepts. To the Babylonians, the stars were attached to a shell, or dome, that rested on mountaintops. The sun and the moon entered and exited the dome through locked gates maintained by a god. The Egyptian concept of the universe was also geocentric, but their universe was surrounded by water. They conceived of stars as lamps hung from the dome and the sun as a disc of fire carried in a boat by the sun god Ra. The idea of stars being attached to an inverted bowl or dome was expanded upon by the ancient Greeks, who believed a *celestial sphere* surrounded the earth. The stars do seem fixed on a celestial sphere, and the earth's rotation provides the illusion that the entire sphere turns while the earth stands still. The Greek concept of a celestial sphere turning around a fixed earth, like the Babylonian and Egyptian concepts, was geocentric.

The early Babylonian observers of the night sky noted that the Sun, the Moon, and the five planets known at the time (Mercury, Venus, Mars, Jupiter, and Saturn) moved across the dome of the sky only along a certain path. The stars followed the motions of these seven celestial bodies along the path but kept the same position relative to each other as if they were fixed on the dome. The seven celestial bodies moved independently of the stars and only within a narrow band across the celestial sphere, with the sun's movement in the center of this band (figure 18.3). This path along which the sun appears to move among the stars is today called the *ecliptic*. As viewed from the earth, the sun appears to move completely around the ecliptic each year. To keep track of the sun in its travels, the Babylonians imagined the arrangements of certain stars to be the shapes of some mythical god, object, or animal. These imagined patterns of stars, or **constellations,** were used to identify twelve equal divisions of the ecliptic through which the sun passed in monthly succession. Thus there are twelve constellations around the ecliptic, and all twelve are called the *zodiac,* which means "circle of animals." The sun moves across a constellation, or "sign," each month, so there are twelve *signs of the zodiac* (figure 18.4). Table 18.1 gives the names and original meanings of these zodiac signs.

Table 18.1

The zodiac: The object, person, or animal (sign) and its original meaning

Ram (Aries)
The Babylonians sacrificed rams during the first month of their year, so the position of the sun in this constellation identified the first month.

Bull (Taurus)
Ancient association of the sun with a bull. The sun rose in this constellation on the first day of spring.

Twins (Gemini)
After an ancient legend of a wolf raising twins who built Rome.

Crab (Cancer)
The sun retreated this month and a crab moves backward.

Lion (Leo)
The lion was an ancient symbol for hot, which was the weather for the month that the sun was in this constellation.

Virgin (Virgo)
After Babylonian myth of Ishtar.

Balance (Libra)
Day and night were the same length this month, so they were balanced when the sun was in this constellation.

Scorpion (Scorpio)
A scorpion comes out at night, or with darkness, and the sun in this constellation marked the approach of winter with longer nights.

Archer (Sagittarius)
The Babylonian god of war.

Goat (Capricorn)
After the ancient legend of the god of the sun who was nursed by a goat when young.

Water Bearer (Aquarius)
This was the month when floods began on the Nile so the sun in this constellation marked the time of floods.

Fishes (Pisces)
This was the spring month when people returned to work after a winter of darkness; the fish was the ancient symbol for life after death, and the sun in this constellation marked the time when everything returned to life.

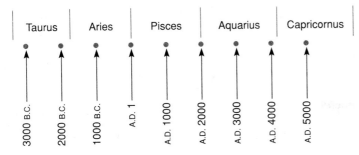

Figure 18.5

During the time of the Babylonians, the sun rose with the constellation Taurus on the first day of spring. Today, however, the sun rises with the constellation Pisces on the first day of spring. The earth's precession will continue to change the position of the sun during a particular month, and 25,780 years after the time of the Babylonians, the sun will again rise with the constellation Taurus on the first day of spring.

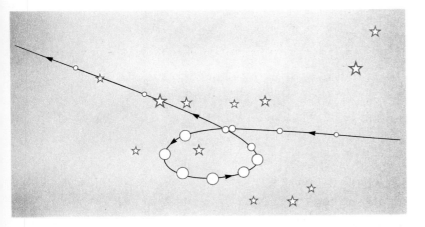

Figure 18.6

The apparent position of Mars against the background of stars as it goes through retrograde motion. Each position is observed approximately two weeks after the previous position.

The early Babylonians used the zodiac as a basis for keeping time. The twelve signs of the zodiac identified the twelve months, with each new month beginning with the new moon. Seven days to a week originated from the motion of the seven celestial bodies according to how long it took each to move across the sky. Thus "Saturn's day" became Saturday, "Sun's day" became Sunday, and "Moon's day" became Monday. The days of Tuesday, Wednesday, Thursday, and Friday are still named after the other four planets in the French, Italian, and Spanish languages, but other names are used today in English and German. By using a vertical rod as a sundial, the Babylonians also divided a day into hours, minutes, and seconds.

The later Babylonians, or Chaldeans, maintained astronomical observatories, but from their beliefs in the gods of nature they also developed astrology, the belief that the stars influence humans. Fully developed by 540 B.C., astrology became the art of studying the stars to guide human affairs. Some people today still have confidence in astrology, even though it is a fanciful extension of beliefs in mythical gods and an animistic conception of nature.

The zodiac signs of three thousand years ago are not the zodiac signs of today. The earth's precession (to be discussed in the next chapter) has shifted the constellations of the zodiac to the west. Thus three thousand years ago the sun entered the "house" (constellation) of Virgo in August. The astrological "forecasts" of today are still based on the sun being in the constellation of Virgo in August. But today, the sun is in the "house of Leo" in August. Because of the earth's precession, the sun will circuit through the entire zodiac in 25,780 years (figure 18.5).

The ancient Greek civilization, beginning about 600 B.C., added observations, reasoned theories, and generally removed the contrived influence of gods from the study of astronomy. They observed, for example, that over time the seven celestial bodies moving on the ecliptic did not hold the same relative positions and sometimes moved backward, making a loop over a period of several months. They called the five bodies that did this the Greek word for wanderer, "planetes," from which the word *planet* is derived. The Greek theory of the universe (which was discussed in chapter 2) had the earth fixed and motionless at the center. In general, the sun, moon, planets, and stars were thought to move around the earth attached to spherical shells. Attempting to account for the movements of the sun, moon, and planets required a model that eventually had fifty-five concentric shells, all centered on the earth and all moving at different rates. The model could not, however, explain why a planet such as Mars would move eastward across the ecliptic one month, slow down during the next month, stop, then move backward (westward) for the next two months, then resume its normal eastward motion. Today, this looping change of motion is called *retrograde motion* (figure 18.6). The time required for retrograde motion ranges from 34 days for Mercury to 139 days for Saturn.

Later Greek astronomers attempted to explain retrograde motion as the result of a planet being fixed on the outer edge of a wheel as it rolled across the sky. Thus they established a means of ranking the distances of the planets from the earth. Since Saturn took the longest time period to go through retrograde motion, the Greeks reasoned that it must be farthest away. Mercury must be the closest planet according to this theory. The moon was reasoned to be the closest of all because it occasionally moved in front of the planets. The Sun required a period of time that was between Venus and Mars, so the Sun was reasoned to be in that position. Ptolemy recorded this ancient Greek geocentric model of the universe in a seven-volume work. The earth-centered Ptolematic system (figure 18.7) became the accepted model of the universe when, along with Aristotle's theories, it became a part of Church doctrine. How the sun-centered model became established will be discussed with the Solar System in the next chapter.

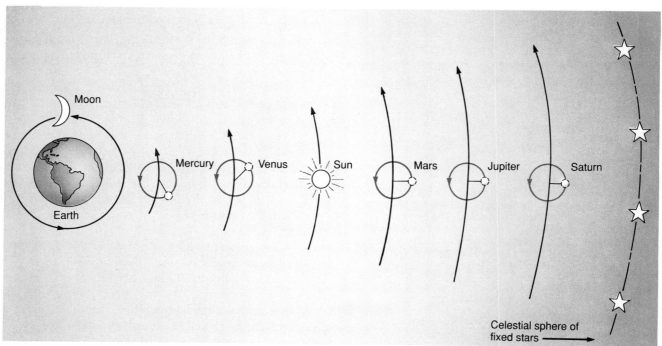

Figure 18.7

A schematic representation of the geocentric Ptolemaic system.

The Night Sky

Away from city lights, dust, and pollution you can make observations of the stars and planets as the ancient astronomers did without any special equipment. Light from the stars and planets must pass through the earth's atmosphere to reach you, however, and the atmosphere affects the light. Stars appear as point *sources* of light, and each star generates its own light. The stars seem to twinkle because density differences in the atmosphere refract the point of starlight one way, then the other, as the air moves. The result is the slight dancing about and change in intensity called twinkling. The points of starlight are much steadier when viewed on a calm night or when viewed from high in the mountains where there is less atmosphere for the starlight to pass through. Astronauts outside the atmosphere see no twinkling, and the stars appear as steady point sources of light.

Back at ground level, within the atmosphere, the *reflected* light from a planet does not seem to twinkle. A planet appears as a disk of light rather than a point source, so refraction from moving air of different densities does not affect the image as much. Sufficient air movement can cause planets to appear to shimmer, however, just as a road appears to shimmer on a hot summer day.

Celestial Location

To locate the ecliptic, planets, or anything else in the sky, you need something to refer to, a referent system. A referent system is easily established by first imagining the sky to be a *celestial sphere* just as the ancient Greeks did (figure 18.8). A coordinate system of lines can be visualized on this celestial sphere

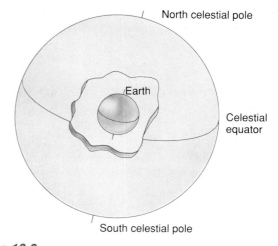

Figure 18.8

The celestial sphere with the celestial equator directly above the earth's equator, and the celestial poles directly above the earth's poles.

just as you think of the coordinate system of latitude and longitude lines on the earth's surface. Imagine that you could inflate the earth until its surface touches the celestial sphere. If you now transfer the latitude and longitude lines to the celestial sphere, you will have a system of sky coordinates. The line of the equator of the earth on the celestial sphere is called the **celestial equator.** The North Pole of the earth touches the celestial sphere at a point called the **north celestial pole.** From the surface of the earth, you can see that the celestial equator is a line on the celestial sphere directly above the earth's equator, and

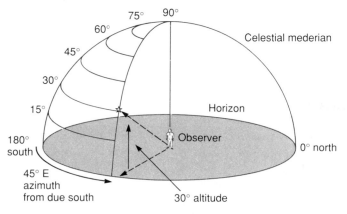

Figure 18.9

Once you have established the celestial equator, the celestial poles, and the celestial meridian, you can use a two-coordinate horizon system to locate positions in the sky. One popular method of using this system identifies the altitude angle (in degrees) from the horizon up to an object on the celestial sphere and the azimuth angle (again in degrees) the object on the celestial sphere is east or west of due south, where the celestial meridian meets the horizon. The illustration shows an altitude of 30° and an azimuth of 45° east of due south.

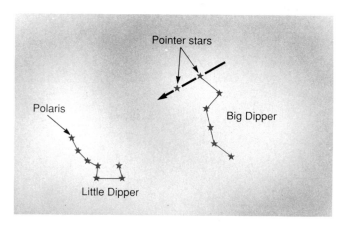

Figure 18.10

The North Star, or Polaris, is located by using the pointer stars of the Big Dipper constellation.

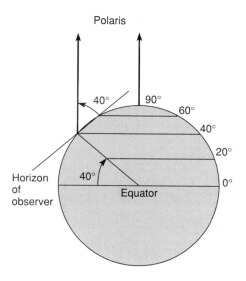

Figure 18.11

The altitude of Polaris above the horizon is approximately the same as the observer's latitude in the Northern Hemisphere.

the north celestial pole is a point directly above the North Pole of the earth. Likewise, the **south celestial pole** is a point directly above the South Pole of the earth.

You can only see half of the overall celestial sphere from any one place on the surface of the earth. Imagine a point on the celestial sphere directly above where you are located. An imaginary line that passes through this point, then passes north through the north celestial pole, continuing all the way around through the south celestial pole and back to the point directly above you makes a big circle called the **celestial meridian** (figure 18.9). Note that the celestial meridian location is determined by where *you* are on the earth. The celestial equator and the celestial poles, on the other hand, are always in the same place no matter where you are.

Overall, the celestial sphere appears to spin, turning on an axis through the celestial poles. A photograph made by pointing a camera at the north celestial pole, then leaving the shutter open for several hours, will show the apparent motion of the celestial sphere with star trails (see figure 18.1a). The moderately bright star near the center is the North Star, *Polaris*. Polaris is almost, but not exactly, at the north celestial pole. You can locate Polaris by finding the constellation named the *Big Dipper*. The two stars on the end of the dipper opposite the handle are called the pointers. Imagine a line moving from the bottom of the dipper upward through the two pointers. The first bright star that this line meets is Polaris (figure 18.10). The angle that you see Polaris above the horizon is your approximate latitude in the Northern Hemisphere. Figure 18.11 shows the geometric relationships between your latitude and the angle of Polaris above the horizon.

If you observe the constellations night after night, you will see that the stars maintain their positions relative to one another as they turn counterclockwise around Polaris. Those near Polaris pivot around it and are called "circumpolar." Those farther out rise in the east, move in an arc, then set in the west.

Celestial Distance

When you look at Polaris, or any other star, it seems impossible to know anything about the distance to such a point of light. The distance *between* stars on the celestial sphere is measured by a technique that was first developed by the Babylonians. The Babylonians considered a circle, such as the celestial meridian, to have 360 divisions, or degrees. Thus 1° is 1/360 of the circle,

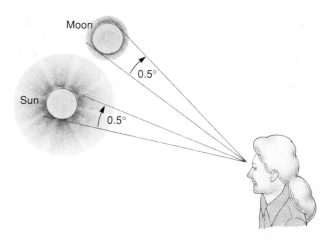

Figure 18.12

The moon and the sun both have an angular size of 0.5°, but the sun is much farther away. The observed angular size depends on distance and the true size of an object.

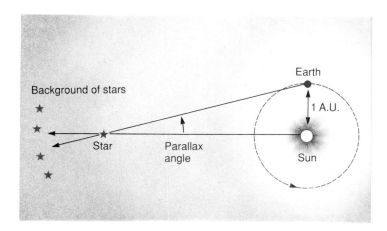

Figure 18.13

The angle that a star seems to move against the background of stars between two observations that are 1 A.U. apart defines the parallax angle. The distance to relatively nearby stars can be determined from the parallax angle (see example 18.1).

which is referred to as 1° *of arc* so it is not confused with temperature. An arc is a segment of the circumference of a circle, and the length of a segment can be used to indicate how far apart things are on the celestial sphere. The entire celestial meridian, for example, contains 360° of arc, but only 180° of arc are visible from one horizon to the other. A star directly overhead is 90° of arc above the horizon. The pointer stars in the Big Dipper are about 5° of arc apart. Polaris is 1° of arc away from the north celestial pole. For smaller measurements of arc, each degree is divided into 60 minutes (60') and each minute is divided into 60 seconds (60''). These subdivisions are called "arc minutes" and "arc seconds" so they are not confused with time measurements.

The apparent size of an object is measured by the angle an object makes, which is called the angular size or *angular diameter.* This is an apparent size because the angular diameter of an object decreases as its distance increases. Both the sun and the moon, for example, have an angular diameter of about 0.5° of arc when viewed from the earth, but the sun is much farther away. Thus the angular size of an object depends both on its distance and its true size (figure 18.12).

The distance to a star can be measured by *parallax,* the apparent shift of position by closer objects against a background when viewed from different observation points. As shown in figure 18.13, a relatively nearby star viewed from the earth in two different parts of the earth's orbit will appear in different locations against the background of stars. The radius of the earth's orbit is called an **astronomical unit** (A.U.). One A.U. is about 1.5×10^8 km (about 9.3×10^7 mi). The baseline of 1 A.U. defines the astronomical unit of distance called a **parsec.** A parsec is the distance at which the angle made from a 1 A.U. baseline is 1 arc second.

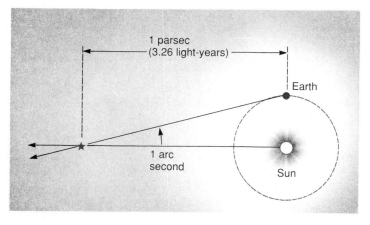

Figure 18.14

A parsec is defined as the distance at which the parallax angle is 1 arc second. A parsec is approximately 3.26 light-years.

The parsec is the distance unit of choice of astronomers since it historically provided a means of measuring distances with a telescope. The distance to a star in parsecs is simply the reciprocal of the parallax angle in seconds of arc, distance = 1/seconds of arc. The other unit of astronomical distance is the **light-year** (ly), which is the distance that light travels in one year, about 9.5×10^{12} km (about 6×10^{12} mi). One parsec is approximately equal to 3.26 light-years (figure 18.14). There is an obvious limit to using the parallax measurement since the angle becomes smaller and smaller with greater distances. At the present time this limit is about 500 light-years from the sun. Distances beyond this require other measurement techniques. Standard referent units of length, such as kilometers or miles have little meaning in astronomy since there are no referent

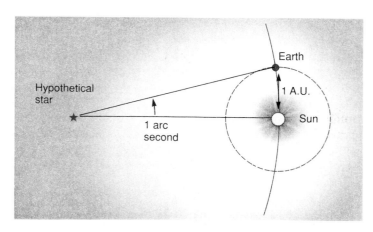

Figure 18.15

points of comparison. Thus the light-year measures distance in terms of time and the parsec measures distance in terms of angles.

Example 18.1

To understand how parallax can be used to find the distance to a star, imagine a hypothetical star that has a parallax of exactly 1.0 arc second (figure 18.15). Reversing your point of view, you could consider the star to be in the center of a huge circle that has a circumference going through the sun and the earth. The angular separation between the earth and the sun, as seen from the hypothetical star, is 1.0 arc second. The distance between the earth and the sun is 1 A.U., so 1 A.U. along the circumference of the circle as seen from the star must make an angle of 1 arc second. One A.U. is about 93 million mi, or more precisely 9.2956×10^7 mi, so the distance around the whole circumference is

$$(60') \times (60'') \times (360°) \times (9.2956 \times 10^7) = 1.2047 \times 10^{14} \text{ mi}$$

The radius of this circle can be found from $C = 2\pi r$, and

$$r = \frac{1.2 \times 10^{14}}{2 \times 3.14} = 1.91 \times 10^{13} \text{ mi}$$

One light-year is 6×10^{12} mi, so the hypothetical star is

$$\frac{1.91 \times 10^{13}}{5.86 \times 10^{12}} = 3.26 \text{ ly}$$

No star is close enough to earth to have a parallax of 1 arc second as the hypothetical star example. One of the nearest stars is Alpha Centauri, with a parallax of 0.8 arc second, which means that Alpha Centauri is about 4.5 light-years away.

Stars

If you could travel by spaceship a few hundred light-years from the earth, you would observe the sun shrink to a bright point of light among the billions and billions of other stars. The sun is just an ordinary star with an average brightness. As the other stars, the sun is a massive, dense ball of gases with a surface heated to incandescence by energy released from fusion reactions deep within. Since the sun is an average star it can be used as a reference for understanding all the other stars.

Origin of Stars

Theoretically, a star with the mass of the sun was born from swirling clouds of hydrogen gas in the deep space between other stars. Such interstellar clouds are called **nebulae.** Most nebula cannot be seen without a telescope, but one in the constellation Orion can be seen on a clear winter night in the Northern Hemisphere. Such clouds have a random, swirling motion, with gas atoms passing by one another mostly unaffected because they are not massive enough for gravity to exert much of a force. By chance, however, the number of atoms in a given swirl may sometimes be large enough that their mutual gravitational attraction begins to pull them together in a cluster. The cluster grows as more atoms are pulled in, which increases the mass and thus the gravitational attraction, and still more atoms are pulled in from farther away. Staggeringly huge numbers of atoms must be present to create the mutual attractions necessary to hold the atoms together in a cluster. Theoretical calculations indicate that on the order of 1×10^{57} atoms are necessary, all within a distance of several trillion miles. When these conditions do occur, the cloud of gas atoms begins to condense by gravitational attraction to a **protostar,** an accumulation of gases that will become a star.

Example 18.2

Compared to the 10^{19} molecules/cm³ of air on the earth, an average concentration of 1,000 hydrogen atoms/cm³ in the Orion Nebula does not seem very dense. However, considering that the Orion Nebula is about 20 light-years across (20×10^{18} cm), a sphere with a volume of 4.1×10^{57} cm³ would enclose the Orion Nebula, and it would contain

$$\frac{1,000 \text{ atoms}}{\text{cm}^3} \times (4.1 \times 10^{57} \text{ cm}^3) = 4.1 \times 10^{60} \text{ atoms}$$

This is a sufficient number of hydrogen atoms to produce

$$\frac{4.1 \times 10^{60} \text{ atoms}}{1 \times 10^{57} \text{ atoms/star}} = 4.1 \times 10^3 \text{ stars}$$

Thus there is a sufficient number of hydrogen atoms in the Orion Nebula to produce 4,100 average stars like the sun.

Gravitational attraction pulls the average protostar from a cloud with a diameter of trillions of miles down to a dense sphere with a diameter of 1.5 million miles or so. As gravitational attraction accelerates the atoms toward the center, they gain kinetic energy and the interior temperature increases. Over a period of some ten million years of contracting and heating,

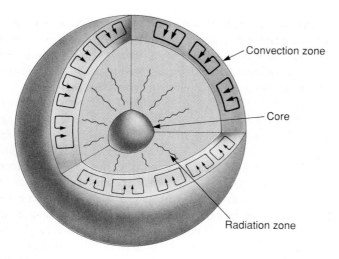

Figure 18.16

The structure of an average, mature star such as the sun. Hydrogen fusion reactions occur in the core, releasing gamma and X-ray radiation. This radiation moves through the radiation zone from particle to particle, eventually heating gases at the bottom of the convection zone. Convection cells carry energy to the surface, where it is emitted to space as visible light, ultraviolet radiation, and infrared radiation.

the temperature and density conditions at the center of the protostar are sufficient to start nuclear fusion reactions. Pressure from hot gases and energy from increasing fusion reactions begin to balance the gravitational attraction over the next seventeen million years, and the newborn, average star begins its stable life, which will continue for the next ten billion years.

The interior of an average star, such as the sun, is modeled after the theoretical pressure, temperature, and density conditions that would be necessary to produce the observed energy and light from the surface. This model describes the interior as a set of three shells: (1) the core, (2) a radiation zone, and (3) the convection zone (figure 18.16).

The **core** is the dense, very hot region where nuclear fusion reactions release gamma and X-ray radiation. The pressure at the core is about 3×10^{11} atmospheres with a central temperature of $1.5 \times 10^{7°}$ C. At this pressure, the density of the core is about 150 g/cm^3, or about twelve times that of solid lead. Because of the plasma conditions, however, the core remains in a gaseous state even at this density. The core contains about half the total mass of an average star but reaches only one-fourth of the distance to the surface.

The **radiation zone** is less dense than the core, having a density of about 1 g/cm^3 (which is the density of water) halfway from the center to the surface. Energy in the form of gamma and X rays from the core is absorbed and reemitted by collisions with atoms in this zone. The radiation slowly diffuses outward because of the countless collisions over a distance comparable to the distance between the earth and the moon. Each collision red-shifts the radiation toward ultraviolet radiation, and it could take millions of years before this radiation finally escapes the radiation zone.

The **convection zone** begins about nine-tenths of the way to the surface, and the density of the gases here is much less, about 1 percent the density of water. Gases at the bottom of this zone are heated by radiation from the zone below, expand from the heating, and are buoyed to the surface by convection. At the surface, the gases emit energy in the form of visible light, ultraviolet radiation, and infrared radiation, which moves out into space. The now cooler gases contract in volume and sink back to the radiation zone to become heated again, continuously carrying energy from the radiation zone to the surface in convection cells. The surface is continuously heated by the convection cells as it gives off energy to space, maintaining a temperature of about 5,800 K (about 5,500° C).

As an average star, the sun converts about 1.4×10^{17} kg of matter to energy every year as hydrogen nuclei are fused to produce helium. The sun was born about five billion years ago and has sufficient hydrogen in the core to continue shining for another four or five billion years. Other stars, however, have masses that are much greater or much less than the mass of the sun so they have different life spans. More massive stars generate higher temperatures in the core because they have a greater gravitational contraction from their greater masses. Higher temperatures mean increased kinetic energy, which results in increased numbers of collisions between hydrogen nuclei with the end result an increased number of fusion reactions. Thus a more massive star uses up its hydrogen more rapidly than a less massive star. On the other hand, stars that are less massive than the sun use their hydrogen at a slower rate so they have longer life spans. The life spans of the stars range from a few million years for large, massive stars, to ten billion years for average stars like the sun, to trillions of years for small, less massive stars.

Brightness of Stars

Stars generate their own energy and light, but some stars appear brighter than others in the night sky. As you can imagine, this difference in brightness could be related to (1) the amount of energy and light produced by the stars, (2) the size of each star, or (3) the distance to a particular star. A combination of these factors is responsible for the brightness of a star as it appears to you in the night sky. A classification scheme for different levels of brightness that you see is called the **apparent magnitude** scale. The apparent magnitude scale is based on a system established by a Greek astronomer over two thousand years ago. Hipparchus made a catalog of the stars he could see and assigned a numerical value to each to identify its relative brightness. The brightness values ranged from one to six, with the number one assigned to the brightest star and the number six assigned to the faintest star that could be seen. Stars assigned the number one came to be known as first-magnitude stars, those a little dimmer as second-magnitude stars, and so on to the faintest stars visible, the sixth-magnitude stars.

When technological developments in the nineteenth century made it possible to measure the brightness of a star, Hipparchus's system of brightness values acquired a precise, quantitative meaning. Today, a first-magnitude star is defined as one that is 100 times brighter than a sixth-magnitude star, with five uniform multiples of decreasing brightness on a scale from the first magnitude to the sixth magnitude. Each magnitude is about 2.51 times fainter as the next highest magnitude number. Thus a first-magnitude star is 2.51 times brighter than a second-magnitude star, $(2.51)^2$ or 6.31 times brighter than a third magnitude star, and so on. Table 18.2 shows the brightness

Table 18.2

Brightness comparisons for apparent magnitude differences

Difference of Two Apparent Magnitudes	Ratio of Brightness between the Two Stars
0.0	1.0
0.5	1.6
1.0	2.5
1.5	4.0
2.0	6.3
2.5	10
3.0	16
3.5	25
4.0	40
4.5	63
5.0	100
5.5	160
6.0	250
6.5	400
7.0	630

ratio that can be used to compare the brightness of two stars based on apparent magnitude differences. When using this table, recall that the *brighter* star has the *lower* magnitude number. This seems backward, but this convention has been followed ever since the first scale was devised by Hipparchus. Also note that some stars were found, by measurement, to be brighter than the apparent magnitude of $+1$, which extends the scale into negative numbers. The brightest star in the night sky is Sirius, for example, with an apparent magnitude of -1.42.

The apparent magnitude of a star depends on how far away stars are in addition to differences in the stars themselves. Stars at a farther distance will appear fainter and those closer will appear brighter just as any other source of light. To compensate for distance differences, astronomers calculate the brightness that stars would appear to have if they were all at a defined, standard distance. The standard distance is defined as 10 parsecs (32.6 light-years), and the brightness of a star at this distance is called the **absolute magnitude.** The sun, for example, is the closest star and has an apparent magnitude of -26.7 at an average distance from the earth. When viewed from the standard distance of 10 parsecs, the sun would have an absolute magnitude of $+4.8$, which is about the brightness of a faint star (figure 18.17).

The absolute magnitude is an expression of **luminosity,** the total amount of energy radiated into space each second from the surface of a star. The sun, for example, radiates 4×10^{26} joules per second from the surface. The luminosity of stars is often compared to the sun's luminosity, with the sun considered to have a luminosity of one unit. When this is done the luminosity of the stars ranges from a low of 10^{-6} sun units for the dimmest stars up to a high of 10^5 sun units. Thus the sun is somewhere in the middle of the range of star luminosity.

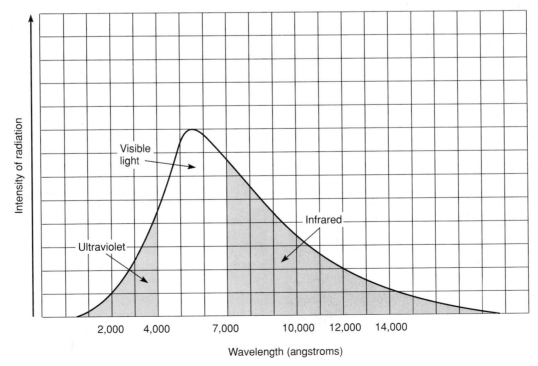

Figure 18.17

Not all energy from a star goes into visible light. The graph shows the distribution of radiant energy emitted from the sun, which has an absolute magnitude of $+4.8$.

The Universe

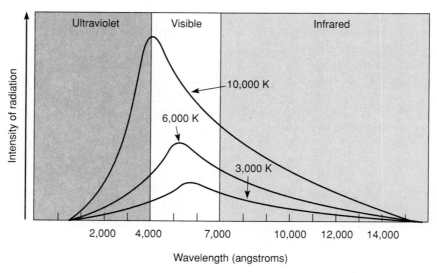

Figure 18.18

The distribution of radiant energy emitted is different for stars with different surface temperatures. Note that the peak radiation of a cooler star is more toward the red part of the spectrum, and the peak radiation of a hotter star is more toward the blue part of the spectrum.

Star Temperature

If you observe the stars on a clear night you will notice that some are brighter than others, but you will also notice some color differences. Some stars have a reddish color, some have a bluish white color, and others have a yellowish color. This color difference is understood to be a result of the relationship that exists between the color and the temperature of an incandescent object (see chapter 10). The colors of the various stars are thus a result of the temperatures of the stars. You see a cooler star as reddish in color and comparatively hotter stars as bluish white. Stars with in-between temperatures, such as the sun, appear to have a yellowish color (figure 18.18).

Astronomers analyze starlight to measure the temperature and luminosity as well as the chemical composition of a star. Light from a star is focused through a telescope on a solid-state photocell device, and the amount of radiation measured is used to calculate the luminosity of the star. By using light filters, the intensity of blue light from a star can be compared to the intensity of red light. Comparison of the two intensities indicates the temperature, since more intense blue light means higher temperatures and more intense red light means cooler temperatures. When the starlight is analyzed in a spectroscope, specific elements can be identified from the unique set of spectral lines that each element emits (see chapter 11).

Astronomers use information about the star temperature and spectra as the basis for a star classification scheme. Originally, the classification scheme was based on sixteen categories according to the strength of the hydrogen line spectra. The groups were identified alphabetically with A for the group with the strongest hydrogen line spectrum, B for slightly weaker lines, and on to the last group with the faintest lines. Later, astronomers realized that the star temperature was the important variable, so they rearranged the categories according to decreasing temperatures. The original letter categories were retained, however, resulting in classes of stars with the hottest temperature first and the coolest last with the sequence O B A F G K M. Table 18.3 compares the color, temperature ranges, and other features of the stellar spectra classification scheme.

The H-R Diagram

In 1910, Henry Russell in the United States and Ejnar Hertzsprung in Denmark independently developed a scheme to classify stars with a temperature-luminosity graph. Today, the graph is called the **Hertzsprung-Russell diagram,** or the *H-R diagram* for short. The diagram is a plot with temperature indicated by spectral types and the true brightness indicated by absolute magnitude. The diagram, as shown in figure 18.19, plots temperature by spectral types sequenced O through M types, so the temperature decreases from left to right. The hottest, brightest stars are thus located at the top left of the diagram and the coolest, faintest stars are located at the bottom right.

Each dot is a data point representing the surface temperature and brightness of a particular star. The sun, for example, is a type G star with an absolute magnitude of about +5, which places the data point for the sun almost in the center of the diagram. This means that the sun is an ordinary, average star with respect to both surface temperature and true brightness.

Most of the stars plotted on an H-R diagram fall in or close to a narrow band that runs from the top left to the lower right. This band is made up of **main sequence stars.** Stars along

Table 18.3
Major stellar spectral types and temperatures

Type	Color	Temperature (K)	Comment
O	Bluish	30,000–80,000	Spectrum with ionized helium and hydrogen but little else; short-lived and rare stars
B	Bluish	10,000–30,000	Spectrum with neutral helium, none ionized
A	Bluish	7,500–10,000	Spectrum with no helium, strongest hydrogen, some magnesium and calcium
F	White	6,000–7,500	Spectrum with ionized calcium, magnesium, neutral atoms of iron
G	Yellow	5,000–6,000	The spectral type of the sun. Spectrum shows sixty-seven elements in the sun
K	Orange-red	3,500–5,000	Spectrum packed with lines from neutral metals
M	Reddish	2,000–3,500	Band spectrum of molecules, e.g., titanium oxide; other related spectral types (R, N, and S) are based on other molecules present in each spectral type

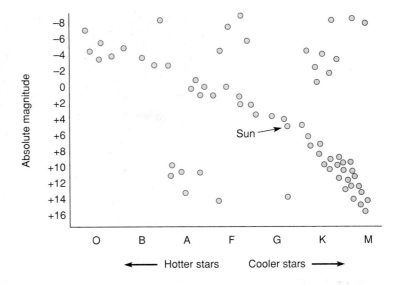

Figure 18.19
The Hertzsprung-Russell diagram.

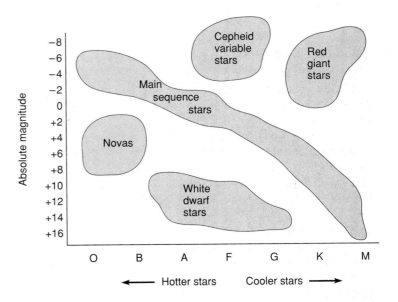

Figure 18.20
The region of the cepheid variable, red giant, main sequence, and white dwarf stars and novas on the H-R diagram.

the main sequence band are normal, mature stars that are using their nuclear fuel at a steady rate. Those stars on the upper left of the main sequence are the brightest, bluest, and most massive stars on the sequence. Those at the lower right are the faintest, reddest, and least massive of the stars on the main sequence. In general, most of the main sequence stars have masses that fall between a range from ten times greater than the mass of the sun (upper left) to one-tenth the mass of the sun (lower right). The extremes, or ends, of the main sequence range from about sixty times more massive than the sun to one-twenty-fifth of the sun's mass. It is the *mass* of a main sequence star that determines its brightness, its temperature, and thus its location on the H-R diagram.

As shown in figure 18.20, there are two groups of stars that have a different set of properties than the main sequence stars. The **red giant stars** are bright, but low-temperature, giants. These reddish stars are enormously bright for their temperature because they are very large, with an enormous surface area giving off light. A red giant might be one hundred times larger but have the same mass as the sun. These low-density red giants are located in the upper right part of the H-R diagram. The **white dwarf stars,** on the other hand, are located at the lower left because they are faint, white-hot stars. A white dwarf is

faint because it is small, perhaps twice the size of the earth. It is also very dense, with a mass approximately equal to the sun. Red giants and white dwarfs are believed to be stages that aging stars pass through.

Stars such as the sun emit a steady light because the force of gravitational contraction is balanced by the outward flow of energy. Other stars vary in brightness over a period of time because the star expands, then collapses periodically during one stage of its life. Some of these *variable stars* have luminosities that vary by small amounts at irregular intervals, and these variables are found in eight regions scattered over the H-R diagram. The **cepheid variables,** on the other hand, have a regular variation in brightness over a period of time. The cepheid variables appear on the H-R diagram in a region close to the main sequence and red giants. They were named after the first star of this type to be discovered, *Delta Cephei.* The cepheid variables are important because they mean something about the life cycle of a star. They are also important because of a relationship between the cycle of brightness variation and the average luminosity of a star. This relationship has given astronomers a means of measuring the distance to other galaxies that could not otherwise have been measured.

Sometimes a star undergoes a violent flare-up, increasing in brightness by some fifteen magnitudes for a short period of time. Such an irregularly flaring star is called a **nova** after the Latin for "new." Many novas occur as *binary stars,* two stars that are circling around a common center of mass under the influence of gravity. The nova may be an aging binary star, perhaps a white dwarf, that pulls matter from the companion binary star, then undergoes a periodic violent explosion as a result of the newly acquired matter upsetting the balance of pressures. A *supernova,* on the other hand, is an extremely violent explosion that spreads the outer layers of a massive star into space. The cause and consequences of a supernova are discussed in the next section.

The Life of a Star

A star is born in a gigantic cloud of gas and dust in interstellar space, then settles to a period of millions or billions of years of calmly shining while it fuses hydrogen nuclei in the core. How long a star shines and what happens to it when it uses up the hydrogen in the core depends on the mass of the star. Some stars, such as the sun, will slowly expand to a bloated red giant, then violently blow off their outer shells to become a white dwarf star. Other, more massive stars become a collapsed core of nuclear matter called a *neutron star.* The collapsed core of still more massive stars might collapse into a *black hole* in space. Of course no one has observed a life cycle of over millions or billions of years. The life cycle of a star is a theoretical outcome of computations concerning relationships between the measured values of the mass of a star, its surface temperature, and luminosity with models based on what is known about nuclear reactions and the theoretical changes that would occur as the fuel for the nuclear reactions is used up. The resulting model of the life cycle of a star and the predicted outcomes seem to agree with observations of stars today, with different groups of stars that can be plotted on the H-R diagram. Thus the groups of stars on the diagram—main sequence, red giants, and white dwarfs, for example—are understood to be stars in various stages of their lives.

The first stage in the theoretical model of the life cycle of a star is the formation of the protostar. As gravity pulls the gas of a protostar together the density, pressure, and temperature

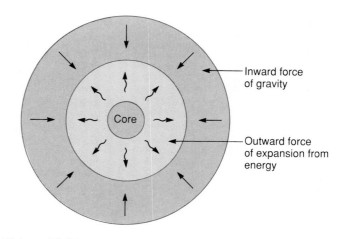

Figure 18.21

A star becomes stable when the outward forces of expansion from the energy released in nuclear fusion reactions balances the inward forces of gravity.

increase from the surface down to the center. Eventually, the conditions are right for nuclear fusion reactions to begin in the core, which requires a temperature of 10 million Kelvins. The initial fusion reaction essentially combines four hydrogen nuclei to form a helium nucleus with the release of much energy. This energy heats the core beyond the temperature reached by gravitational contraction, eventually to 16 million Kelvins. Since the star is a gas, the increased temperature expands the volume of the star. The outward pressure of expansion balances the inward pressure from gravitational collapse, and the star settles down to a balanced condition of calmly converting hydrogen to helium in the core, radiating the energy released into space (figure 18.21). The theoretical time elapsed from the initial formation and collapse of the protostar to the main sequence is about fifty million years.

Where the star is located on the main sequence and what happens to it next depend only on how massive it is. The more massive stars have higher core temperatures and use up their hydrogen more rapidly as they shine at higher surface temperatures (O type stars). Less massive stars shine at lower surface temperatures (M type stars) as they use their fuel at a slower rate. The overall life span on the main sequence ranges from millions of years for O type stars to trillions of years for M type stars. An average one-solar mass star will last about ten billion years.

The next stage in the theoretical life of a star begins as much of the hydrogen in the core has been fused into helium. With fewer hydrogen fusion reactions, less energy is released and less outward balancing pressure is produced, so the star begins to gravitationally collapse. The collapse heats the now helium core and the surrounding shell where hydrogen still exists. The hydrogen in the shell begins to undergo fusion from the increased temperatures, and the increased energy released now expands the outer layers of the star. With an increased surface area the amount of radiation emitted per unit area is less,

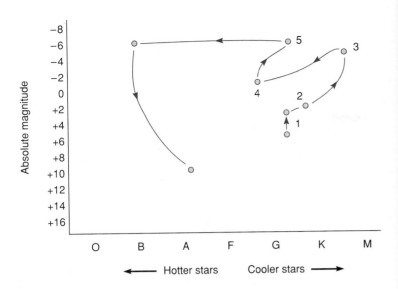

Figure 18.22

The evolution of a star of solar mass as it depletes hydrogen in the core (1), fuses hydrogen in the shell to become a red giant (2 to 3), becomes hot enough to produce helium fusion in the core (3 to 4), then expands to a red giant again as helium and hydrogen fusion reactions move out into the shells (4 to 5). It eventually becomes unstable and blows off the outer shells to become a white dwarf star.

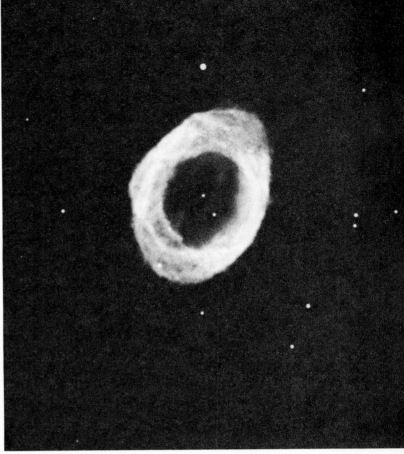

Figure 18.23

The blown-off outer shells of a star form a ringlike structure called a planetary nebula.

and the star acquires the properties of a brilliant red giant. Its data point position on the H-R diagram changes since it now has different luminosity and temperature properties. (The star has not physically *moved,* as is often said. The changing properties move its temperature-luminosity data point, not the star, to a new position.)

After about five hundred million years as a red giant, the star now has a surface temperature of about 4,000 Kelvins compared to its main sequence surface temperature of 6,000 Kelvins. The radius of the red giant is now a thousand times greater, a distance that will engulf the earth when the sun reaches this stage. Even though the surface temperature has decreased from the expansion, the helium core is continually heating and eventually reaches a temperature of 100 million Kelvins, the critical temperature necessary for the helium nuclei to undergo fusion to produce carbon nuclei. The red giant now has helium fusion reactions in the core and hydrogen fusion reactions in a shell around the core. This changes the radius, the surface temperature, and the luminosity with the overall result depending on the composition of the star. In general, the radius and luminosity decreases when this stage is reached, moving the star back toward the main sequence (figure 18.22).

After millions of years of helium fusion reactions, the core is gradually converted to a carbon core and helium fusion begins in the shell surrounding the core. The core reactions decrease as the star now has a helium fusing shell surrounded by a second, hydrogen fusing shell. This releases additional energy, and the star again expands to a red giant for the second time. A star the

size of the sun or less massive may cool enough at this point that nuclei at the surface become neutral atoms rather than a plasma. As neutral atoms they can absorb radiant energy coming from within the star, heating the outer layers. Changes in temperature produce changes in pressure, which change the balance between the temperature, pressure, and the internal energy generation rate. The star begins to expand outward from heating. The expanded gases are cooled by the expansion process, however, and are pulled back to the star by gravity only to be heated and expand outward again. In other words, the outer layers of the star begin to pulsate in and out. Finally, a violent expansion blows off the outer layers of the star, leaving the hot core. Such blown-off outer layers of a star form circular nebula called *planetary nebula* (figure 18.23). The nebula continues moving away from the core, eventually adding to the dust and gases between the stars. The remaining carbon core and helium-fusing shell begins gravitationally to contract to a small, dense *white dwarf* star. A star with the original mass of the sun, or less, slowly cools from white, to red, then to a black lump of carbon in space.

A more massive star will have a different theoretical ending than the slow cooling of a white dwarf. A massive star will contract, just as the less massive stars, after blowing off its outer shells. In a more massive star, however, heat from the contraction may reach the critical temperature of 600 million Kelvins to begin carbon fusion reactions. Thus a more massive star may go through a carbon fusing stage and other fusion reaction stages that will continue to produce new elements until the element iron is reached. After iron, energy is no longer released by the fusion process (see chapter 17), and the star has used up all of its energy sources. Lacking an energy source, the star is no longer able to maintain its internal temperature. The star loses the outward pressure of expansion from the high temperature, which had previously balanced the inward pressure from gravitational attraction. The star thus collapses, then rebounds like a compressed spring into a catastrophic explosion called a **supernova.** A supernova produces a brilliant light in the sky that may last for months before it begins to dim as the new elements that were created during the life of the star diffuse into space. These include all the elements up to iron that were produced by fusion reactions during the life of the star and heavier elements that were created during the instant of the explosion. All the elements heavier than iron were created as some less massive nuclei disintegrated in the explosion, joining with each other and with lighter nuclei to produce the nuclei of the elements from iron to uranium. These fusion-produced and supernova-produced elements are scattered through space by the supernova explosion. As you will see in the next chapter, these newly produced, scattered elements will later become the building blocks for new stars and planets such as the sun and the earth.

The remains of the compressed core after the supernova have still yet another fate if they have (1) a mass greater than 1.4 solar masses or (2) a mass greater than about 3 solar masses or more. Both of these solar masses represent a crucial point in what happens to the remaining core. Matter is fantastically compressed in a white dwarf star, so tightly that there is practically no space between the electrons. The mutual repulsion of electrons forced so close together at such a great density balances the contracting forces of gravity, and the white dwarf has a stable size as it cools, with a density equivalent to 10^3 kg/cm^3. If the core of a supernova has a remaining mass greater than 1.4 solar masses, the gravitational forces on the remaining matter, together with the compressional forces of the supernova explosion, are great enough to collapse nuclei, forcing protons and electrons together into neutrons, forming the core of a **neutron star.** A neutron star is the very small (10 to 20 km diameter), superdense (10^{11} kg/cm^3 or greater) remains of a supernova with a center core of pure neutrons.

Because it is a superconcentrated form of matter, the neutron star also has an extremely powerful magnetic field. This magnetic field of a rotating neutron star is believed to be the most likely source of regular, equally spaced pulsating radio signals from space called **pulsars.** Pulsars were a mystery when first discovered. For a time extraterrestrial life was considered as the source of the signals, so they were jokingly identified as LGM (for Little Green Men) signals. Since that time over 150 pulsars have been identified. There are several different models about how a neutron star emits the regularly spaced, pulsed radio signals, but there are still questions about how the radio beam is generated.

Another theoretical limit occurs if the remaining core has a mass of about three solar masses or more. At this limit the force of gravity overwhelms *all* nucleon forces, including the repulsive forces between like charged particles. If this theoretical limit is reached, nothing can stop the collapse, and the collapsed star will become so dense that even light cannot escape. The star is now a **black hole** in space. Since nothing can stop the collapsing star, theoretically a black hole would continue to collapse to a pinpoint and then to a zero radius called a *singularity.* This event seems contrary to anything that can be directly observed in the physical universe, but it does agree with the general theory of relativity and concepts about the curvature of space produced by such massively dense objects. Black holes are theoretical and none has been seen, of course, because a black hole theoretically pulls in radiation of all wavelengths and emits nothing. Evidence for the existence of a black hole is sought by studying binary systems and X rays that would be given off by matter as it is accelerated into a black hole. There are many uncertainties in such theoretical calculations and deductions, however, and more conclusive evidence may be provided when X-ray telescopes are available outside the earth's atmosphere, perhaps on a space station or perhaps on a later modification of the Hubble Space Telescope.

Galaxies

Stars are associated with other stars on many different levels, from double stars that orbit a common center of mass, to groups of tens or hundreds of stars that have gravitational links and a common origin, to the billions and billions of stars that form the basic unit of the universe, a **galaxy.** The sun is but one of an estimated one hundred billion stars that are held together by gravitational attraction in the Milky Way galaxy. The numbers of stars and vastness of the Milky Way galaxy alone seem almost beyond comprehension, but there is more to come. The Milky Way is but one of *billions* of galaxies that are associated with other galaxies in clusters, and these clusters are associated with one another in superclusters. Through a large telescope you can see more galaxies than individual stars in any direction, each galaxy with its own structure of billions of stars. Yet, there are similarities that point to a common origin. Some of the similarities and associations of stars will be introduced in this section along with the Milky Way galaxy, the vast, flat, spiraling arms of stars, gas, and dust where the sun is located (figure 18.24).

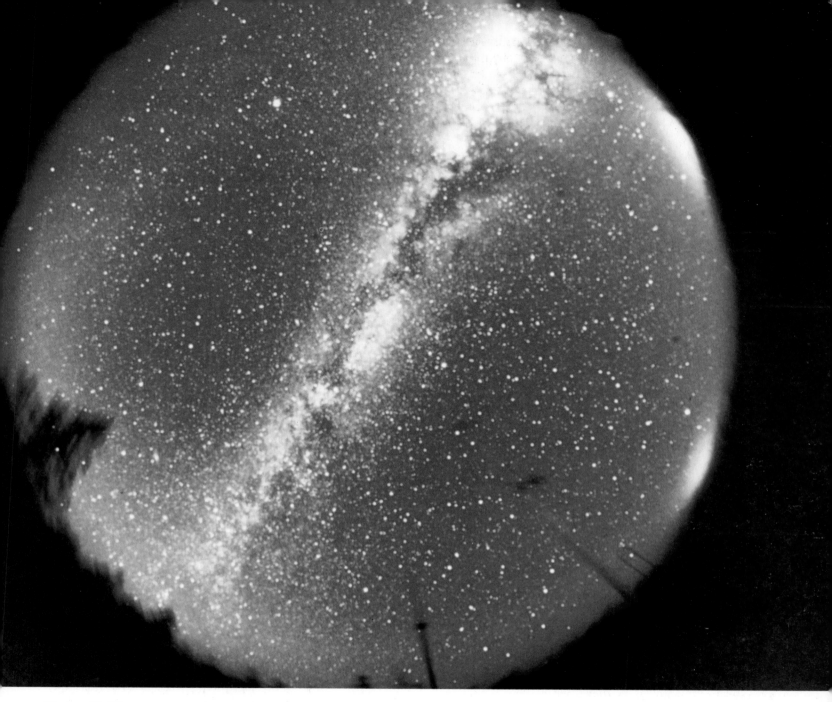

Figure 18.24

A wide-angle view toward the center of the Milky Way galaxy. Parts of the white, milky band are obscured from sight by gas and dust clouds in the Galaxy.

The Milky Way Galaxy

Away from city lights, you can clearly see the faint, luminous band of the Milky Way galaxy on a moonless night. Through a telescope or a good pair of binoculars, you can see that the luminous band is made up of countless numbers of stars. You may also be able to see the faint glow of nebula, concentrations of gas and dust. There are dark regions in the Milky Way that also give an impression of something blocking starlight, such as dust. You can also see small groups of stars called **galactic clusters.** Galactic clusters are gravitationally bound subgroups of as many as one thousand stars that move together within the Galaxy. Other clusters are more symmetrical and tightly packed, containing as many as a million stars, and are known as **globular clusters.**

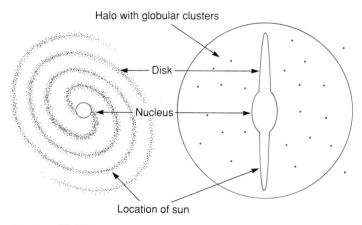

Figure 18.25

The structure of the Milky Way galaxy.

Viewed from a distance in space, the Milky Way would appear to be a huge, flattened cloud of spiral arms radiating out from the center. There are three distinct parts: (1) the spherical concentration of stars at the center of the disk called the *galactic nucleus;* (2) the rotating *galactic disk,* which contains most of the bright, blue stars along with much dust and gas; and (3) a spherical *galactic halo,* which contains some 150 globular clusters located outside the galactic disk (figure 18.25). The sun is located in one of the arms of the galactic disk, some 33,000 light-years from the center. The galactic disk rotates, and the sun completes one full rotation every 200 million years.

The diameter of the *galactic disk* is about 100,000 light-years, or the distance that light traveling at 186,000 miles per second would cover in 100,000 years (5.86×10^{12} miles/year $\times$ 100,000, or 5.86×10^{17} miles). Yet, in spite of the one hundred billion stars in the Milky Way, it is mostly full of emptiness. By way of analogy, imagine reducing the size of the Milky Way disk until stars like the sun are reduced to the size of tennis balls. The distance between two of these tennis-ball-sized stars would now compare to the distance across the state of Texas. The space between the stars is not actually empty since it contains a thin concentration of gas, dust, and molecules of chemical compounds. The gas particles outnumber the dust particles about 10^{12} to 1. The gas is mostly hydrogen, and the dust is mostly solid iron, carbon, and silicon compounds. Over forty different chemical molecules have been discovered in the space between the stars, including many organic molecules. Some nebula consist of clouds of molecules with a maximum density of about 10^6 molecules/cm^3. The gas, dust, and chemical compounds make up about half of the mass of the galactic disk, and the stars in the disk make up the other half. The gas plays an important role in the formation of new stars, and the dust and chemical compounds play an important role in the formation of planets.

The spherical *galactic nucleus* is hidden from the earth by the clouds of dust in the central plane of the galactic disk. Studies of infrared and radio waves from the nucleus and studies of the nuclei of galaxies similar to the Milky Way raise many questions about the nucleus. Studies of other galaxies indicate that the nucleus contains old, red stars with little interstellar dust or gas, with the stars clustered close together within a radius of some 5,000 light-years. Studies of infrared and radio waves indicate that the nucleus has a central part with a one or two light-year radius that is emitting enormous amounts of energy, with matter both streaming out and falling in. The source of this energy and matter is not known.

The *halo* around the Galaxy consists mostly of groups of massive, old red stars in globular clusters, a few individual stars, and not much gas or dust. The stars in these globular clusters are much closer together than those in the disk, and they are believed to be the oldest stars in the Galaxy.

The stars in the *galactic disk* occur in clusters of tens to hundreds of stars gravitationally linked in galactic clusters or binary stars that revolve around a common center of mass, or they occur as individual stars. A galactic cluster (also called an open cluster) of the galactic disk contains a much lower concentration of stars than the globular clusters of the halo. Like the globular clusters, the stars of a galactic cluster are linked by gravity and have a common origin, age, and composition. The stars of a galactic cluster generally fall on the main sequence on an H-R diagram, meaning that they are young to mature stars of various masses. Since the galactic clusters and the bulk of the dust and gas are found in the galactic disk, this is the active part of the Galaxy where young O and B type stars are forming from the gas and dust. These stars shine relatively briefly, then explode to replenish the dust and gas of the spiral arms. Periodically, a longer-lived phenomenon, an average size star such as the sun and its planets, form from the dust and gas. The time scale involved in all of these cycles greatly exceeds the age of the earth.

Other Galaxies

Outside the Milky Way is a vast expanse of emptiness, lacking even the few molecules of gas and dust spread thinly through the galactic nucleus. There is only the light from faraway galaxies and the time that it takes for this light to travel across the vast vacuum of intergalactic space. How far away is the nearest galaxy? Recall that the Milky Way is so large that it takes light 100,000 years to travel the length of its diameter. The nearest galactic neighbor similar to the Milky Way is Andromeda, about 2 million light-years away.

The Andromeda galaxy is similar to the Milky Way in size and shape, with about one hundred billion stars, gas, and dust turning in a giant spiral pinwheel (figure 18.26). Other galaxies have other shapes and other characteristics. The American astronomer Edwin Hubble developed a classification scheme for the structure of galaxies based on his 1926 study of

Figure 18.26

The Andromeda galaxy, which is believed to be similar in size, shape, and structure to the Milky Way galaxy.

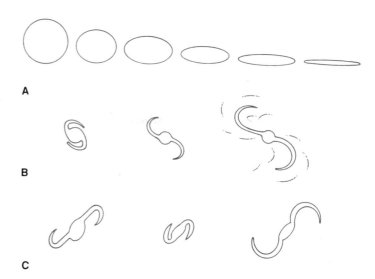

Figure 18.27

Different subgroups in the Hubble classification scheme: (*a*) elliptical galaxies, (*b*) spiral galaxies, and (*c*) barred galaxies.

some six hundred different galaxies. The basic galactic structures were identified as elliptical, spiral, barred, and irregular (figure 18.27).

Elliptical galaxies appear to be spherical to flattened elliptical shapes with rotational symmetry. Hubble identified eight subclasses of elliptical galactic shapes based on the degree of flatness. This shape made up about 20 percent of Hubble's sample, and all galaxies with this shape have properties in common. Elliptical galaxies contain only old stars with no O or B types and very little gas or dust. Elliptical galaxies are large and massive, with about 10^{12} solar masses. All of the evidence indicates that elliptical galaxies are old.

Spiral galaxies have a small spherical nucleus with usually two but sometimes more spiral arms radiating out from the nucleus in a flattened disk. Spiral galaxies made up about half of Hubble's sample. They contain stars of all ages, but the young stars are found in the spirals and the older stars occur in the globular clusters of the halo. Spiral galaxies are slightly less massive than elliptical galaxies, with about 10^{11} solar masses.

Barred galaxies have the shape of a bar across the nucleus with spiral arms radiating from the ends of the bars. About 30 percent of Hubble's sample were barred galaxies. As you would expect, barred galaxies have properties very similar to spiral galaxies.

Irregular galaxies are those that Hubble could not fit into the other categories because they lacked symmetry. The irregular galaxies are relatively few in number, the least massive with only 10^9 solar masses, and contain mostly young stars with the greatest accumulation of dust and gas of all the galaxies.

The Life of a Galaxy

Hubble's classification of galaxies into distinctly different categories of shape was an exciting accomplishment because it suggested that some relationship or hidden underlying order may exist in the shapes. Finding underlying order is important because it leads to the discovery of the physical laws that govern the universe. Soon after Hubble published his classification results in 1926, two models of galactic evolution were proposed. One model, which was suggested by Hubble, had extremely slowly spinning spherical galaxies forming first, which gradually flattened out as their rate of spin increased while they condensed. Thus this is a model of spherical galaxies flattening out to increasingly elliptical shapes, eventually spinning off spirals until they finally broke up into irregular shapes over a long period of time.

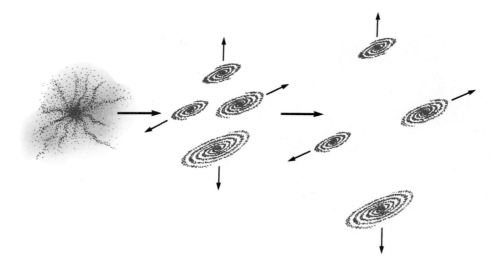

Figure 18.28

The big bang theory of the universe has the space between galaxies expanding as the dust and gas in galaxies become locked up in white dwarf stars, neutron stars, and black holes.

Another model of galactic evolution had the galaxies beginning as irregular shapes that were condensing into the spherical shape. In this model the center part would not rotate as rapidly as the outer parts, so gravity would pull the rapidly moving outer regions into a spherical shape. Thus this is a model of irregular galaxies collapsing to spiral galaxies, which eventually condensed to spherical shapes over a long period of time.

When these two opposing models of galactic evolution were formulated, astronomers did not know about the life of a star and technology had not yet developed to the stage where they could know the spectra types of stars in the classes of galaxies. When this knowledge and information did become available, it was clear that both models were unacceptable. The flattening-out elliptical-to-spiral model is unacceptable because it is inconsistent with the ages of stars observed in these galaxies. Elliptical galaxies contain only old, red stars with no young O or B types and very little gas or dust. Thus they could not evolve into spiral galaxies with half their masses in hydrogen gas and dust and half in stars, many of which are new young stars. Furthermore, the winding-up spiral-to-elliptical model is also unacceptable because spiral galaxies contain old stars, with little gas or dust in the globular clusters of their halos. Since these old stars are just as old as the old stars in the elliptical galaxies, both types of galaxies must be of about the same age.

Today, the current model of galactic evolution is based on the **big bang theory** of the creation of the universe. Evidence for the big bang theory comes from (1) present-day microwave radiation from outer space, (2) current data on the expansion of the universe, (3) the age of the oldest stars in the Milky Way galaxy, and (4) the measured ratios of the amount of radioactive decay products to the radioactive materials they decayed from. All of these different astronomical and physical "clocks" indicate that the universe was created in a "big bang" some

twenty billion years ago, expanding as an intense and brilliant explosion from a primeval fireball with a temperature of some 10^{12} Kelvins. An often-used analogy for the movement of galaxies after the big bang is to consider galaxies as spots on the surface of a balloon that is being inflated. As the balloon expands the distances between all the spots increase, with the result that all the spots find themselves farther and farther apart. In other words, the galaxies are not expanding into space that is already there. It is space itself that is expanding, and the galaxies move with the expanding space (figure 18.28). This concept agrees with the theory of general relativity, that space is generated by the presence of matter and that the contour of space is determined by the distribution of matter (see the Feature at the end of this chapter).

The temperature and the intensity of the initial fireball of radiation, according to the theory, diminished as the universe expanded, and some of the remaining radiation should be measurable today. The presence of the remaining radiation was a theoretical topic of discussion among astronomers, but it was actually discovered in 1965 when scientists with Bell Telephone Laboratories were puzzled by radiation detected by a new radio antenna they were testing. The radiation was a consistent, uniform signal that seemed to come from every direction in space, and the signal intensity was equivalent to a blackbody radiation of about 3 Kelvins. Later, it was determined that they had found the remnant of the original primeval fireball. This measured temperature allowed scientists to trace back the evolution of the universe as its temperature cooled. To cool to a temperature of about 3 Kelvins requires the universe to have formed about twenty billion years ago.

The second set of evidence for the big bang theory came from Edwin Hubble and his pioneer work with galaxies. Hubble had determined the distances to some of the galaxies that had red-shifted spectra. From the Doppler effect (see chapter 7), it

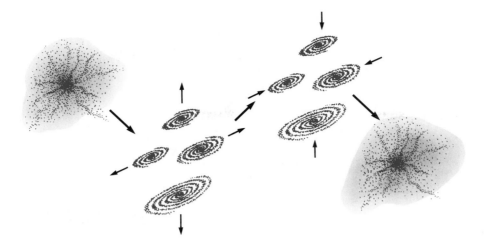

Figure 18.29

The oscillating theory of the universe has the space between the galaxies expanding as does the big bang theory, but the oscillating theory has the galaxies gradually coming back together to begin all over in another big bang.

was known that these galaxies were moving away from the Milky Way. Hubble found a relationship between the distance to a galaxy and the velocity with which it was moving away. He found the velocity to be directly proportional to the distance, that is, the greater the distance to a galaxy the greater the velocity. This means that a galaxy twice as far from the Milky Way as a second galaxy is moving away from the Milky Way at twice the speed as the second galaxy. Since this relationship was seen in all directions, it meant that the universe is expanding uniformly. The same effect would be viewed from any particular galaxy, that is, all the other galaxies are moving away with a velocity proportional to the distance to the other galaxies. This points to a common beginning, a time when all matter in the universe was together. Assuming that the universe has expanded uniformly, Hubble's law of expansion can be used to estimate how far back the expansion began. The result of this calculation gives the answer that the universe began expanding twenty billion years ago.

Current theoretical ideas and models about the universe can lead to interesting speculative ideas about how galaxies formed and what may eventually happen to them. For example, consider a universe that initially consisted of a very hot gas of photons that cooled as the universe expanded. As it cooled, the contents—protons, electrons, and other particles and antiparticles—changed over time, forming other particles. After about a million years of cooling the temperature reached about 3,000 Kelvins and clouds of hydrogen gas began to condense, pulled together in a swirling supercloud. As more hydrogen atoms condensed their mutual gravitational attraction pulled in more atoms and distinct swirls of hydrogen gas were separated. Such clouds of hydrogen gas are called **protogalaxies.** Smaller pockets of hydrogen gas continually form in eddies of the protogalaxy

and eventually form protostars, then stars, as was discussed in the previous section. Different rates of swirling gas clouds probably resulted in the shapes of galaxies observed today.

The first stars formed were probably very massive and short lived, ending as supernova. The supernova blasted out heavy elements, which condensed in the disk as the dust and gas that would form the next generation of stars. With each generation of stars, more and more heavy elements accumulate and more material is tied up in neutron stars. Thus less material is available for the next generation, and the next generation are stars of lower mass, which, in turn, do not experience a supernova ending. Eventually, only less massive, slowly reacting red stars will remain with no gas or dust to create another generation of young hot stars. Only cooling white dwarfs, neutron stars, and black holes will remain. Eventually, everything cools to the same temperature with all matter now locked in cold, black dwarfs and black holes.

A model of the universe that begins with a big bang and ends with a little sigh does not seem to agree with the symmetry that is observed in nature. Since symmetry is usually observed, perhaps the universe will eventually contract, collapsing into another big bang. This model is called the **oscillating theory** of the universe (figure 18.29). Whether or not the big bang or oscillating theory of the universe is ultimately accepted seems to depend on estimates of the density of matter in the universe. If the density is sufficiently great, gravitational attraction will pull the expanding universe together to rebound anew. If the density is not sufficient, the universe will continue expanding. At the present time, measurements and estimates indicate that the density of the universe is *not* sufficient to pull it together. Yet the symmetry of an oscillating universe seems more appealing.

The General Theory of Relativity and Superstrings

Ever since Newton formulated the laws of motion and the universal law of gravitation, it has been possible to predict the movement and positions of objects in the Solar System, the Milky Way galaxy, and other galaxies with an accuracy that reaches the limits of measurement instruments used to test the predictions. But there were small problems of precision that remained unexplained until Einstein formulated the general theory of relativity. This theory made it possible to refine predictions of movement with an astonishing degree of accuracy, the most notable of which was a correction in the orbit of Mercury by forty-three seconds of arc every hundred years. Einstein's theory did not seem to affect scientific predictions greatly on the everyday level, but the basic difference between the Newtonian view of the universe and the view presented by relativity was revolutionary.

Newtonian physics considers the force of gravitational interaction between two objects to depend on the mass of the objects and the space between them. Thus the space between objects was considered to have an affect on matter, but no consideration was given to the affect of the matter on space. Einstein's general theory of relativity has a different interpretation of the relationship between matter and space, that they are equitably connected and space affects matter, but matter also affects space. Einstein's general theory of relativity, which is really his field theory of gravity, considers matter to *change* the local geometry of space; a gravity field is a curved space. The earth, for example, is attracted to the sun because the mass of the sun distorts, or curves, the geometry of space. Thus the force of gravity between the earth and the sun is a consequence of the motion of the earth following the geometry of the underlying curved space. Therefore space affects matter and matter affects space, and the two are regarded as two parts of a single, dynamic whole.

Einstein's theory is a mathematical set of field equations that was verified by three separate but classic tests: (1) the correct, precise description of the orbit of Mercury, which deviates forty-three seconds of arc per century because it is so close to the massive sun; (2) the bending of starlight as it passes near the massive sun; and (3) the gravitational blue-shift of starlight as it enters earth's gravitational field and the gravitational red-shift of starlight as it leaves a massive white dwarf star (these gravitational shifts are unrelated to Doppler shifts). In addition to these experimental verifications of the affects of matter on space, the general theory of relativity seems consistent with theoretical predictions concerning black holes, the expansion of the universe, and the possible existence of gravitational waves.

The general theory of relativity was revolutionary, but the physics is not yet settled. There were problems, which Einstein recognized, since matter still had to be introduced into the field equations separately. Matter, and especially matter in the form of elementary particles, is considered as mathematical points, having no dimensions, which results in calculations yielding infinites. Infinites are not mathematically traceable, yielding the same kind of answers you obtain when you divide by zero, so the mathematical analysis failed. Einstein spent the last thirty years of his life trying to develop a "unified field theory" that would not need to consider matter as a singularity in the field. A unified theory of a continuous field would predict the properties of both matter and space.

In the 1980s, physicists were still struggling with the concept of gravity, now trying to integrate gravity with quantum mechanics (see chapter 11). The problem was to integrate the then-understood four fundamental forces of nature (gravitation, electromagnetic, strong nuclear, and weak nuclear) into a single unified field theory. The problem, physicists believe, is the consideration of quarks, leptons, electrons, and other elementary particles as mathematical points. The location of a point in space is usually identified by three coordinates. But quantum mechanics goes beyond three coordinates to describe the state of elementary particles such as the electron. An electron also has the quality known as spin, and spin is described by special numerical coordinates that are added to the usual three dimensions of space and one of time. The result is a mathematical model of a new kind of space and time called *superspace*. Superspace is described by a new kind of geometry called *supergeometry*. According to this model, spin can generate the conventional dimensions of space, and this may have happened when the universe was created.

A new mathematical model of the primordial universe identifies a beginning with nine-dimensional space locked up in the tiny space of a point. At the birth of the universe three of the dimensions expanded, leaving the other six dimensions curled up in a little ball too small to be observed today. The physical object in this space is not a point but is more like a smoke ring called a *superstring*. When the features of superstrings are enlarged a billion, billion times they mathematically begin to have the properties of elementary particles such as the electron.

The superstring mathematical model has attracted attention because it can mathematically describe all of the fundamental forces without producing the result of infinity as the general theory of relativity does. The superstring theory considers matter in the form of elementary particles to be one-dimensional strings rather than points. These strings are very small, averaging about 10^{-33} cm. How they vibrate determines which elementary particle is represented. The fundamental forces are determined by how they divide into two strings or come together to form one string. All of this is a mathematical model, but a successful model that proposes how all the elementary particles and all the fundamental forces can originate from a single source.

The superstring theory is a mathematically consistent theory that seems to unify space, matter, and the properties of matter. However, the existence of six hidden extra dimensions and the symmetry requirement of invisible matter sounds much like the makings of a good science fiction plot. On the other hand, invisible matter could mean invisible stars, galaxies, and planets, which would explain the unaccounted bending of light that is sometimes observed as it passes through seemingly empty space. Also, perhaps the invisible matter is the "missing" matter that is needed for the oscillating theory of the universe to be accepted. Nonetheless, the existence of invisible matter and of tiny smoke rings of matter that are about 10^{-33} cm long will be very difficult to verify. Energies that can only be provided in the next generation of atom accelerators, such as the Superconducting Supercollider, will be necessary to verify some of the predictions of the superstring theory. Verification would produce a major revolution in physics and astronomy, of course, perhaps solving the mysteries of the universe within the next century.

Perhaps some new, future observations will conclusively favor either the oscillation or continued expansion model. Or perhaps new information will lead to a whole new model, which is how science progresses as humans continually attempt to understand the universe and their place in it.

Summary

The early conception of the universe was *geocentric,* and the ancient Greeks pictured the stars and planets as being attached to a *celestial sphere* that rotated around the earth. Today, the idea of a celestial sphere is still used to locate stars and plot how they move.

The distance between the stars as they appear on the celestial sphere is measured by a segment of the circumference on the sphere. There are 360° in a circle and 1/360 of the circumference is called *1 degree (1°) of arc,* with subdivisions called *arc minutes* and *arc seconds.* The apparent size of an object is measured by the angle between the sides of the object, which is called the *angular size* or the *angular diameter.*

The distance to a relatively close star can be measured by *parallax,* the apparent shift of position of an object against a background. The radius of the earth's orbit is called one *astronomical unit,* or A.U. When a baseline of 1 A.U. is used and the apparent shift of position is 1 arc second, the distance is defined as the astronomical unit of distance called a *parsec.* Another unit of astronomical distance is the *light-year,* the distance that light travels in one year.

Stars are theoretically born in clouds of hydrogen gas and dust in the space between other stars. Gravity pulls huge masses of hydrogen gas together into a *protostar,* a mass of gases that will become a star. The protostar contracts, becoming increasingly hotter at the center, eventually reaching a temperature high enough to start *nuclear fusion* reactions between hydrogen atoms. Pressure from hot gases balances the gravitational contraction, and the average newborn star will shine quietly billions of years. The average star has a dense, hot *core* where nuclear fusion releases radiation, a less dense *radiation zone* where radiation moves outward, and a thin *convection zone* that is heated by the radiation at the bottom, then moves to the surface to emit light to space.

The brightness of a star is related to the amount of energy and light it is producing, the size of the star, and the distance to the star. The *apparent magnitude* is the brightness of a star as it appears to you. To compensate for differences in brightness due to distance, astronomers calculate the brightness that stars would have at 10 parsecs (32.6 light-years). This standard-distance brightness is called the *absolute magnitude.* Absolute magnitude is an expression of *luminosity,* the total amount of energy radiated into space each second from the surface of a star.

Stars appear to have different colors because they have different surface temperatures. A graph of temperature by spectral types and brightness by absolute magnitude is called the *Hertzsprung-Russell diagram,* or H-R diagram for short. Such a graph shows that normal, mature stars fall on a narrow band called the *main sequence* of stars. Where a star falls on the main sequence is determined by its brightness and temperature, which in turn are determined by the mass of the star. Other groups of stars on the H-R diagram have different sets of properties that are determined by where they are in their evolution.

The life of a star consists of several stages, the longest of which is the *main sequence* stage after a relatively short time as a *protostar.* After using up the hydrogen in the core, a star with an average mass expands to a *red giant,* then blows off the outer shell to become a *white dwarf star,* which slowly cools to a black dwarf. The blown-off outer shell forms a *planetary nebula,* which disperses over time to become the gas and dust of interstellar space. More massive stars collapse into a *neutron star* or a *black hole* after a violent *supernova* explosion. Neutron stars are believed to the the source of *pulsars,* regular equally spaced radio signals from space.

Galaxies are the basic units of the universe. The Milky Way galaxy has three distinct parts: (1) the *galactic nucleus,* (2) a rotating *galactic disk,* and (3) a *galactic halo.* The galactic disk contains subgroups of stars that move together as *galactic clusters.* The halo contains symmetrical and tightly packed clusters of millions of stars called *globular clusters.*

All the billions of galaxies can be classified into groups of four structures: *elliptical, spiral, barred,* and *irregular.* Evidence from four different astronomical and physical "clocks" indicates that the galaxies formed some twenty billion years ago, expanding ever since from a common origin in a *big bang.* The *big bang theory* predicts that the universe will slowly die out as more and more matter becomes locked up in cooling white dwarfs and black holes. The *oscillating theory* of the universe predicts that the universe will be drawn together to eventually begin again with another big bang. Measurements and estimates currently indicate that the density of matter in the universe is not sufficient to draw the universe back together from its current expansion.

Key Terms

absolute magnitude (p. 365)	light-year (p. 362)
apparent magnitude (p. 364)	luminosity (p. 365)
astronomical unit (p. 362)	main sequence stars (p. 366)
big bang theory (p. 374)	nebulae (p. 363)
black hole (p. 370)	neutron star (p. 370)
celestial equator (p. 360)	north celestial pole (p. 360)
celestial meridian (p. 361)	nova (p. 368)
cepheid variables (p. 368)	oscillating theory (p. 375)
constellations (p. 358)	parsec (p. 362)
convection zone (p. 364)	protogalaxies (p. 375)
core (p. 364)	protostar (p. 363)
galactic clusters (p. 371)	pulsars (p. 370)
galaxy (p. 370)	radiation zone (p. 364)
geocentric (p. 358)	red giant stars (p. 367)
globular clusters (p. 371)	south celestial pole (p. 361)
Hertzsprung-Russell diagram (p. 366)	supernova (p. 370)
	white dwarf stars (p. 367)

Applying the Concepts

1. A planet that is undergoing retrograde motion appears to stop moving when the earth, in its orbit,
   a. is catching up with the other planet in its orbit.
   b. has drawn even with the other planet in its orbit.
   c. has passed the other planet in its orbit.
   d. none of the above.

2. Stars twinkle and planets do not twinkle because
   a. planets shine by reflected light and stars produce their own light.
   b. all stars are pulsing light sources.
   c. stars appear as point sources of light and planets are disc sources.
   d. all of the above.

3. How much of the celestial meridian can you see from any given point on the surface of the earth?
   a. one-fourth
   b. one-half
   c. three-fourths
   d. all of it

4. Which of the following of the coordinate system of lines depends on where you are on the surface of the earth?
   a. celestial meridian
   b. celestial equator
   c. north celestial pole
   d. none of the above

5. The angle that you see Polaris, the North Star, above the horizon is about the same as your approximate location on
   a. the celestial meridian.
   b. the celestial equator.
   c. a northern longitude.
   d. a northern latitude.

6. If you were at the north celestial pole looking down on the earth, how would it appear to be moving? (Use a globe if you wish.)
   a. clockwise
   b. counterclockwise
   c. it would not appear to move from this location

7. Your answer to question 6 means that the earth turns
   a. from the west toward the east.
   b. from the east toward the west.
   c. not at all.

8. Your answer to question 7 means that the moon, sun, and stars that are not circumpolar appear to rise in the
   a. west, move in an arc, then set in the east.
   b. east, move in an arc, then set in the west.

9. The basic property of a main sequence star that determines most of its other properties, including its location on the H-R diagram, is
   a. brightness.
   b. color.
   c. temperature.
   d. mass.

10. All the elements that are more massive than the element iron were formed in
   a. a nova.
   b. a white dwarf.
   c. a supernova.
   d. the birth of a star.

Answers

1. b 2. c 3. b 4. a 5. d 6. b 7. a 8. b 9. d
10. c

Questions for Thought

1. Describe briefly how the people of ancient civilizations viewed the universe.

2. What is a constellation? For what purpose are constellations used?

3. What was the origin of the zodiac? What was the origin, in general, of the meaning of the signs of the zodiac?

4. How did the present system of twelve months to a year and seven days to a week originate? Suggest a reason why the number seven is considered by some people today to be a mystical, lucky number.

5. Explain how the ancient Greeks reasoned the distances to the seven celestial bodies to create their model of the celestial bodies moving on concentric shells that turned around the earth.

6. Explain why a geocentric model of the universe was so common among the ancient civilizations.

7. Would you ever observe the sun to move along the celestial meridian? Explain.

8. What is the meaning of a distance between stars that is expressed in degrees of arc? What are arc minutes and arc seconds?

9. What is a parsec and how is it defined? What is a light-year and how is it defined?

10. Why are astronomical distances not measured with standard referent units of distance such as kilometers or miles?

11. Explain why a protostar heats up internally as it gravitationally contracts.

12. About how much time is required for the energy released from nuclear fusion reactions in the center part of an average star such as the sun to reach the surface to be emitted as light?

13. Describe in general the structure and interior density, pressure, and temperature conditions of an average star such as the sun.

14. Which size of star has the longest life span, a star sixty times more massive than the sun, one just as massive as the sun, or a star that has a mass of one-twenty-fifth of the sun? Explain.

15. What is the difference between apparent magnitude and absolute magnitude?

16. What does the color of a star indicate about the surface temperature of the star? What is the relationship between the temperature of a star and the spectrum of the star? Describe in general the spectral classification scheme based on temperature and stellar spectra.

17. What is the Hertzsprung-Russell diagram a diagram of? What is the significance of the diagram?

18. What is meant by the main sequence of the H-R diagram? What one thing determines where a star is plotted on the main sequence?

19. Describe in general the life history of a star with an average mass like the sun.

20. What, if anything, is the meaning of the Hubble classification scheme of the galaxies?

21. What is a nova? What is a supernova?

22. Describe the theoretical physical circumstances that lead to the creation of (a) a white dwarf star, (b) a red giant, (c) a neutron star, (d) a black hole, (e) a supernova.

23. Describe the two forces that keep a star in a balanced, stable condition while it is on the main sequence. Explain how these forces are able to stay balanced for a period of billions of years or longer.

24. What is the source of all the elements in the universe that are more massive than helium but less massive than iron? What is the source of all the elements in the universe that are more massive than iron?

25. Why must the internal temperature of a star be hotter for helium fusion reactions than for hydrogen fusion reactions?

26. When does a protostar become a star? Explain.

27. What is a red giant star? Explain the conditions that lead to the formation of a red giant. How can a red giant become brighter than it was as a main sequence star if it now has a lower surface temperature?

28. Why is an average star like the sun unable to have carbon fusion reactions in its core?

29. Describe the structure of the Milky Way galaxy. Where are new stars being formed in the Milky Way? Explain why they are formed in this part of the structure and not elsewhere.

30. If the universe is expanding are the galaxies becoming larger? Explain.

31. What is the evidence that supports a big bang theory of the universe?

32. How is the oscillating theory of the universe different from the big bang theory? What evidence is needed before the oscillating theory of the universe will be acceptable?

Chapter 19

The Solar System

Figure 19.1
This view of the rising earth was seen by the *Apollo 11* astronauts after they entered orbit around the moon. The earth is just above the lunar horizon in this photograph.

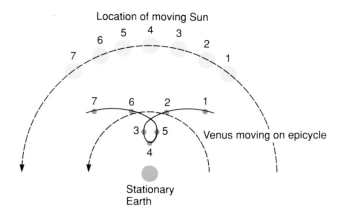

Figure 19.2
The paths of Venus and the Sun at equal time intervals according to the Ptolemaic system. The combination of epicycle and Sun movement explains retrograde motion with a stationary Earth.

FOR generations people have observed the sky in awe, wondering about the bright planets moving across the background of stars, but they could do no more than wonder. You are among the first generations on the earth to see close-up photographs of the planets and to see earth as it appears from space. Spacecrafts have now made thousands of photographs of the other planets and their moons, measured properties of the planets, and in some cases, studied their surfaces with landers. Astronauts have left earth and visited the moon, bringing back rock samples, data, and photographs of earth as seen from the moon (figure 19.1). All of these photographs and findings have given your generation a unique new perspective of the earth, the planets, and the moons, comets, and asteroids that make up the Solar System.

Viewed from the moon, earth is a spectacular blue globe with expanses of land and water covered by huge changing patterns of white clouds. Viewed from a spacecraft, the planets present a very different picture, each unique in its own way. Mercury has a surface that is covered with craters, looking very much like the surface of Earth's Moon. Venus is covered with clouds of sulfuric acid over an atmosphere of mostly carbon dioxide, which is under great pressures with surface temperatures hot enough to melt lead. The surface of Mars has great systems of canyons, inactive volcanos, and surprisingly, dry riverbeds and tributaries. The giant planets of Jupiter and Saturn have orange, red, and white bands of organic and sulfur compounds and storms with gigantic lightning discharges compared to anything ever seen on Earth. One moon of Jupiter has active volcanos spewing out molten sulfur and sulfur dioxide. The outer giant planets of Uranus and Neptune have moons and particles in rings that appear to be covered with powdery, black carbon.

These and many more findings, some fascinating surprises and some expected, have stimulated the imagination as well as added to our comprehension of the frontiers of space. The new information about the sun's impressive system of planets, moons, comets, and asteroids has also added to speculations and theories about the planets and how they evolved over time in space. This information, along with the theories and speculations, will be presented in this chapter to give you a picture of the earth's immediate neighborhood, the Solar System.

Ideas about the Solar System

If you observe the daily motion of the sun and the nightly motion of the moon and stars over a period of time, you can easily convince yourself that all the heavenly bodies revolve around a fixed, motionless earth. The sun, the moon, and the planets do appear to move from east to west across the sky, and the stars seem to be fixed on a turning sphere, maintaining the same positions relative to one another as they move as units on the turning sphere. To consider that the earth moves rather than the sun, the moon, and the planets seems contrary to all these observations.

There is a problem with the observed motion of the planets, however, one troublesome observation that spoils this model of a motionless earth with everything moving around it. When the planets are observed over a period of a year or more, they are observed not to hold the same relative positions as would be expected of this model. The stars and the planets both rise in the east and set in the west, but the planets do not maintain the same positions relative to the background of stars. Over time, the planets appear to move across the background of stars, sometimes slowing, then reversing their direction in a loop before resuming their normal motion. This *retrograde*, or reverse, motion for the planet Mars is shown in figure 18.6.

The Geocentric Model

Early Greek astronomers and philosophers had attempted to explain the observed motions of the sun, moon, and stars with a geometric model, a model of perfect geometrical spheres with attached celestial bodies rotating around a fixed earth in perfect circles. To explain the occasional retrograde motion of the planets with this model, later Greek astronomers had to modify it by assuming a secondary motion of the planets. The modification required each planet to move in a secondary circular orbit as it moved along with the turning sphere. The small circular orbit, or **epicycle,** was centered on the surface of the sphere as it turned around the earth. The planet was understood to move around the epicycle once as the sphere turned around the earth. The combined motion of the movement of the planet around the epicycle as the sphere turned resulted in a loop with retrograde motion (figure 19.2). Thus the earlier model of the Solar System, modified with epicycles, was able to explain all that was known about the movement of the stars and the seven heavenly bodies and all that was observed during the days of the later ancient Greek civilization.

A version of the explanation of retrograde motion—perfectly circular epicycle motion on perfectly spherical turning spheres—was published by Ptolemy in the second century A.D. and came to be known as the **Ptolemaic system.** This system did

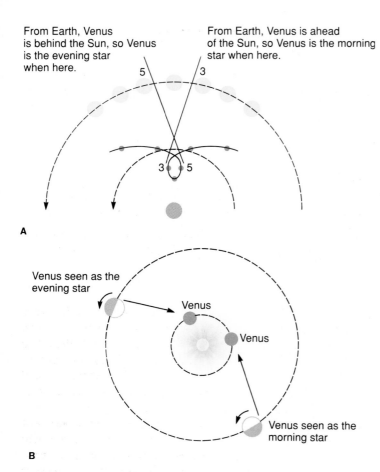

From Earth, Venus is behind the Sun, so Venus is the evening star when here.

From Earth, Venus is ahead of the Sun, so Venus is the morning star when here.

A

Venus seen as the evening star

Venus

Venus

Venus seen as the morning star

B

Figure 19.3

(a) The Ptolemaic system explanation of Venus as a morning star and evening star. (b) The heliocentric system explanation of Venus as a morning star and evening star.

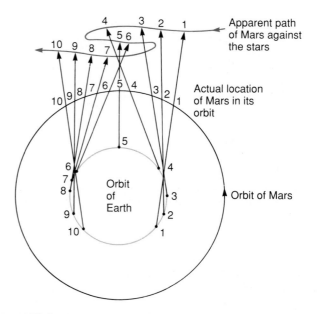

Apparent path of Mars against the stars

Actual location of Mars in its orbit

Orbit of Earth

Orbit of Mars

Figure 19.4

The heliocentric system explanation of retrograde motion.

account for the facts that were known about the Solar System at that time. Not only did the system describe retrograde motion with complex paths as shown in figure 19.2, but it also explained other observations such as why Venus is sometimes observed for a short time near the Sun at sunrise (the morning "star") and other times observed near the Sun for a short time at sunset (the evening "star") (figure 19.3).

Over the years, newly discovered inconsistencies of observed and predicted positions of the planets were discovered, and epicycles were added to epicycles in an attempt to fit the system with observations. The system became increasingly complicated and unmanageable, but it did seem to agree with other ideas of that time. The agreements were that (1) humans were at the center of a universe that was created for them and (2) heaven was a perfect place, so it would naturally be a place of perfectly circular epicycles moving on perfect spheres, which were in turn moving in perfect circles. Thus the Ptolematic

system of a geometric, geocentric Solar System came to be supported by the Church, and this was the accepted understanding of the Solar System for the next fourteen centuries.

The Heliocentric Model

The idea that the earth revolves around the sun rather than the sun moving around the earth was proposed by a Polish astronomer, Nicolas Copernicus, in a book published in 1543. In his book, Copernicus pointed out that the observed motions of the planets could be explained by a model of the earth and other planets revolving around the sun as well as by the Ptolemaic system. In this model each planet moved around the sun in perfect circles at different distances, moving at faster speeds in orbits closer to the sun. When viewed from a moving earth, the other planets would appear to undergo retrograde motion because of the combined motions of the earth and the planets. The Earth, for example, moves along its inner orbit about twice as fast as Mars moves along its outer orbit, which is about one and one-half times farther from the Sun. Since the Earth is moving faster in the inside orbit it will move even with, then pass, Mars in the outer orbit (figure 19.4). As this happens Mars will appear to slow to a stop, move backward, stop again, then move on. This combined motion is similar to what you observe when you pass a slower moving car, which appears to move backward against the background of the landscape as you pass it. In an outer circular path the car would appear to slow, move backward, then forward again as you pass it.

The **Copernican system** of a heliocentric, or sun-centered, Solar System provided a simpler explanation for retrograde motion than the Ptolemaic system, but it was only an alternative way to consider the Solar System. The Copernican system offered no compelling reasons why the alternative Ptolematic

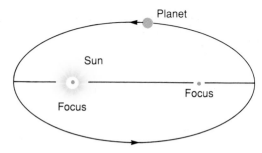

Figure 19.5

Kepler's first law describes the shape of a planetary orbit as an ellipse, which is exaggerated in this figure. The sun is located at one focus of the ellipse.

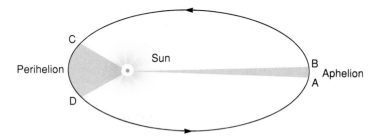

Figure 19.6

Kepler's second law. A line from the sun to a planet at point A sweeps over a certain area as the planet moves to point B in a given time interval. A line from the sun to a planet at point C will sweep over the same area as the planet moves to point D during the same time interval. The time required to move from point A to point B is the same as the time required to move from point C to point D, so the planet moves faster in its orbit at perihelion.

system should be rejected. Furthermore, Copernicus had retained the old Greek idea of planets moving in perfect circles, so there were inconsistencies in predicted and observed motions with this model, too. Clear-cut evidence for rejecting the Ptolemaic system would have to await the detailed measurements of planetary motions made by Tycho Brahe and analysis of those measurements by Johannes Kepler.

Tycho Brahe was a Danish nobleman who constructed highly accurate observatories for his time, which was before the time of the telescope. From his observatory on a little island about twenty miles from Copenhagen, Brahe spent about twenty years (1576–1597) making systematic, uninterrupted measurements of the sun, moon, planets, and stars. His skilled observations resulted in the first precise, continuous record of planetary position. In 1600 Brahe hired a young German, Johannes Kepler, as an assistant. When Brahe died in 1601, Kepler was promoted to Brahe's position and was given access to the vast collection of observation records. Kepler spent the next twenty-five years analyzing the data to find if planets followed circular paths or if they followed the paths of epicycles. Using the careful observations of Tycho Brahe, Kepler found that the planets did not move in epicycles nor did they move in perfect circles. Planets move in the path of an *ellipse* of certain dimensions. He published his findings in 1609 and 1619, establishing the actual paths of planetary movement. Kepler had found the first evidence that the Ptolemaic system of complicated epicycles was unnecessary and unacceptable as a model of the Solar System. His findings also required adjustments of the heliocentric Copernican system, which are described by his three laws of planetary motion. Today, his findings are called **Kepler's laws of planetary motion.**

Kepler's first law states that each planet moves in an orbit that has the shape of an ellipse, with the sun located at one focus (figure 19.5). **Kepler's second law** states that an imaginary line between the sun and a planet moves over equal areas of the ellipse during equal time intervals (figure 19.6). This means that the orbital velocity of a planet varies with where the planet is in the orbit, since the distance from the focus to a given position

varies around the ellipse. The point at which an orbit comes closest to the sun is called the **perihelion,** and the point at which an orbit is farthest from the sun is called the **aphelion.** The shortest line from a planet to the sun at perihelion means that the planet moves most rapidly when here. The short line and rapidly moving planet would sweep out a certain area in a certain time period, for example, one day. The longest line from a planet to the sun at aphelion means that the planet moves most slowly at aphelion. The long line and slowly moving planet would sweep out the same area in one day as was swept out at perihelion. The earth travels fastest in its orbit at perihelion on about January 3 and slowest at aphelion on about July 1.

Kepler's third law states that the time for a planet to complete one orbit (the period) is related to the radius of the major axis (the longest axis of the ellipse) according to $t^2 \propto d^3$. When the time is expressed in the earth units of one year for a revolution and a radius of one astronomical unit, the distance to a planet can be determined by observing the period of revolution and comparing the orbit of the planet with that of the earth. For example, suppose a planet is observed to require eight earth years to complete one orbit. Then,

$$\frac{t \text{ (planet)}^2}{t \text{ (earth)}^2} = \frac{d \text{ (planet)}^3}{d \text{ (earth)}^3}$$

$$\frac{(8)^2}{1} = \frac{(\text{distance})^3}{1}$$

$$64 = (\text{distance})^3$$

$$\text{distance} = \sqrt[3]{64} \text{ A.U.}$$

$$\text{distance} = 4 \text{ A.U.}$$

Thus a planet that takes eight times as long to complete an orbit is four times as far from the sun as the earth. In general, Kepler's third law means that the more distant a planet is from the sun, the longer the time required to complete one orbit.

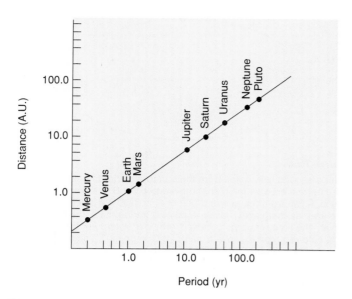

Figure 19.7

Kepler's third law describes a relationship between the time required for a planet to move around the sun and its average distance from the sun. The relationship is that the time squared is proportional to the distance cubed.

Figure 19.7 shows this relationship for the planets of the Solar System. Kepler's third law applies to moons, satellites, and comets in addition to the planets. In the case of moons and satellites, the distance is to the planet the moon or satellite is orbiting, not to the sun.

Kepler's laws were empirically derived from the data collected by Tycho Brahe, and the reason why planets followed these relationships would not be known or understood until Isaac Newton published the law of gravitation some sixty years later (see chapter 3). In the meantime, Galileo constructed his telescope and added observational support to the heliocentric theory. By the time Newton derived the law of gravitation and then improved the accuracy of Kepler's third law, the Solar System was understood to consist of planets that move around the sun in elliptical orbits, paths that could be predicted by applying the law of gravitation. The heliocentric model of the Solar System had evolved to a conceptual model that both explained and predicted what was observed.

Origin of the Solar System

Kepler empirically deduced the geometric laws of planetary motion from the continuous record of precise measurements made by Brahe. Later, these same laws were derived theoretically from Newton's law of gravitation. Thus the heliocentric model of the Solar System was formulated from (1) *observations and measurements* of the movement of the planets over a period of time, (2) *analysis* of these measurements, and (3) empirical deductions, then *verification* of the model from independent and theoretically based calculations that fit the observed measurements. Thus the heliocentric model is accepted because the model can be independently tested and verified by observation.

The theoretical model of the life of a star (see chapter 18), such as the sun, presents a different kind of problem in testing and verification. The life of a star takes place over billions of years, so it is not possible to make observations and measurements to test the model. However, there are billions of stars in various stages of their life cycles that can be observed and measured to test the model of the life cycle. Thus the model of the life of the sun is acceptable because the model can be verified by observing other stars in different stages of their life cycles.

A model of how the Solar System originated presents still another kind of problem in testing and verification. This problem is that the Solar System originated a long time ago, some five billion years ago according to a number of different independent sources of evidence, and that there are no other planetary systems that can be observed, either in existence or in the process of being formed. Other planetary systems are believed to exist, and periodically evidence is found that implies their existence, but the fact remains that other planetary systems are just too far away and too faint to be visible.

Models about the origin of the Solar System are not clear-cut and accepted by everyone because of the difficulties of collecting data and verification. The sun-centered model of the Solar System enjoys general acceptance because data can be collected to verify and support this model. The model of the life of a star, on the other hand, is generally accepted because data and observations support the model even though questions still exist because some of the details are difficult to observe. The most widely accepted theory of the origin of the Solar System is called the **protoplanet nebular model**. A *protoplanet* is the earliest stage in the formation of a planet. The model can be considered in stages, which are not really a part of the model but are simply a convenient way to organize the total picture (figure 19.8).

The first important event in the formation of the Solar System involves stars that disappeared billions of years ago, long before the sun was born. The earth, the other planets, and all the members of the Solar System are composed of elements that were manufactured by these former stars. In a sequence of nuclear reactions, which was described in chapter 18, hydrogen fusion in the core of large stars results in the formation of the elements up to iron. Elements heavier than iron are formed in rare supernova explosions of dying massive stars. Thus *stage A* of the formation of the Solar System consisted of the formation of elements heavier than hydrogen in many, many previously existing stars, including the supernovas of more massive stars. Many stars had to live out their life cycles to provide the raw materials of the Solar System. The death of each star, including supernovas, added newly formed elements to the accumulating gases and dust in interstellar space. Over a long period of time these elements began to concentrate in one region of space as dust, gases, and chemical compounds, but hydrogen was still the most abundant element in the nebula that was destined to become the Solar System.

During *stage B*, the hydrogen gas, dust, elements, and chemical compounds from former stars began to form a large, slowly rotating nebula that was much, much larger than the present Solar System. Under the influence of gravity, the large but diffuse, slowly rotating nebula began to contract, increasing its rate of spin. The largest mass pulled together in the center, contracting to the protostar, which eventually would become the sun. The remaining gases, elements, and dusts formed an enormous, fat, bulging disk called an **accretion disk,** which would eventually form the planets and smaller bodies. The fragments of dust and other solid matter in the disk began to stick together in larger and larger accumulations from numerous collisions. All of the present-day elements of the planets must have been present in the nebula along with the most abundant elements of hydrogen and helium. The elements and the familiar chemical compounds accumulated into basketball-sized chunks of matter or larger. You could speculate about how and when the chemical compounds were formed and also about how the chunks of matter could come together to form what would become the planets and smaller bodies of the Solar System. Logical arguments could be made for completely different speculations. You could speculate, for example, that if the chunks of matter were cold and icy, they would be more likely to stick when pulled together by gravity. Thus over a period of time, perhaps 100 million years or so, huge accumulations of frozen water, frozen ammonia, and frozen crystals of methane began to accumulate, together with silicon, aluminium, and iron oxide plus other metals in the form of rock and mineral grains. Such a slushy mixture would no doubt have been surrounded by an atmosphere of hydrogen, helium, and other vapors thinly interspersed with smaller rocky grains of dust. Local concentrations of certain minerals might have occurred throughout the whole accretion disk, with a greater concentration of iron, for example, in the disk where the protoplanet Mars was forming compared to where the protoplanet Earth was forming. All of the protoplanets, however, might have started out somewhat similarly as huge accumulations of a slushy mixture with an atmosphere of hydrogen and helium gases. Gravitational attraction must have compressed the protoplanets as well as the protosun. During this period of contraction and heating gravitational adjustments continue, and about a fifth of the disk nearest to the protosun must have been pulled into the central body of the protosun, leaving a larger accumulation of matter in the outer part of the accretion disk.

During *stage C*, the warming protosun became established as a star, perhaps undergoing an initial flare-up that has been observed today in other newly forming stars. Such a flare-up might have been of such a magnitude that it blasted away the hydrogen and helium atmospheres of the interior planets out past Mars, but it did not reach far enough out to disturb the hydrogen and helium atmospheres of the outer planets. The innermost of the outer planets, Jupiter and Saturn, might have acquired some of the matter blasted away from the inner planets, becoming the giants of the Solar System by comparison. This

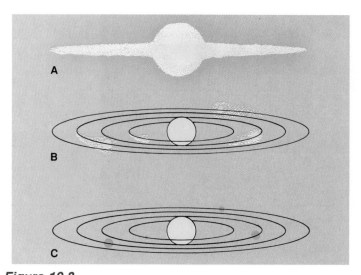

Figure 19.8

Formation of the Solar System according to the protoplanet nebular model, not drawn to scale. (*a*) The process starts with a nebula of gas, dust, and chemical elements from previously existing stars. (*b*) The nebula is pulled together by gravity, collapsing into the protosun and protoplanets. (*c*) As the planets form, they revolve around the sun in orbits that are described by Kepler's laws of planetary motion.

is just speculation, however, and the two giants may have simply formed from greater concentrations of matter in that part of the accretion disk.

The evidence shows that the protoplanets, some more than others, underwent heating early in their formation. Much of the heating may have been provided by gravitational contraction, the same process that gave the protosun sufficient heat to begin its internal nuclear fusion reactions. Heat was also provided from radioactive decay processes inside the protoplanets, and the initial greater heating from the sun may have played a role in the protoplanet heating process. Larger bodies were able to retain this heat better than smaller ones, which radiated it to space more readily. Thus the larger bodies underwent a more thorough heating and melting, perhaps becoming completely molten early in their history. In the larger bodies the heavier elements, such as iron, were pulled to the center of the now molten mass, leaving the lighter elements near the surface. The overall heating and cooling process took millions of years as the planets and smaller bodies were formed. Gases from the hot interiors formed secondary atmospheres of water vapor, carbon dioxide, and nitrogen on the larger interior planets. How these initial secondary atmospheres evolved into the present-day atmospheres of Mars, Venus, and Earth will be discussed in the next section.

In addition to the planets, the existence of other, smaller bodies of the Solar System is explained by the protoplanet nebular model. **Comets,** for example, are believed today to originate from clusters of condensed ices and solids that remain from

Table 19.1

Distances from the sun to planets known in the 1790s

Planet	n	Distance Predicted by (d + 4)/10 (A.U.)	Actual Distance (A.U.)
Mercury	0	0.4	0.39
Venus	3	0.7	0.72
Earth	6	1.0	1.0
Mars	12	1.6	1.5
(Asteroid belt)	24	2.8	—
Jupiter	48	5.2	5.2
Saturn	96	10.0	9.5
Uranus	192	19.6	19.2

the outer regions of the original condensed nebula. These clusters did not undergo melting as did the larger bodies and exist today as a cloud of icy, snowlike bodies that slowly move around the outer limits of the Solar System in a spherical shell. How these icy bodies become comets and what eventually happens to them is discussed in a later section of this chapter.

Between the orbits of Mars and Jupiter is a belt of thousands of small rocky bodies called **asteroids.** Interestingly, the belt of asteroids was discovered from a prediction made by the German astronomer Bode at the end of the eighteenth century. Bode had noticed a pattern of regularity in the spacing of the planets that were known at the time. He found that by expressing the distances of the planets from the sun in astronomical units, these distances could be approximated by the relationship (n + 4)/10, where n is a number in the sequence 0, 3, 6, 12, and so on where each number (except the first) is doubled in succession. When these calculations were done the distances turned out to be very close to the distances of all the planets known at that time, but the numbers also predicted a planet between Mars and Jupiter where there was none. Later a belt of asteroids was found where the Bode numbers predicted there should be a planet. This suggested to some people that a planet had existed between Mars and Jupiter in the past, and somehow this planet was broken into pieces, perhaps by a collision with another large body (table 19.1).

Such patterns of apparent regularity in the spacing of planetary orbits were of great interest because, if a true pattern existed, it may hold meaning about the mechanism that determined the location of planets at various distances from the sun. Many attempts have been made to explain the mechanism of planetary spacing and why a belt of asteroids exists where the Bode numbers predict there should be a planet. The most successful explanations concern Jupiter and the influence of its gigantic gravitational field on the formation of clumps of matter at certain distances from the Sun. In other words, a planet does not exist today between Mars and Jupiter because there never was a planet there. The gravitational influence of Jupiter prevented the clumps of matter from joining together to form a planet, and a belt of asteroids formed instead.

The protoplanet nebular model seems to account for the origin and nature of the planets observed today, the origin of comets, and the existence of the asteroid belt. Yet, there are gaps in the model, with missing explanations of how, for example, the earth's moon formed. On the other hand, relatively recent information from spacecraft missions to the planets has revealed much information about the planets that seems to agree with the model. This information is included with the survey of the planets other than the earth, which follows in the next section.

The Planets

Today, the Solar System consists of a middle-aged main sequence G type star called the sun with nine planets, nearly fifty moons, thousands of asteroids, and many comets all revolving around it (figure 19.9). The sun has an average mass compared to the other main sequence stars, about halfway between the most massive and the least massive. It is, however, 333,400 times more massive than the earth and 1,000 times more massive than Jupiter, the largest planet. The sun, in fact, has 700 times the mass of all the planets, moons, and minor members of the Solar System together. It is the force of gravitational attraction between the comparatively massive sun and the rest of the Solar System that holds it all together. This force of gravitational attraction varies with the product of the mass of each planet and the sun and the distance between them (see chapter 3).

The motion of a planet as it orbits the sun is called a **revolution,** and it takes the earth one year to make one revolution around the sun. If you imagine the earth cutting out a large circle of paper as it revolves around the sun you will have a nearly circular sheet that is an approximate average of 149.6 million km (92.96 million mi) from the edge to the sun. This distance is called one astronomical unit, or A.U. The surface of the imagined near-circle of paper is the *plane of the orbit*. The plane of the earth's orbit is called the *ecliptic*. All of the planets have orbital planes that are within a few degrees of the ecliptic except Mercury and Pluto (the innermost and outermost planets).

The planets also have a second kind of motion because they turn, or spin, like a top as they revolve around the sun. The earth turns completely around every twenty-four hours, and the daily spinning is called **rotation.** The earth rotates around an imaginary line that passes through the geographic North and South Poles. The imaginary line is called the **axis,** and the earth rotates around its axis once a day (figure 19.10).

Table 19.2 compares the basic properties of the nine planets with properties of the orbital plane, period of revolution, and mass of each compared to earth. From this table you can see that the planets can be classified into two major groups based on size, density, and atmospheric chemical composition. The interior planets of Mercury, Venus, and Mars have similar densities and similar compositions that resemble the Earth, so these planets, along with the Earth, are known as the **terrestrial planets.**

Figure 19.9

The order of the planets out from the sun. The orbits and the planet sizes are not drawn to scale.

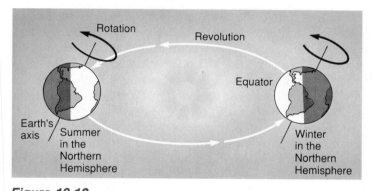

Figure 19.10

The earth's rotation around its axis results in day and night. One year is required for one revolution.

Outside the orbit of Mars are four **giant planets,** which have a similar density and chemical composition. The terrestrial planets are mostly composed of rocky materials and iron with a density range of about 4 to 5.5 g/cm³. The giant planets of Jupiter, Saturn, Uranus, and Neptune, on the other hand, are massive giants mostly composed of hydrogen, helium, and methane with a density of less than 2 g/cm³. The density of the giant planets suggests the presence of rocky materials and iron as the terrestrial planets have, but in this case as a core surrounded by a deep layer of compressed gases beneath a deep atmosphere of vapors and gases.

The terrestrial planets are separated from the giant planets by the asteroid belt. The outermost planet, Pluto, has properties that do not fit with either the terrestrial planets or the giant planets and has very strange orbital properties. This has led to the suggestion that perhaps Pluto is not a true planet at all, but it may be a captured body that orbits the sun.

As you found in the previous section, the protoplanet nebular model leads to speculative ideas about how the terrestrial planets came to be so different from the giant planets. In general, these ideas concern (1) the loss of matter from the interior planets to the sun early in the nebular stage, (2) the loss of the primary, or first, atmospheres of the interior planets as they were "blasted away" by an early flare-up of the newly forming sun, and (3) different initial amounts of melting in the larger bodies compared to the smaller bodies because of different amounts of heating from compression and radioactive decay compared to the loss of heat radiated to space. When rock materials are heated to melting, gases and water vapor are expelled from the molten mixture by a process called **degassing.** This process is still going on today on the earth as volcanoes emit much degassed water and carbon dioxide along with molten lava and ash. The idea here is that this same degassing process occurred in the past. The amount of melting was thus related to the size of a planet, or body and greater amounts of melting would have resulted in greater amounts of degassing. When this idea is coupled with the observation that more massive bodies have greater

Table 19.2

Properties of the planets

	Mercury	Venus	Earth	Mars	Jupiter	Saturn	Uranus	Neptune	Pluto
Average Distance from the Sun:									
in 10^6 km	58	108	150	228	778	1,427	2,870	4,497	5,900
in 10^6 mi	36	67	93	142	483	886	1,782	2,793	3,664
Inclination to Ecliptic	7°	3.4°	0°	1.9°	1.3°	2.5°	0.8°	1.8°	17.2°
Revolution Period (Earth Years)	0.24	0.62	1.00	1.88	11.86	29.46	84.01	164.8	247.7
Rotation Period (Earth Days, hr, min, and sec)	59 days	−243 days	23 hr 56 min 4 sec	24 hr 37 min 23 sec	9 hr 50 min 30 sec	10 hr 14 min	−0.46 days	0.67 days	6.38 days
Mass (Earth = 1)	0.05	0.82	1.00	0.11	317.9	95.2	14.6	17.2	(?)
Equatorial Diameter:									
in km	4,880	12,104	12,756	6,787	142,800	120,000	51,800	49,500	5,000(?)
in mi	3,030	7,517	7,921	4,251	88,680	75,000	32,000	31,000	3,000(?)
Density (g/cm³)	5.4	5.2	5.5	3.9	1.3	0.7	1.2	1.7	(?)
Atmosphere (Major Compounds)	None	CO_2	N_2, O_2	CO_2	H_2, He	H_2, He	H_2, He, CH_4	H_2, He, CH_4	CH_4(?)
Solar Energy Received (cal/cm²/sec)	13.4	3.8	2.0	0.86	0.08	0.02	0.006	0.002	0.001

gravity, and thus a greater ability to hold on to degassed vapors and gases, you have all the information you need to speculate how the secondary atmospheres formed on the terrestrial planets and why these atmospheres are so different today.

Smaller bodies, such as the planet Mercury and the Earth's Moon, did not produce much water through degassing since they did not undergo long-term, thorough melting. The water and gases that were released through the process were lost to space because of their low gravities, so not much of an atmosphere, if any, is found on the planet Mercury and the Earth's Moon today. The planet Mars, being more massive, did produce more water and carbon dioxide gas and was able to hold on to some of both as it formed a secondary atmosphere.

The Earth and Venus are similar in size, the two largest of the terrestrial planets, so both produced appreciable amounts of carbon dioxide and water vapor through the degassing processes. There are complications involving water and carbon dioxide on these two planets, however, because (1) both planets receive intense, energetic ultraviolet solar radiation that dissociates water molecules and (2) carbon dioxide in the atmosphere absorbs infrared radiation emitted from the surface to increase the temperature through the greenhouse effect (see chapter 23). On Venus, the ultraviolet radiation dissociated water molecules to hydrogen and oxygen gas. The less massive hydrogen gas eventually escaped to space from Venus, and the oxygen combined with other elements, becoming locked up in the surface rocks. Thus Venus was left with a secondary atmosphere consisting mostly of carbon dioxide. Meanwhile, the

Earth received less intense ultraviolet radiation than Venus since it was farther from the Sun. Much of the water vapor on the earth was able to condense, forming the oceans, which began recycling through the hydrologic cycle. Much of the carbon dioxide in the earth's atmosphere dissolved in its surface waters, reacting with minerals to become locked up as calcium carbonate (limestone). Most of the original carbon dioxide was thus removed from the early secondary atmosphere of earth. Plant life, in the meantime, was able to grow profusely in the oceans since the waters protected it from the deadly ultraviolet radiation. Plant growth slowly increased the oxygen content of the earth's atmosphere. Gradually an ozone layer accumulated in the upper atmosphere, which now absorbed the deadly ultraviolet radiation from the sun. New life forms on the land as well as the water supply were now protected from the ultraviolet radiation, and the atmosphere continued to evolve into its present-day form, consisting mostly of nitrogen and oxygen with traces of carbon dioxide and water vapor (figure 19.11).

Today, carbon dioxide is found on the earth (1) in the atmosphere in trace amounts, (2) dissolved in the oceans as a gas, (3) locked up in carbonate rocks of the surface, and (4) locked up in calcium carbonate seashells and other plant and animal structures. When all this carbon dioxide is added together, the total amount is about equal to the amount now found in the atmosphere of Venus. The formation of the earth's secondary atmosphere and its abundance of water is thus an outcome of earth (1) being sufficiently massive to generate water and carbon dioxide through degassing; (2) being sufficiently massive to keep the water vapor from escaping to space; (3) being at a distance

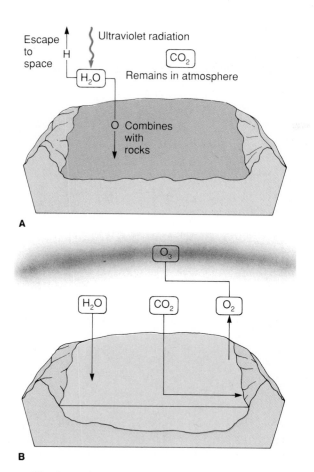

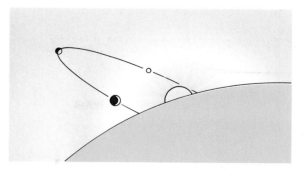

Figure 19.12
Mercury is close to the Sun and is visible only briefly before or after sunrise or sunset, showing phases. Mercury actually appears much smaller in an orbit that is not tilted as much as shown in this figure.

Figure 19.11

(a) The early secondary atmosphere of Venus lost hydrogen to space and oxygen became combined with rocks as ultraviolet radiation decomposed water molecules. (b) On Earth, water formed the oceans, carbon dioxide was removed, and plants released oxygen, which in turn formed an ozone layer in the atmosphere, protecting the water and life below.

from the sun so the water was not immediately dissociated and could form oceans that removed carbon dioxide from the atmosphere through the hydrologic cycle, which permitted it to become locked up in rock compounds. Thus with the carbon dioxide removed, the ultrahigh temperatures that occur on Venus from the greenhouse effect did not happen on the Earth. The overall outcome is that Earth has cooler temperatures than Venus, it is watery, and life occurs on it. There is no evidence of life on the other planets, and why this is so should become apparent as the other planets are considered in more detail.

Mercury

Mercury is the innermost planet, moving rapidly in a highly elliptical orbit that averages about 58 million km (about 36 million mi) from the Sun. This average distance is about 0.4 astronomical unit, or about 0.4 of the average distance of the earth from the sun. Mercury is also the smallest planet with a diameter of less than 5,000 km (about 3,000 mi), which means

that it is slightly larger than the Earth's Moon. Mercury is very bright because it is so close to the Sun, but it is difficult to observe because it only appears briefly for a few hours immediately after sunset or before sunrise. This appearance, low on the horizon, means that Mercury must be viewed through more of the Earth's atmosphere, making the study of such a small object difficult at best (figure 19.12).

In accord with Kepler's third law, as the innermost planet Mercury has the shortest period of revolution around the Sun. Mercury's period of revolution is about three Earth months at eighty-eight days, so Mercury has the shortest "year" of all the planets. With the highest orbital velocity of all the planets, Mercury was appropriately named after the mythical Roman messenger of speed. Oddly, however, this speedy planet has a rather long day in spite of its very short year. With respect to the stars, Mercury rotates once every fifty-nine days. This means that Mercury rotates three times every two orbits.

The long Mercury day with a nearby large, hot Sun in the sky means high temperatures. High temperatures mean higher gas kinetic energies, and with a low gravity, gases easily escape from Mercury so it has only trace gases for an atmosphere. The lack of an atmosphere to even the heat gains from the long days and heat losses from the long nights results in some very large temperature differences. The temperature of the surface of Mercury ranges from a high of about 500° C (about 930° F) in the sunlight to a low of about −180° C (about −350° F) on the dark side. These extreme temperatures are above the melting point of lead (328° C) and below the boiling point of liquid oxygen (−183° C).

Much of what is presently known about Mercury has been provided by radar beams bounced off the planet from radio telescopes on the Earth and by spacecraft missions to the small planet. The American spacecraft *Mariner 10* was launched in late 1973 on a trajectory that took it past Venus, then into orbit around the Sun after passing by Mercury. The spacecraft passed by Mercury three times in 1974 and 1975, transmitting data

The Solar System

Figure 19.13

A photomosaic of Mercury made from pictures taken by the *Mariner 10* spacecraft. The surface of Mercury is heavily cratered, looking much like the surface of Earth's Moon. All the interior planets and the Moon were bombarded early in the life of the Solar System.

and thousands of photographs back to Earth. The photographs transmitted by *Mariner 10* revealed that the surface of Mercury is covered with craters and the planet has a surface that very much resembles the surface of the Earth's Moon. There are large craters, small craters, superimposed craters, and craters with lighter colored rays coming from them just as craters on the moon. Also as on the earth's moon, there are hills and smooth areas with light and dark colors that were covered by lava in the past, sometimes after most of the impact craters were formed (figure 19.13).

The cratering on Mercury is believed to be related to the cratering observed on the Earth's Moon and, in fact, to the cratering that appears on all the interior planets and their satellites. Almost all of this cratering on the planets and their satellites took place about the same time in the past. Studies of the moon's craters by Apollo astronauts indicate that the bombardment took place about four billion years ago. The evidence from radioactive dating of moon rocks and from rates of erosion of the craters by micrometeorites fits with other evidence that

the craters are impact craters formed from collisions with left-over bits and pieces of rocks moving throughout the newly formed Solar System. The cratering was produced on all the planets early in the history of the Solar System. The craters on the surface of the earth were long ago obliterated by the action of water and other erosional agents that are lacking on the moon and other planets. During this same time period the rocky crust of Mercury has been weathered by solar wind and tiny micrometeorites. The surface of Mercury, as the surface of the Earth's Moon, is thus covered by a thin layer of fine dust.

Mercury has no natural satellites, or moons, it has a weak magnetic field, and it has an average density more similar to Venus or Earth than to the Moon. The presence of the magnetic field and the relatively high density for such a small body must mean that Mercury has a relatively large core of molten iron with fewer of the less dense rocks near the surface. This could mean that Mercury lost much of its less dense, outer layer of rock materials sometime during its formation.

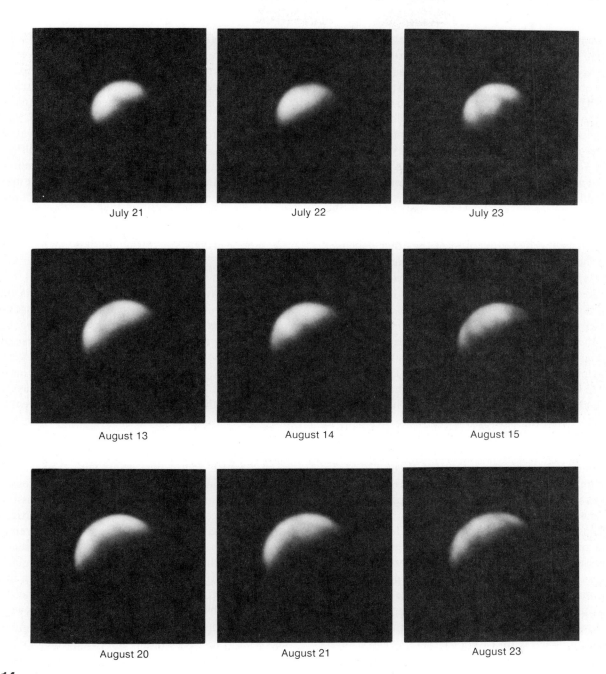

Figure 19.14

Photographs of Venus in ultraviolet light, showing cloud markings over a one-month period of time.

Venus

Venus is the brilliant evening and morning "star" that appears near sunrise or sunset, sometimes shining so brightly that you can see it while it is still daylight. Venus orbits the Sun at an average distance of about 108 million km (about 67 million mi), or 0.7 A.U. in a 225-day orbit. This means that Venus is sometimes to the left of the Sun, appearing as the evening star, and sometimes to the right of the Sun, appearing as the morning star. As figure 19.14 shows, Venus also has phases just as the Moon does. When Venus is in the full phase, however, it is small and farthest away from the Earth. A crescent Venus appears much larger and thus the brightest when it is closest to the Earth. You can see the phases of Venus with a good pair of binoculars.

Venus shines brightly because it is covered with clouds that reflect about 80 percent of the sunlight, making it the brightest object in the sky after the Sun and Moon. These same clouds prevented any observations of the surface of Venus until the early 1960s, when radar astronomers were able to penetrate the clouds and measure the planet's rotation rate. Venus was found to rotate in 243 days with respect to the stars. Since Venus has a 225-day rate of revolution, this means that each day on Venus is longer than a Venus year! Also a surprise, Venus was found to rotate in the *opposite* direction of its direction of revolution. On Venus, you would observe the Sun to rise in the west and set in the east, if you could see it, that is, through all the clouds. The backward rotation is called *retrograde rotation,* since it is opposite the rotation experienced by most of the other planets.

In addition to early studies by radio astronomers, Venus has been the target of many American and Soviet spacecraft probes. The Soviet Venera spacecraft series has landed on the surface of Venus, sending back measurements of temperature, pressure, radioactivity, and chemical data in addition to photographs of the nearby landscape. The American Mariner spacecraft series has collected data on the environment of Venus and sent back photographs of the cloud layers. The American *Pioneer* orbiter spacecraft sent probes through the atmosphere to the surface and mapped much of the surface by radar.

Venus has long been called the sister planet to the Earth since its mass, size, and density are very similar. Venus has an equatorial diameter of about 12,104 km (about 7,517 mi) compared to the earth's equatorial diameter of about 12,756 km (about 7,921 mi). Size, mass, and density, however, are where the similarities with the earth end. Spacecraft probes found a hot, dry surface under tremendous atmospheric pressure. The atmosphere consists mostly of carbon dioxide, a few percent of nitrogen, and traces of water vapor and other gases. The atmospheric pressure at the surface of Venus is almost one hundred times the pressure at the surface of the Earth, a pressure many times what a human could tolerate. The average surface temperature of about 480° C (about 900° F) is comparable to the surface temperature on Mercury, which is hot enough to melt lead. The comparably hot temperature on Venus, which is nearly twice the distance from the Sun as Mercury, is a result of the greenhouse effect. Sunlight filters through the atmosphere of Venus, warming the surface. The surface reemits the energy in the form of infrared radiation, which is absorbed by the almost pure carbon dioxide atmosphere. Carbon dioxide molecules absorb the infrared radiation, increasing their kinetic energy and thus the temperature.

The tremendously hot surface and crushing atmospheric pressure have taken their toll of space probes sent to the surface of Venus. None of the probes has operated for longer than about an hour before it ceased to function. The probes did verify the extreme temperature and pressure measurements before they gave in to the pressure, temperature, or perhaps the clouds and rain consisting of sulfuric acid. A Soviet *Venera* lander was able to transmit a few photographs before it failed.

The surface of Venus is mostly a flat, rolling plane but there are several raised areas, or "continents," on about 5 percent of the surface, a mountain larger than Mount Everest, a great valley deeper and wider than the Grand Canyon, and many large, old impact craters. In general, the surface of Venus appears to have evolved much as the surface of the Earth, but without the erosion by ice, rain, and running water.

Venus, as Mercury, does not have any satellites. Venus does not have a magnetic field as expected. Two conditions seem to be necessary in order for a planet to generate a magnetic field, a molten center part and a relatively rapid rate of rotation. With a 243-day rate of rotation, Venus rotates the slowest of all the planets and thus does not have a magnetic field even if some of the interior of Venus is still liquid like Earth.

Mars

Mars has always attracted attention because of its unique, bright reddish color and its swift retrograde motion against the background of stars. The properties and surface characteristics have also attracted attention, particularly since Mars seems to have similarities to the Earth. It orbits the sun at an average distance of about 228 million km (about 142 million mi), or about 1.5 A.U. It makes a complete orbit every 687 days, about twice the time that the earth takes. Mars rotates in twenty-four hours, thirty-seven minutes, so the length of a day on Mars is about the same as the length of a day on the Earth. The observations that Mars has an atmosphere, light and dark regions that appear to be greenish and change colors with the seasons, and white polar caps that grow and shrink with the seasons led to early speculations (and many fantasies!) about the possibilities of life on Mars. These speculations increased dramatically in 1877 when Schiaparelli, an Italian astronomer, reported seeing "channels" on the Martian surface. Other astronomers began interpreting the dark greenish regions as vegetation and the white polar caps as ice caps as the earth has. In the early part of the twentieth century, the American astronomer who founded the Lowell Observatory in Arizona, Percival Lowell, published a series of popular books showing a network of hundreds of canals on Mars. Lowell, and other respectable astronomers, interpreted what they believed to be canals as evidence of intelligent life on Mars. Other astronomers, however, interpreted the greenish colors and the canals to be illusions, imagined features of astronomers working with the limited telescopes of that time. Since canals never appeared in photographs, said the skeptics, the canals were the result of the human tendency to see patterns in random markings where no patterns actually exist.

Speculation about Martian canals, vegetation, polar caps of ice, and intelligent life on Mars ended in the late 1960s and early 1970s with extensive studies and probes by spacecraft. Limited photographs by Mariner flybys in 1965 and 1969 had provided some evidence that the surface of Mars was much like the Moon, with no canals, vegetation, or much of anything else. Then in 1971, *Mariner 9* became the first spacecraft to orbit Mars, photographing the entire surface as well as making extensive measurements of the Martian atmosphere, temperature

Figure 19.15
Surface picture of Mars taken by the *Viking 1* lander found reddish, fine-grained material, rocks coated with a reddish stain, and groups of blue-black volcanic rocks.

ranges, and chemistry. For about a year *Mariner 9* sent a flood of new and surprising information about Mars back to the Earth (see figure 19.15 for a photo of the surface taken by *Viking 1*).

Mariner 9 found the surface of Mars not to be a crater-pitted surface as is found on the Moon. Mars has had a geologically active past and has four provinces, or regions, of related surface features. There are (1) volcanic regions with inactive volcanos, one larger than any found on the earth, (2) regions with systems of canyons, some larger than any found on the earth, (3) regions of terraced plateaus near the poles, and (4) flat regions pitted with impact craters (figure 19.16). There are also regions of sand dunes and powdery, salmon-colored dust that is whipped into monstrous dust storms by seasonal winds. It was the bright dust periodically covering dark rock surfaces that produced the surface color changes visible from the earth and thought to be seasonal vegetation changes. The white polar caps were found to be mostly dry ice, solid carbon dioxide that covers some underlying water ice. Surprisingly, dry channels suggesting former water erosion were discovered near the cratered regions. These are not the straight, long channels that were

imagined by earlier astronomers but are sinuous, dry riverbed features with dry tributaries. At one time Mars must have had an abundance of liquid water.

Abundant liquid water may have been present on Mars in the past, but none is to be found today. The atmosphere of Mars is very thin, exerting an average pressure at the surface that is only 0.6 percent of the average atmospheric pressure on the Earth's surface. Moreover, this thin Martian atmosphere is about 95 percent carbon dioxide, and 20 percent of this freezes as dry ice at the Martian South Pole every winter. The average Martian surface temperature is $-53°$ C (about $-63°$ F), and any water present is frozen in permafrost or layered sheets, perhaps forming the terraced plateaus near the poles. Today, Mars is much too cold for liquid water to exist.

If liquid water cannot exist on Mars, what formed the dry riverbed features that appear to have been formed by water? In theory, Mars at one time had an atmosphere of much denser concentrations of carbon dioxide and water vapor. A denser Martian atmosphere of carbon dioxide and water vapor could

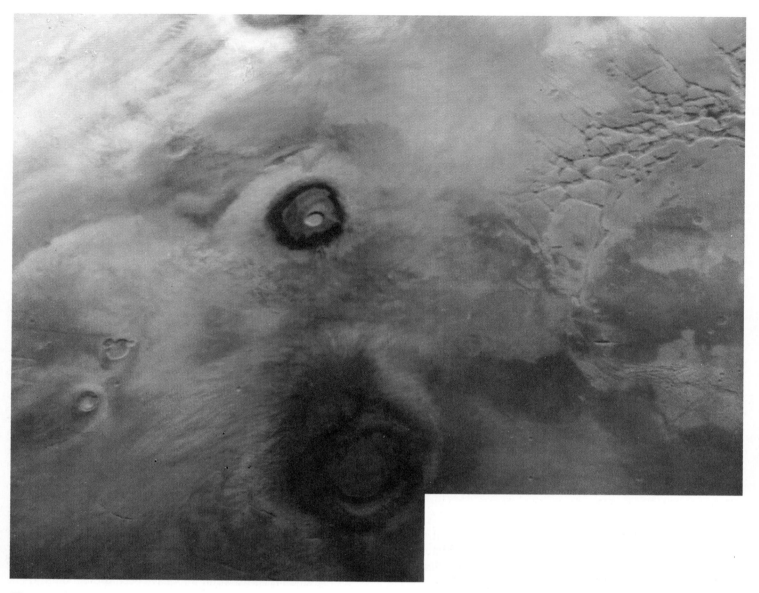

Figure 19.16

A view of the surface of Mars taken by the *Viking Orbiter 1* cameras. The scene shows three volcanoes that rise an average of 17 km (about 11 mi) above the top of a 10 km high (about 6 mi) ridge. Clouds can be seen in the upper portion of the photograph and haze is present in the valleys at the lower right.

have heated the atmosphere of Mars to much warmer temperatures through the greenhouse effect, even though it is half again as far from the Sun as the Earth. According to this theory, both the Earth and Mars had similar secondary atmospheres of carbon dioxide and water that were originally degassed from molten rock. Mars, however, is about half the size of the Earth. It is the size of a planet—or a body such as the moon—that determines the extent of internal melting and how rapidly the internal molten rock cools back to the solid state. Thus a small body such as the moon has less internal melting and cools more rapidly since it is smaller. Evidence from the moon, for example,

points to molten rock and volcanic activity for only about the first 700 million years. Evidence from Mars, which is twice as large as the Moon, points to molten rock and volcanic activity during the first billion years. The Earth, on the other hand, is twice as large as Mars and has had molten rocks and volcanic activity for 4.6 billion years.

The carbon dioxide concentration on Mars was reduced, perhaps by the formation of carbonate rocks as on the Earth and perhaps by some other mechanism. As the volcanic activity on Mars abated, the planet lost its source of water vapor and the carbon dioxide in the atmosphere, so the greenhouse effect also declined. The planet then cooled to its present harsh, cold conditions. The water-carved channels observed today were carved when Mars was much warmer with an active greenhouse effect.

The weather and climate conditions on Mars were monitored for about four years by two *Viking* orbiter spacecraft and their landers, which were parachuted to the surface in 1976. The Viking mission found that Mars has global wind systems similar to those on the Earth and regular dust storms on the surface that are coupled with the seasons. The landers also provided photographs of the immediate surface and analyzed Martian soil samples. The red color of Mars is from the iron oxides and iron-rich rocks on the surface. No evidence of life was found on this mission. The latest American mission to Mars is called the *Mars Observer*.

Mars was named for a mythical god of war, so the two satellites that circle Mars were named Deimos and Phobos after the two companions of the Roman god. Both satellites are small, irregularly shaped, and highly cratered. Phobos is the largest of the two, about 22 km (about 14 mi) across the longest dimension, and Deimos is about 13 km (about 8 mi) across. Both satellites reflect light poorly and have a much lower density than Mars. They are assumed to be captured asteroids rather than naturally occurring moons.

Jupiter

Jupiter is the largest of all the planets, with a mass equivalent to some three hundred Earths and, in fact, contains three-fourths of the mass of all the planets in the Solar System. This massive planet is located an average 778 million km (about 483 million mi), or 5 A.U., from the sun in an orbit that takes about twelve years for one revolution. The internal heating from gravitational contraction was tremendous when this giant formed, and today it still radiates twice the energy that it receives from the sun. Even with its tremendous mass, however, it has only about 1/1,000 of the mass of the sun. Jupiter would have to be almost twice as massive to create the internal temperatures needed to start nuclear fusion reactions, or in other words, to become a star itself. Nonetheless, the giant Jupiter and its system of sixteen satellites seems almost like a smaller version of a planetary system within the Solar System.

Jupiter has been observed, photographed, and probed by two *Pioneer* spacecraft in 1972 and 1974 and by two *Voyager* spacecraft in 1979. Information from these four spacecraft, together with Earth-based observations, make it possible to derive a model of the interior structure of Jupiter. Measurement of the motion of its satellites, for example, provided the needed data to calculate the mass and thus the density of the planet. Jupiter has an average density of 1.3 g/cm³, which is about a quarter of the density of Earth. This low density indicates that Jupiter is mostly made of light elements, such as hydrogen and helium, but does contain a percentage of heavier rocky substances. The model of Jupiter's interior (figure 19.17) is derived from this and other information from spectral studies, studies of rotation

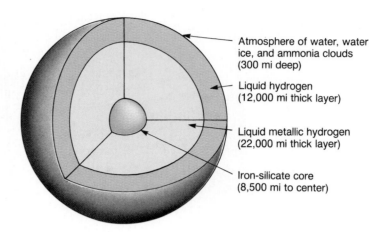

Atmosphere of water, water ice, and ammonia clouds (300 mi deep)

Liquid hydrogen (12,000 mi thick layer)

Liquid metallic hydrogen (22,000 mi thick layer)

Iron-silicate core (8,500 mi to center)

Figure 19.17
The interior structure of Jupiter.

rates, and measurements of heat flow. The model indicates a solid, rocky core with a radius of about 14,000 km (about 8,500 mi). This rocky core is thus more than twice the size of the earth. Above this core is an approximate 35,000 km (about 22,000 mi) thick layer of liquid hydrogen, compressed so tightly by millions of atmospheres of pressure that it is able to conduct electric currents. Liquid hydrogen with this property is called *metallic hydrogen* because it has the conductive ability of metals. Above the layer of metallic hydrogen is a 20,000 km (about 12,000 mi) thick layer of ordinary liquid hydrogen, which is under less pressure. The outer layer, or atmosphere, of Jupiter is a 500 km or so (about 300 mi) zone with water, water ice, ammonia gas, crystalline compounds, and ammonia clouds. It is the uppermost ammonia clouds, perhaps mixed with sulphur and organic compounds, that form the bright orange, white, and yellow bands around the planet. The banding is believed to be produced by atmospheric convection, in which bright, hot gases are forced to the top where they cool, darken in color, and sink back to the surface.

Jupiter's famous Great Red Spot is located near the equator. This permanent, deep, red oval feature was first observed by Robert Hooke in the 1600s and has generated much speculation over the years. The red oval, some 40,000 km long (about 25,000 mi) has been hypothesized to be created by deflected gases above an island or some structure below the atmosphere. Another popular hypothesis is that the Great Red Spot is a giant hurricane that has lasted for some three hundred years or longer. The energy source for such a huge, long-lasting hurricane is unknown (figure 19.18).

Jupiter has sixteen satellites, and the four brightest and largest can be seen from the Earth with a good pair of binoculars. These four are called the *Galilean moons* because they were discovered by Galileo in 1610. The Galilean moons are named Io, Europa, Ganymede, and Callisto (figure 19.19).

A

Figure 19.18

Photos of Jupiter taken by *Voyager 1.* (*a*) From a distance of about 36 million km (about 22 million mi). (*b*) A closer view, from the Great Red Spot to the south pole, showing organized cloud patterns. In general, dark features are warmer and light features are colder. The Great Red Spot soars about 25 km (about 15 mi) above the surrounding clouds and is the coldest place on the planet.

Observations by the *Pioneer* and *Voyager* spacecrafts revealed some fascinating and intriguing information about the moons of Jupiter. Io, for example, was discovered to have powerful, active volcanos that eject enormous plumes of molten sulfur and sulfur dioxide. This is the only volcanic activity that has ever been observed in the Solar System other than on the earth. Europa is covered with smooth water ice, which has a network of long, straight, dark ridges. Ganymede has valleys, ridges, folded mountains, and other evidence of an active geologic history. Callisto, the most distant of the Galilean moons, was found to be the most heavily cratered object in the Solar System. In addition to all the new information about Jupiter's satellites, *Voyager* discovered that Jupiter has a system of rings between the satellites and the planet. These rings are much smaller than the well-known rings of Saturn.

B

Figure 19.19
The four Galilean moons pictured by *Voyager 1*. Clockwise from upper left, Io, Europa, Ganymede, and Callisto. Io and Europa are about the size of Earth's Moon; Ganymede and Callisto are larger than Mercury.

Saturn

Saturn is slightly smaller and substantially less massive than Jupiter, and it has similar surface features, but it is readily identified by its unique, beautiful system of rings. Saturn is about 9.5 A.U. from the Sun, but its system of rings is readily identified with a good pair of binoculars. Saturn also has the lowest average density of any of the planets, 0.7 g/cm³, which is less than the density of water.

The surface of Saturn, as Jupiter's surface, has bright and dark bands that circle the planet parallel to the equator. Saturn also has a smaller version of Jupiter's Great Red Spot, but in general the bands and spot are not as highly contrasted or brightly colored as they are on Jupiter.

Saturn's rings were found by *Voyager 1* and *Voyager 2* to consist of thousands of narrow rings made up of particles, some rings with particles large enough to be measured in meters and some rings with particles that are dust-sized. Rings with small particles were observed to have waves, and the center ring had radial streaks or "spokes." The waves are thought to be a result of collisions between particles that are caused by the gravity of a nearby moon, or perhaps they are electrically charged particles interacting with Saturn's magnetic field. The changing pattern of "spokes," which rotate with the ring, are thought to be related to Saturn's magnetic field and electrical discharges that are up to 100,000 times more energetic than lightning on the

Figure 19.20

A part of Saturn's system of rings, pictured by *Voyager 2* from a distance of about 3 million km (about 2 million mi). More than sixty bright and dark ringlets are seen here; different colors indicate different surface compositions.

Earth. The wealth of information collected on the rings of Saturn by the *Voyager* spacecraft is still being analyzed and interpreted (figure 19.20).

Before the *Voyager* spacecraft mission Saturn was known to have ten satellites: Janus, Mimas, Enceladus, Tethys, Dione, Rhea, Titan, Hyperion, Iapetus, and Phoebe. The *Voyager* observations added eight more smaller satellites to the list. All the ten known moons were observed to be icy and heavily cratered except Titan, which is covered with clouds and impossible to observe. Titan is the only moon in the Solar System that has a substantial atmosphere. Titan is larger than the planet Mercury

and is covered with a 240 km (about 150 mi) deep layer of reddish clouds consisting of methane, nitrogen, carbon dioxide, and organic molecules. The pressure on the surface is 1.5 atmospheres, but with a surface temperature of about $-180°$ C (about $-290°$ F) it is doubtful that life has developed on Titan.

Uranus, Neptune, and Pluto

Uranus and Neptune are two more giant planets that are far, far away from the Earth. Uranus revolves around the Sun at an average distance of over 19 A.U., taking about 84 years to circle the Sun once. Neptune is an average 30 A.U. from the Sun and takes about 165 years for one complete orbit. Thus Uranus is

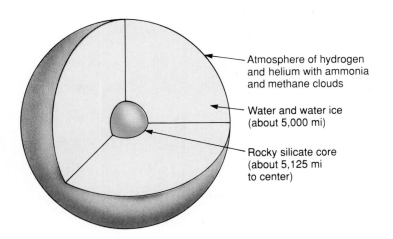

Figure 19.21

The interior structure of Uranus and Neptune.

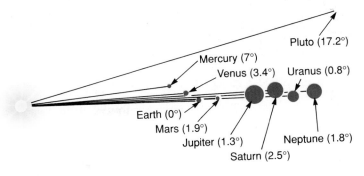

Figure 19.22

The orbital planes of the planets. Notice the exceptional tilt of the orbit of Pluto.

about twice as far away from the Sun as Saturn and Neptune is three times as far away. To give you an idea of these tremendous distances, consider the time required for radio signals to travel from the *Voyager 2* spacecraft to the earth. *Voyager* photographed and made observations of Uranus in 1986, sending this information back to Earth in the form of radio signals. If you consider Uranus to be 2,720 million km from the Earth, a radio signal traveling through space at the speed of light would require 2.72×10^{12} m divided by 3.00×10^8 m/sec, or 9,067 sec, which is more than 2.5 hours! It would be most difficult to carry on a conversation by radio with someone such a distance away. Even farther away, a radio signal from a transmitter near Neptune would require over 4 hours to reach Earth, which means 8 hours would be required for two people to just say, "Hello" to each other!

Uranus and Neptune are more similar to each other than Saturn is to Jupiter. Both are massive, with a diameter of about 50,000 km (about 30,000 mi), which is about a third the mass of Jupiter. Both planets are thought to have similar interior structures (figure 19.21), which is water and water ice surrounding a rocky core with an atmosphere of hydrogen and helium. Because of their great distances from the sun, both have very low average surface temperatures, $-210°$ C (about $-350°$ F) for Uranus and $-220°$ C (about $-360°$ F) for Neptune.

In addition to retrograde rotation, Uranus has an odd orientation of its axis to the plane of its orbit. Most planets have their axes of rotation tilted less than 30° from a vertical line to the plane of the orbit (earth's tilt is 23.5°). Uranus's tilt, however, is 82°, meaning that it is practically on its side. Moving in its orbit around the sun on its side, one pole receives direct sunlight for twenty-one years, then the equator receives direct sunlight for the next twenty-one years, then the other pole receives the direct sunlight for the next twenty-one years, and so on. This would produce some interesting climatic patterns on Uranus, that is if the weak sunlight is able to penetrate the atmosphere.

Voyager 2 discovered ten new satellites, in addition to the five that were previously known, and new rings in the system of rings around Uranus for a total of ten narrow rings and a number of dusty bands. Interestingly, the rock particles making up the rings and the newly discovered moons are dark colored, reflecting less than 5 percent of the sunlight falling on them. By comparison, the earth's moon has a reflectivity of about 10 percent. One explanation of the dark colors is that the surfaces were originally covered with frozen methane, which has been decomposed by radiation, leaving a layer of black carbon.

In 1989, *Voyager 2* visited Neptune and discovered six new satellites, making a total of eight, and a system of rings. *Voyager* found that Neptune has a churning, turbulent atmosphere and that the largest moon, Triton, is about 2,800 km in diameter (about 1,740 mi). Triton rotates in the opposite direction from the planet, unlike other moons in the Solar System. This has led to speculation that Triton could have been a comet that was captured by Neptune.

Pluto, the outermost planet, is a small, cold body, probably of frozen methane, that is about twice the size of Neptune's moon of Triton. It is so small and far away at its approximate 40 A.U. average distance from the sun that little is known about the strange planet. The plane of its orbit is tilted about 17° from the ecliptic, the plane of the earth's orbit, while all the other planets have orbital planes within 2° or 3° of the ecliptic. In addition, the orbit of Pluto is the most eccentric, that is, the least circular, of all the planetary orbits (figure 19.22). Pluto's orbit is so eccentric that it is sometimes nearer the Sun than Neptune and sometimes farther out than Neptune. It has, in fact, been closer to the Sun than Neptune since 1978. It reached perihelion in 1989 and has been moving back around its 248-year orbit since. After the year 2000 Pluto will again be farther from the Sun than Neptune. The two planets do not come close enough together to collide, but because their orbits cross, it might be an escaped moon of Neptune rather than a true planet. No other planets have such strange, crossed orbits, and Pluto remains a puzzle.

The Solar System

Pluto's moon, Charon, was discovered in 1978. It is about half the size of the planet, with the largest ratio of the size of a moon to its planet in the Solar System. Since it has the same period of revolution as the period of rotation for Pluto, Charon would appear stationary from the surface of the planet.

Small Bodies of the Solar System

In the earlier description of how the planets formed from the accretion disk, they were described as forming from protoplanets, icy collections of slushy materials, gases, and rocky particles. Not all of the materials in the accretion disk ended up in the planets, however; some ended up in smaller bodies of the Solar System. Bodies of some of the original icy collections that were too small to form planets themselves and were not incorporated into the planets or the sun still exist today, sometimes orbiting close enough to the sun to become illuminated. These illuminated masses of icy materials may form a tail and are called *comets*. Other leftover materials are more solid, with a rocky or iron composition. These *asteroids* or *minor planets* are thought to be the stuff of a planet that never quite made it to the final stage before being pulled apart. Rare collisions between asteroids break off small pieces, some of which fall to the earth as *meteorites*. Comets, asteroids, and meteorites are the leftovers from the formation of the sun and planets. Presently, the total mass of all these leftovers in and around the Solar System is probably less than the mass of the earth's moon. It must have been much greater in the past, however, as evidenced by the intense bombardment that took place on the moon and other planets some four billion years ago.

Comets

Based on calculations of their observed paths, *comets* are estimated to originate some 50,000 A.U. from the Sun, far beyond the orbit of Pluto. Here, according to other calculations and estimates, is a cloud of billions and billions of small objects that make a spherical halo around the entire Solar System. The objects are made up of porous aggregates of water ice, frozen methane, frozen ammonia, dry ice, and the dust of mineral grains. This cloud of icy, dusty aggregates is called the **Oort cloud,** the birthplace of comets (figure 19.23).

The current theory of the origin of comets was developed by the Dutch astronomer Jan Oort in 1950. According to the theory, the huge cloud of icy, dusty aggregates consists of leftovers from the formation of the Solar System and has been slowly circling the Solar System ever since it formed. Something, perhaps a gravitational nudge from a passing star, moves one of the icy bodies enough that it is pulled toward the sun in what will become an extremely elongated elliptical orbit. When in the Oort cloud the body had a temperature of about −260° C (about −440° F). As it moves toward the sun it warms from the increasing intense solar radiation. Somewhere between Jupiter and

Figure 19.23

The Oort cloud is the birthplace of comets. The cloud surrounds the Solar System, with the bodies of ice and dust slowly circling.

Mars the ice and frozen gases begin to vaporize, forming a hazy head around the comet called a *coma*. The coma grows larger with increased vaporization, perhaps several hundred times larger than the body of the comet, which is now called the *nucleus*. As the comet nears the sun solar wind and solar radiation ionize gases and push particles from the coma, pushing both into the familiar visible *tail* of the comet. The tail can be small or stretch over the entire horizon as seen from the earth, and it might have two parts, one consisting of solid dust particles and one consisting of ionized gases. The solid particles are visible because of reflected sunlight from the dust particles. The ionized gases are fluorescent, emitting visible light because they are excited by ultraviolet radiation from the sun. The tail always points away from the sun, so the tail follows the comet as it approaches the sun but leads the comet as it moves away from the sun (figure 19.24).

Comets are not very massive or solid, and the porous, snowlike mass has a composition more similar to the giant planets than to the terrestrial planets in comparison. Each time a comet passes near the sun, it loses some of its mass through evaporation of gases and loss of dust to the solar wind. After passing the sun the surface forms a thin, fragile crust covered with carbon and other dust particles. Each pass by the sun means a loss of matter and the coma and tail are dimmer with each succeeding path. About 20 percent of the approximately six hundred comets that are known have orbits that return them to the sun within a two-hundred-year period, some of which return as often as every five or ten years. The other 80 percent have long elliptical orbits that return them at intervals exceeding two hundred years. The famous Halley's comet has a smaller elliptical orbit and returns about every seventy-six years. Halley's comet, as all other comets, may eventually break up into a trail of gas and dust particles that orbit the sun.

Figure 19.24
Comet West in 1976. As a comet nears the sun it grows brighter, with the tail always pointing away from the sun.

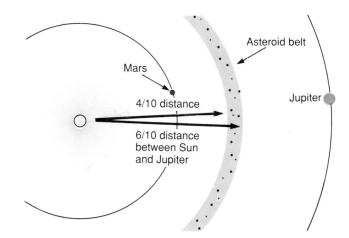

Figure 19.25
Most of the asteroids in the asteroid belt are about halfway between the Sun and Jupiter.

Asteroids

The asteroid belt between Mars and Jupiter was introduced earlier (figure 19.25). This belt contains thousands of *asteroids* that range in size from 1 km or less up to the largest asteroid, named Ceres, which has a diameter of about 1,000 km (over 600 mi). The asteroids are thinly distributed in the belt, 1 million km or so apart (about 600,000 mi), but there is evidence of collisions occurring in the past. Most asteroids larger than 50 km (about 30 mi) have been studied by analyzing the sunlight reflected from their surfaces. These spectra provide information about the composition of the asteroids. Asteroids on the inside of the belt are made of stony materials, and those on the outside of the belt are dark with carbon minerals. Still other asteroids are metallic, containing iron and nickel. These spectral composition studies, analyses of the orbits of asteroids, and studies of meteorites that have fallen to the earth all indicate that the asteroids are not the remains of a planet or planets that

were broken up. The asteroids are now believed to have formed some 4.6 billion years ago from the original solar nebula. During their formation, or shortly thereafter, their interiors were partly melted, perhaps from the heat of short-lived radioactive decay reactions. Their location close to Jupiter, with its gigantic gravitational field, prevented the slow gravitational clumping-together process that would have formed a planet.

Jupiter's gigantic gravitational field also captured some of the asteroids, pulling them into its orbit. Today there are two groups of asteroids, called the *Trojan asteroids,* which lead and follow Jupiter in its orbit. They lead and follow at a distance where the gravitational forces of Jupiter and the Sun balance to keep them in the orbit. A third group of asteroids, called the *Apollo asteroids,* has orbits that cross the orbit of the earth. It is possible that one of the Apollo asteroids could collide with the earth. One theory about what happened to the dinosaurs is based on evidence that such a collision indeed did occur some sixty-five million years ago. This theory will be discussed in a later chapter. The chemical and physical properties of the two satellites of Mars, Phobos and Deimos, are more similar to the asteroids than to Mars. There is thus the likely possibility that the Martian satellites are captured asteroids.

Meteors and Meteorites

Comets leave trails of dust and rock particles after encountering the heat of the sun, and collisions between asteroids in the past have ejected fragments of rock particles into space. In space, the remnants of comets and asteroids are called **meteoroids.** When a meteoroid encounters the earth moving through space, it accelerates toward the surface with a speed that depends on its direction of travel and the relative direction that the earth is moving. It soon begins to heat from air friction in the upper atmosphere, melting into a visible trail of light and smoke. The streak of light and smoke in the sky is called a **meteor.** The

Table 19.3

Some annual meteor showers

Name	Date of Maximum	Hour Rate
Quadrantid	January 3	30
Aqaurid	May 4	5
Perseid	August 12	40
Orionid	October 22	15
Taurids	November 1, 16	5
Leonid	November 17	5
Geminid	December 12	55

A

B

Figure 19.26

(*a*) A stony meteorite. The smooth black surface was melted by friction with the atmosphere. (*b*) An iron meteorite that has been cut, polished, and etched with acid. The pattern indicates that the original material cooled from a molten material over millions of years.

"falling star" or "shooting star" is a meteor. Most meteors burn up or evaporate completely within seconds after reaching an altitude of about 100 km (about 60 mi) because they are nothing more than a speck of dust. A **meteor shower** occurs when the earth passes through a stream of particles left by a comet in its orbit. The earth might meet the stream of particles concentrated in such an orbit on a regular basis as it travels around the sun, resulting in predictable meteor showers (table 19.3). In the third week of October, for example, the earth crosses the orbital path of Halley's comet, resulting in a shower of some ten to fifteen meteors per hour. Meteor showers are named for the constellation in which they appear to originate. The October meteor shower resulting from an encounter with the orbit of Halley's comet, for example, is called the Orionid shower because it appears to come from the constellation Orion.

If a meteoroid survives its fiery trip through the atmosphere to strike the surface of the earth, it is called a **meteorite.** Most meteors are from fragments of comets, but most meteorites generally come from particles that resulted from collisions between asteroids that occurred long ago. Meteorites are classified into three basic groups according to their composition: (1) **iron meteorites,** (2) **stony meteorites,** and (3) **stony-iron meteorites** (figure 19.26). The most common meteorites are stony, composed of the same minerals that make up rocks on the earth. The stony meteorites are further subdivided into two groups according to their structure, the **chondrites** and the **achondrites.** Chondrites have a structure of small spherical lumps of silicate minerals or glass, called **chondrules,** held together by a fine-grained cement. The achondrites do not have the chondrules, as their name implies, but have a homogeneous texture more like volcanic rocks such as basalt that cooled from molten rock.

The iron meteorites are about half as abundant as the stony meteorites. They consist of variable amounts of iron and nickel, with traces of other elements. In general, there is proportionally much more nickel than is found in the rocks of the earth. When cut, polished, and etched, beautiful crystal patterns are observed on the surface of the iron meteorite. The patterns mean that the iron was originally molten, then cooled very slowly over millions of years as the crystal patterns formed.

A meteorite is not, as is commonly believed, a ball of fire that burns up the landscape where it lands. The iron or rock has been in the deep freeze of space for some time, and it travels rapidly through the earth's atmosphere. The outer layers become hot enough to melt, but there is insufficient time for this heat to be conducted to the inside. Thus a newly fallen iron meteorite

Strange Meteorites

Meteorites are a valuable means of learning about the Solar System since they provide direct evidence about objects away from the earth in space. Over two thousand meteorites have been collected, classified, and analyzed for this purpose. Most appear to have remained unchanged since they were formed in the asteroid belt between Jupiter and Mars along with the rest of the Solar System some five billion years ago. Yet a few meteorites are very unusual or strange, not fitting in with the rest. The mystery of two types of these strange meteorites is the topic of this Feature.

The first example of a meteorite mystery is the case of the carbonaceous chondrites, which are soft, black stony meteorites with a high carbon content. Fewer than twenty-five of these soft and sooty black fragments have been recovered since the first one was identified in France in 1806. The carbonaceous chondrite is similar to the ordinary chondrite except that it contains carbon compounds, including amino acids. Amino acids are organic compounds basic to life. The mystery of these meteorites is the meaning of the presence of amino acids. Is it the remains of some former life? Or is it a precursor to life, an organic compound produced naturally when the Solar System formed? If this is true, what is the relationship between the amino acids and life on the earth, and what are the implications for elsewhere? The answer to these questions will probably not be available until more of these strange meteorites are available for study.

The second example of a meteorite mystery is the case of the shergotties, which are meteorites of volcanic origin with strange properties. The shergotties are named after a town in India where the first one identified was observed to fall in 1865. A second shergottie fell in Africa in 1962, and a third was recently found on an ice sheet in Antarctica. The shergotties are known to be meteorites because the first two were actually observed to fall and all three have the typical glassy crust that forms from partial melting during the fall.

The strangeness of the shergotties comes from their age, chemical composition, and physical structure as compared to all the other meteorites. Radioactive dating shows that the shergotties are the youngest meteorites known, cooling from a molten state somewhere between 650 million and 1 billion years ago. All of the other meteorites were originally formed some 4.6 billion years ago in the asteroid belt. In addition, the shergotties do not have the same composition as the asteroids according to spectral analysis. The conclusion so far is that they are not from the asteroid belt. Their chemical composition is more similar to that of a planet such as the Earth or Mars than it is to an asteroid.

One interesting feature of the shergotties is the special kind of glass that comprises about half of their masses. This special glass forms only when certain minerals are subjected to tremendous pressures by impact shock. Evidently these meteorites were subjected to a tremendous shock sometime after cooling from the molten state.

The last piece of puzzle to the mystery comes from the quantitative analysis of the soil on Mars that was performed by the *Viking* landers. The analysis showed that the chemical composition of the shergotties is more similar to Mars than it is to the Earth. Large volcanos were observed on Mars by the *Viking* spacecraft, and calculations indicate that Mars was volcanically active when the shergotties were formed.

Considering all the evidence, would you consider Mars as a credible source of the shergottie meteorites? If so, what kind of astounding event could possibly have ejected rocks from the surface of Mars into space? The answers to these questions are unknown, and the mystery of these strange meteorites continues.

---

will be hot since metals are good heat conductors, but it will not be hot enough to start a fire. A stone meteorite is a poor conductor of heat so it will be merely warm.

According to the radioactive "clock," which is read by determining radioactive decay ratios, meteorites formed some 4.6 billion years ago, which fits with other evidence about the age of the Solar System. Since cosmic rays in outer space cause isotopic changes in materials, physicists are able to measure how long a meteoroid has existed as a small body, that is, when it was broken off from a larger asteroid. This data indicates that the meteoroids were broken off in a few major collisions as opposed to a series of many small collisions.

Summary

The ancient Greeks explained observations of the sky with a *geocentric model* of a motionless earth surrounded by turning geometrical spheres with attached celestial bodies. To explain the observed *retrograde motion* of the planets, later Greek astronomers added small circular orbits, or *epicycles,* to the model, which is called the *Ptolemaic system.* A *heliocentric,* or sun-centered, model of the Solar System was proposed by Nicolas Copernicus. The *Copernican system* considered the planets to move around the sun in circular orbits. Tycho Brahe made the first precise, continuous record of planetary motions. His assistant, Johannes Kepler, analyzed the data to find if planets followed circular paths or if they followed the paths of epicycles. Today, his findings are called *Kepler's laws of planetary motion. Kepler's first law* describes the orbit of a planet as having the shape of an *ellipse* with the sun at one focus. *Kepler's second law* describes the motion of the planet along this ellipse, that a line from the planet to the sun will move over equal areas of the orbital plane during equal periods of time. The point of the orbit closest to the sun is called the *perihelion,* and the point of the orbit farthest from the sun is called the *aphelion. Kepler's third law* states a relationship between the time required for a planet to complete one orbit and its distance from the sun. In general, the farther away from the sun, the longer the period of time to complete an orbit.

The *protoplanet nebular model* is the most widely accepted theory of the origin of the Solar System, and this theory can be considered as a series of events, or stages. *Stage A* is the creation of all the elements heavier than hydrogen in previously existing stars. *Stage B* is the formation of a nebula from the raw materials created in stage A. The nebula contracts from gravitational attraction, forming the *protosun* in the center with a fat, bulging *accretion disk* around it. The sun will form from the protosun, and the planets will form in the accretion disk. *Stage C* begins as the protosun becomes established as a star. The icy remains of the original nebula are the birthplace of *comets. Asteroids* are other remains that did undergo some melting.

The motion of a planet as it orbits the sun is called *revolution,* and its spinning motion is called *rotation.* The plane of the earth's orbit is called the *ecliptic.* The planets can be classified into two major groups: (1) the *terrestrial planets* of Mercury, Venus, Mars, and Earth and (2) the *giant planets* of Jupiter, Saturn, Uranus, and Neptune.

The present-day atmospheres of the terrestrial planets are believed to be secondary atmospheres derived from rocks by a *degassing* process, the expulsion of carbon dioxide, other gases, and water vapor from molten rock. The amount released is related to how much melting occurred on a particular planet or body. The size of the planet or body is also related to its ability to hold on to a degassed atmosphere. The present-day atmospheres of Mars, Venus, and Earth all evolved from the same kind of degassed carbon dioxide and water vapor atmospheres.

Comets are porous aggregates of water ice, frozen methane, frozen ammonia, dry ice, and dust. The Solar System is surrounded by a sphere of clusters of these objects called the *Oort cloud.* Something nudges one of the icy bodies from the Oort cloud and it falls into a long elliptical orbit around the sun. As it approaches the sun increased radiation evaporates ices and pushes ions and dust into a long visible tale. *Asteroids* are rocky or metallic bodies that are mostly located in a belt between Mars and Jupiter. The remnants of comets, fragments of asteroids, and dust are called *meteoroids.* A meteoroid that falls through the earth's atmosphere and melts to a visible trail of light and smoke is called a *meteor.* A meteoroid that survives the trip through the atmosphere to strike the surface of the earth is called a *meteorite.* Most meteors are fragments and pieces of dust from comets. Most meteorites are fragments that resulted from collisions between asteroids.

Key Terms

accretion disk (p. 385)	Kepler's second law (p. 383)
achondrites (p. 402)	Kepler's third law (p. 383)
aphelion (p. 383)	meteor (p. 401)
asteroids (p. 386)	meteorite (p. 402)
axis (p. 386)	meteoroids (p. 401)
chondrites (p. 402)	meteor shower (p. 402)
chondrules (p. 402)	Oort cloud (p. 400)
comets (p. 385)	perihelion (p. 383)
Copernican system (p. 382)	protoplanet nebular model
degassing (p. 387)	(p. 384)
epicycle (p. 381)	Ptolemaic system (p. 381)
giant planets (p. 387)	revolution (p. 386)
iron meteorites (p. 402)	rotation (p. 386)
Kepler's first law (p. 383)	stony-iron meteorites (p. 402)
Kepler's laws of planetary	stony meteorites (p. 402)
motion (p. 383)	terrestrial planets (p. 386)

Applying the Concepts

1. A model of a motionless earth with the sun, moon, and planets moving around it will explain all observations except
   a. stars maintaining the same relative positions.
   b. the sun, moon, and planets rise in the east and set in the west.
   c. the motion of the planets over a year.
   d. none of the above.

2. The earth, other planets, and all the members of the Solar System
   a. have always existed.
   b. formed thousands of years ago from elements that have always existed.
   c. formed millions of years ago, when elements and each body were created at the same time.
   d. formed billions of years ago from elements that were created in many previously existing stars.

3. The atmosphere that is found on the earth today
   a. formed when the earth formed.
   b. is the secondary atmosphere that formed from degassing.
   c. is a secondary atmosphere that has been modified over time.
   d. has always been present just as it is today.

4. The belt of asteroids between Mars and Jupiter is probably
   a. the remains of a planet that exploded.
   b. clumps of matter that condensed from the accretion disk but never got together as a planet.
   c. the remains of two planets that collided.
   d. the remains of a planet that collided with an asteroid or comet.

5. Comparing the amount of interior heating and melting that has taken place over time, the smaller planets underwent
   a. more heating and melting because they are smaller.
   b. the same heating and melting as larger planets.
   c. less heating and melting for a shorter period of time.

6. Which of the following planets would be mostly composed of hydrogen, helium, and methane and have a density of less than 2 g/cm^3?
   a. Uranus
   b. Mercury
   c. Mars
   d. Venus

7. Which of the following planets probably still has its original atmosphere?
   a. Mercury
   b. Venus
   c. Mars
   d. Jupiter

8. Venus appears the brightest when it is in the
   a. full phase.
   b. half phase.
   c. quarter phase.
   d. crescent phase.

9. The small body with a composition and structure closest to the materials that condensed from the accretion disk is a(n)
   a. asteroid.
   b. meteorite.
   c. comet.
   d. none of the above.

10. A small body from space that falls on the surface of the earth is
    a. a meteoroid.
    b. a meteor.
    c. a meteor shower.
    d. a meteorite.

Answers

1. c   2. d   3. c   4. b   5. c   6. a   7. d   8. d   9. c
10. d

Questions for Thought

1. Describe the one observation that is difficult to explain with a simple geocentric model of the Solar System. Explain how the Ptolemaic system took care of this difficulty.

2. Describe the Copernican system and how it accounted for the motions of the planets. Briefly discuss at least three reasons why the Copernican system was not immediately accepted over the Ptolemaic system.

3. Discuss the contributions each of the following made to the eventual acceptance of a heliocentric model of the Solar System: (a) Nicolas Copernicus, (b) Tycho Brahe, (c) Johannes Kepler, (d) Galileo, and (e) Isaac Newton.

4. Describe Kepler's three laws of planetary motion, using diagrams as needed.

5. What are the perihelion and the aphelion? Compare the orbital velocity of a planet at perihelion and at aphelion.

6. Evaluate the comparative difficulties in creating, testing, and verifying models of (a) the structure of the Solar System, (b) the life of a star, and (c) the origin of the Solar System.

7. Describe the protoplanet nebular model of the origin of the Solar System. Which part or parts of this model seem least credible to you? Explain. What information could you look for today that would cause you to accept or modify this least credible part of the model?

8. What are the basic differences between the terrestrial planets and the giant planets? Describe how the protoplanet nebular model accounts for these differences.

9. Identify at least three properties of the terrestrial planets that theoretically determined what kind of atmosphere is present on the terrestrial planets today. Explain the role of each of these planetary properties in determining the atmosphere of a planet.

10. Compare the atmospheres and surface conditions on the two hot planets of Mercury and Venus. Provide reasons why these differences exist.

11. Explain (a) why Venus and Earth are believed to have had similar atmospheres at two different times during their history and (b) why the atmospheres are so different today.

12. Describe the surface and atmospheric conditions on Mars.

13. What evidence exists that Mars at one time had abundant liquid water? If Mars did have liquid water at one time, what happened to it and why?

14. Describe the internal structure of Jupiter and Saturn.

15. What are the rings of Saturn? Name other planets that have ring structures.

16. Describe some of the unusual features found on the moons of Jupiter, Saturn, and Neptune.

17. What are the similarities and the differences between the Sun and Jupiter?

18. Give one idea about why the Great Red Spot exists on Jupiter. Does the existence of a similar spot on Saturn support or not support this idea? Explain.

19. What is so unusual about the motions and orbits of Venus, Uranus, and Pluto?

20. What evidence exists today that the number of rocks and rock particles floating around in the Solar System was much greater in the past soon after the planets formed?

21. What was the source of the water found today in the earth's oceans? Explain the reasoning behind your answer.

22. Explain why carbon dioxide is a major component of the terrestrial planets of Mars and Venus but not of (a) Mercury, (b) Earth, and (c) the giant planets.

23. Explain why oxygen is a major component of the earth's atmosphere but not the atmospheres of Venus or Mars.

24. Using the properties of the planets other than earth, discuss the possibilities of life on each.

25. What is a "shooting star"? Where do they come from? Where do they go?

26. What is an asteroid? What evidence indicates that asteroids are parts of a broken-up planet? What evidence indicates that asteroids are not parts of a broken-up planet?

27. Where do comets come from? Why are astronomers so interested in studying the physical and chemical structure of a comet?

28. What is a meteor? What is the most likely source of meteors?

29. What is a meteorite? What is the most likely source of meteorites?

30. Technically speaking, what is wrong with calling a rock that strikes the surface of the moon a meteorite? Again speaking technically, what should you call a rock that strikes the surface of the moon (or any other planet)?

31. Describe the physical structure and composition of the two main kinds of meteorites, including any subdivisions. What is the meaning of the structure, composition, and texture of a meteorite?

32. If a comet is an icy, dusty body, explain why it appears brightly in the night sky.

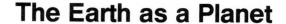

The Earth as a Planet

Artist's concept of the solar system. Shown are the orbits of the planets, the earth being the third planet from the sun, and the other planets and their relative sizes and distances from each other and to the sun. Also shown is the solar system as seen looking toward the earth from the moon.

THE earth is a common object in the Solar System, one of nine planets that goes around the sun once a year in an almost circular orbit. Earth is the third planet out from the sun, it is fifth in mass and fifth in diameter, and it has the greatest density of all the planets. The earth is unique because of its abundant supply of liquid water, its strong magnetic field, and its atmospheric composition. In addition to these physical properties, the earth has a unique set of natural motions that humans have used for thousands of years as a frame of reference to mark time and to identify the events of their lives. These references to the earth's motions are called the day, the month, and the year.

Eventually, about three hundred years ago, people began to understand that their references for time came from an earth that spins like a top as it circles the sun. It was still difficult, however, for them to understand the earth's place in the universe. The problem was not unlike that of a person trying to comprehend the motion of a distant object while riding a moving merry-go-round being pulled by a cart. Actually, the combined motions of the earth are much more complex than a simple moving merry-go-round being pulled by a cart. Imagine trying to comprehend the motion of a distant object while undergoing a combination of the earth's more conspicuous motions, which are as follows:

1. A daily rotation of 1,609 km/hr (about 1,000 mi/hr) at the equator and less at higher latitudes.
2. A monthly revolution of earth around the earth-moon center of gravity at about 50 km/hr (about 30 mi/hr).
3. A yearly revolution around the sun at about an average 106,000 km/hr (about 66,000 mi/hr).
4. A motion of the Solar System compared to other, faraway stars at about 48,000 km/hr (about 30,000 mi/hr).
5. A motion of the local star group that contains the sun as compared to other star clusters of about 1,000,000 km/hr (about 700,000 mi/hr).
6. Movement of the entire Milky Way galaxy relative to other, remote galaxies at about 580,000 km/hr (about 360,000 mi/hr).
7. Minor motions such as cycles of change in (a) the size and shape of the earth's orbit, (b) the tilt of earth's axis, and (c) the time of perihelion. In addition to these slow changes, there is a gradual slowing of the rate of earth's daily rotation.

Basically the earth is moving through space at fantastic speeds, following the sun in a spiral path of a giant helix as it spins like a top. This ceaseless and complex motion in space is relative to various frames of reference, however, and the limited perspective from the earth's surface can result in some very different ideas about the earth and its motions (figure 20.1). This chapter is about the more basic, or fundamental, motions of the earth and its moon. In addition to conceptual understandings and evidences for the motions, some practical human uses of the motions of the planet earth will be discussed.

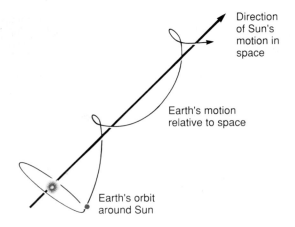

Figure 20.1

The earth undergoes many different motions as it moves through space. There are seven more conspicuous motions, three of which are more obvious on the surface. The earth follows the path of a gigantic helix, moving at fantastic speeds as it follows the sun and the Galaxy through space.

Shape and Size of the Earth

The most widely accepted theory about how the Solar System formed pictures the planets forming in a disk-shaped nebula with a turning, swirling motion. The planets formed from separate accumulations of materials within this disk-shaped, turning nebula, so the orbit of each planet was established along with its rate of rotation as it formed. Thus all the planets move around the sun in the same direction in elliptical orbits that are nearly circular. The flatness of the Solar System results in the observable planets moving in, or near, the plane of the earth's orbit, which is called the plane of the ecliptic.

Today, almost everyone has seen pictures of the earth from space, and it is difficult to deny that the earth has a rounded shape. During the fifth and sixth century B.C., the ancient Greeks decided that the earth must be round because (1) philosophically, they considered the sphere to be the perfect shape and they considered the earth to be perfect, so therefore the earth must be a sphere, (2) the earth was observed to cast a circular shadow on the moon during a lunar eclipse, and (3) ships were observed to slowly disappear below the horizon as they sailed off into the distance. More abstract evidence of a round earth was found in the observation that the altitude of the North Star above the horizon appeared to increase as a person traveled northward. This established that the earth's surface was curved, at least, which seemed to fit with other evidence and philosophical reasonings.

One of the first known attempts to calculate the size of the earth from actual data was made by the Greek astronomer Eratosthenes in about 250 B.C. Eratosthenes lived in Alexandria, Egypt. He learned, perhaps from travelers, that an interesting thing happened every first day of summer (June 21) in

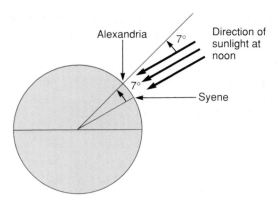

Figure 20.2
Eratosthenes calculated the size of the earth's circumference after learning that the sun's rays were vertical at Syene at noon on the same day they made an angle of a little over 7° at Alexandria. He reasoned that the difference was due to the earth's curved surface. Since 7° is about 1/50 of 360°, then the size of the earth's circumference had to be fifty times the distance between the two towns. (The angle is exaggerated in the diagram for clarity.)

the town of Syene (now Aswan). On that day when the sun reached its highest point in the sky (noon), sunlight would move straight down a deep well without making any shadows on the sides of the well. Eratosthenes decided to check this observation at noon on the first day of summer and found the sun over Alexandria was slightly over 7° from the vertical. Eratosthenes considered the earth to be a sphere and, as the Babylonians before him, divided the circumference of a sphere (a circle) into 360°. Eratosthenes reasoned that since the sun was directly over Syene *at the same instant* that it was a little over 7° from the vertical at Alexandria, the difference must be a consequence of the earth's curved surface. Further, 7° is about one-fifth of 360°, a complete circle, so the distance between Alexandria and Syene must be one-fiftieth of the earth's circumference. Thus the earth's circumference could be calculated by measuring the distance between Alexandria and Syene. Eratosthenes' calculation was made in the length units used at that time, the stadia. The distance between Syene and Alexandria was 5,000 stadia, so the earth's circumference was calculated to be 5,000 stadia times 50, or 250,000 stadia. The exact length of the ancient Greek stadium unit is unknown today. The unit was based on the length of the track in the local stadium, and this was a time long before standard units were established. The shortest of these tracks were about 185 m (about 607 ft), so Eratosthenes' measurement was probably at least 15 to 20 percent too large. Today, the earth's equatorial circumference is measured at 40,075 km (about 24,886 mi). Nonetheless, Eratosthenes' calculation showed that the ancient Greeks had a pretty good idea about the earth's size as well as its shape (figure 20.2).

Today, the shape and size of the earth has been precisely measured by artificial satellites circling the earth. These measurements have found that the earth is not a perfectly round sphere as believed by the ancient Greeks. The earth is flattened

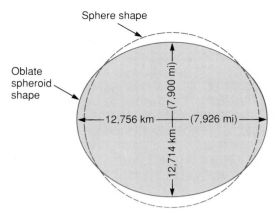

Figure 20.3
The earth has an irregular, slightly lopsided, slightly pear-shaped form. In general, it is considered to have the shape of an oblate spheroid, departing from a perfect sphere as shown here.

at the poles and has an equatorial bulge as do many other planets. In fact, you can observe through a telescope that both Jupiter and Saturn are considerably flattened at the poles. A shape that is flattened at the poles has a greater distance through the equator than through the poles, which is described as an *oblate* shape. The earth, like a water-filled, round balloon resting on a table, has an oblate shape. The earth is not perfectly symmetrically oblate, however, since the North Pole is slightly higher and the South Pole is slightly lower than the average surface. In addition, the earth is not perfectly circular around the equator, with a lump in the Pacific and a depression in the Indian Ocean. The shape of the earth is a slightly pear-shaped, slightly lopsided **oblate spheroid.** All the elevations and depressions are less than 85 m (about 280 ft), however, which is practically negligible compared to the size of the earth. Thus the earth is very close to, but not exactly, an oblate spheroid. The significance of this shape will become apparent when the earth's motions are discussed next (figure 20.3).

Motions of the Earth

Ancient civilizations had a fairly accurate understanding of the size and shape of the earth but had difficulty accepting the idea that the earth moves. The geocentric theory of a motionless earth with the sun, moon, planets, and stars circling the earth was discussed in the previous chapter. Ancient people had difficulty with anything but a motionless earth for at least two reasons: (1) they could not sense any motion of the earth and (2) they had ideas about being at the center of a universe that was created for them. Thus it was not until the 1700s that the concept of an earth in motion became generally accepted. Today, the earth is understood to move a number of different ways, seven of which were identified in the introduction to this chapter. Three of these motions are independent of motions of the sun and the Galaxy. These are (1) a yearly revolution around the sun, (2) a daily rotation on its axis, and (3) a slow clockwise wobble of its axis.

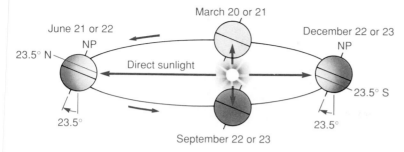

Figure 20.4

The consistent tilt and orientation of the earth's axis as it moves around its orbit is the cause of the seasons. The North Pole is pointing toward the sun during the summer solstice and away from the sun during the winter solstice.

Revolution

The earth moves constantly around the sun in a slightly elliptical orbit that requires an average of one year for one complete circuit. Recall that the movement around the sun is called a *revolution* and that all points of the earth's orbit lie in a plane called the *plane of the ecliptic*. The average distance between the earth and the sun is about 150 million km (about 93 million mi).

The earth's orbit is slightly elliptical, so the earth moves with a speed that varies according to Kepler's laws of planetary motion (see chapter 19). It moves fastest when it is closer to the sun in January, at perihelion, and moves slowest when it is farthest away from the sun in early July, at aphelion. The earth is about 2.5 million km (about 1.5 million mi) closer to the sun in January and about the same distance farther away in July than it would be if the orbit were a circle. This total difference of about 5 million km (about 3 million mi) results in a January sun with an apparent diameter that is 3 percent larger than the July sun, and the earth as a whole receives about 6 percent more solar energy in January. The effect of this closer sun is much less than the effect of some directional relationships, and winter occurs in the Northern Hemisphere when the earth is closest to the sun. Likewise, summer occurs in the Northern Hemisphere when the sun is at its greatest distance from the earth (figure 20.4).

The important directional relationships that override the effect of the earth's distance from the sun involve the daily *rotation,* or spinning, of the earth around an imaginary line through the geographic poles called the earth's *axis.* The important directional relationships are a *constant inclination* of the earth's axis to the plane of the ecliptic and a *constant orientation* of the earth's axis to the stars. The **inclination of the earth's axis** to the plane of the ecliptic is about 66.5° (or 23.5° from a line perpendicular to the plane). This relationship between the plane of the earth's orbit and the tilt of the earth's axis is considered to be the same day after day throughout the year, even though small changes do occur in the inclination over time. Likewise, the **orientation of the earth's axis** to the stars is

considered to be the same throughout the year as the earth moves through its orbit. Again, small changes do occur in the orientation over time. Thus, in general, the axis points in the same direction, remaining essentially parallel to its position during any day of the year. The essentially constant orientation and inclination of the axis results in the earth's axis pointing toward the sun as it moves in one part of its orbit, then pointing away from the sun six months later. The generally constant inclination and orientation of the axis, together with the earth's rotation and revolution, combine to produce three related effects: (1) days and nights that vary in length, (2) changing seasons, and (3) climates that vary with latitude.

Figure 20.4 shows how the North Pole of the earth points toward the sun on June 21 or 22, then away from the sun on December 22 or 23 as it maintains its orientation to the stars. When the North Pole is pointed toward the sun it receives sunlight for a full twenty-four hours, and the South Pole is in the earth's shadow for a full twenty-four hours. This is summer in the Northern Hemisphere with the longest daylight periods and the sun at its maximum noon height in the sky. Six months later, on December 22 or 23, the orientation is reversed with winter in the Northern Hemisphere, the shortest daylight periods, and the sun at its lowest noon height in the sky.

The beginning of a season can be recognized from any one of the three related observations: (1) the length of the daylight period, (2) the altitude of the sun in the sky at noon, or (3) the length of a shadow from a vertical stick at noon. All of these observations vary with changes in the direction of the earth's axis of rotation relative to the sun. On about June 22 and December 22 the sun reaches its highest and lowest noon altitudes as the earth moves to point the North Pole directly toward the sun (June 21 or 22) and directly away from the sun (December 22 or 23). Thus the sun appears to stop increasing or decreasing its altitude in the sky, stop, then reverse its movement twice a year. These times are known as **solstices** after the Latin meaning "sun stand still." The Northern Hemisphere's **summer solstice** occurs on about June 22 and identifies the beginning of the summer season. At the summer solstice the sun at noon has the highest altitude and the shadow from a vertical stick is shorter than any other day of the year. The Northern Hemisphere's **winter solstice** occurs on about December 22 and identifies the beginning of the winter season. At the winter solstice the sun at noon has the lowest altitude and the shadow from a vertical stick is longer than any other day of the year (figure 20.5).

As the earth moves in its orbit between pointing its North Pole toward the sun on about June 22 and pointing it away on about December 22, there are two times when it is halfway between. At these times the axis of the earth is perpendicular to a line between the center of the sun and the earth, and daylight and night are of equal length. These are called the **equinoxes** after the Latin meaning "equal nights." The **spring equinox** (also

The Earth as a Planet

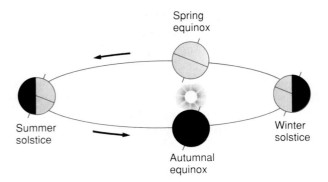

Figure 20.5

The length of daylight during each season is determined by the relationship of the earth's shadow to the tilt of the axis. At the equinoxes, the shadow is perpendicular to the latitudes, and day and night are of equal length everywhere. At the summer solstice, the North Pole points toward the sun and is completely out of the shadow for a twenty-four-hour day. At the winter solstice, the North Pole is in the shadow for a twenty-four-hour night. The situation is reversed for the South Pole.

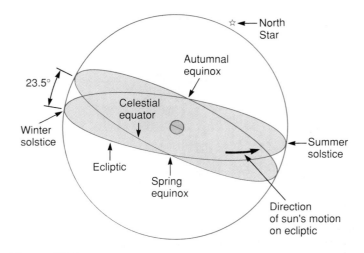

Figure 20.6

The position of the sun on the celestial sphere at the solstices and the equinoxes.

called the **vernal equinox**) occurs on about March 21 and identifies the beginning of the spring season. The **autumnal equinox** occurs on about September 23 and identifies the beginning of the fall season.

The relationship between the apparent path of the sun on the celestial sphere and the seasons is shown in figure 20.6. Recall that the celestial equator is a line on the celestial sphere directly above the earth's equator. The equinoxes are the points on the celestial sphere where the ecliptic, the path of the sun, crosses the celestial equator. Note also that the summer solstice occurs when the ecliptic is 23.5° north of the celestial equator, and the winter solstice occurs when it is 23.5° south of the celestial equator.

Rotation

Observing the apparent turning of the celestial sphere once a day and seeing the east to west movement of the sun, moon, and stars, it certainly seems as if it is the heavenly bodies and not the earth doing the moving. You cannot sense any movement, and there is little apparent evidence that the earth indeed moves. Evidence of a moving earth comes from at least three different observations: (1) the observation that the other planets and the sun rotate, (2) the observation of the changing plane of a long, heavy pendulum at different latitudes on earth, and (3) the observation of the direction of travel of something moving across, but above, the earth's surface such as a rocket.

Other planets, such as Jupiter, and the Sun can be observed to rotate by keeping track of features on the surface such as the Great Red Spot on Jupiter and sunspots on the Sun. While such observations are not direct evidence that the earth also rotates, it does show that other members of the Solar System spin on their axes. As described earlier, Jupiter is also observed to

be oblate, flattened at its poles with an equatorial bulge. Since the earth is also oblate, this is again indirect evidence that the earth rotates too.

The most easily obtained and convincing evidence about the earth's rotation comes from a **Foucault pendulum,** a heavy mass swinging from a long wire. This pendulum is named after the French physicist Jean Foucault, who first used a long pendulum in 1851 to prove that the earth rotates. Foucault started a long, heavy pendulum moving just above the floor, marking the plane of its back-and-forth movement. Over some period of time, the pendulum appeared to slowly change its position, smoothly shifting its plane of rotation. Science museums often show this shifting plane of movement by setting up small objects for the pendulum to knock down. Foucault demonstrated that the pendulum actually maintains its plane of movement in space (inertia) while the earth rotates eastward (counterclockwise) under the pendulum. It is the earth that turns under the pendulum, causing the pendulum to appear to change its plane of rotation. It is difficult to imagine the pendulum continuing to move in a fixed direction in space while the earth, and everyone on it, turns under the swinging pendulum.

Figure 20.7 illustrates the concept of the Foucault pendulum. A pendulum is attached to a support on a stool that is free to rotate. If the stool is slowly turned while the pendulum is swinging, you will observe that the pendulum maintains its plane of rotation while the stool turns under it. If you were much smaller and now on the stool below the pendulum, it would appear to turn as you rotate with the turning stool. This is what happens on the earth. Such a pendulum at the North Pole would make a complete turn in about twenty-four hours. Moving south from the North Pole, the change decreases with latitude until, at the equator, the pendulum would not appear to turn at all. At higher latitudes, the plane of the pendulum appears to move clockwise in the Northern Hemisphere and counterclockwise in the Southern Hemisphere.

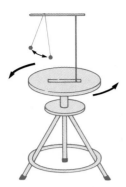

Figure 20.7
The Foucault pendulum swings back and forth in the same plane while a stool is turned beneath it. Likewise, a Foucault pendulum on the earth's surface swings back and forth in the same plane while the earth turns beneath it. The amount of turning observed depends on the latitude of the pendulum.

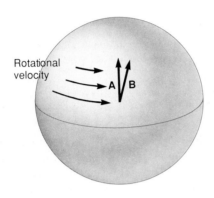

Rotational velocity

Figure 20.8
The earth has a greater rotational velocity at the equator and less toward the poles. As an object moves north or south (A), it passes over land with a different rotational velocity, which produces a deviation to the right in the Northern Hemisphere (B) and to the left in the Southern Hemisphere.

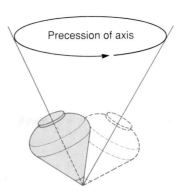

Precession of axis

Figure 20.9
A spinning top wobbles as it spins, and the axis of the top traces out a small circle. The wobbling of the axis is called *precession*.

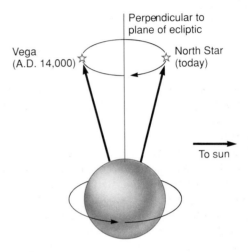

Perpendicular to plane of ecliptic

Vega (A.D. 14,000)

North Star (today)

To sun

Figure 20.10
The slow, continuous precession of the earth's axis results in the North Pole pointing around a small circle over a period of about 26,000 years.

More evidence that the earth rotates is provided by objects that move above and across the earth's surface. As shown in figure 20.8, the earth has a greater rotational velocity at the equator than at the poles. As an object leaves the surface and moves north or south, the surface has a different rotational velocity, so it rotates beneath the object as it proceeds in a straight line. This gives the moving object an apparent deflection to the right of the direction of movement in the Northern Hemisphere and to the left in the Southern Hemisphere. The apparent deflection caused by the earth's rotation is called the **Coriolis effect.** The Coriolis effect will explain the earth's prevailing wind systems as well as the characteristic direction of wind in areas of high pressure and areas of low pressure (see chapter 23).

Precession

If the earth were a perfect spherically shaped ball, its axis would always point to the same reference point among the stars. The reaction of the earth to the gravitational pull of the moon and the sun on its equatorial bulge, however, results in a slow wobbling of the earth as it turns on its axis. This slow wobble of the earth's axis, called **precession,** causes it to swing in a slow circle like the wobble of a spinning top (figure 20.9). It takes the earth's axis about 26,000 years to complete one turn, or wobble. Today, the axis points very close to the North Star, Polaris, but is slowly moving away to point to another star. In about 12,000 years the star Vega will appear to be in the position above the North Pole, and Vega will be the new North Star. The moving pole also causes changes over time in which particular signs of the zodiac appear with the spring equinox (see chapter 18). Because of precession the occurrence of the spring equinox has been moving backward (westward) through the zodiac constellations at about 1 degree every 72 years. Thus after about 26,000 years the spring equinox will have moved through all the constellations and will again approach the constellation of Aquarius for the next "age of Aquarius" (figure 20.10).

The Earth as a Planet

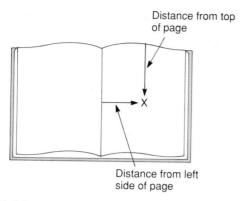

Figure 20.11

Any location on a flat, two-dimensional surface is easily identified with two references from two edges. This technique does not work on a motionless sphere because there are no reference points.

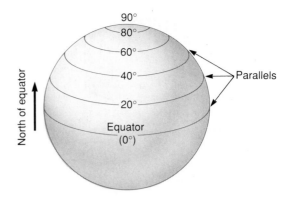

Figure 20.12

A circle that is parallel to the equator is used to specify a position north or south of the equator. A few of the possibilities are illustrated here.

Place and Time

The continuous rotation and revolution of the earth establishes an objective way to determine direction, location, and time on the earth. If the earth were an unmoving sphere there would be no side, end, or point to provide a referent for direction and location. The earth's rotation, however, defines an axis of rotation, which serves as a reference point for determination of direction and location on the entire surface. The earth's rotation and revolution together define cycles, which define standards of time. The following describes how the earth's movements are used to identify both place and time.

Identifying Place

A system of two straight lines can be used to identify a point, or position, on a flat, two-dimensional surface. The position of the letter *X* on this page, for example, can be identified by making a line a certain number of measurement units from the top of the page and a second line a certain number of measurement units from the left side of the page. Where the two lines intersect will identify the position of the letter *X*, which can be recorded or communicated to another person (figure 20.11).

A system of two straight lines can also be used to identify a point, or position, on a sphere except this time the lines are circles. The reference point for a sphere is not as simple as in the flat, two-dimensional case, however, since a sphere does not have a top or side edge. The earth's axis provides the north-south reference point. The equator is a big circle around the earth that is exactly halfway between the two ends, or poles, of the rotational axis. An infinite number of circles are imagined to run around the earth parallel to the equator as shown in figure 20.12. The east and west running parallel circles are called **parallels.** Each parallel is the same distance between the equator and one of the poles all the way around the earth. The distance from the equator to a point on a parallel is called the **latitude** of that point. Latitude tells you how far north or south a point

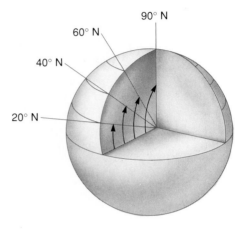

Figure 20.13

If you could see to the earth's center, you would see that latitudes run from 0° at the equator north to 90° at the North Pole (or to 90° south at the South Pole).

is from the equator by telling you the parallel the point is located on. The distance is measured northward from the equator (which is 0°) to the North Pole (90° north) or southward from the equator (0°) to the South Pole (90° south) (figure 20.13). If you are somewhere at a latitude of 35° north, you are somewhere on the earth on the 35° latitude line north of the equator.

Since a parallel is a circle, a location of 40° N latitude could be anyplace on that circle around the earth. To identify a location you need another line, this time one that runs pole to pole and perpendicular to the parallels. These north-south running arcs that intersect at both poles are called **meridians** (figure 20.14). There is no naturally occurring, identifiable meridian that can be used as a point of reference such as the equator serves for parallels, so one is identified as the referent by international agreement. The referent meridian is the one that passes through the Greenwich Observatory near London, England, and this meridian is called the **prime meridian.** The distance from

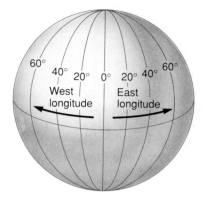

Figure 20.14

Meridians run pole to pole perpendicular to the parallels and provide a reference for specifying east and west directions.

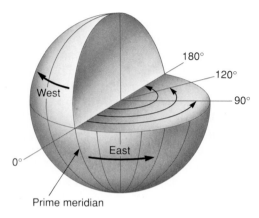

Figure 20.15

If you could see inside the earth, you would see 360° around the equator and 180° of longitude east and west of the prime meridian.

the prime meridian east or west is called the **longitude.** The degrees of longitude of a point on a parallel is measured to the east or to the west from the prime meridian up to 180° (figure 20.15). New Orleans, Louisiana, for example, has a latitude of about 30° N of the equator and a longitude of about 90° W of the prime meridian. The location of New Orleans is therefore described as 30° N, 90° W.

Locations identified with degrees of latitude north or south of the equator and degrees of longitude east or west of the prime meridian are more precisely identified by dividing each degree of latitude into subdivisions of 60 minutes (60′) per degree, and each minute into 60 seconds (60″). On the other hand, latitudes near the equator are sometimes referred to in general as the *low latitudes,* and those near the poles are sometimes called the *high latitudes.*

In addition to the equator (0°) and the poles (90°), the parallels of 23.5° N and 23.5° S from the equator are important references for climatic consideration. The parallel of 23.5° N is called the **tropic of Cancer,** and 23.5° S is called the **tropic of Capricorn.** These two parallels identify the limits

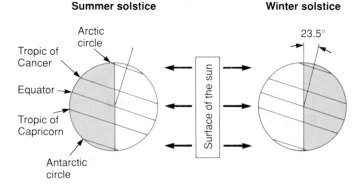

Figure 20.16

At the summer solstice, the noon sun appears directly overhead at the tropic of Cancer (23.5° N), and twenty-four hours of daylight occurs north of the arctic circle (66.5° N). At the winter solstice, the noon sun appears overhead at the tropic of Capricorn (23.5° S) and twenty-four hours of daylight occurs south of the antarctic circle (66.5° S).

toward the poles of where the sun appears directly overhead during the course of a year. The parallel of 66.5° N is called the **arctic circle,** and the parallel of 66.5° S is called the **antarctic circle.** These two parallels identify the limits toward the equator of where the sun appears above the horizon all day for six months during the summer (figure 20.16).

Measuring Time

Standards of time are determined by intervals between two successive events that repeat themselves in a regular way. Since ancient civilizations, many of the repeating events used to mark time have been recurring cycles associated with the rotation of the earth on its axis and its revolution around the sun. Thus the day, month, season, and year are all measures of time based on recurring natural motions of the earth. All other measures of time are based on other events or definitions of events. There are, however, several different ways to describe the day, month, and year, and each depends on a different set of events. These events are described in the following section.

Daily Time

The technique of using astronomical motions for keeping time originated some four thousand years ago with the Babylonian culture. The Babylonians marked the yearly journey of the sun against the background of the stars, which was divided into twelve periods, or months, after the signs of the zodiac. Based on this system, the Babylonian year was divided into twelve months with a total of 360 days. In addition, the Babylonians invented the week and divided the day into hours, minutes, and seconds. The week was identified as a group of seven days, each based on one of the seven heavenly bodies that were known at the time. The hours, minutes, and seconds of a day were determined from the movement of the shadow around a straight, vertical rod.

Figure 20.17

A sundial indicates the apparent local solar time at a given instant in a given location. The time read from a sundial is based on an average solar time usually different from the time read from a clock.

As seen from a place in space above the North Pole, the earth rotates counterclockwise turning toward the east. On the earth, this motion causes the sun to appear to rise in the east, travel across the sky, and set in the west. The changing angle between the tilt of the earth's axis and the sun produces an apparent shift of the sun's path across the sky, northward in the summer season and southward in the winter season. The apparent movement of the sun across the sky was the basis for the ancient as well as the modern standard of time known as the day.

Today, everyone knows that the earth turns as it moves around the sun, but it is often convenient to regard space and astronomical motions as the ancient Greeks did, as a celestial sphere that turns around a motionless earth. Recall that the celestial meridian is a great circle on the celestial sphere that passes directly overhead where you are and continues around the earth through both celestial poles. The movement of the sun across the celestial meridian identifies an event of time called **noon.** As the sun appears to travel west it crosses meridians that are farther and farther west, so the instant identified as noon moves west with the sun. The instant of noon at any particular longitude is called the **apparent local noon** for that longitude because it identifies noon from the apparent position of the sun in the sky. The morning hours before the sun crosses the meridian are identified as *ante meridiem* (A.M.) hours, which is Latin for "before meridian." Afternoon hours are identified as *post meridiem* (P.M.) hours, which is Latin for "after the meridian."

There are several ways that the movement of the sun across the sky can be measured. The ancient Babylonians, for example, used a vertical rod called a *gnomon* to make and measure a shadow that moved as a result of the apparent changes of the sun's position. The gnomon eventually evolved into a *sundial,* a vertical or slanted gnomon with divisions of time marked on a horizontal plate beneath the gnomon. The shadow from the gnomon indicates the **apparent local solar time** at a given place and a given instant from the apparent position of the sun in the sky. If you have ever read the time from a sundial, you know that it usually does not show the same time as a clock or a watch (figure 20.17). In addition, sundial time is nonuniform, fluctuating throughout the course of a year, sometimes running ahead of clock time and sometimes running behind clock time.

A sundial shows the apparent local solar time, but clocks are set to measure a uniform standard time based on **mean solar time.** Mean solar time is a uniform time averaged from the apparent solar time. The apparent solar time is nonuniform, fluctuating because (1) the earth moves sometimes faster and sometimes slower in its elliptical orbit around the sun and (2) the equator of the earth is inclined to the ecliptic. The combined consequences of these two effects is a variable, nonuniform sundial time as compared to the uniform mean solar time, otherwise known as clock time.

A day is defined as the length of time required for the earth to rotate once on its axis. There are different ways to measure this rotation, however, which result in different definitions of the day. A **sidereal day** is the interval between two consecutive crossings of the celestial meridian by a particular star ("sidereal" is a word meaning "star"). This interval of time depends only on the time the earth takes to rotate 360° on its axis. One sidereal day is practically the same length as any other sidereal day since the earth's rate of rotation is constant for all practical purposes.

An **apparent solar day** is the interval between two consecutive crossings of the celestial meridian by the sun, for example, from one local solar noon to the next solar noon. Since the earth is moving in orbit around the sun, the earth must turn a little bit farther to compensate for its orbital movement, bringing the sun back to local solar noon (figure 20.18). As a consequence, the apparent solar day is about four minutes longer than the sidereal day. This additional time accounts for the observation that the stars and constellations of the zodiac rise about four minutes earlier every night, appearing higher in the sky at the same clock time until they complete a yearly cycle. A sidereal day is twenty-three hours, fifty-six minutes, and four seconds long. A **mean solar day** is twenty-four hours long, averaged from the mean solar time to keep clocks in closer step with the sun than would be possible using the variable apparent solar day. Just how out of synchronization the apparent solar day can become with a clock can be illustrated with another ancient way of keeping track of the sun's motions in the sky, the "hole in the wall" sun calendar and clock.

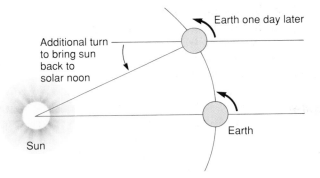

Earth one day later

Additional turn
to bring sun
back to
solar noon

Earth

Sun

Figure 20.18

Because the earth is moving in orbit around the sun, it must rotate an additional distance each day, requiring about four minutes to bring the sun back across the celestial meridian (local solar noon). This explains why the stars and constellations rise about four minutes earlier every night.

Variations of the "hole in the wall" sun calendar were used all over the world by many different ancient civilizations, including the early Native Americans of the American Southwest. More than one ancient Native American ruin has small holes in the western wall aligned in such a way to permit sunlight to enter a chamber only on the longest and shortest days of the year. This established a basis for identifying the turning points in the yearly cycle of seasons.

A hole in the roof can be used as a sun clock, but it will require a whole year to establish the meaning of a beam of sunlight shining on the floor. Imagine a beam of sunlight passing through a small hole to make a small spot of light on the floor. For a year you mark the position of the spot of light on the floor *each day* when your clock tells you the *mean solar time is noon.* You trace out an elongated, lopsided figure eight with the small end pointing south and the larger end pointing north (figure 20.19a). Note by following the monthly markings shown in figure 20.19b that the figure-eight shape is actually traced out by the spot of sunlight making two S shapes as the sun changes its apparent position in the sky. Together, the two S shapes make the shape of the figure eight.

Why did the sunbeam trace out a figure eight over a year? The two extreme north-south positions of the figure are easy to understand because by December, the earth is in its orbit with the North Pole tilted away from the sun. At this time the direct rays of the sun fall on the tropic of Capricorn (23.5° south of the equator), and the sun appears low in the sky as seen from the Northern Hemisphere. Thus on this date, the winter solstice, a beam of sunlight strikes the floor at its northernmost position beneath the hole. By June, the earth has moved halfway around its orbit and the North Pole is now tilted toward the sun. The direct rays of the sun now fall on the tropic of Cancer (23.5° north of the equator), and the sun appears high in the sky as

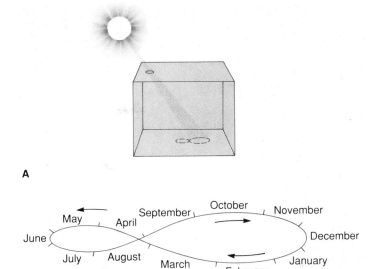

A

May   April   September   October   November

June   December

July   August   March   February   January

B

Figure 20.19

(a) During a year, a beam of sunlight traces out a lopsided figure eight on the floor if the position of the light is marked at noon every day. (b) The location of the point of light on the figure eight during each month.

seen from the Northern Hemisphere (figure 20.20). Thus on this date, the summer solstice, a beam of sunlight strikes the floor at its southernmost position beneath the hole.

If everything else were constant, the path of the spot would trace out a straight line between the northernmost and southernmost positions beneath the hole. The east and west movements of the point of light as it makes an S shape on the floor must mean, however, that the sun crosses the celestial meridian (noon) earlier one part of the year and later the other part. This early and late arrival is explained in part by the earth moving at different speeds in its orbit. Recall that the earth moves faster when at perihelion, when it is closest to the sun during the winter season. Starting from perihelion, the faster moving earth travels farther in its orbit as it completes one rotation. This means that the earth must rotate farther to bring the sun back across the celestial meridian. As a result, the sun appears to move more slowly across the sky. By April the apparent local noon occurs almost eight minutes after the mean solar time as shown by a clock. Six months later the earth is moving from aphelion, so it appears to move across the sky more rapidly. In October the apparent local noon occurs almost eight minutes before the mean solar time as shown by a clock. Figure 20.21 is a graph of these differences in time, showing how many minutes the apparent local solar time (sundial time) is ahead or behind the mean solar time (clock time) for each month. Note that the two clocks are together at the times of perihelion and aphelion of the earth in its orbit.

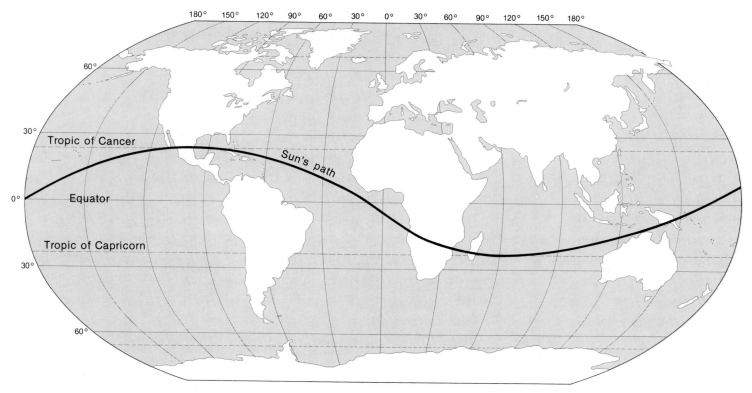

Figure 20.20

The path of the sun's direct rays during a year. The sun is directly over the tropic of Cancer at the summer solstice and high in the Northern Hemisphere sky. At the winter solstice, the sun is directly over the tropic of Capricorn and low in the Northern Hemisphere sky.

If changes in orbital speed were the only reason that the sun does not cross the sky at the same rate during the year, the spot of sunlight on the floor would trace out an oval rather than a figure eight. The plane of the ecliptic, however, does not coincide with the plane of the earth's equator, so the sun appears at different angles in the sky and this makes it appear to change its speed during different times of the year. As shown in figure 20.22, this effect changes the length of the apparent solar day by making the sun up to ten minutes later or earlier than the mean solar time four times a year between the solstices and equinoxes.

The two effects add up to a cumulative variation between the apparent local solar time (sundial time) and the mean solar time (clock time) (figure 20.23). This cumulative variation is known as the **equation of time,** which shows how many minutes sundial time is faster or slower than clock time during different days of the year. The equation of time is often shown on globes in the figure-eight shape called an *analemma,* which also can be used to determine the latitude of direct solar radiation for any day of the year.

Since the local mean time varies with longitude, every place on an east-west line around the earth could possibly have clocks that were a few minutes ahead of those to the west and a few minutes behind those to the east. To avoid the confusion that would result from many clocks set to local mean solar time, the earth's surface is arbitrarily divided into one-hour **standard time zones** (figure 20.24). Since there are 360° around the earth and 24 hours in a day, this means that each time zone is 360° divided by 24, or 15° wide. These 15° zones are adjusted so that whole states are in the same time zone, or the zones are adjusted for other political reasons. The time for each zone is defined as the mean solar time at the middle of each zone. When you cross a boundary between two zones, the clock is set ahead one hour if you are traveling east and back one hour if you are traveling west. Most states adopt **daylight saving time** during the summer, setting clocks ahead one hour in the spring and back one hour in the fall ("spring ahead and fall back"). Daylight saving time results in an extra hour of daylight during summer evenings.

The 180° meridian is arbitrarily called the **international date line,** an imaginary line established to compensate for cumulative time zone changes (figure 20.25). A traveler crossing

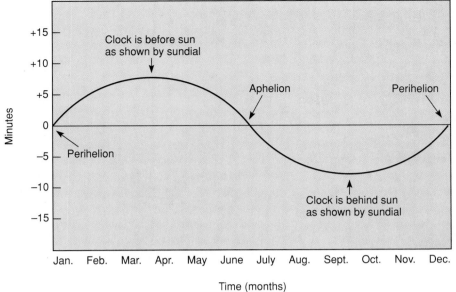

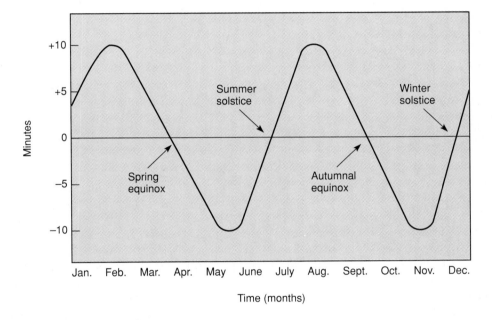

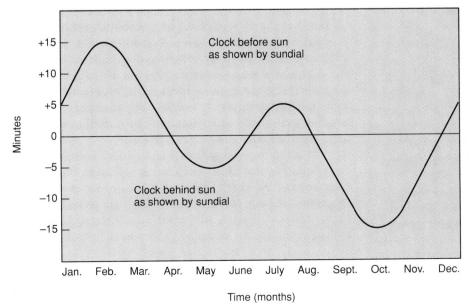

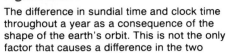

Figure 20.21

The difference in sundial time and clock time throughout a year as a consequence of the shape of the earth's orbit. This is not the only factor that causes a difference in the two clocks (see figures 20.22 and 20.23).

Figure 20.22

The difference in sundial time and clock time throughout a year as a consequence of the angle between the plane of the ecliptic and the plane of the equator (see also figures 20.21 and 20.23).

Figure 20.23

The equation of time, which shows how many minutes sundial time is faster or slower than clock time during different months of the year. The correction comes from a combination of two factors shown in figures 20.21 and 20.22.

The Earth as a Planet

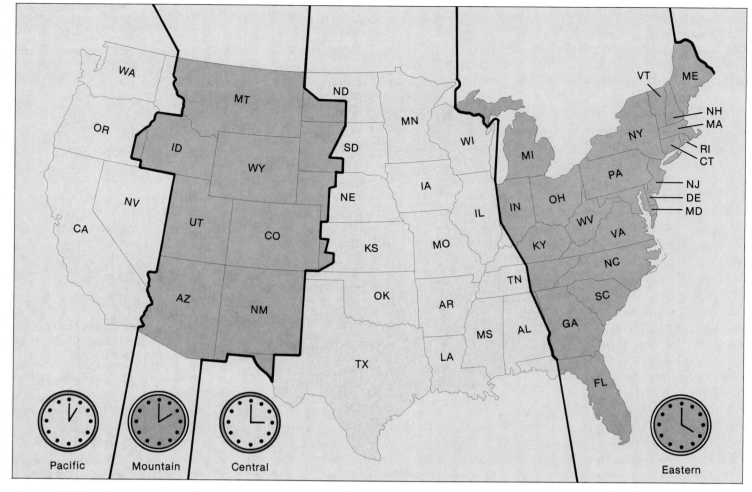

Figure 20.24

The standard time zones. Hawaii and most of Alaska are two hours earlier than Pacific Standard Time.

the date line gains or loses a day just as crossing a time zone boundary results in the gain or loss of an hour. A person moving across the line while traveling westward gains a day, for example, Tuesday is followed by Thursday. A person crossing the line while traveling eastward repeats a day, for example, Tuesday is followed by Tuesday.

Yearly Time

A *year* is generally defined as the interval of time required for the earth to make one complete revolution in its orbit. As was the case for definitions of a day, there are different definitions of what is meant by a year. The most common definition of a year is the interval between two consecutive spring equinoxes, which is known as the **tropical year** (*trope* is Greek for "turning"). The tropical year is 365 days, 5 hours, 48 minutes, and 46 seconds, or 365.24220 mean solar days.

A **sidereal year** is defined as the interval of time required for the earth to move around its orbit so the sun is again in the same position relative to the stars. The sidereal year is slightly longer than the tropical year because the earth rotates more than 365.25 times during one revolution. Thus the sidereal year is 365.25636 mean solar days, which is about 20 minutes longer than the tropical year.

The tropical and sidereal years would be the same interval of time if the earth's axis pointed in a consistent direction. The precession of the axis, however, results in the axis pointing in a slightly different direction with time. This shift of direction over the course of a year moves the position of the spring equinox westward, and the equinox is observed 20 minutes before the orbit has been completely circled. The position of the spring equinox against the background of the stars thus moves westward by some 50 minutes of arc per year.

It is the *tropical year* that is used as a standard time interval to determine the calendar year. The earth does not complete an exact number of turns on its axis while completing one trip around the sun, so it becomes necessary to periodically adjust the calendar so it stays in step with the seasons. The calendar system that was first designed to stay in step with the seasons

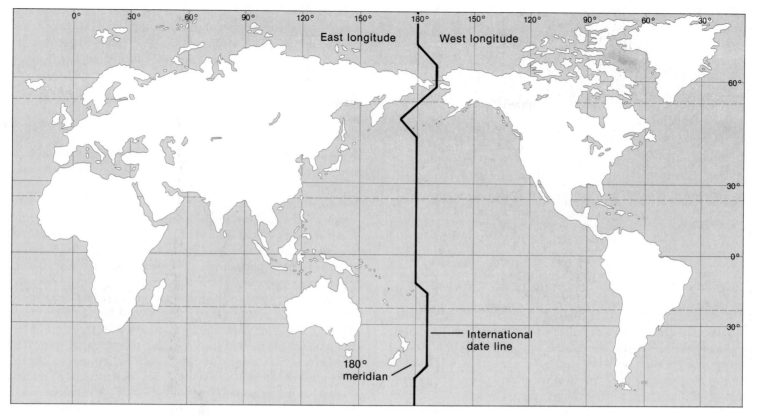

Figure 20.25
The international date line follows the 180° meridian but is arranged in a way that land areas and island chains have the same date.

was devised by the ancient Romans. Julius Caesar reformed the calendar, beginning in 46 B.C., to have a 365 day year with a 366 day year (leap year) every fourth year. Since the tropical year of 365.24220 mean solar days is very close to 365¼ days, the system, called a *Julian calendar,* accounted for the ¼ day by adding a full day to the calendar every fourth year. The Julian calendar was very similar to the one presently used, except the year began in March, the month of the spring equinox. The month of July was named in honor of Julius Caesar, and the following month was later named after his successor, Augustus.

There was a slight problem with the Julian calendar since it was longer than the tropical year by 365.25 minus 365.24220, or 0.0078 day per year. This small interval (which is 11 minutes, 14 seconds) does not seem significant when compared to the time in a whole year. But over the years the error of minutes and seconds grew to an error of days. By 1582, when Pope Gregory XIII revised the calendar, the error had grown to 10 days. This revision resulted in the *Gregorian calendar,* which is the system used today. Since the accumulated error of 0.0078 day per year is almost 0.75 day per century, it follows that four centuries will have 0.75 times 4, or 3 days of error. The Gregorian system corrects for the accumulated error by dropping the additional leap year day three centuries out of every four. Thus the century

year of 2000 will be a leap year with 366 days, but the century years of 2100, 2200, and 2300 will not be leap years. You will note that this approximation still leaves an error of 0.0003 day per century, so another calender revision will be necessary in a few thousand years to keep the calendar in step with the seasons.

Monthly Time
In ancient times, people often used the moon to measure time intervals that were longer than a day but shorter than a year. The word "month," in fact, has its origins in the word "moon" and its period of revolution. The moon revolves around the earth in an orbit that is inclined to the plane of the earth's orbit, the plane of the ecliptic, by about 5°. The moon is thus never more than about ten apparent diameters from the ecliptic. It revolves in this orbit in about 27⅓ days as measured by two consecutive crossings of any star. This period is called a **sidereal month.** The moon rotates in the same period as the time of revolution, so the sidereal month is also the time required for one rotation. Because the rotation and revolution rates are the same, you always see the same side of the moon from the earth.

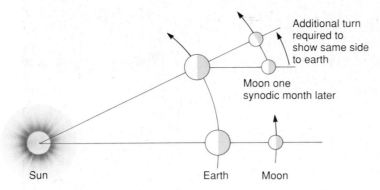

Figure 20.26

As the moon moves in its orbit around the earth, it must revolve a greater distance to bring the same part to face the earth. The additional turning requires about 2.2 days, making the synodic month longer than the sidereal month.

The ancient concept of a month was based on the **synodic month,** the interval of time from new moon to new moon (or any two consecutive identical phases). The synodic month is longer than a sidereal month at a little more than 29½ days. The moon's phases (see next section) are determined by the relative positions of the earth, moon, and sun. As shown in figure 20.26, the moon moves with the earth in its orbit around the sun. During one sidereal month the moon has to revolve a greater distance before the same phase is observed on the earth and this greater distance requires 2.2 days. This makes the synodic month about 29½ days long, only a little less than 1/12 of a year, or the period of time the present calendar identifies as a "month."

The Moon

Next to the sun, the moon is the largest, brightest object in the sky. The moon is the earth's nearest neighbor at an average distance of 380,000 km (about 238,000 mi), and surface features can be observed with the naked eye. With the aid of a telescope or a good pair of binoculars, you can see light-colored mountainous regions called the **lunar highlands,** smooth dark areas called **maria,** and many sizes of craters, some with bright streaks extending from them (figure 20.27). The smooth dark areas are called maria after a Latin word meaning "sea." They acquired this name from early observers who thought the dark areas were oceans and the light areas were continents. Today, the maria are understood to have formed from ancient floods of molten lava that poured across the surface and solidified to form the "seas" of today. There is no water and no atmosphere on the moon.

Many facts about the moon were established during the Apollo mission, the first human exploration of a place away from the earth. A total of twelve Apollo astronauts walked on the moon, taking thousands of photographs, conducting hundreds of experiments, and returning to the earth with over 380 kg (about 840 lb) of moon rocks. In addition, instruments were left on the moon that continued to radio data back to the earth after the Apollo program ended in 1972. As a result of the Apollo mission many questions were answered about the moon, but unanswered questions still remain. Among others, the big question that remains to be answered is how the moon formed in the first place.

Composition and Features

The Apollo astronauts found that the surface of the moon is covered by a 3 m (about 10 ft) layer of fine gray dust that contains microscopic glass beads. The dust and beads were formed from millions of years of continuous bombardment of micrometeorites. These very small meteorites generally burn up in the earth's atmosphere. The moon does not have an atmosphere that protects its surface, so they have continually fragmented and pulverized the surface in a slow, steady rain. The glass beads are believed to have formed when larger meteorite impacts melted part of the surface, which was immediately forced into a fine spray that cooled rapidly while above the surface.

The rocks on the surface of the moon were found to be mostly *basalts,* a type of rock formed on the earth from the cooling and solidification of molten lava. The dark-colored rocks from the maria are similar to earth's basalts but contain greater amounts of titanium and iron oxides. The light-colored rocks from the highlands are mostly *breccias,* a kind of rock made up of rock fragments that have been compacted together. On the moon, the compacting was done by meteorite impacts. The rocks from the highlands contain more aluminum and less iron than the maria basalts and thus have a lower density (2.9 g/cm^3) than the darker mare rocks (3.3 g/cm^3).

All the moon rocks contained a substantial amount of radioactive elements, which made it possible to precisely measure the age of the rocks. The light-colored rocks from the highlands were formed some 4 billion years ago. The dark-colored rocks from the maria were much younger, with ages ranging from 3.1 to 3.8 billion years. This indicates a period of repeated volcanic eruptions and lava flooding over a 700 million year period that ended about 3 billion years ago.

Seismometers left on the moon by Apollo astronauts detected only very weak moonquakes, so weak that they would not be felt by a person. These moonquakes are thought to be produced by the nearby impact of larger meteoroids or by a slight cracking of the crust from gravitational interactions with the earth and the sun. The movement of these seismic waves through the moon's interior suggests that the moon has an internal structure. The outer layer of solid rock, or crust, is about 65 km (about 40 mi) thick on the side that always faces the earth and is about twice as thick on the far side. The data also suggest a small, partly molten iron core at a depth of about 900 km (about 600 mi) beneath the surface. A small core would account for the moon's low density (3.34 g/cm^3) as compared to the earth's average (5.5 g/cm^3) and would account for the observation that the moon has no general magnetic field.

Figure 20.27

You can easily see the light-colored lunar highlands, smooth and dark maria, and many craters on the surface of earth's nearest neighbor in space.

The Earth as a Planet

History of the Moon

The moon rocks brought back to earth, the results of the lunar seismographs, and all the other data gathered through the Apollo missions have increased our knowledge about the moon, leading to new understandings of how it formed. This model pictures the present moon developing through four distinct stages of events.

The *origin stage,* the first stage in the history of the moon, describes how the moon originally formed. There is still uncertainty about how the moon formed, and there are three kinds of theories concerning its origin:

1. The *fission* theory—the moon formed from a part of the earth that broke away early in the earth's formation.
2. The *condensation* theory—the moon and the earth formed at the same time in neighboring parts of the original solar nebula.
3. The *capture* theory—the moon formed elsewhere, independently of the earth, and was later captured by the earth's gravitational field when it moved by the earth.

All three theories about how the moon formed have strong and weak points, and the findings of the Apollo mission did not rule out any one theory. The condensation theory, or some version of it, seems most likely, with the earth and moon condensing from the solar nebula some 4.6 billion years ago.

The *molten surface stage,* the second stage, occurred during the first 200 million years. Heating from a number of sources could have been involved in the melting of the entire lunar surface 100 km (about 60 mi) or so deep. The heating needed to melt the surface is believed to have generally accumulated from the impacts of rock fragments, which were leftover debris from the formation of the Solar System that intensely bombarded the moon. As the bombardment subsided the molten outer layer cooled and solidified to solid rock. Between 4.2 to 3.9 billion years ago the subsiding meteorite bombardment on the cooling crust formed most of the craters that are seen on the moon today.

The *molten interior stage,* the third stage in the development of the moon, involved the melting of the interior of the moon. Radioactive decay had been slowly heating the interior, and 3.8 billion years ago, or about a billion years after the moon formed, sufficient heat accumulated to melt the interior. The light and heavier rock materials separated during this period, perhaps producing a small iron core. Molten lava flowed into basins on the surface during this period, forming the smooth, darker maria seen today. The lava flooding continued for about 700 million years, ending about 3.1 billion years ago.

The *cold and quiet stage,* the fourth and last stage in the development of the moon, began 3.1 billion years ago as the last lava flow cooled and solidified. Since that time the surface of the moon has been continually bombarded by micrometeorites and a few larger meteorites. With the exception of a few new craters, the surface of the moon has changed little in the last three billion years.

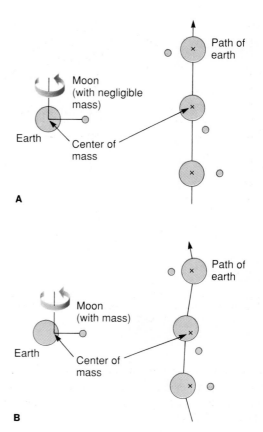

Figure 20.28

(*a*) If the moon had a negligible mass, the center of gravity between the moon and earth would be the earth's center and the earth would follow a smooth orbit around the sun. (*b*) The actual location of the center of mass between the earth and moon results in a slightly in and out, or wavy, path around the sun.

The Earth-Moon System

The earth and its moon are unique in the Solar System because of the size of the moon. It is not the largest satellite, but the ratio of its mass to the earth's mass is greater than the mass ratio of any other moon to its planet. This comparison excludes the Pluto-Charon pair, which really appears to be more like a binary planet than a planet and its satellite. The moon has a diameter of 3,476 km (about 2,159 mi), which is about one-fourth the diameter of the earth, and a mass of about 1/81 of the earth's mass. This is a small fraction of the earth's mass, but it is enough to effect the earth's motion as it revolves around the sun.

If the moon had a negligible mass it would circle the earth with a center of rotation (center of mass) located at the center of the earth. In this situation the center of the earth would follow a smooth path around the sun (figure 20.28a). The mass of the moon, however, is great enough to move the center of rotation away from the earth's center toward the moon. As a result, both bodies act as a system, moving around a center of mass. The

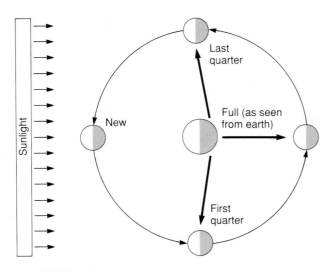

Figure 20.29

Half of the moon is always lighted by the sun and half is always in the shadow. The moon phases result from the view of the lighted and dark parts as the moon revolves around the earth.

center of mass between the earth and moon follows a smooth orbit around the sun. The earth follows a slightly wavy path around the sun as it slowly revolves around the common center of mass (figure 20.28b).

Phases of the Moon

The phases of the moon are a result of the changing relative positions of the earth, moon, and sun as the earth-moon system moves around the sun. Sunlight always illuminates half of the moon and half is always in shadow. As the moon's path takes it between the earth and sun, then to the dark side of the earth, you see different parts of the illuminated half called *phases* (figure 20.29). When the moon is on the dark side of the earth you see the entire illuminated half of the moon called the **full moon** (or the full phase). Halfway around the orbit, the lighted side of the moon now faces away from the earth and the unlighted side now faces the earth. This dark appearance is called the **new moon** (or the new phase). In the new phase the moon is not *directly* between the earth and the sun, so it does not produce an eclipse (see the next section).

As the moon moves from the new phase in its orbit around the earth, you will eventually see half the lighted surface, which is known as the **first quarter.** Often the unlighted part of the moon shines with a dim light of reflected sunlight from the earth called *earthshine*. Note that the division between the lighted and unlighted part of the moon's surface is curved in an arc. A straight line connecting the ends of the arc is perpendicular to the direction of the sun (figure 20.30). After the first quarter the moon moves to its full phase, then to the **last quarter** (see figure 20.29). The period of time between two consecutive phases, such as new moon to new moon, is the synodic month, or about 29.5 days.

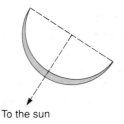

Figure 20.30

The cusps, or horns, of the moon always point away from the sun. A line drawn from the tip of one cusp to the other is perpendicular to a straight line between the moon and the sun.

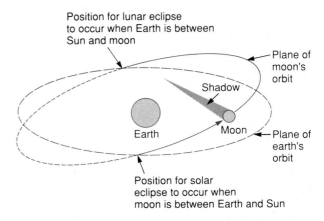

Figure 20.31

The plane of the moon's orbit is inclined to the plane of the earth's orbit by about 5°. An eclipse occurs only where the two planes intersect and the earth, moon, and sun are in a line.

Eclipses of the Sun and Moon

Sunlight is not visible in the emptiness of space because there is nothing to reflect the light, so the long conical shadow behind each spherical body is not visible either. One side of the earth and one side of the moon are always visible because they reflect sunlight. The shadow from the earth or from the moon becomes noticeable only when it falls on the illuminated surface of the other body. This event of the earth or moon shadow falling on the other body is called an **eclipse.** Most of the time eclipses do not occur because the plane of the moon's orbit is inclined to the earth's orbit about 5° (figure 20.31). As a result, the shadow from the moon or the shadow from the earth usually falls above or below the other body, too high or too low to produce an eclipse. An eclipse occurs only when the sun, moon, and earth are in a line with each other.

The shadow from the earth and the shadow from the moon are long cones that point away from the sun. Both cones have two parts, an inner cone of a complete shadow called the **umbra** and an outer cone of partial shadow called the **penumbra.** When and where the umbra of the moon's shadow falls on the earth, people see a **total solar eclipse.** During a total solar eclipse the new moon completely covers the disk of the sun. The total solar

The Earth as a Planet

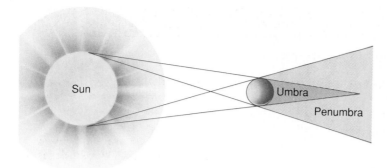

Figure 20.32

People in a location where the tip of the umbra falls on the surface of the earth see a total eclipse. People in locations where the penumbra fall see a partial eclipse.

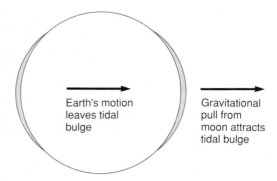

Figure 20.33

Two forces are important in the production of tides, the gravitational attraction that pulls on earth's waters on the side facing the moon and the centripetal reaction that pulls the earth out of a straight-line path, leaving a tidal bulge on the side of the earth opposite the moon.

eclipse is preceded and followed by a partial eclipse, which is seen when the observer is in the penumbra. If the observer is in a location where only the penumbra passes, then only a partial eclipse will be observed (figure 20.32). More people see partial than full solar eclipses because the penumbra covers a larger area. The occurrence of a total solar eclipse is a rare event in a given location, occurring once every several hundred years and then lasting for less than seven minutes.

The moon's cone-shaped shadow averages a length of 375,000 km (about 233,000 mi), which is less than the average distance between the earth and the moon. The moon's elliptical orbit brings it sometimes closer to and sometimes farther from the earth. A total solar eclipse occurs only when the moon is close enough so at least the tip of its umbra reaches the surface of the earth. If the moon's umbra fails to reach the earth, an **annular eclipse** occurs. Annular means "ring-shaped," and during this eclipse the edge of the sun is seen to form a bright ring around the moon. As before, people located in the area where the penumbra falls will see a partial eclipse. The annular eclipse occurs more frequently than the total solar eclipse.

When the moon is full and the sun, moon, and earth are lined up so the earth's shadow falls on the moon a **lunar eclipse** occurs. The earth's shadow is much larger than the moon's diameter, so a lunar eclipse is visible to everyone on the night side of the earth. This larger shadow also means a longer eclipse that may last for hours. As the umbra moves over the moon the darkened part takes on a reddish, somewhat copper-colored glow from light refracted and scattered into the umbra by earth's atmosphere. This light passes through the thickness of the earth's atmosphere on its way to the eclipsed moon, and it acquires the reddish color for the same reason that a sunset is red, that is, that much of the blue light has been removed by scattering in the earth's atmosphere.

Tides

If you live near or have ever visited a coastal area of the ocean, you are familiar with the periodic rise and fall of sea level known as **tides.** The relationship between the motions of the moon and the magnitude and timing of tides has been known and studied since very early times. These relationships are that (1) the greatest range of tides occurs at full and new moon phases,

(2) the least range of tides occurs at quarter moon phases, and (3) in most oceans the time between two high tides or between two low tides is an average of twelve hours and twenty-five minutes. The period of twelve hours and twenty-five minutes is half the average time interval between consecutive passages of the moon across the celestial meridian. A location on the surface of the earth is directly under the moon when it crosses the meridian and directly opposite it on the far side of the earth an average twelve hours and twenty-five minutes later. There are two *tidal bulges* that follow the moon as it moves around the earth, one on the side facing the moon and one on the opposite side. In general, tides are a result of these bulges moving westward around the earth.

A simplified explanation of the two tidal bulges involves two basic factors, the gravitational attraction of the moon and the motion of the earth-moon system (figure 20.33). Water on the earth's surface is free to move, and the moon's gravitational attraction pulls the water to the tidal bulge on the side of the earth facing the moon. This tide-raising force directed toward the moon bulges the water in mid-ocean some .75 m (about 2.5 ft), but it also bulges the land, producing a land tide. Since the earth is much more rigid than water, the land tide is much smaller at about 12 cm (about 4.5 in). Since all parts of the land bulge together this movement is not evident without measurement by sensitive instruments.

The tidal bulge on the side of the earth opposite to the moon is explained by movement of the earth-moon system. Recall that the moon does not simply revolve around the earth, but both the moon and earth revolve around a center of gravity between the pair. Since the earth is much more massive than the moon the center of gravity of the pair is much closer to the earth, located within the earth opposite the moon. The center of the moon revolves around this center of gravity in a large circle, and the center of the earth revolves around it in a small circle. Water, being a fluid free to move, follows a straight-line path (in accord with Newton's first law of motion) as the earth moves in its small circle around the earth-moon center of gravity. Earth's gravity, however, pulls the water out of the straight-line path. This forms a tidal bulge on the *outside* of earth's path of revolution about the common center of gravity, that is, on the side of the earth opposite from the moon.

Earth's Motions and Ice Ages

The evidence is abundant and widespread that the earth has undergone repeated cycles of glaciation in the past, periods when glaciers and ice sheets covered much of the land. There have been at least ten of these periods of glaciation, or *ice ages* as they are called, in the past million years. The most recent ice age maximum occurred about 18,000 years ago when about one-third of the earth's land was covered with ice. A great sheet of ice, perhaps several kilometers (about 1.25 mi) thick in places, covered all of Canada into the United States as far south as Long Island, along the Ohio and Missouri Rivers, and along a line to Oregon. Half the states in the United States were partly or completely covered by ice during this ice age. At that time, average temperatures were roughly 8° C (about 14° F) lower than the average today in temperate climates. About 10,000 years ago the average temperature began to increase and the North American ice sheet melted back, retreating from its maximum advance. The average temperature then increased gradually until some 5,000 years ago, cooling slightly since. Other temperature fluctuations have followed, most recently causing the rapid retreat of glaciers in Alaska and Canada.

Is a new ice age coming in the future or is the climate becoming warmer? To make such a long-range climatic forecast requires some explanation of why ice ages occur. There are different theories about the cause of ice ages, ranging from variations in solar energy output to the movement of continents, but recent evidence seems to support a theory concerning the earth's motions around the sun. The theory is called the Milankovitch theory after the geologist who originally proposed it in 1920. This theory proposes that long-term variations in the earth's climate are a result of a combination of three factors: (1) slow changes in the size and shape of the

earth's orbit around the sun, (2) slow changes in the inclination of earth's axis, and (3) slow changes in the direction of earth's axis with respect to the stars. These three factors are seen to result in changes in the amount of solar energy received at different latitudes during different seasons of the year.

The changes in the basic earth motions of (1) orbital size and shape, (2) tilt, or inclination of axis, and (3) precession are produced by small gravitational attractions of the moon and the giant planets of Jupiter and Saturn. They produce nearly periodic changes that can be calculated with great accuracy for periods of thousands of years.

The size and shape of the earth's orbit varies with a period of about 100,000 years and changes the solar energy received by the earth by a factor of about 0.3 percent over a million years. The tilt of the earth's axis, on the other hand, varies between 22.1° and 24.5° with a period of about 40,000 years. This period of changes of tilt does not alter the total amount of solar energy received by the earth, but it does vary the amount of sunlight received during the summers and winters of both hemispheres by as much as 20 percent. This would be important because in order for ice to accumulate, growing into an ice sheet, snow must survive the warmer summer months. Less sunlight than normal in the Northern Hemisphere could result in a growing ice sheet, moving the climate toward an ice age.

The earth's precession, with its 26,000 year period, couples with the rotation of the earth's elliptical orbit around the sun to determine the time of perihelion. The time of perihelion has a period of 20,000 years, presently occurring during the summer of the Southern Hemisphere. As a whole, the earth receives about 6 percent more solar energy at perihelion than six months later. In about another 12,000 years, precession will have

moved the time of perihelion so it occurs during the summer of the Northern Hemisphere.

The combinations of changes in the earth's orbit do correlate with the times of the ice ages according to geologic evidence as well as computer analysis. Computer programs analyzed the amount of solar energy received at summertime high latitude locations in the Northern Hemisphere according to variations in the earth's orbit, axial tilt, and time of perihelion. The analysis predicted major ice ages occurring in a 100,000 year cycle with minor fluctuations at periods of 20,000 and 40,000 years. This prediction correlates roughly with the evidence of past ice ages according to the geologic evidence. The computer model also indicates that the earth will reach another cold period in 60,000 years after some periods of lesser variations.

There are apparently connections between variations in the earth's orbit and the time of ice ages, but it is nonetheless difficult to use this information alone to predict the next ice age. Other major factors are important as well as changes in the earth's orbit. Volcanic eruptions or even a collision with a large meteorite could inject major amounts of dust into the atmosphere. On a small scale, a single volcano in recent times was observed to inject a sufficient amount of dust into the atmosphere to scatter and reflect sunlight, lowering the average summer temperature by several degrees. On the other hand, the continued burning of fossil fuels and destruction of rain forests are increasing the present-day carbon dioxide concentration of the atmosphere. There is a possibility of an increasing greenhouse effect, or warming trend, from such human activities. All of the variations, fluctuations, and possibilities emphasize the difficulty of climatic forecasting, even with understandings about the role of changes in the earth's orbit and the ice ages.

The tides do not actually appear as alternating bulges that move around the earth. There are a number of factors that influence the making and moving of the bulges in complex interactions that determine the timing and size of a tide at a given time in a given location. Some of these factors include (1) the relative positions of the earth, moon, and sun, (2) the elliptical orbit of the moon, which sometimes brings it closer to the earth, and (3) the size, shape, and depth of the basin holding the water.

The relative positions of the earth, moon, and sun determine the size of a given tide because the sun, as well as the moon, produces a tide-raising force. The sun is much more massive than the moon, but it is so far away that its tide-raising force is about half that of the closer moon. Thus the sun basically modifies lunar tides rather than producing distinct tides of its own. For example, the earth, moon, and sun are nearly in

The Earth as a Planet

line during the full and new moon phases. At these times the lunar and solar tide-producing forces act together, producing tides that are unusually high and corresponding low tides that are unusually low. The period of these unusually high and low tides are called periods of **spring tides.** Spring tides occur every two weeks and have nothing to do with the spring season. When the moon is in its quarter phases the sun and moon are at right angles to one another and the solar tides occur between the lunar tides, causing unusually less pronounced high and low tides called **neap tides.** The period of neap tides also occurs every two weeks.

The size of the lunar-produced tidal bulge varies as the moon's distance from the earth changes. The moon's elliptical orbit brings it closest to the earth at a point called **perigee** and at a farthest point called **apogee.** At perigee the moon is about 44,800 km (about 28,000 mi) closer to the earth than at apogee, so its gravitational attraction is much greater. When perigee coincides with a new or full moon, especially high spring tides result.

The open basins of oceans, gulfs, and bays are all connected but have different shapes and sizes and have bordering land masses in all possible orientations to the westward moving tidal bulges. Water in each basin responds differently to the tidal forces, responding as periodic resonant oscillations that move back and forth much like the water in a bowl shifts when carried. Thus coastal regions on open seas may experience tides that range between about 1 and 3 m (about 3 to 10 ft), but mostly enclosed basins such as the Gulf of Mexico have tides less than about 1/3 m (about 1 ft). The Gulf of Mexico, because of its size, depth, and limited connections with the open ocean, responds only to the stronger tidal attractions and has only one high and one low tide per day. Even lakes and ponds respond to tidal attractions, but the result is too small to be noticed. Other basins, such as the Bay of Fundy in Nova Scotia, are funnel-shaped and undergo an unusually high tidal range. The Bay of Fundy has experienced as much as a 15 m (about 50 ft) tidal range.

As the tidal bulges are pulled against a rotating earth, friction between the moving water and the ocean basin tends to slow the earth's rotation over time. This is a very small slowing effect that is increasing the length of each day by about 1.5 seconds per 100,000 years. Evidence for this slowing comes from a number of sources including records of ancient solar eclipses. The solar eclipses of 2,000 years ago occurred 3 hours earlier than would be expected by using today's time but were on the mark if a lengthening day is considered. Fossils of a certain species of coral still living today provide further evidence of a lengthening day. This particular coral adds daily growth rings, and 500-million-year-old fossils show that the day was about 21 hours long at that time. Finally, the moon is moving away from

the earth at a rate of about 4 cm (about 1.5 in) per year. This movement out to a larger orbit is a necessary condition to conserve angular momentum as the earth slows. As the moon moves from the earth the length of the month increases. Some time in the distant future, millions of years from now, both the day and the month will be equal, about 50 of the present days long.

Summary

The earth is an *oblate spheroid* that undergoes three basic motions: (1) a yearly *revolution* around the sun, (2) a daily *rotation* on its axis, and (3) a slow wobble of its axis called *precession.*

As the earth makes its yearly *revolution* around the sun, it maintains a generally *constant inclination of its axis* to the *plane of the ecliptic* of 66.5°, or 23.5° from a line perpendicular to the plane. In addition, the earth maintains a generally constant *orientation of its axis* to the stars, which always points in the same direction. The constant inclination and orientation of the axis, together with the earth's rotation and revolution, produce three effects: (1) days and nights that vary in length, (2) seasons that change during the course of a year, and (3) climates that vary with latitude. When the earth is at a place in its orbit so the axis points toward the sun, the Northern Hemisphere experiences the longest days and the summer season. This begins on June 21 or 22, which is called the *summer solstice.* Six months later, the axis points away from the sun and the Northern Hemisphere experiences the shortest days and the winter season. This begins on December 22 or 23 and is called the *winter solstice.* On March 20 or 21 the earth is halfway between the solstices and has days and nights of equal length, which is called the *spring (or vernal) equinox.* On September 22 or 23 the *autumnal equinox,* another period of equal nights and days, identifies the beginning of the fall season.

Precession is a slow wobbling of the axis as the earth spins. Precession is produced by the gravitational tugs of the sun and moon on the earth's equatorial bulge.

Lines around the earth that are parallel to the equator are circles called *parallels.* The distance from the equator to a point on a parallel is called the *latitude* of that point. North and south arcs that intersect at the poles are called *meridians.* The meridian that runs through the Greenwich Observatory is a reference line called the *prime meridian.* The distance of a point east or west of the prime meridian is called the *longitude* of that point.

The event of time called *noon* is the instant the sun appears to move across the celestial meridian. The instant of noon at a particular location is called the *apparent local noon.* The time at a given place that is determined by a sundial is called the *apparent local solar time.* It is the basis for an averaged, uniform standard time called the *mean solar time.* Mean solar time is the time used to set clocks.

A *sidereal day* is the interval between two consecutive crossings of the celestial meridian by a star. An *apparent solar day* is the interval between two consecutive crossings of the celestial meridian by the sun, from one apparent solar noon to the next. A *mean solar day* is twenty-four hours as determined from mean solar time. The *equation of time* shows how the local solar time is faster or slower than the clock time during different days of the year.

The earth's surface is divided into one-hour *standard time zones* that are about 15° of meridian wide. The *international date line* is the 180° meridian; you gain a day if you cross this line while traveling westward and repeat a day if you are traveling eastward.

A *tropical year* is the interval between two consecutive spring equinoxes. A *sidereal year* is the interval of time between two consecutive crossings of a star by the sun. It is the tropical year that is used as a standard time interval for the calendar year. A *sidereal month* is the interval of time between two consecutive crossings of a star by the moon. The *synodic month* is the interval of time from a new moon to the next new moon. The synodic month is about 29 1/2 days long, which is about 1/12 of a year.

The surface of the moon has light-colored mountainous regions called *highlands*, smooth dark areas called *maria*, and many sizes of *craters* and is covered by a layer of fine *dust*. Samples of rocks returned to the earth by Apollo astronauts revealed that the highlands are composed of basalt breccias that were formed some 4 billion years ago. The maria are basalts that formed from solidified lava some 3.1 to 3.8 billion years ago. This, and other data, indicate that the moon developed through four stages.

The earth and moon act as a system, with both bodies revolving around a common center of mass located under the earth's surface. This combined motion around the sun produces three phenomena: (1) as the earth-moon system revolves around the sun different parts of the illuminated moon surface, called *phases*, are visible from the earth; (2) a *solar eclipse* is observed where the moon's shadow falls on the earth, and a *lunar eclipse* is observed where the earth's shadow falls on the moon; and (3) the *tides*, a periodic rising and falling of sea level, are produced by gravitational attractions of the moon and sun and by the movement of the earth-moon system.

Key Terms

annular eclipse (p. 424)
antarctic circle (p. 413)
apogee (p. 426)
apparent local noon (p. 414)
apparent local solar time
   (p. 414)
apparent solar day (p. 414)
arctic circle (p. 413)
autumnal equinox (p. 410)
Coriolis effect (p. 411)
daylight saving time (p. 416)
eclipse (p. 423)
equation of time (p. 416)
equinoxes (p. 409)
first quarter (p. 423)
Foucault pendulum (p. 410)
full moon (p. 423)
inclination of earth's axis
   (p. 409)
international date line (p. 416)
last quarter (p. 423)
latitude (p. 412)
longitude (p. 413)
lunar eclipse (p. 424)
lunar highlands (p. 420)
maria (p. 420)
mean solar day (p. 414)
mean solar time (p. 414)
meridians (p. 412)

neap tides (p. 426)
new moon (p. 423)
noon (p. 414)
oblate spheroid (p. 408)
orientation of earth's axis
   (p. 409)
parallels (p. 412)
penumbra (p. 423)
perigee (p. 426)
precession (p. 411)
prime meridian (p. 412)
sidereal day (p. 414)
sidereal month (p. 419)
sidereal year (p. 418)
solstices (p. 409)
spring equinox (p. 409)
spring tides (p. 426)
standard time zones (p. 416)
summer solstice (p. 409)
synodic month (p. 420)
tides (p. 424)
total solar eclipse (p. 423)
tropical year (p. 418)
tropic of Cancer (p. 413)
tropic of Capricorn (p. 413)
umbra (p. 423)
vernal equinox (p. 410)
winter solstice (p. 409)

Applying the Concepts

1. If you are located at 20° N latitude the sun will appear directly overhead
   a. never.
   b. once a year.
   c. twice a year.
   d. four times a year.

2. The time as read from a sundial is the same as the time read from a clock
   a. all the time.
   b. only once a year.
   c. twice a year.
   d. four times a year.

3. You are traveling west by jet and cross three time zone boundaries. If your watch reads 3:00 P.M. when you arrive, you should reset it to
   a. 12:00 noon.
   b. 6:00 P.M.
   c. 12:00 midnight.
   d. 6:00 A.M.

4. If it is Sunday when you cross the international date line while traveling westward the next day is
   a. Wednesday.
   b. Sunday.
   c. Tuesday.
   d. Saturday.

5. What has happened to the surface of the moon during the last three billion years?
   a. heavy meteorite bombardment produced craters
   b. widespread lava flooding from the interior
   c. both widespread lava flooding and meteorite bombardment
   d. not much

6. If you see a full moon, an astronaut on the moon looking back at the earth would see a
   a. full earth.
   b. new earth.
   c. first quarter earth.
   d. last quarter earth.

7. A lunar eclipse can occur only during the moon phase of
   a. full moon.
   b. new moon.
   c. first or last quarter.
   d. any of the above.

8. A total solar eclipse can occur only during the moon phase of
   a. full moon.
   b. new moon.
   c. first or last quarter.
   d. any of the above.

9. A lunar eclipse does not occur every month because
   a. the plane of the moon's orbit is inclined to the ecliptic.
   b. of precession.
   c. the earth moves faster in its orbit when closest to the sun.
   d. the earth's axis is tilted with respect to the sun.

The Earth as a Planet

10. The least range between high and low tides occurs during
    a. full moon.
    b. new moon.
    c. quarter moon phases.
    d. an eclipse.

Answers

1. c 2. d 3. a 4. a 5. d 6. b 7. a 8. b 9. a
10. c

Questions for Thought

1. Briefly describe the more conspicuous of the earth's motions. Identify which of these motions are independent of the Sun and the Galaxy.

2. Describe some evidences that (a) the earth is shaped like a sphere and (b) that the earth moves.

3. Use sketches with brief explanations to describe how the constant inclination and constant orientation of the earth's axis produces (a) a variation in the number of daylight hours and (b) a variation in seasons throughout a year.

4. Where on the earth are you if you observe the following at the instant of apparent local noon on September 23? (a) The shadow from a vertical stick points northward. (b) There is no shadow on a clear day. (c) The shadow from a vertical stick points southward.

5. What is the meaning of the word "solstice"? What causes solstices? On about what dates do solstices occur?

6. What is the meaning of "equinox"? What causes equinoxes? On about what dates do equinoxes occur?

7. What is precession?

8. Briefly describe how the earth's axis is used as a reference for a system that identifies locations on the earth's surface.

9. Use a map or a globe to identify the latitude and longitude of your present location.

10. The tropic of Cancer, tropic of Capricorn, arctic circle, and antarctic circle are parallels that are identified with specific names. What parallels do the names represent? What is the significance of each?

11. What is the meaning of (a) noon, (b) A.M., and (c) P.M.?

12. Explain why the time shown by a sundial does not usually agree with the time shown by a clock. Describe how the sundial time can be corrected to clock time.

13. Explain why standard time zones were established. In terms of longitude, how wide is a standard time zone? Why was this width chosen?

14. When it is 12:00 noon in Texas, what time is it (a) in Jacksonville, Florida? (b) in Bakersfield, California? (c) at the North Pole?

15. On what date is the earth closest to the sun? What season is occurring in the Northern Hemisphere at this time? Explain this apparent contradiction.

16. Explain why a lunar eclipse is not observed once a month.

17. Use a sketch and briefly describe the conditions necessary for a total eclipse of the sun.

18. Using sketches, briefly describe the position of the earth, moon, and sun during each of the major moon phases.

19. If you were on the moon as people on the earth observed a full moon, in what phase would you observe the earth?

20. Briefly describe three theories about the origin of the moon.

21. What are the smooth, dark areas that can be observed on the face of the moon? When did they form?

22. What made all the craters that can be observed on the moon? When did this happen?

23. What phase is the moon in if it rises at sunset? Explain your reasoning.

24. Why doesn't an eclipse of the sun occur at each new moon when the moon is between the earth and the sun?

25. Is the length of time required for the moon to make one complete revolution around the earth the same length of time required for a complete cycle of moon phases? Explain.

26. What is an annular eclipse? Which is more common, an annular eclipse or a total solar eclipse? Why?

27. Does an eclipse of the sun occur during any particular moon phase? Explain.

28. Identify the moon phases that occur with (a) a spring tide and (b) a neap tide.

29. What was the basic problem with the Julian calendar? How does the Gregorian calendar correct this problem?

30. What is the source of the dust found on the moon?

31. Describe the four stages in the moon's history.

32. Explain why everyone on the dark side of the earth can see a lunar eclipse, but only a limited few ever see a solar eclipse on the lighted side.

33. Explain why there are two tidal bulges on opposite sides of the earth.

34. Describe how the Foucault pendulum provides evidence that the earth turns on its axis.

35. Why are consecutive high tides commonly twelve hours and twenty-five minutes apart.

Section IV

The land surface, the ocean, and the atmosphere of the earth appear to be separate parts of the earth. Recently developed theories and ideas about the earth have found that all parts are connected and are engaged in an ongoing cycling of materials. Thus, changes in one part will affect other parts.

Earth Sciences

Some of the most recent and revolutionary ideas of science have been developed in the earth sciences. The revolution began with the understanding that parts of the earth's surface are continually moving. This understanding led to the theory of plate tectonics, which brought together the dynamics of moving continents, ocean basins that are constantly being created and destroyed, and a crust subjected to volcanoes and earthquakes. The revolution of ideas continued as scientists began to understand that the separate parts of the earth—the interior, the surface, the atmosphere, and the water on the surface—are all engaged in an ongoing cycling of materials. Compared to other planets, this ongoing cycling between the different parts of the earth makes it unique in a number of ways. Not the least of the results of this unique cycling is the presence of life on the earth.

The fascinating story of earth's unique, unified cycles begins with a chapter on rocks, the basic materials that make up the earth. Evidence for a number of sources is used to describe how the earth's interior is arranged as well as how rocks and rock materials cycle into and out of the interior. Evidence for the oceans leads to the development of the plate tectonic theory,

which is basic to understanding the other cycles. Next, you will learn how rocks are folded, faulted, and sculptured to produce the landscape you see on the earth today. Understanding the sculpturing processes will enable you to interpret the landscape for meaning about the earth's past and perhaps its future.

The story of earth's cycles continues with the atmosphere, the thin layer of air around the earth. Causes of changing weather, predicting the weather, and climate are considered along with interactions between the atmosphere and earth's ocean. Finally, earth's waters are considered as another part of planet earth, a part that has its own unique patterns of circulation and cycles.

As you can see, the earth is not made of isolated, independent parts. All parts of this dynamic earth are related, and changes in one part affect the other parts. Yet, in spite of all the changes and cycles, each part of the earth has remained amazingly stable for millions of years. The study of the earth sciences is thus viewed in a broad context, a context that will help you to understand the earth on which you live as well as present-day environmental concerns.

The Dynamic Earth

Figure 21.1

An aerial view from the south of the eruption of Mount St. Helens volcano on May 18, 1980.

UNDERSTANDINGS about the earth come from many branches of science, each with specialized scientists who consider different scales of space and time. *Astronomers* study the motions of earth in space, its place in the universe, and how the earth formed and evolved over time. *Oceanographers* are primarily concerned with the composition and motions of earth's oceans. *Meteorologists* are concerned with the composition of the atmosphere and how it changes over time. *Geologists* are concerned primarily with the rocks, landscapes, and history of earth through time. Each of these branches of science has its own subdisciplines. For example, there are geologists who study rocks (petrology), geologists who study earthquakes and the earth's interior (seismology), and geologists who study the early history of the earth (paleontology). Together, all the scientists who study the earth are known as *earth scientists.*

The separation of the earth sciences into independent branches and subdisciplines was traditionally done for convenience in studying a large and complex earth. In the past, scientists in each branch studied their field without considering the earth as an interacting whole. Today, most earth scientists consider changes in the earth as taking place in an overall dynamic system. The parts of the earth's interior, the rocks on the surface, the oceans, and the atmosphere are today understood to be parts of a complex, interacting system with a cyclic movement of materials from one part to another.

How can materials cycle through changes from the interior, to the surface, and to the atmosphere and back? As you will see in this and the following chapters, the answer to this question is found in the unique combination of fluids of the earth. No other planet has earth's combination of (1) an atmosphere consisting mostly of nitrogen and oxygen, (2) a surface that is mostly covered with liquid water, and (3) an interior that is partly fluid, partly semifluid, and partly solid. Earth's atmosphere is unique both in terms of its composition and in terms of interactions with the liquid water surface. These interactions have cycled materials, such as carbon dioxide, from the atmosphere to the land and oceans of the earth. The internal flow of rock materials, on the other hand, produces the large-scale motion of the earth's continents and the associated phenomena of earthquakes and volcanoes. Volcanoes cycle carbon dioxide back into the atmosphere (figure 21.1), and the movement of land cycles rocks from the earth's interior to the surface and back to the interior again. Altogether, the earth's atmosphere, liquid water, and motion of its land masses make up a dynamic cycling system that is found only on the planet earth.

Earth also appears to be unique because there is life on it but not on the other planets. The cycling of atmospheric gases and vapors, waters of the surface, and flowing interior rock materials sustains a wide diversity of life on the earth, with some two million different species of plants and animals. Yet, there is no evidence of even one species of life existing outside the earth. The existence of life on the earth must be directly related to the earth's unique, dynamic system of interacting fluids.

This and the remaining chapters of this book are about the dynamic nature of planet earth. This chapter is concerned with earth materials, the internal structure of the earth, and the large-scale motion of the continents of earth. The remaining chapters will consider changes that occur on the surface, in the atmosphere, and in the ocean.

Solid Earth Materials

The earth, as all other solid matter in the universe, is made up of atoms of the chemical elements. The elements are not distributed equally throughout the mass of the earth, however, nor are they equally abundant. As you shall see in the next section, there is evidence that the earth was molten during an early stage in its development. During this molten stage the heavier abundant elements, such as iron and nickel, apparently sank to the deep interior of the earth, leaving a thin layer of lighter elements on the surface. This thin layer is called the *crust.* The rocks and rock materials that you see on the surface and the materials sampled in even the deepest mines and well holes are all materials of the earth's crust. The bulk of the earth's mass lies below the crust and has not been directly sampled.

Chemical analysis of literally thousands of rocks from the earth's surface found that only eight elements make up about 98.6 percent of the crust with all the other elements making up the remaining 1.4 percent. Oxygen is the most abundant element, making up about 50 percent of the weight of the crust. Silicon makes up over 25 percent, so oxygen and silicon alone make up about 75 percent of the earth's solid surface.

Figure 21.2 shows the eight most abundant elements that occur as elements or combine to form the chemical compounds of the earth's crust. They make up the solid materials of the earth's crust that are known as minerals and rocks. In general, a *mineral* is a naturally occurring element or chemical compound with a crystalline structure. This definition will be expanded later. For now, consider a mineral to be a solid material of the earth that has both a known chemical composition and a crystalline structure that is unique to that mineral. About 2,500 minerals are known to exist, but only about 20 are common in the crust. Examples of these common minerals are quartz, calcite, and gypsum.

Minerals are the fundamental building blocks of the rocks making up the earth's crust. A *rock* is a solid aggregation of minerals or mineral materials that has been cohesively brought together by a rock-forming process. There are many possibilities of different kinds of rocks that could exist from many different variations of mineral mixtures. Within defined ranges of composition, however, there are only about 20 common rocks making up the crust of the earth. Examples of common rocks are sandstone, limestone, and granite.

How the atoms of elements combine depends on the number and arrangement of electrons around the atoms. How they form a mineral with specific properties and a specific crystalline structure depends on the electrons and the size of the ions as well. A discussion of the chemical principles that determine the structure and properties of chemical compounds is found in chapters 12 and 13. You may wish to review these principles before continuing with the discussion of common minerals and rocks.

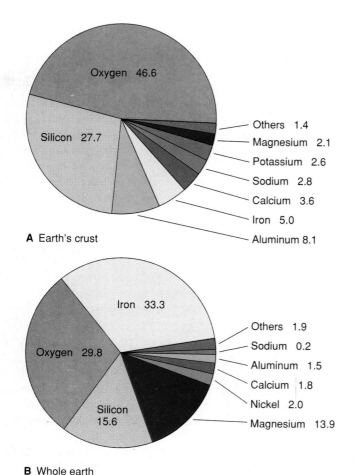

A Earth's crust

B Whole earth

Figure 21.2

(a) The percentage by weight of the elements that make up the earth's crust. (b) The percentage by weight of the elements that make up the whole earth.

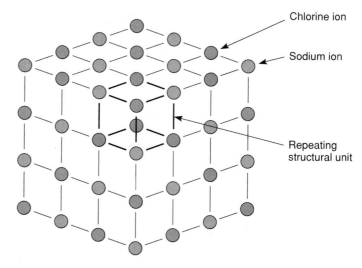

Figure 21.3

A crystal is composed of a structural unit that is repeated in three dimensions. This is the basic structural unit of a crystal of sodium chloride, the mineral halite.

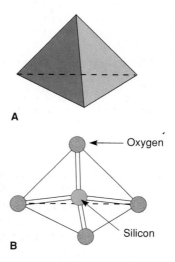

Figure 21.4

(a) The geometric shape of a tetrahedron with four equal sides. (b) A silicon and four oxygen atoms are arranged in the shape of a tetrahedron with the silicon in the center. This is the basic building block of all silicate minerals.

Minerals

In everyday usage the word "mineral" can have several different meanings. It can mean something your body should have (vitamins and minerals), something a fertilizer furnishes for a plant (nitrogen, potassium, and phosphorus) or sand, rock, and coal taken from the earth for human use (mineral resources). In the earth sciences, a **mineral** is defined as a naturally occurring, inorganic solid element or compound with a crystalline structure (figure 21.3). This definition means that the element or compound cannot be synthetic (thus, naturally occurring), must not be produced by a living organism (thus, inorganic), and must have atoms arranged in a regular, repeating pattern (a crystal structure). Note that the crystal structure of a mineral can be present on the microscopic scale and it is not necessarily obvious to the unaided eye.

The crystalline structure of a mineral can be made up of atoms of one or more kinds of elements. The diamond, for example, is a mineral with only carbon atoms in a strong crystalline structure. Mica, on the other hand, is a mineral with atoms of seven different elements in a sheetlike crystalline structure.

However many kinds of atoms are present, each mineral has its own defined chemical composition or range of chemical compositions. A range of chemical compositions is possible because the composition of some minerals can vary with the substitution of chemically similar elements. For example, some atoms of magnesium might be substituted for some chemically similar atoms of calcium. Such substitutions might slightly alter some properties but not enough to make a different mineral.

Silicon and oxygen are the most abundant elements in the earth's crust, and, as you would expect, the most common minerals contain these two elements. All minerals are classified on the basis of whether or not the mineral structure contains these

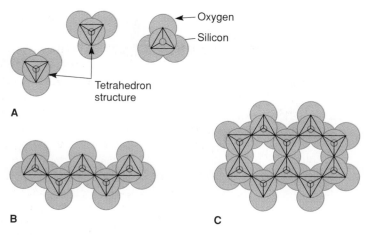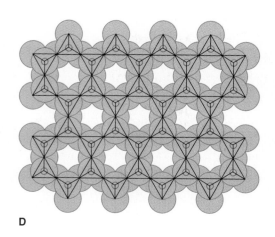

Figure 21.5

(a) Isolated silicon-oxygen tetrahedra do not share oxygens. This structure occurs in the mineral olivine. (b) Single chains of tetrahedra are formed by each silicon ion having two oxygens all to itself and sharing two with other silicons at the same time. This structure occurs in hornblende. (c) Double chains of tetrahedra are formed by silicon ions sharing either two or three oxygens. This structure occurs in augite. (d) The sheet structure in which each silicon shares three oxygens occurs in the micas, resulting in layers that pull off easily because of cleavage between the sheets.

two elements. The two main groups are thus called the *silicates* and the *nonsilicates*. Note, however, that the silicates can contain some other elements in addition to silicon and oxygen. The silicate minerals are by far the most abundant, making up about 92 percent of the earth's crust. When an atom of silicon (Si^{+4}) combines with four oxygen atoms (O^{-2}), a tetrahedral structure of $(SiO_4)^{+4}$ forms (figure 21.4). All **silicates** have this basic silicon-oxygen tetrahedral unit either isolated or joined together in the crystal structure. The structure has four unattached electrons on the oxygen atoms that can combine with metallic ions such as iron or magnesium. They can also combine with the silicon atoms of *other* tetrahedral units. Some silicate minerals are thus made up of single tetrahedral units combined with metallic ions. Other silicate minerals are combinations of tetrahedral units combined in chains, sheets, or an interlocking framework (figure 21.5).

The silicate minerals that form rocks can be conveniently subdivided into two groups based on the presence of iron and magnesium. The basic tetrahedron structure joins with ions of iron, magnesium, calcium, and other elements in the **ferromagnesian silicates.** Examples of ferromagnesian silicates are *olivine, augite, hornblende,* and *biotite.* They have a greater density and a darker color than the other silicates because of the presence of the metal ions. Augite, hornblende, and biotite are very dark in color, practically black, and olivine is light green.

The **nonferromagnesian silicates** do not contain iron or magnesium ions. These minerals have a light color and a low density compared to the ferromagnesians. This group includes the minerals of *muscovite (white mica), the feldspars,* and *quartz.*

The silicate minerals can also be classified according to the structural differences, or arrangements of the basic tetrahedra structures. There are four major arrangements of the units: (1) isolated tetrahedra, (2) chain silicates, (3) sheet silicates, and (4) framework silicates. Other structures are possible, but are not as common. The impact of meteorites, for example, creates sufficiently high temperatures and pressures to form a silicate structure with six oxygens around each silicon atom rather than the typical four. A similar structure is believed to exist deep within the earth. Another interesting less-common structure is found in the asbestos minerals. These minerals have a sheet silicate structure that is rolled into fiberlike strands, resulting in a wide variety of silicates that are fibrous.

The remaining 8 percent of minerals making up the earth's crust, which do not have silicon-oxygen tetrahedra in their crystal structure, are called **nonsilicates.** There are six subgroups of nonsilicates: (1) carbonates, (2) sulfates, (3) oxides, (4) halides, (5) sulfides, and (6) native elements. Some common nonsilicates are identified in four of the larger groups listed in table 21.1. The carbonates are the most abundant of the nonsilicates, but others are important as fertilizers, sources of metals, and sources of industrial chemicals.

Each rock-forming mineral has its own set of physical properties because each mineral has (1) a chemical composition and (2) a crystal structure that are unlike any other mineral. The exact composition and crystal structure of an unknown mineral can be determined by using laboratory procedures, and this type of analysis is necessary when the crystal structures are

The Dynamic Earth

Table 21.1

Classification scheme of some common minerals

Minerals				
Silicates		**Nonsilicates**		
Ferromagnesian	*Nonferromagnesian*	*Clay Minerals*	*Carbonates*	*Sulfates*
Biotite	Quartz	Kaolinite	Calcite	Gypsum
Hornblende	Muscovite (white mica)	Illite	Dolomite	Anhydrite
Augite	Feldspars (aluminum silicates)	Chlorite	*Oxides*	*Sulfides*
Olivine	Plagioclase		Hematite	Pyrite
	(sodium feldspar)		Magnetite	Sphalerite
	(calcium feldspar)		Ilmenite	Galena
	Orthoclase (potassium feldspar)		Aluminum oxides	Chalcopyrite

too small to be visible to the unaided eye. If a particular mineral sample happens to have formed large crystals with well-developed shapes, it is often possible to tell one mineral from another through identifying characteristics. There are about eight characteristics, or physical properties, that are useful in identifying minerals. These are the characteristics of color, streak, hardness, crystal form, cleavage, fracture, luster, and density.

The *color* of a mineral is an obvious characteristic, but it is often not very useful for identification. While some minerals always seem to appear the same color, many will vary from one specimen to the next. Variation in color is usually caused by the presence of small amounts of chemical impurities in the mineral that have nothing to do with its basic composition. The mineral quartz, for example, is colorless in its pure form but other samples may appear milky white, rose pink, golden yellow, or purple.

A more consistent characteristic of a mineral is *streak,* the color of the mineral when it is finely powdered. Streak is tested by rubbing a mineral across a piece of unglazed tile or porcelain, which leaves a line of powdered mineral on the tile. Surprisingly, the streak of a mineral is more consistent than is the color of the overall sample. The streak of the same mineral usually shows the same color even though different samples of the mineral may have different colors. Hematite, for example, always leaves a red-brown streak even though the colors of different samples may range from gray to black in overall color.

Hardness is the resistance of a mineral to being scratched. Classically, hardness is measured by using the Mohs Hardness Scale, which is a list of ten minerals in order of hardness. The softest mineral is talc, which is assigned a hardness of 1. The hardest mineral is diamond, which is assigned a hardness of 10. Altogether, there are ten minerals on the scale that can be used as test minerals (table 21.2). A hardness test is made by trying to scratch an unknown mineral or by using the unknown mineral to try to scratch one of the test minerals. If the unknown

Table 21.2

The Mohs Hardness Scale

Mineral	Assigned Hardness
Talc	1
Gypsum	2
Calcite	3
Fluorite	4
Apatite	5
Orthoclase	6
Quartz	7
Topaz	8
Corundum	9
Diamond	10

Note: The hardness of some common objects is sometimes used for comparisons rather than an actual test mineral. Here are some of the common objects and their approximate hardnesses on the same scale: fingernail, 2.5; copper penny, 3.5; ordinary glass, 5 to 6; pocketknife blade, 5 to 6.

mineral scratches a test mineral, the unknown mineral is harder than the test mineral. If the unknown mineral is scratched by the test mineral, the unknown mineral is not as hard as the test mineral. If both minerals are scratched by each other, they have the same hardness. However, the hardness test yields only approximate findings since there are many minerals of a particular hardness.

The *crystal form,* or shape, of a well-developed crystal of a mineral is often a useful clue to its identity. The crystal form is related to the internal geometric arrangement of the atoms making up the crystal structure. The ions of sodium chloride are arranged in a cubic structure, and table salt tends to crystallize in the shape of cubes (see figure 21.3). Thus halite, a mineral composed of sodium chloride, occurs with a cubic structure. There are six basic groups of crystal forms, each with a characteristic symmetry of the flat surfaces, or faces, of the crystal. A crystal with a cubic structure, for example, belongs to the isometric group, which has three equal-length crystal axes at right angles to each other.

Another property that is controlled by the internal crystal structure is *cleavage,* the tendency of minerals to break along smooth planes. Where the cleavage occurs depends on zones of weakness in the crystal structure. Mica, for example, will break along zones of weakness into very thin sheets. Calcite and halite will break in three directions, and if you hit either mineral with a hammer, it will shatter into little pieces with angles like the original one.

If a mineral does not have a well-defined zone of weakness, it may show *fracture* rather than cleavage. In fracture the broken surface is irregular and not in the smooth plane of a cleavage. A distinctive type of fracture is the conchoidal fracture of volcanic glass, quartz, and a few other minerals. Conchoidal fracture breaks along curved surfaces like a shell.

The *luster* of a mineral describes the surface sheen, that is, the way the mineral reflects light. Minerals that have the surface sheen of a metal are described as being *metallic.* Other descriptions of luster include *pearly* (like a pearl), *vitreous* (like glass), and *earthy.*

Density is a ratio of the mass of a mineral to its volume, or the compactness of the matter making up the mineral. Often a mineral density is expressed as *specific gravity,* which is a ratio of the mineral density to the density of water. In the metric system the density of water is 1 g/cm^3, so specific gravity will have the same numerical value as its density (the units cancel). The specific gravity of a mineral will depend on two factors, the kind of atoms of which it is composed and the way the atoms are packed in the crystal lattice. A diamond, for example, has carbon atoms arranged in a close-packed structure and has a specific gravity of 3.5. Graphite, however, has carbon atoms in a loose-packed structure and has a specific gravity of 2.2. To obtain an exact specific gravity a mineral sample must be pure and without cracks, bubbles, or substitutions of chemically similar elements. These conditions are difficult to meet, and a range of specific gravities is sometimes specified for certain minerals.

There are a few other properties and tests that can be used for a few minerals, such as taste, feel, melting point, reaction to a magnet, and so forth. Some of these special properties, such as the double image seen through a calcite crystal, might identify an unknown mineral in an instant. Otherwise, an analysis of the other, more general properties is needed. In general, the properties of minerals are used to find out what an unknown mineral is not. Using a mineral chart, a specific characteristic is selected for testing. For example, suppose you start with the streak test and find that the mineral leaves a white streak. This test would eliminate all the minerals that do not leave a white streak. Using the table of properties, you would then perform a second test on the minerals that leave a white streak, and the

Figure 21.6

Granite is a coarse-grained igneous rock composed mostly of light-colored, light-density nonferromagnesian minerals. The earth's continental areas are dominated by granite and by rocks with the same mineral composition of granite.

second test would eliminate still more possibilities. Eventually, you will have a good idea of what an unknown sample is by finding out what it is not.

Rocks

Elements are *chemically* combined to make minerals. Minerals are *physically* combined to make rocks. A **rock** is defined as an aggregation of one or more minerals and perhaps mineral materials that have been brought together into a cohesive solid. Mineral materials include volcanic glass, a silicate that is not considered a mineral because it lacks a crystalline structure. Thus a rock can consist of one or more kinds of minerals, all held together in a solid with other materials such as glass. Most rocks are composed of silicate minerals, as you might expect since most minerals are silicates. Granite, for example, is a rock that is primarily three silicate minerals: quartz, mica, and feldspar. You can see the grains of these three minerals in a freshly broken surface of most samples of granite (figure 21.6).

Rocks can be described and classified by many different characteristics, such as mineral composition, color, density, or texture. There is a broader classification scheme that is first used, however, and this scheme is based on the way the rocks were formed. There are three main groups in this scheme: (1) *igneous rocks,* which formed as a hot, molten mass of rock materials cooled and solidified; (2) *sedimentary rocks,* which formed from particles or dissolved materials from previously existing rocks; and (3) *metamorphic rocks,* which formed from igneous or sedimentary rocks that were subjected to high temperatures and pressures that deformed or recrystallized the rock without complete melting.

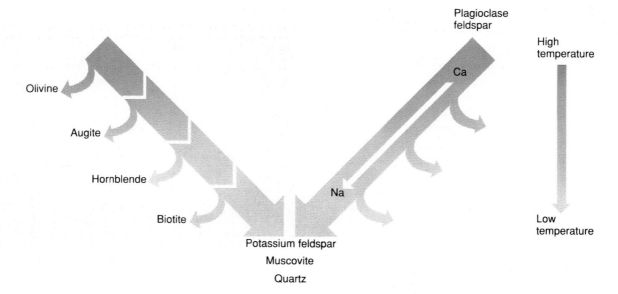

Olivine

Augite

Hornblende

Biotite

Potassium feldspar

Muscovite

Quartz

Plagioclase
feldspar

Ca

Na

High
temperature

Low
temperature

Figure 21.7

Bowen's reaction series. Minerals at the top of the series (olivine, augite, and calcium-rich plagioclase) crystallize at higher temperatures, leaving the magma enriched in silica. Later, the residual magma cools and lighter-colored, less dense minerals (orthoclase feldspar, quartz, and white mica) crystallize. Thus, granitelike rocks can form from a magma that would have produced basaltic rocks had it cooled quickly.

Igneous Rocks

The word *igneous* comes from the Latin *ignis,* which means "fire." This is an appropriate name for **igneous rocks,** which are defined as rocks that formed from a hot, molten mass of melted rock materials. Some very high temperatures, high enough to melt rocks, had to occur as the first step in forming these rocks. The molten rock material from which igneous rocks form is known as **magma.** Magma may cool and crystallize to solid igneous rock either below or on the surface of the earth. Magma that is forced out to the surface is also called **lava.** Lava is the familiar molten rock associated with an erupting volcano. The earth has had a history of molten materials, and all rocks of the earth were at one time igneous rocks. Today, about two-thirds of the outer layer, or crust, is made up of igneous rocks. This is not apparent in many locations because the surface is covered by other kinds of rocks and rock materials (sand, soil, etc.).

As a magma cools, atoms in the melt begin to lose kinetic energy and come together to form the orderly array of a crystal structure. How rapidly the cooling takes place determines the *texture* of the igneous rock being formed. In general, a *coarse-grained* texture means that you can see mineral crystals with the unaided eye. If you need a lens or a microscope to see the crystals the texture is said to be *fine-grained*. The presence of a fine-grained or coarse-grained texture tells you something about the cooling history of a particular igneous rock.

How rapidly a magma cools and hardens is generally determined by its location. Magma that cools slowly deep below the surface produces coarse-grained **intrusive igneous rocks.**

Below the surface the magma loses heat slowly, and the atoms have sufficient time to produce large crystals. Lava that cools rapidly above the surface produces fine-grained **extrusive igneous rocks.** Rapid cooling does not result in sufficient time for large crystals to form so extrusive rocks are fine-grained. Very rapid cooling results in no time for *any* crystals to form, and a volcanic glass is produced as a result (glass does not have an orderly arrangement of atoms and is therefore not a crystal).

Differences in the rate of cooling also affect the combinations of minerals that are present in a particular igneous rock. Each mineral has its own range of freezing-point temperatures, that is, its own temperature range in which it begins to form a solid material from a melt. The crystallization sequence that occurs as a result is called **Bowen's reaction series.** The series, as shown in figure 21.7, is arranged with minerals at the top that crystallize at higher temperatures and minerals at the bottom that crystallize at lower temperatures. Minerals at the top of the series are ferromagnesian silicates, minerals that are rich in iron and magnesium. Minerals at the bottom of the series are nonferromagnesian silicates, minerals that are rich in silicon and generally lack iron and magnesium. Since the ferromagnesian silicates crystallize at a higher temperature, these minerals are the first to form in a cooling magma. If a magma crystallizes directly, it will contain the ferromagnesian minerals listed at the top of the series. On the other hand, if it cools slowly, perhaps far below the surface, the minerals containing the iron and magnesium will form crystals that sink to the bottom of the liquid magma. Thus the remaining magma, and the minerals that crystallize later, will become progressively richer in silicon as more and more iron and magnesium are removed.

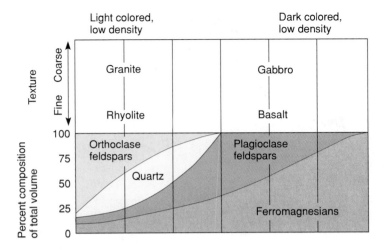

Figure 21.8
Igneous rock classification scheme based on mineral composition and texture. There are other blends of minerals with various textures, many of which have a specific name.

Table 21.3
A simplified classification scheme for clastic sediments and rocks

Sediment Name	Size Range	Rock
Boulder	Over 256 mm (10 in)	Conglomerate or breccia*
Gravel	2 to 256 mm (0.08–10 in)	Sandstone
Sand	1/16 to 2 mm (0.025–0.08 in)	Sandstone
Silt (or dust)	1/256 to 1/16 mm (0.00015–0.025 in)	Siltstone**
Clay (or dust)	Less than 1/256 mm (less than 0.00015 in)	Claystone**

*Conglomerate has a rounded fragment; breccia has an angular fragment.
**Both also known as mudstone; called shale if it splits along parallel planes.

A general classification scheme for igneous rocks is given in figure 21.8. Igneous rocks are various mixtures of minerals, and this scheme names the rocks according to their mineral composition and their texture. Note that the mineral composition changes continuously from one side of the chart to the other. There are many blendings and intermediate types of igneous rocks possible, but this chart identifies only the most important ones.

Igneous rocks on the left side of figure 21.8 are blends of the nonferromagnesian minerals, so they are comparatively light in density and color, appearing to be light grey, white, or ivory colored. The most common igneous rock with the minerals on the left side of the chart is **granite.** If you look closely at the surface of a freshly broken piece of granite you will note that it is *coarse-grained* with noticeable particles of different size, shape, and color. The vitreous, white particles are probably orthoclase feldspar, which makes up about 45 percent of the particles. The clear, glassy looking particles are probably quartz crystals, which make up about 25 percent of the total sample. The remaining particles of black specks are ferromagnesian minerals.

Rocks with the chemical composition of granite make up the bulk of the earth's continents, and granite is the most common intrusive rock in the continental crust. As shown in the chart, **rhyolite** and **obsidian** are the chemical equivalents of granite, except they have a different texture. Rhyolite is fine-grained and obsidian is a translucent volcanic glass.

Igneous rocks on the right side of the chart usually have a greater density than rocks with the granite chemical composition and are very dark in color. The most common example of these dark, relatively high-density igneous rocks is **basalt.** Basalt is the dark, *fine-grained* igneous rock that you probably associate with cooled and hardened lava. Basalt is fine-grained so you cannot see any mineral particles, and a freshly broken surface looks sugary. As shown in the chart, basalt is about half plagioclase feldspars and about half ferromagnesian minerals.

Basaltic rocks, meaning rocks with the chemical composition of basalt, make up the ocean basins and much of the earth's interior. Basalt is the most common extrusive rock found on the earth's surface. The coarse-grained chemical equivalent of basalt is called **gabbro.**

Sedimentary Rocks

Sedimentary rocks are rocks that formed from particles or dissolved materials from previously existing rocks. Igneous rocks at the earth's surface, for example, are exposed to conditions very different from the environment in which they formed. Chemical reactions with air and water tend to break down and dissolve some minerals, freeing more chemically stable particles and grains in the process. The more stable particles are mechanically broken down by the action of moving water, wind, and ice. Many such processes are at work altering rocks through a process known as *weathering.* Weathering is discussed in detail in the next chapter. For now, weathering is of interest because of the role it plays in providing the rocks and mineral grains and dissolved materials that will become sedimentary rocks. The weathered materials are transported by moving water, wind, and ice and deposited as sediments. **Sediments** are accumulations of silt, sand, or gravel that settle out of the atmosphere or out of water. There are actually two sources of sediments, weathered rock fragments and dissolved rock materials.

Weathered rock fragments are called **clastic sediments** after a Greek word meaning "broken." Clastic sediments accumulated from rocks that are in various stages of being broken down, so there is a wide range of sizes of clastic sediments (table 21.3). The largest of the clastic sediments, boulders and gravel, are the raw materials for the sedimentary rock that is called **conglomerate** or **breccia,** depending on if the fragments are well-rounded or angular. **Sandstone,** as the name implies, is a sedimentary rock formed from sand that has been consolidated into solid rock. The smallest clastic sediments, silt and clay, are consolidated into solid **siltstone** and **claystone.** If either of these

The Dynamic Earth

Table 21.4

A classification scheme for sedimentary rocks

Sediment Type	Particle or Composition	Rock
Clastic	Larger than sand	Conglomerate or breccia
Clastic	Sand	Sandstone
Clastic	Silt and clay	Siltstone, claystone, or shale
Chemical	Calcite	Limestone
Chemical	Dolomite	Dolomite
Chemical	Gypsum	Gypsum
Chemical	Halite (sodium chloride)	Salt

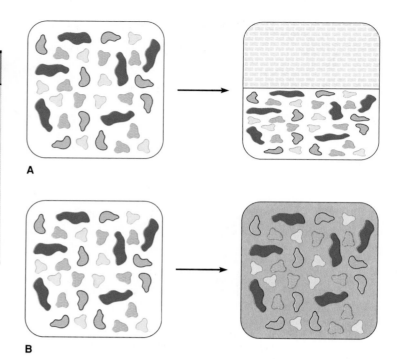

A

B

Figure 21.9

(*a*) In compaction, the sediment grains are packed more tightly together, often by overlying sediments, as represented by the bricks. (*b*) In cementation, fluids contain dissolved minerals that are precipitated in the space between the grains, cementing them together into a rigid, solid mass.

sedimentary rocks tends to break along planes into flat pieces they are called **shale.** Note that when a clastic sediment is referred to as "clay," it means a sediment size (less than 1/256 mm) and not the name of the clay mineral. When deposited from the air, clay- and silt-sized particles are commonly called "dust."

Dissolved rock materials form **chemical sediments** that are removed from solution to form sedimentary rocks. The dissolved materials are ions from minerals and rocks that have been completely broken down. Once they are transported to lakes or oceans, the dissolved ions are available to make sediments through one of three paths: (1) chemical precipitation from solution, (2) crystallization from evaporating water, or (3) as biological sediments. The most abundant chemical sedimentary rocks are the carbonates and evaporates. The carbonates are **limestone** and **dolomite.** Limestone is composed of calcium carbonate, the mineral called calcite. Dolomite probably formed from limestone by replacement of calcium ions with magnesium ions (both belong to the same chemical family). Limestone is precipitated directly from fresh or salt water or indirectly by the actions of plants and animals that form shells of calcium carbonate (table 21.4).

Most sediments are deposited as many separate particles that accumulate in certain environments as loose sediments. Such accumulations of rock fragments, chemical deposits, or animal shells must become consolidated into a solid, coherent mass to become sedimentary rock. There are two main parts to this *lithification,* or rock forming process, compaction and cementation (figure 21.9).

The weight of an increasing depth of overlying sediments causes an increasing pressure on the sediments below. This pressure squeezes the deeper sediments together, gradually reducing the pore space between the individual grains. This **compaction** of the grains reduces the thickness of a sediment deposit, squeezing out water as the grains are packed more

tightly together. Compaction alone is usually not enough to make an unconsolidated deposit into solid rock. Cementation is needed to hold the compacted grains together.

In **cementation** the spaces between the sediment particles are filled with a chemical deposit. The chemical deposit binds the particles together into the rigid, cohesive mass of a sedimentary rock. Compaction and cementation may occur at the same time, but the cementing agent must have been introduced before compaction restricts the percolation of the fluid through the open spaces. Many soluble minerals can serve as cementing agents, and calcite (calcium carbonate) and silica (silicon dioxide) are common.

Metamorphic Rocks

The third group of rocks is called metamorphic. **Metamorphic rocks** are previously existing rocks that have been changed by heat, pressure, or hot solutions into a distinctly different rock. The heat, pressure, or hot solutions that produced the changes are associated with (1) geologic events of movement of the crust, which will be discussed soon, and (2) heating and hot solutions from the intrusion of a magma. Pressures from movement of the crust can change the rock texture by flattening, deforming, or realigning mineral grains. Temperatures from an intruded magma must be just right to produce a metamorphic rock. They must be high enough to disrupt the crystal structures to cause them to recrystallize but not high enough to melt the rocks and form igneous rocks.

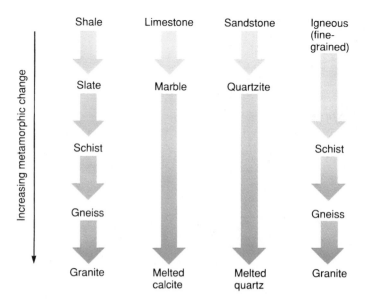

Figure 21.10
Increasing metamorphic change occurs with increasing temperatures and pressures. If the melting point is reached, the change is no longer metamorphic and igneous rocks are formed.

The exact changes caused by heat and pressure depend on the mineral composition of the parent rock and the extent of the pressure, temperature, and hot solutions that may or may not be present to induce chemical changes. In general, pressure on parent rocks with flat crystal flakes (such as clays and mica) tends to align the flakes in parallel sheets. This new crystal alignment is called **foliation** after the Latin for "leaf" (as in the leaves, or pages, of a closed book). Foliation gives a metamorphic rock the property of breaking along the planes between the aligned mineral grains, a characteristic known as *rock cleavage*. The extent of foliation is determined by the extent of the metamorphic changes. For example, **slate** is a metamorphic rock formed from the sedimentary rock shale. Slate is fine-grained with no crystals visible to the unaided eye. Alignment of the microscopic crystals results in a tendency of slate to split into flat sheets. Greater heat and pressure can cause more metamorphic change, resulting in larger crystals and increased foliation. The metamorphic rock called **schist** can be produced from slate by further metamorphism. In schist the cleavage surfaces are now visible and coarser mica crystals are visible to the unaided eye. Still further metamorphism of schist may break down the mica crystals and produce alternating bands of light and dark minerals. These bands are characteristic of the metamorphic rock **gneiss** (pronounced "nice"). Gneiss can also be produced by strong metamorphism of other rock types such as granite. Slate, schist, and gneiss are but three examples of a continuous transition that can take place from the metamorphism of shale all the way until it is completely melted to become an igneous rock (figure 21.10).

Table 21.5
A classification scheme for metamorphic rocks

Metamorphic Texture	Metamorphic Rock
Nonfoliated	Quartzite and marble
Very finely foliated	Slate
Finely foliated	Schist
Coarsely foliated	Gneiss

Some metamorphic rocks are nonfoliated because they consist mainly of one mineral and the grains are not aligned into sheets. When a quartz-rich sandstone is metamorphosed, the new rock has recrystallized, tightly locking grains. The resulting metamorphic rock is the tough, hard rock called **quartzite. Marble** is another nonfoliated metamorphic rock that forms from recrystallized limestone (table 21.5).

The Rock Cycle
Earth is a dynamic planet with a constantly changing surface and interior. As you will see in the next sections, internal changes alter the earth's surface by moving the continents and, for example, building mountains that are soon worn away by weathering and erosion. Seas advance and retreat over the continents as materials are cycled from the atmosphere to the land and from the surface to the interior of the earth and then back again. Through all this continual change rocks are transformed from one type to another. There is not a single rock on the earth's surface today that has remained unchanged through the earth's long history. The concept of continually changing rocks through time is called the **rock cycle** (figure 21.11). The rock cycle concept views an igneous, a sedimentary, or a metamorphic rock as the present but temporary stage in the ongoing transformation of rocks to new types. Any particular rock sample today has gone through countless transformations in the 4.6 billion year history of the earth and will continue to do so in the future.

The Earth's Interior

Many of the properties and characteristics of earth, including the structure of its interior, can be explained from how it formed and evolved. Theories and ideas about how the earth and the rest of the Solar System formed were discussed in detail in chapter 19. In general, the earth is considered to have formed about 4.6 billion years ago in a rotating disk of particles and grains that had condensed around the central protosun. The condensed rock, iron, and mineral grains were pulled together by gravity, growing eventually to a planet-sized mass with the same force of gravitational attraction that is present today. Not all the bits and pieces of matter in the original solar nebula were incorporated into the newly formed planets, however, and they were soon being pulled by gravity to the newly born planets and

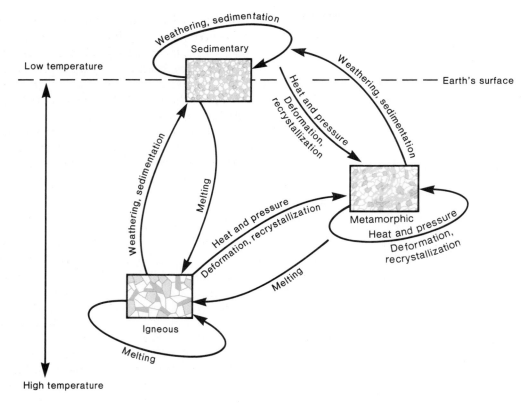

Figure 21.11
A schematic diagram of the rock cycle concept, which states that geologic processes act continuously to produce new rocks from old ones.

their satellites. All sizes of these leftover bits and pieces of matter thus began bombarding the planets and their moons. The bombardment was so intense that the heat generated by impact after impact increased the surface temperature to the melting point. Evidence visible on the moon and other planets today indicates that the bombardment was substantial, continuing for several hundred million years. Calculations of the heating that resulted from this tremendous bombardment indicate that sufficient heat was liberated to melt the entire surface of the earth to a layer of glowing, molten lava. Thus the early earth had a surface of molten lava that eventually cooled and crystallized to solid igneous rocks as the bombardment gradually subsided, then stopped.

The earth then began to undergo a second melting, this time from the inside. The interior slowly accumulated heat from the radioactive decay of uranium, thorium, and potassium isotopes (see chapter 17). Heat conducts slowly through great thicknesses of rock and rock materials. After about a billion years of accumulating heat, parts of the interior became hot enough to melt to pockets of magma. Iron and other metals were pulled from the magma toward the center of the earth, leaving less dense rocks toward the surface. The melting probably did not occur all at one time throughout the interior, but rather it occurred in local pockets of magma. Each magma became molten, cooled to a solid, and perhaps repeated the cycle numerous times.

With each cyclic melting the heavier abundant elements were pulled by gravity toward the center of the earth and additional heat was generated by the release of gravitational energy. Today, the earth's interior still contains an outer core of molten material that is predominantly iron. The environment of the center of earth today is extreme, with estimates of pressures up to 3.5 million atmospheres (3.5 million times the pressure of the atmosphere at the surface, 14.7 lb/in²). Recent estimates of the temperatures at the earth's core are about the same as the temperature of the surface of the sun, about 6,000° C (11,000° F).

The melting and flowing of iron to the earth's center were the beginnings of *differentiation,* the separation of materials that gave the earth its present-day stratified, or layered, interior. The different freezing-point temperatures of the basic minerals, as illustrated in Bowen's reaction series, further differentiated the materials of the earth into a total of three basic divisions, the *core,* the *mantle,* and the *crust* (figure 21.12).

The Core

Information about the nature of the earth's **core,** the center part of the earth, comes from studies of three sources of information: (1) seismological data, (2) the nature of meteorites, and (3) geological data at the surface of the earth. Seismological data is information from the study of earthquake waves that pass through the earth. This data indicates that the earth's core

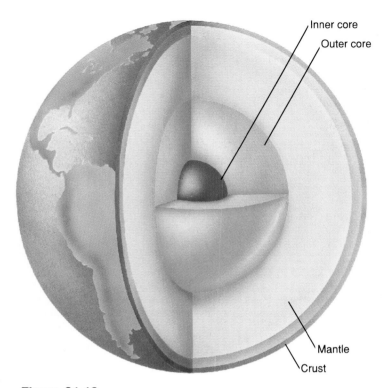

Figure 21.12
The structure of the earth's interior.

begins at a depth of 2,900 km (about 1,800 mi) below the surface. Since the earth's center is 6,370 km (about 3,960 mi) below the surface, the core has a radius of 3,470 km (about 2,160 mi). Seismological data also indicates that the inner core is solid and that the outer core is liquid. The solid inner core has a radius of 1,200 km (about 750 mi). Overall, the core makes up about 15 percent of the earth's total volume and about one-third of its mass.

Evidence from the nature of meteorites indicates that the earth's core is mostly iron. The earth has a strong magnetic field that has its sources in the turbulent flow of the liquid part of the earth's core. In order to do this, the materials of the core would have to be an electrical conductor, that is, a metal such as iron. There are two general kinds of meteorites that fall to the earth, stony meteorites, which are made of silicate minerals, and iron meteorites, which are made of iron or of a nickel-iron alloy. Since the earth has a silicate-rich crust and mantle, by analogy the earth's core must consist of iron or a nickel and iron alloy.

Geological studies of surface rocks also provide evidence about the nature of the earth's core. Iron minerals, such as magnetite (Fe_3O_4), crystallize as particles in a cooling magma and become magnetized and oriented to the earth's magnetic field like tiny compass needles. When the rock crystallizes to a solid, these tiny compass needles become frozen in the orientations they had at the time. Such rocks thus provide evidence of the direction and distance to earth's ancient magnetic poles. The

study of ancient magnetism, called *paleomagnetics,* provides some very interesting information about the earth's magnetic field. Paleomagnetic studies indicate that the earth's magnetic field has been reversed many times in the past as the north magnetic pole has become the south magnetic pole and vice versa (more on this later). Paleomagnetic studies, along with the determination of the ages of rocks containing the magnetic minerals, indicate that the earth's magnetic field was first formed about 3.5 billion years ago, or about a billion years after the earth formed. This fits nicely with other understandings about how the earth formed and the calculated time required for radioactive heating to melt the earth's interior.

The Mantle

The middle part of the earth's interior is called the **mantle.** The mantle is a 2,870 km (about 1,780 mi) thick shell that surrounds the metallic core. This shell takes up about 80 percent of the total volume of the earth and accounts for about two-thirds of the earth's total mass. It accounts for less of the mass because it is composed of silicate minerals that are less dense than the metallic core.

Information about the composition and nature of the mantle comes from (1) studies of seismological data, (2) studies of the nature of meteorites, and (3) studies of materials from the mantle that have been ejected to the earth's surface. The evidence from these separate sources all indicates that the mantle is composed of silicates, predominantly the ferromagnesian silicate *olivine.* Meteorites, as mentioned earlier, are basically either iron meteorites or stony meteorites. The bulk of the stony meteorites are the carbonaceous chondrites, silicates with a composition that would produce the chemical composition of olivine if they were melted and the heavier elements removed by differentiation. This chemical composition also agrees closely with the composition of basalt, the most common volcanic rock found on the surface of the earth.

Laboratory studies of the wave transmission characteristics of olivine under increasing conditions of pressure and temperature agree closely with the observed velocity of earthquake waves in the mantle. Both sources of information suggest a mantle that is not uniform throughout but that changes character at depths of about 400 km and 700 km (about 250 mi and 430 mi). At these depths there is a very rapid increase in the velocity of seismic waves. Laboratory studies suggest that the pressure and temperature conditions found at these depths correspond to changes in the structure of the olivine mineral.

In the upper 400 km (about 250 mi) the dominant olivine mineral structure is that of a typical silicate, that is, a tetrahedral structure of a silicon atom surrounded by four oxygen atoms. Then, at pressures and temperatures corresponding to a depth of about 400 km, the structure collapses to a closer-packed arrangement that increases the density. The increased density would transmit seismic waves at a greater velocity, which is exactly what is observed in seismic data.

443

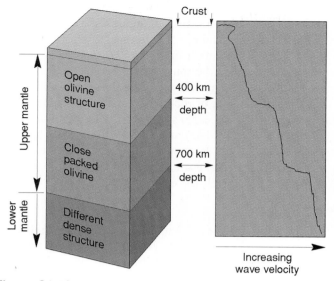

Figure 21.13

Seismic wave velocities increase in velocity at depths of about 400 km and 700 km (about 250 mi and 430 mi). This finding agrees closely with laboratory studies of changes in the character of mantle materials that would occur at these depths from increases in temperature and pressure.

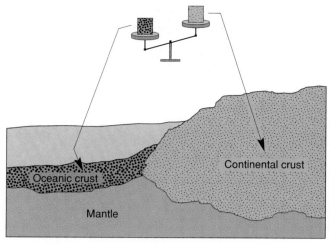

Figure 21.14

Continental crust is less dense, granite-type rock, while the oceanic crust is more dense basaltic rock. Both types of crust behave as if they were floating on the mantle, which is more dense than either type of crust.

At pressures and temperatures corresponding to a depth of about 700 km (about 430 mi) the close-packed structure is changed further, this time to a new arrangement with six oxygen atoms around each silicon atom. Again, this information from laboratory studies agrees closely with changes in seismic wave velocities that are observed at that depth (figure 21.13). Since it is also observed that earthquakes always occur above 700 km (about 430 mi), this depth is identified as the boundary between the *upper mantle* and the *lower mantle*. The lower mantle is composed of the six oxygen, one silicon structure silicate and the upper mantle is composed of zones consisting of open- and close-packed four oxygen, one silicon tetrahedral structures.

The Crust

The outer shell of the earth, the **crust,** rests on top of the mantle. The boundary between the crust and the mantle was discovered in 1909 by the Yugoslavian scientist Mohorovicic. The boundary is marked by a sharp increase in the velocity of seismic waves as they pass from the crust to the mantle. Today, this boundary is called the **Mohorovicic discontinuity,** or the "Moho" for short. The boundary is actually a zone 1 or 2 km thick (about 1,000 or 2,000 yards) where compressional seismic waves increase in velocity because of changes in the composition of the materials. Note that changes in seismic wave velocities in the mantle are a result of structural changes in minerals of the same composition. Seismic waves increase in velocity at the Moho because the composition on both sides of the boundary is different. The mantle is richer in ferromagnesian minerals and poorer in silicon than the crust.

Studies of the Moho show that the crust varies in thickness around the earth's surface. It is thicker under the continents, varying from 10 km (about 6 mi) to more than 70 km (about 40 mi) beneath mountain chains. The crust under the oceans is much thinner, ranging from 5 km to 8 km (about 3 mi to 5 mi) thick. The continental crust is also much older, with parts up to 3.8 million years old compared to the less than 200 million year age for the oldest oceanic crust.

In general, continental crust is less dense, granite-type rock with a density of about 2.7 g/cm³. Oceanic crust, on the other hand, is made up of basaltic rock with a density of about 3.0 g/cm³. The less dense crust behaves as if it were floating on the mantle, much as less dense ice floats on water. There are explainable exceptions, but in general the thicker, less dense continental crust "floats" on the mantle above sea level and the thin, dense oceanic crust "floats" on the mantle far below sea level (figure 21.14). The concept of less dense rock "floating" on more dense rock is called **isostasy.** The isostasy concept explains that mountains have a thick crustal structure for the same reason that most of an iceberg is below the water level. Thus, by the analogy of isostasy, most of a mountain is below the surrounding continental crust because it sinks deeper into the mantle.

A Different Structure

There is strong evidence that the earth has a layered structure with a core, mantle, and crust as described previously. This description of the structure is important for historical reasons and for understanding how the earth evolved over time. There is another structure that can be described, however, and this structure is far more important in understanding the past history and present appearance of the earth's surface, including the phenomena of earthquakes and volcanoes (figure 21.15).

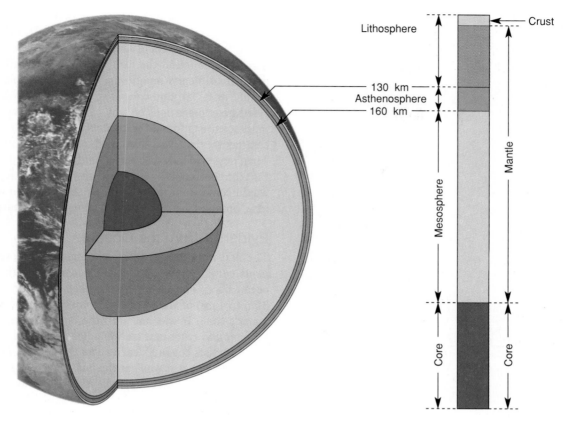

Figure 21.15

The earth's interior showing the weak, plastic layer called the *asthenosphere*. The rigid, solid layer above the asthenosphere is called the *lithosphere*. The lithosphere is broken into plates that move on the asthenosphere like giant tabular ice sheets floating on water. This arrangement is the foundation for plate tectonics, which explains many changes that occur on the earth's surface such as earthquakes, volcanoes, and mountain building.

The important part of this different structural description of the earth's interior was first identified from seismic data. There is a thin zone in the mantle, ranging in thickness from a depth of 130 km (81 mi) to 160 km (100 mi), in which seismic waves undergo a sharp *decrease* in velocity. This low-velocity zone is evidently a hot, elastic, semiliquid layer that extends unbroken around the entire earth. It is called the **asthenosphere** after the Greek for "weak shell." The asthenosphere is weak because it is plastic and mobile and yields to stresses. In some regions the asthenosphere is completely liquid, containing pockets of magma.

The rocks above and below the asthenosphere are rigid, solid, and brittle. The solid layer above the asthenosphere is called the **lithosphere** after the Greek for "stone shell." The lithosphere is also known as the "strong layer" in contrast to the "weak layer" of the asthenosphere. The lithosphere includes the entire crust, the Moho, and the upper part of the mantle. The rest of the solid, dense mantle below the asthenosphere is called the **mesosphere,** or "middle shell." As you will see in the next section, the asthenosphere is one important source of

magma that reaches the earth's surface. It is also a necessary part of the mechanism involved in the movement of the crust. The lithosphere is made up of comparatively rigid plates that are moving, floating on the asthenosphere like giant tabular ice sheets floating on the ocean.

Plate Tectonics

If you observe the shape of the continents on a world map or a globe, you will notice that some of the shapes look as if they would fit together like the pieces of a puzzle. The most obvious is the eastern edge of North and South America, which seems to fit the western edge of Europe and Africa in a slight S-shaped curve. Such patterns between continental shapes seem to suggest that the continents were at one time together, breaking apart and moving to their present positions some time in the past. Impressed by these patterns, Antonio Snider published a sketch in 1855 showing Africa and South America fitting together as one landmass. This sketch gave rise to the bold speculation that the continents had at one time been part of a single landmass (figure 21.16).

The Dynamic Earth

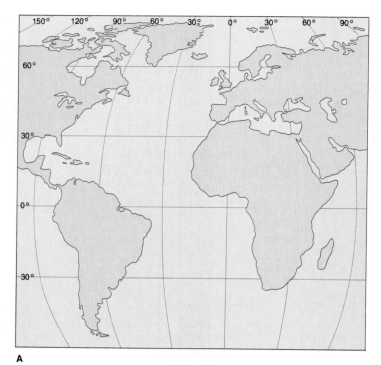

A

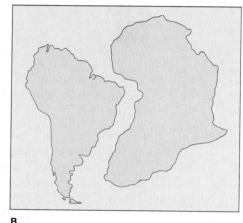

B

Figure 21.16

(a) Normal position of the continents on a world map. (b) A sketch of South America and Africa suggesting that they once might have been joined together and then subsequently separated by continental drift.

In the early 1900s a German geologist named Alfred Wegener became enamored with the idea that the continents had shifted positions and published papers on the subject for nearly two decades. Wegener supposed that at one time there was a single large landmass, which he called "Pangaea," from the Greek meaning "all lands." He pointed out that similar fossils found in landmasses on both sides of the Atlantic Ocean today must be from animals and plants that lived in Pangaea, which later broke up and split into smaller continents. Wegener's concept came to be known as **continental drift,** the idea that individual continents could shift positions on the earth's surface. Some people found the idea of continental drift plausible, but most had difficulty imagining how a huge and massive continent could "drift" around on a solid earth. Since Wegener had provided no explanations of why or how continents might do this, most scientists found the concept unacceptable. The concept of continental drift was dismissed as an interesting but odd idea. Then new evidence discovered in the 1950s and 1960s indicated that the continents have indeed moved. The first of this evidence would come from the bottom of the ocean and would lead to a new, broader theory about movement of the earth's crust.

Evidence from the Ocean

The first important studies concerning the movement of continents came from studies of the ocean basin, the bottom of the ocean floor. The basins are covered by 4 km to 6 km (about 3 mi to 4 mi) of water and were not easily observed during Wegener's time. It was not until the development and refinement of sonar and other new technologies during the 1940s and 1950s that scientists began to learn about the nature of the ocean basin. They found that it was not the flat, featureless plain that many had imagined. There are valleys, hills, mountains, and mountain ranges. Long, high, and continuous chains of mountains that seem to run clear around the earth were discovered, and these chains are called **oceanic ridges.** The *Mid-Atlantic Ridge* is one such oceanic ridge, located in the center of the Atlantic Ocean basin. The Mid-Atlantic Ridge divides the Atlantic Ocean into two nearly equal parts. Where it is high enough to reach sea level, it makes oceanic islands such as Iceland and the Azores (figure 21.17). The basins also contain **oceanic trenches.** These trenches are long, narrow, and deep troughs with steep sides. Oceanic trenches always run parallel to the edge of continents, a preferred orientation that seems to invite some kind of explanation.

Studies of the Mid-Atlantic Ridge during the late 1950s found at least three related groups of data and observations that also seemed to invite some kind of explanation. Among these related groups were the following: (1) submarine earthquakes were discovered and measured, but the earthquakes were all observed to occur mostly in a narrow band under the crest of the Mid-Atlantic Ridge; (2) a long, continuous, and deep valley was observed to run along the crest of the Mid-Atlantic Ridge for its length; this continuous, crest-running valley is called a **rift**; and, (3) a large amount of heat was found to be escaping from the rift. One explanation of the related groups of findings is that the rift might be a crack in the earth's crust, a fracture through which basaltic lava flowed to build up the ridge. The evidence of excessive heat flow, earthquakes along the crest of the ridge, and the very presence of the ridge all led to a **seafloor spreading** hypothesis. This hypothesis explained that hot, molten rock moved up from the interior of the earth to emerge along the rift,

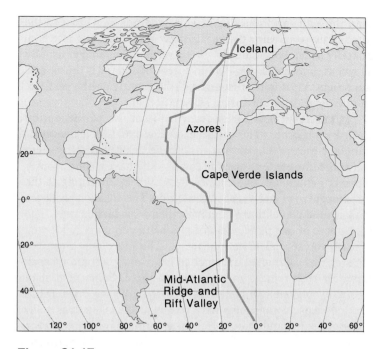

Figure 21.17

The Mid-Atlantic Ridge divides the Atlantic Ocean into two nearly equal parts. Where the ridge reaches above sea level, it makes oceanic islands such as Iceland and the Azores.

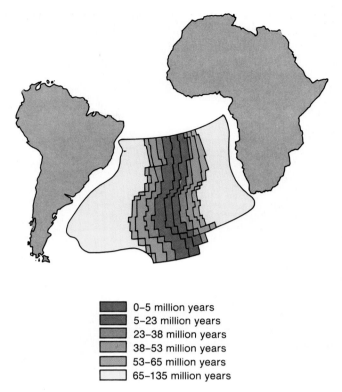

0–5 million years
5–23 million years
23–38 million years
38–53 million years
53–65 million years
65–135 million years

Figure 21.18

The pattern of seafloor ages on both sides of the Mid-Atlantic Ridge reflects seafloor spreading activity. Younger rocks are found closer to the ridge.

Source: From a map by W. C. Pitman III, R. L. Larson, and E. M. Herron, 1974, *Geological Society of America.*

flowing out in both directions to create new rocks along the ridge. The creation of new rock like this would tend to spread the seafloor in both directions, so thus the name. The test of this hypothesis would come from further studies, this time on the ages and magnetic properties of the seafloor along the ridge (figure 21.18).

Evidence of the age of sections of the seafloor was obtained by drilling into the ocean floor from a research ship. From this drilling, scientists were able to obtain samples of fossils and sediments at progressive distances outward from the Mid-Atlantic Ridge. They found thin layers of sediments near the ridge that became progressively thicker toward the continents. This is a pattern you would expect if the seafloor was spreading because older layers would have more time to accumulate greater depths of sediments. The fossils and sediments in the bottom of the layer were also progressively older at increasing distances from the ridge. The oldest, which were about 150 million years old, were near the continents. This would seem to indicate that the Atlantic Ocean did not exist until 150 million years ago. At that time a fissure formed between Africa and South America, and new materials have been continuously forced into the fissure, forcing the two continents apart.

More convincing evidence of seafloor spreading came from the paleomagnetic discovery of patterns of magnetic strips in the basaltic rocks of the ocean floor. The earth's magnetic field has been reversed many times in the last 150 million years. The periods of time between each reversal were not equal, ranging from thousands to millions of years. Since iron minerals in molten basalt formed, became magnetized, and then froze in the orientation they had when the rock cooled, they made a record of reversals in the earth's ancient magnetic field (figure 21.19). Analysis of the magnetic pattern in the rocks along the Mid-Atlantic Ridge found identical patterns of magnetic bands on both sides of the ridge, just what you would expect if molten rock were forced out of the rift, cooled to solid basalt, then moved away from the rift on both sides. The pattern of magnetic bands also matched patterns of reversals measured elsewhere, providing a means of determining the age of the basalt. This showed that the oceanic crust is like a giant conveyor belt that is moving away from the Mid-Atlantic Ridge in both directions. It is moving at an average 5 cm (about 2 in) a year, which is about how fast your fingernails grow. This means that in 50 years the seafloor will have moved 5 cm/yr × 50 yr, or 2.5 m (about 8 ft). This slow rate is why most people do not recognize that the seafloor—and the continents, as you will see—move.

Lithosphere Plates and Boundaries

The strong evidence for seafloor spreading soon led to the development of a new theory called **plate tectonics.** According to plate tectonics the lithosphere is broken into a number of fairly rigid plates that move on the asthenosphere. Some plates, as

The Dynamic Earth

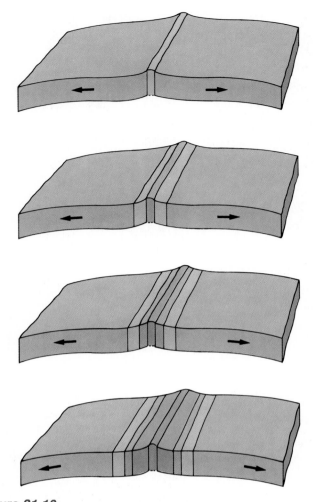

Figure 21.19

Formation of magnetic strips on the seafloor. As each new section of seafloor forms at the ridge, iron minerals become magnetized in a direction that depends on the orientation of the earth's field at that time. This makes a permanent record of reversals of the earth's magnetic field.

shown in figure 21.20, contain continents and part of an ocean basin, while other plates contain only ocean basins. The plates move, and the movement helps to explain why mountains form where they do, the occurrence of earthquakes, the occurrence of volcanoes, and in general, the entire changing surface of the earth.

Earthquakes, volcanoes, and most rapid changes in the earth's crust occur at the edge of a plate, which is called a *plate boundary*. There are three general kinds of plate boundaries that describe how one plate moves relative to another: (1) divergent boundaries, (2) convergent boundaries, and (3) transform boundaries.

Divergent boundaries occur between two plates moving away from each other. Magma forms as the plates separate, decreasing pressure on the mantle below. This molten material

from the asthenosphere rises, cools, and adds new crust to the edges of the separating plates. The new crust tends to move horizontally from both sides of the divergent boundary, usually known as an oceanic ridge. A divergent boundary is thus a **new crust zone.** Most new crust zones are presently on the seafloor, producing seafloor spreading (figure 21.21).

Convergent boundaries occur between two plates moving toward each other. The creation of new crust at a divergent boundary means that old crust must be destroyed somewhere else at the same rate or else the earth would have a continuously expanding diameter. Old crust is destroyed by returning to the asthenosphere at convergent boundaries. The collision produces an elongated belt of down-bending called a **subduction zone.** The lithosphere of one plate, which contains the crust, is subducted beneath the second plate and melts. The more dense components of this subducted and melted crust may become igneous materials that remain in the asthenosphere. Some of it may eventually migrate to a spreading ridge to make new crust again. The less dense components may return to the surface as a silicon, potassium, and sodium rich lava, forming volcanoes on the upper plate or it may cool below the surface to form a body of granite. Thus the oceanic lithosphere is being recycled through this process, which explains why ancient seafloor rocks do not exist. Convergent boundaries produce related characteristic geologic features depending on the nature of the materials in the plates. There are three general possibilities: (1) converging continental and oceanic plates, (2) converging oceanic plates, and (3) converging continental plates.

As an example of *ocean-continent plate convergence,* consider the plate containing the South American continent (the South American Plate) and its convergent boundary with an oceanic plate (the Nazca Plate) along its western edge. Continent-oceanic plate convergence produces a characteristic set of geologic features as the oceanic plate of denser basaltic material is subducted beneath the less dense granite-type continental plate (figure 21.22a). The subduction zone is marked by an oceanic trench (the Peru-Chile Trench), deep-seated earthquakes, and volcanic mountains on the continent (the Andes Mountains). The trench is formed from the down-bending associated with subduction, the volcanic mountains are formed from subducted and melted crust that rises up through the overlying plate to the surface, and the earthquakes are associated with the movement of the subducted crust under the overlying crust.

Ocean-ocean plate convergence produces another set of characteristic and related geologic features (figure 21.22b). The northern boundary of the oceanic Pacific Plate, for example, converges with the oceanic part of the North American Plate near the Bering Sea. The Pacific Plate is subducted, forming the Aleutian oceanic trench with a zone of earthquakes that are shallow near the trench and progressively more deep-seated toward the continent. The deeper earthquakes are associated with the movement of more deeply subducted crust into the mantle. The Aleutian Islands are typical **island arcs,** which are

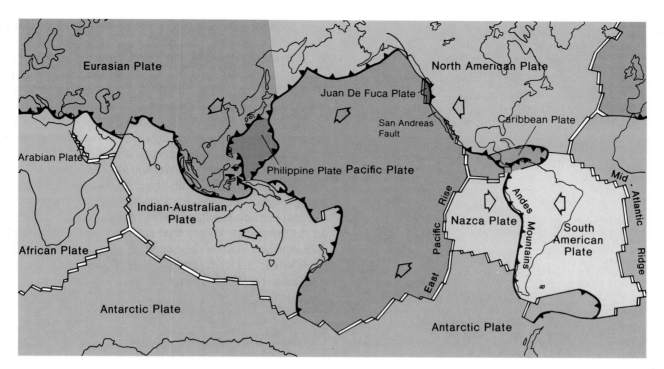

Figure 21.20

The major plates of the lithosphere that move on the asthenosphere.

Source: After W. Hamilton, U.S. Geological Survey.

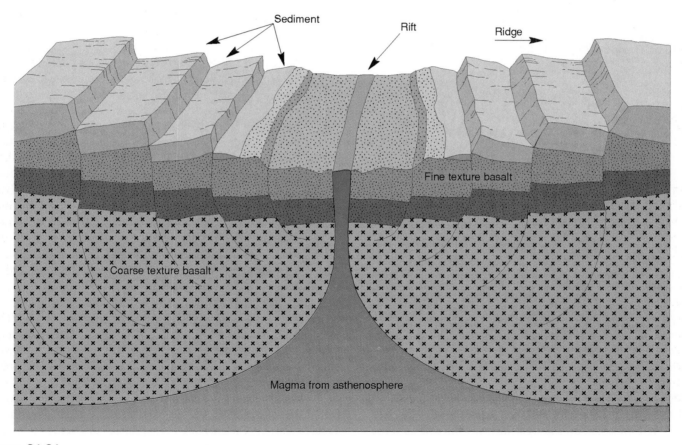

Figure 21.21

A divergent boundary is a new crust zone where molten magma from the asthenosphere rises, cools, and adds new crust to the edges of the separating plates. Magma that cools at deeper depths forms a coarse-grained basalt while surface lava cools to a fine-grained basalt. Note that deposited sediment is deeper farther from the spreading rift.

The Dynamic Earth

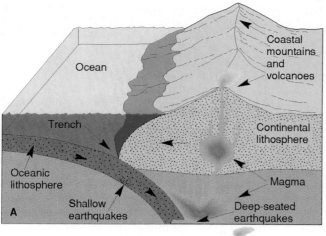

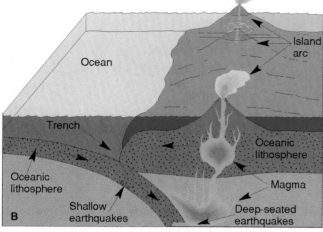

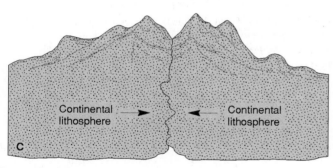

Figure 21.22

(a) Ocean-continent plate convergence. This type of plate boundary accounts for shallow and deep-seated earthquakes, an oceanic trench, volcanic activity, and mountains along the coast. (b) Ocean-ocean plate convergence. This type of plate convergence accounts for shallow and deep-seated earthquakes, an oceanic trench, and a volcanic arc above the subducted plate. (c) Continent-continent plate convergence. Rocks are deformed and some lithosphere thickening occurs, but neither plate is subducted to any great extent.

curving chains of volcanic islands that occur over the belt of deep-seated earthquakes. These islands form where the melted subducted material rises up through the overriding plate above sealevel. The Japanese, Mariana, and Indonesian islands are similar groups of arc islands associated with converging oceanic-oceanic plate boundaries.

During *continent-continent plate convergence* subduction does not occur since the less dense, granite-type materials tend to resist subduction (figure 21.22c). Instead, the colliding plates pile up into a deformed and thicker crust of the lighter material. Such a collision produced the thick, elevated crust known as the Tibetan Plateau and the Himalayan Mountains.

Transform boundaries occur between two plates sliding by each other. Crust is neither created nor destroyed at transform boundaries; one plate slides horizontally past another along a long, vertical fault. The movement is neither smooth nor equal along the length of the fault, however, since short segments move independently with sudden jerks that are separated by periods without motion. The Pacific Plate, for example, is moving slowly to the northwest, sliding past the North American Plate. The San Andreas Fault is one boundary along the California coastline. Vibrations from plate movements along this boundary are the famous California earthquakes.

Present-Day Understandings

The theory of plate tectonics, developed during the late 1960s and early 1970s, is new compared to most major scientific theories. Measurements are still being made, evidence is still being gathered and evaluated, and the exact number of plates and their boundaries are yet to be determined with certainty. The major question that remains to be answered is what drives the plates, forcing them apart, together, and by each other? The most widely favored explanation is that slowly turning **convective cells** in the plastic asthenosphere drive the plates. According to this hypothesis, hot fluid materials rise at the diverging boundaries. Some of the material escapes to form new crust, but most of it spreads out beneath the lithosphere. As it moves beneath the lithosphere, it drags the overlying plate with it. Eventually, it cools and sinks back inward under a subduction zone.

There is uncertainty about the existence of convective cells in the asthenosphere and their possible role because of a lack of clear evidence. Seismic data is not refined enough to show convective cell movement beneath the lithosphere. In addition, deep-seated earthquakes occur to depths of about 700 km (about 400 mi), which means that descending fluids—or parts of a subducted plate—must extend to that depth. This could mean that a convective cell might operate all the way down to the core-mantle boundary some 2,900 km (about 1,800 mi) below the surface. This presents another kind of problem because little is known about the lower mantle and how it interacts with the asthenosphere. Theorizing without information is called speculation, and that is the best that can be done with existing data. The full answer may include the role of heat and the role of gravity, but the mechanism of what drives the plates will probably have to await a better understanding of how the lower mantle behaves.

What is generally accepted about the plate tectonic theory is that the solid materials of the earth are engaged in a continual cycle of change. Oceanic crust is subducted, melted, then partly returned to the crust as volcanic igneous rocks in island arcs and along continental plate boundaries. Other parts of the subducted crust become mixed with the asthenosphere, returning

Earth's Hot Spots

According to plate tectonics, the entire lithosphere of the earth is made up of separate plates that are in a state of slow, continuous, and complex motion. Not everything is moving, however, since there are relatively stationary features called hot spots. *Hot spots* are sites where plumes of hot rock materials rise from deep within the mantle. There are some forty active hot spots under continental and oceanic plates as well as under diverging plate boundaries. The sites of these upwelling plumes are marked by broad bulges in the plates and by volcanic activity that occurs far from plate boundaries.

Hot spots are relatively fixed at a certain latitude and longitude because they originate deep within the mantle. As a plate moves over such a fixed hot spot, a trail of volcanic features marks places where the plate has been over the hot spot, then moved on. Active volcanoes and features of a magma intrusion, such as geysers and hot springs, mark where the hot spot is presently located at the end of the trail. The Hawaiian Island chain, for example, is part of an even longer chain of volcanic features that extends northwesterly across the Pacific seafloor to the mostly submerged Midway Islands, then turns northward in the entirely submerged Emperor chain of extinct volcanoes. The oldest rocks are located at the northern-most extinct volcanoes of the Emperor chain. The rocks then become progressively younger across the Emperors to the southeast end of the Hawaiian Island chain, the island of Hawaii. The hot spot is now under Hawaii, the site of active volcanic activity. The evidence seems clear that the Emperor extinct volcanoes and then the Hawaiian Islands were created one after another as the Pacific Plate moved over a hot spot (box figure 21.1).

As shown in box figure 21.2, the trail of the hot spot currently under the Yellowstone region of Wyoming shows that this spot moved across the Snake River plain as the North American Plate moved west by southwest.

Chemical analyses of materials erupted from hot-spot volcanoes have found a composition that suggests a deep-mantle source, perhaps from near the core. The presence of high amounts of radioactive decay isotopes suggests that the upwelling plumes are driven by heat released by the decay of radioactive elements. Since hot-spot volcanoes erupt intermittently, the plumes probably rise in a series of blobs rather than in a continuous stream.

The substantial evidence for deep-mantle plumes has suggested a "plume model" origin of the asthenosphere. In this model the decay of radioactive elements deep in the mantle heats local materials to the point that they flow readily. This hot, easily flowing mantle material rises in blobs to the asthenosphere, where it collects in a layer that goes around the entire earth. The asthenosphere then becomes the source of new crust at divergent boundaries and serves as the plastic layer that crustal plates move across. Thus the ultimate energy source of seafloor spreading, plate movement, and the related features of oceanic trenches, earthquakes, and volcanoes is the decay of radioactive elements deep in the mantle.

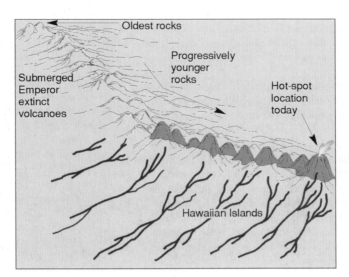

Box Figure 21.1

The volcanic rocks are progressively younger from the northwest end of the Emperor chain of extinct volcanoes, through the Hawaiian Island chain, and to the volcanically active island at the end of the chain. This is one piece of evidence that the Pacific Plate has slowly moved over an enduring hot spot, a plumelike rising of molten rock from deep within the earth's interior.

Box Figure 21.2

Hot spots leave a trail of extinct volcanoes, magma intrusions, and an elevated lithosphere where the plate has moved over the upwelling plume. The present hot springs and geysers of Yellowstone National Park are accounted for by the present location of a hot spot that left a trail of lava flows across the Snake River plain.

as new crust at diverging boundaries. The materials of the crust and the mantle are thus cycled back and forth in a mixing that may include the deep mantle and the core as well. There is more to this story of a dynamic earth that undergoes constant change. The story continues in the next chapters with different cycles to consider.

Summary

The elements silicon and oxygen make up 75 percent of all the elements in the outer layer, or *crust*, of the earth. The elements combine to make crystalline chemical compounds called minerals. A *mineral* is defined as a naturally occurring, inorganic solid element or compound with a crystalline structure.

About 92 percent of the minerals of the earth's crust are composed of silicon and oxygen, the *silicate minerals*. The basic unit of the silicates is a *tetrahedral structure* that combines with positive metallic ions or with other tetrahedral units to form chains, sheets, or an interlocking framework.

The *ferromagnesian silicates* are tetrahedral structures combined with ions of iron, magnesium, calcium, and other elements. The ferromagnesian silicates are darker in color and more dense than other silicates. The *nonferromagnesian silicates* do not have iron or magnesium ions, and they are lighter in color and less dense than the ferromagnesians. The *nonsilicates minerals* do not contain silicon and are carbonates, sulfates, oxides, halides, sulfides, and native elements.

A *rock* is defined as an aggregation of one or more minerals that have been brought together into a cohesive solid. *Igneous rocks* formed as hot, molten *magma* cooled and crystallized to firm, hard rocks. Magma that cools slowly produces coarse-grained *intrusive igneous rocks*. Magma that cools rapidly produces fine-grained *extrusive igneous rocks*. The most abundant igneous rocks are the ferromagnesian rich *basalt* and the silicon rich and ferromagnesian poor *granite*.

Sedimentary rocks are formed from *sediments*, accumulations of weathered rock materials that settle out of the atmosphere or out of water. Sedimentary rocks from *clastic sediments*, or rock fragments, are named according to the size of the sediments making up the rock: *conglomerate, sandstone,* and *shale* in decreasing sediment size. Chemical sediments form from precipitation, crystallization, or the action of plants and animals. *Limestone* is the most common sedimentary rock from chemical sediments. Sediments become sedimentary rocks through *lithification,* a rock-forming process that involves both the *compaction* and *cementation* of the sediments.

Metamorphic rocks are previously existing rocks that have been changed by heat, pressure, or hot solution into a different kind of rock without melting. Increasing metamorphism can change the sedimentary rock shale to *slate,* which is then changed to *schist,* which can then be changed to *gneiss.* Each of these stages has a characteristic crystal size and alignment known as *foliation. Quartzite* and *marble* are examples of two nonfoliated metamorphic rocks.

The *rock cycle* concept views an igneous, a sedimentary, or a metamorphic rock as the temporary stage in the ongoing transformation of rocks to new types.

The earth has a layered interior that formed as the earth's materials underwent *differentiation,* the separation of materials while in the molten state. The center part, or *core,* is predominantly iron with a solid inner part and a liquid outer part. The core makes up about 15 percent of the earth's total volume and about a third of its total mass.

The *mantle* is the middle part of the earth's interior, which accounts for about two-thirds of the earth's total mass and about 80 percent of its total volume. The mantle is predominantly composed of the ferromagnesian silicate *olivine,* which undergoes structural changes at two depths from the increasing heat and pressure. The outer layer, or *crust,* of the earth is separated from the mantle by the *Mohorovicic discontinuity.* The crust of the *continents* is composed mostly of less dense granite-type rocks. The crust of the *ocean basins* is composed mostly of the more dense basaltic rocks. Both types of crust seem to "float" on the mantle according to the concept of *isostasy.*

Another way to consider the earth's interior structure is to consider the weak layer in the upper mantle, the *asthenosphere,* which extends unbroken around the entire earth. The rigid, solid, and brittle layer above the asthenosphere is called the *lithosphere.* The lithosphere includes the entire crust, the Moho, and the upper part of the mantle. Another rigid, solid layer below the asthenosphere is the *mesosphere.*

The shapes of the continents suggested to some that the continents were together at one time and have shifted positions to their present locations. This idea, first developed in the early 1900s, came to be known as *continental drift* and was generally dismissed by most scientists. Evidence from the ocean floor that was gathered in the 1950s and 1960s revived interest in the idea that continents could move. The evidence for *seafloor spreading* came from related observations concerning oceanic ridge systems, from sediment and fossil dating of materials outward from the ridge, and from magnetic patterns of seafloor rocks. Confirmation of seafloor spreading led to the *plate tectonic theory.* According to plate tectonics, new basaltic crust is added at *diverging boundaries* of plates and old crust is *subducted* at *converging boundaries.* Mountain building, volcanoes, and earthquakes are seen as *related geologic features* that are caused by plate movements. The force behind the movement of plates is uncertain, but it may involve *convection* in the deep mantle.

Key Terms

asthenosphere (p. 445)
basalt (p. 439)
Bowen's reaction series (p. 438)
breccia (p. 439)
cementation (p. 440)
chemical sediments (p. 440)
clastic sediments (p. 439)
claystone (p. 439)
compaction (p. 440)
conglomerate (p. 439)
continental drift (p. 446)
convective cells (p. 450)
convergent boundaries (p. 448)
core (p. 442)
crust (p. 444)
divergent boundaries (p. 448)
dolomite (p. 440)
extrusive igneous rocks (p. 438)
ferromagnesian silicates (p. 435)
foliation (p. 441)
gabbro (p. 439)
gneiss (p. 441)
granite (p. 439)

igneous rocks (p. 438)
intrusive igneous rocks (p. 438)
island arcs (p. 448)
isostasy (p. 444)
lava (p. 438)
limestone (p. 440)
lithosphere (p. 445)
magma (p. 438)
mantle (p. 443)
marble (p. 441)
mesosphere (p. 445)
metamorphic rocks (p. 440)
mineral (p. 434)
Mohorovicic discontinuity (p. 444)
new crust zone (p. 448)
nonferromagnesian silicates (p. 435)
nonsilicates (p. 435)
obsidian (p. 439)
oceanic ridges (p. 446)
oceanic trenches (p. 446)
plate tectonics (p. 447)

quartzite (p. 441)
rhyolite (p. 439)
rift (p. 446)
rock (p. 437)
rock cycle (p. 441)
sandstone (p. 439)
schist (p. 441)
seafloor spreading (p. 446)

sedimentary rocks (p. 439)
sediments (p. 439)
shale (p. 440)
silicates (p. 435)
siltstone (p. 439)
slate (p. 441)
subduction zone (p. 448)
transform boundaries (p. 450)

Applying the Concepts

1. The earth's mantle has a chemical composition that agrees closely with the composition of
   a. basalt.
   b. iron and nickle.
   c. granite.
   d. gneiss.

2. From seismological data, the earth's shadow zone indicates that part of the earth's interior is
   a. liquid.
   b. a semiconductor.
   c. a hollow, empty shell.
   d. solid throughout.

3. Mohorovicic discontinuity is a change in seismic wave velocity that takes place because of
   a. structural changes in minerals of the same composition.
   b. changes in the composition on both sides of the boundary.
   c. a shift in the density of minerals of the same composition.
   d. changes in the temperature with depth.

4. The oldest rocks are found in
   a. the continental crust.
   b. the oceanic crust.
   c. neither, since both are the same age.

5. The least dense rocks are found in
   a. the continental crust.
   b. the oceanic crust.
   c. neither, since both are the same density.

6. The idea of seafloor spreading along the Mid-Atlantic Ridge was supported by evidence from
   a. changes in magnetic patterns and ages of rocks away from the ridge.
   b. faulting and volcanoes on the continents.
   c. the observation that the continents fit together like a puzzle.
   d. all of the above.

7. According to the plate tectonics theory, seafloor spreading takes place at a
   a. convergent boundary between plates.
   b. subduction zone.
   c. divergent boundary between plates.
   d. transform boundary between plates.

8. The presence of an oceanic trench, a chain of volcanic mountains along the continental edge, and deep-seated earthquakes are characteristics of a(n)
   a. ocean-ocean plate convergence.
   b. ocean-continent plate convergence.
   c. continent-continent plate convergence.
   d. none of the above.

9. The presence of an oceanic trench with shallow earthquakes and island arcs with deep-seated earthquakes are characteristics of a(n)
   a. ocean-ocean plate convergence.
   b. ocean-continent plate convergence.
   c. continent-continent plate convergence.
   d. none of the above.

10. The ongoing occurrence of earthquakes without seafloor spreading, oceanic trenches, or volcanoes is most characteristic of a
    a. convergent boundary between plates.
    b. subduction zone.
    c. divergent boundary between plates.
    d. transform boundary between plates.

Answers
1. a 2. a 3. b 4. a 5. a 6. a 7. c 8. b
9. a 10. d

Questions for Thought

1. What are the characteristics that make a mineral different from other solid materials of the earth? Is ice a mineral? Explain.

2. Describe the silicate minerals in terms of (a) structural arrangement and (b) composition.

3. Explain why each mineral has its own unique set of physical properties.

4. Identify at least eight physical properties that are useful in identifying minerals. From this list identify two properties that are probably the most useful and two that are probably the least useful in identifying an unknown mineral. Give reasons for your choices.

5. Explain how the identity of an unknown mineral is determined by finding out what the mineral is not.

6. What is a rock?

7. Describe the concept of the rock cycle.

8. Describe the concept of isostasy.

9. Briefly explain the basic differences between the three major kinds of rocks based on the way they are formed.

10. Which major kind of rock, based on the way they are formed, would you expect to find most of in the earth's crust? Explain.

11. What is the difference between magma and lava?

12. What is meant by the texture of an igneous rock? What does the texture of an igneous rock tell you about its cooling history?

13. What are the basic differences between basalt and granite, the two most common igneous rocks of the earth's crust? In what part of the earth's crust are basalt and granite most common? Explain.

14. Explain why a cooled and crystallized magma might have ferromagnesian silicates in the lower part and nonferromagnesian silicates in the upper part.

15. Is the igneous rock basalt *always* fine-grained? Explain.

16. What are clastic sediments? How are they classified and named?

17. Briefly describe the rock-forming process that changes sediments into solid rock.

The Dynamic Earth

18. What are metamorphic rocks? What limits the maximum temperatures possible in metamorphism? Explain.

19. Describe what happens to the minerals as shale is metamorphosed to slate, then schist, then to gneiss. Is it possible to metamorphose shale directly to gneiss or must it go through the slate and schist sequence first? Explain.

20. Describe one theory of how the earth came to have a core composed mostly of iron. What evidence provides information about the nature of the earth's core?

21. Briefly describe the internal composition and structure of the (a) core, (b) mantle, and (c) crust of the earth.

22. What is the asthenosphere? Why is it important in modern understandings of the earth?

23. Describe the parts of the earth included in the (a) lithosphere, (b) asthenosphere, (c) crust, and (d) mantle.

24. What is continental drift? How is it different from plate tectonics?

25. Rocks, sediments, and fossils around an oceanic ridge have a pattern concerning their ages. What is the pattern? Explain what the pattern means.

26. Describe the origin of the magnetic strip patterns found in the rocks along an oceanic ridge.

27. Explain why ancient rocks are not found on the seafloor.

28. Describe the three major types of plate boundaries and what happens at each.

29. What is an island arc? Where are they found? Explain why they are found at this location.

30. Briefly describe a model that explains how the earth developed a layered internal structure.

31. Briefly describe the theory of plate tectonics and how it accounts for the existence of certain geologic features.

32. What is an oceanic trench? What is their relationship to major plate boundaries? Explain this relationship.

33. Describe the probable source of all the earthquakes that occur in southern California.

34. The northwestern coast of the United States has a string of volcanoes running along the coast. According to plate tectonic theory, what does this mean about this part of the North American Plate? What geologic feature would you expect to find on the seafloor off the northwestern coast? Explain.

35. Explain how the crust of the earth is involved in a dynamic, ongoing, recycling process.

Chapter 22

▼

The Surface of the Earth

Figure 22.1
This famous natural bridge is an example of a landform created by the sculpturing power of weathering and erosion. It is Rainbow Bridge in the Rainbow Bridge National Monument, Utah.

THE central idea of plate tectonics, which was discussed in chapter 21, is that the earth's surface is made up of rigid plates that are moving. On the average, this motion is at a rate of several centimeters a year. Since the plates, and the continents riding on them, are in constant motion, any given map of the world is actually only a snapshot that shows the relative positions of the continents at the present time. The continents occupied a different position in the distant past. They will occupy different positions in the distant future. The surface of the earth, which seems so solid and stationary, is in fact mobile and moving.

Plate tectonics has changed the accepted way of thinking about the solid, stationary nature of the earth's surface and ideas about the permanence of the surface as well. The surface of the earth is no longer viewed as having a permanent nature but is understood to be involved in an ongoing cycle of destruction and renewal. Old crust is destroyed as it is plowed back into the mantle through subduction, it melts, and it becomes mixed with the mantle. New crust is created as molten materials move from the mantle through seafloor spreading and volcanoes. Over time, much of the crust must cycle into and out of the mantle.

The mantle-crust cycle is but one of many cycles involving earth materials. Crust exposed to the atmosphere is involved in another kind of cycle, which was introduced in the previous chapter as the rock cycle. The atmosphere and rocks of the crust interact, and rocks are weathered to sediments, which are buried and formed into sedimentary rock. Metamorphosis or melting may take place as forces from the movement of plates deform and elevate rocks into mountains. Here, they are again weathered to sediments and the rock cycling process begins again. You might begin to wonder if anything about the earth is fixed or unchanging.

The movement of plates, the crust-mantle cycle, and the igneous-sedimentary-metamorphic rock cycle all combine to produce a constantly changing surface. This chapter is about these changes and how they account for the varied and interesting scenery that you can observe across the countryside. There are basically two types of surface changes: (1) changes that originate within the earth, resulting in a building up of the surface, and (2) changes that occur from rocks being exposed to the atmosphere and water, resulting in a sculpturing and tearing down of the surface (figure 22.1). The building up and tearing down of the land has left fascinating evidence of what happened in the past, and you will learn how to read some of this evidence. From a knowledge of this type of evidence and its meaning, you will find something of far more interest and significance on the earth's surface than the scenic aspect. The chapter concludes with a discussion of how history can be read from rocks, from the origin of the earth some 4.6 billion years ago to the present.

Shaping the Landscape

Many early thoughts about the earth's landscape were rather strange by today's standards because the vastness of geologic time was not appreciated. Geologic time was not appreciated because changes that occur in the landscape are so slow that a person observes little change in a single lifetime. Thus it was commonly believed that continents and their plains, mountains, and canyons were formed by a single event in the past. The Grand Canyon, for example, was not interpreted as the result of incomprehensibly slow river erosion but as the result of a giant crack or rip that appeared in the surface. The canyon that you see today was thus interpreted as forming when the earth split open and the Colorado River fell into the split. Early ideas about the earth's past were based on a concept of catastrophes like this that created the landscape all at once, with little or no change occurring since that time. This interpretation seemed to fit with the observation of no changes in a person's lifetime.

About two hundred years ago the idea of unchanging, catastrophically formed landscapes was challenged by James Hutton, a Scottish physician. Hutton, who is known today as the founder of modern geology, traveled widely throughout the British Isles. Hutton was a keen observer of rocks, rock structures, and other features of the landscape. He noted that sandstone, for example, was made up of rock fragments that appeared to be (1) similar to the sand being carried by rivers and (2) similar to the sand making up the beaches next to the sea. He also noted fossil shells of sea animals in sandstone on the land, while the living relatives of these animals were found in the shallow waters of the sea. This and other evidence led Hutton to realize that rocks were ground into fragments, then carried by rivers to the sea. He surmised that these particles would be reformed into rocks later, then lifted and shaped into the hills and mountains of the land. He saw all this as quiet, orderly change that required only *time* and the ongoing work of the water and other forces to make the sediments back into rocks. With Hutton's logical conclusion came the understanding that the earth's history could be interpreted by tracing it backward, from the present to the past. This tracing required a frame of reference of slow, uniform change, not the catastrophic frame of reference of previous thinkers. The frame of reference of uniform change is today called the **principle of uniformity.** The principle of uniformity is often represented by a statement that "the present is the key to the past." This statement means that the same geologic processes you see changing rocks today are the very same processes that changed them in the ancient past, although not necessarily at the same rate. This has been the basic, guiding frame of reference used by geologists since the time of Hutton. The concept of how the constant change occurs has evolved with the development of plate tectonics, but the basic frame of reference is the same. You will see how the principle of uniformity is applied as you make a fascinating trip into the history of a landscape. The trip begins by first considering what can happen to rocks and rock layers that are deeply buried.

Rock Stress

Any solid material responds to pressure in a way that depends on the extent of the pressure, the nature of the material, and other variables such as the temperature. Consider, for example, what happens if you apply an increasing pressure to the outside metal of a car door. With increasing pressure you can observe at least four different and separate responses: (1) At first, the metal successfully resists a slight pressure and nothing happens. (2) At a somewhat greater pressure you will be able to deform, or bend, the metal into a concave surface. The metal will return to its original shape, however, when the pressure is removed. This is called an *elastic deformation* since the metal was able to spring back into its original shape. (3) At a still greater pressure the metal is deformed into a concave surface, but this time the metal does not return to its original shape. This means the *elastic limit* of the metal has been exceeded, and it has now undergone a *plastic deformation*. Plastic deformation permanently alters the shape of a material. (4) Finally, at some great pressure the metal will rupture, resulting in a *break* in the material. Many materials, including rocks, respond to increasing pressures in this way, showing no change, an elastic change with recovery, and a plastic change with no recovery, then finally breaking from the pressure.

Rocks in the earth's solid outer crust are subjected to forces as the plates move into, away from, or alongside each other. The plates are subjected to stress by these forces. A **stress** is a force that tends to compress, pull apart, or deform a rock. Just as the metal in your car door, a rock is able to withstand lower stress up to a limit, but then it may undergo elastic deformation, plastic deformation, or breaking with progressively greater pressures. Exactly how the rock responds depends on (1) the nature of the rock, (2) the temperature of the rock, (3) how slowly or quickly the stress is applied over time, and (4) the confining pressure on the rock, which is a result of how deeply it is buried. In general, rocks are able to withstand compressional rather than pulling-apart stresses. Cold rocks are more likely to break than warm rocks, which tend to undergo plastic deformation. In addition, a stress that is applied quickly tends to break the rock, whereas stress applied more slowly over time, perhaps thousands of years, tends to result in plastic deformation.

In general, rocks at great depths are under great pressure at higher temperatures. These rocks tend to undergo elastic deformation, then elastic flow, so rocks at great depths are bent and deformed extensively. Rocks at lesser depths can also bend, but they have a lower elastic limit and break more readily (figure 22.2). Rock deformation often results in recognizable surface features called folds and faults, the topics of the following sections.

Folding

Sediments that form most sedimentary rocks are deposited in nearly flat, horizontal layers at the bottom of a body of water. Over time, conditions change on the land and different mixtures of sediments are deposited in distinct layers of varying thickness. Thus most sedimentary rocks occur naturally as structures of horizontal layers, or beds. A sedimentary rock layer that is not horizontal has been subjected to some kind of stress. The source of the stress could be from colliding plates of the lithosphere, from the intrusion of a magma, or from a plate moving

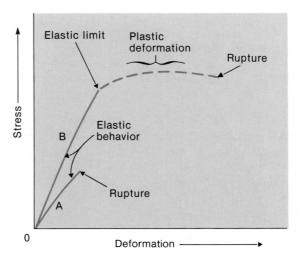

Figure 22.2

Stress and deformation relationships for deeply buried, warm rocks under high pressure (A) and cooler rocks near the surface (B). Breaking occurs when stress exceeds rupture strength.

over a hot spot (see chapter 21). Stress on buried layers of horizontal rocks can result in plastic deformation, resulting in a wrinkling of the layers into **folds.** Folds of all shapes and sizes can occur from plastic deformation, depending generally on the regional or local nature of the stress and other factors. Widespread, regional horizontal stress on deeply buried sedimentary rock layers can produce symmetrical up-and-down folds shaped like waves on water. A vertical, upward stress, on the other hand, can produce a large upwardly bulging fold called a **dome.** A corresponding downwardly bulging fold is called a **basin.** When the stress is great and extensive, complex overturned folds can result.

The most common regional structures from deep plastic deformation are arch-shaped and trough-shaped folds. In general, an arch-shaped fold is called an **anticline** (figure 22.3). The corresponding trough-shaped fold is called a **syncline** (figure 22.4). Anticlines and synclines sometimes alternate across the land like waves on water. You can imagine that great compressional forces must have been involved over a wide region to wrinkle the land like this.

Anticlines, synclines, and other types of folds are not always visible as such on the earth's surface. The ridges of anticlines are constantly being weathered into sediments. The sediments, in turn, tend to collect in the troughs of synclines, filling them in. The Appalachian Mountains have ridges of rocks that are more resistant to weathering, forming hills and mountains (figure 22.5). The San Joaquin Valley, on the other hand, is a very large syncline in California. A syncline of great extent is called a **geosyncline.** The prefix "geo" is from the Greek meaning "earth." Thus a geosyncline is a trough-shaped fold in the earth, implying a very large syncline.

The Surface of the Earth

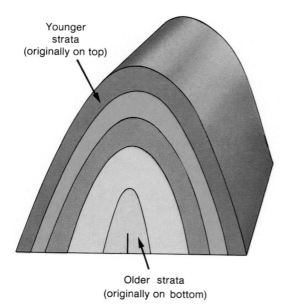

Younger strata (originally on top)

Older strata (originally on bottom)

Figure 22.3

An anticline, or arching fold, in layered sediments. Note that the oldest strata are at the center.

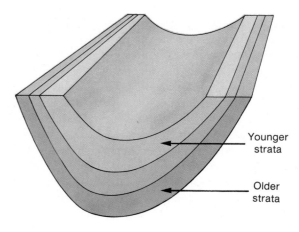

Younger strata

Older strata

Figure 22.4

A syncline showing the reverse age pattern.

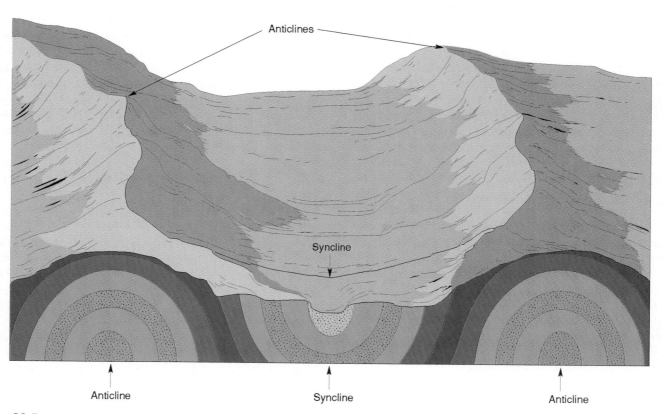

Anticlines

Syncline

Anticline

Syncline

Anticline

Figure 22.5

Parallel mountain ridges often reflect the underlying structure of anticlines and synclines of folded rock beds.

Figure 22.6
The basalt in the Devil's Post Pile, San Joaquin River, California, has developed columnar jointing. This type of jointing forms at right angles to the surface as the basalt cools. The Devil's Tower in Wyoming also shows this type of jointing as do many basalt structures.

Note that any kind of rock can be folded. Sedimentary rocks are usually the best example of folding, however, since the fold structures of rock layers are easy to see and describe. Folding is much harder to see in igneous or metamorphic rocks that are blends of minerals without a layered structure.

Faulting

Rock layers do not always respond to stress by folding. Rocks near the surface are cooler and under less pressure so they tend to be more brittle. A sudden stress on these rocks may reach the rupture point, resulting in a cracking and breaking of the rock structure. If there is breaking of rock without a relative

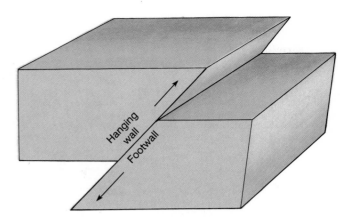

Figure 22.7
Identification of the hanging wall and footwall of a fault.

movement on either side of the break, the crack is called a **joint.** Joints are common in rocks exposed at the surface of the earth. They can be produced from compressional stresses, but they are also formed by other processes such as the contraction of an igneous rock while cooling. Basalt often develops joints from the contraction of cooling magma. The joints are parallel and evenly spaced in basalt, resulting in the appearance of columns. The Devil's Post Pile in California and Devil's Tower in Wyoming are classic examples of columnar joints in basalt (figure 22.6).

When there is relative movement between the rocks on either side of a fracture the crack is called a **fault.** When faulting occurs the rocks on one side move relative to the rocks on the other side along the surface of the fault, which is called the **fault plane.** Faults are generally described in terms of (1) the steepness of the fault plane, that is, the angle between the plane and an imaginary horizontal plane, and (2) the direction of relative movement. This description has meaning about the directions from which the deforming stresses have come to produce the fault.

One classification scheme for faults is based on an orientation referent borrowed from mining (many ore veins are associated with fault planes). Imagine a mine with a fault plane running across a horizontal shaft. Unless the plane is perfectly vertical, a miner would stand on the mass of rock below the fault plane and look up at the mass of rock above. Hence the mass of rock below is called the *footwall* and the mass of rock above is called the *hanging wall* (figure 22.7). How the footwall and hanging wall have moved relative to one another describes three basic classes of faults, the normal, reverse, and thrust faults. A **normal fault** is one in which the hanging wall has moved downward relative to the footwall. This seems "normal" in the sense that you would expect an upper block to slide *down* a lower block along a slope. The opposite situation, where the hanging wall has moved upward relative to the footwall is called a **reverse fault.** A reverse fault with a low-angle fault plane is also called a **thrust fault** (figure 22.8).

The Surface of the Earth

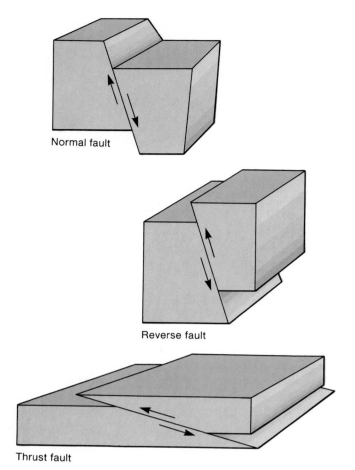

Figure 22.8

Faults are subdivided on the basis of the sense of motion and the angle of the fault plane into normal, reverse, and thrust faults.

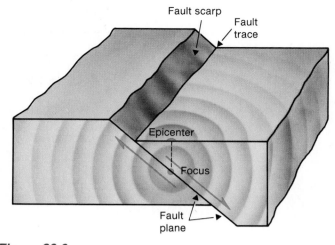

Figure 22.9

Simplified diagram of a fault, illustrating component parts and associated earthquake terminology.

The relative movement of blocks of rocks along a fault plane provides information about the stresses that produced the movement. Reverse and thrust faulting results from compressional stress in the direction of the movement. Normal faulting, on the other hand, results from a pulling-apart stress, which might be associated with diverging plates, for example, or with the stretching and bulging up of the crust over a hot spot.

Earthquakes

Major earthquakes occur along existing fault planes or when a new fault is formed by the fracturing of rock. In either case, earthquakes occur along a fault plane when there is displacement of one side relative to the other. The sudden movement of blocks of rock produces vibrations that move out as waves throughout the earth, and the vibrations are called an *earthquake.*

Almost all earthquakes occur near plate boundaries, and occurrences elsewhere are few and far between. The actual place where an earthquake originates beneath the surface is called the **earthquake focus.** The point on the earth's surface directly

above the focus is called the **earthquake epicenter** (figure 22.9). Along subduction zones shallow-focus earthquakes, meaning those originating within the upper 50 km (about 30 mi), have epicenters near where the lithosphere is first subducted. The deep-focus earthquakes, meaning those originating between 300 km and 700 km (about 200 mi to 400 mi), have epicenters that are progressively farther from the trench with an increasing depth of focus. This is interpreted as meaning that the plunging lithosphere is moving deeper and deeper into the mantle and the earthquakes are a result of this movement.

Most earthquakes have a shallow focus, originating within the upper 50 km (about 30 mi) of the earth's surface. These shallow-focus earthquakes are typical of those that occur at the boundary between the North American Plate, which is moving along against the Pacific Plate. In California, the boundary between these two plates is known as the San Andreas Fault. The two plates are tightly pressed against each other, and friction between the rocks along the fault prevents them from moving easily. Stress continues to build along the entire fault as one plate attempts to move along the other. Some elastic deformation does occur from the stress, but eventually the rupture strength of the rock (or the friction) is overcome. The stressed rock, now released of the strain, snaps suddenly into new positions in a phenomenon known as **elastic rebound** (figure 22.10). The rocks are displaced to new positions on either side of the fault, and the vibrations from the sudden movement are felt as an earthquake. The elastic rebound and movement tends to occur along short segments of the fault at different times rather than along long lengths. Thus the resulting earthquake tends to be a localized phenomenon rather than a regional one.

Earthquakes are detected and measured by an instrument called a seismograph. The seismograph detects three kinds of waves in the solid earth, a longitudinal (compressional) wave called a *primary wave* (or P-wave), a transverse (shear) wave called a *secondary wave* (or S-wave), and an up-and-down

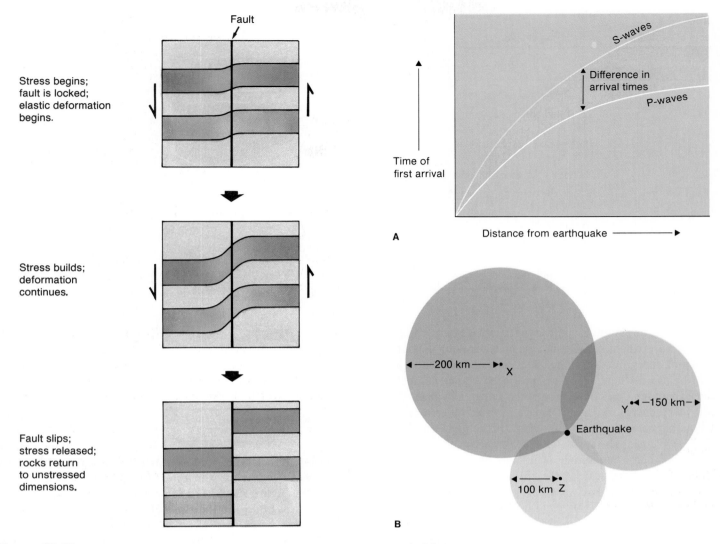

Figure 22.10

The phenomenon of elastic rebound along fault zones. The rocks deform elastically under stress until failure, then snap back to their original, undeformed condition after the earthquake.

Figure 22.11

Use of seismic waves in locating earthquakes. (*a*) Difference in times of first arrivals of P-waves and S-waves is a function of the distance from the focus. (*b*) Triangulation using data from several seismograph stations allows location of the earthquake.

(crests and troughs) wave that travels across the surface called a *long wave* (or L-wave) (figure 22.11). S-waves and P-waves move out from the focus and through the solid earth's interior. L-waves are much like a water wave that moves across the solid surface of the earth. S-waves and P-waves provide information about the location and magnitude of an earthquake as well as information about the earth's interior. P-waves are the fastest and travel through solid and liquid materials. S-waves are next in speed but do not travel through liquids since liquids do not have the cohesion necessary to transmit a shear, or side-to-side, motion. L-waves are the slowest and occur where S-waves or P-waves reach the surface.

To locate an epicenter, at least three recording stations measure the time lag between the arrival of the P-waves and the slower S-waves. The difference in the speed between the two

waves is a constant, so the farther they travel the greater the time lag between the arrival of the faster P-waves and the slower S-waves. Knowing the speed of the two waves, it is possible to calculate the distance to their source by measuring the time lag. However, the calculated distance provides no information about the direction or location of the source of the waves. The location is found by first using the calculated distance as the radius of a circle drawn on a map. Where the circles of the three recording stations intersect identifies the location of their source.

Earthquakes occur in a wide range of intensities, from the many that are barely detectable to the few that cause widespread destruction. The energy of the vibrations, or motion of the land, associated with an earthquake is called its magnitude. Earthquake magnitude is most often measured on the **Richter**

The Surface of the Earth

Table 22.1

Effects of earthquakes of various magnitudes

Richter Magnitudes	Description
0–2	Smallest detectable earthquake
2–3	Detected and measured but not generally felt
3–4	Felt as small earthquake but no damage occurs
4–5	Minor earthquake with local damage
5–6	Moderate earthquake with structural damage
6–7	Strong earthquake with destruction
7–8	Major earthquake with extensive damage and destruction
8–9	Great earthquake with total destruction

Table 22.2

Earthquake safety rules from the National Oceanic and Atmospheric Administration

During the Shaking
1. *Don't panic.*
2. If you are indoors, stay there. Seek protection under a table or desk or in a doorway. Stay away from glass. Don't use matches, candles, or any open flame; douse all fires.
3. If you are outside, move away from buildings and power lines and stay in the open. Don't run through or near buildings.
4. If you are in a moving car, bring it to a stop as quickly as possible but stay in it. The car's springs will absorb some of the shaking and it will offer you protection.

After the Shaking
1. Check, but do *not* turn on, utilities. If you smell gas, open windows, shut off the main valve, and leave the building. Report the leak to the utility and don't reenter the building until it has been checked out. If water mains are damaged, shut off the main valve. If electrical wiring is shorting, close the switch at the main meter box.
2. Turn on radio or television (if possible) for emergency bulletins.
3. Stay off the telephone except to report an emergency.
4. Stay out of severely damaged buildings that could collapse in aftershocks.
5. Don't go sightseeing; you will only hamper the efforts of emergency personnel and repair crews.

Source: National Oceanic and Atmospheric Administration.

Scale. This scale assigns a number that increases with the magnitude of an earthquake. The numbers have meaning about the severity of the ground-shaking vibrations and the energy released by the earthquake. In general, each higher number indicates about ten times more ground movement and about thirty times more energy released than the preceding number. An earthquake measuring below three on this scale is usually not felt by people near the epicenter. The largest earthquake measured so far had a magnitude near nine, but there is actually no upper limit to the scale (table 22.1).

Origin of Mountains

Most of the interesting features of the earth's surface have been created by folding and faulting, and the most prominent of these features is *mountains*. Mountains are elevated parts of the earth's crust that rise abruptly above the surrounding surface. Most mountains do not occur in isolation but in groups that are associated in chains or belts. These long, thin belts are generally found along the edges of continents rather than in the continental interior. There are a number of complex processes involved in the origin of mountains and mountain chains, and no two mountains are exactly alike. For convenience, however, mountains can be classified according to three basic origins: (1) folding, (2) faulting, and (3) volcanic activity.

Folding

The major mountain ranges of the earth—the Appalachian, Rocky, and Himalayan mountains, for example—have a great vertical relief that involves complex folding on a very large scale. The crust was thickened in these places as compressional forces produced tight, almost upright folds. Thus folding is a major feature of the major mountain ranges, but faulting and igneous intrusions are invariably also present. Differential weathering of different rock types produced the parallel features of the Appalachian Mountains, which are so prominent in satellite photographs (figure 22.12). The folded sedimentary rocks of the Rockies are evident in the almost upright beds along the flanks of the front range.

A broad arching fold, which is called a dome, produced the Black Hills of South Dakota. The sedimentary rocks from the top of the dome have been weathered away, leaving a somewhat circular area of more resistant granite hills surrounded by upward-tilting sedimentary beds (figure 22.13). The Adirondack Mountains of New York are another example of this type of mountain formed from folding, called domed mountains.

Faulting

Compression and relaxation of compressional forces on a regional scale can produce large-scale faults, shifting large crustal blocks up or down relative to one another. Huge blocks of rocks can be thrust to mountainous heights, creating a series of fault block mountains. Fault block mountains rise sharply from the surrounding land along the steeply inclined fault plane. The mountains are not in the shape of blocks, however, because weathering has carved them into their familiar mountainlike shapes (figure 22.14). The Teton Mountains of Wyoming and the Sierra Nevadas of California are classic examples of fault block mountains that rise abruptly from the surrounding land. The various mountain ranges of Nevada, Arizona, Utah, and southeastern California have large numbers of fault block mountains that generally trend north and south.

Figure 22.12

The folded structure of the Appalachian Mountains is obvious in this *Skylab* photograph of the Virginia-Tennessee-Kentucky boundary area. The clouds are over the Blue Ridge Mountains.

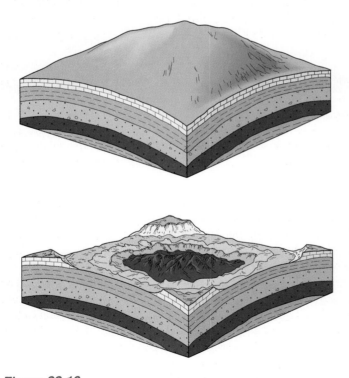

Figure 22.13

A domed mountain begins as a broad, upwarped fold, or dome. The overlying rock of the dome is eroded away, leaving more resistant underlying rock as hills and mountains in a somewhat circular shape. These mountains are surrounded by layers of the rock that formerly covered the dome.

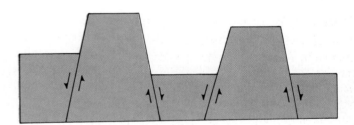

Original sharply bounded
fault blocks softened by erosion
and sedimentation

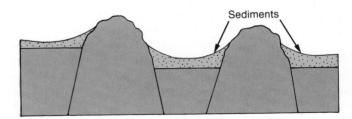

Sediments

Figure 22.14

Fault block mountains are weathered and eroded as they are elevated, resulting in a rounded shape and sedimentation rather than sharply edged fault blocks.

The Surface of the Earth

Volcanic Activity

Lava and other materials from volcanic vents can pile up to mountainous heights on the surface. These accumulations can form local volcano-formed mountains near mountains produced by folding or faulting. Such mixed-origin mountains are common in northern Arizona, New Mexico, and in western Texas. The Cascade Mountains of Washington and Oregon are a series of towering volcanic peaks, most of which are not active today. As a source of mountains, volcanic activity has an overall limited impact on the continents. The major mountains built by volcanic activity are the midoceanic ridges formed at diverging plate boundaries.

The origin of mountain systems and belts of mountains involves a complex mixture and history of folding, faulting, and intrusive igneous activity. Individual mountains can be identified as having an origin of more faulting or more folding, but the overall picture is best seen from generalizations about the long, thin belts of mountains that have grown along the edge of continents at converging boundaries. Such boundaries are seen as places of folding, faulting and associated earthquakes, and the volcanic activity that would be expected with the forces and igneous activities that would thicken the crust. Thus plate tectonics explains that new mountain belts are built as the crust thickens at a convergent boundary. This belt is slowly weathered and worn down as the next belt of mountains begins to build at the new continental edge. Mountains are thus involved in their own cycle of creation and destruction on the ever-changing surface of our dynamic earth.

Sculpturing the Landscape

A mountain of solid granite would appear to be a very stable, permanent structure. A large mass of granite deep within the earth's crust might endure eons without change. But granite at the surface is constantly exposed to air, water, and other agents of change. The granite is altered in appearance and in composition, slowly crumbling into rock fragments. Rock fragments are moved downhill by gravity and other agents, exposing more granite, which was previously deeply buried. The process continues and ultimately the mountain of solid granite is reduced to a mass of loose rock fragments. Figure 22.15 is actually a snapshot of a mountain in a stage between its formation and its eventual destruction to rock fragments.

The changes that result in the breaking up, crumbling, and destruction of any kind of solid rock are called **weathering.** The term implies changes in rocks from the action of the weather, but it actually encompasses a variety of chemical, physical, and biological processes. These weathering processes are important and necessary in (1) the rock cycle, (2) the formation of soils, and (3) the movement of rock materials over the earth's surface. Weathering is important in the rock cycle (see chapter 21) because it produces sediment, the raw materials for new rocks. It is important in the formation of soils because soil is a surface

Figure 22.15

The piles of rocks and rock fragments around a mass of solid rock is evidence that the solid rock is slowly crumbling away. This solid rock that is crumbling to rock fragments is in the Grand Canyon, Arizona.

accumulation of rock fragments and organic matter. Weathering is also important in the rock cycle and in the further weathering of a rock structure because it prepares the rock materials for movement. Before the process of weathering the rock is mostly confined to one location as a solid mass.

The products of weathering are moved by the influence of gravity, water, and other agents. The process of moving these products is called **erosion.** Weathering prepares the way for erosion by attacking the breaking solid rock into fragments. The fragments are eroded downhill by gravity and carried away by erosional agents such as running water, ice, and wind. The combined action of weathering and erosion thus wears away and lowers the surface of the earth. The land surface has been sculptured into its present shape by weathering and erosion.

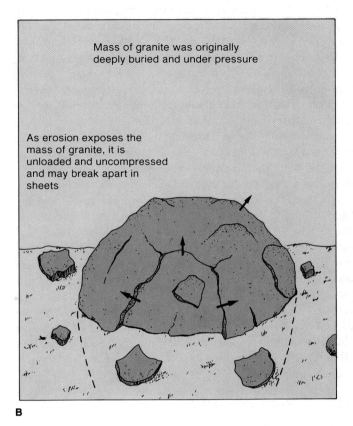

Figure 22.16

(*a*) Frost wedging and (*b*) exfoliation are two examples of mechanical weathering, or disintegration, of solid rock.

Weathering

There are two basic kinds of weathering that act to break down rocks, mechanical weathering and chemical weathering. **Mechanical weathering** is the physical breaking up of rocks without any changes in their chemical composition. Mechanical weathering results in the breaking up of rocks into smaller and smaller pieces, so it is also called *disintegration*. **Chemical weathering** is the breakdown of minerals by chemical reactions with water, gases of the atmosphere, or solutions. Chemical weathering results in the dissolving or breaking down of the minerals in rocks, so it is also called *decomposition*.

Examples of mechanical weathering include the disintegration of rocks caused by (1) wedging effects and (2) the effects of reduced pressure. Wedging effects are often caused by the repeated freezing and thawing of water in the pores and small cracks of otherwise solid rock. As water in a pore or a crack freezes, it expands, exerting a pressure on the walls of the pore or crack and making it slightly larger. The ice melts and the enlarged pore or crack becomes filled with water for another cycle of freezing and thawing. As the process is repeated many

times, small pores and cracks become larger and larger, eventually forcing pieces of rock to break off. This process is called **frost wedging** (figure 22.16a). Frost wedging is an important cause of mechanical weathering in mountains and other locations where repeated cycles of water freezing and thawing occur. The roots of trees and shrubs can also mechanically wedge rocks apart as they grow into cracks. You may have noticed the work of roots when trees or shrubs have grown next to a sidewalk for some period of time.

As more and more weathered materials are removed from the surface the downward pressure from the weight of the material on the rock below becomes less and less. The rock below begins to expand upward, fracturing into concentric sheets from the effect of reduced pressure. These curved, sheetlike plates fall away later in the mechanical weathering process called **exfoliation** (figure 22.16b). Granite commonly weathers by exfoliation, producing characteristic dome-shaped hills and rounded boulders. Stone Mountain, Georgia, is a well-known

The Surface of the Earth

example of an exfoliation-shaped dome. The onionlike structure of exfoliated granite is a common sight in the Sierras, Adirondacks, and any mountain range where older exposed granite is at the surface.

Examples of chemical weathering include (1) oxidation, (2) carbonation, and (3) hydration. **Oxidation** is a reaction between oxygen and the minerals making up rocks. The ferromagnesian minerals, for example, contain iron, magnesium, and other metal ions in a silicate structure (see chapter 21). Iron can react with oxygen to produce several different iron oxides, each with its own characteristic color. The most common iron oxide (hematite) has a deep red color. Other oxides of iron are brown to yellowish-brown. It is the presence of such iron oxides that color many sedimentary rocks and soils. The red soils of Oklahoma, Georgia, and many other places are colored by the presence of iron oxides produced by chemical weathering.

Carbonation is a reaction between carbonic acid and the minerals making up rocks. Rainwater is naturally somewhat acidic because it dissolves carbon dioxide from the air. This forms a weak acid known as carbonic acid (H_2CO_3). Carbonic acid falls on the land, seeping into cracks and crevices, where it reacts with minerals. Limestone, for example, is easily weathered to a soluble form by carbonic acid. The limestone caves of Missouri, Kentucky, New Mexico, and elsewhere were produced by the chemical weathering of limestone by carbonation. Minerals containing calcium, magnesium, sodium, potassium, and iron are chemically weathered by carbonation to produce salts that are soluble in water.

Hydration is a reaction between water and the minerals that make up rocks. The process of hydration includes (1) the dissolving of a mineral and (2) the combining of water directly with a mineral. Some minerals, for example, halite (which is sodium chloride), dissolve in water to form a solution. The carbonates formed from carbonation are mostly soluble, so they are easily leached from a rock by dissolving. Water also combines directly with some minerals to form new, different minerals. The feldspars, for example, undergo carbonation to produce (1) water-soluble potassium carbonate, (2) a chemical product that combines with water to produce a clay mineral, and (3) silica. The silica, which is silicon dioxide (SiO_2), may appear as a suspension of finely divided particles or in solution.

Mechanical and chemical weathering are interrelated, working together to break up and decompose solid rocks of the earth's surface. In general, mechanical weathering results in cracks in solid rocks and broken-off coarse fragments. Chemical weathering results in finely pulverized materials and ions in solution, the ultimate decomposition of a solid rock. Consider, for example, a mountain of solid granite, the most common rock found on continents. In general, granite is made up of 65 percent feldspars, 25 percent quartz, and about 10 percent ferromagnesian minerals. Mechanical weathering begins the destruction process as exfoliation and frost wedging create cracks

in the solid mass of granite. Rainwater, with dissolved oxygen and carbon dioxide, flows and seeps into the cracks and reacts with ferromagnesian minerals to form soluble carbonates and metal oxides. Feldspars undergo carbonation and hydration, forming clay minerals and soluble salts, which are washed away. Quartz is less susceptible to chemical weathering and remains mostly unchanged to form sand grains. The end products of the complete weathering of granite are quartz sand, clay minerals, metal oxides, and soluble salts.

Erosion

Weathering prepares the way for erosion to carry away the fragments, clays, and solutions that have been produced from solid rock. These materials can be moved to a lower elevation by the direct result of gravity acting alone. They can also be moved to a lower elevation by gravity acting through some intermediate agent, such as running water, glaciers, or wind. The erosion of weathered materials as a result of gravity alone will be considered first.

Gravity

Gravity acts on every mass of materials on the surface of the earth, constantly pulling parts of elevated regions toward lower levels. Rocks in the elevated regions are temporarily able to resist this constant pull through their cohesiveness with a main rock mass or by the friction of the rock on a slope. Whenever anything happens to reduce the cohesiveness or to reduce the friction, gravity pulls the freed material to a lower elevation. Thus gravity acts directly on individual rock fragments, and it acts directly on large amounts of surface materials as a mass, pulling all to a lower elevation. Erosion caused by gravity acting directly is called **mass movement.** Mass movement can be so slow that it is practically imperceptible. **Creep,** the slow downhill movement of soil down a steep slope, for example, is detectable only from the peculiar curved growth patterns of trees growing in the slowly moving soil (figure 22.17). At the other extreme, mass movement can be as sudden and swift as a single rock bounding and clattering down a slope from a cliff. Either slow or sudden, mass movement is a small victory for gravity in the ongoing process of leveling the land mass of the earth.

In general, mass movement can be classified according to (1) the nature and amount of material moved and (2) the nature and rate of the movement. The nature of the materials could include finely sorted materials such as soil, sand, or snow to large blocks or sheets of rocks. **Landslide** is a general term for rapid movement of any type or mass of materials, from the short slump of a hillside to the avalanche of a whole mountainside. A short description of some of the more specific mass movement processes follows.

A *rock fall* is the rapid tumbling, bouncing, and/or free fall of individual rock fragments from a cliff or steep slope (figure 22.18). Rocks that are part of the cliff or high up on the slope are weakened and broken up by weathering, lose their support, and fall to the base of the slope. Accumulations of rock frag-

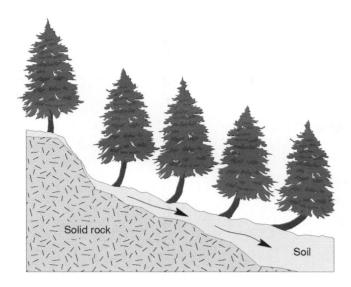

Figure 22.17
The slow creep of soil is evidenced by the strange growth pattern of these trees.

Figure 22.18
Rockfall (schematic).

ments are called **talus.** The weathering of talus continues at a greater rate because the rocks now have a greater surface area exposed than they did as part of the exterior of a solid rock mass.

A *rock slide* is a sudden, rapid movement of a coherent unit of rock along a clearly defined surface or plane. Rock slides occur most often by movement of sedimentary rock beds down a plane of bedding weakness. *Slump,* on the other hand, is the movement of rocks or a soil mass over a short distance. A mass of soil, for example, often falls a short distance as a coherent unit from a stream bank into the stream.

A *flow* is a mass movement of a variety of materials such as soil, rocks, and water with a thick, fluidlike flow. *Mudflows* are common flows that occur in the hills of semiarid areas. This

happens because the clays and other weathering products accumulate over the long dry periods of the semiarid climate. When a heavy rain finally does occur, the accumulated clays and sands form a thick mud that flows downhill, often moving houses in the process. A *debris avalanche* is a mass movement of a wide variety of materials such as rocks, trees, soils, and so forth in a single chaotic flow. Avalanches that occur from deep accumulations of snow high in the mountains often include trees, rocks, and other materials in such a chaotic flow.

Running Water
Of gravity and all the erosional agents of gravity that remove rock materials to lower levels, running water is the most important. Streams and major rivers are at work, for the most part, twenty-four hours a day every day of the year moving rock fragments and dissolved materials from elevated land masses to the oceans. Anytime you see mud, clay, and sand being transported by a river, you know that the river is at work moving mountains, bit by bit, to the ocean. It has been estimated that rivers remove enough dissolved materials and sediments to lower the whole surface of the United States about one meter (about a yard) in slightly more than 27,000 years. At that rate of erosion (and with no uplift), the entire continent could be eroded flat in little over 10 million years, a very short time compared to the 4.6 billion year age of the earth.

Erosion by running water begins with rainfall. Each raindrop impacting the soil moves small rock fragments about, but it also begins to dissolve some of the soluble products of weathering. If the rainfall is heavy enough, a shallow layer, or sheet, of water forms on the surface, moving small fragments and dissolved materials across the surface. This *sheet erosion* moves fragments and dissolved material to small streams at lower levels (figure 22.19). The small streams move to larger channels, and the running water transports materials three different ways: (1) as dissolved rock materials carried in solution, (2) as clay minerals and small grains carried in suspension, and (3) as sand and larger rock fragments that are rolled, bounced, and slid along the bottom of the streambed. Just how much material is eroded and carried by the stream depends on the volume of water, its velocity, and the load that it is already carrying.

In addition to transporting materials that were weathered and eroded by other agents of erosion, streams do their own erosive work. Streams can dissolve soluble materials directly from rocks and sediments. They also quarry and pluck fragments and pieces of rocks from beds of solid rock by hydraulic action. Most of the erosion accomplished directly by streams, however, is done by the more massive fragments that are rolled, bounced, and slid along the streambed and against each other. This results in a grinding and filing action on the fragments and a wearing away of the streambed.

As a stream cuts downward into its bed, other agents of erosion such as mass movement begin to widen the channel as materials slump into the moving water. The load that the stream carries is increased by this slumping, which slows the stream.

The Surface of the Earth

Figure 22.19
Moving streams of water slowly erode the land as they carry away
dissolved materials and sediments.

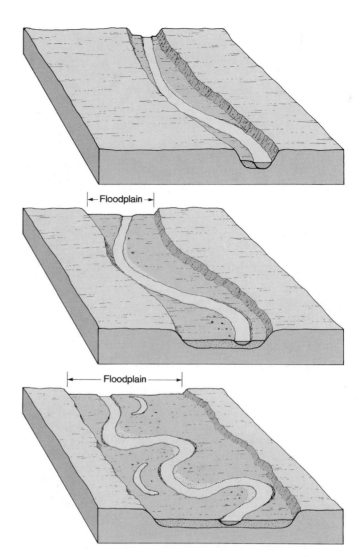

Figure 22.20
A floodplain is broadened by meandering.

As the stream slows, it begins to develop bends, or **meanders,**
along the channel. Meanders have a dramatic effect on stream
erosion because the water moves faster downstream around an
outside bank than it does around the inside bank. This difference in stream velocity means that the stream has a greater erosion ability on the outside, downstream side and less on the
sheltered area inside the curves. The stream begins to widen the
floor of the valley through which it runs by eroding on the outside of the meander, then depositing the eroded material on the
inside of another bend downstream. The stream thus begins to
erode laterally, slowly working its way across the land. Sometimes two bends in the stream meet, forming a cutoff meander
called an **oxbow lake.**

A stream, along with mass movement, develops a valley
on a widening floodplain (figure 22.20). A **floodplain** is the wide,
level floor of a valley built by a stream. It is called a floodplain
because this is where the stream floods when it spills out of its
channel. The development of a stream channel into a widening
floodplain seems to follow a general, idealized aging pattern
(figure 22.21). When a stream is on a recently uplifted land
mass, it has a steep gradient, a vigorous, energetic ability to

erode the land, and characteristic features known as the stage
of youth. *Youth* is characterized by a steep gradient, a V-shaped
valley without a floodplain, and the presence of features that
interrupt its smooth flow such as boulders in the streambed,
rapids, and waterfalls. Stream erosion during youth is predominantly downward. The stream eventually erodes its way into
maturity by eroding away the boulders, rapids, and waterfalls
and in general smoothing and lowering the stream gradient.
During maturity meanders form over a wide floodplain that now
occupies the valley floor. The higher elevations are now more
sloping hills at the edge of the wide floodplain rather than steep-
sided walls close to the river channel. *Old age* is marked by a
very low gradient in extremely broad, gently sloping valleys. The
stream now flows slowly in broad meanders over the wide floodplain. Floods are more common in old age since the stream is
carrying a full load of sediments and flows sluggishly.

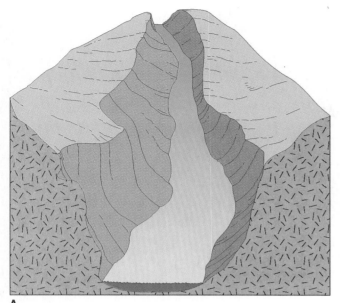

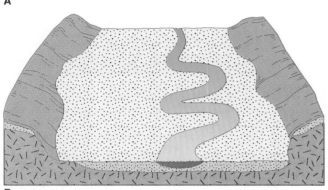

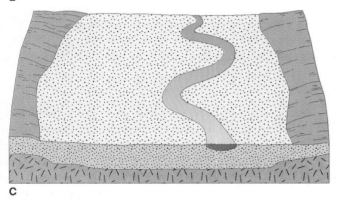

Figure 22.21

Three stages in the aging and development of a stream valley: (a) youth, (b) maturity, and (c) old age.

Many assumptions are made in any generalized scheme of the erosional aging of a stream. Streams and rivers are dynamic systems that respond to local conditions, so it is possible to find an "old-age feature" such as meanders in an otherwise youthful valley. This is like finding a gray hair on an 18-year-old youth—the presence of the gray hair does not mean old age.

In general, old-age characteristics are observed near the *mouth* of a stream where it flows into an ocean, lake, or another stream. Youthful characteristics are observed at the *source,* where the water collects first to form the stream channel. As the stream slowly lowers the land, the old-age characteristics will move slowly but surely toward the source.

When the stream flows into the ocean or a lake it loses all of its sediment-carrying ability. It drops the sediments, forming a deposit at the mouth called a **delta.** Large rivers such as the Mississippi River have large and extensive deltas that actually extend the land mass more and more over time. In a way, you could think of the Mississippi Delta as being formed from pieces and parts of the Rocky Mountains, the Ozark Mountains, and other elevated land masses that the Mississippi has carried there over time.

Glaciers

Glaciers presently cover only about 10 percent of the earth's continental land area and much of this is at higher latitudes, so it might seem that glaciers would not have much of an overall effect in eroding the land. However, ice has sculptured much of the present landscape, and features attributed to glacial episodes are found over about three-quarters of the continental surface. Only a few tens of thousands of years ago, sheets of ice covered major portions of North America, Europe, and Asia. Today, the most extensive glaciers in the United States are those of Alaska, which cover about 3 percent of the state's land area. Less extensive glacier ice is found in the mountainous regions of Washington, Montana, California, Colorado, and Wyoming.

A **glacier** is a mass of ice on land that moves under its own weight. Glacier ice forms gradually from snow, but the quantity of snow needed to form a glacier does not fall in a single winter. Glaciers form in cold climates where some snow and ice persist throughout the year. The amount of winter snowfall must exceed the summer melting to accumulate a sufficient mass of snow to form a glacier. As the snow accumulates, it is gradually transformed into ice. The weight of the overlying snow packs it down, driving out much of the air and causing it to recrystallize into a coarser, denser mass of interlocking ice crystals that appears to have a blue to deep blue color. Complete conversion of snow into glacial ice may take from 5 years to 3,500 years, depending on such factors as climate and rate of snow accumulation at the top of the pile. Eventually the mass of ice will become large enough that it begins to flow, spreading out from the accumulated mass. Glaciers that form at high elevations in mountainous regions, which are called **alpine glaciers,** tend to flow downhill through a valley, so they are also called *valley glaciers* (figure 22.22). Glaciers that cover a large area of a continent are called **continental glaciers.** Continental glaciers can cover whole continents and reach thicknesses of a kilometer or more. Today, the remaining continental glaciers are found on Greenland and in the Antarctic.

Glaciers move slowly and unpredictably, spreading like a huge blob of putty under the influence of gravity. As an alpine

Figure 22.22
An alpine glacier—Knik Glacier, Alaska—as seen from an airplane.
Note the streaks of sediment created where two flows of ice come
together.

glacier moves downhill through a V-shaped valley the sides and
bottom of the valley are eroded wider and deeper. When the
glacier later melts, the V-shaped valley is now a U-shaped valley
that has been straightened and deepened by the the glacial erosion. The glacier does its erosional work using three different
techniques: (1) bulldozing, (2) abrasion, and (3) plucking.
Bulldozing, as the term implies, is the pushing along of rocks,
soil, and sediments by the leading edge of an advancing glacier.
Deposits of bulldozed rocks and other materials that remain after
the ice melts are called **moraines.** *Plucking* occurs as water seeps
into cracked rocks and freezes, becoming a part of the solid glacial mass. As the glacier moves on, it pulls the fractured rock
apart and plucks away chunks of it. The process is accelerated
by the frost-wedging action of the freezing water. Plucking at
the upper-most level of an alpine glacier, combined with weathering of the surrounding rocks, produces a rounded or bowl-like
depression known as a **cirque.** *Abrasion* occurs as the rock fragments frozen into the moving glacial ice scratch, polish, and
grind against surrounding rocks at the base and along the valley
walls. The result of this abrasion is the pulverizing of rock into
ever finer fragments, eventually producing a powdery, silt-sized
sediment called **rock flour.** Suspended rock flour in meltwater
from a glacier gives the water a distinctive gray to blue-gray
color.

Glaciation is continuously at work eroding the landscape
in Alaska and many mountainous regions today. The glaciation
that formed the landscape features in the Rockies, the Sierras,
and across the northeastern United States took place thousands
of years ago. (See the Feature at the end of chapter 20 for a
discussion of a possible cause of the ice ages and the possibility
of new ice ages in the future.)

Wind

Like running water and ice, wind acts as an agent shaping the
surface of the land. It can erode, transport, and deposit materials. However, wind is considerably less efficient than ice or
water in modifying the surface. Wind is much less dense and
does not have the eroding or carrying power of the more dense
water or ice. In addition, a stream generally flows most of the
time, but the wind blows only occasionally in most locations.
Thus on a worldwide average winds move only a few percent as
much material as do streams. Wind also lacks the ability to
attack rocks chemically such as water does through carbonation
and other processes, and the wind cannot carry dissolved sediments in solution. Even in many deserts, more sediment is moved
during the brief periods of intense surface runoff following an
occasional rainstorm than is moved by wind during the prolonged dry periods.

Flowing air and moving water do have much in common
as agents of erosion since both are fluids. Both can move larger

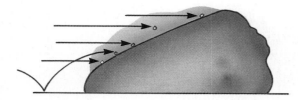

If wind is predominantly from one direction, rocks will be planed off or flattened on the upwind side.

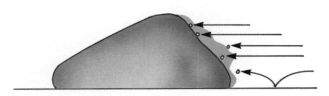

With a persistent shift in wind direction, additional facets are cut in the rock.

Figure 22.23
Ventifact formation by abrasion from one or several directions.

particles by rolling them along the surface and finer particles by carrying them in suspension. Both can move larger and more massive particles with increased velocities. Water is denser and more viscous than air, so it is more efficient at transporting quantities of material than is the wind, but the processes are quite similar.

Two major processes of wind erosion are called abrasion and deflation. **Wind abrasion** is a natural sandblasting process that occurs when the particles carried along by the wind break off small particles and polish what they strike. Generally, the harder mineral grains such as quartz sand accomplish this best near the ground where the wind is bouncing them along. Wind abrasion can strip paint from a car exposed to the moving particles of a dust storm, eroding the paint along with rocks on the surface. Rocks and boulders exposed to repeated windstorms where the wind blows consistently from one or a few directions may be planed off from the repeated action of this natural sandblasting. Rocks sculptured by wind abrasion are called **ventifacts,** after the Latin meaning "wind-made" (figure 22.23).

Deflation, named after the Latin meaning "to blow away," is the widespread removal of loose materials from the surface. Deflation is naturally most active where winds are unobstructed and the materials are exposed and not protected by vegetation. These conditions are often found on deserts, beaches, and on unplanted farmland between crops. During the 1930s several years of drought killed the native vegetation in the Plains states during a period of increased farming activity. Unusually strong winds eroded the unprotected surface, removing and transporting hundreds of millions of tons of soil. This period of prolonged drought, dust storms, and general economic disaster for farmers in the area is known as the Dust Bowl.

The most common windblown deposits are dunes and loess. A **dune** is a low mound or ridge of sand or other sediments. Dunes form when sediment-bearing wind encounters an obstacle that reduces the wind velocity. With a slower velocity the wind cannot carry as large a load, so sediments are deposited on the surface. This creates a windbreak, which results in a growing obstacle, a dune. Once formed, a dune tends to migrate, particularly if the winds blow predominantly from one direction. Dunes are commonly found in semiarid areas or near beaches.

Another common wind deposit is called **loess.** Loess is a very fine dust, or silt, that has been deposited over a large area. One such area is located in the central part of the United States, particularly to the east sides of the major rivers of the Mississippi basin. Apparently this deposit originated from the rock flour produced during the last great ice age. The rock flour was probably deposited along the major river valleys and later moved eastward by the prevailing westerly winds. Since rock flour is produced by the mechanical grinding action of glaciers, it has not been chemically broken down. Thus the loess deposit contains many minerals that were not leached out of the deposit as typically occurs with chemical weathering. It also has an open, porous structure since it does not have as much of the chemically produced clay minerals. The good moisture-holding capacity from this open structure, together with the presence of minerals that serve as plant nutrients, make farming on the deposit particularly productive.

Development of Landscapes

The landscape provides interesting scenery with a variety of features such as mountains, valleys, and broad, rolling hills. The features of the earth's surface are called **landforms.** Landforms include (1) *broad features,* such as a mountain, plain, or plateau, and (2) *minor features,* such as a hill, valley, or canyon. Broad or minor, all landforms are temporary expressions between the forces that elevate the land and the weathering and erosion that level it.

No two landforms are identical because each has been produced and sculptured by a variety of processes. Thus there is no exact way to describe how a particular landform came to be, but generalizations are possible. Generalizations are based on three factors: (1) rock structure, (2) weathering and erosion processes, and (3) stages of erosion.

The *structure* of the rocks determines the shape of the minor landforms. Structure refers to the *type* of rock (igneous, metamorphic, or sedimentary) and the *attitude,* that is, if the rocks have been disturbed by faulting or folding. The type of rock determines how well a rock resists weathering. The sedimentary rock limestone, for example, is highly susceptible to

Figure 22.24

This melting glacier is the source for a stream that flows through a valley in the youth stage.

chemical weathering, while the metamorphic rock quartzite is highly resistant to chemical weathering. The attitude of the rock also determines how well a rock resists weathering and erosion. Limestone beds that have been faulted and folded, for example, are more easily eroded than flat-lying beds of limestone.

The *processes* of weathering and erosion that attack the rock structure are influenced and controlled by other factors such as climate and elevation. Chemical weathering, for example, is more dominant in warm, moist climates, and mechanical weathering is more dominant in dry climates. Thus landforms

in warm, moist climates tend to have softer, rounded outlines from the accumulations of clay minerals, sand, and other finely divided products of chemical weathering. The landforms in dry climates, on the other hand, tend to have sharp angular outlines from the mass movement of rock material from vertical cliffs.

The *stage* of landform development describes how effective the processes have been in attacking the rock structure. Stage describes the extent to which the processes have completed their work, that is, the amount of the original surface that remains. Mountains, for example, are said to be *youthful* when the processes of weathering and erosion have not had time to do much of their work. Youthful mountains are characterized by a prominent relief of steep peaks and narrow, steep valleys. The steep peaks may be from cirques produced by glaciers, and the steep valleys have been cut by streams, but neither process has yet greatly altered the original structure (figure 22.24). By *maturity* the original structure has been worn down to rounded forms and slopes. Eventually, even the mightiest mountain is worn down to nearly flat rolling plains during *old age*. The nearly flat surface is called a **peneplain,** which means a region that is "almost" a plain. Often hills of resistant rock called **monadnocks** exist on the peneplain during the last stages of old age. Theoretically, the monadnocks and peneplain will be reduced to the lowest level possible, which approximates sea level. More than likely, however, the land will be uplifted before this happens, causing **rejuvenation** of the erosion processes. Rejuvenation renews the effectiveness of the weathering and erosion processes, and the cycle begins again with youthful landform structures being superimposed on the old-age structures.

Rocks and Time

There are many different fascinating stories that can be read from a landscape if you know what to look for. Overall, there is a story about the present stage of erosion of the landscape and what it means about the history of the region. The structures in the landscape, such as hills and valleys, also tell a story about folding, faulting, and erosional processes in the past. The story reaching back the farthest in time is told by individual rocks. Each rock has its own combination of minerals that began to change the moment the rock was created. The story of the individual rock, the landscape feature, and the entire region describes the history of the region and how it came to be what it is today. Most knowledge about the history of the earth has come from the direct observation and interpretation of clues found in the rocks of the earth's crust.

Reading Rocks

Reading history from the rocks of the earth's crust requires an understanding of how different kinds of rocks are formed, how rocks are weathered, and the environmental conditions that must have existed for the erosion and deposition of sediments. Thus the mineral composition, texture, and sedimentary structure of rocks are clues about past events. The clues are interpreted within a frame of reference that can be described with several basic principles of geology. The following is a summary of these basic guiding principles.

The *principle of uniformity,* as described earlier, is sometimes stated as "the present is the key to the past." This statement means that the geologic features that you see today have been formed in the past by the same processes of crustal movement, erosion, and deposition that are observed today. By studying the processes now shaping the earth, you can thus understand how it has evolved through time. This principle establishes a frame of reference that the surface of the earth has been continuously and gradually modified over the immense span of geologic time.

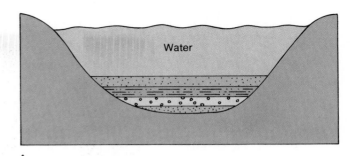

A

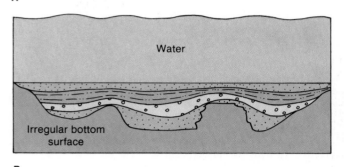

B

Figure 22.25

The principle of original horizontality. (*a*) Sediments tend to be deposited in horizontal layers, (*b*) Even where the sediments are draped over an irregular surface, they tend toward the horizontal.

According to the *principle of original horizontality,* old rocks are continually being changed to new ones in the continuous processes of crustal movement, erosion, and deposition (figure 22.25). As sediments are deposited in a basin of deposition, such as a lake or ocean, they accumulate in essentially horizontal stratified layers. Thus any layer of sedimentary rocks that is not horizontal has been subjected to forces that have deformed the earth's surface.

The *principle of superposition* states that layers of sediments are deposited in succession in horizontal layers, which later are compacted and cemented into sedimentary rock (figure 22.26). In an undeformed sequence, these horizontal layers are arranged in chronological order with the oldest layers at the bottom. Each consecutive layer will be younger than the one below it. This is true, of course, only if the layers have not been turned over by deforming forces.

According to the *principle of crosscutting relationships,* any geologic feature that cuts across or is intruded into a rock mass must be younger than the rock mass (figure 22.27). Thus if a fault cuts across a layer of sedimentary rocks, the fault is the youngest feature. Faults, folds, and igneous intrusions are always younger than the rocks they are found in.

Shifting sites of erosion and deposition: The principle of uniformity states that the earth processes going on today have always been occurring. This does not mean, however, that they always occur in the same place. As erosion wears away the rock layers at a site, the sediments produced are deposited someplace

The Surface of the Earth

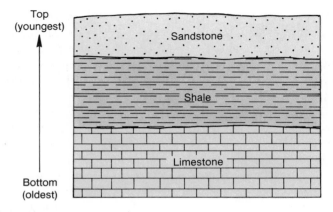

Figure 22.26

The principle of superposition. In an undisturbed sedimentary sequence, the rocks on the bottom were deposited first, and the depositional ages become younger as you progress to the top of the pile.

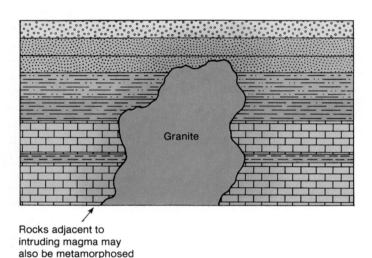

Figure 22.27

A granite intrusion cutting across older rocks.

Rocks adjacent to intruding magma may also be metamorphosed by its heat.

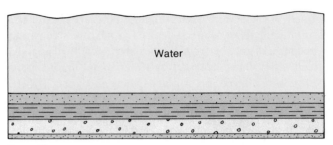

Deposition

Erosional surface

Rocks tilted, eroded

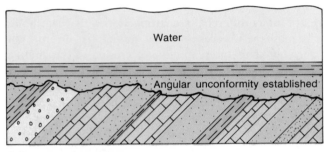

Water

Angular unconformity established

Subsequent deposition

Figure 22.28

Angular unconformity. Development involves some deformation and erosion before sedimentation is resumed.

else. Later, the sites of erosion and deposition may shift, and the sediments are deposited on top of the eroded area. When the new sediments later are formed into new sedimentary rocks there will be a time lapse between the top of the eroded layer and the new layers. A time break in the rock record is called an **unconformity.** An unconformity may occur within a sedimentary sequence of the same kind or between different kinds of rocks. The most obvious kind of unconformity to spot is an **angular unconformity.** An angular unconformity, as illustrated in figure 22.28, is one in which the bedding planes above and below the unconformity are not parallel. An angular unconformity usually implies some kind of tilting or folding followed by a significant period of erosion, which in turn was followed by a period of deposition.

The principle of superposition, the principle of cross-cutting relationships, and the presence of an unconformity all have meaning about the order of geologic events that have occurred in the past. This order can be used to unravel a complex sequence of events such as the one shown in figure 22.29. The presence of fossils can help too. The *principle of faunal succession* recognizes that life forms have changed through time. Old life forms disappear from the fossil record and new ones appear, but the same form is never exactly duplicated independently at two different times in history. This principle implies that the same type of fossil organisms that lived only a brief geologic time should occur only in rocks that are the same age. Distinctive fossils of organisms that lived only a brief time are called **index fossils.** Index fossils, together with the other principles used in reading rocks, make it possible to compare the ages of rocks exposed in two different locations. This is called **correlation** between rock units. Correlations of exposed rock units separated by a few kilometers are easier to do, but correlations

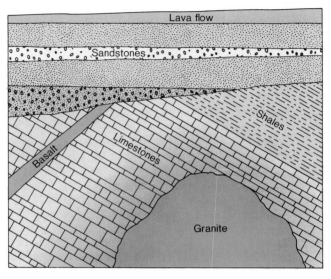

Figure 22.29

Deciphering a complex rock sequence. The limestones must be oldest (law of superposition), followed by the shales. The granite and basalt must both be younger than the limestone they crosscut (note the metamorphosed zone around the granite). It is not possible to tell whether the igneous rocks predate or postdate the shales or to determine whether the sedimentary rocks were tilted before or after the igneous rocks were emplaced. After the limestones and shales were tilted, they were eroded and then the sandstones were deposited on top. Finally, the lava flow covered the entire sequence.

have been done with exposed rock units that are separated by an ocean. Correlation allows the ordering of geologic events according to age. Since this process can only determine the age of a rock unit or geologic event relative to some other unit or event, it is called **relative dating.** Dates with numerical ages are determined by means different from correlation.

Geologic Time

Most human activities are organized by time intervals of minutes, hours, days, months, and years. These time intervals are based on the movements of the earth and moon, as discussed in chapter 20. Short time intervals are measured in minutes and hours, which are typically tracked by watches and clocks. Longer time intervals are measured in days, months, and years, which are typically tracked by calendars. But how do you measure and track time intervals for something as old as the earth? First, you would need to know the age of the earth, then you would need some consistent, measurable events to divide the overall age into intervals. Questions about the age of the earth have puzzled humans for thousands of years, dating back at least to the time of the ancient Greek philosophers. Many people have attempted to answer this question and understand geologic time but with little success until the last few decades.

One early estimate of the age of the earth was attempted by seventeenth-century biblical scholars. Based on the generations of people that could be counted in biblical history, it was determined that the formation of the earth occurred in the year 4004 B.C. On the authority of the biblical scholars, this date was generally accepted for the next century or so. The date of 4004 B.C. meant that the earth and all of its surface features had formed over a period of about 6,000 years. This required a model of great cataclysm to explain how all the earth's features could possibly have formed over this span.

Near the end of the eighteenth century, John Hutton reasoned out the principle of uniformity, and people began to assume a much older earth than they had formerly. The problem then became one of finding some uniform change or process that could serve as a geologic clock to measure the age of the earth. To serve as a geologic clock a process or change would need to meet three criteria: (1) the process must have been operating since the earth began, (2) the process must be uniform or at least subject to averaging, and (3) the process must be measurable.

During the nineteenth century many attempts were made to find earth processes that would meet these criteria. Among others, the processes explored were (1) the rate that salt is being added to the ocean, (2) the rate that sediments are being deposited, and (3) the rate that the earth is cooling. Comparing the load of salts being delivered to the ocean by all the rivers, and assuming the ocean was initially pure water, it was calculated that about 100 million years would be required for the present salinity to be reached. The calculations did not consider the amount of materials being removed from the ocean by organisms and by chemical sedimentation, however, so this technique was considered to be unacceptable. Even if the amount of materials removed was known, it would actually result in the age of the ocean, not the age of the earth.

A number of separate and independent attempts were made to measure the rate of sediment deposition, and then compare that rate to the thickness of sedimentary rocks found on the earth. Dividing the total thickness by the rate of deposition resulted in estimates of an earth age that ranged from about 20 to 1,500 million years. The wide differences occurred because there are gaps in many sedimentary rock sequences, periods when sedimentary rocks were being eroded away to be deposited elsewhere as sediments again. There were just too many unknowns for this technique to be acceptable.

The idea of measuring the rate that the earth is cooling for use as a geologic clock assumed that the earth was initially a molten mass and has been cooling ever since. Calculations estimating the temperature that the earth must have been to be molten were compared to the earth's present rate of cooling. This resulted in an estimated age of 20 to 40 million years. These calculations were made back in the nineteenth century before it was understood that natural radioactivity is adding heat to the earth's interior, so it has required much longer to cool the earth to its present temperature.

Soon after the beginning of the twentieth century the discovery of the radioactive decay process in the elements of minerals and rocks led to the development of a new, accurate geologic clock. This clock finds the **radiometric age** of rocks in years by

Earth's Past and Future

A fascinating picture of the history of earth has resulted from the development of plate tectonics, the exploration of the Solar System by spacecraft, and visits to the moon by astronauts. This picture of history begins about 4.6 billion years ago as earth formed from the condensation and accretion of a solid mass from the solar nebula of gas and dust. Heating from the contraction of this solid mass, heating from the early impacts of planetary matter, and heating from abundant radioactive matter resulted in widespread melting of the early earth. This melting made possible a global differentiation of earth materials, producing a basic structure of an iron-rich core, a silicon-rich mantle, and a less dense silicon-rich crust. This early earth had a primitive atmosphere that was altered by carbon dioxide, water vapor, and other gases from the widespread magma.

Evidence about the early earth is found in rocks that radioactively date back to 3.8 billion years. Metamorphic rocks that formed from previously existing sedimentary rocks some 3.8 billion years ago indicate that the geologic processes operating before this time were much as the geologic processes operating today. The earliest igneous rocks provide evidence that they were also formed much as they are being formed today, but at much higher temperatures. The ancient rock records indicate that the earth before 3.8 billion years ago had much smaller continental land masses and had an atmosphere and ocean that lacked today's relative abundance of free oxygen.

About 2.5 billion years ago the rock record indicates a period of magmatic differentiation that produced much granite in the crust, resulting in extensive elevated continental land masses. About a billion years ago this land mass existed as an early supercontinent, that is, one continuous continental landmass. During this time plate tectonics must have been operational, much as it is today, as the supercontinent was broken into separate landmasses that moved apart into separate, smaller continental masses. The rock record also indicates that the atmospheric oxygen content began to increase at this time as algae and other organisms that produce oxygen through photosynthesis increased in number. The chemical composition of the atmosphere and the oceans thus changed, reaching their present state about 570 million years ago. The fossil shells of marine organisms also appear in rocks of this age, marking the beginning of the Paleozoic era, the time of ancient life.

The earth's landmasses continued to move, propelled by convection from radioactive sources within the interior. Over a period of some 200 million years the landmasses were gradually joined together by plate motions. The Appalachian Mountains were formed as North America collided with Europe and Africa during part of this convergence process. The landmasses came completely together about 225 million years ago, forming the supercontinent of Pangaea with a superocean surrounding it. Soon, however, this second supercontinent began breaking up as the first one did with smaller continental masses moving apart. They have moved slowly ever since to their present-day locations.

The breaking up of Pangaea and formation of today's continental landmasses can be traced along with the development of higher organisms according to the rock and fossil records. The time line is from radioactive dating, but only approximate dates are given here for a generalized picture. Fossils of early shelled organisms, fish, land plants, and early reptiles are found in progressively older rocks that formed between 240 and 570 million years ago when Pangaea began to break apart. Part of the breaking-up process included the opening of the present Atlantic Ocean. The Atlantic Ocean formed just before the fossils of early birds and mammals appeared in the fossil records about 200 million years ago. The Rocky Mountains and the Alps were elevated about the same time as the extinction of dinosaurs about 65 million years ago. Australia and Antarctica became separated about the same time as the fossils of early horses appeared in the record some 50 million years ago. Stone-Age humans appeared within the last 2 million years before a period of worldwide glaciation, generally known as the great ice age. The origin of modern humans is still uncertain, but the best-known ancestral candidate is the Cro-Magnon people who developed during the last stages of the ice age between 35,000 to 8,000 B.C.

What is the next stage in earth's ongoing history? Eventually, the earth's internal supply of radioactive elements will be depleted through radioactive decay. The interior regions of partial melting, the asthenosphere, and the interior of the earth will cool, solidifying to a solid mass as the heat released from radioactive decay diminishes along with the supply. Without the source of heat, there will be no convective currents and therefore no forces to move the continents. If the continents do not move converging plate boundaries will slow to a stop, ending uplift, mountain building, and volcanic activity. Hot spots will die out, ending this source of volcanic activity and elevated landmasses. Weathering and erosion will continue, however, slowly wearing down the land. Sediments will accumulate in the ocean basin, no longer driven back into the mantle at subduction zones. The surface will be slowly reduced to a flat, featureless plane with an elevation that approximates sea level. Rocks will no longer be cycled into and out of the earth's interior and the present-day chemical cycles will end, changing the composition of the atmosphere, the ocean, and the soils that presently support plant life. The resulting chemical state, and how well it will support the present forms of life, are not yet detailed. Discussions of the present-day linked cycles involving the earth's interior, the atmosphere, and the ocean are discussed in the following chapters. These cycles may provide some clues for forecasting the consequences of changes in the dynamic, complex earth system.

---

measuring the radioactive decay of unstable elements within the crystals of certain minerals. Since radioactive decay occurs at a constant, known rate, the ratio of the remaining amount of an unstable element to the amount of decay products present can be used to calculate the time that the unstable element has been a part of that crystal (see chapter 17). Potassium, uranium, and thorium are radioactive isotopes that are often included in the minerals of rocks, so they are often used as radioactive clocks.

A recently developed geologic clock is based on the magnetic orientation of magnetic minerals. These minerals become aligned with the earth's magnetic field when the igneous rock crystallizes, making a record of the field at that time. The earth's magnetic field is global and has undergone a number of reversals in the past (see chapter 9). A **geomagnetic time scale** has been established from the number and duration of magnetic field reversals occurring during the past 6 million years. Combined with radiometric age dating, the geomagnetic time scale is making possible a worldwide geologic clock that can be used to determine local chronologies.

The Geological Time Scale

A yearly calendar helps you keep track of events over time by dividing the year into months, weeks, and days. In a similar way, the **geological time scale** helps you keep track of events that have occurred in the earth's geologic history. The first development of this scale came from the work of William Smith, an English surveyor. Smith discovered that certain rock layers in England occurred in the same order, top to bottom, wherever they were located. He also found that he could correlate and identify each layer by the kinds of fossils in the rocks of the layers. In 1815, he published a geologic map of England, identifying the rock layers in a sequence from oldest to youngest. Smith's work was followed by extensive geological studies of the rock layers in other countries. Soon it was realized that similar, distinctive index fossils appeared in rocks of the same age when the principle of superposition was applied. For example, the layers at the bottom contained fossils of trilobites, but trilobites were not found in the upper levels (figure 22.30a). On the other hand, fossil shells of ammonites appeared in the middle levels but not the lower or upper levels of the rocks (figure 22.30b). The top layer was found to contain the fossils of animals identified as still living today. The early appearance and later disappearance of fossils in progressively younger rocks are explained by organic evolution and extinction, events that could be used to mark the time boundaries of the earth's geologic history.

The major blocks of time in earth's geologic history are called **eras,** each identified by the appearance and disappearance of particular fossils in the sedimentary rock record. There are four main eras: (1) **Cenozoic,** which refers to the time of recent life. Recent life means that the fossils for this time period are identical to the life found on earth today. (2) **Mesozoic,** which refers to the time of middle life. Middle life means that some of the fossils for this time period are similar to the life found on earth today but many are different from anything living today. (3) **Paleozoic,** which refers to the time of ancient life. Ancient life means that the fossils for this time period are very different from anything living on the earth today. (4) **Precambrian,** which refers to the time before ancient life. This means that the rocks for this time period contain very few fossils. The eras were divided into blocks of time called **periods** and the periods were further subdivided into smaller blocks of time called **epochs** (figure 22.31).

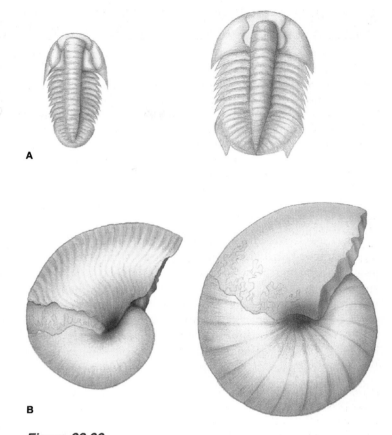

Figure 22.30

(a) Fossil trilobites from Cambrian rocks. (b) Fossil ammonites from Permian rocks.

Era	Period	Epoch	Time of start (millions of years ago)
Cenozoic	Quaternary	Holocene	0.1
		Pleistocene	2
	Tertiary	Pliocene	5
		Miocene	24
		Oligocene	37
		Eocene	58
		Paleocene	66
Mesozoic	Cretaceous		144
	Jurassic		208
	Triassic		245
Paleozoic	Permian		286
	Carboniferous:		
	Pennsylvanian		320
	Mississippian		360
	Devonian		408
	Silurian		438
	Ordovician		505
	Cambrian		570
Precambrian			4,600

Figure 22.31

The divisions of the geological time scale.

The Surface of the Earth

The geological time scale developed in the 1800s was without dates because geologists did not yet have a way to measure the length of the eras, periods, or epochs. This was a *relative time scale,* based on the superposition of rock layers and fossil records of organic evolution. With the development of radiometric dating, it became possible to attach numbers to the time scale. When this was accomplished it became apparent that geologic history spanned very long periods of time. In addition, the part of the time scale with evidence of life makes up less than 15 percent of all of earth's history. This does not mean that life appeared suddenly in the Cambrian period. The fossil record is incomplete since it is the hard parts of animals or plants that form fossils, usually after rapid burial. Thus the soft-bodied life forms that existed during the Precambrian era would make Precambrian fossils exceedingly rare. The Precambrian fossils that have been found are chiefly those of deposits from algae, a few fungi, and the burrow holes of worms. Another problem in finding ancient fossils of soft-bodied life forms is that heat and pressure have altered many of the ancient rocks over time, destroying any fossil evidence that may have been present.

In general, the earliest abundant fossils are found in rocks from the Cambrian period at the beginning of the Paleozoic era. They represent an abundance of oceanic life with over a thousand different species of animals. There is no fossil evidence of life of any kind living on the land during the Cambrian period. The dominant life form of the Cambrian ocean were trilobites and brachiopods (see figure 22.30). The trilobites, now extinct, made up more than half the life population during the Cambrian period.

More and more species of brachiopods appear in the rocks of the Ordovician and Silurian periods, reaching their climax in the Devonian period. Primitive forms of ammonites, shelled organisms that later become more important, first appear in the Devonian period. Fossils of fish first appear in the Ordovician period, then become abundant and diversified by the Devonian period. Sharks were common as was a primitive form of an air-breathing fish. Primitive evergreen and fernlike trees appeared on the land at this time according to the Devonian fossils.

The Carboniferous age, which is made up of the Mississippian and Pennsylvanian periods in the United States, was a time of fast-growing, soft-wooded forests. These were not the modern trees of today but were spore-bearing giants that later would form great coal deposits. Fossils of the first reptiles and the first winged insects are found in rocks from this age. The Paleozoic era closed with the extinction of many types of plant and animal life of that time.

The dinosaurs first appeared in the Triassic period, outnumbering all the other reptiles until the close of the Mesozoic period. Fossils of the first birds, the first mammals, the first flowering plants, and the first deciduous trees appeared in the rocks of this era. Like the close of the Paleozoic era, the Mesozoic era ended with a great dying of land and marine life, which resulted in the extinction of many species, including the dinosaurs.

Throughout the Mesozoic era the reptiles dominated life on the earth. As the Cenozoic era opened the dinosaurs were extinct and the mammals became the dominant life form. The Cenozoic era is thus called "The Age of the Mammals."

You may wonder about the major extinctions of life on the earth during the close of the Paleozoic and Mesozoic eras. There are several theories about what could possibly cause such worldwide extinctions. Based on evidence of a thin clay layer marking the boundary between the Cretaceous and Tertiary periods, one theory proposes that a huge (16 km diameter and 10^{15} kg mass) meteorite struck the earth. The impact would have thrown a tremendous amount of dust into the atmosphere, obscuring the sun and significantly changing the climate and thus the conditions of life on the earth. The resulting colder climate may have led to the extinction of many plant and animal species, including the dinosaurs. This theory is based on the clay layer, which theoretically formed as the dust settled, and its location in the rock record at the time of the extinctions. The layer is enriched with a rare metal, iridium, which is not found on the earth in abundance but occurs in certain meteorites in greater abundance. Perhaps the great extinctions of the past occurred with heavy meteorite bombardments every 100 million years or so. There are other theories as to what could cause such apparently periodic extinctions.

Summary

The *principle of uniformity* is the frame of reference that the same geologic processes you see changing rocks today are the same processes that changed them in the past.

Rocks respond to *stress* by (1) withstanding the stress without change, (2) undergoing *elastic deformation,* (3) undergoing *plastic deformation,* or (4) *breaking.* Exactly how a particular rock responds to stress depends on the nature of the rock, the temperature, and how quickly the stress is applied over time.

Deeply buried rocks are at a higher temperature and tend to undergo *plastic deformation,* resulting in a wrinkling of the layers into *folds.* The most common are an arch-shaped fold called an *anticline* and a trough-shaped fold called a *syncline.* Anticlines and synclines are most easily observed in sedimentary rocks because they have bedding planes, or layers.

Rocks near the surface tend to break from a sudden stress. A break without movement of the rock is called a *joint.* When movement does occur between the rocks on one side of a break relative to the other side the break is called a *fault.*

The vibrations that move out as waves from the movement of rocks are called an *earthquake.* The actual place where an earthquake originates is called its *focus.* The place on the surface directly above a focus is called an *epicenter.* There are three kinds of waves that travel through the earth, *S-waves, P-waves,* and *L-waves.* The magnitude of earthquake waves is measured on the *Richter Scale.*

Folding and faulting produce prominent features on the surface called *mountains.* Mountains can be classified as having an origin of *folding, faulting,* or *volcanic origin.* In general, mountains that occur in long narrow belts called *ranges* have an origin that can be explained by *plate tectonics.*

Weathering is the breaking up, crumbling, and destruction of any kind of solid rock. The movement of weathered rock materials is called *erosion*. The combined action of weathering and erosion wears away and lowers the surface of the earth.

The physical breaking up of rocks is called *mechanical weathering*. The breakdown of minerals by chemical reactions is called *chemical weathering*. The end products of the complete weathering of rocks are removed by direct erosion by gravity, called *mass movement*, or through the agents of *running water, glaciers,* or *wind*.

Landforms are temporary expressions between the forces that elevate the land and the weathering and erosion that is tearing it down. Landforms and the rocks that make them up contain clues about their history. These clues are interpreted within a frame of reference based on (1) the *principle of uniformity*, (2) the *principle of original horizontality*, (3) the *principle of superposition*, (4) the *principle of crosscutting relationships*, (5) the observation that *sites of past erosion and deposition have shifted* over time (shifting sites produce an *unconformity*, or break, in the rock record when erosion removes part of the rocks), and (6) the *principle of faunal succession*.

Geologic time is measured through the radioactive decay process, determining the *radiometric age* of rocks in years. A *geomagnetic time scale* has been established from the number and duration of reversals in the magnetic field of the earth's past.

Correlation and the determination of the numerical ages of rocks and events has led to the development of a *geological time scale*. The major blocks of time on this calendar are called *eras*. The eras are the (1) *Cenozoic*, the time of recent life, (2) *Mesozoic*, the time of middle life, (3) *Paleozoic*, the time of ancient life, and (4) *Precambrian*, the time before ancient life. The eras are divided into smaller blocks of time called *periods*, and the periods are further subdivided into *epochs*. The fossil record is seen to change during each era, ending with *great extinctions* of plant and animal life.

Key Terms

alpine glaciers (p. 469)	folds (p. 457)
angular unconformity (p. 474)	frost wedging (p. 465)
anticline (p. 457)	geological time scale (p. 477)
basin (p. 457)	geomagnetic time scale (p. 477)
carbonation (p. 466)	geosyncline (p. 457)
Cenozoic (p. 477)	glacier (p. 469)
chemical weathering (p. 465)	hydration (p. 466)
cirque (p. 470)	index fossils (p. 474)
continental glaciers (p. 469)	joint (p. 459)
correlation (p. 474)	landforms (p. 471)
creep (p. 466)	landslide (p. 466)
deflation (p. 471)	loess (p. 471)
delta (p. 469)	mass movement (p. 466)
dome (p. 457)	meanders (p. 468)
dune (p. 471)	mechanical weathering (p. 465)
earthquake epicenter (p. 460)	Mesozoic (p. 477)
earthquake focus (p. 460)	monadnocks (p. 473)
elastic rebound (p. 460)	moraines (p. 470)
epochs (p. 477)	normal fault (p. 459)
eras (p. 477)	oxbow lake (p. 468)
erosion (p. 464)	oxidation (p. 466)
exfoliation (p. 465)	Paleozoic (p. 477)
fault (p. 459)	peneplain (p. 473)
fault plane (p. 459)	periods (p. 477)
floodplain (p. 468)	Precambrian (p. 477)

principle of uniformity (p. 456)	syncline (p. 457)
radiometric age (p. 475)	talus (p. 467)
rejuvenation (p. 473)	thrust fault (p. 459)
relative dating (p. 475)	unconformity (p. 474)
reverse fault (p. 459)	ventifacts (p. 471)
Richter Scale (p. 461)	weathering (p. 464)
rock flour (p. 470)	wind abrasion (p. 471)
stress (p. 457)	

Applying the Concepts

1. The basic difference in the frame of reference called the principle of uniformity and the catastrophic frame of reference used by previous thinkers is
   a. the energy requirements for catastrophic changes is much less.
   b. the principle of uniformity requires more time for changes to take place.
   c. catastrophic changes have a greater probability of occurring.
   d. none of the above.

2. When subjected to stress, rocks buried at great depths are under great pressure at high temperatures so they tend to undergo
   a. no change because of the pressure.
   b. elastic deformation because of the high temperature.
   c. plastic deformation.
   d. breaking or rupture.

3. Earthquakes that occur at the boundary between two tectonic plates moving against each other occur along
   a. the entire length of the boundary at once.
   b. short segments of the boundary at different times.
   c. the entire length of the boundary at different times.
   d. none of the above.

4. Other than igneous activity, all mountain ranges have an origin resulting from
   a. folding.
   b. faulting.
   c. stresses.
   d. sedimentation.

5. Of any given series of rock weathering events, the last one to occur would probably be
   a. exfoliation.
   b. frost wedging.
   c. carbonation.
   d. disintegration.

6. Which kind of erosion has had the greatest overall effect in eroding the land?
   a. continental glaciers
   b. alpine glaciers
   c. wind
   d. running water

7. A likely source of loess is
   a. rock flour.
   b. a cirque.
   c. a terminal moraine.
   d. an accumulation of ventifacts.

The Surface of the Earth

8. The landscape in a dry climate tends to be more angular because the dry climate
   a. has more winds.
   b. lacks as much chemical weathering.
   c. has less rainfall.
   d. has stronger rock types.

9. You would expect to find the least number of fossils in rocks from which era?
   a. Cenozoic
   b. Mesozoic
   c. Paleozoic
   d. Precambrian

10. You would expect to find relatively more fossils, but fossils of life very different from anything living today, in rocks from which era?
    a. Cenozoic
    b. Mesozoic
    c. Paleozoic
    d. Precambrian

Answers

1. b 2. c 3. b 4. c 5. c 6. d 7. a 8. b 9. d
10. c

Questions for Thought

1. What is the principle of uniformity? What are the underlying assumptions of this principle?

2. Describe the responses of rock layers to increasing compressional stress when (a) it increases slowly on deeply buried, warm layers, (b) it increases slowly on cold rock layers, and (c) it is applied quickly to rock layers of any temperature.

3. Describe the difference between a syncline and an anticline, using sketches as necessary.

4. What does the presence of folded sedimentary rock layers mean about the geologic history of an area?

5. Describe the conditions that would lead to faulting as opposed to folding of rock layers.

6. How would plate tectonics explain the occurrence of (a) normal faulting? (b) reverse faulting?

7. What is an earthquake? What produces an earthquake?

8. Where would the theory of plate tectonics predict that earthquakes would occur?

9. Describe how the location of an earthquake is identified by a seismic recording station.

10. Briefly explain (a) how and where folded mountains form and (b) how fault block mountains form.

11. Compare and contrast mechanical and chemical weathering.

12. The magnitude of an earthquake is measured on the Richter Scale. What does each higher number mean about an earthquake?

13. Identify three areas of probable active volcanic activity today in the United States. Explain your reasoning or reasons for selecting these three areas.

14. Granite is the most common rock found on continents. What are the end products after granite has been completely weathered? What happens to these weathering products?

15. What other erosion processes are important as a stream of running water carves a valley in the mountains? Explain.

16. Describe how a stream transports the products of weathering.

17. What is a floodplain?

18. Describe the characteristic features associated with stream erosion as the stream valley passes through the stages of youth, maturity, and old age.

19. What is a glacier? How does a glacier erode the land?

20. What is the principle of superposition?

21. What is the geological time scale? What is the meaning of the eras?

22. Why does the rock record only go back 3.8 billion years? If this missing record were available, what do you think it would show? Explain.

23. Do igneous, metamorphic, or sedimentary rocks provide the most information about earth history? Explain.

24. What major event marked the end of the Paleozoic and Mesozoic eras according to the fossil record? Describe one theory that accounts for this.

25. Briefly describe the principles and assumptions that form the basis of interpreting earth's history from the rocks.

26. A meteorite strikes the earth's surface and creates a new landform of a huge crater. Does this violate the principle of uniformity? Explain.

27. Briefly describe the processes that (a) build up the earth's surface and (b) reduce the elevation of the earth's surface.

28. Describe the two major deposits of wind erosion.

29. Describe the sequence of geologic events represented by an angular unconformity. Begin with the deposition of the sediments that formed the oldest rocks represented.

30. Discuss the basic source of energy that produces the earthquakes in southern California.

31. Could a stream erode the land lower than sea level? Explain.

32. Explain why glacial erosion produces a U-shaped valley but stream erosion produces a V-shaped valley.

33. Describe how the principles of superposition, horizontality, and faunal succession are used in the relative dating of sedimentary rock layers.

34. How are the numbers of the ages of eras and other divisions of the geological time scale determined?

35. What is rock flour and how is it produced?

Chapter 23

The Atmosphere of Earth

Figure 23.1
This tree was uprooted by a storm that was predicted, but the couple could do nothing to stop or slow the storm. Understanding the atmosphere and weather phenomena may help in predicting changes, but it is doubtful that weather will ever be controlled on a large scale.

ALMOST all planets in our Solar System have an *atmosphere,* a shell of gases that surrounds the solid mass of the planet. In general, the planets located in the outer part of the Solar System have an atmosphere consisting mostly of hydrogen and helium, or a mixture of these two gases with methane. Earth is located between two planets, Venus and Mars, that have atmospheres of mostly carbon dioxide. Mercury, which is closest to the Sun, is the only planet without an atmosphere. The Earth is also unusual since it is the only planet with an atmosphere of mostly nitrogen and oxygen. An explanation of why these differences exist between the planets was discussed in chapter 19.

The earth's atmosphere has a unique composition because of the cyclic flow of materials that takes place between different parts of the earth. These cycles, which do not exist on the other planets, involve the movement of materials between the surface and the interior (see chapter 21) and the building-up and tearing-down cycles on the surface (see chapter 22). Materials also cycle in and out of the earth's atmosphere. Carbon dioxide, for example, is the major component of the atmospheres around Venus and Mars, and the early Earth had a similar atmosphere. Today, carbon dioxide is a very minor part of Earth's atmosphere. It has been maintained as a minor component in a mostly balanced state for about the past 570 million years, cycling into and out of the atmosphere.

Water is also involved in a global cyclic flow between the atmosphere and the surface. Water on the surface is mostly in the ocean, with lesser amounts in lakes, streams, and underground. Not much water is found in the atmosphere at any one time on a worldwide basis, but billions of tons are constantly evaporating into the atmosphere each year and returning as precipitation in an ongoing cycle.

The cycling of carbon dioxide and water to and from the atmosphere takes place in a dynamic system that is energized by the sun. Radiant energy from the sun heats parts of the earth more than other parts. Winds redistribute this energy with temperature changes, rain, snow, and other changes that are generally referred to as the *weather.*

Understanding and predicting the weather is the subject of **meteorology.** Meteorology is the science of the atmosphere and weather phenomena, from understanding everyday rain and snow to predicting not so common storms and tornadoes (figure 23.1). Understanding weather phenomena depends on a knowledge of the atmosphere and the role of radiant energy on a rotating earth that is revolving around the sun. This chapter is concerned with understanding the atmosphere of the earth, its cycles, and the influence of radiant energy on the atmosphere. This understanding will be put to use in the next chapter, which is concerned with weather and climate.

The Atmosphere

The atmosphere is a relatively thin shell of gases that surrounds the solid earth. If you could see the molecules making up the atmosphere, you would see countless numbers of rapidly moving particles, all undergoing a terrific, chaotic jostling from the billions and billions of collisions occurring every second. Since this

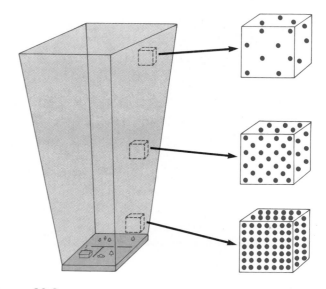

Figure 23.2

At greater altitudes, the same volume contains fewer molecules of the gases that make up the air. This means that the density of air decreases with increasing altitude.

jostling mass of tiny particles is pulled toward the earth by gravity, more are found near the surface than higher up. Thus the atmosphere thins rapidly with increasing distance above the surface, gradually merging with the very diffuse medium of outer space.

To understand how rapidly the atmosphere thins with altitude, imagine a very tall stack of open boxes. At any given instant each consecutively higher box would contain fewer numbers of the jostling molecules than the box below it. Molecules in the lowest box on the surface, at sea level, might be able to move a distance of only 1×10^{-8} m (about 3×10^{-6} in) before colliding with another molecule. A box moved to an altitude of 80 km (about 50 mi) above sea level would have molecules that could move perhaps 10^{-2} m (about 1/2 in) before colliding with another molecule. At 160 km (about 100 mi) the distance traveled would be about 2 m (about 7 ft). As you can see, the distance between molecules increases geometrically with increasing altitude. Since air density is defined by the number of molecules in a unit volume, the density of the atmosphere decreases geometrically with increasing altitude (figure 23.2).

It is often difficult to imagine a distance above the surface because there is nothing visible in the atmosphere for comparison. Imagine a stack of boxes from the surface upward, each with progressively fewer molecules per unit volume. Imagine that this stack of boxes is so tall that it reaches from the surface past the top of the atmosphere. Now imagine that this tremendously tall stack of boxes is tipped over and carefully laid out horizontally on the surface of the earth. How far would you have to move along these boxes to reach the box that was in outer space, outside of the atmosphere? From the bottom box, you would cover a distance of only 5.6 km (about 3.5 mi) to reach the box that was above 50 percent of the mass of the earth's

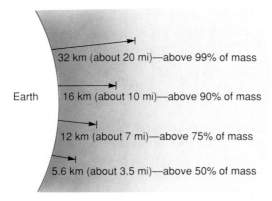

Figure 23.3

The earth's atmosphere thins rapidly with increasing altitude and is much closer to the earth than most people realize.

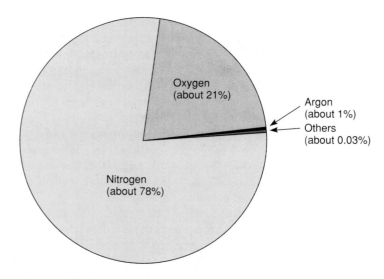

Figure 23.4

The earth's atmosphere has a unique composition of gases when compared to the other planets in the Solar System.

atmosphere. At 12 km (about 7 mi), you would reach the box that was above 75 percent of the earth's atmosphere. At 16 km (about 10 mi), you will reach the box that was above about 90 percent of the atmosphere. And, after only 32 km (about 20 mi), you will reach the box that was above 99 percent of the earth's atmosphere. The significance of these distances might be better appreciated if you can imagine the distances to some familiar locations, for example, from your campus to a store 16 km (about 10 mi) away would place you above 90 percent of the atmosphere if you were to travel this same distance straight up.

Since the average radius of the solid earth is about 6,373 km (3,960 mi), you can see that the atmosphere is a very thin shell with 99 percent of the mass within 32 km (about 20 mi) by comparison. The atmosphere is much closer to the earth than most people realize (figure 23.3).

Composition of the Atmosphere

A sample of pure, dry air is colorless, odorless, and mostly the molecules of just three gases, nitrogen (N_2), oxygen (O_2), and argon (Ar). Nitrogen is the most abundant (about 78 percent of the total volume), followed by oxygen (about 21 percent), then argon (about 1 percent). The molecules of these three gases are well mixed, and this composition is nearly constant everywhere near the earth's surface (figure 23.4). Nitrogen does not readily enter into chemical reactions with rocks, so it has accumulated in the atmosphere. Some nitrogen is involved in a nitrogen cycle, however, as it is removed from the atmosphere by certain bacteria in the soil and by lightning. Both form nitrogen compounds that are essential to the growth of plants. These nitrogen compounds are absorbed by plants and consequently utilized throughout the food chain. Eventually the nitrogen is returned to the atmosphere through the decay of plant and animal matter. Overall, these processes of nitrogen removal and release must be in balance since the amount of nitrogen in the atmosphere is essentially constant over time.

Oxygen gas also cycles into and out of the atmosphere in balanced processes of removal and release. Oxygen is removed by (1) living organisms as food units are oxidized to carbon dioxide and water and (2) chemical weathering of rocks as metals and other elements combine with oxygen to form oxides. Oxygen is released by green plants as a result of the photosynthesis process, and the amount released balances the amount removed by organisms and weathering. So oxygen, as well as nitrogen, is maintained in a state of constant composition through balanced chemical reactions.

The third major component of the atmosphere, argon, is inert and does not enter into any chemical reactions or cycles. It is produced as a product of radioactive decay and once released, remains in the atmosphere as an inactive filler.

In addition to the relatively fixed amounts of nitrogen, oxygen, and argon, the atmosphere contains variable amounts of water vapor. Water vapor is the invisible, molecular form of water in the gaseous state, which should not be confused with fog or clouds. Fog and clouds are tiny droplets of liquid water, not water in the single molecular form of water vapor. The amount of water vapor in the atmosphere can vary from a small fraction of a percent composition by volume in cold, dry air to about 4 percent in warm, humid air. This small, variable percentage of water vapor is essential in maintaining life on the earth. It enters the atmosphere by evaporation, mostly from the ocean, and leaves the atmosphere as rain or snow. The continuous cycle of evaporation and precipitation is called the **hydrologic cycle.** The hydrologic cycle is considered in detail in chapters 24 and 25.

Apart from the variable amounts of water vapor, the relatively fixed amounts of nitrogen, oxygen, and argon make up about 99.97 percent of the volume of a sample of dry air. The remaining 0.03 percent is mostly carbon dioxide (CO_2) and traces of the inert gases neon, helium, krypton, and xenon, along with less than 5 parts per million of free hydrogen, methane, and nitrous oxide. The carbon dioxide content varies locally near cities from the combustion of fossil fuels and from the respiration and decay of organisms and materials produced by organisms. The overall atmospheric concentration of carbon dioxide is regulated (1) by removal from the atmosphere through the photosynthesis process of green plants, (2) by massive exchanges of carbon dioxide between the ocean and the atmosphere, and (3) by chemical reactions between the atmosphere and rocks of the surface, primarily limestone.

The ocean contains some fifty times more carbon dioxide than the atmosphere in the form of carbonate ions and in the form of a dissolved gas. The ocean seems to serve as an equilibrium buffer, absorbing more if the atmospheric concentration increases and releasing more if the atmospheric concentration decreases. Limestone rocks contain an amount of carbon dioxide that is equal to about twenty times the mass of all of earth's present atmosphere. If all this chemically locked-up carbon dioxide were released, the atmosphere would have a concentration of carbon dioxide similar to the present atmosphere of Venus. This amount of carbon dioxide would result in a tremendous increase in the atmospheric pressure and temperatures on earth. Overall, however, equilibrium exchange processes with the ocean, rocks, and living things regulate the amount of carbon dioxide in the atmosphere. Measurements have indicated a yearly increase of about 1 part per million of carbon dioxide in the atmosphere over the last several decades. This increase is believed to be a result of the destruction of tropical rain forests along with increased fossil fuel combustion. The possible consequences of a continuing increase in the carbon dioxide composition of the atmosphere will be discussed soon in another section.

In addition to gases and water vapor, the atmosphere contains particles of dust, smoke, salt crystals, and tiny solid or liquid particles called *aerosols*. The particles of an aerosol are typically 0.005 micrometers in diameter (a micrometer is 10^{-6} meter). These particles become suspended and are dispersed throughout the molecules of the atmospheric gases. Aerosols are produced by combustion, often resulting in air pollution (see the feature at the end of this chapter). Aerosols are also produced by volcanoes and forest fires. Volcanoes, smoke from combustion, and the force of the wind lifting soil and mineral particles into the air all contribute to dust particles larger than aerosols in the atmosphere. These larger particles, which range in size up to 500 micrometers, are not suspended as the aerosols are and they soon settle out of the atmosphere as dust and soot.

Tiny particles of salt crystals that are suspended in the atmosphere come from the mist created by ocean waves and the surf. This mist forms an atmospheric aerosol of seawater that evaporates, leaving the solid salt crystals suspended in the air.

The aerosol of salt crystals and dust becomes well mixed in the lower atmosphere around the globe, playing a large and important role in the formation of clouds.

Atmospheric Pressure

The atmosphere exerts a pressure that decreases with increasing altitude above the surface. Atmospheric pressure can be understood in terms of two different frames of reference that will be useful for different purposes. These two frames of reference are called (1) a hydrostatic frame of reference and (2) a molecular frame of reference. *Hydrostatics* is a consideration of the pressure exerted by a fluid at rest. In this frame of reference atmospheric pressure is understood to be produced by the mass of the atmosphere being pulled to the earth's surface by gravity. In other words, atmospheric pressure is the pressure from the weight of the atmosphere above you. The atmosphere is deepest at the earth's surface, so the greatest pressure is found at the surface. Pressure is less at higher altitudes because there is less air above you and the air is thinner. The *molecular* frame of reference is a consideration of the number of molecules (nitrogen, oxygen, etc.) and the force with which they strike a surface. Air pressure in this frame of reference is understood to be the result of the composite bombardment of air molecules. Atmospheric pressure is greatest at the earth's surface because there are more molecules at lower levels. At higher altitudes in the atmosphere, fewer molecules are present per unit volume, so the pressure they exert is less.

At the earth's surface (sea level), the atmosphere exerts a force of about 10.0 newtons on each square centimeter (14.7 lb/sq in). As you go to higher altitudes above sea level the pressure geometrically decreases with increasing altitude. At an altitude of about 5.6 km (about 3.5 mi) the air pressure is about half of what it is at sea level, about 5.0 newtons/cm² (7.4 lb/in²). At 12 km (about 7 mi) the air pressure is about 2.5 newton/cm² (3.7 lb/in²). Compare this decreasing air pressure at greater elevations to figure 23.3. Again, you can see that most of the atmosphere is very close to the earth, and it thins rapidly with increasing altitude. Even a short elevator ride takes you high enough that the atmospheric pressure on your eardrum is reduced. You equalize the pressure by opening your mouth, allowing the air under greater pressure inside the eardrum to move through the eustachian tube. This makes a "pop" sound that most people associate with changes in air pressure.

Atmospheric pressure is measured by an instrument called a **barometer.** The mercury barometer was invented in 1643 by an Italian named Torricelli. He closed one end of a glass tube, then filled it with mercury. The tube was then placed, open end down, in a bowl of mercury while holding the mercury in the tube with a finger. When Torricelli removed his finger with the open end below the surface in the bowl, a small amount of mercury moved into the bowl leaving a vacuum at the top end of the tube. The mercury remaining in the tube was supported by the atmospheric pressure on the surface of the mercury in the

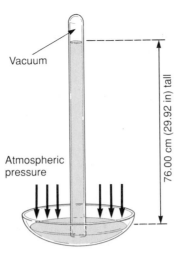

Figure 23.5
The mercury barometer measures the atmospheric pressure from the balance between the pressure exerted by the weight of the mercury in a tube and the pressure exerted by the atmosphere. As the atmospheric pressure increases and decreases, the mercury rises and falls with the changes in pressure. This sketch shows the average height of the column at sea level.

bowl. The pressure exerted by the weight of the mercury in the tube thus balanced the pressure exerted by the atmosphere. At sea level, Torricelli found that atmospheric pressure balanced a column of mercury about 76.00 cm (29.92 in) tall (figure 23.5).

As the atmospheric pressure increases and decreases the height of the supported mercury column moves up and down. Atmospheric pressure can be expressed in terms of the height of such a column of mercury. Public weather reports give the pressure by referring to such a mercury column, for example, "The pressure is 30 in (about 76 cm) and rising." Meteorologists use a measure of atmospheric pressure called a **millibar.** A millibar of pressure is defined as 1,000 dynes per cm² (a dyne is a unit of force equal to 10^{-5} newton; see chapter 3).

If the atmospheric pressure at sea level is measured many times over long periods of time, an average value of 76.00 cm (29.92 in) of mercury is obtained. This average measurement is called the **standard atmospheric pressure** and is sometimes referred to as the *normal pressure*. It is also called *one atmosphere of pressure*. In meteorology, the normal sea level pressure is known as 1,013.25 millibars (or 760.0 mm) of mercury.

Warming the Atmosphere

Temperature, heat, and the heat transfer processes were discussed in chapters 5 and 6. The temperature of a substance is related to the kinetic energy of the molecules making up that substance. The higher the temperature, the greater the kinetic energy. Thus a high temperature means high molecular kinetic energy and low temperature means low molecular kinetic energy. The relationship between the temperature of an object and the nature and amount of radiant energy it emits was discussed in

chapter 10. In general, all objects with a temperature above absolute zero emit radiant energy. The higher the temperature the greater the amount of all wavelengths emitted but with greater proportions of shorter wavelengths. You know, for example, that a "red hot" piece of metal is not as hot as an "orange hot" piece of metal and that the bluish white light from a welder's torch means a very hot temperature. The wavelengths are progressively shorter from red to orange to blue, indicating a progressively higher temperature in each case. Since the surface of the sun is very hot (6,000 K), it radiates much of its energy at short wavelengths. The cooler surface of the earth (300 K) does more of its radiating at longer wavelengths, wavelengths too long to be detected by the human eye. The wavelengths that the earth emits are in the infrared range of the electromagnetic spectrum. Humans cannot see infrared radiation, but they can feel warmth when this radiation is absorbed. Thus it is often referred to as "heat radiation."

Of the total energy radiated by the sun, about 40 percent is in the visible light (violet to red) part of the electromagnetic spectrum, about 9 percent is in the invisible ultraviolet part, and about 51 percent is in the invisible infrared part of the spectrum. This is the radiation from the sun that reaches the outermost part of the earth's atmosphere. Here, when the sunlight is perpendicular to the outer edge and the earth is at an average distance from the sun, it produces about 1,370 watts per square meter. This amount has been called the **solar constant** because the quantity was believed to remain essentially constant. Recent measurements by spacecraft and satellites outside the earth's atmosphere have found, however, that the amount of solar energy received outside the earth's atmosphere decreases slightly, up to a tenth of a percent, when large numbers of sunspots are present on the surface of the sun. As more measurements are taken over longer periods of time, the solar constant may be found not to be so constant after all.

Radiation from the sun must first pass through the atmosphere before reaching the earth's surface. The atmosphere filters, absorbs, and reflects incoming solar radiation as shown in figure 23.6. On the average, the earth as a whole reflects about 30 percent of the total radiation back into space, with two-thirds of the reflection occurring from clouds. The amount reflected at any one time depends on the extent of cloud cover, the amount of dust in the atmosphere, and the extent of snow and vegetation on the surface. Substantial changes in any of these influencing variables could increase or decrease the reflectivity, leading to increased heating or cooling of the atmosphere.

As figure 23.6 shows, only about one-half of the incoming solar radiation reaches the earth's surface. The reflection and selective filtering by the atmosphere allow a global average of about 240 watts per square meter to reach the surface. The value of solar radiation received by the surface is less than half of the solar constant because it is a worldwide average that has been selectively filtered. Wide variations from the average occur with latitude as well as with the season.

The Atmosphere of Earth

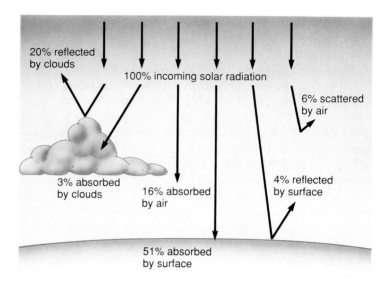

20% reflected
by clouds

100% incoming solar radiation

6% scattered
by air

3% absorbed
by clouds

16% absorbed
by air

4% reflected
by surface

51% absorbed
by surface

Figure 23.6

On the average, the earth's surface absorbs only 51 percent of the incoming solar radiation after it is filtered, absorbed, and reflected. This does not include the radiation emitted back to the surface from the greenhouse effect, which is equivalent to ninety-three units if the percentages in this figure are considered as units of energy.

The incoming solar radiation that does reach the earth's surface is absorbed. Rocks, soil, water, and the ground become warmer as a result. These materials emit the absorbed solar energy as infrared radiation, wavelengths longer than the visible part of the electromagnetic spectrum. This longer wavelength infrared radiation has a frequency that matches some of the natural frequencies of vibration of carbon dioxide and water molecules. This match means that carbon dioxide and water molecules readily absorb infrared radiation that is emitted from the surface of the earth. The absorbed infrared energy shows up as an increased kinetic energy, which is indicated by an increase in temperature. Carbon dioxide and water vapor molecules in the atmosphere now emit infrared radiation of their own, this time in all directions. Some of this reemitted radiation is again absorbed by other molecules in the atmosphere, some is emitted to space, and significantly, some is absorbed by the surface to start the process all over again. The net result is that less of the energy from the sun escapes immediately to space after being absorbed and emitted as infrared. It is retained through the process of being redirected to the surface, increasing the surface temperature more than it would have otherwise been. The more carbon dioxide that is present in the atmosphere, the more energy that will be bounced around and redirected back toward the surface, increasing the temperature near the surface. The process of heating the atmosphere in the lower parts by the absorption and reemission of infrared radiation is called the **greenhouse effect.** It is called the greenhouse effect because greenhouse glass allows the short wavelengths of solar radiation to pass into the greenhouse but does not allow

the longer infrared radiation to leave. This analogy is misleading, however, because carbon dioxide and water vapor molecules do not "trap" infrared radiation, but they are involved in a dynamic absorption and downward reemission process that increases the surface temperature. The more carbon dioxide molecules that are involved in this dynamic process, the more the temperature will increase. More layers of glass on a greenhouse will not increase the temperature significantly. The significant heating factor in a real greenhouse is the blockage of convection by the glass, a process that does not occur from the presence of carbon dioxide and water vapor in the atmosphere.

Structure of the Atmosphere

Convection currents and the repeating absorption and reemission processes of the greenhouse effect tend to heat the atmosphere from the ground up. In addition, the higher altitude parts of the atmosphere lose radiation to space more readily than the lower altitude parts. Thus the lowest part of the atmosphere is warmer, and the temperature decreases with increasing altitude. On the average the temperature decreases about 6.5° C for each kilometer of altitude (3.5° F/1,000 ft). This change of temperature with altitude is called the **observed lapse rate.** The observed lapse rate applies only to air that is not rising or sinking, and the actual change with altitude can be very different from this average value. For example, a stagnate mass of very cold air may settle over an area, producing colder temperatures near the surface than in the air layers above. Such a layer where the temperature *increases* with height is called an **inversion** (figure 23.7). Inversions often occur on calm winter days after the arrival of a cold front. They also occur on calm, clear, and cool nights ("C" nights) when the surface rapidly loses radiant energy to space. In either case, the situation results in a "cap" of cooler, more dense air overlying the warmer air beneath. This often leads to an increase of air pollution because the inversion prevents dispersion of the pollutants.

Temperature decreases with height at the observed lapse rate until an average altitude of about 11 km (about 6.7 mi), where it then begins to remain more or less constant with increasing altitude. The layer of the atmosphere from the surface up to where the temperature stops decreasing with height is called the **troposphere.** Almost all weather occurs in the troposphere, which is named after the Greek meaning "turning layer." The upper boundary of the troposphere is called the **tropopause.** The tropopause is identified by the altitude where the temperature stops decreasing and remains constant with increasing altitude. This altitude varies with latitude and with the season. In general, the tropopause is nearly one and one-half times higher than the average over the equator and about half the average altitude over the poles. It is also higher in the summer than in the winter at a given latitude. Whatever its altitude, the tropopause marks the upper boundary of the atmospheric turbulence and the weather that occurs in the troposphere. The average temperature at the tropopause is about −60° C (about −80° F).

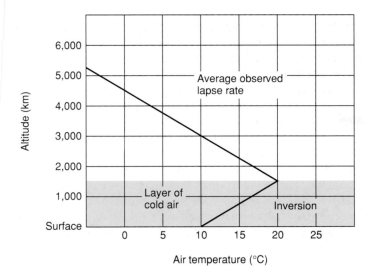

Figure 23.7

On the average, the temperature decreases about 6.5° C/1,000 km, which is known as the observed lapse rate. An inversion is a layer of air in which the temperature increases with height.

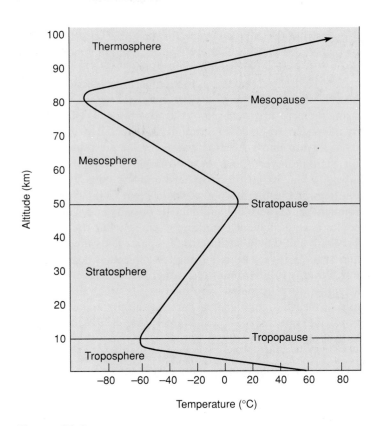

Figure 23.8

The structure of the atmosphere based on temperature differences.

Above the tropopause is the second layer of the atmosphere called the **stratosphere.** This layer is named after the Greek for "stratified layer." It is stratified, or layered, because the temperature increases with height. Cooler air below means that consecutive layers of air are denser on the bottom, which leads to a stable situation rather than the turning turbulence of the troposphere below. The stratosphere contains little moisture or dust and lacks convective turbulence, making it a desirable altitude to fly. Temperature in the lower stratosphere increases gradually with increasing altitude to a height of about 48 km (about 30 mi), where it reaches a maximum of about 10° C (about 50° F). This altitude marks the upper boundary of the stratosphere, the **stratopause** (figure 23.8).

The temperature increases in the stratosphere as a result of interactions between high-energy ultraviolet radiation and ozone. Ozone is triatomic oxygen (O_3) that is concentrated mainly in the lower portions of the stratosphere. Diatomic molecules of oxygen (O_2) are concentrated in the troposphere and monatomic molecules of oxygen (O) are found in the outer edges of the atmosphere. Although the amount of ozone present in the stratosphere is not great, its presence is vital to life on the earth's surface. Without the **ozone shield** ultraviolet radiation would reach the surface of the earth at a level sufficient to damage both skin and eyes even more readily than it now does. It is believed that the incidence of skin cancer would rise dramatically without the protection offered by the ozone.

Here is how the ozone shield works. The ozone concentration is not static because there is an ongoing process of ozone formation and destruction. For ozone to form, diatomic oxygen must first be broken down into the monatomic form. Short ultraviolet (UV) radiation is absorbed by ordinary oxygen, which breaks it down into the single atom (O) form. This reaction is

significant because of (1) the high-energy ultraviolet that is removed from the sunlight and (2) the monatomic oxygen that is formed, which will combine to make ozone that will absorb even more ultraviolet radiation. This initial reaction is

$$O_2 + UV \rightarrow O + O$$

When the O molecule collides with an O_2 molecule and any third, neutral molecule (NM), the following reaction takes place:

$$O_2 + O + NM \rightarrow O_3 + NM$$

When O_3 is exposed to ultraviolet the ozone absorbs the UV radiation and breaks down to two forms of oxygen in the following reaction,

$$O_3 + UV \rightarrow O_2 + O$$

The monatomic molecule that is produced combines with an ozone molecule to produce two diatomic molecules,

$$O + O_3 \rightarrow 2 O_2$$

and the process starts all over again. There is much concern about Freon and other similar chemicals that make their way to the stratosphere. These chemicals are broken down by UV radiation, releasing chlorine, which reacts with the oxygen. This forms a compound that is stable, so it effectively removes the

The Atmosphere of Earth

oxygen from the ongoing ozone forming and UV radiation absorbing process. Many believe that all such chemicals will be banned by international agreement about the end of this decade.

Above the stratopause, the temperature decreases again, just as in the stratosphere, then increases with altitude. The rising temperature is caused by the absorption of solar radiation by molecular fragments present at this altitude.

Layers above the stratopause are the **mesosphere** (Greek for "middle layer") and the **thermosphere** (Greek for "warm layer"). The name thermosphere and the high temperature readings of the thermosphere would seem to indicate an environment that is actually not found at this altitude. The gas molecules here do have a high kinetic energy, but the air here is very thin and the molecules are far apart. Thus the average kinetic energy is very high, but the few molecules do not transfer much energy to a thermometer. A thermometer here would show a temperature far below zero for this reason, even though the same average kinetic energy back at the surface would result in a temperature beyond any temperature ever recorded in the hottest climates.

The **exosphere** (Greek for "outer layer") is the outermost layer where the molecules merge with the diffuse vacuum of space. Molecules of this layer that have sufficient kinetic energy are able to escape and move off into space. The thermosphere and upper mesosphere are sometimes called the **ionosphere** because of the free electrons and ions at this altitude. The electrons and ions here are responsible for reflecting radio waves around the earth and for the northern lights.

The Winds

The troposphere is heated from the bottom up as the surface of the earth absorbs incoming solar radiation and emits infrared radiation. The infrared radiation is absorbed and reemitted numerous times by carbon dioxide and water molecules as the energy works its way back to space. The overall result of this ongoing absorption and reemission process is the observed lapse rate, the decrease of temperature upward in the troposphere. The observed lapse rate is an *average* value, which means that the actual condition at a given place and time is probably higher or lower than this value. The composition of the surface varies from place to place, consisting of many different types and forms of rock, soil, water, ice, snow, and so forth. These various materials absorb and emit energy at various rates, which results in an uneven heating of the surface. You may have noticed that different materials vary in their abilities to absorb and emit energy if you have ever walked barefooted across some combination of grass, concrete, asphalt, and dry sand on a hot sunny day.

Uneven heating of the earth's surface sets the stage for *convection* (see chapter 5). As a local region of air becomes heated, the increased kinetic energy of the molecules expands the mass of air, reducing its density. This less dense air is buoyant

and is pushed upward by nearby cooler, more dense air. This results in three general motions of air: (1) the upward movement of air over a region of greater heating, (2) the sinking of air over a cooler region, and (3) a horizontal air movement between the cooler and warmer regions. In general, a horizontal movement of air is called **wind,** and the direction of a wind is defined as the direction from which it blows.

Air in the troposphere rises, moves as the wind, and sinks. All three of these movements are related and all occur at the same time in different places. During a day with gentle breezes on the surface, the individual, fluffy clouds you see are forming over areas where the air is moving upward. The clear air between the clouds is over areas where the air is moving downward. On a smaller scale, air can be observed moving from a field of cool grass toward an adjacent asphalt parking lot on a calm, sunlit day. Soap bubbles or smoke will often reveal the gentle air movement of this localized convection.

Depending on local surface conditions, which are discussed in the next section, the wind usually averages about 16 km/hr (about 10 mi/hr) and has an average rising and sinking velocity of about 2 km/hr (about 1 mi/hr). These normal, average values are greatly exceeded during storms and severe weather events. A hurricane has winds that exceed 120 km/hr (about 75 mi/hr), and a thunderstorm can have updrafts and downdrafts between 50 and 100 km/hr (about 30 to 60 mi/hr). The force exerted by such winds can be very destructive to structures on the surface. An airplane unfortunate enough to be caught in a thunderstorm can be severely damaged as it is tossed about by the updrafts and downdrafts.

Local Wind Patterns

Considering average conditions, there are two factors that are important for a generalized model to help you understand local wind patterns. These factors are (1) the relationship between air temperature and air density and (2) the relationship between air pressure and the movement of air. The relationship between air temperature and air density was discussed in the introduction to this section. This relationship is that cool air has a greater density than warm air because a mass of warming air expands and a mass of cooling air contracts. Warm, less dense air is buoyed upward by cooler, more dense air, which results in the upward, downward, and horizontal movement of air called a convection cell.

The upward and downward movement of air leads to the second part of the generalized model, that (1) the upward movement produces a "lifting" effect on the surface that results in an area of lower atmospheric pressure and (2) the downward movement produces a "piling up" effect on the surface that results in an area of higher atmospheric pressure. On the surface, air is seen to move from the "piled up" area of higher pressure horizontally to the "lifted" area of lower pressure (figure 23.9). In other words, air generally moves from an area of higher pressure to an area of lower pressure. The movement of air and the pressure differences occur together, and neither is the cause of

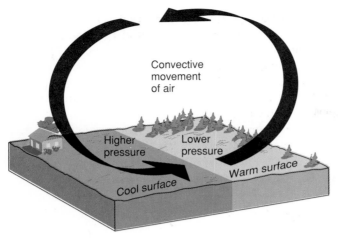

Figure 23.9

A model of the relationships between differential heating, the movement of air, and pressure difference in a convective cell. Cool air pushes the less dense, warm air upward, reducing the surface pressure. As the uplifted air cools and becomes more dense, it sinks, increasing the surface pressure.

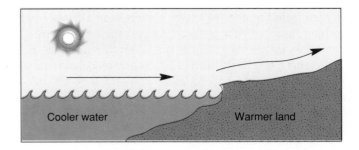

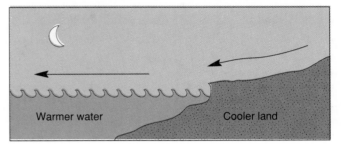

Figure 23.10

The land warms and cools more rapidly than an adjacent large body of water. During the day, the land is warmer and air over the land expands and is buoyed up by cooler, more dense air from over the water. During the night, the land cools more rapidly than the water and the direction of the breeze is reversed.

the other. This is an important relationship in a working model of air movement that can be observed and measured on a very small scale, such as between an asphalt parking lot and a grass field. It can also be observed and measured for local, regional wind patterns and for worldwide wind systems.

Adjacent areas of the surface can have different temperatures because of different heating or cooling rates. The difference is very pronounced between adjacent areas of land and water. Under identical conditions of incoming solar radiation, the temperature changes experienced by the water will be much less than the changes experienced by the adjacent land. There are three principal reasons for this difference: (1) The specific heat of water is about twice the specific heat of soil (see chapter 5). This means that it takes more energy to increase the temperature of water than it does for soil. Equal masses of soil and water exposed to sunlight will result in the soil heating about 1° C while the water heats 1/2° C from absorbing the same amount of solar radiation. (2) Water is a transparent fluid that is easily mixed, so the incoming solar radiation warms a body of water throughout, spreading out the heating effect. Incoming solar radiation on land, on the other hand, warms a relatively thin layer on the top, concentrating the heating effect. (3) The water is cooled by evaporation, which helps keep a body of water at a lower temperature than an adjacent landmass under identical conditions of incoming solar radiation.

A local wind pattern may result from the resulting temperature differences between a body of water and adjacent landmasses. If you have ever spent some time along a coast, you may have observed that a cool, refreshing gentle breeze blows from the water toward the land during the summer. During the day the temperature of the land increases more rapidly than the water temperature. The air over the land is therefore heated

more, expands, and becomes less dense. Cool, dense air from over the water moves inland under the air over the land, buoying it up. The air moving from the sea to the land is called a **sea breeze.** The sea breeze along a coast may extend inland several miles during the hottest part of the day in the summer. The same pattern is sometimes observed around the Great Lakes during the summer, but this breeze usually does not reach more than several city blocks inland. During the night the land surface cools more rapidly than the water and a breeze blows from the land to the sea (figure 23.10).

Another pattern of local winds develops in mountainous regions. If you have ever visited a mountain in the summer, you may have noticed that there is usually a breeze or wind blowing up the mountain slope during the afternoon. This wind pattern develops because the air over the mountain slope is heated more than the air in a valley. As shown in figure 23.11, the air over the slope becomes warmer because it receives more direct sunlight than the valley floor. Sometimes this air movement is so gentle that it would be unknown except for the evidence of clouds that form over the peaks during the day and evaporate at night. During the night the air on the slope cools as the land loses radiant energy to space. As the air cools, it becomes denser and flows downslope, forming a reverse wind pattern to the one observed during the day.

The Atmosphere of Earth

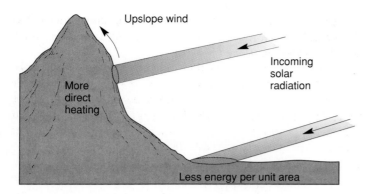

Figure 23.11

Incoming solar radiation falls more directly on the side of a mountain, which results in differential heating. The same amount of sunlight falls on the areas shown in this illustration, with the valley floor receiving a more spread-out distribution of energy per unit area. The overall result is an upslope mountain breeze during the day. During the night, dense cool air flows downslope for a reverse wind pattern.

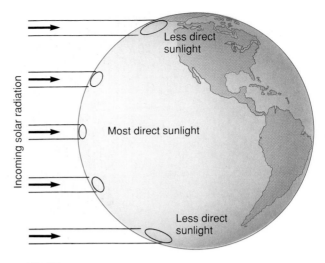

Figure 23.12

On a global, yearly basis, the equatorial region of the earth receives more direct incoming solar radiation than the higher latitudes. As a result, average temperatures are higher in the equatorial region and decrease with latitude toward both poles. This sets the stage for worldwide patterns of prevailing winds, high and low areas of atmospheric pressure, and climatic patterns.

During cooler seasons cold, dense air may collect in valleys or over plateaus, forming a layer or "puddle" of cold air. Such an accumulation of cold air often results in some very cold nighttime temperatures for cities located in valleys, temperatures that are much colder than anywhere in the surrounding region. Some weather disturbance, such as an approaching front, can disturb such an accumulation of cold air and cause it to pour out of its resting place and through canyons or lower valleys. Air moving from a higher altitude like this becomes compressed as it moves to lower elevations under increasing atmospheric pressure. Compression of air, or any gas for that matter, increases the temperature by increasing the kinetic energy of the molecules (see chapter 6). This creates a wind called a **Chinook,** which is common to mountainous and adjacent regions. A Chinook is a wind of compressed air with sharp temperature increases that can melt away any existing snow cover in a single day. The *Santa Ana* is a well-known compressional wind that occurs in southern California.

Global Wind Patterns

Local wind patterns tend to mask the existence of the overall global wind pattern that is also present. The global wind pattern is not apparent if the winds are observed and measured for a particular day, week, or month. It does become apparent when the records for a long period of time are analyzed. These records show that the earth has a large-scale pattern of atmospheric circulation that varies with latitude. There are belts in which the winds average an overall circulation in one direction, belts of higher atmospheric pressure averages, and belts of lower atmospheric pressure averages. This has led to a generalized pattern of atmospheric circulation and a global atmospheric model. This model, as you will see, today provides the basis for the daily weather forecast for local and regional areas.

As with local wind patterns, it is temperature imbalances that drive the global circulation of the atmosphere. The earth receives more direct solar radiation in the equatorial region than it does at higher latitudes (figure 23.12). As a result, the temperatures of the lower troposphere are generally higher in the equatorial region, decreasing with latitude toward both poles. The lower troposphere from 10° N to 10° S of the equator is heated, expands, and becomes less dense. Hot air rises in this belt around the equator, known as the **intertropical convergence zone.** The rising air cools because it expands as it rises, resulting in heavy average precipitation. The tropical rain forests of the earth occur in this zone of high temperatures and heavy rainfall. As the now dry, rising air reaches the upper parts of the troposphere it begins to spread toward the north and toward the south, sinking back toward the earth's surface (figure 23.13). The descending air reaches the surface to form a high-pressure belt that is centered about 30° N and 30° S of the equator. Air moving on the surface away from this high-pressure belt produces the prevailing northeast trade winds and the prevailing westerly winds of the Northern Hemisphere. The great deserts of the earth are also located in this high-pressure belt of descending dry air.

Poleward of the belt of high pressure the atmospheric circulation is controlled by a powerful belt of wind near the top of the troposphere called a **jet stream.** Jet streams are sinuous, meandering loops of winds that tend to extend all the way around the earth, moving generally from the west in both hemispheres at speeds of 160 km/hr (about 100 mi/hr) or more. A jet stream may occur as a single belt, or loop, of wind but sometimes it divides into two or more parts. The jet stream develops north

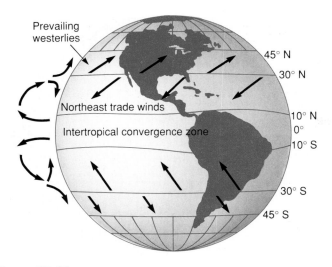

Figure 23.13
Part of the generalized global circulation pattern of the earth's atmosphere. The scale of upward movement of air above the intertropical convergence zone is exaggerated for clarity. The troposphere over the equator is thicker than elsewhere, reaching a height of about 20. km (about 12 mi).

and south loops of waves much like the waves you might make on a very long rope. These waves vary in size, sometimes beginning as a small ripple but then growing slowly as the wave moves eastward. Waves that form on the jet stream bulge toward the poles (called a crest) or toward the equator (called a trough). Warm air masses move toward the poles ahead of a trough and cool air masses move toward the equator behind a trough as it moves eastward. The development of a wave in the jet stream is understood to be one of the factors that influences the movement of warm and cool air masses, a movement that results in weather changes on the surface.

The intertropical convergence zone, the 30° belt of high pressure, and the northward and southward migration of a meandering jet stream all shift toward or away from the equator during the different seasons of the year. The troughs of the jet stream influence the movement of alternating cool and warm air masses over the belt of the prevailing westerlies, resulting in frequent shifts of fair weather to stormy weather, then back again. The average shift during the year is about 6° of latitude, which is sufficient to control the overall climate in some locations. The influence of this shift of the global circulation of the earth's atmosphere will be considered as a climatic influence after considering the role of water and air masses in frequent weather changes.

Water and the Atmosphere

Water exists on the earth in all three states: (1) as a liquid when the temperature is generally above the freezing point of 0° C (32° F), (2) as a solid in the form of ice, snow, or hail when the temperature is generally below the freezing point, and (3) as the invisible, molecular form of water in the gaseous state, which is called *water vapor.*

Over 98 percent of all the water on the earth exists in the liquid state, mostly in the ocean, and only a small, variable amount of water vapor is in the atmosphere at any given time. Since so much water seems to fall as rain or snow at times, it may be a surprise that the overall atmosphere really does not contain very much water vapor. If the average amount of water vapor in the earth's atmosphere were condensed to liquid form, the vapor *and* all the droplets present in clouds would form a uniform layer around the earth only 3 cm (about 1 in) thick. Nonetheless, it is this small amount of water vapor that is eventually responsible for (1) contributing to the greenhouse effect, which helps make the earth a warmer planet, (2) serving as one of the principal agents in the weathering and erosion of the land, which creates soils and sculptures the landscape, and (3) maintains life, for almost all plants and animals cannot survive without water. It is the ongoing cycling of water vapor into and out of the atmosphere that makes all this possible. Understanding this cycling process and the energy exchanges involved is also closely related to understanding the earth's weather patterns.

Evaporation and Condensation

Water tends to undergo a liquid-to-gas or a gas-to-liquid phase change at any temperature. The phase change can occur in either direction at any temperature. The temperature of liquid water and the temperature of water vapor are associated with the *average* kinetic energy of the water molecules. The word *average* implies that some of the molecules have a greater kinetic energy and some have less. If a molecule of water that has an exceptionally high kinetic energy is near the surface, and is headed in the right direction, it may overcome the attractive forces of the other water molecules and escape the liquid to become a gas. This is the process of evaporation. A supply of energy must be present to maintain the process of evaporation, and the water robs this energy from the surroundings. This explains why water at a higher temperature evaporates more rapidly than water at a lower temperature. More energy is available at higher temperatures to maintain the process at a faster rate.

Water molecules that evaporate move about in all directions and some will strike the liquid surface. The same forces that it escaped from earlier now capture the molecule, returning it to the liquid state. This is called the process of condensation. Condensation is the opposite of evaporation. In *evaporation,* more molecules are leaving the liquid state than are returning. In *condensation,* more molecules are returning to the liquid state than are leaving. This is a dynamic, ongoing process with molecules leaving and returning continuously (figure 23.14). If the air were perfectly dry and still, more molecules would leave (evaporate) the liquid state than would return (condense). Eventually, however, an equilibrium would be reached with as

The Atmosphere of Earth

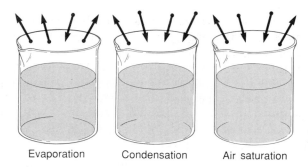

Evaporation   Condensation   Air saturation

Figure 23.14

Evaporation and condensation are occurring all the time. If the number of molecules leaving the liquid state exceeds the number returning, the water is evaporating. If the number of molecules returning to the liquid state exceeds the number leaving, the water vapor is condensing. If both rates are equal, the air is saturated, that is, the relative humidity is 100 percent.

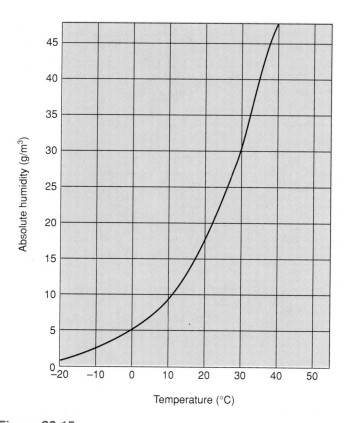

Figure 23.15

The maximum amount of water vapor that can be in the air at different temperatures. The amount of water vapor in the air at a particular temperature is called the absolute humidity.

many molecules returning to the liquid state per unit of time as are leaving. An equilibrium condition between evaporation and condensation occurs in **saturated air.** Saturated air occurs when the processes of evaporation and condensation are in balance.

Air will remain saturated as long as (1) the temperature remains constant and (2) the processes of evaporation and condensation remain balanced. Temperature influences the equilibrium condition of saturated air because increases or decreases in the temperature mean increases or decreases in the kinetic energy of water vapor molecules. Water vapor molecules usually undergo condensation when attractive forces between the molecules can pull them together into the liquid state. Lower temperature means lower kinetic energies, and slow-moving water vapor molecules spend more time close to one another and close to the surface of liquid water. Spending more time close together means an increased likelihood of attractive forces pulling the molecules together. On the other hand, higher temperature means higher kinetic energies, and higher kinetic energy molecules are less likely to be pulled together. As the temperature increases, there is therefore less tendency for water molecules to return to the liquid state. If the temperature is increased in an equilibrium condition, more water vapor must be added to the air to maintain the saturated condition. Warm air can therefore hold more water vapor than cooler air. In fact, warm air on a typical summer day can hold five times as much water vapor as cold air on a cold winter day.

Humidity

The amount of water vapor in the air is referred to generally as **humidity.** Damp, moist air is more likely to have condensation than evaporation, and this air is said to have a *high humidity.* Dry air is more likely to have evaporation than condensation, on the other hand, and this air is said to have a *low humidity.* A measurement of the amount of water vapor in the air at a

particular time is called the **absolute humidity** (figure 23.15). At room temperature, for example, humid air might contain 15 grams of water vapor in each cubic meter of air. At the same temperature air of low humidity might have an absolute humidity of only 2 grams per cubic meter. The absolute humidity can range from near zero up to a maximum that is determined by the temperature at the time. Since the temperature of the water vapor present in the air is the same as the temperature of the air, the maximum absolute humidity is usually said to be determined by the air temperature. What this really means is that the maximum absolute humidity is determined by the temperature of the water vapor, that is, the average kinetic energy of the water vapor.

The relationship between the *actual* absolute humidity at a particular temperature and the *maximum* absolute humidity that can occur at that temperature is called the **relative humidity.** Relative humidity is a ratio between (1) the amount of water vapor in the air and (2) the amount of water vapor needed to saturate the air at that temperature. The relationship is

$$\frac{\text{actual absolute humidity at present temperature}}{\text{maximum absolute humidity at present temperature}} \times 100\%$$
$$= \frac{\text{relative}}{\text{humidity}}$$

For example, suppose a measurement of the water vapor in the air at 10° C (50°F) finds an absolute humidity of 5.0 g/m³. According to figure 23.15, the maximum amount of water vapor that can be in the air when the temperature is 10° C is 10. g/m³. The relative humidity is then

$$\frac{5.0 \text{ g/m}^3}{10. \text{ g/m}^3} \times 100\% = 50\%$$

If the absolute humidity had been 10. g/m³, then the air would have all the water vapor it could hold and the relative humidity would be 100 percent. A humidity of 100 percent means that the air is saturated at the present temperature.

The important thing to understand about relative humidity is that the capacity of air to hold water vapor changes with the temperature. Cold air cannot hold as much water vapor, and warming the air will increase its capacity. With the capacity *increased* and the same amount of water in the air, the relative humidity *decreases* because you can now add more water vapor to the air than you could before. Just warming the air, for example, can reduce the relative humidity from 50 percent to 3 percent. Lower relative humidity results because warming the air increases the capacity of air to hold water vapor. This explains the need to humidify a home in the winter. Evaporation occurs very rapidly when the humidity is low. Evaporation is a cooling process since the higher kinetic energy molecules are the ones to escape, lowering the average kinetic energy as they evaporate. Dry air will therefore cause you to feel cool even though the air temperature is fairly high. Adding moisture to the air will enable you to feel warmer at lower air temperatures, and thus lower your fuel bill. The relationship between the capacity of air to hold water vapor and temperature also explains why the relative humidity increases in the evening after the sun goes down. A cooler temperature means less capacity of air to hold water vapor. With the same amount of vapor in the air, a reduced capacity means a higher relative humidity.

Relative humidity is closely associated with your comfort. When the wind blows evaporation is increased because the moving air also moves evaporated water vapor molecules away from your body, reducing the possibility that they could condense and release the latent heat of vaporization. Thus wind makes the temperature *seem* much lower. In the winter, the cooling power of the wind is called the **wind chill factor.** For example, a temperature of 2° C (about 35° F) in air moving at 16 km/hr (about 10. mi/hr) will have the cooling equivalent of still air with a temperature of about −7° C (about 20° F).

You have probably heard the expression, "It's not the heat, it's the humidity." A higher humidity will lessen the tendency for evaporation to exceed condensation, slowing the cooling process of evaporation. You often fan your face or use an electric fan to move evaporated water vapor molecules away from the surface of your skin, increasing the net rate of evaporation. The use of a fan will not help cool you at all if the relative humidity is 100 percent. At 100 percent relative humidity the air is saturated and the rate of evaporation is equal to the rate of condensation so there can be no net cooling.

Evaporation occurs at a rate that is proportional to the absolute humidity, ranging from a maximum rate when the air is driest to no net evaporation when the air is saturated. Since evaporation is a cooling process, it is possible to use a thermometer to measure humidity. An instrument called a **psychrometer** has two thermometers, one of which has a damp cloth wick around its bulb end. As air moves past the two thermometer bulbs, the ordinary thermometer (the dry bulb) will measure the present air temperature. Water will evaporate from the wet wick (the wet bulb) until an equilibrium is reached between water vapor leaving the wick and water vapor returning to the wick from the air. Since evaporation lowers the temperature, the depression of the temperature of the wet bulb thermometer is an indirect measure of the water vapor present in the air. The relative humidity can be determined by obtaining the dry and wet bulb temperature readings and referring to a relative humidity table such as the one found in appendix C. If the humidity is 100 percent, no net evaporation will take place from the wet bulb, and both wet and dry bulb temperatures will be the same. The lower the humidity, the greater the difference in the temperature reading of the two thermometers.

You can also use the relative humidity table in appendix C to find out how much cooling occurs by evaporation at a given relative humidity and air temperature. For this purpose, you use the table "inside out." Locate the row for the present (dry bulb) temperature, then find the present relative humidity on that row inside the table. The relative humidity reading will be located in the column that has the wet bulb depression at the top of the column. The wet bulb depression is the maximum amount of cooling that can occur by evaporation at that relative humidity and air temperature.

You may have noticed that your hair becomes more or less curly in humid weather. The human hair absorbs moisture from the air, becoming longer in humid air and shorter in dry air. Since the change of length is proportional to changes in humidity it is possible to use human hair to measure the humidity. A **hair hygrometer** is an instrument that measures the humidity from changes in the length of hair. A bundle of the hair is held under tension by a spring, and changes in the length moves a pointer on a dial that is calibrated to read the relative humidity.

The Condensation Process

In still air under a constant pressure the rate of evaporation depends primarily on three factors: (1) the surface area of the liquid that is exposed to the atmosphere, (2) the air and water temperature, and (3) the amount of water vapor in the air at the time, that is, the relative humidity. The opposite process, condensation, depends primarily on two factors: (1) the relative humidity and (2) the temperature of the air, or more directly, the kinetic energy of the water vapor molecules. During condensation molecules of water vapor join together to produce liquid water on the surface as dew or in the air as the droplets of water making up fog or clouds. Water molecules may also join together to produce solid water in the form of frost or snow.

The Atmosphere of Earth

Before condensation can occur, however, the air must be saturated, which means that the relative humidity must be 100 percent with the air containing all the vapor it can hold at the present temperature. A parcel of air can become saturated as a result of (1) water vapor being added to the air from evaporation, (2) cooling, which reduces the capacity of the air to hold water vapor and therefore increases the relative humidity, or (3) a combination of additional water vapor with cooling.

The process of condensation of water vapor explains a number of common observations. You are able to "see your breath" on a cold day, for example, because the high moisture content of your exhaled breath is condensed into tiny water droplets by cold air. The small fog of water droplets evaporates as it spreads into the surrounding air with a lower moisture content. The white trail behind a high-flying jet aircraft is also a result of condensation of water vapor. Water is one of the products of combustion, and the white trail is condensed water vapor, a trail of tiny droplets of water in the cold upper atmosphere. The trail of water droplets is called a *contrail* after "condensation trail." Back on the surface, a cold glass of beverage seems to "sweat" as water vapor molecules near the outside of the glass are cooled, moving more slowly. Slowly moving water vapor molecules spend more time closer together, and the molecular forces between the molecules pull them together, forming a thin layer of liquid water on the outside of the cold glass. This same condensation process sometimes results in a small stream of water from the cold air conditioning coils of an automobile or home mechanical air conditioner.

As air is cooled, its capacity to hold water vapor is reduced to lower and lower levels. Even without water vapor being added to the air a temperature will eventually be reached at which saturation, 100 percent humidity, occurs. Further cooling below this temperature will result in condensation. The temperature at which condensation begins is called the **dew point temperature.** If the dew point is above 0° C (32° F) the water vapor will condense on surfaces as a liquid called **dew.** If the temperature is at or below 0° C the vapor will condense on surfaces as a solid called **frost.** Note that dew and frost form on the tops, sides, and bottoms of objects. Dew and frost condense directly on objects and do not "fall out" of the air. Note also that the temperature that determines if dew or frost forms is the temperature of the object where they condense. This temperature near the open surface can be very different from the reported air temperature, which is measured more at eye level in a sheltered instrument enclosure.

Observations of where and when dew and frost form can lead to some interesting things to think about. Dew and frost, for example, seem to form on "C" nights, nights that can be described by the three "C" words of clear, calm, and cool. Dew and frost also seem to form more (1) in open areas rather than under trees or other shelters, (2) on objects such as grass rather than on the flat, bare ground, and (3) in low-lying areas before they form on slopes or the sides of hills. What is the meaning of these observations?

Dew and frost are related to clear nights and open areas because these are the conditions best suited for the loss of infrared radiation. Air near the surface becomes cooler as infrared radiation is emitted from the grass, buildings, streets, and everything else that absorbed the shorter wavelength radiation of incoming solar radiation during the day. Clouds serve as a blanket, keeping the radiation from escaping to space so readily. So a clear night is more conducive to the loss of infrared radiation and therefore to cooling. On a smaller scale, a tree serves the same purpose, holding in radiation and therefore retarding the cooling effect. Thus, an open area on a clear, calm night would have cooler air near the surface than would be the case on a cloudy night or under the shelter of a tree.

The observation that dew and frost form on objects such as grass before forming on flat, bare ground is also related to loss of infrared radiation. Grass has a greater exposed surface area than the flat, bare ground. A greater surface area means a greater area from which infrared radiation can escape, so grass blades cool more rapidly than the flat ground. Other variables, such as specific heat, may be involved, but overall frost and dew are more likely to form on grass and low-lying shrubs before they form on the flat, bare ground.

Dew and frost form in low-lying areas before forming on slopes and the sides of hills because of the density differences of cool and warm air. Cool air is more dense than warm air and is moved downhill by gravity, pooling in low-lying areas. You may have noticed the different temperatures of low-lying areas if you have ever driven across hills and valleys on a clear, calm, and cool evening. Citrus and other orchards are often located on slopes of hills rather than on valley floors because of the gravity drainage of cold air.

It is air near the surface that is cooled first by the loss of radiation from the surface. Calm nights favor dew or frost formation because the wind mixes the air near the surface that is being cooled with warmer air above the surface. If you have ever driven near a citrus orchard, you may have noticed the huge, airplanelike fans situated throughout the orchard on poles. These fans are used on "C" nights when frost is likely to form to mix the warmer, upper layers of air with the cooling air in the orchard (figure 23.16).

Condensation occurs on the surface as frost or dew when the dew point is reached. When does condensation occur in the air? Water vapor molecules in the air are constantly colliding and banging into each other, but they do not just join together to form water droplets, even if the air is saturated. The water molecules need something to condense upon. Condensation of water vapor into fog or cloud droplets takes place on tiny particles present in the air. The particles are called **condensation nuclei.** There are hundreds of tiny dust, smoke, soot, and salt crystals suspended in each cubic centimeter of the air that serve as condensation nuclei. Tiny salt crystals, however, are particularly effective condensation nuclei because salt crystals attract water molecules. You may have noticed that salt in a salt shaker becomes moist on a humid day because of the way it attracts

Figure 23.16

Fans like this one are used to mix the warmer, upper layers of air with the cooling air in the orchard on nights when frost is likely to form.

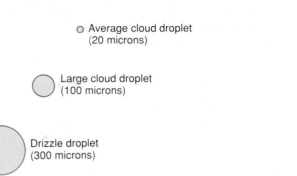

- Condensation nucleus (0.2 microns)

○ Average cloud droplet (20 microns)

Large cloud droplet (100 microns)

Drizzle droplet (300 microns)

Average raindrop (2,000 microns)

Figure 23.17

This figure compares the size of the condensation nuclei to the size of typical condensation droplets. Note that 1 micron is 1/1,000 mm.

water molecules. Tiny salt crystals suspended in the air act the same way, serving as nuclei that attract water vapor into tiny droplets of liquid water.

After water vapor molecules begin to condense on a condensation nucleus, other water molecules will join the liquid water already formed and the tiny droplet begins to increase in volume. The water droplets that makes up a cloud are about 1,500 times larger than a condensation nuclei, and these droplets can condense out of the air in a matter of minutes. As the volume increases, however, the process slows, and hours and days are required to form the even larger droplet and drops. For comparison to the sizes shown in figure 23.17, consider that the average human hair is about 100 microns in diameter. This is about the same diameter as the large cloud droplet of water. Large raindrops have been observed falling from clouds that formed only a few hours previously, so it must be some process or processes other than the direct condensation of raindrops that form precipitation. These processes are discussed in the next chapter.

Fog and Clouds

Fog and clouds are both accumulations of tiny droplets of water that have been condensed from the air. These water droplets are very small, on the order of 0.02 to 0.1 mm (about 8×10^{-4} to 4×10^{-3} in) in diameter, and a very slight upward movement of the air will keep them from falling. If they do fall they usually evaporate. Fog is sometimes described as a cloud that forms at or near the surface. A fog, as a cloud, forms because air containing water vapor and condensation nuclei has been cooled to the dew point. Some types of fog form under the same "C" night conditions favorable for dew or frost to form, that is, on clear, cool, and calm nights when the relative humidity is high. Sometimes this type of fog forms only in valleys and low-lying areas where cool air accumulates (figure 23.18). This type of fog is typical of inland fogs, those that form away from bodies of water. Other types of fog may form somewhere else, such as in the humid air over an ocean, then move inland. Many fogs that occur along coastal regions were formed over the ocean and then carried inland by breezes. A third type of fog looks much like steam rising from melting snow on a street, steam rising over a body of water into cold air, or steam rising over streets after a summer rain shower. These are examples of a temporary fog that forms as a lot of water vapor is added to cool air. This is a cool fog, like other fogs, and is not hot as the steamlike appearance may lead you to believe.

Sometimes a news report states something about the sun "burning off" a fog. A fog does not burn, of course, because it is made up of droplets of water. What the reporter really means is that the sun's radiation will increase the temperature, which increases the air capacity to hold water vapor. With an increased capacity to hold water the relative humidity drops, and the fog simply evaporates back to the state of invisible water vapor molecules.

Clouds, like fogs, are made up of tiny droplets of water that have been condensed from the air. Luke Howard, an English weather observer, made one of the first cloud classification

The Atmosphere of Earth

A

B

C

Figure 23.18
Some types of fog form in cool, moist air on clear nights. Since the cool air is dense, the fog is often close to the ground and in depressions and low-lying areas.

schemes in 1803. He used the Latin terms *cirrus* (curly), *cumulus* (piled up), and *stratus* (spread out) to identify the basic shapes of clouds (figure 23.19). The clouds usually do not occur just in these basic cloud shapes but in combinations of the different shapes. Later, Howard's system was modified by expanding the different shapes of clouds into ten classes by using the basic cloud shapes and altitude as criteria. Clouds give practical hints about the approaching weather. The relationship between the different cloud shapes and atmospheric conditions and what clouds can mean about the coming weather is discussed in the next chapter.

Figure 23.19
The three basic shapes of clouds: (*a*) cirrus, (*b*) cumulus, and (*c*) stratus.

Pollution of Earth's Atmosphere

The earth's atmosphere is never "pure" in the sense of being just a mixture of gases, because the atmosphere naturally contains a mixture of suspended particles of dust, soot, and salt crystals. These are not pollutants, however, since they occur naturally as part of the atmosphere. An *air pollutant* is defined as a human-produced foreign substance that reduces visibility, irritates the senses, or is someway detrimental to humans or their surroundings. The primary source of air pollutants today is the waste products released into the air from the exhaust of (1) internal combustion engines and (2) boilers and furnaces of industries, power plants, and homes.

Engines, boilers, and furnaces do not always produce air pollution just because they are operating. The atmosphere can usually absorb a limited amount of waste products from exhaust gases without detrimentally changing the environment. Whether the waste products become pollutants or not depends on a number of factors, such as the population density, the number of automobiles, the concentration of industries, and the immediate weather conditions. The right combination of population density, automobiles, and industries can produce air pollution under certain weather conditions, such as an inversion. Under other weather conditions the same amount of waste products can be released into the air without producing air pollution. Many people realize, however, that the limit of the atmosphere to absorb waste products is being rapidly reached in many locations. This Feature is about what the waste products are, how they can be harmful to people and their environment, and how they can be limited.

All fuels used today are basically made up of hydrocarbon molecules (hydrogen and carbon), mixtures of other chemicals, and impurities. Coal also contains carbon in an uncombined state. If the fuels were just pure hydrocarbons or uncombined carbon burned completely at "ordinary" temperatures, the reaction would yield just water and carbon dioxide with the energy released. Of these products, carbon dioxide can be a pollutant when produced in sufficient quantities. The complete reaction rarely happens, however, because (1) incomplete combustion often occurs, (2) some boilers and furnaces often operate at some very high temperatures, and (3) fuels often contain impurities, some more than others. Incomplete combustion can yield carbon monoxide, unburned particles of carbon, and mixtures of various hydrocarbon chemical by-products. All of these can be toxic, irritating, and vision-obscuring pollutants under the right conditions. High-combustion temperatures can eliminate these products of incomplete combustion, but high temperatures produce their own set of possible pollutants. High temperatures can oxidize nitrogen in the air to form different nitrogen oxide compounds with the general formula of NO_x. Compounds with this general formula are called "knox." The subscript x of knox means that several oxides are possible, nitrogen oxide (NO) and nitrogen dioxide (NO_2), for example. The general formula NO_x implies the presence of any or all of the nitrogen oxides. The NO_x compounds are associated with the dirty brown haze of air pollution and are partly responsible for the acid rain problem. So far, there are three separate sets of pollutants that could possibly be produced by (1) complete combustion at normal temperatures, (2) incomplete combustion, and (3) combustion at high temperatures. To each of these possibilities you can add the pollutants produced by the ever-present fuel impurities. Troublesome fuel impurities include (1) nitrogen compounds, which can produce NO_x compounds at lower combustion temperatures, (2) sulfur compounds, which can produce sulfur dioxide (SO_2), another compound implicated in the acid rain problem, and (3) various solid impurities, which can become tiny particles called *particulates*.

After water vapor, carbon dioxide is the second most abundant waste product from burning and combustion of fuels. The use of fossil fuels (petroleum and coal) releases about twenty billion tons of carbon dioxide into the atmosphere each year. The total amount of carbon dioxide has been slowly increasing over the years, apparently from the combined effect of increased fossil fuel consumption and the clearing and burning of the world's tropical rain forests. Scientists are concerned that increases in the amount of carbon dioxide will eventually lead to increases in the average temperatures of the world through the greenhouse effect. Opinions differ about when or if the increased warming will happen and about what the consequences will be if it does happen. Some believe that an average temperature increase will melt some of the polar ice cap, flooding coastal cities and changing the global climate patterns. Others believe that increased carbon dioxide and warmer climate patterns would produce lush vegetation, therefore benefiting humans through increased agricultural production. Still other scientists have the opinion that no predictions can be made because of the number of other variables involved. Increased water vapor and dust in the atmosphere, for example, could reflect more and more sunlight. This could cancel any warming from the greenhouse effect or, in fact, could lead to an overall cooling trend. Scientists do seem to agree on at least two points about carbon dioxide, water vapor, and the other products of combustion. That agreement is that human activity has now reached the level that (1) pollution is becoming an influencing part of the atmosphere and (2) any resulting changes in the earth's climate will last a long time.

Carbon monoxide is an odorless, colorless, and poisonous gas. Exposure to concentrations of 100 parts per million produces symptoms of breathing difficulties and headaches. Exposure to concentrations of 1,000 parts per million can be fatal. Carbon monoxide is the most abundant poisonous pollutant in the air of most large cities. It is mostly produced by automobile engines in need of a tune-up, but all engines produce some carbon monoxide. Even if a tuned engine only produces between 0.01 and 0.1 percent carbon monoxide in its exhaust, too many cars in too little space during certain types of weather and atmospheric conditions can lead to unacceptable levels of carbon monoxide.

Sulfur is released to the atmosphere from natural sources, such as decaying organic matter, and from metal smelting, petroleum refining, and the burning of fuels such as petroleum and coal, which contain sulfur impurities. There are different forms of sulfur oxides, but the most common is sulfur dioxide (SO_2). A study of all sources of SO_2 released in the eastern parts of Canada and the United States found that an average of 85,000 tons of sulfur emissions were released per day. About 60 percent of this came from coal-fired power plants.

Sulfur dioxide gas can injure plants and plant materials. If inhaled in sufficient quantities, it can make breathing difficult because it causes constriction of the air tubes in human lungs. Sulfur dioxide has also been implicated as an acid rain precursor, one of the substances that is later converted to acid by atmospheric processes.

Nitrogen oxides in the atmosphere are oxidized to nitrogen dioxide, which absorbs ultraviolet radiation from the sun. This breaks the compound down into nitric oxide and monatomic oxygen (O). Monatomic oxygen combines with diatomic oxygen (O_2) to produce ozone (O_3). Ozone is strongly reactive and causes the cracking and deterioration of rubber and other materials. In sufficient quantities, it is also a health hazard.

Ozone and nitrogen oxide combine with the hydrocarbons from incomplete fossil fuel combustion in a photochemical reaction. In the presence of sunlight they combine to form peroxyacyl nitrate (PAN), a principal component of smog that irritates eyes and damages mucous membranes. Photochemical smog usually occurs in large cities, but it can occur anywhere that a mixture of hydrocarbons and nitrogen oxides is exposed to sunlight.

Particulates include the larger dust, ash, and smoke particles and the much smaller suspended aerosols. About 80 percent of the particulates in the atmosphere are from windblown dust, sea mist, volcanoes, and other natural sources. Automobiles, incinerators, and a variety of industrial and power plant burning of fuels contribute to the human-added 20 percent. Particulates are of concern because of potential health problems and aesthetics. Smaller particles can lodge in the lower respiratory tract of humans, leading to asthma, emphysema, or lung cancer. In the past, cities were covered with black soot from the particulates released by coal burning. The increased use of cleaner-burning petroleum and the control technologies required with the use of coal have reduced this problem to one of reduced visibility.

The readily definable sources of air pollution, in order of greatest to least emissions, are (1) automobiles, (2) major industries, (3) power plants, and (4) home space heating. The automobile is the prime polluter, contributing almost all of the carbon monoxide and, along with industries, most of the hydrocarbons. Automobiles and power plants contribute most of the nitrogen oxides and power plants and industries contribute most of the sulfur oxides.

The problem of air pollution is more acute in cities, where most people live. Large numbers of people use the same air, which receives the waste products from automobiles and other sources. Stringent state and federal automobile emission standards have reduced the total automobile emissions other than CO_2 and water vapor. The problem continues, however, because of too many cars in too little space. So, we will continue to hear about "acceptable levels" of air pollution and warnings about dangerous levels when they occur. Many newspaper weather reports include information about carbon monoxide, ozone, and particulates and the "acceptable" or "health hazard" levels. The Clean Air Act of 1971 established tight federal standards on the combustion emissions of industries and power plants. All are required to employ the "latest technology" of pollution controls, no matter what the expense. The cost of pollution control equipment is now often up to one-fourth of the total cost of building a new coal-fired power plant. This increased construction cost, and the associated operational costs, has resulted in a 10 to 15 percent increase in a total monthly electric bill. As new, modern pollution control equipment is developed through research, more options for emission control will become available. The question may change from What can we do about air pollution? to How clean do you want the air . . . and how much are you willing to pay for it?

Summary

The earth's *atmosphere* thins rapidly with increasing altitude. Pure, dry air is mostly *nitrogen, oxygen,* and *argon* with traces of *carbon dioxide* and other gases. Atmospheric air also contains a variable amount of *water vapor*. Water vapor cycles into and out of the atmosphere through the *hydrologic cycle* of evaporation and precipitation.

Atmospheric pressure is measured with a *mercury barometer*. At sea level, the atmospheric pressure will support a column of mercury about 76.00 cm (about 29.92 in) tall. This is the average pressure at sea level, and it is called the *standard atmospheric pressure, normal pressure,* or *one atmosphere of pressure*.

Sunlight is absorbed by the materials on the earth's surface, which then emit infrared radiation. Infrared is absorbed by carbon dioxide and water molecules in the atmosphere, which then reemit the energy many times before it reaches outer space again. The overall effect warms the lower atmosphere from the bottom up in a process called the *greenhouse effect*.

The layer of the atmosphere from the surface up to where the temperature stops decreasing with height is called the *troposphere*. The *stratosphere* is the layer above the troposphere. Temperatures in the stratosphere increase because of the interaction between ozone (O_3) and ultraviolet radiation from the sun. Other layers of the atmosphere are the *mesosphere, thermosphere, exosphere,* and the *ionosphere*.

The surface of the earth is not heated uniformly by sunlight. This results in a *differential heating*, which sets the stage for *convection*. The horizontal movement of air on the surface from convection is called *wind*. A generalized model for understanding why the wind blows involves (1) the relationship between *air temperature and air density* and (2) the relationship between *air pressure and the movement of air*. This model explains local wind patterns and wind patterns observed on a global scale.

The amount of water vapor in the air at a particular time is called the *absolute humidity*. The *relative humidity* is a ratio between the amount of water vapor that is in the air and the amount needed to saturate the air at the present temperature.

When the air is saturated, condensation can take place. The temperature at which this occurs is called the *dew point temperature*. If the dew point temperature is above freezing, *dew* will form. If the temperature is below freezing, *frost* will form. Both dew and frost form directly on objects and do not fall from the air.

Water vapor condenses in the air on *condensation nuclei*. If this happens near the ground, the accumulation of tiny water droplets is called a *fog*. *Clouds* are accumulations of tiny water droplets in the air above the ground. In general, there are three basic shapes of clouds, *cirrus*, *cumulus*, and *stratus*. These basic cloud shapes have meaning about the atmospheric conditions and about the coming weather conditions.

Key Terms

absolute humidity (p. 492)	meteorology (p. 482)
barometer (p. 484)	millibar (p. 485)
Chinook (p. 490)	observed lapse rate (p. 486)
condensation nuclei (p. 494)	ozone shield (p. 487)
dew (p. 494)	psychrometer (p. 493)
dew point temperature (p. 494)	relative humidity (p. 492)
exosphere (p. 488)	saturated air (p. 492)
frost (p. 494)	sea breeze (p. 489)
greenhouse effect (p. 486)	solar constant (p. 485)
hair hygrometer (p. 493)	standard atmospheric pressure
humidity (p. 492)	(p. 485)
hydrologic cycle (p. 483)	stratopause (p. 487)
intertropical convergence zone	stratosphere (p. 487)
(p. 490)	thermosphere (p. 488)
inversion (p. 486)	tropopause (p. 486)
ionosphere (p. 488)	troposphere (p. 486)
jet stream (p. 490)	wind (p. 488)
mesosphere (p. 488)	wind chill factor (p. 493)

Applying the Concepts

1. If it were not for the ocean, the earth's atmosphere would probably be mostly
   a. nitrogen.
   b. carbon dioxide.
   c. oxygen.
   d. argon.

2. Your ear makes a "pop" sound as you go up in an elevator because air
   a. is moving to the backside of your eardrum.
   b. is moving from the backside of your eardrum.
   c. moves into your outer ear, striking the eardrum.
   d. pressure increases with increasing altitude.

3. The greenhouse effect results in warmer temperatures near the surface because
   a. clouds trap infrared radiation near the surface.
   b. some of the energy is reradiated back toward the surface.
   c. carbon dioxide molecules do not permit the radiation to leave.
   d. carbon dioxide and water vapor both trap infrared radiation.

4. On a clear, calm, cool night you would expect the air temperature over a valley floor to be what temperature compared to the air temperature over a slope to the valley?
   a. cooler
   b. warmer
   c. the same temperature
   d. sometimes warmer and sometimes cooler

5. Air moving down a mountain slope is often warm because
   a. it has been closer to the sun.
   b. cool air is more dense and settles to lower elevations.
   c. it is compressed as it moves to lower elevations.
   d. this only occurs during the summertime.

6. Considering earth's overall atmosphere, you would expect more rainfall to occur in a zone of
   a. high atmospheric pressure.
   b. low atmospheric pressure.
   c. prevailing westerly winds.
   d. prevailing trade winds.

7. Considering earth's overall atmosphere, you would expect to find a desert located in a zone of
   a. high atmospheric pressure.
   b. low atmospheric pressure.
   c. prevailing westerly winds.
   d. prevailing trade winds.

8. Without adding or removing any water vapor a sample of air experiencing an increase in temperature will have
   a. a higher relative humidity.
   b. a lower relative humidity.
   c. the same relative humidity.
   d. a changed absolute humidity.

9. Cooling a sample of air results in a(n)
   a. increased capacity to hold water vapor.
   b. decreased capacity to hold water vapor.
   c. unchanged capacity to hold water vapor.

10. On a clear, calm, and cool night, dew or frost is most likely to form
    a. under trees or other shelters.
    b. on bare ground on the side of a hill.
    c. under a tree on the side of a hill.
    d. on grass in an open, low-lying area.

Answers

1. b 2. b 3. b 4. a 5. c 6. b 7. a 8. b 9. b
10. d

Questions for Thought

1. Briefly describe how the atmosphere thins with altitude by describing the mass contained within different altitudes.

2. Do density boundaries exist between the atmosphere at the earth's surface and the vacuum of outer space? Explain.

3. What are the major and minor gases that make up the earth's atmosphere? In addition to the gases, what else is always found in the atmosphere?

4. Describe how (a) nitrogen and (b) oxygen are cycled into and out of the atmosphere.

5. What is the hydrologic cycle? Why is it important?

6. Describe how carbon dioxide is cycled into and out of the atmosphere.

7. Describe how a mercury barometer measures the atmospheric pressure, using sketches as needed.

8. What is the meaning of "normal" atmospheric pressure?

9. Explain how it is possible to use a barometer as an altimeter.

10. What is the solar constant?

11. Describe what happens to sunlight after it enters the earth's atmosphere. What happens to it when it reaches the surface?

12. Explain the greenhouse effect. Is a greenhouse a good analogy for the earth's atmosphere? Explain.

13. Describe the temperature layers of the earth's atmosphere from the surface upward.

14. What is a temperature inversion? Why does it increase air pollution?

15. Describe how the ozone shield protects living things on the earth's surface. Why is there some concern about this shield?

16. An unprotected person in the thermosphere (warm layer) would quickly freeze. Explain this apparent contradiction.

17. Explain why the temperature in the troposphere decreases with increasing altitude.

18. What is wind? What is the energy source for wind? Explain.

19. Explain the relationship between air temperature and air density.

20. Why does heated air rise?

21. Explain the relationship between air pressure and the movement of air.

22. Locations near large bodies of water seem to have more even temperatures than locations away from the water. Explain why this is so.

23. Explain why a breeze blows from the sea to the land during hot days of the summer but not during the cool days or nights.

24. How is it possible to have a warm wind blowing from a mountainous region during the winter months?

25. What is the intertropical convergence zone? Why do tropical rain forests occur in this zone?

26. Provide an explanation for the observation that an airplane flight at the top of the troposphere takes several hours longer to fly from the east coast to the west coast than it does to make the return trip.

27. If evaporation cools the surroundings does condensation warm the surroundings? Explain.

28. Explain why warm air can hold more water vapor than cool air.

29. Compare and contrast absolute humidity and relative humidity.

30. Explain why a cooler temperature can result in a higher relative humidity when no water vapor was added to the air.

31. What is the meaning of the expression, "It's not the heat, it's the humidity."

32. Explain how a psychrometer measures the relative humidity.

33. What is the meaning of the dew point temperature?

34. Explain why frost is more likely to form on a clear, calm, and cool night than on nights with other conditions.

35. Explain why the presence of dust, smoke, and other particles in the atmosphere are necessary in order for clouds to form.

36. How are a fog and a cloud alike? How are they different?

37. Describe the three basic shapes of clouds.

Chapter 24

Weather and Climate

Figure 24.1
Weather is a description of the changeable aspects of the atmosphere, that is, the temperature, rainfall, pressure, and so forth, at a particular time. These changes usually affect your daily life one way or another, but some of these changes seem more inconvenient than others.

As you learned in chapter 23, the earth's atmosphere can be described by its mass, density, composition, and structure. These terms describe the condition, or state, of the atmosphere at a given time. They describe aspects of the atmosphere that are not expected to change much over a short period of time. Weather, on the other hand, is described by terms that indicate both the present conditions and predicted changes. Some of these terms, in fact, carry with them an expectation of change. Examples of these terms are overcast, rainy, cold front, stormy, and windy. Other weather terms might or might not mean change, for example, clear, cold, hot, humid, dry, and so forth. Sometimes weather changes are slow and gradual, but other times they can be rapid and violent. Usually you listen to or read a weather forecast to find out what types of changes are going to occur (figure 24.1).

Sometimes you hear weather forecasts with combinations of terms that link together in familiar patterns. For example, during certain times of the year hot and humid weather is often followed by windy and stormy weather, which is then followed by clear and cool weather. When patterns of combinations occur like this it means that there are underlying reasons, that is, cause and effect relationships that are producing the observed patterns. These are important relationships that you will want to know, for knowledge about the relationships will help you understand what is happening to the weather at the present time as well as what may happen next.

What causes weather and what causes changes in the weather are the subjects of this chapter. The chapter begins with relationships associated with the weather terms of clear, cloudy, overcast, and rainy. Cloud-forming processes and the origins of precipitation will be explained, with descriptions of why and how clouds and precipitation form. In this section you will learn the type of precipitation that you might or might not expect to receive from the basic types of clouds. You will also learn the atmospheric conditions necessary for clouds to form.

Sometimes the weather changes slowly over several days, but other times rapid changes occur. Some weather terms used to describe predicted rapid changes are windy, rainy, snowstorms, and severe storm warnings. Weather in North America is the most changeable in the world, and the section on weather producers will explain why these changes take place, including the major changes of thunderstorms, tornadoes, and hurricanes.

The section on weather forecasting describes how computers are used to predict the coming weather, problems with making such predictions, and why any prediction beyond three to five days is usually no better than a wild guess. Finally, the average weather for a location, the climate, will be described along with what factors are responsible for producing the climate where you live.

Clouds and Precipitation

Water cycles continuously into and out of the atmosphere through the processes of evaporation, condensation, and precipitation. When water evaporates, individual water molecules leave the liquid and enter the atmosphere as the gas called *water vapor*.

While in the liquid state, water molecules are held together by attractive molecular forces. Water molecules have a wide range of kinetic energies, and occasionally the more energetic ones are able to overcome the attractive forces, breaking away. When they do escape, water vapor molecules carry the latent heat of vaporization with them as discussed in chapter 6. Because water vapor molecules take energy with them, evaporation is a cooling process. If incoming solar radiation did not supply energy, the surface and the ocean would soon become cooler and cooler from the continuous evaporation that takes place. The sun supplies the energy that maintains surface temperatures, which allows the ongoing evaporation of water. Thus it is the sun that supplies the energy required to evaporate water.

Water vapor in the atmosphere does not remain for more than several weeks, but during this time it is transported by the winds of the earth. Eventually, the air becomes cooled and the relative humidity increases to 100 percent. The water vapor in the saturated air now condenses to form the tiny droplets of clouds. The water returns to the surface as precipitation that falls from the clouds. Each year, on the average, about 97 cm (about 38 in) of water evaporate from the earth's ocean, but only 90 cm (about 35 in) are returned by precipitation. The deficit is made up, on the average, by the return of 7 cm (about 3 in) per year by streams flowing from the continents into the oceans. If rivers and streams did not cycle water back to the ocean, it would be lowered each year by a depth of 7 cm (about 3 in). On the other hand, if the atmosphere did not cycle water vapor back over the land, there would eventually be no water on the land. Both streams and precipitation are part of a never-ending series of events involving the ocean and lands of the earth. The series of events is called the **hydrologic cycle.** Overall, the hydrologic cycle can be considered to have four main events: (1) *evaporation* of water from the ocean, (2) *transport* of water vapor through the atmosphere, (3) *condensation and precipitation* of water on the lands and, (4) the *return of water* to the ocean by rivers and streams (figure 24.2). This definition of the hydrologic cycle involves only the ocean and the lands, but water vapor also evaporates from the land and may condense and precipitate back to the land without ever returning to the ocean. This can be considered as a small subcycle within the overall hydrologic cycle. The ocean-land exchange is the major cycle, and there are actually many small subcycles that also exist. The following section is about the part of the hydrologic cycle that returns water to the earth's surface. The cloud-forming condensation processes will be considered first, followed by a discussion of the processes that result in precipitation falling from the clouds.

Cloud-Forming Processes

Clouds form when a mass of air above the surface is cooled to its dew point temperature. In general, the mass of air is cooled because something has given it an upward push, moving it to higher levels in the atmosphere. There are three major causes of upward air movement: (1) *convection* resulting from differential heating, (2) mountain ranges that serve as *barriers* to

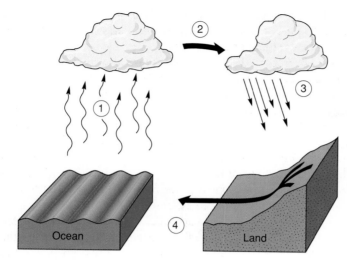

Figure 24.2

The main events of the hydrologic cycle are: (1) The evaporation of water from the ocean, (2) the transport of water vapor through the atmosphere, (3) condensation and precipitation of water on the land, and (4) return of water to the ocean by rivers and streams.

moving air masses, and (3) the meeting of *moving air masses* with different densities, for example, a cold, dense mass of air meeting a warm, less dense mass of air.

The white, puffy cumulus clouds that typically form during a late summer afternoon are a result of differential heating on the surface. Air moves upward over the warmer surface in a vertical column called a *thermal* and back toward the ground in the spaces between the clouds. Hawks and other birds that soar, making lazy circles in the sky, are using the uplifting thermals to gain altitude. Cumulus clouds form over thermals, and the base of a cumulus cloud identifies the altitude at which the rising air has been cooled to the dew point. Sea breezes also produce cumulus clouds over the land surface during the day and over the water surface during the night (see figure 23.10). The formation of cumulus clouds around the tops of mountain peaks is evidence of the upward movement of a mountain breeze. Convection is often associated with the phrase "heated air rises," which actually means that the less dense air is forced upward by cooler, more dense air. Thus in addition to the clouds that form as a result of differential heating, cumulus clouds sometimes form over forest fires from convective processes.

Moving air that runs into a mountain range is uplifted, resulting in cooling and the formation of clouds. In the zone of prevailing westerlies the western slopes of mountain ranges have more clouds and receive more precipitation than the eastern slopes. Air forced upward on the western slope is cooled to the dew point and condensation results in the clouds. The third major cause of upward air movement is the meeting of two masses of air with different densities. Warm air is less dense than cold air and is forced upward by the cooler, more dense air mass. This sometimes results in widespread areas covered by spread-out stratus clouds.

The three major causes of uplifted air sometimes result in clouds, but just as often they do not. Whether clouds form or not depends on the condition of the atmosphere at the time. As a parcel of warm air is moved upward by convection, mountain barriers, or air masses of different densities, the parcel tends to stay together, mixing very little with the surrounding air. It might be helpful to think of this parcel of rising air as the air that stays together inside a huge balloon. But the air parcel stays together without the balloon. As the parcel is forced upward it expands because the atmospheric pressure decreases with altitude. The parcel thus expands as it rises and does work in expanding outward against the surrounding air. Doing work requires the expenditure of energy, and the parcel of air becomes cooler because it is expanding. You are familiar with cooling from expansion if you have ever noticed that high-pressure air released from a tire or compressed gases released from an aerosol can are cool. The decrease in temperature of an expanding gas is called **adiabatic cooling.** Adiabatic cooling takes place without heat flowing out of the gas, and the cooling results only from the energy lost by expansion. Similarly, when work is done on a parcel of gas by compressing it, the temperature increases from the compression. **Adiabatic heating** takes place without heat flowing into the gas, and the heating results only from the energy gained by compression. Rising air cools adiabatically since the cooling results from work performed by the molecules of gases in the air. Descending air warms adiabatically since the warming results from work done on the molecules of gases in the air. The rate of cooling or warming in the absence of condensation is called the **dry adiabatic lapse rate.** This rate is about 10° C for each increase or decrease of 1 km of altitude (about 5.5° F/1,000 ft).

In the last chapter you learned that the *average* temperature decrease with altitude is about 6.5° C for each increase of 1 km in altitude (3.5° F/1,000 ft). Recall that this average lapse rate refers to the temperature lapse (decrease) that you would expect to find, on the average, for air that is not rising or falling at various altitudes above the surface. Figure 24.3 compares the average temperature of air that is not moving (the average lapse rate) with the temperature of air that is being forced upward (the dry adiabatic lapse rate) at various altitudes.

The *actual* lapse rate of the atmosphere varies with the altitude at a rate determined by the weather conditions at the time the measurement was taken. It is the actual lapse rate and the dry adiabatic lapse rate that will determine if a parcel of air that has been given an upward shove will continue to rise, fall back to the surface, or stay put. This, in turn, determines if clouds will form and what happens to the clouds. In other words, part of the overall weather condition is determined by what happens to a parcel of air that is shoved upward by differential heating, mountains, or masses of air with different densities.

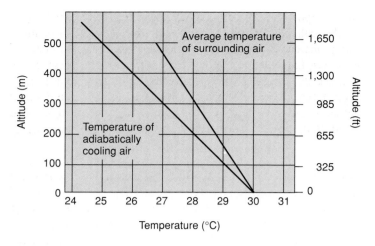

Figure 24.3

As a parcel of dry air is moved upward, it expands and cools according to the dry adiabatic lapse rate. This graph compares the temperature of a rising and adiabatically cooling parcel of dry air with the average temperature of the surrounding atmosphere when the temperature at the surface is 30° C (86° F).

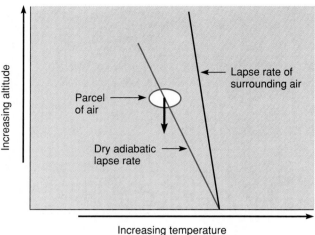

Figure 24.4

If a parcel of air is moved up or down, it will cool or warm according to the dry adiabatic lapse rate. In a state of atmospheric stability, the parcel of air will always be cooler, and therefore more dense, than the surrounding air at any altitude. It will, therefore, return to the original level when the upward force is removed.

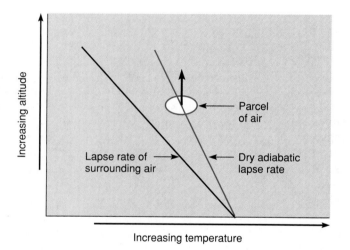

Figure 24.5

In a state of atmospheric instability, a parcel of air will always be warmer, and therefore less dense, than the surrounding air at any altitude. The parcel will, therefore, continue on in the direction pushed when the upward force is removed.

What happens to a rising parcel of air depends upon the difference in density between the parcel and the surrounding air that it has been shoved into. Since the density of a parcel of air is determined by the temperature of the air, the temperature is an indication of density. By comparing the actual lapse rate (which is measured by instruments attached to a weather balloon) with the dry adiabatic lapse rate, the state of **atmospheric stability** can be determined. There are many different states of atmospheric stability, and different states can exist at different atmospheric levels. The following is a simplified description of just a few of the possible states, first considering dry air only.

The atmosphere is in a state of *stability* when the atmospheric lapse rate is less than the dry adiabatic lapse rate. If a parcel of air is moved upward, it expands and cools according to the dry adiabatic lapse rate. It will have a density that is more, less, or equal to the density of the surrounding air. In stability, it will be cooler and therefore more dense than the surrounding air at any altitude it is lifted to. To verify that it is cooler, place your pencil point on the dry adiabatic lapse rate line in figure 24.4. As you follow the line upward, you will see that it is always to the left of the air temperature of the surroundings at any altitude (the actual lapse rate). Being cooler, the lifted parcel of air will be more dense than the surroundings. If a parcel of air is moved up to a higher level and released in a stable atmosphere, it will move back to its former level. A displaced parcel of air always returns to its original level when the atmosphere is stable. Any clouds that do develop in a stable atmosphere are usually arranged in the horizontal layers of stratus-type clouds.

The atmosphere is in a state of *instability* when the atmospheric lapse rate is greater than the dry adiabatic lapse rate. Again, an uplifted parcel of air will expand and cool according to the dry adiabatic lapse rate. In instability, it will be warmer and therefore less dense than the surrounding air at any altitude it is lifted to. As you follow the cooling line of the uplifted air in figure 24.5, you will see that it is always to the right of the surrounding air temperature at any altitude. If a parcel of air is moved up to a higher level in an unstable atmosphere, it will continue in the direction moved after the uplifting force is removed. Cumulus clouds usually develop in an unstable atmosphere, and the rising parcels of air, called thermals, can result in a very bumpy airplane ride.

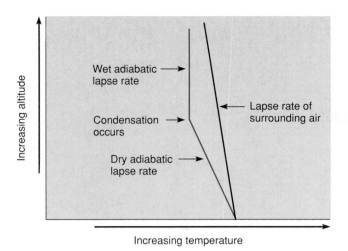

Figure 24.6

When the dew point temperature is reached in a rising parcel of air, the latent heat of vaporization is released as water vapor condenses. This release of heat warms the air, decreasing the density and accelerating the ascent. The new lapse rate from the release of latent heat is called the wet adiabatic lapse rate.

So far only dry air has been considered. As air moves upward and cools from expansion, sooner or later the dew point is reached and the air becomes saturated. As some of the water vapor in the rising parcel condenses to droplets, the latent heat of vaporization is released. The rising parcel of air now cools at a slower rate because of the release of this latent heat of vaporization, a rate that is now called the **wet adiabatic lapse rate** (figure 24.6). The release of latent heat warms the air in the parcel and decreases the density even more, accelerating the ascent. This leads to further condensation and the formation of towering cumulus clouds. The wet adiabatic lapse rate is variable since air at different temperatures can hold different amounts of water vapor. Different amounts of water vapor mean that different amounts of heating can occur through condensation and the release of the latent heat of vaporization.

The state of the atmosphere and the moisture content of an uplifted parcel of air are not the only variables affecting whether or not clouds form. A parcel of air, given an upward push by convection, mountains, or a dense air mass, expands as it reaches higher and higher altitudes under less atmospheric pressure. It cools adiabatically from the expansion, and as it cools the relative humidity increases. When the dew point temperature is reached the water vapor in the air tends to condense, but now it requires the help of tiny microscopic particles called *condensation nuclei*. Without condensation nuclei, water vapor condenses to tiny droplets that are soon torn apart by collisions with other water vapor molecules. Thus without condensation nuclei further cooling can result in the air parcel becoming **supersaturated,** containing more than its normal saturation amount of water vapor. Condensation nuclei permit the condensation of water vapor into tiny droplets that are not torn apart by molecular collisions. Soon the stable droplets grow on a condensation nucleus to a diameter of about one-tenth the diameter of a human hair, but a wide range of sizes may be present. An average cloud may contain hundreds of such droplets per cubic centimeter with an average density of 1 gram of liquid water per cubic meter of cloud. This accumulation of tiny droplets can easily remain suspended in the air from the slightest air movement.

A single water droplet in a cloud is so tiny that it would be difficult to see with the unaided eye. When they accumulate in huge numbers in clouds, what you see depends on the size of the droplets making up the cloud and the relative position of you, the cloud, and the sun. You will see a white cloud, for example, if you are between the cloud and the sun so that you see reflected sunlight from the cloud. The same cloud will appear to be gray if it is between you and the sun, positioned so that it filters the sunlight coming to you.

Origin of Precipitation

Water that returns to the surface of the earth, in either the liquid or solid form, is called **precipitation** (figure 24.7). Note that dew and frost are not classified as precipitation because they form directly on the surface and do not fall through the air. Precipitation seems to form in clouds by one of two processes: (1) the *coalescence* of cloud droplets or (2) the *growth of ice crystals.* It would appear difficult for cloud droplets to merge, or coalesce, with one another since any air movement would seem to move them all at the same time, not bringing them together. Condensation nuclei come in different sizes, however, and cloud droplets of many different sizes form on these different-sized nuclei. Larger cloud droplets are slowed less by air friction as they drift downward, and they collide and merge with smaller droplets as they fall. They may merge, or coalesce, with a million other droplets before they fall from the cloud as raindrops. This **coalescence process** of forming precipitation is thought to take place in warm cumulus clouds that form near the ocean in the tropics. These clouds contain giant salt condensation nuclei and have been observed to produce rain within about twenty minutes after forming.

Clouds at middle latitudes, away from the ocean, also produce precipitation, so there must be a second way that precipitation forms. The **ice-crystal process** of forming precipitation is important in clouds that extend high enough in the atmosphere to be above the freezing point of water. Water molecules are more strongly bonded to each other in an ice crystal than in liquid water. Thus an ice crystal can capture water molecules and grow to a larger size while neighboring water droplets are evaporating. As they grow larger and begin to drift toward the surface, they may coalesce with other ice crystals or droplets of water, soon falling from the cloud. During the summer, they fall

Figure 24.7

Precipitation is water in the liquid or solid form that returns to the surface of the earth. The tiny droplets in clouds become precipitation by merging to form larger droplets or by the growth of ice crystals. The fog you see here is made up of tiny droplets, much like the tiny droplets that form precipitation in clouds.

through warmer air below and reach the ground as raindrops. During the winter, they fall through cooler air below and reach the ground as snow.

Tiny water droplets do not freeze as readily as a larger mass of liquid water, and many droplets do not freeze until the temperature is below about −40° C (−40° F). Water that is still in the liquid state when the temperature is below the freezing temperature is said to be **supercooled.** Supercooled clouds of water droplets are common between the temperatures of −40° C and 0° C (−40° F and 32° F), a range of temperatures that is often found in the upper atmosphere. The liquid droplets at these temperatures need a solid particle called an **ice-forming nuclei** to freeze upon. Generally, dust from the ground serves as ice-forming nuclei that start the ice-crystal process of forming precipitation. Artificial rainmaking has been successful by (1) dropping crushed dry ice, which is cooler than −40° C, on top of a supercooled cloud, and (2) by introducing

"seeds" of ice-forming nuclei in supercooled clouds. Tiny crystals from the burning of silver iodide are effective ice-forming nuclei, producing ice crystals at temperatures as high as −4.0° C (about 25° F). Attempts at ground-based cloud seeding with silver iodide in the mountains of the western United States have suggested up to 15 percent more snowfall, but it is difficult to know how much snowfall would have resulted without the seeding.

In general, the basic form of a cloud has meaning about the general type of precipitation that can occur as well as the coming weather. Cumulus clouds usually produce showers or thunderstorms that last only brief periods of time. Longer periods of drizzle, rain, or snow usually occur from stratus clouds. Cirrus clouds do not produce precipitation of any kind, but they may have meaning about the coming weather, which is discussed in the following section.

Weather Producers

The general circulation of the earth's atmosphere, which was discussed in chapter 23, is useful as a model to help you understand why the weather changes as it does on the earth. The general circulation of the atmosphere described is a simplified, idealized model based on averages of a featureless, uniform earth. A more detailed model would require an application of physics that cannot be developed in a short chapter in a physical science textbook. Also, note that the winds and the belts of high and low pressure described in this model are generalized, average values that are different from the everyday changes of winds and pressures that are shown on daily weather maps.

The idealized model of the general atmospheric circulation starts with the poleward movement of warm air from the tropics. The region between 10° N and 10° S of the equator receives more direct radiation, on the average, than other regions of the earth's surface. The air over this region is heated more, expands, and becomes less dense as a consequence of the heating. This less dense air is buoyed up by convection to heights up to 20 km (about 12 mi) as it is cooled by radiation to less than −73° C (about −110° F). This accumulating mass of cool, dry air spreads north and south toward both poles (see figure 23.13), then sinks back toward the surface at about 30° N and 30° S. The descending air is warmed by adiabatic heating and is warm and dry by the time it reaches the surface. Part of the sinking air then moves back toward the equator across the surface, completing a large convective cell. This giant cell has a low-pressure belt over the equator and high-pressure belts over the subtropics near latitudes of 30° N and 30° S. The other part of the sinking air moves poleward across the surface, producing belts of westerly winds in both hemispheres to latitudes of about 60°. On an earth without landmasses next to bodies of water, a belt of low pressure would probably form around 60° in both hemispheres and a high-pressure region would form at both poles.

The overall pattern of pressure belts and belts of prevailing winds is seen to shift north and south with the seasons, resulting in a seasonal shift in the types of weather experienced at a location. This shift of weather is related to three related weather producers: (1) the movement of large bodies of air, called *air masses,* that have acquired the temperature and moisture conditions where they have been located, (2) the leading *fronts* of air masses when they move, and (3) the local *high-* and *low-pressure* patterns that are associated with air masses and fronts. These are the features shown on almost all daily weather maps, and they are the topics of this section.

Air Masses

An **air mass** is defined as a large, more or less uniform body of air with nearly the same temperature and moisture conditions. An air mass forms when a large body of air, perhaps covering millions of square kilometers, remains over a large area of land or water for an extended period of time. While it is stationary it acquires the temperature and moisture characteristics of the land or water through the heat transfer processes of conduction, convection, and radiation and through the moisture transfer processes of evaporation and condensation. For example, a large body of air that remains over the cold, dry, snow-covered surface of Siberia for some time will become cold and dry. A large body of air that remains over a warm tropical ocean, on the other hand, will become warm and moist. Knowledge about the condition of air masses is important because they tend to retain the acquired temperature and moisture characteristics when they finally break away, sometimes moving long distances. An air mass that formed over Siberia can bring cold, dry air to your location while an air mass that formed over a tropical ocean will bring warm, moist air.

Air masses are classified according to the temperature and moisture conditions where they originate. There are two temperature extreme possibilities, a **polar air mass** from a cold region and a **tropical air mass** from a warm region. There are also two moisture extreme possibilities, a moist **maritime air mass** from over the ocean and a generally dry **continental air mass** from over the land. Thus there are four main types of air masses that can influence the weather where you live: (1) continental polar, (2) maritime polar, (3) continental tropical, and (4) maritime tropical. Figure 24.8 shows the general direction in which these air masses usually move over the mainland United States.

Once an air mass leaves its source region it can move at speeds of up to 800 km (about 500 mi) per day while mostly retaining the temperature and moisture characteristics of the source region. If it slows and stagnates over a new location, however, the air may again begin to acquire a new temperature and moisture equilibrium with the surface. When a location is under the influence of an air mass, the location is having a period of **air mass weather.** This means that the weather conditions will generally remain the same from day to day with slow, gradual changes. Air mass weather will remain the same until a new air mass moves in or until the air mass acquires the conditions of the new location. This process may take days or several weeks, and the weather conditions during this time depend on the conditions of the air mass and conditions at the new location. For example, a polar continental air mass arriving over a cool, dry land area may produce a temperature inversion with the air colder near the surface than higher up. When the temperature increases with height the air is stable and cloudless, and cold weather continues with slow, gradual warming. The temperature inversion may also result in hazy periods of air pollution in some locations. A continental air mass arriving over a generally warmer land area, on the other hand, results in a condition of instability. In this situation each day will start clear and cold, but differential heating during the day develops cumulus clouds in the unstable air. After sunset the clouds evaporate, and a clear night results because the thermals during the day carried away the dust and air pollution. Thus a dry, cold air mass can bring different weather conditions, each depending on the properties of the air mass and the land it moves over.

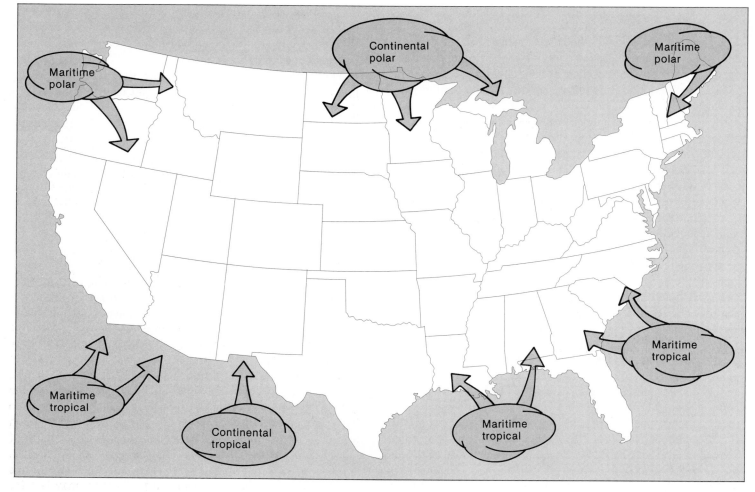

Figure 24.8

The general movement of the four main types of air masses that
influence the weather over the continental United States. The tropical
air masses visit most often in the summer and the polar air masses visit
most often during the winter. During other times, the polar and tropical
air masses battle back and forth over the land.

Weather Fronts

The boundary between air masses of different temperatures is
called a **front.** A front is actually a thin transition zone between
two air masses that ranges from about 5 to 30 km thick (about
3 to 20 mi), and the air masses do not mix other than in this
narrow zone. The density differences between the two air masses
prevent any general mixing since the warm, less dense air mass
is forced upward by the cooler, more dense air moving under it.
You may have noticed on a daily weather map that fronts are
usually represented with a line bulging outward in the direction
of cold air mass movement. A cold air mass is much like a huge,
flattened bubble of air that moves across the land (figure 24.9).
The line on a weather map represents the place where the leading
edge of this huge, flattened bubble of air touches the surface of
the earth.

A **cold front** is formed when a cold air mass moves into
warmer air, displacing it in the process. A cold front is generally
steep, and when it runs into the warmer air it forces it to rise
quickly. If the warm air is moist, it is quickly cooled adiabatically to the dew point temperature, resulting in large, towering
cumulus clouds and thunderclouds along the front (figure 24.10).
You may have observed that thunderstorms created by an advancing cold front often form in a line along the front. These
thunderstorms can be intense but are usually over quickly, soon
followed by a rapid drop in temperature from the cold air mass
moving past your location. The passage of the cold front is also
marked by a rapid shift in the wind direction and a rapid increase in the barometric pressure. Before the cold front arrives,
the wind is generally moving toward the front as warm, less dense
air is forced upward by the cold, more dense air. The lowest

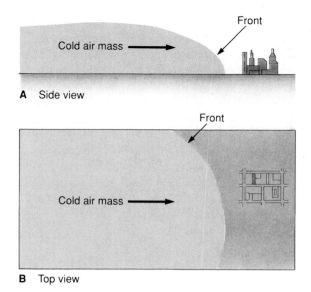

A Side view

B Top view

Figure 24.9

(a) A cold air mass is similar to a huge, flattened bubble of cold air that moves across the land. The front is the boundary between two air masses, a narrow transition zone of mixing. (b) A front is represented by a line on a weather map, which shows the location of the front at ground level.

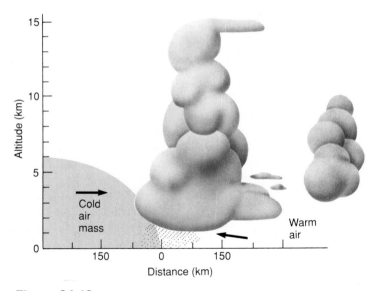

Figure 24.10

An idealized cold front showing the types of clouds that might occur when an unstable cold air mass moves through unstable warm air. Stable air would result in more stratus clouds rather than cumulus clouds.

barometric pressure reading is associated with the lifting of the warm air at the front. After the front passes your location, you are in the cooler, more dense air that is settling outward, so the barometric pressure increases and the wind shifts with the movement of the cold air mass.

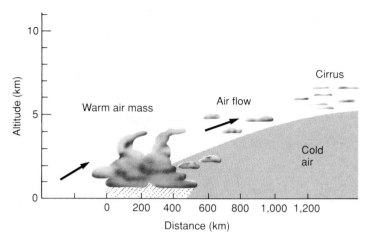

Figure 24.11

An idealized warm front showing a warm air mass overriding and pushing cold air in front of it. Notice that the overriding warm air produces a predictable sequence of clouds far in advance of the moving front.

A **warm front** forms when a warm air mass advances over a mass of cooler air. Since the advancing warm air is less dense than the cooler air it is displacing, it generally overrides the cooler air forming a long, gently sloping front. Because of this, the overriding warm air may form clouds far in advance of the ground-level base of the front (figure 24.11). This may produce high cirrus clouds a day or more in advance of the front, which are followed by thicker and lower stratus clouds as the front advances. Usually these clouds result in a broad band of drizzle, fog, and the continuous light rain usually associated with stratus clouds. This light rain (and snow in the winter) may last for days as the warm front passes.

Sometimes the forces influencing the movement of a cold or warm air mass lessen or become balanced, and the front stops advancing. When this happens a stream of cold air moves along the north side of the front and a stream of warm air moves along the south side in an opposite direction. This is called a **stationary front** because the edge of the front is not advancing. A stationary front may sound as if it is a mild frontal weather maker because it is not moving. Actually, a stationary front represents an unstable situation that can result in a major atmospheric storm. This type of storm is discussed in the following section.

Waves and Cyclones

A slowly advancing cold front and a stationary front often develop a bulge, or *wave,* in the boundary between cool and warm air moving in opposite directions (figure 24.12b). The wave grows as the moving air is deflected, forming a warm front moving northward on the right side and a cold front moving southward on the left side. Cold air is more dense than warm air, and the cold air moves faster than the slowly moving warm front. As the faster moving cold air catches up with the slower

Weather and Climate

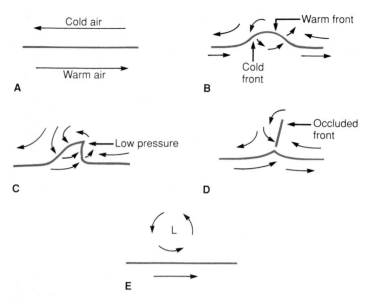

Figure 24.12

The development of a low-pressure center, or cyclonic storm, along a stationary front as seen from above. (a) A stationary front with cold air on the north side and warm air on the south side. (b) A wave develops, producing a warm front moving northward on the right side and a cold front moving southward on the left side. (c) The cold front lifts the warm front off the surface at the apex, forming a low-pressure center. (d) When the warm front is completely lifted off the surface, an occluded front is formed. (e) The cyclonic storm is now a fully developed low-pressure center.

moving warm air, the cold air underrides the warm air, lifting it upward. This lifting action produces a low-pressure area at the point where the two fronts come together (figure 24.12c). The lifted air expands, cools adiabatically, and reaches the dew point. Clouds form and precipitation begins from the lifting and cooling action. Within days after the wave first appears the cold front completely overtakes the warm front forming an occlusion (figure 24.12d). An **occluded front** is one that has been lifted completely off the ground into the atmosphere. The disturbance is now a *cyclonic storm* with a fully developed low-pressure center. Since its formation, this low-pressure cyclonic storm has been moving, taking its associated stormy weather with it in a generally easterly direction. Such cyclonic storms usually follow principal tracks along a front (figure 24.13). Since they are observed generally to follow these same tracks, it is possible to predict where the storm might move to next.

A **cyclone** is defined as a low-pressure center where the winds move into the low-pressure center and are forced upward. As air moves in toward the center, the Coriolis effect (see chapter 20) and friction with the ground cause the moving air to veer to the right of the direction of motion. In the Northern Hemisphere this rightward veering of moving air produces a counterclockwise circulation pattern of winds around the low-pressure center (figure 24.14). The upward movement associated with the low-pressure center of a cyclone cools the air adiabatically, resulting in clouds, precipitation, and stormy conditions.

Air is sinking in the center of a region of high pressure, producing winds that move outward. In the Northern Hemisphere, the Coriolis effect and frictional forces deflect this wind to the right, producing a clockwise circulation (figure 24.14). A high-pressure center is called an **anticyclone,** or simply a **high.** Since air in a high-pressure zone sinks, it is warmed adiabatically and the relative humidity is lowered. Thus clear, fair weather is usually associated with a high. By observing the barometric pressure, you can watch for decreasing pressure, which can mean the coming of a cyclone and its associated stormy weather. You can also watch for increasing pressure, which means a high and its associated fair weather are coming. Consulting a daily weather map makes such projections a much easier job, however.

Major Storms

There is a wide range of weather changes that can take place as a front passes because there is a wide range of possible temperature, moisture, stability, and other conditions between the new air mass and the air mass that it is displacing. The changes that accompany some fronts may be so mild that they go unnoticed. Others are noticed only as a day with breezes or gusty winds. Still other fronts are accompanied by a rapid and violent weather change called a **storm.** A snowstorm, for example, is a rapid weather change that may happen as a cyclonic storm moves over a location. The most rapid and violent changes occur with three kinds of major storms: (1) thunderstorms, (2) tornadoes, and (3) hurricanes.

Thunderstorms

A **thunderstorm** is a brief but intense storm with rain, lightning and thunder, gusty and often strong winds, and sometimes hail. Thunderstorms usually develop in warm, very moist, and unstable air. These conditions set the stage for a thunderstorm to develop when something lifts a parcel of air, starting it moving upward. This is usually accomplished by the same three general causes that produce cumulus clouds: (1) differential heating, (2) mountain barriers, or (3) along an occluded or cold front. Thunderstorms that occur from differential heating usually occur during warm, humid afternoons after the sun has had time to establish convective thermals. In the Northern Hemisphere the most of these convective thunderstorms occur during the month of July. Frontal thunderstorms, on the other hand, can occur any month and any time of the day or night that a front moves through warm, moist, and unstable air.

Frontal thunderstorms generally move with the front that produced them. Thunderstorms that developed in mountains or over flat lands from differential heating can move miles after they form, sometimes appearing to wander aimlessly across the land. These storms are not just one big rain cloud but are sometimes made up of cells that are born, grow to maturity, then die out in less than an hour. The thunderstorm, however, may last longer than an hour because new cells are formed as old ones die out. Each cell is about 2 to 8 km (about 1 to 5 mi) in diameter and goes through three main stages in its life: (1) cumulus, (2) mature, and (3) final (figure 24.15).

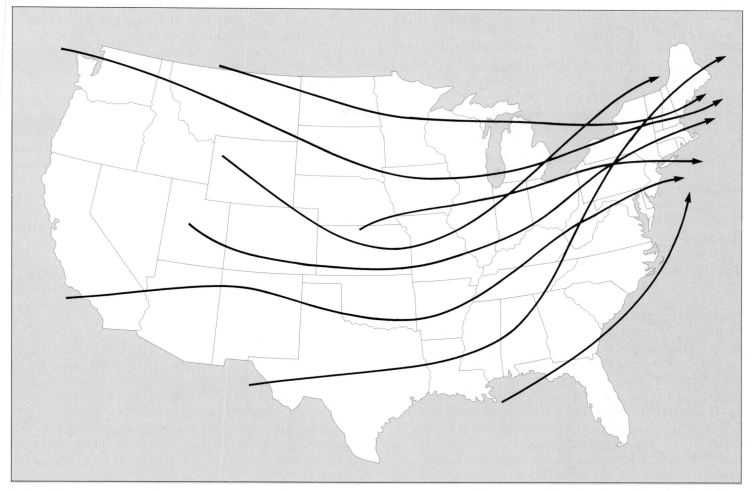

Figure 24.13

Cyclonic storms usually follow principal storm tracks across the continental United States in a generally easterly direction. This makes it possible to predict where the low-pressure storm might move to next.

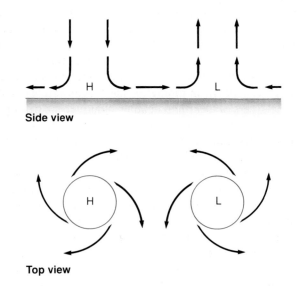

Side view

Top view

Figure 24.14

Air sinks over a high-pressure center and moves away from the center on the surface, veering to the right in the northern hemisphere to create a clockwise circulation pattern. Air moves toward a low-pressure center on the surface, rising over the center. As air moves toward the low-pressure center on the surface, it veers to the right in the northern hemisphere to create a counterclockwise circulation pattern.

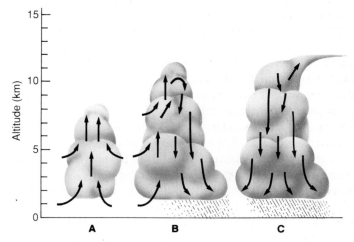

Figure 24.15

Three stages in the life of a thunderstorm cell. (*a*) The cumulus stage begins as warm, moist air is lifted in an unstable atmosphere. All the air movement is upward in this stage. (*b*) The mature stage begins when precipitation reaches the ground. This stage has updrafts and downdrafts side by side, which create violent turbulence. (*c*) The final stage begins when all the updrafts have been cut off and only downdrafts exist. This cuts off the supply of moisture, and the rain decreases as the thunderstorm dissipates. The anvil-shaped top is a characteristic sign of this stage.

Weather and Climate

Damage from a thunderstorm is usually caused by the associated lightning, strong winds, or hail. As illustrated in figure 24.15, the first stage of a thunderstorm begins as convection, mountains, or a dense air mass slightly lifts a mass of warm, moist air in an unstable atmosphere. The lifted air mass expands and cools adiabatically to the dew point temperature and a cumulus cloud forms. The latent heat of vaporization released by the condensation process accelerates the upward air motion, called an *updraft,* and the cumulus cloud continues to grow to towering heights. Soon the upward-moving, saturated air reaches the freezing level and ice crystals and snowflakes begin to form. When they become too large to be supported by the updraft, they begin to fall toward the surface, melting into raindrops in the warmer air they fall through. When they reach the surface, this marks the beginning of the mature stage. As the raindrops fall through the air, friction between the falling drops and the cool air produces a downdraft in the region of the precipitation. The cool air accelerates toward the surface at speeds up to 90 km/hr (about 55 mi/hr), spreading out on the ground when it reaches the surface. In regions where dust is raised by the winds, this spreading mass of cold air from the thunderstorm has the appearance of a small cold front with a steep, bulging leading edge. This miniature cold front may play a role in lifting other masses of warm, moist air in front of the thunderstorm, leading to the development of new cells. This stage in the life of a thunderstorm has the most intense rainfall, winds, and possibly hail. As the downdraft spreads throughout the cloud, the supply of new moisture from the updrafts is cut off and the thunderstorm enters the final, dissipating stage. The entire life cycle, from cumulus cloud to the final stage, lasts for about an hour as the thunderstorm moves across the surface. During the mature stage of powerful updrafts, the top of the thunderstorm may reach all the way to the top of the troposphere, forming a cirrus cloud that is spread into an anvil shape by the strong winds at this high altitude.

The updrafts, downdrafts, and falling precipitation separate tremendous amounts of electric charges that accumulate in different parts of the thundercloud. Large drops of water tend to carry negative charges, and cloud droplets tend to lose them. The upper part of the thunderstorm develops an accumulation of positive charges as cloud droplets are uplifted, and the middle portion develops an accumulation of negative charges from larger drops that fall. There are many other charging processes at work, such as induction (see chapter 8), and the lower part of the thundercloud develops both negative and positive charges. The voltage of these charge centers builds to the point that the electrical insulating ability of the air between them is overcome and a giant electrical discharge called *lightning* occurs (figure 24.16). Lightning discharges occur from the cloud to the ground, from the ground to a cloud, from one part of the cloud to another part, or between two different clouds. The discharge takes place in a fraction of a second and may actually consist of a number of strokes rather than one big discharge. The discharge produces an extremely high temperature around the channel, which

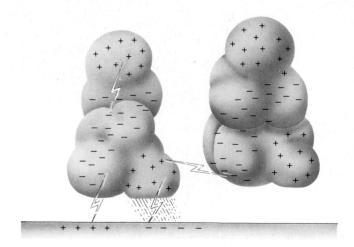

Figure 24.16

Different parts of a thunderstorm cloud develop centers of electric charge. Lightning is a giant electric spark that discharges the accumulated charges.

may be only 6 cm or so wide (about 2 in). The air it travels through is heated quickly, expanding into a sudden pressure wave that you hear as *thunder.* A nearby lightning strike produces a single, loud crack. Farther away strikes sound more like a rumbling boom as the sound from the separate strokes become separated over distance. Echoing of the thunder produced at farther distances also adds to the rumbling sounds. The technique of estimating the distance to a lightning stroke by measuring the interval between the flash of the lightning and the boom of the thunder is discussed in chapter 7. Lightning can present a risk for people in the open, near bodies of water, or under a single, isolated tree during a thunderstorm. The safest place to be during a thunderstorm is inside a car or a building with a metal frame.

Updrafts are also responsible for **hail,** a frozen form of precipitation that can be very destructive to crops, automobiles, and other property. Hailstones can be irregular, somewhat spherical, or flattened forms of ice that range from the size of a BB to the size of a softball. Most hailstones, however, are less than 2 cm (about 1 in) in diameter. The larger hailstones have alternating layers of clear and opaque, cloudy ice. These layers are believed to form as the hailstone goes through cycles of falling then being returned to the upper parts of the thundercloud by updrafts. The clear layers are believed to form as the hailstone moves through heavy layers of supercooled water droplets, which accumulate quickly on the hailstone but freeze slowly because of the release of the latent heat of fusion. The cloudy layers are believed to form as the hailstone accumulates snow crystals or moves through a part of the cloud with less supercooled water droplets. In either case, rapid freezing traps air bubbles, which result in the opaque, cloudy layer. Thunderstorms with hail are most common during the month of May in the states of Colorado, Kansas, and Nebraska.

Figure 24.17
A tornado might be small, but it is the most violent storm that occurs on the earth. This tornado, moving across an open road, eventually struck Dallas, Texas.

Tornadoes

A **tornado** is the smallest, most violent weather disturbance that occurs on the earth (figure 24.17). Tornadoes occur with intense thunderstorms, resembling a long, narrow funnel or ropelike structure that drops down from a thundercloud and may or may not touch the ground. This ropelike structure is a rapidly whirling column of air, usually 100 to 400 m (about 330 to 1,300 ft) in diameter. The bottom of the column moves across the surface, sometimes skipping into the air, then back down again at speeds that average about 50 km/hr (about 30 mi/hr). The speed of the whirling air in the column has been estimated to be well over 300 km/hr (about 200 mi/hr), leaving a path of destruction wherever the column touches the surface. The destruction is produced by the powerful winds, the sudden drop in atmospheric pressure that occurs at the center of the funnel, and the debris that is flung through the air like projectiles. A passing tornado sounds like very loud, continuous rumbling thunder with cracking and hissing noises that are punctuated by the crashing of debris projectiles.

On the average, several hundred tornadoes are reported in the United States every year. These occur mostly during spring and early summer afternoons over the Great Plains states. The states of Texas, Oklahoma, Kansas, and Iowa have such a high occurrence of tornadoes that the region is called "tornado alley." During the spring and early summer, this region has maritime tropical air from the Gulf of Mexico at the surface. Above this warm, moist layer is a layer of dry, unstable air that has just crossed the Rocky Mountains, moved along rapidly by the jet stream. The stage is now set for some event, such as a cold air mass moving in from the north, to shove the warm, moist air upward, and the result will be violent thunderstorms with tornadoes.

Hurricanes

A **hurricane** is a large, violent circular storm that is born over the warm, tropical ocean near the equator. A hurricane is a **tropical cyclone** with winds that exceed 120 km/hr (75 mi/hr). It is a cyclone because it is a low-pressure center with counterclockwise winds in the Northern Hemisphere. It is called a tropical cyclone because it is born in the tropics. The same type of tropical cyclone in the western Pacific is called a **typhoon.** Both typhoons and hurricanes, however, are tropical cyclones. The wind speed of 120 km/hr (75 mi/hr) is the defined difference between a tropical storm and a tropical cyclone. A storm with winds less than this defined value is a tropical storm. A storm with winds at or above 120 km/hr is a tropical cyclone, which is called a hurricane in some places and a typhoon in other places.

A tropical cyclone is similar to the wave cyclone of the mid-latitudes because both have low-pressure centers with a counterclockwise circulation in the Northern Hemisphere. They are different because a wave cyclone is usually about 2,500 km (about 450 mi) wide, has moderate winds, and receives its energy from the temperature differences between two air masses. A tropical cyclone, on the other hand, is often less than 200 km (about 125 mi) wide, has very strong winds, and receives its energy from the latent heat of vaporization released during condensation.

A fully developed hurricane has heavy bands of clouds, showers, and thunderstorms that rapidly rotate around a relatively clear, calm eye (figure 24.18). As a hurricane approaches a location the air seems unusually calm as a few clouds appear, then thicken as the wind begins to gust. Over the next six hours or so the overall wind speed increases as strong gusts and intense rainshowers occur. Thunderstorms, perhaps with tornadoes, and the strongest winds occur just before the winds suddenly die down and the sky clears with the arrival of the eye of the hurricane. The eye is an average of 10 to 15 km (about 6 to 9 mi) across, and it takes about an hour or so to cross a location. When the eye passes the intense rainshowers, thunderstorms, and hurricane-speed winds begin again, this time

Weather and Climate

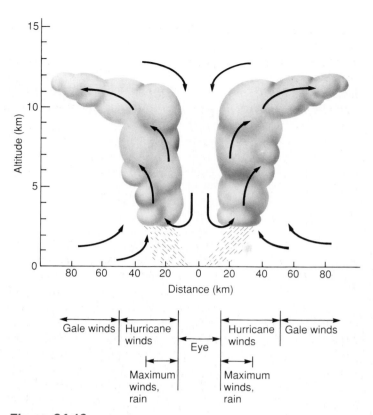

Figure 24.18

Cross section of a hurricane.

Figure 24.19

Supercomputers make routine weather forecasts possible by solving mathematical equations that describe changes in a mathematical model of the atmosphere.

blowing from the opposite direction. The whole sequence of events may be over in a day or two, but hurricanes are unpredictable and sometimes stall in one location for days. In general, they move at a rate of 15 to 50 km/hr (about 10 to 30 mi/hr).

Most of the damage from hurricanes results from strong winds, flooding, and the occasional tornado. Flooding occurs from the intense, heavy rainfall but also from the increased sea level that results from the strong, constant winds blowing seawater toward the shore. The sea level can be raised some 5 m (about 16 ft) above normal, with storm waves up to 15 m (about 50 ft) high on top of this elevated sea level. Overall, large inland areas can be flooded with extensive property damage. A single hurricane moving into a populated coastal region has caused billions of dollars of damage and the loss of hundreds of lives in the past. Today, the National Weather Service tracks hurricanes by weather satellites. Warnings of hurricanes, tornadoes, and severe thunderstorms are broadcast locally over special weather alert stations located across the country.

Weather Forecasting

Today, weather predictions are based on information about the characteristics, location, and rate of movement of air masses and associated fronts and pressure systems. This information is summarized as average values, then fed into a computer model of the atmosphere. The model is a scaled-down replica of the real atmosphere, and changes in one part of the model result in changes in another part of the model just as they do in the real atmosphere. Underlying the computer model are the basic scientific laws concerning solar radiation, heat, motion, and the gas laws. All these laws are written as a series of mathematical equations, which are applied to thousands of data points in a three-dimensional grid that represents the atmosphere. The computer is given instructions about the starting conditions at each data point, that is, the average values of temperature, atmospheric pressure, humidity, wind speed, and so forth. The computer is then instructed to calculate the changes that will take place at each data point, according to the scientific laws, within a very short period of time. This requires billions of mathematical calculations when the program is run on a worldwide basis. The new calculated values are then used to start the process all over again and it is repeated some 150 times to obtain a one-day forecast (figure 24.19).

Figure 24.20
The climate determines what types of plants and animals live in a location, the types of houses that people build, and the life-styles of people. This orange tree, for example, requires a climate that is relatively frost free, yet it requires some cool winter nights to produce a sweet fruit.

A problem with the computer model of the atmosphere is that small-scale events are inadequately treated, and this introduces errors that grow when predictions are attempted for farther and farther into the future. Small eddies of air, for example, or gusts of wind in a region have an impact on larger-scale atmospheric motions such as those larger than a cumulus cloud. But all of the small eddies and gusts cannot be observed without filling the atmosphere with measuring instruments. This lack of ability to observe small events that can change the large-scale events introduces uncertainties in the data, which, over time, will increasingly affect the validity of a forecast. If the uncertainties were cut in half by doubling the ability to measure the small-scale events, a three-day precision forecast might be extended by a day or two. More data points in the computer model, however, would require an enormous number of calculations to keep track of all the changes at each point. On a worldwide basis, a supercomputer must now do some 504 billion mathematical calculations just to make a forecast for next week. For forecasts of more than five days, the number of calculations and the effect of uncertainties increase greatly. It has been estimated that the reductions of observational errors could increase

the range of accurate forecasting up to two weeks. The ultimate range of accurate forecasting will require a better understanding—and thus an improved model—of patterns of changes that occur in the ocean as well as in the atmosphere. All of this increased understanding and reduction of errors lead to an estimated ultimate future forecast of three weeks, beyond which any pinpoint forecast would be only slightly better than a wild guess. In the meantime, regional and local daily weather forecasts are fairly accurate, and computer models of the atmosphere now provide the basis for extending the forecasts for up to about a week.

Climate

Changes in the atmospheric condition over a brief period of time, such as a day or a week, are referred to as changes in the *weather*. Weather changes follow a yearly pattern of seasons that are referred to as winter weather, summer weather, and so on. All of these changes are part of a composite, larger pattern called **climate**. Climate is the general pattern of the weather that occurs for a region over a number of years. Among other things, the climate determines what types of vegetation grow in a particular region, resulting in characteristic groups of plants associated with the region (figure 24.20). For example, orange, grapefruit, and palm trees grow in a region that has a climate with warm monthly temperatures throughout the year. On the other hand, blueberries, aspen, and birch trees grow in a region that has cool temperature patterns throughout the year. Climate determines what types of plants and animals live in a location, the types of houses that people build, and the life-styles of people. Climate also influences the processes that shape the landscape, the type of soils that form, the suitability of the region for different types of agriculture, and how productive the agriculture will be in a region. This section is about climate, what determines the climate of a region, and how climate patterns are classified.

Major Climate Groups

Climate is determined by the same basic three elements that determine the weather, that is, temperature, moisture, and the movement of air. As you learned earlier, incoming solar radiation determines the air temperature, provides the energy that drives the hydrologic cycle, and is the ultimate cause of nearly all motion of the atmosphere. It is the uneven distribution of the incoming solar radiation that results in the great variety of temperature conditions, moisture patterns, and general circulation of the atmosphere at different latitudes of the earth. Since the climate varies so much from one location to the next, even at the same latitude, it is evident that other climatic influences are also at work. These climatic influences will be considered after first looking at the most important factor in different climates, incoming solar radiation.

The earth's atmosphere is heated directly by incoming solar radiation and by absorption of infrared radiation from the surface as discussed in chapter 23. The amount of heating at any particular latitude on the surface depends primarily on two factors: (1) the *intensity* of the incoming solar radiation, which is determined by the angle at which the radiation strikes the surface, and (2) the *time* that the radiation is received at the surface, that is, the number of daylight hours compared to the number of hours of night.

The earth is so far from the sun that all rays of incoming solar radiation reaching the earth are essentially parallel. The earth, however, has a mostly constant orientation of its axis with respect to the stars as it moves around the sun in its orbit. Since the inclined axis points toward the sun part of the year and away from the sun the other part, radiation reaches different latitudes at different angles during different parts of the year. The orientation of the earth's axis to the sun during different parts of the year also results in days and nights of nearly equal length in the equatorial region but increasing differences at increasing latitudes to the poles. During the polar winter months, the night is twenty-four hours long, which means no solar radiation is received at all. The equatorial region receives more solar radiation during a year, and the amount received decreases toward the poles as a result of (1) yearly changes in intensity and (2) yearly changes in the number of daylight hours (see chapter 20).

In order to generalize about the amount of radiation received at different latitudes, some means of organizing, or grouping, the latitudes is needed (figure 24.21). For this purpose the latitudes are organized into three groups: (1) the **low latitudes,** those that some time of the year receive *vertical* solar radiation at noon, (2) the **high latitudes,** those that some time of the year receive *no* solar radiation at noon, and (3) the **middle latitudes,** which are between the low and high latitudes. This definition of low, middle, and high latitudes means that the low latitudes are between the tropics of Cancer and Capricorn (between 23 1/2° N and 23 1/2° S latitudes) and that the high latitudes are above the arctic and antarctic circles (above 66 1/2° N and above 66 1/2° S latitudes).

In general, (1) the *low latitudes* receive a high amount of incoming solar radiation that varies little during a year. Temperatures are high throughout the year, varying little from month to month. (2) The *middle latitudes* receive a higher amount of incoming radiation during one part of the year and a lower amount during the other part. Overall temperatures are cooler than in the low latitudes and have a wide seasonal variation. (3) The *high latitudes* receive a maximum amount of radiation during one part of the year and none during another part. Overall temperatures are low, with the highest range of annual temperatures.

The low, middle, and high latitudes provide a basic framework for describing the earth's climates. These climates are associated with the low, middle, and high latitudes illustrated in figure 24.21, but they are defined in terms of yearly temperature averages. It is necessary to define the basic climates in terms

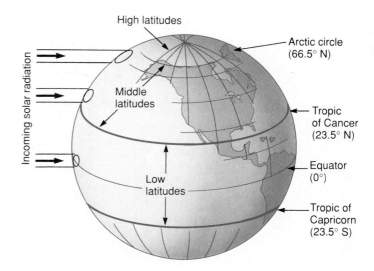

Figure 24.21

Latitude groups based on incoming solar radiation. The low latitudes receive vertical solar radiation at noon some time of the year, the high latitudes receive no solar radiation at noon during some time of the year, and the middle latitudes are in-between.

of temperature because land and water surfaces react differently to incoming solar radiation, creating a different temperature. Temperature and moisture are the two most important climate factors, and temperature will be considered first.

The principal climate zones are defined in terms of yearly temperature averages, which occur in broad regions (figure 24.22). They are (1) the **tropical climate zone** of the low latitudes, (2) the **polar climate zone** of the high latitudes, and (3) the **temperate climate zone** of the middle latitudes. The tropical climate zone is near the equator and receives the greatest amount of sunlight throughout the year. Overall, the tropical climate zone is hot. Average monthly temperatures stay above 18° C (64° F), even during the coldest month of the year. The other extreme is found in the polar climate zone, where the sun never sets during some summer days and never rises during some winter days. Overall, the polar climate zone is cold. Average monthly temperatures stay below 10° C (50° F), even during the warmest month of the year. The temperate climate zone is between the polar and tropical zones, with average temperatures that are neither very cold nor very hot. Average monthly temperatures stay between 10° C and 18° C (50° F and 64° F) throughout the year.

General patterns of precipitation and winds are also associated with the low, middle, and high latitudes. An idealized model of the global atmospheric circulation and pressure patterns was described in chapter 23. Recall that this model described a huge convective movement of air in the low latitudes, with air being forced upward over the equatorial region. This air expands, cools to the dew point, and produces abundant rainfall throughout most of the year. On the other hand, air is slowly sinking over 30° N and 30° S of the equator, becoming warm

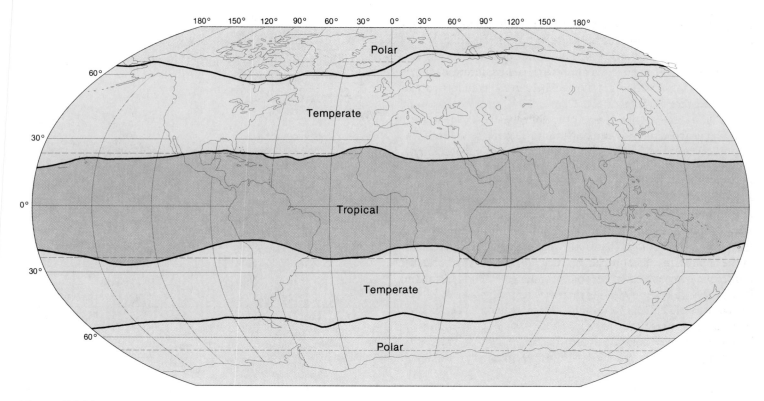

Figure 24.22

The principal climate zones are defined in terms of yearly temperature averages, which are determined by the amount of solar radiation received at the different latitude groups.

and dry as it is compressed. Most of the great deserts of the world are near 30° N or 30° S latitude for this reason. There is another wet zone near 60° latitudes and another dry zone near the poles. These wet and dry zones are shifted north and south during the year with the changing seasons. This results in different precipitation patterns in each season. Figure 24.23 shows where the wet and dry zones are in winter and in summer seasons.

Local Climatic Influence

Latitude determines the basic tropical, temperate, and polar climatic zones, and the wet and dry zones move back and forth over the latitudes with the seasons. If these were the only factors influencing the climate, you would expect to find the same climatic conditions at all locations with the same latitude. This is not what is found, however, because there are four major factors that affect a regional climate. These are (1) altitude, (2) mountains, (3) large bodies of water, and (4) ocean currents. The following describes how these four factors modify the climate of a region.

The first of the four regional climate factors is *altitude*. The atmosphere is warmed mostly by the greenhouse effect from the surface upward, and air at higher altitudes increasingly radiates more and more of its energy to space. Average air temperatures therefore decrease with altitude, and locations with

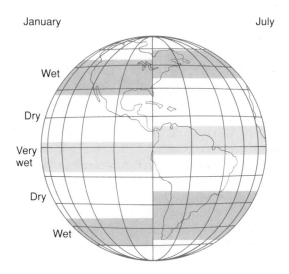

Figure 24.23

The idealized general rainfall patterns over the earth shift with seasonal shifts in the wind and pressure areas of earth's general atmospheric circulation patterns.

higher altitudes will have lower average temperatures. This is why the tops of mountains are often covered with snow when none is found at lower elevations. St. Louis, Missouri, and Denver, Colorado, are located almost at the same latitude (within 1° of 39° N), so you might expect the two cities to have about the same average temperature. Denver, however, has an altitude of 1,609 m (5,280 ft) and the altitude of St. Louis is 141 m (465 ft). The yearly average temperature for Denver is about 10° C (about 50° F), and for St. Louis it is about 14° C (about 57° F). In general, higher altitude means lower average temperature.

The second of the regional climate factors is *mountains*. In addition to the temperature change caused by the altitude of the mountain, mountains also affect the conditions of a passing air mass. The western United States has mountainous regions along the coast. When a moist air mass from the Pacific meets these mountains it is forced upward and cools adiabatically. Water vapor in the moist air mass condenses, clouds form, and the air mass loses much of its moisture as precipitation falls on the western side of the mountains. Air moving down the eastern slope is compressed and becomes warm and dry. As a result the western slopes of these mountains are moist and have forests of spruce, redwood, and fir trees. The eastern slopes are dry and have grassland or desert vegetation.

The third of the regional climate factors is a large body of *water*. Water, as discussed previously, has a higher specific heat than land material, it is transparent, and it loses energy through evaporation. All of these affect the temperature of a landmass located near a large body of water, making the temperatures more even from day to night and from summer to winter. San Diego, California, and Dallas, Texas, for example, are at about the same latitude (both almost 33° N), but San Diego is at a seacoast and Dallas is inland. Because of its nearness to water, San Diego has an average summer temperature about 7° C (about 13° F) cooler and an average winter temperature about 5° C (about 9° F) warmer than the average temperatures in Dallas. Nearness to a large body of water keeps the temperature more even at San Diego. This relationship is observed to occur for the earth as a whole as well as locally. The Northern Hemisphere is about 39 percent land and 61 percent water and has an average yearly temperature range of about 14° C (about 25° F). On the other hand the Southern Hemisphere is about 19 percent land and 81 percent water and has an average yearly temperature range of about 7° C (about 13° F). There is little doubt about the extent to which nearness to a large body of water influences the climate.

The fourth of the regional climate factors is *ocean currents*. In addition to the evenness brought about by being near the ocean, currents in the ocean can bring water that has a different temperature than the land. For example, currents can move warm water northward or they can move cool water southward (figure 24.24). This can influence the temperatures of air masses that move from the water to the land and thus the

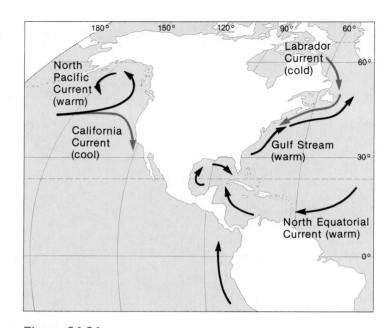

Figure 24.24
Ocean currents can move large quantities of warm or cool water to influence the air temperatures of nearby landmasses.

temperatures of the land. For example, the North Pacific current brings warm waters to the western coast of North America, which results in warmer temperatures for cities near the coast.

Describing Climates

Describing the earth's climates presents a problem because there are no sharp boundaries that exist naturally between two adjacent regions with different climates. Even if two adjacent climates are very different, one still blends gradually into the other. For example, if you are driving from one climate zone to another you might drive for miles before becoming aware that the vegetation is now different than it was a hour ago. Since the vegetation is very different from what it was before, you know that you have driven from one regional climate zone to another.

Actually, no two places on the earth have exactly the same climate. Some plants will grow on the north or south side of a building, for example, but not on the other side. The two sides of the building could be considered as small, local climate zones within a larger, major climate zone. The following is a general description of the regional climates of North America, followed by descriptions of how some local climates develop.

Major Climate Zones

Major climate zones are described by first considering the principal polar, temperate, and tropical climate zones, then looking at subdivisions within each that result from differences in moisture. Within one of the major zones, for example, there may be an area near the ocean that is influenced by air masses from the ocean. If so, the area has a **marine climate.** Because of the influence of the ocean, areas with marine climates have mild winters and cool summers compared to areas farther inland. In

Weather, Climate, and the Ocean

The atmosphere is a system that responds to incoming solar radiation, the spinning earth, and other factors such as the amount of water vapor present. The ocean is also a system that responds to incoming solar radiation, the spinning earth, and other factors that are discussed in chapter 25. The two systems also undergo changes by interacting with each other, most visibly in the tropical cyclone. The ocean system supplies water vapor, latent heat, and condensation nuclei, which are the essential elements of a tropical cyclone as well as everyday weather changes and climate. The atmosphere, on the other hand, drives the ocean with prevailing winds, moving warm or cool water to locations where they affect the climate on the land. There is a complex, interdependent relationship between the ocean and the atmosphere, and it is probable that even small changes in one system can lead to changes in the other. For example, small changes in the temperature of seawater could lead to changes in the weather, creating more intense or less intense tropical cyclones. Changes in the temperature of seawater could also affect the climate of lands the water is moved by. Scientists are engaged in meteorological research designed to understand, among other things, changes in the weather or changes in the climate of a region that result from changes in the ocean. In the late 1970s an apparent link was found between changes in the atmosphere and changes in the ocean, a series of events known as *El Niño.*

To describe the series of events called El Niño, it is first necessary to describe the usually normal conditions of the Pacific Ocean along the equator in July. This area has systems of prevailing wind belts, pressure systems, and ocean currents that are normally associated with the intertropical convergence zone. This means a high pressure system over the ocean west of North America and a second high pressure system over the ocean west of South America. The wind patterns around these two high-pressure systems push the surface seawater offshore, westward along the

equator. Along the coast, water from below the surface moves to the surface to replace the water moved seaward. This process, called upwelling, brings cool water with abundant dissolved phosphates, nitrates, and other nutrients from the deep, darker water to the surface. Here, with the more sufficient light of the surface, tiny marine plants use the nutrients to grow in abundance. The abundant supply of tiny marine plants, in turn, supports an extensive population of fish. Thus it is the upwelling that is responsible for the large fishing industry off the coast of Ecuador and northern Peru. Periodic failures of the upwelling, and associated catastrophic declines in the yearly fish catch, received international attention in the early 1980s when it was believed that an apparent link had been found between these periodic events and global weather changes.

It is normal for the waters in the Peruvian fishing area to become slightly warmer each year as the earth's prevailing wind belts and associated ocean currents shift southward with the season. In late December each year, the southward migration moves a warmer southward current past Ecuador and northern Peru, reducing the upwelling and warming the sea water 1° C to 2° C (about 2° F to 4° F). Since this event occurred every year about Christmas, the local population called the event El Niño, which is Spanish for the Christ child. Normally, the slightly warmer waters and the reduced upwelling lasts for about six months before resuming to the nutrient-rich upwelling.

About every six or seven years or so, the El Niño becomes unusually intense with an 8° C to 10° C (about 14° F to 18° F) warming that lasts for a year or more. No upwelling occurs during this intense period, cutting off the source of nutrients and eventually the food supply for the fish, birds, humans, and other organisms in the food chain. It is the intense event that is generally referred to as an El Niño by meteorologists, even though the local population calls each smaller, yearly event by the same name.

Meteorologists became interested in El Niño when certain weather patterns seem to fluctuate with its occurrence. For example, the difference between high and low atmospheric pressure systems along the equator seemed to become even greater during the intense event. In addition, the trade winds seemed to diminish, sometimes reversing directions completely when an especially intense El Niño occurred. Other apparent links found during the early 1980s included droughts in normally wet climates and heavy rainfall in normally dry climates, such as in southern California. Some of these apparently linked changes were global in nature, affecting places as far away as India, Africa, and the Soviet Union. It appeared that the atmosphere and the ocean were somehow linked, with a change in one system influencing a change in the other. The major problem in this linking was a lack of understanding of what caused the links and a lack of consistency in the links themselves. For example, southern California did not always have an unusually wet season every time an El Niño occurred and in fact experienced a drought during one event.

Scientists have continued to study the El Niño since the mid-1980s, searching for patterns that will reveal consistent cause-and-effect links. Part of the problem may be that other factors, such as a volcanic eruption, may influence part of the linkage but not another part. Another part of the problem may be the influence of unknown factors, such as the circulation of water deep beneath the surface, the track taken by tropical cyclones, or the energy released by tropical cyclones one year compared to the next. The results so far have indicated that atmosphere-ocean interactions are much more complex than early theoretical models had predicted. Sometimes a new model will predict some weather changes that occur with an El Niño, but no model is yet consistently correct in predicting the conditions that lead to the event and the weather patterns that result. All this may someday lead to an understanding of how the ocean and the atmosphere interact on this dynamic planet.

addition to the moderate temperatures, areas with a marine climate have abundant precipitation with an average 50 to 75 cm (20 to 30 in) yearly precipitation. The western coast of Canada and the northwest coasts of Washington, Oregon, and northern California have a marine climate. These regions are covered with forests of spruce, fir, and other conifers.

Also within a major polar, temperate, or tropical zone there may be an area that is far from an ocean, influenced mostly by air masses from large land areas. If so, the area has a **continental climate.** A continental climate does not have an even temperature as does the marine climate because the land heats and cools rapidly. Thus summers are hot and winters are cold.

Climates can be further classified as being **arid,** which means dry, or **humid,** which means moist. An area with an arid climate is defined as one that receives less than 25 cm (10 in) of precipitation per year. An area with a humid climate is defined as one that receives 50 cm (20 in) or more precipitation per year. An area that receives between 25 and 50 cm (10 and 20 in) precipitation per year is defined as **semiarid.**

Table 24.1 describes the principal climate types of North America, and figure 24.25 gives the general location of these climates. Both are based on average temperatures and amounts of precipitation as described in the table. Recall that the actual climate in a given location may not agree with the general description because a local climate factor may change the climate. Also recall that the climates blend gradually from one location to the next and do not change suddenly as you move across one of the lines.

Local Climates

The spread of cities, construction of high-rise buildings, paving of roads, and changes in the natural vegetation and landscape can change the local climate. Concrete, metal, stone, and glass react differently to incoming solar radiation than the natural vegetation and soils they replaced. High-rise buildings not only have a greater area exposed to solar radiation but also are capable of slowing and channeling the wind. Concrete and asphalt streets, roads, and parking lots also change the local climate because they are better absorbers of incoming solar radiation than natural vegetation and become heat sources for increased convection. They also make it impossible for precipitation to soak into the ground, increasing the likelihood of flooding. Large cities make a greater contribution to what has been called the "heat island" or the "heat dome" effect. In addition to causing the small changes in the actual air temperature, the buildings, concrete, and asphalt in large cities emit much more infrared radiation than is given off from the landscape in the country. This increased radiation causes people to feel much warmer in the city during the day and especially at night. The overall feeling of being warmer is also influenced by the decreased wind speed that occurs because large buildings block the wind, as well as the increased amounts of humidity in the city air.

Table 24.1

Principal climate types

Climate Type	Description
Polar	Long, very cold winters, cold summers, dry
Humid Continental (Subarctic)	Long, cold winters, cool summers, moderate precipitation
Humid Continental (Middle Latitudes)	Cold winters, moderate summers, moderate precipitation
Humid Continental (Low Latitudes)	Mild winters, hot summers, moderate precipitation
Humid Subtropical	Short, mild winters, humid summers, moderate precipitation
Tropical Wet	Hot and humid all year with heavy precipitation
Tropical Wet/Dry (Subtropical)	Hot all year with alternating wet and dry seasons
Semiarid	Varying temperatures with low precipitation
Desert	Hot days and cool nights, arid
Marine	Moderate, rainy summers, mild winters
Mediterranean	Hot, dry summers with short, mild and wet winters
Highlands	Conditions vary with altitude

Changes in the local climate are not restricted to large cities. A local pattern of climate is called a **microclimate.** A large city creates a new microclimate. Certain plants will grow within one microclimate but not another. For example, in some locations lichens or mosses will grow on one side of a tree but not the other. A different microclimate exists on each side of the tree. The planting or cutting down of trees around a house can change the microclimate around the house. Air pollution also creates a new microclimate (see the Feature in chapter 23). Dust, particulates, and smog also contribute to the "heat dome" of a large city by holding in radiation at night and reflecting incoming solar radiation during the day. This reduces convection, making the air pollution and smog last longer.

Summary

The *cloud-forming process* begins when something gives a parcel of air an upward push. The three major causes of upward air movement are (1) *convection,* (2) *barriers* to moving air masses, and (3) the *meeting of moving air masses.* As a parcel of air is pushed upward, it comes under less atmospheric pressure and expands. An expanding parcel of air does work on the surroundings and becomes cooler through the process known as *adiabatic cooling.* The rate of cooling as a parcel of dry air is lifted in the atmosphere is called the *dry adiabatic lapse rate.* The dry adiabatic lapse rate is 10° C/km (about 5.5° F/1,000

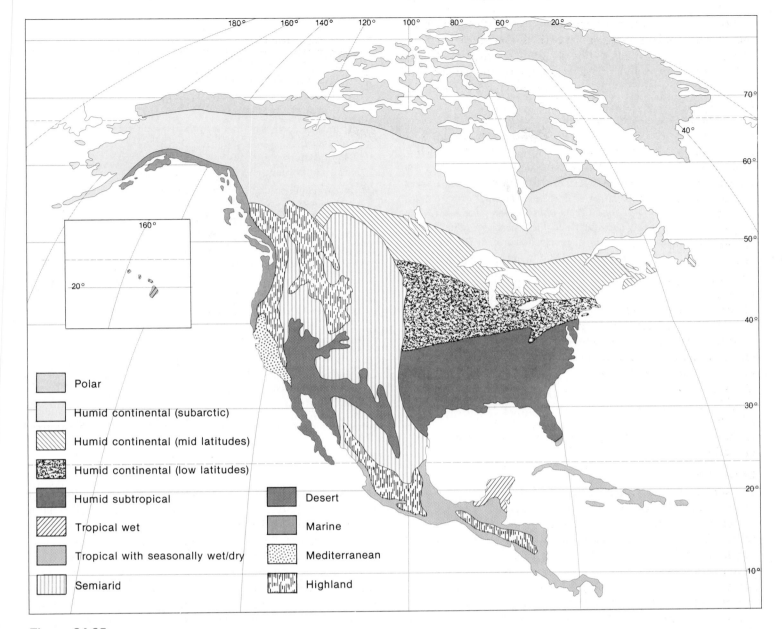

Figure 24.25

This map highlights the approximate location of the major types of climates in North America. See table 24.1 for a description of these climates.

Legend:

- Polar
- Humid continental (subarctic)
- Humid continental (mid latitudes)
- Humid continental (low latitudes)
- Humid subtropical
- Tropical wet
- Tropical with seasonally wet/dry
- Semiarid
- Desert
- Marine
- Mediterranean
- Highland

ft). The relationship between the measured lapse rate and the dry adiabatic lapse rate describes if a state of *atmospheric stability* or *atmospheric instability* exists.

Condensation nuclei act as centers of condensation as water vapor forms tiny droplets around the microscopic particles. The accumulation of large numbers of tiny droplets is what you see as a cloud.

Water that returns to the earth in liquid or solid form falls from the clouds as *precipitation*. Precipitation forms in clouds through two processes: (1) the *coalescence* of cloud droplets or (2) the *growth of ice crystals* at the expense of water droplets.

Weather changes are associated with the movement of large bodies of air called *air masses,* the leading *fronts* of air masses when they move, and local *high-and low-pressure* patterns that accompany air masses or fronts. Examples of air masses include (1) *continental polar,* (2) *maritime polar,* (3) *continental tropical,* and (4) *maritime tropical.*

When a location is under the influence of an air mass the location is having *air mass weather* with slow, gradual changes. More rapid changes take place when the *front,* a thin transition zone between two air masses, passes a location.

A *stationary front* often develops a bulge, or *wave,* that forms into a moving cold front and a moving warm front. The faster moving cold front overtakes the warm front, lifting it into the air to form an *occluded front.* The lifting process forms a low-pressure center called a *cyclone.* Cyclones are associated with heavy clouds, precipitation, and stormy conditions because of the lifting action.

A *thunderstorm* is a brief, intense storm with rain, lightning and thunder, gusty and strong winds, and sometimes hail. A *tornado* is the smallest, most violent weather disturbance the occurs on the earth. A *hurricane* is a *tropical cyclone,* a large, violent circular storm that is born over warm tropical waters near the equator.

The general pattern of the weather that occurs for a region over a number of years is called *climate.* The three principal climate zones are (1) the *tropical climate zone,* (2) the *polar climate zone,* and (3) the *temperate climate zone.* The climate in these zones is influenced by four factors that determine the local climate: (1) *altitude,* (2) *mountains,* (3) *large bodies of water,* and (4) *ocean currents.* The climate for a given location is described by first considering the principal climate zone, then looking at subdivisions within each that result from local influences.

Key Terms

adiabatic cooling (p. 503)	low latitudes (p. 516)
adiabatic heating (p. 503)	marine climate (p. 518)
air mass (p. 507)	maritime air mass (p. 507)
air mass weather (p. 507)	microclimate (p. 520)
anticyclone (p. 510)	middle latitudes (p. 516)
arid (p. 520)	occluded front (p. 510)
atmospheric stability (p. 504)	polar air mass (p. 507)
climate (p. 515)	polar climate zone (p. 516)
coalescence process (p. 505)	precipitation (p. 505)
cold front (p. 508)	semiarid (p. 520)
continental air mass (p. 507)	stationary front (p. 509)
continental climate (p. 520)	storm (p. 510)
cyclone (p. 510)	supercooled (p. 506)
dry adiabatic lapse rate (p. 503)	supersaturated (p. 505)
front (p. 508)	temperate climate zone (p. 516)
hail (p. 512)	thunderstorm (p. 510)
high (p. 510)	tornado (p. 513)
high latitudes (p. 516)	tropical air mass (p. 507)
humid (p. 520)	tropical climate zone (p. 516)
hurricane (p. 513)	tropical cyclone (p. 513)
hydrologic cycle (p. 502)	typhoon (p. 513)
ice-crystal process (p. 505)	warm front (p. 509)
ice-forming nuclei (p. 506)	wet adiabatic lapse rate (p. 505)

Applying the Concepts

1. A thunderstorm that occurs at 3:00 A.M. over a flat region of the country was probably formed by
   a. convection.
   b. a barrier to moving air.
   c. the meeting of moving air masses.
   d. any of the above.

2. White, puffy cumulus clouds that form over a flat region of the country during the late afternoon of a clear, warm day are probably the result of
   a. convection.
   b. a barrier to moving air.
   c. the meeting of moving air masses.
   d. none of the above.

3. Cumulus clouds usually mean an atmospheric state of
   a. stability.
   b. instability.
   c. equilibrium.

4. A parcel of air with a relative humidity of 50 percent is given an upward shove into the atmosphere. What is necessary before cloud droplets form in this air?
   a. adiabatic cooling
   b. saturation
   c. condensation nuclei
   d. all of the above

5. When water vapor in the atmosphere condenses to liquid water
   a. dew falls to the ground.
   b. rain or snow falls to the ground.
   c. a cloud forms.
   d. any of the above, depending on conditions.

6. Longer periods of drizzle, rain, or snow are usually associated with which basic form of a cloud?
   a. stratus
   b. cumulus
   c. cirrus
   d. none of the above

7. Brief periods of showers are usually associated with which type of cloud?
   a. stratus
   b. cumulus
   c. cirrus
   d. none of the above

8. The type of air mass weather that results from the arrival of polar continental air is
   a. frequent snowstorms with rapid changes.
   b. clear and cold with gradual changes.
   c. unpredictable but with frequent and rapid changes.
   d. much the same from day to day, the conditions depending on the air mass and the local conditions.

9. The appearance of high cirrus clouds, followed by thicker and lower stratus clouds, then continuous light rain over several days probably means which of the following air masses has moved into your area?
   a. continental polar
   b. maritime tropical
   c. continental tropical
   d. maritime polar

10. Most of the great deserts of the world are located
    a. near the equator.
    b. 30° N or 30° S latitude.
    c. 60° N or 60° S latitude.
    d. anywhere, as there is no pattern to their location.

Answers

1. c 2. a 3. b 4. d 5. c 6. a 7. b 8. d 9. b
10. b

Questions for Thought

1. What is the hydrologic cycle?

2. Describe and explain the energy exchange processes that take place during the hydrologic cycle.

3. What is a cloud? Describe how a cloud forms.

4. What is the role of convection, mountains, and moving air masses in forming a cloud?

5. What is special about the air temperature at the base of a cumulus cloud that is forming? Explain your reasoning.

6. Give a reason for the observation that regions near the ocean during the summer have cumulus clouds over the land during the day and over the water during the night.

7. A very young person states that clouds are around mountain peaks because they get caught on the peaks. What is your explanation for the observation that clouds are observed around mountain peaks but not elsewhere?

8. Why does a rising parcel of air become cooler as it rises? What is the name of this process?

9. What is atmospheric stability? What does this have to do with what kind of clouds may form on a given day?

10. What are the atmospheric conditions for a stable atmosphere? What are the atmospheric conditions for atmospheric instability?

11. Explain why the formation of a cloud is usually accompanied by an acceleration of a rising parcel of air.

12. Why are condensation nuclei important in the cloud-forming process?

13. Describe two different methods of artificial rainmaking, explaining how each method produces rain.

14. Describe two ways that precipitation may form from the water droplets of a cloud.

15. What is an air mass and how is it different from ordinary air?

16. What is a supercooled cloud? Why do they form?

17. Describe a model that explains why (a) the region around the equator is hot and humid and (b) the regions around 30° N and 30° S are warm and dry.

18. Describe the four main types of air masses that can influence the weather.

19. What is a front?

20. What kinds of clouds and weather changes are usually associated with the passing of (a) a warm front? (b) a cold front?

21. Describe how a low-pressure center forms along a stationary front.

22. Describe the wind direction, pressure, and weather conditions that are usually associated with (a) low-pressure centers and (b) high-pressure centers.

23. Compare and contrast a frontal cyclone with a tropical cyclone.

24. In which of the four basic types of air masses would you expect to find afternoon thunderstorms? Explain.

25. Describe the three main stages in the life of a thunderstorm cell, identifying the events that mark the beginning and end of each stage.

26. Describe how each of the following events of a thunderstorm develop: (a) hail, (b) lightning, (c) thunder, and (d) strong gusty winds.

27. What is a tornado? When and where do tornadoes usually form?

28. What is a hurricane? Describe how the weather conditions change as a hurricane approaches, passes directly over, then moves away from a location.

29. What probably produced a thunderstorm that occurred at 3:00 A.M.? Explain.

30. Explain why the characteristic anvil-shaped cloud usually develops during the last stage of a thunderstorm.

31. Explain how a computer is used in weather forecasting.

32. What are the problems associated with computer-based weather forecasting?

33. How is climate different from the weather?

34. Explain why the average air temperature decreases with increasing (a) latitude and (b) altitude.

35. Describe the average conditions found in the three principal climate zones.

36. Identify the four major factors that influence the climate of a region and explain how each does its influencing.

37. What is a microclimate? Describe a microclimate that exists where you live.

38. Explain why an area near the ocean usually has a different climate than farther inland.

39. Describe the differences between climates that are described with the terms (a) arid, (b) humid, and (c) semiarid.

40. Since heated air rises, why is snow found on top of a mountain and not at lower elevations?

Earth's Water

Figure 25.1

This is one of the water canals of the present-day system in Phoenix, Arizona. These canals were reconstructed from a system built by Native Americans five hundred years ago and then abandoned. The name "Phoenix" is from a mythical bird by the same name that was consumed by fire and then arose from its ashes.

THROUGHOUT history humans have diverted rivers and reshaped the land to ensure a supply of fresh water. There is evidence, for example, that ancient civilizations along the Nile River diverted water for storage and irrigation some five thousand years ago. The ancient Greeks and Romans built systems of aqueducts to divert streams to their cities some two thousand years ago. Some of these aqueducts are still standing today. More recent water diversion activities were responsible for the name of Phoenix, Arizona. Phoenix was named after a mythical bird that arose from its ashes after being consumed by fire. The city was given this name because it is built on a system of canals that were first designed and constructed by ancient Native Americans, then abandoned hundreds of years before settlers reconstructed the ancient canal system (figure 25.1). Water is and always has been an essential resource. Where water is in short supply, humans have historically turned to extensive diversion and supply projects to meet their needs.

Precipitation is the basic source of the water supply found today in streams, lakes, and beneath the earth's surface. Much of the precipitation that falls on the land, however, evaporates back into the atmosphere before it has a chance to become a part of this supply. The water that does not evaporate mostly moves directly to rivers and streams, flowing back to the ocean, but some soaks into the land. The evaporation of water, condensation of water vapor, and the precipitation-making processes were introduced in chapter 24 as important weather elements. They are also part of the generalized *hydrologic cycle* of evaporation from the ocean, transport through the atmosphere by moving air masses, precipitation on the land, and movement of water back to the ocean. Only part of this cycle was considered previously, however, and this was the part from evaporation through precipitation. This chapter is concerned with the other parts of the hydrologic cycle, that is, what happens to the water that falls on the land and that makes it back to the ocean. The chapter begins with a discussion of how water is distributed on the earth and a more detailed look at the hydrologic cycle. Then the travels of water across and into the land will be considered as streams, wells, springs, and other sources of usable water are discussed as limited resources. The tracing of the hydrologic cycle will be completed as the water finally makes it back to the ocean. This last part of the cycle will consider the nature of the ocean floor, the properties of seawater, and how waves and currents are generated. The water is now ready to evaporate, starting another one of earth's never-ending cycles.

Water on the Earth

Some water is tied up in chemical bonds deep in the earth's interior, but free water is the most abundant chemical compound near the surface. Water is five or six times more abundant than the most abundant mineral in the outer 6 km (about 4 mi) of the earth, so it should be no surprise that water covers about 70 percent of the surface. On the average, about 98 percent of this water exists in the liquid state in depressions on the surface and in sediments. Of the remainder, about 2 percent exists in the solid state as snow and ice on the surface in colder locations.

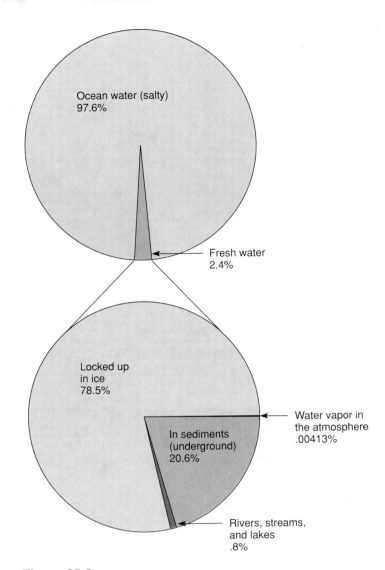

Figure 25.2
The distribution of all the water found on the earth's surface.

Only a fraction of a percent exists as a variable amount of water vapor in the atmosphere at a given time. Water is continually moving back and forth between these "reservoirs," but the percentage found in each is assumed to be essentially constant.

As shown in figure 25.2, over 97 percent of the earth's water is stored in the earth's oceans. This water contains a relatively high level of dissolved salts, which will be discussed in a later section. This dissolved salt makes ocean water unfit for human consumption and unfit for most agricultural purposes. All other water, which is fit for human consumption and agriculture, is called **fresh water.** About two-thirds of the earth's freshwater supply is locked up in the ice caps of Greenland and the Antarctic and in glaciers. This leaves less than 1 percent of all the water found on the earth as available fresh water. There is a generally abundant supply, however, because the freshwater supply is continually replenished by the hydrologic cycle.

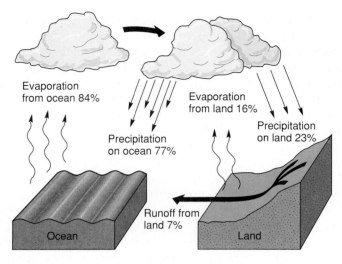

Evaporation from ocean 84%

Evaporation from land 16%

Precipitation on land 23%

Precipitation on ocean 77%

Runoff from land 7%

Ocean

Land

100% is based on a global average of 85 cm/yr precipitation.

Figure 25.3

On the average, more water is evaporated from the ocean than is returned by precipitation. More water is precipitated over the land than evaporates. The difference is returned to the ocean by rivers and streams.

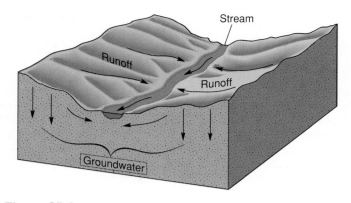

Stream

Runoff

Runoff

Groundwater

Figure 25.4

Some of the precipitation soaks into the ground to become groundwater. Groundwater slowly moves underground, and some of it emerges in a streambed to keep the stream running during dry spells.

Evaporation of water from the ocean is an important process of the hydrologic cycle because (1) water vapor leaves the dissolved salts behind, forming precipitation that is fresh water, and (2) the gaseous water vapor is easily transported in the atmosphere from one part of the earth to another. Over a year this natural desalination process produces and transports enough fresh water to cover the entire earth with a layer about 85 cm (about 33 in) deep. Precipitation is not evenly distributed like this, of course, and some places receive much more while other places receive almost none. Considering global averages, more water is evaporated from the ocean than returns by precipitation. On the other hand, more water is precipitated over the land than evaporates back to the atmosphere. The net amount evaporated and precipitated over the land and over the ocean is balanced by the return of water to the ocean by streams and rivers (figure 25.3). This fresh water returning on and under the land is the source of fresh water. What happens to the fresh water during its return to the ocean is discussed in the following section.

Fresh Water

The basic source of fresh water is precipitation, but not all precipitation ends up as part of the freshwater supply. Liquid water is always evaporating, even as it falls. In arid climates, rain sometimes evaporates completely before reaching the surface, even from a fully developed thunderstorm. Evaporation continues from the water that does reach the surface. Puddles and standing water on the hard surface of city parking lots and streets, for example, gradually evaporate back to the atmosphere after a rain and the surface is soon dry. There are many

factors that determine how much of a particular rainfall evaporates, but in general more than two-thirds of the rain eventually returns to the atmosphere. The remaining amount either (1) flows downhill across the surface of the land toward a lower place or (2) soaks into the ground. Water moving across the surface is called **runoff.** Runoff begins as rain accumulates in thin sheets of water that move across the surface of the land. These sheets collect into a small body of running water called a **stream.** A stream is defined as any body of water that is moving across the land, from one so small that you could step across it to the widest river. Water that soaks into the ground moves down to a saturated zone where it is called *groundwater.* Groundwater moves through sediments and rocks beneath the surface, slowly moving downhill. Streams carry the runoff of a recent rainfall or melting snow, but otherwise most of the flow comes from groundwater that has seeped into the stream channel. This explains how a permanent stream is able to continue flowing when it is not being fed by runoff or melting snow (figure 25.4). Where or when the source of groundwater is in low supply a stream may flow only part of the time, and it is designated as an *intermittent stream.*

The amount of a rainfall that becomes runoff or groundwater depends on a number of factors, including (1) the type of soil on the surface, (2) how dry the soil is, (3) the amount and type of vegetation, (4) the steepness of the slope, and (5) if the rainfall is a long, gentle one or a cloudburst. Different combinations of these factors can result in from 5 percent to almost 100 percent of given rainfall running off, with the rest evaporating or soaking in the ground. On the average, however, about 70 percent of all precipitation evaporates back into the atmosphere, about 30 percent becomes runoff, and less than 1 percent soaks into the ground.

Surface Water

If you could follow the smallest of streams downhill, you would find that it eventually merges with other streams until you reach a major river. The land area drained by a stream is known as

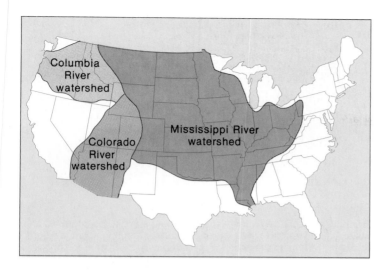

Figure 25.5
The approximate watersheds of the Columbia River, the Colorado River, and the Mississippi River.

the stream's drainage basin or **watershed.** Each stream has its own watershed, but the watershed of a large river includes all the watersheds of the smaller streams that feed into the larger river. Figure 25.5 shows the watersheds of the Columbia River, the Colorado River, and the Mississippi River. Note that the water from the Columbia River and the Colorado River watersheds empties into the Pacific Ocean. The Mississippi River watershed drains into the Gulf of Mexico, which is part of the Atlantic Ocean.

Two adjacent watersheds are separated by a line called a **divide.** Rain that falls on one side of a divide flows into one watershed and rain that falls on the other side flows into the other watershed. A *continental divide* separates river systems that drain into opposite sides of a continent. The North American continental divide trends northwestward through the Rocky Mountains. Imagine standing over this line with a glass of water in each hand, then pouring the water to the ground. The water from one glass will eventually end up in the Atlantic Ocean, and the water from the other glass will end up in the Pacific Ocean. Sometimes the Appalachian Mountains are considered to be an eastern continental divide, but water from both sides of this divide ends up on the same side of the continent, in the Atlantic Ocean.

Water moving downhill is sometimes stopped by a depression in a watershed, a depression where water temporarily collects as a standing body of fresh water. A smaller body of standing water is usually called a **pond,** and one of much larger size is called a **lake.** A lake is supposed to be deep enough in some place or places that sunlight does not reach the bottom, but this definition is not always followed when lakes and ponds are named. Part of the problem is the interpretation of what is meant by "sunlight." In general, only blue-green light remains at a depth of about 10 m (about 33 ft), and there is insufficient light for plant life at a depth of about 80 m (about 260 ft).

A pond or lake can occur naturally in a depression or it can be created by building a dam on a stream. A natural pond, a natural lake, or a pond or lake created by building a dam is called a **reservoir** if it is used for (1) water storage, (2) flood control, or (3) generating electricity. A reservoir can be used for one or two of these purposes but not generally for all three. A reservoir built for water storage, for example, is kept as full as possible to store water. This use is incompatible with use for flood control, which would require a low water level in the reservoir in order to catch runoff, preventing waters from flooding the land. In addition, extensive use of reservoir water to generate electricity requires the release of water, which could be incompatible with water storage. The water of streams, ponds, lakes, and reservoirs is collectively called *surface water,* and all serve as sources of fresh water. The management of surface water, as you can see, can present some complicated problems.

Groundwater

Precipitation soaks into the ground, or *percolates* slowly downward until it reaches an area, or zone, where the open spaces between rock and soil particles are completely filled with water. Water from such a saturated zone is called **groundwater.** There is a tremendous amount of water stored as groundwater, which makes up a supply about thirty times larger than all the surface water on the earth. Groundwater is an important source of fresh water for human consumption and for agriculture. Groundwater is often found within 100 m (about 330 ft) of the surface, even in arid regions where little surface water is found. Groundwater is the source of water for wells in addition to being the source that keeps streams flowing during dry periods.

Water is able to percolate down to a zone of saturation because sediments contain open spaces between the particles called *pore spaces.* The more pore space a sediment has, the more water it will hold. The total amount of pore spaces in a given sample of sediment is a measure of its **porosity.** Sand and gravel sediments, for example, have grains that have large pore spaces, so these sediments have a high porosity. In order for water to move through a sediment, however, the pore spaces must be connected. The ability of a given sample of sediment to transmit water is a measure of its **permeability.** Sand and gravel have a high permeability because the grains do not fit tightly together, allowing water to move from one pore space to the next. Sand and gravel sediments thus have a high porosity as well as a high permeability. Clay sediments, on the other hand, have small, flattened particles that fit tightly together. Clay thus has a low porosity, and when saturated or compressed, clay becomes *impermeable,* meaning water cannot move through it at all (figure 25.6).

The amount of groundwater available in a given location depends on a number of factors, such as the present and past climate, the slope of the land, and the porosity and permeability of the sediments beneath the surface. Generally sand and gravel sediments, along with solid sandstone, have the best porosity and permeability for transmitting groundwater. Other solid

Earth's Water

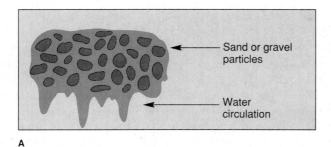

Sand or gravel particles

Water circulation

A

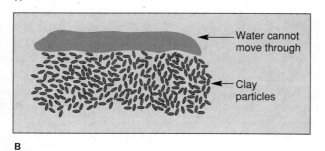

Water cannot move through

Clay particles

B

Figure 25.6

(*a*) Sand and gravel have large, irregular particles with large pore spaces, so they have a high porosity. Water can move from one pore space to the next, so they also have a high permeability. (*b*) Clay has small, flat particles, so it has a low porosity and is practically impermeable because water cannot move from one pore to the next.

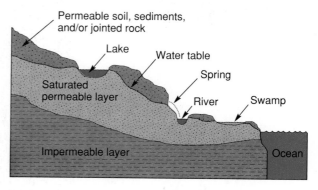

Permeable soil, sediments, and/or jointed rock

Lake

Water table

Spring

Saturated permeable layer

River

Swamp

Impermeable layer

Ocean

Figure 25.7

Groundwater from below the water table seeps into lakes, streams, and swamps and returns to the surface naturally at a spring. Groundwater eventually returns to the ocean, but the trip may take hundreds of years.

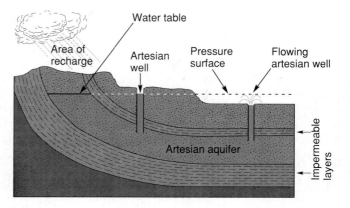

Water table

Area of recharge

Artesian well

Pressure surface

Flowing artesian well

Artesian aquifer

Impermeable layers

Figure 25.8

An artesian aquifer has groundwater that is under pressure from the water at higher elevations. The pressure will cause the water to rise above the water table in a well drilled into the aquifer, becoming a flowing well if the pressure is sufficiently high.

rocks, such as granite, can also transmit groundwater if they are sufficiently fractured by joints and cracks. In any case, groundwater will percolate downward until it reaches an area where pressure and other conditions have eliminated all pores, cracks, and joints. Above this downward limit it collects in all available spaces to form a **zone of saturation.** Water from the zone of saturation is considered to be groundwater. Water from the zone above is not considered to be groundwater. The surface of the boundary between the zone of saturation and the zone above is called the **water table.** The surface of a water table is not necessarily horizontal, but it tends to follow the topography of the surface in a humid climate. A hole that is dug or drilled through the earth to the water table is called a well. The part of the well that is below the water table will fill with groundwater, and the surface of the water in the well is generally at the same level as the water table.

Precipitation falls on the land and percolates down to the zone of saturation, then begins to move laterally, or sideways, to lower and lower elevations until it finds its way back to the surface. This surface outflowing could take place at a stream, pond, lake, swamp, or spring (figure 25.7). Groundwater flows gradually and very slowly through the tiny pore spaces, moving at a rate that ranges from kilometers (miles) per day to meters (feet) per year. Surface streams, on the other hand, move much faster at rates up to about 30 km per hour (about 20 mi/hr).

An **aquifer** is a layer of sand, gravel, sandstone, or other highly permeable material beneath the surface that is capable of producing water. In some places an aquifer carries water from a higher elevation, resulting in a pressure on water trapped by

impermeable layers at lower elevations. Groundwater that is under such a confining pressure is in an **artesian** aquifer. "Artesian" refers to the pressure, and groundwater from an artesian well rises above the top of the aquifer but not necessarily to the surface. Some artesian wells are under sufficient pressure to produce a fountainlike flow or spring (figure 25.8). Some people call groundwater from any deep well "artesian water," which is technically incorrect.

Fresh Water as a Resource

Water is an essential resource, not only because it is required for life processes but also because of its role in a modern industrialized society. Water is used in the home for drinking, cooking, and cleaning, as a carrier to remove wastes, and for maintaining lawns and gardens. These domestic uses lead to an equivalent consumption of about 570 liters per person each day (about 150 gal/person/day), but this is only about 10 percent

Figure 25.9

The filtering beds of a city water treatment facility. Surface water contains more sediments, bacteria, and other suspended materials because it is active on the surface and is exposed to the atmosphere. This means that surface water must be filtered and treated when used as a domestic resource, a processing not required when groundwater is used as the resource.

of the total consumed. Average daily use of water in the United States amounts to some 5,700 liters per person each day (about 1,500 gal/person/day), or about enough water to fill a small swimming pool once a week. The bulk of the water is used by agriculture (about 40 percent), for the production of electricity (about 40 percent), and for industrial purposes (about 10 percent). These overall percentages of use vary from one region of the country to another, depending on (1) the relative proportions of industry, agriculture, and population, (2) the climate of the region, (3) the nature of the industrial or agricultural use, and (4) other variables. In an arid climate with a high proportion of farming and fruit growing, for example, up to two-thirds of the available water might be used for agriculture.

Most of the water supply is obtained from the surface water resources of streams, lakes, and reservoirs. Surface water contains more sediments, bacteria, and possible pollutants than groundwater because it is more active and is directly exposed to the atmosphere. This means that surface water requires filtering to remove suspended particles, treatment to kill bacteria, and sometimes processing to remove pollution. In spite of the additional processing and treatment costs, surface water is less costly as a resource than groundwater. Groundwater is naturally filtered as it moves through the pore spaces of an aquifer, so it is usually relatively free of suspended particles and bacteria. Thus the processing or treatment of groundwater is usually not necessary (figure 25.9). But groundwater, on the other

Figure 25.10

This is groundwater pumped from the ground for irrigation. In some areas, groundwater is being removed from the ground faster than it is being replaced by precipitation, resulting in a water table that is falling deeper and deeper. It is thus possible that the groundwater resource will soon become depleted in some areas.

hand, will cost more to use as a resource because it must be pumped to the surface. The energy required for this pumping can be very expensive. In addition, groundwater generally contains more dissolved minerals (hard water), which may require additional processing or chemical treatment to remove the troublesome minerals.

The use of surface water as a source of fresh water means that the supply depends on precipitation. When a drought occurs, low river and lake resources may require curtailing water consumption. The curtailing of consumption occurs more often when a drought lasts for a longer period of time and when smaller lakes and reservoirs make the supply sensitive to rainfall amounts. In some parts of the western United States, such as the Colorado River watershed, *all* of the surface water is already being used, with certain percentages allotted for domestic, industrial, and irrigation uses. Groundwater is also used in this watershed, and in some locations it is being pumped from the ground faster than it is being replenished by precipitation (figure 25.10). As the population grows and new industries develop, more and more demands are placed on the surface water supply, which has already been committed to other uses, and on the diminishing supply of groundwater. This raises some very controversial issues about how fresh water should be divided among agriculture, industries, and city domestic use. Agricultural interests claim they should have the water because they produce the food and fibers that people must have. Industrial interests claim they should have the water because they create the jobs and the products that people must have. Cities, on the other hand, claim that domestic consumption is the most important because people cannot survive without water. Who should have the first priority for water use in such cases?

Some have suggested that people should not try to live and grow food in areas that have a short water supply, that plenty of fresh water is available elsewhere. Others have suggested that humans have historically moved rivers and reshaped the land to obtain water, so perhaps one answer to the problem is to find new sources of fresh water. Possible sources include the recycling of waste water and turning to the largest supply of water in the world, the ocean. About 90 percent of the water used by industries is presently dumped as a waste product. In some areas city waste water is already being recycled for use in power plants and for watering parks. A practically limitless supply of fresh water could be available by desalting ocean water, something which occurs naturally in the hydrologic cycle. The ocean, and the nature of seawater, is the topic of the following section. The treatment of seawater to obtain a new supply of fresh water is presently too expensive because of the cost of energy to accomplish the task. New technologies, perhaps ones that use solar energy, may make this more practical in the future. In the meantime, the best sources of extending the supply of fresh water appear to be the control of pollution, the recycling of waste water, and conservation of the existing supply.

Seawater

More than 70 percent of the surface of the earth is covered by seawater, with an average depth of 3,800 m (about 12,500 ft). The land areas cover 30 percent, less than a third of the surface, with an average elevation of only about 830 m (about 2,700 ft). With this comparison, you can see that humans live on and fulfill most of their needs by drawing from a small part of the total earth. As populations continue to grow and as resources of the land continue to diminish, the ocean will be looked at more as a resource rather than a convenient place for dumping wastes. The ocean already provides some food and is a source of some minerals, but it can possibly provide fresh water, new sources of food, new sources of important minerals, and new energy sources in the future. There are vast deposits of phosphorite and manganese nodules on the ocean bottom, for example, that can provide valuable minerals. Phosphate is an important fertilizer needed in agriculture, and the land supplies are becoming depleted. Manganese nodules, which occur in great abundance on the ocean bottom, can be a source of manganese, iron, copper, cobalt, and nickel. Seawater contains enough deuterium to make it a feasible source of energy. One gallon of seawater contains about a spoonful of deuterium, with the energy equivalent of three hundred gallons of gasoline. It has been estimated there is sufficient deuterium in the oceans to supply power at one hundred times the present consumption for the next ten billion years. The development of controlled nuclear fusion is needed,

Figure 25.11
These watermen are using hydraulic tongs to gather oysters from the Chesapeake Bay. In the future, the ocean may provide new sources of food through extensive farming of the sea, a practice called aquaculture.

however, to utilize this potential energy source. The sea may provide new sources of food through *aquaculture,* the farming of the sea the way that the land is presently farmed. Some aquaculture projects have already started with the farming of oysters, clams, and certain fishes, but these projects have barely begun to utilize the full resources that are possible (figure 25.11).

Part of the problem of utilizing the ocean is that the ocean has remained mostly unexplored and a mystery until recent times. Only now are scientists beginning to understand the complex patterns of the circulation of ocean waters, the nature of the chemical processes at work in the ocean, and the interactions of the ocean and the atmosphere and to chart the topography of the ocean floor. The following section will briefly describe some of these findings about the nature of seawater, how the oceans move, and what lies beneath the vast, watery surface of the earth.

Oceans and Seas

The vast body of salt water that covers more than 70 percent of the earth's surface is usually called the *ocean* or the *sea.* Although there is really only one big ocean on the earth, specific regions have been given names for convenience in describing locations. For this purpose, three principal regions are recognized, the (1) Atlantic Ocean, (2) Indian Ocean, and (3) Pacific Ocean. As shown in figure 25.12, these are not separate, independent bodies of salt water but are actually different parts of earth's single, continuous ocean. In general, the **ocean** is a single, continuous body of salt water on the surface of the earth. Specific regions (Atlantic, Indian, and Pacific) are often subdivided further into North Atlantic Ocean, South Atlantic Ocean, and so on.

Earth's Water

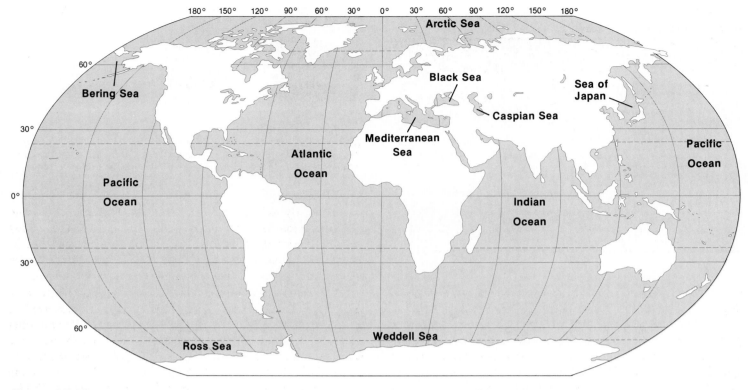

Figure 25.12

Distribution of the oceans and major seas on the earth's surface. There is really only one ocean; for example, where is the boundary between the Pacific, Atlantic, and Indian oceans in the southern hemisphere?

A **sea** is usually a smaller part of the ocean, a region with different characteristics that distinguish it from the larger ocean of which it is a part. However, the term "sea" is also used in the name of certain inland bodies of salty water, but not always. The ancient term "Seven Seas" is today synonymous with "ocean," meaning all parts of earth's huge body of salt water.

The Pacific Ocean is the largest of the three principal ocean regions. It has the largest surface area, covering 180 million km² (about 70 million mi²), and has the greatest average depth of 3.9 km (about 2.4 mi). The Pacific is circled by active converging plate boundaries, so it is sometimes described as being circled by a "rim of fire." It is called this because of the volcanoes associated with the converging plates. The "rim" also has the other associated features of converging plate boundaries such as oceanic trenches, island arcs, and earthquakes. The Atlantic Ocean is second in size, with a surface area of 107 million km² (about 41 million mi²), and the shallowest average depth of only 3.3 km (about 2.1 mi). The Atlantic Ocean is bounded by nearly parallel continental margins with a diverging plate boundary between. It lacks the trench and island arc features of the Pacific, but it does have islands, such as the Azores and Iceland, that are a part of the Mid-Atlantic Ridge at the plate boundary. The shallow seas of the Atlantic, such as the Mediterranean, Caribbean, and Gulf of Mexico, contribute to the shallow average depth of the Atlantic. The Indian Ocean has the smallest surface area, with 74 million km² (about 29 million mi²), and an average depth of 3.8 km (about 2.4 mi).

As mentioned earlier, a "sea" is usually a part of an ocean that is identified because some characteristic sets it apart. For example, the Mediterranean, Gulf of Mexico, and Caribbean seas are bounded by land, and they are located in a warm, dry climate. Evaporation of seawater is greater than usual at these locations, which results in the seawater being saltier. Being bounded by land and having saltier seawater characterizes these locations as being different from the rest of the Atlantic. The Sargasso Sea, on the other hand, is a part of the Atlantic that is not bounded by land and has a normal concentration of sea salts. This sea is characterized by having an abundance of floating brown seaweeds that accumulate in this region because of the global wind and ocean current patterns. The Arctic Sea, which is also sometimes called the Arctic Ocean, is a part of the North Atlantic Ocean that is less salty. Thus the terms "ocean" and "sea" are really arbitrary terms that are used to describe different parts of the earth's one continuous ocean.

The Nature of Seawater

According to one theory, the ocean is an ancient feature of the earth's surface, forming at least three billion years ago as the earth cooled from its early molten state. The seawater, and much

Table 25.1

Major dissolved materials in seawater

Ion	Percent (by weight)
Chloride (Cl^-)	55.05
Sodium (Na^+)	30.61
Sulfate (SO_4^{-2})	7.68
Magnesium (Mg^{+2})	3.69
Calcium (Ca^{+2})	1.16
Potassium (K^+)	1.10
Bicarbonate (HCO_3^-)	0.41
Bromine (Br^-)	0.19
Total	99.89

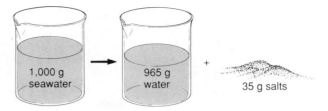

Figure 25.13

Salinity is defined as the mass of salts dissolved in 1.0 kg of seawater. Thus, if a sample of seawater has a salinity of 35‰, a 1,000 g sample would evaporate 965 g of water and leave 35 g of sea salts behind.

of the dissolved materials, is believed to have formed from the degassing of water vapor and other gases from molten rock materials. The degassed water vapor soon condensed, and over a period of time it began collecting as a liquid in the depression of the early ocean basin. Ever since, seawater has continuously cycled through the hydrologic cycle, returning water to the ocean through the world's rivers. For millions of years, these rivers have carried large amounts of suspended and dissolved materials to the ocean. These dissolved materials, including salts, stay behind in the seawater as the water again evaporates, condenses, falls on the land, and then brings more dissolved materials much like a continuous conveyor belt.

You might wonder why the ocean basin has not become filled in by the continuous supply of sediments and dissolved materials that would accumulate over millions of years. The basin has not filled in because (1) accumulated sediments have been recycled to the earth's interior through plate tectonics and (2) dissolved materials are removed by natural processes just as fast as they are supplied by the rivers. Some of the dissolved materials, such as calcium and silicon, are removed by plants and animals to make solid shells, bones, and other hard parts. Other dissolved materials, such as iron, magnesium, and phosphorous, form solid deposits directly and also make sediments that settle to the ocean floor. Hard parts of plants and animals and solid deposits are cycled to the earth's interior along with suspended sediments that have settled out of the seawater. Studies of fossils and rocks indicate that the composition of seawater has changed little over the past 600 million years.

The dissolved materials of seawater are present in the form of ions because of the strong dissolving ability of water molecules. Almost all of the chemical elements are present, but only six ions make up more than 99 percent of any given sample of seawater. As shown in table 25.1, chlorine and sodium are the most abundant ions. These are the elements of sodium chloride, or common table salt. As a sample of seawater evaporates the positive metal ions join with the different negative ions to form a complex mixture of ionic compounds known as *sea salt*. Sea

salt is mostly sodium chloride, but it also contains salts of the four metal ions (sodium, magnesium, calcium, and potassium) combined with the different negative ions of chlorine, sulfate, bicarbonate, and so on. Note that this mixture also includes magnesium sulfate, a strong laxative known as Epsom salt.

The amount of dissolved salts in seawater is measured as **salinity.** Salinity is defined as the mass of salts dissolved in 1.0 kg, or 1,000 g of seawater. Since the salt content is reported in parts per thousand, the symbol ‰ is used (% means parts per hundred). Thus 35‰ means that 1,000 g of seawater contains 35 g of dissolved salts (and 965 g of water). This is the same concentration as a 3.5 percent salt solution (figure 25.13). Oceanographers use the salinity measure because the mass of a sample of seawater does not change with changes in the water temperature. Other measures of concentration are based on the volume of a sample, and the volume of a liquid does vary as it expands and contracts with changes in the temperature. Thus by using the salinity measure any corrections due to temperature differences are eliminated.

The average salinity of seawater is about 35‰, but the concentration varies from a low of about 32‰ in some locations up to a high of about 36‰ in other locations. The salinity of seawater in a given location is affected by factors that tend to increase or decrease the concentration. The concentration is increased by two factors, evaporation and the formation of sea ice. Evaporation increases the concentration because it is water vapor only that evaporates, leaving the dissolved salts behind in a greater concentration. Ice that forms from freezing seawater increases the concentration because when ice forms the salts are excluded from the crystal structure. Thus sea ice is fresh water and the removal of this water leaves the dissolved salts behind in a greater concentration. The salinity of seawater is decreased by three factors, heavy precipitation, the melting of ice, and the addition of fresh water by a large river. All three of these factors tend to dilute seawater with fresh water, which lowers the concentration of salts.

Note that increases or decreases in the salinity of seawater are brought about by the addition or removal of fresh water. This changes only the amount of water present in the solution. The *kind* or *proportion* of the ions present (table 25.1)

Figure 25.14
Air will dissolve in water, and cooler water will dissolve more air than warmer water. The bubbles you see here are bubbles of carbon dioxide that came out of solution as the soda became warmer.

would expect, however, as seawater contains a large amount of this gas. More carbon dioxide can dissolve in seawater because it reacts with water to form carbonic acid, H_2CO_3, the same acid that is found in a bubbly cola. In seawater, carbonic acid breaks down into bicarbonate and carbonate ions, which tend to remain in solution. Water temperature and the salinity have an influence on how much gas can be dissolved in seawater, and increasing either, or both, will reduce the amount of gases that can be dissolved. Cold, lower salinity seawater in colder regions will dissolve more gases than the warm, higher salinity seawater in tropical locations. Abundant plant life in the upper, sunlit water tends to reduce the concentration of carbon dioxide and increase the concentration of dissolved oxygen through the process of photosynthesis. With increasing depth, less light penetrates the water, and below about 80 m (about 260 ft) there is insufficient light for photosynthesis. Thus more plant life and more dissolved oxygen are found above this depth. Below this depth there are no plants, more dissolved carbon dioxide, and less dissolved oxygen. The oxygen-poor, deep ocean water does eventually circulate back to the surface, but the complete process may take several thousand years.

Movement of Seawater

Consider the enormity of the earth's ocean with a surface area of some 361 million km² (about 139 million mi²) and a volume of 1,370 million km³ (about 328 million mi³) of seawater. There must be a terrific amount of stirring in such an enormous amount of seawater to produce the well-mixed, uniform chemical composition that is found in seawater throughout the world. The amount of mixing required is more easily imagined if you consider the long history of the ocean, the very long period of time over which the mixing has occurred. Based on investigations of the movement of seawater, it has been estimated that there is a complete mixing of all the earth's seawater about every 2,000 years or so. With an assumed age of 3 billion years, this means that the earth's seawater has been mixed 3,000,000,000 ÷ 2,000, or 1.5 million times. With this much mixing, you would be surprised if seawater were *not* identical all around the earth.

How does seawater move to accomplish such a complete mixing? Seawater is in a constant state of motion, both on the surface and deep within. The surface has two types of motion: (1) *waves,* which have been produced by some disturbance, such as the wind, and (2) *currents,* which move water from one place to another. Waves travel across the surface as a series of wrinkles that range from a few centimeters high to more than 30 m (100 ft) high. Waves crash on the shore as booming breakers and make the surf. This produces local currents as water moves along the shore and back out to sea. There are also permanent, worldwide currents that move ten thousand times more water across the ocean than all the water moving in all the large rivers on the land. Beneath the surface there are currents that move water up in some places and move it down in other places. Finally, there are enormous deep ocean currents that move tremendous volumes of seawater. The overall movement of many of the currents on the surface and their relationship to the deep

in seawater does not change with increased or decreased amounts of fresh water. The same proportion, meaning the same chemical composition, is found in seawater of any salinity of any sample taken from any location anywhere in the world, from any depth of the ocean, or taken anytime of the year. Seawater has a remarkably uniform composition that varies only in concentration. This means that the ocean is well mixed and thoroughly stirred around the entire earth. How seawater becomes so well mixed and stirred on a worldwide basis is discussed in the next section.

If you have ever allowed a glass of tap water to stand for a period of time, you may have noticed tiny bubbles collecting as the water warms. These bubbles are atmospheric gases, such as nitrogen and oxygen, that were dissolved in the water (figure 25.14). Seawater also contains dissolved gases in addition to the dissolved salts. Near the surface, seawater contains mostly nitrogen and oxygen, in similar proportions to the mixture that is found in the atmosphere. There is more carbon dioxide than you

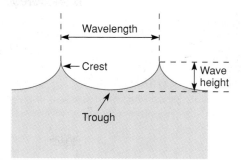

Figure 25.16

The simplest form of ocean waves showing some basic characteristics. Most waves do not look like this representation because most are complicated mixtures of superimposed waves with a wide range of sizes and speeds.

Figure 25.15

The surface of the ocean is rarely, if ever, still. These waves were formed by a local wind. In the open, such waves could end up as swell that travels all the way to the other side of the ocean.

ocean currents are not yet fully mapped or understood. The surface waves are better understood. The general trend and cause of permanent, worldwide currents in the ocean can also be explained. The following is a brief description and explanation of waves, currents, and the deep ocean movements of seawater.

Waves

Any slight disturbance will create ripples that move across a water surface. For example, if you gently blow on the surface of water in a glass you will see a regular succession of small ripples moving across the surface. These ripples, which look like small, moving wrinkles, are produced by the friction of the air moving across the water surface. The surface of the ocean is much larger, but a gentle wind produces patches of ripples in a similar way. These patches appear, then disappear as the wind begins to blow over calm water. If the wind continues to blow, larger and longer-lasting ripples are made, and the moving air can now push directly on the side of the ripples. A ripple may

eventually grow into an **ocean wave,** a moving disturbance that travels across the surface of the ocean. In its simplest form, each wave has a ridge, or mound, of water called a **crest,** which is followed by a depression called a **trough.** Ocean waves are basically repeating series of these crests and troughs that move across the surface like wrinkles (figure 25.15).

The simplest form of an ocean wave can be described by measurements of three distinct characteristics: (1) the **wave height,** which is the vertical distance between the top of a crest and the bottom of the next trough, (2) the **wavelength,** which is the horizontal distance between two successive crests (or other successive parts of the wave), and (3) the **wave period,** which is the time required for two successive crests (or other successive parts) of the wave to pass a given point (figure 25.16).

The characteristics of an ocean wave formed by the wind depends on three factors: (1) the wind speed, (2) the length of time that the wind blows, and (3) the *fetch,* which is the distance the wind blows across the open ocean. As you can imagine, larger waves are produced by strong winds that blow for a longer time over a long fetch. In general, longer-blowing, stronger winds produce waves with greater wave heights, longer wavelengths, and longer periods, but a given wind produces waves with a wide range of sizes and speeds. In addition, the wind does not blow in just one direction, and shifting winds produce a chaotic pattern of waves of many different heights and wavelengths. Thus the surface of the ocean in the area of a storm or strong wind has a complicated mixture of many sizes and speeds of superimposed waves. The smaller waves soon die out from friction within the water, and the larger ones grow as the wind pushes against the crest. Ocean waves range in height from a few centimeters up to more than 30 m (about 100 ft), but giant waves more than 15 m (about 50 ft) are extremely rare.

The larger waves of the chaotic, superimposed mixture of waves in a storm area last longer than the winds that formed them, and they may travel for hundreds or thousands of kilometers from their place of origin. The longer wavelength waves travel faster and last longer than the shorter wavelength waves,

Earth's Water

so the longer wavelength waves tend to outrun the shorter wave-length waves as they die out from energy losses to water friction. Thus the irregular, superimposed waves created in the area of a storm become transformed as they travel away from the area. They become regular groups of low-profile, long wavelength waves that are called **swell.** The regular waves of swell that you might observe near a shore may have been produced by a storm that occurred days before, thousands of kilometers across the ocean.

The regular, low-profile crests and troughs of swell carry energy across the ocean, but they do not transport water across the open ocean. If you have ever been in a boat that is floating in swell, you know that you move in a regular pattern of up and forward on each crest, then backward and down on the following trough. The boat does not move along with the waves unless it is moved along by a wind or by some current. Likewise, a particle of water on the surface moves upward and forward with each wave crest, then backward and down on the following trough, tracing out a nearly circular path through this motion. The particle returns to its initial position, without any forward movement while tracing out the small circle. Note that the diameter of the circular path is equal to the wave height (figure 25.17). Water particles farther below the surface also trace out circular paths as a wave passes. The diameters of these circular paths below the surface are progressively smaller with increasing depth. Below a depth equal to about half the wavelength there is no circular movement of the particles. Thus you can tell how deeply the passage of a wave disturbs the water below if you measure the wavelength.

As swell moves from the deep ocean to the shore the waves pass over shallower and shallower water depths. When a depth is reached that is equal to about half the wavelength, the circular motion of the water particles begin to reach the ocean bottom. The water particles now move across the ocean bottom, and the friction between the two results in the waves moving slower as the wave height increases. These important modifications result in a change in the direction of travel and in an increasingly unstable situation as the wave height increases.

Most waves move toward the shore at some angle. As the wave crest nearest the shore starts to slow, the part still over deep water continues on at the same velocity. The slowing at the shoreward side *refracts,* or bends, the wave so it is more parallel to the shore. Thus waves always appear to approach the shore head-on, arriving at the same time on all parts of the shore.

After the waves reach water that is less than one-half the wavelength, friction between the bottom and the circular motion of the water particles progressively slow the bottom part of the wave. The wave front becomes steeper and steeper as the top overruns the bottom part of the wave. When the wave front becomes too steep the top part breaks forward and the wave is now called a **breaker.** In general, this occurs where the water depth is about one and one-third times the wave height. The zone where the breakers occur is called **surf** (figure 25.18).

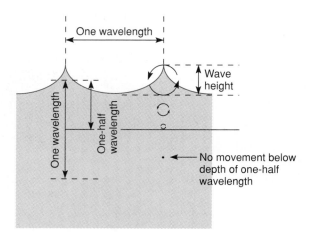

Figure 25.17

Water particles are moved in a circular motion by a wave passing in the open ocean. On the surface, a water particle traces out a circle with a diameter that is equal to the wave height. The diameters of the circles traced out by water particles decrease with depth to a depth that is equal to one-half the wavelength of the ocean wave.

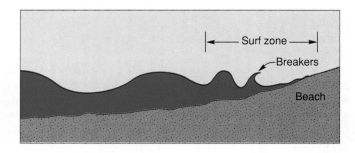

Figure 25.18

As a pattern of swell approaches a gently sloping beach, friction between the circular motion of the water particles and the bottom slows the wave and the wave front becomes steeper and steeper. When the depth is about one and one-third times the wave height, the wave breaks forward, moving water toward the beach.

Waves break in the foamy surf, sometimes forming smaller waves that then proceed to break in progressively shallower water. The surf may have several sets of breakers before the water is finally thrown on the shore as a surging sheet of seawater. The turbulence of the breakers in the surf zone and the final surge expend all the energy that the waves may have brought from thousands of kilometers away. Some of the energy does work in eroding the shoreline, breaking up rock masses into the sands that are carried by local currents back to the ocean. The rest of the energy goes into the kinetic energy of water molecules, which appears as a temperature increase.

Swell does not transport water with the waves over a distance, but small volumes of water are moved as a growing wave is pushed to greater heights by the wind over the open ocean.

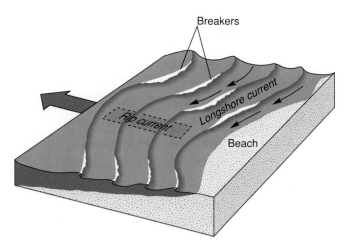

Figure 25.19
Breakers result in a buildup of water along the beach that moves as a longshore current. Where it finds a shore bottom that allows it to return to the sea, it surges out in a strong flow called a rip current.

A strong wind can topple such a wave on the open ocean, producing a foam-topped wave known as a *whitecap.* In general, whitecaps form when the wind is blowing at 30 km/hr (about 20 mi/hr) or more.

Waves do transport water where breakers occur in the surf zone. When a wave breaks, it tosses water toward the shore where the water begins to accumulate. This buildup of water tends to move away in currents, or streams, as the water returns to a lower level. Some of the water might return directly to the sea by moving beneath the breakers. This direct return of water forms a weak current known as **undertow.** Other parts of the accumulated water might be pushed along by the waves, producing a **longshore current** that moves parallel to the shore in the surf zone. This current moves parallel to the shore until it finds a lower place or a channel that is deeper than the adjacent bottom. Where the current finds such a channel, it produces a **rip current,** a strong stream of water that bursts out against the waves and returns water through the surf to the sea (figure 25.19). The rip current usually extends beyond the surf zone, then diminishes. A rip current, or where rip currents are occurring, can usually be located by looking for the combination of (1) a lack of surf, (2) darker looking water, which means a deeper channel, and (3) a turbid, or muddy, streak of water that extends seaward from the channel indicated by the darker water that lacks surf.

In addition to waves created by winds, important waves are created by earthquakes (see chapter 22) and by tides (see chapter 20). Movement of the earth's crust or underwater landslides can produce giant destructive waves called **tsunamis.** Tsunamis are sometimes incorrectly called "tidal waves" (they are in no way associated with tides or tide-making processes). Tsunamis can be the largest of all ocean waves, but their tremendous, destructive energy is often not apparent until they

reach the shore. A tsunami traveling across the deep ocean can have wavelengths up to 200 km (about 120 mi) and wave heights up to 0.5 m (about 1.5 ft) while traveling in an impulse-produced series of five to ten or so waves. Because of the small wave height, a tsunami often moves unnoticed across the deep ocean. But the tsunami forms taller waves as it slows in the shallow water near the shore, waves that can reach over 8 m (about 25 ft) tall. Waves of this size can result in tremendous property damage and loss of life along low coastal areas, especially if the waves make giant breakers. Many low coastal areas of Japan, Alaska, Hawaii, Chile, and the Philippines have experienced the destructive effects of tsunamis.

Tides are basically produced by the gravitational pull of the moon and the rotation of the earth-moon system as discussed in chapter 20. The basic pattern of two high tides and two low tides every twenty-four hours and fifty minutes can produce strong **tidal currents** in narrow bays. These are reversing currents, moving up the bay with the rising tide and out to sea with the falling tide. If the bay is long and narrow, such as some in Alaska and Nova Scotia, a **tidal bore** may be produced. A tidal bore is a wave that moves rapidly up the bay as the tide rises. Sometimes the tidal bore is easily observed, and when the conditions are right, the bore is large enough to carry a person on a surfboard for a great distance.

Ocean Currents

Waves generated by the winds, earthquakes, and tidal forces keep the surface of the ocean in a state of constant motion. Local, temporary currents associated with this motion, such as rip currents or tidal currents, move seawater over a short distance. Seawater also moves in continuous **ocean currents,** streams of water that stay in about the same path as they move through other seawater over large distances. Ocean currents can be difficult to observe directly since they are surrounded by water that looks just like the water in the current. Wind is likewise difficult to observe directly since the moving air looks just like the rest of the atmosphere. Unlike the wind, an ocean current moves *continuously* in about the same path, often carrying water with different chemical and physical properties than the water it is moving through. Thus an ocean current can be identified and tracked by measuring the physical and chemical characteristics of the current and the surrounding water. This shows where the current is coming from and where in the world it is going. In general, ocean currents are produced by (1) density differences in seawater and (2) winds that blow persistently in the same direction.

Density Currents. The density of seawater is influenced by three factors: (1) the water temperature, (2) salinity, and (3) suspended sediments. Cold water is generally more dense than warm water, thus sinking and displacing warmer water. Seawater of a high salinity has a higher relative density than less salty water, so it sinks and displaces water of less salinity. Likewise, seawater with a larger amount of suspended sediments has a higher relative density than clear water, so it sinks

Earth's Water

and displaces clear water. The following describes how these three ways of changing the density of seawater result in the ocean current known as a **density current,** which is an ocean current that flows because of density differences.

The earth receives more incoming solar radiation in the tropics than it does at the poles, which establishes a temperature difference between the tropical and polar ocean. The surface water in the polar ocean is often at or below the freezing point of fresh water, while the surface water in the tropical ocean averages about 26° C (about 79° F). Seawater freezes at a temperature below that of fresh water because the salt content lowers the freezing point. Seawater does not have a set freezing point, however, because as it freezes the salinity is increased as salt is excluded from the ice structure. Increased salinity lowers the freezing point more, so the more ice that freezes from seawater, the lower the freezing point for the remaining seawater. Cold seawater near the poles is therefore the densest, sinking and creeping slowly as a current across the ocean floor toward the equator. Where and how such a cold, dense bottom current moves is influenced by the shape of the ocean floor, the rotation of the earth, and other factors. The size and the distance that cold bottom currents move can be a surprise. Cold, dense water from the arctic, for example, moves in a 200 m (about 660 ft) diameter current on the ocean bottom between Greenland and Iceland. This current carries an estimated 5 million cubic meters per second (about 177 million cubic ft/sec) of seawater to the 3.5 km (about 2.1 mi) deep water of the North Atlantic Ocean. This is a flow rate about 250 times larger than that of the Mississippi River. At about 30° N, the cold arctic waters meet even denser water that has moved in currents all the way from the antarctic to the deepest part of the North Atlantic Basin (figure 25.20).

A second type of density current results because of differences in salinity. The water in the Mediterranean, for example, has a high salinity because it is mostly surrounded by land in a warm, dry climate. The Mediterranean seawater, with its higher salinity, is more dense than the seawater in the open Atlantic Ocean. This density difference results in two separate currents that flow in opposite directions between the Mediterranean and the Atlantic. The greater density seawater flows from the bottom of the Mediterranean into the Atlantic, while the less dense Atlantic water flows into the Mediterranean near the surface. The dense Mediterranean seawater sinks to a depth of about 1,000 m (about 3,300 ft) in the Atlantic, where it spreads over a large part of the North Atlantic Ocean. This increases the salinity of this part of the ocean, making it one of the more saline areas in the world.

The third type of density current occurs when underwater sediments on a slope slide toward the ocean bottom, producing a current of muddy or turbid water called a **turbidity current.** Turbidity currents are believed to be a major mechanism that moves sediments from the continents to the ocean basin. They

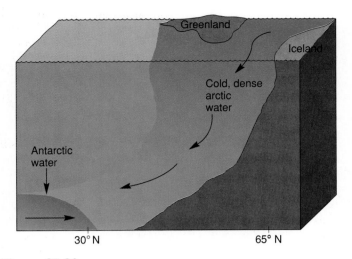

Figure 25.20

A cold-density current carries about 250 times more water than the Mississippi River from the arctic and between Greenland and Iceland to the deep Atlantic Ocean. At about 30° N latitude, it meets water that has moved by cold-density currents all the way from the antarctic.

may also be responsible for some undersea features, such as submarine canyons. Turbidity currents are believed to occur only occasionally, however, and none has ever been directly observed or studied. There is thus no data or direct evidence of how they form or what effects they have on the ocean floor.

Surface Currents. There are broad and deep-running ocean currents that slowly move tremendous volumes of water relatively near the surface. As shown in figure 25.21, each current is actually part of a worldwide system, or circuit, of currents. This system of ocean currents is very similar to the worldwide system of prevailing winds (see chapter 23). This similarity exists because it is the friction of the prevailing winds on the seawater surface that drives the ocean currents. The currents are modified by other factors, such as the rotation of the earth and the shape of the ocean basins, but they are basically maintained by the wind systems.

Each ocean has a great system of moving water called a **gyre** that is centered in the mid-latitudes. The gyres rotate to the right in the Northern Hemisphere and to the left in the Southern Hemisphere because of the Coriolis effect (see chapter 20). The movement of water around these systems, or gyres, plus some smaller systems, form the surface circulation system of the world ocean. Each part of the system has a separate name, usually based on its direction of flow. All are called "currents" except one that is called a "stream" (the Gulf Stream) and those that are called "drifts." Both the Gulf Stream and the drifts are currents that are part of the connected system.

The major surface currents are like giant rivers of seawater that move through the ocean near the surface. You know that all the currents are connected, for a giant river of water cannot just start moving in one place, then stop in another. The

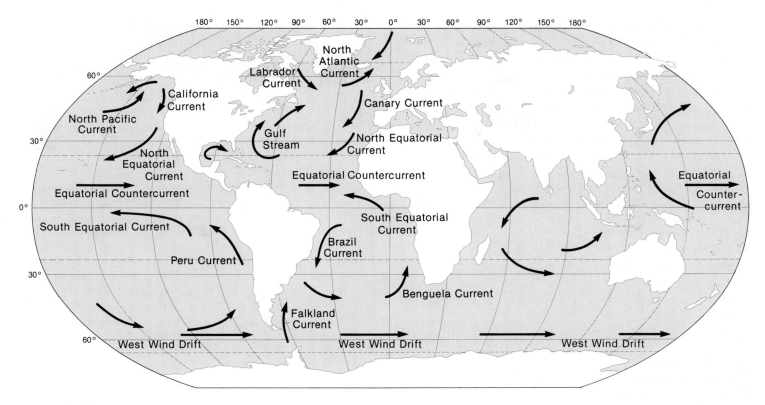

Figure 25.21
The earth's system of ocean currents.

Gulf Stream, for example, is a current about 100 km (about 60 mi) wide that may extend to a depth of 1 km (about 0.6 mi) below the surface, moving more that 75 million cubic meters of water per second (about 2.6 billion cubic ft/sec). The Gulf Stream carries more than 370 times more water than the Mississippi River. The California Current is weaker and broader, carrying cool water southward at a relatively slow rate. The flow rate of all the currents must be equal, however, since all the ocean basins are connected and the sea level is changing very little, if at all, over long periods of time.

The Ocean Floor

Some of the features of the ocean floor were discussed in chapter 21 because they were important in developing the theory of plate tectonics. Many features of the present ocean basins were created from the movement of large crustal plates, according to plate tectonics theory, and in fact some ocean basins are thought to have originated with the movement of these plates. There is also evidence that some features of the ocean floor were modified during the ice ages of the past. During an ice age, much water becomes locked up in glacial ice, which lowers the sea level. The sea level dropped as much as 140 m (about 460 ft) during the most recent major ice age, exposing the margins of the continents to erosion. Today, these continental margins are flooded with seawater, forming a zone of relatively shallow water

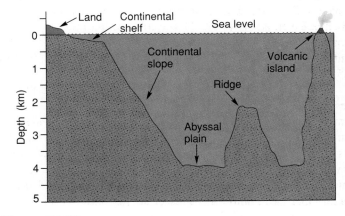

Figure 25.22
Some features of the ocean floor. Note that the inclination of the features is not as steep as this vertical exaggeration would suggest.

called the **continental shelf** (figure 25.22). The continental shelf is considered to be a part of the continent and not the ocean, even though it is covered with an average depth of about 130 m (about 425 ft) of seawater. The shelf slopes gently away from the shore for an average of 75 km (about 47 mi), but they are much wider on the edge of some parts of continents than other parts.

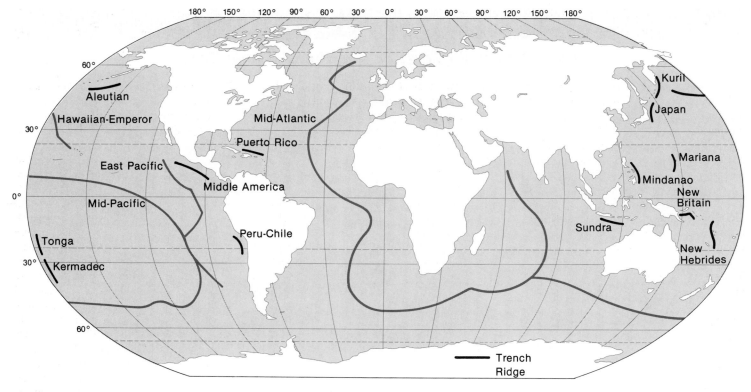

Figure 25.23
The location of some of the larger ridges and trenches of the earth's ocean floor.

The continental shelf is a part of the continent that happens to be flooded by seawater at the present time. It still retains some of the general features of the adjacent land that is above water, such as hills, valleys, and mountains, but these features were smoothed off by the eroding action of waves when the sea level was lower. Today, a thin layer of sediments from the adjacent land covers these smoothed-off features.

Beyond the gently sloping continental shelf is a steeper feature called the **continental slope.** The continental slope is the transition between the continent and the deep ocean basin. The water depth at the top of the continental slope is about 120 m (about 390 ft), then plunges to a depth of about 3,000 m (about 10,000 ft) or more. The continental slope is generally 20 to 40 km (about 12 to 25 mi) wide, so the inclination is similar to that encountered driving down a steep mountain road on an interstate highway. At various places around the world the continental slopes are cut by long, deep, and steep-sided **submarine canyons.** Some of these canyons extend from the top of the slope and down the slope to the ocean basin. Such a submarine canyon can be similar in size and depth to the Grand Canyon on the Colorado River of Arizona. Submarine canyons are believed to have been eroded by turbidity currents, which were discussed earlier.

Beyond the continental slope is the bottom of the ocean floor, the **ocean basin.** Ocean basins are the deepest part of the ocean, covered by about 4 to 6 km (about 2 to 4 mi) of seawater.

The basin is mostly a practically level plain called the **abyssal plain** and long, rugged mountain chains called **ridges** that rise thousands of meters above the abyssal plain. The Atlantic Ocean and Indian Ocean basins have ridges that trend north and south near the center of the basin. The Pacific Ocean basin has its ridge running north and south near the eastern edge. The Pacific Ocean basin also has more **trenches** than the Atlantic Ocean or the Indian Ocean basins (figure 25.23). A trench is a long, relatively narrow, steep-sided trough that occurs along the edges of the ocean basins. Trenches range in depth from about 8 to 11 km (about 5 to 7 mi) deep below sea level. The origin of ridges and trenches was discussed in chapter 21.

The ocean basin and ridges of the ocean cover more than half of the earth's surface, accounting for more of the total surface than all the land of the continents. The plain of the ocean basin alone, in fact, covers an area about equal to the area of the land. Scattered over the basin are more than ten thousand steep volcanic peaks called **seamounts.** By definition, seamounts rise more than 1 km (about 0.6 mi) above the ocean floor, sometimes higher than the sea level surface of the ocean. A seamount that sticks above the water level makes an island. The Hawaiian Islands are examples of such giant volcanoes that have formed islands. Most seamount-formed islands are in the Pacific Ocean. Most islands in the Atlantic, on the other hand, are the tops of volcanoes of the Mid-Atlantic Ridge.

Deep Ocean Exploration and the Dynamic Earth

According to the theory of plate tectonics, the earth's surface is made up of a dozen or so rigid plates that are in motion, moving about 5 to 15 cm (about 2 to 6 in) per year with respect to one another. The plates are moving apart at the mid-ocean ridges where molten rock moves up from the mantle below, cools, and forms new crust. The size of the earth does not increase from the addition of the new crust, however, as old crust is incorporated back into the mantle at subduction zones. The subduction zone is where one plate is being forced under another plate. The spreading and the subduction boundaries have associated volcanic activity, faulting, and earthquakes, but the type of geologic activity associated with each boundary is very different.

The continents are generally viewed as less dense materials that "float" higher than the more dense materials making up the ocean crust. Thus continents are moved away on plates as new ocean crust forms at ridges, then moves steadily away from the ridges and eventually sinks in a subduction zone. The continental plates sometimes collide and merge with other continental plates in the process. In addition, continental plates have split apart with a new ocean basin forming between the parts. In fact, the geologic evidence indicates that the continents were originally all together in one huge supercontinent at least twice in the past. The original supercontinent split apart to form new ocean basins and continents, then moved together to close the ocean basins and form a supercontinent for a second time. The second supercontinent split into separate continents, then moved apart for the second time to form the continents and ocean basins of today.

More evidence indicates that the opening and closing of the Atlantic Ocean basin might be periodic, operating on a 500 million year cycle. Around the Pacific Ocean, on the other hand, there is no evidence of opening and closing. Another difference is observed in that the ocean crust of the Pacific is being subducted under the continents around it, while the ocean crust of the Atlantic generally moves with the continents. This could mean that the Pacific basin is the original, permanent feature and the Atlantic basin is a temporary feature that opens and closes with movements of the continents.

If there is a pattern to the observations and if there is regularity to the opening and closing of the Atlantic Ocean, and not the Pacific, it could have meaning about changes in the climate of the entire earth. Do such patterns of change exist, and if they do, what does it mean about the future of the earth? One of the problems of seeking answers to such questions has been that the mid-ocean ridges and the subduction zones are at the bottom of the ocean, far beyond the reach of scientists. Only since the 1970s have advances in technology opened the way for study and direct observation of this important part of the earth's crust.

The investigation of the features of the ocean floor was made possible by the development of highly sensitive but durable cameras, high-resolution sonar scanners, and sensing devices to measure the magnetization, electrical, and seismic wave properties of crustal rocks. In addition to these instruments, which could be lowered and towed underwater by a surface ship, deep-diving submarine vehicles were developed to take research scientists directly to the deepest parts of the ocean. Both the instrument packages and a submarine vehicle named *Alvin* were used to first study the Atlantic and Pacific ocean ridges in the 1970s. Some amazing discoveries were made during these expeditions.

The *Alvin* made several explorations in different locations of the ridge in the Pacific, in addition to an exploration of the Atlantic Ocean ridge and other studies. During the Pacific studies, which lowered three scientists to a depth of 2.6 km (about 1.6 mi), something unexpected was found on the ridge.

There were young flows of basalt lava, as was expected, but on the lava were mounds of chemical precipitates and organic debris. The mounds had funnel-shaped structures that were up to 10 m (about 33 ft) tall and 0.5 m (about 2 ft) wide. Hot, black fluids were billowing from these structures, vents that released water with temperatures up to 350° C (about 660° F) into the near-freezing seawater. The black color was from tiny particles of iron, copper, and zinc sulfides. Dense colonies of white crabs, large clams, and clusters of giant white tube worms with bright red plumes occupied an area about the size of a football field around the hydrothermal vents. These animals, previously unknown, were discovered in 1977 living in total darkness under a pressure that is 275 times greater than the pressure at the surface. Later studies found that these animals represented a whole new ecosystem, one based on sulfide-oxidizing bacteria rather than on green plants that utilize sunlight. The energy source for this ecosystem is chemical rather than sunlight. The hydrothermal vents released hot seawater, which seeped downward to the magma under the ridge, where it became superheated, dissolved sulfide minerals, then discharged back into the ocean. Such hydrothermal vents, and the sulfide-oxidizing ecosystems, were found in two separate locations some 3 km (about 2 mi) apart in the Pacific.

The mid-ocean ridge in the Pacific was found to be very different from the mid-ocean ridge in the Atlantic, indicating that perhaps the Pacific and the Atlantic ridges are created by different mechanisms. How or why these mechanisms are different is not yet known. Further deep ocean exploration will be needed to understand the processes that are involved in creating deep ocean ridges and what the differences might mean about the future of the earth. Who knows what other deep ocean surprises await the scientists who are trying to understand this complex earth and its cycles!

Summary

Precipitation that falls on the land either evaporates, flows across the surface, or soaks into the ground. Water moving across the surface is called *runoff*. Water that moves across the land as a small body of running water is called a *stream*. A stream drains an area of land known as the stream drainage basin or *watershed*. The watershed of one stream is separated from the watershed of another by a line called a *divide*. Water that collects as a small body of standing water is called a *pond*, and a larger body is called a *lake*. A *reservoir* is a natural pond, a natural lake, or a lake or pond created by building a dam for water management or control. The water of streams, ponds, lakes, and reservoirs is collectively called *surface water*.

Precipitation that soaks into the ground *percolates* downward until it reaches a *zone of saturation*. Water from the saturated zone is called *groundwater*. The amount of water that a material will hold depends on its *porosity*, and how well the water can move through the material depends on its *permeability*. The surface of the zone of saturation is called the *water table*.

The *ocean* is the single, continuous body of salt water on the surface of the earth. A *sea* is a smaller part of the ocean with different characteristics. The dissolved materials in seawater are mostly the ions of six substances, but sodium ions and chlorine ions are the most abundant. *Salinity* is a measure of the mass of salts dissolved in 1,000 grams of seawater.

An *ocean wave* is a moving disturbance that travels across the surface of the ocean. In its simplest form a wave has a ridge called a *crest* and a depression called a *trough*. Waves have a characteristic *wave height*, *wavelength*, and *wave period*. The characteristics of waves made by the wind depend on the wind *speed*, the *time* the wind blows, and the *fetch*. Regular groups of low-profile, long-wavelength waves are called *swell*. When swell approaches a shore, the wave slows and increases in wave height. This slowing *refracts*, or bends, the waves so they approach the shore head-on. When the wave height becomes too steep the top part breaks forward forming *breakers* in the *surf zone*. Water accumulates at the shore from the breakers and returns to the sea as *undertow*, as *longshore currents*, or in *rip currents*.

Earthquakes or undersea landslides produce large, destructive waves called *tsunamis*. Tsunamis do not have a large wave height at open sea, but they can do tremendous damage when they reach a low coastal area. A tide moving in or out of a narrow bay can produce a wave called a *tidal bore*.

Ocean currents are streams of water that move through other seawater over large distances. Some ocean currents are *density currents*, which are caused by differences in *water temperature, salinity,* or *suspended sediments*. Each ocean has a great system of moving water called a *gyre* that is centered in mid-latitudes. Different parts of a gyre are given different names such as the *Gulf Stream* or the *California Current*.

The ocean floor is made up of the *continental shelf,* the *continental slope,* and the *ocean basin*. The ocean basin has two main parts, the *abyssal plain* and mountain chains called *ridges*.

Key Terms

abyssal plain (p. 540)
aquifer (p. 528)
artesian (p. 528)
breaker (p. 536)

continental shelf (p. 539)
continental slope (p. 540)
crest (p. 535)
density current (p. 538)

divide (p. 527)
fresh water (p. 525)
groundwater (p. 527)
gyre (p. 538)
lake (p. 527)
longshore current (p. 537)
ocean (p. 531)
ocean basin (p. 540)
ocean currents (p. 537)
ocean wave (p. 535)
permeability (p. 527)
pond (p. 527)
porosity (p. 527)
reservoir (p. 527)
ridges (p. 540)
rip current (p. 537)
runoff (p. 526)
salinity (p. 533)
sea (p. 532)

seamounts (p. 540)
stream (p. 526)
submarine canyons (p. 540)
surf (p. 536)
swell (p. 536)
tidal bore (p. 537)
tidal currents (p. 537)
trenches (p. 540)
trough (p. 535)
tsunamis (p. 537)
turbidity current (p. 538)
undertow (p. 537)
watershed (p. 527)
water table (p. 528)
wave height (p. 535)
wavelength (p. 535)
wave period (p. 535)
zone of saturation (p. 528)

Applying the Concepts

1. In a region of abundant rainfall a layer of extensively cracked but otherwise solid granite could serve as a limited source of groundwater because it has
   a. limited permeability and no porosity.
   b. average porosity and average permeability.
   c. no permeability and no porosity.
   d. limited porosity and no permeability.

2. How many different oceans are actually on the earth's surface?
   a. fourteen
   b. seven
   c. three
   d. one

3. The Gulf of Mexico is a shallow sea of the
   a. Atlantic Ocean.
   b. Pacific Ocean.
   c. Indian Ocean.
   d. none of the above.

4. Measurement of the salts dissolved in seawater taken from various locations throughout the world show that seawater has a
   a. uniform chemical composition and a variable concentration.
   b. variable chemical composition and a variable concentration.
   c. uniform chemical composition and a uniform concentration.
   d. variable chemical composition and a uniform concentration.

5. The salinity of seawater is *increased* locally by
   a. the addition of water from a large river.
   b. heavy precipitation.
   c. the formation of sea ice.
   d. none of the above.

6. Considering only the available light and the dissolving ability of gases in seawater, more abundant life should be found in a
   a. cool, relatively shallow ocean.
   b. warm, very deep ocean.
   c. warm, relatively shallow ocean.
   d. cool, very deep ocean.

7. The regular, low-profile waves called swell are produced from
   a. constant, prevailing winds.
   b. small, irregular waves becoming superimposed.
   c. longer wavelengths outrunning and outlasting shorter wavelengths.
   d. all wavelengths becoming transformed by gravity as they travel any great distance.

8. If the wavelength of swell is 10.0 m, then you know that the fish below the surface feel the waves to a depth of
   a. 5.0 m.
   b. 10.0 m.
   c. 20.0 m.
   d. however deep it is to the bottom.

9. In general, a breaker forms where the water depth is about one and one-third times the wave
   a. period.
   b. length.
   c. height.
   d. width.

10. The greatest volume of water is moved by the
    a. Mississippi River.
    b. California Current.
    c. Gulf Stream.
    d. Colorado River.

Answers

1. a 2. d 3. a 4. a 5. c 6. a 7. c 8. a 9. c
10. c

Questions for Thought

1. How are the waters of the earth distributed as a solid, a liquid, and a gas at a given time? How much of the water is salt water and how much is fresh water?

2. Describe the hydrologic cycle. Why is the hydrologic cycle important in maintaining a supply of fresh water? Why is the hydrologic cycle called a cycle?

3. Describe in general all the things that happen to the water that falls on the land.

4. Explain how a stream can continue to flow even during a dry spell.

5. Explain the relationship between runoff, streams, and a watershed.

6. What is the continental divide? How do the Rocky Mountains act as the continental divide?

7. What determines if a particular zone of wet sediments is called groundwater or not?

8. For what purposes are reservoirs constructed? Can they be used for all purposes at the same time? Explain.

9. What determines how much groundwater a given sediment will hold? What determines how well groundwater will move through a given sediment?

10. What determines how deep groundwater will soak below the surface? What happens to the water above this lower limit?

11. What is the water table? What is the relationship between the depth to the water table and the depth that a well must be drilled? Explain.

12. What is an aquifer? What is special about an artesian aquifer?

13. Compare the advantages and disadvantages of using (a) surface water, and (b) groundwater as a source of fresh water.

14. Prepare arguments for (a) agriculture, (b) industries, and (c) cities each having first priority in the use of a limited water supply. Identify one of these arguments as being the "best case" for first priority, then justify your choice.

15. Discuss some possible ways of extending the supply of fresh water.

16. How many oceans are found on the surface of the earth? Explain how you arrived at this number.

17. What is the theoretical origin of all the ocean waters?

18. The world's rivers and streams carry millions of tons of dissolved materials to the ocean each year. Explain why this does not increase the salinity of the ocean.

19. Describe what makes seawater salty.

20. What is salinity? What processes tend to increase salinity? What processes tend to reduce salinity?

21. Explain how the salinity of seawater can vary from one location to another while the composition remains the same everywhere.

22. With all other factors the same, more abundant plant life is found in the upper, sunlit part of cold ocean water than in warm ocean water. Suggest a reason for this observation.

23. Explain why seawater does not have a fixed freezing point.

24. Describe three ways that the density of seawater is changed.

25. Is it possible for an ocean current to exist alone, not as a part of a larger system of circulation? Explain.

26. Compare the volume of water transported by ocean currents and the volume transported by the rivers and streams of the land.

27. Explain why the waves of swell start to refract when the water depth is equal to about one-half a wavelength.

28. What is the evidence that the world ocean is well mixed? Explain.

29. Describe how the wind forms ocean waves.

30. What is swell and how does it form?

31. Why do waves always seem to approach the shore head-on?

32. What is a tsunami? Why are they often not noticed on the open sea?

33. What factors determine the size of an ocean wave made by the wind?

34. Describe what is found beneath the vast, watery surface of the earth called the ocean.

35. Describe how a breaker forms from swell. What is surf?

36. Describe what you would look for to avoid where rip currents occur at a beach.

Appendix A

Mathematical Review

Working with Equations

Many of the problems of science involve an equation, a shorthand way of describing patterns and relationships that are observed in nature. Equations are also used to identify properties and to define certain concepts, but all uses have well-established meanings, symbols that are used by convention, and allowed mathematical operations. This appendix will assist you in better understanding equations and the reasoning that goes with the manipulation of equations in problem-solving activities.

Background

In addition to a knowledge of rules for carrying out mathematical operations, an understanding of certain quantitative ideas and concepts can be very helpful when working with equations. Among these helpful concepts are (1) the meaning of inverse and reciprocal, (2) the concept of a ratio, and (3) fractions.

The term *inverse* means the opposite, or reverse, of something. For example, addition is the opposite, or inverse, of subtraction, and division is the inverse of multiplication. A *reciprocal* is defined as an inverse multiplication relationship between two numbers. For example, if the symbol n represents any number (except zero), then the reciprocal of n is $1/n$. The reciprocal of a number $(1/n)$ multiplied by that number (n) always gives a product of 1. Thus the number multiplied by 5 to give 1 is $1/5$ ($5 \times 1/5 = 5/5 = 1$). So $1/5$ is the reciprocal of 5 and 5 is the reciprocal of $1/5$. Each number is the *inverse* of the other.

The fraction $1/5$ means 1 divided by five, and if you carry out the division it gives the decimal 0.2. Calculators that have a $1/x$ key will do the operation automatically. If you enter 5, then press the $1/x$ key, the answer of 0.2 is given. If you press the $1/x$ key again, the answer of 5 is given. Each of these numbers is a reciprocal of the other.

A *ratio* is a comparison between two numbers. If the symbols m and n are used to represent any two numbers, then the ratio of the number m to the number n is the fraction m/n. This expression means to divide m by n. For example, if m is 10 and n is 5, the ratio of 10 to 5 is $10/5$, or 2:1.

Working with *fractions* is sometimes necessary in problem-solving exercises, and an understanding of these operations is needed to carry out unit calculations. It is helpful in many of these operations to remember that a number (or a unit) divided by itself is equal to 1, for example,

$$\frac{5}{5} = 1 \qquad \frac{\text{inch}}{\text{inch}} = 1 \qquad \frac{5 \text{ inches}}{5 \text{ inches}} = 1$$

When one fraction is divided by another fraction, the operation commonly applied is to "invert the denominator and multiply." For example, 2/5 divided by 1/2 is

$$\frac{\frac{2}{5}}{\frac{1}{2}} = \frac{2}{5} \times \frac{2}{1} = \frac{4}{5}$$

What you are really doing when you invert the denominator of the larger fraction and multiply is making the denominator (1/2) equal to 1. Both the numerator (2/5) and the denominator (1/2) are multiplied by 2/1, which does not change the value of the overall expression. The complete operation is

$$\frac{\frac{2}{5}}{\frac{1}{2}} \times \frac{\frac{2}{1}}{\frac{2}{1}} = \frac{\frac{2}{5} \times \frac{2}{1}}{\frac{1}{2} \times \frac{2}{1}} = \frac{\frac{4}{5}}{\frac{2}{2}} = \frac{\frac{4}{5}}{1} = \frac{4}{5}$$

Symbols and Operations

The use of symbols seems to cause confusion for some students because it seems different from their ordinary experiences with arithmetic. The rules are the same for symbols as they are for numbers, but you cannot do the operations with the symbols until you know what values they represent. The operation signs, such as $+$, $\div$, $\times$, and $-$, are used with symbols to indicate the operation that you *would* do if you knew the values. Some of the mathematical operations are indicated several ways. For example, $a \times b$, $a \cdot b$, and ab all indicate the same thing, that a is to be multiplied by b. Likewise, $a \div b$, a/b, and $a \times 1/b$ all indicate that a is to be divided by b. Since it is not possible to carry out the operations on symbols alone, they are called *indicated operations*.

Operations in Equations

An equation is a shorthand way of expressing a simple sentence with symbols. The equation has three parts: (1) a left side, (2) an equal sign (=), which indicates the equivalence of the two sides, and (3) a right side. The left side has the same value and units as the right side, but the two sides may have a very different appearance. The two sides may also have the symbols that indicate mathematical operations (+, −, ×, and so forth) and may be in certain forms that indicate operations (a/b, ab, and so forth). In any case, the equation is a complete expression that states the left side has the same value and units as the right side.

Equations may contain different symbols, each representing some unknown quantity. In science, the term "solve the equation" means to perform certain operations with one symbol (which represents some variable) by itself on one side of the equation. This single symbol is usually, but not necessarily, on the left side and is not present on the other side. For example, the equation $F = ma$ has the symbol F on the left side. In science, you would say that this equation is solved for F. It could also be solved for m or for a, which will be considered shortly. The equation $F = ma$ is solved for F, and the *indicated operation* is to multiply m by a because they are in the form ma, which means the same thing as $m \times a$. This is the only indicated operation in this equation.

A solved equation is a set of instructions that has an order of indicated operations. For example, the equation for the relationship between a Fahrenheit and Celsius temperature, solved for °C, is $°C = 5/9(F − 32°)$. A list of indicated operations in this equation is as follows:

1. Subtract 32° from the given Fahrenheit temperature.
2. Multiply the result of (1) by 5.
3. Divide the result of (2) by 9.

Why are the operations indicated in this order? Because the bracket means 5/9 of the *quantity* $(F − 32°)$. In its expanded form, you can see that $5/9(F − 32°)$ actually means $5/9(F) − 5/9(32°)$. Thus you cannot multiply by 5 or divide by 9 until you have found the quantity of $(F − 32°)$. Once you have figured out the order of operations, finding the answer to a problem becomes almost routine as you complete the needed operations on both the numbers and the units.

Solving Equations

Sometimes it is necessary to rearrange an equation to move a different symbol to one side by itself. This is known as solving an equation for an unknown quantity. But you cannot simply move a symbol to one side of an equation. Since an equation is a statement of equivalence, the right side has the same value as the left side. If you move a symbol, you must perform the operation in a way that the two sides remain equivalent. This is accomplished by "canceling out" symbols until you have the unknown on one side by itself. One key to understanding the canceling operation is to remember that a fraction with the same number (or unit) over itself is equal to 1. For example, consider the equation $F = ma$, which is solved for F. Suppose you are considering a problem in which F and m are given, and the unknown is a. You need to solve the equation for a so it is on one side by itself. To eliminate the m, you do the *inverse* of the indicated operation on m, dividing both sides by m. Thus,

$$F = ma$$

$$\frac{F}{m} = \frac{ma}{m}$$

$$\frac{F}{m} = a$$

Since m/m is equal to 1, the a remains by itself on the right side. For convenience, the whole equation may be flipped to move the unknown to the left side,

$$a = \frac{F}{m}$$

Thus a quantity that indicated a multiplication (ma) was removed from one side by an inverse operation of dividing by m.

Consider the following inverse operations to "cancel" a quantity from one side of an equation, moving it to the other side:

If the Indicated Operation of the Symbol You Wish to Remove Is:	Perform This Inverse Operation on Both Sides of the Equation:
multiplication	division
division	multiplication
addition	subtraction
subtraction	addition
squared	square root
square root	square

Example ▼

The equation for finding the kinetic energy of a moving body is $KE = 1/2mv^2$. You need to solve this equation for the velocity, v.

Solution

The order of indicated operations in the equation is as follows:

1. Square v.
2. Multiply v^2 by m.
3. Divide the result of (2) by 2.

To solve for v, this order is *reversed* as the "canceling operations" are used:

Step 1: Multiply both sides by 2

$$KE = \frac{1}{2}mv^2$$

$$2KE = \frac{2}{2}mv^2$$

$$2KE = mv^2$$

Step 2: Divide both sides by m

$$\frac{2KE}{m} = \frac{mv^2}{m}$$

$$\frac{2KE}{m} = v^2$$

Step 3: Take the square root of both sides

$$\sqrt{\frac{2KE}{m}} = \sqrt{v^2}$$

$$\sqrt{\frac{2KE}{m}} = v$$

or

$$v = \sqrt{\frac{2KE}{m}}$$

The equation has been solved for v, and you are now ready to substitute quantities and perform the needed operations (see example 1.3 in chapter 1 for information on this topic).

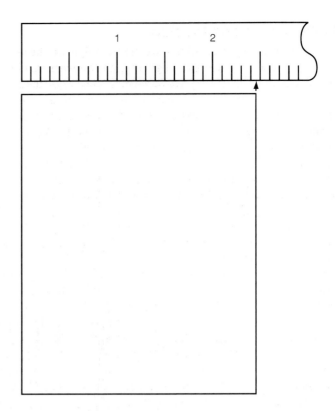

The numerical value of any measurement will always contain some uncertainty. Suppose, for example, that you are measuring one side of a square piece of paper as shown in figure A.1. You could say that the paper is *about* 2.5 cm wide and you would be correct. This measurement, however, would be unsatisfactory for many purposes. It does not approach the true value of the length and contains too much uncertainty. It seems clear that the paper width is larger than 2.4 cm but shorter than 2.5 cm. But how much larger than 2.4 cm? You cannot be certain if the paper is 2.44, 2.45, or 2.46 cm wide. As your best estimate, you might say that the paper is 2.45 cm wide. Everyone would agree that you can be certain about the first two numbers (2.4) and they should be recorded. The last number (0.05) has been estimated and is not certain. The two certain numbers, together with one uncertain number, represents the greatest accuracy possible with the ruler being used. The paper is said to be 2.45 cm wide.

A *significant figure* is a number that is believed to be correct with some uncertainty only in the last digit. The value of the width of the paper, 2.45 cm, represents three significant figures. As you can see, the number of significant figures can be determined by the degree of accuracy of the measuring instrument being used. But suppose you need to calculate the area of the paper. You would multiply 2.45 cm $\times$ 2.45 cm and the product for the area would be 15.24635 cm². This is a greater accuracy than you were able to obtain with your measuring instrument. The result of a calculation can be no more accurate than the values being treated. Because the measurement had only three significant figures (two certain, one uncertain), then the answer can have only three significant figures. The area is correctly expressed as 15.2 cm².

There are a few simple rules that will help you determine how many significant figures are contained in a reported measurement:

1. All digits reported as a direct result of a measurement are significant.
2. Zero is significant when it occurs between nonzero digits. For example, 607 has three significant figures, and the zero is one of the significant figures.
3. In figures reported as *larger than the digit one,* the digit zero is not significant when it follows a nonzero digit to indicate place. For example, in a report that "23,000 people attended the rock concert," the digits 2 and 3 are significant but the zeros are not significant. In this situation the 23 is the measured part of the figure and the three zeros tell you an estimate of how many attended the concert, that is, 23 thousand. If the figure is a measurement rather than an estimate, then it is written *with a decimal point after the last zero* to indicate that the zeros *are* significant. Thus 23,000 has *two* significant figures (2 and 3), but 23,000. has *five* significant figures. The figure 23,000 means "about 23 thousand," but 23,000. means 23,000. and not 22,999 or 23,001.
4. In figures reported as *smaller than the digit one,* zeros after a decimal point that come before nonzero digits *are not* significant and serve only as place holders. For example, 0.0023 has two significant figures, 2 and 3. Zeros alone after a decimal point or zeros after a nonzero digit indicate a measurement, however, so these zeros *are* significant. The figure 0.00230, for example, has three significant figures since the 230 means 230 and not 229 or 231. Likewise, the figure 3.000 cm has four significant figures because the presence of the three zeros means that the measurement was actually 3.000 and not 2.999 or 3.001.

Multiplication and Division

When multiplying or dividing measurement figures, the answer may have no more significant figures than the *least* number of significant figures in the figures being multiplied or divided. This simply means that an answer can be no more accurate than the least accurate measurement entering into the calculation. You cannot improve the accuracy of a measurement by doing a calculation. For example, in multiplying 54.2 mi/hr $\times$ 4.0 hr to find out the total distance traveled, the first figure (54.2) has three significant figures but the second (4.0) has only two significant figures. The answer can contain only two significant figures since this is the weakest number of those involved in the calculation. The correct answer is therefore 220 mi, not 216.8 mi. This may seem strange since multiplying the two numbers together gives the answer of 216.8 mi. This answer, however, means a greater accuracy than is possible and the accuracy cannot be improved over the weakest number involved in the calculation. Since the weakest number (4.0) has only two significant figures the answer must also have only two significant figures, which would be 220 mi.

The result of a calculation is *rounded off* to have the same least number of significant figures as the least number of a measurement involved in the calculation. When rounding numbers the last significant figure is increased by one if the number after it is five or larger. If the number after the last significant figure is four or less, the nonsignificant figures are simply dropped. Thus, if two significant figures are called for in an answer as in the preceding example, 216.8 is rounded up to 220 because the last number after the two significant figures is 6 (a number larger than 5). If the calculation result had been 214.8, the rounded number would be 210 miles.

Note that *measurement figures* are the only figures involved in the number of significant figures in the answer. Numbers used as a part of a counting process are not included in the determination of significant figures in the answer. In dividing by 2 to find an average, for example, the 2 is not a measurement so it is ignored when considering the number of significant figures.

Addition and Subtraction

Addition and subtraction operations involving measurements, as multiplication and division, cannot result in an answer that implies greater accuracy than the measurements had before the calculation. Recall that the last digit to the right in a measurement is uncertain, that is, it is the result of an estimate. The answer to an addition or subtraction calculation can have this uncertain number *no farther from the decimal place than it was in the weakest number involved in the calculation.* Thus when 8.4 is added to 4.926, the weakest number is 8.4 and the uncertain number is .4, one place to the right of the decimal. The sum of 13.326 is therefore rounded to 13.3, reflecting the placement of this weakest doubtful figure.

In an example concerning percentage error, an experimental result of 511 hz was found for a tuning fork with an accepted frequency value of 522 hz. The error calculation is

$$\frac{(522 \text{ hz} - 511 \text{ hz})}{522 \text{ hz}} \times 100\% = 2.1\%$$

Since 522 − 511 is 11, the least number of significant figures of measurements involved in this calculation is *two*. Note that the "100" does not enter into the determination since it is not a measurement number. The calculated result (from a calculator) is 2.1072797, which is rounded off to have only two significant figures, so the answer is recorded as 2.1%.

Conversion of Units

The measurement of most properties results in both a numerical value and a unit. The statement that a glass contains 50 cm³ of a liquid conveys two important concepts—the numerical value of 50 and the referent unit of cubic centimeters. Both the numerical value and the unit are necessary to communicate correctly the volume of the liquid.

When working with calculations involving measurement units, *both* the numerical value and the units are treated mathematically. As in other mathematical operations, there are general rules to follow.

1. Only properties with *like units* may be added or subtracted. It should be obvious that adding quantities such as 5 dollars and 10 dimes is meaningless. You must first convert to like units before adding or subtracting.
2. Like or unlike units may be multiplied or divided and treated in the same manner as numbers. You have used this rule when dealing with area (length $\times$ length = length², for example, cm $\times$ cm = cm²) and when dealing with volume (length $\times$ length $\times$ length = length³, for example, cm $\times$ cm $\times$ cm = cm³).

You can use these two rules to create a *conversion ratio* that will help you change one unit to another. Suppose you need to convert 2.300 kg to grams. First, write the relationship between kilograms and grams:

$$1{,}000. \text{ g} = 1.000 \text{ kg}$$

Next, divide both sides by what you wish to convert *from* (kilograms in this example):

$$\frac{1{,}000. \text{ g}}{1.000 \text{ kg}} = \frac{1.000 \text{ kg}}{1.000 \text{ kg}}$$

One kilogram divided by 1 kg equals 1, just as 10 divided by 10 equals 1. Therefore, the right side of the relationship becomes 1:

$$\frac{1{,}000. \text{ g}}{1.000 \text{ kg}} = 1$$

The 1 is usually understood, that is, not stated and the operation is called *canceling*. Canceling leaves you with the fraction 1,000. g/1.000 kg, which is a conversion ratio that can be used to convert from kilograms to grams. You simply multiply the conversion ratio by the numerical value and unit you wish to convert:

$$= 2.300 \text{ kg} \times \frac{1,000. \text{ g}}{1.000 \text{ kg}}$$

$$= \frac{2.300 \times 1,000. \text{ kg} \times \text{g}}{1.000 \quad \text{kg}}$$

$$= \boxed{2,300 \text{ g}}$$

The kilogram units cancel. Showing the whole operation with units only, you can see how you end up with the correct unit of grams:

$$\text{kg} \times \frac{\text{g}}{\text{kg}} = \frac{\text{kg} \cdot \text{g}}{\text{kg}} = \text{g}$$

Since you did obtain the correct unit, you know that you used the correct conversion ratio. If you had blundered and used an inverted conversion ratio, you would obtain

$$2.300 \text{ kg} \times \frac{1.000 \text{ kg}}{1,000. \text{ g}} = .0023 \frac{\text{kg}^2}{\text{g}}$$

which yields the meaningless, incorrect units of kg^2/g. Carrying out the mathematical operations on the numbers and the units will always tell you whether or not you used the correct conversion ratio.

Example

A distance is reported as 100. km and you want to know how far this is in miles.

Solution

First, you need to obtain a *conversion factor* from a textbook or reference book, which usually lists the conversion factors by properties in a table. Such a table will show two conversion factors for kilometers and miles: (1) 1.00 km = 0.621 mi and (2) 1.00 mi = 1.609 km. You select the factor that is in the same form as your problem, for example, your problem is 100. km = ? mi. The conversion factor in this form is 1.00 km = 0.621 mi.

Second, you convert this conversion factor into a *conversion ratio* by dividing the factor by what you wish to convert *from:*

conversion factor:	1.00 km = 0.621 mi
divide factor by what you want to convert from:	$\dfrac{1.00 \text{ km}}{1.00 \text{ km}} = \dfrac{0.621 \text{ mi}}{1.00 \text{ km}}$
resulting conversion ratio:	$\dfrac{0.621 \text{ mi}}{\text{km}}$

Note that if you had used the 1.00 mi = 1.609 km factor, the resulting units would be meaningless. The conversion ratio is now multiplied by the numerical value *and unit* you wish to convert:

$$100. \text{ km} \times \frac{0.621 \text{ mi}}{\text{km}}$$

$$(100.)(0.621) \frac{\text{km} \cdot \text{mi}}{\text{km}}$$

$$62.1 \text{ mi}$$

Example

A service station sells gasoline by the liter and you fill your tank with 72.0 liters. How many gallons is this? (Answer: 19.0 gal)

Scientific Notation

Most of the properties of things that you might measure in your everyday world can be expressed with a small range of numerical values together with some standard unit of measure. The range of numerical values for most everyday things can be dealt with by using units (1s), tens (10s), hundreds (100s), or perhaps thousands (1,000s). But the actual universe contains some objects of incredibly large size that require some very big numbers to describe. The sun, for example, has a mass of about 1,970,000,000,000,000,000,000,000,000,000 kg. On the other hand, very small numbers are needed to measure the size and parts of an atom. The radius of a hydrogen atom, for example, is about 0.00000000005 m. Such extremely large and small numbers are cumbersome and awkward since there are so many zeros to keep track of, even if you are successful in carefully counting all the zeros. A method does exist to deal with extremely large or small numbers in a more condensed form. The method is called *scientific notation,* but it is also sometimes called *powers of ten* or *exponential notation* since it is based on exponents of 10. Whatever it is called, the method is a compact way of dealing with numbers that not only helps you keep track of zeros but provides a simplified way to make calculations as well.

In algebra you save a lot of time (as well as paper) by writing ($a \times a \times a \times a \times a$) as a^5. The small number written to the right and above a letter or number is a superscript called an *exponent*. The exponent means that the letter or number is to be multiplied by itself that many times, for example, a^5 means "a" multiplied by itself five times, or $a \times a \times a \times a \times a$. As you can see, it is much easier to write the exponential form of this operation than it is to write it out in the long form. Scientific notation uses an exponent to indicate the power of the base 10.

The exponent tells how many times the base, 10, is multiplied by itself. For example,

$$10,000. = 10^4$$
$$1,000. = 10^3$$
$$100. = 10^2$$
$$10. = 10^1$$
$$1. = 10^0$$
$$0.1 = 10^{-1}$$
$$0.01 = 10^{-2}$$
$$0.001 = 10^{-3}$$
$$0.0001 = 10^{-4}$$

This table could be extended indefinitely, but this somewhat shorter version will give you an idea of how the method works. The symbol 10^4 is read as "ten to the fourth power" and means $10 \times 10 \times 10 \times 10$. Ten times itself four times is 10,000, so 10^4 is the scientific notation for 10,000. It is also equal to the number of zeros between the 1 and the decimal point, that is, to write the longer form of 10^4 you simply write 1, then move the decimal point four places to the *right;* ten to the fourth power is 10,000.

The power of ten table also shows that numbers smaller than one have negative exponents. A negative exponent means a reciprocal:

$$10^{-1} = \frac{1}{10} = 0.1$$

$$10^{-2} = \frac{1}{100} = 0.01$$

$$10^{-3} = \frac{1}{1000} = 0.001$$

To write the longer form of 10^{-4}, you simply write 1 then move the decimal point four places to the *left;* ten to the negative fourth power is 0.0001.

Scientific notation usually, but not always, is expressed as the product of two numbers: (1) a number between 1 and 10 that is called the *coefficient* and (2) a power of ten that is called the *exponent.* For example, the mass of the sun that was given in long form earlier is expressed in scientific notation as

$$1.97 \times 10^{30} \text{ kg}$$

and the radius of a hydrogen atom is

$$5.0 \times 10^{-11} \text{ m}$$

In these expressions, the coefficients are 1.97 and 5.0 and the power of ten notations are the exponents. Note that in both of these examples, the exponent tells you where to place the decimal point if you wish to write the number all the way out in the long form. Sometimes scientific notation is written without a coefficient, showing only the exponent. In these cases the coefficient of 1.0 is understood, that is, not stated. If you try to enter a scientific notation in your calculator, however, you will need to enter the understood 1.0 or the calculator will not be able to function correctly. Note also that 1.97×10^{30} kg and the expressions 0.197×10^{31} kg and 19.7×10^{29} kg are all correct expressions of the mass of the sun. By convention, however, you will use the form that has one digit to the left of the decimal.

Example

What is 26,000,000 in scientific notation?

Solution
Count how many times you must shift the decimal point until one digit remains to the left of the decimal point. For numbers larger than the digit 1, the number of shifts tells you how much the exponent is increased, so the answer is

$$2.6 \times 10^7$$

which means the coefficient 2.6 is multiplied by 10 seven times.

Example

What is 0.000732 in scientific notation? (Answer: 7.32×10^{-4})

It was stated earlier that scientific notation provides a compact way of dealing with very large or very small numbers, but it provides a simplified way to make calculations as well. There are a few mathematical rules that will describe how the use of scientific notation simplifies these calculations.

To *multiply* two scientific notation numbers, the coefficients are multiplied as usual and the exponents are *added* algebraically. For example, to multiply (2×10^2) by (3×10^3), first separate the coefficients from the exponents,

$$(2 \times 3) \times (10^2 \times 10^3),$$

then multiply the coefficients and add the exponents,

$$6 \times 10^{(2+3)} = 6 \times 10^5$$

Adding the exponents is possible because $10^2 \times 10^3$ means the same thing as $(10 \times 10) \times (10 \times 10 \times 10)$, which equals $(100) \times (1,000)$, or 100,000, which is expressed as 10^5 in scientific notation. Note that two negative exponents add algebraically, for example $10^{-2} \times 10^{-3} = 10^{[(-2)+(-3)]} = 10^{-5}$. A negative and a positive exponent also add algebraically, as in $10^5 \times 10^{-3} = 10^{[(+5)+(-3)]} = 10^2$.

If the result of a calculation involving two scientific notation numbers does not have the conventional one digit to the left of the decimal, move the decimal point so it does, changing the exponent according to which way and how much the decimal point is moved. Note that the exponent increases by one number for each decimal point moved to the left. Likewise, the exponent decreases by one number for each decimal point moved to the right. For example, $938. \times 10^3$ becomes 9.38×10^5 when the decimal point is moved two places to the left.

To *divide* two scientific notation numbers, the coefficients are divided as usual and the exponents are *subtracted*. For example, to divide (6×10^6) by (3×10^2), first separate the coefficients from the exponents,

$$(6 \div 3) \times (10^6 \div 10^2)$$

then divide the coefficients and subtract the exponents,

$$2 \times 10^{(6-2)} = 2 \times 10^4$$

Note that when you subtract a negative exponent, for example, $10^{[(3)-(-2)]}$, you change the sign and add, $10^{(3+2)} = 10^5$.

Example

Solve the following problem concerning scientific notation:

$$\frac{(2 \times 10^4) \times (8 \times 10^{-6})}{8 \times 10^4}$$

Solution

First, separate the coefficients from the exponents,

$$\frac{2 \times 8}{8} \times \frac{10^4 \times 10^{-6}}{10^4}$$

then multiply and divide the coefficients and add and subtract the exponents as the problem requires,

$$2 \times 10^{\{[(4)+(-6)]-[4]\}}$$

solving the remaining additions and subtractions of the coefficients gives

$$2 \times 10^{-6}$$

Appendix B

Solubilities Chart

	Acetate	Bromide	Carbonate	Chloride	Fluoride	Hydroxide	Iodide	Nitrate	Oxide	Phosphate	Sulfate	Sulfide
Aluminum	S	S	—	S	s	i	S	S	i	i	S	d
Ammonium	S	S	S	S	S	S	S	S	—	S	S	S
Barium	S	S	i	S	s	S	S	S	S	i	i	d
Calcium	S	S	i	S	i	s	S	S	s	i	s	d
Copper (I)	—	s	i	s	i	—	i	—	i	—	d	i
Copper (II)	S	S	i	S	S	i	S	S	i	i	S	i
Iron (II)	S	S	i	S	s	i	S	S	i	i	S	i
Iron (III)	S	S	i	S	s	i	S	S	i	i	S	d
Lead	S	s	i	s	i	i	s	S	i	i	i	i
Magnesium	S	S	i	S	i	i	S	S	i	i	S	d
Mercury (I)	s	i	i	i	d	d	i	S	i	i	i	i
Mercury (II)	S	s	i	S	d	i	i	S	i	i	i	i
Potassium	S	S	S	S	S	S	S	S	S	S	S	i
Silver	s	i	i	i	S	—	i	S	i	i	i	i
Sodium	S	S	S	S	S	S	S	S	d	S	S	S
Strontium	S	S	s	S	i	s	S	S	—	i	i	i
Zinc	S	S	i	S	S	i	S	S	i	i	S	i

S—soluble
i—insoluble
s—slightly soluble
d—decomposes

Relative Humidity (%)

Dry-Bulb Temperature (°C)	Difference between Wet-Bulb and Dry-Bulb Temperatures (°C)																			
	1	2	3	4	5	6	7	8	9	10	11	12	13	14	15	16	17	18	19	20
0	81	64	46	29	13															
1	83	66	49	33	17															
2	84	68	52	37	22	7														
3	84	70	55	40	26	12														
4	86	71	57	43	29	16														
5	86	72	58	45	33	20	7													
6	86	73	60	48	35	24	11													
7	87	74	62	50	38	26	15													
8	87	75	63	51	40	29	19	8												
9	88	76	64	53	42	32	22	12												
10	88	77	66	55	44	34	24	15	6											
11	89	78	67	56	46	36	27	18	9											
12	89	78	68	58	48	39	29	21	12											
13	89	79	69	59	50	41	32	23	15	7										
14	90	79	70	60	51	42	34	26	18	10										
15	90	80	71	61	53	44	36	27	20	13	6									
16	90	81	71	63	54	46	38	30	23	15	8									
17	90	81	72	64	55	47	40	32	25	18	11									
18	91	82	73	65	57	49	41	34	27	20	14	7								
19	91	82	74	65	58	50	43	36	29	22	16	10								
20	91	83	74	66	59	51	44	37	31	24	18	12	6							
21	91	83	75	67	60	53	46	39	32	26	20	14	9							
22	92	83	76	68	61	54	47	40	34	28	22	17	11	6						
23	92	84	76	69	62	55	48	42	36	30	24	19	13	8						
24	92	84	77	69	62	56	49	43	37	31	26	20	15	10	5					
25	92	84	77	70	63	57	50	44	39	33	28	22	17	12	8					
26	92	85	78	71	64	58	51	46	40	34	29	24	19	14	10	5				
27	92	85	78	71	65	58	52	47	41	36	31	26	21	16	12	7				
28	93	85	78	72	65	59	53	48	42	37	32	27	22	18	13	9	5			
29	93	86	79	72	66	60	54	49	43	38	33	28	24	19	15	11	7			
30	93	86	79	73	67	61	55	50	44	39	35	30	25	21	17	13	9	5		
31	93	86	80	73	67	61	56	51	45	40	36	31	27	22	18	14	11	7		
32	93	86	80	74	68	62	57	51	46	41	37	32	28	24	20	16	12	9	5	
33	93	87	80	74	68	63	57	52	47	42	38	33	29	25	21	17	14	10	7	
34	93	87	81	75	69	63	58	53	48	43	39	35	30	28	23	19	15	12	8	5
35	94	87	81	75	69	64	59	54	49	44	40	36	32	28	24	20	17	13	10	7

Appendix D

Solutions for Group A Chapter Exercises

Chapter 1

1. Answers will vary but should have the relationship of 100 cm = 1.00 m, for example, 178 cm = 1.78 m.

2. Since mass density is given by the relationship $\rho = m/V$, then

$$\rho = \frac{m}{V} = \frac{272.0 \text{ g}}{20.0 \text{ cm}^3}$$

$$= \frac{272.0}{20.0} \frac{\text{g}}{\text{cm}^3}$$

$$= \boxed{13.6 \ \frac{\text{g}}{\text{cm}^3}}$$

The units g/cm³ are already as simplified as they can be.

3. Solving the relationship $\rho = m/V$ for volume gives $V = m/\rho$. In order to have the same units in the equation, 1.500 kg is *first* converted to 1,500. g (because the density is given in grams), and

$$V = \frac{m}{\rho} = \frac{1,500. \text{ g}}{3.00 \ \frac{\text{g}}{\text{cm}^3}}$$

$$= \frac{1,500.}{3.00} \frac{\text{g}}{1} \times \frac{\text{cm}^3}{\text{g}}$$

$$= 500. \ \frac{\text{g} \cdot \text{cm}^3}{\text{g}}$$

$$= \boxed{500. \text{ cm}^3}$$

4. (a) Units must be converted first so they are the same. One way to express this ratio is 1,800.0 cm³/3,600 cm³, so the motorcycle engine is 1/2 the size of the automobile engine. (b) The ratio does not have units. The units cancel when the relationship is simplified by division.

$$\frac{1,800.0 \text{ cm}^3}{3,600 \text{ cm}^3} = \frac{1,800.0}{3,600} \frac{\text{cm}^3}{1} \times \frac{1}{\text{cm}^3}$$

$$= \frac{1,800.0}{3,600} \frac{\text{cm}^3}{\text{cm}^3}$$

$$= \boxed{\frac{1}{2}} \text{ (or 1:2)}$$

5. The problem asks for a mass and gives a volume, so you need a relationship between mass and volume. A reference table gives the mass density of water as 1.0 g/cm³, which is a density that

is easily remembered. The volume is given in L, which should first be converted to cm³ because this is the unit in which density is expressed. The relationship of $\rho = m/V$ solved for mass is ρV, so the solution is

$$m = \rho V = \left(1.0 \ \frac{\text{g}}{\text{cm}^3} \right) (40,000 \text{ cm}^3)$$

$$= 1.0 \times 40,000 \ \frac{\text{g}}{\text{cm}^3} \times \text{cm}^3$$

$$= 40,000 \ \frac{\text{cm}^3 \cdot \text{g}}{\text{cm}^3}$$

$$= 40,000 \text{ g} = \boxed{40. \text{ kg}}$$

6. From a table of densities, the mass density of aluminum is given as 2.7 g/cm³. Converting the given mass to the same units, 2.0 kg = 2,000 g. Solving $\rho = m/V$ for the volume gives

$$V = \frac{m}{\rho} = \frac{2,000 \text{ g}}{2.7 \ \frac{\text{g}}{\text{cm}^3}}$$

$$= \frac{2,000}{2.7} \frac{\text{g}}{1} \times \frac{\text{cm}^3}{\text{g}}$$

$$= 740.74 \ \frac{\text{g} \cdot \text{cm}^3}{\text{g}}$$

$$= 740.74 \text{ cm}^3 = \boxed{740 \text{ cm}^3}$$

7. The length of one side of the box is 0.1 m. Reasoning: Since the density of water is 1.00 g/cm³, then the volume of 1,000 g of water is 1,000 cm³. A cubic box with a volume of 1,000 cm³ is 10 cm (since 10 × 10 × 10 = 1,000). Converting 10 cm to m units, the cube is 0.1 m on each edge.

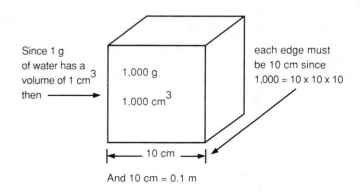

Since 1 g of water has a volume of 1 cm³ then → 1,000 g 1,000 cm³ each edge must be 10 cm since 1,000 = 10 x 10 x 10

10 cm

And 10 cm = 0.1 m

8. A ratio is one number compared to another, usually simplified by dividing. Dividing cancels the units, for example,

$$\frac{\text{specific}}{\text{gravity}} = \frac{\text{density of substance}}{\text{density of water}}$$

$$= \frac{0.92 \frac{g}{cm^3}}{1.00 \frac{g}{cm^3}} = \frac{0.92}{1.00} \frac{g}{cm^3} \times \frac{cm^3}{g}$$

$$= \frac{0.92}{1.00} \frac{\cancel{g} \cdot \cancel{cm^3}}{\cancel{g} \cdot \cancel{cm^3}}$$

$$= \boxed{0.92}$$

9. The volume of the tank in ft^3 is $(25.0 \text{ gal})(0.134 \text{ ft}^3/\text{gal})$ = 3.35 ft³. The weight of the gasoline can be obtained from the relationship $D = w/V$ by solving for the weight, or $W = DV$. Thus,

$$W = DV = \left(42.0 \frac{lb}{ft^3}\right)(3.35 \text{ ft}^3)$$

$$= (42.0)(3.35) \frac{lb}{ft^3} \times ft^3$$

$$= 140 \frac{\cancel{ft^3} \cdot lb}{\cancel{ft^3}}$$

$$= \boxed{140. \text{ lb}}$$

10. Since the fresh water has a weight of 31.2 lb, the container must have a volume of

$$D = \frac{W}{V} \therefore V = \frac{W}{D} = \frac{31.2 \text{ lb}}{62.4 \frac{lb}{ft^3}}$$

$$= \frac{31.2}{62.4} \text{ lb} \times \frac{ft^3}{lb}$$

$$= 0.500 \frac{\cancel{lb} \cdot ft^3}{\cancel{lb}}$$

$$= 0.500 \text{ ft}^3$$

The same volume of seawater would have a weight of

$$D = \frac{W}{V} \therefore W = DV = \left(64.0 \frac{lb}{ft^3}\right)(0.500 \text{ ft}^3)$$

$$= (64.0)(0.500) \frac{lb}{ft^3} \times ft^3$$

$$= 32.0 \frac{\cancel{ft^3} \cdot lb}{\cancel{ft^3}}$$

$$= \boxed{32.0 \text{ lb}}$$

Chapter 2

1. The distance (d) and the time (t) quantities are given in the problem, and

$$\bar{v} = \frac{d}{t}$$

$$= \frac{250.0 \text{ mi}}{5.0 \text{ hr}}$$

$$= \frac{250.0}{5.0} \frac{mi}{hr}$$

$$= \boxed{50. \frac{mi}{hr}}$$

The units cannot be simplified further. Note that the decimal after the 50 identifies *two* significant figures in the answer, which is the least number of significant figures of the measurements involved in the division operation.

2. One of the units (hr or min) must be changed so they are the same. Since such speeds are usually given as mi/hr or km/hr, convert the minutes to hours. This is easily done since 20 minutes is 1/3 of an hour. Solving the average velocity-distance-time relationship for distance and using the given quantities:

$$\bar{v} = \frac{d}{t} \therefore d = \bar{v}t$$

$$= \left(18.0 \frac{km}{hr}\right)\left(\frac{1}{3} \text{ hr}\right)$$

$$= 18.0 \times \frac{1}{3} \frac{\cancel{hr} \cdot km}{\cancel{hr}}$$

$$= \boxed{6 \text{ km}}$$

3. Average speed is represented by the symbol $\bar{v}$, and the relationship between the quantities is

(a) $\bar{v} = \dfrac{d}{t} = \dfrac{400.0 \text{ mi}}{8.0 \text{ hr}} = \boxed{50. \text{ mi/hr}}$

(b) $\bar{v} = \dfrac{d}{t} = \dfrac{400.0 \text{ mi}}{7.0 \text{ hr}} = \boxed{57 \text{ mi/hr}}$

4. The bicycle averages a velocity equal to the *total* distance covered divided by the *total* time for the trip. Thus:

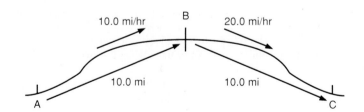

A to B: $t = \dfrac{d}{\bar{v}} = \dfrac{10.0 \text{ mi}}{10.0 \dfrac{\text{mi}}{\text{hr}}}$

$= \dfrac{10.0}{10.0} \text{ mi} \times \dfrac{\text{hr}}{\text{mi}}$

$= \dfrac{10.0}{10.0} \dfrac{\text{mi} \cdot \text{hr}}{\text{mi}} = 1.00 \text{ hr}$

B to C: $= \dfrac{10.0 \text{ mi}}{20.0 \dfrac{\text{mi}}{\text{hr}}}$

$= \dfrac{10.0}{20.0} \text{ mi} \times \dfrac{\text{hr}}{\text{mi}}$

$= \dfrac{10.0}{20.0} \dfrac{\text{mi} \cdot \text{hr}}{\text{mi}} = 0.500 \text{ hr}$

A to C totals: $t = 1.50 \text{ hr}$ and $d = 20.0 \text{ mi}$

$\therefore \bar{v} = \dfrac{d}{t} = \dfrac{20.0 \text{ mi}}{1.50 \text{ hr}} = \boxed{13.3 \dfrac{\text{mi}}{\text{hr}}}$

5. (a) The term "at rest" means an initial velocity of $v_i = 0$ and an acceleration to a certain speed in a certain time period means (in this problem) $v_f = 60.0$ ft/sec. Acceleration is a change of velocity in a unit of time, so

$\bar{a} = \dfrac{v_f - v_i}{t} = \dfrac{60.0 \dfrac{\text{ft}}{\text{sec}} - 0}{10.0 \text{ sec}}$

$= \dfrac{60.0}{10.0} \dfrac{\text{ft}}{\text{sec}} \times \dfrac{1}{\text{sec}}$

$= \dfrac{60.0}{10.0} \dfrac{\text{ft}}{\text{sec} \cdot \text{sec}}$

$= \boxed{6.00 \dfrac{\text{ft}}{\text{sec}^2}}$

(b) Lacking further information, you would have to assume that this question is asking for the velocity after 15.0 sec of acceleration from rest ($v_i = 0$), so

$\bar{a} = \dfrac{v_f - v_i}{t} \quad \therefore v_f = \bar{a}t + v_i$

$= \left(6.00 \dfrac{\text{ft}}{\text{sec}^2}\right)(15.0 \text{ sec})$

$= (6.00)(15.0) \dfrac{\text{ft}}{\text{sec}^2} \times \text{sec}$

$= 90.0 \dfrac{\text{ft} \cdot \text{sec}}{\text{sec} \cdot \text{sec}}$

$= \boxed{90.0 \dfrac{\text{ft}}{\text{sec}}}$

6. The question is asking for a final velocity (v_f). The acceleration ($\bar{a} = 9.0$ ft/sec²) and time ($t = 8.0$ sec) are given. The initial velocity (v_i) is zero since the car accelerates from "rest." Therefore the question involves a relationship between acceleration ($\bar{a}$), final velocity (v_f), and time (t), so

$\bar{a} = \dfrac{v_f - v_i}{t} \quad \therefore v_f = \bar{a}t + v_i$

$= \left(9.0 \dfrac{\text{ft}}{\text{sec}^2}\right)(8.0 \text{ sec})$

$= 9.0 \times 8.0 \dfrac{\text{ft}}{\text{sec}^2} \times \text{sec}$

$= 72 \dfrac{\text{ft} \cdot \text{sec}}{\text{sec} \cdot \text{sec}}$

$= \boxed{72 \text{ ft/sec}}$

7. For this type of problem, note that the velocity, distance, and time relationship calls for $\bar{v}$, or *average* velocity. The quantity of 88.0 ft/sec is the *initial* velocity that the car had before stopping. Therefore the average velocity must be obtained before solving for the time required:

(a) $\bar{v} = \dfrac{d}{t} \quad \therefore t = \dfrac{d}{\bar{v}}$

$\bar{v} = \dfrac{v_f + v_i}{2} = \dfrac{0 + 88.0 \text{ ft/sec}}{2} = 44.0 \text{ ft/sec}$

$t = \dfrac{100.0 \text{ ft}}{44.0 \text{ ft/sec}}$

$= \dfrac{100.0}{44.0} \text{ ft} \times \dfrac{\text{sec}}{\text{ft}}$

$= 2.27 \dfrac{\text{ft} \cdot \text{sec}}{\text{ft}}$

$= \boxed{2.27 \text{ sec}}$

(Note: The 2 is not considered in significant figures since it is not the result of a measurement.)

(b) $\bar{a} = \dfrac{v_f - v_i}{t} = \dfrac{0 - 88.0 \text{ ft/sec}}{2.27 \text{ sec}}$

$= \dfrac{-88.0}{2.27} \dfrac{\text{ft}}{\text{sec}} \times \dfrac{1}{\text{sec}}$

$= -38.8 \dfrac{\text{ft}}{\text{sec} \cdot \text{sec}}$

$= \boxed{-38.8 \dfrac{\text{ft}}{\text{sec}^2}}$

(Note: The negative sign simply means that the car is slowing, or decelerating, from some given speed.)
One g is 32.0 ft/sec², so the deceleration was

$\dfrac{38.8 \text{ ft/sec}^2}{32.0 \text{ ft/sec}^2} = \dfrac{38.8 \text{ ft/sec}^2}{32.0 \text{ ft/sec}^2} = \boxed{1.21 \text{ g's}}$

8. A ball thrown straight up decelerates to a velocity of zero then accelerates back to the surface just as a dropped ball would do from the height reached. Thus the time required to decelerate upward is the same as the time required to accelerate downward, and the ball returns to the surface with the same velocity with which it was thrown (neglecting friction). Therefore:

$$\bar{a} = \frac{v_f - v_i}{t} \therefore v_f = \bar{a}t + v_i$$

$$= \left(9.80 \ \frac{m}{sec^2}\right)(3.0 \ sec)$$

$$= (9.80)(3.0) \ \frac{m}{sec^2} \times sec$$

$$= 29 \ \frac{m \cdot sec}{sec \cdot sec}$$

$$= \boxed{29 \ m/sec}$$

(or 96 ft/sec in English units)

9. These three questions are easily answered by using the three sets of relationships, or equations, that were presented in this chapter:

(a) $v_f = \bar{a}t = \left(9.80 \ \frac{m}{sec^2}\right)(4.00 \ sec)$

$$= 9.80 \times 4.00 \ \frac{m}{sec^2} \times sec$$

$$= 39.2 \ \frac{m \cdot sec}{sec \cdot sec}$$

$$= \boxed{39.2 \ m/sec}$$

(b) $\bar{v} = \frac{v_f + v_i}{2} = \frac{39.2 \ m/sec + 0}{2} = \boxed{19.6 \ m/sec}$

(c) $\bar{v} = \frac{d}{t} \therefore d = \bar{v}t = \left(19.6 \ \frac{m}{sec}\right)(4.00 \ sec)$

$$= 19.6 \times 4.00 \ \frac{m}{sec} \times sec$$

$$= 78.4 \ \frac{m \cdot sec}{sec}$$

$$= \boxed{78.4 \ m}$$

10. Note that this problem can be solved with a series of three steps as in the previous problem, or it can be solved by the equation that combines all the relationships into one step. Either method

is acceptable, but the following example of a one-step solution reduces the possibility of error since fewer calculations are involved:

$$d = \frac{1}{2} \ gt^2 = \frac{1}{2}\left(9.80 \ \frac{m}{sec^2}\right)(5.00 \ sec)^2$$

$$= \frac{1}{2}\left(9.80 \ \frac{m}{sec^2}\right)(25.0 \ sec^2)$$

$$= \left(\frac{1}{2}\right)(9.80)(25.0) \ \frac{m}{sec^2} \times sec^2$$

$$= 4.90 \times 25.0 \ \frac{m \cdot sec^2}{sec^2}$$

$$= \boxed{123 \ m}$$

11. Note that the question asks for a unit of time and that the acceleration ($\bar{a}$), initial velocity (v_i), and the final velocity (v_f) are given. The *first step,* before anything else is done, is to write the equation as a relationship between these quantities and then solve the equation for the unknown (time in this problem). Since unit conversions are not necessary, quantities are then substituted for the symbols and mathematical operations are performed on both the numbers and units:

$$\bar{a} = \frac{v_f - v_i}{t} \therefore t = \frac{v_f - v_i}{\bar{a}} = \frac{50.0 \ \frac{m}{sec} - 10.0 \ \frac{m}{sec}}{4.0 \ \frac{m}{sec^2}}$$

$$= \frac{40.0}{4.0} \ \frac{m}{sec} \times \frac{sec^2}{m}$$

$$= 10. \ \frac{m \cdot sec \cdot sec}{m \cdot sec}$$

$$= \boxed{10. \ sec}$$

12.

$$\bar{a} = g \therefore g = \frac{v_f - v_i}{t} \therefore v_f = gt$$

$$= \left(9.80 \ \frac{m}{sec^2}\right)(3.0 \ sec)$$

$$= (9.80)(3.0) \ \frac{m}{sec^2} \times sec$$

$$= 29.4 \ \frac{m \cdot sec}{sec \cdot sec}$$

$$= 29.4 \ \frac{m}{sec} = \boxed{29 \ \frac{m}{sec}}$$

Chapter 3

1. (a) Weight (W) is a downward force from the acceleration of gravity (g) on the mass (m) of an object. This relationship is the same as Newton's second law of motion, $F = ma$, and

$$W = mg = (1.25 \text{ kg})\left(9.80 \frac{m}{sec^2}\right)$$

$$= (1.25)(9.80) \text{ kg} \times \frac{m}{sec^2}$$

$$= 12.25 \frac{kg \cdot m}{sec^2}$$

$$= \boxed{12.3 \text{ N}}$$

(b) The relationship between force (F), mass (m), and acceleration (a) is given by Newton's second law of motion and

$$F = ma \therefore a = \frac{F}{m} = \frac{10.0 \frac{kg \cdot m}{sec^2}}{1.25 \text{ kg}}$$

$$= \frac{10.0}{1.25} \frac{kg \cdot m}{sec^2} \times \frac{1}{kg}$$

$$= 8.00 \frac{kg \cdot m}{kg \cdot sec^2}$$

$$= \boxed{8.00 \frac{m}{sec^2}}$$

(Note how the units were treated mathematically in this solution and why it is necessary to show the units for a newton of force. The resulting unit in the answer *is* a unit of acceleration, which provides a check that the problem was solved correctly.)

2.
$$F = ma = (1.25 \text{ kg})\left(5.00 \frac{m}{sec^2}\right)$$

$$= (1.25)(5.00) \text{ kg} \times \frac{m}{sec^2}$$

$$= 6.25 \frac{kg \cdot m}{sec^2}$$

$$= \boxed{6.25 \text{ N}}$$

(Note that the solution is correctly reported in *newton* units of force rather than kg · m/sec².)

3. The bicycle tire exerts a backward force on the road and the equal and opposite reaction force of the road on the bicycle produces the forward motion. (The motion is always in the direction of the applied force.) Therefore,

$$F = ma = (70.0 \text{ kg})\left(2.0 \frac{m}{sec^2}\right)$$

$$= (70.0)(2.0) \text{ kg} \times \frac{m}{sec^2}$$

$$= 140 \frac{kg \cdot m}{sec^2}$$

$$= \boxed{140 \text{ N}}$$

4. The question requires finding a force in the metric system, which is measured in newtons of force. Since newtons of force are defined in kg, m, and sec, unit conversions are necessary and these should be done first. There are 1,000. m in 1.000 km and 3,600. sec in 1.000 hr, and

$$1.000 \frac{km}{hr} = \frac{1,000. \text{ m}}{3,600. \text{ sec}} = 0.2778 \frac{m}{sec}$$

Dividing both sides of this conversion factor by what you are converting *from* gives the conversion ratio of

$$\frac{0.2778 \frac{m}{sec}}{\frac{km}{hr}}$$

Multiplying this conversion ratio times the two velocities in km/hr will convert them to m/sec as follows:

$$\left(0.2778 \frac{\frac{m}{sec}}{\frac{km}{hr}}\right)\left(80.0 \frac{km}{hr}\right)$$

$$= (0.2778)(80.0) \frac{m}{sec} \times \frac{hr}{km} \times \frac{km}{hr}$$

$$= 22.2 \frac{m}{sec}$$

$$\left(0.2778 \frac{\frac{m}{sec}}{\frac{km}{hr}}\right)\left(44.0 \frac{km}{hr}\right)$$

$$= (0.2778)(44.0) \frac{m}{sec} \times \frac{hr}{km} \times \frac{km}{hr}$$

$$= 12.2 \frac{m}{sec}$$

Now you are ready to find the appropriate relationship between the quantities involved. This involves two separate equations: Newton's second law of motion and the relationship of quantities involved in acceleration. These may be combined as follows:

$$F = ma \text{ and } a = \frac{v_f - v_i}{t} \therefore F = m\left(\frac{v_f - v_i}{t}\right)$$

Now you are ready to substitute quantities for the symbols and perform the necessary mathematical operations:

$$= (1{,}500 \text{ kg})\left(\frac{22.2 \text{ m/sec} - 12.2 \text{ m/sec}}{10.0 \text{ sec}}\right)$$

$$= (1{,}500 \text{ kg})\left(\frac{10.0 \text{ m/sec}}{10.0 \text{ sec}}\right)$$

$$= 1{,}500 \times 1.00 \ \frac{\text{kg} \cdot \dfrac{\text{m}}{\text{sec}}}{\text{sec}}$$

$$= 1{,}500 \ \frac{\text{kg} \cdot \text{m}}{\text{sec}} \times \frac{1}{\text{sec}}$$

$$= 1{,}500 \ \frac{\text{kg} \cdot \text{m}}{\text{sec} \cdot \text{sec}}$$

$$= 1{,}500 \ \frac{\text{kg} \cdot \text{m}}{\text{sec}^2}$$

$$= 1{,}500 \text{ N} = \boxed{1.5 \times 10^3 \text{ N}}$$

5. A unit conversion is needed as in the previous problem:

$$\left(90.0 \ \frac{\text{km}}{\text{hr}}\right)\left(0.2778 \ \frac{\dfrac{\text{m}}{\text{sec}}}{\dfrac{\text{km}}{\text{hr}}}\right) = 25.0 \text{ m/sec}$$

(a) $F = ma \therefore m = \dfrac{F}{a}$ and $a = \dfrac{v_f - v_i}{t}$, so

$$m = \frac{F}{\dfrac{v_f - v_i}{t}} = \frac{5{,}000.0 \ \dfrac{\text{kg} \cdot \text{m}}{\text{sec}^2}}{\dfrac{25.0 \text{ m/sec} - 0}{5.0 \text{ sec}}}$$

$$= \frac{5{,}000.0 \ \dfrac{\text{kg} \cdot \text{m}}{\text{sec}^2}}{5.0 \ \dfrac{\text{m}}{\text{sec}^2}}$$

$$= \frac{5{,}000.0}{5.0} \ \frac{\text{kg} \cdot \text{m}}{\text{sec}^2} \times \frac{\text{sec}^2}{\text{m}}$$

$$= 1{,}000 \ \frac{\text{kg} \cdot \text{m} \cdot \text{sec}^2}{\text{m} \cdot \text{sec}^2}$$

$$= \boxed{1.0 \times 10^3 \text{ kg}}$$

(b)

$$W = mg$$

$$= (1.0 \times 10^3 \text{ kg})\left(9.80 \ \frac{\text{m}}{\text{sec}^2}\right)$$

$$= (1.0 \times 10^3)(9.80) \ \text{kg} \times \frac{\text{m}}{\text{sec}^2}$$

$$= 9.8 \times 10^3 \ \frac{\text{kg} \cdot \text{m}}{\text{sec}^2}$$

$$= \boxed{9.8 \times 10^3 \text{ N}}$$

6.

$$W = mg$$

$$= (70.0 \text{ kg})\left(9.80 \ \frac{\text{m}}{\text{sec}^2}\right)$$

$$= 70.0 \times 9.80 \ \text{kg} \times \frac{\text{m}}{\text{sec}^2}$$

$$= 686 \ \frac{\text{kg} \cdot \text{m}}{\text{sec}^2}$$

$$= \boxed{686 \text{ N}}$$

7. The pound unit is an English unit of weight, which is a force (not a mass). The mass of the motorcycle is

$$W = mg \therefore m = \frac{W}{g} = \frac{320.0 \text{ lb}}{32.0 \text{ ft/sec}^2}$$

which can be used as a ratio for mass (instead of slug units) as follows

$$F = ma = (m)\left(\frac{v_f - v_i}{t}\right)$$

$$= \left(\frac{320.0 \text{ lb}}{32.0 \text{ ft/sec}^2}\right)\left(\frac{88.0 \text{ ft/sec} - 44.0 \text{ ft/sec}}{4.0 \text{ sec}}\right)$$

$$= \left(\frac{320.0}{32.0}\right)\left(\frac{44.0}{4.0}\right) \frac{\text{lb}}{1} \times \frac{\text{sec}^2}{\text{ft}} \times \frac{\text{ft}}{\text{sec}} \times \frac{1}{\text{sec}}$$

$$= (10.0)(11) \ \frac{\text{lb} \cdot \text{sec} \cdot \text{sec} \cdot \text{ft}}{\text{sec} \cdot \text{sec} \cdot \text{ft}}$$

$$= \boxed{110 \text{ lb}}$$

8. Again, the English unit of pound is a unit of weight (a force), not mass. The mass of the person is the ratio of

$$W = mg \therefore m = \frac{W}{g} = \frac{160.0 \text{ lb}}{32.0 \text{ ft/sec}^2} \text{ and}$$

$$F = ma = \left(\frac{160.0 \text{ lb}}{32.0 \text{ ft/sec}^2}\right)\left(23.0 \ \frac{\text{ft}}{\text{sec}^2}\right)$$

$$= \frac{160.0}{32.0} \times 23.0 \ \frac{\text{lb}}{1} \times \frac{\text{sec}^2}{\text{ft}} \times \frac{\text{ft}}{\text{sec}^2}$$

$$= (5.00)(23.0) \ \frac{\text{lb} \cdot \text{sec}^2 \cdot \text{ft}}{\text{sec}^2 \cdot \text{ft}}$$

$$= \boxed{115 \text{ lb}}$$

9. The cable must support the weight of the elevator and occupants (weight is a force) *plus* the force involved in the acceleration.

$$F = mg + ma$$

$$= (750.0 \text{ kg})(9.80 \text{ m/sec}^2) + (750.0 \text{ kg})(2.00 \text{ m/sec}^2)$$

$$= 7{,}350 \text{ N} + 1{,}500 \text{ N}$$

$$= \boxed{8{,}850 \text{ N}}$$

10. $F = ma$ and $a = \dfrac{v_f - v_i}{t}$ $\therefore$ $F = m\left(\dfrac{v_f - v_i}{t}\right)$ and

$$v_f = \dfrac{Ft}{m} + v_i = \dfrac{\left(1,000.0 \, \dfrac{kg \cdot m}{sec^2}\right)(10.0 \, sec)}{1,000.0 \, kg}$$

$$= \dfrac{(1,000.0)(10.0)}{1,000.0} \, \dfrac{kg \cdot m}{sec^2} \times \dfrac{sec}{1} \times \dfrac{1}{kg}$$

$$= 10.0 \, \dfrac{kg \cdot m \cdot sec}{kg \cdot sec \cdot sec}$$

$$= \boxed{10.0 \, \dfrac{m}{sec}}$$

11.
$$p = mv$$
$$= (50.0 \, kg)(2.00 \, m/sec)$$
$$= \boxed{100. \, kg \cdot m/sec}$$

(Note the lowercase p is the symbol used for momentum. This is one of the few cases where the English letter does not provide a clue about what it stands for. The units for momentum are also somewhat unusual for metric units since they do not have a name or single symbol to represent them.)

12.
$$F_c = \dfrac{mv^2}{r}$$

$$= \dfrac{(0.20 \, kg)\left(3.0 \, \dfrac{m}{sec}\right)^2}{1.5 \, m}$$

$$= \dfrac{(0.20 \, kg)\left(9.0 \, \dfrac{m^2}{sec^2}\right)}{1.5 \, m}$$

$$= \dfrac{0.20 \times 9.0}{1.5} \, \dfrac{kg \cdot m^2}{sec^2} \times \dfrac{1}{m}$$

$$= 1.2 \, \dfrac{kg \cdot m \cdot m}{sec^2 \cdot m}$$

$$= \boxed{1.2 \, N}$$

13. Note that unit conversions are necessary since the definition of the newton unit of force involves different units:

$$F_c = \dfrac{mv^2}{r} \therefore v = \sqrt{\dfrac{F_c r}{m}} = \sqrt{\dfrac{\left(1.0 \, \dfrac{kg \cdot m}{sec^2}\right)(0.50 \, m)}{0.10 \, kg}}$$

$$= \sqrt{\dfrac{1.0 \times 0.50}{0.10} \, \dfrac{kg \cdot m^2}{sec^2} \times \dfrac{1}{kg}}$$

$$= \sqrt{5.0 \, m^2/sec^2}$$

$$= \boxed{2.2 \, m/sec}$$

14. (a) $F_c = \dfrac{mv^2}{r} = \dfrac{(1,000.0 \, kg)(10.0 \, m/sec)^2}{20.0 \, m}$

$$= \dfrac{(1,000.0 \, kg)(100.0 \, m^2/sec^2)}{20.0 \, m}$$

$$= \dfrac{(1,000.0)(100.0)}{20.0} \, \dfrac{kg \cdot m^2}{sec^2} \times \dfrac{1}{m}$$

$$= 5,000 \, \dfrac{kg \cdot m \cdot m}{sec^2 \cdot m}$$

$$= \boxed{5.00 \times 10^3 \, N}$$

(b) The centripetal force that keeps a car on the road while moving around a curve is the frictional force between the tires and the road.

15. Newton's laws of motion consider the resistance to a change of motion, or mass, and not weight. The astronaut's mass is

$$W = mg \therefore m = \dfrac{W}{g} = \dfrac{1,960.0 \, \dfrac{kg \cdot m}{sec^2}}{9.80 \, \dfrac{m}{sec^2}}$$

$$= \dfrac{1,960.0}{9.80} \, \dfrac{kg \cdot m}{sec^2} \times \dfrac{sec^2}{m} = 200. \, kg$$

From Newton's second law of motion, you can see that the 100 N rocket gives the 200 kg astronaut an acceleration of

$$F = ma \therefore a = \dfrac{F}{m} = \dfrac{100 \, \dfrac{kg \cdot m}{sec^2}}{200 \, kg}$$

$$= \dfrac{100}{200} \, \dfrac{kg \cdot m}{sec^2} \times \dfrac{1}{kg} = 0.5 \, m/sec$$

An accleration of 0.5 m/sec² for 2.0 sec will result in a final velocity of

$$a = \dfrac{v_f - v_i}{t} \therefore v_f = at + v_i$$

$$= (0.5 \, m/sec^2)(2.0 \, sec)$$

$$= \boxed{1 \, m/sec}$$

Chapter 4

1. (a)
$$W = Fd$$
$$= (10.0 \, lb)(15.0 \, ft)$$
$$= (10.0)(15.0) \, ft \times lb$$
$$= \boxed{150. \, ft \cdot lb}$$

(b) The distance of the bookcase from some horizontal reference level did not change so the gravitational potential energy does not change.

2. The force (F) needed to lift the book is equal to the weight (W) of the book, or F = W. Since W = mg, then F = mg. Work is defined as the product of a force moved through a distance, or W = Fd. The work done in lifting the book is therefore W = mgd, and:

(a)

$$W = mgd$$

$$= (2.0 \text{ kg})(9.80 \text{ m/sec}^2)(2.00 \text{ m})$$

$$= (2.0)(9.80)(2.00) \ \frac{\text{kg} \cdot \text{m}}{\text{sec}^2} \times \text{m}$$

$$= 39.2 \ \frac{\text{kg} \cdot \text{m}^2}{\text{sec}^2}$$

$$= 39.2 \text{ J} = \boxed{39 \text{ J}}$$

(b)

$$PE = Mgh = \boxed{39 \text{ J}}$$

(c)

$$PE_{\text{lost}} = KE_{\text{gained}} = mgh = \boxed{39 \text{ J}}$$

(or)

$$v = \sqrt{2gh} = \sqrt{(2)(9.80 \text{ m/sec}^2)(2.00 \text{ m})}$$

$$= \sqrt{39.2 \text{ m}^2/\text{sec}^2} \longleftarrow \quad \text{(Note)}$$

$$= 6.26 \text{ m/sec}$$

$$KE = \frac{1}{2} \ mv^2 = \left(\frac{1}{2}\right)(2.0 \text{ kg})(6.26 \text{ m/sec})^2$$

$$= \left(\frac{1}{2}\right)(2.0 \text{ kg})(39.2 \text{ m}^2/\text{sec}^2)$$

$$= (1.0)(39.2) \ \frac{\text{kg} \cdot \text{m}^2}{\text{sec}^2}$$

$$= \boxed{39 \text{ J}}$$

3. Note that the gram unit must be converted to kg to be consistent with the definition of a newton-meter, or joule unit of energy:

$$KE = \frac{1}{2} \ mv^2 = \left(\frac{1}{2}\right)(0.15 \text{ kg})(30.0 \text{ m/sec})^2$$

$$= \left(\frac{1}{2}\right)(0.15 \text{ kg})(900. \text{ m}^2/\text{sec}^2)$$

$$= \left(\frac{1}{2}\right)(0.15 \text{ kg})(900.) \ \frac{\text{kg} \cdot \text{m}^2}{\text{sec}^2}$$

$$= 67.5 \text{ J} = \boxed{68 \text{ J}}$$

4. The km/hr unit must first be converted to m/sec before finding the kinetic energy. Note also that the work done to put an object in motion is equal to the energy of motion, or kinetic energy, that it has as a result of the work. The work needed to bring the object to a stop is also equal to the kinetic energy of the moving object:

Unit conversion:

$$1 \ \frac{\text{km}}{\text{hr}} = 0.2778 \ \frac{\frac{\text{m}}{\text{sec}}}{\frac{\text{km}}{\text{hr}}} = \left(90.0 \ \frac{\text{km}}{\text{hr}}\right)\left(0.2778 \ \frac{\frac{\text{m}}{\text{sec}}}{\frac{\text{km}}{\text{hr}}}\right) = 25.0 \text{ m/sec}$$

(a)

$$KE = \frac{1}{2} \ mv^2 = \frac{1}{2}(1,000.0 \text{ kg})(25.0 \text{ m/sec})^2$$

$$= \frac{1}{2} \ (1,000.0 \text{ kg})(625 \text{ m}^2/\text{sec}^2)$$

$$= \frac{1}{2} \ (1,000.0)(625) \ \frac{\text{kg} \cdot \text{m}^2}{\text{sec}^2}$$

$$= 312.5 \text{ kJ} = \boxed{313 \text{ kJ}}$$

(b)

$$W = Fd = KE = \boxed{313 \text{ kJ}}$$

(c)

$$KE = W = Fd = \boxed{313 \text{ kJ}}$$

5. $KE = \frac{1}{2}mv^2$

$$= \frac{1}{2} \ (60.0 \text{ kg})(3.00 \text{ m/sec})^2$$

$$= (30.0 \text{ kg})(9.00 \text{ m}^2/\text{sec}^2)$$

$$= 30.0 \times 9.00 \ \text{kg} \times \frac{\text{m}^2}{\text{sec}^2}$$

$$= \boxed{270. \text{ J}}$$

$$KE = \frac{1}{2} \ mv^2$$

$$= \frac{1}{2} \ (60.0 \text{ kg})(6.00 \text{ m/sec})^2$$

$$= (30.0 \text{ kg})(36.0 \text{ m}^2/\text{sec}^2)$$

$$= 30.0 \times 36.0 \ \text{kg} \times \frac{\text{m}^2}{\text{sec}^2}$$

$$= \boxed{1,080 \text{ J}}$$

Thus, doubling the speed results in a fourfold increase in kinetic energy.

6.

$$KE = \frac{1}{2} \ mv^2$$

$$= \frac{1}{2} \ (70.0 \text{ kg})(6.00 \text{ m/sec})^2$$

$$= (35.0 \text{ kg})(36.0 \text{ m}^2/\text{sec}^2)$$

$$= 35.0 \times 36.0 \ \text{kg} \times \frac{\text{m}^2}{\text{sec}^2}$$

$$= \boxed{1,260 \text{ J}}$$

$$KE = \frac{1}{2} mv^2$$

$$= \frac{1}{2} (140.0 \text{ kg})(6.00 \text{ m/sec})^2$$

$$= (70.0 \text{ kg})(36.0 \text{ m}^2/\text{sec}^2)$$

$$= 70.0 \times 36.0 \text{ kg} \times \frac{m^2}{sec^2}$$

$$= \boxed{2,520 \text{ J}}$$

Thus, doubling the mass results in a doubling of the kinetic energy.

7. **(a)** The force needed is equal to the weight of the student. The English unit of a lb is a force unit, so

$$W = Fd$$

$$= (170.0 \text{ lb})(25.0 \text{ ft})$$

$$= \boxed{4,250 \text{ ft} \cdot \text{lb}}$$

(b) Work (W) is defined as a force (F) moved through a distance (d), or $W = Fd$. Power (P) is defined as work (W) per unit of time (t), or $P = W/t$. Therefore

$$P = \frac{Fd}{t}$$

$$= \frac{(170.0 \text{ lb})(25.0 \text{ ft})}{10.0 \text{ sec}}$$

$$= \frac{(170.0)(25.0)}{10.0} \frac{ft \cdot lb}{sec}$$

$$= 425 \frac{ft \cdot lb}{sec}$$

1 hp is defined as $550. \dfrac{ft \cdot lb}{sec}$ and

$$\frac{425 \text{ ft} \cdot \text{lb/sec}}{550. \text{ ft} \cdot \text{lb/sec}} = \boxed{0.773 \text{ hp}}$$

Note that the student's power rating (425 ft·lb/sec) is less than the power rating defined as 1 hp (550. ft·lb/sec). Thus the student's horsepower must be *less* than 1 hp. A simple analysis such as this will let you know if you inverted the ratio or not.

8. **(a)** The force (F) needed to lift the elevator is equal to the weight of the elevator. Since the work (W) = Fd and power (P) = W/t, then

$$P = \frac{Fd}{t} \therefore t = \frac{Fd}{P}$$

$$= \frac{(2,000.0 \text{ lb})(20.0 \text{ ft})}{\left(550. \dfrac{\dfrac{ft \cdot lb}{sec}}{hp} \right)(20.0 \text{ hp})}$$

$$= \frac{40,000 \text{ ft} \cdot \text{lb}}{11,000 \dfrac{ft \cdot lb}{sec}} \times \frac{1}{hp} \times hp$$

$$= \frac{40,000 \text{ ft} \cdot \text{lb}}{11,000} \times \frac{sec}{1} \times \frac{sec}{ft \cdot lb}$$

$$= 3.64 \frac{ft \cdot lb \cdot sec}{ft \cdot lb}$$

$$= \boxed{3.64 \text{ sec}}$$

(b)

$$\bar{v} = \frac{d}{t}$$

$$= \frac{20.0 \text{ ft}}{3.64 \text{ sec}}$$

$$= \boxed{5.49 \text{ ft/sec}}$$

9. **(a)** 15.0 ft − 10.0 ft = 5.0 ft change of position by the brick falling to the lower ledge.

$$PE = mgh \text{ and } W = mg \therefore PE = Wh$$

$$= (10.0 \text{ lb})(5.0 \text{ ft})$$

$$= \boxed{50. \text{ ft} \cdot \text{lb loss}}$$

(b)

$$PE = Wh$$

$$= (10.0 \text{ lb})(10.0 \text{ ft})$$

$$= \boxed{100. \text{ ft} \cdot \text{lb}}$$

10. Note that some unit conversions are needed in this problem.
(a) A force of 1,300.0 N is applied over a distance of 20.0 m.

$$W = Fd$$

$$= \left(1,300.0 \frac{kg \cdot m}{sec^2} \right)(20.0 \text{ m})$$

$$= 1,300.0 \times 20.0 \text{ m} \times \frac{kg \cdot m}{sec^2}$$

$$= 26,000 \frac{kg \cdot m^2}{sec^2}$$

$$= \boxed{26.0 \text{ kJ}}$$

(b) The force needed in this case is the weight of the furniture.

$$W = Fd$$

$$= \left(6{,}500.0 \ \frac{kg \cdot m}{sec^2} \right)(4.00 \ m)$$

$$= 6{,}500.0 \times 4.00 \ m \times \frac{kg \cdot m}{sec^2}$$

$$= 26{,}000 \ \frac{kg \cdot m^2}{sec^2}$$

$$= \boxed{26.0 \ kJ}$$

(c) Less force is needed to push the furniture up the ramp (disregarding friction), but the work done is the same in either case.

11. Neglecting air resistance, the ball will rise to a maximum potential energy, then fall back to the surface with an increasing kinetic energy. Since the potential energy lost as the ball falls is equal to the kinetic energy gained, the kinetic energy at the surface is equal to the potential energy at the top of the ascent. The problem can therefore be solved by using the relationships of quantities in potential energy (note that the gram unit must be converted to kg):

$$PE = mgh \ \therefore \ h = \frac{PE}{mg}$$

$$= \frac{40.0 \ \dfrac{kg \cdot m^2}{sec^2}}{(0.200 \ kg)\left(9.80 \ \dfrac{m}{sec^2} \right)}$$

$$= \frac{40.0}{(0.200)(9.80)} \ \frac{kg \cdot m^2}{sec^2} \times \frac{1}{kg} \times \frac{sec^2}{m}$$

$$= \frac{40.0}{1.96} \ \frac{kg \cdot m \cdot m \cdot sec^2}{kg \cdot m \cdot sec^2}$$

$$= \boxed{20.4 \ m}$$

12. Since $PE_{lost} = KE_{gained}$ then $mgh = \frac{1}{2} \ mv^2$. Solving for v,

$$v = \sqrt{2gh} = \sqrt{(2)(32.0 \ ft/sec^2)(9.80 \ ft)}$$

$$= \sqrt{(2)(32.0)(9.80) \ ft^2/sec^2}$$

$$= \sqrt{627 \ ft^2/sec^2}$$

$$= \boxed{25.0 \ ft/sec}$$

13.

$$KE = \frac{1}{2}mv^2 \ \therefore \ v = \sqrt{\frac{2KE}{m}}$$

$$= \sqrt{\frac{(2)\left(200{,}000 \ \dfrac{kg \cdot m^2}{sec^2} \right)}{1{,}000.0 \ kg}}$$

$$= \sqrt{\frac{400{,}000}{1{,}000.0} \ \frac{kg \cdot m^2}{sec^2} \times \frac{1}{kg}}$$

$$= \sqrt{\frac{400{,}000}{1{,}000.0} \ \frac{kg \cdot m^2}{kg \cdot sec^2}}$$

$$= \sqrt{400 \ m^2/sec^2}$$

$$= \boxed{20 \ m/sec}$$

14. The maximum velocity occurs at the lowest point with a gain of kinetic energy equivalent to the loss of potential energy in falling 3.0 in (which is 0.25 ft), so

$$KE_{gained} = PE_{lost}$$

$$\frac{1}{2}mv^2 = mgh$$

$$v = \sqrt{2gh}$$

$$= \sqrt{(2)(32.0 \ ft/sec^2)(0.25 \ ft)}$$

$$= \sqrt{(2)(32.0)(0.25) \ ft/sec^2 \times ft}$$

$$= \sqrt{16 \ ft^2/sec^2}$$

$$= \boxed{4.0 \ ft/sec}$$

15. (a) $W = Fd$ and the force F that is needed to lift the load upward is mg, so $W = mgh$. Power is W/t, so

$$P = \frac{mgh}{t}$$

$$= \frac{(250 \ kg)(9.80 \ m/sec^2)(80.0 \ m)}{39.2 \ sec}$$

$$= \frac{(250.0)(9.80)(80.0)}{39.2} \ \frac{kg}{1} \times \frac{m}{sec^2} \times \frac{m}{1} \times \frac{1}{sec}$$

$$= \frac{196{,}000}{39.2} \ \frac{kg \cdot m^2}{sec^2} \times \frac{1}{sec}$$

$$= 5{,}000 \ \frac{J}{sec}$$

$$= \boxed{5.00 \ kw}$$

(b) There are 746 w per horsepower, so

$$\frac{5{,}000 \ w}{746 \ \dfrac{w}{hp}} = \frac{5{,}000}{746} \ \frac{w}{1} \times \frac{hp}{w}$$

$$= 6.70 \ \frac{w \cdot hp}{w}$$

$$= \boxed{6.70 \ hp}$$

Chapter 5

1. (a)

$$T_C = \frac{5}{9} \ (T_F - 32°)$$

$$= \frac{5}{9} \ (20.0 - 32°)$$

$$= \frac{5}{9} \ (-12.0°)$$

$$= -\frac{60.0}{9}$$

$$= \boxed{-6.67° \ C}$$

(b)

$$T_C = \frac{5}{9}(T_F - 32°)$$

$$= \frac{5}{9}(68.0° - 32°)$$

$$= \frac{5}{9}(36.0°)$$

$$= \frac{180.°}{9}$$

$$= \boxed{20.0° \text{ C}}$$

(c)

$$T_C = \frac{5}{9}(T_F - 32°)$$

$$= \frac{5}{9}(98.6° - 32°)$$

$$= \frac{5}{9}(66.6°)$$

$$= \frac{333°}{9}$$

$$= \boxed{37.0° \text{ C}}$$

(d)

$$T_C = \frac{5}{9}(T_F - 32°)$$

$$= \frac{5}{9}(118.0° - 32°)$$

$$= \frac{5}{9}(86.00°)$$

$$= \frac{430.0°}{9}$$

$$= \boxed{47.78° \text{ C}}$$

2. (a)

$$T_F = \frac{9}{5}T_C + 32°$$

$$= \frac{9}{5} \times 20.0° + 32°$$

$$= \frac{180.°}{5} + 32°$$

$$= 36.0° + 32°$$

$$= \boxed{68.0° \text{ F}}$$

(b)

$$T_F = \frac{9}{5}T_C + 32°$$

$$= \frac{9}{5} \times 100.0° + 32°$$

$$= \frac{900.0°}{5} + 32°$$

$$= 180.0° + 32°$$

$$= \boxed{212.0° \text{ F}}$$

(c)

$$T_F = \frac{9}{5}T_C + 32°$$

$$= \frac{9}{5} \times 50.0° + 32°$$

$$= \frac{450.°}{5} + 32°$$

$$= 90.0° + 32°$$

$$= \boxed{122° \text{ F}}$$

(d)

$$T_F = \frac{9}{5}T_C + 32°$$

$$= \frac{9}{5} \times 0.0° + 32°$$

$$= \frac{0.0°}{5} + 32°$$

$$= 0.0° + 32°$$

$$= \boxed{32° \text{ F}}$$

3. (a)

$$T_K = T_C + 273 \therefore T_C = T_K - 273$$

$$= 0 - 273$$

$$= \boxed{-273° \text{ C}}$$

(b)

$$T_F = \frac{9}{5}T_C + 32°$$

$$= \frac{9}{5}(-273°) + 32°$$

$$= \frac{-2457}{5} + 32°$$

$$= -491.4 + 32$$

$$= \boxed{-459° \text{ F}}$$

4. (a)

$$T_C = T_K - 273$$

$$= 4.00 - 273$$

$$= -269° \text{ C}$$

$$T_F = \frac{9}{5}T_C + 32°$$

$$= \frac{9}{5}(-269°) + 32°$$

$$= \boxed{-452° \text{ F}}$$

(b)

$$T_C = T_K - 273$$

$$= 100.0 - 273$$

$$= -173° \text{ C}$$

$$T_F = \frac{9}{5}T_C + 32°$$

$$= \frac{9}{5}(-173°) + 32°$$

$$= \boxed{-279° \text{ F}}$$

(c)

$$T_C = T_K - 273$$

$$= 300.0 - 273$$

$$= 27.00°\ C$$

$$T_F = \frac{9}{5}\ (27.00°) + 32°$$

$$= \frac{243.0}{5} + 32°$$

$$= \boxed{80.6°\ F}$$

(d)

$$T_C = T_K - 273$$

$$= 6,000.0 - 273$$

$$= 5,727.0°\ C$$

$$T_F = \frac{9}{5}\ (5727.0°) + 32°$$

$$= \frac{51,543.0°}{5} + 32°$$

$$= \boxed{10,340°\ F}$$

5.

$$Q = mc\Delta T$$

$$= (200.0\ g)\left(0.093\ \frac{cal}{g \cdot °C}\right)(40.0°\ C - 20.0°\ C)$$

$$= (200.0)(0.093)(20.0)\ g \times \frac{cal}{g \cdot °C} \times °C$$

$$= 372\ \frac{g \cdot cal \cdot °C}{g \cdot °C}$$

$$= \boxed{372\ cal}$$

6. First, you need to know the energy of the moving bike and rider. Since the speed is given as 36.00 km/hr, convert to m/sec by multiplying by 0.2778 m/sec per km/hr. Then,

$$KE = \frac{1}{2}\ mv^2$$

$$= \frac{1}{2}\ (100.0\ kg)(10.00\ m/sec)^2$$

$$= \frac{1}{2}\ (100.0\ kg)(100.0\ m^2/sec^2)$$

$$= \frac{1}{2}\ (100.0)(100.0)\ \frac{kg \cdot m^2}{sec^2}$$

$$= 5,000.\ J$$

Second, this energy is converted to the calorie heat unit through the mechanical equivalent of heat relationship, that 1.0 kcal = 4,185 J, or that 1.0 cal = 4.185 J. Thus,

$$\frac{1.0\ cal}{4.185\ J} = 0.2389\ \frac{cal}{J}$$

$$= (0.2389)(5,000)\ \frac{cal}{J} \times J$$

$$= 1,195\ \frac{cal \cdot J}{J}$$

$$= \boxed{1.195\ kcal}$$

7. First, you need to find the energy of the falling bag. Since the potential energy lost equals the kinetic energy gained, the energy of the bag just as it hits the ground can be found from

$$PE = mgh$$

$$= (10.0\ kg)(9.80\ m/sec^2)(5.00\ m)$$

$$= (10.0)(9.80)(5.00)\ \frac{kg \cdot m}{sec^2} \times m$$

$$= 490.\ J$$

In calories, this energy is equal to

$$(0.2389\ cal/J)(490.\ J) = 117\ cal$$

Second, the temperature change can be calculated from the equation giving the relationship between a quantity of heat (Q), mass (m), specific heat of the substance (c), and the change of temperature:

$$Q = mc\Delta T \ \therefore\ \Delta T = \frac{Q}{mc}$$

$$= \frac{0.117\ kcal}{(10.0\ kg)\left(0.250\ \frac{kcal}{kg \cdot °C}\right)}$$

$$= \frac{0.117}{(10.0)(0.250)}\ \frac{kcal}{1} \times \frac{1}{kg} \times \frac{kg \cdot °C}{kcal}$$

$$= 0.0468\ \frac{kcal \cdot kg \cdot °C}{kcal \cdot kg}$$

$$= \boxed{0.0468°\ C}$$

8. The Calorie used by dietitians is a kilocalorie, so 250.0 Cal is 250.0 kcal. The mechanical energy equivalent is 1 kcal = 4,185 J, so (250.0 kcal)(4,185 J/kcal) = 1,046,250 J.

Since $W = Fd$ and the force needed is equal to the weight (mg) of the person, $W = mgh = (75.0\ kg)(9.8\ m/sec^2)(10.0\ m)$

= 7,350 J for each stairway climb.

A total of 1,046,250 J would require $\frac{1,046,250\ J}{7,350\ J}$ per climb, or 142 trips up the stairs.

9. For unit consistency,

$$T_C = \frac{5}{9}\ (T_F - 32°) = \frac{5}{9}\ (68° - 32°) = \frac{5}{9}\ (36°) = 20°\ C$$

$$= \frac{5}{9}\ (32° - 32°) = \frac{5}{9}\ (0°) = 0°\ C$$

glass bowl

$$Q = mc\Delta T$$

$$= (0.5\ kg)\left(0.2\ \frac{kcal}{kg \cdot °C}\right)(20°\ C)$$

$$= (0.5)(0.2)(20)\ \frac{kg}{1} \times \frac{kcal}{kg \cdot °C} \times \frac{°C}{1}$$

$$= \boxed{2\ kcal}$$

iron pan

$$Q = mc\Delta T$$

$$= (0.5 \text{ kg})\left(0.11 \frac{\text{kcal}}{\text{kg} \cdot {}^\circ\text{C}}\right)(20^\circ \text{ C})$$

$$= (0.5)(0.11)(20) \text{ kg} \times \frac{\text{kcal}}{\text{kg} \cdot {}^\circ\text{C}} \times {}^\circ\text{C}$$

$$= \boxed{1 \text{ kcal}}$$

10. Note that a specific heat expressed in cal/g has the same numerical value as a specific heat expressed in kcal/kg because you can cancel the k units. Of course you could convert 896 cal to 0.896 kcal, but one of the two methods is needed for unit consistency.

$$Q = mc\Delta T \therefore m = \frac{Q}{c\Delta T}$$

$$= \frac{896 \text{ cal}}{\left(0.056 \frac{\text{cal}}{\text{g} \cdot {}^\circ\text{C}}\right)(80.0^\circ \text{ C})}$$

$$= \frac{896}{(0.056)(80.0)} \frac{\text{cal}}{1} \times \frac{\text{g} \cdot {}^\circ\text{C}}{\text{cal}} \times \frac{1}{{}^\circ\text{C}}$$

$$= 200 \text{ g}$$

$$= \boxed{0.200 \text{ kg}}$$

11. Since a watt is defined as a J/sec, finding the total energy in J will tell the time:

$$Q = mc\Delta T$$

$$= (250.0 \text{ g})\left(1.0 \frac{\text{cal}}{\text{g} \cdot {}^\circ\text{C}}\right)(60.0^\circ \text{ C})$$

$$= (250.0)(1.0)(60.0) \text{ g} \times \frac{\text{cal}}{\text{g} \cdot {}^\circ\text{C}} \times {}^\circ\text{C}$$

$$= 15,000 \text{ cal}$$

This energy in J is $(15,000 \text{ cal})\left(4.185 \frac{\text{J}}{\text{cal}}\right) = 62,775 \text{ J}$

A 300.0 w heater uses energy at a rate of $300.0 \frac{\text{J}}{\text{sec}}$ so $\frac{62,775}{300.0}$

$= 209.25$ sec is required, which is $\dfrac{209.25 \text{ sec}}{60 \dfrac{\text{sec}}{\text{min}}} = 3.49$ min, or

$$\boxed{\text{about } 3\frac{1}{2} \text{ min}}$$

12. $$Q = mc\Delta T \therefore c = \frac{Q}{m\Delta T}$$

$$= \frac{60 \text{ cal}}{(100.0 \text{ g})(20.0^\circ \text{ C})}$$

$$= \frac{60}{(100.0)(20.0)} \frac{\text{cal}}{\text{g} \cdot {}^\circ\text{C}}$$

$$= \boxed{0.03 \frac{\text{cal}}{\text{g} \cdot {}^\circ\text{C}}}$$

Chapter 6

1. Since the problem specified a solid changing to a liquid without a temperature change, you should recognize that this is a question about a phase change only. The phase change from solid to liquid (or liquid to solid) is concerned with the latent heat of fusion, and the relationship is given in equation 6.1. For water, the latent heat of fusion is given as 80.0 cal/g, and

$$m = 250.0 \text{ g} \qquad\qquad Q = mL_f$$

$$L_{f \text{ (water)}} = 80.0 \text{ cal/g} \qquad\qquad = (250.0 \text{ g})\left(80.0 \frac{\text{cal}}{\text{g}}\right)$$

$$Q = ?$$

$$= 250.0 \times 80.0 \frac{\text{g} \cdot \text{cal}}{\text{g}}$$

$$= 20,000 \text{ cal} = \boxed{20.0 \text{ kcal}}$$

2. To change water at 80.0° C to steam at 100.0° C requires two separate quantities of heat, which can be called Q_1 and Q_2: Q_1 is the amount of heat needed to warm the water from 80.0° C to the boiling point, 100.0° C ($\Delta T = 20.0^\circ$ C). You learned about this relationship in the previous chapter, $Q = mc\Delta T$. Q_2 is the amount of heat needed to take 100.0° C water through the phase change to steam (water vapor) at 100.0° C. The phase change from a liquid to a gas (or gas to liquid) is concerned with the latent heat of vaporization and the relationship is given in equation 6.2. For water, the latent heat of vaporization is given as 540.0 cal/g.

$$m = 250.0 \text{ g}$$

$$L_{v \text{ (water)}} = 540.0 \text{ cal/g}$$

$$Q = ?$$

$$Q_1 = mc\Delta T$$

$$= (250.0 \text{ g})\left(1.00 \frac{\text{cal}}{\text{g} \cdot {}^\circ\text{C}}\right)(20.0^\circ \text{ C})$$

$$= (250.0)(1.00)(20.0) \text{ g} \times \frac{\text{cal}}{\text{g} \cdot {}^\circ\text{C}} \times {}^\circ\text{C}$$

$$= 5,000 \frac{\text{g} \cdot \text{cal} \cdot {}^\circ\text{C}}{\text{g} \cdot {}^\circ\text{C}}$$

$$= 5,000 \text{ cal}$$

$$= 5.00 \text{ kcal}$$

$$Q_2 = mL_v$$

$$= (250.0 \text{ g})\left(540.0 \frac{\text{cal}}{\text{g}}\right)$$

$$= 250.0 \times 540.0 \frac{\text{g} \cdot \text{cal}}{\text{g}}$$

$$= 135,000 \text{ cal}$$

$$= 135.0 \text{ kcal}$$

$$Q_{\text{total}} = Q_1 + Q_2$$

$$= 5.00 \text{ kcal} + 135.0 \text{ kcal}$$

$$= \boxed{140.0 \text{ kcal}}$$

3. To change 20.0° C water to steam at 125.0° C requires three separate quantities of heat: Q_1, an amount of heat needed to warm the water from 20.0° C to 100.0° C ($\Delta T = 80.0$° C); Q_2, an amount of heat needed to take 100.0° C water to steam at 100.0° C, and; Q_3, an amount of heat needed to warm the steam from 100.0° C to 125.0° C. According to table 5.1, the c for steam is 0.48 cal/g·°C.

$$m = 100.0 \text{ g}$$

$$\Delta T_{water} = 80.0° \text{ C}$$

$$\Delta T_{steam} = 25.0° \text{ C}$$

$$L_{v \text{ (water)}} = 540.0 \text{ cal/g}$$

$$c_{steam} = 0.48 \text{ cal/g·°C}$$

$$Q_1 = mc\Delta T$$

$$= (100.0 \text{ g})\left(1.00 \ \frac{cal}{g·°C}\right)(80.0° \text{ C})$$

$$= (100.0)(1.00)(80.0) \text{ g} \times \frac{cal}{g·°C} \times °C$$

$$= 8,000 \ \frac{g·cal·°C}{g·°C}$$

$$= 8,000 \text{ cal}$$

$$= 8.00 \text{ kcal}$$

$$Q_2 = mL_v$$

$$= (100.0 \text{ g})\left(540.0 \ \frac{cal}{g}\right)$$

$$= 100.0 \times 540.0 \ \frac{g·cal}{g}$$

$$= 54,000 \text{ cal}$$

$$= 54.00 \text{ kcal}$$

$$Q_3 = mc\Delta T$$

$$= (100.0 \text{ g})\left(0.48 \ \frac{cal}{g·°C}\right)(25.0° \text{ C})$$

$$= (100.0)(0.48)(25.0) \text{ g} \times \frac{cal}{g·°C} \times °C$$

$$= 1,200 \ \frac{g·cal·°C}{g·°C}$$

$$= 1,200 \text{ cal}$$

$$= 1.2 \text{ kcal}$$

$$Q_{total} = Q_1 + Q_2 + Q_3$$

$$= 8.00 \text{ kcal} + 54.00 \text{ kcal} + 1.2 \text{ kcal}$$

$$= \boxed{63.2 \text{ kcal}}$$

4. (a) **Step 1:** Cool water from 18.0° C to 0° C.

$$Q_1 = mc\Delta T$$

$$= (400.0 \text{ g})\left(1.00 \ \frac{cal}{g·°C}\right)(18.0° \text{ C})$$

$$= (400.0)(1.00)(18.0) \text{ g} \times \frac{cal}{g·°C} \times °C$$

$$= 7,200 \ \frac{g·cal·°C}{g·°C}$$

$$= 7,200 \text{ cal}$$

$$= 7.20 \text{ kcal}$$

Step 2: Find the energy needed for the phase change of water at 0° C to ice at 0° C.

$$Q_2 = mL_f$$

$$= (400.0 \text{ g})\left(80.0 \ \frac{cal}{g}\right)$$

$$= 400.0 \times 80.0 \ \frac{g·cal}{g}$$

$$= 32,000 \text{ cal}$$

$$= 32.0 \text{ kcal}$$

Step 3: Cool the ice from 0° C to ice at −5.00° C.

$$Q_3 = mc\Delta T$$

$$= (400.0 \text{ g})\left(0.50 \ \frac{cal}{g·°C}\right)(5.00° \text{ C})$$

$$= 400.0 \times 0.50 \text{ g} \times 5.00 \ \frac{cal}{g·°C} \times °C$$

$$= 1,000 \ \frac{g·cal·°C}{g·°C}$$

$$= 1,000 \text{ cal}$$

$$= 1.0 \text{ kcal}$$

$$Q_{total} = Q_1 + Q_2 + Q_3$$

$$= 7.20 \text{ kcal} + 32.0 \text{ kcal} + 1.0 \text{ kcal}$$

$$= \boxed{40.2 \text{ kcal}}$$

(b)

$$Q = mL_v, \therefore m = \frac{Q}{L_v}$$

$$= \frac{40,200 \text{ cal}}{40.0 \ \frac{cal}{g}}$$

$$= \frac{40,200}{40.0} \ \frac{cal}{1} \times \frac{g}{cal}$$

$$= 1,005 \ \frac{cal·g}{cal}$$

$$= \boxed{1,010 \text{ g}}$$

5. The relationship found in equation 6.8 is needed for this problem, but first the temperature must be converted to absolute values,

$$T_K = \frac{5}{9}(T_F - 32) + 273 \therefore$$

$$T_H = \frac{5}{9}(4{,}937 - 32) + 273 = 2{,}998 \text{ K}$$

$$T_L = \frac{5}{9}(2{,}832 - 32) + 273 = 1{,}829 \text{ K}$$

Applying these Kelvin temperatures in equation 6.8,

$$e = 1 - \frac{T_L}{T_H} \times 100\%$$

$$= 1 - \frac{1{,}829 \text{ K}}{2{,}998 \text{ K}} \times 100\%$$

$$= 1 - 0.6100 \times 100\%$$

$$= 0.3900 \times 100\%$$

$$= \boxed{39.00\%}$$

6. The relationship between the given quantities is found in equation 6.6,

$$e = \frac{W}{JQ_H} \times 100\%$$

$$= \frac{2{,}092 \text{ J}}{\left(4{,}184 \ \frac{\text{J}}{\text{kcal}}\right)(5.00 \text{ kcal})} \times 100\%$$

$$= \frac{2{,}092 \text{ J}}{4{,}184 \times 5.00 \ \frac{\text{J} \cdot \text{kcal}}{\text{kcal}}} \times 100\%$$

$$= \frac{2{,}092 \text{ J}}{20{,}920 \text{ J}} \times 100\%$$

$$= 0.100 \times 100\%$$

$$= \boxed{10.0\%}$$

7. (a) From equation 6.5, note that the heat values (or the value of J) must be converted for unit consistency.

$$W = J(Q_H - Q_L)$$

$$= 4{,}184 \ \frac{\text{J}}{\text{kcal}}(0.3000 \text{ kcal} - 0.2000 \text{ kcal})$$

$$= 4{,}184 \ \frac{\text{J}}{\text{kcal}} \times 0.1000 \text{ kcal}$$

$$= 4{,}184 \times 0.1000 \ \frac{\text{J} \cdot \text{kcal}}{\text{kcal}}$$

$$= \boxed{418.4 \text{ J}}$$

(b) From equation 6.7,

$$e = 1 - \frac{Q_L}{Q_H} \times 100\%$$

$$= 1 - \frac{200.0 \text{ cal}}{300.0 \text{ cal}} \times 100\%$$

$$= 1 - 0.6667 \times 100\%$$

$$= 0.3333 \times 100\%$$

$$= \boxed{33.33\%}$$

8. Equation 6.8 requires absolute temperatures, so these are converted first,

$$T_H = \frac{5}{9}(80.0 - 32) + 273 = 300. \text{ K}$$

$$T_L = \frac{5}{9}(40.0 - 32) + 273 = 277 \text{ K}$$

$$e = 1 - \frac{T_L}{T_H} \times 100\%$$

$$= 1 - \frac{277 \text{ K}}{300. \text{ K}} \times 100\%$$

$$= 1 - 0.923 \times 100\%$$

$$= 0.077 \times 100\%$$

$$= \boxed{7.7\%}$$

9. From equation 6.5,

$$W = J(Q_H - Q_L)$$

$$= 4{,}184 \ \frac{\text{J}}{\text{kcal}}(55.0 \text{ kcal} - 40.0 \text{ kcal})$$

$$= \left(4{,}184 \ \frac{\text{J}}{\text{kcal}}\right)(15.0 \text{ kcal})$$

$$= 4{,}184 \times 15.0 \ \frac{\text{J} \cdot \text{kcal}}{\text{kcal}}$$

$$= 62{,}760 \text{ J}$$

$$= \boxed{62.8 \text{ kJ}}$$

10. The graph in figure 6.10 gives the maximum amount of water vapor that can be in the air at various temperatures. The graph has temperatures in ° C, so temperature conversions are required (or remember that 32.0° F = 0.00° C and that 68.0° F, which is defined as "room temperature," is 20.0° C).

First, find 0° C on the graph and trace up and down the 0° C line to find the maximum water vapor content at that temperature. It intersects at about 5 g/m³. Since the relative humidity is given as 30 percent, this means that the air has an actual moisture content of 5 g/m³ × 0.30, or 1.5 g/m³.

Second, find the maximum water vapor content for air at 20° C. According to the graph, this value is about 18 g/m³. Now you are ready to use equation 6.3 to find the relative humidity,

$$\text{relative humidity} = \frac{g/m^3_{(present)}}{g/m^3_{(max)}} \times 100\%$$

$$= \frac{1.50 \ g/m^3}{18.0 \ g/m^3} \times 100\%$$

$$= 0.0833 \times 100\%$$

$$= \boxed{8.33\%}$$

Chapter 7

1.

$$v = f\lambda$$

$$= \left(10.0 \ \frac{1}{sec}\right)(0.500 \ m)$$

$$= \boxed{5.00 \ \frac{m}{sec}}$$

2. The distance between two *consecutive* condensations (or rarefactions) is one wavelength, so $\lambda = 3.00$ m and

$$v = f\lambda$$

$$= \left(112.0 \ \frac{1}{sec}\right)(3.00 \ m)$$

$$= \boxed{336 \ \frac{m}{sec}}$$

3. (a) One complete wave every 4.00 sec means that T = 4.00 sec. (Note that the symbol for the *time of a cycle* is T. Do not confuse this symbol with the symbol for temperature.)

(b)

$$f = \frac{1}{T}$$

$$= \frac{1}{4.00 \ sec}$$

$$= \frac{1}{4.00} \ \frac{1}{sec}$$

$$= 0.250 \ \frac{1}{sec}$$

$$= \boxed{0.250 \ hz}$$

4. The distance from one condensation to the next is one wavelength, so

$$v = f\lambda \ \therefore \ \lambda = \frac{v}{f}$$

$$= \frac{330. \ \dfrac{m}{sec}}{260. \ \dfrac{1}{sec}}$$

$$= \frac{330. \ m}{260. \ sec} \times \frac{sec}{1}$$

$$= \boxed{1.27 \ m}$$

5. (a)

$$v = f\lambda = \left(256 \ \frac{1}{sec}\right)(1.34 \ m) = \boxed{343 \ m/sec}$$

(b)

$$= \left(440. \ \frac{1}{sec}\right)(0.780 \ m) = \boxed{343 \ m/sec}$$

(c)

$$= \left(750. \ \frac{1}{sec}\right)(0.457 \ m) = \boxed{343 \ m/sec}$$

(d)

$$= \left(2,500. \ \frac{1}{sec}\right)(0.1372 \ m) = \boxed{343 \ m/sec}$$

6. The speed of sound at 0.0° C is 1,087 ft/sec, and

(a)

$$V_{T_F} = V_0 + \left(\frac{2.0 \ ft/sec}{°C}\right)(T_F)$$

$$= 1,087 \ ft/sec + \left(\frac{2.0 \ ft/sec}{°C}\right)(0.0° \ C)$$

$$= 1,087 + (2.0)(0.0) \ ft/sec + \frac{ft/sec}{°C} \times °C$$

$$= 1,087 \ ft/sec + 0.0 \ ft/sec$$

$$= \boxed{1,087 \ ft/sec}$$

(b)

$$V_{20°} = 1,087 \ ft/sec + \left(\frac{2.0 \ ft/sec}{°C}\right)(20.0° \ C)$$

$$= 1,087 \ ft/sec + 40.0 \ ft/sec$$

$$= \boxed{1,127 \ ft/sec}$$

(c)

$$V_{40°} = 1,087 \ ft/sec + \left(\frac{2.0 \ ft/sec}{°C}\right)(40.0° \ C)$$

$$= 1,087 \ ft/sec + 80.0 \ ft/sec$$

$$= \boxed{1,167 \ ft/sec}$$

(d)

$$V_{80°} = 1,087 \ ft/sec + \left(\frac{2.0 \ ft/sec}{°C}\right)(80.0° \ C)$$

$$= 1,087 \ ft/sec + 160. \ ft/sec$$

$$= \boxed{1,247 \ ft/sec}$$

7. For consistency with the units of the equation given, 43.7° F is first converted to 6.50° C. The velocity of sound in this air is

$$V_{T_F} = V_0 + \left(\frac{2.0 \ ft/sec}{°C}\right)(T_F)$$

$$= 1,087 \ ft/sec + \left(\frac{2.0 \ ft/sec}{°C}\right)(6.50° \ C)$$

$$= 1,087 \ ft/sec + 13 \ ft/sec$$

$$= 1,100 \ ft/sec$$

The distance that a sound with this velocity travels in the given time is

$$v = \frac{d}{t} \therefore d = vt$$

$$= (1{,}100 \text{ ft/sec})(4.80 \text{ sec})$$

$$= (1{,}100)(4.80) \; \frac{\text{ft} \cdot \text{sec}}{\text{sec}}$$

$$= 5{,}280 \text{ ft}$$

$$= \frac{5{,}280 \text{ ft}}{2}$$

$$= \boxed{2{,}640 \text{ ft}}$$

Since the sound traveled from the rifle to the cliff and then back, the cliff must be 5,280 feet/2 = 2,640 feet, or about one-half mile away.

8. This problem requires three steps, (1) conversion of the F temperature value to C, (2) calculating the velocity of sound in air at this temperature, and (3) calculating the distance from the calculated velocity and the given time:

$$V_{T_F} = V_0 + \left(\frac{2.0 \text{ ft/sec}}{^\circ C} \right)(T_F)$$

$$= 1{,}087 \text{ ft/sec} + \left(\frac{2.0 \text{ ft/sec}}{^\circ C} \right)(26.67^\circ \text{ C})$$

$$= 1{,}087 \text{ ft/sec} + 53 \text{ ft/sec} = 1{,}140. \text{ ft/sec}$$

$$v = \frac{d}{t} \therefore d = vt$$

$$= (1{,}140. \text{ ft/sec})(4.63 \text{ sec})$$

$$= \boxed{5{,}280 \text{ ft (about 1 mi)}}$$

9. A tube closed at one end must have a minimum of 1/4 wavelength in order to have a node at the closed end and a resonant antinode at the open end. Therefore,

$$L = \frac{1}{4}\lambda \therefore \lambda = 4 \text{ L}$$

$$= (4)(0.25 \text{ m})$$

$$= 1.0 \text{ m}$$

$$v = f\lambda$$

$$= \left(340.0 \; \frac{1}{\text{sec}} \right)(1.0 \text{ m})$$

$$= (340.0)(1.0) \; \frac{1}{\text{sec}} \times \text{m}$$

$$= \boxed{340 \; \frac{\text{m}}{\text{sec}}}$$

10. (a)
$$v = f\lambda \therefore \lambda = \frac{v}{f}$$

$$= \frac{1{,}125 \; \frac{\text{ft}}{\text{sec}}}{440 \; \frac{1}{\text{sec}}}$$

$$= \frac{1{,}125}{440} \; \frac{\text{ft}}{\text{sec}} \times \frac{\text{sec}}{1}$$

$$= 2.56 \; \frac{\text{ft} \cdot \text{sec}}{\text{sec}}$$

$$= \boxed{2.6 \text{ ft}}$$

(b)
$$v = f\lambda \therefore \lambda = \frac{v}{f}$$

$$= \frac{5{,}020}{440} \; \frac{\text{ft}}{\text{sec}} \times \frac{\text{sec}}{1}$$

$$= 11.4 \text{ ft} = \boxed{11 \text{ ft}}$$

11. The wavelength can be found from the relationship between the length of the air column and λ (see the solution to problem number 9): $\lambda = 4L = 4(0.240 \text{ m}) = 0.960 \text{ m}$. The velocity can be determined from the given air temperature:

$$v_{T_P} = v_0 + \left(\frac{0.60 \text{ m/sec}}{^\circ C} \right)(T_P)$$

$$= 331 \text{ m/sec} + \left(\frac{0.60 \text{ m/sec}}{^\circ C} \right)(20.0^\circ \text{ C})$$

$$= 331 \text{ m/sec} + 12 \text{ m/sec} = 343 \text{ m/sec}$$

From the wave equation,

$$v = f\lambda \therefore f = \frac{v}{\lambda}$$

$$= \frac{343 \text{ m/sec}}{0.960 \text{ m}}$$

$$= \frac{343}{0.960} \; \frac{\text{m}}{\text{sec}} \times \frac{1}{\text{m}}$$

$$= 357 \; \frac{1}{\text{sec}}$$

$$= \boxed{357 \text{ hz}}$$

12. The fundamental frequency is found from the relationship:

$$f_n = \frac{nv}{4L} \text{ where } n = 1 \text{ is the fundamental and } n = 3, 5, 7, \text{ and}$$

so on are the possible harmonics.

Assuming room temperature, since you have to assume some temperature to find the velocity (room temperature is defined to be 20.0° C), the velocity of sound is 343 m/sec. Thus

fundamental frequency (n = 1):

$$f = \frac{nv}{4L} = \frac{(1)(343 \text{ m/sec})}{(4)(0.700 \text{ m})}$$

$$= \frac{343}{2.80} \frac{m}{sec} \times \frac{1}{m}$$

$$= 122.5 \frac{1}{sec}$$

$$= \boxed{123 \text{ hz}}$$

first overtone above the fundamental (n = 3):

$$f = \frac{nv}{4L} = \frac{(3)(343 \text{ m/sec})}{(4)(0.700)}$$

$$= \frac{1,029}{2.80} \frac{m}{sec} \times \frac{1}{m}$$

$$= 367.5 \frac{1}{sec}$$

$$= \boxed{368 \text{ hz}}$$

Chapter 8

1. The fundamental charge (q) of a single (n = 1) electron (e) is 1.60×10^{-19} C. The negative charge means an excess of electrons and

$$q = ne, \therefore n = \frac{q}{e}$$

$$= \frac{1.00 \times 10^{-14} \text{ C}}{1.60 \times 10^{-19} \frac{C}{electron}}$$

$$= \frac{1.00 \times 10^{-14}}{1.60 \times 10^{-19}} \frac{C}{1} \times \frac{electron}{C}$$

$$= 6.25 \times 10^4 \frac{C \cdot electron}{C}$$

$$= \boxed{6.25 \times 10^4 \text{ electron}}$$

2. (a) The force is away from each other since both charges are negative.

(b)

$$F = \frac{kq_1q_2}{d^2}$$

$$= \frac{(9.00 \times 10^9 \text{ N·m}^2/\text{C}^2)(3.00 \times 10^{-14} \text{ C})(2.00 \times 10^{-12} \text{ C})}{(2.00 \times 10^{-2} \text{ m})^2}$$

$$= \frac{(9.00 \times 10^9)(3.00 \times 10^{-14})(2.00 \times 10^{-12})}{4.00 \times 10^{-4}} \frac{\frac{N \cdot m^2}{C^2} \times C \times C}{m^2}$$

$$= \frac{5.40 \times 10^{-16}}{4.00 \times 10^{-4}} \frac{N \cdot m^2}{C^2} \times C^2 \times \frac{1}{m^2}$$

$$= \boxed{1.35 \times 10^{-12} \text{ N}}$$

3.

$$\text{potential difference} = \frac{work}{charge}$$

or

$$V = \frac{W}{q}$$

$$= \frac{7.50 \text{ J}}{5.00 \text{ C}}$$

$$= 1.50 \frac{J}{C}$$

$$= \boxed{1.50 \text{ V}}$$

4.

$$\text{electric current} = \frac{charge}{time}$$

or

$$I = \frac{q}{t}$$

$$= \frac{6.00 \text{ C}}{2.00 \text{ sec}}$$

$$= 3.00 \frac{C}{sec}$$

$$= \boxed{3.00 \text{ A}}$$

5. A current of 1.00 amp is defined as 1.00 C/sec. Since the fundamental charge of the electron is 1.60×10^{-19} C/electron

$$\frac{1.00 \frac{C}{sec}}{1.60 \times 10^{-19} \frac{C}{electron}}$$

$$= 6.30 \times 10^{18} \frac{C}{sec} \times \frac{electron}{C}$$

$$= \boxed{6.30 \times 10^{18} \frac{electrons}{sec}}$$

6.

$$\text{resistance} = \frac{potential \ difference}{electric \ current}$$

or

$$R = \frac{V}{I}$$

$$= \frac{120. \text{ V}}{4.00 \text{ A}}$$

$$= 30.0 \frac{V}{A}$$

$$= \boxed{30.0 \ \Omega}$$

7.

$$R = \frac{V}{I} \therefore I = \frac{V}{R}$$

$$= \frac{120.\ V}{60.0\ \dfrac{V}{A}}$$

$$= \frac{120.}{60.0}\ V \times \frac{A}{V}$$

$$= \boxed{2.00\ A}$$

8. (a)

$$R = \frac{V}{I} \therefore V = IR$$

$$= (1.20\ A)\left(10.0\ \frac{V}{A}\right)$$

$$= \boxed{12.0\ V}$$

(b)

power = (current)(potential difference)

or

$$P = IV$$

$$= \left(1.20\ \frac{C}{sec}\right)\left(12.0\ \frac{J}{C}\right)$$

$$= (1.20)(12.0)\ \frac{C}{sec} \times \frac{J}{C}$$

$$= 14.4\ \frac{J}{sec}$$

$$= \boxed{14.4\ w}$$

9. Note that there are two separate electrical units that are rates: the amp (coulomb/sec), and the watt (joule/sec). The question asked for a rate of using energy, which is measured in joules, so you are looking for the power of the radio in watts:

$$I = \frac{V}{R}$$

$$= \frac{3.00\ V}{15.0\ \dfrac{V}{A}}$$

$$= 0.200\ A$$

$$P = IV$$

$$= (0.200\ C/sec)(3.00\ J/C)$$

$$= \boxed{0.600\ w}$$

10.

$$\frac{(1,200\ w)(0.25\ hr)}{1,000.\ w/kw} = 0.30\ kwhr$$

$$(0.30\ kwhr)(\$0.10/kwhr) = \boxed{\$0.03\ (3¢)}$$

11. The relationship between power (P), current (I), and volts (V) will provide a solution. Since the relationship considers power in watts the first step is to convert horsepower to watts. One horsepower is equivalent to 746 watts, so:

$$(746\ w/hp)(2.00\ hp) = 1,492\ w$$

$$P = IV \therefore I = \frac{P}{V}$$

$$= \frac{1,492\ \dfrac{J}{sec}}{12.0\ \dfrac{J}{C}}$$

$$= \frac{1,492}{12.0}\ \frac{J}{sec} \times \frac{C}{J}$$

$$= 124.3\ \frac{C}{sec}$$

$$= \boxed{124\ A}$$

12. (a) The rate of using energy is J/sec, or the watt. Since 1.00 hp = 746 w,

inside motor: (746 w/hp)(1/3 hp) = 249 w

outside motor: (746 w/hp)(1/3 hp) = 249 w

compressor motor: (746 w/hp)(3.70 hp) = 2,760 w

249 w + 249 w + 2,760 w = $\boxed{3,258\ w}$

(b)

$$\frac{(3,258\ w)(1.00\ hr)}{1,000.\ w/kw} = 3.26\ kwhr$$

$$(3.26\ kwhr)(\$0.10/kwhr) = \boxed{\$0.33\ per\ hr}$$

(c) ($0.33/hr)(12 hr/day)(30. day/mo) = $118.80 = $\boxed{\$120}$

13. The solution is to find how much current each device draws and then to see if the total current is less or greater than the breaker rating:

$$\text{toaster: } I = \frac{V}{R} = \frac{120.\ V}{15.0\ V/A} = 8.00\ A$$

motor: (0.20 hp)(746 w/hp) = 1.2 A

three 100 w bulbs = (3)(100. w) = 300. w

$$I = \frac{P}{V} = \frac{300.\ J/sec}{120.\ J/C} = 2.50\ A$$

$$\text{iron: } I = \frac{600.\ J/sec}{120.\ J/C} = 5.00\ A$$

The sum of the currents from each of the devices is 16.7 A, so the total current is greater than 15.0 amp and the circuit breaker will trip.

Chapter 9

1. $L = 150.$ cm
$= 1.50$ m
$B = 0.400$ tesla
$v = 30.0$ cm/sec
$= 0.300$ m/sec
$V = ?$

$V = BLv$

$= \left(0.400 \ \frac{N}{A \cdot m}\right)(1.50 \text{ m})\left(0.300 \ \frac{m}{sec}\right)$

$= (0.400)(1.50)(0.300) \ \frac{N}{\frac{C}{sec} \cdot m} \times m \times \frac{m}{sec}$

$= 0.180 \ \frac{N \cdot m}{C}$

$= 0.180 \ \frac{J}{C}$

$= \boxed{0.180 \text{ V}}$

2. (a) $V_P = 1{,}200.$ V
$N_P = 1$ loop
$N_S = 200$ loops
$V_S = ?$

$\frac{V_P}{N_P} = \frac{V_s}{N_s} \ \therefore \ V_s = \frac{V_P N_s}{N_P}$

$V_s = \frac{(1{,}200. \text{ V})(200 \text{ loops})}{1 \text{ loop}}$
$= \boxed{240{,}000 \text{ V}}$

(b) $I_P = 40.0$ A $V_P I_P = V_s I_s \ \therefore \ I_s = \frac{V_P I_P}{V_s}$
$I_s = ?$

$I_s = \frac{1{,}200. \text{ V} \times 40.0 \text{ A}}{240{,}000 \text{ V}}$

$= \frac{1{,}200 \times 40.0}{240{,}000} \ \frac{V \cdot A}{V}$

$= \boxed{0.20 \text{ A}}$

3. $B = 2.00 \times 10^{-5}$ tesla
$L = 5.00$ m
$v = 1.00$ m/sec
$V = ?$

$V = BLv$

$= \left(2.00 \times 10^{-5} \ \frac{N}{A \cdot m}\right)(5.00 \text{ m})\left(1.00 \ \frac{m}{sec}\right)$

$= (2.00 \times 10^{-5})(5.00)(1.00) \ \frac{N}{\frac{C}{sec} \cdot m} \times m \times \frac{m}{sec}$

$= 1.00 \times 10^{-4} \ \frac{N \cdot m}{C}$

$= 1.00 \times 10^{-4} \ \frac{J}{C}$

$= \boxed{1.00 \times 10^{-4} \text{ V}}$

4. $L = 100.$ m
$B = 2.00 \times 10^{-5}$ tesla
$P = 48.0$ kw
$E_P = 240{,}000$ V
$I = ?$
$F = ?$

Step 1:

$P = IV \ \therefore \ I = \frac{P}{V}$

$I = \frac{48{,}000 \ \frac{J}{sec}}{240{,}000 \ \frac{J}{C}}$

$= \frac{48{,}000}{240{,}000} \ \frac{J}{sec} \times \frac{C}{J}$

$= 0.20 \ \frac{J \cdot C}{J \cdot sec}$

$= 0.20$ A

Step 2:

$F_{max} = ILB$

$= \left(0.20 \ \frac{C}{sec}\right)(100. \text{ m})\left(2.00 \times 10^{-5} \ \frac{N}{\frac{C}{sec} \cdot m}\right)$

$= (0.20)(100.)(2.00 \times 10^{-5}) \ \frac{C}{sec} \times \frac{m}{1} \times \frac{N}{1} \times \frac{sec}{C} \times \frac{1}{m}$

$= \boxed{4.0 \times 10^{-4} \text{ N}}$

5. $L = 500.$ cm $F_{max} = ILB \ \therefore \ B = \frac{F_{max}}{IL}$
$= 5.00$ m
$F = 0.400$ N $B = \frac{0.400 N}{(8.00 \text{ A})(5.00 \text{ m})}$
$I = 8.00$ A
$B = ?$ $= \frac{0.400}{(8.00)(5.00)} \ \frac{N}{A \cdot m}$
$= \boxed{0.010 \text{ tesla}}$

6. (a) For this and other problems where you need the charge of a single particle, recall that the fundamental charge of an electron (and a proton) is 1.60×10^{-19} C.

$B = 0.500$ tesla

$v = 6.00 \times 10^{-5}$ m/sec $\qquad B = \dfrac{F_{max}}{qv} \therefore F_{max} = Bqv$

$m = 9.00 \times 10^{-32}$ kg

$q = 1.60 \times 10^{-19}$ C

$F = ?$

$$F_{max} = \left(0.500 \ \dfrac{N}{\frac{C}{sec} \cdot m}\right)(1.60 \times 10^{-19} \ C)\left(6.00 \times 10^{-5} \ \dfrac{m}{sec}\right)$$

$$= (0.500)(1.60 \times 10^{-19})(6.00 \times 10^{-5})$$

$$\dfrac{N}{1} \times \dfrac{sec}{C} \times \dfrac{1}{m} \times \dfrac{C}{1} \times \dfrac{m}{sec}$$

$$= \boxed{4.80 \times 10^{-24} \ N}$$

(b) $\qquad\qquad F = ma \therefore a = \dfrac{F}{m}$

$$a = \dfrac{4.80 \times 10^{-24} \ \dfrac{kg \cdot m}{sec^2}}{9.00 \times 10^{-32} \ kg}$$

$$= \dfrac{4.80 \times 10^{-24}}{9.00 \times 10^{-32}} \ \dfrac{kg \cdot m}{sec^2} \times \dfrac{1}{kg}$$

$$= \boxed{5.33 \times 10^7 \ \dfrac{m}{sec^2}}$$

7. $V = 765{,}000$ V

$I = 0.260$ A

$d = 60.0$ m

$k = 1.00 \times 10^{-5}$ N/A²

$B = ?$

$$B = k \dfrac{2I}{d}$$

$$= \left(1.00 \times 10^{-5} \ \dfrac{N}{A^2}\right) \dfrac{2 \times 0.260 \ A}{60.0 \ m}$$

$$= (1.00 \times 10^{-5}) \dfrac{2 \times 0.260}{60.0} \ \dfrac{N}{A \cdot A} \times \dfrac{A}{1} \times \dfrac{1}{m}$$

$$= \dfrac{5.20 \times 10^{-6}}{60.0} \ \dfrac{N}{A \cdot m}$$

$$= \boxed{8.67 \times 10^{-8} \ \text{tesla}}$$

8. (a) According to the left-hand rule used for electron current, pointing your thumb to the east will result in your fingers pointing south horizontally across the floor.

(b) $I = 5.00$ A

$d = 1.00$ m

$k = 1.00 \times 10^{-7}$ N/A²

$B = ?$

$$B = k \dfrac{2 \ I}{d}$$

$$= \left(1.00 \times 10^{-7} \ \dfrac{N}{A^2}\right) \dfrac{2 \times 5.00 \ A}{1.00 \ m}$$

$$= \dfrac{(1.00 \times 10^{-7})(2)(5.00)}{1.00} \ \dfrac{N}{A \cdot A} \times \dfrac{A}{m}$$

$$= 1.00 \times 10^{-6} \ \dfrac{N}{A \cdot m}$$

$$= \boxed{1.00 \times 10^{-6} \ \text{tesla}}$$

9. $d = 10.0$ cm

$\quad\ = 0.100$ m

$B = 2.00 \times 10^{-5}$ tesla $\qquad B = k \dfrac{2 \ I}{d} \therefore I = \dfrac{B \ d}{2 \ k}$

$k = 1.00 \times 10^{-7}$ N/A²

$I = ?$

$$I = \dfrac{\left(2.00 \times 10^{-5} \ \dfrac{N}{A \cdot m}\right)(0.100 \ m)}{(2)\left(1.00 \times 10^{-7} \ \dfrac{N}{A^2}\right)}$$

$$= \dfrac{(2.00 \times 10^{-5})(0.100)}{(2)(1.00 \times 10^{-7})} \ \dfrac{N}{A \cdot m} \times m \times \dfrac{A \cdot A}{N}$$

$$= \dfrac{2.00 \times 10^{-6}}{2.00 \times 10^{-7}} \ A$$

$$= \boxed{10.0 \ A}$$

10. (a) $V_s = 12.0$ V $\qquad \dfrac{V_P}{N_P} = \dfrac{V_s}{N_s} \therefore \dfrac{N_P}{N_s} = \dfrac{V_P}{V_s}$

$I_s = 0.500$ A

$V_P = 120.$ V $\qquad \dfrac{N_P}{N_s} = \dfrac{120. \ V}{12.0 \ V} = \dfrac{10}{1}$

$\dfrac{N_P}{N_s} = ?$

or

$\boxed{\text{10 primary to 1 secondary}}$

(b) $I_P = ?$ $\qquad V_P I_P = V_s I_s \therefore I_P = \dfrac{V_s I_s}{V_P}$

$$I_P = \dfrac{(12.0 \ V)(0.500 \ A)}{120. \ V}$$

$$= \dfrac{12.0 \times 0.500}{120.} \ \dfrac{V \cdot A}{V}$$

$$= \boxed{0.05 \ A}$$

(c) $P_s = ?$ $P_s = I_s V_s$

$$= (0.500 \text{ A})(12.0 \text{ V})$$

$$= 0.500 \times 12.0 \; \frac{\cancel{C}}{\text{sec}} \times \frac{J}{\cancel{C}}$$

$$= 6.00 \; \frac{J}{\text{sec}}$$

$$= \boxed{6.00 \text{ w}}$$

11. (a) $V_P = 120. \text{ V}$

$N_P = 50 \text{ loops}$

$N_s = 150 \text{ loops}$

$I_P = 5.00 \text{ A}$

$V_s = ?$

$$\frac{V_P}{N_P} = \frac{V_s}{N_s} \; \therefore \; V_s = \frac{V_P N_s}{N_P}$$

$$V_s = \frac{120. \text{ V} \times 150 \text{ loops}}{50 \text{ loops}}$$

$$= \frac{120. \times 150}{50} \; \frac{\text{V} \cdot \cancel{\text{loops}}}{\cancel{\text{loops}}}$$

$$= \boxed{360. \text{ V}}$$

(b) $I_s = ?$ $V_P I_P = V_s I_s \; \therefore \; I_s = \frac{V_P I_P}{V_s}$

$$I_s = \frac{(120. \text{ V})(5.00 \text{ A})}{360. \text{ V}}$$

$$= \frac{120. \times 5.00}{360.} \; \frac{\cancel{\text{V}} \cdot \text{A}}{\cancel{\text{V}}}$$

$$= \boxed{1.67 \text{ A}}$$

(c) $P_s = ?$ $P_s = I_s V_s$

$$= \left(1.67 \; \frac{C}{\text{sec}}\right)\left(360. \; \frac{J}{C}\right)$$

$$= 1.67 \times 360. \; \frac{\cancel{C}}{\text{sec}} \times \frac{J}{\cancel{C}}$$

$$= 601.2 \; \frac{J}{\text{sec}}$$

$$= \boxed{601 \text{ w}}$$

12. $B = 6.00 \times 10^{-2} \text{ tesla}$

$v = 3.00 \times 10^7 \text{ m/sec}$

$q = 1.60 \times 10^{-19} \text{ C}$

$F_{max} = ?$

$$B = \frac{F_{max}}{qv} \; \therefore \; F_{max} = Bqv$$

$$F_{max} = \left(6.00 \times 10^{-2} \; \frac{N}{A \cdot m}\right)(1.60 \times 10^{-19} \text{ C})\left(3.00 \times 10^7 \; \frac{m}{\text{sec}}\right)$$

$$= (6.00 \times 10^{-2})(1.60 \times 10^{-19})(3.00 \times 10^7)$$

$$\frac{N}{\dfrac{C}{\text{sec}} \cdot m} \times C \times \frac{m}{\text{sec}}$$

$$= 2.88 \times 10^{-13} \; N \times \frac{\text{sec}}{\cancel{C}} \times \frac{1}{\cancel{m}} \times \cancel{C} \times \frac{\cancel{m}}{\text{sec}}$$

$$= \boxed{2.88 \times 10^{-13} \text{ N}}$$

Chapter 10

1. The relationship between the speed of light in a transparent material (v), the speed of light in a vacuum ($c = 3.00 \times 10^8$ m/sec) and the index of refraction (n) is n = c/v. According to table 10.1, the index of refraction for water is n = 1.33 and ice is n = 1.31.

(a) $c = 3.00 \times 10^8 \text{ m/sec}$

$n = 1.33$

$v = ?$

$$n = \frac{c}{v} \; \therefore \; v = \frac{c}{n}$$

$$v = \frac{3.00 \times 10^8 \text{ m/sec}}{1.33}$$

$$= \boxed{2.26 \times 10^8 \text{ m/sec}}$$

(b) $c = 3.00 \times 10^8 \text{ m/sec}$

$n = 1.31$

$v = ?$

$$v = \frac{3.00 \times 10^8 \text{ m/sec}}{1.31}$$

$$= \boxed{2.29 \times 10^8 \text{ m/sec}}$$

2. $d = 1.50 \times 10^8 \text{ km}$

$\quad = 1.50 \times 10^{11} \text{ m}$

$c = 3.00 \times 10^8 \text{ m/sec}$

$t = ?$

$$v = \frac{d}{t} \; \therefore \; t = \frac{d}{v}$$

$$t = \frac{1.50 \times 10^{11} \text{ m}}{3.00 \times 10^8 \; \dfrac{m}{\text{sec}}}$$

$$= \frac{1.50 \times 10^{11}}{3.00 \times 10^8} \; m \times \frac{\text{sec}}{m}$$

$$= 5.00 \times 10^2 \; \frac{\cancel{m} \cdot \text{sec}}{\cancel{m}}$$

$$= \frac{5.00 \times 10^2 \text{ sec}}{60.0 \; \dfrac{\text{sec}}{\text{min}}}$$

$$= \frac{5.00 \times 10^2}{60.0} \; \cancel{\text{sec}} \times \frac{\text{min}}{\cancel{\text{sec}}}$$

$$= \boxed{8.33 \text{ min}}$$

3. $d = 6.00 \times 10^9 \text{ km}$

$\quad = 6.00 \times 10^{12} \text{ m}$

$c = 3.00 \times 10^8 \text{ m/sec}$

$t = ?$

$$v = \frac{d}{t} \quad \therefore t = \frac{d}{v}$$

$$t = \frac{6.00 \times 10^{12} \text{ m}}{3.00 \times 10^8 \frac{m}{sec}}$$

$$= \frac{6.00 \times 10^{12}}{3.00 \times 10^8} \cancel{m} \times \frac{sec}{\cancel{m}}$$

$$= 2.00 \times 10^4 \text{ sec}$$

$$= \frac{2.00 \times 10^4 \text{ sec}}{3,600. \frac{sec}{hr}}$$

$$= \frac{2.00 \times 10^4}{3.600 \times 10^3} \cancel{sec} \times \frac{hr}{\cancel{sec}}$$

$$= \boxed{5.56 \text{ hr}}$$

4. From equation 10.1, note that both angles are measured from the normal and that the angle of incidence (θ_i) equals the angle of reflection (θ_r), or

$$\theta_i = \theta_r \quad \therefore \quad \boxed{\theta_i = 10°}$$

5. $v = 2.20 \times 10^8$ m/sec

$c = 3.00 \times 10^8$ m/sec

$n = ?$

$$n = \frac{c}{v}$$

$$= \frac{3.00 \times 10^8 \frac{m}{sec}}{2.20 \times 10^8 \frac{m}{sec}}$$

$$= 1.36$$

According to table 10.1, the substance with an index of refraction of 1.36 is $\boxed{\text{ethyl alcohol.}}$

6. **(a)** From equation 10.3,

$\lambda = 6.00 \times 10^{-7}$ m $\qquad c = \lambda f \therefore f = \frac{c}{\lambda}$

$c = 3.00 \times 10^8$ m/sec

$f = ?$

$$f = \frac{3.00 \times 10^8 \frac{m}{sec}}{6.00 \times 10^{-7} \text{ m}}$$

$$= \frac{3.00 \times 10^8}{6.00 \times 10^{-7}} \frac{\cancel{m}}{sec} \times \frac{1}{\cancel{m}}$$

$$= 5.00 \times 10^{14} \frac{1}{sec}$$

$$= \boxed{5.00 \times 10^{14} \text{ hz}}$$

(b) From equation 10.5,

$f = 5.00 \times 10^{14}$ hz

$h = 6.63 \times 10^{-34}$ J·sec

$E = ?$

$$E = hf$$

$$= (6.63 \times 10^{-34} \text{ J·sec})\left(5.00 \times 10^{14} \frac{1}{sec}\right)$$

$$= (6.63 \times 10^{-34})(5.00 \times 10^{14}) \text{ J·}\cancel{sec} \times \frac{1}{\cancel{sec}}$$

$$= \boxed{3.32 \times 10^{-19} \text{ J}}$$

7. First, you can find the energy of one photon of the peak intensity wavelength (5.60×10^{-7} m) by using equation 10.3 to find the frequency, then equation 10.5 to find the energy:

Step 1: $c = \lambda f \therefore f = \frac{c}{\lambda}$

$$= \frac{3.00 \times 10^8 \frac{m}{sec}}{5.60 \times 10^{-7} \text{ m}}$$

$$= 5.36 \times 10^{14} \text{ hz}$$

Step 2: $E = hf$

$$= (6.63 \times 10^{-34} \text{ J·sec})(5.36 \times 10^{14} \text{ hz})$$

$$= 3.55 \times 10^{-19} \text{ J}$$

Step 3: Since one photon carries an energy of 3.55×10^{-19} J and the overall intensity is 1,000.0 w, for each square meter there must be an average of

$$\frac{1,000.0 \frac{J}{sec}}{3.55 \times 10^{-19} \frac{J}{photon}}$$

$$\frac{1.000 \times 10^3}{3.55 \times 10^{-19}} \frac{\cancel{J}}{sec} \times \frac{photon}{\cancel{J}}$$

$$\boxed{2.82 \times 10^{21} \frac{photon}{sec}}$$

8. **(a)** $f = 4.90 \times 10^{14}$ hz $\qquad c = \lambda f \therefore \lambda = \frac{c}{f}$

$c = 3.00 \times 10^8$ m/sec

$\lambda = ?$

$$\lambda = \frac{3.00 \times 10^8 \frac{m}{sec}}{4.90 \times 10^{14} \frac{1}{sec}}$$

$$= \frac{3.00 \times 10^8}{4.90 \times 10^{14}} \frac{m}{\cancel{sec}} \times \frac{\cancel{sec}}{1}$$

$$= \boxed{6.12 \times 10^{-7} \text{ m}}$$

(b) According to table 10.2, this is the frequency and wavelength of orange.

9. $f = 5.00 \times 10^{20}$ hz

$h = 6.63 \times 10^{-34}$ J·sec

$E = ?$

$$E = hf$$

$$= (6.63 \times 10^{-34} \text{ J·sec})\left(5.00 \times 10^{20} \frac{1}{\text{sec}}\right)$$

$$= (6.63 \times 10^{-34})(5.00 \times 10^{20}) \text{ J·sec} \times \frac{1}{\text{sec}}$$

$$= \boxed{3.32 \times 10^{-13} \text{ J}}$$

10. $\lambda = 1.00$ mm

$= 0.001$ m

$f = ?$

$c = 3.00 \times 10^8$ m/sec

$h = 6.63 \times 10^{-34}$ J·sec

$E = ?$

Step 1: $c = \lambda f \therefore f = \dfrac{v}{\lambda}$

$$f = \frac{3.00 \times 10^8 \frac{\text{m}}{\text{sec}}}{1.00 \times 10^{-3} \text{ m}}$$

$$= \frac{3.00 \times 10^8}{1.00 \times 10^{-3}} \frac{\text{m}}{\text{sec}} \times \frac{1}{\text{m}}$$

$$= 3.00 \times 10^{11} \text{ hz}$$

Step 2:

$$E = hf$$

$$= (6.63 \times 10^{-34} \text{ J·sec})\left(3.00 \times 10^{11} \frac{1}{\text{sec}}\right)$$

$$= (6.63 \times 10^{-34})(3.00 \times 10^{11}) \text{ J·sec} \times \frac{1}{\text{sec}}$$

$$= \boxed{1.99 \times 10^{-22} \text{ J}}$$

Chapter 11

1. $m = 1.68 \times 10^{-27}$ kg

$v = 3.22 \times 10^3$ m/sec

$h = 6.63 \times 10^{-34}$ J·sec

$\lambda = ?$

$$\lambda = \frac{h}{mv}$$

$$= \frac{6.63 \times 10^{-34} \text{ J·sec}}{(1.68 \times 10^{-27} \text{ kg})\left(3.22 \times 10^3 \frac{\text{m}}{\text{sec}}\right)}$$

$$= \frac{6.63 \times 10^{-34}}{(1.68 \times 10^{-27})(3.22 \times 10^3)} \frac{\text{J·sec}}{\text{kg} \times \frac{\text{m}}{\text{sec}}}$$

$$= \frac{6.63 \times 10^{-34}}{5.41 \times 10^{-24}} \frac{\frac{\text{kg·m}^2}{\text{sec·sec}} \times \text{sec}}{\text{kg} \times \frac{\text{m}}{\text{sec}}}$$

$$= 1.23 \times 10^{-10} \frac{\text{kg·m·m}}{\text{sec}} \times \frac{1}{\text{kg}} \times \frac{\text{sec}}{\text{m}}$$

$$= \boxed{1.23 \times 10^{-10} \text{ m}}$$

2. **(a)**

$n = 6$

$E_1 = -13.6$ eV

$E_6 = ?$

$$E_n = \frac{E_1}{n^2}$$

$$E_6 = \frac{-13.6 \text{ eV}}{6^2}$$

$$= \frac{-13.6 \text{ eV}}{36}$$

$$= \boxed{-0.378 \text{ eV}}$$

(b)

$$= (-0.378 \text{ eV})\left(1.60 \times 10^{-19} \frac{\text{J}}{\text{eV}}\right)$$

$$= (-0.378)(1.60 \times 10^{-19}) \text{ eV} \times \frac{\text{J}}{\text{eV}}$$

$$= \boxed{-6.05 \times 10^{-20} \text{ J}}$$

3. **(a)** Energy is related to the frequency and Planck's constant in equation 11.1, $E = hf$. From equation 11.6,

$$hf = E_h - E_1 \therefore E = E_h - E_1$$

$E_h = (n = 2) = -5.44 \times 10^{-19}$ J

$E_1 = (n = 6) = -6.05 \times 10^{-20}$ J

$E = ?$ J

$$E = E_h - E_1$$

$$= (-6.05 \times 10^{-20} \text{ J}) - (-5.44 \times 10^{-19} \text{ J})$$

$$= \boxed{4.84 \times 10^{-19} \text{ J}}$$

(b) $E_h = -0.377$ eV*

$E_1 = -3.40$ eV*

$E = ?$ eV

$$E = E_h - E_1$$

$$= (-0.377 \text{ eV}) - (-3.40 \text{ eV})$$

$$= \boxed{3.02 \text{ eV}}$$

*From figure 11.13

4. $(n = 6) = -6.05 \times 10^{-20}$ J

$(n = 2) = -5.44 \times 10^{-19}$ J

$h = 6.63 \times 10^{-34}$ J·sec

$f = ?$

$$hf = E_h - E_l \therefore f = \frac{E_h - E_l}{h}$$

$$f = \frac{(-6.05 \times 10^{-20} \text{ J}) - (-5.44 \times 10^{-19} \text{ J})}{6.63 \times 10^{-34} \text{ J·sec}}$$

$$= \frac{4.84 \times 10^{-19} \text{ J}}{6.63 \times 10^{-34} \text{ J·sec}}$$

$$= 7.29 \times 10^{14} \frac{1}{\text{sec}}$$

$$= \boxed{7.29 \times 10^{14} \text{ hz}}$$

5. $(n = 1) = -13.6$ eV $E_n = \dfrac{E_l}{n^2}$

$E = ?$

$$= \frac{-13.6 \text{ eV}}{1^2}$$

$$= -13.6 \text{ eV}$$

Since the energy of the electron is -13.6 eV, it will require 13.6 eV (or 2.18×10^{-19} J) to remove the electron.

6. $q/m = -1.76 \times 10^{11}$ C/kg

$q = -1.60 \times 10^{-19}$ C

$m = ?$

$$\text{mass} = \frac{\text{charge}}{\text{charge/mass}}$$

$$= \frac{-1.60 \times 10^{-19} \text{ C}}{-1.76 \times 10^{11} \dfrac{\text{C}}{\text{kg}}}$$

$$= \frac{-1.60 \times 10^{-19}}{-1.76 \times 10^{11}} \text{ C} \times \frac{\text{kg}}{\text{C}}$$

$$= \boxed{9.09 \times 10^{-31} \text{ kg}}$$

7. $\lambda = -1.67 \times 10^{-10}$ m

$m = 9.11 \times 10^{-31}$ kg

$v = ?$

$$\lambda = \frac{h}{mv} \therefore v = \frac{h}{m\lambda}$$

$$v = \frac{6.63 \times 10^{-34} \text{ J·sec}}{(9.11 \times 10^{-31} \text{ kg})(1.67 \times 10^{-10} \text{ m})}$$

$$= \frac{6.63 \times 10^{-34}}{(9.11 \times 10^{-31})(1.67 \times 10^{-10})} \frac{\text{J·sec}}{\text{kg·m}}$$

$$= \frac{6.63 \times 10^{-34}}{1.52 \times 10^{-40}} \frac{\dfrac{\text{kg·m}^2}{\text{sec·sec}} \times \text{sec}}{\text{kg·m}}$$

$$= 4.36 \times 10^6 \frac{\text{kg·m·m}}{\text{sec}} \times \frac{1}{\text{kg}} \times \frac{1}{\text{m}}$$

$$= \boxed{4.36 \times 10^6 \frac{\text{m}}{\text{sec}}}$$

8. (a) Boron: $1s^2 2s^2 2p^1$
(b) Aluminum: $1s^2 2s^2 2p^6 3s^2 3p^1$
(c) Potassium: $1s^2 2s^2 2p^6 3p^6 4s^1$

9. (a) Boron is atomic number 5 and there are 5 electrons.
(b) Aluminum is atomic number 13 and there are 13 electrons.
(c) Potassium is atomic number 19 and there are 19 electrons.

10. (a) Argon: $1s^2 2s^2 2p^6 3s^2 3p^6$
(b) Zinc: $1s^2 2s^2 2p^6 3s^2 3p^6 4s^2 3d^{10}$
(c) Bromine: $1s^2 2s^2 2p^6 3s^2 3p^6 4s^2 3d^{10} 6s^2$

Chapter 12

1. The answers to this question are found in the list of elements and their symbols, which is located on the inside back cover of this text.

(a) Silicon: Si
(b) Silver: Ag
(c) Helium: He
(d) Potassium: K
(e) Magnesium: Mg
(f) Iron: Fe

2. Atomic weight is the weighted average of the isotopes as they occur in nature. Thus

Lithium-6:   6.01512 u $\times$ 0.0742 = 0.446 u

Lithium-7:   7.016 u $\times$ 0.9258 = 6.4054 u

Lithium-6 contributes 0.446 u of the weighted average and lithium-7 contributes 6.4954 u. The atomic weight of lithium is therefore

$$\begin{array}{r} 0.446 \text{ u} \\ + 6.4954 \text{ u} \\ \hline 6.941 \text{ u} \end{array}$$

3. Recall that the subscript is the atomic number, which identifies the number of protons. In a neutral atom, the number of protons equals the number of electrons so the atomic number tells you the number of electrons, too. The superscript is the mass number, which identifies the number of neutrons and the number of protons in the nucleus. The number of neutrons is therefore the mass number minus the atomic number.

	Protons	Neutrons	Electrons
(a)	6	6	6
(b)	1	0	1
(c)	18	22	18
(d)	1	1	1
(e)	79	118	79
(f)	92	143	92

4.

		Period	Family
(a)	Radon (Rn)	6	VIIIA
(b)	Sodium (Na)	3	IA
(c)	Copper (Cu)	4	IB
(d)	Neon (Ne)	2	VIIIA
(e)	Iodine (I)	5	VIIA
(f)	Lead (Pb)	6	IVA

5. Recall that the number of outer shell electrons is the same as the family number for the representative elements:

(a) Li: 1
(b) N: 5
(c) F: 7
(d) Cl: 7
(e) Ra: 2
(f) Be: 2

6. The same information that was used in question 5 can be used to draw the dot notation:

(a) $\overset{\textstyle .}{\underset{\textstyle .}{B}} \cdot$

(b) $\cdot \overset{\textstyle ..}{\underset{\textstyle ..}{Br}} \cdot$

(c) Ca:

(d) K ·

(e) $\cdot \overset{\textstyle ..}{\underset{\textstyle ..}{O}} \cdot$

(f) $\cdot \overset{\textstyle ..}{\underset{\textstyle ..}{S}} \cdot$

7. The charge is found by identifying how many electrons are lost or gained in achieving the noble gas structure:

(a) Boron 3+
(b) Bromine 1−
(c) Calcium 2+
(d) Potassium 1+
(e) Oxygen 2−
(f) Sulfur 2−

8. Metals have one, two, or three outer electrons and are located in the left two-thirds of the periodic table. Semiconductors are adjacent to the line that separates the metals and nonmetals. Look at the periodic table on the inside back cover and you will see:

(a) Krypton—nonmetal
(b) Cesium—metal
(c) Silicon—semiconductor
(d) Sulfur—nonmetal
(e) Molybdenum—metal
(f) Plutonium—metal

9. (a) Bromine gained an electron to acquire a 1− charge, so it must be in family VIIA (the members of this family have seven electrons and need one more to acquire the noble gas structure).

(b) Potassium must have lost one electron, so it is in IA.
(c) Aluminum lost three electrons, so it is in IIIA.
(d) Sulfur gained two electrons, so it is in VIA.
(e) Barium lost two electrons, so it is in IIA.
(f) Oxygen gained two electrons, so it is in VIA.

10. (a) $^{16}_{8}O$
(b) $^{23}_{11}Na$
(c) $^{3}_{1}H$
(d) $^{35}_{17}Cl$

Chapter 13

1.

(a) K · + $\cdot \overset{\textstyle ..}{\underset{\textstyle ..}{I}}$: → $K^{+} \left(\overset{\textstyle ..}{\underset{\textstyle ..}{:I:}} \right)^{-}$ (or KI)

(b) Sr : + $\cdot \overset{\textstyle ..}{\underset{\textstyle ..}{S}}$: → $Sr^{2+} \left(\overset{\textstyle ..}{\underset{\textstyle ..}{:S:}} \right)^{2-}$ (or SrS)

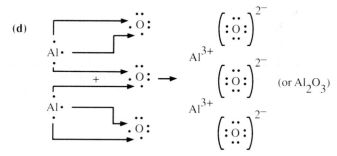

(c) (or Na_2O)

(d) (or Al_2O_3)

2. (a) Sulfur is in family VIA, so sulfur has six valence electrons and will need two more to achieve a stable outer structure like the noble gases. Two more outer shell electrons will give the sulfur atom a charge of 2−. Copper^{2+} will balance the 2− charge of sulfur, so the name is copper (II) sulfide. Note the "-ide" ending for compounds that have only two different elements.

(b) Oxygen is in family VIA, so oxygen has six valence electrons and will have a charge of 2−. Using the crossover technique in reverse, you can see that the charge on the oxygen is 2− and the charge on the iron is 3−. Therefore the name is iron (III) oxide.

(c) From information in (a) and (b), you know that oxygen has a charge of 2−. The chromium ion must have the same charge to make a neutral compound as it must be, so the name is chromium (II) oxide. Again, note the "-ide" ending for a compound with two different elements.

(d) Sulfur has a charge of 2−, so the lead ion must have the same positive charge to make a neutral compound. The name is lead (II) sulfide.

3. The name of some common polyatomic ions are in table 13.4. Using this table as a reference, the names are

(a) hydroxide
(b) sulfite
(c) hypochlorite
(d) nitrate
(e) carbonate
(f) perchlorate

4. The Roman numeral tells you the charge on the variable-charge elements. The charges for the polyatomic ions are found in table 13.4. The charges for metallic elements can be found in tables 13.1 and 13.2. Using these resources and the crossover technique, the formulas are as follows:

(a) $Fe(OH)_3$
(b) $Pb_3(PO_4)_2$
(c) $ZnCO_3$
(d) NH_4NO_3
(e) $KHCO_3$
(f) K_2SO_3

5. Table 13.6 has information about the meaning of prefixes and stem names used in naming covalent compounds. (a), for example, asks for the formula of carbon tetrachloride. Carbon has no prefixes, so there is one carbon atom, and it comes first in the formula because it comes first in the name. The "tetra-" prefix means four, so there are four chlorine atoms. The name ends in "-ide" so you know there are only two elements in the compound. The symbols can be obtained from the list of

elements on the inside back cover of this text. Using all this information from the name, you can think out the formula for carbon tetrachloride. The same process is used for the other compounds and formulas:

(a) CCl_4 (d) SO_3
(b) H_2O (e) N_2O_4
(c) MnO_2 (f) As_2S_5

6. Again using information from table 13.6, this question requires you to reverse the thinking procedure you learned in question 5.

(a) carbon monoxide
(b) carbon dioxide
(c) carbon disulfide
(d) dinitrogen monoxide
(e) tetraphosphorus trisulfide
(f) dinitrogen trioxide

7. The types of bonds formed are predicted by using the electronegativity scale in figure 13.7 and finding the absolute difference. On this basis:

(a) Difference = 1.0, which means polar covalent bond
(b) Difference = 0, which means nonpolar covalent
(c) Difference = 0, which means nonpolar covalent
(d) Difference = 0.4, which means nonpolar covalent
(e) Difference = 1.6, which means polar covalent and almost ionic

Chapter 14

1. (a) $MgCl_2$ is an ionic compound, so the formula has to be empirical.
   (b) C_2H_2 is a covalent compound, so the formula might be molecular. Since it is not the simplest whole number ratio (which would be CH), then the formula is molecular.
   (c) BaF_2 is ionic; the formula is empirical.
   (d) C_8H_{18} is not the simplest whole number ratio of a covalent compound, so the formula is molecular.
   (e) CH_4 is covalent, but the formula might or might not be molecular (?).
   (f) S_8 is a nonmetal bonded to a nonmetal (itself); this is a molecular formula.

2. (a) $CuSO_4$

   1 of Cu = 1 × 35.5 u =   35.5 u
   1 of S = 1 × 32.0 u =   32.0 u
   4 of O = 4 × 16.0 u =   64.0 u
                               101.5 u

   (b) CS_2

   1 of C = 1 × 12.0 u = 12.0 u
   2 of S = 2 × 32.0 u = 64.0 u
                           76.0 u

   (c) $CaSO_4$

   1 of Ca = 1 × 40.1 u =   40.1 u
   1 of S = 1 × 32.0 u =   32.0 u
   4 of O = 4 × 16.0 u =   64.0 u
                            136.1 u

   (d) Na_2CO_3

   2 of Na = 2 × 23.0 u =   46.0 u
   1 of C = 1 × 12.0 u =   12.0 u
   3 of O = 3 × 16.0 u =   48.0 u
                            106.0 u

3. (a) FeS_2

   For Fe: $\dfrac{(55.9\ u\ Fe)(1)}{119.9\ u\ FeS_2} \times 100\%\ FeS_2 = 46.6\%$ Fe

   For S: $\dfrac{(32.0\ u\ S)(1)}{119.9\ u\ FeS_2} \times 100\%\ FeS_2 = 53.4\%$ S

   or (100% FeS_2) − (46.6% Fe) = 53.4% S

   (b) H_3BO_3

   For H: $\dfrac{(1.0\ u\ H)(3)}{61.8\ u\ H_3BO_3} \times 100\%\ H_3BO_3 = 4.85\%$ H

   For B: $\dfrac{(10.8\ u\ B)(1)}{61.8\ u\ H_3BO_3} \times 100\%\ H_3BO_3 = 17.5\%$ B

   For O: $\dfrac{(61.8\ u\ O)(3)}{61.8\ u\ H_3BO_3} \times 100\%\ H_3BO_3 = 77.7\%$ O

   (c) $NaHCO_3$

   For Na: $\dfrac{(23.0\ u\ Na)(1)}{84.0\ u\ NaHCO_3} \times 100\%\ NaHCO_3 = 27.4\%$ Na

   For H: $\dfrac{(1.0\ u\ H)(1)}{84.0\ u\ NaHCO_3} \times 100\%\ NaHCO_3 = 1.2\%$ H

   For C: $\dfrac{(12.0\ u\ C)(1)}{84.0\ u\ NaHCO_3} \times 100\%\ NaHCO_3 = 14.3\%$ C

   For O: $\dfrac{(16.0\ u\ O)(3)}{84.0\ u\ NaHCO_3} \times 100\%\ NaHCO_3 = 57.1\%$ O

   (d) $C_9H_8O_4$

   For C: $\dfrac{(12.0\ u\ C)(9)}{180.0\ u\ C_9H_8O_4} \times 100\%\ C_9H_8O_4 = 60.0\%$ C

   For H: $\dfrac{(1.0\ u\ H)(8)}{180.0\ u\ C_9H_8O_4} \times 100\%\ C_9H_8O_4 = 4.4\%$ H

   For O: $\dfrac{(16.0\ u\ O)(4)}{180.0\ u\ C_9H_8O_4} \times 100\%\ C_9H_8O_4 = 35.6\%$ O

4. (a) $2\ SO_2 + O_2 \rightarrow 2\ SO_3$
   (b) $4\ P + 5\ O_2 \rightarrow 2\ P_2O_5$
   (c) $2\ Al + 6\ HCl \rightarrow 2\ AlCl_3 + 3\ H_2$
   (d) $2\ NaOH + 6\ HCl \rightarrow Na_2SO_4 + 2\ H_2O$
   (e) $Fe_2O_3 + 3\ CO \rightarrow 2\ Fe + 3\ CO_2$
   (f) $3\ Mg(OH)_2 + 2\ H_3PO_4 \rightarrow Mg_3(PO_4)_2 + 6\ H_2O$

5. (a) General form of $XY + AZ \rightarrow XZ + AY$ with precipitate formed: Ion exchange reaction.
   (b) General form of $X + Y \rightarrow XY$: Combination reaction.
   (c) General form of $XY \rightarrow X + Y + \ldots$: Decomposition reaction.
   (d) General form of $X + Y \rightarrow XY$: Combination reaction.
   (e) General form of $XY + A \rightarrow AY + X$: Replacement reaction.
   (f) General form of $X + Y \rightarrow XY$: Combination reaction.

6. (a) $C_5H_{12}(g) + 8\ O_2(g) \rightarrow 5\ CO_2(g) + 6\ H_2O(g)$
   (b) $HCl(aq) + NaOH(aq) \rightarrow NaCl(aq) + H_2O(l)$
   (c) $2\ Al(s) + Fe_2O_3(s) \rightarrow Al_2O_3(s) + 2\ Fe(l)$
   (d) $Fe(s) + CuSO_4(aq) \rightarrow FeSO_4(aq) + Cu(s)$
   (e) $MgCl(aq) + Fe(NO_3)_2(aq) \rightarrow$ No reaction (all possible compounds are soluble and no gas or water was formed).
   (f) $C_6H_{10}O_5(s) + 6\ O_2(g) \rightarrow 6\ CO_2(g) + 5\ H_2O(g)$

7. (a)
$$2 \text{ KClO}_3(s) \xrightarrow{\Delta} 2 \text{ KCl}(s) + 3 \text{ O}_2\uparrow$$

(b)
$$2 \text{ Al}_2\text{O}_3(l) \xrightarrow{elec} 4 \text{ Al}(s) + 3 \text{ O}_2\uparrow$$

(c)
$$\text{CaCO}_3(s) \xrightarrow{\Delta} \text{CaO}(s) + \text{CO}_2\uparrow$$

8. (a) $2 \text{ Na}(s) + 2 \text{ H}_2\text{O}(l) \rightarrow 2 \text{ NaOH}(aq) + \text{H}_2\uparrow$

(b) $\text{Au}(s) + \text{HCl}(aq) \rightarrow$ No reaction (gold is below hydrogen in the activity series)

(c) $\text{Al}(s) + \text{FeCl}_3(aq) \rightarrow \text{AlCl}_3(aq) + \text{Fe}(s)$

(d) $\text{Zn}(s) + \text{CuCl}_2(aq) \rightarrow \text{ZnCl}_2(aq) + \text{Cu}(s)$

9. (a) $\text{NaOH}(aq) + \text{HNO}_3(aq) \rightarrow \text{NaNO}_3(aq) + \text{H}_2\text{O}(l)$

(b) $\text{CaCl}_2(aq) + \text{KNO}_3(aq) \rightarrow$ No reaction

(c) $3 \text{ Ba(NO}_3)_2(aq) + 2 \text{ Na}_3\text{PO}_4(aq) \rightarrow 6 \text{ NaNO}_3(aq) + \text{Ba}_3(\text{PO}_4)_2\downarrow$

(d) $2 \text{ KOH}(aq) + \text{ZnSO}_4(aq) \rightarrow \text{K}_2\text{SO}_4(aq) + \text{Zn(OH)}_2\downarrow$

10. One mole of oxygen combines with 2 moles of acetylene, so 0.5 mole of oxygen would be needed for 1 mole of acetylene. Therefore, 1 L of C_2H_2 requires 0.5 L of O_2.

Chapter 15

1. $m_{\text{solute}} = 1.75$ g

$m_{\text{solution}} = 50.0$ g

$\%_{\text{weight}} = ?$

$$\% \text{ solute} = \frac{m_{\text{solute}}}{m_{\text{solution}}} \times 100\% \text{ solution}$$

$$= \frac{1.75 \text{ g NaCl}}{50.0 \text{ g solution}} \times 100\% \text{ solution}$$

$$= \boxed{3.50\% \text{ NaCl}}$$

2. $m_{\text{solution}} = 103.5$ g

$m_{\text{solute}} = 3.50$ g

$\%_{\text{weight}} = ?$

$$\% \text{ solute} = \frac{m_{\text{solute}}}{m_{\text{solution}}} \times 100\% \text{ solution}$$

$$= \frac{3.50 \text{ g NaCl}}{103.5 \text{ g solution}} \times 100\% \text{ solution}$$

$$= \boxed{3.38\% \text{ NaCl}}$$

3. Since ppm is defined as the weight unit of solute in 1,000,000 weight units of solution, the percent by weight can be calculated just like any other percent. The weight of the dissolved sodium and chlorine ions is the part, and the weight of the solution is the whole, so

$$\% = \frac{\text{part}}{\text{whole}} \times 100\%$$

$$= \frac{30,113 \text{ g NaCl ions}}{1,000,000 \text{ g seawater}} \times 100\% \text{ seawater}$$

$$= \boxed{3.00\% \text{ NaCl ions}}$$

4. $m_{\text{solution}} = 250$ g

$\% \text{ solute} = 3.0\%$

$m_{\text{solute}} = ?$

$$\% \text{ solute} = \frac{m_{\text{solute}}}{m_{\text{solution}}} \times 100\% \text{ solution}$$

$$\therefore$$

$$m_{\text{solute}} = \frac{(m_{\text{solution}})(\% \text{ solute})}{100\% \text{ solution}}$$

$$= \frac{(250 \text{ g})(3.0\%)}{100\%}$$

$$= \boxed{7.5 \text{ g}}$$

5. $\% \text{ solution} = 12\%$ solution

$V_{\text{solution}} = 200$ mL

$V_{\text{solute}} = ?$

$$\% \text{ solution} = \frac{V_{\text{solute}}}{V_{\text{solution}}} \times 100\% \text{ solution}$$

$$\therefore$$

$$V_{\text{solute}} = \frac{(\% \text{ solution})(V_{\text{solution}})}{100\% \text{ solution}}$$

$$= \frac{(12\% \text{ solution})(200 \text{ mL})}{100\% \text{ solution}}$$

$$= \boxed{24 \text{ mL alcohol}}$$

6. $\% \text{ solution} = 40\%$

$V_{\text{solution}} = 50$ mL

$V_{\text{solute}} = ?$

$$\% \text{ solution} = \frac{V_{\text{solute}}}{V_{\text{solution}}} \times 100\% \text{ solution}$$

$$\therefore$$

$$V_{\text{solute}} = \frac{(\% \text{ solution})(V_{\text{solution}})}{100\% \text{ solution}}$$

$$= \frac{(40\% \text{ solution})(50 \text{ mL})}{100\% \text{ solution}}$$

$$= \boxed{20 \text{ mL alcohol}}$$

7. (a)
$$\% \text{ concentration} = \frac{\text{ppm}}{1 \times 10^4}$$

$$= \frac{5}{1 \times 10^4}$$

$$= \boxed{0.0005\% \text{ DDT}}$$

(b)
$$\% \text{ part} = \frac{\text{part}}{\text{whole}} \times 100\% \text{ whole}$$

$$\therefore$$

$$\text{whole} = \frac{(100\%)(\text{part})}{\% \text{ part}}$$

$$= \frac{(100\%)(17.0 \text{ g})}{0.0005\%} = \boxed{34,000 \text{ g or 34 kg}}$$

8. (a) $HC_2H_3O_2(aq) + H_2O(l) \rightleftarrows H_3O^+(aq) + C_2H_5O_2^-(aq)$
      acid   base

(b) $C_6H_6NH_2(l) + H_2O(l) \rightleftarrows C_6H_5NH_3^+(aq) + OH^-(aq)$
      base   acid

(c) $HClO_4(aq) + HC_2H_3O_2(aq) \rightleftarrows H_2C_2H_3O_2^+ (aq)$
      acid   base
  $+ ClO_4^-(aq)$

(d) $H_2O(l) + H_2O(l) \rightleftarrows H_3O^+(aq) + OH^-(aq)$
      base   acid

Chapter 16

1.

(a)

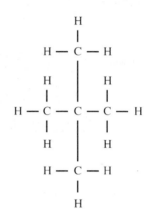

(b)

(c) 2,2-dimethylpropane

2.

n-pentane

3-methylpentane

2-methylpentane

2,2-dimethylbutane

2,3-dimethylbutane

3.

(a)

(b)

(c)

$$H - \overset{\overset{\displaystyle H}{|}}{\underset{\underset{\displaystyle H}{|}}{C}} - \overset{\overset{\displaystyle H}{|}}{\underset{\underset{\displaystyle H}{|}}{C}} - C \equiv C - \overset{\overset{\displaystyle CH_3}{|}}{\underset{\underset{\displaystyle CH_3}{|}}{C}} - \overset{\overset{\displaystyle H}{|}}{\underset{\underset{\displaystyle H}{|}}{C}} - \overset{\overset{\displaystyle H}{|}}{\underset{\underset{\displaystyle H}{|}}{C}} - H$$

4. (a) 2-chloro-4-methylpentane
   (b) 2-methyl-1-pentene
   (c) 2-methyl-3-ethyl-3-hexene

5. The 2,2,3-trimethylbutane is more highly branched, so it will have the higher octane rating.

2,2,3-trimethylbutane

$$H - \overset{\overset{\displaystyle H}{|}}{\underset{\underset{\displaystyle H}{|}}{C}} - \overset{\overset{\displaystyle CH_3}{|}}{\underset{\underset{\displaystyle CH_3}{|}}{C}} - \overset{\overset{\displaystyle CH_3}{|}}{\underset{\underset{\displaystyle H}{|}}{C}} - \overset{\overset{\displaystyle H}{|}}{\underset{\underset{\displaystyle H}{|}}{C}} - H$$

2,2-dimethylpentane

$$H - \overset{\overset{\displaystyle H}{|}}{\underset{\underset{\displaystyle H}{|}}{C}} - \overset{\overset{\displaystyle CH_3}{|}}{\underset{\underset{\displaystyle CH_3}{|}}{C}} - \overset{\overset{\displaystyle H}{|}}{\underset{\underset{\displaystyle H}{|}}{C}} - \overset{\overset{\displaystyle H}{|}}{\underset{\underset{\displaystyle H}{|}}{C}} - \overset{\overset{\displaystyle H}{|}}{\underset{\underset{\displaystyle H}{|}}{C}} - H$$

6. (a) alcohol
   (b) amide
   (c) ether
   (d) ester
   (e) organic acid

Chapter 17

1. (a) cobalt-60: 27 protons, 33 neutrons
   (b) potassium-40: 19 protons, 21 neutrons
   (c) neon-24: 10 protons, 14 neutrons
   (d) lead-204: 82 protons, 122 neutrons

2. (a) $^{60}_{27}Co$ (c) $^{24}_{10}Ne$
   (b) $^{40}_{19}K$ (d) $^{204}_{82}Pb$

3. (a) cobalt-60: Radioactive because odd numbers of protons (27) and odd numbers of neutrons (33) are usually unstable.
   (b) potassium-40: Radioactive, again having an odd number of protons (19) and an odd number of neutrons (21).
   (c) neon-24: Stable, because even numbers of protons and neutrons are usually stable.
   (d) lead-204: Stable, because even numbers of protons and neutrons *and* because 82 is a particularly stable number of nucleons.

4. (a) $^{56}_{26}Fe \rightarrow \, ^{0}_{-1}e + \, ^{56}_{27}Co$
   (b) $^{7}_{4}Be \rightarrow \, ^{0}_{-1}e + \, ^{7}_{5}B$
   (c) $^{64}_{29}Cu \rightarrow \, ^{0}_{-1}e + \, ^{64}_{30}Zn$
   (d) $^{24}_{11}Na \rightarrow \, ^{0}_{-1}e + \, ^{24}_{12}Mg$
   (e) $^{214}_{82}Pb \rightarrow \, ^{0}_{-1}e + \, ^{214}_{83}Bi$
   (f) $^{32}_{15}P \rightarrow \, ^{0}_{-1}e + \, ^{32}_{16}S$

5. (a) $^{235}_{92}U \rightarrow \, ^{4}_{2}He + \, ^{231}_{90}Th$
   (b) $^{226}_{88}Ra \rightarrow \, ^{4}_{2}He + \, ^{222}_{86}Rn$
   (c) $^{239}_{94}Pu \rightarrow \, ^{4}_{2}He + \, ^{235}_{92}U$
   (d) $^{214}_{83}Bi \rightarrow \, ^{4}_{2}He + \, ^{210}_{81}Tl$
   (e) $^{230}_{90}Th \rightarrow \, ^{4}_{2}He + \, ^{226}_{88}Ra$
   (f) $^{210}_{84}Po \rightarrow \, ^{4}_{2}He + \, ^{206}_{82}Pb$

6. Thirty-two days is four half-lives. After the first half-life (8 days), 1/2 oz will remain. After the second half-life (8 + 8, or 16 days), 1/4 oz will remain. After the third half-life (8 + 8 + 8, or 24 days), 1/8 oz will remain. After the fourth half-life (8 + 8 + 8 + 8, or 32 days), 1/16 oz will remain, or 6.3×10^{-2} oz.

7. The relationship between half-life and the radioactive decay constant is found in equation 17.2, which is first solved for the decay constant:

$$t_{1/2} = \frac{0.693}{k} \; \therefore \; k = \frac{0.693}{t_{1/2}}$$

A decay constant calculation requires units of seconds, so 27.6 years is next converted to seconds:

$$(27.6 \text{ yr})(3.15 \times 10^7 \text{ sec/yr}) = 8.69 \times 10^8 \text{ sec}$$

This value in seconds is now used in equation 17.2 solved for k:

$$k = \frac{0.693}{8.69 \times 10^8 \text{ sec}} = 7.97 \times 10^{-10}/\text{sec}$$

8. The relationship between the radioactive decay constant, the decay rate, and the number of nuclei is found in equation 17.1, which can be solved for the rate:

$$k = \frac{\text{rate}}{n} \therefore \text{rate} = kn$$

The number of nuclei in a molar mass of strontium-90 (87.6 g) is Avogadro's number, 6.02×10^{23}, so

$$\text{rate} = (7.97 \times 10^{-10}/\text{sec})(6.02 \times 10^{23} \text{ nuclei})$$

$$= 4.80 \times 10^{14} \text{ nuclei/sec}$$

9. A curie (Ci) is defined as 3.70×10^{10} disintegrations/sec, so a molar mass that is decaying at a rate of 4.80×10^{14} nuclei/sec has a curie activity of

$$\frac{4.80 \times 10^{14}}{3.70 \times 10^{10}} = 1.30 \times 10^4 \text{ Ci}$$

10. The Fe-56 nucleus has a mass of 55.9206 u, but the individual masses of the nucleons are

$$
\begin{array}{ll}
26 \text{ protons} \times 1.00728 \text{ u} & = 26.1893 \text{ u} \\
30 \text{ neutrons} \times 1.00867 \text{ u} & = \underline{30.2601 \text{ u}} \\
& 56.4494 \text{ u}
\end{array}
$$

The mass defect is thus

$$
\begin{array}{r}
56.4494 \text{ u} \\
- 55.9206 \text{ u} \\
\hline
0.5288 \text{ u}
\end{array}
$$

The mass defect is equivalent to the binding energy according to $E = mc^2$. For a molar mass of Fe-56, the mass defect is

$$E = (5.29 \times 10^{-4} \text{ kg})\left(3.00 \times 10^8 \, \frac{m}{\text{sec}}\right)^2$$

$$= (5.29 \times 10^{-4} \text{ kg})\left(9.00 \times 10^{16} \, \frac{m^2}{\text{sec}^2}\right)$$

$$= 4.76 \times 10^{13} \, \frac{\text{kg} \cdot m^2}{\text{sec}^2}$$

$$= 4.76 \times 10^{13} \text{ J}$$

For a single nucleus,

$$\frac{4.76 \times 10^{13} \text{ J}}{6.02 \times 10^{23} \text{ nuclei}} = 7.90 \times 10^{-11} \text{ J/nuclei}$$

Glossary

A

absolute humidity
the amount of water vapor in the air at a particular time

absolute magnitude
brightness of stars as if they were all at a defined, standard distance from earth

absolute scale
temperature scale whose scale is set so that zero is at absolute zero; the Kelvin temperature scale is an absolute scale

absolute zero
the lowest temperature attainable; occurs when all random motion of molecules has ceased

abyssal plain
the level plain of the ocean floor

acceleration
a change in velocity per change in time; this change can be in either speed, direction, or some combination of speed and direction

accretion disk
fat bulging disk of gas and dust from the remains of the gas cloud that forms around a protostar

achondrites
homogeneously textured stony meteorites

acid
any substance that is a proton donor when dissolved in water

acid-base indicator
a vegetable dye used to distinguish acid and base solutions by a color change

adiabatic cooling
the decrease in temperature of an expanding gas that involves no heat flowing out of the gas

adiabatic heating
the increase in temperature of compressed gas that involves no heat flowing into the gas

air mass
a large, more or less uniform body of air with nearly the same temperature and moisture conditions throughout

air mass weather
the weather experienced within a given air mass; characterized by gradual changes from day to day

alcohol
an organic compound with a general formula of ROH

aldehyde
an organic molecule with the general formula RCHO

alkali metals
shiny, low density metals that can be cut with a knife and that react violently with water to form an alkaline solution; members of family IA of the periodic table

alkaline earth metals
soft, reactive metals that are less reactive than alkali metals; members of family IIA of the periodic table

alkanes
hydrocarbons with single covalent bonds between the carbon atoms

alkenes
hydrocarbons with a double covalent carbon-carbon bond

alkyne
hydrocarbon with a carbon-carbon triple bond

allotropic forms
elements that can have several different structures

alpha particle
the nucleus of a helium atom emitted as radiation from a decaying heavy nucleus

alpine glaciers
glaciers that form at high elevations in mountainous regions

alternating current
current that first moves one direction, then the opposite direction with a regular frequency

altimeter
a device that measures altitude

amino acids
organic functional groups that form polypeptides and proteins

amp
unit of electric current; equivalent to C/sec

ampere
full name of the unit amp

amplitude
the extent of displacement from the equilibrium condition; the size of a wave

aneroid barometer
a device that measures atmospheric pressure by comparing it to a reference pressure

angle of incidence
angle of the incident ray measured from a line perpendicular to the boundary at the point of reflection

angle of reflection
angle of the reflected ray measured from a line perpendicular to the boundary at the point of reflection

angular momentum quantum number
used to indicate the energy sublevels of electrons within the main energy levels of an atom

angular unconformity
a boundary in rock where the bedding planes above and below the time interruption unconformity are not parallel

annular eclipse
occurs when the penumbra reaches the surface of the earth; as seen from earth, the sun forms a bright ring around the disk of the new moon

Antarctic Circle
parallel identifying the limit toward the equator where the sun appears above the horizon all day for six months during the summer; located at 66 1/2° S latitude

anticline
an arch-shaped fold in rock strata

anticyclone
a high pressure center with winds flowing away from the center; associated with clear weather

antinodes
regions of maximum vibration in a standing wave

aphelion
point at which an orbit is farthest from the sun

apogee
when the moon's orbit is farthest from earth

apparent local noon
the instant of the sun crossing the celestial meridian at any particular longitude

apparent local solar time
the time found from the position of the sun in the sky; the shadow of the gnomon on a sundial

apparent magnitude
classification scheme for different levels of stars' brightness that you see

apparent solar day
the interval between two consecutive crossings of the celestial meridian by the sun

aquifer
a layer of sand, gravel, or other highly permeable material that is filled with water beneath the surface

Arctic Circle
parallel identifying the limit toward the equator where the sun appears above the horizon all day for six months during the summer; located at 66 1/2° N latitude

area
the extent of a surface

arid
dry climate classification; receives less than 25 cm (10 in) precipitation per year

aromatic hydrocarbons
organic compounds based on the benzene ring structure

artesian
term describing the condition where pressure forces groundwater from a well to rise above the aquifer

asteroids
small rocky bodies left over from the formation of the solar system; most are accumulated in a zone between the orbits of Mars and Jupiter

asthenosphere
hot, elastic, semiliquid layer of the earth located in the mantle; ranges in thickness from a depth of 130 km to 160 km

astronomical unit
radius of the earth's orbit

atmospheric stability
the condition of a parcel of air found by relating the parcel's actual lapse rate to the dry adiabatic lapse rate

atom
the smallest unit of an element that can exist alone or in combination with other elements

atomic mass unit
unit of mass of atomic nuclei based on the carbon-12 atom that is defined to mass 12 amu

atomic number
the number of protons in the nucleus of an atom

atomic weight
weighted average of the masses of stable isotopes of an element as they occur in nature

autumnal equinox
equinox that occurs about September 23

Avogadro's number
the number of C-12 atoms in exactly 12.0 g of C-12; 6.02×10^{23} atoms or particles

axis
the imaginary line about which a planet or other object rotates

B

background radiation
ionizing radiation from natural sources

Balmer series
a set of line spectra emitted by hydrogen atom electrons as they drop from excited states to the ground state

band of stability
a graph of the number of neutrons versus the number of protons in nuclei; nuclei that have the neutron to proton ratios located in this band do not undergo radioactive decay

barometer
an instrument that measures atmospheric pressure

basalt
most common of the dark, high density igneous rocks; it is fine grained and about half plagioclase feldspars and half ferromagnesian materials

base
any substance that is a proton acceptor when dissolved in water

basin
a large, depressed landform into which streams drain; a circular syncline

beats
regularly spaced increases and decreases of volume due to alternating constructive and destructive interference between two sound waves

beta particle
high energy electron emitted as ionizing radiation from a decaying nucleus

big bang theory
current model of galactic evolution in which the universe was created from an intense and brilliant explosion from a primeval fireball

binding energy
the energy required to break a nucleus into its constituent protons and neutrons

blackbody radiation
radiation emitted by an ideal material that perfectly absorbs and emits radiation

black hole
the theoretical remaining core of a supernova collapsed so densely that even light cannot escape

Bohr model
model of the structure of the atom that attempted to correct the deficiencies of the solar system model and account for the Balmer series

boiling
the process in a liquid where, throughout the entire liquid, molecules are entering the gaseous state; occurs when the vapor pressure equals the atmospheric pressure on the liquid

boiling point
the temperature at which a liquid boils and changes into a gas or vapor

boundary
the division between two regions of differing physical properties

Bowen's reaction series
crystallization series that occurs as a result of the different freezing point temperatures of various minerals present in magma

breaker
a wave whose front has become so steep that the top part has broken forward of the wave

breccia
a rock with highly angular, large fragments

British thermal unit
abbreviated Btu; the amount of energy or heat needed to increase the temperature of one pound of water one degree Fahrenheit

buffer solution
a solution consisting of a weak acid and a salt that has the same negative ion as the acid

C

caloric theory of heat
old, discarded theory that considered heat to be an invisible, weightless, self-repulsive fluid that flowed from one object to another

calorie
the amount of energy or heat needed to increase the temperature of one gram of water one degree Celsius

Calorie
the dieter's "calorie"; equivalent to one kilocalorie

carbohydrates
organic compounds that include sugars, starches, and cellulose; are used by plants and animals for structure, protection, and food

carbonation
a chemical reaction that occurs naturally between carbonic acid and rock minerals

Carnot cycle
an idealized way of describing the transformations between heat and work that occur in a heat engine

cathode rays
negatively charged particles (electrons) that are emitted from a negative terminal in an evacuated glass tube

celestial equator
line of the equator of the earth on the celestial sphere

celestial meridian
an imaginary line in the sky directly above you that runs north through the north celestial pole, south through the south celestial pole, and back around the other side to make a big circle around the earth

cellulose
a polysaccharide abundant in plants that forms the fibers in cell walls that preserves the structure of plant materials

Celsius scale
referent scale of temperature with 100 degrees between the two reference points; based on the freezing and boiling points of water at normal atmospheric pressure

cementation
process by which spaces between sediment particles are filled with binding chemical deposits

Cenozoic
geologic era of recent life

centigrade
alternate name for the Celsius scale

centrifugal force
an apparent outward force on an object following a circular path; this apparent force seems to pull an object out of the circular path into the object's natural straight-line motion

centripetal force
the force required for an object to follow a circular path instead of straight-line motion

cepheid variables
stars that have a regular variation in brightness over a period of time

chain reaction
reaction where the products of a nuclear reaction are able to produce more nuclear reactions in a self-sustaining series

chemical bond
an attractive force that holds atoms together in a compound

chemical change
identity of matter is altered and new substances are formed

chemical energy
a form of energy involved in chemical reactions that results in changes in internal potential energy

chemical equation
concise way of describing what happens in a chemical reaction

chemical equilibrium
occurs when two opposing reactions happen at the same time and at the same rate

chemical reaction
a change in matter where different chemical substances are created by forming or breaking chemical bonds

chemical sediments
dissolved rock materials that have precipitated from solution

chemical weathering
the breakdown of minerals in rocks by chemical reactions with water, gases of the atmosphere, or solutions

chemistry
the science concerned with the study of matter and the changes it undergoes

Chinook
a warm wind consisting of air that has been warmed by compression

chondrites
subdivision of stony meteorites containing small spherical lumps of silicate minerals or glass

chondrules
small spherical lumps of silicate minerals or glass found in some meteorites

cirque
a bowl-like depression formed by glacial erosion

clastic sediments
weathered rock fragments

claystone
rock formed from clay consolidated into solid rock

climate
the general pattern of weather that occurs in a region over a number of years

coalescence process
the process by which raindrops form from the merging of millions of tiny water droplets

cold front
the front that is formed as a cold air mass moves into warmer air

combination chemical reaction
a synthesis reaction in which two or more substances combine to form a single compound

comets
celestial objects originating from the outer edges of the solar system that move about the sun in highly elliptical orbits; solar heating and pressure from the solar wind form a tail on the comet that points away from the sun

compaction
the process of pressure squeezing sediments together and squeezing water out

compound
a pure substance that can be decomposed by a chemical change into simpler substances with a fixed mass ratio

concentration
the relative amounts of solute and solvent in a solution

concept
a generalized mental image of an object or idea

condensation (sound)
increased density and pressure of gas molecules in a wave

condensation (water vapor)
when where more vapor or gas molecules are returning to the liquid state than are evaporating

condensation nuclei
tiny particles suspended in the air on which water condenses

condensation point
the temperature at which a gas or vapor condenses into a liquid; the dew point

conduction
the transfer of heat from molecule to molecule

conglomerate
sedimentary rock formed from large, rounded clastic sediments

constellations
imagined patterns of stars

constructive interference
the condition in which two waves arriving at the same place at the same time in phase add amplitudes to create a new, higher wave crest

continental air mass
dry air masses from over land

continental climate
a climate influenced by air masses from large land areas

continental drift
the idea that continents can shift positions on the earth's surface

continental glaciers
glaciers that cover a large area of a continent

continental shelf
the flooded margins of the continents that form a zone of relatively shallow water

continental slope
steep slope forming the transition between the continental shelf and the ocean basin

control rods
rods inserted between fuel rods in a nuclear reactor to absorb neutrons and thus control the rate of the nuclear chain reaction

convection
transfer of heat by the movement of high energy molecules or the transfer of a fluid from one place to another

convection zone (of a star)
least dense star shell where surface gases emit energy in the form of visible light, ultraviolet radiation, and infrared radiation

convective cells
complete convective circulation pattern; slowly turning regions in the plastic asthenosphere that might drive the motion of plate tectonics

conventional current
the direction of flow if current is considered to be carried by positive charges, for example, from the positive terminal to the negative terminal of a battery

convergent boundaries
boundaries that occur between two plates moving toward each other

coordinate covalent bond
formed when the shared electron pair is contributed by one atom

Copernican system
heliocentric, or sun centered solar system theory

core (of a star)
dense, very hot region of a star where nuclear fusion reactions release gamma and X-ray radiation

core (of earth)
the center part of the earth

Coriolis effect
the apparent deflection due to the rotation of the earth; it is to the right in the Northern Hemisphere

correlation
the determination of the equivalence in geologic age by comparing the rocks in two separate locations

coulomb
unit that measures electric charge by identifying the equivalent of how many electrons have been transferred onto or away from an object

Coulomb's law
relationship between charge, distance, and magnitude of the electrical force between two bodies

covalent bond
chemical bond formed by the sharing of a pair of electrons

covalent compound
compound held together by covalent bonds

creep
the slow downhill movement of soil down a steep slope

crest
the high mound of water that is part of a wave

critical angle
limit to the angle of incidence when all light rays are reflected internally

critical mass
mass of fissionable material needed to sustain a chain reaction

crude oil
petroleum pumped from the ground that has not yet been refined into usable groups

crust
the thin, solid layer of rock that rests on top of the Mohorovicic discontinuity

Curie
unit of nuclear activity defined as 3.70×10^{10} nuclear disintegrations per second

cycle
a complete vibration

cyclone
a low-pressure center where the winds move into the low-pressure center and are forced upward

D

data
measurement information used to describe something

data points
points that may be plotted on a graph to represent simultaneous measurements of two related variables

Daylight Saving Time
setting clocks ahead one hour during the summer to more effectively utilize the longer days of summer

decibel scale
a nonlinear scale of loudness based on the ratio of the intensity level of a sound to the intensity at the threshold of hearing

decomposition chemical reaction
a reaction in which a compound is broken down into the elements that make up the compound, into simpler compounds, or into elements and simpler compounds

deflation
the widespread removal of base materials from the surface by the wind

degassing
process where gases and water vapor were released from rocks heated to melting during the early stages of the formation of a planet

degradation of energy
transformation of energy from high quality to low quality

delta
sediment deposits at the mouth of a river

density
the ratio of mass per volume

density current
an ocean current that flows because of density differences

destructive interference
the condition in which two waves arriving at the same point at the same time out of phase add amplitudes to create zero total disturbance

dew
condensation on surfaces

dew point temperature
the temperature at which condensation begins

diffraction
the bending of light around the edge of an opaque object

diffuse reflection
light rays reflected in many random directions

direct current
constant electrical current in one direction

direct proportion
when two variables increase or decrease together in the same ratio (at the same rate)

disaccharides
two monosaccharides joined together with the loss of a water molecule

dispersion
the effect of spreading colors of light into a spectrum with a material that has an index of refraction that varies with wavelength

divergent boundaries
boundaries that occur between two plates moving away from each other

divide
line separating two adjacent watersheds

dolomite
sedimentary rock formed from limestone by replacement of calcium ions with magnesium ions

dome
landform of roughly symmetrical upfold; a smooth mountain summit

doppler effect
the apparent shift in the frequency of sound or light due to relative motion between the source of the sound or light and the observer

double bond
covalent bond formed when two pairs of electrons are shared by two atoms

dry adiabatic lapse rate
the rate of adiabatic cooling or warming of air in the absence of condensation or evaporation

dune
a low mound or ridge of sand or other sediments

E

earthquake epicenter
point on the earth's surface directly above an earthquake focus

earthquake focus
place where an earthquake originates

echo
a reflected sound that can be distinguished from the original sound

eclipse
when the shadow of a celestial body falls on the surface of another celestial body

elastic rebound
the sudden snap of stressed rock into new positions; the recovery from elastic strain

electrical conductors
materials that have electrons that are free to move throughout the material, e.g., metals

electrical force
a push or a pull that is the result of electric charges interacting

electrical insulators
electrical nonconductors, or materials that obstruct the flow of electric current

electrical nonconductors
materials that do not have electrons that are free to move throughout the material (plastic, wood, etc.)

electrical resistance
the property of opposing or reducing electric current

electric circuit
a device used to perform work from electrical energy; it contains three parts: a voltage source, a continuous path for the current, and a device to perform the work

electric current
the flow of electric charge

electric energy
the form of energy from electromagnetic interactions

electric field
force field produced by an electrical charge

electric field lines
model of the action at a distance of electrical force; gives the direction of the electric force on a positive charge

electric field strength
size of the force on a unit charge located in an electric field

electric generator
a device that utilizes electromagnetic induction to produce electricity

electric potential energy
potential energy due to the position of a charge near other charges

electrolytes
solutions of ionic substances that will conduct an electric current

electromagnet
a magnet formed by a solenoid that can be turned on and off by turning the current on and off

electromagnetic force
the force of attraction or repulsion between two charged particles

electromagnetic induction
process in which current is induced by moving a loop of wire in a magnetic field or by changing the magnetic field

electromagnetic spectrum
the complete range of frequencies of electromagnetic waves

electromagnetic wave
a pulse of rapidly regenerating electric and magnetic fields moving through space

electron
subatomic particle that has the smallest negative charge possible; usually found in an orbital of an atom

electron configuration
the arrangement of electrons in orbitals and suborbitals about the nucleus of an atom

electron current
opposite to conventional current; the direction of flow of electrons, for example, electrons moving from a negative terminal to a positive terminal of a battery

electron dot notation
short way to show an element and its outer shell electrons

electronegativity
the comparative ability of atoms of an element to attract bonding electrons

electron pair
a pair of electrons with different spin quantum numbers that may occupy a suborbital

electrons
negatively charged particles present in all matter

electron volt
the energy gained by an electron moving across a potential difference of one volt

electrostatic charge
an accumulated electric charge at rest

element
a pure substance that cannot be broken down into anything simpler by chemical or physical means

empirical formula
identifies the elements present in a compound and describes the simplest whole number ratio of atoms of these elements with subscripts

energy
the ability to do work

English system
a system of measurement that originally used sizes of parts of the human body as referents

entropy
the thermodynamic measure of disorder

epicycle
in the geocentric model, the secondary circular orbit that explained the occasional retrograde motion of the planets

epochs
subdivisions of geologic periods

equation
a statement that describes a relationship where quantities on one side of the equal sign are identical to quantities on the other side

Equation of Time
the cumulative variation between the apparent local solar time and the mean solar time

equinoxes
Latin meaning "equal nights"; time when the axis of the earth is perpendicular to a line between the centers of the sun and earth

eras
the major blocks of time in the earth's geologic history

erosion
the process of moving the products of weathering

esters
class of organic compounds that make up fats, oils, and waxes; certain esters give fruit and flowers their taste and odor

ether
class of organic compounds with the general formula ROR', mostly used as industrial and laboratory solvents

evaporation
when more molecules are leaving a liquid for the gaseous state than are returning from the gas to the liquid

excited states
energy states of an atom that are above the minimum energy state for that atom

exfoliation
the fracturing and breaking away of curved, sheetlike plates from bare rock surfaces via physical or chemical weathering

exosphere
the outermost layer of the atmosphere

extrusive igneous rocks
fine-grained igneous rocks formed by lava that cools rapidly above the surface

F

Fahrenheit scale
referent scale of temperature with 180 degrees between the two reference points; based on the freezing and boiling points of water

family
vertical columns on the elements table that have similar properties

fats
organic compounds of esters formed from glycerol; also called triglycerides

fault
a fracture along which relative movement has occurred between the rocks on either side

fault plane
the surface of a fault

ferromagnesian silicates
silicates that contain iron and magnesium

ferromagnetic properties
describes substances attracted to magnets

first law of motion
every object remains at rest or in motion in a straight line unless acted on by an unbalanced force

first law of thermodynamics
the relationship between the change of the internal energy state of a system, the heat added, and the work done by the system; it is the thermodynamic statement of conservation of energy

first quarter
the moon phase that occurs after a new moon when one half of the moon disk is lighted; since the lighted part is one fourth of the whole moon, this is called the first quarter after a new moon

floodplain
the wide, level floor of a valley built by a stream

fluids
matter that has the ability to flow; the individual molecules are able to move freely

folds
wrinkling of rock layers due to stress

foliation
flat crystal flakes aligned in parallel sheets

force
a push or a pull that can change the motion of an object

force field
a model describing action at a distance by giving the magnitude and direction of force on a unit particle

formula
describes what elements are in a compound and in what proportions

formula weight
the sum of the atomic weights of all the atoms in a chemical formula

fossil fuels
organic fuels that contain the stored radiant energy of the sun converted to chemical energy by plants that lived millions of years ago

Foucault pendulum
a heavy mass swinging from a long wire that illustrates the rotation of the earth

free fall
when objects fall toward the earth with no forces acting upward, air resistance is neglected

freezing point
the temperature at which a liquid freezes into a solid

frequency
the number of cycles of a wave occurring in one second

fresh water
water fit for human consumption, having less than 2,000 ppm salinity

front
the boundary between air masses of different temperatures

frost
frozen condensation on surfaces

frost wedging
the process of ice freezing inside cracks and pores and the wedging apart of the rocks

fuel rod
long zirconium alloy tubes containing fissionable material for use in a nuclear reactor

full moon
when the entire side of the moon facing the earth is illuminated

functional group
the atom or group of atoms in an organic molecule that is responsible for the chemical properties of a particular class or group of organic chemicals

fundamental charge
smallest common charge known

fundamental frequency
the lowest frequency that can set up standing waves in an air column or on a string

fundamental properties
a property that cannot be defined in simpler terms other than to describe how it is measured

G

g
the acceleration of an object in free fall due to the force of gravity; its magnitude is 9.80 m/s² (32.0 ft/s²)

gabbro
the coarse grained chemical equivalent of basalt

galactic clusters
gravitationally bound subgroups of as many as 1,000 stars that move together within the galaxy

galaxy
group of billions and billions of stars that form the basic unit of the universe

gamma ray
very short wavelength electromagnetic radiation emitted by decaying nuclei

gases
a state of matter whose constituent molecules have weak cohesive forces acting between them; neither the shape nor the volume of the substance is fixed

gasohol
solution of ethanol and gasoline

gauss
measure of magnetic field strength; smaller than a tesla

Geiger counter
device that counts ions produced by the passage of ionizing radiation as a means of measuring the radiation

geocentric
earth-centered universe

Geologic Time Scale
a "calendar" of geologic history

geomagnetic time scale
time scale established from the number and duration of magnetic field reversals during the past 6 million years

geosyncline
a syncline of great extent

giant planets
the large outer planets that all have similar densities and compositions

glacier
a mass of ice on land that moves under its own weight

globular clusters
symmetrical and tightly packed cluster of as many as a million stars

glycerol
an alcohol with three hydroxyl groups per molecule

glycogen
starchlike polysaccharide synthesized by the human body and stored in the muscles and liver

glycol
an alcohol with two hydroxyl groups per molecule

gneiss
a metamorphic rock that has alternating bands of light and dark minerals

gram-atomic weight
the mass in grams of one mole of an element; numerically equal to its atomic weight

gram-formula weight
the mass in grams of one mole of the compound that is numerically equal to its formula weight

gram-molecular weight
the gram-formula weight of a molecular compound

granite
light colored, coarse-grained igneous rock common on continents; igneous rocks formed by blends of quartz and feldspars with small amounts of micas, hornblende, and other minerals

greenhouse effect
the process of heating in the lower parts of the atmosphere through the absorption and re-emission of infrared radiation by certain gases in the atmosphere

ground state
the lowest energy state of an atom

groundwater
water from a saturated zone beneath the surface

gyre
the great circular systems of moving water in each ocean

H

hail
a frozen form of precipitation

hair hygrometer
a device that measures humidity from changes in the length of hair

half-life
the time required for one-half of the unstable nuclei in a radioactive substance to decay

halogen
very reactive nonmetalic elements common in salt compounds; elements of the family VIIA in the period table

hard water
water that contains relatively high concentrations of dissolved salts of calcium and magnesium

harmonics
frequencies higher than the fundamental frequency that can set up standing waves in an air column or on a string

heat
energy flow resulting from a temperature difference; energy that increases the internal energy of a substance, that is, the energy of the individual molecules forming a substance

heat death
the term referring to the state of the universe when all the high quality energy has been degraded to low quality energy

heat engine
device that can use heat flow between regions of different temperature to perform work

heat of formation
energy released in a chemical reaction

heat pump
device that moves heat from one place to another against the natural flow of heat

Heisenberg uncertainty principle
you cannot measure both the exact momentum and the exact position of a subatomic particle at the same time; when the more exact of the two is known, the less certain you are of the value of the other

hertz
unit used to measure frequency; equivalent to one cycle per second

Hertzsprung-Russell diagram
diagram to classify stars with a temperature-luminosity graph

high latitudes
latitudes close to the poles; those that sometime receive no solar radiation at noon

high pressure center
another term for anticyclone

horsepower
measurement of power equivalent to 550 ft·lb/sec

humid
moist climate classification; receives more than 50 cm (20 in) precipitation per year

humidity
the amount of water vapor in the air; see relative humidity

hurricane
large, violent circular storms born over warm tropical oceans near the equator

hydration
the attraction of water molecules for ions; a reaction that occurs between water and minerals that make up rocks

hydrocarbon
an organic compound consisting of only the two elements, hydrogen and carbon

hydrocarbon derivatives
organic compounds that can be thought of as forming when one or more hydrogen atoms on a hydrocarbon have been replaced by some element or group of elements other than hydrogen

hydrogen bond
a weak to moderate bond between the hydrogen end (+) of a polar molecule and the negative end (−) of a second polar molecule

hydrologic cycle
the cycle of continuous evaporation and precipitation of water in the earth's atmosphere

hydronium ion
a molecule of water with an attached hydrogen ion

hypothesis
a tentative explanation of a phenomenon that provides a framework for understanding and that is compatible with the data describing that phenomenon

I

ice-forming nuclei
small, solid particles suspended in air; ice can form on the suspended particles

igneous rocks
rocks that formed from a hot, molten mass

incandescent
matter emitting visible light as a result of high temperatures

incident ray
line representing the direction of motion of light approaching a boundary

inclination of earth axis
tilt of the earth's axis measured from the plane of the ecliptic; considered to be the same throughout the year

index fossils
distinctive fossils of organisms that lived only a brief time; used to date rock

index of refraction
the ratio of the speed of light in a vacuum to the speed of light in a material

inertia
the property of matter where an object will remain in unchanging motion or at rest in the absence of an unbalanced force

infrasonic
sound waves having too low a frequency to be heard by the human ear; sound having a frequency of less than 20 Hz

inorganic chemistry
the study of all compounds and elements in which carbon is not the principal element

insulators
materials that do not conduct heat well; they usually have many small air spaces that make it more difficult to transfer molecular vibrations from one molecule to another

intensity
a measure of the energy carried by a wave

interference
phenomenon of light where the relative phase difference between two light waves produces light or dark spots; is a result of light's wavelike nature

intermolecular forces
forces of interaction between molecules

internal energy
total sum of all the energy of all the molecules forming an object

international date line
an imaginary line established to compensate for cumulative time zone changes

intertropical convergence zone
the belt around the equator where air heats and rises

intrusive igneous rocks
coarse-grained igneous rocks formed by magma that cools slowly deep below the surface

inverse proportion
the relationship where the value of one variable increases while the value of the second variable decreases at the same rate (in the same ratio)

inversion
a layer of air where temperature increases with height

ion
an atom or particle that has a net charge because of the gain or loss of electrons; polyatomic ions are groups of banded atoms that have a net charge

ion exchange reaction
a reaction that takes place when the ions of one compound interact with the ions of another forming 1) a solid that comes out of solution, 2) a gas, or 3) water

ionic bond
chemical bond of electrostatic attraction between negative and positive ions

ionic compounds
compounds held together by ionic bonds

ionization
process of forming ions from molecules

ionization counter
a device that counts ions produced by radiation

ionized
an atom or particle that has a net charge because it has gained or lost electrons

ionosphere
the thermosphere and upper mesosphere where free electrons and ions are present

iron meteorites
meteorite classification group whose members are composed mainly of iron

island arcs
curving chains of volcanic islands that occur over belts of deep-seated earthquakes

isomers
compounds with the same molecular formula but with different structures

isostasy
the concept of less dense rock "floating" in more dense rock

isotope
atoms of an element with identical chemical properties but with different masses

J

jet stream
a powerful belt of wind near the top of the troposphere

joint
a break in rock without a relative movement of the rock on either side of the break

joule
metric unit used to measure work and energy; can also be used to measure heat; equivalent to newton·meter

K

Kelvin scale
an absolute temperature scale whose unit size is based on the Celsius scale

Kepler's first law
each planet moves in an elliptical orbit, with the sun located at one focus

Kepler's laws of planetary motion
the three laws describing the motion of the planets

Kepler's second law
an imaginary line between the sun and a planet moves over equal areas of the ellipse during equal time intervals

Kepler's third law
square of the period of an orbit is directly proportional to the cube of the radius of the major axis of the orbit

ketone
an organic compound with the general formula RCOR′

kilocalorie
the amount of energy required to increase the temperature of one kilogram of water one degree Celsius; equivalent to 1,000 calories

kilogram
the fundamental unit of mass in the metric system of measurement

kinetic energy
the energy associated with motion

kinetic molecular theory
the collection of assumptions about the tiny atoms and molecules that form every substance

L

lake
a large body of standing water

landforms
the features of the surface of the earth

landslide
general term for rapid movement of any type or mass of materials

last quarter
the moon phase that occurs after a full moon when one half of the moon disk is lighted; since the lighted part is one fourth of the whole moon, this is called the last quarter before a new moon

latent heat
refers to the heat "hidden" in the changes of state

latent heat of fusion
the heat required to convert one gram of a substance from the solid to the liquid phase, or the heat released by one gram of a substance when changing from the liquid phase to the solid phase

latent heat of vaporization
the heat required to convert one gram of a substance from the liquid phase to the gaseous phase, or the heat released when one gram of gas changes from the gaseous phase to the liquid phase

latitude
the angular distance from the equator to a point on a parallel

lava
magma that is forced to the surface

law of conservation of energy
energy can never be created or destroyed—it can only change form

law of conservation of mass
matter is neither created nor destroyed in a chemical reaction

law of conservation of momentum
in the absence of external forces, the total momentum of a group of interacting objects remains constant

light ray
model using lines to show the direction of motion of light to describe the travels of light

light year
the distance that light travels in one year

limestone
sedimentary rock formed chiefly from calcium carbonate

linear scale
a scale, generally on a graph, where equal intervals represent equal changes in the value of the variable

lines of force
lines drawn to make an electric field strength map

line spectrum
narrow lines of color in an otherwise dark spectrum; these lines can be used as "fingerprints" to identify gases

liquids
a state of matter whose molecules have interactions that are stronger than in a gas but not strong enough to keep the molecules near fixed positions; liquids have definite volume but no permanent shape

liter
a metric system unit of volume usually used for liquids

lithosphere
solid layer above the lithosphere that includes the crust, the Moho, and the upper part of the mantle

loess
very fine dust or silt deposits

longitude
angular distance east or west from the prime meridian

longitudinal wave
a wave in which the displacements back and forth forming the wave are in the same direction as the wave is traveling

longshore current
accumulated water pushed along by waves that produces a current that moves parallel to the shore

low latitudes
latitudes close to the equator; those that sometimes receive vertical solar radiation at noon

luminosity
the total amount of energy radiated into space each second from the surface of a star

luminous
something producing light

lunar eclipse
occurs when the shadow of the earth falls on the moon

lunar highlands
light colored mountainous regions of the moon

M

macromolecule
very large molecule that is a combination of many smaller, similar molecules joined together in a chainlike structure

magma
molten rock material from which igneous rocks form

magnetic domain
tiny regions in permanent magnets, each with their own north and south poles

magnetic field
model used to describe how magnetic forces on moving charges act at a distance

magnetic poles
the two ends of a magnet about which the force of attraction is concentrated

magnetic quantum number
quantum number of an electron related to the energy of its orbital as the orbital is oriented in space by an external magnetic field

magnitude
the size of a measurement or a vector

main sequence stars
normal, mature stars that use their nuclear fuel at a steady rate

manipulated variable
in an experiment, a quantity that can be controlled or manipulated; also known as the independent variable

mantle
middle part of the earth's interior

marble
nonfoliated metamorphic rock formed from recrystallized limestone

maria
smooth, dark areas on the moon

marine climate
a climate influenced by air masses from over an ocean

maritime air mass
a moist air mass that forms over the ocean

mass
a measure of inertia

mass defect
the difference between the sum of the masses of the individual nucleons forming a nucleus and the actual mass of that nucleus

mass movement
erosion caused by the direct action of gravity

mass number
sum of the number of protons and neutrons in a nucleus

matter
anything that occupies space and has mass

matter waves
term for the wavelike properties of subatomic particles

meanders
sweeping bends in a river channel

mean solar day
twenty-four hour day averaged from the mean solar time

mean solar time
a uniform time averaged from the apparent solar time

measurement
the process of comparing a property of an object to a well-defined and agreed-upon referent

mechanical energy
the form of energy associated with machines and other familiar objects

mechanical weathering
the physical breaking up of rocks without any changes in their chemical composition

melting point
the temperature at which a solid melts into a liquid

meridians
north-south running arcs that intersect at both poles

mesosphere
means middle layer; the atmospheric layer above the stratosphere; the solid, dense mantle below the asthenosphere

Mesozoic
geologic era referring to the time of middle life

metal
matter having the physical properties of conductivity, malleability, ductility, and luster

metamorphic rocks
previously existing rocks that have been changed into a distinctly different rock by heat, pressure, or hot solutions

meteor
the streak of light and smoke in the sky left by a meteoroid falling toward the earth

meteorite
small celestial bodies that fall and strike the surface of the earth

meteoroids
in space, remnants of comets and asteroids

meteorology
the science of understanding and predicting weather

meteor shower
event when many meteorites fall in a short period of time

meter
the fundamental metric unit of length

metric system
a system of referent units based upon invariable referents of nature that have been defined as standards

microclimate
a local, small scale pattern of climate

middle latitudes
latitudes equally far from the poles and equator; between the high and low latitudes

millibar
a measure of atmospheric pressure equivalent to 1,000 dynes per cm^2

mineral
a naturally occurring, inorganic solid element or chemical compound with a crystalline structure

miscible fluids
fluids that can mix in any proportion

mixture
matter made of unlike parts that have a variable composition and can be separated into their component parts by physical means

model
a mental or physical representation of something that cannot be observed directly; used as an aid to understanding

moderator
a substance in a nuclear reactor that slows fast neutrons so the neutrons can participate in nuclear reactions

Mohorovicic discontinuity
boundary between the crust and mantle

mole
an amount of a substance that contains Avogadro's number of atoms, ions, molecules, or any other chemical unit

molecular formula
a chemical formula that identifies the actual numbers of atoms in a molecule

molecular weight
the formula weight of a molecular substance

molecule
chemical point of view: a particle composed of two or more atoms held together by an attractive force called a chemical bond; kinetic point of view: smallest particle of a compound or a gaseous element that can exist and still retain the characteristic properties of a substance

momentum
the product of the mass of an object times its velocity

monadnocks
hills of resistant rock that are found on peneplains

monosaccharides
simple sugars that are mostly 6-carbon molecules

moraines
deposits of bulldozed rocks and other materials left behind by a melted glacier

N

natural frequency
the frequency of vibration of an elastic object that depends on the size, composition, and shape of the object

neap tide
period of less pronounced high and low tides

nebulae
interstellar clouds of hydrogen gas that may develop into a star

negative electric charge
one of the two types of electric charge; repels other negative charges and attracts positive charges

negative ion
neutral atom that picks up an electron and hence has a negative charge

net force
the resulting force after all vector forces have been added

neutralized
acid or base properties have been lost through a chemical reaction

neutron
neutral subatomic particle usually found in the nucleus of an atom

neutron star
very small super-dense remains of a supernova with a center core of pure neutrons

new crust zone
zone near a divergent boundary made up of new crust formed from magma upwelling at the boundary

new moon
when the entire side of the moon facing the earth is dark

newton
the unit defining the proportion between mass, acceleration, and force; it is equivalent to the following product of units: $kg \cdot m/sec^2$

noble gas
colorless, odorless, nonreactive gases

node
regions on a standing wave that do not oscillate

noise
sounds made up of groups of waves of random frequency and intensity

nonelectrolytes
covalent compounds that form molecular solutions; they cannot conduct an electric current

nonferromagnesian silicates
silicates that do not contain iron or magnesium ions

nonmetal
matter that doesn't have a metallic luster, is a poor conductor of heat and electricity, and is not malleable or ductile

nonsilicates
minerals that do not contain the silicon and oxygen tetrahedral structure in their crystal structure

noon
when the sun moves across the celestial meridian

normal
a line perpendicular to a boundary

normal fault
a fault where the hanging wall has moved downward with respect to the foot wall

north celestial pole
the point where the north pole of the earth touches the celestial sphere

north pole
north-seeking pole of a magnet

nova
irregularly flaring star

nuclear energy
the form of energy obtained from reactions involving the nuclei of atoms

nuclear fission
nuclear reaction of splitting a massive nucleus into more stable, less massive nuclei with an accompanying release of energy

nuclear force
a strong force of attraction that operates over very short distances; this force overcomes the electric repulsion of protons in a nucleus and binds the nucleus together

nuclear fusion
nuclear reaction of low mass nuclei fusing together to form more stable and more massive nuclei with an accompanying release of energy

nuclear reactor
steel vessel in which a controlled chain reaction of fissionable materials releases energy

nucleons
the protons and neutrons in the nucleus of an atom

nucleus
tiny, relatively massive positively charged center of an atom containing protons and neutrons

nucleus
small dense center of an atom

numerical constant
a constant without units; a number

O

objects
things that can be seen or touched

oblate spheroid
the shape of the earth—a somewhat squashed spherical shape

observed lapse rate
the rate of change in temperature compared to change in altitude

obsidian
name of volcanic glass

occluded front
a front that has been lifted off the ground into the atmosphere

ocean
the single, continuous body of salt water on the surface of the earth

ocean basin
the deep bottom of the ocean floor

ocean currents
streams of water within the ocean that stay in about the same path as they move over large distances

oceanic ridges
long, high, continuous, suboceanic mountain chains that seem to run completely around the earth

oceanic trenches
long, narrow, deep troughs with steep sides

ocean wave
a moving disturbance that travels across the surface of the ocean

octet rule
atoms attempt to acquire an outer orbital with eight electrons

ohm
unit of resistance; equivalent to volts/amps

Ohm's law
the electric potential difference is directly proportional to the product of the current times the resistance

oil field
petroleum accumulated in and under rock layers

oils
long-term storage of energy in plants

Oort cloud
in the outer reaches of the solar system, the cloud of icy, dusty aggregates from which comets originate

opaque
materials that do not allow the transmission of any light

orbital
the region of space around the nucleus of an atom where an electron is likely to be found

organic acids
acids derived from organisms; organic compounds with a general formula of RCOOH

organic chemistry
the study of compounds in which carbon is the principal element

orientation of earth's axis
direction that the earth's axis points; considered to be the same throughout the year

origin
the only point on a graph where both the x and y variables have a value of zero at the same time

oscillating theory
that states since symmetry is observed in nature, the universe will end in another big bang after contracting and collapsing

overtones
the sounds made by the harmonics of a musical instrument

oxbow lake
a meander cut off from a river to form a lake

oxidation
the process of a substance losing electrons during a chemical reaction

oxidation number
the positive or negative charge of the atoms of an element in a compound

oxidation-reduction reaction
a chemical reaction in which electrons are transferred from one atom to another

oxidized
substance that has lost electrons in a chemical reaction

oxidizing agents
substances that take electrons from other substances

ozone shield
layer of atmosphere in which ultraviolet absorbing chemical reactions occur

P

Paleozoic
geologic era referring to the time of ancient life

parallels
east and west running circles that are parallel to the equator on a globe

parsec
astronomical unit of distance where the distance at which the angle made from 1 AU baseline is 1 arc second

parts per billion
concentration ratio of parts of solute to parts of solution

parts per million
concentration ratio of parts of solute to parts of solution

Pauli exclusion principle
no two electrons in an atom can have the same four quantum numbers

peneplain
a nearly flat landform that is the end result of the weathering and erosion of the land surface

penumbra
the zone of partial darkness in a shadow

percent by volume
the volume of solute in 100 volumes of solution

percent by weight
the weight of solute in 100 weight units of solution

perigee
when the moon's orbit is closest to the earth

perihelion
point at which an orbit comes closest to the sun

period in periodic table
horizontal rows of elements with increasing atomic numbers; runs from left to right on the element table

period of wave
the time required for one complete cycle of a wave

periodic law
similar physical and chemical properties recur periodically when the elements are listed in order of increasing atomic number

periods
subdivisions of geologic eras

permeability
the ability of a sediment to transmit water

petroleum
mixture of alkanes, cyclo-alkanes, and some aromatic hydrocarbons

phase change
when a substance changes from one state of matter to another

photoelectric effect
the movement of electrons as a result of energy acquired from absorbed light

photons
a quanta of energy in light wave; the particle associated with light

pH scale
scale that measures the acidity of a solution with numbers below seven representing acids, seven representing neutral, and numbers above seven representing bases

physical change
a change of the state of a substance but not the identity of the substance

pitch
the frequency of a sound wave

Planck's constant
proportionality constant in the relationship between the energy of vibrating molecules and their frequency of vibration

plasma
a state of matter; a very hot gas consisting of electrons and atoms that have been stripped of their electrons because of high kinetic energies

plate tectonics
the theory that the earth's crust is made of rigid plates that float on the asthenosphere

polar air mass
cold air mass that forms in cold regions

polar climate zone
climate zone of the high latitudes; average monthly temperatures stay below 10° C (50° F)

polar covalent bond
a covalent bond in which there is an unequal sharing of bonding electrons

polarized
light whose constituent transverse waves are all vibrating in the same direction

Polaroid
a film that transmits only polarized light

polyatomic ion
ion made up of many atoms

polymer
long chain of repeating monomer molecules

polymers
huge, chainlike molecules made of hundreds or thousands of smaller, repeating molecular units called monomers

polysaccharides
polymers consisting of monosaccharide units joined together in straight or branched chains; starches, glycogen, or cellulose

pond
a small body of standing water

porosity
the ratio of pore space to the total volume of a rock or soil sample, expressed as a percentage

positive electric charge
one of the two types of electric charge; repels other positive charges and attracts negative charges

positive ion
atom or particle that has a net positive charge due to an electron or electrons being torn away

potential energy
energy associated with changes in position or shape

power
the rate at which energy is transferred or the rate at which work is performed

Precambrian
geologic era referring to the time of ancient life

precession
the slow wobble of the axis of the earth similar to the wobble of a spinning top

precipitation
water that returns to the surface of the earth in the solid or liquid form

pressure
the ratio of force to unit area

primary coil
coil of wire connected to a source of alternating current that is part of a transformer

primary loop
part of nuclear power plant; the closed pipe system that carries heated water from the reactor to a steam generator

prime meridian
referent meridian that passes through the Greenwich Observatory in England

principle of uniformity
states that the processes changing rocks today are the processes that changed them in the past

principle quantum number
used to describe the main energy level of an electron in terms of its most probable distance from the nucleus

proof
double the ethanol concentration by volume of an alcoholic beverage

properties
qualities or attributes that, taken together, are usually unique or peculiar to an object

proportionality constant
a constant applied to a proportionality statement that transforms the statement into an equation

proteins
macromolecular polymers made of smaller molecules of amino acids

protogalaxies
collection of gas, dust, and young stars in the process of forming a galaxy

proton
subatomic particle that has the smallest possible positive charge, usually found in the nucleus of an atom

protoplanet nebular model
model of the formation of the solar system that states that the planets formed from gas and dust leftover from the formation of the sun

protostar
an accumulation of gases that will become a star

psychrometer
a two-thermometer device used to measure humidity

Ptolemaic system
geocentric model of the structure of the solar system that uses epicycles to explain retrograde motion

pulsars
the source of regular, equally spaced pulsating radio signals believed to be the result of the magnetic field of a rotating neutron star

pure substance
materials that are the same throughout and have a fixed definite composition

pure tone
sound made by very regular repeating vibrations

Q

quad
one quadrillion Btu; used to describe very large amounts of heat energy

quanta
fixed amounts; usually referring to fixed amounts of energy absorbed or emitted by matter

quantities
measured properties; includes the numerical value of the measurement and the unit used in the measurement

quantum mechanics
model of the atom based on the mechanics of the wavelike properties of electrons and other subatomic particles

quantum numbers
numbers that describe the energy state of an electron bound to an atom

quartzite
quartz rich metamorphic rock formed from sandstone

R

rad
a measure of radiation received by a material (*r*adiation *a*bsorbed *d*ose)

radiant energy
energy that can travel through space; see radiation

radiation
energy in a form that can travel through space with no medium required to conduct it

radiation zone
less dense region than the star's core where gamma and X-rays from the core are absorbed and reemitted by collisions with atoms

radioactive decay
the natural spontaneous disintegration or decomposition of a nucleus

radioactive decay constant
a specific constant for a particular isotope that is the ratio of the rate of nuclear disintegration per unit of time to the total number of radioactive nuclei

radioactive decay series
series of decay reactions that begins with one radioactive nucleus that decays to a second nucleus that decays to a third nucleus and so on until a stable nucleus is reached

radioactivity
spontaneous emission of particles or energy from an atomic nucleus as it disintegrates

radiometric age
age determined by measuring the radioactive decay of unstable elements

rarefaction
decreased density and pressure of gas molecules in a wave

ratio
a relationship between two numbers, one divided by the other, such as the ratio of distance per time is speed

real image
an image generated by a lens or mirror that can be projected onto a screen

red giant stars
bright, low temperature giant stars that are enormously bright for their temperature due to their size

redox reaction
short name for oxidation-reduction reaction

reduced
substance that has gained electrons in a chemical reaction

reducing agents
supplies electrons to the substance being reduced in a chemical reaction

reduction
the process of gaining electrons in a chemical reaction

referent
referring to or thinking of a property in terms of another, more familiar object

reflected ray
a line representing direction of motion of light reflected from a boundary

reflection
the change when light, sound, or other waves bounce backwards off a boundary

refraction
a change in the direction of travel of light, sound, or water waves crossing a boundary

refrigerant
a volatile fluid used in heat pumps that undergoes a liquid-gas phase change at low temperature

rejuvenation
process of uplifting land that renews the effectiveness of weathering and erosion processes

relative dating
dating the age of a rock unit or geological event relative to some other unit or event

relative humidity
the ratio of how much water vapor the air holds to the maximum amount of water vapor the air could hold at that temperature

rem
measure of radiation that considers the biological effects of different kinds of ionizing radiation

replacement chemical reaction
reaction in which an atom or polyatomic ion is replaced in a compound by a different atom or polyatomic ion

representative elements
members of the A group families of the periodic table

reservoir
a natural or artificial pond or lake used to store water, control floods, or generate electricity

resonance
when the frequency of an external force matches the natural frequency and standing waves are set up

responding variable
the variable that responds to changes in the manipulated variable, also known as the dependent variable because its value depends upon the value of the manipulated variable

reverberation
apparent increase in volume caused by reflections

reverse fault
a fault where the hanging wall has moved upward with respect to the foot wall

revolution
the motion of a planet as it orbits the sun

rhyolite
a fine-grained igneous rock that is the chemical equivalent of granite

Richter Scale
a scale that measures the severity of earthquakes with each higher number indicating 10 times more ground movement and about 30 times more energy

ridges
long, rugged mountain chains rising thousands of meters above the abyssal plains of the ocean basin

rift
a break in the crust of the earth

rip current
strong, brief current that runs against the surf and out to sea

rock
a solid aggregation of minerals or mineral materials that have been brought together into a cohesive solid

rock cycle
process of rocks continually changing from one type to another

rock flour
rock pulverized by a glacier into a powdery, silt-sized sediment

rotation
the spinning of a planet

runoff
water moving across the surface of the earth

S

salinity
the mass of salts dissolved in 1,000 g of solution

salt
any ionic compound except one with hydroxide or oxide ions

sandstone
sedimentary rock formed from sand and consolidated into solid rock

saturated air
air in which an equilibrium exists between evaporation and condensation

saturated molecule
an organic molecule that has the maximum number of hydrogen atoms

saturated solution
a state of equilibrium that exists between dissolving solute and solute coming out of solution

scalars
measurements that have magnitude only

schist
a metamorphic rock produced from slate by further metamorphism

scientific law
a relationship between quantities usually described by an equation in the physical sciences

scientific principle
a relationship between quantites concerned with a specific, narrow range of observations and behavior

scintillation counter
device that measures the flashes of light produced when radiation strikes a phosphor

sea
a smaller part of the ocean with characteristics that distinguish it from the larger ocean

sea breeze
air moving from sea to land

seafloor spreading
the process by which hot, molten rock moves up from the interior of the earth to emerge along mid-oceanic rifts, flowing out in both directions to create new rocks and spread the seafloor apart

seamounts
steep, submerged volcanic peaks on the abyssal plain

second
the standard unit of time in both the metric and English systems of measurement

secondary coil
coil of wire in a transformer linked to the primary coil via electromagnetic induction

secondary loop
part of nuclear power plant; the closed pipe system that carries steam from a steam generator to the turbines, then back to the steam generator as feedwater

second law of motion
the acceleration of an object is directly proportional to the net force acting on that object and inversely proportional to the mass of the object

second law of thermodynamics
states that heat only flows from warmer to cooler objects and that it is impossible to convert heat completely into mechanical energy

sedimentary rocks
rocks formed from particles or dissolved minerals from previously existing rocks

sediments
accumulations of silt, sand, or gravel that settle out of the atmosphere or out of water

semiarid
climate classification between arid and humid; receives between 25 and 50 cm (10 and 20 in) precipitation per year

semiconductors
elements that sometimes act like metals, sometimes act like nonmetals, and sometimes act like both

shale
siltstone or claystone that tends to break along planes into flat pieces

shell model of the nucleus
considers that protons and neutrons move in energy levels, or shells, in the nucleus (like the shell structure of electrons in an atom)

shells
refers to the layers that electrons occupy around the nucleus

shock wave
an intense disturbance caused by wave fronts, such as sound waves, piling atop one another

sidereal day
the interval between two consecutive crossings of the celestial meridian by a particular star

sidereal month
the time interval between two consecutive crossing of the moon across any star

sidereal year
the time interval required for the earth to move around its orbit so that the sun is in the same position against the stars

silicates
minerals that contain silicon-oxygen tetrahedra either isolated or joined together

siltstone
sedimentary rock formed from silt (1/16 to 1/256 mm diameter sediment) that has been consolidated into solid rock

simple harmonic motion
the vibratory motion that occurs when there is a restoring force opposite to and proportional to a displacement

single bond
covalent bond in which a single pair of electrons is shared by two atoms

slate
a metamorphic rock formed from shale, a sedimentary rock

slope
the ratio of changes in the y variable to changes in the x variable or how fast the y-value increases as the x-value increases

solar constant
the averaged solar power received by the outermost part of the earth's atmosphere—about 1,370 watts per square meter

solenoid
many loops of wire formed into a cylindrical coil that becomes magnetic when a current runs through it

solids
a state of matter whose substances have definite shapes and volumes because the molecules remain close to fixed positions due to strong interactions between the molecules

solstices
time when the sun is at its maximum or minimum altitude in the sky

solubility
the concentration that is reached in a saturated solution at a particular temperature

solute
the components that dissolve in the solvent

solution
a homogeneous mixture of ions or molecules of two or more substances

solvent
the component of a solution present in the larger amount

sonic boom
sound waves that pile up into a shock wave when a source is traveling at or faster than the speed of sound

sound quality
characteristic of the sound produced by a musical instrument; determined by the presence and relative strengths of the overtones produced by the instrument

south celestial pole
the point where the south pole of the earth touches the celestial sphere

south pole
the south-seeking pole of a magnet

specific heat
the amount of energy or heat needed to increase the temperature of one gram of a substance one degree Celsius; each substance has its own specific heat

speed
the rate of change in position per change in time; this does not include the direction of change

spin quantum number
describes the spin orientation of an electron relative to an external magnetic field

spring equinox
equinox that occurs on about March 21

spring tide
high and low tides that occur every two weeks because of the relative positions of the earth, moon, and sun

standard atmospheric pressure
the average atmospheric pressure at sea level; the standard pressure is 29.92 inches or 760.0 mm of mercury (1,013.25 millibar)

standard time zones
15° wide zones defined to have the same time throughout the zone

standard unit
a unit established as the standard among its type of referent units upon which the value of the other referent units of the same type are based

standing waves
condition where two waves of equal frequency traveling in opposite directions meet and form stationary regions of maximum displacement due to constructive interference and stationary regions of zero displacement due to destructive interference

starch
group of complex carbohydrates that plants use as a stored food source

states of matter
the forms matter can take—solid, liquid, and gas

stationary front
a front that is not advancing

steam generator
part of nuclear power plant; the heat exchanger that heats feedwater from the secondary loop to steam with the very hot water from the primary loop

step-down transformer
a transformer that decreases the voltage of a current

step-up transformer
a transformer that increases the voltage of a current

stony-iron meteorites
meteorites composed of silicate minerals and metallic iron

stony meteorites
rock-forming meteorites composed mostly of silicate minerals

storm
a rapid and violent weather change

stratopause
the upper boundary of the stratosphere

stratosphere
the layer of the atmosphere above the troposphere where temperature increases with height

stream
a large or small body of running water

stress
a force that tends to compress, pull apart, or deform rock

strong acid
acids that ionize completely in water

strong base
completely ionic in solution and has hydroxide ions

subduction zone
the region near a convergent boundary where the crust forming one plate is forced under the crust of another plate into the interior of the earth

sublimation
the phase change of a solid directly into a vapor or gas

submarine canyons
deep, steep-sided canyons that cut through the continental slopes

summer solstice
in the Northern Hemisphere, the time when the sun reaches its maximum altitude in the sky

superconductors
conductors that conduct electric current with no resistance

supercooled
water in the liquid state when the temperature is below the freezing point

supernova
a star that collapses due to inward pressure and then rebounds like a compressed spring into a catastrophic explosion

supersaturated
containing more than the normal saturation amount of water

surf
the zone where breakers occur

swell
regular groups of low profile, long wavelength waves

syncline
a trough-shaped fold in rock strata

synodic month
the interval of time from new moon to new moon

T

talus
accumulations of rock fragments at the base of a slope

temperate climate zone
climate zone of the middle latitudes; average monthly temperatures stay between 10° C and 18° C (50° F and 64° F)

temperature
a measure of the average kinetic energy of all the molecules that make up a substance

terrestrial planets
planets that have similar densities and compositions that resemble the earth

tesla
magnetic field strength unit equivalent to N/Am

theory
a broad, detailed explanation that guides the development of hypotheses and interpretations of experiments in a field of study

thermal efficiency
the ratio of work accomplished by a heat engine to the heat it absorbs

thermodynamics
the branch of physical science concerned with the study of heat and its relationship to mechanical energy

thermometer
a device used to measure the temperature of a substance

thermosphere
atmospheric layer of the earth that is above the mesosphere

third law of motion
whenever two objects interact, the force exerted on the first object by the second object is equal in size and opposite in direction to the force exerted by the second object on the first object

thrust fault
a reverse fault with a low-angle fault plane

thunderstorm
a brief, intense storm with rain, lightning, thunder, strong winds, and sometimes hail

tidal bore
a strong tidal current produced in very long, very narrow bays

tidal currents
a wavelike rise in the water level produced in narrow bays by the tides

tide
periodic rise and fall of the level of the sea from the gravitational attraction of the moon and sun and from the motion of the earth-moon system

tornado
smallest, most violent weather disturbance that occurs on earth

total internal reflection
condition where all light is reflected back from a boundary between a material with a lower index of refraction and a material with a higher index of refraction

total solar eclipse
eclipse that occurs when the umbra of the moon falls on the surface of the earth; the new moon completely covers the disk of the sun

transform boundaries
in plate tectonics, boundaries that occur between two plates sliding by each other

transformer
a device that uses electromagnetic induction to increase or decrease voltage

transition elements
members of the B group families of the periodic table

transparent
term describing materials that allow the transmission of light

transverse wave
a wave in which the displacements back and forth forming the wave are perpendicular to the direction of the motion of the wave

trenches
a long, relatively narrow, steep-sided trough that occurs along the edges of the ocean basins

triglyceride
ester formed from three fatty acids and glycerol

triple bond
covalent bond formed when three pairs of electrons are shared by two atoms

Tropic of Cancer
parallel identifying the Northern limit where the sun appears directly overhead; located at 23 1/2° N latitude

Tropic of Capricorn
parallel identifying the Southern limit where the sun appears directly overhead; located at 23 1/2° S latitude

tropical air mass
a warm air mass from warm regions

tropical climate zone
climate zone of the low latitudes; average monthly temperatures stay above 18° C (64° F)

tropical cyclone
a hurricane

tropical year
the time interval between two consecutive spring equinoxes; used as standard for the calendar year

tropopause
the upper boundary of the troposphere

troposphere
layer of the atmosphere from the surface to where the temperature stops decreasing with height

trough
the depression in the surface of the ocean following a wave crest

tsunamis
giant destructive waves created by undersea earthquakes or underwater landslides

turbidity current
a muddy current produced by underwater landslides

typhoon
the name for hurricanes in the Western Pacific

Glossary

U

ultrasonic
sound waves too high in frequency to be heard by the human ear; frequencies above 20,000 Hz

umbra
the inner core of a complete shadow

unconformity
a time break in the rock record

undertow
the weak current produced by the direct return of water from the shore to the sea

unit
in measurement, a well-defined and agreed-upon referent

universal law of gravitation
every object in the universe is attracted to every other object with a force directly proportional to the product of their masses and inversely proportional to the square of the distance between them

unpolarized light
light consisting of transverse waves vibrating in all conceivable random directions

unsaturated
an organic molecule that does not contain the maximum number of hydrogen atoms

V

valence
the number of covalent bonds an atom can form

valence electrons
outer electrons that determine the chemical properties of an atom and that participate in chemical bonding

Van Allen belts
belts of radiation caused by cosmic ray particles becoming trapped and following the earth's magnetic field lines between the poles

van der Walls force
general term for weak attractive intermolecular forces

vapor
the gaseous state of a substance that is normally in the liquid state

variables
changing quantities usually represented by a letter or symbol

vectors
quantities that require both a magnitude and direction to describe them

velocity
the speed and direction of a moving object

ventifacts
rocks sculpted by wind abrasion

vernal equinox
another name for the spring equinox

vibration
a back and forth motion that repeats itself

virtual image
an image where light rays appear to originate from a mirror or lens; this image cannot be projected on a screen

volt
unit of potential difference equivalent to J/C

voltage drop
the electric potential difference across a resistor or other part of a circuit that consumes power

voltage source
source of electric power in an electric circuit that maintains a constant voltage supply to the circuit

volume
how much space something occupies

W

warm front
the front that forms when a warm air mass advances over a cool air mass

watershed
the land area drained by a stream

water table
the boundary between the zone of saturation and the zone above

watt
metric unit for power; equivalent to J/sec

wave
a disturbance that moves through a medium

wave equation
the relationship of the velocity of a wave to the product of the wavelength and frequency of the wave

wave front
a region of maximum displacement in a wave; a condensation in a sound wave

wave height
the vertical distance of an ocean wave between the top of the wave crest and the bottom of the next trough

wavelength
the horizontal distance between successive wave crests or other successive parts of the wave

wave mechanics
alternate name for quantum mechanics derived from the wavelike properties of subatomic particles

wave period
the time required for two successive crests or other successive parts of the wave to pass a given point

weak acid
acids that only partially ionize because of an equilibrium reaction with water

weak base
a base only partially ionized because of an equilibrium reaction with water

weathering
the changes that result in the breaking up, crumbling, and destruction of any kind of solid rock

wet adiabatic lapse rate
the adiabatic cooling or warming rate of air with condensation or evaporation of water vapor

white dwarf stars
faint, white-hot stars that are very small and dense

wind
horizontal movement of air

wind abrasion
the natural sand-blasting process that occurs when wind particles break off small particles of rock and polish the rock they strike

wind chill factor
in the winter, the cooling power of wind

winter solstice
in the sky in the Northern Hemisphere, the time when the sun reaches its minimum altitude

work
the magnitude of applied force times the distance through which the force acts; can be thought of as the process by which one form of energy is transformed to another

Z

zone of saturation
zone of sediments beneath the surface in which water has collected in all available spaces

Credits

Photographs

Section Openers
Section 1: © M. Delesantro/Visuals Unlimited; **Section 2:** © Carolina Biological Supply/Visuals Unlimited; **Section 3, 4:** © Science NASA/Visuals Unlimited.

Chapter 1
Figure 1.1, 1.2: Courtesy Keith Jennings/News Bureau, Arizona State University; **1.3:** Courtesy Bill W. Tillery; **1.13:** Courtesy Keith Jennings/News Bureau, Arizona State University; **1.18:** Courtesy Bill W. Tillery.

Chapter 2
Figure 2.1: Courtesy News Bureau, Arizona State University; photo by Conley Photography, Inc.; **2.8:** Courtesy Smithsonian Institution (Photo No. 46843); **2.12:** Courtesy Smithsonian Institution (Photo No. 36170); **Box 2.1 all:** Courtesy General Motors; **2.20:** Courtesy Keith Jennings/News Bureau, Arizona State University.

Chapter 3
Figure 3.1: Courtesy Beech Aircraft Corporation; **3.2:** Courtesy Smithsonian Institution (Photo No. 30823A); **3.5:** Courtesy News Bureau, Arizona State University; photo by Conley Photography, Inc.; **3.9:** Courtesy Bill W. Tillery; **3.11, Box 3.1:** NASA.

Chapter 4
Figure 4.1: Courtesy Salt River Project; **4.14, 4.15:** Courtesy Bill W. Tillery; **4.16, 4.18, Box 4.1:** Courtesy Arizona Public Service Company.

Chapter 5
Figure 5.1: Courtesy Bethlehem Steel Corporation; **5.5:** Courtesy News Bureau, Arizona State University; photo by John C. Wheatly; **5.10:** Courtesy Honeywell, Inc.; **5.18:** Courtesy of Manville Company.

Chapter 6
Figure 6.1: Courtesy General Motors; **6.2:** HREM, Arizona State University; **6.9, 6.11:** Courtesy Bill W. Tillery.

Chapter 7
Figure 7.1: Courtesy Keith Jennings/News Bureau, Arizona State University.

Chapter 8
Figure 8.1: Arizona Public Service Company; **8.2:** Courtesy Bill W. Tillery; **8.15:** Arizona Public Service Company; **8.19 left:** Courtesy Bill W. Tillery; **right:** GTE Electrical Products; **8.20:** Courtesy Bill W. Tillery.

Chapter 9
Figure 9.1: Arizona Public Service Company; **9.2:** Courtesy Bill W. Tillery; **9.23, Box 9.1:** Arizona Public Service Company.

Chapter 10
Figure 10.1: Bureau of Land Management; **10.7 both:** 1987 Libbey-Owens-Ford Company.

Chapter 11
Figure 11.1: Courtesy News Bureau, Arizona State University; **11.3:** Courtesy Bill W. Tillery; **11.9:** Courtesy News Bureau, Arizona State University, Charles R. Carley.

Chapter 12
Figure 12.1, 12.2, 12.5: Courtesy Bill W. Tillery.

Chapter 13
Figure 13.1: Stewart Lindsay-Physics Department, Arizona State University; **13.9, 13.11:** Courtesy Bill W. Tillery.

Chapter 14
Figure 14.1: © Science/John Deere/Visuals Unlimited; **14.3, 14.4, 14.7, 14.8:** Courtesy Bill W. Tillery; **14.9:** Courtesy Keith Jennings/News Bureau, Arizona State University; **14.10, 14.12:** Courtesy Bill W. Tillery.

Chapter 15
Figure 15.1: Courtesy City of Tempe, Arizona; **15.2:** Courtesy Bill W. Tillery.

Chapter 16
Figure 16.1: Courtesy the American Petroleum Institute.

Chapter 17
Figure 17.1: Arizona Public Service Company; **17.2 both:** © Fundamental Photographs; **17.10:** Courtesy Bill W. Tillery.

Chapter 18
Figure 18.1 A: Lick Observatory Photographs; **18.2:** © H. Sommer/Visuals Unlimited; **18.23:** Lick Observatory Photographs; **18.24:** National Optical Astronomy Observatories; **18.26:** Lick Observatory Photographs.

Chapter 19
Figure 19.1, 19.13: NASA; **19.14:** Lick Observatory Photographs; **19.15, 19.16, 19.18 both, 19.19, 19.20:** NASA; **19.24:** Lick Observatory Photographs; **19.26 both:** Center for Meteorite Studies, Arizona State University.

Chapter 20
Page 406: NASA; **Figure 20.17:** Courtesy Bill W. Tillery; **20.27:** Lick Observatory Photographs.

Chapter 21
Figure 21.1: Courtesy A. Post, USGS Photo Library, Denver, CO; **21.6:** Courtesy Bill W. Tillery.

Chapter 22
Figure 22.1: Courtesy W. R. Hansen, USGS Photo Library, Denver, CO; **22.6:** National Park Service. Photo by Cecil W. Stoughton; **22.12:** NASA; **22.15:** National Park Service. Photo by Wm. Belknap, Jr.; **22.19:** © William J. Weber/Visuals Unlimited; **22.22:** Courtesy R. D. Miller, USGS Photo Library, Denver, CO; **22.24:** Courtesy Bill W. Tillery.

Chapter 23
Figure 23.1: © Nick Noyes/Visuals Unlimited; **23.16:** Courtesy Bill W. Tillery; **23.18:** © Bruce Berg/Visuals Unlimited; **23.19:** NOAA.

Chapter 24
Figure 24.1: © Bruce S. Cushing/Visuals Unlimited; **24.7, 24.17:** NOAA; **24.19:** Courtesy Conley Photography/News Bureau, Arizona State University; **24.20:** Courtesy Bill W. Tillery.

Chapter 25
Figure 25.1: Courtesy Bill W. Tillery; **25.10:** Salt River Project; **25.11, 25.14:** Courtesy Bill W. Tillery; **25.15:** Courtesy Michael J. Reber, Chesapeake Biological Laboratory.

Line Art

Chapter 21
Figures 21.11 and 21.19: From Carla W. Montgomery, *Physical Geology,* 2d ed. Copyright © 1990 Wm. C. Brown Publishers, Dubuque, Iowa. All Rights Reserved. Reprinted by permission.

Chapter 22
Figures 22.2, 22.3, 22.4, 22.7, 22.8, 22.10, 22.11, 22.14, 22.16, 22.18, 22.20, 22.23, 22.25, 22.26, 22.27, 22.28, and 22.29: From Carla W. Montgomery, *Physical Geology,* 2d ed. Copyright © 1990 Wm. C. Brown Publishers, Dubuque, Iowa. All Rights Reserved. Reprinted by permission.

Illustrator Credits

John Foerster
Figures 21.13, 21.14, 21.15, 21.21, 21.22 A–C, Box Figure 21.1, Box Figure 21.2, Figures 22.5, 22.13, 22.17, 22.21 A–C, 22.31, 23.9, 23.10, 23.11, 24.2, 25.3, 25.4, 25.6, 25.7, 25.8, 25.18, 25.19, 25.20, 25.22.

McCullough Graphics (text art)

TA 11.1, TA 13.1, TA 13.2, TA 13.3, TA 13.4, TA 13.5, TA 13.6, TA 13.7, TA 13.8, TA 13.9, TA 13.10, TA 13.11, TA 13.12, TA 13.13, TA 13.14, TA 13.15, TA 13.16, TA 13.17, TA 13.18, TA 13.19, TA 13.20, TA 13.21, TA 13.22, TA 13.23, TA 13.24, TA 13.25, TA 13.26, TA 13.27, TA 13.28, TA 13.29, TA 13.30, TA 13.31, TA 13.32, TA 13.33, TA 13.34, TA 13.35, TA 13.36, TA 13.37, TA 13.38, TA 13.39, TA 13.40, TA 13.41, TA 13.42, TA 13.43, TA 15.1, TA 15.2, TA 15.3, TA 15.4, TA 16.1, TA 16.2, TA 16.3, TA 16.4, TA 16.5, TA 16.6, TA 16.7, TA 16.8, TA 16.9, TA 16.10, TA 16.11, TA 16.12, TA 16.13, TA 16.14, TA 16.15, TA 16.16, TA 16.17, TA 16.18, TA 16.19, TA 16.20, TA 16.21, TA 16.22, TA 16.23, TA 16.24, TA A.1, TA A.2, TA A.3, TA A.4, TA A.5, TA A.6, TA A.7, TA A.8, TA A.9.

Rolin Graphics

Figures 1.4, 1.5, 1.6, 1.7, 1.8, 1.9, 1.10, 1.11, 1.12, 1.14, 1.15, 1.16, 1.17, 2.2, 2.3, 2.4, 2.5, 2.6 A–B, 2.7, 2.9, 2.10, 2.11, 2.13, 2.14, 2.15, 2.16, 2.17, 2.18, 2.19, 3.3 A–B, 3.4, 3.6, 3.7, 3.8, 3.10, 3.12, 3.13, 3.14, 3.15, 4.2, 4.3, 4.4, 4.5, 4.6, 4.7, 4.8, 4.9, 4.10, 4.11, 4.12, 4.13, 4.17, 4.19, 4.20, 4.21, 4.22, 4.23, 5.2, 5.3, 5.4, 5.6, 5.7, 5.8, 5.9, 5.11, 5.12, 5.13, 5.14, 5.15, 5.16, 5.17, 5.19, 5.20, Box Fig. 5.1, Box Fig. 5.2, Box Fig. 5.3, Figures 6.3, 6.4, 6.5, 6.6, 6.7, 6.8, 6.10, 6.12, 6.13, 6.14, 6.15, 6.16, 6.17, 6.18, 6.19, Box Fig. 6.1, Figures 7.2, 7.3, 7.4, 7.5, 7.6, 7.7 A–B, 7.8, 7.9, 7.10, 7.11, 7.12, 7.13, 7.14, 7.15, 7.16, 7.17, 7.18, 7.19, 7.20, 7.21, 7.22, 7.23, 7.24, 7.25, 8.3, 8.4, 8.5, 8.6, 8.7, 8.8, 8.9 A–B, 8.10, 8.11, 8.12, 8.13, 8.14, 8.16, 8.17, 8.18, 9.3, 9.4, 9.6, 9.7, 9.8, 9.9, 9.10, 9.11, 9.12, 9.13, 9.14, 9.15, 9.16, 9.17, 9.18, 9.19, 9.20, 9.21, 9.22, 9.24, Box Fig. 9.2, Figures 10.2, 10.3, 10.4, 10.5, 10.6, 10.8, 10.9, 10.10, 10.11, 10.12, 10.13, 10.14, 10.15, 10.16, 10.17, 10.18, 10.19, 10.20, 10.21, 10.22, 10.23, 10.24, 10.25, 10.26, 11.2, 11.4, 11.5, 11.6, 11.7, 11.8, 11.10, 11.11, 11.12, 11.13, 11.14, 11.15, 11.16, 11.17, 11.18, 12.3, 12.4, 12.6, 12.7, 12.8, 12.9, 12.10, 12.11, 12.12, 12.13, 12.14, 12.15 A–B, 12.16, 12.17, 12.18, 12.19, 12.20, 13.2, 13.3, 13.4, 13.5, 13.6, 13.7, 13.8, 13.10, 13.12, Box Figure 13.1, Figures 14.2, 14.5, 14.6, 14.11, 14.13, 14.14, 14.15 A–C, 15.3 A–C, 15.4, 15.5, 15.6, 15.7, 15.8, 15.9, 15.10, 15.11, 15.12, 15.13, 15.14, 15.15, 15.16, 15.17, 15.18, 16.2, 16.3, 16.4, 16.5, 16.6, 16.7, 16.8, 16.9, 16.10, 16.11, 16.12, 16.13, 16.14, 16.15, 16.16, 16.17, 16.18, 16.19, 16.20, 16.21, 16.21A, Box Figure 16.1, Figures 17.3, 17.4, 17.5, 17.6, 17.7, 17.8, 17.9, 17.10, 17.12, 17.13, 17.14, 17.15, 17.16, 17.17, 17.18, 17.19, 18.1, 18.3, 18.4, 18.5, 18.6, 18.7, 18.8, 18.9, 18.10, 18.11, 18.12, 18.13, 18.14, 18.15, 18.16, 18.17, 18.18, 18.19, 18.20, 18.21, 18.22, 18.25, 18.27, 18.28, 18.29, 19.2, 19.3 A–B, 19.4, 19.5, 19.6, 19.7, 19.8, 19.9, 19.10, 19.11, 19.12, 19.17, 19.21, 19.22, 19.23, 19.25, 20.1, 20.2, 20.3, 20.4, 20.5, 20.6, 20.7, 20.8, 20.9, 20.10, 20.11, 20.12, 20.13, 20.14, 20.15, 20.16, 20.18, 20.19 A–B, 20.21, 20.22, 20.23, 20.26, 20.28, 20.29, 20.30, 20.31, 20.32, 20.33, 21.2, 21.3, 21.4, 21.5, 21.7, 21.8, 21.9, 21.10, 21.12, 22.30, 23.2, 23.3, 23.4, 23.5, 23.6, 23.7, 23.8, 23.12, 23.13, 23.14, 23.15, 23.17, 24.3, 24.4, 24.5, 24.6, 24.9, 24.10, 24.11, 24.12, 24.14, 24.15, 24.16, 24.18, 24.21, 24.23, 25.2, 25.13, 25.16, 25.17.

Alice Theide

Figures 9.5, 20.20, 20.24, 20.25, 21.16, 21.17, 24.8, 24.13, 24.22, 24.24, 24.25, 25.5, 25.12, 25.21, 25.23.

Index

Neptune, properties of the planet, 399
Net and unbalanced forces, 45–46
Net force, 32
Neutralized, acid and base reaction, 299
Neutrino, 221
Neutrons, 145, 211
Neutron star, 368, 370
Newcomen, Thomas, 112, 229
Newton, Issac, 43, 192, 194–95
Newton's
   law of gravitation, 53–56
   laws of motion, 43–50
Newton unit of force, 46–47
Nitroglycerine, 318
Nobel, Alfred, 319
Noble gases, 239
Nodes, 135
Noise, 135
Nonelectrolytes, 295
Nonferromagnesian silicates, 435
Nonmetals, properties of, 226
Nonpolar molecule, 255
Noon, 414
Normal, between angles of incidence and
     reflection, 190
Normal fault, 459
North celestial pole, 360
Northern lights, 171
Nova, 368
Nuclear
   equation, 334–35
   fission, 345–46
   force, 335
   fusion, 345, 348–49
   power plants, 346–48
   reactor, 346
Nuclear energy, 70, 343–50
   as viewed by electric utilities, 75
Nucleons, 334
Nucleus, 145
   discovery of, 210–11
   nature of, 335
   shell model of, 335
   size of compared to size of atom, 211

O

Oblate spheroid, 408
Observed lapse rate, 486
Obsidian, 439
Occluded front, 510
Ocean, 531–32
   basin, 540
   currents, 537
   exploration, 541
   floor, 539–40
   waves, 535
Oceanic
   ridges, 446
   trenches, 446
Ocean thermal energy conversion (OTEC), 76
Octane number, 315
Octet rule, 249
Odor and taste of fruits and flowers, 319–20
Oersted, Hans Christian, 166, 172, 175
Ohm's law, 155
Ohm unit of electrical resistance, 155
Oil field, 314
One-way mirrors, 188

Oort cloud, 400
Opaque materials, 188
Open tube vibrating air column, 137
Orbital, 218
Organic acids, 319
Organic chemistry
   definition of, 308
   system of naming isomers, 311–12
Organic molecules, and heredity, 327
Origin of stars, 363
Oscillating theory of universe, 375
Out of phase waves, 132
Overtones, 136
Oxbow lake, 468
Oxidation, 274
   number, 274
   reactions, 273–75
Oxidation, of rocks, 466
Oxidized substance, 274
Oxidizing agent, 274
Oxygen, discovery of, 230
Ozone layer
   and chemical pollution, 487
   as a shield, 487

P

Paleozoic era, 477
Parallels, 412
Parsec, 362
Parts per billion, 289
Parts per million, 289
Passive solar house, 76
   design, 95
Pauli, Wolfgang, 219
Pauli exclusion principle, 219
Peneplain, 473
Penumbra, 423
Percent
   composition of compounds, 268
   by volume, 290
   by weight, 290
Perigee, 426
Perihelion, 383
Period
   of periodic table, 236
   of vibration, 124
   of a wave, 128
Periodic
   law, 235–36
   motion, 124
   vibration, 123
Periods of geologic time, 477
Permeability, of sediment, 527
Petroleum
   how formed, 73–74
   major uses, 74
   nature of, 314
   short history of use, 314–15
Petroleum ether, 315
Petroleum products, table of, 315
pH
   scale, 300
   table of common substances, 301
Phase change, 103–5
Phenyl, 314
Phlogiston, 229
Photoelectric effect, 198
Photons, 199, 212, 221
Physical change, 228

Pioneer spacecraft, 392, 395, 396
Pitch, 127
Planck, Max, 199, 211
Planck's constant, 199
Plane of the ecliptic, 409
Planetary motion, laws of, 383
Planetary nebula, 369
Planet properties, table, 388
Planets, and Greek astronomy, 359
Plasma, 85
Plastic deformation, 457
Plate tectonics, 445–52
   unanswered questions, 450
Pluto, properties of the planet, 399
Polar air mass, 507
Polar covalent bond, 255
Polaris, 361
Polarized light, 197
Polar molecule, 255
Polaroid, 197
Polyatomic ion, 254
Polymers, 317, 324
Polysaccharides, 322
Pond, 527
Porosity, of sediment, 527
Positive ion, 145
Positron, 221
Post meridiem (P.M.), 414
Potential energy, 65
   zero reference point, 66
Power, 63
   units of, 64
Precambrian era, 477
Precession, of earth's axis, 411
Precipitation, 505
Pressure, 106
Priestley, Joseph, 230
Primary loop, in pressurized water nuclear
     reactor, 347
Prime meridian, 412
Principal quantum number, 218
Principle, definition of scientific, 15
Principle of
   crosscutting relationships, 473
   original horizonality, 473
   superposition, 473
   uniformity, 456, 473
Problem solving
   approach and procedures, 17
   format, 12
   working with, 544
Products, in chemical equation, 270
Projectiles, 38
Proof, of an alcohol beverage, 318
Propane, 315
Properties, 2–4. See also metric standard units,
     e.g., meter
   fundamental properties, 5
Properties of
   acids and bases, theory about, 299
   salts, 301
Proportional relationships, 11
Proteins, 320–22
Protogalaxies, 375
Proton, 145
   discovery of, 211
Protostar, 363

Table of Atomic Weights

Element	Symbol	Atomic Number	Atomic Weight	Element	Symbol	Atomic Number	Atomic Weight
Actinium	Ac	89	227.0278	Mercury	Hg	80	200.59
Aluminum	Al	13	26.98154	Molybdenum	Mo	42	95.94
Americium	Am	95	(243)	Neodymium	Nd	60	144.24
Anitmony	Sd	51	121.75	Neon	Ne	10	20.179
Argon	Ar	18	39.948	Neptunium	Np	93	237.0482
Arsenic	As	33	74.9216	Nickel	Ni	28	58.69
Astatine	At	85	(210)	Niobium	Nb	41	92.9064
Barium	Ba	56	137.33	Nitrogen	N	7	14.0067
Berkelium	Bk	97	(247)	Nobelium	No	102	(259)
Beryllium	Be	4	9.01218	Osmium	Os	76	190.2
Bismuth	Bi	83	208.9804	Oxygen	O	8	15.9994
Boron	B	5	10.811	Palladium	Pd	46	106.42
Bromine	Br	35	79.704	Phosphorus	P	15	30.97376
Cadmium	Cd	48	112.41	Platinum	Pt	78	195.08
Calcium	Ca	20	40.078	Plutonium	Pu	94	(244)
Californium	Cf	98	(251)	Polonium	Po	84	(209)
Carbon	C	6	12.011	Potassium	K	19	39.0983
Cerium	Ce	58	140.12	Praseodymium	Pr	59	140.9077
Cesium	Cs	55	132.9054	Promethium	Pm	61	(145)
Chlorine	Cl	17	35.453	Protactinium	Pa	91	231.0359
Chromium	Cr	24	51.9961	Radium	Ra	88	226.0254
Cobalt	Co	27	58.9332	Radon	Rn	86	(222)
Copper	Cu	29	63.546	Rhenium	Re	75	186.207
Curium	Cm	96	(247)	Phodium	Rh	45	102.9055
Dysprosium	Dy	66	162.50	Rubidium	Rb	37	85.4678
Einsteinium	Es	99	(252)	Ruthenium	Ru	44	101.07
Erbium	Er	68	167.26	Samarium	Sm	62	150.36
Europium	Eu	63	151.96	Scandium	Sc	21	44.95591
Fermium	Fm	100	(257)	Selenium	Se	34	78.96
Fluorine	F	9	18.998403	Silicon	Si	14	28.0855
Francium	Fr	87	(223)	Silver	Ag	47	107.8682
Gadolinium	Gd	64	157.25	Sodium	Na	11	22.98977
Gallium	Ga	31	69.723	Strontium	Sr	38	87.62
Germainum	Ge	32	72.59	Sulfur	S	16	32.066
Gold	Au	79	196.9665	Tantalum	Ta	73	180.9479
Hafnium	Hf	72	178.49	Technetium	Tc	43	(98)
Helium	He	2	4.002602	Tellurium	Te	52	127.60
Holmium	Ho	67	164.9304	Terbium	Tb	65	158.9254
Hydrogen	H	1	1.00794	Thallium	Tl	81	204.383
Indium	In	49	114.82	Thorium	Th	90	232.0381
Iodine	I	53	126.9045	Thulium	Tm	69	168.9342
Iridium	Ir	77	192.22	Tin	Sn	50	118.710
Iron	Fe	26	55.847	Titanium	Ti	22	47.88
Krypton	Kr	36	83.80	Tungsten	W	74	183.85
Lanthanum	La	57	138.9055	Uranium	U	92	238.0289
Lawrencium	Lr	103	(260)	Vanadium	V	23	50.9415
Lead	Pb	82	207.2	Xenon	Xe	54	131.29
Lithium	Li	3	6.941	Ytterbium	Yb	70	173.04
Lutetium	Lu	71	174.967	Yttrium	Y	39	88.9059
Magnesium	Mg	12	24.305	Zinc	Zn	30	65.39
Manganese	Mn	25	54.9380	Zirconium	Zr	40	91.224
Mendelevium	Md	101	(258)				